T0258213

Introduction to Many-Body Physics

A modern, graduate-level introduction to many-body physics in condensed matter, this textbook explains the tools and concepts needed for a research-level understanding of the correlated behavior of quantum fluids.

Starting with an operator-based introduction to the quantum field theory of many-body physics, this textbook presents the Feynman diagram approach, Green's functions, and finite-temperature many-body physics before developing the path integral approach to interacting systems. Special chapters are devoted to the concepts of Landau–Fermi-liquid theory, broken symmetry, conduction in disordered systems, superconductivity, local moments and the Kondo effect, and the physics of heavy-fermion metals and Kondo insulators.

A strong emphasis on concepts and numerous exercises make this an invaluable course book for graduate students in condensed matter physics. It will also interest students in nuclear, atomic, and particle physics.

Piers Coleman is a Professor at the Center for Materials Theory at the Serin Physics Laboratory at Rutgers, The State University of New Jersey. He also holds a part-time position as Professor of Theoretical Physics, Royal Holloway, University of London. He studied as an undergraduate at the University of Cambridge and later as a graduate student at Princeton University. He has held positions at Trinity College, University of Cambridge, and the Kavli Institute for Theoretical Physics, University of California, Santa Barbara.

Professor Coleman is an active contributor to the theory of interacting electrons in condensed matter. He believes deeply that the great frontier of science lies in understanding the remarkable emergent properties of matter that develop when quantum particles come together. He is currently co-director of I2CAM, the International Institute for Complex Adaptive Matter. He invented the slave boson approach to strongly correlated electron systems and believes that new mathematical methods open the door to major new discoveries.

Professor Coleman has a long-standing interest in highly correlated d- and f-electron materials, novel forms of superconductivity, topological matter, and the unsolved problem of quantum criticality in metals. He is also interested in science outreach, and co-produced *Music of the Quantum* with his brother, the musician Jaz Coleman.

Introduction to Many-Body Physics

PIERS COLEMAN

Rutgers, The State University of New Jersey

CAMBRIDGE
UNIVERSITY PRESS

CAMBRIDGE
UNIVERSITY PRESS

University Printing House, Cambridge CB2 8BS, United Kingdom

One Liberty Plaza, 20th Floor, New York, NY 10006, USA

477 Williamstown Road, Port Melbourne, VIC 3207, Australia

4843/24, 2nd Floor, Ansari Road, Daryaganj, Delhi – 110002, India

79 Anson Road, #06–04/06, Singapore 079906

Cambridge University Press is part of the University of Cambridge.

It furthers the University's mission by disseminating knowledge in the pursuit of education, learning and research at the highest international levels of excellence.

www.cambridge.org
Information on this title: www.cambridge.org/9780521864886

© Piers Coleman 2015

Cover illustration © Swarez Modern Art Ltd. 2015

This publication is in copyright. Subject to statutory exception and to the provisions of relevant collective licensing agreements, no reproduction of any part may take place without the written permission of Cambridge University Press.

First published 2015
3rd printing 2019

Printed in Singapore by Markono Print Media Pte Ltd

A catalogue record for this publication is available from the British Library

Library of Congress Cataloguing in Publication data
Coleman, Piers, 1958 – author.
Introduction to many body physics / Piers Coleman, Rutgers University, New Jersey.
pages   cm
Includes bibliographical references and index.
ISBN 978-0-521-86488-6
1. Many-body problem.   2. Condensed matter.   I. Title.
QC174.17.P7C64   2015
530.4′1–dc23
2015001415

ISBN 978-0-521-86488-6 Hardback

Cambridge University Press has no responsibility for the persistence or accuracy of URLs for external or third-party internet websites referred to in this publication, and does not guarantee that any content on such websites is, or will remain, accurate or appropriate.

To Ronald Coleman, who shared a love of invention and experiment.

Contents

Preface

This book could not have been written without the inspiration of my family, mentors, friends, colleagues and students, whom I thank from the bottom of my heart. Certain folk deserve a very special mention. As a kid, my mother inspired me with a love of mathematics and science. My father taught me that theory is fine, but amounts to nought without down-to-earth pragmatism and real-world experiment. I owe so much to Gil Lonzarich at Cambridge who inspired and introduced me to the beauty of condensed matter physics, and to Phil Anderson, who introduced me, as a graduate student, to the idea of emergence and the notion that deep new physics is found within simple yet elegant concepts.

I particularly want to thank my wife and physics colleague at Rutgers, Premi Chandra, who not only encouraged and shared ideas with me, but kept it real by constantly reminding me to think about my audience. At Rutgers, my colleagues Elihu Abrahams, Natan Andrei, Lev Ioffe and Gabi Kotliar receive my special appreciation, who over the years have kept physics exciting and real by sharing with me their ideas and questions, and listening to my own. I also want to especially note my Russian friends and collaborators, who have provided constant input and new insight, especially the late Anatoly Larkin and my two close friends, Andrey Chubukov and Alexei Tsvelik, each of whom has shared with me the wonderful Russian ideal of "kitchen table" physics. Throughout the book, there are various references to the history of many-body physics – here I have benefited immensely over the years, especially through discussion with David Pines, David Khmelnitskii, Lev Gor'kov and Igor Dzyaloshinskii. My apologies to you for any innaccuracies you find in this aspect of the text. I also wish to thank Andy Schofield, who has shared with me his ideas about presenting basic many-body physics. Many others have read and corrected the book with a critical eye, providing wonderful suggestions, including Annica Black-Schaffer, Eran Lebanon, Anna Posazhennikova and my former student Revaz Ramazashvili. Finally, my deep thanks to many students and postdocs who have listened to my lectures over the years, helping to improve the course and presentation.

I particularly want to thank the National Science Foundation and the Department of Energy, who over the years have supported my research in condensed matter theory at Rutgers University. Special thanks in this context goes to my program managers at the National Science Foundation and the Department of Energy, Daryl Hess and James Davenport, respectively. We are so fortunate in the United States to have scientifically adept and involved program managers. The final stages of this book were finished with the support of National Science Foundation grant DMR 1309929 and the support of a Simons Fellowship while at the Kavli Institute for Theoretical Physics, Santa Barbara. Finally, I want to thank Simon Capelin at Cambridge University Press for his constant patience and encouragement during the long decade it has taken to write this book.

Introduction

This book is written with the graduate student in mind. I had in mind to write a text that would introduce my students to the basic ideas and concepts behind many-body physics. At the same time, I felt very strongly that I would like to share my excitement about this field, for without feeling the thrill of entering uncharted territory I do not think one has the motivation to learn and to make the passage from learning to research.

Traditionally, as physicists we ask "what are the microscopic laws of nature?", often proceeding with the brash certainty that, once revealed, these laws will have such profound beauty and symmetry that the properties of the universe at large will be self-evident. This basic philosophy can be traced from the earliest atomistic philosophy of Democritus to the most modern quests to unify quantum mechanics and gravity.

The dreams and aspirations of many-body physics interwine the reductionist approach with a complementary philosophy: that of *emergent phenomena*. In this view, fundamentally new kinds of phenomena emerge within complex assemblies of particles which cannot be anticipated from an *a priori* knowledge of the microscopic laws of nature. Many-body physics aspires to synthesize, from the microscopic laws, new principles that govern the macroscopic realm, asking:

> *What emergent principles and laws develop as we make the journey from the microscopic to the macroscopic?*

This is a comparatively modern and far less familiar scientific philosophy. Charles Darwin was perhaps the first to seek an understanding of emergent laws of nature. Following in his footsteps, Ludwig Boltzmann and James Clerk Maxwell were among the first physicists to appreciate the need to understand how emergent principles are linked to microscopic physics. From Boltzmann's biography [1], we learn that he was strongly influenced and inspired by Charles Darwin. In more modern times, a strong advocate of this philosophy has been Philip W. Anderson, who first introduced the phrase "emergent phenomenon" into physics. In an influential article entitled "More is different," written in 1967 [2], he captured the philosophy of emergence, writing:

> The behavior of large and complex aggregations of elementary particles, it turns out, is not to be understood in terms of a simple extrapolation of the properties of a few particles. Instead, at each level of complexity entirely new properties appear, and the understanding of the new behaviors requires research which I think is as fundamental in its nature as any other.

In an ideal world, I would hope that from this short course your knowledge of many-body techniques will grow hand-in-hand with an appreciation of the motivating philosophy.

In many ways, this dual track is essential, for often one needs both inspiration and overview to steer one lightly through the formalism, without getting bogged down in mathematical quagmires.

I have tried in the course of the book to mention aspects of the history of the field. We often forget that the act of discovering the laws of nature is a very human and very passionate one. Indeed, the act of creativity in physics research is very similar to the artistic process. Sometimes, scientific and artistic revolutions even go hand-in-hand, for the desire for change and revolution often crosses between art and the sciences [3]. I think it is important for students to gain a feeling of this passion behind the science, and for this reason I have often included a few words about the people and the history behind the ideas that appear in this text. There are, unfortunately, very few texts that tell the history of many-body physics. Abraham Pais' book *Inward Bound* [4] has some important chapters on the early stages of many-body physics. A few additional references are included at the end of this chapter [5–7].

There are several texts that can be used as reference books in parallel with this book, of which a few deserve special mention. The student reading this book will need to consult standard references on condensed matter and statistical mechanics. Among these let me recommend *Statistical Physics, Part 2* by Lifshitz and Pitaevksii [8]. For a conceptual underpining of the field, I recommend Anderson's classic *Basic Notions of Condensed Matter Physics* [9]. For an up-to-date perspective on solid state physics from a many-body physics perspective, may I refer you to *Advanced Solid State Physics* by Philip Phillips [10]. Among the classic references to many-body physics let me also mention *Methods of Quantum Field Theory* by Abrikosov, Gor'kov and Dzyaloshinskii ("AGD") [11]. This is the text that drove the quantum many-body revolution of the 1960s and 1970s, yet it is still very relevant today, if rather terse. Other many-body texts which introduce the reader to the Green's function approach to many-body physics include *Many-Particle Physics* by G. Mahan [12], notable for the large number of problems he provides, *Green's Functions for Solid State Physicists* by Doniach and Sondheimer [13] and a very light introduction to the subject, *A Guide to Feynman Diagrams in the Many-Body Problem* by Richard Mattuck [14]. Among the more recent treatments, let me note Alexei Tsvelik's *Quantum Field Theory in Condensed Matter Physics* [15], which provides a wonderful introduction to bosonization and conformal field theory, and *Condensed Matter Field Theory* by Alexander Altland and Ben Simons [17], a perfect companion volume to my own work. As a reference to the early developments of many-body physics, I recommend *The Many-Body Problem* by David Pines [16], which contains a compilation of the classic early papers in the field. Lastly, let me direct the reader to numerous excellent online reference sources; in addition to the online physics archive http://arXiv.org, let me mention the lecture notes on solid state physics and many-body theory by Chetan Nayak [18].

Here is a brief summary of the book:

- Scales and complexity: the gulf of time (T), length scale (L), particle number (N), and complexity that separates the microscopic from the macroscopic (Chapter 1)
- Second quantization: the passage from the wavefunction to the field operator, and an introduction to the excitation concept (Chapters 2–4)

- The Green's function: the fundamental correlator of quantum fields (Chapter 5)
- Landau Fermi-liquid theory: an introduction to Landau's phenomenological theory of interacting fermions and his concept of the quasiparticle (Chapter 6)
- Feynman diagrams: an essential tool for visualizing and calculating many-body processes (Chapter 7)
- Finite temperature and imaginary time: by replacing $it \rightarrow \tau$, $e^{-iHt} \rightarrow e^{-H\tau}$, quantum field theory is extended to finite temperature, where we find an intimate link between fluctuations and dissipation (Chapters 8–9)
- Electron transport theory: using many-body physics to calculate the resistance of a metal; second-quantized treatment of weakly disordered metals: the Drude metal and the derivation of Ohm's law (Chapter 10)
- The concepts of broken symmetry and generalized rigidity (Chapter 11)
- Path integrals: using the coherent state to link the partition function and the S-matrix by an integral over all possible time-evolved paths of the many-body system ($Z = \int_{PATH} e^{-S/\hbar}$); using path integrals to study itinerant magnetism (Chapters 12–13)
- Superconductivity and BCS theory, including anisotropic pairing (Chapters 14–15)
- Introduction to the physics of local moments and the Kondo effect (Chapter 16)
- Introduction to the physics of heavy electrons and mixed valence using the large-N and slave boson approaches (Chapters 17–18).

Finally, some notes on conventions. This book uses standard SI notation, which means abandoning some of the notational elegance of cgs units, but brings the book into line with international standards.

Following early Russian texts on physics and many-body physics, and Mahan's *Many Particle Physics* [12], I use the convention that the charge on the electron is

$$e = -1.602\ldots \times 10^{-19}\,\mathrm{C}.$$

In other words, $e = -|e|$ denotes the magnitude *and* the sign of the electron charge. This convention minimizes the number of minus signs required. The magnitude of the electron charge is denoted by $|e|$ in formulae such as the electron cyclotron frequency $\omega_c = \frac{|e|B}{m}$. With this notation, the Hamiltonian of an electron in a magnetic field is given by

$$H = \frac{(\mathbf{p} - e\mathbf{A})^2}{2m} + e\phi,$$

where $\mathbf{A}$ is the vector potential and ϕ the electric potential. We choose the notation ϕ for electric potential in order to avoid confusion with the frequent use of V for potential and also for hybridization.

Following a tradition started in the Landau and Lifshitz series, this book uses the notation

$$F = E - TS - \mu N$$

for the Landau free energy – the grand canonical version of the traditional Helmholtz free energy ($E - TS$), which for simplicity will be referred to as the free energy.

One of the more difficult choices in the book concerns the notation for the density of states of a Fermi gas. To deal with the different conventions used in Fermi-liquid theory, in superconductivity and in local-moment physics, I have adopted the notation

$$\mathcal{N}(0) \equiv 2N(0)$$

to denote the total density of states at the Fermi energy, where $N(0)$ is the density of states per spin. The alternative notation $N(0) \equiv \rho$ is used in Chapters 16, 17, and 18, in keeping with traditional notation for the Kondo effect.

References

[1] E. Broda and L. Gray, *Ludwig Boltzmann: Man, Physicist, Philosopher*, Woodbridge, 1983.

[2] P. W. Anderson, More is different, *Science*, vol. 177, p. 393, 1972.

[3] R. March, *Physics for Poets*, McGraw-Hill, 1992.

[4] A. Pais, *Inward Bound: Of Matter and Forces in the Physical World*, Oxford University Press, 1986.

[5] L. Hoddeson, G. Baym, and M. Eckert, The development of the quantum mechanical electron theory of metals: 1928–1933, *Rev. Mod. Phys.*, vol. 59, p. 287, 1987.

[6] M. Riordan and L. Hoddeson, *Crystal Fire*, W.W. Norton, 1997.

[7] L. Hoddeson and V. Daitch, *True Genius: The Life and Science of John Bardeen*, National Academy Press, 2002.

[8] E.M.L. Lifshitz and L. P. Pitaevksii, *Statistical Mechanics, Part 2: Theory of the Condensed State*, Landau and Lifshitz Course on Theoretical Physics, vol. 9, trans. J.B. Sykes and M.J. Kearsley, Pergamon Press, 1980.

[9] P. W. Anderson, *Basic Notions of Condensed Matter Physics*, Benjamin Cummings, 1984.

[10] P. Phillips, *Advanced Solid State Physics*, Cambridge University Press, 2nd edn., 2012.

[11] A. A. Abrikosov, L. P. Gor'kov, and I. E. Dzyaloshinski, *Methods of Quantum Field Theory in Statistical Physics*, Dover, 1977.

[12] G. D. Mahan, *Many-Particle Physics*, Plenum, 3rd edn., 2000.

[13] S. Doniach and E. H. Sondheimer, *Green's Functions for Solid State Physicists*, Imperial College Press, 1998.

[14] R. Mattuck, *A Guide to Feynman Diagrams in the Many-Body Problem*, Dover, 2nd edn., 1992.

[15] A. Tsvelik, *Quantum Field Theory in Condensed Matter Physics*, Cambridge University Press, 2nd edn., 2003.

[16] D. Pines, *The Many-Body Problem*, Wiley Advanced Book Classics, 1997.

[17] A. Altland and B. Simons, *Condensed Matter Field Theory*, Cambridge University Press, 2006.

[18] C. Nayak, *Quantum Condensed Matter Physics*, http://stationq.cnsi.ucsb.edu/nayak/courses.html, 2004.

Scales and complexity

We do in fact know the microscopic physics that governs all metals, chemistry, materials and possibly life itself. In principle, all can be determined from the many-particle wavefunction

$$\Psi(\vec{x}_1, \vec{x}_2, \ldots \vec{x}_N, t), \tag{1.1}$$

which in turn is governed by the Schrödinger equation [1, 2], written out for identical particles as

$$\left\{ -\frac{\hbar^2}{2m} \sum_{j=1}^{N} \nabla_j^2 + \sum_{i<j} V(\vec{x}_i - \vec{x}_j) + \sum_j U(\vec{x}_j) \right\} \Psi = i\hbar \frac{\partial \Psi}{\partial t}. \tag{1.2}$$

There are of course a few details that we have omitted. For instance, if we're dealing with electrons, then

$$V(\vec{x}) = \frac{e^2}{4\pi \epsilon_0} \frac{1}{|\vec{x}|} \tag{1.3}$$

is the Coulomb interaction potential, where $e = -|e|$ is the charge on the electron. In an electromagnetic field we must also *gauge* the derivatives $\nabla \rightarrow \nabla - i(e/\hbar)\mathbf{A}$ and $U(\vec{x}) \rightarrow U(\vec{x}) + e\phi(\vec{x})$, where $\vec{A}$ is the vector potential and $\phi(\vec{x})$ is the electric potential. Furthermore, to be complete we must discuss spin and the antisymmetry of Ψ under particle exchange; moreoover, if we are to treat the vibrations of the lattice, we must include the nuclear locations in the wavefunction. But with these various provisos, we have every reason to believe that the many-body Schrödinger equation contains the essential microscopic physics that governs the macroscopic behavior of materials.

Unfortunately this knowledge is only the beginning. Why? Because, at the most pragmatic level, we are defeated by the sheer complexity of the problem. Even the task of solving the Schrödinger equation for modest multi-electron atoms proves insurmountable without bold approximations. The problem facing the condensed matter physicist, with systems involving 10^{23} atoms, is qualitatively more severe. The amount of storage required for numerical solution of Schrödinger equation grows exponentially with the number of particles, so with a macroscopic number of interacting particles this becomes far more than a technical problem: it becomes one of *principle*. Indeed, we believe that the gulf between the microscopic and the macroscopic is something qualitative and fundamental, so much so that new properties emerge in macroscopic systems that we cannot anticipate a priori by using brute-force analysis of the Schrödinger equation.

The *Hitchhiker's Guide to the Galaxy* [3] describes a super-computer called Deep Thought that, after millions of years spent calculating "the answer to the ultimate question of life and the universe", reveals it to be "42." Douglas Adams' cruel parody of reductionism holds a certain sway in physics today. Our "42," is Schrödinger's many-body equation: a set of relations whose complexity grows so rapidly that we can't trace its full consequences to macroscopic scales. All is fine, provided we wish to understand the workings of isolated atoms or molecules up to sizes of about a nanometer, but, between the nanometer and the micron, wonderful things start to occur that severely challenge our understanding. Physicists have coined the term "emergence" from evolutionary biology to describe these phenomena [4–7].

The pressure of a gas is an example of emergence: it's a cooperative property of large numbers of particles which cannot be anticipated from the behavior of one particle alone. Although Newton's laws of motion account for the pressure in a gas, 180 years elapsed before James Clerk Maxwell developed the statistical description of atoms needed to understand pressure.

Let us dwell a little more on this gulf of complexity that separates the microscopic from the macroscopic. We can try to describe this gulf using four main catagories of scale:

- T: time
- L: length
- N: number of particles
- C: complexity.

1.1　T: Time scale

We can make an estimate of the characteristic quantum time scale by using the uncertainty principle, $\Delta \tau \Delta E \sim \hbar$, so that

$$\Delta \tau \sim \frac{\hbar}{[1 \, \text{eV}]} \sim \frac{\hbar}{10^{-19} \, \text{J}} \sim 10^{-15} \text{s}. \qquad (1.4)$$

Although we know the physics on this time scale, in our macroscopic world the characteristic time scale ~ 1s, so that

$$\frac{\Delta \tau_{\text{macroscopic}}}{\Delta \tau_{\text{quantum}}} \sim 10^{15}. \qquad (1.5)$$

To link quantum and macroscopic time scales, we must make a leap comparable with an extrapolation from the time scale of a heartbeat to the age of the universe (10 billion years $\sim 10^{17}$s).

1.2　L: length scale

An approximate measure for the characteristic length scale in the quantum world is the de Broglie wavelength of an electron in a hydrogen atom,

$$L_{\text{quantum}} \sim 10^{-10} \text{m}, \qquad (1.6)$$

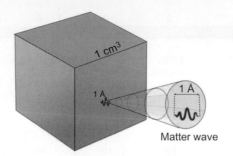

Matter wave

The typical size of a de Broglie wave is 10^{-10} m, to be compared with a typical scale of 1 cm for a macroscopic crystal.

Fig. 1.1

so

$$\frac{L_{\text{macroscopic}}}{L_{\text{quantum}}} \sim 10^8 \tag{1.7}$$

(see Figure 1.1). At the beginning of the twentieth century, a leading philosopher-physicist of the era, Ernst Mach, argued to Boltzmann that the atomic hypothesis was metaphysical, for, he argued, one could simply not envisage any device with the resolution to detect or image something so small. Today, this incredible gulf of scale is routinely spanned in the lab with scanning tunneling microscopes, able to view atoms at sub-angstrom resolution.

1.3 *N*: particle number

To visualize the number of particles in a single mole of a substance, it is worth reflecting that a crystal containing a mole of atoms occupies a volume of roughly $1\,\text{cm}^3$. From the quantum perspective, this is a cube with approximately 100 million atoms along each edge. Avagadro's number,

$$N_{\text{macroscopic}} = 6 \times 10^{23} \sim (100 \text{ million})^3, \tag{1.8}$$

is placed in perspective by reflecting that the number of atoms in a grain of sand is roughly comparable with the number of sand-grains in a 1-mile-long beach. But in quantum matter, the sand-grains, which are electrons, quantum mechanically interfere with one another, producing a state that is much more than the simple sum of its constituents.

1.4 *C*: complexity and emergence

Real materials are like "macroscopic atoms," where the quantum interference among the constituent particles gives rise to a range of complexity and diversity that constitutes the largest gulf of all. We can attempt to quantify the "complexity" axis by considering the

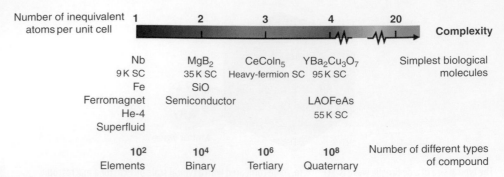

Number of inequivalent atoms per unit cell ... Complexity

Fig. 1.2 Condensed matter of increasing complexity. As the number of inequivalent atoms per unit cell grows, the complexity of the material and the potential for new types of behavior grows. In the labels, "X K SC" denotes a superconductor with a transition temperature $T_c = X$; thus MgB_2 has a 35 K transition temperature.

number of atoms per unit cell of a crystal. Whereas there are roughly 100 stable elements, there are roughly 100^2 stable binary compounds. The number of stable tertiary compounds is conservatively estimated at more than 10^6, of which still only a tiny fraction have been explored experimentally. At each step the range of diversity increases (see Figure 1.2), and there is reason to believe that, at each level of complexity, new types of phenomena begin to emerge.

But it is really the confluence of the length and time scales, particle number and complexity that provides the canvas for "more is different." While classical matter develops new forms of behavior on large scales, the potential for quantum matter to develop emergent properties is far more startling. Let us take an example given by Anderson [8]. Consider similar atoms of niobium and gold: at the angstrom level, there is really nothing to distinguish them, yet as crystals on the macroscale one is a lustrous metal, the other a superconductor. Up to about 30 nm, there is little to distinguish copper and niobium, yet at longer length scales everything changes. Electrons can roam freely across gold crystals, forming the conducting fluid that gives it its lustrous metallic properties. But in niobium, beyond 30 nm the electrons pair up into "Cooper pairs". By the time we reach the micron scale, these pairs congregate by the billions into a pair condensate, transforming the crystal into an entirely new metallic state: a type II superconductor,[1] which conducts without resistance, excludes magnetic fields and has the ability to levitate magnets.

Niobium is an elemental superconductor, with a transition temperature $T_c = 9.2$ K that is pretty typical of conventional low-temperature superconductors. When experimenters began to explore the properties of quaternary compounds in the 1980s, they came across the completely unexpected phenomenon of high-temperature superconductivity. Even today, three decades later, research has only begun to explore the vast universe of quaternary compounds, and the pace of discovery has not slackened. In the two years preceeding publication of this book, superconductivity has been tentatively observed at 108 K in a single-layer FeSe superconductor, and at 190 K in H_2S at high pressure. I'd like to think

---

[1] Niobium, together with vanadium and technitium, is one of only three elements to exhibit type II superconductivity, which allows magnetic fields to penetrate, forming a vortex lattice. See Section 11.5.4.

that before this book goes out of print, many more families of superconductor will have been discovered.

Superconductivity is only a beginning. First of all, it is only one of a large number of broken-symmetry states that can develop in "hard" quantum matter. But in assemblies of softer organic molecules, a tenth of a micron is already enough for the emergence of life. Self-sustaining microbes little more than 200 nm in size have recently been discovered. While we understand, more or less, the principles that govern the superconductor, we do not yet understand those that govern the emergence of life on roughly the same spatial scale [9].

References

[1] E. Schrödinger, Quantisierung als Eigenwertproblem I (Quantization as an eigenvalue problem), *Ann. Phys.*, vol. 79, p. 361, 1926.

[2] E. Schrödinger, Quantisierung als Eigenwertproblem IV (Quantization as an eigenvalue problem), *Ann. Phys.*, vol. 81, p. 109, 1926.

[3] D. Adams, *The Hitchhiker's Guide to the Galaxy*, Pan Macmillan, 1979.

[4] P. W. Anderson, More is different, *Science*, vol. 177, p. 393, 1972.

[5] R. B. Laughlin, D. Pines, J. Schmalian, B. P. Stojkovic, and P. Wolynes, The middle way, *Proc. Natl. Acad. Sci. U.S.A.*, vol. 97, 2000.

[6] R. B. Laughlin, *A Different Universe*, Basic Books, 2005.

[7] P. Coleman, The frontier at your fingertips, *Nature*, vol. 446, 2007.

[8] P. W. Anderson, *Basic Notions of Condensed Matter Physics*, Benjamin Cummings, 1984.

[9] J. C. Séamus Davis, *Music of the Quantum*, http://musicofthequantum.rutgers.edu, 2005.

2 Quantum fields

2.1 Overview

At the heart of quantum many-body theory lies the concept of the quantum field. Like a classical field $\phi(x)$, a quantum field is a continuous function of position, except that now this variable is an operator $\hat{\phi}(x)$. Like all other quantum variables, the quantum field is in general a strongly fluctuating degree of freedom that only becomes sharp in certain special eigenstates; its function is to add or subtract particles to the system. The appearance of particles or *quanta* of energy $E = \hbar\omega$ is perhaps the greatest single distinction between quantum and classical fields.

This astonishing feature of quantum fields was first recognized by Albert Einstein, who in 1905 and 1907 made the proposal that the fundamental excitations of continuous media – the electromagnetic field and crystalline matter in particular – are carried by quanta [1–4], with energy

$$E = \hbar\omega.$$

Einstein made this bold leap in two stages, first showing that Planck's theory of black-body radiation could be reinterpreted in terms of photons [1, 2], and one year later generalizing the idea to the vibrations inside matter [3] which, he reasoned, must also be made up of tiny wavepackets of sound that we now call *phonons*. From his phonon hypothesis Einstein was able to explain the strong temperature dependence of the specific heat in diamond – a complete mystery from a classical standpoint. Yet despite these early successes, it took a further two decades before the machinery of quantum mechanics gave Einstein's ideas a concrete mathematical formulation.

Quantum fields are intimately related to the idea of second quantization. First quantization permits us to make the jump from the classical world to the simplest quantum systems. The classical momentum and position variables are replaced by operators such as

$$E \rightarrow i\hbar\partial_t$$
$$p \rightarrow \hat{p} = -i\hbar\partial_x, \tag{2.1}$$

while the Poisson bracket which relates canonical conjugate variables is now replaced by the quantum commutator [5, 6]:

$$[x, p] = i\hbar. \tag{2.2}$$

The commutator is the key to first quantization, and it is the non-commuting property that leads to quantum fluctuations and the Heisenberg uncertainty principle. (See Examples

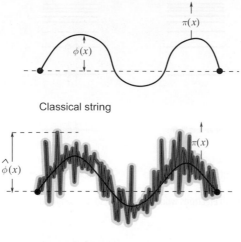

Classical string

Quantum string

Contrasting classical and quantum strings.

Fig. 2.1

2.1 and 2.2.) Second quantization permits us to take the next step, extending quantum mechanics to

- macroscopic numbers of particles
- an *excitation* or *quasiparticle* description of low-energy physics
- the dynamical response and internal correlations of large systems
- collective behavior and broken-symmetry phase transitions.

In its simplest form, second quantization elevates classical fields to the status of operators. The simplest example is the quantization of a classical string, as shown in Figure 2.1. Classically, the string is described by a smooth field $\phi(x)$ which measures the displacement from equilibrium, plus the conjugate field $\pi(x)$ which measures the transverse momentum per unit length. The classical Hamiltonian is

$$H = \int dx \left[\frac{T}{2} \left(\nabla_x \phi(x) \right)^2 + \frac{1}{2\rho} \pi(x)^2 \right], \tag{2.3}$$

where T is the tension in the string and ρ the mass per unit length. In this case, second quantization is accomplished by imposing the canonical commutation relation

$$[\phi(x), \pi(y)] = i\hbar \delta(x - y). \qquad \text{canonical commutation relation} \qquad (2.4)$$

In this respect, second quantization is no different from conventional quantization, except that the degrees of freedom are defined continuously throughout space. The basic method I have just described works for collective fields such as sound vibrations, or for the electromagnetic field, but we also need to know how to develop the field theory of identical particles, such as an electron gas in a metal or a fluid of identical helium atoms.

For particle fields, the process of second quantization is more subtle, for here the underlying fields have no strict classical counterpart. Historically, the first steps to dealing with

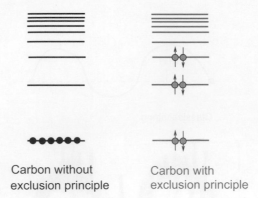

Carbon without
exclusion principle

Carbon with
exclusion principle

Fig. 2.2 Without the exclusion principle, all electrons would occupy the same atomic orbital. There would be no chemistry, no life.

such many-particle systems were made in atomic physics. In 1925 Wolfgang Pauli proposed his famous *exclusion principle* [7] to account for the diversity of chemistry, and for the observation that atomic spectra could be understood only if one assumed there was no more than one electron per quantum state (Figure 2.2). A year later, Paul Dirac and Enrico Fermi examined the consequences of this principle for a gas of particles, which today we refer to as *fermions*. Dirac realized that the two fundamental varieties of particle, fermions and bosons, could be related to the parity of the many-particle wavefunction under particle exchange [8]:

$$\Psi(\text{particle 1 at A, particle 2 at B}) = e^{i\Theta}\Psi(\text{particle 1 at B, particle 2 at A}). \qquad (2.5)$$

If one exchanges the particles twice, the total phase is $e^{2i\Theta}$. If we are to avoid a many-valued wavefunction, then we must have

$$e^{2i\Theta} = 1 \Rightarrow e^{i\Theta} = \begin{cases} +1 \text{ for bosons} \\ -1 \text{ for fermions} \end{cases}. \qquad (2.6)$$

The choice of $e^{i\Theta} = -1$ leads to a wavefunction which is completely antisymmetric under particle exchange, which immediately prevents more than one particle in a given quantum state.[1]

In 1927, Pascual Jordan and Oskar Klein realized that to cast the physics of a many-body system into a more compact form, one needs to introduce an operator for the particle itself – the field operator. With their innovation, it proves possible to unshackle ourselves from the many-body wavefunction. The particle field operator

$$\hat{\psi}(x) \qquad (2.7)$$

can be very loosely regarded as a quantization of the one-body Schrödinger wavefunction. Jordan and Klein [9] proposed that the particle field and its complex conjugate are conjugate variables. With this insight, the second quantization of bosons is achieved by

---

[1] In dimensions below three, it is possible to have wavefunctions with several Reimann sheets, which gives rise to the concept of fractional statistics and "anyons."

introducing a non-zero commutator between the particle field and its complex conjugate. The new *quantum fields* that emerge play the roles of creating and destroying particles:

$$\underbrace{\psi(x), \psi^*(x)}_{\text{one-particle wavefunction}} \xrightarrow{[\psi(x), \psi^\dagger(y)] = \delta(x - y)} \underbrace{\hat{\psi}(x), \hat{\psi}^\dagger(x).}_{\text{destruction/creation operator}} \text{ bosons} \tag{2.8}$$

For fermions, the existence of an antisymmetric wavefunction means that particle fields must *anticommute*, i.e.,

$$\psi(x)\psi(y) = -\psi(y)\psi(x), \tag{2.9}$$

a point first noted by Jordan and then developed by Jordan and Wigner [10]. The simplest example of anticommuting operators is provided by the Pauli matrices: we are now going to have to get used to a whole continuum of such operators! Jordan and Wigner realized that the second quantization of fermions requires that the non-trivial commutator between conjugate particle fields must be replaced by an anticommutator:

$$\underbrace{\psi(x), \psi^*(x)}_{\text{one-particle wavefunction}} \xrightarrow{\{\psi(x), \psi^\dagger(y)\} = \delta(x - y)} \underbrace{\hat{\psi}(x), \hat{\psi}^\dagger(x).}_{\text{destruction/creation operator}} \text{ fermions} \tag{2.10}$$

The operation $\{a, b\} = ab + ba$ denotes the anticommutator. Remarkably, just as bosonic physics derives from commutators, fermionic physics derives from an algebra of anticommutators.

How real is a quantum field and what is its physical significance? To get a feeling for its meaning, let us look at some key properties. The transformation from wavefunction to operator also extends to more directly observable quantities. Consider, for example, the electron probability density $\rho(x) = \psi^*(x)\psi(x)$ of a one-particle wavefunction $\psi(x)$. By elevating the wavefunction to the status of a field operator, we obtain

$$\rho(x) = |\psi(x)|^2 \longrightarrow \hat{\rho}(x) = \hat{\psi}^\dagger(x)\hat{\psi}(x), \tag{2.11}$$

which is the density *operator* for a many-body system. Loosely speaking, the squared magnitude of the quantum field represents the density of particles.

Another aspect of the quantum field we have to understand is its relationship to the many-body wavefunction. This link depends on a new concept, the *vacuum*. This unique state, denoted by $|0\rangle$, is devoid of particles, and for this reason it is the only state for which there is no amplitude to destroy a particle, so

$$\psi(x)|0\rangle = 0. \qquad \text{the vacuum} \tag{2.12}$$

We shall see that, as a consequence of the canonical algebra, the creation operator $\hat{\psi}^\dagger(x)$ increments the number of particles by one (see Figure 2.3), *creating* a particle at x, so that

$$|x_1\rangle = \psi^\dagger(x_1)|0\rangle \tag{2.13}$$

is a single particle at x_1,

$$|x_1, \ldots x_N\rangle = \psi^\dagger(x_N) \cdots \psi^\dagger(x_1)|0\rangle \tag{2.14}$$

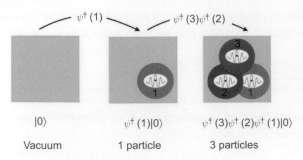

$$\psi^\dagger(1) \qquad \psi^\dagger(3)\psi^\dagger(2)$$

$$|0\rangle \qquad\qquad \psi^\dagger(1)|0\rangle \qquad\qquad \psi^\dagger(3)\psi^\dagger(2)\psi^\dagger(1)|0\rangle$$

Vacuum   1 particle   3 particles

Fig. 2.3   Action of creation operator on vacuum to create a one-particle and a three-particle state.

is the N-particle state with particles located at $x_1, \ldots x_N$ and

$$\langle x_1, \ldots x_N| = \langle 0|[\psi^\dagger(x_N) \cdots \psi^\dagger(x_1)]^\dagger = \langle 0|\psi(x_1) \cdots \psi(x_N) \tag{2.15}$$

is its conjugate "bra" vector. The wavefunction of an N-particle state $|N\rangle$ is given by the overlap of $\langle x_1, \ldots x_N|$ with $|N\rangle$:

$$\psi(x_1, \ldots x_N) = \langle x_1, \ldots x_N|N\rangle = \langle 0|\psi(x_1) \cdots \psi(x_N)|N\rangle. \tag{2.16}$$

So, many-body wavefunctions correspond to matrix elements of the quantum fields. From this link we can see that the exchange symmetry under particle exchange is directly linked to the exchange algebra of the field operators. For bosons and fermions, respectively, we have

$$\langle 0| \cdots \psi(x_r)\psi(x_{r+1}) \cdots |N\rangle = \pm\langle 0| \cdots \psi(x_{r+1})\psi(x_r) \cdots |N\rangle \tag{2.17}$$

(where $+$ refers to bosons, $-$ to fermions), so that

$$\psi(x_r)\psi(x_{r+1}) = \pm\psi(x_{r+1})\psi(x_r). \tag{2.18}$$

From this we see that bosonic operators commute, but fermionic operators must *anticommute*. Thus it is the exchange symmetry of identical quantum particles that dictates the commuting or anticommuting algebra of the associated quantum fields.

Unlike classical fields, quantum fields are in a state of constant fluctuation. This applies both to collective fields, as in the example of the string in Figure 2.1, and to quantum fluids. Just as the commutator between position and momentum gives rise to the uncertainty principle, $[x, p] = i\hbar \longrightarrow \Delta x \Delta p \gtrsim \hbar$, the canonical commutation or anticommutation relations give rise to a similar relation between the amplitude and phase of the quantum field. Under certain conditions the fluctuations of a quantum field can be eliminated, and in these extreme limits the quantum field begins to take on a tangible classical existence. In a Bose superfluid, for example, the quantum field becomes a sharp variable, and we can ascribe a meaning to the expectation of the quantum field,

$$\langle \psi(x) \rangle = \sqrt{\rho_s}e^{i\theta}, \tag{2.19}$$

where ρ_s measures the density of particles in the superfluid condensate. We shall see that there is a parallel relation between the uncertainity in phase and density of quantum fields:

$$\Delta N \Delta \theta \gtrsim 1, \tag{2.20}$$

where θ is the average phase of a condensate and N the number of particles it contains. When N is truly macroscopic, the uncertainty in the phase may be made arbitrarily small, so that in a Bose superfluid the phase becomes sufficiently well defined that it is possible to observe interference phenomena! Similar situations arise inside a laser, where the phase of the electromagnetic field becomes well defined, or in a superconductor, where the phase of the pair electrons in the pair condensate becomes well defined.

In the next two chapters we will return to see how all these features appear systematically in the context of free-field theory. We will begin with collective bosonic fields, which behave as a dense ensemble of coupled harmonic oscillators, and continue in the next chapter to consider conserved particles, where the exchange symmetry of the wavefunction leads to the commutation and anticommutation algebra of Bose and Fermi fields. We shall see how this information enables us to completely solve the properties of a non-interacting Bose or Fermi fluid.

It is the non-commuting properties of quantum fields that generate their intrinsic "graininess." Quantum fields, though nominally continuous degrees of freedom, can always be decomposed in terms of a discrete particular content. The action of a collective field involves the creation of a wavepacket centered at x by both the creation and destruction of quanta; schematically:

$$\phi(x) = \sum_{\mathbf{k}} \left[\begin{array}{c} \text{boson creation} \\ \text{momentum } -\mathbf{k} \end{array} + \begin{array}{c} \text{boson destruction} \\ \text{momentum } \mathbf{k} \end{array} \right] e^{-i\mathbf{k}\cdot\mathbf{x}}. \tag{2.21}$$

Examples include quanta of sound (phonons) and quanta of radiation (photons). In a similar way, the action of a particle-creation operator creates a particle localized at x; schematically:

$$\psi^{\dagger}(x) = \sum_{\mathbf{k}} \left[\begin{array}{c} \text{particle creation} \\ \text{momentum } \mathbf{k} \end{array} \right] e^{i\mathbf{k}\cdot\mathbf{x}}. \tag{2.22}$$

When the underlying particles develop coherence, the quantum field or certain combinations of the quantum fields start to behave as classical collective fields. It is the ability of quantum fields to describe continuous classical behavior *and* discrete particulate behavior in a unified way that makes them so very special.

Example 2.1   By considering the positivity of the quantity $\langle A(\lambda)^{\dagger} A(\lambda) \rangle$, where $\hat{A} = \hat{x} + i\lambda\hat{p}$ and λ is a real number, prove the Heisenberg uncertainty relation $\Delta x \Delta p \geq \frac{\hbar}{2}$.

Solution

Let us expand

$$\langle A^{\dagger} A \rangle = \langle (\hat{x} - i\lambda\hat{p})(\hat{x} + i\lambda\hat{p}) \rangle = \langle \hat{x}^2 \rangle + i\lambda \overbrace{\langle [\hat{x}, \hat{p}] \rangle}^{i\hbar} + \lambda^2 \langle \hat{p}^2 \rangle$$

$$= \langle \hat{x}^2 \rangle - \lambda\hbar + \lambda^2 \langle \hat{p}^2 \rangle > 0. \tag{2.23}$$

Completing the square on this result, we have

$$\langle A^\dagger A \rangle = \langle \hat{p}^2 \rangle \left(\lambda - \frac{\hbar}{2\langle \hat{p}^2 \rangle} \right)^2 + \left(\langle \hat{x}^2 \rangle - \frac{\hbar^2}{4\langle \hat{p}^2 \rangle} \right) > 0. \tag{2.24}$$

For this term to be positive definite, it follows that

$$\langle x^2 \rangle \langle \hat{p}^2 \rangle \geq \left(\frac{\hbar}{2} \right)^2. \tag{2.25}$$

Now this isn't quite what we wanted. However, let us replace $\hat{A} \to \hat{B} = \hat{A} - \langle \hat{A} \rangle$, $\hat{p} \to \Delta \hat{p} = \hat{p} - \langle \hat{p} \rangle$ and $\hat{x} \to \Delta x = \hat{x} - \langle \hat{x} \rangle$. Since the commutator $[\Delta x, \Delta p] = i\hbar$ is unchanged, we can repeat the same arguments for $\langle \hat{B}^\dagger \hat{B} \rangle \geq 0$, with the result that

$$\Delta x \Delta p = \sqrt{\langle \Delta x^2 \rangle \langle \Delta \hat{p}^2 \rangle} \geq \frac{\hbar}{2}, \tag{2.26}$$

proving the Heisenberg uncertainty relation.

Example 2.2
(a) How does the uncertainty principle prevent the collapse of the hydrogen atom?
(b) Is the Heisenberg uncertainty principle enough to explain the stability of matter?

Solution

(a) The stability of a hydrogen atom can be indeed be understood as a qualitative consequence of the Heisenberg uncertainty principle. The energy in the ground state is

$$E \sim -\frac{e^2}{4\pi \epsilon_0} \left\langle \frac{1}{r} \right\rangle + \frac{\langle p^2 \rangle}{2m}. \tag{2.27}$$

In a classical hydrogen atom, as $\langle r \rangle \to 0$ the Coulomb energy is unbounded below and the atom would be unstable. In a quantum atom, we can imagine variationally adjusting a trial ground-state wavefunction, varying its characteristic radius $\langle r \rangle$ and following how the potential and kinetic energy depend on $\langle r \rangle$. The uncertainty principle ensures that

$$\frac{\langle p^2 \rangle}{2m} \sim \frac{\hbar^2}{2m\langle r^2 \rangle}, \tag{2.28}$$

so that

$$E \sim -\frac{e^2}{4\pi \epsilon_0} \left\langle \frac{1}{r} \right\rangle + \frac{\hbar^2}{2m\langle r^2 \rangle}.$$

As the characteristic radius of the wavefunction decreases, the decreasing uncertainty in position leads to an increasing uncertainty in momentum, which in turn causes the kinetic energy to grow faster than the Coulomb energy with the reduction in $\langle r \rangle$. The compromise between the Coulomb and uncertainty energy ultimately stabilizes the atom at a radius $\langle r \rangle \sim a_B$ of the order of the Bohr radius a_B.

(b) No: the uncertainty relation is not sufficient to account for the stability of electronic matter. Although the uncertainty relation stabilizes the hydrogen atom, in multi-electron systems it is the Pauli exclusion principle that stabilizes electronic matter and makes it "hard." Were electrons bosons, they would collapse into the lowest $1s$ state of the atom, giving rise to a much denser and far more bland version of matter. Instead, the exclusion principle prevents multiple occupancy of atomic orbitals, so that, as atoms are brought together, the atomic orbitals overlap; and to avoid double occupancy, the exclusion principle forces electrons into higher-energy orbitals, creating a rapid increase in the energy and creating a strong repulsive interaction that is the origin of the hardness of matter.

2.2 Collective quantum fields

Here, we will begin to familiarize ourselves with quantum fields by developing the field theory of a free bosonic field. It is important to realize that a bosonic quantum field is nothing more than a set of linearly coupled oscillators. So long as the system is linear, the modes of oscillation can always be decomposed into a linear sum of independent normal modes, each one a simple harmonic oscillator. The harmonic oscillater is thus the basic building block for bosonic field theories.

Our basic strategy for quantizing collective bosonic fields thus consists of two basic parts. First, we reduce the Hamiltonian to its normal modes. For translationally invariant systems, this is just a matter of Fourier transforming the field and its conjugate momenta. Second, we quantize the normal-mode Hamiltonian as a sum of independent harmonic oscillators:

$$H(\phi, \pi) \xrightarrow{\text{Fourier transform}} \text{normal coordinates} \xrightarrow{\phi_q \sim (a_q + a^\dagger_{-q})} H = \sum_q \hbar\omega_q(n_q + \tfrac{1}{2}).$$

(2.29)

The first part of this procedure is essentially identical for both quantum and classical oscillators. The second stage is nothing more than the quantization of a single harmonic oscillator. Consider the family of lattices shown in Figure 2.4. We shall start with a single oscillator at one site. We shall then graduate to one- and higher-dimensional chains of oscillators, as shown in Figure 2.4.

2.3 Harmonic oscillator: a zero-dimensional field theory

Although the Schrödinger approach is most widely used in first quantization, it is the Heisenberg approach [5, 11] that opens the door to second quantization. In the Schrödinger approach, one solves the wave equation

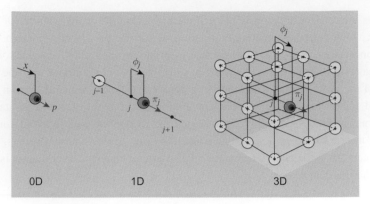

Fig. 2.4   Family of zero-, one- and three-dimensional harmonic crystals.

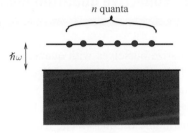

Fig. 2.5   Illustrating the excitation picture for a single harmonic oscillator.

$$\left(\frac{-\hbar^2 \partial_x^2}{2m} + \frac{1}{2}m\omega^2 x^2 \right) \psi_n = E_n \psi_n, \tag{2.30}$$

from which one finds that the energy levels are evenly spaced according to

$$E_n = (n + \frac{1}{2})\hbar\omega, \tag{2.31}$$

where ω is the frequency of the oscillator.

The door to second quantization is opened by reinterpreting these evenly spaced energy levels in terms of quanta, each of energy $\hbar\omega$. The nth excited state corresponds to the addition of n quanta to the ground state (see Figure 2.5). We shall now see how we can put mathematical meat on these words by introducing an operator $a^\dagger$ that creates these quanta, so that the nth excited state is obtained by acting n times on the ground state with the creation operator:

$$|n\rangle = \frac{1}{\sqrt{n!}}(a^\dagger)^n|0\rangle. \tag{2.32}$$

Let us now see how this works. The Hamiltonian for this problem involves conjugate position and momentum operators, as follows:

$$H = \frac{p^2}{2m} + \frac{1}{2}m\omega^2 x^2,$$

$$[x, p] = i\hbar. \tag{2.33}$$

quantum harmonic oscillator

In the ground state, the particle in the harmonic potential undergoes zero-point motion, with uncertainties in position and momentum Δp and Δx which satisfy $\Delta x \Delta p \sim \hbar$. Since the zero-point kinetic and potential energies are equal, $\Delta p^2/2m = m\omega^2 \Delta x^2/2$, so

$$\Delta x = \sqrt{\frac{\hbar}{m\omega}}, \quad \Delta p = \sqrt{m\omega\hbar} \tag{2.34}$$

define the scale of zero-point motion. It is useful to define dimensionless position and momentum variables by factoring out the scale of zero-point motion:

$$\xi = \frac{x}{\Delta x}, \quad p_\xi = \frac{p}{\Delta p}. \tag{2.35}$$

One quickly verifies that $[\xi, p_\xi] = i$ are still canonically conjugate, and that now

$$H = \frac{\hbar\omega}{2}\left[\xi^2 + p_\xi^2\right]. \tag{2.36}$$

Next, introduce the *creation* and *annihilation* operators

$$a^\dagger = \frac{1}{\sqrt{2}}(\xi - ip_\xi) \quad \text{creation operator}$$

$$a = \frac{1}{\sqrt{2}}(\xi + ip_\xi). \quad \text{annihilation operator} \tag{2.37}$$

Since $[a, a^\dagger] = \frac{-i}{2}\left([\xi, p_\xi] - [p_\xi, \xi]\right) = 1$, these operators satisfy the algebra

$$\left.\begin{aligned} [a, a] = [a^\dagger, a^\dagger] &= 0 \\[2mm] [a, a^\dagger] &= 1. \end{aligned}\right\} \quad \text{canonical commutation rules} \tag{2.38}$$

It is this algebra which lies at the heart of bosonic physics, enabling us to interpret the creation and annihilation operators as the objects which add and remove quanta of vibration to and from the system.

To follow the trail further, we rewrite the Hamiltonian in terms of a and $a^\dagger$. Since $\xi = (a + a^\dagger)/\sqrt{2}$ and $p_\xi = (a - a^\dagger)/\sqrt{2}\,i$, the core of the Hamiltonian can be rewritten as

$$\xi^2 + p_\xi^2 = a^\dagger a + aa^\dagger. \tag{2.39}$$

But $aa^\dagger = a^\dagger a + 1$, from the commutation rules, so that

$$H = \hbar\omega\left[a^\dagger a + \frac{1}{2}\right]. \tag{2.40}$$

This has a beautifully simple interpretation. The second term is just the zero-point energy $E_0 = \hbar\omega/2$. The first term contains the *number operator*

$$\hat{n} = a^\dagger a, \qquad \text{number operator} \tag{2.41}$$

which counts the number of vibrational quanta added to the ground state. Each of these quanta carries energy $\hbar\omega$.

To see this, we need to introduce the concept of the vacuum, defined as the unique state such that

$$a|0\rangle = 0. \tag{2.42}$$

From (2.40), this state is clearly an eigenstate of H, with energy $E = \hbar\omega/2$. We now assert that the state

$$|N\rangle = \frac{1}{\lambda_N}(a^\dagger)^N|0\rangle, \tag{2.43}$$

where λ_N is a normalization constant, contains N quanta.

To verify that $\hat{n}$ counts the number of bosons, we use commutation algebra to show that $[\hat{n}, a^\dagger] = a^\dagger$ and $[\hat{n}, a] = -a$, or

$$\hat{n}a^\dagger = a^\dagger(\hat{n} + 1)$$
$$\hat{n}a = a(\hat{n} - 1), \tag{2.44}$$

which means that when $a^\dagger$ or a act on a state, they respectively add or remove one quantum of energy. Suppose that

$$\hat{n}|N\rangle = N|N\rangle \tag{2.45}$$

for some N. Then, from (2.44),

$$\hat{n}\, a^\dagger|N\rangle = a^\dagger(\hat{n} + 1)|N\rangle = (N + 1)\, a^\dagger|N\rangle, \tag{2.46}$$

so that $a^\dagger|N\rangle \equiv |N + 1\rangle$ contains $N + 1$ quanta. Since (2.45) holds for $N = 0$, it holds for all N. To complete the discussion, let us fix λ_N by noting that, from the definition of $|N\rangle$,

$$\langle N - 1|aa^\dagger|N - 1\rangle = \left(\frac{\lambda_N}{\lambda_{N-1}}\right)^2 \langle N|N\rangle = \left(\frac{\lambda_N}{\lambda_{N-1}}\right)^2. \tag{2.47}$$

But since $aa^\dagger = \hat{n} + 1$, $\langle N - 1|aa^\dagger|N - 1\rangle = N\langle N - 1|N - 1\rangle = N$. Comparing these two expressions, it follows that $\lambda_N/\lambda_{N-1} = \sqrt{N}$, and since $\lambda_0 = 1$, $\lambda_N = \sqrt{N!}$.

Summarizing the discussion:

$$H = \hbar\omega(\hat{n} + \tfrac{1}{2})$$

$$\hat{n} = a^\dagger a \qquad\qquad \text{number operator} \tag{2.48}$$

$$|N\rangle = \frac{1}{\sqrt{N!}}(a^\dagger)^N|0\rangle. \qquad\qquad \text{N-boson state}$$

Using these results, we can quickly learn many things about the quantum fields a and $a^\dagger$. Let us look at a few examples. First, we can transform all time dependence from the states to the operators by moving to a Heisenberg representation, writing

$$a(t) = e^{iHt/\hbar} a e^{-iHt/\hbar}. \qquad \text{Heisenberg representation} \qquad (2.49)$$

This transformation preserves the canonical commutation algebra and the form of H. The equation of motion for $a(t)$ is given by

$$\frac{da(t)}{dt} = \frac{i}{\hbar}[H, a(t)] = -i\omega a(t), \qquad (2.50)$$

so that the Heisenberg operators are given by

$$a(t) = e^{-i\omega t} a$$
$$a^\dagger(t) = e^{i\omega t} a^\dagger. \qquad (2.51)$$

Using these results, we can decompose the original momentum and displacement operators as follows:

$$\hat{x}(t) = \Delta x \xi(t) = \frac{\Delta x}{\sqrt{2}}\left(a(t) + a^\dagger(t)\right) = \sqrt{\frac{\hbar}{2m\omega}}\left(ae^{-i\omega t} + a^\dagger e^{i\omega t}\right)$$

$$\hat{p}(t) = \Delta p p_\xi(t) = -i\sqrt{\frac{m\hbar\omega}{2}}\left(ae^{-i\omega t} - a^\dagger e^{i\omega t}\right). \qquad (2.52)$$

Notice how the displacement operator – a priori a continuous variable – has the action of creating and destroying discrete quanta.

We can use this result to compute the correlation functions of the displacement.

Example 2.3 Calculate the autocorrelation function $S(t - t') = \frac{1}{2}\langle 0|\{x(t), x(t')\}|0\rangle$ and the response function $R(t - t') = (i/\hbar)\langle 0|[x(t), x(t')]|0\rangle$ in the ground state of the quantum harmonic oscillator.

Solution

We may expand the correlation function and response function as follows:

$$S(t_1 - t_2) = \frac{1}{2}\langle 0|x(t_1)x(t_2) + x(t_2)x(t_1)|0\rangle$$

$$R(t_1 - t_2) = (i/\hbar)\langle 0|x(t_1)x(t_2) - x(t_2)x(t_1)|0\rangle. \qquad (2.53)$$

But we may expand $x(t)$ as given in (2.52). The only term which survives in the ground state is the term proportional to $aa^\dagger$, so that

$$\langle 0|x(t)x(t')|0\rangle = \frac{\hbar}{2m\omega}\langle 0|aa^\dagger|0\rangle e^{-i\omega(t-t')}. \qquad (2.54)$$

Now, using (2.53) we obtain

$$\frac{1}{2}\langle 0|\{x(t), x(t')\}|0\rangle = \frac{\hbar}{2m\omega}\cos\left[\omega(t - t')\right] \qquad \text{correlation function}$$

$$-i\langle 0|[x(t), x(t')]|0\rangle = \frac{1}{m\omega}\sin\left[\omega(t - t')\right]. \qquad \text{response function}$$

We shall later see that $R(t - t')$ gives the response of the ground state to an applied force $F(t')$, so that at a time t the displacement is given by

$$\langle x(t) \rangle = \int_{-\infty}^{t} R(t - t')F(t')dt'. \tag{2.55}$$

Remarkably, the response function is identical with a classical harmonic oscillator.

Example 2.4 Calculate the number of quanta present in a harmonic oscillator with characteristic frequency ω, at temperature T.

Solution

To calculate the expectation value of any operator at temperature T, we need to consider an ensemble of systems in different quantum states $|\Psi\rangle = \sum_n c_n |n\rangle$. The expectation value of operator $\hat{A}$ in state $|\Psi\rangle$ is then

$$\langle \hat{A} \rangle = \langle \Psi | A | \Psi \rangle = \sum_{m,n} c_m^* c_n \langle m | \hat{A} | n \rangle. \tag{2.56}$$

In a position basis (assuming $\langle x | A | x' \rangle = A(x)\delta(x - x')$ is local), this is

$$\langle \hat{A} \rangle = \sum_{m,n} c_m^* c_n \int dx \psi_m^*(x) A(x) \psi_m(x). \tag{2.57}$$

But now we have to average over the typical state $|\Psi\rangle$ in the ensemble, which gives

$$\overline{\langle \hat{A} \rangle} = \sum_{m,n} \overline{c_m^* c_n} \langle m | \hat{A} | n \rangle = \sum_{m,n} \rho_{mn} \langle m | \hat{A} | n \rangle, \tag{2.58}$$

where $\rho_{mn} = \overline{c_m^* c_n}$ is the *density matrix*. If the ensemble is in equilibrium with an incoherent heat bath at temperature T, quantum statistical mechanics asserts that there are no residual phase correlations between the different energy levels, which acquires a Boltzmann distribution

$$\rho_{mn} = \overline{c_m^* c_n} = p_n \delta_{n,m}, \tag{2.59}$$

where $p_n = e^{-\beta E_n}/Z$ is the Boltzmann distribution, with $\beta = 1/k_B T$, and k_B is Boltzmann's constant. Let us now apply this to our problem, where

$$\hat{A} = \hat{n} = a^\dagger a \tag{2.60}$$

is the number operator. In this case,

$$\langle \hat{n} \rangle = \sum_n (e^{-\beta E_n}/Z)\langle n | \hat{n} | n \rangle = \frac{1}{Z} \sum_n n e^{-\beta E_n}. \tag{2.61}$$

To normalize the distribution, we must have $\sum_n p_n = 1$, so that

$$Z = \sum_n e^{-\beta E_n}. \tag{2.62}$$

Finally, since $E_n = \hbar\omega(n + \frac{1}{2})$,

$$\langle \hat{n} \rangle = \frac{\sum_n e^{-\beta\hbar\omega(n+\frac{1}{2})} n}{\sum_n e^{-\beta\hbar\omega(n+\frac{1}{2})}} = \frac{\sum_n e^{-\lambda n} n}{\sum_n e^{-\lambda n}} \qquad (\lambda = \beta\hbar\omega). \qquad (2.63)$$

The sum in the denominator is a geometric series,

$$\sum_n e^{-\lambda n} = \frac{1}{1 - e^{-\lambda}}, \qquad (2.64)$$

and the numerator is given by

$$\sum_n e^{-\lambda n} n = -\frac{\partial}{\partial\lambda} \sum_n e^{-\lambda n} = \frac{e^{-\lambda}}{(1 - e^{-\lambda})^2}, \qquad (2.65)$$

so that

$$\langle \hat{n} \rangle = \frac{1}{e^\lambda - 1} = \frac{1}{e^{\beta\hbar\omega} - 1}, \qquad (2.66)$$

which is the famous Bose–Einstein distribution function.

2.4 Collective modes: phonons

We now extend the discussion of the previous section from zero to higher dimensions. Let us go back to the lattice shown in Figure 2.4 . To simplify our discussion, let imagine that at each site there is a single elastic degree of freedom. For example, consider the longitudinal displacement of an atom along a one-dimensional chain that runs in the x-direction. For the jth atom,

$$x_j = x_j^0 + \phi_j. \qquad (2.67)$$

If π_j is the conjugate momentum to x_j, then the two variables must satisfy canonical commutation relations

$$[\phi_i, \pi_j] = i\hbar\delta_{ij}. \qquad (2.68)$$

Notice how variables at different sites are fully independent. We'll imagine that our one-dimensional lattice has $\mathcal{N}_s$ sites, and we shall make life easier by working with periodic boundary conditions, so that $\phi_{j+\mathcal{N}_s} \equiv \phi_j$ and $\pi_j \equiv \pi_{j+\mathcal{N}_s}$. Suppose nearest neighbors are connected by a "spring," in which case the total total energy is then a sum of kinetic and potential energy:

$$\hat{H} = \sum_{j=1,\mathcal{N}_s} \left[\frac{\pi_j^2}{2m} + \frac{m\omega^2}{2}(\phi_j - \phi_{j+1})^2 \right], \qquad (2.69)$$

where m is the mass of an atom.

Now the great simplifying feature of this model is that that it possesses *translational symmetry*, so that under the translation

$$\pi_j \to \pi_{j+1}, \qquad \phi_j \to \phi_{j+1} \tag{2.70}$$

the Hamiltonian and commutation relations remain unchanged. If we shrink the size of the lattice to zero, this symmetry will become a continuous translational symmetry. The generator of these translations is the *crystal momentum* operator, which must therefore commute with the Hamiltonian. Because of this symmetry, it makes sense to transform to operators that are diagonal in momentum space, so we'll Fourier transform all fields as follows:

$$\left. \begin{array}{l} \phi_j = \frac{1}{\sqrt{\mathcal{N}_s}} \sum_q e^{iqR_j} \phi_q \\[2mm] \pi_j = \frac{1}{\sqrt{\mathcal{N}_s}} \sum_q e^{iqR_j} \pi_q \end{array} \right\} \qquad R_j = ja. \tag{2.71}$$

The periodic boundary conditions $\phi_j = \phi_{j+\mathcal{N}_s}$, $\pi_j = \pi_{j+\mathcal{N}_s}$ mean that the values of q entering in this sum must satisfy $qL = 2\pi n$, where $L = \mathcal{N}_s a$ is the length of the chain and n is an integer. Thus the wavevectors

$$q \equiv q_n = \frac{2\pi}{L} n \qquad (n \in [1, \mathcal{N}_s]) \tag{2.72}$$

must be discrete multiples of $\frac{2\pi}{L}$. The range of q is thus $q \in [0, \frac{2\pi}{a}]$. As in any periodic structure, the crystal momentum is only defined modulo a reciprocal lattice vector, which in this case is $2\pi/a$, so that $q + \frac{2\pi}{a} \equiv q$ (since $\exp[i(q + \frac{2\pi}{a})R_j] = \exp[i(qR_j + 2\pi j)] = \exp[iqR_j]$). This means that the integer n in q_n is only defined modulo the number of sites $\mathcal{N}_s$.

The functions $\frac{1}{\sqrt{\mathcal{N}_s}} e^{iqR_j} \equiv \langle j|q \rangle$ form a complete orthogonal basis, and we may use Fourier's theorem to show that

$$\langle q_r | q_s \rangle \equiv \sum_{j=1}^{\mathcal{N}_s} \langle q_r | j \rangle \langle j | q_s \rangle \equiv \frac{1}{\mathcal{N}_s} \sum_{j=1}^{\mathcal{N}_s} e^{i(q_s - q_r)R_j} = \delta_{rs} \qquad \text{orthogonality} \tag{2.73}$$

vanishes unless $q_r = q_s$ (see Exercise 2.2). This result is immensely useful, and we shall use it time and time again. Applying the orthogonality relation to (2.71), we can check that the inverse transformations are

$$\phi_q = \frac{1}{\sqrt{\mathcal{N}_s}} \sum_j e^{-iqR_j} \phi_j$$

$$\pi_q = \frac{1}{\sqrt{\mathcal{N}_s}} \sum_j e^{-iqR_j} \pi_j. \tag{2.74}$$

Notice that, since ϕ_j and π_j are Hermitian operators, it follows that $\phi_q^\dagger = \phi_{-q}$ and $\pi_q^\dagger = \pi_{-q}$. Using orthogonality, we can verify that the transformed commutation relations are

$$[\phi_q, \pi_{-q'}] = \frac{1}{\mathcal{N}_s} \sum_{i,j} e^{-i(qR_i - q'R_j)} \overbrace{[\phi_i, \pi_j]}^{i\hbar\delta_{ij}}$$

$$= \frac{i\hbar}{\mathcal{N}_s} \sum_j e^{-i(q-q')R_j} = i\hbar \delta_{qq'}. \tag{2.75}$$

We shall now see that π_q and ϕ_q are quantized versions of *normal coordinates* which bring the hamiltonian back into the standard harmonic oscillator form. To check that the Hamiltonian is truly diagonal in these variables, we

1. expand ϕ_j and π_j in terms of their Fourier components
2. regroup the sums so that the summation over momenta is on the outside
3. eliminate all but one summation over momentum by carrying out the internal sum over site variables. This will involve terms like $\mathcal{N}_s^{-1} \sum_j e^{i(q+q')R_j} = \delta_{q+q'}$, which constrains $q' = -q$ and eliminates the sum over q'.

With a bit of practice, these steps can be carried out very quickly. In transforming the potential energy, it is useful to rewrite it in the form

$$V = \frac{m\omega^2}{2} \sum_j \phi_j (2\phi_j - \phi_{j+1} - \phi_{j-1}).$$

(2.76)

The term in brackets can be Fourier transformed as follows:

$$\omega^2 (2\phi_j - \phi_{j+1} - \phi_{j-1}) = \frac{1}{\sqrt{\mathcal{N}_s}} \sum_q \overbrace{\omega^2 [2 - e^{iqa} - e^{-iqa}]}^{4\omega^2 \sin^2(qa/2) \equiv \omega_q^2} \times \phi_q\, e^{iqR_j}$$

$$\equiv \frac{1}{\sqrt{\mathcal{N}_s}} \sum_q \omega_q^2\, \phi_q\, e^{iqR_j},$$

(2.77)

where we have defined $\omega_q = 2\omega |\sin(qa/2)|$. Inserting this into (2.76), we obtain

$$V = \frac{m}{2} \sum_{q,q'} \omega_q^2\, \phi_{-q'}\phi_q \overbrace{\mathcal{N}_s^{-1} \sum_j e^{i(q-q')R_j}}^{\delta_{q,q'}}$$

$$= \sum_q \frac{m\omega_q^2}{2} \phi_{-q}\phi_q.$$

(2.78)

Carrying out the same procedure on the kinetic energy, we obtain

$$H = \sum_q \left(\frac{1}{2m} \pi_q \pi_{-q} + \frac{m\omega_q^2}{2} \phi_q \phi_{-q} \right),$$

(2.79)

which expresses the Hamiltonian in terms of normal coordinates ϕ_q and π_q. So far, all of the transformations have preserved the ordering of the operators, so it is no surprise that the quantum and classical expressions for the Hamiltonian in terms of normal coordinates are formally identical.

It is useful to note that at $q = 0$, $\omega_q = 0$, so that there is no contribution to the potential energy from the $q = 0$ mode. This mode corresponds to the center-of-mass motion of the entire system, and it is useful to separate it from the oscillatory modes, writing

$$H = \overbrace{\frac{1}{2m}\pi_0^2}^{H_{CM}} + \sum_{q \neq 0} \left(\frac{1}{2m}\pi_q \pi_{-q} + \frac{m\omega_q^2}{2}\phi_q\phi_{-q} \right).$$

(2.80)

The first term is just the center-of-mass energy.

The next step merely repeats the procedure carried out for the single harmonic oscillator. For $q \neq 0$ we define a set of conjugate creation and annihilation operators

$$\left.\begin{array}{l} a_q = \sqrt{\frac{m\omega_q}{2\hbar}}\left(\phi_q + \frac{i}{m\omega_q}\pi_q\right) \\[2mm] a_q^\dagger = \sqrt{\frac{m\omega_q}{2\hbar}}\left(\phi_{-q} - \frac{i}{m\omega_q}\pi_{-q}\right) \end{array}\right\} \quad [a_q, a_{q'}^\dagger] = \frac{-i}{2\hbar}\Big[[\phi_q, \pi_{-q'}] - [\pi_q, \phi_{-q'}]\Big] = \delta_{q,q'}.$$

$$(2.81)$$

The second expression for $a_q^\dagger$ is obtained by taking the complex conjugate of a_q, remembering that $\phi_q^\dagger = \phi_{-q}$ and $\pi_q^\dagger = \pi_{-q}$, since the underlying fields are real.

The inversion of these expressions is

$$\pi_q = -i\sqrt{\frac{m\omega_q \hbar}{2}}\left(a_q - a_{-q}^\dagger\right)$$
$$\phi_a = \sqrt{\frac{\hbar}{2m\omega_q}}\left(a_q + a_{-q}^\dagger\right).$$

$$(2.82)$$

Notice how the Fourier component of the field at wavevector q either destroys a phonon of momentum q or creates a phonon of momentum $-q$. Both processes reduce the total momentum by q.

It follows that

$$\pi_q \pi_{-q} = \frac{m\omega_q \hbar}{2}\left(a_{-q}^\dagger a_{-q} + a_q a_q^\dagger - a_{-q}^\dagger a_q^\dagger - a_q a_{-q}\right)$$

$$\phi_q \phi_{-q} = \frac{\hbar}{2m\omega_q}\left(a_{-q}^\dagger a_{-q} + a_q a_q^\dagger + a_{-q}^\dagger a_q^\dagger + a_q a_{-q}\right).$$

$$(2.83)$$

Adding the two terms inside the Hamiltonian (2.80) then gives

$$H = H_{CM} + \frac{1}{2}\sum_{q\neq 0} \hbar\omega_q\left(a_q a_q^\dagger + a_{-q}^\dagger a_{-q}\right).$$

$$(2.84)$$

Now we can replace $-q \to q$ inside the summation and use the commutation relations $a_q a_q^\dagger = a_q^\dagger a_q + 1$, so that

$$H = H_{CM} + \sum_{q\neq 0} \hbar\omega_q\left(a_q^\dagger a_q + \frac{1}{2}\right).$$

$$(2.85)$$

Since each set of a_q and $a_q^\dagger$ obey canonical commutation relations, we can immediately identify $n_q = a_q^\dagger a_q$ as the number operator for quanta in the qth momentum state. Remarkably, the system of coupled oscillators can be reduced to a sum of independent harmonic oscillators with characteristic frequency ω_q, energy $\hbar\omega_q$ and momentum q. Each normal mode of the original classical system corresponds to a particular phonon excitation. From now on we will drop the center-of-mass term H_{CM} with the understanding that summations over q do not include $q = 0$.

We can immediately generalize all of our results from a single harmonic oscillator. For example, the general state of the system will now be an eigenstate of the phonon occupancies:

$$|\Psi\rangle = |n_{q_1}, n_{q_2}, \ldots n_{q_N}\rangle = \prod_{\otimes} |n_{q_i}\rangle = \left[\prod_i \frac{(a_{q_i}^\dagger)^{n_{q_i}}}{\sqrt{n_{q_i}!}}\right]|0\rangle,$$

$$(2.86)$$

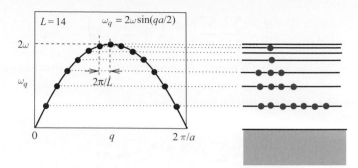

Illustrating the excitation picture for a chain of coupled oscillators, length $L = 14$.

Fig. 2.6

where the vacuum is the unique state that is annihilated by all of the a_q. In this state, the occupation numbers n_q are diagonal, so this is an energy eigenstate with energy

$$E = E_0 + \sum_q n_q \hbar \omega_q, \tag{2.87}$$

where $E_0 = \frac{1}{2} \sum_q \hbar \omega_q$ is the zero-point energy (see Figure 2.6).

Remarks

- The quantized displacements of a crystal are called *phonons*. Quantized fluctuations of magnetization in a magnet are *magnons*.
- We can easily transform to a Heisenberg representation, whereupon $a_q(t) = a_q e^{-i\omega_q t}$.
- We can expand the local field entirely in terms of phonons. Using (2.82), we obtain

$$\phi_j(t) = \frac{1}{\sqrt{N_s}} \sum_q \phi_q e^{iqR_j}$$

$$= \phi_{CM}(t) + \frac{1}{\sqrt{N_s}} \sum_{q \neq 0} \sqrt{\frac{\hbar}{2m\omega_q}} \left[a_q(t) + a^\dagger_{-q}(t) \right] e^{iqR_j}, \tag{2.88}$$

where $\phi_{CM} = \frac{1}{N_s} \sum_j \phi_j$ is the center-of-mass displacement.

- The transverse displacements of the atoms can be readily included by simply upgrading the displacement and momentum ϕ_j and π_j to vectors. For springs, the energies associated with transverse and longitudinal displacements are not the same because the stiffness associated with transverse displacements depends on the tension. Nevertheless, the Hamiltonian has an identical form for the one longitudinal and two transverse modes, provided one inserts a different stiffness for the transverse modes. The initial Hamiltonian is then simply a sum over three degenerate polarizations $\lambda \in [1, 3]$:

$$\hat{H} = \sum_{\lambda=1,3} \sum_{j=1, N_s} \left[\frac{\pi^2_{j\lambda}}{2m} + \frac{m\omega^2_\lambda}{2}(\phi_{j\lambda} - \phi_{j+1\lambda})^2 \right], \tag{2.89}$$

where $\omega^2_1 = \omega^2$ for the longitudinal mode and $\omega^2_{2,3} = T/a$, where T is the tension in the spring, for the two transverse modes. By applying the same procedure to all three modes, the final Hamiltonian then becomes

$$H = \sum_{\lambda=1,3} \sum_q \hbar\omega_{q\lambda} \left(a_{q\lambda}^\dagger a_{q\lambda} + \frac{1}{2} \right),$$

where $\omega_{q\lambda} = 2\omega_\lambda \sin(qa/2)$. Of course, in more realistic crystal structures the energies of the three modes will no longer be degenerate.

- We can generalize all of this discussion to a two- or three-dimensional square lattice by noting that the orthogonality relation becomes

$$\mathcal{N}_s^{-1} \sum_j e^{-i(\mathbf{q}-\mathbf{q}')\cdot\mathbf{R}_j} = \delta_{\mathbf{q}-\mathbf{q}'}, \tag{2.90}$$

where now

$$\mathbf{q} = \frac{2\pi}{L}(i_i, i_2, \ldots i_D) \tag{2.91}$$

and $\mathbf{R}_j$ is a site on the lattice. The general form for the potential energy is slightly more complicated, but one can still cast the final Hamiltonian in terms of a sum over longitudinal and transverse modes.

- The zero-point energy $E_0 = \frac{1}{2}\sum_q \hbar\omega_q$ is very important in ^{3}He and ^{4}He crystals, where the lightness of the atoms gives rise to such large phonon frequencies that the crystalline phase is unstable at ambient pressure, melting under the influence of quantum zero-point motion. The resulting *quantum fluids* exhibit the remarkable property of superfluidity.

2.5 The thermodynamic limit: $L \rightarrow \infty$

In the previous section, we examined a system of coupled oscillators on a finite lattice. By restricting a system to a finite lattice, we impose a restriction on the *maximium* wavelength, and hence the excitation spectrum. This is known as an "infrared cut-off". When we take $L \rightarrow \infty$, the allowed momentum states become closer and closer together, and we now have a continuum in momentum space.

What happens to the various momentum summations in the thermodynamic limit, $L \rightarrow \infty$? When the allowed momenta become arbitrarily close together, the discrete summations over momentum must be replaced by continuous integrals. For each dimension, the increment in momentum appearing inside the discrete summations is

$$\Delta q = \frac{2\pi}{L}, \tag{2.92}$$

so that $L\frac{\Delta q}{2\pi} = 1$ (see Figure 2.7). Thus in one dimension, the summation over the discrete values of q can be formally rewritten as

$$\sum_{q_j} \{\ldots\} = L \sum_{q_j} \frac{\Delta q}{2\pi} \{\ldots\}, \tag{2.93}$$

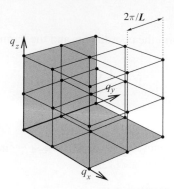

Fig. 2.7

Illustrating the grid of allowed momenta for a three-dimensional crystal of dimensions L^3. In the limit $L \to \infty$, the grid becomes a continuum, with $(L/2\pi)^3$ points per unit volume of momentum space.

where $q_j = 2\pi \frac{j}{L}$ and $j \in [1, \mathcal{N}_s]$. When we take $L \to \infty$, q becomes a continuous variable $q \in [0, 2\pi/a]$, where $a = L/\mathcal{N}_s$ is the lattice spacing, so that the summation can now be replaced by a continuous integral:

$$\sum_q \{\ldots\} \to L \int_0^{2\pi/a} \frac{dq}{2\pi} \{\ldots\}. \qquad (2.94)$$

Similarly, in D dimensions, we can regard the D-dimensional sum over momentum as a sum over tiny hypercubes, each of volume

$$(\Delta q)^D = \frac{(2\pi)^D}{L^D}, \qquad (2.95)$$

so that $L^D \frac{(\Delta q)^D}{(2\pi)^D} = 1$ and

$$\sum_q \{\ldots\} = L^D \sum_q \frac{(\Delta q)^D}{(2\pi)^D} \{\ldots\} \to L^D \int_{0<q_i<2\pi/a} \frac{d^D q}{(2\pi)^D} \{\ldots\}, \qquad (2.96)$$

where the integral is over a hypercube in momentum space, with sides of length $2\pi/a$.

Example 2.5   Re-express the Hamiltonian $\hat{H}$ of a simplified three-dimensional harmonic crystal in terms of phonon number operators and calculate the zero-point energy, where

$$H = \sum_j \frac{\pi_j^2}{2m} + \sum_{j, \mathbf{a} = (\hat{x}, \hat{y}, \hat{z})} \frac{m\omega_0^2}{2} (\Phi_j - \Phi_{j+\mathbf{a}})^2, \qquad (2.97)$$

where $\phi_j \equiv \phi(x_j)$ and $\pi_j \equiv \pi(x_j)$ denote canonically conjugate (scalar) displacement and momenta at site j, and $\hat{a} = (\hat{x}, \hat{y}, \hat{z})$ denotes the unit vector separating nearest-neighbor atoms.

Solution

First we must Fourier transform the coordinates and the harmonic potential. The potential can be rewritten as

$$\hat{V} = \frac{1}{2} \sum_{i,j} V_{i-j} \phi_i \phi_j, \tag{2.98}$$

where

$$V_{\mathbf{R}} = m\omega_0^2 \sum_{\mathbf{a}=(\hat{x},\hat{y},\hat{z})} (2\delta_{\mathbf{R}} - \delta_{\mathbf{R}-\mathbf{a}} - \delta_{\mathbf{R}+\mathbf{a}}). \tag{2.99}$$

The Fourier transform of this expression is

$$V_{\mathbf{q}} = \sum_{\mathbf{R}} V_{\mathbf{R}} e^{-i\mathbf{q}\cdot\mathbf{R}}$$

$$= m\omega_0^2 \sum_{\mathbf{a}=(\hat{x},\hat{y},\hat{z})} (2 - e^{-i\mathbf{q}\cdot\mathbf{a}} - e^{i\mathbf{q}\cdot\mathbf{a}})$$

$$= 2m\omega_0^2 \sum_{l=x,y,z} [1 - \cos(q_l a)], \tag{2.100}$$

so that, writing $V_{\mathbf{q}} = m(\omega_{\mathbf{q}})^2$, it follows that the normal-mode frequencies are given by

$$\omega_{\mathbf{q}} = 2\omega_0 [\sin^2(q_x a/2) + \sin^2(q_y a/2) + \sin^2(q_z a/2)]^{\frac{1}{2}}. \tag{2.101}$$

Fourier transforming the fields

$$\phi_j = \frac{1}{\sqrt{\mathcal{N}_s}} \sum_{\mathbf{q}} \phi_{\mathbf{q}} e^{i\mathbf{q}\cdot\mathbf{x}}$$

$$\pi_j = \frac{1}{\sqrt{\mathcal{N}_s}} \sum_{\mathbf{q}} \pi_{\mathbf{q}} e^{i\mathbf{q}\cdot\mathbf{x}}, \tag{2.102}$$

where $\mathbf{q} = \frac{2\pi}{L}(i,j,k)$ are the discrete momenta of a cubic crystal of volume L^3 with periodic boundary conditions, we find

$$H = \sum_{\mathbf{q}} \left[\frac{\pi_{\mathbf{q}} \pi_{-\mathbf{q}}}{2m} + \frac{m\omega_{\mathbf{q}}^2}{2} \phi_{\mathbf{q}} \phi_{-\mathbf{q}} \right]. \tag{2.103}$$

Defining the creation and annihilation operators

$$b_{\mathbf{q}} = \sqrt{\frac{m\omega_{\mathbf{q}}}{2\hbar}} \left(\phi_{\mathbf{q}} + \frac{i}{m\omega_{\mathbf{q}}} \pi_{\mathbf{q}} \right), \qquad b_{\mathbf{q}}^\dagger = \sqrt{\frac{m\omega_{\mathbf{q}}}{2\hbar}} \left(\phi_{-\mathbf{q}} - \frac{i}{m\omega_{\mathbf{q}}} \pi_{-\mathbf{q}} \right), \tag{2.104}$$

we reduce the Hamiltonian to its standard form,

$$H = \sum_{\mathbf{q}} \hbar\omega_{\mathbf{q}} \left(\hat{n}_{\mathbf{q}} + \frac{1}{2} \right), \tag{2.105}$$

where $\hat{n}_{\mathbf{q}} = b_{\mathbf{q}}^\dagger b_{\mathbf{q}}$ is the phonon number operator.

In the ground state, $n_{\mathbf{q}} = 0$, so that the zero-point energy is

$$E_0 = \sum_{\mathbf{q}} \frac{\hbar \omega_{\mathbf{q}}}{2} \longrightarrow V \int \frac{d^3 q}{(2\pi)^3} \frac{\hbar \omega_{\mathbf{q}}}{2}, \tag{2.106}$$

where $V = L^3$. Substituting for $\omega_{\mathbf{q}}$, we obtain

$$E_0 = V \prod_{l=1,3} \int_0^{2\pi/a} \frac{dq_l}{2\pi} \hbar \omega_0 \sqrt{\sum_{l=1,3} \sin^2(q_l a/2)}$$

$$= \mathcal{N}_s \hbar \omega_0 I_3, \tag{2.107}$$

where

$$I_3 = \int_{0 < u_1, u_2, u_3 < \pi} \frac{d^3 u}{\pi^3} \sqrt{\sum_{l=1,3} \sin^2(u_l)} = 1.19 \tag{2.108}$$

and $\mathcal{N}_s$ is the number of sites.

Remarks

- The zero-point energy per unit cell of the crystal is $\hbar \omega_0 I_3$, a finite number.
- Were we to take the *continuum limit*, taking the lattice separation to zero, the zero-point energy would diverge, due to the profusion of ultraviolet modes.

2.6 The continuum limit: $a \to 0$

In contrast to the thermodynamic limit, when we take the continuum limit we remove the discrete character of the problem, allowing fluctuations of arbitrarily small wavelength and hence arbitrarily large energy. For a discrete system with periodic boundary conditions, the momentum in any one direction cannot exceed $2\pi/a$. By taking a to zero, we send the ultraviolet cut-off to infinity.

In the continuum limit, it is natural to normalize our plane-wave basis per unit volume, writing

$$\langle \mathbf{x} | \mathbf{q} \rangle \longrightarrow e^{i \mathbf{q} \cdot \mathbf{x}}. \tag{2.109}$$

The orthogonality condition is now

$$\langle \mathbf{q}' | \mathbf{q} \rangle = \int d^D x \, e^{i(\mathbf{q} - \mathbf{q}') \cdot \mathbf{x}} = L^D \delta_{\mathbf{q}, \mathbf{q}'}, \tag{2.110}$$

where $\delta_{\mathbf{q}, \mathbf{q}'}$ is the discrete delta function on the grid of allowed wavevectors. If we now also take the thermodynamic limit, the discrete delta function is replaced by the D-dimensional Dirac delta function as follows:

$$L^D \delta_{\mathbf{q}, \mathbf{q}'} = (2\pi)^D \frac{\delta_{\mathbf{q}, \mathbf{q}'}}{(\Delta q)^D} \longrightarrow (2\pi)^D \delta^D(\mathbf{q} - \mathbf{q}'), \tag{2.111}$$

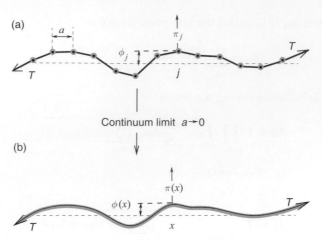

Fig. 2.8 Illustrating (a) a discrete and (b) a continuous string. By taking the length between units in the string to zero, and maintaining the density per unit length and the tension, we arrive at the continuum limit.

so the orthogonality becomes

$$\int d^D x\, e^{i(\mathbf{q}-\mathbf{q}')\cdot\mathbf{x}} = (2\pi)^D \delta^D(\mathbf{q} - \mathbf{q}'). \tag{2.112}$$

As a simple example, we shall consider a one-dimensional string. The important lesson that we shall learn is that both the discrete model and the continuum model have the same long-wavelength physics. Their behavior will only differ at very short distances, at high frequencies and over short times. This is a very simple example of the concept of renormalization. Provided we are interested in low-energy properties, the details of the string at short – distances – whether it is discrete or continuous don't matter.

In many respects, the continuum model is more satisfying and elegant. We shall see, however, that we always have to be careful in going to the continuum limit because this introduces quantum fluctuations on arbitrarily short length scales. These fluctuations don't affect the low-energy excitations, but they do mean that the zero-point fluctuations of the field become arbitrarily large.

Let us start out with a discrete string, as shown in Figure 2.8. For small displacements, the Hamiltonian for this discrete string is identical to that of the previous section, as we can see by the following argument. If a string is made up of point particles of mass m separated by a distance a, with a tensile force T acting between them, then, for small transverse displacements ϕ_j, the link between the jth and $(j + 1)$th particle is expanded by an amount $\Delta s_j = (\phi_j - \phi_{j+1})^2/2a$, raising the potential energy by an amount $T\Delta s_j$. The Hamiltonian is then

$$\hat{H} = \sum_{j=1,\mathcal{N}_s} \left[\frac{\pi_j^2}{2m} + \frac{T}{2a}(\phi_j - \phi_{j+1})^2 \right], \tag{2.113}$$

which reverts to (2.69) with the replacement $T/a \to m\omega^2$.

To take the continuum limit, we let $a \rightarrow 0$, preserving $\rho = m/a$. In this limit, we may replace

$$a \sum_j \rightarrow \int dx$$

$$\frac{(\phi_j - \phi_{j+1})^2}{a^2} \rightarrow (\nabla_x \phi(x))^2. \tag{2.114}$$

Making the replacement

$$\pi_j/a \rightarrow \pi(x_j), \tag{2.115}$$

we obtain

$$H = \int dx \left[\frac{T}{2} (\nabla_x \phi)^2 + \frac{1}{2\rho} \pi(x)^2 \right]. \tag{2.116}$$

On the discrete lattice we have the commutation relations

$$[\phi(x_i), \pi(x_j)] = i\hbar \tilde{\delta}(x_i - x_j), \tag{2.117}$$

where $\tilde{\delta}(x_i - x_j) = a^{-1} \delta_{ij}$. In the limit $a \rightarrow 0$, $\tilde{\delta}(x_i - x_j)$ behaves as a Dirac delta function, so

$$[\phi(x), \pi(y)] = i\hbar \delta(x - y). \tag{2.118}$$

Notice that while we have taken $a \rightarrow 0$ to obtain the continuum limit, in principle we can still retain the finite length L of the string. If we preserve the same periodic boundary conditions $\phi(x) = \phi(x + L)$, the momenta $q = \frac{2\pi}{L} n$ will still be discrete, despite the continuous position variables.

We shall, however, skip this step and go directly to the thermodynamic limit. Transforming to momentum space, we obtain

$$\phi_q = \int dx \phi(x) e^{-iqx}$$

$$\pi_q = \int dx \pi(x) e^{-iqx}, \tag{2.119}$$

from which we immediately derive the canonical commutation relation

$$[\phi_q, \pi_{-q'}] = \int dx dx' e^{-i(qx - q'x')} \overbrace{[\phi(x), \pi(x')]}^{i\hbar \delta(x-x')}$$

$$= i\hbar \overbrace{\int dx e^{-i(q - q')x}}^{\langle q | q' \rangle = 2\pi \delta(q - q')} = i\hbar \times 2\pi \delta(q - q'). \tag{2.120}$$

Now it is just a question of repeating the steps of the previous section, but for the continuous fields ϕ_q and π_q. Using the inverse Fourier transforms,

$$\phi(x) = \int \frac{dq}{2\pi} \phi_q e^{iqx}$$

$$\pi(x) = \int \frac{dq}{2\pi} \pi_q e^{iqx}. \tag{2.121}$$

When we transform the Hamiltonian, we obtain

$$H = \int \frac{dq}{2\pi} \left[\frac{\pi_q \pi_{-q}}{2\rho} + \frac{Tq^2}{2} \phi_q \phi_{-q} \right],$$

(2.122)

where the main effect is the replacement of the gradient by its Fourier transform, $|\nabla_x \phi|^2 \rightarrow q^2 |\phi_q|^2$. We can write this in a form reminiscent of the harmonic oscillator:

$$H = \int \frac{dq}{2\pi} \left[\frac{\pi_q \pi_{-q}}{2\rho} + \frac{\rho \omega_q^2}{2} \phi_q \phi_{-q} \right],$$

(2.123)

where now $\omega_q = c|q|$, and $c = \sqrt{T/\rho}$ is the velocity of sound.

Now the only problem with the above procedure is that, by going to the continuum, the spectrum $\omega_q = c|q|$ stretches to infinity, with an unbounded number of high-frequency or ultraviolet modes. To carry out actual calculations, we still need some kind of momentum cut-off. A crude way would be simply to cut off the momentum integrals. A slightly more elegant way is to introduce a small exponential convergence factor into the Fourier transform, *defining* the real-space fields by

$$\phi(x) = \int \frac{dq}{2\pi} \phi_q e^{iqx} e^{-\epsilon|q|/2}, \qquad \pi(x) = \int \frac{dq}{2\pi} \pi_q e^{iqx} e^{-\epsilon|q|/2}.$$

(2.124)

If we repeat the calculation of the commutation relation, we now find

$$[\phi(x), \pi(y)] = \int \frac{dq dq'}{(2\pi)^2} e^{i(qx - q'x')} \overbrace{[\phi_q, \pi_{-q'}]}^{2\pi i\hbar\delta(q-q')} e^{-\frac{\epsilon}{2}(|q|+|q'|)}$$

$$= i\hbar \int \frac{dq}{(2\pi)} e^{iq(x-x') - |q|\epsilon} = i\hbar \times \left[\int_0^\infty e^{-(\epsilon - i(x-x'))q} + \int_{-\infty}^0 e^{(\epsilon + i(x-x'))q} \right]$$

$$= \frac{i\hbar}{2\pi} \left[\frac{1}{\epsilon - i(x - x')} + \frac{1}{\epsilon + i(x - x')} \right] = i\hbar \times \frac{1}{\pi} \overbrace{\left(\frac{\epsilon}{\epsilon^2 + (x - x')^2} \right)}^{\delta_\epsilon(x-x')},$$

(2.125)

showing that the removal of the ultraviolet modes smears the delta function into a Lorentzian of width ϵ.

The regulated Hamiltonian now takes the form

$$H = \int \frac{dq}{2\pi} \left[\frac{\pi_q \pi_{-q}}{2\rho} + \frac{\rho \omega_q^2}{2} \phi_{\mathbf{q}} \phi_{-\mathbf{q}} \right] e^{-\epsilon|q|}.$$

(2.126)

Notice how this has almost exactly the same form as the the discrete lattice, but now the high-momentum modes are cut off by the exponential factor, rather than the finite size of the Brillouin zone.

Defining the creation and annihilation operators by the relations

$$\phi_q = \sqrt{\frac{\hbar}{2\rho\omega_q}} [a_q + a_{-q}^\dagger]$$

$$\pi_q = -i\sqrt{\frac{\hbar\rho\omega_q}{2}} [a_q - a_{-q}^\dagger],$$

(2.127)

we find that the creation and annihilation operators satisfy

$$[a_q, a^\dagger_{q'}] = 2\pi \delta(q - q'). \qquad (2.128)$$

We may now rewrite the Hamiltonian as

$$H = \int_{-\infty}^{\infty} \frac{dq}{2\pi} \frac{\hbar\omega_\mathbf{q}}{2} \left(a^\dagger_q a_q + a_{-q} a^\dagger_{-q} \right) e^{-\epsilon|q|}. \qquad (2.129)$$

If we reorder the boson operators, we obtain

$$H = \int_{\infty}^{\infty} \frac{dq}{2\pi} \hbar\omega_\mathbf{q} \left(a^\dagger_q a_q + \overbrace{2\pi \delta(0)}^{L} \frac{1}{2} \right) e^{-\epsilon|q|} \qquad (2.130)$$

The first term corresponds to the excitations of the string, and we recognize the last term as the zero-point energy of the string. Had we started out on a finite but long string, the term $2\pi \delta(0)$ would be replaced by L, which is merely a statement that the zero-point energy scales with the length:

$$E_{zp} = L \int_0^{\infty} \frac{dq}{2\pi} \hbar c |q| e^{-\epsilon|q|} = \frac{L\hbar c}{2\pi \epsilon^2}. \qquad (2.131)$$

Once we remove the momentum cut-off, the momentum sum is unbounded and the zero-point energy per unit length becomes infinite in the continuum limit. It often proves convenient to remove this nasty infinity by introducing the concept of *normal ordering*. If we take any operator A, then we denote its normal-ordered counterpart by the symbol $: A :$. The operator $: A :$ is the same as A, except that all the creation operators have been ordered to the left of all of the annihilation operators. All commutators associated with the ordering are neglected. The normal-ordered Hamiltonian

$$: H := \int_{-\infty}^{\infty} \frac{dq}{2\pi} \hbar\omega_q a^\dagger_q a_q \qquad (\omega_q = c|q|) \qquad (2.132)$$

measures the excitation energy above the ground state.

Finally, let us look at the displacement of the string. The fields in coordinate space are given by

$$\phi(x, t) = \int \frac{dq}{2\pi} \sqrt{\frac{\hbar}{2\rho\omega_q}} \left[a_q(t) + a^\dagger_{-q}(t) \right] e^{iqx} e^{-\epsilon|q|/2}, \qquad (2.133)$$

where, as in the case of the harmonic oscillator,

$$a_q(t) = a_q e^{-i\omega_q t}, \qquad a^\dagger_q(t) = a_q e^{i\omega_q t}. \qquad (2.134)$$

Note:

- The generalization of the quantum string to higher dimensions is written

$$H = \int d^d x \left[\frac{T}{2} (\nabla\phi)^2 + \frac{1}{2\rho} \pi(x)^2 \right]$$

$$[\phi(x), \pi(y)] = i\hbar \delta^d(x - y). \qquad (2.135)$$

Sometimes it is useful to rescale $\phi(x) \to \phi(x)/\sqrt{\rho}$, $\pi(x) \to \pi(x)\sqrt{\rho}$, so that

$$H = \frac{1}{2} \int d^d x \left[(c\nabla\phi)^2 + \pi^2 \right]. \tag{2.136}$$

In two dimensions, this describes a fluctuating quantum membrane.

- In particle physics, the "massive" version of the above model, written as

$$H = \frac{1}{2} \int d^d x \left[\phi \left(-c^2 \nabla^2 + \left(\frac{mc^2}{\hbar} \right)^2 \right) \phi + \pi^2 \right], \tag{2.137}$$

where c is the speed of light, is called the *Klein–Gordon Hamiltonian*. In this model, the elementary quanta have energy $E_q = \sqrt{(\hbar cq)^2 + (mc^2)^2}$. This also corresponds to a string where a uniform displacement ϕ costs an energy proportional to $m^2\phi^2$.

Example 2.6   Calculate the the equal-time ground-state correlation function

$$S(x) = \frac{1}{2} \langle 0 | (\phi(x) - \phi(0))^2 | 0 \rangle \tag{2.138}$$

for a one-dimensional string.

Solution

Let us begin by rewriting

$$S(x) = \langle 0 | (\phi(0)^2 - \phi(x)\phi(0)) | 0 \rangle, \tag{2.139}$$

where we have used translational and inversion symmetry to replace $\langle 0 | \phi(x)^2 | 0 \rangle = \langle 0 | \phi(0)^2 | 0 \rangle$ and $\langle 0 | \phi(x)\phi(0) | 0 \rangle = \langle 0 | \phi(0)\phi(x) | 0 \rangle$.

When we expand $\phi(x)$ and $\phi(0)$ in terms of creation and annihilation operators, only terms of the form $\langle 0 | a_q a_{-q'}^\dagger | 0 \rangle = \langle 0 | [a_q, a_{-q'}^\dagger] | 0 \rangle = (2\pi)\delta(q - q')$ will survive. Let us write this out explicitly:

$$\begin{aligned}
S(x) &= \int \frac{dq\,dq'}{(2\pi)^2} \frac{\hbar}{2\rho c \sqrt{|q||q'|}} \langle 0 | [a_q + a_{-q}^\dagger][a_{-q'} + a_{q'}^\dagger] | 0 \rangle (1 - e^{iqx}) e^{-|q|\epsilon} \\
&= \frac{\hbar}{2\rho c} \int \frac{dq}{2\pi} e^{-|q|\epsilon} \left(\frac{1 - e^{iqx}}{|q|} \right) \\
&= \left(\frac{\hbar}{\rho c} \right) \left[\frac{1}{4\pi} \ln \left(\frac{\epsilon^2 + x^2}{\epsilon^2} \right) \right],
\end{aligned} \tag{2.140}$$

where, to obtain the last step, we first calculate

$$\begin{aligned}
\frac{dS}{dx} &= -\frac{i\hbar}{2\rho c} \int \frac{dq}{2\pi} e^{iqx - q|\epsilon|} \mathrm{sgn}(q) \\
&= -\frac{i\hbar}{2\rho c} \left[\int_0^\infty \frac{dq}{2\pi} e^{-[\epsilon - ix]q} - \int_{-\infty}^0 \frac{dq}{2\pi} e^{[\epsilon + ix]q} \right] \\
&= -\frac{i\hbar}{4\pi\rho c} \left[\frac{1}{\epsilon - ix} - \frac{1}{\epsilon + ix} \right] = \frac{\hbar}{2\pi\rho c} \mathrm{Im} \left(\frac{1}{\epsilon - ix} \right)
\end{aligned} \tag{2.141}$$

and then integrate the answer on x, noting that $S(0) = 0$, to get

$$S(x) = \frac{\hbar}{2\pi\rho c} \mathrm{Im} \int_0^x \frac{1}{\epsilon - ix'} dx' = \frac{\hbar}{2\pi\rho c} \mathrm{Re} \ln\left(\frac{\epsilon - ix}{\epsilon}\right) = \frac{\hbar}{4\pi\rho c} \ln\left(\frac{\epsilon^2 + x^2}{\epsilon^2}\right). \quad (2.142)$$

Remarks

- Were we to send the cut-off $\epsilon \to 0$, the fluctuations at a given distance x would diverge logarithmically with ϵ, because the number of short-wavelength (ultraviolet) fluctuations becomes unbounded.
- We could have also obtained this result by working with a discrete string and taking $a \to 0$ at the end of the calculation. Had we done this, we would have found that

$$S(x) = \frac{\hbar}{2m} \sum_q \left(\frac{1 - e^{iqx}}{\omega_q}\right), \quad (2.143)$$

which has the same long-wavelength behavior.

- Had we repeated this calculation in D dimensions, the integral over q would become a D-dimensional integral. In this case,

$$S(x) \sim \int d^D q \left(\frac{1 - e^{iqx}}{|q|}\right) \sim \int_{1/x}^{1/\epsilon} \frac{d^D q}{|q|} \sim \frac{1}{x^{D-1}} \quad (2.144)$$

at short distances. Notice that the e^{iqx} term averages to zero at large q, but cuts off the integral for $q \lesssim 1/x$. In higher dimensions, the phase space for number of short-wavelength fluctuations grows as q^D, which leads to stronger fluctuations at short distances.

Exercises

Exercise 2.1 In 1906, in what is arguably the first paper in theoretical condensed matter physics [3], Albert Einstein postulated that vibrational excitations of a solid are quantized with energy $\hbar\omega$, just like photons in the vacuum. Repeat his calculation for diamond: calculate the energy $E(T)$ of one mole of simple harmonic oscillators with characteristic frequency ω at temperature T and show that the specific heat capacity is

$$C_V(T) = \frac{dE}{dT} = RF\left(\frac{\hbar\omega}{k_B T}\right),$$

where

$$F(x) = \left(\frac{x/2}{\sinh(x/2)}\right)^2$$

and $R = N_{AV}k_B$, the product of Avagadro's number N_{AV} and Boltzmann's constant k_B. Plot $C(T)$ and show that it deviates from Dulong and Petit's law, $C_V = (R/2)$ per quadratic degree of freedom, at temperatures $T \ll \hbar\omega/k_B$.

Exercise 2.2 Consider the orthogonality relation in (2.73):

$$\sum_j \langle q_m|j\rangle\langle j|q_n\rangle \equiv \frac{1}{N_s}\sum_j e^{i(q_n - q_m)R_j} = \delta_{nm}, \tag{2.145}$$

where $q_n = n\frac{2\pi}{L}$, $q = n\frac{2\pi}{L} = n\frac{2\pi}{N_s a}$ are the discrete wavevectors, $\mathcal{N}_s = L/a$ is the number of sites in the chain and a is the lattice spacing. By substituting $R_j = ja$ and treating this expression as a geometric series, show that

$$\sum_j \langle q_m|j\rangle\langle j|q_n\rangle \equiv \frac{1}{N_s}\sum_{j=0}^{\mathcal{N}_s - 1} e^{i(q_n - q_m)ja} = \frac{(-1)^{(N-1)(n-m)}}{\mathcal{N}_s}\frac{\sin[\pi(n-m)]}{\sin[\frac{\pi}{\mathcal{N}_s}(n-m)]} \equiv \delta_{nm},$$

thereby proving orthogonality.

Exercise 2.3 For the harmonic oscillator $H = \hbar\omega[a^\dagger a + \frac{1}{2}]$, we know that

$$\langle \hat{n}\rangle = n(\omega) = \frac{1}{e^{\beta\hbar\omega} - 1}, \tag{2.146}$$

where $\beta = 1/(k_B T)$ and $\hat{n} = a^\dagger a$ is the number operator. In the ground state, using the equations of motion for the creation and annihilation operators, we showed that the zero-point fluctuations in position were described by the correlation function

$$\frac{1}{2}\langle\{x(t), x(0)\}\rangle = \frac{\hbar}{2m\omega}\cos\omega t. \tag{2.147}$$

Generalize this result to finite temperatures. You should find that there are two terms in the correlation function. Please give them a physical interpretation.

Exercise 2.4 (a) Show that, if a is a canonical Bose operator, the canonical transformation

$$b = ua + va^\dagger$$
$$b^\dagger = ua^\dagger + va \tag{2.148}$$

(where u and v are real) preserves the canonical commutation relations, provided $u^2 - v^2 = 1$.

(b) Using the results of (a), diagonalize the Hamiltonian

$$H = \omega\left(a^\dagger a + \frac{1}{2}\right) + \frac{1}{2}\Delta(a^\dagger a^\dagger + aa) \tag{2.149}$$

by transforming it into the form $H = \tilde{\omega}(b^\dagger b + \frac{1}{2})$. Find $\tilde{\omega}$, u and v in terms of ω and Δ. What happens when $\Delta = \omega$?

(c) The Hamiltonian in (b) has a boson pairing term. Show that the ground state of H can be written as a coherent condensate of paired bosons, given by

$$|\tilde{0}\rangle = e^{-\alpha(a^\dagger a^\dagger)}|0\rangle.$$

Calculate the value of α in terms of u and v. (Hint: $|\tilde{0}\rangle$ is the vacuum for b, i.e. $b|\tilde{0}\rangle = (ua + va^\dagger)|\tilde{0}\rangle = 0$. Calculate the commutator of $[a, e^{-\alpha a^\dagger a^\dagger}]$ by expanding the exponential as a power series. Find a value of α that guarantees that b annihilates the vacuum $|\tilde{0}\rangle$.)

Exercise 2.5 (a) Find the classical normal-mode frequencies and normal coordinates for the one-dimensional chain with Hamiltonian

$$H = \sum_j \left[\frac{p_j^2}{2m_j} + \frac{k}{2}(\phi_j - \phi_{j-1})^2 \right], \tag{2.150}$$

where at even sites $m_{2j} = m$ and at odd sites $m_{2j+1} = M$. Please sketch the dispersion curves.

(b) What is the gap in the excitation spectrum?

(c) Write the diagonalized Hamiltonian in second-quantized form and discuss how you might arrive at your final answer. You will now need two types of creation operator.

Exercise 2.6 According to the *Lindeman criterion*, a crystal melts when the RMS displacement of its atoms exceeds one-third of the average separation of the atoms. Consider a three-dimensional crystal with separation a, atoms of mass m and a nearest-neighbor quadratic interaction $V = \frac{m\omega^2}{2}(\vec{\Phi}_R - \vec{\Phi}_{R+a})^2$.

(a) Estimate the amplitude of zero-point fluctuations using the uncertainty principle, to show that, if

$$\frac{\hbar}{m\omega a^2} > \zeta_c, \tag{2.151}$$

where ζ_c is a dimensionless number of order one, the crystal will be unstable, even at absolute zero, and will melt due to zero-point fluctuations. (*Hint*: what would the answer be for a simple harmonic oscillator?)

(b) Calculate ζ_c in the above model. If you like, to start out imagine that the atoms only move in one direction, so that Φ is a scalar displacement at the site with equilibrium position $\mathbf{R}$. Calculate the RMS zero-point displacement of an atom, $\sqrt{\langle 0|\Phi(x)^2|0\rangle}$. Now generalize your result to take account of the fluctuations in three orthogonal directions.

(c) Suppose $\hbar\omega/k_B = 300\,\text{K}$ and the atom is a helium atom. Assuming that ω is independent of atom separation a, estimate the critical atomic separation a_c at which the solid becomes unstable to quantum fluctuations. Note that in practice ω is dependent on a, and rises rapidly at short distances, with $\omega \sim a^{-\alpha}$, where $\alpha > 2$. Is the solid stable for $a < a_c$ or for $a > a_c$?

Exercise 2.7 Find the transformation that diagonalizes the Hamiltonian

$$H = \sum_j \left\{ J_1(a_{i+1}^\dagger a_i + \text{H.c.}) + J_2(a_{i+1}^\dagger a_i^\dagger + \text{H.c.}) \right\}, \tag{2.152}$$

where the ith site is located at $R_j = aj$, and "H.c." denotes the Hermitian conjugate. You may find it helpful to (i) transform to momentum space, writing $a_j = \frac{1}{N^{1/2}} \sum_q e^{iqR_j} a_q$, and (ii) carry out a canonical transformation of the form $b_q = u_q a_q + v_q a_{-q}^\dagger$, where $u^2 - v^2 = 1$. What happens when $J_1 = J_2$?

Exercise 2.8 This problem sketches the proof that the displacement of the quantum harmonic oscillator, originally in its ground state (in the distant past), is given by

$$\langle x(t) \rangle = \int_{-\infty}^{\infty} R(t - t')f(t')dt', \tag{2.153}$$

where

$$R(t - t') = \frac{i}{\hbar}\langle 0|[x(t), x(t')]|0\rangle \tag{2.154}$$

is the response function and $x(t)$ is the position operator in the Heisenberg representation of H_0. A more detailed discussion can be found in Chapter 9.

An applied force $f(t)$ introduces an additional forcing term to the harmonic oscillator Hamiltonian

$$\hat{H}(t) = H_0 + V(t) = \hat{H}_0 - f(t)\hat{x}, \tag{2.155}$$

where $H_0 = \hbar\omega(a^\dagger a + \frac{1}{2})$ is the unperturbed Hamiltonian. To compute the displacement of the harmonic oscillator, it is convenient to work in the *interaction representation*, which is the Heisenberg representation for H_0. In this representation, the time evolution of the wavefunction is due to the force term. The wavefunction of the harmonic oscillator in the interaction representation $|\psi_I(t)\rangle$ is related to the Schrödinger state $|\psi_S(t)\rangle$ by the relation $|\psi_I(t)\rangle = e^{iH_0t/\hbar}|\psi_S(t)\rangle$.

(a) By using the equation of motion for the Schrödinger state, $i\hbar\partial_t|\psi_S(t)\rangle = (H_0 + V(t))|\psi_S(t)\rangle$, show that the time evolution of the wavefunction in the interaction representation is

$$i\hbar\partial_t|\psi_I(t)\rangle = V_I(t)|\psi_I(t)\rangle = -f(t)\hat{x}(t)|\psi_I(t)\rangle, \tag{2.156}$$

where $V_I(t) = e^{iH_0t/\hbar}\hat{V}(t)e^{-iH_0t/\hbar} = -x(t)f(t)$ is the force term in the interaction representation.

(b) Show that, if $|\psi(t)\rangle = |0\rangle$ at $t = -\infty$, then the leading-order solution to the above equation of motion is

$$|\psi_I(t)\rangle = |0\rangle + \frac{i}{\hbar}\int_{-\infty}^{t} dt'f(t')\hat{x}(t')|0\rangle + O(f^2), \tag{2.157}$$

so that

$$\langle\psi_I(t)| = \langle 0| - \frac{i}{\hbar}\int_{-\infty}^{t} dt'f(t')\langle 0|\hat{x}(t') + O(f^2). \tag{2.158}$$

(c) Using the results just derived, expand the expectation value $\langle\psi_I(t)|x(t)|\psi_I(t)\rangle$ to linear order in f, obtaining the above cited result.

References

[1] A. Einstein, Concerning an heuristic point of view towards the emission and transformation of light, *Ann. Phys. (Leipzig)*, vol. 17, p. 132, 1905.

[2] A. B. Arons and M. B. Peppard, Einstein's proposal of the photon concept: a translation of the Annalen der Physik paper of 1905, *Am. J. Phys.*, vol. 33, p. 367, 1965.

[3] A. Einstein, Planck's theory of radiation and the theory of the specific heat, *Ann. Physik*, vol. 22, p. 180, 1907.

[4] A. Pais, *Subtle is the Lord: the Science and the Life of Albert Einstein*, Oxford University Press, 1982.

[5] M. Born and P. Jordan, Zur Quantenmechanik (On quantum mechanics), *Z. Phys.*, vol. 34, p. 858, 1925.

[6] P. A. M. Dirac, The fundamental equations of quantum mechanics, *Proc. R. Soc. A*, vol. 109, p. 642, 1925.

[7] W. Pauli, Die Quantumtheorie und die Rotverschiebung der Spektralien (Quantum theory and the red shift of spectra), *Z. Phys.*, vol. 26, p. 765, 1925.

[8] P. A. M. Dirac, On the theory of quantum mechanics, *Proc. R. Soc. A*, vol. 112, p. 661, 1926.

[9] P. Jordan and O. Klein, Zum Mehrkörperproblem der Quantentheorie (On the many-body problem of quantum theory), *Z. Phys.*, vol. 45, p. 751, 1927. The second quantization condition for bosons appears in equation (14) of this paper.

[10] P. Jordan and E. Wigner, Uber das Paulische Aquivalenzverbot (On the Pauli exclusion principle), *Z. Phys.*, vol. 47, p. 631, 1928.

[11] W. Heisenberg, Uber quanten theoretische Umdeutung kinematischer und mechanischer Beziehungen (Quantum theoretical reinterpretation of kinematic and mechanical relations), *Z. Phys.*, vol. 33, p. 879, 1925.

3 Conserved particles

From Einstein and Planck's quantum hypothesis, we learned that the collective excitations of a field are discrete "massless" particles. This leads us naturally to reverse the logic, and to ask – in the case where we have conserved particles, particles with mass, such as a gas of ^{4}He atoms, or an electron gas inside a metal – do those particles also arise from a quantum field?

$$\text{fields} \rightleftharpoons \text{particles}$$

This is the motivation for this chapter: to find a way to rewrite Schrödinger's many-body quantum mechanics in terms of a quantum field that encodes both the particles and their statistics.

First-quantized quantum mechanics *can* deal with many-body physics, through the introduction of a many-particle wavefunction. This is the approach used in quantum chemistry, where the number of electrons is large but not macroscopic. The quantum chemistry approach revolves around the many-body wavefunction. For N particles, this a function of $3N$ variables and N spins. The Hamiltonian is then an operator expressed in terms of these coordinates:

$$\psi \longrightarrow \psi(x_1, x_2, \ldots x_N, t)$$
$$H \longrightarrow \sum_j \left[-\frac{\hbar^2}{2m} \nabla_j^2 + U(x_j) \right] + \frac{1}{2} \sum_{i<j} V(x_i - x_j). \tag{3.1}$$

With a few famous exceptions this method is cumbersome, and ill-suited to macroscopically large systems. The most notable exceptions occur in low-dimensional problems, where wavefunctions of macroscopically large ensembles of interacting particles have been obtained. Examples include:

- Bethe ansatz solutions to interacting one-dimensional and impurity problems [1–4]
- Laughlin's wavefunction for interacting electrons in high magnetic fields, at commensurate filling factors [5, 6].

Second quantization provides a general way of approaching many-body systems in which the wavefunction plays a minor role. As we mentioned in Chapter 2, the essence of second quantization is a process of raising the Schrödinger wavefunction to the level of an operator which satisfies certain *canonical commutation* or *canonical anticommutation* algebras. In first-quantized physics, physical properties of a quantum particle such as its density, kinetic energy and potential energy can be expressed in terms of the one-particle wavefunction. Second quantization elevates each of these quantities to the status of an operator by replacing the one-particle wavefuncion by its corresponding field operator:

$$
\left.
\begin{array}{ccc}
\psi(x,t) & \longrightarrow & \hat{\psi}(x,t) \\
\text{one-particle wavefunction} & & \text{field operator} \\
& & \\
O(\psi^*, \psi) & \longrightarrow & \hat{O}(\hat{\psi}^\dagger, \hat{\psi}).
\end{array}
\right\} \quad \text{second quantization} \qquad (3.2)
$$

For example, Born's famous expression for the one-particle (probability) density becomes an operator as follows:

$$
\rho(x) = |\psi(x)|^2 \longrightarrow \hat{\rho}(x) = \hat{\psi}^\dagger(x)\hat{\psi}(x), \qquad (3.3)
$$

so that the potential energy associated with an external potential is

$$
\hat{V} = \int d^3x\, U(x)\hat{\rho}(x). \qquad (3.4)
$$

Similarly, the kinetic energy in first quantization

$$
T[\psi^*,\ \psi] = \int d^3x\, \psi^*(x)\left[-\frac{\hbar^2}{2m}\nabla^2\right]\psi(x) \qquad (3.5)
$$

becomes the operator

$$
\hat{T} = \int d^3x\, \hat{\psi}^\dagger(x)\left[-\frac{\hbar^2}{2m}\nabla^2\right]\hat{\psi}(x). \qquad (3.6)
$$

Finally,

$$
H = \int d^3x\, \hat{\psi}^\dagger(x)\left[-\frac{\hbar^2}{2m}\nabla^2 + U(x)\right]\hat{\psi}(x) + \frac{1}{2}\int d^3x\, d^3x'\, V(x-x') : \hat{\rho}(x)\hat{\rho}(x') : \qquad (3.7)
$$

is the complete many-body Hamiltonian in second-quantized form. Here $V(x - x')$ is the interaction potential between the particles, and the symbol ":" reflects the fact that order of the operators counts. The *normal ordering* operator ": ... :" indicates that all creation operators between the two colons must be ordered to lie to the left of all destruction operators.

3.1 Commutation and anticommutation algebras

In 1928, Jordan and Wigner [7] proposed that the microscopic field operators describing identical particles divide up into two types. This is an axiom of quantum field theory. For bosons, field operators satisfy a commutation algebra, whereas fermions satisfy an anticommutation algebra. Since we will be dealing with many of their properties in parallel, it useful to introduce a unified notation for commutators and anticommutators, as follows:

$$
\begin{aligned}
\{a, b\} &= ab + ba \equiv [a, b]_+ \\
[a, b] &= ab - ba \equiv [a, b]_-
\end{aligned} \qquad (3.8)
$$

so that

$$
[a, b]_\mp = ab \mp ba. \qquad (3.9)
$$

Table 3.1 First and second quantization treatment of conserved particles.

	First quantization	Second quantization
Wavefunction $\longrightarrow$ field operator	$\psi(x) = \langle x|\psi\rangle$	$\hat{\psi}(x)$
Commutator	$[x,p] = i\hbar$	$[\hat{\psi}(x), \hat{\psi}^{\dagger}(x')]_{\mp} = \delta^{D}(x - x')$
Density	$\rho(x) = |\psi(x)|^2$	$\hat{\rho}(x) = \hat{\psi}^{\dagger}(x)\hat{\psi}(x)$
Arbitrary basis	$\psi_{\lambda} = \langle\lambda|\psi\rangle$	$\hat{\psi}_{\lambda}$
Change of basis	$\langle\bar{s}|\psi\rangle = \sum_{\lambda}\langle\bar{s}|\lambda\rangle\langle\lambda|\psi\rangle$	$\hat{a}_s = \sum_{\lambda}\langle\bar{s}|\lambda\rangle\hat{\psi}_{\lambda}$
Orthogonality	$\langle\lambda|\lambda'\rangle = \delta_{\lambda\lambda'}$	$[\psi_{\lambda}, \psi_{\lambda'}^{\dagger}]_{\mp} = \delta_{\lambda\lambda'}$
One-particle energy	$\frac{p^2}{2m} + U$	$\int_x \hat{\psi}^{\dagger}(x)\left(-\frac{\hbar^2}{2m}\nabla^2 + U(x)\right)\hat{\psi}(x)$
Interaction	$\sum_{i<j} V(x_i - x_j)$	$\hat{V} = \frac{1}{2}\int_{x,x'} V(x - x') : \hat{\rho}(x)\hat{\rho}(x') :$ $= \frac{1}{2}\sum V(\mathbf{q})c_{\mathbf{k+q}}^{\dagger}c_{\mathbf{k'-q}}^{\dagger}c_{\mathbf{k'}}c_{\mathbf{k}}$
Many-body wavefunction	$\Psi(x_1, x_2, \ldots x_N)$	$\langle 0|\hat{\psi}(x_1)\cdots\hat{\psi}(x_N)|\Psi\rangle$
Schrödinger equation	$\left(\sum\mathcal{H}_i + \sum_{i<j} V_{ij}\right)\Psi = i\hbar\dot{\Psi}$	$\left[\mathcal{H}^{(0)} + \int_{x'}\hat{\rho}(x')V(x'-x)\right]\hat{\psi}(x)$ $= i\hbar\dot{\hat{\psi}}(x)$

We shall adopt the $-/+$ subscript notation in this chapter while discussing both bosons and fermions together.

The algebra of field operators is then

$$[\psi(1), \psi(2)]_{\mp} = [\psi^{\dagger}(2), \psi^{\dagger}(1)]_{\mp} = 0 \quad\left.\right\}\text{ bosons/fermions} \qquad (3.10)$$

$$[\psi(1), \psi^{\dagger}(2)]_{\mp} = \delta(1-2).$$

When spin is involved, $1 \equiv (x_1, \sigma_1)$ and $\delta(1-2) = \delta^{(D)}(x_1 - x_2)\delta_{\sigma_1\sigma_2}$. We shall motivate these axioms in two ways: (i) by showing, in the case of bosons, that they are a natural result of trying to quantize the one-particle wavefunction; (ii) by showing that they lead to the first-quantized formulation of many-body physics, naturally building the particle exchange statistics into the mathematical framework.

Table 3.1 summarizes the main points of second quantization that we shall now discuss in detail.

3.1.1 Heuristic derivation for bosons

The name "second quantization" derives from the notion that many-body physics can be obtained by quantizing the one-particle wavefunction. Philosophically, this is very tricky, for surely the wavefunction is already a quantum object? Let us imagine, however, a thought experiment in which we prepare a huge number of non-interacting particles in such a way that they are all in precisely the same quantum state. The feasibility of this does not worry us here, but note that it can actually be done for a large ensemble of bosons by condensing them into a single quantum state. In this circumstance, every single particle lies in the same one-particle state. If we time-evolve the system, we can think of the single-particle wavefunction as a semiclassical variable describing the collective behavior of the condensate.

Let us briefly recall one-particle quantum mechanics. If the particle is in a state $|\psi\rangle$, then we can always expand the state in terms of a complete basis $\{|n\rangle\}$, as follows:

$$|\psi(t)\rangle = \sum_n |n\rangle \overbrace{\langle n|\psi(t)\rangle}^{\psi_n(t)} = \sum_n |n\rangle \psi_n(t), \qquad (3.11)$$

so that $|\psi_n(t)|^2 = p_n(t)$ gives the probability of being in state n. Applying Schrödinger's equation, $\hat{H}|\psi\rangle = i\hbar\partial_t|\psi\rangle$, gives

$$i\hbar\dot{\psi}_n(t) = \sum_m \langle n|H|m\rangle \psi_m(t)$$

$$i\hbar\dot{\psi}_n^*(t) = -\sum_m \langle m|H|n\rangle \psi_m^*(t). \qquad (3.12)$$

Now if we write the ground-state energy as a functional of the $\psi_m(t)$, we get

$$H(\psi, \psi^*) = \langle H\rangle = \sum_{m,n} \psi_m^* \psi_n \langle m|H|n\rangle. \qquad (3.13)$$

We see that the equations of motion can be written in Hamiltonian form

$$\dot{\psi}_m = \frac{\partial H(\psi, \psi^*)}{i\hbar\partial\psi_m^*} \qquad \left(\text{cf. } \dot{q} = \frac{\partial H}{\partial p}\right)$$

$$i\hbar\dot{\psi}_m^* = -\frac{\partial H(\psi, \psi^*)}{\partial\psi_m} \qquad \left(\text{cf. } \dot{p} = -\frac{\partial H}{\partial q}\right), \qquad (3.14)$$

so we can identify

$$\{\psi_n, i\hbar\psi_n^*\} \equiv \{q_n, p_n\} \qquad (3.15)$$

as the canonical position and momentum coordinates.

But suppose we don't have a macroscopic number of particles in a single state. In this case, the amplitudes $\psi_n(t)$ are expected to undergo quantum fluctuations. Let us examine what happens if we *second-quantize* these variables, making the replacement

$$[q_n, p_m] = i\hbar\delta_{nm} = i\hbar[\psi_n, \psi_m^\dagger] \qquad (3.16)$$

or

$$\begin{aligned}
[\psi_n, \psi_m] &= [\psi_n^\dagger, \psi_m^\dagger] = 0, \\
[\psi_n, \psi_m^\dagger] &= \delta_{nm}.
\end{aligned} \qquad (3.17)$$

In terms of these operators, our second-quantized Hamiltonian becomes

$$H = \sum_{m,l} \hat{\psi}_m^\dagger \hat{\psi}_l \langle m|H|l\rangle. \qquad (3.18)$$

If we now use this to calculate the time evolution of the quantum fields, we obtain

$$- i\hbar \partial_t \psi_j = [\hat{H}, \ \psi_j] = \sum_{m,l} \langle m|H|l\rangle \overbrace{[\psi_m^\dagger \psi_l, \psi_j]}^{-\delta_{mj}\psi_l}. \tag{3.19}$$

Eliminating the sum over m, we obtain

$$- i\hbar \partial_t \psi_j = - \sum_l \langle j|H|l\rangle \psi_l$$

$$- i\hbar \partial_t \psi_j^\dagger = [\hat{H}, \ \psi_j^\dagger] = \sum_l \psi_l^\dagger \langle l|H|j\rangle, \tag{3.20}$$

where the complex conjugated expression gives the time evolution of $\psi_l^\dagger$. Remarkably, the equations of motion of the operators match the time evolution of the one-particle amplitudes. But now that we have operators, we have all the new physics associated with quantum fluctuations of the particle fields.

3.2 What about fermions?

Remarkably, as Jordan and Wigner first realized, we recover precisely the same time evolution if we second-quantize the operators using anticommutators [7] rather than commutators, and it is this choice that gives rise to fermions and the exclusion principle. But for fermions we cannot offer a heuristic argument, because they don't condense: as far as we know, there is no situation in which individual Fermi field operators behave semiclassically (although of course, in a superconductor, pairs of fermions behave semiclassically).

In fact, all of the operations we carried out above work equally well with either canonical commutation or canonical anticomutation relations:

$$[\psi_n, \psi_m]_\mp = [\psi_n^\dagger, \psi_m^\dagger]_\mp = 0$$

$$[\psi_n, \psi_m^\dagger]_\mp = \delta_{nm} \tag{3.21}$$

where the $\mp$ refers to bosons/fermions, respectively.

To evaluate the equation of motion of the field operators, we need to know the commutator $[H, \psi_n]$. Using the relation

$$[ab, c] = a[b, c]_\mp \pm [a, c]_\mp b \tag{3.22}$$

we may verify that

$$[\psi_m^\dagger \psi_l, \psi_j] = \psi_m^\dagger \overbrace{[\psi_l, \psi_j]_\mp}^{0} \pm \overbrace{[\psi_m^\dagger, \psi_j]_\mp}^{-\delta_{mj}} \psi_l$$

$$= -\delta_{mj}\psi_l, \tag{3.23}$$

so that

$$-i\hbar\partial_t\psi_j = [\hat{H},\ \psi_j] = \sum_{m,l}\langle m|H|l\rangle \overbrace{[\psi_m^\dagger\psi_l,\psi_j]}^{-\delta_{mj}\psi_l}$$
$$= -\sum_l \langle j|H|l\rangle\psi_l, \tag{3.24}$$

independently of whether we use an anticommuting or commuting algebra.

Let us now go on, and look at some general properties of second-quantized operators that hold for both bosons and fermions.

3.3 Field operators in different bases

Let us first check that our results don't depend on the one-particle basis we use. To do this, we must confirm that the commutation or anticommutation algebra of bosons or fermions is basis-independent. Let us recall that, in quantum mechanics, if Hilbert space is spanned by an orthonormal and complete basis of states $\{|r\rangle\}$, then this basis satisfies

$$\langle r|s\rangle = \delta_{rs} \qquad \text{orthonormality}$$

$$\sum_r |r\rangle\langle r| = 1. \qquad \text{completeness} \tag{3.25}$$

The completeness identity is particularly valuable – one of my favourite equations – because we can always insert the identity $1 = \sum_r |r\rangle\langle r|$, permitting operator expressions to be decomposed in terms of matrix elements with a particular basis.

Suppose we have two complete, orthonormal bases of *one-particle* states: the $\{|r\rangle\}$ basis and a new $\{|\tilde{s}\rangle\}$ basis, where

$$|\psi\rangle = \sum_r |r\rangle\psi_r = \sum_s |\tilde{s}\rangle a_s, \tag{3.26}$$

where $\langle\tilde{s}|\psi\rangle = a_s$ and $\langle r|\psi\rangle = \psi_r$. Introducing the completeness relation $1 = \sum_r |r\rangle\langle r|$ into the first expression, we obtain

$$\overbrace{\langle\tilde{s}|\psi\rangle}^{a_s} = \sum_r \langle\tilde{s}|r\rangle \overbrace{\langle r|\psi\rangle}^{\psi_r}. \tag{3.27}$$

If this is how the one-particle states transform between the two bases, then we must use the same unitary transformation to relate the field operators that destroy particles in the two bases:

$$\hat{a}_s = \sum_r \langle\tilde{s}|r\rangle\hat{\psi}_r. \tag{3.28}$$

The commutation algebra of the new operators is now

$$[\hat{a}_s, \hat{a}_p^\dagger]_\pm = \sum_{l,m} \langle \tilde{s}|l \rangle \overbrace{[\hat{\psi}_l, \hat{\psi}_m^\dagger]_\pm}^{\delta_{lm}} \langle m|\tilde{p} \rangle. \tag{3.29}$$

This is just the pre- and post-multiplication of a unit operator by the unitary matrix $U_{sl} = \langle \tilde{s}|l \rangle$ and its conjugate $U_{mp}^\dagger = \langle m|\tilde{p} \rangle$. The final result is unity, as expected:

$$[\hat{a}_s, \hat{a}_p^\dagger]_\pm = \sum_r \langle \tilde{s}|r \rangle \langle r|\tilde{p} \rangle = \langle \tilde{s}|\tilde{p} \rangle = \delta_{sp}. \tag{3.30}$$

In other words, the canonical commutation algebra is preserved by unitary transformations of basis.

A basis of particular importance is the position basis. The one-particle wavefunction can always be decomposed in a discrete basis, as follows:

$$\psi(x) = \langle x|\psi \rangle = \sum_n \langle x|n \rangle \psi_n, \tag{3.31}$$

where $\langle x|n \rangle = \phi_n(x)$ is the wavefunction of the nth state. We now define the corresponding destruction operator,

$$\hat{\psi}(x) = \sum_n \langle x|n \rangle \hat{\psi}_n, \tag{3.32}$$

which defines the field operator in real space. Using completeness of the one-particle eigenstates $1 = \int d^D x |x\rangle \langle x|$, we can expand the orthogonality relation $\delta_{nm} = \langle n|m \rangle$ as

$$\delta_{nm} = \langle n| \overbrace{\hat{1}}^{1=\int d^D x|x\rangle\langle x|} |m \rangle = \int d^D x \langle n|x \rangle \langle x|m \rangle.$$

By integrating (3.32) over x with $\langle n|x \rangle$, we can then invert this equation to obtain

$$\psi_n = \int d^D x \langle n|x \rangle \psi(x), \qquad \psi_n^\dagger = \int d^D x \psi^\dagger(x) \langle x|n \rangle. \tag{3.33}$$

You can see by now that, so far as transformation laws are concerned, $\psi_n \sim \langle n|$ and $\psi(x) \sim \langle x|$ transform like "bra" vectors, while their conjugates transform like "kets."

By moving to a real-space representation, we have traded a discrete basis for a continuous basis. The corresponding "unit" operator appearing in the commutation algebra now becomes a delta function:

$$\begin{aligned}
[\psi(x), \psi^\dagger(y)]_\pm &= \sum_{n,m} \langle x|n \rangle \langle m|y \rangle \overbrace{[\psi_n, \psi_m^\dagger]_\pm}^{\delta_{nm}} \\
&= \sum_n \langle x|n \rangle \langle n|y \rangle = \langle x|y \rangle \\
&= \delta^3(x-y),
\end{aligned} \tag{3.34}$$

where we have assumed a three-dimensional system.

Another basis of importance is that provided by the one-particle energy eigenstates. In this basis $\langle l|H|m\rangle = E_l\delta_{lm}$, so the Hamiltonian becomes diagonal:

$$H = \sum_l E_l\psi_l^\dagger\psi_l = \sum E_l\hat{n}_l. \tag{3.35}$$

The Hamiltonian of the non-interacting many-body system thus divides up into a set of individual components, each one describing the energy associated with the occupancy of a given one-particle eigenstate. The eigenstates of the many-body Hamiltonian are thus labeled by the occupancy of the lth one-particle state. Of course, in a real-space basis the Hamiltonian becomes more complicated. Formally, if we transform this back to the real-space basis, we find that

$$H = \int d^Dx\, d^Dx'\, \psi^\dagger(x)\langle x|H|x'\rangle\psi(x'). \tag{3.36}$$

For free particles in space, the one-particle Hamiltonian is

$$\langle x|H|x'\rangle = \left[-\frac{\hbar^2}{2m}\nabla^2 + U(x)\right]\delta^D(x-x'), \tag{3.37}$$

so that the Hamiltonian becomes

$$H = \int d^Dx\psi^\dagger(x)\left[-\frac{\hbar^2}{2m}\nabla^2 + U(x)\right]\psi(x), \tag{3.38}$$

which, despite its formidable appearance, is just a transformed version of the diagonalized Hamiltonian (3.35).

Example 3.1   By integrating by parts, taking care with the treatment of surface terms, show that the second-quantized Hamiltonian (3.38) can be rewritten in the form

$$H = \int d^Dx \left(\frac{\hbar^2}{2m}|\nabla\psi(x)|^2 + U(x)|\psi(x)|^2\right), \tag{3.39}$$

where we have taken a notational liberty common in field theory, denoting $|\nabla\psi(x)|^2 \equiv \vec{\nabla}\psi^\dagger(x)\cdot\vec{\nabla}\psi(x)$ and $|\psi(x)|^2 \equiv \psi^\dagger(x)\psi(x)$.

Solution

Let us concentrate on the kinetic energy term in the Hamiltonian, writing $H = T + U$, where

$$T = \int d^Dx\psi^\dagger(x)\left(-\frac{\hbar^2}{2m}\nabla^2\right)\psi(x). \tag{3.40}$$

Integrating this term by parts, we can split it into "bulk" and "surface" terms, as follows:

$$T = \frac{\hbar^2}{2m}\int d^Dx\vec{\nabla}\psi^\dagger(x)\cdot\vec{\nabla}\psi(x) - \frac{\hbar^2}{2m}\overbrace{\int d^Dx\vec{\nabla}\cdot\left(\psi^\dagger(x)\vec{\nabla}\psi(x)\right)}^{T_S}. \tag{3.41}$$

Using the divergence theorem, we can rewrite the total derivative as a surface integral:

$$T_S = -\frac{\hbar^2}{2m} \int d\vec{S} \cdot \left(\psi^\dagger(x)\vec{\nabla}\psi(x) \right). \tag{3.42}$$

Now it is tempting to just drop this term as a surface term that "vanishes at infinity." However, here we are dealing with operators, so this brash step requires a little contemplation before we take it for granted. One way to deal with this term is to use periodic boundary conditions. In this case there really are no boundaries, or, more strictly speaking, opposite boundaries cancel ($\int_R dS + \int_L dS = 0$), so the surface term is zero. But suppose we had used hard-wall boundary conditions, what then?

Well, in this case we can decompose the field operators in terms of the one-particle eigenstates of the cavity. Remembering that, under change of bases, $\psi(x) \sim \langle x|$ and $\psi^\dagger(x) \sim |x\rangle$ behave as "bras" and "kets" respectively, we write

$$\psi(x) = \sum_n \overbrace{\langle x|n\rangle}^{\phi_n(x)} \psi_n, \qquad \psi^\dagger(x) = \sum_n \psi_n^\dagger \overbrace{\langle n|x\rangle}^{\phi_n^*(x)}.$$

Substituting these expressions into (3.42), the surface term becomes

$$T_S = \sum_{n,m} t_{nm}^{(S)} \psi_n^\dagger \psi_m$$

$$t_{nm}^S = -\frac{\hbar^2}{2m} \int d\vec{S} \cdot \phi_n^*(x)\vec{\nabla}\phi_m(x). \tag{3.43}$$

Provided $\phi_n(x) = 0$ on the surface, it follows that the matrix elements $t_{nm}^S = 0$, so that $\hat{T}_S = 0$.

Thus, whether we use hard-wall or periodic boundary conditions, we can drop the surface contribution to the kinetic energy in (3.41), enabling us to write

$$T = \frac{\hbar^2}{2m} \int d^D x |\vec{\nabla}\psi(x)|^2,$$

and when we add in the potential term, we obtain (3.39).

3.4　Fields as particle creation and annihilation operators

By analogy with collective fields, we now interpret the quantity $\hat{n}_l = \psi_l^\dagger \psi_l$ as the number operator, counting the number of particles in the one-particle state l. The total particle number operator is then

$$N = \sum_l \psi_l^\dagger \psi_l. \tag{3.44}$$

Using relation (3.22), it is easy to verify that, for both fermions and bosons,

$$[\hat{N}, \psi_l] = [\hat{n}_l, \psi_l] = -\psi_l, \qquad [\hat{N}, \psi_l^\dagger] = [\hat{n}_l, \psi_l^\dagger] = \psi_l^\dagger. \tag{3.45}$$

In other words, $\hat{N}\psi_l^\dagger = \psi_l^\dagger(\hat{N}+1)$, so that $\psi_l^\dagger$, adds a particle to state l. Similarly, since $\hat{N}\psi_l = \psi_l(\hat{N}-1)$, ψ_l *destroys* a particle from state l.

There is, however, a vital difference between bosons and fermions. For bosons, the number of particles n_l in the lth state is unbounded, but for fermions, since

$$\psi_l^{\dagger 2} = \frac{1}{2}\{\psi_l^\dagger, \psi_l^\dagger\} = 0, \tag{3.46}$$

the amplitude for adding more than one particle to a given state is always *zero*. We can never add more than one particle to a given state: in other words, the *exclusion principle* follows from the algebra! The occupation number bases for bosons and fermions are given by

$$|n_1, n_2, \ldots n_l, \ldots\rangle = \prod_l \frac{(\psi_l^\dagger)^{n_l}}{\sqrt{n_l!}}|0\rangle \qquad (n_r = 0, 1, 2, \ldots) \quad \text{bosons}$$
$$|n_1, n_2, \ldots n_r\rangle = (\psi_r^\dagger)^{n_r}\cdots(\psi_1^\dagger)^{n_1}|0\rangle \quad (n_r = 0, 1), \qquad \text{fermions} \tag{3.47}$$

A specific example for fermions is

$$\overset{1\,2\,3\,4\,5\,6}{|101101\rangle} = \psi_6^\dagger \psi_4^\dagger \psi_3^\dagger \psi_1^\dagger |0\rangle, \tag{3.48}$$

which contains particles in the 1st, 3rd, 4th and 6th one-particle states. Notice how the *order* in which we add the particles affects the sign of the wavefunction, so exchanging particles 4 and 6 gives

$$\psi_4^\dagger \psi_6^\dagger \psi_3^\dagger \psi_1^\dagger |0\rangle = -\psi_6^\dagger \psi_4^\dagger \psi_3^\dagger \psi_1^\dagger |0\rangle = -\overset{1\,2\,3\,4\,5\,6}{|101101\rangle}. \tag{3.49}$$

By contrast, a bosonic state is symmetric; for example,

$$\overset{1\,2\,3\,4\,5\,6}{|805241\rangle} = \frac{1}{\sqrt{4!\,2!\,5!\,8!}}\psi_6^\dagger(\psi_5^\dagger)^4(\psi_4^\dagger)^2(\psi_3^\dagger)^5(\psi_1^\dagger)^8|0\rangle. \tag{3.50}$$

To get further insight, let us transform the number operator to a real-space basis by writing

$$\hat{N} = \int d^D x\, d^D y \sum_l \psi^\dagger(x) \overset{\delta^D(x-y)}{\overbrace{\langle x|l\rangle\langle l|y\rangle}} \psi(y), \tag{3.51}$$

so that

$$\hat{N} = \int d^D x\, \psi^\dagger(x)\psi(x). \tag{3.52}$$

From this expression, we are immediately led to identify

$$\rho(x) = \psi^\dagger(x)\psi(x) \tag{3.53}$$

as the density operator. Furthermore, since

$$[\rho(y), \psi(x)] = \mp[\psi^\dagger(y), \psi(x)]_{\pm}\psi(y) = -\delta^3(x-y)\psi(y), \tag{3.54}$$

we can we can identify $\psi(x)$ as the operator which annihilates a particle at x.

Example 3.2   Using the result (3.54) that, if

$$\hat{N}_{\mathcal{R}} = \int_{y \in \mathcal{R}} d^3 y \rho(\vec{y})$$

(3.55)

measures the number of particles in some region $\mathcal{R}$,

$$[\hat{N}_{\mathcal{R}}, \psi(x)] = \begin{cases} -\psi(x), & (x \in \mathcal{R}) \\ 0 & (x \notin \mathcal{R}). \end{cases}$$

(3.56)

By localizing the region $\mathcal{R}$ around x_0, use this to prove that $\psi(x_0)$ annihilates a particle at position x_0.

Solution

By directly commuting $\hat{N}_{\mathcal{R}}$ with $\psi(x)$, we obtain

$$[\hat{N}_{\mathcal{R}}, \psi(x)] = \int_{y \in \mathcal{R}} [\rho(y), \psi(x)] = -\int_{y \in \mathcal{R}} \delta^3(x-y)\psi(y) = \begin{cases} -\psi(x) & (x \in \mathcal{R}) \\ 0 & (x \notin \mathcal{R}). \end{cases}$$

Suppose $|n_{\mathcal{R}}\rangle$ is a state with a definite number $n_{\mathcal{R}}$ of particles inside $\mathcal{R}$. If the region $\mathcal{R}$ is centered around x_0, then it follows that

$$\hat{N}_{\mathcal{R}} \psi(x_0)|n_{\mathcal{R}}\rangle = \psi(x_0)(\hat{N}_{\mathcal{R}} - 1)|n_{\mathcal{R}}\rangle = (n_{\mathcal{R}} - 1)\psi(x_0)|n_{\mathcal{R}}\rangle$$

contains one less particle. In this way, we see that $\psi(x)$ annihilates a particle from inside region $\mathcal{R}$, no matter how small that region is made, proving that $\psi(x)$ annihilates a particle at position x_0.

Example 3.3   Suppose $b_{\mathbf{q}}$ destroys a boson in a cubic box of side length L, where $\mathbf{q} = \frac{2\pi}{L}(i, j, k)$ is the momentum of the boson. Express the field operators in real space, and show that they satisfy canonical commutation relations. Write down the Hamiltonian in both bases.

Solution

The field operators in momentum space satisfy $[b_{\mathbf{q}}, b_{\mathbf{q}'}^{\dagger}] = \delta_{\mathbf{q}\mathbf{q}'}$. We may expand the field operator in real space as follows:

$$\psi(x) = \sum_{\mathbf{q}} \langle \mathbf{x}|\mathbf{q}\rangle b_{\mathbf{q}}.$$

(3.57)

Now

$$\langle \mathbf{x}|\mathbf{q}\rangle = \frac{1}{L^{3/2}} e^{i\mathbf{q}\cdot\mathbf{x}}$$

(3.58)

is the one-particle wavefunction of a boson with momentum $\mathbf{q}$. Calculating the commutator between the fields in real space, we obtain

$$[\psi(\mathbf{x}), \psi^\dagger(\mathbf{y})] = \sum_{q,q'} \langle \mathbf{x}|\mathbf{q}\rangle \langle \mathbf{q}'|\mathbf{y}\rangle \overbrace{[b_\mathbf{q}, b_{\mathbf{q}'}^\dagger]}^{\delta_{\mathbf{qq'}}} = \sum_q \langle \mathbf{x}|\mathbf{q}\rangle \langle \mathbf{q}|\mathbf{y}\rangle$$

$$= \frac{1}{L^3} \sum_q e^{i\mathbf{q}\cdot(\mathbf{x}-\mathbf{y})} = \delta^{(3)}(\mathbf{x} - \mathbf{y}). \tag{3.59}$$

The last two steps could have been carried out by noting that $\sum_\mathbf{q} |\mathbf{q}\rangle \langle \mathbf{q}| = 1$, so that $[\psi(\mathbf{x}), \psi^\dagger(\mathbf{y})] = \langle \mathbf{x}|\mathbf{y}\rangle = \delta^3(\mathbf{x} - \mathbf{y})$.

The Hamiltonian for the bosons in a box is

$$H = -\frac{\hbar^2}{2m} \int d^3x \, \psi^\dagger(x) \nabla^2 \psi(x). \tag{3.60}$$

We now Fourier transform this, writing

$$\psi^\dagger(x) = \frac{1}{L^{3/2}} \sum_q e^{-i\mathbf{q}\cdot x} b_\mathbf{q}^\dagger$$

$$\nabla^2 \psi(x) = -\frac{1}{L^{3/2}} \sum_q q^2 e^{i\mathbf{q}\cdot x} b_\mathbf{q}. \tag{3.61}$$

Substituting into the Hamiltonian, we obtain

$$H = \frac{1}{L^3} \sum_{q,\,q'} \epsilon_\mathbf{q} b_{\mathbf{q}'}^\dagger b_\mathbf{q} \overbrace{\int d^3x \, e^{i\mathbf{q}-\mathbf{q}'\cdot x}}^{L^3 \delta_{\mathbf{q}-\mathbf{q}'}} = \sum_q \epsilon_\mathbf{q} b_\mathbf{q}^\dagger b_\mathbf{q} \tag{3.62}$$

where

$$\epsilon_\mathbf{q} = \left(\frac{\hbar^2 q^2}{2m} \right) \tag{3.63}$$

is the one-particle energy.

3.5 The vacuum and the many-body wavefunction

We are now in a position to build up the many-body wavefunction. Once again, of fundamental importance is the notion of the vacuum, the unique state $|0\rangle$ which is annihilated by all field operators. If we work in the position basis, we can add a particle at site x to make the one-particle state

$$|x\rangle = \psi^\dagger(x)|0\rangle. \tag{3.64}$$

Notice that the overlap between two one-particle states is

$$\langle x|x'\rangle = \langle 0|\psi(x)\psi^\dagger(x')|0\rangle. \tag{3.65}$$

By using the (anti) commutation algebra to move the creation operator in the above expression to the right-hand side, where it annihilates the vacuum, we obtain

$$\langle 0|\psi(x)\psi^\dagger(x')|0\rangle = \overbrace{\langle 0|[\psi(x),\ \psi^\dagger(x')]_\mp|0\rangle}^{\delta^{(3)}(x-x')} = \delta^{(3)}(x - x'). \tag{3.66}$$

We can equally well add many particles, forming the N-particle state:

$$|x_1, x_2, \ldots x_N\rangle = \psi^\dagger(x_N)\cdots\psi^\dagger(x_2)\psi^\dagger(x_1)|0\rangle. \tag{3.67}$$

Now the corresponding "bra" state is given by

$$\langle x_1, x_2, \ldots x_N| = \langle 0|\psi(x_1)\psi(x_2)\cdots\psi(x_N). \tag{3.68}$$

The wavefunction of the N-particle state $\Psi_S(t)$ is the overlap with this state:

$$\Psi_S(x_1, x_2, \ldots x_N, t) = \langle x_1, x_2, \ldots x_N|\Psi_S(t)\rangle = \langle 0|\psi(x_1)\psi(x_2)\cdots\psi(x_N)|\Psi_S(t)\rangle. \tag{3.69}$$

Remarks

- In the above expression, the time dependence of the wavefunction lies in the "ket" vector $|\Psi(t)\rangle$. We can alternatively write the wavefunction in terms of the time-dependent Heisenberg field operators $\psi(x, t) = e^{iHt/\hbar}\psi(x)e^{-iHt/\hbar}$ and the stationary Heisenberg "ket" vector $|\Psi_H\rangle = e^{iHt/\hbar}|\Psi_S(t)\rangle$, as follows:

$$\Psi(x_1, x_2, \ldots x_N, t) = \langle 0|\psi(x_1, t)\psi(x_2, t)\cdots\psi(x_N, t)|\Psi_H\rangle. \tag{3.70}$$

- The commutation/anticommutation algebra guarantees that the symmetry of this wavefunction under particle exchange is positive for bosons and negative for fermions, so that, if we permute the particles $(1\,2\cdots N) \to (P_1 P_2 \cdots P_N)$,

$$\langle 0|\psi(x_{P_1})\psi(x_{P_2})\cdots\psi(x_{P_N})|\Psi_S(t)\rangle = (\pm 1)^P\langle 0|\psi(x_1)\psi(x_2)\cdots\psi(x_N)|\Psi(t)\rangle, \tag{3.71}$$

where P is the number of pairwise permutations involved in making the permutation. Notice that, for fermions, this hard-wires the Pauli exclusion principle into the formalism, and guarantees a node at locations where any two position (and spin) coordinates coincide.

Example 3.4   Two spinless fermions are added to a cubic box with sides of length L, in momentum states k_1 and k_2, forming the state

$$|\Psi\rangle = |\mathbf{k}_1, \mathbf{k}_2\rangle = c^\dagger_{\mathbf{k}_2}c^\dagger_{\mathbf{k}_1}|0\rangle. \tag{3.72}$$

Calculate the two-particle wavefunction

$$\Psi(x_1, x_2) = \langle x_1, x_2|\Psi\rangle. \tag{3.73}$$

Solution

Written out explicitly, the wavefunction is

$$\Psi(x_1, x_2) = \langle 0|\psi(x_1)\psi(x_2)c^\dagger_{\mathbf{k}_2}c^\dagger_{\mathbf{k}_1}|0\rangle. \tag{3.74}$$

To evaluate this quantity, we commute the two destruction operators to the right, until they annihilate the vacuum. Each time a destruction operator passes a creation operator, we generate a "contraction" term,

$$\{\psi(x), c^\dagger_{\mathbf{k}}\} = \int d^3y \overbrace{\{\psi(x), \psi^\dagger(y)\}}^{\delta^3(x-y)}\langle y|\mathbf{k}\rangle = \langle x|\mathbf{k}\rangle = L^{-3/2}e^{i\mathbf{k}\cdot\mathbf{x}}. \tag{3.75}$$

Carrying out this procedure, we generate a sum of pairwise contractions, as follows:

$$
\begin{aligned}
\langle 0|\psi(x_1)\psi(x_2)c^\dagger_{\mathbf{k}_2}c^\dagger_{\mathbf{k}_1}|0\rangle &= \langle x_1|\mathbf{k}_1\rangle\langle x_2|\mathbf{k}_2\rangle - \langle x_1|\mathbf{k}_2\rangle\langle x_2|\mathbf{k}_1\rangle \\
&= \begin{vmatrix} \langle x_1|\mathbf{k}_1\rangle & \langle x_1|\mathbf{k}_2\rangle \\ \langle x_2|\mathbf{k}_1\rangle & \langle x_2|\mathbf{k}_2\rangle \end{vmatrix} \\
&= \frac{1}{L^3}\left[e^{i(\mathbf{k}_1\cdot\mathbf{x}_1+\mathbf{k}_2\cdot\mathbf{x}_2)} - e^{i(\mathbf{k}_1\cdot\mathbf{x}_2+\mathbf{k}_2\cdot\mathbf{x}_1)} \right].
\end{aligned}
$$

Note: the determinantal expression for the two-particle wavefunction is an example of a *Slater determinant*. The N-dimensional generalization can be used to define the wavefunction of the corresponding N-particle state.

3.6 Interactions

Second quantization is easily extended to interactions. Classically, the interaction potential energy of a continuous plasma of particles is given by

$$V = \frac{1}{2}\int d^3x\, d^3x'\, V(x-x')\rho(x)\rho(x'), \tag{3.76}$$

so we might expect that the corresponding second-quantized expression is

$$\frac{1}{2}\int d^3x\, d^3x'\, V(x-x')\hat{\rho}(x)\hat{\rho}(x'). \tag{3.77}$$

This is *wrong*, because we have not been careful about the ordering of operators. Were we to use (3.77), then a one-particle state would interact with itself! We require that the action of the potential on the vacuum, or a one-particle state, gives zero:

$$\hat{V}|0\rangle = \hat{V}|x\rangle = 0. \tag{3.78}$$

To guarantee this, we need to be careful that we normal order the field operators, by permuting them so that all destruction operators are on the right-hand side. All additional terms that are generated by permuting the operators are dropped, but the signs associated with the permutation process are preserved. We denote the normal-ordering process by two colons. Thus,

$$: \rho(x)\rho(y) := \; : \psi^\dagger(x)\psi(x)\psi^\dagger(y)\psi(y) :$$
$$= \mp\psi^\dagger(x)\psi^\dagger(y)\psi(x)\psi(y) = \psi^\dagger(y)\psi^\dagger(x)\psi(x)\psi(y), \quad (3.79)$$

where we can remove the normal-ordering signs once all destruction operators are on the right-hand side. The correct expression for the interaction potential is then

$$V = \frac{1}{2}\int d^3x\, d^3x'\, V(x-x') : \hat{\rho}(x)\hat{\rho}(x') :$$
$$= \sum_{\alpha,\beta} \frac{1}{2}\int d^3x\, d^3x'\, V(x-x')\psi_\alpha^\dagger(x')\psi_\beta^\dagger(x)\psi_\beta(x)\psi_\alpha(x'), \quad (3.80)$$

where we have written a more general expression for fields with spin α, $\beta \in \pm\frac{1}{2}$.

Example 3.5   Show that the action of the operator V on the many-body state $|x_1, \ldots x_N\rangle$ is given by

$$\hat{V}|x_1, x_2, \ldots x_N\rangle = \sum_{i<j} V(x_i - x_j)|x_1, x_2, \ldots x_N\rangle. \quad (3.81)$$

Solution

First, note that this result is very natural, since the state $|x_1, \ldots, x_N\rangle$ is simultaneously an eigenstate of each of the N particle positions. Nevertheless, we'd like to confirm this with the machinery of the operators.

To prove this, we first prove the intermediate result

$$[\hat{V}, \psi^\dagger(x)] = \int d^3y\, V(x-y)\psi^\dagger(x)\rho(y). \quad (3.82)$$

This result can be obtained by expanding the commutator as follows:

$$[\hat{V}, \psi^\dagger(x)] = \frac{1}{2}\int_{y,y'} V(y-y')\psi^\dagger(y)\psi^\dagger(y') \overbrace{[\psi(y')\psi(y), \psi^\dagger(x)]}^{\delta(y-x)\psi(y')\pm\delta(y'-x)\psi(y)}$$

$$= \psi^\dagger(x)\frac{1}{2}\int_{y'} V(x-y')\rho(y') \pm \frac{1}{2}\int_y V(y-x) \overbrace{\psi^\dagger(y)\psi^\dagger(x)}^{\pm\psi^\dagger(x)\psi^\dagger(y)} \psi(y)$$

$$= \int_y V(x-y)\psi^\dagger(x)\rho(y), \quad (3.83)$$

where the lower sign choice on the second line is for fermions.

We now calculate

$$\hat{V}|x_1, \ldots x_N\rangle = \hat{V}\psi^\dagger(x_N)\cdots\psi^\dagger(x_1)|0\rangle \quad (3.84)$$

by commuting $\hat{V}$ successively to the right until it annihilates with the vacuum. Each time we hop $\hat{V}$ to the right over a field operator, we generate a remainder term. Commuting $\hat{V}$ past the jth creation operator,

$$\psi^\dagger(x_N) \cdots \hat{V}\psi^\dagger(x_j) \cdots \psi^\dagger(x_j)|0\rangle = \psi^\dagger(x_N) \cdots \psi^\dagger(x_j)\hat{V} \cdots \psi^\dagger(x_1)|0\rangle + \mathcal{R}_j,$$

$$(3.85)$$

we generate the remainder

$$\mathcal{R}_j = \int d^3 y \psi^\dagger(x_N) \cdots V(y - x_j)\psi^\dagger(x_j)\rho(y) \cdots \psi^\dagger(x_1)|0\rangle.$$

$$(3.86)$$

To write $\mathcal{R}_j$ in a more usable form, we use $\rho(y)\psi^\dagger(x_i) = \psi^\dagger(x_i)\rho(y) + \psi^\dagger(x_i)\delta(y - x_i)$ to commute the density operator to the right until it annihilates the vacuum. The terms generated by this procedure can be written

$$\mathcal{R}_j = \sum_{i=1}^{j-1} V(x_i - x_j)\psi^\dagger(x_N) \cdots \psi^\dagger(x_j) \cdots \psi^\dagger(x_i) \cdots \psi^\dagger(x_1)|0\rangle$$

$$= \sum_{i=1}^{j-1} V(x_i - x_j)|x_1, x_2, \ldots x_N\rangle.$$

$$(3.87)$$

Our final answer is the sum of the remainders $\mathcal{R}_j$:

$$\hat{V}\psi^\dagger(x_N) \cdots \psi^\dagger(x_1)|0\rangle = \sum_{j=2,N} \mathcal{R}_j = \sum_{i<j} V(x_i - x_j)|x_1, x_2, \ldots x_N\rangle.$$

$$(3.88)$$

In other words, the state $|x_1, \ldots x_N\rangle$ is an eigenstate of the interaction operator, with eigenvalue given by the classical interaction potential energy.

To get another insight into the interaction, we shall now rewrite it in the momentum basis. This is very useful in translationally invariant systems, where momentum is conserved in collisions. Let us imagine we are treating fermions with spin. The transformation to a momentum basis is then

$$\psi_\sigma(x) = \int_{\mathbf{k}} c_{\mathbf{k}\sigma} e^{i(\mathbf{k}\cdot\mathbf{x})}$$

$$\psi_\sigma^\dagger(x) = \int_{\mathbf{k}} c_{\mathbf{k}\sigma}^\dagger e^{-i(\mathbf{k}\cdot\mathbf{x})},$$

$$(3.89)$$

where $\{c_{\mathbf{k}\sigma}, c_{\mathbf{k}'\sigma'}^\dagger\} = (2\pi)^3 \delta^3(\mathbf{k} - \mathbf{k}')\delta_{\sigma\sigma'}$ are canonical fermion operators in momentum space and we have used the shorthand notation

$$\int_{\mathbf{k}} = \int \frac{d^3 k}{(2\pi)^3}.$$

$$(3.90)$$

We shall also Fourier transform the interaction

$$V(x - x') = \int_{\mathbf{q}} V(\mathbf{q})e^{i\mathbf{q}\cdot(\mathbf{x}-\mathbf{x}')}.$$

$$(3.91)$$

When we substitute these expressions into the interaction, we need to regroup the Fourier terms so that the momentum integrals are on the outside and the spatial integrals are on the inside. Doing this, we obtain

$$\hat{V} = \frac{1}{2} \sum_{\sigma\sigma'} \int_{\mathbf{k}_{1,2,3,4}} V(\mathbf{q}) \times c_{\mathbf{k}_4\sigma}^\dagger c_{\mathbf{k}_3\sigma'}^\dagger c_{\mathbf{k}_2\sigma'} c_{\mathbf{k}_1\sigma} \times \text{spatial integrals},$$

$$(3.92)$$

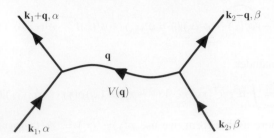

Fig. 3.1 Scattering of two particles, showing transfer of momentum $\mathbf{q}$.

where the spatial integrals take the form

$$\int d^3x\, d^3x'\, e^{i(\mathbf{k}_1-\mathbf{k}_4+\mathbf{q})\cdot\mathbf{x}}\, e^{i(\mathbf{k}_2-\mathbf{k}_3-\mathbf{q})\cdot\mathbf{x}'} = (2\pi)^6\delta^{(3)}(\mathbf{k}_4 - \mathbf{k}_1 - \mathbf{q})\delta^{(3)}(\mathbf{k}_3 - \mathbf{k}_2 + \mathbf{q}), \quad (3.93)$$

which impose momentum conservation at each scattering event. Using the spatial integrals to eliminate the integrals over $\mathbf{k}_3$ and $\mathbf{k}_4$, the final result is

$$\hat{V} = \frac{1}{2}\sum_{\alpha,\beta}\int_{\mathbf{k}_1,\mathbf{k}_2,\mathbf{q}} V(\mathbf{q})c^{\dagger}_{\mathbf{k}_1+\mathbf{q}\alpha}c^{\dagger}_{\mathbf{k}_2-\mathbf{q}\beta}c_{\mathbf{k}_2\beta}c_{\mathbf{k}_1\alpha}. \quad (3.94)$$

In other words, when the particles scatter at positions x and x', momentum is conserved. Particle 1 comes in with momentum $\mathbf{k}_1$, and gains momentum $\mathbf{q}$ from particle 2:

$$\begin{array}{lll} \text{particle 1:} & \mathbf{k}_1 & \longrightarrow \quad \mathbf{k}_1 + \mathbf{q} \\ \text{particle 2:} & \mathbf{k}_2 & \longrightarrow \quad \mathbf{k}_2 - \mathbf{q} \end{array} \quad (3.95)$$

as illustrated in Figure 3.1. The matrix element associated with this scattering process is merely the Fourier transform of the potential $V(\mathbf{q})$.

Example 3.6 Particles interact via a delta-function interaction, $V(x) = Ua^3\delta^{(3)}(x)$. Write down the second-quantized interaction in a momentum space representation.

Solution

The Fourier transform of the interaction is

$$V(q) = \int d^3x\, Ua^3\delta(x)e^{-i\mathbf{q}\cdot\mathbf{x}} = Ua^3, \quad (3.96)$$

so the interaction in momentum space is

$$\hat{V} = \sum_{\alpha,\beta}\frac{Ua^3}{2}\int_{\mathbf{k}_1,\mathbf{k}_2,\mathbf{q}} c^{\dagger}_{\mathbf{k}_1-\mathbf{q}\alpha}c^{\dagger}_{\mathbf{k}_2+\mathbf{q}\beta}c_{\mathbf{k}_2\beta}c_{\mathbf{k}_1\alpha}. \quad (3.97)$$

Example 3.7 A set of fermions interact via a screened Coulomb (Yukawa) potential,

$$V(r) = \frac{Ae^{-\lambda r}}{r}. \quad (3.98)$$

Write down the interaction in momentum space.

Solution

The interaction in momentum space is given by

$$\hat{V} = \frac{1}{2} \sum_{\alpha, \beta} \int_{\mathbf{k}_1, \mathbf{k}_2, \mathbf{q}} V(\mathbf{q}) c^\dagger_{\mathbf{k}_1+\mathbf{q}\alpha} c^\dagger_{\mathbf{k}_2-\mathbf{q}\beta} c_{\mathbf{k}_2\beta} c_{\mathbf{k}_1\alpha}, \tag{3.99}$$

where

$$V(\mathbf{q}) = \int d^3x \frac{Ae^{-\lambda r}}{r} e^{-i\mathbf{q}\cdot\mathbf{x}}. \tag{3.100}$$

To carry out this integral, we use polar coordinates, with the z-axis aligned along the direction $\hat{\mathbf{q}}$. Writing $\mathbf{q} \cdot \mathbf{x} = qr\cos\theta$, then $d^3x = r^2 d\phi d\cos\theta \to 2\pi r^2 d\cos\theta$, so that

$$V(\mathbf{q}) = \int 4\pi r^2 dr V(r) \underbrace{\frac{1}{2} \int_{-1}^{1} d\cos\theta e^{-iqr\cos\theta}}_{\langle e^{-i\mathbf{q}\cdot\mathbf{x}} \rangle = \frac{\sin qr}{qr}}, \tag{3.101}$$

so that, for an arbitrary spherically symmetric potential,

$$V(q) = \int_0^\infty 4\pi r^2 dr V(r) \left(\frac{\sin qr}{qr} \right). \tag{3.102}$$

In this case,

$$V(q) = \frac{4\pi A}{q} \int_0^\infty dr e^{-\lambda r} \sin(qr) = \frac{4\pi A}{q^2 + \lambda^2}. \tag{3.103}$$

Notice that the Coulomb interaction

$$V(r) = \frac{e^2}{4\pi\epsilon_0 r} \tag{3.104}$$

is the infinite-range limit of the Yukawa potential, with $\lambda = 0$, $A = e^2/4\pi\epsilon_0$, so that, for the Coulomb interaction,

$$V(q) = \frac{e^2}{q^2\epsilon_0}. \tag{3.105}$$

Example 3.8 If one transforms to a new one-particle basis, writing $\psi(x) = \sum_s \Phi_s(x)c_s$, show that the interaction becomes

$$\hat{V} = \frac{1}{2} \sum_{lmnp,\alpha\beta} \langle lm|V|pn \rangle c^\dagger_{l\alpha} c^\dagger_{m\beta} c_{n\beta} c_{p\alpha}, \tag{3.106}$$

where

$$\langle lm|V|pn \rangle = \int_{x,y} (\Phi^*_l(\mathbf{x})\Phi_p(\mathbf{x}))(\Phi^*_m(\mathbf{y})\Phi^*_n(\mathbf{y}))V(\mathbf{x} - \mathbf{y}) \tag{3.107}$$

is the interaction matrix element for scattering $|pn\rangle \to |lm\rangle$.

Solution

Substituting $\psi_\alpha^\dagger(x) = \sum_l \Phi_l^*(x)c_{l\alpha}^\dagger$, $\psi_\beta^\dagger(y) = \sum_m \Phi_m^*(y)c_{m\beta}^\dagger$ etc. into

$$\hat{V} = \frac{1}{2}\sum_{\alpha,\beta}\int_{\mathbf{x},\mathbf{y}}\left[\psi_\alpha^\dagger(\mathbf{x})\psi_\beta^\dagger(\mathbf{y})\psi_\beta(\mathbf{y})\psi_\alpha(\mathbf{x})\right]V(\mathbf{x}-\mathbf{y}), \tag{3.108}$$

we obtain

$$\hat{V} = \frac{1}{2}\sum_{l,m,n,p,\alpha,\beta}\overbrace{\left[\int_{\mathbf{x},\mathbf{y}}\Phi_l^*(\mathbf{x})\Phi_m^*(\mathbf{y})\Phi_n(\mathbf{y})\Phi_p(\mathbf{x})V(\mathbf{x}-\mathbf{y})\right]}^{\langle lm|\hat{V}|pn\rangle}c_{l\alpha}^\dagger c_{m\beta}^\dagger c_{n\beta}c_{p\alpha}$$

$$= \frac{1}{2}\sum_{lmnp,\alpha\beta}\langle lm|V|pn\rangle c_{l\alpha}^\dagger c_{m\beta}^\dagger c_{n\beta}c_{p\alpha}. \tag{3.109}$$

3.7   Equivalence with the many-body Schrödinger equation

In this section, we establish that our second-quantized version of the many-body Hamiltonian is indeed equivalent to the many-body Schrödinger equation. Let us start with the Hamiltonian for an interacting gas of charged particles,

$$H = \overbrace{\sum_\sigma \int_x \psi_\sigma^\dagger\left[-\frac{\hbar^2\nabla^2}{2m}+U(x)-\mu\right]\psi_\sigma(x)}^{H_0} + \overbrace{\frac{1}{2}\int_{x,x'}V(x-x'):\hat{\rho}(x)\hat{\rho}(x'):}^{\hat{V}}, \tag{3.110}$$

where $\int_x \equiv \int d^3x$ and, by convention, we work in the grand canonical ensemble, subtracting the term μN from the Schrödinger Hamiltonian H_S: $H = H_S - \mu N$. For a Coulomb interaction,

$$V(x-x') = \frac{e^2}{4\pi\epsilon_0|x-x'|}, \tag{3.111}$$

but the interaction might take other forms, such as the *hard-core* interaction between neutral atoms in liquid ^{3}He and ^{4}He.

The Heisenberg equation of motion of the field operator is

$$i\hbar\frac{\partial\psi_\sigma}{\partial t} = [\psi_\sigma, H]. \tag{3.112}$$

Using the relations

$$[\psi_\sigma(x), \psi_{\sigma'}^\dagger(x')O_{x'}\psi_{\sigma'}(x')] = \delta_{\sigma\sigma'}\delta^3(x-x')O_x\psi_\sigma(x),$$
$$[\psi_\sigma(x), :\rho(x_1)\rho(x_2):] = \delta^3(x_1-x)\rho(x_2)\psi_\sigma(x) + \delta^3(x_2-x)\rho(x_1)\psi_\sigma(x),$$

we can see that the commutators of the one- and two-particle parts of the Hamiltonian with the field operator are

$$[\psi_\sigma(x), H_0] = \left[-\frac{\hbar^2\nabla^2}{2m} + U(x) - \mu\right]\psi_\sigma(x)$$

$$[\psi_\sigma(x), V] = \int d^3x' V(x' - x)\rho(x')\psi_\sigma(x). \tag{3.113}$$

The final equation of motion of the field operator thus resembles a one-particle Schrödinger equation:

$$i\hbar\frac{\partial\psi_\sigma}{\partial t} = \left[-\frac{\hbar^2\nabla^2}{2m} + U(x) - \mu\right]\psi_\sigma(x) + \int d^3x' V(x' - x)\rho(x')\psi_\sigma(x). \tag{3.114}$$

If we now apply this to the many-body wavefunction, we obtain

$$i\hbar\frac{\partial\Psi(1, 2, \ldots N)}{\partial t} = i\hbar\sum_{j=1,N}\langle 0|\psi(1)\cdots\frac{\partial\psi(j)}{\partial t}\cdots\psi(N)|\Psi\rangle$$

$$= \sum_j\left[-\frac{\hbar^2\nabla_j^2}{2m} + U(x_j) - \mu\right]\Psi$$

$$+ \sum_j\int d^3x' V(x' - x_j)\langle 0|\psi(1)\cdots\rho(x')\psi_\sigma(x_j)\cdots\psi(N)|\Psi\rangle.$$

By commuting the density operator to the left until it annihilates with the vacuum, we find that

$$\langle 0|\psi(1)\cdots\rho(x')\psi_\sigma(x_j)\cdots\psi(N)|\Psi\rangle = \sum_{l<j}\delta^3(x' - x_l)\langle 0|\psi(1)\cdots\psi(N)|\Psi\rangle, \tag{3.115}$$

so that the final expression for the time evolution of the many-body wavefunction is precisely the same as we obtain in a first-quantized approach:

$$i\hbar\frac{\partial\Psi}{\partial t} = \left(\sum_j\mathcal{H}_j^{(0)} + \sum_{l<j}V_{lj}\right)\Psi. \tag{3.116}$$

The second-quantized approach has the advantages that it builds in the exchange statistics and does not need an explicit reference to the many-body wavefunction.

Notice, finally, that instead of using the equation of motion of the fields, we could have alternatively proved that $|\Psi\rangle$ satisfies the many-body Schrödinger equation using the results

$$\langle x_1, \ldots x_N|H_0 = \sum_j\mathcal{H}_j^{(0)}\langle x_1, \ldots x_N| \tag{3.117}$$

and

$$\langle x_1, \ldots x_N|\hat{V} = \sum_{i<j}V(x_i - x_j)\langle x_1, \ldots x_N|, \tag{3.118}$$

so that

$$i\hbar\frac{\partial\Psi}{\partial t} = \langle x_1, \ldots x_N|\hat{H}_0 + \hat{V}|\Psi\rangle = \left(\sum_j\mathcal{H}_j^{(0)} + \sum_{l<j}V_{lj}\right)\langle x_1, \ldots x_N|\Psi\rangle. \tag{3.119}$$

3.8 Identical conserved particles in thermal equilibrium

3.8.1 Generalities

By quantizing the particle field, we have been led to a version of quantum mechanics with a vastly expanded Hilbert space which includes the vacuum and all possible states with an arbitrary number of particles. An exactly parallel development occurs in statistical thermodynamics in making the passage from a canonical to a grand canonical ensemble, where systems are considered to be in equilibrium with a heat and particle bath. Not surprisingly, then, second quantization provides a beautiful way of treating a grand canonical ensemble of identical particles.

When we come to treat conserved particles in thermal equilibrium, we have to take into the account the conservation of two independent quantities:

- energy, E
- particle number, N.

Statistical mechanics usually begins with an ensemble of identical systems of definite particle number N and energy E (more precisely, particle number and energy lying in the narrow ranges $[N, N + dN]$ and $[E, E + dE]$, respectively). Such an ensemble is called a *microcanonical ensemble*. This is a confusing name, because it suggests something *small*, yet typically a microcanonical ensemble is an ensemble of identical macroscopic systems that play the role of a heat bath [8–11]. The ergodic hypothesis of statistical mechanics assumes that, in such an ensemble, all accesible quantum states within this narrow band of allowed energies and particle numbers are equally probable (equal *a priori* probability).

Now suppose we divide the system into two parts: a vast heat bath and a tiny sub-system, exchanging energy and particles as shown in Figure 3.2, until they reach a state of thermal equilibrium. In the vast heat and particle bath, the energy levels are so close together that they form a continuum. The density of states per unit energy and particle number is taken to be $W(E', N')$, where E' is the energy and N' the number of particles in the bath. When the tiny sub-system is in a quantum state $|\lambda\rangle$ with energy E_λ and particle number N_λ, the large system has energy $E' = E - E_\lambda$ and particle number $N' = N - N_\lambda$.

Assuming equal *a priori* probability, the probability that the small system is in state $|\lambda\rangle$ is proportional to the number of states $W(E, N)$ of the heat bath with energy $E - E_\lambda$ and particle number $N - N_\lambda$:

$$p(E_\lambda, N_\lambda) \propto W(E - E_\lambda, N - N_\lambda) = e^{\ln W(E - E_\lambda, N - N_\lambda)}. \tag{3.120}$$

Following Boltzmann, we can tentatively identify $W(E, N)$ with the entropy $S(E, N)$ of the heat bath (see Exercise 3.6), according to the famous formula

$$S_B(E, N) = k_B \ln W(E, N), \tag{3.121}$$

where we have included the subscript B to indicate the heat bath. It follows that

$$p(E_\lambda, N_\lambda) \propto \exp\left[\frac{1}{k_B} S_B(E - E_\lambda, N - N_\lambda)\right]. \tag{3.122}$$

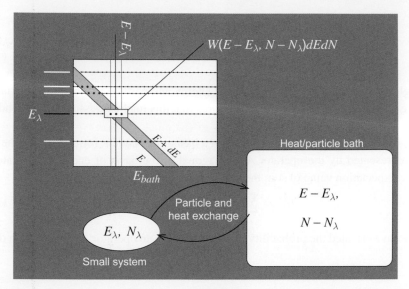

$W(E - E_\lambda, N - N_\lambda)dEdN$

Heat/particle bath

$E - E_\lambda,$

$N - N_\lambda$

Particle and
heat exchange

E_λ, N_λ

Small system

Illustrating equilibrium between a small system and a large heat bath. The diagram at top left illustrates how the number of states with energy E_λ and particle number N_λ is proportional to the density of states in the big system.

Fig. 3.2

Now E_λ and N_λ are tiny perturbations to the total energy and particle number of the heat bath, so we may approximate $S(E - E_\lambda, N - N_\lambda)$ by a linear expansion:

$$S_B(E - E_\lambda, N - N_\lambda) = S_B(E, N) - E_\lambda \frac{\partial S_B}{\partial E} - N_\lambda \frac{\partial S_B}{\partial N} + \cdots . \qquad (3.123)$$

Now according to thermodynamics, $dE = TdS + \mu dN$, where T and μ are the temperature and chemical potential, respectively, so that $dS_B = \frac{1}{T}dE - \frac{\mu}{T}dN$, allowing us to identify

$$\frac{1}{k_B}\frac{\partial S_B}{\partial E} = \frac{\partial \ln W}{\partial E} = \frac{1}{k_B T} \equiv \beta$$
$$\frac{1}{k_B}\frac{\partial S_B}{\partial N} = \frac{\partial \ln W}{\partial N} = -\frac{\mu}{k_B T} \equiv -\mu\beta. \qquad (3.124)$$

These are the Lagrange multipliers associated with the conservation of energy and of particle number.[1] Once we have made this expansion, it follows that the probability of being in state $|\lambda\rangle$ is

$$p_\lambda = \frac{1}{Z} e^{-\beta(E_\lambda - \mu N_\lambda)}, \qquad (3.125)$$

where the normalizing partition function is $Z = \sum_\lambda e^{-\beta(E_\lambda - \mu N_\lambda)}$.

To recast statistical mechanics in the language of many-body theory, we need to rewrite the above expression in terms of operators. Let us begin with the partition function, which we may rewrite as

---

[1] Incidentally, if you are uncomfortable with the use of classical thermodynamics to identify these quantities in terms of the temperature and chemical potential, you may regard these assignments as tentative, pending calculations of physical properties that allow us to definitively identify them in terms of temperature and chemical potential.

$$Z = \sum_\lambda e^{-\beta(E_\lambda - \mu N_\lambda)}$$
$$= \sum_\lambda \langle \lambda | e^{-\beta(\hat{H} - \mu \hat{N})} | \lambda \rangle = \text{Tr}[e^{-\beta(\hat{H} - \mu \hat{N})}]. \quad (3.126)$$

Although we started with the eigenstates of energy and particle number, the invariance of the trace under unitary transformations ensures that this final expression is independent of the many-body basis.

Next, we cast the expectation value $\langle \hat{A} \rangle$ in a basis-independent form. Suppose the quantity A, represented by the operator $\hat{A}$, is diagonal in the basis of energy eigenstates $|\lambda\rangle$. Then the expectation value of A in the ensemble is

$$\langle A \rangle = \sum_\lambda p_\lambda \langle \lambda | \hat{A} | \lambda \rangle = \text{Tr}[\hat{\rho} \hat{A}]. \quad (3.127)$$

Here we have elevated the probability distribution p_λ to an operator, the Boltzmann density matrix:

$$\hat{\rho} = \sum_\lambda |\lambda\rangle p_\lambda \langle \lambda | = Z^{-1} e^{-\beta(\hat{H} - \mu \hat{N})}. \quad (3.128)$$

The derivation of (3.127) assumed that $\hat{A}$ could be simultaneously diagonalized with the energy and particle number. However, quantum statistical mechanics makes the radical assertion that (3.127) holds for all quantum operators $\hat{A}$ representing observables, *even when the operator $\hat{A}$ does not commute with $\hat{H}$ or $\hat{N}$* and is thus not diagonal in the energy and particle number basis.

3.8.2 Identification of the free energy: key thermodynamic properties

There are a number of key thermodynamic quantities of great interest: the energy E, the particle number N, the entropy S and the free energy $F = E - ST - \mu N$. One of the key relations from elementary thermodynamics is that

$$dE = TdS - \mu dN - PdV. \quad (3.129)$$

By putting $F = E - TS - \mu N$ and $dF = dE - dTS - SdT - \mu dN - Nd\mu$, one can also derive

$$dF = -SdT - Nd\mu - PdV, \quad (3.130)$$

a relationship of great importance.

The energy and particle number can easily be written in the language of second quantization as

$$E = \text{Tr}[\hat{H}\hat{\rho}]$$
$$N = \text{Tr}[\hat{N}\hat{\rho}], \quad (3.131)$$

but what about the entropy? From statistical mechanics, we know that the general expression for the entropy is

$$S = -k_B \sum_\lambda p_\lambda \ln p_\lambda. \quad (3.132)$$

Now since the diagonal elements of the density matrix are p_λ, we can rewrite this expression as

$$S = -k_B \text{Tr}[\hat{\rho} \ln \hat{\rho}]. \tag{3.133}$$

If we substitute $\ln \hat{\rho} = -\beta(\hat{H} - \mu \hat{N}) - \ln Z$ into this expression, we obtain

$$S = \frac{1}{T} \text{Tr} \hat{\rho}(H - \mu N) + k_B \ln Z$$

$$= \frac{1}{T}(E - \mu N) + k_B \ln Z, \tag{3.134}$$

i.e. $-k_B T \ln Z = E - ST - \mu N$, from which we identify

$$F = -k_B T \ln Z \tag{3.135}$$

as the free energy.

Summarizing these key thermodynamic relationships, we have:

$F = -k_B T \ln Z$	free energy
$Z = \text{Tr}[e^{-\beta(\hat{H} - \mu \hat{N})}]$	partition function
$\hat{\rho} = \dfrac{e^{-\beta(\hat{H} - \mu \hat{N})}}{Z}$	density matrix
$N = \text{Tr}[\hat{N}\hat{\rho}] = -\dfrac{\partial F}{\partial \mu}$	particle number
$S = -k_B \text{Tr}[\hat{\rho} \ln \hat{\rho}] = -\dfrac{\partial F}{\partial T}$	entropy
$P = -\dfrac{\partial F}{\partial V}$	pressure
$E - \mu N = \text{Tr}[(\hat{H} - \mu \hat{N})\hat{\rho}] = -\dfrac{\partial \ln Z}{\partial \beta}$	energy

Notice how all the key thermodynamic properties can be written as appropriate derivatives of the free energy.

Example 3.9 (a) Enumerate the energy eigenstates of a single fermion Hamiltonian,

$$H = E c^\dagger c, \tag{3.136}$$

where $\{c, c^\dagger\} = 1$, $\{c, c\} = \{c^\dagger, c^\dagger\} = 0$.
(b) Calculate the number of fermions at temperature T.

Solution

(a) The states of this problem are the vacuum state and the one-particle state:

$$\begin{array}{lll} |0\rangle & E_0 & = 0 \\ |1\rangle = c^\dagger |0\rangle & E_1 & = E. \end{array} \tag{3.137}$$

(b) The number of fermions at temperature T is given by

$$\langle \hat{n} \rangle = \text{Tr}[\hat{\rho} \hat{n}], \tag{3.138}$$

where $\hat{n} = c^\dagger c$,

$$\rho = e^{-\beta(\hat{H} - \mu \hat{N})}/Z \tag{3.139}$$

is the density matrix, and

$$Z = \text{Tr}[e^{-\beta(H-\mu N)}] \qquad (3.140)$$

is the partition function. For this problem, we can write out the matrices explicitly:

$$e^{-\beta(H-\mu N)} = \begin{bmatrix} 1 & 0 \\ 0 & e^{-\beta(E-\mu)} \end{bmatrix}, \qquad \hat{n} = \begin{bmatrix} 0 & 0 \\ 0 & 1 \end{bmatrix}, \qquad (3.141)$$

so that

$$Z = 1 + e^{-\beta(E-\mu)} \qquad (3.142)$$

and

$$\text{Tr}[\hat{n}e^{-\beta(H-\mu N)}] = e^{-\beta(E-\mu)}. \qquad (3.143)$$

The final result is thus

$$\langle \hat{n} \rangle = \frac{e^{-\beta(E-\mu)}}{1 + e^{-\beta(E-\mu)}} = \frac{1}{e^{\beta(E-\mu)} + 1}, \qquad (3.144)$$

which is the famous Fermi–Dirac function for the number of fermions in a state of energy E and chemical potential μ.

3.8.3 Independent particles

In a system of independent particles with many energy levels E_λ, each energy level can be regarded as an independent member of a microcanonical ensemble. Formally, this is because the Hamiltonian is a sum of independent Hamiltonians

$$H - \mu N = \sum_\lambda (E_\lambda - \mu)\hat{n}_\lambda, \qquad (3.145)$$

so that the partition function is then a product of the individual partition functions:

$$Z = \text{Tr}\left[\prod_{\lambda \otimes} e^{-\beta(E_\lambda-\mu)\hat{n}_\lambda}\right]. \qquad (3.146)$$

Since the trace of an (exterior) product of matrices, this is equal to the product of their individual traces ($\text{Tr} \prod_{\lambda\otimes} = \prod_\lambda \text{Tr}$),

$$Z = \prod_\lambda \text{Tr}[e^{-\beta(E_\lambda-\mu)\hat{n}_\lambda}] = \prod_\lambda Z_\lambda. \qquad (3.147)$$

Since

$$Z_\lambda = \begin{cases} 1 + e^{-\beta(E_\lambda-\mu)} & \text{fermions} \\ 1 + e^{-\beta(E_\lambda-\mu)} + e^{-2\beta(E_\lambda-\mu)} + \cdots = (1 - e^{-\beta(E_\lambda-\mu)})^{-1}, & \text{bosons} \end{cases} \qquad (3.148)$$

the corresponding free energy is given by

$$F = \mp k_B T \sum_\lambda \ln[1 \pm e^{-\beta(E_\lambda - \mu)}]. \qquad \left\{ \begin{array}{l} + \text{ for fermions} \\ - \text{ for bosons} \end{array} \right. \qquad (3.149)$$

The occupancy of the lth level is independent of all the other levels, and is given by

$$\langle \hat{n}_l \rangle = \text{Tr}[\hat{\rho} \hat{n}_l] = \text{Tr}\left[(\prod_\otimes \hat{\rho}_\lambda) \hat{n}_l \right]$$

$$= \prod_{\lambda \neq l} \overbrace{\text{Tr}[\rho_\lambda]}^{=1} \times \text{Tr}[\rho_l \hat{n}_l] = \frac{1}{e^{\beta(E_l - \mu)} \pm 1}, \qquad (3.150)$$

where $(+)$ refers to fermions and $(-)$ to bosons.

In the next chapter, we shall examine the consequences of these relationships.

Exercises

Exercise 3.1 In this question $c_i^\dagger$ and c_i are fermion creation and annihilation operators and the states are fermion states. Use the convention $|11111000\cdots\rangle = c_5^\dagger c_4^\dagger c_3^\dagger c_2^\dagger c_1^\dagger |\text{vacuum}\rangle$.

(a) Evaluate $c_3^\dagger c_6 c_4 c_6^\dagger c_3 |111111000\cdots\rangle$.

(b) Write $|1101100100\cdots\rangle$ in terms of excitations about the "filled Fermi sea" $|1111100000\cdots\rangle$. Interpret your answer in terms of electron and hole excitations.

(c) Find $\langle \psi | \hat{N} | \psi \rangle$, where $|\psi\rangle = A|100\rangle + B|111000\rangle$, $\hat{N} = \sum_i c_i^\dagger c_i$.

Exercise 3.2 Consider two fermions, a_1 and a_2.

(a) Show that the Bogoliubov transformation

$$c_1 = u a_1 + v a_2^\dagger$$
$$c_2^\dagger = -v a_1 + u a_2^\dagger, \qquad (3.151)$$

where u and v are real, preserves the canonical anticommutation relations if $u^2 + v^2 = 1$.

(b) Use this result to show that the Hamiltonian

$$H = \epsilon(a_1^\dagger a_1 - a_2 a_2^\dagger) + \Delta(a_1^\dagger a_2^\dagger + \text{H.c.}) \qquad (3.152)$$

can be diagonalized in the form

$$H = \sqrt{\epsilon^2 + \Delta^2}(c_1^\dagger c_1 + c_2^\dagger c_2 - 1). \qquad (3.153)$$

(c) What is the ground-state energy of this Hamiltonian?

(d) Write out the ground-state wavefunction in terms of the original operators $c_1^\dagger$ and $c_2^\dagger$ and their corresponding vacuum $|0\rangle$ $(c_{1,2}|0\rangle = 0)$.

Exercise 3.3 This is an alterative derivation of the normal-ordered interaction.

(a) Show that, for a system of N identical classical particles with density $\rho(\mathbf{x}) = \sum_{j=1,N} \delta^{(3)}(\mathbf{x} - \mathbf{x}_j)$, the interaction energy, corrected to avoid self-interaction, is given by

$$V = \sum_{i<j} V(\mathbf{x}_i - \mathbf{x}_j) = \frac{1}{2} \int d^3x d^3y V(\mathbf{x} - \mathbf{y})\rho(\mathbf{x})\rho(\mathbf{y}) - \frac{1}{2}V(0)N.$$

(b) Second-quantize the above result and show that it leads directly to the normal-ordered form of the interaction, i.e. that

$$\hat{V} = \frac{1}{2} \int d^3x d^3y V(\mathbf{x} - \mathbf{y})\hat{\rho}(\mathbf{x})\hat{\rho}(\mathbf{y}) - \frac{1}{2} \int d^3x V(0)\hat{\rho}(x)$$
$$= \frac{1}{2} \int d^3x d^3y V(\mathbf{x} - \mathbf{y}) : \hat{\rho}(\mathbf{x})\hat{\rho}(\mathbf{y}) : .$$

Exercise 3.4 Consider a system of fermions or bosons created by the field $\psi^\dagger(\mathbf{r})$ interacting under the potential

$$V(r) = \begin{cases} U & (r < R) \\ 0 & (r > R). \end{cases} \tag{3.154}$$

(a) Write the interaction in second-quantized form.

(b) Switch to the momentum basis, where $\psi(\mathbf{r}) = \int \frac{d^3k}{(2\pi)^3} c_{\mathbf{k}} e^{i\mathbf{k}\cdot\mathbf{r}}$. Verify that $[c_{\mathbf{k}}, c_{\mathbf{k}'}^\dagger]_\pm = (2\pi)^3 \delta^{(3)}(\mathbf{k} - \mathbf{k}')$ and write the interaction in this new basis. Sketch the form of the interaction in momentum space.

Exercise 3.5

(a) Show that, for a general system of conserved particles at chemical potential μ, the total particle number in thermal equilibrium can be written as

$$N = -\partial F/\partial \mu, \tag{3.155}$$

where

$$F = -k_B T \ln Z$$
$$Z = \text{Tr}[e^{-\beta(\hat{H}-\mu N)}]. \tag{3.156}$$

(b) Apply this to a single bosonic energy level, where

$$H - \mu N = (\epsilon - \mu)a^\dagger a \tag{3.157}$$

and $\hat{a}^\dagger$ creates either a fermion or a boson, to show that

$$\langle \hat{n} \rangle = \frac{1}{e^{\beta(\epsilon-\mu)} - 1}. \tag{3.158}$$

Why does μ have to be negative for bosons?

Exercise 3.6 (Equivalence of the microcanonical and Gibbs ensembles for large systems.) In a microcanonical ensemble, the density matrix can be written

$$\hat{\rho}_M = \frac{1}{W}\delta(E - \hat{H})\delta(N - \hat{N}),$$

where E and N are the energy and particle number, respectively, while

$$W \equiv W(E, N) = \text{Tr}\left[\delta(E - \hat{H})\delta(N - \hat{N})\right]$$

is the density of states at energy E, particle number N. This normalizing quantity plays a role similar to the partition function in the Gibbs ensemble.

(a) By rewriting the delta functions inside the above trace W as inverse Laplace transforms, such as

$$\delta(x - \hat{H}) = \int_{\beta_0 - i\infty}^{\beta_0 + i\infty} \frac{d\beta}{2\pi i} e^{-\beta(x - \hat{H})},$$

and evaluating the resulting integrals at the saddle point of the integrand, show that, for a large system, W is related to the entropy by Boltzmann's relation,

$$S(E, N) = k_B \ln W(E, N).$$

(b) Using your result, show that, in a large system, the expectation value of an operator is the same for corresponding Gibbs and microcanonical ensembles, namely

$$\langle A \rangle = \text{Tr}[\rho_M \hat{A}] = \text{Tr}[\rho_B \hat{A}],$$

where $\hat{\rho}_B = Z^{-1} e^{-\beta(H - \mu \hat{N})}|_{\beta = \beta_0, \mu = \mu_0}$ is the Boltzmann density matrix evaluated at the saddle point values of β_0 and μ_0,

$$\beta_0 = \frac{\partial \ln W}{\partial E}, \qquad \mu_0 = \beta_0^{-1} \frac{\partial \ln W}{\partial N}.$$

References

[1] H. Bethe, Zur Theorie der Metalle: I. Eigenwerte und Eigenfunktionen der linearer Atomkerte *(On the theory of metals: I. eigenvalues and eigenfunctions of the linear atom chain)*, Z. Phys., vol. 71, p. 205, 1931.

[2] M. Takahashi, *Thermodynamics of One-Dimensional Solvable Models*, Cambridge University Press, 1999.

[3] N. Andrei, K. Furuya, and J. H. Lowenstein, Solution of the Kondo problem, *Rev. Mod. Phys.*, vol. 55, p. 331, 1983.

[4] A. Tsvelik and P. Wiegman, The exact results for magnetic alloys, *Adv. Phys.*, vol. 32, p. 453, 1983.

[5] R. B. Laughlin, Anomalous quantum Hall effect: an incompressible quantum fluid with fractionally charged excitations, *Phys. Rev. Lett.*, vol. 50, no. 18, p. 1395, 1983.

[6] S. Girvin, Quantum Hall effect: novel excitations and broken symmetries, in *Topological Aspects of Low-Dimensional Systems*, ed. A. Comtet, T. Jolicoeur, S. Ouvry, F. David, pp. 53–175, Springer-Verlag and Les Editions de Physique, 2000.

[7] P. Jordan and E. Wigner, Uber das Paulische Aquivalenzverbot (On the Pauli exclusion principle), *Z. Phys.*, vol. 47, p. 631, 1928.

[8] F. Reif, *Fundamentals of Statistical and Thermal Physics*, McGraw-Hill, 1965.

[9] N. Saito, M. Toda, and R. Kubo, *Statistical Physics I*, Springer, 1998.

[10] R. P. Feynman, *Statistical Mechanics: A Set of Lectures*, Westview Press, 1998.

[11] M. Kardar, *Statistical Physics of Particles*, Cambridge University Press, 2007.

Simple examples of second quantization   4

In this chapter, we give three examples of the application of second quantization, mainly to non-interacting systems.

4.1 Jordan–Wigner transformation

A non-interacting gas of fermions is still highly correlated: the exclusion principle introduces a *hard-core* interaction between fermions in the same quantum state. This feature is exploited in the Jordan–Wigner representation of spins. A classical spin is represented by a vector pointing in a specific direction. Such a representation is fine for quantum spins with extremely large spin S, but once the spin S becomes small, spins behave as very new kinds of object. Now their spin becomes a quantum variable, subject to its own zero-point motions. Furthermore, the spectrum of excitations becomes discrete or grainy.

Quantum spins are notoriously difficult objects to deal with in many-body physics, because they do not behave as canonical fermions or bosons. In one dimension, however, it turns out that spins with $S = \frac{1}{2}$ actually behave like fermions. We shall show this by writing the quantum spin-$\frac{1}{2}$ Heisenberg chain as an interacting one-dimensional gas of fermions, and we shall actually solve the limiting case of the one-dimensional spin-$\frac{1}{2}$ x-y model, in which the Ising (z) component of the interaction is set to zero.

Jordan and Wigner observed [1] that the "down" and "up" states of a single spin can be thought of as empty and singly occupied fermion states (Figure 4.1.), enabling them to make the mapping (see Figure 4.1)

$$| \uparrow \rangle \equiv f^{\dagger}|0\rangle, \qquad | \downarrow \rangle \equiv |0\rangle. \tag{4.1}$$

An explicit representation of the spin-raising and spin-lowering operators is then

$$S^{+} = f^{\dagger} = \begin{bmatrix} 0 & 1 \\ 0 & 0 \end{bmatrix}$$

$$S^{-} = f \equiv \begin{bmatrix} 0 & 0 \\ 1 & 0 \end{bmatrix}. \tag{4.2}$$

The z component of the spin operator can be written

$$S_z = \frac{1}{2}\left[| \uparrow \rangle\langle \uparrow | - | \downarrow \rangle\langle \downarrow | \right] \equiv f^{\dagger}f - \frac{1}{2}. \tag{4.3}$$

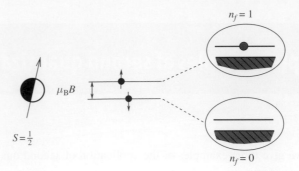

Fig. 4.1   The "up" and "down" states of a spin $\frac{1}{2}$ can be treated as a one-particle state which is either full or empty.

We can also reconstruct the transverse spin operators

$$
S_x = \frac{1}{2}(S^+ + S^-) = \frac{1}{2}(f^\dagger + f)
$$
$$
S_y = \frac{1}{2i}(S^+ - S^-) = \frac{1}{2i}(f^\dagger - f). \tag{4.4}
$$

The explicit matrix representation of these operators makes it clear that they satisfy the same algebra:

$$
[S_a, S_b] = i\epsilon_{abc}S_c. \tag{4.5}
$$

Curiously, due to a hidden supersymmetry, they also satisfy an anticommuting algebra:

$$
\{S_a, S_b\} = \frac{1}{4}\{\sigma_a, \sigma_b\} = \frac{1}{2}\delta_{ab}. \tag{4.6}
$$

In this way, the Pauli spin operators provided Jordan and Wigner with an elementary model of a fermion.

Unfortunately the represeentation needs to be modified if there is more than one spin, for independent spin operators commute but independent fermions anticommute. Jordan and Wigner discovered a way to fix this difficulty in one dimension by attaching a phase factor called a *string* to the fermions [1]. For a chain of spins in one dimension, the Jordan–Wigner representation of the spin operator at site j is defined as

$$
S_j^+ = f_j^\dagger e^{i\phi_j}, \tag{4.7}
$$

where the phase operator ϕ_j contains the sum over all fermion occupancies at sites to the left of j:

$$
\phi_j = \pi \sum_{l<j} n_l. \tag{4.8}
$$

The operator $e^{i\hat{\phi}_j}$ is known as a *string operator*.

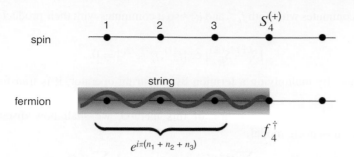

Fig. 4.2

Illustrating the Jordan–Wigner transformation. The spin-raising operator at site $j = 4$ is decomposed into a product of a fermion operator and a string operator.

The complete transformation is then

$$
\left.
\begin{aligned}
S_j^z &= f_j^\dagger f_j - \tfrac{1}{2} \\
S_j^+ &= f_j^\dagger e^{i\pi \sum_{l<j} n_l} \\
S_j^- &= f_j e^{-i\pi \sum_{l<j} n_l} .
\end{aligned}
\right\} \quad \text{Jordan–Wigner transformation} \qquad (4.9)
$$

(Notice that $e^{i\pi n_j} = e^{-i\pi n_j}$ is a Hermitian operator so that the overall sign of the phase factors can be reversed without changing the spin operator.) In words (Figure 4.2):

$$\text{spin} = \text{fermion} \times \text{string}.$$

The important property of the string is that it *anticommutes* with any fermion operator to the left of its free end. To see this, note first that the operator $e^{i\pi n_j}$ anticommutes with f_j. This follows because f_j reduces n_j from unity to zero so that, acting to the right of f_j, since $n_j = 1$, $e^{i\pi n_j} = -1$ and hence $f_j e^{i\pi n_j} = -f_j$; whereas, acting to the left of f_j, since $n_j = 0$, $e^{i\pi n_j} f_j = f_j$. It follows that

$$\{e^{i\pi n_j}, f_j\} = e^{i\pi n_j} f_j + f_j e^{i\pi n_j} = f_j - f_j = 0 \qquad (4.10)$$

and similarly, from the conjugate of this expression, $\{e^{i\pi n_j}, f_j^\dagger\} = 0$. Now the phase factor $e^{i\pi n_l}$ at any other site $l \neq j$ commutes with f_j and $f_j^\dagger$, so that the string operator $e^{i\hat{\phi}_j}$ anticommutes with all fermions at all sites l to the left of j, i.e. $l < j$:

$$\{e^{i\phi_j}, f_l^{(\dagger)}\} = 0 \qquad (l < j),$$

while commuting with fermions at all other sites $l \geq j$:

$$[e^{i\phi_j}, f_l^{(\dagger)}] = 0 \qquad (l \geq j).$$

We can now verify that the transverse spin operators satisfy the correct commutation algebra. Suppose $j < k$; then $e^{i\phi_j}$ commutes with fermions at sites j and k, so that

$$[S_j^{(\pm)}, S_k^{(\pm)}] = [f_j^{(\dagger)} e^{i\phi_j}, f_k^{(\dagger)} e^{i\phi_k}] = e^{i\phi_j} [f_j^{(\dagger)}, f_k^{(\dagger)} e^{i\phi_k}].$$

But $f_j^{(\dagger)}$ antcommutes with both $f_k^{(\dagger)}$ and $e^{i\phi_k}$ so it commutes with their product $f_k^{(\dagger)}e^{i\phi_k}$, and hence

$$[S_j^{(\pm)}, S_k^{(\pm)}] \propto [f_j^{(\dagger)}, f_k^{(\dagger)}e^{i\phi_k}] = 0. \tag{4.11}$$

So we see that by multiplying a fermion by the string operator, it is transformed into a boson.

As an example of the application of this method, we shall now discuss the one-dimensional Heisenberg model,

$$H = -J\sum[S_j^x S_{j+1}^x + S_j^y S_{j+1}^y] - J_z \sum_j S_j^z S_{j+1}^z. \tag{4.12}$$

In real magnetic systems, local moments can interact via ferromagnetic or antiferromagnetic interactions. Ferromagnetic interactions generally arise as a result of *direct exchange*, in which the Coulomb repulsion energy is lowered when electrons are in a triplet state because the wavefunction is then spatially antisymmetric. Antiferromagnetic interactions are generally produced by the mechanism of *super exchange*, in which electrons on neighboring sites that form singlets (antiparallel spin) lower their energy through virtual quantum fluctuations into high-energy states in which they occupy the same orbital. Here we have written the model as if the interactions are ferromagnetic.

For convenience, the model can be rewritten as

$$H = -\frac{J}{2}\sum[S_{j+1}^+ S_j^- + \text{H.c.}] - J_z \sum_j S_j^z S_{j+1}^z, \tag{4.13}$$

where "H.c." denotes the Hermitian conjugate. To "fermionize" the first term, we note that all terms in the strings cancel except for $e^{i\pi n_j}$, which has no effect:

$$\frac{J}{2}\sum_j S_{j+1}^+ S_j^- = \frac{J}{2}\sum_j f_{j+1}^\dagger e^{i\pi n_j} f_j = \frac{J}{2}\sum_j f_{j+1}^\dagger f_j, \tag{4.14}$$

so that the transverse component of the interaction induces a "hopping" term in the fermionized Hamiltonian. Notice that the string terms would enter if the spin interaction involved next-nearest neighbors. The z component of the Hamiltonian becomes

$$-J_z \sum_j S_{j+1}^z S_j^z = -J_z \sum_j \left(n_{j+1} - \frac{1}{2}\right)\left(n_j - \frac{1}{2}\right). \tag{4.15}$$

Notice how the ferromagnetic interaction means that spin fermions attract one another. The transformed Hamiltonian is then

$$H = -\frac{J}{2}\sum_j (f_{j+1}^\dagger f_j + f_j^\dagger f_{j+1}) + J_z \sum_j n_j - J_z \sum_j n_j n_{j+1}. \tag{4.16}$$

Interestingly enough, the pure x-y model has no interaction term in it, so this case can be mapped onto a non-interacting fermion problem, a discovery made by Lieb, Schulz and Mattis in 1961 [2].

To write out the fermionized Hamiltonian in its most compact form, let us transform to momentum space, writing

$$f_j = \frac{1}{\sqrt{N}} \sum_k s_k e^{ikR_j}, \tag{4.17}$$

where $s_k^\dagger$ creates a spin excitation in momentum space, with momentum k. In this case, the one-particle terms become

$$J_z \sum_j n_j = J_z \sum_k s_k^\dagger s_k$$

$$-\frac{J}{2} \sum_j (f_{j+1}^\dagger f_j + \text{H.c.}) = -\frac{J}{2N_s} \sum_k (e^{-ika} + e^{ika}) s_k^\dagger s_{k'} \overbrace{\sum_j e^{-i(k-k')R_j}}^{N\delta_{kk's}}$$

$$= -J \sum_k \cos(ka) s_k^\dagger s_k. \tag{4.18}$$

The anisotropic Heisenberg Hamiltonian can thus be written

$$H = \sum_k \omega_k s_k^\dagger s_k - J_z \sum_j n_j n_{j+1}, \tag{4.19}$$

where

$$\omega_k = (J_z - J\cos ka) \tag{4.20}$$

defines a *magnon excitation energy*. We can also cast the second term in momentum space by noticing that the interaction is a function of $i - j$, which is $-J_z/2$ for $i - j = \pm 1$ but zero otherwise. The Fourier transform of this short-range interaction is $V(q) = -J_z \cos qa$, so that Fourier transforming the interaction term gives

$$H = \sum_k \omega_k s_k^\dagger s_k - \frac{J_z}{N_s} \sum_{k,k',q} \cos(qa) \, s_{k-q}^\dagger s_{k'+q}^\dagger s_{k'} s_k. \tag{4.21}$$

This transformation holds for both the ferromagnet and the antiferromagnet. In the former case, the fermionic spin excitations correspond to the magnons of the ferromagnet. In the latter case, the fermionic spin excitations are often called *spinons*.

To see what this Hamiltonian means, let us first neglect the interactions. This is a reasonable thing to do in the limiting cases of (i) the Heisenberg ferromagnet, $J_z = J$, and (ii) the x-y model, $J_z = 0$.

- *Heisenberg ferromagnet, $J_z = J$* (Figure 4.3)
  In this case, the spectrum

$$\omega_k = 2J \sin^2(ka/2) \tag{4.22}$$

is always positive, so that there are no magnons present in the ground state. The ground state can thus be written

$$|0\rangle = |\downarrow\downarrow\downarrow \cdots\rangle, \tag{4.23}$$

corresponding to a state with a spontaneous magnetization $M = -N_s/2$.

Curiously, since $\omega_{k=0} = 0$, it costs no energy to add a magnon of arbitrarily long wavelength. This is an example of a *Goldstone mode*, and the reason it arises is that the

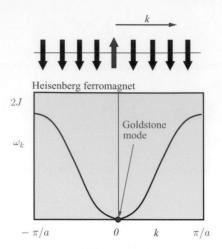

Heisenberg ferromagnet

Fig. 4.3 Excitation spectrum of the one-dimensional Heisenberg ferromagnet.

spontaneous magnetization could actually point in any direction. Suppose we want to rotate the magnetization through an infinitesimal angle $\delta\theta$ about the x-axis; then the new state is given by

$$|\psi\rangle' = e^{i\delta\theta S_x}|\downarrow\downarrow\cdots\rangle$$

$$= |\downarrow\downarrow\cdots\rangle + i\frac{\delta\theta}{2}\sum_j S_j^+|\downarrow\downarrow\cdots\rangle + O(\delta\theta^2). \tag{4.24}$$

The change in the wavefunction is proportional to the state

$$S_{TOT}^+|\downarrow\downarrow\cdots\rangle \equiv \sum_j f_j^\dagger e^{i\phi_j}|0\rangle$$

$$= \sum_j f_j^\dagger|0\rangle = \sqrt{N_s}\, s_{k=0}^\dagger|0\rangle. \tag{4.25}$$

In other words, the action of adding a single magnon at $q = 0$ rotates the magnetization infinitesimally upwards. Rotating the magnetization should cost no energy, and this is the reason why the $k = 0$ magnon is a zero-energy excitation.

- *x-y ferromagnet* (Figure 4.4)

  As J_z is reduced from J, the spectrum develops a negative part, and magnon states with negative energy will become occupied. For the pure x-y model, where $J_z = 0$, the interaction identically vanishes, and the excitation spectrum of the magnons is given by $\omega_k = -J\cos ka$, as sketched in Figure 4.4. All the negative-energy fermion states with $|k| < \pi/2a$ are occupied, so the ground state is given by

$$|\Psi_g\rangle = \prod_{|k|<\pi/2a} s_k^\dagger|0\rangle. \tag{4.26}$$

The band of magnon states is thus precisely half-filled, so that

$$\langle S_z\rangle = \left\langle n_f - \frac{1}{2}\right\rangle = 0, \tag{4.27}$$

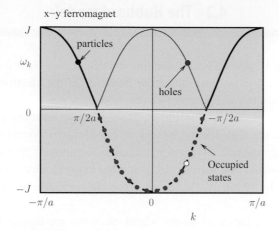

x–y ferromagnet

Fig. 4.4

Excitation spectrum of the one-dimensional x-y ferromagnet, showing how the negative energy states are filled. The negative-energy dispersion curve is "folded over" to describe the positive hole excitation energy.

so that, remarkably, there is no ground-state magnetization. We may interpret this loss of ground-state magnetization as a consequence of the growth of quantum spin fluctuations in going from the Heisenberg to the x-y ferromagnet.

Excitations of the ground state can be made, either by adding a magnon at wavevectors $|k| > \pi/2a$ or by annihilating a magnon at wavevectors $|k| < \pi/2a$, to form a *hole*. The energy to form a hole is $-\omega_k$. To represent the hole excitations, we make a *particle–hole transformation* for the occupied states, writing

$$\tilde{s}_k = \begin{cases} s_k & (|k| > \pi/2a) \\ s_{-k}^\dagger, & (|k| < \pi/2a). \end{cases} \tag{4.28}$$

These are the "physical" excitation operators. Since $s_k^\dagger s_k = 1 - s_k s_k^\dagger$, the Hamiltonian of the pure x-y ferromagnet can be written

$$H_{xy} = \sum_k J|\cos ka|(\tilde{s}_k^\dagger \tilde{s}_k - \frac{1}{2}). \tag{4.29}$$

Notice that, unlike the Heisenberg ferromagnet, the magnon excitation spectrum is now linear. The ground-state energy is evidently

$$E_g = -\frac{1}{2} \sum_k J|\cos ka|$$

$$= -N_s a \int_{-\pi/2a}^{\pi/2a} \frac{dk}{2\pi} J\cos(ka) = -N_s \frac{J}{\pi}. \tag{4.30}$$

But if there is no magnetization, why are there zero-energy magnon modes at $q = \pm\pi/a$? Although there is no true long-range order, it turns out that the spin correlations in the x-y model display power-law correlations with an infinite spin correlation length, generated by the gapless magnons in the vicinity of $q = \pm\pi/a$.

4.2 The Hubbard model

In a real electronic system such as a metallic crystal, at first sight it might appear to be a task of hopeless complexity to model the behavior of the electron fluid. Fortunately, even in complex systems, at low energies only a certain subset of the electronic degrees of freedom are excited. This philosophy is closely tied up with the idea of renormalization – the idea that the high-energy degrees of freedom in a system can be successively eliminated or "integrated out" to reveal an effective Hamiltonian that describes the important low-energy physics. One such model which has enjoyed great success is the Hubbard model, introduced by Philip W. Anderson, John Hubbard, Martin Gutziller and Junjiro Kanamori [3–6].

Suppose we have a lattice of atoms where electrons are almost localized in atomic orbitals at each site. In this case we can use a basis of atomic orbitals. The operator which creates a particle at site j is

$$c_{j\sigma}^\dagger = \int d^3x \, \psi^\dagger(\mathbf{x})_\sigma \, \Phi(\mathbf{x} - \mathbf{R}_j), \tag{4.31}$$

where $\Phi(\mathbf{x})$ is the wavefunction of a particle in the localized atomic orbital. In this basis, the Hamiltonian governing the motion of and interactions between the particles can be written quite generally as

$$H = \sum_{i,j} \langle i|H_0|j\rangle c_{i\sigma}^\dagger c_{j\sigma} + \frac{1}{2} \sum_{lmnp} \langle lm|V|pn\rangle c_{l\sigma}^\dagger c_{m\sigma'}^\dagger c_{n\sigma'} c_{p\sigma}, \tag{4.32}$$

where $\langle i|H_0|j\rangle$ is the one-particle matrix element between states i and j, and $\langle lm|V|pn\rangle$ is the interaction matrix element between two-particle states $|lm\rangle$ and $|pn\rangle$ (see Figure 4.5).

Let us suppose that the energy of an electron in this state is ϵ. If this orbital is highly localized, then the amplitude for it to tunnel or "hop" between sites will decay exponentially with the distance between the sites, and to a good approximation we can eliminate all but nearest-neighbor hopping. In this case, the one-particle matrix elements which govern the motion of electrons between sites are

$$\langle j|H^{(0)}|i\rangle = \begin{cases} \epsilon & (j = i) \\ -t & (i, j \text{ nearest neighbors}) \\ 0 & (\text{otherwise}). \end{cases} \tag{4.33}$$

The hopping matrix element between neighboring states is generally given by an overlap integral between the wavefunctions and the negative crystalline potential, and for this reason is taken to be be negative. Now of this matrix element the matrix element of the interaction between electrons at different sites will be given by

$$\langle lm|V|pn\rangle = \int_{x,x'} \Phi_l^*(x)\Phi_p(x)\Phi_m^*(x')\Phi_n^*(x')V(x - x'), \tag{4.34}$$

but, in practice, if the states are well localized, this will be dominated by the onsite interaction between two electrons in a single orbital, so that we may approximate

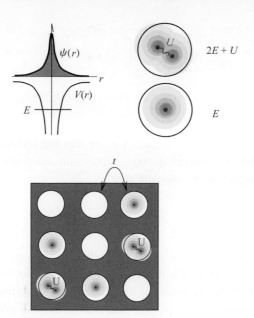

Fig. 4.5

Illustrating the Hubbard model. When two electrons of opposite spin occupy a single atom, this gives rise to a Coulomb repulsion energy U. The amplitude to hop from site to site in the crystal is t.

$$\langle lm|V|pn\rangle = \begin{cases} U & (l = p = m = n) \\ 0 & \text{(otherwise)}. \end{cases} \tag{4.35}$$

In this situation, the interaction term in (4.32) simplifies to

$$\frac{U}{2} \sum_{j,\sigma\sigma'} c_{j\sigma}^{\dagger} c_{j\sigma'}^{\dagger} c_{j\sigma'} c_{j\sigma} = U \sum_{j} n_{j\uparrow} n_{j\downarrow}, \tag{4.36}$$

where the exclusion principle ($c_{j\sigma}^{2} = 0$) means that the interaction term vanishes unless σ and σ' are opposite spins. The Hubbard model can be thus be written

$$H = -t \sum_{j,\hat{a},\sigma} [c_{j+\hat{a}\sigma}^{\dagger} c_{j\sigma} + \text{H.c.}] + \epsilon \sum_{j\sigma} c_{j\sigma}^{\dagger} c_{j\sigma} + U \sum_{j} n_{j\uparrow} n_{j\downarrow}, \tag{4.37}$$

where $n_{j\sigma} = c_{j\sigma}^{\dagger} c_{j\sigma}$ represents the number of electrons of spin σ at site j. For completeness, let us rewrite this in momentum space, putting

$$c_{j\sigma} = \frac{1}{\sqrt{N_s}} \sum_{\mathbf{k}} c_{\mathbf{k}\sigma} e^{i\mathbf{k}\cdot\mathbf{R}_j}, \tag{4.38}$$

whereupon

$$H = \sum_{\mathbf{k}\sigma} \epsilon_{\mathbf{k}} c_{\mathbf{k}\sigma}^{\dagger} c_{\mathbf{k}\sigma} + \frac{U}{N_s} \sum_{\mathbf{q},\mathbf{k},\mathbf{k'}} c_{\mathbf{k}-\mathbf{q}\uparrow}^{\dagger} c_{\mathbf{k'}+\mathbf{q}\downarrow}^{\dagger} c_{\mathbf{k'}\downarrow} c_{\mathbf{k}\uparrow}, \tag{4.39}$$

Hubbard model

where

$$\epsilon_{\mathbf{k}} = \sum_i \langle j + \mathbf{R}_i | H_0 | j \rangle e^{i\mathbf{k} \cdot \mathbf{R}_i}$$

$$= -2t(\cos k_x + \cos k_y + \cos k_z) + \epsilon \qquad (4.40)$$

is recognized as the kinetic energy of the electron excitations, which results from their *coherent* hopping motion from site to site. We see that the Hubbard model describes a band of electrons with kinetic energy $\epsilon_{\mathbf{k}}$ and a momentum-independent "point" interaction of strength U between particles of opposite spin.

Remark

- This model has played a central part in the theory of magnetism, metal–insulator transitions and, most recently, the description of electron motion in high-temperature superconductors. With the exception of one-dimensional physics, we do not yet have a complete understanding of the physics that this model can give rise to. One established prediction of the Hubbard model is that, under certain circumstance, if interactions become too large the electrons become localized to form what is called a *Mott insulator*. This occurs when the band is half-filled, with one electron per site. What is unclear at the present time is what happens to the Mott insulator when it is doped. There are many who believe that a complete understanding of the doped Mott insulator will enable us to understand high-temperature superconductivity.

4.3 Non-interacting particles in thermal equilibrium

Before we start to consider the physics of the interacting problem, let us go back and look at the ground-state properties of free particles (see Figure 4.6). What is not commonly recognized is that the ground state of non-interacting but identical particles is in fact a *highly correlated* many-body state. For this reason, the non-interacting ground state has a robustness that does not exist in its classical counterpart. In the next chapter, we shall embody some of these thoughts by considering the action of turning on the interactions adiabatically. For the moment, however, we shall content ourselves with looking at a few of the ground-state properties of non-interacting gases of identical particles.

In practice, quantum effects will influence a fluid of identical particles at the point where their characteristic wavelength is comparable with the separation between particles. At a temperature T the RMS momentum of particles is given by $p_{RMS}^2 = 3mk_BT$, so that the characteristic de Broglie wavelength is given by

$$\lambda_T = \frac{h}{\sqrt{p_{RMS}^2}} = \frac{h}{\sqrt{3mk_BT}}. \qquad (4.41)$$

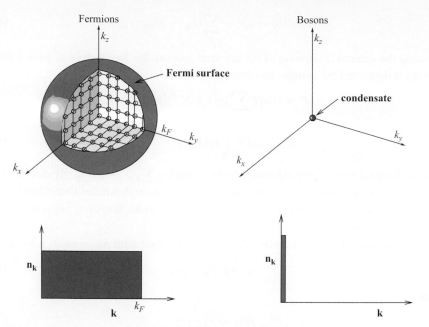

Contrasting the ground states of non-interacting fermions and non-interacting bosons. Fermions form a degenerate Fermi gas, with all one-particle states below the Fermi energy individually occupied. Bosons form a Bose–Einstein condensate, with a macroscopic number of bosons in the zero-momentum state.

Fig. 4.6

When $\lambda_T \sim \rho^{-1/3}$, the characteristic temperature is of order

$$k_B T^* \sim \frac{\hbar^2 \rho^{2/3}}{2m}. \tag{4.42}$$

Below this temperature, identical particles start to interfere with one another, and a quantum mechanical treatment of the fluid becomes necessary. In a Fermi fluid exclusion statistics tends to keep particles apart, enhancing the pressure, whereas for a Bose fluid the correlated motion of particles in the condensate tends to lower the pressure, ultimately causing it to vanish at the Bose–Einstein condensation temperature. In electron fluids inside materials, this characteristic temperature is two orders of magnitude larger than room temperature, which makes electricity one of the most dramatic examples of quantum physics in everyday phenomena!

4.3.1 Fluid of non-interacting fermions

The thermodynamics of a fluid of fermions leads to the concept of a *degenerate Fermi liquid*, important in a wide range of physical situations such as

- the ground state and excitations of metals
- the low-energy physics of liquid ^{3}He
- the degenerate Fermi gas of neutrons, electrons and protons that lies within a neutron star.

The basic physics of each of these cases can, to a first approximation, be described by a fluid of non-interacting fermions, with Hamiltonian

$$H = H_S - \mu N = \sum_{\sigma} (E_{\mathbf{k}} - \mu) c_{\mathbf{k}\sigma}^{\dagger} c_{\mathbf{k}\sigma}. \tag{4.43}$$

Following the general discussion of the previous section, the free energy of such a fluid of fermions is described by a single free energy functional:

$$F = -k_B T \sum_{\mathbf{k}\sigma} \ln[1 + e^{-\beta(E_{\mathbf{k}} - \mu)}]$$

$$= -2k_B T V \int_{\mathbf{k}} \ln[1 + e^{-\beta(E_{\mathbf{k}} - \mu)}], \tag{4.44}$$

where we have taken the thermodynamic limit, replacing $\sum_{\mathbf{k}\sigma} \to 2V \int_{\mathbf{k}}$. By differentiating F with respect to volume, temperature and chemical potential, we can immediately derive the pressure, entropy and particle density of this fluid. Let us, however, begin with a more physical discussion.

In thermal equilibrium the number of fermions in a state with momentum $\mathbf{p} = \hbar\mathbf{k}$ is

$$n_{\mathbf{k}} = f(E_{\mathbf{k}} - \mu), \tag{4.45}$$

where

$$f(x) = \frac{1}{e^{\beta x} + 1} \tag{4.46}$$

is the Fermi–Dirac function. At low temperatures, this function resembles a step, with a jump in occupancy spread over an energy range of order $k_B T$ around the chemical potential. At absolute zero, $f(x) \to \theta(-x)$, so that the occupancy of each state is given by

$$n_{\mathbf{k}} = \theta(\mu - E_{\mathbf{k}}). \tag{4.47}$$

This is a step function with an abrupt change in occupation when $\epsilon = \mu$, corresponding to the fact that states with $E_{\mathbf{k}} < \mu$ are completely occupied and states above this energy are empty. The zero-temperature value of the chemical potential is often called the *Fermi energy*. In momentum space, the occupied states form a sphere whose radius in momentum space, k_F, is often referred to as the *Fermi momentum*.

The ground state corresponds to a state where all fermion states with momentum $k < k_F$ are occupied:

$$|\psi_g\rangle = \prod_{|\mathbf{k}| < k_F, \, \sigma} c_{\mathbf{k}\sigma}^{\dagger} |0\rangle. \tag{4.48}$$

Excitations above this ground state are produced by the addition of particles at energies above the Fermi wavevector, or the creation of *holes* beneath the Fermi wavevector. To describe these excitations, we make the following *particle–hole* transformation:

$$a_{\mathbf{k}\sigma}^{\dagger} = \begin{cases} c_{\mathbf{k}\sigma}^{\dagger} & (k > k_F) & \text{particle} \\ \text{sgn}(\sigma) c_{-\mathbf{k}-\sigma} & (k < k_F). & \text{hole} \end{cases} \tag{4.49}$$

Beneath the Fermi surface, we must replace $c_{\mathbf{k}\sigma}^{\dagger} c_{\mathbf{k}\sigma} \to 1 - a_{\mathbf{k}\sigma}^{\dagger} a_{\mathbf{k}\sigma}$, so that in terms of particle and hole excitations, the Hamiltonian can be rewritten

$$H - \mu N = \sum_{\mathbf{k}\sigma} |(E_{\mathbf{k}} - \mu)| a_{\mathbf{k}\sigma}^{\dagger} a_{\mathbf{k}\sigma} + F_g, \tag{4.50}$$

where

$$F_g = \sum_{|\mathbf{k}|<k_F,\sigma} (E_\mathbf{k} - \mu) = 2V \int_{|\mathbf{k}|<k_F} (E_\mathbf{k} - \mu) \tag{4.51}$$

is the ground-state free energy and E_g and N are respectively the ground-state energy and particle number. Notice that:

- To create a hole with momentum $\mathbf{k}$ and spin σ, we must destroy a fermion with momentum $-\mathbf{k}$ and spin $-\sigma$. (The additional multiplying factor of sgn(σ) in the hole definition is a technical feature, required so that the particle and hole have the same spin operators.)
- The excitation energy of a particle or hole is given by $\epsilon_\mathbf{k}^* = |E_\mathbf{k} - \mu|$, corresponding to "reflecting" the excitation spectrum of the negative energy fermions about the Fermi energy.

The ground-state density of a Fermi gas is given by the volume of the Fermi surface, as follows:

$$\langle \hat{\rho} \rangle = \frac{1}{V} \sum_{\mathbf{k}\sigma} \langle c_{\mathbf{k}\sigma}^\dagger c_{\mathbf{k}\sigma} \rangle = 2 \int_{k<k_F} \frac{d^3k}{(2\pi)^3} = \frac{2}{(2\pi)^3} V_{FS}, \tag{4.52}$$

where

$$V_{FS} = \frac{4\pi}{3} k_F^3 = \left(\frac{4\pi}{3} \right) \left(\frac{2m\epsilon_F}{\hbar^2} \right)^{3/2} \tag{4.53}$$

is the volume of the Fermi surface. The relationship between the density of particles, the Fermi wavevector and the Fermi energy is thus

$$\left\langle \frac{\hat{N}}{V} \right\rangle = \frac{1}{3\pi^2} k_F^3 = \frac{1}{3\pi^2} \left(\frac{2m\epsilon_F}{\hbar^2} \right)^{3/2}. \tag{4.54}$$

In an electron gas, where the characteristic density is $N/V \sim 10^{29} \text{m}^{-3}$ the characteristic Fermi energy is of order $1\,\text{eV} \sim 10\,000\,\text{K}$. In other words, the characteristic energy of an electron is two orders of magnitude larger than would be expected classically. This is a stark and dramatic consequence of the exchange interference between identical particles, and it is one of the great early triumphs of quantum mechanics to have understood this basic piece of physics.

Let us briefly look at finite temperatures. Here, by differentiating the free energy with respect to volume and chemical potential, we obtain

$$P = -\frac{\partial F}{\partial V} = \frac{-F}{V} = 2k_B T \int_\mathbf{k} \ln[1 + e^{-\beta(E_\mathbf{k}-\mu)}]$$

$$N = -\frac{\partial F}{\partial \mu} = 2 \int_\mathbf{k} f(E_\mathbf{k} - \mu). \tag{4.55}$$

The second equation *defines* the chemical potential in terms of the particle density at a given temperature. The first equation shows that, apart from a minus sign, the pressure is simply the free energy density. These two equations can be solved parametrically as a function of chemical potential. At high temperatures the pressure reverts to the ideal gas law, $PV = Nk_BT$, but at low temperatures the pressure is determined by the Fermi energy

$$P = 2 \int_{|\mathbf{k}|<k_F} (\mu - E_{\mathbf{k}}) = \frac{2N}{5V} \epsilon_F. \qquad (4.56)$$

The final result is obtained by noting that the first term in this expression is $\mu(N/V)$. The first term contains an integral over $d^3k \sim k^2 dk \to k_F^3/3$, whereas the second term contains an integral over $E_k d^3k \sim k^4 dk \to k_F^5/5$, so the second term is $3/5$ of the first term. Not surprisingly, this quantity is basically the density of fermions times the Fermi energy – a pressure that is hundreds of times larger than the classical pressure in a room-temperature electron gas.

Remarks

- At first sight, one might reasonably doubt whether the the key features of the degenerate Fermi gas would survive once interactions are present. In particular, one is tempted to wonder whether the Fermi surface is blurred by particle–particle interactions. Yet, remarkably, for modest repulsive interactions the Fermi surface is believed stable in dimensions bigger than one. This is because electrons at the Fermi surface have no phase space for scattering. This is the basis of Landau's *Fermi-liquid theory* of interacting fermions.
- In a remarkable result, due to Luttinger and Ward, in interacting Fermi liquids the jump in the occupancy at the Fermi wavevector $Z_{\mathbf{k_F}}$ remains finite, although reduced from unity ($Z_{\mathbf{k_F}} < 1$).

4.3.2 Fluid of bosons: Bose–Einstein condensation

Bose–Einstein condensation (BEC) was predicted in 1924, the outcome of Einstein's extension of Bose's pioneering calculations on the statistics of a gas of identical bosons. However, it was not until 70 years later, in 1995, that the group of Eric Cornell and Carl Wieman [7] at the National Institute the University of Colorado, Boulder and independently that of Wolfgang Ketterle [8] at Massachesetts Institute of Technology succeeded in cooling a low-density gas of atoms – initially rubidium and sodium atoms – through the Bose–Einstein transition temperature. The closely related phenomenon of superfluidity was first observed in the late 1930s by Donald Misener and Jack Allen, working in Toronto and Cambridge [9], and Piotr Kapitza in Moscow [10]. Superfluidity results from a kind of Bose–Einstein condensation, in a dense quantum fluid where interactions between the particles become important. In the modern context, ultra-cold, ultra-dilute gases of alkali atoms are contained inside a magnetic atom trap in which the Zeeman energy of the atoms, spin-aligned with the magnetic field, confines them to the region of highest field [11]. Lasers are used to precool a small quantity of atoms inside a magnetic trap using a method known as *Doppler cooling*, in which the tiny blue shift of the laser light seen by atoms moving towards a laser causes them to selectively absorb photons, which are then re-emitted in random directions, a process which gradually slows them down, reducing their average

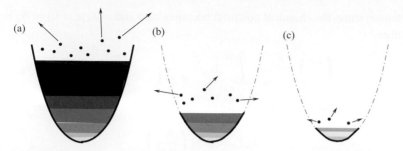

Illustrating evaporative cooling in an atom trap. (a) Atoms are held within a magnetic potential. (b) As the height of the potential well is dropped, the most energetic atoms evaporate from the well, progressively reducing the temperature. (c) A Bose–Einstein condensate, with a finite fraction of the gas in a single momentum state, forms when the temperature drops below the condensation temperature.

Fig. 4.7

temperature. Doppler pre-cooling cools the atoms to about $10-100\,\mu\mathrm{K}$. The second stage involves *evaporative cooling*, a process in which the most energetic atoms are allowed to evaporate out of the well while systematically lowering the height of the well. As the well height drops, the temperature of the gas plummets to the nano-Kelvin range required to produce Bose–Einstein condensation (or Fermi liquid formation) in these gases (see Figure 4.7).

To understand the phenomenon of BEC, consider the density of a gas of bosons, which at a finite temperature takes almost precisely the same form as for fermions:

$$\rho = \int_{\mathbf{k}} \frac{1}{e^{\beta(E_{\mathbf{k}}-\mu)} - 1}, \tag{4.57}$$

where we have written the expression for spinless bosons, as would be the case for a gas of liquid ^{4}He or ultra-dilute potassium atoms, for instance. But there is a whole world of physics in the innocent minus sign in the denominator! Whereas for fermions the chemical potential is positive, the chemical potential for bosons is negative. For a gas at fixed volume, (4.57) thus defines the chemical potential $\mu(T)$. By changing variables, writing

$$x = \beta E_{\mathbf{k}} = \beta\frac{\hbar^2 k^2}{2m}, \quad \left(\frac{m}{\beta\hbar^2}\right) dx = k\,dk, \quad \frac{d^3k}{(2\pi)^3} \rightarrow \frac{4\pi k^2\,dk}{(2\pi)^3}$$

$$= \frac{1}{\sqrt{2}\pi^2}\left(\frac{m}{\beta\hbar^2}\right)^{3/2}\sqrt{x}\,dx, \tag{4.58}$$

we can rewrite the boson density in the form

$$\rho = \frac{2}{\sqrt{\pi}\tilde{\lambda}_T^3}\int_0^\infty dx\sqrt{x}\frac{1}{e^{x-\beta\mu} - 1}, \tag{4.59}$$

where

$$\tilde{\lambda}_T = \left(\frac{2\pi\hbar^2}{mk_BT}\right)^{1/2} \tag{4.60}$$

is a convenient definition of the thermal de Broglie wavelength. In order to maintain a fixed density as one lowers the temperature, the chemical potential $\mu(T)$ must rise. At a certain temperature, the chemical potential becomes zero and $\rho(T, \mu = 0) = N/V$. At this temperature,

$$\left(\frac{\tilde{\lambda}_T}{a}\right)^3 = \frac{2}{\sqrt{\pi}} \int_0^\infty dx \sqrt{x} \frac{1}{e^x - 1} = \zeta(\frac{3}{2}) = 2.61, \tag{4.61}$$

where $a = \rho^{-1/3}$ is the interparticle spacing. The corresponding temperature

$$k_B T_o = 3.31 \left(\frac{\hbar^2}{ma^2}\right) \tag{4.62}$$

is the *Bose–Einstein* condensation temperature.

Below this temperature, the number of bosons in the $\mathbf{k} = 0$ state becomes macroscopic, i.e.

$$n_{\epsilon=0} = \frac{1}{e^{-\beta\mu} - 1} = N_0(T) \tag{4.63}$$

becomes a finite fraction of the total particle number. Since $N_0(T)$ is macroscopic, it follows that

$$\frac{\mu}{k_B T} = -\frac{1}{N_0(T)} \tag{4.64}$$

is infinitesimally close to zero. For this reason, we must be careful to split off the $\mathbf{k} = 0$ contribution to the particle density, writing

$$N = N_0(T) + \sum_{\mathbf{k} \neq \mathbf{0}} n_{\mathbf{k}} \tag{4.65}$$

and *then* taking the thermodynamic limit of the second term. For the density, this gives

$$\rho = \frac{N}{V} = \rho_0(T) + \int_{\mathbf{k}} \frac{1}{e^{\beta(E_{\mathbf{k}})} - 1}. \tag{4.66}$$

The second term is proportional to $\tilde{\lambda}_T^{-3} \propto T^{3/2}$. Since the first term vanishes at $T = T_0$, it follows that, below the Bose–Einstein condensation temperature, the density of bosons in the condensate is given by

$$\rho_0(T) = \rho \left[1 - \left(\frac{T}{T_0}\right)^{3/2}\right]. \tag{4.67}$$

Remarks

- Bose–Einstein condensation is an elementary example of a second-order phase transition.
- Bose–Einstein condensation is an example of a broken-symmetry phase transition. It turns out that the same phenomenon survives, in a more robust form, if repulsive interactions between the bosons are present. In the interacting Bose–Einstein condensate, the field operator $\psi(x)$ for the bosons actually acquires a macroscopic expectation value

$$\langle \psi(x) \rangle = \sqrt{\rho_0} e^{i\phi(x)}. \tag{4.68}$$

In a non-interacting Bose condensate, the phase $\phi(x)$ lacks rigidity, and does not have a well-defined meaning. However, in an interacting condensate, the phase $\phi(x)$ becomes a smooth function of position and gradients of the phase result in a *superflow* of particles – a flow of atoms which is completely free of viscosity.

Example 4.1 In a laser-cooled atom trap, atoms are localized in a region of space through the Zeeman energy of interaction between the atomic spin and the external field. As the field changes direction, the "up" and "down" spin atoms adiabatically evolve their orientations to remain parallel with the magnetic field, and the trapping potential of the "up" spin atoms is determined by the magnitude of the Zeeman energy $V(x) = g\mu_B JB(x)$, which has a parabolic form:

$$V(x) = \frac{m}{2}\left[\omega_x^2 x^2 + \omega_y^2 y^2 + \omega_z^2 z^2\right].$$

Show that the fraction of bosons condensed in the atom trap is now given by

$$\frac{N_0(T)}{N} = 1 - \left(\frac{T}{T_{BE}}\right)^3.$$

Solution

In the atom trap, one-particle states of the atoms are harmonic oscillator states with energy $E_{lmn} = \hbar(l\omega_x + m\omega_y + n\omega_z)$ (where the constant has been omitted). In this case, the number of particles in the trap is given by

$$N = \sum_{l,m,n} \frac{1}{e^{\beta E_{lmn}} - 1}.$$

The summation over the single-particle quantum numbers can be converted to an integral over energy, provided the condensate fraction is split off the sum, so that

$$\sum_{lmn} \frac{1}{e^{\beta E_{lmn}} - 1} = N_0(T) + \int dE \rho(E)\frac{1}{e^{\beta E} - 1},$$

where N_0 is the number of atoms in the condensate and

$$\rho(E) = \sum_{lmn\ (E_{lmn}\neq 0)} \delta(E - E_{lmn})$$

is the density of states. By converting this sum to an integral, we obtain

$$\rho(E) = \int dl\,dm\,dn\,\delta(E - E_{lmn})$$

$$= \int \frac{dE_x dE_y dE_z}{\hbar\omega_x \hbar\omega_y \hbar\omega_z}\delta(E_x + E_y + E_z - E)$$

$$= \frac{1}{(\hbar\tilde{\omega})^3}\int_0^E dE_x \int_0^{E_x} dE_y = \frac{E^2}{2(\hbar\tilde{\omega})^3}. \qquad (\tilde{\omega} = (\omega_x\omega_y\omega_z)^{1/3})$$

The quadratic dependence of this function on energy replaces the square-root dependence of the corresponding quantity for free bosons. The number of particles outside the condensate is proportional to T^3:

$$\int dE \rho(E) \frac{1}{e^{\beta E} - 1} = \frac{T^3}{2(\hbar\tilde{\omega})^3} \overbrace{\int dx \frac{x^2}{e^x - 1}}^{2\zeta_3} = N \left(\frac{T}{T_{BE}}\right)^3,$$

where $k_B T_{BE} = \hbar\tilde{\omega}(N/\zeta_3)^{1/3}$, so that the condensate fraction is now given by

$$\frac{N_0(T)}{N} = 1 - \left(\frac{T}{T_{BE}}\right)^3.$$

Example 4.2   Using the results of Section 4.3, show that the ideal gas law is modified by the interference between identical particles, so that

$$P = nk_B T \mathcal{F}^{\pm}(\mu/k_B T), \tag{4.69}$$

where n is the number density of particles, $\mathcal{F}^{\pm}(z) = g^{\pm}(z)/h^{\pm}(z)$ and

$$g^{\pm}(z) = \pm \int_0^{\infty} dx \sqrt{x} \ln[1 \pm e^{-(x-z)}]$$

$$h^{\pm}(z) = \int_0^{\infty} dx \sqrt{x} \frac{1}{e^{(x-z)} \pm 1}, \tag{4.70}$$

where the upper sign refers to fermions, the lower sign to bosons. Sketch the dependence of pressure on temperature for a gas of identical bosons and for a gas of identical fermions with the same density.

Solution

Let us begin by deriving an explicit expression for the free energy of a free gas of fermions or bosons. We start with

$$F = \mp(2S + 1)k_B TV \int_{\mathbf{k}} \ln[1 \pm e^{-\beta(E_{\mathbf{k}} - \mu)}], \tag{4.71}$$

where S is the spin of the particle. Making the change of variables,

$$x = \beta E_{\mathbf{k}} = \beta \frac{\hbar^2 k^2}{2m},$$

$$\frac{d^3 k}{(2\pi)^3} \rightarrow \frac{2}{\tilde{\lambda}_T^3 \sqrt{\pi}} \sqrt{x} dx, \tag{4.72}$$

where $\tilde{\lambda}_T = \sqrt{2\pi\hbar^2/(mk_B T)}$ is the rescaled thermal de Broglie wavelength, we obtain

$$F = \mp(2S + 1)k_B T \frac{V}{\tilde{\lambda}_T^3} \frac{2}{\sqrt{\pi}} \int dx \sqrt{x} \ln[1 \pm e^{-(x+\mu\beta)}]. \tag{4.73}$$

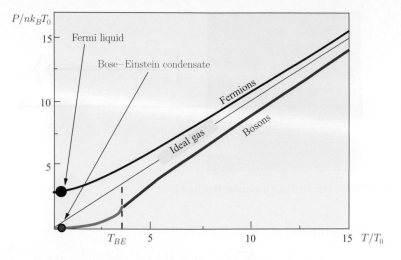

Pressure dependence in a Fermi or Bose gas, where temperature is measured in units of $k_B T_0 = \hbar^2/ma^2$.

Fig. 4.8

Taking derivatives with respect to volume and chemical potential, we obtain the following results for the pressure and the particle density:

$$P = -\frac{\partial F}{\partial V} = \pm(2S+1)\frac{k_B T}{\tilde{\lambda}_T^3}\frac{2}{\sqrt{\pi}}\int dx\sqrt{x}\ln[1 \pm e^{-(x-\mu\beta)}]$$

$$n = -\frac{\partial F}{V\partial\mu} = \frac{(2S+1)}{\tilde{\lambda}_T^3}\frac{2}{\sqrt{\pi}}\int dx\sqrt{x}\frac{1}{e^{(x-\mu\beta)} \pm 1}. \tag{4.74}$$

Dividing the pressure by the density, we obtain the quoted result for the ideal gas.

To plot these results, it is convenient to rewrite the temperature and pressure in the form

$$T = T_0[h^{\pm}(\mu\beta)]^{-2/3}$$

$$\frac{P}{nk_B T_0} = \frac{g^{\pm}(\mu\beta)}{[h^{\pm}(\mu\beta)]^{5/3}}, \tag{4.75}$$

where $k_B T_0 = \frac{\hbar^2}{ma^2}$, permitting both the pressure and the temperature to be plotted parametrically as a function of $\mu\beta$. Figure 4.8 shows the results of such a plot.

Exercises

Exercise 4.1

(a) Use the Jordan–Wigner transformation to show that the one-dimensional anisotropic x–y model,

$$H = -\sum_j [J_1 S_x(j)S_x(j+1) + J_2 S_y(j)S_y(j+1)], \tag{4.76}$$

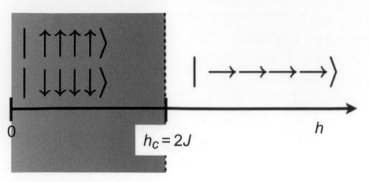

Fig. 4.9   Phase diagram of transverse-field Ising model. See Exercise 4.2.

can be written as

$$H = -\sum_j [t(d_{j+1}^\dagger d_j + \text{H.c.}) + \Delta(d_{j+1}^\dagger d_j^\dagger + \text{H.c.})], \qquad (4.77)$$

where $t = \frac{1}{4}(J_1 + J_2)$ and $\Delta = \frac{1}{4}(J_2 - J_1)$.

(b) Calculate the excitation spectrum for this model and sketch your results. Comment specifically on the two cases $J_1 = J_2$ and $J_2 = 0$.

Exercise 4.2 The transverse one-dimensional field Ising model provides the simplest example of a *quantum phase transition*: a phase transition induced by quantum zero-point motion (Figure 4.9). This model is written

$$H = -J\sum_j S_z(j)S_z(j+1) - h\sum_j S_x(j), \qquad (4.78)$$

where S_z is the z component of a spin $\frac{1}{2}$, while the the magnetic field h acts in the transverse (x) direction. (For convenience, one can assume periodic boundary conditions with N_s sites, so that $j \equiv j \,\text{mod}(N_s)$.) At $h = 0$, the model describes a one-dimensional Ising model, with long-range ferromagnetic order associated with a two-fold degenerate ferromagnetic ground state:

$$|\Psi_\uparrow\rangle = |\uparrow_1\rangle|\uparrow_2\rangle \cdots |\uparrow_{N_s}\rangle \qquad (4.79)$$

or

$$|\Psi_\downarrow\rangle = |\downarrow_1\rangle|\downarrow_2\rangle \cdots |\downarrow_{N_s}\rangle. \qquad (4.80)$$

A finite transverse field mixes "up" and "down" states, and, for infinitely large h, the system has a single ground state, with the spins all pointing in the x direction:

$$|\Psi_\rightarrow\rangle = \prod_{j=1,N_s} \left(\frac{|\uparrow_j\rangle + |\downarrow_j\rangle}{\sqrt{2}}\right).$$

In other words, there is a *quantum phase transition* – a phase transition driven by quantum fluctuations between the doubly degenerate ferromagnet at small h and a singly degenerate state polarized in the x direction at large h.

(a) By rotating the above model so that the magnetic field acts in the $+x$ direction and the Ising interaction acts on the spins in the x direction, show that the transverse-field Ising model can be rewritten as

$$H = -J \sum_j S_x(j)S_x(j+1) - h \sum_j S_z(j).$$

(b) Use the Jordan–Wigner transformation to show that the "fermionized" version of this Hamiltonian can be written

$$H = \frac{J}{4} \sum_j (f_j - f_j^\dagger)(f_{j+1} + f_{j+1}^\dagger) - h \sum_j f_j^\dagger f_j. \tag{4.81}$$

(c) Writing $f_j = \frac{1}{\sqrt{N_s}} \sum_k d_k e^{ikR_j}$, where $R_j = aj$, show that H can be rewritten in momentum space as

$$H = \sum_{k \in [0,\pi/a]} \left[\epsilon_k(d_k^\dagger d_k - d_{-k}d_{-k}^\dagger) - i\Delta_k(d_k^\dagger d_{-k}^\dagger - d_{-k}d_k) \right], \tag{4.82}$$

where the sum over $k = \frac{2\pi}{N_s a}(1, 2, \ldots N_s/2) \in [0, \frac{\pi}{a}]$ is restricted to half the Brillouin zone, while $\epsilon_k = -\frac{J}{2}\cos ka - h$ and $\Delta_k = \frac{J}{2}\sin ka$.

(d) Using the results of Exercise 4.1, show that the spectrum of the excitations is described by Dirac fermions with a dispersion

$$E_k = \sqrt{\epsilon_k^2 + \Delta_k^2} = \sqrt{2Jh \sin^2(ka/2) + (h - J/2)^2},$$

so that the gap in the excitation spectrum closes at $h = h_c = J/2$. What is the significance of this field?

Exercise 4.3 Consider the non-interacting Hubbard model for next-nearest-neighbor hopping on a two-dimensional lattice:

$$H - \mu N = -t \sum_{j,\hat{a}=\hat{x},\hat{y},\sigma} [c_{j+\hat{a}\sigma}^\dagger c_{j\sigma} + \text{H.c.}] - \mu \sum_{j\sigma} c_{j\sigma}^\dagger c_{j\sigma},$$

where $n_{j\sigma} = c_{j\sigma}^\dagger c_{j\sigma}$ represents the number of electrons of spin component $\sigma = \pm\frac{1}{2}$ at site j.

(a) Show that the dispersion of the electrons in the absence of interactions is given by

$$\epsilon(\vec{k}) = -2t(\cos k_x a + \cos k_y a) - \mu,$$

where a is the distance between sites and $\vec{k} = (k_x, k_y)$ is the wavevector.

(b) Derive the relation between the number of electrons per site, n_e, and the area of the Fermi surface.

(c) Sketch the Fermi surface when
   (i) $n_e < 1$
   (ii) $n_e = 1$, corresponding to "half-filling."

(d) The corresponding interacting Hubbard model, with an interaction term $Un_\uparrow n_\downarrow$ at each site, describes a class of material called *Mott insulators*, which includes the mother compounds for high-temperature superconductors. What feature of

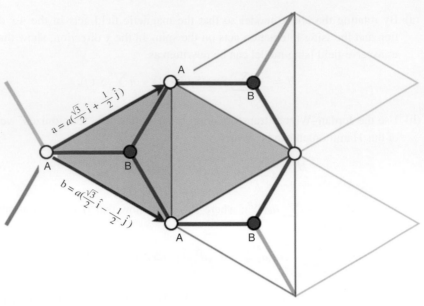

$$a = a(\tfrac{\sqrt{3}}{2}\hat{i} + \tfrac{1}{2}\hat{j})$$

$$b = a(\tfrac{\sqrt{3}}{2}\hat{i} - \tfrac{1}{2}\hat{j})$$

Fig. 4.10　　Honeycomb structure of graphene. See Exercise 4.4.

the Fermi surface at half-filling makes the non-interacting ground state unstable to spin density wave formation and the development of a gap around the Fermi surface?

(e) Derive the dispersion for the case when, in the one-particle Hamiltonian, there is an additional next-nearest-neighbor hopping matrix element of strength $-t'$, across the diagonal. (*Hint*: use the Fourier transform of $t(R)$, given by $t(\vec{k}) = \sum_{\vec{R}} t(\vec{R}) e^{-i\vec{k}\cdot\vec{R}}$.) How does this affect the dispersion at half-filling?

Exercise 4.4 Electrons in a monolayer of graphene move on a honeycomb lattice as shown in Figure 4.10. The vertices of each unit cell form a triangular lattice of side length a, located at positions $\mathbf{r}_i = m\mathbf{a} + n\mathbf{b}$, where $\mathbf{a} = a\left(\tfrac{\sqrt{3}}{2}\hat{i} + \tfrac{1}{2}\hat{j}\right)$ and $\mathbf{b} = a\left(\tfrac{\sqrt{3}}{2}\hat{i} - \tfrac{1}{2}\hat{j}\right)$ are the lattice vectors. There are two atoms per unit cell, labeled A and B. In a simplified model of graphene, electrons can occupy π orbitals at either the A or the B site, with a tight-binding hopping matrix element $-t$ between neighboring sites.

(a) Construct a tight-binding model for graphene. For simplicity, ignore the spin of the electron. Suppose the creation operator for an electron in the A or B orbital in the ith cell is $\psi_{A,B}^{\dagger}(\mathbf{r}_i)$. Show that the tight-binding Hamiltonian can be written in the form

$$H = -t \sum_{j} \left\{ \left[\psi_B^{\dagger}(\mathbf{r}_i) + \psi_B^{\dagger}(\mathbf{r}_i - \mathbf{a}) + \psi_B^{\dagger}(\mathbf{r}_i - \mathbf{b}) \right] \psi_A(\mathbf{r}_i) + \text{H.c.} \right\}$$
$$+ (\epsilon - \mu) \sum_{i} (n_A(i) + n_B(i)),$$

where ϵ is the energy of a localized orbital.

(b) By transforming to momentum space, writing

$$\psi_\lambda^\dagger(\mathbf{r}_j) = \frac{1}{\sqrt{N_s}} \sum_{\mathbf{k}} c_{\mathbf{k}\lambda}^\dagger e^{-i\mathbf{k}\cdot\mathbf{r}_j} \qquad (\lambda = \mathrm{A}, \mathrm{B}),$$

with N_s the number of unit cells in the crystal, show that the Hamiltonian can be written

$$H = \sum_{\mathbf{k}} \left(c_{\mathbf{k}B}^\dagger, c_{\mathbf{k}A}^\dagger \right) \begin{bmatrix} \epsilon - \mu & \Delta(\mathbf{k}) \\ \Delta^*(\mathbf{k}) & \epsilon - \mu \end{bmatrix} \begin{pmatrix} c_{\mathbf{k}B} \\ c_{\mathbf{k}A} \end{pmatrix},$$

where

$$\Delta(\mathbf{k}) = -t(1 + e^{i\mathbf{k}\cdot\mathbf{a}} + e^{i\mathbf{k}\cdot\mathbf{b}}),$$

with energy eigenstates

$$\epsilon(\mathbf{k}) = \pm|\Delta(\mathbf{k})| + (\epsilon - \mu).$$

(c) Show that $\Delta(\mathbf{k}) = 0$ at two points in the Brillouin zone where $\mathbf{k} \cdot \mathbf{a} = -\mathbf{k} \cdot \mathbf{b} = \pm\frac{2\pi}{3}$, given by

$$\mathbf{k} = \pm\mathbf{K},$$

where $\mathbf{K} = \frac{4\pi}{3a}\hat{\mathbf{j}}$.

(d) By expanding around $\mathbf{k} = \pm\mathbf{K} + \mathbf{p}$, show that, when $\mathbf{p}$ is small, $\Delta_{\mathbf{p}\pm\mathbf{K}} = \pm\tilde{c}(p_y \pm ip_x)$, where $\tilde{c} = \frac{\sqrt{3}}{2}at$ is a "renormalized" speed of light. By defining a spinor for the two cones

$$\psi_{\mathbf{p}+} = \begin{pmatrix} c_{\mathbf{p}+\mathbf{K}B} \\ c_{\mathbf{p}+\mathbf{K}A} \end{pmatrix} \qquad \psi_{\mathbf{p}-} = \begin{pmatrix} c_{\mathbf{p}-\mathbf{K}A} \\ -c_{\mathbf{p}-\mathbf{K}B} \end{pmatrix},$$

show that the low-energy Hamiltonian can be written as a Dirac equation,

$$H = \sum_{\mathbf{p}\lambda=\pm} \psi_{\mathbf{p}\lambda}^\dagger \left((\vec{\sigma} \times \mathbf{p}) + (\epsilon - \mu)\mathbb{1} \right) \psi_{\mathbf{p}\lambda},$$

where $\vec{\sigma}$ is a Pauli pseudo-spin matrix acting in the two-component sublattice space, so that when $\epsilon - \mu = 0$ the excitation spectrum is defined by two Dirac cones with $E(\mathbf{p}) = \pm\tilde{c}p$.

References

[1] P. Jordan and E. Wigner, Über das Paulische Äquivalenzverbot (On the Pauli exclusion principle), *Z. Phys.*, vol. 47, p. 631, 1928.

[2] E. Lieb, T. Schultz, and D. Mattis, Two soluble models of an antiferromagnetic chain, *Ann. Phys.*, vol. 16, no. 3, p. 407, 1961.

[3] J. Hubbard, Electron correlations in narrow energy bands, *Proc. R. Soc. A*, vol. 276, p. 238, 1963.

[4] M. C. Gutzwiller, Effect of correlation on the ferromagnetism of transition metals, *Phys. Rev. Lett.*, vol. 10, no. 5, p. 159, 1963. The Hubbard model was written down independently by Gutzwiller in equation (11) of this paper.

[5] J. Kanamori, Electron correlation and ferromagnetism of transition metals, *Prog. Theor. Phys.*, vol. 30, no. 3, p. 275, 1963. The Hubbard model, with the modern notation "U" for the interaction, was independently introduced by Kanamori in equation (1) of this paper.

[6] P. W. Anderson, New approach to the theory of superexchange interactions, *Phys. Rev.*, vol. 115, p. 2, 1959.

[7] M. H. Anderson, J. R. Ensher, M. R. Matthews, C. E. Wieman, and E. A. Cornell, Observation of Bose–Einstein condensation in a dilute atomic vapor, *Science*, vol. 269, p. 198, 1995.

[8] K. B. Davis, M.-O. Mewes, M. R. Andrews, N. J. van Druten, D. S. Durfee, D. M. Kurn, and W. Ketterle, Bose–Einstein condensation in a gas of sodium atoms, *Phys. Rev. Lett.*, vol. 75, p. 3969, 1995.

[9] J. F. Allen and A. D. Misener, The l-phenomenon of liquid helium and the Bose–Einstein degeneracy, *Nature*, vol. 75, p. 141, 1938.

[10] P. Kapitza, Viscosity of liquid helium below the l-point, *Nature*, vol. 141, p. 74, 1938.

[11] C. J. Pethick and H. Smith, Bose-Einstein Condensation in Dilute Gases, Cambridge University Press, 2008.

Green's functions

Ultimately, we are interested in more than just free systems. We would like to understand what happens to our system as we dial up the interaction strength from zero to its full value. We also want to know the response of our complex system to external perturbations, such as an electromagnetic field. We have to recognize that we cannot in general expect to diagonalize the problem of interest. We do not even need interactions to make the problem complex: a case in point is a disordered metal, where averaging over typically disordered configurations introduces effects reminiscent of interactions, and can even lead to new kinds of physics, such as electron localization. We need some general way of examinining the change of the system in response to these effects even though we can't diagonalize the Hamiltonian.

In general we will be considering problems where we introduce new terms to a non-interacting Hamiltonian, represented by V. The additional term might be due to

- external electromagnetic fields, which modify the kinetic energy in the Hamiltonian as follows:

$$-\frac{\hbar^2}{2m}\nabla^2 \rightarrow -\frac{\hbar^2}{2m}\left(\nabla - i\frac{e}{\hbar}\mathbf{A}\right)^2 \tag{5.1}$$

- interactions between particles:

$$\hat{V} = \frac{1}{2}\int d\mathbf{1}d\mathbf{2}V(1-2)\psi^\dagger(\mathbf{1})\psi^\dagger(\mathbf{2})\psi(\mathbf{2})\psi(\mathbf{1}) \tag{5.2}$$

- a random potential:

$$\hat{V} = \int d\mathbf{1}V(\mathbf{1})\rho(\mathbf{1}), \tag{5.3}$$

where $V(x)$ is a random function of position.

We would like to examine what happens when the new term V in the Hamiltonian is small enough to be considered a perturbation. Even if the term of interest is not small, we can still try to make it small by writing

$$H = H_0 + \lambda\hat{V}. \tag{5.4}$$

This is a useful excercise, for it enables us to consider the effect of adiabatically dialing up the strength of the additional term in the Hamiltonian from zero to its full value, as illustrated in Figure 5.1.

This is a dangerous procedure, but sometimes it really works. Life is interesting, because in macroscopic systems the perturbation of interest often leads to an instability. This can sometimes occur for arbitrarily small λ. At other times, the instability occurs when the

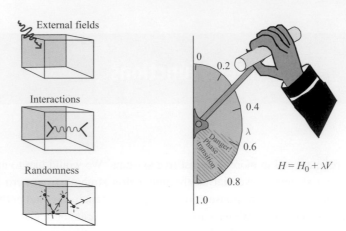

External fields

Interactions

Randomness

$H = H_0 + \lambda V$

Fig. 5.1 "Dialing up the interaction": treating perturbations to a non-interacting Hamiltonian by adjusting the strength of the perturbation.

strength of the new term reaches some critical value λ_c. When this happens, the ground state can change. If the change is a continuous one, then the point where the instability develops is a *quantum critical point*, a point of great interest. Beyond this point, for $\lambda > \lambda_c$, if we are lucky we can find some new starting $H'_0 = H_0 + \Delta H$, $\hat{V}' = \hat{V} - \Delta H$. If H'_0 is a good description of the ground state, then we can once again apply this adiabatic procedure, writing

$$H = H'_0 + \lambda' \hat{V}'. \tag{5.5}$$

If a phase transition occurs, then H'_0 will in all probability display a spontaneous *broken symmetry*. The region of Hamiltonian space where $H \sim H'_0$ describes a new phase of the system, and H'_0 is closely associated with the notion of a *fixed-point* Hamiltonian.

This motivates us to develop a general perturbative approach to many-body systems, and this rapidly leads us into the realm of Green's functions and Feynman diagrams. A Green's function describes the elementary correlations and responses of a system. Feynman diagrams are a way of graphically displaying the scattering processes that result from a perturbation.

5.1 Interaction representation

So far we have known two representations of quantum theory the Schrödinger representation, where it is the wavefunction that evolves, and the Heisenberg representation, where the operators evolve and the states are stationary. We are interested in observable quantities more than wavefunctions, and so we aspire to the Heisenberg representation. In practice, however, we always want to know what happens if we change the Hamiltonian a little. If we change H_0 to $H_0 + V$ but stick to the Heisenberg representation for H_0, then we are now using the *interaction representation* (see Table 5.1, where the subscripts S, H and I indicate

Table 5.1 Representations		
Representation	States	Operators
Schrödinger	Change rapidly $i\frac{\partial}{\partial t}|\psi_S(t)\rangle = H|\psi_S(t)\rangle$	Constant
Heisenberg	Constant	Evolve $-i\frac{\partial O_H(t)}{\partial t} = [H, O_H(t)]$
Interaction $H = H_0 + V$	States change slowly $i\frac{\partial}{\partial t}|\psi_I(t)\rangle = V_I(t)|\psi_I(t)\rangle$	Evolve according to H_0 $-i\frac{\partial O_I(t)}{\partial t} = [H_0, O_I(t)]$

that operators and states are to be evaluated in the Schrödinger, Heisenberg or interaction representations, respectively).

Let us examine the interaction representation in greater detail. In the discussion that follows, we simplify the notation by taking taking $\hbar = 1$. We begin by writing the Hamiltonian in two parts, $H = H_0 + V$. States and operators in this representation are defined as

$$\left.\begin{array}{l} |\psi_I(t)\rangle = e^{iH_0 t}|\psi_S(t)\rangle \\ \\ O_I(t) = e^{iH_0 t} O_S e^{-iH_0 t}. \end{array}\right\} \text{ removes rapid state evolution due to } H_0 \qquad (5.6)$$

The evolution of the wavefunction is thus

$$|\psi_I(t)\rangle = U(t)|\psi_I(0)\rangle$$
$$U(t) = e^{iH_0 t} e^{-iHt}, \qquad (5.7)$$

or, more generally,

$$|\psi_I(t)\rangle = S(t, t')|\psi_I(t')\rangle$$
$$S(t, t') = U(t)U^\dagger(t'). \qquad (5.8)$$

The time evolution of $U(t)$ can be derived as follows:

$$i\frac{\partial U}{\partial t} = i\left(\frac{\partial e^{iH_0 t}}{\partial t}\right)e^{-iHt} + ie^{iH_0 t}\left(\frac{\partial e^{-iHt}}{\partial t}\right)$$
$$= e^{iH_0 t}(-H_0 + H)e^{-iHt}$$
$$= [e^{iH_0 t} V e^{-iH_0 t}]U(t)$$
$$= V_I(t)U(t), \qquad (5.9)$$

so that

$$i\frac{\partial S(t_2, t_1)}{\partial t_2} = V(t_2)S(t_2, t_1), \qquad (5.10)$$

where, from now on, all operators are implicitly assumed to be in the interaction representation.

Now we would like to exponentiate this time-evolution equation, but unfortunately the operator $V(t)$ is not constant, and furthermore $V(t)$ at one time does not commute with $V(t')$ at another time. To overcome this difficulty, Schwinger invented a device called the *time-ordering operator*:

Time-ordering operator   Suppose $\{O_1(t_1), O_2(t_2), \ldots O_N(t_N)\}$ is a set of operators at different times $\{t_1, t_2, \ldots t_N\}$. If P is the permutation that orders the times, so that $t_{P_1} > t_{P_2}, \ldots t_{P_N}$, then if the operators are entirely bosonic, containing an even number of fermionic operators, the time-ordering operator is defined as

$$T\big[O_1(t_1)O_2(t_2)\cdots O_N(t_N)\big] = O_{P_1}(t_{P_1})O_{P_2}(t_{P_2})\cdots O_{P_N}(t_{P_N}). \qquad (5.11)$$

For later use we note that, if the operator set contains fermionic operators composed of an odd number of fermionic operators, then

$$T\big[F_1(t_1)F_2(t_2)\cdots F_N(t_N)\big] = (-1)^P F_{P_1}(t_{P_1})F_{P_2}(t_{P_2})\cdots F_{P_N}(t_{P_N}). \qquad (5.12)$$

where P is the number of pairwise permutations of fermions involved in the time-ordering process.

Suppose we divide the time interval $[t_1, t_2]$, where $t_2 > t_1$, into N identical segments of period $\Delta t = (t_2 - t_1)/N$, where the time at the midpoint of the nth segment is $\tau_n = t_1 + (n - \frac{1}{2})\Delta t$. The S-matrix can be written as a product of S-matrices over each intermediate time segment, as follows (see Figure 5.2):

$$S(t_2, t_1) = S(t_2, \tau_N - \tfrac{\Delta t}{2})S(\tau_{N-1} + \tfrac{\Delta t}{2}, \tau_{N-1} - \tfrac{\Delta t}{2})\cdots S(\tau_1 + \tfrac{\Delta t}{2}, t_1). \qquad (5.13)$$

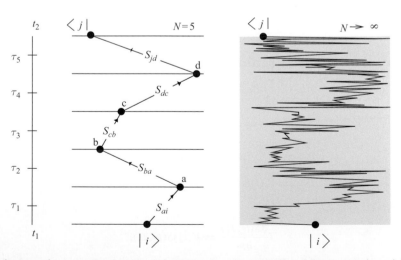

Fig. 5.2   Each contribution to the time-ordered exponential corresponds to the amplitude to follow a particular path in state space. The S-matrix is given by the limit of the process in which the number of time segments goes to infinity.

Provided N is large, then over the short time interval Δt we can approximate

$$S(\tau + \tfrac{\Delta t}{2}, \tau - \tfrac{\Delta t}{2}) = e^{-iV(\tau)\Delta t} + O(1/N^2), \tag{5.14}$$

so that we can write

$$S(t_2, t_1) = e^{-iV(\tau_N)\Delta t} e^{-iV(\tau_{N-1})\Delta t} \cdots e^{-iV(\tau_1)\Delta t} + O(1/N). \tag{5.15}$$

Using the time-ordering operator, this can be written

$$S(t_2, t_1) = T\Big[\prod_{j=1}^{N} e^{-iV(\tau_j)\Delta t}\Big] + O(1/N). \tag{5.16}$$

The beauty of the time-ordering operator is that, even though $A(t_1)$ and $A(t_2)$ don't commute, we can treat them as commuting operators so long as we always time-order them. This means that we can write

$$T[e^{A(t_1)} e^{A(t_2)}] = T[e^{A(t_1)+A(t_2)}] \tag{5.17}$$

because, in each time-ordered term in the Taylor expansion, we never have to commute operators, so the algebra is the same as for regular complex numbers. With this trick, we can write

$$S(t_2, t_1) = \lim_{N\to\infty} T\big[e^{-i\sum_j V(\tau_j)\Delta t}\big]. \tag{5.18}$$

The limiting value of this time-ordered exponential is written as

$$S(t_2, t_1) = T\left[\exp\left\{-i\int_{t_1}^{t_2} V(t)dt\right\}\right]. \qquad \text{time-ordered exponential} \tag{5.19}$$

This is the famous *time-ordered exponential* of the interaction representation.

Remarks

- The time-ordered exponential is intimately related to Feynman's notion of the path integral. The time-evolution operator $\langle p|S(\tau_j + \Delta\tau/2, \tau_j - \Delta\tau/2)|q\rangle = S_{pq}(\tau_j)$ across each segment of time is a matrix that takes one from state q to state p. The total time-evolution operator is just a sum of matrix products over each intermediate time segment. Thus the amplitude to go from state i at time t_1 to state f at time t_2 is given by

$$S_{fi}(t_2, t_2) = \sum_{\text{path}=\{p_1,\ldots p_{N-1}\}} S_{f,p_{N-1}}(\tau_N) \cdots S_{p_2 p_1}(\tau_2) S_{p_1 i}(\tau_1). \tag{5.20}$$

Each term in this sum is the amplitude to go along the path of states

$$\text{path } i \;\to\; f : i \to p_1 \to p_2 \to \cdots p_{N-1} \to f. \tag{5.21}$$

The limit in which the number of segments goes to infinity is a path integral.

- One can formally expand the time-ordered exponential as a power series, writing

$$S(t_2, t_1) = \sum_{n=0,\infty} \frac{(-i)^n}{n!} \int_{t_1}^{t_2} d\tau_1 \cdots d\tau_n T[V(\tau_1) \cdots V(\tau_n)].$$ (5.22)

The nth term in this expansion can be simply interpreted as the amplitude to go from the initial to the final state, scattering n times off the perturbation V. This form of the S-matrix is very useful in making a perturbation expansion. By explicitly time-ordering the nth term, one obtains $n!$ identical terms, so that

$$S(t_2, t_1) = \sum_{n=0,\infty} (-i)^n \int_{t_1, \{\tau_n > \tau_{n-1} \ldots > \tau_1\}}^{t_2} d\tau_1 \cdots d\tau_n V(\tau_n) \cdots V(\tau_1).$$ (5.23)

This form for the S-matrix is obtained by iterating the equation of motion:

$$S(t_2, t_1) = 1 - i \int_{t_1}^{t_2} d\tau\, V(\tau) S(\tau, t_1),$$ (5.24)

which provides an alternative derivation of the time-ordered exponential.

5.1.1 Driven harmonic oscillator

To illustrate the concept of the time-ordered exponential, we shall show how it is possible to evaluate the S-matrix for a driven harmonic oscillator, where $H = H_0 + V(t)$,

$$H_0 = \omega \left(b^\dagger b + \frac{1}{2} \right)$$

$$V(t) = \bar{z}(t)b + b^\dagger z(t).$$ (5.25)

Here the forcing terms are written in their most general form: $z(t)$ and $\bar{z}(t)$ are forces which create and annihilate quanta, respectively. A conventional force in the Hamiltonian $H = H_0 - f(t)\hat{x}$ gives rise to a particular case in which $\bar{z}(t) = z(t) = f(t)/(2m\omega)^{1/2}$. We shall show that if the forcing terms are zero in the distant past and distant future and the system is initially in the ground state, the amplitude to stay in this state is

$$S[\bar{z}, z] = \langle 0|Te^{-i\int_{-\infty}^{\infty} dt[\bar{z}(t)b(t) + b^\dagger(t)z(t)]}|0\rangle = \exp\left[-i \int_{-\infty}^{\infty} dt dt'\, \bar{z}(t) G(t - t') z(t') \right],$$ (5.26)

where $G(t-t') = -i\theta(t-t')e^{-i\omega(t-t')}$ is our first example of a one-particle *Green's function*. The importance of this result is that we have a precise algebraic result for the response of the ground state to an arbitrary force term. Once we know the response to an arbitrary force, we can, as we shall see, deduce the nth-order moments or correlation functions of the Bose fields.

Proof: To prove this result, we need to evaluate the time-ordered exponential

$$\langle 0|T \exp\left[-i \int_{-\tau}^{\tau} dt[\bar{z}(t)b(t) + b^\dagger(t)z(t)] \right] |0\rangle,$$ (5.27)

where $b(t) = b e^{i\omega t}$ and $b^\dagger(t) = b^\dagger e^{i\omega t}$. To evaluate this integral, we divide up the interval $t \in (t_1, t_2)$ into N segments $t \in (\tau_j - \Delta\tau/2, \tau_j + \Delta\tau_j)$ of width $\Delta\tau = 2\tau/N$ and write down the discretized time-ordered exponential as

$$S_N = e^{A_N - A_N^\dagger} \cdots e^{A_r - A_r^\dagger} \cdots e^{A_1 - A_1^\dagger}, \tag{5.28}$$

where we have used the shorthand notation

$$A_r = -i\bar{z}(\tau_r)b(\tau_r)\Delta\tau$$
$$A_r^\dagger = i b^\dagger(\tau_r)z(\tau_r)\Delta\tau. \tag{5.29}$$

To evaluate the ground state expectation of this exponential, we need to *normal-order* the exponential, bringing the terms involving annihilation operators e^{A_r} to the right-hand side of the expression. To do this, we use the result[1]

$$e^{\hat{\alpha} + \hat{\beta}} = e^{\hat{\beta}} e^{\hat{\alpha}} e^{[\hat{\alpha}, \hat{\beta}]/2} \tag{5.30}$$

and the related result that follows by equating $e^{\hat{\alpha} + \hat{\beta}} = e^{\hat{\beta} + \hat{\alpha}}$,

$$e^{\hat{\alpha}} e^{\hat{\beta}} = e^{\hat{\beta}} e^{\hat{\alpha}} e^{[\hat{\alpha}, \hat{\beta}]}. \tag{5.31}$$

These results hold if $[\hat{\alpha}, \hat{\beta}]$ commutes with $\hat{\alpha}$ and $\hat{\beta}$. We use these relations to separate $e^{A_r - A_r^\dagger} \rightarrow e^{-A_r^\dagger} e^{A_r} e^{-[A_r, A_r^\dagger]/2}$ and commute the e^{A_r} to the right, past terms of the form $e^{-A_s^\dagger}$, $e^{A_r} e^{-A_s^\dagger} = e^{-A_s^\dagger} e^{A_r} e^{-[A_r, A_s^\dagger]}$. We observe that in our case,

$$[A_r, A_s^\dagger] = \Delta\tau^2 \bar{z}(\tau_r)z(\tau_s)e^{-i\omega(\tau_r - \tau_s)} \tag{5.32}$$

is a c-number, so we can use the above theorem. We first normal-order each term in the product, writing $e^{A_r - A_r^\dagger} = e^{-A_r^\dagger} e^{A_r} e^{-[A_r, A_r^\dagger]/2}$ so that

$$S_N = e^{-A_N^\dagger} e^{A_N} \cdots e^{-A_1^\dagger} e^{A_1} e^{-\sum_r [A_r, A_r^\dagger]/2}. \tag{5.33}$$

Now we move the general term e^{A_r} to the right-hand side, picking up the residual commutators along the way, to obtain

$$S_N = \overbrace{e^{-\sum_r A_r^\dagger} e^{\sum_r A_r}}^{:S_N:} \exp\Big[-\sum_{r \geq s} [A_r, A_s^\dagger](1 - \tfrac{1}{2}\delta_{rs})\Big], \tag{5.34}$$

where the δ_{rs} term is present because, by (5.33), we get half a commutator when $r = s$. The vacuum expectation value of the first term is unity, so that

---

[1] To prove this result, consider $f(x) = e^{x\hat{\alpha}} e^{x\hat{\beta}}$. Differentiating $f(x)$, we obtain $\frac{df}{dx} = e^{x\hat{\alpha}}(\hat{\alpha} + \hat{\beta})e^{x\hat{\beta}}$. Now if $[\hat{\alpha}, \hat{\beta}]$ commutes with $\hat{\alpha}$ and $\hat{\beta}$, then $[\hat{\alpha}^n, \hat{\beta}] = n[\hat{\alpha}, \hat{\beta}]\hat{\alpha}^{n-1}$, so that the commutator $[e^{x\hat{\alpha}}, \hat{\beta}] = x[\hat{\alpha}, \hat{\beta}]e^{x\hat{\alpha}}$. It follows that $\frac{df}{dx} = (\hat{\alpha} + \hat{\beta} + x[\hat{\alpha}, \hat{\beta}])f(x)$. We can integrate this expression to obtain $f(x) = \exp[x(\hat{\alpha} + \hat{\beta}) + \frac{x^2}{2}[\hat{\alpha}, \hat{\beta}]]$. Setting $x = 1$ then gives $e^{\hat{\alpha}} e^{\hat{\beta}} = e^{\hat{\alpha} + \hat{\beta}} e^{\frac{1}{2}[\hat{\alpha}, \hat{\beta}]}$. If we interchange α and β, we obtain $e^{\hat{\beta}} e^{\hat{\alpha}} = e^{\hat{\alpha} + \hat{\beta}} e^{-\frac{1}{2}[\hat{\alpha}, \hat{\beta}]}$. Combining the two expressions, $e^{\alpha} e^{\beta} = e^{\beta} e^{\alpha} e^{[\alpha, \beta]}$.

$$S(t_2, t_1) = \lim_{\Delta\tau \to 0} \exp\left[-\sum_{s \leq r} \Delta\tau^2 \bar{z}(\tau_r) z(\tau_s) e^{-i\omega(\tau_r - \tau_s)} \left(1 - \frac{1}{2}\delta_{rs}\right)\right]$$

$$= \exp\left[-\int_{-\tau}^{\tau} d\tau d\tau' \bar{z}(\tau)\theta(\tau - \tau')e^{-i\omega(\tau - \tau')}z(\tau')\right], \tag{5.35}$$

where the δ_{rs} term contributes a term of order $\Delta\tau$ to the exponent that vanishes in the limit $\Delta\tau \to 0$. So, placing $G(t - t') = -i\theta(\tau - \tau')e^{-i\omega(\tau - \tau')}$,

$$S(t_2, t_1) = \exp\left[-i\int_{-\tau}^{\tau} d\tau d\tau' \bar{z}(\tau) G(t - t') z(\tau')\right]. \tag{5.36}$$

Finally, taking the limits of the integral to infinity ($\tau \to \infty$), we obtain the quoted result.

Example 5.1 A charged particle of charge q and mass m is in the ground state of a harmonic potential of characteristic frequency ω. Show that, after exposure to an electric field E for a time T, the probability that it remains in the ground state is given by

$$p = \exp[-4g^2 \sin^2(\omega T/2)], \tag{5.37}$$

where the coupling constant

$$g^2 = \frac{V_{spring}}{\hbar\omega} \tag{5.38}$$

is the ratio of the potential energy $V_{spring} = q^2 E^2/(2m\omega^2)$ stored in a classical spring stretched by a force to qE the quantum of energy $\hbar\omega$.

Solution

The probability $p = |S(T, 0)|^2$ of remaining in the ground state is the square of the amplitude

$$S(T, 0) = \langle\phi|Te^{-\frac{i}{\hbar}\int_0^T V(t)dt}|\phi\rangle. \tag{5.39}$$

Notice that, since we explicitly reintroduced $\hbar \neq 1$, we must now use

$$\frac{V(t)}{\hbar} = -\frac{qE(t)}{\hbar}x(t) \tag{5.40}$$

in the time-ordered exponential, where $E(t)$ is the electric field. Writing $x = \sqrt{\frac{\hbar}{2m\omega}}(b + b^\dagger)$, we can recast V in terms of boson creation and annihilation operators as $V(t)/\hbar = \bar{z}(t)b(t) + b^\dagger(t)z(t)$, where

$$z(t) = \bar{z}(t) = -\frac{1}{\hbar}\sqrt{\frac{\hbar}{2m\omega}}qE(t) = -\sqrt{\frac{V\omega}{\hbar}}\theta(t). \tag{5.41}$$

Here $V = \frac{q^2 E^2}{2m\omega^2}$ is the potential energy of the spring in a constant field E. Using the relationship derived in (5.36), we deduce that

$$S(T, 0) = e^{-iA},$$

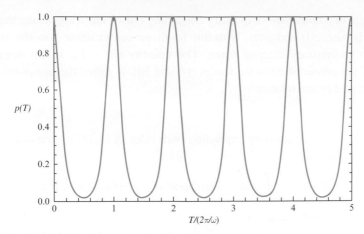

Probability $p(T)$ for an oscillator to remain in its ground state after exposure to an electric field for time T, illustrated for the case $V/\hbar\omega = 1$.

Fig. 5.3

where the phase term

$$A = \int_0^T dt_1 dt_2 \bar{z}(t_1) G(t_1 - t_2) z(t_2)$$

and $G(t) = -ie^{-i\omega t}\theta(t)$ is the Green's function. Carrying out the integral, we obtain

$$A = -i\frac{V\omega}{\hbar} \int_0^T dt \int_0^t dt' e^{-i\omega(t-t')} = -i\frac{V\omega}{\hbar} \int_0^T dt \frac{1}{i\omega}[1 - e^{-i\omega t}]$$

$$= -\frac{VT}{\hbar} + \frac{2V}{\hbar\omega} e^{-i\omega T/2} \sin\frac{\omega T}{2}$$

$$= -\frac{VT}{\hbar}\left[1 - \frac{\sin(\omega T)}{\omega T}\right] - i\frac{2V}{\hbar\omega}\sin^2\left(\frac{\omega T}{2}\right). \tag{5.42}$$

The real part of A contains a term that grows linearly in time (Re $A \sim -VT/\hbar$), giving rise to uniform growth in the phase of $S(T) \sim e^{iVT/\hbar}|S(T,0)|$ that we recognize as a consequence of the shift in the ground state energy of the oscillator $E_g \to \frac{\hbar\omega}{2} - V$ in the applied field. The imaginary part determines the probability of remaining in the ground state, which is given by

$$p = |S(T,0)|^2 = e^{2\text{Im}[A]} = \exp\left(-\frac{4V}{\hbar\omega}\sin^2\frac{\omega T}{2}\right),$$

demonstrating the oscillatory amplitude for remaining in the ground state (Figure 5.3).

5.1.2 Wick's theorem and generating functionals

The time-ordered exponential in the generating function

$$S[\bar{z}, z] = \langle 0|Te^{-i\int_{-\infty}^{\infty} dt[\bar{z}(t)b(t)+b^\dagger(t)z(t)]}|0\rangle = \exp\left[-i\int_{-\infty}^{\infty} dt dt' \bar{z}(t)G(t - t')z(t')\right] \tag{5.43}$$

is an example of a *functional*: a quantity containing one or more arguments that are functions (in this case, $z(t)$ and $\bar{z}(t)$). With this result we can examine how the ground state responds to an arbitrary external force. The quantity $G(t - t')$, which determines the response of the ground state to the forces $z(t)$ and $\bar{z}(t)$, is called the *one-particle Green's function*, defined by the relation

$$G(t - t') = -i\langle 0|Tb(t)b^{\dagger}(t')|0\rangle. \tag{5.44}$$

We may confirm this relation by expanding both sides of (5.43) to first order in $\bar{z}$ and z. The left-hand side gives

$$1 + (-i)^2 \int dt dt' \bar{z}(t)\langle 0|Tb(t)b^{\dagger}(t')|0\rangle.z(t') + O(\bar{z}^2, z^2), \tag{5.45}$$

whereas the right-hand side gives

$$1 - i \int dt dt' \bar{z}(t)G(t - t')z(t') + O(\bar{z}^2, z^2). \tag{5.46}$$

Comparing the coefficients, we confirm (5.44).

Order-by-order in z and $\bar{z}$, the relationships between the left-hand and right-hand sides of the expansion (5.43) of the generating functional $S[\bar{z}, z]$ provide an expansion for all the higher-order correlation functions of the harmonic oscillator in terms of the elementary Green's function $G(t - t')$, an expansion, known as *Wick's theorem*. From the left-hand side of (5.43), we see that each time we differentiate the generating functional we bring down operators $b(1)$ and $b^{\dagger}(1')$ inside the Green's function according to the relations

$$i\frac{\delta}{\delta\bar{z}(1)} \to \hat{b}(1), \qquad i\frac{\delta}{\delta z(1')} \to \hat{b}(1'), \tag{5.47}$$

where we have used the shorthand $1 \equiv t$, $1' \equiv t'$. For example,

$$\frac{i}{S}\frac{\delta S}{\delta\bar{z}(1)} = \frac{\langle 0|\hat{S}b(1)|0\rangle}{\langle 0|\hat{S}|0\rangle} = \langle b(1)\rangle = \int d1'G(1 - 1')z(1'), \tag{5.48}$$

so if there is a force present, the boson field develops an expectation value, which in the original oscillator corresponds to a state with a finite displacement or momentum. If we differentiate this expression again and set the source terms to zero, we get the two-particle Green's function,

$$i^2\frac{\delta^2 S}{\delta z(1')\delta\bar{z}(1)}\bigg|_{\bar{z}, z=0} = \langle 0|Tb(1)b^{\dagger}(1')|0\rangle = iG(1 - 1'). \tag{5.49}$$

If we take a $2n$th-order derivative, we obtain the n-particle Green's function:

$$i^{2n}\frac{\delta^{2n}S[\bar{z}, z]}{\delta z(1')\delta z(2')\cdots\delta\bar{z}(2)\delta\bar{z}(1)}\bigg|_{\bar{z}, z=0} = \langle 0|Tb(1)\cdots b(n)b^{\dagger}(n')\cdots b^{\dagger}(1')|0\rangle. \tag{5.50}$$

We define the quantity

$$G(1, \ldots n; 1', \ldots n') = (-i)^n\langle 0|Tb(1)\cdots b(n)b^{\dagger}(n')\cdots b^{\dagger}(1')|0\rangle$$

$$= i^n\frac{\delta^{2n}S[\bar{z}, z]}{\delta z(1')\delta z(2')\cdots\delta\bar{z}(2)\delta\bar{z}(1)}\bigg|_{\bar{z}, z=0} \tag{5.51}$$

as the *n-particle Green's function*. Now we can obtain an expansion for this quantity by differentiating the right-hand side of (5.43). After the first n differentiations we get

$$i^n \frac{\delta^n S}{\delta \bar{z}(n) \cdots \delta \bar{z}(1)} = S[\bar{z}, z] \int \prod_{s=1}^{n} ds' \, G(s - s') z(s').$$ (5.52)

Now there are $n!$ permutations P of the $z(s')$, so that when we carry out the remaining n differentiations, ultimately setting the source terms to zero, we obtain

$$i^n \frac{\delta^{2n} S}{\delta z(1') \cdots \delta z(n') \delta \bar{z}(n) \cdots \delta \bar{z}(1)} = \sum_P \prod G(r - P'_r),$$ (5.53)

where P_r is the rth component of the permutation $P = (P_1 P_2 \cdots P_n)$. Comparing relations (5.51) and (5.53), we obtain

$$G(1, \ldots n; 1', \ldots n') = \sum_P \prod_r G(r - P'_r).$$ (5.54)

Wick's theorem

It is a remarkable property of non-interacting systems that the n-particle Green's functions are determined entirely in terms of the one-particle Green's functions. In (5.54) each destruction event at time $t_r \equiv r$ is paired up with a corresponding creation event at time $t'_{P_r} \equiv P'_r$. The connection between these two events is often called a *contraction*, denoted as follows:

$$(-i)^n \langle 0|T \cdots b(r) \cdots b^{\dagger}(P'_r) \cdots |0\rangle = G(r - P'_r) \times (-i)^{n-1} \langle 0|T \cdots |0\rangle.$$ (5.55)

Notice that, since particles are conserved, we can only contract a creation operator with a destruction operator. According to Wick's theorem, the expansion of the n-particle Green's function in (5.50) is carried out as a sum over all possible contractions, denoted as follows:

$$G(1, \ldots n') = \sum G(1 - P'_1) G(2 - P'_2) \cdots G(r - P'_r) \cdots$$

$$= \sum_P (-i)^n \langle 0|Tb(1) b(2) \cdots b(r) \cdots b^{\dagger}(P'_r) \cdots b^{\dagger}(P'_1) \cdots b^{\dagger}(P'_2) \cdots |0\rangle.$$ (5.56)

Physically, this result follows from the identical nature of the bosonic quanta or particles. When we take the n particles out at times $t_1, \ldots t_n$, there is no way to know the order in which we are taking them out. The net amplitude is the sum of all possible ways of taking out the particles – this is the meaning of the sum over permutations P.

Finally, notice that the generating functional result can be generalized to an arbitrary number of oscillators by replacing $(z, \bar{z}) \rightarrow (z_r, \bar{z}_r)$, whereupon

$$\langle 0|T \exp\left[-i\int_{-\infty}^{\infty} dt\left(\bar{z}_r(t)b_r(t) + b_r^\dagger(t)z_r(t)\right)\right]|0\rangle$$

$$= \exp\left[-i\int_{-\infty}^{\infty} dtdt'\bar{z}_r(t)G_{rs}(t - t')z_s(t')\right], \qquad (5.57)$$

where now $G_{rs}(t - t') = -i\langle 0|Tb_r(t)b_s^\dagger(t')|0\rangle = -i\delta_{rs}\theta(t - t')e^{-i\omega_r(t-t')}$, and summation over repeated indices is implied. This provides the general basis for Wick's theorem. The concept of a generating functional can also be generalized to fermions, with the proviso that now we must replace $(z, \bar{z})$ by anticommuting numbers $(\eta, \bar{\eta})$, a point we return to later.

5.2 Green's functions

Green's functions are the elementary response functions of a many-body system. The one-particle Green's function is defined as

$$G_{\lambda\lambda'}(t - t') = -i\langle\phi|T\psi_\lambda(t)\psi_{\lambda'}^\dagger(t')|\phi\rangle, \qquad (5.58)$$

where $|\phi\rangle$ is the many-body ground state, $\psi_\lambda(t)$ is the field in the Heisenberg representation and

$$T\psi_\lambda(t)\psi_{\lambda'}^\dagger(t') = \begin{cases} \psi_\lambda(t)\psi_{\lambda'}^\dagger(t') & (t > t') \\ \pm\psi_{\lambda'}^\dagger(t')\psi_\lambda(t) & (t < t') \end{cases} \quad \begin{cases} \text{bosons}: + \\ \text{fermions}: - \end{cases} \qquad (5.59)$$

defines the time-ordering for fermions and bosons. Diagrammatically, this quantity is represented as follows

$$G_{\lambda\lambda'}(t - t') = \quad \overset{\lambda,t}{\underset{\longleftarrow}{\rule{5cm}{0pt}}} \overset{\lambda',t'}{} . \qquad (5.60)$$

Quite often, we shall be dealing with translationally invariant systems, where λ denotes the momentum and spin of the particle $\lambda \equiv \mathbf{p}\sigma$. If spin is a good quantum number (no magnetic field, no spin–orbit interactions), then

$$G_{\mathbf{k}\sigma,\mathbf{k}'\sigma'}(t - t') = \delta_{\sigma\sigma'}\delta_{\mathbf{k}\mathbf{k}'}G(\mathbf{k}, t - t') \qquad (5.61)$$

is diagonal (where in the continuum limit $\delta_{\mathbf{k}\mathbf{k}'} \to (2\pi)^D\delta^{(D)}(\mathbf{k} - \mathbf{k}')$). In this case, we denote

$$G(\mathbf{k}, t - t') = -i\langle\phi|T\psi_{\mathbf{k}\sigma}(t)\psi_{\mathbf{k}\sigma}^\dagger(t')|\phi\rangle = \quad t\overset{\mathbf{k}}{\underset{\longleftarrow}{\rule{4cm}{0pt}}}t'. \qquad (5.62)$$

We can also define Green's functions in coordinate space:

$$G(\mathbf{x} - \mathbf{x}', t - t') = -i\langle\phi|T\psi_\sigma(\mathbf{x}, t)\psi_\sigma^\dagger(\mathbf{x}', t')|\phi\rangle, \qquad (5.63)$$

which we denote diagrammatically by

$$G(\mathbf{x} - \mathbf{x}', t - t') = (x,t)\overset{}{\underset{\longleftarrow}{\rule{4cm}{0pt}}}(x', t'). \qquad (5.64)$$

By writing $\psi_\sigma(\mathbf{x}, t) = \int_\mathbf{k} \psi_{\mathbf{k}\sigma} e^{i(\mathbf{k}\cdot\mathbf{x})}$, we see that the coordinate-space Green's function is just the Fourier transform of the momentum-space Green's function:

$$G(\mathbf{x} - \mathbf{x}', t) = \int_{\mathbf{k},\mathbf{k}'} e^{i(\mathbf{k}\cdot\mathbf{x} - \mathbf{k}'\cdot\mathbf{x}')} \overbrace{-i\langle\phi|T\psi_{\mathbf{k}\sigma}(t)\psi_{\mathbf{k}'\sigma}^\dagger(0)|\phi\rangle}^{\delta_{\mathbf{k}\mathbf{k}'}G(\mathbf{k}, t-t')}$$

$$= \int \frac{d^3k}{(2\pi)^3} G(\mathbf{k}, t) e^{i\mathbf{k}\cdot(\mathbf{x}-\mathbf{x}')}. \tag{5.65}$$

It is also often convenient to Fourier transform in time:

$$G(\mathbf{k}, t) = \int_{-\infty}^{\infty} \frac{d\omega}{2\pi} G(\mathbf{k}, \omega) e^{-i\omega t}. \tag{5.66}$$

The quantity

$$G(\mathbf{k}, \omega) = \int_{-\infty}^{\infty} dt G(\mathbf{k}, t) e^{i\omega t} = \underset{\xleftarrow{\hspace{3cm}}}{\overset{k, \omega}{}} \tag{5.67}$$

is known as the *propagator*. We can then relate the Green's function in coordinate space to its propagator, as follows:

$$-i\langle\phi|T\psi_\sigma(\mathbf{x}, t)\psi_\sigma^\dagger(\mathbf{x}', t')|\phi\rangle = \int \frac{d^3k\, d\omega}{(2\pi)^4} G(\mathbf{k}, \omega) e^{i[(\mathbf{k}\cdot(\mathbf{x}-\mathbf{x}') - \omega(t-t')]}. \tag{5.68}$$

5.2.1 Green's function for free fermions

As a first example, let us calculate the Green's function of a degenerate Fermi liquid of non-interacting fermions in its ground state. We shall take the heat bath into account, using a Heisenberg representation where the heat bath's contribution to the energy is subtracted away, so that

$$H = \hat{H}_0 - \mu N = \sum_\sigma \epsilon_\mathbf{k} c_{\mathbf{k}\sigma}^\dagger c_{\mathbf{k}\sigma} \tag{5.69}$$

is the Hamiltonian used in the Heisenberg representation and $\epsilon_\mathbf{k} = \frac{\hbar^2 k^2}{2m} - \mu$. We will frequently reserve use of c for the creation operator of fermions in momentum space. The ground state for a fluid of fermions is given by

$$|\phi\rangle = \prod_{\sigma|\mathbf{k}|<k_f} c_{\mathbf{k}\sigma}^\dagger |0\rangle. \tag{5.70}$$

In the Heisenberg representation, $c_{\mathbf{k}\sigma}^\dagger(t) = e^{i\epsilon_\mathbf{k} t} c_{\mathbf{k}\sigma}^\dagger$, $c_{\mathbf{k}\sigma}(t) = e^{-i\epsilon_\mathbf{k} t} c_{\mathbf{k}\sigma}$. For forward time propagation, it is only possible to add a fermion above the Fermi energy:

$$\langle\phi|c_{\mathbf{k}\sigma}(t)c_{\mathbf{k}'\sigma'}^\dagger(t')|\phi\rangle = \delta_{\sigma\sigma'}\delta_{\mathbf{k}\mathbf{k}'}e^{-i\epsilon_\mathbf{k}(t-t')}\langle\phi|c_{\mathbf{k}\sigma}c_{\mathbf{k}\sigma}^\dagger|\phi\rangle$$

$$= \delta_{\sigma\sigma'}\delta_{\mathbf{k}\mathbf{k}'}(1 - n_\mathbf{k})e^{-i\epsilon_\mathbf{k}(t-t')}, \tag{5.71}$$

where $n_\mathbf{k} = \theta(|k_F| - |\mathbf{k}|)$. For backwards time propagation, it is only possible to destroy a fermion, creating a hole, below the Fermi energy:

$$\langle\phi|c_{\mathbf{k}'\sigma'}^\dagger(t')c_{\mathbf{k}\sigma}(t)|\phi\rangle = \delta_{\sigma\sigma'}\delta_{\mathbf{k}\mathbf{k}'}n_\mathbf{k}e^{-i\epsilon_\mathbf{k}(t-t')}, \tag{5.72}$$

so that

$$G(\mathbf{k}, t) = -i[(1 - n_{\mathbf{k}})\theta(t) - n_{\mathbf{k}}\theta(-t)]e^{-i\epsilon_{\mathbf{k}}t} \tag{5.73}$$

can be expanded as

$$G(\mathbf{k}, t) = \begin{cases} -i\theta_{|\mathbf{k}|-|\mathbf{k}_F|}e^{-i\epsilon_{\mathbf{k}}t} & (t > 0) \quad \text{particles} \\[2mm] i\theta_{|\mathbf{k}_F|-|\mathbf{k}|}e^{-i\epsilon_{\mathbf{k}}t} & (t < 0). \quad \text{holes: particles moving backwards in time} \end{cases} \tag{5.74}$$

The unification of particle and hole excitations in a single function is one of the great utilities of the time-ordered Green's function.[2]

Next, let us calculate the Fourier transform of the Green's function. This is given by

$$G(\mathbf{k}, \omega) = -i \int_{-\infty}^{\infty} dt e^{i(\omega - \epsilon_{\mathbf{k}})t} \overbrace{e^{-|t|\delta}}^{\text{convergence factor}} \left[\theta_{k-k_F}\theta(t) - \theta_{k_F-k}\theta(-t) \right]$$

$$= -i \left[\frac{\theta_{k-k_F}}{\delta - i(\omega - \epsilon_{\mathbf{k}})} - \frac{\theta_{k_F-k}}{\delta + i(\omega - \epsilon_{\mathbf{k}})} \right] = \frac{1}{\omega - \epsilon_{\mathbf{k}} + i\delta_{\mathbf{k}}}, \tag{5.75}$$

where $\delta_{\mathbf{k}} = \delta \operatorname{sgn}(k - k_F)$. The free fermion propagator is then

$$G(\mathbf{k}, \omega) = \frac{1}{\omega - \epsilon_{\mathbf{k}} + i\delta_{\mathbf{k}}} = \frac{\mathbf{k}, \omega}{\longleftarrow}. \tag{5.76}$$

The Green's function contains both static and dynamical information about the motion of particles in the many-body system. For example, we can use it to calculate the density of particles in a Fermi gas

$$\langle \hat{\rho}(x) \rangle = \sum_{\sigma} \langle \psi_{\sigma}^{\dagger} \psi_{\sigma} \rangle = -\sum_{\sigma} \langle \phi | T \psi_{\sigma}(x, 0^-) \psi_{\sigma}^{\dagger}(x, 0) | \phi \rangle$$

$$= -i(2S + 1)G(\mathbf{x}, 0^-)|_{\mathbf{x}=0}, \tag{5.77}$$

where S is the spin of the fermion. We can also use it to calculate the kinetic energy density:

$$\langle \hat{T}(x) \rangle = -\frac{\hbar^2}{2m} \sum_{\sigma} \langle \psi_{\sigma}^{\dagger}(x) \nabla_x^2 \psi_{\sigma}(x) \rangle = \frac{\hbar^2 \nabla_x^2}{2m} \sum_{\sigma} \langle \phi | T \psi_{\sigma}(x, 0^-) \psi_{\sigma}^{\dagger}(\vec{x}', 0) | \phi \rangle \Big|_{\mathbf{x}-\mathbf{x}'=0}$$

$$= i(2S + 1)\frac{\hbar^2 \nabla^2}{2m} G(\mathbf{x}, 0^-) \Big|_{\mathbf{x}=0}. \tag{5.78}$$

---

[2] According to an aprocryphal story, the relativistic counterpart of this notion, that positrons are electrons traveling backwards in time, was invented by Richard Feynman while a graduate student of John Wheeler at Princeton. Wheeler was strict, allowing his graduate students precisely half an hour of discussion a week, employing a chess clock as a timer at the meeting. Wheeler treated Feynman no differently and when the alloted time was up, he stopped the clock and announced that the session was over. At their second meeting, Feynman apparently arrived with his own clock, and at the end of the half hour, Feynman stopped his own clock to announce that Wheeler's time was up. During this meeting they discussed the physics of positrons and Feynman came up with the idea that a positron was an electron traveling backwards in time and that there might only be one electron in the whole universe, threading backwards and forwards in time. To mark the discovery, at the third meeting Feynman arrived with a modified clock which he had fixed to start at 30 minutes and run backwards to zero!

Example 5.2 By relating the particle density and kinetic energy density to the one-particle Green's function, calculate the particle and kinetic energy density of particles in a degenerate Fermi liquid.

Solution

We begin by writing $\langle \hat{\rho}(x) \rangle = -i(2S + 1)G(\vec{0}, 0^-)$. Writing this out explicitly, we obtain

$$\langle \rho(x) \rangle = (2S + 1) \int \frac{d^3k}{(2\pi)^3} \left[\int \frac{d\omega}{2\pi i} e^{i\omega\delta} \frac{1}{\omega - \epsilon_{\mathbf{k}} + i\delta_{\mathbf{k}}} \right], \tag{5.79}$$

where the convergence factor appears because we are evaluating the Green's function at a small negative time $-\delta$. We have explicitly separated out the frequency and momentum integrals. The poles of the propagator are at $\omega = \epsilon_{\mathbf{k}} - i\delta$ if $k > k_F$, but at $\omega = \epsilon_{\mathbf{k}} + i\delta$ if $k < k_F$, as illustrated in Figure 5.4. The convergence factor means that we can calculate the complex integral using Cauchy's theorem by completing the contour in the upper half complex plane, where the integrand dies away exponentially. The pole in the integral will only pick up those poles associated with states below the Fermi energy, so that

$$\int \frac{d\omega}{2\pi i} e^{i\omega\delta} \frac{1}{\omega - \epsilon_{\mathbf{k}} + i\delta_{\mathbf{k}}} = \theta_{k_F - |\mathbf{k}|} \tag{5.80}$$

and hence

$$\rho = (2S + 1) \int_{k < k_F} \frac{d^3k}{(2\pi)^3} = (2S + 1) \frac{V_F}{(2\pi)^3}. \tag{5.81}$$

In a similar way, the kinetic energy density is written

$$\langle T(x) \rangle = (2S + 1) \int \frac{d^3k}{(2\pi)^3} \frac{\hbar^2 k^2}{2m} \left[\int \frac{d\omega}{2\pi i} e^{i\omega\delta} \frac{1}{\omega - \epsilon_{\mathbf{k}} + i\delta_{\mathbf{k}}} \right]$$

$$= (2S + 1) \int_{k < k_F} \frac{d^3k}{(2\pi)^3} \frac{\hbar^2 k^2}{2m} = \frac{3}{5} \epsilon_F \rho. \tag{5.82}$$

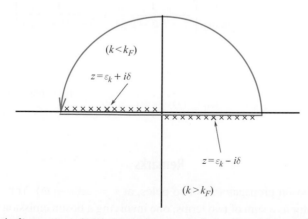

Showing how the path of integration in (5.80) picks up the pole contributions from the occupied states beneath the Fermi surface.

Fig. 5.4

5.2.2 Green's function for free bosons

As a second example, let us examine the Green's function of a gas of non-interacting bosons, described by

$$H = \sum_{\mathbf{q}} \omega_{\mathbf{q}} \left[b_{\mathbf{q}}^{\dagger} b_{\mathbf{q}} + \frac{1}{2} \right], \tag{5.83}$$

where the physical field operator $\phi(x)$ is related to a sum of creation and annihilation operators:

$$\phi(x) = \int_q \phi_{\mathbf{q}} e^{i\mathbf{q}\cdot\mathbf{x}}$$

$$\phi_{\mathbf{q}} = \sqrt{\frac{\hbar}{2m\omega_{\mathbf{q}}}} [b_{\mathbf{q}} + b_{-\mathbf{q}}^{\dagger}]. \tag{5.84}$$

Since there are no bosons present in the ground state, boson destruction operators annihilate the ground state, which in this case is the bosonic vacuum $|\phi\rangle \equiv |0\rangle$. The only terms contributing to the Green's function are then

$$-i\langle\phi|Tb_{\mathbf{q}}(t)b_{\mathbf{q}}^{\dagger}(0)|\phi\rangle = -i\theta(t)e^{-i\omega_{\mathbf{q}}t},$$

$$-i\langle\phi|Tb_{-\mathbf{q}}^{\dagger}(t)b_{-\mathbf{q}}(0)|\phi\rangle = -i\theta(-t)e^{i\omega_{\mathbf{q}}t}, \tag{5.85}$$

so that

$$D(\mathbf{q}, t) = -i\langle\phi|\phi(\mathbf{q}, t)\phi(-\mathbf{q}, 0)|\phi\rangle = -i\frac{\hbar}{2m\omega_{\mathbf{q}}}\left[\theta(t)e^{-i\omega_{\mathbf{q}}t} + \theta(-t)e^{i\omega_{\mathbf{q}}t}\right]. \tag{5.86}$$

If we Fourier transform this quantity, we obtain the boson propagator,

$$D(\mathbf{q}, \nu) = \int_{-\infty}^{\infty} dt e^{-\delta|t|+i\nu t} D(\mathbf{q}, t)$$

$$= -i\frac{\hbar}{2m\omega_{\mathbf{q}}} \left[\frac{1}{\delta - i(\nu - \omega_{\mathbf{q}})} + \frac{1}{\delta + i(\nu + \omega_{\mathbf{q}})} \right] \tag{5.87}$$

or

$$D(\mathbf{q}, \nu) = \frac{\hbar}{2m\omega_{\mathbf{q}}} \left[\frac{2\omega_{\mathbf{q}}}{\nu^2 - (\omega_{\mathbf{q}} - i\delta)^2} \right]. \qquad \text{boson propagator} \tag{5.88}$$

Remarks

- Note that the boson propagator has two poles, at $\nu = \pm(\omega - i\delta)$. You can think of the boson propagator as a sum of two terms, one involving a boson emission that propagates *forwards* in time from the emitter, a second involving boson absorption that propagates *backwards* in time from the absorber:

$$D(\mathbf{q}, \nu) = \frac{\hbar}{2m\omega_{\mathbf{q}}} \left[\overbrace{\frac{1}{\nu - (\omega_{\mathbf{q}} - i\delta)}}^{\text{emission}} + \overbrace{\frac{1}{-\nu - (\omega_{\mathbf{q}} - i\delta)}}^{\text{absorption}} \right]. \tag{5.89}$$

- We shall shortly see that amplitude to absorb and emit bosons by propagating fermions is directly related to the boson propagator. For example, when there is an interaction of the form

$$H_{int} = g \int d^3 x \phi(\mathbf{x}) \rho(\mathbf{x}), \tag{5.90}$$

the exchange of virtual bosons between particles gives rise to *retarded* interactions,

$$V(\mathbf{q}, t - t') = \frac{g^2}{\hbar} D(\mathbf{q}, t - t'), \tag{5.91}$$

whereby a passing fermion produces a potential change in the environment which lasts a characteristic time $\Delta\tau \sim 1/\omega_0$ where ω_0 is the characteristic value of $\omega_{\mathbf{q}}$. From the Fourier transform of this expression, you can see that the time average of this interaction, proportional to $D(\mathbf{q}, \nu = 0) = -\frac{\hbar}{m\omega_q^2}$, is negative, i.e. the virtual exchange of a spinless boson mediates an attractive interaction.

5.3 Adiabatic concept

The adiabatic concept is one of the most valuable concepts in many-body theory. What does it mean to understand a many-body problem when we can never, except in the most special cases, expect to solve the problem exactly? The adiabatic concept provides an answer to this question.

Suppose we are interested in a many-body problem with Hamiltonian H, with ground state $|\Psi_g\rangle$ which we cannot solve exactly. Instead we can often solve a simplified version of the many-body Hamiltonian H_0 where the ground state $|\tilde{\Psi}_g\rangle$ has the same *symmetry* as $|\Psi_g\rangle$. Suppose we start in the ground state $|\tilde{\Psi}_g\rangle$ and now slowly evolve the Hamiltonian from H_0 to H, i.e. if $\hat{V} = H - H_0$, we imagine that the state time-evolves according to the Hamiltonian

$$H(t) = H_0 + \lambda(t)V$$

$$\lambda(t) = e^{-|t|\delta} \tag{5.92}$$

where δ is arbitrarily small.

As we adiabatically evolve the system, the ground state and excited states will evolve, as shown in Figure 5.5. In such an evolution process, the energy levels will typically show *energy level repulsion*. If any two levels get too close together, matrix elements between the two states will cause them to repel one another. However, it is possible for states of different symmetry to cross, because selection rules prevent them from mixing. Sometimes

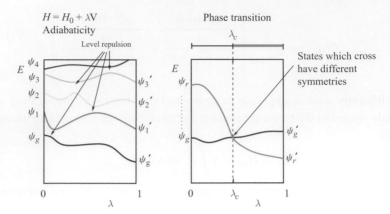

Fig. 5.5 Illustrating the evolution of the Hilbert space as the Hamiltonian is adiabatically evolved. In the first case, the ground state can be adiabatically evolved all the way to $\lambda = 1$. In the second case, a phase transition occurs at $\lambda = \lambda_c$, where a previously excited state, with a different symmetry to the ground state, crosses below the ground state.

such an adiabatic evolution will lead to *level crossing*, whereby at $\lambda = \lambda_c$ some excited state ψ_r with different symmetry to the ground state crosses to a lower energy than the ground state. Such a situation leads to *spontaneous symmetry breaking*. A simple example is when a ferromagnetic ground state becomes stabilized by interactions.

In general, however, if there is no symmetry-changing phase transition as the interaction V is turned on, adiabatic evolution can be used to turn on *interactions* and to evolve the ground state from $\tilde{\Psi}_g$ to Ψ_g.

These ideas play a central role in the development of perturbation theory and Feynman diagrams. They are also, however, of immense qualitative importance, for adiabatic evolution defines an equivalence class of ground states with the same qualitative physics. The adiabatic principle was first employed with great success in the 1950s. Murray Gell-Mann and Francis Low used it to prove their famous relation linking non-interacting and interacting Green's functions [1]. Later in the decade, Landau [2–4] used the adiabatic idea in a brilliantly qualitative fashion to formulate his theory of interacting Fermi liquids, which we examine in detail in the next chapter.

5.3.1 Gell-Mann–Low theorem

Suppose we gradually turn on, and later gradually turn off, an interaction V so that

$$V(t) = e^{-\epsilon |t|} V(0) \tag{5.93}$$

acquires its full magnetitude at $t = 0$ and vanishes in the distant past and in the far future. The quantity $\tau_A = \epsilon^{-1}$ sets the characteristic "switch-on time" for the process. Adiabaticity requires that we ultimately let $\epsilon \to 0$, sending the switch-on time to infinity, $\tau_A \to \infty$. When we start out at $t = -\infty$, the ground state is $|-\infty\rangle$, and the interaction and Heisenberg representations coincide. If we now evolve to the present in the Heisenberg representation, the states do not evolve, so the ground state is unchanged:

$$|\phi\rangle_H \equiv |-\infty\rangle, \tag{5.94}$$

and all the interesting physics of the interaction V is encoded in the operators. We would like to calculate the correlation or Green's functions of a set of observables in the fully interacting system. The Gell-Mann–Low theorem enables us to relate the Green's function of the interacting system to the Green's function of the non-interacting system at $t = -\infty$. The key result is

$$\langle\phi|TA(t_1)B(t_2)\cdots R(t_r)|\phi\rangle_H = \langle+\infty|TS[\infty,-\infty]A(t_1)B(t_2)\cdots R(t_r)|-\infty\rangle_I$$

$$S[\infty,-\infty] = T\exp\left[-i\int_{-\infty}^{\infty} V(t')dt'\right]. \tag{5.95}$$

The state $|+\infty\rangle = S(\infty,-\infty)|-\infty\rangle$ corresponds to the ground state, in the interaction representation in the distant future. If adiabaticity holds, then the process of slowly turning on and then turning off the interaction will return the system to its original state, up to a phase, so that $|+\infty\rangle = e^{2i\delta}|-\infty\rangle$. We can then write $e^{2i\delta} = \langle-\infty|\infty\rangle$, so that

$$\langle+\infty| = e^{-2i\delta}\langle-\infty| = \frac{\langle-\infty|}{\langle-\infty|+\infty\rangle}, \tag{5.96}$$

and the Gell-Mann–Low formula becomes

$$\langle\phi|TA(t_1)B(t_2)\cdots R(t_r)|\phi\rangle_H = \frac{\langle-\infty|TS[\infty,-\infty]A(t_1)B(t_2)\cdots R(t_r)|-\infty\rangle_I}{\langle-\infty|S[\infty,-\infty]|-\infty\rangle}. \tag{5.97}$$

Remarks

- With the Gell-Mann–Low relation, we relate the Green's function of a set of complex operators in an interacting system to the Green's function of a set of simple operators multiplied by the S-matrix.
- The Gell-Mann–Low relation is the starting point for the Feynman diagram expansion of Green's functions. When we expand the S-matrix as a power series in V, each term in the expansion can be written as an integral over Green's functions of the non-interacting problem. Each of these terms corresponds to a particular Feynman diagram.
- When we expand the vacuum expectation value of the S-matrix, we will see that this leads to *linked cluster* diagrams.

Proof: To prove this result, let $U(t) = S(t,-\infty)$ be the time-evolution operator for the interaction representation. Since the interaction and Heisenberg states coincide at $t = -\infty$, and $|\psi_H\rangle$ does not evolve with time,

$$|\psi_I(t)\rangle = U(t)|\psi_H\rangle. \tag{5.98}$$

Since $U(t)A_H(t)|\psi_H\rangle = A_I(t)|\psi_I(t)\rangle = A_I(t)U(t)|\psi_H\rangle$, the relation between operators in the two representations must be

$$A_H(t) = U^\dagger(t)A_I(t)U(t). \tag{5.99}$$

Suppose $t_1 > t_2 > t_3 > \cdots t_r$. Then, using this relation we may write

$$\langle\phi|A(t_1)\cdots R(t_r)|\phi\rangle_H$$

$$= \langle-\infty|U^\dagger(t_1)A_I(t_1)\overbrace{U(t_1)U^\dagger(t_2)}^{S(t_1,t_2)}\cdots\overbrace{U(t_{r-1})U^\dagger(t_r)}^{S(t_{r-1},t_r)}R_I(t_r)U(t_r)|-\infty\rangle,$$

where we have identified $|\phi\rangle_H \equiv |-\infty\rangle$. Now $S(t_1,t_2) = U(t_1)U^\dagger(t_2)$ is the operator that time-evolves the states of the interaction representation, so we may rewrite the above result as

$$\langle 0|A(t_1)\cdots R(t_r)|0\rangle_H = \langle-\infty|\overbrace{U^\dagger(t_1)}^{S^\dagger(t_1,-\infty)}A_I(t_1)S(t_1,t_2)\cdots S(t_{r-1},t_r)R_I(t_r)\overbrace{U(t_r)}^{S(t_r,-\infty)}|-\infty\rangle,$$

where we have replaced $U(t) \to S(t,-\infty)$. Now $S(\infty,t_1)S(t_1,-\infty)|-\infty\rangle = |\infty\rangle$ and, since S is a unitary matrix, $S^\dagger(\infty,t_1)S(\infty,t_1) = 1$, so multiplying both sides by $S^\dagger(\infty,t_1)$, $S(t_1,-\infty)|-\infty\rangle = S^\dagger(\infty,t_1)|\infty\rangle$ and by taking their complex conjugates,

$$\langle-\infty|S^\dagger(t_1,-\infty) = \langle\infty|S(\infty,t_1). \tag{5.100}$$

Inserting this into the above expression gives

$$\langle 0|A(t_1)\cdots R(t_r)|0\rangle_H = \langle+\infty|S(\infty,t_1)A_I(t_1)S(t_1,t_2)\cdots S(t_{r-1},t_r)R_I(t_r)S(t_r,-\infty)|-\infty\rangle$$

$$= \langle+\infty|T\overbrace{S(\infty,t_1)S(t_1,t_2)\cdots S(t_r,-\infty)}^{S(\infty,-\infty)}A_I(t_1)\cdots R_I(t_r)|-\infty\rangle,$$

where we have used the time-ordering operator to separate out the S-matrix terms from the operators. Finally, since we assumed $t_1 > t_2 > \cdots > t_r$, we can write

$$\langle\phi|T[A(t_1)\cdots R(t_r)]|\phi\rangle_H = \langle+\infty|T[S(\infty,-\infty)A_I(t_1)B_I(t_2)\cdots R_I(t_r)]|-\infty\rangle. \tag{5.101}$$

Although we have proved this expression for a particular time-ordering, it is clear that if we permute the operators the time-ordering will always act to time-order both sides, and thus this expression holds for an arbitrary time-ordering of operators.

5.3.2 Generating function for free fermions

The generating function derived for the harmonic oscillator can be generalized to free fermions by the use of *anticommuting* or *Grassman numbers* η and $\bar{\eta}$. The simplest model is

$$\begin{aligned} H &= \epsilon c^\dagger c \\ V(t) &= \bar{\eta}(t)c(t) + c^\dagger(t)\eta(t). \end{aligned} \tag{5.102}$$

The corresponding generating functional is given by

$$S[\bar{\eta}, \eta] = \langle \phi | T \exp \left(-i \int_{-\infty}^{\infty} dt \left[\bar{\eta}(t) c(t) + c^{\dagger}(t) \eta(t) \right] \right) | \phi \rangle$$

$$= \exp \left[-i \int_{-\infty}^{\infty} dt dt' \, \bar{\eta}(t) G(t - t') \eta(t') \right]$$

$$G(t - t') = -i \langle \phi | T c(t) c^{\dagger}(t') | \phi \rangle, \tag{5.103}$$

where $|\phi\rangle$ is the ground state for the non-interacting Hamiltonian. To prove this result, we use the same method as for the harmonic oscillator. As before we split up the S-matrix into N discrete time-slices, writing

$$S_N = e^{A_N - A_N^{\dagger}} \cdots e^{A_r - A_r^{\dagger}} \cdots e^{A_1 - A_1^{\dagger}}, \tag{5.104}$$

where

$$A_r = \bar{\eta}(t_r)(-ice^{-i\epsilon t_r})\Delta t$$

$$A_r^{\dagger} = \eta(t_r)(ic^{\dagger}e^{i\epsilon t_r})\Delta t. \tag{5.105}$$

The next step requires a little care, for when $\epsilon < 0$, $|\phi\rangle = c^{\dagger}|0\rangle$ is the vacuum for holes ($h = c^{\dagger}$), rather than particles, so that in this case we need to *anti-normal-order* the S-matrix. Carrying out the ordering process, we obtain

$$S_N = \begin{cases} e^{-\sum_r A_r^{\dagger}} e^{\sum_r A_r} \exp \left[-\sum_{r \geq s} [A_r, A_s^{\dagger}](1 - \frac{1}{2}\delta_{rs}) \right] & (\epsilon > 0) \\ \\ e^{\sum_r A_r} e^{-\sum_r A_r^{\dagger}} \exp \left[\sum_{r \leq s} [A_r, A_s^{\dagger}](1 - \frac{1}{2}\delta_{rs}) \right] & (\epsilon < 0). \end{cases} \tag{5.106}$$

When we take the expectation value $\langle \phi | S_N | \phi \rangle$, the first term in these expressions gives unity. Calculating the commutators, in the exponent we obtain

$$[A_r, A_s^{\dagger}] = \Delta t^2 [\bar{\eta}(t_r)c, c^{\dagger}\eta(t_s)]e^{-i\epsilon(t_r - t_s)}$$

$$= \Delta t^2 \bar{\eta}(t_r)\{c, c^{\dagger}\}\eta(t_s)e^{-i\epsilon(t_r - t_s)}$$

$$= \Delta t^2 \bar{\eta}(t_r)\eta(t_s)e^{-i\epsilon(t_r - t_s)}. \tag{5.107}$$

(Notice how the anticommuting property of the Grassman variables $\bar{\eta}(t_r)\eta(t_s) = -\eta(t_s)\bar{\eta}(t_r)$ means that we can convert a commutator of $[A_r, A_s]$ into an anticommutator $\{c, c^{\dagger}\}$.) Next, taking the limit $N \to \infty$, we obtain

$$S[\bar{\eta}, \eta] = \begin{cases} \exp \left[-\int_{-\infty}^{\infty} dt dt' \, \bar{\eta}(t)\theta(t - t')\eta(t')e^{-i\epsilon(t - t')} \right] & (\epsilon > 0) \\ \\ \exp \left[\int_{-\infty}^{\infty} d\tau d\tau' \, \bar{\eta}(\tau)\theta(t' - t)\eta(\tau')e^{-i\epsilon(t - t')} \right] & (\epsilon < 0). \end{cases} \tag{5.108}$$

By introducing the Green's function

$$G(t) = -i \left[(1 - f(\epsilon))\theta(t) - f(\epsilon))\theta(-t) \right] e^{-i\epsilon t},$$

we can compactly combine these two results into the final form,

$$S(t_2, t_1) = \exp\left[-i \int_{-\infty}^{\infty} dt\,dt'\, \bar{\eta}(t) G(t - t') \eta(t')\right]. \tag{5.109}$$

A more heuristic derivation, however, is to recognize that derivatives of the generating functional bring down Fermi operators inside the time-ordered exponential:

$$-i\frac{\delta}{\delta\eta(t)} \langle\phi|T\hat{S}\cdots|\phi\rangle = \langle\phi|T\hat{S}c^{\dagger}(t)\cdots|\phi\rangle$$

$$i\frac{\delta}{\delta\bar{\eta}(t)} \langle\phi|T\hat{S}\cdots|\phi\rangle = \langle\phi|T\hat{S}c(t)\cdots|\phi\rangle, \tag{5.110}$$

where $\hat{S} = T\exp\left[-i\int dt'\left(\bar{\eta}(t')c(t') + c^{\dagger}(t')\eta(t')\right)\right]$, so that, inside the expectation value,

$$-i\frac{\delta}{\delta\eta(t)} \equiv c^{\dagger}(t)$$

$$i\frac{\delta}{\delta\bar{\eta}(t)} \equiv c(t) \tag{5.111}$$

and

$$-i\frac{\delta \ln S}{\delta\eta(1)} = \frac{\langle\phi|Tc^{\dagger}(1)\hat{S}|\phi\rangle}{\langle\phi|\hat{S}|\phi\rangle} \equiv \langle c^{\dagger}(1)\rangle, \tag{5.112}$$

where $\hat{S} = T\exp\left[-i\int V(t')dt'\right]$. Here, we have used the Gell-Mann–Low theorem to identify the quotient above as the expectation value for $c^{\dagger}(1)$ in the presence of the source terms. Differentiating one more time,

$$\frac{\delta^2 \ln S[\bar{\eta},\eta]}{\delta\bar{\eta}(2)\delta\eta(1)} = \frac{\langle\phi|Tc(2)c^{\dagger}(1)\hat{S}|\phi\rangle}{\langle\phi|\hat{S}|\phi\rangle} - \frac{\langle\phi|Tc(2)\hat{S}|\phi\rangle}{\langle\phi|\hat{S}|\phi\rangle}\frac{\langle\phi|Tc^{\dagger}(1)\hat{S}|\phi\rangle}{\langle\phi|\hat{S}|\phi\rangle}$$

$$= \langle Tc(2)c^{\dagger}(1)\rangle - \langle c(2)\rangle\langle c^{\dagger}(1)\rangle$$

$$= \langle T\delta c(2)\delta c^{\dagger}(1)\rangle. \tag{5.113}$$

This quantity describes the variance in the fluctuations $\delta c^{(\dagger)}(2) \equiv c^{(\dagger)}(2) - \langle c^{(\dagger)}(2)\rangle$ of the fermion field about their average value. When the source terms η and $\bar{\eta}$ are introduced, they induce a finite (Grassman) expectation value of the fields $\langle c(1)\rangle$ and $\langle c^{\dagger}(1)\rangle$, but the absence of interactions between the modes means they won't change the amplitude of fluctuations about the mean, so that

$$\frac{\delta^2 \ln S[\bar{\eta},\eta]}{\delta\bar{\eta}(2)\delta\eta(1)} = \langle Tc(1)c^{\dagger}(2)\rangle\bigg|_{\eta,\,\bar{\eta}=0} = iG(2-1),$$

and we can then deduce that

$$\ln S[\bar{\eta},\eta] = -i\int d1\,d2\,\bar{\eta}(2)G(2-1)\eta(1). \tag{5.114}$$

There is no constant term, because $S = 1$ when the source terms are removed, and we arrive back at (5.103).

The generalization of the generating functional to a gas of fermions with many one-particle states is just a question of including an appropriate sum over one-particle states, i.e.

$$
\begin{aligned}
H &= \sum_\lambda \epsilon_\lambda c_\lambda^\dagger c_\lambda \\
V(t) &= \sum_\lambda \bar{\eta}_\lambda(t) c_\lambda(t) + c_\lambda^\dagger(t) \eta_\lambda(t).
\end{aligned}
\tag{5.115}
$$

The corresponding generating functional is given by

$$
S[\bar{\eta}, \eta] = \langle \phi | T \exp\left[-i \int d1 \sum_\lambda \bar{\eta}_\lambda(1) c_\lambda(1) + c_\lambda^\dagger(1) \eta_\lambda(1) \right] | \phi \rangle
$$

$$
= \exp\left[-i \sum_\lambda \int d1 d2 \bar{\eta}_\lambda(1) G_\lambda(1-2) \eta_\lambda(2) \right]
$$

$$
G_\lambda(1-2) = -i \langle \phi | T c_\lambda(1) c_\lambda^\dagger(2) | \phi \rangle.
\tag{5.116}
$$

Example 5.3 Show, using the generating function, that, in the presence of a source term,

$$
\langle c_\lambda(1) \rangle = \int d2 G_\lambda(1-2) \eta_\lambda(2).
\tag{5.117}
$$

Solution

Taking the (functional) derivative of (5.116) with respect to η_λ, from the left-hand side of (5.116), we obtain

$$
\frac{\delta S[\bar{\eta}, \eta]}{\delta \bar{\eta}_\lambda(1)} = -i \langle \phi | T c_\lambda(1) \exp\left[-i \int dt V(t) \right] | \phi \rangle,
\tag{5.118}
$$

so that

$$
i \frac{\delta \ln S[\bar{\eta}, \eta]}{\delta \bar{\eta}_\lambda(1)} = \frac{i}{S[\bar{\eta}, \eta]} \frac{\delta S[\bar{\eta}, \eta]}{\delta \bar{\eta}_\lambda(1)} = \frac{\langle \phi | T c_\lambda(1) \exp\left[-i \int dt V(t) \right] | \phi \rangle}{\langle \phi | T \exp\left[-i \int dt V(t) \right] | \phi \rangle} = \langle c_\lambda(1) \rangle.
\tag{5.119}
$$

Now taking the logarithm of the right-hand side of (5.116), we obtain

$$
i \ln S[\bar{\eta}, \eta] = \sum_\lambda \int d1 d2 \bar{\eta}_\lambda(1) G_\lambda(1-2) \eta_\lambda(2),
\tag{5.120}
$$

so that

$$
i \frac{\delta \ln S[\bar{\eta}, \eta]}{\delta \bar{\eta}_\lambda(\tau)} = \int d2 G_\lambda(1-2) \eta_\lambda(2).
\tag{5.121}
$$

Combining (5.119) with (5.121) we obtain the final result,

$$
\langle c_\lambda(1) \rangle = \int d2 G_\lambda(1-2) \eta_\lambda(2).
\tag{5.122}
$$

5.3.3 The spectral representation

In the non-interacting Fermi liquid, we saw that the propagator contained a single pole, at $\omega = \epsilon_{\mathbf{k}}$. What happens to the propagator when we turn on the interactions? Remarkably, it retains its same general analytic structure, except that now the single pole divides into a plethora of poles, each one corresponding to an excitation energy for adding or removing a particle from the ground state. The general result, called the *spectral representation* of the propagator [5, 6], is that

$$G(\mathbf{k}, \omega) = \sum_{\lambda} \frac{|M_{\lambda}(\mathbf{k})|^2}{\omega - \epsilon_{\lambda} + i\delta_{\lambda}}, \tag{5.123}$$

where $\delta_{\lambda} = \delta \, \mathrm{sgn}(\epsilon_{\lambda})$ and the total pole strength

$$\sum_{\lambda} |M_{\lambda}(\mathbf{k})|^2 = 1 \tag{5.124}$$

is unchanged. Notice how the positive-energy poles of the Green's function are below the real axis at $\epsilon_{\lambda} - i\delta$, while the negative-energy poles are *above* the real axis, preserving the pole structure of the non-interacting Green's function.

  If the ground state is an N-particle state, then the state $|\lambda\rangle$ is either an $(N + 1)$ or $(N - 1)$-particle state. The poles of the Green's function are related to the excitation energies $E_{\lambda} - E_g > 0$ according to

$$\epsilon_{\lambda} = \begin{cases} E_{\lambda} - E_g > 0 & (|\lambda\rangle \in |N + 1\rangle) \\ -1 \times (E_{\lambda} - E_g) < 0 & (|\lambda\rangle \in |N - 1\rangle), \end{cases} \tag{5.125}$$

and the corresponding matrix elements are

$$M_{\lambda}(\mathbf{k}) = \begin{cases} \langle \lambda | c_{\mathbf{k}\sigma}^{\dagger} | \phi \rangle & (|\lambda\rangle \in |N + 1\rangle) \\ \langle \lambda | c_{\mathbf{k}\sigma} | \phi \rangle & (|\lambda\rangle \in |N - 1\rangle). \end{cases} \tag{5.126}$$

Notice that the excitation energies $E_{\lambda} - E_g > 0$ are always positive, so $\epsilon_{\lambda} > 0$ measures the energy to add an electron, while $\epsilon_{\lambda} < 0$ measures $(-1) \times$ the energy to create a hole state.

  In practice, the poles in the interacting Green's function blur into a continuum of excitation energies with infinitesimal separations. To deal with this situation, we define a quantity known as the *spectral function*, given by the imaginary part of the Green's function:

$$A(\mathbf{k}, \omega) = \frac{1}{\pi} \mathrm{Im} G(\mathbf{k}, \omega - i\delta). \qquad \text{spectral function} \tag{5.127}$$

By shifting the frequency ω by a small imaginary part which is taken to zero at the end of the calculation, overriding the δ_{λ} in (5.123), all the poles of $G(\mathbf{k}, \omega - i\delta)$ are moved above the real axis. Using Cauchy's principle-part equation, $1/(x - i\delta) = P(1/x) + i\pi \delta(x)$, where P denotes the principal part, we can use the spectral representation (5.123) to write

$$A(\mathbf{k}, \omega) = \sum_{\lambda} |M_{\lambda}(\mathbf{k})|^2 \delta(\omega - \epsilon_{\lambda})$$

$$= \sum_{\lambda} \left[|\langle \lambda | c_{\mathbf{k}\sigma}^{\dagger} | \phi \rangle|^2 \theta(\omega) + |\langle \lambda | c_{\mathbf{k}\sigma} | \phi \rangle|^2 \theta(-\omega) \right] \delta(|\omega| - (E_{\lambda} - E_g)), \quad (5.128)$$

where now the normalization of the pole strengths means that

$$\int_{-\infty}^{\infty} A(\mathbf{k}, \omega) d\omega = \sum_{\lambda} |M_{\lambda}(\mathbf{k})|^2 = 1. \quad (5.129)$$

Since the excitation energies are positive, $E_{\lambda} - E_g > 0$, from (5.125), it follows that ϵ_{λ} is positive for electron states and negative for hole states, so

$$A(\mathbf{k}, \omega) = \theta(\omega)\rho_e(\mathbf{k}, \omega) + \theta(-\omega)\rho_h(\mathbf{k}, -\omega), \quad (5.130)$$

where

$$\rho_e(\omega) = \sum_{\lambda} |\langle \lambda | c_{\mathbf{k}\sigma}^{\dagger} | \phi \rangle|^2 \delta(\omega - (E_{\lambda} - E_g)) \qquad (\omega > 0) \quad (5.131)$$

and

$$\rho_h(\omega) = \sum_{\lambda} |\langle \lambda | c_{\mathbf{k}\sigma} | \phi \rangle|^2 \delta(\omega - (E_g - E_{\lambda})) \qquad (\omega > 0) \quad (5.132)$$

are the spectral functions for adding electrons or holes of energy ω to the system, respectively. To a good approximation, in high-energy spectroscopy $\rho_{e,h}(\mathbf{k}, \omega)$ is directly proportional to the cross-section for adding or removing an electron of energy $|\omega|$. Photo-emission and inverse photoemission experiments can, in this way, be used to directly measure the spectral functions of electronic systems.

To derive this spectral decomposition, we suppose that we know the complete Hilbert space of energy eigenstates $\{|\lambda\rangle\}$. By injecting the completeness relation $\sum |\lambda\rangle\langle\lambda| = 1$ between the creation and annihilation operators in the Green's function, we can expand it as follows:

$$G(\mathbf{k}, t) = -i \left[\langle \phi | c_{\mathbf{k}\sigma}(t) c_{\mathbf{k}\sigma}^{\dagger}(0) | \phi \rangle \theta(t) - \langle \phi | c_{\mathbf{k}\sigma}^{\dagger}(0) c_{\mathbf{k}\sigma}(t) | \phi \rangle \theta(-t) \right]$$

$$= -i \sum_{\lambda} \left[\langle \phi | c_{\mathbf{k}\sigma}(t) \overbrace{|\lambda\rangle\langle\lambda|}^{=1} c_{\mathbf{k}\sigma}^{\dagger}(0) | \phi \rangle \theta(t) - \langle \phi | c_{\mathbf{k}\sigma}^{\dagger}(0) \overbrace{|\lambda\rangle\langle\lambda|}^{=1} c_{\mathbf{k}\sigma}(t) | \phi \rangle \theta(-t) \right].$$

By using energy eigenstates, we are able to write

$$\langle \phi | c_{\mathbf{k}\sigma}(t) | \lambda \rangle = \langle \phi | e^{iHt} c_{\mathbf{k}\sigma} e^{-iHt} | \lambda \rangle = \langle \phi | c_{\mathbf{k}\sigma} | \lambda \rangle e^{i(E_g - E_{\lambda})t}$$

$$\langle \lambda | c_{\mathbf{k}\sigma}(t) | \phi \rangle = \langle \lambda | e^{iHt} c_{\mathbf{k}\sigma} e^{-iHt} | \phi \rangle = \langle \lambda | c_{\mathbf{k}\sigma} | \phi \rangle e^{i(E_{\lambda} - E_g)t}. \quad (5.133)$$

Notice that the first term involves adding a particle of momentum $\mathbf{k}$, spin σ, so that the state $|\lambda\rangle = |N + 1; \mathbf{k}\sigma\rangle$ is an energy eigenstate with $N + 1$ particles, momentum $\mathbf{k}$ and spin σ. Similarly, in the second matrix element, a particle of momentum $\mathbf{k}$, spin σ has been *subtracted*, so that $|\lambda\rangle = |N - 1; -\mathbf{k} - \sigma\rangle$. We can thus write the Green's function in the form

$$G(\mathbf{k}, t) = -i \sum_\lambda \left[|\langle\lambda|c^\dagger_{\mathbf{k}\sigma}|\phi\rangle|^2 e^{-i(E_\lambda - E_g)t}\theta(t) - |\langle\lambda|c_{\mathbf{k}\sigma}|\phi\rangle|^2 e^{-i(E_g - E_\lambda)t}\theta(-t) \right],$$

where we have simplified the expression by writing $\langle\phi|c_{\mathbf{k}\sigma}|\lambda\rangle = \langle\lambda|c^\dagger_{\mathbf{k}\sigma}|\phi\rangle^*$ and $\langle\lambda|c_{\mathbf{k}\sigma}|\phi\rangle = \langle\phi|c^\dagger_{\mathbf{k}\sigma}|\lambda\rangle^*$. This has precisely the same structure as a non-interacting Green's function, except that $\epsilon_{\mathbf{k}} \to E_\lambda - E_g$ in the first term and $\epsilon_{\mathbf{k}} \to E_g - E_\lambda$ in the second term. We can use this observation to carry out the Fourier transform, whereupon

$$G(\mathbf{k}, \omega) = \sum_\lambda \left[\frac{|\langle\lambda|c^\dagger_{\mathbf{k}\sigma}|\phi\rangle|^2}{\omega - (E_\lambda - E_g) + i\delta} + \frac{|\langle\lambda|c_{\mathbf{k}\sigma}|\phi\rangle|^2}{\omega - (E_g - E_\lambda) - i\delta} \right],$$

which is the formal expansion of (5.123).

To show that the total pole strength is unchanged by interactions, we expand the sum over pole strengths, and then use completeness again, as follows:

$$\sum_\lambda |M_\lambda(\mathbf{k})|^2 = \sum_\lambda |\langle\lambda|c^\dagger_{\mathbf{k}\sigma}|\phi\rangle|^2 + |\langle\lambda|c_{\mathbf{k}\sigma}|\phi\rangle|^2$$

$$= \sum_\lambda \langle\phi|c_{\mathbf{k}\sigma} \overbrace{|\lambda\rangle\langle\lambda|}^{=1} c^\dagger_{\mathbf{k}\sigma}|\phi\rangle + \langle\phi|c^\dagger_{\mathbf{k}\sigma} \overbrace{|\lambda\rangle\langle\lambda|}^{=1} c_{\mathbf{k}\sigma}|\phi\rangle$$

$$= \langle\phi| \overbrace{\{c_{\mathbf{k}\sigma}, c^\dagger_{\mathbf{k}\sigma}\}}^{=1} |\phi\rangle = 1. \tag{5.134}$$

Example 5.4 Using the spectral decomposition, show that the momentum distribution function in the ground state of a translationally invariant system of fermions is given by the integral over the "filled" states,

$$\sum_\sigma \langle c^\dagger_{\mathbf{k}\sigma} c_{\mathbf{k}\sigma}\rangle = (2S + 1) \int_{-\infty}^0 d\omega A(\mathbf{k}, \omega).$$

Solution

Let us first write the occupancy in terms of the one-particle Green's function evaluated at time $t = 0^-$:

$$\langle n_{\mathbf{k}\sigma}\rangle = \langle\phi|n_{\mathbf{k}\sigma}|\phi\rangle = -i \times -i\langle\phi|Tc_{\mathbf{k}\sigma}(0^-)c^\dagger_{\mathbf{k}\sigma}(0)|\phi\rangle = -iG(\mathbf{k}, 0^-).$$

Now, using the spectral representation (5.134),

$$\langle n_{\mathbf{k}\sigma}\rangle = -iG(\mathbf{k}, 0^-) = \sum_\lambda |\langle\lambda|c_{\mathbf{k}\sigma}|\phi\rangle|^2 = \sum_\lambda |M_\lambda(\mathbf{k})|^2\theta(-\epsilon_\lambda),$$

since $|M_\lambda(\mathbf{k})|^2 = |\langle\lambda|c_{\mathbf{k}\sigma}|\phi\rangle|^2$ for $\epsilon_\lambda < 0$. This is just the sum over the negative-energy part of the spectral function. Now since $A(\mathbf{k}, \omega) = \sum_\lambda |M_\lambda(\mathbf{k})|^2\delta(\omega - \epsilon_\lambda)$, it follows that, at absolute zero,

$$\int_{-\infty}^{0} d\omega\, A(\mathbf{k}, \omega) = \sum_{\lambda} |M_{\lambda}(\mathbf{k})|^2 \overbrace{\int_{-\infty}^{0} d\omega\, \delta(\omega - \epsilon_{\lambda})}^{\theta(-\epsilon_{\lambda})} = \sum_{\lambda} |M_{\lambda}(\mathbf{k})|^2 \theta(-\epsilon_{\lambda}),$$

so that

$$\sum_{\sigma} \langle n_{\mathbf{k}\sigma} \rangle = (2S + 1) \int_{-\infty}^{0} d\omega\, A(\mathbf{k}, \omega).$$

Example 5.5   Show that the zero-temperature Green's function can be written in terms of the spectral function as follows:

$$G(\mathbf{k}, \omega) = \int d\epsilon\, \frac{1}{\omega - \epsilon(1 - i\delta)} A(\mathbf{k}, \epsilon).$$

Solution

Introduce the relationship $1 = \int d\epsilon\, \delta(\epsilon - (E_{\lambda} - E_g))$ and $1 = \int d\epsilon\, \delta(\epsilon + (E_{\lambda} - E_g))$ into (5.134) to obtain

$$G(\mathbf{k}, \omega) = \int d\epsilon\, \frac{1}{\omega - \epsilon + i\delta} \sum_{\lambda} |\langle \lambda | c_{\mathbf{k}\sigma}^{\dagger} | \phi \rangle|^2 \delta(\epsilon - (E_{\lambda} - E_g))$$

$$+ \int d\epsilon\, \frac{1}{\omega - \epsilon - i\delta} \sum_{\lambda} |\langle \lambda | c_{\mathbf{k}\sigma} | \phi \rangle|^2 \delta(\epsilon + (E_{\lambda} - E_g)). \qquad (5.135)$$

Now in the first term, $\epsilon > 0$, while in the second term, $\epsilon < 0$, enabling us to rewrite this expression as

$$G(\mathbf{k}, \omega) = \int d\epsilon\, \frac{1}{\omega - \epsilon(1 - i\delta)}$$

$$\times \sum_{\lambda} \overbrace{\left[|\langle \lambda | c_{\mathbf{k}\sigma}^{\dagger} | \phi \rangle|^2 \theta(\epsilon) + |\langle \lambda | c_{\mathbf{k}\sigma} | \phi \rangle|^2 \theta(-\epsilon) \right] \delta(|\epsilon| - (E_{\lambda} - E_g))}^{A(\mathbf{k}, \epsilon)},$$

giving the quoted result.

5.4 Many-particle Green's functions

The n-particle Green's function determines the amplitude for n particles to go from one starting configuration to another:

$$\overbrace{\{1', 2', \dots n'\}}^{\text{initial particle positions}} \xrightarrow{G} \overbrace{\{1, 2, \dots n\}}^{\text{final particle positions}}, \qquad (5.136)$$

where $1' \equiv (\mathbf{x}', t')$, etc. and $1 \equiv (\mathbf{x}, t)$, etc. The n-particle Green's function is defined as

$$G(1, 2, \ldots n; 1', 2', \ldots n') = (-i)^n \langle \phi | T \psi(1) \psi(2) \cdots \psi(n) \psi^\dagger(n') \cdots \psi^\dagger(1') | \phi \rangle$$

and represented diagrammatically as

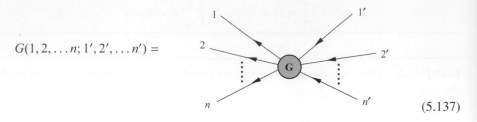

$$G(1, 2, \ldots n; 1', 2', \ldots n') = \tag{5.137}$$

In systems without interactions, the n-body Green's function can always be decomposed in terms of the one-body Green's function using Wick's theorem. This is because particles propagate without scattering off one another. Suppose a particle which ends up at r comes from location P'_r, where P_r is the rth element of a permutation P of $(1, 2, \ldots n)$. The amplitude for this process is

$$G(r - P'_{\mathbf{r}}) \tag{5.138}$$

and the overall amplitude for all n particles to go from locations P'_r to positions r is then

$$\zeta^P G(1 - P'_1) G(2 - P'_2) \cdots G(n - P'_n), \tag{5.139}$$

where $\zeta = +1$ for bosons and -1 for fermions, and p is the number of pairwise permutations required to make the permutation P. This prefactor arises because, for fermions, every time we exchange two of them we pick up a minus sign in the amplitude. Wick's theorem states the physically reasonable result that the n-body Green's function of a non-interacting system is given by the sum of all such amplitudes:

$$G(1, 2, \ldots n; 1', 2', \ldots n') = \sum \zeta^P \prod_{r=1,n} G(r - P'_r). \tag{5.140}$$

For example, the two-body Green's function is given by

$$G(1, 2; 1', 2') = G(1, 1') G(2, 2') \pm G(1, 2') G(2, 1').$$

The process of identifying pairs of initial and final states in the n-particle Green's function is often referred to as a *contraction*. When we contract two field operators inside a Green's function, we associate an amplitude with the contraction, as follows:

$$\langle 0|T[\cdots \psi(1)\cdots \psi^\dagger(2)\cdots]|0\rangle \quad \longrightarrow \quad \langle 0|T[\psi(1)\psi^\dagger(2)]|0\rangle = iG(1-2)$$

$$\langle 0|T[\cdots \psi^\dagger(2)\cdots \psi(1)\cdots]|0\rangle \quad \longrightarrow \quad \langle 0|T[\psi^\dagger(2)\psi(1)]|0\rangle = \pm iG(1-2).$$

Each product of Green's functions in the Wick's theorem expansion of the propagator is a particular contraction of the n-body Green's function, thus:

$$(-i)^n \langle 0|T[\psi(1)\psi(2)\cdots \psi(n)\cdots \psi^\dagger(P_2')\cdots \psi^\dagger(P')\cdots \psi^\dagger(P_n')]|0\rangle$$
$$= \zeta^P G(1-P_1')G(2-P_2')\cdots G(n-P_n'), \tag{5.141}$$

where now P is just the number of times the contraction lines cross one another. Wick's theorem then states that the n-body Green's function is given by the sum over all possible contractions:

$$(-i)^n \langle \phi|\, T\, \psi(1)\psi(2)\cdots \psi^\dagger(n')|\phi\rangle$$

$$= \sum_{\text{all contractions}} (-i)^n \langle 0|T[\psi(1)\psi(2)\cdots \psi(n)\cdots \psi^\dagger(P_2')\cdots \psi^\dagger(P_1')\cdots \psi^\dagger(P_n')]|0\rangle.$$
$$\tag{5.142}$$

Example 5.6 Show how the expansion of the generating functional in the absence of interactions can be used to derive Wick's theorem.

Solution

To derive Wick's theorem, we need to expand both sides of our equality,

$$S[\bar\eta, \eta] = \langle \phi|T\exp\left[-i\int d1(\bar\eta(1)c(1) + c^\dagger(1)\eta(1))\right]|\phi\rangle$$
$$= \exp\left[-i\int d1d2\bar\eta(1)G(1-2)\eta(2)\right], \tag{5.143}$$

where we have lumped all position, time, and quantum number subscripts into a single numeral so that, e.g. $c(1) \equiv c_{\lambda_1}(x_1, t_1)$. Let's start by working on the left-hand side of the generating function equality. Each time we differentiate with respect to a source or sink term, we bring down a creation or annihilation operator, according to the equivalence

$$i\frac{\delta}{\delta\bar\eta(1)} = c(1), \qquad -i\frac{\delta}{\delta\eta(1)} \equiv c^\dagger(1). \tag{5.144}$$

Thus if we differentiate with respect to n source terms $(\eta(1), \ldots \eta(n))$ we obtain

$$(-i)^n \frac{1}{S} \frac{\delta^n S}{\delta \eta(n) \cdots \delta \eta(1)} = \frac{1}{S} \langle 0|Te^{-iV} c^\dagger(n) \cdots c^\dagger(1)|0\rangle$$

$$= \langle Tc^\dagger(n) \cdots c^\dagger(1)\rangle, \tag{5.145}$$

where we have denoted $V = \int d1[\bar{\eta}(1)c(1) + \bar{c}(1)\eta(1)]$. Now, differentiating with respect to the corresponding sink terms,

$$\frac{1}{S} \frac{\delta^{2n} S}{\delta \bar{\eta}(1') \cdots \delta \bar{\eta}(n')\delta \eta(n) \cdots \delta \eta(1)} = \frac{1}{S} \langle 0|Te^{-iV} c(1') \cdots c(n')^\dagger(n) \cdots c^\dagger(1)|0\rangle$$

$$= \langle Tc(1') \cdots c(n')c^\dagger(n) \cdots c^\dagger(1)\rangle. \tag{5.146}$$

Let's now apply the same sequence of differentiations to the right-hand side of the generating function identity. First, differentiating with respect to the source terms, we obtain

$$(-i)^n \frac{1}{S} \frac{\delta^n S}{\delta \eta(n) \cdots \delta \eta(1)} = \int d1'' \cdots dn'' \bar{\eta}(n'')G(n'' - n) \cdots \bar{\eta}(1'')G(1'' - 1)$$

$$= \int d1'' \cdots dn'' \bar{\eta}(n'') \cdots \bar{\eta}(1'')G(n'' - n) \cdots G(1'' - 1), \tag{5.147}$$

where we have denoted the internal dummy integration coordinates by a double prime. When we now differentiate with respect to the sink variables $\bar{\eta}(r')$, we can identify the primed variables $\bar{\eta}(r')$ with any permutation P of the double-primed variables $\bar{\eta}(P_r'')$, and each permutation gives rise to a different contribution, weighted by the sign of the permutation:

$$\frac{\delta^n}{\delta \bar{\eta}(1') \cdots \delta \bar{\eta}(n')} \bar{\eta}(n'') \cdots \bar{\eta}(1'') = \sum_P (-1)^P \delta(1' - P_1'') \cdots \delta(n' - P_n''), \tag{5.148}$$

so that

$$\frac{1}{S} \frac{\delta^{2n} S}{\delta \bar{\eta}(1') \cdots \delta \bar{\eta}(n')\delta \eta(n) \cdots \eta(1)} = i^n \sum_P (-1)^P G(P_n' - n) \cdots G(P_1' - 1). \tag{5.149}$$

Comparing equations (5.146) and (5.149), we obtain

$$(-i)^n \langle Tc(1') \cdots c(n')c^\dagger(n) \cdots c(1)\rangle = \sum_P (-1)^P G(P_n' - n) \cdots G(P_1' - 1), \tag{5.150}$$

which is the Wick decomposition of the non-interacting, n-particle Green's function.

Exercises

Exercise 5.1 A particle with $S = \frac{1}{2}$ is placed in a large magnetic field $\vec{B} = (B_1 \cos \omega t, B_1 \sin \omega t, B_0)$, where $B_0 >> B_1$.
(a) Treating the oscillating part of the Hamiltonian as the interaction, write down the Schrödinger equation in the interaction representation.

(b) Find $U(t) = \text{T}\exp\left[-i\int_{-\infty}^{t} H_{int}(t')dt'\right]$ by whatever method proves most convenient.

(c) If the particle starts out at time $t = 0$ in the state $S_z = -\frac{1}{2}$, what is the probability it is in this state at time t?

Exercise 5.2 (Optional derivation of bosonic generating functional)

Consider the forced harmonic oscillator

$$H(t) = \omega b^\dagger b + \bar{z}(t)b + b^\dagger z(t), \tag{5.151}$$

where $z(t)$ and $\bar{z}(t)$ are arbitrary, independent functions of time. Consider the S-matrix

$$S[z, \bar{z}] = \langle 0|T\hat{S}(\infty, -\infty)|0\rangle = \langle 0|T\exp\left(-i\int_{-\infty}^{\infty} dt[\bar{z}(t)b(t) + b^\dagger(t)z(t)]\right)|0\rangle, \tag{5.152}$$

where $\hat{b}(t)$ denotes $\hat{b}$ in the interaction representation. Consider changing the function $\bar{z}(t)$ by an infinitesimal amount

$$\bar{z}(t) \rightarrow \bar{z}(t) + \Delta\bar{z}(t_0)\delta(t - t_0). \tag{5.153}$$

The quantity

$$\lim_{\Delta\bar{z}(t_0)\to 0} \frac{\Delta S[z, \bar{z}]}{\Delta\bar{z}(t_0)} = \frac{\delta S[z, \bar{z}]}{\delta\bar{z}(t_0)}$$

is called the *functional derivative* of S with respect to $\bar{z}$.

(a) Using the Gell-Man–Low formula

$$\langle\psi(t)|b|\psi(t)\rangle = \frac{\langle 0|T\hat{S}(\infty, -\infty)b(t)|0\rangle}{\langle 0|T\hat{S}(\infty, -\infty)|0\rangle}, \tag{5.154}$$

prove the following identity:

$$i\delta\ln S[z, \bar{z}]/\delta\bar{z}(t) \equiv \tilde{b}(t) = \langle b(t)\rangle = \langle\psi(t)|b|\psi(t)\rangle. \tag{5.155}$$

(b) Use the equation of motion to show that

$$\frac{\partial}{\partial t}\tilde{b}(t) = i\langle[H(t), b(t)]\rangle = -i[\epsilon\tilde{b}(t) + z(t)].$$

(c) Solve the above differential equation to show that the expectation value

$$\tilde{b}(t) = \int_{-\infty}^{\infty} G(t - t')z(t'), \tag{5.156}$$

where $G(t - t') = -i\langle 0|T[b(t)b^\dagger(t')]|0\rangle$ is the free Green's function for the harmonic oscillator.

(d) Use (c) and (b) together to obtain the fundamental result

$$S[z, \bar{z}] = \exp\left[-i\int_{-\infty}^{\infty} dt dt'\bar{z}(t)G(t - t')z(t')\right]. \tag{5.157}$$

Exercise 5.3 Consider a harmonic oscillator with charge e, so that an applied field changes the Hamiltonian $H \rightarrow H_0 - eE(t)\hat{x}$, where $\hat{x}$ is the displacement and $E(t)$ is the field. Let the system initially be in its ground state, and suppose a constant electric field E is applied for a time T.

(a) Rewrite the Hamiltonian in the form of a forced harmonic oscillator,

$$H(t) = \omega b^\dagger b + \bar{z}(t)b + b^\dagger z(t), \tag{5.158}$$

and show that

$$z(t) = \bar{z}(t) = \begin{cases} \omega\alpha & (T > t > 0) \\ 0 & (\text{otherwise}), \end{cases} \tag{5.159}$$

deriving an explicit expression for α in terms of the field E, mass m and frequency ω of the oscillator.

(b) Use the explicit form of $S[\bar{z}, z]$,

$$S[z, \bar{z}] = \exp\left[-i \int_{-\infty}^{\infty} dt dt' \bar{z}(t) G(t - t') z(t')\right], \tag{5.160}$$

where $G(t - t') = -i\langle 0|T[b(t)b^\dagger(t')]|0\rangle$ is the free bosonic Green's function, to calculate the probability $p(T)$ that the system is still in the ground state after time T. Express your result in terms of α, ω and T. Sketch the form of $p(T)$ and comment on your result.

References

[1] M. Gell-Mann and F. Low, Bound states in quantum field theory, *Phys. Rev.*, vol. 84, p. 350 (Appendix), 1951.

[2] L. D. Landau, The theory of a Fermi liquid, *J. Exp. Theor. Phys.*, vol. 3, p. 920, 1957.

[3] L. D. Landau, Oscillations in a Fermi liquid, *J. Exp. Theor. Phys.*, vol. 5, p. 101, 1957.

[4] L. D. Landau, On the theory of the Fermi liquid, *J. Exp. Theor. Phys.*, vol. 8, p. 70, 1959.

[5] G. Källén, On the definition of the renormalization constants in quantum electro-dynamics, *Helv. Phys. Acta.*, vol. 25, p. 417, 1952.

[6] H. Lehmann, Über Eigenschaften von Ausbreitungsfunktionen und Renormierungskonstanten quantisierter Felder (On the properties of propagators and renormalization constants of quantized fields), *Nuovo Cimento*, vol. 11, p. 342, 1954.

Landau Fermi-liquid theory

6.1 Introduction

One of the remarkable features of a Fermi fluid is its robustness against perturbation. In a typical electron fluid inside metals, the Coulomb energy is comparable with the electron kinetic energy, constituting a major perturbation to the electron motions. Yet, remarkably, the non-interacting model of the Fermi gas reproduces many qualitative features of metallic behavior, such as a well-defined Fermi surface, a linear specific heat capacity, and a temperature-independent paramagnetic susceptibility. Such *Landau Fermi-liquid behavior* appears in many contexts – in metals at low temperatures, in the core of neutron stars, in liquid ^{3}He – and, most recently, it has become possible to create Fermi liquids with tunable interactions in atom traps. As we shall see, our understanding of Landau Fermi-liquids is intimately linked with the idea of adiabaticity, introduced in the previous chapter.

In the 1950s, physicists on both sides of the Iron Curtain pondered the curious robustness of Fermi liquid physics against interactions. At Princeton, David Bohm and David Pines carried out the first quantization of the interacting electron fluid, proposing that the effects of long-range interactions are absorbed by a canonical transformation that separates the excitations into a high-frequency plasmon and a low-frequency fluid of renormalized electrons [1]. On the other side of the world, Lev Landau at the Kapitza Low Temperature Institute in Moscow came to the conclusion, that the robustness of the Fermi liquid is linked to the idea of adiabaticity and the Fermi exclusion principle [2].

At first sight, the possibility that an almost free Fermi fluid might survive the effects of interactions seems hopeless. With interactions, a moving fermion decays by emitting arbitrary numbers of low-energy particle–hole pairs, so how can it ever form a stable particle-like excitation? Landau realized that a fermion outside the Fermi surface cannot scatter into an occupied momentum state below the Fermi surface, so the closer it is to the Fermi surface, the smaller the phase space available for decay. We will see that, as a consequence, the inelastic scattering rate grows quadratically with excitation energy ϵ and temperature:

$$\tau^{-1}(\epsilon) \propto (\epsilon^2 + \pi^2 T^2). \tag{6.1}$$

In this way, particles at the Fermi energy develop an infinite lifetime. Landau named these long-lived excitations *quasiparticles*. *Landau Fermi-liquid theory* [2–5] describes the collective physics of a fluid of these quasiparticles.

It was a set of experiments on liquid ^{3}He half a world away from Moscow that helped to crystallize Landau's ideas. In the aftermath of the Second World War, the availability of isotopically pure ^{3}He as a by-product of the Manhattan Project made it possible for the first time to experimentally study this model Fermi liquid. The first measurements were carried at Duke University by Fairbank, Ard, and Walters [6]. While ^{4}He atoms are bosons, atoms of the much rarer isotope, ^{3}He are spin-$\frac{1}{2}$ fermions. These atoms contain a neutron and two protons in the nucleus, neutralized by two orbital electrons in a singlet state, forming a composite, neutral fermion. ^{3}He is a much simpler quantum fluid than the electron fluid of metals:

- Without a crystal lattice, liquid ^{3}He is isotropic and enjoys the full translational and Galilean symmetries of the vacuum.
- ^{3}He atoms are neutral, interacting via short-range interactions, avoiding the complications of a long-range Coulomb interaction in metals.

Prior to Landau's theory, the only available theory of a degenerate Fermi liquid was Sommerfeld's model for non-interacting fermions. A key property of the non-interacting Fermi liquid is the presence of a large, finite density of single-particle excitations at the Fermi energy, given by [1]

$$\mathcal{N}(0) = 2\frac{(4\pi)p^2}{(2\pi\hbar)^3}\frac{dp}{d\epsilon_p}\bigg|_{p=p_F} = \frac{mp_F}{\pi^2\hbar^3}, \tag{6.2}$$

where we use a script $\mathcal{N}(0)$ to distinguish the total density of states from the density of states per spin, $N(0) = \mathcal{N}(0)/2$. The argument of $\mathcal{N}(\epsilon)$ is the energy $\epsilon = E - \mu$, measured relative to the chemical potential μ. A magnetic field splits the "up" and "down" Fermi surfaces, shifting their energy by an amount $-\sigma\mu_F B$, where $\sigma = \pm 1$ and $\mu_F = \frac{g}{2}\frac{e\hbar}{2m}$ is half the product of the Bohr magneton for the fermion and the g-factor associated with its spin. The number of "up" and "down" fermions is thereby changed by an amount $\delta N_\uparrow = -\delta N_\downarrow = \frac{1}{2}\mathcal{N}(0)(\mu_F B)$, inducing a net magnetization $M = \chi B$, where

$$\chi = \mu_F(N_\uparrow - N_\downarrow)/B = \mu_F^2 \mathcal{N}(0) \tag{6.3}$$

is the *Pauli paramagnetic susceptibility*. For electrons, $g \approx 2$ and $\mu_F \equiv \mu_B = \frac{e\hbar}{2m}$ is the Bohr magneton, so the Pauli susceptibility of a free electron gas is $\mu_B^2 \mathcal{N}(0)$.

In a degenerate Fermi liquid, the energy is given by

$$\mathcal{E}(T) = E(T) - \mu N = \sum_{\mathbf{k}\sigma=\pm 1/2} \epsilon_\mathbf{k}\frac{1}{e^{\beta\epsilon_\mathbf{k}} + 1}. \tag{6.4}$$

Here we use the notation $\mathcal{E} = E - \mu N$ to denote the energy measured in the grand canonical ensemble. The variation of this quantity at low temperatures (where, to order T^2, the chemical potential is constant) depends only on the free-particle density of states at the Fermi energy $N(0)$. The low-temperature specific heat

---

[1] Note: in the discussion that follows, we shall normalize all extensive properties per unit volume. Thus the density of states $N(\epsilon)$, the specific heat C_V, and the magnetization M will all refer to those quantities per unit volume.

$$C_V = \frac{d\mathcal{E}}{dT} = \mathcal{N}(0) \int_{-\infty}^{\infty} d\epsilon\, \epsilon \frac{d}{dT} \left(\frac{1}{e^{\beta\epsilon} + 1} \right)$$

$$= \mathcal{N}(0) k_B^2 T \int_{-\infty}^{\infty} dx \underbrace{\frac{x^2}{(e^x + 1)(e^{-x} + 1)}}_{\pi^2/3} = \overbrace{\frac{\pi^2}{3} \mathcal{N}(0) k_B^2\, T}^{=\gamma} \qquad (6.5)$$

is linear in temperature. Since both the specific heat and the magnetic susceptibility are proportional to the density of states, the ratio of these two quantities, $W = \chi/\gamma$, often called the *Wilson ratio* or *Stoner enhancement factor*, is set purely by the size of the magnetic moment:

$$W = \frac{\chi}{\gamma} = 3 \left(\frac{\mu_F}{\pi k_B} \right)^2. \qquad (6.6)$$

Fairbank, Ard, and Walters' experiment confirmed the Pauli paramagnetism of liquid ^{3}He, but the measured Wilson ratio is about three times larger than predicted by Sommerfeld theory. Landau's explanation of these results is based on the idea that one can track the evolution of the properties of the Fermi liquid by adiabatically switching on the interactions. He considered a hypothetical gas of non-interacting helium atoms with no forces of repulsion between, for which Sommerfeld's model would certainly hold. Suppose the interactions are now turned on slowly. Landau argued that since the fermions near the Fermi surface had nowhere to scatter to, the low-lying excitations of the Fermi liquid would evolve adiabatically, in the sense discussed in the previous chapter, so that each quantum state of the fully interacting liquid ^{3}He would be in *precise one-to-one correspondence with the states of the idealized non-interacting Fermi liquid* [4].

6.2 The quasiparticle concept

The *quasiparticle* concept is a triumph of Landau's Fermi-liquid theory, for it enables us to continue using the idea of an independent particle, even in the presence of strong interactions. It also provides a framework for understanding the robustness of the Fermi surface while accounting for the effects of interactions.

A quasiparticle is the adiabatic evolution of the non-interacting fermion into an interacting environment. The conserved quantum numbers of this excitation – its spin, its "charge" and its momentum – are unchanged, but Landau reasoned that its dynamical properties – the effective magnetic moment and mass of the quasiparticle – would be *renormalized* to new values g^* and m^*, respectively. Subsequent measurements on ^{3}He [5, 6] revealed that the quasiparticle mass and enhanced magnetic moment g^* are approximately

$$m^* = (2.8) m_{(^3\text{He})}$$
$$(g^*)^2 = 3.3 (g^2)_{(^3\text{He})}. \qquad (6.7)$$

These renormalizations of the quasiparticle mass and magnetic moment are elegantly accounted for in Landau Fermi-liquid theory in terms of a small set of *Landau parameters* which characterize the interaction, as we now shall see.

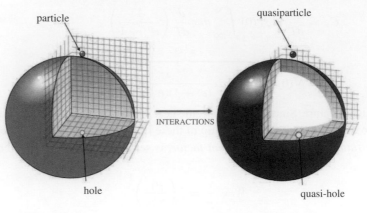

(a) Fermi liquid   (b) Landau Fermi-liquid

Fig. 6.1   (a) In the non-interacting Fermi liquid, a stable particle can be created anywhere outside the Fermi surface, a stable hole excitation anywhere inside the Fermi surface. (b) When the interactions are turned on adiabatically, particle excitations near the Fermi surface adiabatically evolve into quasiparticles, with the same charge, spin and momentum. Quasiparticles and quasiholes are only well defined near the Fermi surface of the Landau Fermi-liquid.

Let us label the momentum of each particle in the original non-interacting Fermi liquid by $\mathbf{p}$ and its spin component $\sigma = \pm\frac{1}{2}$. The number $n_{\mathbf{p}\sigma}$ of fermions of momentum $\mathbf{p}$ and spin component σ is either one or zero. The complete quantum state of the non-interacting system is labeled by these occupancies. We write

$$\Psi = |n_{\mathbf{p}_1\sigma_1}, \, n_{\mathbf{p}_2\sigma_2}, \ldots\rangle. \tag{6.8}$$

In the ground state Ψ_0, all states with momentum p less than the Fermi momentum are occupied, all states above the Fermi surface are empty (see Figure 6.1a)

$$\text{ground state } \Psi_0 \quad : \quad n_{\mathbf{p}\sigma} = \begin{cases} 1 & (p < p_F) \\ 0 & (\text{otherwise}). \end{cases} \tag{6.9}$$

Landau argued that, if one turned on the interactions infinitely slowly, this state would evolve smoothly into the ground state of the interacting Fermi liquid. This is an example of the adiabatic evolution encountered in the previous chapter. For the adiabatic evolution to work, the Fermi liquid ground state has to remain stable. This is a condition that certainly fails when the system undergoes a phase transition into another ground state, a situation that may occur at a certain critical interaction strength. However, up to this critical value, the adiabatic evolution of the ground state can take place. The energy of the final ground state is unknown, but we can call it E_0.

Suppose we now add a fermion above the Fermi surface of the original state. We can repeat the adiabatic switch-on of the interactions, but it is a delicate procedure for an excited state, because away from the Fermi surface, an electron can decay by emitting low-energy particle–hole pairs, which dissipates its energy in an irreversible fashion. To avoid this irreversibility, the lifetime $\tau(\epsilon) \sim \epsilon^{-2}$ of the particle (see Section 6.1) must be longer than the adiabatic switch-on time $\tau_A = \epsilon^{-1}$ encountered in (5.93), and since this time becomes infinite, strict adiabaticity is only possible for excitations that lie on the Fermi surface, where τ_e is infinite. A practical Landau Fermi-liquid theory requires that we consider excitations that are a finite distance away from the Fermi surface, and when we do this, we tacitly ignore the finite lifetime of the quasiparticles. By doing so, we introduce an error of order $\tau^{-1}(\epsilon_{\mathbf{p}})/|\epsilon_{\mathbf{p}}| \propto |\epsilon_{\mathbf{p}}|$. This error can be made arbitrarily small, provided we restrict our attention to small perturbations to the ground state.

Adiabatic evolution conserves the momentum of the quasiparticle state, which will then evolve smoothly into a final state that we can label as:

$$\text{quasiparticle}: \Psi_{\mathbf{p}_0 \sigma_0} \qquad n_{\mathbf{p}\sigma} = \begin{cases} 1 & (p < p_F \text{ or } \mathbf{p} = \mathbf{p}_0, \sigma = \sigma_0) \\ 0 & (\text{otherwise}) \end{cases} \qquad (6.10)$$

(see Figure 6.1b). This state has total momentum $\mathbf{p}_0$, where $|\mathbf{p}_0| > p_F$ and an energy $E(\mathbf{p}_0) > E_0$ larger than the ground state. It is called a *quasiparticle state* because it behaves in almost every respect like a single particle. Notice in particular that the Fermi surface momentum p_F is *preserved* by the adiabatic introduction of interactions. Unlike free particles, however, the Landau quasiparticle is only a well-defined concept close to the Fermi surface. Far from the Fermi surface, quasiparticles develop a lifetime, and once the lifetime is comparable with the quasiparticle excitation energy, the quasiparticle concept loses its meaning.

The energy required to create a single quasiparticle is

$$E_{\mathbf{p}_0}^{(0)} = E(p_0) - E_0, \qquad (6.11)$$

where the superscript (0) denotes a single excitation in the absence of any other quasiparticles. We shall mainly work in the grand canonical ensemble, using $\mathcal{E} = E - \mu N$ in place of the absolute energy, where μ is the chemical potential, enabling us to explore the variation of the energy at constant particle number N. The corresponding quasiparticle excitation energy is then

$$\epsilon_{\mathbf{p}_0}^{(0)} = E_{\mathbf{p}_0}^{(0)} - \mu = \mathcal{E}(p_0) - \mathcal{E}_0. \qquad (6.12)$$

Notice that, since $|\mathbf{p}_0| > p_F$, this energy is positive.

In a similar way, we can also define a *quasihole* state, in which a quasiparticle is removed the Fermi sea:

$$\text{quasihole}: \overline{\Psi}_{\mathbf{p}_0 \sigma_0} \qquad n_{\mathbf{p}\sigma} = \begin{cases} 1 & (p < p_F \text{ except when } \mathbf{p} = \mathbf{p}_0, \sigma = \sigma_0) \\ 0 & (\text{otherwise}), \end{cases}$$

$$(6.13)$$

where the bar is used to denote the hole and now $|\mathbf{p}_0| < p_F$ is beneath the Fermi surface. The energy of this state is $\overline{E}(\mathbf{p}_0) = E_0 - E_{\mathbf{p}_0}$, since we have removed a particle. Now the change in particle number is $\Delta N = -1$, so the excitation energy of a single quasihole, measured in the grand canonical ensemble, is then

$$\overline{\epsilon}_{\mathbf{p}_0}^{(0)} = -E_{\mathbf{p}_0}^{(0)} + \mu = -\epsilon_{\mathbf{p}_0}^{(0)}, \qquad (6.14)$$

i.e. the energy to create a quasihole is the negative of the corresponding quasiparticle energy $\epsilon_{\mathbf{p}_0}$. Of course, when $|\mathbf{p}_0| < p_F$, $\epsilon_{\mathbf{p}_0} < 0$, so that the quasihole excitation energy $\overline{\epsilon}_{\mathbf{p}_0}^{(0)}$ is always positive, as required for a stable ground state. In this way, the energy to create a quasihole or quasiparticle is always given by $|\epsilon_{\mathbf{p}_0}|$, independently of whether $\mathbf{p}_0$ is above or below the Fermi surface.

The quasiparticle concept would be of limited value if it were limited to individual excitations. At a finite temperature, a dilute gas of these particles is excited around the Fermi surface and these particles interact. How can the particle concept survive once one has a finite density of excitations? Landau's appreciation of a very subtle point enabled him to answer this question. He realized that since the phase space for quasiparticle scattering vanishes quadratically with the quasiparticle energy, it follows that the quasiparticle occupancy at a given momentum $\mathbf{p}_F$ on the Fermi surface becomes a constant of the motion. In this way, the Landau Fermi-liquid is characterized by an *infinite* set of conserved quantities $n_{\mathbf{p}\sigma}$, so that, on the Fermi surface,

$$[H, n_{\mathbf{p}_F\sigma}] = 0. \qquad (\mathbf{p}_F \in \text{FS}) \qquad (6.15)$$

It follows that the only residual scattering that remains on the Fermi surface is *forward* scattering, i.e.

$$(\mathbf{p}_1, \mathbf{p}_2) \rightarrow (\mathbf{p}_1 - \mathbf{q}, \mathbf{p}_2 + \mathbf{q}). \qquad (\mathbf{q} = 0 \text{ on Fermi surface }) \qquad (6.16)$$

The challenge is to develop a theory that describes the free energy $F[\{n_{\mathbf{p}\sigma}\}]$ and the slow, long-distance hydrodynamics of these conserved quantities.

Example 6.1   Suppose $|\Psi_0\rangle = \prod_{|\mathbf{p}|<p_F,\sigma} c^\dagger_{\mathbf{p}\sigma}|0\rangle$ is the ground state of a non-interacting Fermi liquid, where $c^\dagger_{\mathbf{p}\sigma}$ creates a "bare" fermion. By considering the process of adiabatically turning on the interaction, time-evolving the one-particle state $c^\dagger_{\mathbf{p}_0\sigma}|FS\rangle$ from the distant past to the present ($t = 0$) in the interaction representation, write down an expression for the ground-state wavefunction $|\psi\rangle$ and the quasiparticle-creation operator of the fully interacting system.

Solution

The time-evolution operator from the distant past in the interaction representation is

$$U = T \exp\left[-i \int_{-\infty}^0 \hat{V}_I(t)dt\right], \qquad (6.17)$$

where $\hat{V}_I(t)$ is the interaction operator, written in the interaction representation. If we add a particle to the filled Fermi sea and adiabatically time-evolve from the distant past to the present, we obtain

$$c^\dagger_{\mathbf{p}\sigma}|\Psi_0\rangle \longrightarrow Uc^\dagger_{\mathbf{p}\sigma}|\Psi_0\rangle = \overbrace{(Uc^\dagger_{\mathbf{p}\sigma}U^\dagger)}^{\text{QP } a^\dagger_{\mathbf{p}\sigma}}\overbrace{U|\Psi_0\rangle}^{|\phi\rangle}. \qquad (6.18)$$

If the adiabatic evolution avoids a quantum phase transition, then

$$|\phi\rangle = U|FS\rangle \qquad (6.19)$$

is the ground state of the fully interacting system. In this case, we may interpret

$$a^\dagger_{\mathbf{p}\sigma} = (Uc^\dagger_{\mathbf{p}\sigma}U^\dagger) \qquad (6.20)$$

as the quasiparticle-creation operator. Note that if we try to rewrite this object in terms of the original creation operator $c_{\mathbf{p}\sigma}^{\dagger}$, it involves combinations of one fermion with particle–hole pairs. See Section 6.8 for a more detailed discussion.

6.3 The neutral Fermi liquid

These physical considerations led Landau to conclude that the energy of a gas of quasiparticles could be expressed as a functional of the quasiparticle occupancies $n_{\mathbf{p}\sigma}$. Following Landau, we shall develop the Fermi liquid concept using an idealized neutral Landau Fermi-liquid, like ^{3}He-, in which the quasiparticles move in free space, interacting isotropically via a short-range interaction, forming a neutral fluid.

If the density of quasiparticles is low, it is sufficient to expand the energy in the small deviations in particle number $\delta n_{\mathbf{p}\sigma} = n_{\mathbf{p}\sigma} - n_{\mathbf{p}\sigma}^{(0)}$ from equilibrium. This leads to the Landau energy functional $\mathcal{E}(\{n_{\mathbf{p}\sigma}\}) = E(\{n_{\mathbf{p}\sigma}\}) - \mu N$, where

$$\mathcal{E} = \mathcal{E}_0 + \sum_{\mathbf{p}\sigma}(E_{\mathbf{p}\sigma}^{(0)} - \mu)\delta n_{\mathbf{p}\sigma} + \frac{1}{2}\sum_{\mathbf{p},\mathbf{p}',\sigma,\sigma'} f_{\mathbf{p}\sigma,\mathbf{p}'\sigma'}\delta n_{\mathbf{p}\sigma}\delta n_{\mathbf{p}'\sigma'} + \cdots . \tag{6.21}$$

The first-order coefficient

$$\epsilon_{\mathbf{p}\sigma}^{(0)} \equiv E_{\mathbf{p}\sigma}^{(0)} - \mu = \frac{\delta\mathcal{E}}{\delta n_{\mathbf{p}\sigma}} \tag{6.22}$$

describes the excitation energy of an isolated quasiparticle. Provided we can ignore spin–orbit interactions, then the total magnetic moment is a conserved quantity, so the magnetic moments of the quasiparticles are preserved by interactions. In this case, $\epsilon_{\mathbf{p}\sigma}^{(0)} = \epsilon_{\mathbf{p}}^{(0)} - \sigma\mu_F B$, where μ_F is the un-renormalized magnetic moment of an isolated fermion.

The quasiparticle energy can be expanded linearly in momentum near the Fermi surface:

$$E_p^{(0)} = v_F(p - p_F) + \mu^{(0)}, \tag{6.23}$$

where v_F is the Fermi velocity at the Fermi energy $\mu^{(0)}$, where $\mu^{(0)}$ is the chemical potential in the ground state. The quasiparticle effective mass m^* is then defined in terms of v_F as

$$v_F = \left.\frac{d\epsilon_p^{(0)}}{dp}\right|_{p=p_F} = \frac{p_F}{m^*}. \tag{6.24}$$

We can use this mass to define a quasiparticle density of states,

$$\mathcal{N}^*(\epsilon) = 2\sum_{\mathbf{p}}\delta(\epsilon - \epsilon_{\mathbf{p}}^{(0)}) = 2\int\frac{4\pi p^2 dp}{(2\pi\hbar)^3}\delta(\epsilon - \epsilon_p^{(0)}) = \frac{p^2}{\pi^2\hbar^3}\frac{dp}{d\epsilon_p^{(0)}}. \tag{6.25}$$

Using (6.24), it follows that

$$\mathcal{N}^*(0) = \frac{m^* p_F}{\pi^2\hbar^3}. \tag{6.26}$$

In this way, the effective mass m^* determines the density of states at the Fermi energy: large effective masses lead to large densities of states.

The second-order coefficients

$$f_{\mathbf{p}\sigma,\mathbf{p}'\sigma'} = \left.\frac{\delta^2 \mathcal{E}}{\delta n_{\mathbf{p}\sigma} \delta n_{\mathbf{p}'\sigma'}}\right|_{\delta n_{\mathbf{p}''\sigma''} = 0} \tag{6.27}$$

describe the interactions between quasiparticles at the Fermi surface. These partial derivatives are evaluated in the presence of an otherwise "frozen" Fermi sea, where all other quasiparticle occupancies are fixed. Landau was able to show that, in an isotropic Fermi liquid, the quasiparticle mass m^* is related to the dipolar component of these interactions, as we shall shortly demonstrate. The Landau interaction can be regarded as an interaction operator that acts on a thin shell of quasiparticle states near the Fermi surface. If $\hat{n}_{\mathbf{p}\sigma} = \psi^\dagger_{\mathbf{p}\sigma} \psi_{\mathbf{p}\sigma}$ is the quasiparticle occupancy, where $\psi^\dagger_{\mathbf{p}\sigma}$ is the quasiparticle-creation operator, then one is tempted to write

$$H_I \sim \frac{1}{2} \sum_{\mathbf{p}\sigma\mathbf{p}'\sigma'} f_{\mathbf{p}\sigma,\mathbf{p}'\sigma'} \hat{n}_{\mathbf{p}\sigma} \hat{n}_{\mathbf{p}'\sigma'}. \tag{6.28}$$

Written this way, we see that the Landau interaction term is a *forward scattering amplitude* between quasiparticles whose initial and final momenta are unchanged. In practice, one has to allow for slowly varying quasiparticle densities $n_{\mathbf{p}\sigma}(\mathbf{x})$, writing

$$H_I \sim \frac{1}{2} \int d^3x \sum_{\mathbf{p}\sigma\mathbf{p}'\sigma'} f_{\mathbf{p}\sigma,\mathbf{p}'\sigma'} \hat{n}_{\mathbf{p}\sigma}(\mathbf{x}) \hat{n}_{\mathbf{p}'\sigma'}(\mathbf{x}), \tag{6.29}$$

where $n_{\mathbf{p}\sigma}(\mathbf{x})$ is the local quasiparticle density. Using the Fourier transformed density operator $\hat{n}_{\mathbf{p}\sigma}(\mathbf{q}) = \psi^\dagger_{\mathbf{p}-\mathbf{q}/2\sigma} \psi_{\mathbf{p}+\mathbf{q}/2\sigma} = \int_{\mathbf{x}} e^{-i\mathbf{q}\cdot\mathbf{x}} n_{\mathbf{p}\sigma}(\mathbf{x})$, a more correct formulation of the Landau interaction is

$$H_I = \frac{1}{2} \sum_{\mathbf{p}\sigma\mathbf{p}'\sigma',|\mathbf{q}|<\Lambda} f_{\mathbf{p}\sigma,\mathbf{p}'\sigma'}(\mathbf{q}) \hat{n}_{\mathbf{p}\sigma}(\mathbf{q}) \hat{n}_{\mathbf{p}'\sigma'}(-\mathbf{q}), \tag{6.30}$$

where Λ is a cut-off that restricts the momentum transfer to values smaller than the thickness of the shell of quasiparticles. The Landau coefficients for the neutral Fermi liquid are then the zero-momentum limit $f_{\mathbf{p}\sigma,\mathbf{p}'\sigma'} = f_{\mathbf{p}\sigma,\mathbf{p}'\sigma'}(\mathbf{q} = 0)$. The existence of such a limit requires that the interaction has a finite range, so that its Fourier transform at $\mathbf{q} = 0$ is well defined. This requirement is met in neutral Fermi liquids; however, the Coulomb interaction does not meet this requirement. The extension of Landau's Fermi liquid concept to charged Fermi liquids requires that we separate out the long-range part of the Coulomb interaction – a point that will be returned to later.

Interactions mean that quasiparticle energies are sensitive to changes in the quasiparticle occupancies. Suppose the quasiparticle occupancies deviate from the ground state as follows: $n_{\mathbf{p}\sigma} \to n_{\mathbf{p}\sigma} + \delta n_{\mathbf{p}\sigma}$. The corresponding change in the total energy is then

$$\frac{\delta \mathcal{E}}{\delta n_{\mathbf{p}\sigma}} = \epsilon_{\mathbf{p}\sigma} \equiv E_{\mathbf{p}\sigma} - \mu = \epsilon_{\mathbf{p}\sigma}^{(0)} + \sum_{\mathbf{p}'\sigma'} f_{\mathbf{p}\sigma,\mathbf{p}'\sigma'} \delta n_{\mathbf{p}'\sigma'}. \tag{6.31}$$

The second term is the change in the quasiparticle energy induced by the polarization of the Fermi sea.

To determine thermodynamic properties of the Landau Fermi-liquid, we also need to know the entropy of the fluid. Fortunately, when we turn on interactions adiabatically the entropy is invariant, so that it must maintain the dependence on particle occupancies that it has in the non-interacting system, i.e.

$$S = -k_B \sum_{\mathbf{p},\sigma} [n_{\mathbf{p}\sigma} \ln n_{\mathbf{p}\sigma} + (1 - n_{\mathbf{p}\sigma}) \ln(1 - n_{\mathbf{p}\sigma})]. \tag{6.32}$$

The full thermodynamics are determined by the free energy $F = \mathcal{E} - TS = E - \mu N - TS$, which is the sum of (6.21) and (6.32).

$$F(\{n_{\mathbf{p}\sigma}\}) = \mathcal{E}_0(\mu) + \sum_{\mathbf{p}\sigma} \epsilon_{\mathbf{p}\sigma}^{(0)} \delta n_{\mathbf{p}\sigma} + \frac{1}{2} \sum_{\mathbf{p},\mathbf{p}',\sigma,\sigma'} f_{\mathbf{p}\sigma,\mathbf{p}'\sigma'} \delta n_{\mathbf{p}\sigma} \delta n_{\mathbf{p}'\sigma'}$$

$$+ k_B T \sum_{\mathbf{p},\sigma} [n_{\mathbf{p}\sigma} \ln n_{\mathbf{p}\sigma} + (1 - n_{\mathbf{p}\sigma}) \ln(1 - n_{\mathbf{p}\sigma})]. \tag{6.33}$$

free energy of Landau Fermi-liquid

6.3.1 Landau parameters

The power of the Landau Fermi-liquid theory lies in its ability to parameterize the interactions in terms of a small number of multipole parameters called Landau parameters. These parameters describe how the original non-interacting Fermi-liquid theory is renormalized by the feedback effect of interactions on quasiparticle energies. Table 6.1 summarizes the key properties of the Landau Fermi-liquid.

In a Landau Fermi-liquid in which spin is conserved, the interaction is invariant under spin rotations and can in general be written in the form [2]

$$f_{\mathbf{p}\sigma,\mathbf{p}'\sigma'} = f_{\mathbf{p},\mathbf{p}'}^s + f_{\mathbf{p},\mathbf{p}'}^a \sigma \sigma'. \tag{6.35}$$

The spin-dependent part of the interaction is the magnetic component of the quasiparticle interaction.

In practice, we are only interested in quasiparticles with a small excitation energy, so we only need to know the values of $f_{\mathbf{p},\mathbf{p}'}^{s,a}$ near the Fermi surface, permitting us to set $\mathbf{p} = p_F \hat{\mathbf{p}}$, $\mathbf{p}' = p_F \hat{\mathbf{p}}'$, where $\hat{\mathbf{p}}$ and $\hat{\mathbf{p}}'$ are the unit vectors on the Fermi surface. In an isotropic Landau

---

[2] To see that this result follows from spin rotation invariance, we need to recognize that the quasiparticle occupancies $n_{\mathbf{p}\sigma}$ we have considered are actually the diagonal elements of a quasiparticle density matrix $n_{\mathbf{p}\alpha\beta}$. With this modification, the interaction becomes a matrix $f_{\mathbf{p}\alpha\beta;\mathbf{p}'\gamma\eta}$ whose most general rotationally invariant form is

$$f_{\mathbf{p}\alpha\beta;\mathbf{p}'\gamma\eta} = f^s(\mathbf{p},\mathbf{p}')\delta_{\alpha\beta}\delta_{\gamma\eta} + f^a(\mathbf{p},\mathbf{p}')\vec{\sigma}_{\alpha\beta} \cdot \vec{\sigma}_{\gamma\eta}. \tag{6.34}$$

The diagonal components of this interaction recover the results of (6.35).

Table 6.1 Key properties of the Fermi liquid.		
Property	Non-interacting	Landau Fermi-liquid
Fermi momentum	p_F	Unchanged
Density of particles	$2\frac{v_{FS}}{(2\pi)^3}$	Unchanged
Density of states	$\mathcal{N}(0) = \frac{mp_F}{\pi^2\hbar^3}$	$\mathcal{N}^*(0) = \frac{m^*p_F}{\pi^2\hbar^3}$
Effective mass	m	$m^* = m(1 + F_1^s)$
Specific heat coefficient $C_V = \gamma T$	$\gamma = \frac{\pi^2}{3}k_B^2\mathcal{N}(0)$	$\gamma = \frac{\pi^2}{3}k_B^2\mathcal{N}^*(0)$
Spin susceptibility	$\chi_s = \mu_F^2\mathcal{N}(0)$	$\chi_s = \mu_F^2\frac{\mathcal{N}^*(0)}{1+F_0^a}$
Charge susceptibility	$\chi_C = \mathcal{N}(0)$	$\chi_C = \frac{\mathcal{N}^*(0)}{1+F_0^s}$
Collective modes	–	Sound ($\omega\tau \ll 1$)
		Zero sound ($\omega\tau \gg 1$)

Fermi liquid, the physics is invariant under spatial rotations, so that interactions on the Fermi surface only depend on the relative angle θ between $\hat{\mathbf{p}}$ and $\hat{\mathbf{p}}'$. We write

$$f_{\mathbf{p},\mathbf{p}'}^{s,a} = f^{s,a}(\cos\theta). \qquad (\cos\theta = \hat{\mathbf{p}}\cdot\hat{\mathbf{p}}') \qquad (6.36)$$

We convert the interaction to a dimensionless function by multiplying it by the quasiparticle density of states $N^*(0)$:

$$F^{s,a}(\cos\theta) = \mathcal{N}^*(0)f^{s,a}(\cos\theta). \qquad (6.37)$$

These functions can now be expanded as a multipole expansion in terms of Legendre polynomials:

$$F^{s,a}(\cos\theta) = \sum_{l=0}^{\infty}(2l+1)F_l^{s,a}P_l(\cos\theta). \qquad (6.38)$$

The coefficients F_l^s and F_l^a are the *Landau parameters*. The spin-symmetric components F_l^s parameterize the non-magnetic part of the interaction, while the spin-antisymmetric F_l^a define the magnetic component of the interaction. These parameters determine how distortions of the Fermi surface are fed back to modify quasiparticle energies.

We can invert (6.38) using the orthogonality relation $\frac{1}{2}\int_{-1}^{1}dc\ P_l(c)P_{l'}(c) = (2l+1)^{-1}\delta_{l,l'}$:

$$F_l^{s,a} = \frac{1}{2}\int_{-1}^{1}dc\ F^{s,a}(c)P_l(c) \equiv \langle F^{s,a}(\hat{\Omega})P_l(\hat{\Omega})\rangle_{\hat{\Omega}}, \qquad (6.39)$$

where $\langle\ldots\rangle_{\hat{\Omega}}$ denotes an average over solid angle. It is useful to rewrite this angular average as an average over the Fermi surface. To do this we note that, since $2\sum_{\mathbf{k}}\delta(\epsilon_{\mathbf{k}}) = \mathcal{N}^*(0)$, the function $\frac{2}{\mathcal{N}^*(0)}\delta(\epsilon_{\mathbf{k}})$ behaves as a normalized *projector* onto the Fermi surface, so that

$$F_l^{s,a} = \langle F^{s,a}(\hat{\Omega})P_l(\hat{\Omega})\rangle_{FS} = \frac{2}{\mathcal{N}^*(0)}\sum_{\mathbf{p}'}F_{\mathbf{p},\mathbf{p}'}^{s,a}P_l(\cos\theta_{\mathbf{p},\mathbf{p}'})\delta(\epsilon_{\mathbf{p}'}), \qquad (6.40)$$

and since $F_{\mathbf{p},\mathbf{p}'}^{s,a} = \mathcal{N}^*(0)f_{\mathbf{p},\mathbf{p}'}^{s,a}$,

$$F_l^{s,a} = 2 \sum_{\mathbf{p'}} f_{\mathbf{p},\mathbf{p'}}^{s,a} P_l(\cos\theta_{\mathbf{p},\mathbf{p'}})\delta(\epsilon_{\mathbf{p'}}). \tag{6.41}$$

This form is very convenient for later calculations.

Example 6.2 Use first-order perturbation theory to calculate the Landau interaction parameters for a fluid of fermions with a weak interaction described by

$$H = \sum_{\mathbf{p}\sigma} E_{\mathbf{p}} n_{\mathbf{p}\sigma} + \frac{\lambda}{2} \sum_{\mathbf{p}\sigma,\mathbf{p'}\sigma',\mathbf{q}} V(q) c_{\mathbf{p}-\mathbf{q}\sigma}^\dagger c_{\mathbf{p'}+\mathbf{q}\sigma'}^\dagger c_{\mathbf{p'}\sigma'} c_{\mathbf{p}\sigma},$$

where $E_{\mathbf{p}}$ is the energy of the non-interacting Fermi gas, $V(q) = \int \frac{d^3q}{(2\pi)^3} e^{-i\mathbf{q}\cdot\mathbf{r}} V(r)$ is the Fourier transform of the interaction potential $V(r)$, and $\lambda \ll 1$ is a very small coupling constant. (*Hint*: use first-order perturbation theory in λ to compute the energy of a state

$$\Psi = |n_{\mathbf{p}_1\sigma_1}, n_{\mathbf{p}_2\sigma_2}, \ldots\rangle \tag{6.42}$$

to leading order in the interaction strength λ, and then read off the terms quadratic in $n_{\mathbf{p}\sigma}$.)

Solution

To leading order in λ, the total energy is given by $E = \langle \Psi | H | \Psi \rangle$, or

$$E = \sum_{\mathbf{p}\sigma} E_{\mathbf{p}} n_{\mathbf{p}\sigma} + \frac{\lambda}{2} \sum_{\mathbf{p}\sigma,\mathbf{p'}\sigma',\mathbf{q}} V(q) \langle \Psi | c_{\mathbf{p}-\mathbf{q}\sigma}^\dagger c_{\mathbf{p'}+\mathbf{q}\sigma'}^\dagger c_{\mathbf{p'}\sigma'} c_{\mathbf{p}\sigma} | \Psi \rangle. \tag{6.43}$$

The matrix element $\langle \Psi | c_{\mathbf{p}-\mathbf{q}\sigma}^\dagger c_{\mathbf{p'}+\mathbf{q}\sigma'}^\dagger c_{\mathbf{p'}\sigma'} c_{\mathbf{p}\sigma} | \Psi \rangle$ in the interaction term vanishes unless the two-quasiparticle state annihilated by the two destruction operators has an overlap with the two-particle state created by the two creation operators, i.e.

$$\langle \Psi | c_{\mathbf{p}-\mathbf{q}\sigma}^\dagger c_{\mathbf{p'}+\mathbf{q}\sigma'}^\dagger c_{\mathbf{p'}\sigma'} c_{\mathbf{p}\sigma} | \Psi \rangle = \langle \mathbf{p}-\mathbf{q},\sigma;\mathbf{p'}+\mathbf{q},\sigma' | \mathbf{p},\sigma;\mathbf{p'}\sigma' \rangle n_{\mathbf{p}\sigma} n_{\mathbf{p'}\sigma'}$$

$$= \left(\delta_{\mathbf{q}=0} - \delta_{\mathbf{p}-\mathbf{q},\mathbf{p'}}\delta_{\sigma,\sigma'} \right) n_{\mathbf{p}\sigma} n_{\mathbf{p'}\sigma'}, \tag{6.44}$$

where the second term occurs when the outgoing state is the *exchange* of the incoming two-quasiparticle state.

Inserting (6.44) into (6.43), we obtain

$$\sum_{\mathbf{p}\sigma} E_{\mathbf{p}} n_{\mathbf{p}\sigma} + \frac{\lambda}{2} \sum_{\mathbf{p}\sigma,\mathbf{p'}\sigma'} [V(0) - V(\mathbf{p}-\mathbf{p'})\delta_{\sigma\sigma'}] n_{\mathbf{p'}\sigma'} n_{\mathbf{p}\sigma}, \tag{6.45}$$

enabling us to read off the Landau interaction as

$$f_{\mathbf{p}\sigma,\mathbf{p'}\sigma'} = \lambda[V(q=0) - V(\mathbf{p}-\mathbf{p'})\delta_{\sigma\sigma'}] + O(\lambda^2). \tag{6.46}$$

It follows that the symmetric and antisymmetric parts of the interaction parameters are

$$f_{\mathbf{p},\mathbf{p'}}^s = \lambda\left[V(q=0) - \frac{1}{2}V(\mathbf{p}-\mathbf{p'})\right] + O(\lambda^2)$$

$$f_{\mathbf{p},\mathbf{p'}}^a = -\frac{\lambda}{2}V(\mathbf{p}-\mathbf{p'}) + O(\lambda^2). \tag{6.47}$$

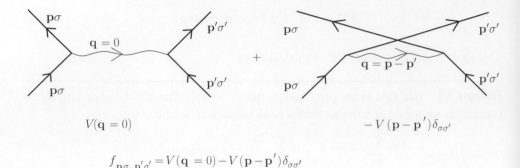

$$f_{\mathbf{p}\sigma, \mathbf{p}'\sigma'} = V(\mathbf{q} = 0) - V(\mathbf{p} - \mathbf{p}')\delta_{\sigma\sigma'}$$

Fig. 6.2 Feynman diagrams for leading-order contributions to the Landau parameter for an interaction $V(\mathbf{q})$. Wavy line represents the interaction between quasiparticles.

Note the following:

- The Landau interaction is only well defined if $V(q = 0)$ is finite, which implies that the interaction is short-range.
- The second term in the interaction corresponds to the "exchange" of identical particles. For a repulsive interaction, this gives rise to an *attractive f^a*. We can represent the interaction term by the Feynman diagrams shown in Figure 6.2.

6.3.2 Equilibrium distribution of quasiparticles

Remarkably, despite interactions, the Landau Fermi-liquid preserves the equilibrium Fermi–Dirac momentum distribution. The key idea here is that, in thermal equilibrium, the free energy (6.33) is stationary with respect to small changes $\delta n_{\mathbf{p}\sigma}$ in quasiparticle occupancies, so that

$$\delta F = \sum_{\mathbf{p}\sigma} \delta n_{\mathbf{p}\sigma} \left[\epsilon_{\mathbf{p}\sigma} + k_B T \ln\left(\frac{n_{\mathbf{p}\sigma}}{1 - n_{\mathbf{p}\sigma}} \right) \right] + O(\delta n_{\mathbf{p}\sigma}^2) = 0. \tag{6.48}$$

Stationarity of the free energy, $\delta F = 0$, enforces the thermodynamic identity $\delta F = \delta \mathcal{E} - T\delta S = 0$, or $d\mathcal{E} = TdS$. This requires that the linear coefficient of $\delta n_{\mathbf{p}\sigma}$ in (6.48) is zero, which implies that the quasiparticle occupancy

$$n_{\mathbf{p}\sigma} = \frac{1}{e^{\beta \epsilon_{\mathbf{p}\sigma}} + 1} = f(\epsilon_{\mathbf{p}\sigma}) \tag{6.49}$$

is determined by the Fermi–Dirac distribution function of its energy. There is a subtlety here, however, for the quantity $\epsilon_{\mathbf{p}\sigma}$ contains the feedback effect of interactions, as given in (6.31):

$$\epsilon_{\mathbf{p}\sigma} = \epsilon_{\mathbf{p}\sigma}^{(0)} + \sum_{\mathbf{p}'\sigma'} f_{\mathbf{p}\sigma, \mathbf{p}'\sigma'} \delta n_{\mathbf{p}'\sigma'}. \tag{6.50}$$

Let us first consider the low-temperature behavior in the absence of a field. In this case, as the temperature is lowered the density of thermally excited quasiparticles will go to zero, and in this limit the quasiparticle distribution function is asymptotically given by

$$n_{\mathbf{p}\sigma} = f(\epsilon_{\mathbf{p}}^{(0)}). \tag{6.51}$$

In the ground state this becomes a step function, $n_{\mathbf{p}\sigma}|_{T=0} = \theta(-\epsilon_{\mathbf{p}}^{(0)}) = \theta(\mu - E_{\mathbf{p}}^{(0)})$, as expected.

To obtain the specific heat, we must calculate $C_V dT = d\mathcal{E} = \sum_{\mathbf{p}} \epsilon_{\mathbf{p}\sigma}^{(0)} \delta n_{\mathbf{p}\sigma}$. At low temperatures, $\delta n_{\mathbf{p}\sigma} = \frac{\partial f(\epsilon_{\mathbf{p}\sigma}^{(0)})}{\partial T} dT$, so that

$$C_V = \sum_{\mathbf{p}\sigma} \epsilon_{\mathbf{p}\sigma}^{(0)} \left(\frac{\partial f(\epsilon_{\mathbf{p}\sigma}^{(0)})}{\partial T} \right) \rightarrow \mathcal{N}^*(0) \int_{-\infty}^{\infty} d\epsilon \, \epsilon \left(\frac{\partial f(\epsilon)}{\partial T} \right), \tag{6.52}$$

where, as in (6.5) the summation is replaced by an integral over the density of states near the Fermi surface. Apart from the renormalization of the energies, this is precisely the same result obtained in (6.5), leading to

$$C_V = \gamma T, \qquad \gamma = \frac{\pi^2 k_B^2}{3} \mathcal{N}^*(0). \tag{6.53}$$

6.4 Feedback effects of interactions

One can visualize the Landau Fermi-liquid as a deformable sphere, like a large water droplet in zero gravity. The Fermi sphere changes shape when the density or magnetization of the fluid is modified, or if a current flows. These deformations act back on the quasi-particles via the Landau interactions, to change the quasiparticle energies. These feedback effects are a generalization of the idea of a Weiss field in magnetism. When the feedback is positive, it can lead to instabilities, such as the development of magnetism. A Fermi surface can also oscillate collectively about its equilibrium shape. In a conventional gas, density oscillations cannot take place without collisions. In a Landau Fermi-liquid, we will see that the interactions play a non-trivial role that gives rise to collisionless collective oscillations of the Fermi surface, called *zero sound* (literally zero-collision sound), that are absent in the free Fermi gas [7].

To examine the feedback effects of interactions, let us suppose an external potential or field is applied to induce a polarization of the Fermi surface, as illustrated in Figure 6.3. There are various kinds of external field we can consider. A simple change in the chemical potential

$$\delta\epsilon_{\mathbf{p}\sigma}^0 = -\delta\mu \tag{6.54}$$

will induce an isotropic enlargement of the Fermi surface. The application of a magnetic field

$$\delta\epsilon_{\mathbf{p}\sigma}^0 = -\sigma\mu_F B \tag{6.55}$$

will induce a spin polarization. We can also consider the application of a vector potential which couples to the quasiparticle current

$$\delta\epsilon_{\mathbf{p}\sigma}^0 = -\mathbf{A} \cdot \frac{e\mathbf{p}}{m} \tag{6.56}$$

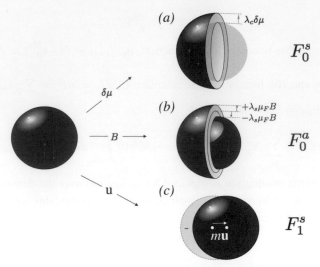

Fig. 6.3 Illustrating the polarization of the Fermi surface by (a) a change in chemical potential to produce an isotropic charge polarization; (b) application of a magnetic field to produce a spin polarization; (c) dipolar polarization of the Fermi surface that accompanies a current of quasiparticles. The Landau parameter governing each polarization is indicated on the right-hand side.

in a translationally invariant system. Notice how, in each case, the applied field couples to a conserved quantity (the particle number, the spin, or the current), which is unchanged by interactions. This means that the energy associated with the application of the external field is unchanged by interactions for any quasiparticle configuration $\{n_{\mathbf{p}\sigma}\}$, which guarantees that the coupling to the external field is identical to that for non-interacting particles. This is the reason for the appearance of the unrenormalized mass in (6.56). For each of these cases, there will of course be a feedback effect of the interactions that we now calculate.

From (6.31), the change in the quasiparticle energy will now contain two terms – one due to direct coupling to the external field, the other derived from the induced polarization $\delta n_{\mathbf{p}\sigma}$ of the Fermi surface:

$$\delta\epsilon_{\mathbf{p}\sigma} = \delta\epsilon_{\mathbf{p}\sigma}^{(0)} + \sum_{\mathbf{p}'\sigma'} f_{\mathbf{p}\sigma\mathbf{p}'\sigma'}\delta n_{\mathbf{p}'\sigma'}. \tag{6.57}$$

In this case, the equilibrium quasiparticle occupancies become

$$n_{\mathbf{p}\sigma} = f(\epsilon_{\mathbf{p}}^{(0)} + \delta\epsilon_{\mathbf{p}\sigma}) = f(\epsilon_{\mathbf{p}}^{(0)}) + f'(\epsilon_{\mathbf{p}}^{(0)})\delta\epsilon_{\mathbf{p}\sigma}. \tag{6.58}$$

As the temperature is lowered to zero, the derivative of the Fermi function evolves into a delta function, $-f'(\epsilon) \sim \delta(\epsilon)$, so that the quasiparticle occupancy is given by

$$n_{\mathbf{p}\sigma} = \overbrace{\theta(-\epsilon_{\mathbf{p}}^{(0)})}^{n_{\mathbf{p}\sigma}^{(0)}} + \overbrace{[-\delta(\epsilon_{\mathbf{p}}^{(0)})\delta\epsilon_{\mathbf{p}\sigma}]}^{\delta n_{\mathbf{p}\sigma}}. \tag{6.59}$$

$\delta n_{\mathbf{p}\sigma} = -\delta(\epsilon_{\mathbf{p}}^{(0)})\delta\epsilon_{\mathbf{p}\sigma}$ represents the polarization of the Fermi surface, which will feed back into the interaction (6.57) as follows:

$$\delta n_{\mathbf{p}\sigma} = -\delta(\epsilon_{\mathbf{p}}^{(0)})\delta\epsilon_{\mathbf{p}\sigma}$$

$$\delta\epsilon_{\mathbf{p}\sigma} = \delta\epsilon_{\mathbf{p}\sigma}^{(0)} + \sum_{\mathbf{p}'\sigma'} f_{\mathbf{p}\sigma\mathbf{p}'\sigma'}\delta n_{\mathbf{p}'\sigma'}. \tag{6.60}$$

The resulting shift in the quasiparticle energies must then satisfy the self-consistency relation:

$$\delta\epsilon_{\mathbf{p}\sigma} = \delta\epsilon_{\mathbf{p}\sigma}^{(0)} - \sum_{\mathbf{p}'\sigma'} f_{\mathbf{p}\sigma\mathbf{p}'\sigma'}\delta(\epsilon_{\mathbf{p}'}^{(0)})\delta\epsilon_{\mathbf{p}'\sigma}. \tag{6.61}$$

This feedback process preserves the symmetry of the external perturbation, but its strength in a given symmetry channel depends on the corresponding Landau paramater. Thus, isotropic charge and spin polarization of the Fermi surface, shown in Figure 6.3(a) and Figure 6.3(b), are fed back via the isotropic charge and magnetic Landau parameters F_0^s and F_0^a. When the quasiparticle fluid is set into motion at velocity $\vec{u}$, this induces a dipolar polarization of the Fermi surface, shown in (Figure 6.3(c)), which is fed back via the dipolar Landau parameter F_1^s. This process is responsible for the renormalization of the effective mass.

Consider a change in the quasiparticle potential that has a particular multipole symmetry, so that the "bare" change in quasiparticle energy is

$$\delta\epsilon_{\mathbf{p}\sigma}^{(0)} = v_l Y_{lm}(\hat{\mathbf{p}}), \tag{6.62}$$

where Y_{lm} is a spherical harmonic. The renormalized response of the quasiparticle energy given by (6.61) must have the same symmetry, but will have a different magnitude t_l:

$$\delta\epsilon_{\mathbf{p}\sigma} = t_l Y_{lm}(\hat{\mathbf{p}}). \tag{6.63}$$

When this is fed back through the interaction, according to (6.61) it produces an additional shift in the quasiparticle energy given by $\sum_{\mathbf{p}'\sigma'} f_{\mathbf{p}\sigma,\mathbf{p}'\sigma'}\delta n_{\mathbf{p}'\sigma'} = -F_l^s t_l Y_{lm}(\hat{\mathbf{p}})$ (see Example 6.3), so that the total change in the energy is given by $\delta\epsilon_{\mathbf{p}\sigma} = (v_l - F_l^s t_l)Y_{lm}(\hat{\mathbf{p}})$. Comparing this result with (6.63), we see that

$$t_l = (v_l - F_l^s t_l). \tag{6.64}$$

This is the symmetry-resolved version of (6.61). Consequently,[3]

$$t_l = \frac{v_l}{1 + F_l^s}. \tag{6.65}$$

We may interpret t_l as the scattering t-matrix associated with the potential v_l. If $F_l^s > 0$, corresponding to a repulsive interaction, negative feedback occurs which causes the response

---

[3] Note: in Landau's original formulation [2], the Landau parameters were defined without the normalizing factor $(2l + 1)$ in (6.73). With such a normalization the F_l are a factor of $2l + 1$ larger and one must replace $F_l^s \rightarrow \frac{1}{2l+1}F_l^s$ in (6.38).

to be suppressed. This is normally the case in the isotropic channel, where repulsive inter-actions tend to suppress the polarizability of the Fermi surface. By contrast, if $F_l^s < 0$, corresponding to an attractive interaction, positive feedback enhances the response. Indeed, if F_l^s drops down to the critical value $F_l^s = -1$, an instability will occur and the Lan-dau Fermi surface becomes unstable to a deformation, a process called a *Pomeranchuk instability*.

A similar calculation can be carried out for a spin polarization of the Fermi surface, where the shifts in the quasiparticle energies are

$$\delta\epsilon_{\mathbf{p}\sigma}^{(0)} = \sigma v_l^a Y_{lm}(\hat{\mathbf{p}}), \qquad \delta\epsilon_{\mathbf{p}\sigma} = \sigma t_l^a Y_{lm}(\hat{\mathbf{p}}). \tag{6.66}$$

Now the spin-dependent polarization of the Fermi surface feeds back via the spin-dependent Landau parameters, so that

$$t_l^a = \frac{v_l^a}{1 + F_l^a}. \tag{6.67}$$

The isotropic response ($l = 0$) corresponds to a simple spin polarization of the Fermi sur-face. If spin interactions grow to the point where $F_0^a = -1$, the Fermi surface becomes unstable to the formation of a spontaneous spin polarization. This is called a *Stoner instability*, and results in ferromagnetism.

Example 6.3   Calculate the response of the quasiparticle energy to a charge or spin polar-ization with a specific multipole symmetry. First, consider a spin-independent polarization of the Fermi surface of the form

$$\delta n_{\mathbf{p}\sigma} = -t_l Y_{lm}(\hat{\mathbf{p}}) \times \delta(\epsilon_{\mathbf{p}}^{(0)}),$$

where $Y_{lm}(\hat{\mathbf{p}})$ is a spherical harmonic. Show that the resulting shift in quasiparticle energies is given by

$$\delta\epsilon_{\mathbf{p}\sigma} = -t_l F_l^s Y_{lm}(\hat{\mathbf{p}}).$$

Determine the corresponding result for a magnetic polarization of the Fermi surface of the form

$$\delta n_{\mathbf{p}\sigma} = -\sigma t_l^a Y_{lm}(\hat{\mathbf{p}}) \times \delta(\epsilon_{\mathbf{p}}^{(0)}).$$

Solution

According to (6.31), the change in quasiparticle energy due to the polarization of the Fermi surface is given by

$$\delta\epsilon_{\mathbf{p}\sigma} = \sum_{\mathbf{p}'\sigma'} f_{\mathbf{p}\sigma,\mathbf{p}'\sigma'}\delta n_{\mathbf{p}'\sigma'}. \tag{6.68}$$

Substituting $\delta n_{\mathbf{p}\sigma} = -t_l Y_{lm}(\hat{\mathbf{p}}) \times \delta(\epsilon_{\mathbf{p}}^{(0)})$, then

$$\delta\epsilon_{\mathbf{p}\sigma} = -t_l \sum_{\mathbf{p}'\sigma'} f_{\mathbf{p}\sigma,\mathbf{p}',\sigma'} Y_{lm}(\hat{\mathbf{p}}') \times \delta(\epsilon_{\mathbf{p}'}^{(0)}). \tag{6.69}$$

Decomposing the interaction into its magnetic and non-magnetic components, $f_{\mathbf{p}\sigma,\mathbf{p}'\sigma'} = f^s(\hat{\mathbf{p}} \cdot \hat{\mathbf{p}}') + \sigma\sigma' f^a(\hat{\mathbf{p}} \cdot \hat{\mathbf{p}}')$, only the non-magnetic component survives the spin summation, so that

$$\delta\epsilon_{\mathbf{p}\sigma} = -2t_l \sum_{\mathbf{p}'} f^s(\hat{\mathbf{p}} \cdot \hat{\mathbf{p}}')Y_{lm}(\hat{\mathbf{p}}') \times \delta(\epsilon_{\mathbf{p}'}^{(0)}). \tag{6.70}$$

Replacing the summation over momentum by an angular average over the Fermi surface,

$$2\sum_{\mathbf{p}'} \delta(\epsilon_{\mathbf{p}'}^{(0)}) \rightarrow \mathcal{N}^*(0) \int \frac{d\Omega_{\hat{\mathbf{p}}'}}{4\pi}, \tag{6.71}$$

we obtain

$$\delta\epsilon_{\mathbf{p}\sigma} = -t_l \times \mathcal{N}^*(0) \int \frac{d\Omega_{\hat{\mathbf{p}}'}}{4\pi} f^s(\hat{\mathbf{p}} \cdot \hat{\mathbf{p}}')Y_{lm}(\hat{\mathbf{p}}')$$

$$= -t_l \int \frac{d\Omega_{\hat{\mathbf{p}}'}}{4\pi} F^s(\hat{\mathbf{p}} \cdot \hat{\mathbf{p}}')Y_{lm}(\hat{\mathbf{p}}'). \tag{6.72}$$

Now we can expand the interaction in terms of Legendre polynomials, which can in turn be decomposed into spherical harmonics:

$$F^s(\cos\theta) = \sum_l (2l+1)F_l^s P_l(\hat{\mathbf{p}} \cdot \hat{\mathbf{p}}') = 4\pi \sum_{l,m} F_l^s Y_{lm}(\hat{\mathbf{p}})Y_{lm}^*(\hat{\mathbf{p}}'). \tag{6.73}$$

When we substitute this into (6.72) we may use the orthogonality of the spherical harmonics to obtain

$$\delta\epsilon_{\mathbf{p}\sigma} = -t_l \sum_{l'm'} F_{l'}^s Y_{l'm'}(\mathbf{p}) \overbrace{\int d\Omega_{\mathbf{p}'} Y_{l'm'}^*(\hat{\mathbf{p}}')Y_{lm}(\mathbf{p}')}^{\delta_{l'l}\delta_{m'm}}$$

$$= -t_l F_l^s Y_{lm}(\hat{\mathbf{p}}). \tag{6.74}$$

For a spin-dependent polarization, $\delta n_{\mathbf{p}\sigma} = -t_l^a \sigma Y_{lm}(\hat{\mathbf{p}})\delta(\epsilon_{\mathbf{p}}^{(0)})$, it is the magnetic part of the interaction that contributes. We can generalize the above result to obtain

$$\delta\epsilon_{\mathbf{p}\sigma} = \sigma t_l^a \times F_l^a Y_{lm}(\hat{\mathbf{p}}). \tag{6.75}$$

6.4.1 Renormalization of paramagnetism and compressibility by interactions

The simplest polarization response functions of a Landau Fermi-liquid are its "charge" and spin susceptibility.

$$\chi_c = \frac{1}{V}\frac{\partial N}{\partial \mu}, \qquad \chi_s = \frac{1}{V}\frac{\partial M}{\partial B}, \tag{6.76}$$

where V is the volume. Here, we use the term "charge" density to refer to the density response function of the neutral Fermi liquid. These responses involve an isotropic polarization of the Fermi surface. In a neutral fluid, the bulk modulus $\kappa = -V\frac{dP}{dV}$ is directly related to the charge susceptibility per unit volume, $\kappa = \frac{n^2}{\chi_c}$, where $n = N/V$ is the particle density. Thus a smaller charge susceptibility implies a stiffer fluid. [4]

When we apply a chemical potential or a magnetic field, the "bare" quasiparticle energies respond isotropically:

$$\delta\epsilon_{\mathbf{p}\sigma}^{(0)} = \delta E_{\mathbf{p}\sigma}^{(0)} - \delta\mu = -\sigma\mu_F B - \delta\mu. \tag{6.77}$$

Feedback via the interactions renormalizes the response of the full quasiparticle energy:

$$\delta\epsilon_{\mathbf{p}\sigma} = -\sigma\lambda_s\mu_F B - \lambda_c\delta\mu. \tag{6.78}$$

Since these are isotropic responses, the feedback is transmitted through the $l = 0$ Landau parameters,

$$\lambda_s = \frac{1}{1 + F_0^a}$$

$$\lambda_c = \frac{1}{1 + F_0^s}. \tag{6.79}$$

When we apply a pure chemical potential shift, the resulting change in quasiparticle number is $\delta N = \lambda_c \mathcal{N}^*(0)\delta\mu$, so the charge susceptibility is given by

$$\chi_c = \lambda_c N^*(0) = \frac{\mathcal{N}^*(0)}{1 + F_0^s}. \tag{6.80}$$

Typically, repulsive interactions cause $F_0^s > 0$, reducing the charge susceptibility, making the fluid "stiffer." In ^{3}He, $F_0^s = 10.8$ at low pressures, which is roughly ten times stiffer than expected, based on its density of states.

A reverse phenomenon occurs in the spin response of Landau Fermi-liquids. In a magnetic field, the change in the number of up and down quasiparticles is $\delta n_\uparrow = -\delta n_\downarrow = \lambda_s\mathcal{N}^*(0)\mu_F B$. The resulting change in magnetization is $\delta M = \mu_F(\delta n_\uparrow - \delta n_\downarrow) = \lambda_s\mu_F^2\mathcal{N}^*(0)B$, so the spin susceptibility is

$$\chi_s = \lambda_s\mu_F^2 N^*(0) = \frac{\mu_F^2\mathcal{N}^*(0)}{1 + F_0^a}. \tag{6.81}$$

There are a number of interesting points to be made here:

- The *Wilson ratio*, defined as the ratio of χ_s/γ in the interacting system to that in the non-interacting system, is given by

$$W = \frac{\left(\frac{\chi}{\gamma}\right)}{\left(\frac{\chi}{\gamma}\right)_0} = \frac{1}{1 + F_0^a}. \tag{6.82}$$

---

[4] In a fluid, where $-\partial F/\partial V = P$, the extensive nature of the free energy guarantees that $F = -PV$, so that the Gibbs free energy $G = F + PV = 0$ vanishes. But $dG = -SdT - Nd\mu + VdP = 0$, so in the ground state $Nd\mu = VdP$ and hence $\kappa = -V\left.\frac{dP}{dV}\right|_N = -N\left.\frac{d\mu}{dV}\right|_N$, but $\mu = \mu(N/V)$ is a function of particle density alone, so that $-N\left.\frac{d\mu}{dV}\right|_N = \frac{N^2}{V}\left.\frac{d\mu}{dN}\right|_V = \frac{n^2}{\chi_c}$ where $n = N/V$. It follows that $\kappa = \frac{n^2}{\chi_c}$.

In the context of ferromagnetism, this quantity is often referred to as the *Stoner enhancement factor*. In Landau Fermi-liquids with strong ferromagnetic exchange interactions between fermions, F_0^a is negative, enhancing the Pauli susceptibility. This is the origin of the enhancement of the Pauli susceptibility in liquid ^{3}He, where $W \sim 3$. In palladium metal (Pd), $W = 10$ is even more substantially enhanced [8].

- When a Landau Fermi-liquid is tuned to the point where $F_0^a \to -1$, $\chi \to \infty$, leading to a ferromagnetic instability, called a *Stoner instability*. This is an example of a ferromagnetic *quantum critical point*, a point where quantum zero-point fluctuations of the magnetization develop infinite-range correlations in space and time. At such a point, the Wilson ratio will diverge.

6.4.2 Mass renormalization

Using this formulation of the interacting Fermi gas, Landau was able to link the renormalization of quasiparticle mass to the dipole component of the interactions F_1^s. As the fermion moves through the medium, the backflow of the surrounding fluid enhances its effective mass according to the relation

$$m^* = m \left(1 + F_1^s\right). \tag{6.83}$$

Another way to understand quasiparticle mass renormalization is to consider the current carried by a quasiparticle. Whether we are dealing with neutral or physically charged quasiparticles, the total number of particles is conserved and we can ascribe a particle current $\mathbf{v}_F = \mathbf{p}_F/m^*$ to each quasiparticle. We can rewrite this current in the form

$$\mathbf{v}_F = \frac{\mathbf{p}_F}{m^*} = \underbrace{\frac{\mathbf{p}_F}{m}}_{\text{bare current}} - \overbrace{\frac{\mathbf{p}_F}{m}\left(\frac{F_1^s}{1 + F_1^s}\right)}^{\text{backflow}}. \tag{6.84}$$

The first term is the bare current associated with the original particle, whereas the second term is backflow of the surrounding Fermi sea (Figure 6.4).

Mass renormalization increases the density of states from $N(0) = \frac{mp_F}{\pi^2} \to \mathcal{N}^*(0) = \frac{m^* p_F}{\pi^2}$, i.e. it has the effect of compressing the spacing between the fermion energy levels, which increases the number of quasiparticles that are excited at a given temperature by a factor m^*/m: this enhances the linear specific heat:

$$C_V^* = \frac{m^*}{m} C_V, \tag{6.85}$$

where C_V is the Sommerfeld value for the specific heat capacity. Experimentally, the specific heat of ^{3}He is enhanced by a factor of 2.8, from which we know that $m^* \approx 3m$.

Landau's original derivation depends on the use of Galilean invariance. Here we use an equivalent derivation, based on the observation that backflow is a feedback response to the dipolar distortion of the Fermi surface which develops in the presence of a current. This enables us to calculate the mass renormalization in an analogous fashion to the

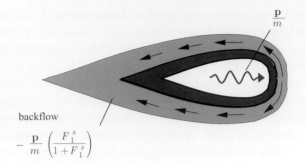

Fig. 6.4 Backflow in the Landau Fermi-liquid. The particle current in the absence of backflow is $\frac{\mathbf{p}}{m}$. Backflow of the Fermi liquid introduces a reverse current $-\left(\frac{F_1^s}{1+F_1^s}\right)\frac{\mathbf{p}}{m}$.

renormalization of the spin susceptibility and compressibility, carried out in Section 6.4.1, except that now we must introduce the conjugate field to current – that is, a vector potential.

To this end, we imagine that each quasiparticle carries a conserved charge $q = 1$ and that the flow of quasiparticles is coupled to a fictitious vector potential $q\mathbf{A} \equiv \mathbf{A}_N$. The microscopic Hamiltonian in the presence of the vector potential is then given by

$$H[\mathbf{A}_N] = \sum_\sigma \int d^3x \frac{1}{2m} \psi_\sigma^\dagger(x) \left[(-i\hbar\nabla - \mathbf{A}_N)^2\right] \psi_\sigma(x) + \hat{V}, \qquad (6.86)$$

where $\hat{V}$ contains the translationally invariant interactions. Notice that the effect of $\mathbf{A}_N$ is to change the momentum of each particle by $-\mathbf{A}_N$, so that $H[\mathbf{A}_N]$ corresponds to the Hamiltonian transformed into a Galilean reference frame moving at speed $\mathbf{u} = \mathbf{A}_N/m$. Landau's original derivation used the Galilean equivalence of the Fermi liquid to compute the mass renormalization.

Since the vector potential $\mathbf{A}_N$ is coupled to a conserved quantity, the momentum, we can treat it in the same way as a chemical potential or magnetic field. The linear term in $\mathbf{A}_N$ in the total energy is $\delta\hat{H} = -\mathbf{A}_N \cdot \frac{\hat{\mathbf{P}}}{m}$, where $\hat{\mathbf{P}}$ is the conserved total momentum operator. For a non-interacting system the change in the total energy for a small vector potential at fixed particle occupancies $n_{\mathbf{p}\sigma}$ is

$$\delta E = \langle \delta H \rangle = -\frac{\langle \mathbf{P} \rangle}{m} \cdot \mathbf{A}_N = -\sum_{\mathbf{p}\sigma} \left(\frac{\mathbf{p}}{m} \cdot \mathbf{A}_N\right) n_{\mathbf{p}\sigma}. \qquad (6.87)$$

Provided the momentum is conserved, this is also the change in the energy of the *interacting* Fermi liquid, at fixed quasiparticle occupancy, i.e. without backflow. In this way, we see that turning on the vector potential changes

$$\epsilon_{\mathbf{p}\sigma}^{(0)} \to \epsilon_{\mathbf{p}\sigma}^{(0)} + \delta\epsilon_{\mathbf{p}\sigma}^{(0)}, \qquad (6.88)$$

where

$$\delta\epsilon_{\mathbf{p}\sigma}^{(0)} = -\frac{\mathbf{p}}{m} \cdot \mathbf{A}_N = -A_N \frac{p_F}{m} \cos\theta. \qquad (6.89)$$

Here, θ is the angle between the vector potential and the quasiparticle momentum. Thus the vector potential introduces a *dipolar potential* around the Fermi surface. Notice how the conservation of momentum guarantees that it is the *bare* mass m that enters into $\delta\epsilon_{\mathbf{p}\sigma}^{(0)}$.

Now when we take account of the feedback effect caused by the redistribution of quasi-particles in response to this potential, the quasiparticle energy becomes $E_{\mathbf{p}-q\mathbf{A}} = \frac{(\mathbf{p}-\mathbf{A}_N)^2}{2m^*}$. Here, the replacement of $\mathbf{p} \to \mathbf{p} - q\mathbf{A}_N = \mathbf{p} - \mathbf{A}_N$ is guaranteed because the quasiparticle carries the same conserved charge $q = 1$ as the original particles. In this way, we see that, in the *presence* of backflow, the change in quasiparticle energy

$$\delta\epsilon_{\mathbf{p}\sigma} = -\frac{\mathbf{p}}{m^*} \cdot \mathbf{A}_N = -A_N \frac{p_F}{m^*} \cos\theta. \tag{6.90}$$

involves the renormalized mass m^*.

Since the vector potential induces a dipolar perturbation to the Fermi surface, using the results from Section 6.4, we conclude that backflow feedback effects involve the spin-symmetric $l = 1$ Landau parameter F_1^s (Equation (6.65)):

$$\delta\epsilon_{\mathbf{p}\sigma} = \left(\frac{1}{1 + F_1^s}\right) \delta\epsilon_{\mathbf{p}\sigma}^{(0)}. \tag{6.91}$$

Inserting (6.89) and (6.90) into this relation, we obtain

$$\frac{m}{m^*} = \frac{1}{1 + F_1^s} \tag{6.92}$$

or $m^* = m(1 + F_1^s)$.

Note the following:

- The Landau mass renormalization formula relies on the conservation of particle current when the interactions are adiabatically turned on. In a crystal lattice, although crystal momentum is still conserved, particle current is not conserved and at present there is no known way of writing down an expression for $\delta\epsilon_{\mathbf{p}\sigma}^{(0)}$ and $\delta\epsilon_{\mathbf{p}\sigma}$ in terms of crystal momentum that would permit the derivation of a mass renormalization formula for electrons in a crystal.
- Since $F_1^s = \mathcal{N}^*(0)f_1^s$ involves the renormalized density of states $\mathcal{N}^*(0) = \frac{m^* p_F}{\pi^2}$, the renormalized mass m^* actually appears on both sides of (6.83). If we use (6.39) to rewrite $F_1^s = \frac{m^*}{m}N(0)f_1^s$, where $N(0) = \frac{mp_F}{\pi^2}$ is the unrenormalized density of states, we can then solve for m^* in terms of m to obtain

$$m^* = \frac{m}{1 - N(0)f_1^s}. \tag{6.93}$$

This expression predicts that $m^* \to \infty$ at $N(0)f_1^s = 1$, i.e. that the quasiparticle density of states and hence the specific heat coefficient will diverge if the interactions become too strong. This possibility was first anticipated by Neville Mott, who predicted that, in presence of large interactions, fermions will localize, a phenomonon now called a *Mott transition*.

There are numerous examples of heavy-electron systems which lie close to such a local-ization transition, in which $m_e^*/m_e \gg 1$. Quasiparticle masses in excess of $1000m_e$ have been observed via specific heat measurements. In practice, the transition where the mass diverges is usually associated with the development of some other sort of order, such as antiferromagnetism or solidification. Since the phase transition occurs at zero temperature in the absence of thermal fluctuations, it is an example of a quantum phase transition. Such mass divergences have been observed in a variety of different contexts in charged electron systems, but they have also been observed as a second-order quantum phase transition, at the Mott localization transition which accompanies the solidification of two-dimensional liquid ^{3}He [9].

6.4.3 Quasiparticle scattering amplitudes

In Section 6.3 we introduced the quasiparticle interactions $f_{\mathbf{p}\sigma,\mathbf{p}'\sigma'}$ as the variation of the quasiparticle energy $\epsilon_{\mathbf{p}\sigma}$ with respect to changes in the quasiparticle occupancy $\delta n_{\mathbf{p}'\sigma'}$, under the condition that the rest of the Fermi sea stays in its ground state:

$$f_{\mathbf{p}\sigma,\mathbf{p}'\sigma'} = \left.\frac{\delta\epsilon_{\mathbf{p}\sigma}}{\delta n_{\mathbf{p}'\sigma'}}\right|_{n_{\mathbf{p}''\sigma''}} = \frac{1}{\mathcal{N}^*(0)}\left[F^s(\hat{\mathbf{p}}\cdot\hat{\mathbf{p}}') + \sigma\sigma' F^a(\hat{\mathbf{p}}\cdot\hat{\mathbf{p}}')\right]. \tag{6.94}$$

The quantity $f_{\mathbf{p}\sigma,\mathbf{p}'\sigma'}$ can be regarded as a bare forward scattering amplitude between the quasiparticles. It proves very useful to define the corresponding quantities when the Fermi sea is allowed to respond to the original change in quasiparticle occupancies, as follows:

$$a_{\mathbf{p}\sigma,\mathbf{p}'\sigma'} = \frac{\delta\epsilon_{\mathbf{p}\sigma}}{\delta n_{\mathbf{p}'\sigma'}} = \frac{1}{\mathcal{N}^*(0)}\left[A^s(\hat{\mathbf{p}}\cdot\hat{\mathbf{p}}') + \sigma\sigma' A^a(\hat{\mathbf{p}}\cdot\hat{\mathbf{p}}')\right]. \tag{6.95}$$

Microscopically, the quantities $a_{\mathbf{p}\sigma\mathbf{p}'\sigma'}$ correspond to the t-matrix for forward scattering of the quasiparticles. These amplitudes can decoupled in precisely the same way as the Landau interaction (6.73):

$$A^\alpha(\cos\theta) = \sum_l (2l+1)A_l^\alpha P_l(\cos\theta)$$

$$= 4\pi\sum_{l,m} A_l^\alpha Y_{lm}(\hat{\mathbf{p}})Y_{lm}^*(\hat{\mathbf{p}}'). \qquad (\alpha = s, a) \tag{6.96}$$

These two sets of parameters are also governed by the feedback effects of interactions:

$$A_l^\alpha = \frac{F_l^\alpha}{1 + F_l^\alpha} \qquad (\alpha = s, a). \tag{6.97}$$

The derivation of this relation closely follows the derivation of relations (6.65) and (6.67); we now repeat the derivation by solving the *Bethe–Salpeter* integral equation that links the scattering amplitudes. The change in the quasiparticle energy is

$$\delta\epsilon_{\mathbf{p}\sigma} = f_{\mathbf{p}\sigma,\mathbf{p}'\sigma'}\delta n_{\mathbf{p}'\sigma'} + \sum_{\mathbf{p}''\sigma''\neq(\mathbf{p}',\sigma')} f_{\mathbf{p}\sigma,\mathbf{p}''\sigma''}\delta n_{\mathbf{p}''\sigma''}, \tag{6.98}$$

where the second term is the induced polarization of the Fermi surface (6.59), $\delta n_{\mathbf{p}''\sigma'} = -\delta(\epsilon_{\mathbf{p}''}^{(0)})\delta\epsilon_{\mathbf{p}''\sigma'}$, so that

$$\delta\epsilon_{\mathbf{p}\sigma} = f_{\mathbf{p}\sigma,\mathbf{p}'\sigma'}\delta n_{\mathbf{p}'\sigma'} - \sum_{\mathbf{p}''\sigma''} f_{\mathbf{p}\sigma,\mathbf{p}''\sigma''}\delta(\epsilon_{\mathbf{p}''}^{(0)})\delta\epsilon_{\mathbf{p}''\sigma'}. \tag{6.99}$$

Substituting $\delta\epsilon_{\mathbf{p}\sigma} = a_{\mathbf{p}\sigma\mathbf{p}'\sigma'}\delta n_{\mathbf{p}'\sigma'}$, then dividing through by $\delta n_{\mathbf{p}'\sigma'}$, we obtain

$$a_{\mathbf{p}\sigma\mathbf{p}\sigma'} = f_{\mathbf{p}\sigma,\mathbf{p}'\sigma'} - \sum_{\mathbf{p}''\sigma''} f_{\mathbf{p}\sigma,\mathbf{p}''\sigma''}\delta(\epsilon_{\mathbf{p}''}^{(0)})a_{\mathbf{p}''\sigma'\mathbf{p}'\sigma'}. \tag{6.100}$$

This integral equation for the scattering amplitudes is a form of the Bethe–Satpeter equation relating the bare scattering amplitude f to the t-matrix described by a.

Now near the Fermi surface, we can decompose the scattering amplitudes using (6.94) and (6.95), while replacing the momentum summation by an angular integral, $\sum_{\mathbf{p}''} \rightarrow \frac{1}{2}\mathcal{N}^*(0)\int d\epsilon'' \int \frac{d\Omega_{\hat{\mathbf{p}}''}}{4\pi}$, so that this equation becomes

$$A^\alpha(\hat{\mathbf{p}}\cdot\hat{\mathbf{p}}') = F^\alpha(\hat{\mathbf{p}}\cdot\hat{\mathbf{p}}') - \int \frac{d\Omega_{\hat{\mathbf{p}}''}}{4\pi} F^\alpha(\hat{\mathbf{p}}\cdot\hat{\mathbf{p}}'')A^\alpha(\hat{\mathbf{p}}''\cdot\hat{\mathbf{p}}'). \tag{6.101}$$

If we decompose F and T in terms of spherical harmonics using (6.73) and (6.96) in the second term, we obtain

$$\int \frac{d\Omega_{\hat{\mathbf{p}}''}}{4\pi} F^\alpha(\hat{\mathbf{p}}\cdot\hat{\mathbf{p}}'')A^\alpha(\hat{\mathbf{p}}''\cdot\hat{\mathbf{p}}') = (4\pi)^2 \sum_{lm,l'm'} F_l^\alpha A_{l'}^\alpha Y_{lm}(\hat{\mathbf{p}})\overbrace{\int \frac{d\Omega_{\hat{\mathbf{p}}''}}{4\pi} Y_{lm}^*(\hat{\mathbf{p}}'')Y_{l'm'}(\hat{\mathbf{p}}'')}^{\delta_{ll'}\delta_{mm'}/(4\pi)} Y_{l'm'}^*(\hat{\mathbf{p}}')$$

$$= (4\pi) \sum_{lm} F_l^\alpha A_l^\alpha Y_{lm}(\hat{\mathbf{p}})Y_{lm}^*(\hat{\mathbf{p}}')$$

$$= \sum_l (2l+1)F_l^\alpha A_l^\alpha P_l(\hat{\mathbf{p}}\cdot\hat{\mathbf{p}}'). \tag{6.102}$$

Extracting coefficients of the Legendre polynomials in (6.101) then gives $A_l^\alpha = F_l^\alpha - F_l^\alpha A_l^\alpha$, from which the result

$$A_l^\alpha = \frac{F_l^\alpha}{1 + F_l^\alpha} \qquad (\alpha = s, a) \tag{6.103}$$

follows. The quasiparticle processes described by these scattering amplitudes involve no momentum transfer between the quasiparticles. Geometrically, scattering processes in which $\mathbf{q} = 0$ correspond to a situation where the momenta of incoming and outgoing quasiparticles lie in the same plane. Scattering processes which involve situations where the plane defined by the outgoing momenta is tipped through an angle ϕ with respect to the incoming momenta, as shown in Figure 6.5, involve a finite momentum transfer $q = 2p_F|\sin\theta/2\sin\phi/2|$. Provided this momentum transfer is very small compared with the Fermi momentum, i.e. $\phi << 1$, then one can extend the t-matrix equation as follows:

$$A_l^\alpha(\mathbf{q}) = \frac{F_l^\alpha(\mathbf{q})}{1 + F_l^\alpha(\mathbf{q})} \qquad (q << p_F). \tag{6.104}$$

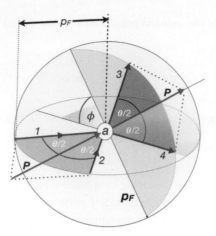

Fig. 6.5 Showing the geometry associated with quasiparticle scattering $1 + 2 \rightarrow 3 + 4$. The momentum transfered in this process is $q = |\mathbf{p}_4 - \mathbf{p}_1| = 2p_F \sin\theta/2 \sin\phi/2$. $\mathbf{P} = \mathbf{p}_1 + \mathbf{p}_2$ is the total incoming momentum. The Landau parameters determine forward scattering processes in which $\phi = 0$.

It is important to realize, however, that Landau Fermi-liquid theory is only really reliable for those processes where ϕ is small.

6.5 Collective modes

The most common collective mode of a fluid or a gas is *sound*. Conventional sound results from collisions among particles which redistribute momentum within the fluid. As such, sound is a low-frequency phenomenon that operates at frequencies much smaller than the typical quasiparticule scattering rate τ^{-1}, i.e. $\omega << \tau^{-1}$ or $\omega\tau << 1$. One of the startling predictions of Landau Fermi-liquid theory is the existence of a collionless collective mode that operates at high frequencies, $\omega\tau >> 1$: *zero sound*. Zero sound is associated with collective oscillations of the Fermi surface and does not involve collisions. Whereas conventional sound travels at a speed below the Fermi velocity, zero sound is "supersonic," traveling at speeds in excess of the Fermi velocity. Historically, the observation of zero sound in liquid ^{3}He clinched Landau Fermi-liquid theory, firmly establishing it as a foundation of fermionic many-body physics.

Let us now contrast zero sound and "first sound." Conventional sound is associated with oscillations in the density of a fluid, and hydrodynamics tells us that

$$u_1^2 = \frac{\kappa}{\rho} = \frac{\kappa}{mn}, \tag{6.105}$$

where $\rho = mn$ is the density of the fluid and $\kappa = -V\frac{\partial P}{\partial V}$ is the bulk modulus. From our previous discussion, $\kappa = \frac{n^2}{\chi_c}$ and $\chi_c = \mathcal{N}^*(0)/(1 + F_0^s)$, so the velocity of first sound in a Fermi liquid is given by

$$u_1^2 = \frac{n}{m\chi_c} = \frac{n}{m\mathcal{N}^*(0)}(1 + F_0^s). \tag{6.106}$$

Replacing $n = \frac{p_F^3}{3\pi^2}$, $\mathcal{N}^*(0) = \frac{m^*p_F}{\pi^2}$, and $m = m^*/(1 + F_1^s)$ we obtain

$$u_1^2 = \frac{v_F^2}{3}(1 + F_0^s)(1 + F_1^s). \tag{6.107}$$

In the non-interacting limit, $u_1 = v_F/\sqrt{3}$ is smaller than the Fermi velocity.

To understand zero sound we need to consider variations in the quasiparticle distribution function $n_{\mathbf{p}}(\mathbf{x}, t)$. Provided that the characteristic frequency ω is much smaller than the Fermi energy, $\omega \ll \epsilon_F$, and the characteristic wavevector is much smaller than the Fermi wavevector, $q \ll k_F$, fluctuations in the quasiparticle occupancy can be treated semiclassically, which leads to a Boltzmann equation,

$$\frac{Dn_{\mathbf{p}\sigma}}{Dt} = I[\{n_{\mathbf{p}\sigma}\}], \tag{6.108}$$

where

$$\frac{Dn_{\mathbf{p}\sigma}}{Dt} = \frac{\partial n_{\mathbf{p}\sigma}}{\partial t} + \dot{\mathbf{x}} \cdot \nabla_{\mathbf{x}} n_{\mathbf{p}\sigma} + \dot{\mathbf{p}} \cdot \nabla_{\mathbf{p}} n_{\mathbf{p}\sigma} \tag{6.109}$$

is the total rate of change of the quasiparticle occupancy $n_{\mathbf{p}\sigma}(\mathbf{x}, t)$, taking into account the movement of quasiparticles through phase space. I is the collision rate. In a semiclassical treatment, the rate of change of momentum and position are determined from Hamilton's equations, $\dot{\mathbf{p}} = -\nabla_{\mathbf{x}}\epsilon_{\mathbf{p}}$ and $\dot{\mathbf{x}} = \nabla_{\mathbf{p}}\epsilon_{\mathbf{p}}$, so that

$$\frac{Dn_{\mathbf{p}\sigma}}{Dt} = \frac{\partial n_{\mathbf{p}\sigma}}{\partial t} + \nabla_{\mathbf{p}}\epsilon_{\mathbf{p}} \cdot \nabla_{\mathbf{x}} n_{\mathbf{p}\sigma} - \nabla_{\mathbf{x}}\epsilon_{\mathbf{p}\sigma} \cdot \nabla_{\mathbf{p}} n_{\mathbf{p}\sigma}. \tag{6.110}$$

We now consider small fluctuations of the Fermi surface defined by

$$n_{\mathbf{p}}(\mathbf{x}, t) = f(\epsilon_{\mathbf{p}}^{(0)}) + e^{i\mathbf{q}\cdot\mathbf{x}-i\omega t}\alpha_{\mathbf{p}\sigma}, \tag{6.111}$$

where $\alpha_{\mathbf{p}\sigma}$ is the amplitude of the fluctuations. The terms contributing to the total rate of change $Dn_{\mathbf{p}\sigma}/Dt$ are of order $O(\omega\delta n)$, whereas the collision term $I[n] \sim O(\tau^{-1}\delta n)$ is of the order of the collision rate τ^{-1}. In the high-frequency limit, $\omega\tau \gg 1$, the collision terms can then be neglected, leading to the collisionless Boltzmann equation:

$$\frac{\partial n_{\mathbf{p}\sigma}}{\partial t} + \nabla_{\mathbf{p}}\epsilon_{\mathbf{p}} \cdot \nabla_{\mathbf{x}} n_{\mathbf{p}\sigma} - \nabla_{\mathbf{x}}\epsilon_{\mathbf{p}\sigma} \cdot \nabla_{\mathbf{p}} n_{\mathbf{p}\sigma} = 0. \tag{6.112}$$

For small periodic oscillations in the Fermi surface, the first two terms in (6.112) can be written

$$\frac{\partial n_{\mathbf{p}\sigma}}{\partial t} + \nabla_{\mathbf{p}}\epsilon_{\mathbf{p}} \cdot \nabla_{\mathbf{x}} n_{\mathbf{p}\sigma} = -i(\omega - \mathbf{v}_F \cdot \mathbf{q})\alpha_{\mathbf{p}\sigma} e^{i\mathbf{q}\cdot\mathbf{x}-i\omega t}. \tag{6.113}$$

In the last term of (6.112), the position dependence of the quasiparticle energies derives from interactions

$$\nabla_{\mathbf{x}}\epsilon_{\mathbf{p}\sigma} = \sum_{\sigma'} \int_{\mathbf{p}'} f_{\mathbf{p}\sigma,\mathbf{p}'\sigma'} \nabla_{\mathbf{x}} n_{\mathbf{p}',\sigma'}$$

$$= i\mathbf{q} e^{i\mathbf{q}\cdot\mathbf{x}-i\omega t} \sum_{\sigma'} \int_{\mathbf{p}'} f_{\mathbf{p}\sigma,\mathbf{p}'\sigma'} \alpha_{\mathbf{p}',\sigma'}. \tag{6.114}$$

Replacing $\nabla_{\mathbf{p}} n_{\mathbf{p}\sigma} = \frac{\partial f}{\partial \epsilon} \mathbf{v}_F$, the collisionless Boltzmann equation becomes

$$(\omega - \mathbf{v}_F \cdot \mathbf{q})\alpha_{\mathbf{p}\sigma} + v_F \cdot \mathbf{q} \left(-\frac{df}{d\epsilon}\right) \sum_{\sigma'} \int_{\mathbf{p}'} f_{\mathbf{p}\sigma,\mathbf{p}'\sigma'}\alpha_{\mathbf{p}',\sigma'} = 0. \qquad (6.115)$$

For a mode propagating at speed u, $\omega = uq$. If we express $\mathbf{v}_F \cdot \mathbf{q} = v_F q \cos\theta_{\mathbf{p}}$, and write the mode velocity as a factor s times the Fermi velocity, $u = sv_F$, then this becomes

$$(s - \cos\theta_{\mathbf{p}})\alpha_{\mathbf{p}\sigma} + \cos\theta_{\mathbf{p}} \left(-\frac{df}{d\epsilon}\right) \sum_{\sigma'} \int_{\mathbf{p}'} f_{\mathbf{p}\sigma,\mathbf{p}'\sigma'}\alpha_{\mathbf{p}',\sigma'} = 0. \qquad (6.116)$$

We see that the fluctuations in occupancy associated with a zero sound mode, $\alpha_{\mathbf{p}\sigma} = \eta_\sigma(\hat{\mathbf{p}})\left(-\frac{df}{d\epsilon}\right)$, are proportional to the energy derivative of the Fermi function, and thus confined to within an energy scale T of the Fermi surface. The function $\eta_\sigma(\hat{\mathbf{p}})$ describes the distribution around the Fermi surface, and this function satisfies the self-consistent relation

$$\eta_\sigma(\mathbf{p}) = \frac{\cos\theta_{\mathbf{p}}}{2(s - \cos\theta_{\mathbf{p}})} \sum_{\sigma'} \int \frac{d\Omega_{\mathbf{p}'}}{4\pi} F_{\mathbf{p}\sigma,\mathbf{p}'\sigma'} \eta_\sigma(\hat{\mathbf{p}}'). \qquad (6.117)$$

For spin-independent zero sound waves, the right-hand side only involves F^s and can be written

$$\eta(\mathbf{p}) = \frac{\cos\theta_{\mathbf{p}}}{(s - \cos\theta_{\mathbf{p}})} \int \frac{d\Omega_{\mathbf{p}'}}{4\pi} F^s_{\mathbf{p},\mathbf{p}'} \eta(\hat{\mathbf{p}}'). \qquad (6.118)$$

To illustrate the solution of this equation, consider the case where the interaction is entirely isotropic and spin-independent, so that the only non-vanishing Landau parameter is F^s_0. In this case, the angular function is spin-independent and given by

$$\eta(\theta) = A\frac{\cos(\theta)}{s - \cos(\theta)}, \qquad (6.119)$$

where A is a constant. Substituting this form into the integral equation, we obtain the following formula for $s = u/v_F$:

$$A = \int_{-1}^{1} \frac{d\cos\theta}{2} \frac{\cos\theta}{s - \cos\theta} AF^s_0 = AF^s_0 \left[-1 + \frac{s}{2}\ln\left(\frac{s+1}{s-1}\right)\right], \qquad (6.120)$$

so that

$$\frac{s}{2}\ln\left(\frac{s+1}{s-1}\right) - 1 = \frac{1}{F^s_0}. \qquad (6.121)$$

For large s, the function on the left-hand side vanishes asymptotically as $1/(3s^2)$, and since the right-hand side vanishes at large interaction F^s_0, it follows that for large interaction strength the zero sound velocity is much greater than v_F:

$$u = sv_F = v_F\sqrt{\frac{F^s_0}{3}} \qquad (F^s_0 \gg 1). \qquad (6.122)$$

For small interaction strength, $s \to 1$ and the zero sound velocity approaches the Fermi velocity.

Experimentally, zero sound has been observed by a variety of methods. Low-frequency zero sound couples directly to vibrations at the wall of the fluid, and can be detected directly as a propagating density mode. Zero sound can also be probed at higher frequencies using neutron and X-ray scattering. Neutron scattering experiments find that, at high frequencies, the zero sound mode re-enters the particle–hole continuum where, as a damped excitation, it acquires a *roton* minimum similar to collective modes in bosonic ^{4}He.

6.6 Charged Fermi liquids: Landau–Silin theory

One of the most useful extensions of Landau Fermi-liquid theory is to charged Fermi liquids, which underpins our understanding of electrons in metals. Charged Fermi liquids present an additional challenge because of the long-range Coulomb interaction. The extension of Landau Fermi-liquid theory to incorporate the long-range part of the Coulomb interaction was originally made by Silin [10, 11]. In neutral Fermi liquids, the existence of well-defined Landau interaction parameters depends on a short-range interaction $V(q)$ with a well-defined zero momentum limit $\mathbf{q} \to \mathbf{0}$ (see also Example 6.2). Yet the long-range Coulomb interaction $V(q) = \frac{e^2}{\epsilon_0 q^2}$ is singular as $\mathbf{q} \to 0$. Charged quasiparticles act as sources for an electric potential which satisfies Gauss' law

$$\nabla^2 \phi_P = \frac{e}{\epsilon_0} \sum_{\mathbf{p}} \delta n_{\mathbf{p}\sigma}(\mathbf{x}). \qquad \text{polarization field} \qquad (6.123)$$

The field $\mathbf{E}_P = -\nabla \phi_P$ that this produces polarizes the surrounding quasiparticle fluid to form a *polarization cloud* around the quasiparticle which screens its charge, so that the net interaction between screened quasiparticles has a finite range. Nevertheless, this poses a subtle technical problem, for screening requires a collective quasiparticle response, yet the Fermi liquid interactions are determined by variation of the quasiparticle energy in response to a change in quasiparticle occupancy against an otherwise frozen (and hence unpolarized) Fermi sea:

$$f_{\mathbf{p}\sigma,\mathbf{p}'\sigma'}(\mathbf{x}, \mathbf{x}') = \frac{\delta \epsilon_{\mathbf{p}\sigma}(\mathbf{x})}{\delta n_{\mathbf{p}'\sigma'}(\mathbf{x}')}\bigg|_{\delta n_{\mathbf{p}''\sigma''}=0}. \qquad (6.124)$$

In a frozen Fermi sea, the quasiparticle interaction must then be unscreened at large distances, forcing it to be singular as $\mathbf{q} \to 0$.

The solution to this problem was proposed by Silin in 1957. Silin proposed splitting the electric potential ϕ produced by charged particles into two parts: the long-range classical polarization field ϕ_P considered above, and a short-range, fluctuating quantum component,

$$\phi(\mathbf{x}) = \phi_P(\mathbf{x}) + \delta \phi_Q(\mathbf{x}). \qquad (6.125)$$

The quantum component is driven by the virtual creation of electron–hole pairs around a charged particle. These processes involve momentum transfer of the order of the Fermi momentum p_F, and are hence localized to within a short distance of the order of the quasiparticle de Broglie wavelength $\lambda \sim h/p_F$ around the quasiparticle. Silin proposed that these

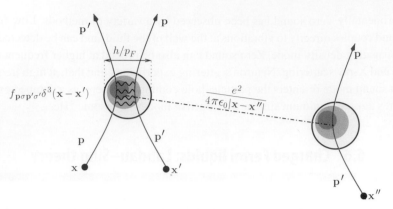

Fig. 6.6 Interactions in a charged Fermi liquid. The short-range part of the interaction results from quantum fluctuations of the polarization field (see Section 7.7.2). The long-range component of the interaction derives from the induced polarization field around the quasiparticle.

virtual fluctuations in the electric potential introduce a second, short-range component to the quasiparticle interactions. Silin's theory isolates the polarization field as a separate term, so that the quasiparticle energy is written

$$\epsilon_{\mathbf{p}\sigma}(\mathbf{x}) = \epsilon_{\mathbf{p}}^{(0)} + e\phi_P(\mathbf{x}) + \sum_{\mathbf{p}'\sigma'} f_{\mathbf{p}\sigma,\mathbf{p}'\sigma'}\delta n_{\mathbf{p}'\sigma'}(\mathbf{x}). \tag{6.126}$$

In momentum space, the change in the quasiparticle energy is given by

$$\delta\epsilon_{\mathbf{p}\sigma}(\mathbf{q}) = e\phi_P(\mathbf{q}) + \sum_{\mathbf{p}'\sigma'} f_{\mathbf{p}\sigma,\mathbf{p}'\sigma'}\delta n_{\mathbf{p}'\sigma'}(\mathbf{q}). \tag{6.127}$$

However, Gauss' law implies that $e\phi(\mathbf{q}) = \frac{e^2}{\epsilon_0 q^2}\sum_{\mathbf{p}'\sigma'}\delta n_{\mathbf{p}'\sigma'}(\mathbf{q})$. Combining these results, we see that

$$\delta\epsilon_{\mathbf{p}\sigma}(\mathbf{q}) = \sum_{\mathbf{p}'\sigma'}\left(\frac{e^2}{\epsilon_0 q^2} + f_{\mathbf{p}\sigma,\mathbf{p}'\sigma'}\right)\delta n_{\mathbf{p}'\sigma'}(\mathbf{q}). \tag{6.128}$$

In other words, the effective interaction takes the form (see Figure 6.6)

$$f_{\mathbf{p}\sigma,\mathbf{p}'\sigma'}(\mathbf{q}) = \overbrace{\frac{e^2}{\epsilon_0 q^2}}^{\substack{\text{Long-range interaction}\\\text{from polarization field}}} + \underbrace{f_{\mathbf{p}\sigma,\mathbf{p}'\sigma'}}_{\text{short-range residual interaction}}. \tag{6.129}$$

There are a number of points to emphasize about Silin's theory:

- When the interaction is decomposed in terms of (q-dependent) Landau parameters, the singular interaction only enters into the $l = 0$, spin-symmetric component; all the other components are determined by $f_{\mathbf{p}\sigma\mathbf{p}'\sigma'}$, so that

$$F_l^s(\mathbf{q}) = \frac{e^2 \mathcal{N}^*(0)}{\epsilon_0 q^2} \delta_{l0} + F_l^s \tag{6.130}$$

and $F_l^a = \tilde{F}_l^a$.

- The Landau–Silin theory can be derived in a Feynman diagram formalism. In such an approach, the short-range part of the interaction is associated with multiple scattering off the Coulomb interaction.
- The short-range interaction $f_{\mathbf{pp'}}$ is a quantum phenomenon, *distinct from classical Thomas–Fermi screening of the quasiparticle charge*, which results from the polarizing effects of the long-range $1/q^2$ component of the interaction.

To illustrate this last point, let us calculate the linear response of the quasiparticle density $\delta\rho(\mathbf{q}) = \chi_c(\mathbf{q})\delta\mu(\mathbf{q})$ to a slowly varying chemical potential $\delta\mu(\mathbf{x}) = \delta\mu(\mathbf{q})e^{i\mathbf{q}\cdot\mathbf{x}}$ where $\chi_c(\mathbf{q})$ is the charge susceptibility. In a neutral Fermi liquid, for $q \ll p_F$ the long-wavelength density response is determined by $\chi_c(\mathbf{q}) \approx \chi_n$, where

$$\chi_n = \frac{\mathcal{N}^*(0)}{1 + F_0^s}, \tag{6.131}$$

as found in (6.80). In the charged Fermi liquid, we replace $F_0^s \to F_0^s(\mathbf{q}) = \frac{e^2 \mathcal{N}^*(0)}{\epsilon_0 q^2} + F_0^s$, which gives

$$\chi_c(\mathbf{q}) = \frac{\mathcal{N}^*(0)}{1 + (\frac{e^2 \mathcal{N}^*(0)}{\epsilon_0 q^2} + F_0^s)} = \frac{\chi_n}{1 + \frac{\kappa^2}{q^2}} = \frac{\chi_n}{1 + \frac{e^2}{\epsilon_0 q^2}\chi_n}, \tag{6.132}$$

where

$$\kappa^2 = \frac{e^2}{\epsilon_0}\chi_n = \frac{e^2}{\epsilon_0}\left(\frac{\mathcal{N}^*(0)}{1 + F_0^s}\right) \tag{6.133}$$

defines a Thomas–Fermi screening length $l_{TF} = \kappa^{-1}$. At large momenta, $q \gg \kappa$ (distances $x \ll l_{TF}$), the response is exactly that of the neutral fluid, but at small momenta, $q \ll \kappa$ (distances $x \gg l_{TF}$), the charge density response is heavily suppressed.

Historically, the Landau–Silin approach changed the way of thinking about metal physics. In early many-body theory of the electron gas, the singular nature of the Coulomb interaction was a primary focus, and many-body physics in the 1950s was in essence the study of quantum plasmas. With Landau–Silin theory, the long-range Coulomb interaction becomes of secondary interest, because this component of the interaction is unrenormalized and can be added in later as an afterthought. This is a major change in philosophy, which shifts our interest to the short-range components of the quasiparticle interactions. In essence, the Landau–Silin observation liberates us from the singular aspects of the Coulomb interaction, and enables us to treat the physics of strongly correlated electrons as a close companion to other neutral Fermi systems.

Example 6.4 Calculate the scattering t-matrix in Landau–Silin theory to display the screening effect of the long-range interaction.

Solution

If we introduce a small modulation in the quaisparticle occupancy at momentum $\mathbf{p}'$, while "freezing" the rest of the Fermi sea, then the quasiparticle energies will acquire a modulated component

$$\delta\epsilon_{\mathbf{p}}^{(0)}(\mathbf{q}) = f_{\mathbf{p},\mathbf{p}'}^{s}(\mathbf{q})\delta n_{\mathbf{p}'}(\mathbf{q}), \qquad (6.134)$$

where $f_{\mathbf{p},\mathbf{p}'}^{s} = \left(\frac{e^2}{\epsilon_0 q^2} + f_{\mathbf{p},\mathbf{p}'}^{s}\right)$ is the spin-symmetric part of the interaction. (For convenience we temporarily drop the spin indices from the subscripts.) If we now allow the quasiparticle sea to polarize, the feedback effect will renormalize the the quasiparticle energies as follows:

$$\delta\epsilon_{\mathbf{p}}(\mathbf{q}) = a_{\mathbf{p},\mathbf{p}'}^{s}(\mathbf{q})\delta n_{\mathbf{p}'}(\mathbf{q}), \qquad (6.135)$$

where a^s is a screened quasiparticle interaction to be calculated. At low momenta $\mathbf{q}$ in an isotropic system, both f and a can be expanded in spherical harmonics, as in (6.73), by writing

$$f_{\mathbf{p},\mathbf{p}'}^{s}(\mathbf{q}) = \frac{4\pi}{\mathcal{N}^*(0)}\sum_{l} F_l^s(\mathbf{q})Y_{lm}(\hat{\mathbf{p}})Y_{lm}^*(\hat{\mathbf{p}}')$$

$$a_{\mathbf{p},\mathbf{p}'}^{s}(\mathbf{q}) = \frac{4\pi}{\mathcal{N}^*(0)}\sum_{l} A_l^s(\mathbf{q})Y_{lm}(\hat{\mathbf{p}})Y_{lm}^*(\hat{\mathbf{p}}'). \qquad (6.136)$$

For very small $\mathbf{q}$, we can solve for the relationship between T_l^s and F_l^s using the methods of Section 6.4.3, which gives

$$A_l^s(\mathbf{q}) = \frac{F_l^s(\mathbf{q})}{1 + F_l^s(\mathbf{q})}. \qquad (6.137)$$

But from (6.130) the $\mathbf{q}$ dependence only enters the $l = 0$ component of the spin-symmetric scattering, where $F_0(\mathbf{q}) = \frac{e^2\mathcal{N}^*(0)}{\epsilon_0 q^2} + F_0$, so that

$$A_0^s(\mathbf{q}) = \frac{\frac{e^2\mathcal{N}^*(0)}{\epsilon_0 q^2} + F_0^s}{1 + \frac{e^2\mathcal{N}^*(0)}{\epsilon_0 q^2} + F_0^s} = \frac{\kappa^2}{(\kappa^2 + q^2)}\left(\frac{1}{1 + F_0^s}\right) + A_0^{s(neutral)}, \qquad (6.138)$$

where $A_0^{s(neutral)} = \frac{F_0^s}{1+F_0^s}$ is the $l = 0$ scattering t-matrix of the equivalent neutral Fermi liquid and $\kappa^2 = \frac{e^2\mathcal{N}^*(0)}{\epsilon_0(1+F_0^s)}$. Since all other components are unchanged by the long-range Coulomb interaction it follows that the interaction t-matrix of the charged Fermi liquid is a sum of the original neutral interaction plus a screened Coulomb correction,

$$a_{\mathbf{p}\sigma,\mathbf{p}'\sigma}(\mathbf{q}) = \frac{1}{(1 + F_0^s)^2}\frac{e^2}{\epsilon_0(q^2 + \kappa^2)} + a_{\mathbf{p}\sigma,\mathbf{p}'\sigma'}^{(neutral)}. \qquad (6.139)$$

Note how the residual Coulomb part of the t-matrix is heavily suppressed when F_0^s becomes large.

6.7 Inelastic quasiparticle scattering

6.7.1 Heuristic derivation

In this section we show how the Pauli exclusion principle limits the phase space for scattering of quasiparticles in a Landau Fermi-liquid, giving rise to a scattering rate with a quadratic dependence on excitation energy and temperature,

$$\frac{1}{\tau} \propto [\epsilon^2 + \pi^2 T^2].$$ (6.140)

The dominant decay mode of a quasiparticle is into three quasiparticles. There are also higher-order processes that involve a quasiparticle decaying into a quasiparticle, and n particle–hole pairs:

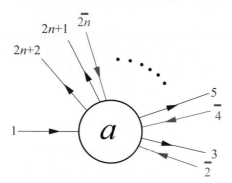

We'll see that the phase space for these higher-order decay processes vanishes with a high power of the energy ($\propto \epsilon^{2n+1}$), allowing us to neglect them relative to the leading process at low temperature and energy. For our discussion, we will denote a hole in the quasiparticle state j as $\bar{j}$, denoting the quasihole energy by $\tilde{\epsilon}_j = -\epsilon_j > 0$. By Fermi's golden rule, the rate of decay into n particle–hole pairs is

$$\Gamma_{2n+1}(\epsilon_1) \sim \frac{2\pi}{\hbar} \sum_{\substack{2,3...2n+2 \\ \tilde{\epsilon}_2, \epsilon_3...>0}} |a(1;\bar{2},3,...2n{+}2)|^2 \delta[\epsilon_1 - (\tilde{\epsilon}_2 + \epsilon_3 + \tilde{\epsilon}_4 + \cdots + \epsilon_{2n+2})],$$ (6.141)

where $a(1;\bar{2},3,...2n+1)$ is the amplitude for the scattering process, $\tilde{\epsilon}_2, \tilde{\epsilon}_4, ... \tilde{\epsilon}_{2n}$ denote the energies of the outgoing quasiholes, and $\epsilon_3, \epsilon_5, ... \epsilon_{2n+1}, \epsilon_{2n+2}$ denote the energies of the outgoing quasiparticles. The energies of the final-state quasiparticles and holes must all be positive, while also summing up to give the initial energy. When the incoming particle is close to the Fermi energy, ϵ and the all final-state energies $\epsilon_1 > \epsilon_i > 0$ must also lie close to the Fermi energy, so we can replace $|a|^2$ by an appropriate Fermi surface average:

$$\langle |a_{2n+1}|^2 \rangle = \sum_{2,3,...2n+2} |a(1;\bar{2},3,...2n+2)|^2 \delta(\tilde{\epsilon}_2) \cdots \delta(\epsilon_{2n+1}),$$ (6.142)

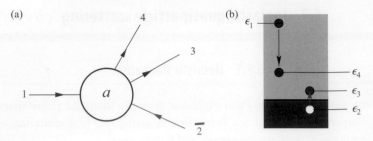

Fig. 6.7   Decay of a quasiparticle into two quasiparticles and a quasihole: (a) scattering process; (b) energies of final states.

to obtain [5]

$$\Gamma_{2n+1}(\epsilon) \sim \frac{2\pi}{\hbar} \langle |a_{2n+1}|^2 \rangle$$

$$\times \int_0^\infty d\tilde{\epsilon}_2 \cdots d\epsilon_{2n+1} \delta[\epsilon - (\tilde{\epsilon}_2 + \cdots + \epsilon_{2n+1})] \propto \frac{\epsilon^{2n}}{(2n)!}. \qquad (6.143)$$

In this way,[6] the phase space for decay into $2n + 1$ quasiparticles vanishes as ϵ_1^{2n}. This means that, near the Fermi surface, quasiparticle decay is dominated by the decay into two quasiparticles and a quasihole, denoted by $\mathbf{1} \rightarrow \bar{\mathbf{2}} + \mathbf{3} + \mathbf{4}$, as illustrated in Figure 6.7.

The decay rate for this process is given by

$$\Gamma(\epsilon) = \frac{2\pi}{\hbar} \langle |a_3|^2 \rangle \frac{\epsilon^2}{2}. \qquad (6.144)$$

On dimensional grounds, we expect the averaged squared matrix element to scale as $\langle |a_3|^2 \rangle \sim \frac{w^2}{\epsilon_F}$, where w is a dimensionless measure of the strength of the scattering, so that $\Gamma \sim \frac{2\pi}{\hbar} \frac{\epsilon^2}{\epsilon_F}$.

6.7.2 Detailed calculation of three-body decay process

We now present a more detailed calculation of quasiparticle decay, deriving a result that was first obtained by Alexei Abrikosov and Isaak Khalatnikov in 1957 [3]. The amplitude to produce an outgoing hole in state $\bar{\mathbf{2}}$ is equal to the amplitude to absorb an incoming particle in state $\mathbf{2}$, so we denote

$$a(\mathbf{1} \rightarrow \bar{\mathbf{2}} + \mathbf{3} + \mathbf{4}) = a(\mathbf{1} + \mathbf{2} \rightarrow \mathbf{3} + \mathbf{4}) \equiv a(\mathbf{1}, \mathbf{2}; \mathbf{3}, \mathbf{4}). \qquad (6.145)$$

---

[5] Formally this is done by inserting $1 = \prod_{i=1}^{2n+1} \int_{-\infty}^\infty d\epsilon_i \delta(\epsilon_i)$ into (6.141),

[6] This last integral can be done by regarding the ϵ_r as the differences $\epsilon_j = s_j - s_{j-1}$ between an ordered set of co-ordinates $s_{2n+1} > s_{2n} > \cdots > s_1$ where $s_0 = 0$, so that

$$\int_0^\infty \underbrace{d\epsilon_1 \cdots d\epsilon_{2n+1}}_{=ds_1 \cdots ds_{2n+1}} \delta[\epsilon - (\epsilon_1 + \epsilon_2 + \cdots + \epsilon_{2n+1})] = \int_0^\infty ds_{2n+1} \delta(\epsilon - \overbrace{s_{2n+1}}^{s_{2n+1}}) \int_0^{s_{2n+1}} ds_{2n} \cdots \int_0^{s_2} ds_1$$

$$= \frac{\epsilon^{2n}}{(2n)!}.$$

Using Fermi's golden rule, the net scattering rate into state **1** is given by

$$I[n_{\mathbf{p}}] = \frac{2\pi}{\hbar} \sum_{2,3,4} |a(\mathbf{1},\mathbf{2};\mathbf{3},\mathbf{4})|^2 \left[\overbrace{(1 - n_2)n_3 n_4(1 - n_1)}^{\bar{\mathbf{2}}+\mathbf{3}+\mathbf{4}\rightarrow\mathbf{1}} - \overbrace{n_1 n_2(1 - n_3)(1 - n_4)}^{\mathbf{1}\rightarrow\bar{\mathbf{2}}+\mathbf{3}+\mathbf{4}} \right]$$

$$\times \; (2\pi\hbar)^3 \delta^{(3)}(\mathbf{p}_1 + \mathbf{p}_2 - \mathbf{p}_3 - \mathbf{p}_4) \, \delta(\epsilon_1 + \epsilon_2 - \epsilon_3 - \epsilon_4), \qquad (6.146)$$

where $\sum_{2} \equiv \int \frac{d^3p}{(2\pi\hbar)^3}$ denotes a sum over final-state momenta, and the delta functions impose the conservation of momentum and energy, respectively. The terms inside the square brackets determine the *a priori* probabilities for the scattering process. For scattering into state **1**, the initial states must be occupied and the final state must be empty, so the *a priori* probability is $(1 - n_2)n_3 n_4 \times (1 - n_1)$, where $(1 - n_2)$ is the probability that the quasihole state $\bar{\mathbf{2}}$ is occupied and $n_3 n_4$ is the probability that **3** and **4** are occupied, while $(1 - n_1)$ is the probability that the final quasiparticle state **1** is empty. The second term in the brackets describes the scattering out of state **1**, and can be understood in a similar way.

In thermal equilibrium, the scattering rate vanishes, $I[n_{\mathbf{p}}^{(0)}] = 0$, and for small deviations from equilibrium we may expand the collision integral to linear order in $\delta n_{\mathbf{p}} = n_{\mathbf{p}} - n_{\mathbf{p}}^{(0)}$, identifying the coefficient as the quasiparticle decay rate as $I[n_{\mathbf{p}}] = -\Gamma \delta n_1 + O(\delta n_{\mathbf{p}}^2)$, where $\Gamma = -\frac{\delta I}{\delta n_1}$, or

$$\Gamma = \frac{2\pi}{\hbar} \sum_{2,3,4} |a(1,2;3,4)|^2 \left[n_2^{(0)}(1 - n_3^{(0)})(1 - n_4^{(0)}) + (1 - n_2^{(0)})n_3^{(0)} n_4^{(0)} \right]$$

$$\times \; (2\pi\hbar)^3 \delta(\epsilon_1 + \epsilon_2 - \epsilon_3 - \epsilon_4)\delta^{(3)}(\mathbf{p}_1 + \mathbf{p}_2 - \mathbf{p}_3 - \mathbf{p}_4). \qquad (6.147)$$

The occupation factors in the square brackets impose the Fermi statistics. (For convenience, we shall drop the superscript "(0)" denoting the equilibrium distribution in subsequent equations.) These terms are easiest to understand at absolute zero, where $n_{\mathbf{p}} = \theta(-\epsilon_{\mathbf{p}})$ restricts $\epsilon_{\mathbf{p}} < 0$ and $1 - n_{\mathbf{p}} = \theta(\epsilon_{\mathbf{p}})$ restricts $\epsilon_{\mathbf{p}} > 0$. The first term, $n_2(1 - n_3)(1 - n_4)$, enforces the constraint that the excitation energies $-\epsilon_2, \epsilon_3, \epsilon_4 > 0$ are all positive. (Recall that the ϵ_j refer to quasiparticle energies, so $-\epsilon_2 = \bar{\epsilon}_2$ is the excitation energy of the outgoing hole in state $\bar{\mathbf{2}}$.) At absolute zero, the second term, $(1 - n_2)n_3 n_4$, is zero unless the excitation energies are negative, and vanishes when $\epsilon_1 > 0$. Now the delta function $\delta(\epsilon_1 + \epsilon_2 - \epsilon_3 - \epsilon_4)$ enforces energy conservation, $\bar{\epsilon}_2 + \epsilon_3 + \epsilon_4 = \epsilon_1$. Together with the requirement that the scattered quasiparticle energies are positive, this term forces all three excitation energies $|\epsilon_{2,3,4}|$ to be smaller than ϵ. In this way we see that, for small ϵ, the final quasiparticle states must lie very close to the Fermi momentum.

With this understanding, at low temperatures we can replace the integrals over three-dimensional momentum by the product of an energy and an angular integral over the direction of the momenta on the Fermi surface:

$$\sum_{\mathbf{p}'} \rightarrow \frac{\mathcal{N}^*(0)}{2} \int \frac{d\Omega_{\hat{\mathbf{p}}'}}{4\pi} \times \int d\epsilon'. \qquad (6.148)$$

This factorization between the energy and momentum degrees of freedom is a hallmark of the Landau Fermi-liquid. Using it, we can factorize (6.147) into two parts:

$$\Gamma = \frac{2\pi}{\hbar} \overbrace{\langle |a_3|^2 \rangle}^{\text{angular average}} \times \overbrace{\left\langle\ n_2(1-n_3)(1-n_4) + (1-n_2)n_3n_4\ \right\rangle}^{\text{energy phase space integral}}_{\epsilon_2,\epsilon_3,\epsilon_4}, \qquad (6.149)$$

where

$$\langle |a_3|^2 \rangle = \left(\frac{\mathcal{N}^*(0)}{2} \right)^3 \int \frac{d\Omega_2 d\Omega_3 d\Omega_4}{(4\pi)^3} |a(1,2;3,4)|^2 (2\pi\hbar)^3 \delta^{(3)}[p_F(\hat{\mathbf{n}}_1 + \hat{\mathbf{n}}_2 - \hat{\mathbf{n}}_3 - \hat{\mathbf{n}}_4)] \qquad (6.150)$$

is the angular average and

$$\langle \ldots \rangle_{\epsilon_2,\epsilon_3,\epsilon_4} = \int d\epsilon_2 d\epsilon_3 d\epsilon_4 \delta(\epsilon_1 + \epsilon_2 - \epsilon_3 - \epsilon_4)[\ldots] \qquad (6.151)$$

is the energy phase space integral. At absolute zero, the argument of the phase space integral restricts the final states to have positive excitation energies, giving $\frac{\epsilon_1^2}{2}$, as obtained from (6.143) for $n = 1$. At finite temperature (see Example 6.6), thermal broadening leads to an additional quadratic temperature dependence in the phase space integral:[7]

$$\left\langle\ n_2(1-n_3)(1-n_4) + (1-n_2)n_3n_4\ \right\rangle_{\epsilon_{2,3,4}} = \frac{1}{2}\left(\epsilon_1^2 + (\pi k_B T)^2 \right). \qquad (6.152)$$

To calculate the average squared matrix element, it is convenient to first ignore the spin of the quasiparticle. To evaluate the angular integral, we need to consider the geometry of the scattering process near the Fermi surface, which is illustrated in Figure 6.5. At low temperatures, all initial and final momenta lie on the Fermi surface, $|\mathbf{p}_j| = p_F$. The total momentum in the particle–particle channel is $\mathbf{P} = \mathbf{p}_1 + \mathbf{p}_2$. Suppose the angle between $\mathbf{p}_1$ and $\mathbf{p}_2$ is θ, so that each of these momenta subtends an angle $\theta/2$ with $\mathbf{P}$, as shown in Figure 6.5; then $|\mathbf{P}| = 2p_F \sin\theta/2$. Now since the total momentum is conserved, $\mathbf{p}_3 + \mathbf{p}_4 = \mathbf{P}$ also, so that $|\mathbf{p}_3 + \mathbf{p}_4| = 2p_F \sin\theta/2$, which means that $\mathbf{p}_3$ and $\mathbf{p}_4$ also subtend an angle $\theta/2$ with $\mathbf{P}$. However, in general the planes defined by $\mathbf{p}_{1,2}$ and $\mathbf{p}_{3,4}$ are not the same, and we denote the angle between them by ϕ. In general, the scattering amplitude $a(\theta, \phi)$ will be a function of the two angles θ and ϕ. In this way, we can parameterize the scattering amplitude by $a(\theta, \phi)$.

A detailed evaluation of the angular integral $\langle |a_3|^2 \rangle$ (see Example 6.5) leads to the result

$$\langle |a_3|^2 \rangle = \frac{1}{2} \times \pi^2 \left(\frac{\mathcal{N}^*(0)\hbar}{2p_F} \right)^3 \left\langle \frac{|a(\theta,\phi)|^2}{2\cos\theta/2} \right\rangle_\Omega, \qquad (6.153)$$

where

$$\left\langle \frac{|a(\theta,\phi)|^2}{2\cos\theta/2} \right\rangle_\Omega \equiv \int \frac{d\cos\theta\, d\phi}{4\pi} \left(\frac{|a(\theta,\phi)|^2}{2\cos\theta/2} \right) \qquad (6.154)$$

denotes a weighted, normalized angular average of the scattering rate over the Fermi surface. For identical spinless particles, the final states with scattering angle ϕ and $\phi + \pi$ are

---

[7] The first term in the phase space integral corresponds to the decay $1 \to \bar{2} + 3 + 4$ of a quasiparticle, while the second term describes the regeneration of quasiparticles via the reverse process, $\bar{2} + 3 + 4 \to 1$. The classic treatment of the quasiparticle decay given by Abrikosov and Khalatnikov [3] and by Morel and Nozières [12], reproduced in Nozières and Pines [4] and in Mahan [13], only includes the first process, which introduces an additional factor $1/(1 + e^{-\beta\epsilon_1})$ into this expression.

are indistinguishable, and the prefactor of $\frac{1}{2}$ is introduced into (6.153) to take account of the overcounting that occurs when we integrate from $\phi = 0$ to $\phi = 2\pi$.

The complete scattering rate for a spinless quasiparticle is then given by

$$\Gamma = \frac{2\pi}{\hbar} \times \left\langle \frac{\frac{1}{2}|a(\theta,\phi)|^2}{2\cos\theta/2} \right\rangle_\Omega \pi^2 \left(\frac{\mathcal{N}^*(0)\hbar}{2p_F} \right)^3 \times \left(\frac{\epsilon^2 + (\pi k_B T)^2}{2} \right). \qquad (6.155)$$

Let us now consider how this answer changes when we reinstate the spin of the quasi-particles. In this case, we must sum over the two spin orientations of quasiparticle 2, corresponding to the case where the spin of 1 and 2 are either parallel ($A_{\uparrow\uparrow}$) or antiparal-lel ($A_{\uparrow\downarrow}$). When the spins of the two quasiparticles are parallel, they are indistinguishable and we must keep the factor of $\frac{1}{2}$, but when the spins are antiparallel, the particles are distinguishable and this factor is omitted. So, to take account of spin, we must replace

$$\frac{1}{2}|a(\theta,\phi)|^2 \rightarrow \frac{1}{2}|a_{\uparrow\uparrow}(\theta,\phi)|^2 + |a_{\uparrow\downarrow}(\theta,\phi)|^2 \qquad (6.156)$$

in (6.155). Following the original convention of Abrikosov and Khalatnikov [3], we denote

$$\frac{2\pi}{\hbar} \left(|a_{\uparrow\downarrow}(\theta,\phi)|^2 + \frac{1}{2}|a_{\uparrow\uparrow}(\theta,\phi)|^2 \right) = 2W(\theta,\phi). \qquad (6.157)$$

Applying these substutions to (6.155), and writing $\mathcal{N}^*(0) = m^* p_F/(\pi^2 \hbar^3)$, we obtain

$$\Gamma = \frac{(m^*)^3}{8\pi^4 \hbar^6} \left\langle \frac{W(\theta,\phi)}{2\cos\theta/2} \right\rangle_\Omega \times (\epsilon^2 + (\pi k_B T)^2). \qquad (6.158)$$

This result was originally obtained by Abrikosov and Khalatnikov in 1957 [3]. An alter-native way to rewrite this expression is to identify the normalized scattering amplitudes $\mathcal{N}^*(0)a_{\alpha\beta}(\theta,\phi) = A_{\alpha\beta}(\theta,\phi) \equiv A_{\alpha\beta}(\mathbf{q})$ with the dimensionless t-matrix introduced in Section 6.4.3. From this we see that the average matrix elements can be written in terms of a dimensionless parameter w^2:

$$w^2 = \left\langle \frac{|A_{\uparrow\downarrow}(\theta,\phi)|^2 + \frac{1}{2}|A_{\uparrow\uparrow}(\theta,\phi)|^2}{2\cos\theta/2} \right\rangle_\Omega. \qquad (6.159)$$

In many strongly interacting systems, w is close to unity. Using this notation, the scattering rate (6.158) can be written in the form

$$\Gamma = \frac{2\pi}{\hbar} \overbrace{\left(\frac{w^2}{16\epsilon_F} \right)}^{\langle |a_3|^2 \rangle} \left[\frac{\epsilon^2 + (\pi k_B T)^2}{2} \right]. \qquad (6.160)$$

Apart from the factor of 16 in the denominator, this is what we guessed on dimensional grounds.

There are two important regimes of behavior to note:

- $|\epsilon_\mathbf{p}| \ll \pi k_B T$: $\Gamma \propto T^2$. Near the Fermi surface, quasiparticles are thermally excited, with a T^2 scattering rate that is independent of energy.
- $|\epsilon_\mathbf{p}| \gg \pi k_B T$: $\Gamma \propto \epsilon_\mathbf{p}^2$. For higher-energy quasiparticles, the scattering rate is quadratically dependent on energy.

Example 6.5   Calculate the angular average of the scattering amplitude

$$\left\langle |a_3|^2 \right\rangle = \left(\frac{\mathcal{N}^*(0)}{2} \right)^3 \int \frac{d\Omega_2 d\Omega_3 d\Omega_4}{(4\pi)^3} |a(1,2;3,4)|^2 (2\pi\hbar)^3 \delta^{(3)}[p_F(\hat{\mathbf{n}}_1 + \hat{\mathbf{n}}_2 - \hat{\mathbf{n}}_3 - \hat{\mathbf{n}}_4)]$$

(6.161)

in the dominant quasiparticle decay processes.

Solution

We first replace $\delta^{(3)}[p_F(\hat{\mathbf{n}}_1 + \hat{\mathbf{n}}_2 - \hat{\mathbf{n}}_3 - \hat{\mathbf{n}}_4)] \rightarrow \frac{1}{p_F^3}\delta^{(3)}[\hat{\mathbf{n}}_1 + \hat{\mathbf{n}}_2 - \hat{\mathbf{n}}_3 - \hat{\mathbf{n}}_4]$, so that

$$\left\langle |a_3|^2 \right\rangle = \left(\frac{\mathcal{N}^*(0)\hbar}{4p_F} \right)^3 \int d\Omega_2 d\Omega_3 d\Omega_4 \delta^{(3)}[\hat{\mathbf{n}}_1 + \hat{\mathbf{n}}_2 - \hat{\mathbf{n}}_3 - \hat{\mathbf{n}}_4]|a(1,2;3,4)|^2. \quad (6.162)$$

To carry out the angular integral, we use polar co-ordinates for $\hat{\mathbf{n}}_2 \equiv (\theta, \phi_2)$, $\hat{\mathbf{n}}_3 \equiv (\theta_3, \phi_3)$, and $\hat{\mathbf{n}}_4 = (\theta_4, \phi_4)$ (as illustrated in Figure 6.8), where θ and ϕ_2 are the polar angles of $\mathbf{n}_2$ relative to $\mathbf{n}_1$, $\theta_{3,4}$ are the angles between $\hat{\mathbf{n}}_{3,4}$ and the direction of the total momentum $\hat{\mathbf{P}}$, while ϕ_3 is the azimuthal angle of $\mathbf{n}_3$ measured relative to the plane defined by $\hat{\mathbf{n}}_1$ and $\hat{\mathbf{n}}_2$ and ϕ_4 is azimuthal angle of $\mathbf{n}_4$ measured relative to the common plane of $\hat{\mathbf{n}}_3$ and $\hat{\mathbf{P}}$. The delta function in the integral will force $\hat{\mathbf{n}}_3$ and $\hat{\mathbf{n}}_4$ to lie in a place, so that ultimately, we only need to know the dependence of the amplitude $a(\theta, \phi_3)$ on θ and ϕ_3.

Taking the z-axis to lie along $\hat{\mathbf{P}}$ and choosing the y-axis to lie along $\hat{\mathbf{P}} \times \hat{\mathbf{n}}_3$, then in this coordinate system $\hat{\mathbf{n}}_1 + \hat{\mathbf{n}}_2 = (0, 0, 2\cos\theta/2)$, $\hat{\mathbf{n}}_3 = (\sin\theta_3, 0, \cos\theta_3)$, and $\hat{\mathbf{n}}_4 = (\sin\theta_4\cos\phi_4, \sin\theta_4\sin\phi_4, \cos\theta_4)$, so that

$$\hat{\mathbf{n}}_3 + \hat{\mathbf{n}}_4 - \hat{\mathbf{n}}_1 - \hat{\mathbf{n}}_2$$
$$= (\sin\theta_3 + \sin\theta_4\cos\phi_4, \sin\theta_4\sin\phi_4, \cos\theta_3 + \cos\theta_4 - 2\cos(\theta/2)).$$

Factorizing the three-dimensional delta function into its x, y and, z components gives

$$\delta^{(3)}[(\hat{\mathbf{n}}_1 + \hat{\mathbf{n}}_2 - \hat{\mathbf{n}}_3 - \hat{\mathbf{n}}_4)]$$
$$= \delta[\sin\theta_3 + \sin\theta_4\cos\phi_4]\delta[\sin\theta_4\sin\phi_4]\delta[\cos\theta_3 + \cos\theta_4 - 2\cos(\theta/2)].$$

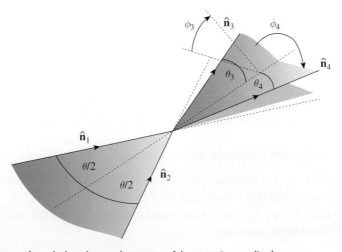

Fig. 6.8   Co-ordinate system used to calculate the angular average of the scattering amplitude.

Integrating over $d\Omega_4 = \sin\theta_4 d\theta_4 d\phi_4$ forces $\phi_4 = \pi$ and $\theta_4 = \theta_3$ (note that $\phi_4 = 0$ satisfies the second delta function, but this then requires that $\sin\theta_3 = -\sin\theta_4$, which is not possible when $\theta_{3,4} \in [0, \pi)$). Resolving the delta functions around these points, we may write

$$\delta[\sin\theta_3 + \sin\theta_4 \cos\phi_4]\,\delta[\sin\theta_4 \sin\phi_4] = \frac{\delta(\theta_3 - \theta_4)}{\cos\theta_4} \frac{\delta(\phi_4 - \pi)}{\sin\theta_4}.$$

When we carry out the integral over $d\Omega_4 = \sin\theta_4 d\theta_4 d\phi_4$, we then obtain

$$\int d\Omega_4 \delta^{(3)}[\hat{\mathbf{n}}_1 + \hat{\mathbf{n}}_2 - \hat{\mathbf{n}}_3 - \hat{\mathbf{n}}_4]|a(\theta, \phi_3)|^2 = \frac{1}{\cos\theta_3}\delta[2\cos\theta_3 - 2\cos(\theta/2)]|a(\theta, \phi_3)|^2.$$

Integrating over $d\Omega_3 = d\phi_3 d\cos\theta_3$ imposes $\theta_3 = \theta/2$, so that

$$\int d\Omega_3 d\Omega_4 \delta^{(3)}[\hat{\mathbf{n}}_1 + \hat{\mathbf{n}}_2 - \hat{\mathbf{n}}_3 - \hat{\mathbf{n}}_4]|a(\theta, \phi_3)|^2 = \int \frac{d\phi_3}{2\cos\theta/2}|a(\theta, \phi_3)|^2.$$

The azimuthal angle ϕ_2 of $\hat{\mathbf{n}}_2$ about $\mathbf{n}_1$ does not enter into the integral, so we may integrate over this angle and write the measure $d\Omega_2 \equiv 2\pi d\cos\theta$. The complete angular integral is then

$$\int d\Omega_2 d\Omega_3 d\Omega_4 \delta^{(3)}[\hat{\mathbf{n}}_1 + \hat{\mathbf{n}}_2 - \hat{\mathbf{n}}_3 - \hat{\mathbf{n}}_4]|a(\theta, \phi_3)|^2 = 2\pi \int \frac{d\phi_3 d\cos\theta}{2\cos\theta/2}|a(\theta, \phi_3)|^2.$$

Substituting this result into (6.162), the complete angular average is then

$$\left\langle |a_3|^2 \right\rangle = \pi^2 \left(\frac{\mathcal{N}^*(0)\hbar}{2p_F}\right)^3 \int \frac{d\cos\theta d\phi}{4\pi} \frac{|a(\theta, \phi)|^2}{2\cos\theta/2},$$

where we have relabeled ϕ_3 as ϕ. Notice (i) that the weighted angular average is normalized, so that if $|a(\theta, \phi)|^2 = |a|^2$ is constant, $\langle |a_3|^2 \rangle = \pi^2 \left(\frac{\mathcal{N}^*(0)\hbar}{2p_F}\right)^3 |a|^2$, and (ii) that since the denominator in the average vanishes for $\theta = \pi$, the angular average contributing to the quasiparticle decay is weighted towards large-angle scattering events in which the outgoing quasiparticles have opposite momenta $\mathbf{p}_3 = -\mathbf{p}_4$. This feature is closely connected with the Cooper pair instability discussed in Chapter 13.

Example 6.6 Compute the energy phase space integral

$$I(\epsilon, T) = \int_{-\infty}^{\infty} d\epsilon_2 d\epsilon_3 d\epsilon_4 \delta(\epsilon + \epsilon_2 - \epsilon_3 - \epsilon_4)\left[n_2(1 - n_3)(1 - n_4) + (1 - n_2)n_3 n_4\right],$$

where $n_i \equiv f(\epsilon_i) = 1/(e^{\beta\epsilon} + 1)$ denotes the Fermi function evaluated at energy ϵ_i.

Solution

As a first step, we make the change of variable $\epsilon_2 \to -\epsilon_2$, so that the integral becomes

$$I(\epsilon, T) = \int_{-\infty}^{\infty} d\epsilon_2 d\epsilon_3 d\epsilon_4 \delta(\epsilon - (\epsilon_2 + \epsilon_3 + \epsilon_4))\left[(1 - n_2)(1 - n_3)(1 - n_4) + n_2 n_3 n_4\right]$$

$$= \int_{-\infty}^{\infty} d\epsilon_2 d\epsilon_3 d\epsilon_4 \delta(\epsilon - (\epsilon_2 + \epsilon_3 + \epsilon_4))\left[n_2 n_3 n_4 + \{\epsilon \leftrightarrow -\epsilon\}\right].$$

Next, we rewrite the delta function as a Fourier transform, $\delta(x) = \int \frac{d\alpha}{2\pi} e^{i\alpha x}$, so that $I(\epsilon, T) = I_1(\epsilon, T) + I_1(-\epsilon, T)$, where

$$I_1(\epsilon, T) = \frac{1}{2\pi} \int d\alpha \, d\epsilon_2 \, d\epsilon_3 \, d\epsilon_4 e^{i\alpha[\epsilon - (\epsilon_2 + \epsilon_3 + \epsilon_4)]} [n_2 n_3 n_4].$$

By carrying out a contour integral around the poles of the Fermi function $f(z)$ at $z = i\pi T(2n + 1)$ in the lower half-plane, we may deduce that

$$\int_{-\infty}^{\infty} d\epsilon \, e^{-i(\alpha + i\delta)\epsilon} f(\epsilon) = 2\pi i T \sum_{n=0}^{\infty} e^{-(\alpha + i\delta)\pi T(2n+1)} = \frac{\pi i T}{\sinh(\alpha + i\delta)\pi T},$$

where a small imaginary part has been added to α to guarantee convergence. This enables us to carry out the energy integrals in $I_1(\epsilon, T)$, obtaining

$$I_1(\epsilon, T) = \int \frac{d\alpha}{2\pi} e^{i\alpha\epsilon} \left(\frac{\pi i T}{\sinh(\alpha + i\delta)\pi T} \right)^3.$$

Now, to carry out this integral, we need to distort the contour into the upper half complex plane. The function $1/\sinh(\alpha + i\delta)\pi T$ has poles at $\alpha = in/T - i\delta$, so the distorted contour wraps around the poles with $n \geq 0$. The cube of this function has both triple and simple poles at these locations. To evaluate the residues of these poles, we expand $\sinh \alpha \pi T$ to third order in $\delta\alpha = (\alpha - \frac{in}{T})$ about the poles, to obtain

$$\sinh \alpha \pi T = (-1)^n \pi T \delta\alpha \left(1 + \frac{(\pi T)^2}{3!} \delta\alpha^2 \right) + \cdots,$$

so that, near the poles,

$$\left(\frac{i\pi T}{\sinh \alpha \pi T} \right)^3 = -i \frac{(-1)^n}{\delta\alpha^3} \left(1 - \frac{(\pi T)^2}{2} \delta\alpha^2 \right)$$

$$= -i(-1)^n \left(\frac{1}{\delta\alpha^3} - \frac{(\pi T)^2}{2\delta\alpha} \right).$$

The complete contour integral becomes

$$I_1(\epsilon, T) = \sum_{n=1}^{\infty} (-1)^n \oint \frac{d\alpha}{2\pi i} \left[\frac{1}{(\alpha - \frac{in}{T})^3} - \frac{1}{2} \frac{(\pi T)^2}{(\alpha - \frac{in}{T})} \right] e^{i\alpha\epsilon}$$

$$= -\sum_{n=1}^{\infty} (-1)^n \oint \frac{d\alpha}{2\pi i} \frac{1}{(\alpha - \frac{in}{T})} \left[\frac{\epsilon^2}{2} + \frac{(\pi T)^2}{2} \right] e^{i\alpha\epsilon}$$

$$= -\left[\frac{\epsilon^2}{2} + \frac{(\pi T)^2}{2} \right] \sum_{n=1}^{\infty} (-1)^n e^{-n\epsilon/T} = \frac{1}{1 + e^{\epsilon/T}} \left[\frac{\epsilon^2}{2} + \frac{(\pi T)^2}{2} \right].$$

Finally, adding $I_1(\epsilon, T) + I_1(-\epsilon, T)$ gives

$$I(\epsilon, T) = \frac{1}{2} [\epsilon^2 + (\pi T)^2].$$

6.7.3 Kadowaki-Woods ratio and "local Fermi liquids"

Heuristic discussion

One of the direct symptoms of Landau Fermi-liquid behavior in a metal is a T^2 temperature dependence of resistivity at low temperatures:

$$\rho(T) = \rho_0 + AT^2. \tag{6.163}$$

Here ρ_0 is the *residual resistivity* due to the scattering of electrons off impurities. The quadratic temperature dependence of the resistivity is a direct reflection of the quadratic scattering rate $\Gamma \propto T^2$ expected in Landau Fermi-liquids. Evidence that this term is directly related to electron–electron scattering is provided by a remarkable scaling relation between the A coefficient of the resistivity and the square of the zero-temperature linear coefficient of the specific heat, $\gamma = C_V/T|_{T\to 0}$:

$$\frac{A}{\gamma^2} = \alpha \approx 1 \times 10^{-5} \mu\Omega\,\text{cm}\,(\text{K mol/mJ})^2. \tag{6.164}$$

The constancy of the ratio A/γ^2 was first discovered in transition metals by Michael Rice [15], and extended to the broad class of intermetallic *heavy-fermion metals* by Kadowaki and Woods, after whom it is now named [16]. The units employed for the Kadowaki–Woods ratio are $\mu\Omega\,\text{cm}$ for the resistivity and $\text{mJ}/(\text{mol K}^2)$ for the specific heat coefficient. In practice, the Kadowaki–Woods ratio is only constant within a given class of compounds: in transition metals, $\alpha \sim 10^{-5}\mu\Omega\,\text{cm}\,(\text{K mol/mJ})^2$ [15], but in the intermetallic heavy-electron metals, $\alpha \sim 1 \times 10^{-5}\,\mu\Omega\,\text{cm}\,(\text{K mol/mJ})^2$ over a very wide range of mass renormalizations (Figure 6.9).

To understand Kadowaki–Woods scaling, we need to keep track of how A and γ depend on the Fermi energy. In the previous section, we found that the electron–electron scattering rate is set by the Fermi energy, $\tau^{-1} \sim T^2/\epsilon_F$. If we insert this into the Drude scattering formula for the resistivity $\rho = m^*/(ne^2\tau)$, since $m^* \propto 1/\epsilon_F$ we deduce that $\rho \sim (T^2/\epsilon_F^2)$, i.e. $A \propto 1/\epsilon_F^2$. By contrast, the specific heat coefficient, $\gamma \propto m^* \propto 1/\epsilon_F$, is inversely proportional to the Fermi energy, so that

$$A \propto \left(\frac{1}{\epsilon_F}\right)^2, \qquad \gamma \propto \frac{1}{\epsilon_F} \Rightarrow \frac{A}{\gamma^2} \sim \text{constant}. \tag{6.165}$$

In strongly correlated metals, the Fermi energy varies from eV to meV scales, so the A coefficient can vary over eight orders of magnitude. This strong dependence of A on the Fermi energy of the Landau Fermi-liquid is canceled by γ^2.

Estimate of the Kadowaki–Woods ratio

To obtain an estimate of the coefficient A, it is useful to regard a metal as a stack of 2D layers with separation a, so that $\rho = a\rho_{2D} = a/\sigma_{2D}$, where σ_{2D} is the dimensionless conductivity per layer. If we use the Drude formula for the conductivity in two dimensions, $\sigma_{2D} = ne^2\tau/m$, putting $n = 2 \times \pi k_F^2/(2\pi)^2$ and $\hbar/\tau = \Gamma$ we obtain

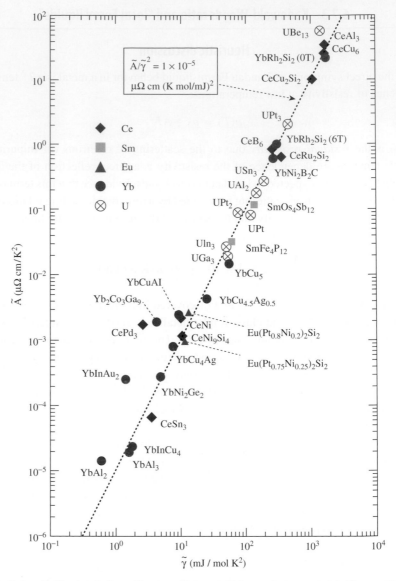

Fig. 6.9 Showing the Kadowaki–Woods ratio for a wide range of intermetallic heavy-electron materials. The quantities $\tilde{A} = A/(N(N-1))$ and $\tilde{\gamma} = \gamma/(N(N-1))$ have been normalized to account for their dependence on the ground-state degeneracy N of the magnetic scattering ion. Reprinted with permission from H. Tsujii, *et al.*, *Phys. Rev. Lett.*, vol. 94, p. 057201, 2005. Copyright 2005 by the American Physical Society.

$$\rho = a \overbrace{\left(\frac{h}{2e^2}\right)}^{\rho_\square = 12.9\,\mathrm{k}\Omega} \left(\frac{\Gamma}{2\epsilon_F}\right). \tag{6.166}$$

In the previous section, we found that $\Gamma = 2\pi(w/4)^2(\pi k_B T)^2/\epsilon_F$. Putting this together then gives

$$\rho = (a\rho_\square)\pi \left(\frac{w}{4}\right)^2 \left(\frac{\pi k_B T}{\epsilon_F}\right)^2. \tag{6.167}$$

(The prefactor $a\rho_\square$ is sometimes called the *unitary resistance*, and corresponds to the resistivity of a metal in which the scattering rate is of the order of the Fermi energy. If we put $a \sim 1 - 4\overset{\circ}{A}$, $\rho_\square \sim 13k\Omega$, we obtain $a\rho_\square \sim 100 - 500\,\mu\Omega\,cm$.) It follows that

$$A \approx (a\rho_\square)\pi^3 \left(\frac{w}{4}\right)^2 \times \left(\frac{1}{T_F}\right)^2, \tag{6.168}$$

where $T_F = \epsilon_F/k_B$ is the Fermi temperature.

Now, using (6.53), the specific heat coefficient per unit volume is $\gamma = \frac{1}{3}\pi^2 k_B^2 \mathcal{N}^*(0) = \frac{\pi^2 k_B^2}{2\epsilon_F}n$, where n is the number of electrons per unit volume. Thus the specific heat coefficient per electron is simply $\gamma_e = \frac{\pi^2 k_B^2}{2\epsilon_F}$ and the specific heat per mole of electrons is $\gamma_M = \frac{1}{2}\pi^2 R\frac{1}{T_F}$, where $R = k_B N_{AV}$ is the gas constant and N_{AV} is Avagadro's number. So if there are n_e electrons per unit cell,

$$\gamma_M^2 \sim \frac{\pi^4 R^2}{4}\frac{(n_e)^2}{T_F^2}, \tag{6.169}$$

giving

$$\alpha = \frac{A}{\gamma^2} \sim \left(\frac{w^2}{4\pi}\right)\left(\frac{\rho_\square}{R^2}\right) \times \frac{a}{(n_e)^2}. \tag{6.170}$$

If we take $\rho_\square = 13 \times 10^9\,\mu\Omega$, $R = 8.3 \times 10^3\,mJ/(mol\,K)$ and $w^2/(4\pi) \sim 1$, to obtain

$$\alpha \sim 2 \times 10^{-5} \times \left(\frac{a[nm]}{(n_e)^2}\right)\mu\Omega\,cm\,(K\,mol/mJ)^2, \tag{6.171}$$

giving a number of the right order of magnitude. Kadowaki and Woods found that $\alpha \approx 10^{-5}\,\mu\,\Omega\,cm\,(K\,mol/mJ)^2$ in a wide range of intermetallic heavy-fermion compounds. In transition metal compounds, $\alpha \approx 10^{-6}\,\mu\Omega\,cm\,(K\,mol/J)^2$ has a considerably smaller value, related to the higher carrier density.

Local Fermi liquids

A fascinating aspect of this estimate is that we needed to put $w^2/(4\pi) \sim 1$ to get an answer comparable with measurements. The tendency of w to be of order unity is a feature of a broad class of *strongly correlated* metals. Although Landau Fermi-liquid theory does not give us information on the detailed angular dependence of the scattering amplitude $A(\theta, \phi)$, we can make a great deal of progress by assuming that the scattering t-matrix is local. In fact, this is a reasonable assumption in systems where the important Coulomb interactions lie within core states of an atom, as in transition metal and rare-earth atoms. In this case,

$$a_{\sigma\sigma'}(\theta, \phi) = a^s + a^a \sigma\sigma' \tag{6.172}$$

is approximately independent of the quasiparticle momenta and momentum transfer. This is the local approximation to the Landau Fermi-liquid. When "up" quasiparticles scatter, the antisymmetry of scattering amplitudes under particle exchange guarantees that

$a_{\uparrow\uparrow}(\theta, \phi) = -a_{\uparrow\uparrow}(\theta, \phi+\pi)$. But if a is independent of scattering amplitude, then it follows that $a_{\uparrow\uparrow} = a^s + a^a = 0$, so that

$$a_{\sigma\sigma'}(\theta, \phi) = a^s(1 - \sigma\sigma') \tag{6.173}$$

in a local Landau Fermi-liquid.

Now we can relate the $a_{\sigma\sigma'} = A_{\sigma\sigma'}/N^{*(0)}$ to the dimensionless scattering amplitudes introduced in Section 6.4.3. By (6.80), the charge susceptibility is given by

$$\chi_c = \mathcal{N}^*(0) \times \left(\frac{1}{1+F_0^s}\right) = \mathcal{N}^*(0) \times \left(1 - \frac{F_0^s}{1+F_0^s}\right) = N^*(0) \times (1 - A_0^s). \tag{6.174}$$

In strongly interacting electron systems the density of states is highly renormalized, so that $\mathcal{N}^*(0) >> N(0)$, but the charge susceptibility is basically unaffected by interactions, given by $\chi_c = N(0) << \mathcal{N}^*(0)$. This implies that $A_0^s \approx 1$, so that $a^s = 1/\mathcal{N}^*(0)$, which in turn implies that the dimensionless ratio w introduced in the previous section is close to $w = 1$.

6.8 Microscopic basis of Fermi-liquid theory

Although Landau's Fermi-liquid theory is a phenomenological theory, based on physical arguments, it translates naturally into the language of diagrammatic many-body theory. The Landau school played a major role in the adaptation of Feynman diagrammatic approaches to many-body physics. However, Feynman diagrams do not appear until the third of Landau's three papers on Fermi-liquid theory [17]. The classic microscopic treatments of Fermi-liquid theory are based on the analysis of many-body perturbation theory to infinite order carried out in the late 1950s and early 1960s.

Victor M. Galitskii [18], at the Kapitza Institute, Moscow, gave the first formulation of Landau's theory in terms of diagrammatic many-body theory. Shortly thereafter, Joaquin Luttinger, John Ward, and Philippe Nozières developed the detailed diagrammatic many-body framework for Landau Fermi-liquid theory by considering the analytic properties of inifinite-order perturbation theory [19, 20]. Here we end with a brief discussion of some of the key results of these analyses.

From the outset, it was understood that the Landau Fermi-liquid is always potentially unstable to superconductivity. By the late 1960s it also became clear that Landau Fermi-liquid theory does not apply in one-dimensional conductors, where the phase space scattering arguments used to support the idea of the Landau quasiparticle no longer apply. In one dimension, the Landau quasiparticle becomes unstable, breaking up into collective modes that independently carry spin and charge degrees of freedom. We call such a fluid a *Luttinger liquid* [21]. However, with this exception, few questioned the robustness of Landau Fermi-liquid theory until the 1980s. In 1986, the discovery of high-temperature superconductors led to a resurgence of interest in this topic, for in the normal state these materials cannot be easily understood in terms of Landau Fermi-liquid theory. For example, these materials display a linear resistivity up to high temperatures; at this time-remains an unsolved mystery. This has led to speculatation that in two or three dimensions, Landau

Fermi-liquid theory might break down into a higher-dimensional analogue of the one-dimensional Luttinger liquid. In the wake of this interest, Landau Fermi-liquid theory was re-examined from the perspective of the *renormalization group* [22, 23]. The conclusion of these analyses is that, unlike in one dimension, Fermi liquids are not generically unstable in two and higher dimensions. While this does not rule out the possibility of new kinds of metallic behavior, Landau Fermi-liquid theory continues to provide the bedrock for our understanding of basic metals in two or three dimensions.

As we noted in the previous chapter, the process of adiabatically "switching on" interactions can be understood as a unitary transformation of the original states of the non-interacting Fermi sea. Thus the ground state and the one-quasiparticle state are given by

$$|\phi\rangle = U|\Psi_0\rangle$$
$$|\widetilde{\mathbf{k}\sigma}\rangle = U|\mathbf{k}\sigma\rangle, \tag{6.175}$$

where $|\Psi_0\rangle$ is the filled Fermi sea of the non-interacting system and $\mathbf{k}$ is a momentum very close to the Fermi surface. In fact, using the results of Section 5.1, we can write U as a time-ordered exponential,

$$U = T\left[\exp\left\{-i\int_{-\infty}^{0} V(t)dt\right\}\right], \tag{6.176}$$

where $\hat{V}$ is the interaction, written in the interaction representation. Now since $|\mathbf{k}\sigma\rangle = c_{\mathbf{k}\sigma}^{\dagger}|\Psi_0\rangle$, where $c_{\mathbf{k}\sigma}^{\dagger}$ is the particle-creation operator for the non-interacting Hamiltonian, it follows that

$$|\widetilde{\mathbf{k}\sigma}\rangle = \overbrace{Uc_{\mathbf{k}\sigma}^{\dagger}U^{\dagger}}^{a_{\mathbf{k}\sigma}^{\dagger}}|\phi\rangle, \tag{6.177}$$

so that the quasiparticle-creation operator is given by

$$a_{\mathbf{k}\sigma}^{\dagger} = Uc_{\mathbf{k}\sigma}^{\dagger}U^{\dagger}. \tag{6.178}$$

From this line of reasoning, we can see that the operator that creates the one-quasiparticle state is nothing more than the original creation bare creation operator, unitarily time-evolved from the distant past to the present in the interaction representation.

While this formal procedure can always be carried out, the existence of the Landau Fermi-liquid requires that, in the thermodynamic limit, the resulting state preserves a finite overlap with the state formed by adding a bare particle to the ground state, i.e.

$$Z_{\mathbf{k}} = |\langle\widetilde{\mathbf{k}\sigma}_0|c_{\mathbf{k}\sigma}^{\dagger}|\phi\rangle|^2 > 0. \qquad \text{wavefunction renormalization} \tag{6.179}$$

This overlap is called the *wavefunction renormalization constant*, and so long as this quantity is finite on the Fermi surface, the Landau Fermi-liquid is alive and well.

In general, near the Fermi energy the electron creation operator will have an expansion as a sum of states containing one, three, five and any odd number of quasiparticle and hole states, each with the same total spin, charge, and momentum of the initial bare particle:

$$c^\dagger_{\mathbf{k}\sigma} = \sqrt{Z_{\mathbf{k}}}\, a^\dagger_{\mathbf{k}\sigma} + \sum_{\mathbf{k}_4+\mathbf{k}_3=\mathbf{k}_2+\mathbf{k}} A(\mathbf{k}_4\sigma_4, \mathbf{k}_3\sigma_3; \mathbf{k}_2\sigma_2, \mathbf{k}\sigma) a^\dagger_{\mathbf{k}_4\sigma_4} a^\dagger_{\mathbf{k}_3\sigma_3} a_{\mathbf{k}_2\sigma_2} + \cdots. \quad (6.180)$$

There are three important consequences that follow:

- **Sharp quasiparticle peak in the spectral function** When a particle is added to the ground state, it excites a continuum of states $|\lambda\rangle$ with energy distribution described by the spectral function (5.128):

$$A(\mathbf{k}, \omega) = \frac{1}{\pi} \mathrm{Im}\, G(\mathbf{k}, \omega - i\delta) = \sum_\lambda |M_\lambda|^2 \delta(\omega - \epsilon_\lambda), \quad (6.181)$$

where the squared amplitude $|M_\lambda|^2 = |\langle \lambda | c^\dagger_{\mathbf{k}\sigma} |\phi\rangle|^2$. In a Landau Fermi-liquid, the spectral function retains a sharp *quasiparticle pole* at the Fermi energy (see Figure 6.10). If we split off the $\lambda \equiv \mathbf{k}\sigma$ contribution to the summation in (6.181) we then get

$$A(\mathbf{k}, \omega) = \frac{1}{\pi} \mathrm{Im}\, G(\mathbf{k}, \omega - i\delta) = \overbrace{Z_{\mathbf{k}\sigma}\, \delta(\omega - \epsilon_{\mathbf{k}})}^{\text{qp peak}} + \overbrace{\sum_{\lambda \neq \mathbf{k}\sigma} |M_\lambda|^2 \delta(\omega - \epsilon_\lambda)}^{\text{continuum}}. \quad (6.182)$$

- **Sudden jump in the momentum distribution** In a non-interacting Fermi liquid, the particle momentum distribution function exhibits a sharp Fermi distribution function which is preserved by the *quasiparticles* in Landau Fermi-liquid theory:

$$\langle \phi | (\hat{n}_{\mathbf{k}\sigma}) \mathrm{qp} | \phi \rangle = \theta(\mu - E_{\mathbf{k}}), \quad (6.183)$$

where here $(\hat{n}_{\mathbf{k}\sigma})\mathrm{qp} = \tilde{c}^\dagger_{\mathbf{k}\sigma} \tilde{c}_{\mathbf{k}\sigma}$ is the quasiparticle occupancy. Remarkably, part of this jump survives interactions (see Figure 6.11). To see this effect, we write the momentum distribution function of the particles as

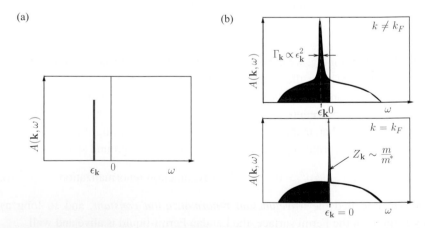

(a)　　　　　　　　(b)

Fig. 6.10 (a) In a non-interacting Fermi system, the spectral function is a sharp delta function at $\omega = \epsilon_{\mathbf{k}}$. (b) In an interacting Fermi liquid for $k \neq k_F$, the quasiparticle forms a broadened peak of width $\Gamma_{\mathbf{k}}$ at $\omega_{\mathbf{k}}$. If $k = k_F$, this peak becomes infinitely sharp, corresponding to a long-lived quasiparticle on the Fermi surface. The weight in the quasiparticle peak is $Z_{\mathbf{k}} \sim m/m^*$, where m^* is the effective mass.

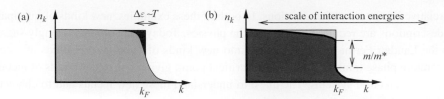

(a) In a non-interacting Fermi liquid, a temperature T that is smaller than the Fermi energy slightly "blurs" the Fermi surface. (b) In a Landau Fermi-liquid, the exclusion principle stabilizes the jump in occupancy at the Fermi surface, even though the bare interaction energy is far greater than the Fermi energy.

Fig. 6.11

$$\langle \hat{n}_{\mathbf{k}\sigma} \rangle = \langle \phi | c^{\dagger}_{\mathbf{k}\sigma} c_{\mathbf{k}\sigma} | \phi \rangle = \int_{-\infty}^{0} d\omega A(\mathbf{k}, \omega), \qquad (6.184)$$

where we have used the results of Section 5.3.3 to relate the particle number to the integral over the spectral function below the Fermi energy. When we insert (6.182) into this expression, the contribution from the quasiparticle peak vanishes if $\epsilon_{\mathbf{k}} > 0$, but gives a contribution $Z_{\mathbf{k}}$ if $\epsilon_{\mathbf{k}} < 0$, so that

$$\langle \hat{n}_{\mathbf{k}\sigma} \rangle = Z_{\mathbf{k}} \theta(-\epsilon_{\mathbf{k}}) + \text{smooth background}. \qquad (6.185)$$

This is a wonderful illustration of the organizing power of the Pauli exclusion principle. One might have expected interactions to have the same effect as temperature, which smears the Fermi distribution by an amount of order $k_B T$. Although interactions do smear the momentum distribution, the jump continues to survive in reduced form so long as the Landau Fermi-liquid is intact.

- **Luttinger sum rule** In the Landau Fermi-liquid, the Fermi surface volume measures the particle density n_F. Since the Fermi surfaces of the quasiparticles and of the unrenormalized particles coincide, it follows that the Fermi surface volume must be an adiabatic invariant when the interactions are turned on:

$$n_F = (2S + 1) \frac{v_{FS}}{(2\pi)^3}. \qquad \text{Luttinger sum rule} \qquad (6.186)$$

The demonstration of this conservation law within infinite-order perturbation theory was first derived by Joaquin Luttinger in 1962, and it is known as the *Luttinger sum rule*. In interacting fermion systems, the conservation of particle number leads to a set of identities between different many-body Green's functions called *Ward identities*. Luttinger showed how these identities can be used to relate the Fermi surface volume to the particle density.

Today, more than a half-century after Landau's original idea, Landau Fermi-liquid theory continues to be a mainstay of our understanding of interacting metals. However, increasingly physicists are asking when and how the Landau Fermi-liquid breaks down, and what new types of fermion fluids may form instead. We know that the Landau Fermi-liquid does not survive in one-dimensional conductors, where quasiparticles break up into collective spin and charge excitations, or in high magnetic fields, where the formation of widely spaced Landau levels effectively quenches the kinetic energy of the particles, enhancing the relative importance of interactions. In both of these examples, new kinds of quasiparticle descriptions are required to describe the physics. Today, experiments strongly suggest that the Landau Fermi-liquid breaks up into new kinds of "non-Fermi" fluids at a zero-temperature phase transition or quantum critical point, giving rise to new kinds of metallic behavior in electron systems. The quest to understand these new metals and to characterize their excitation spectra is one of the great open problems of modern condensed matter theory.

Exercises

Exercise 6.1 Calculate the Landau parameters to leading order in $\lambda_{1,2}$ for a Fermi liquid with the contact interactions:

(a) $V(x - x') = \lambda_1 \delta^{(3)}(x - x')$.

(b) $V(x - x') = -\lambda_2 \nabla^2 \delta^3(x - x')$ (so that $V(q) = \lambda_1 q^2$ in Fourier space).

(c) Taking the results of (a) and (b) literally, sketch the regions of the λ_1, λ_2 phase diagram where the Fermi surface becomes unstable.

Exercise 6.2 (a) Show that, if we ignore both the $\mathbf{q}$ dependence of the Landau $f_{\mathbf{p}\sigma \mathbf{p}'\sigma'}(\mathbf{q}) = f_{\mathbf{p}\sigma \mathbf{p}'\sigma'}$ and the cut-off on momentum transfer, so that

$$H_I = \frac{1}{2} \sum_{\mathbf{p}\sigma \mathbf{p}'\sigma'} f_{\mathbf{p}\sigma, \mathbf{p}'\sigma'} \hat{n}_{\mathbf{p}\sigma}(\mathbf{q}) \hat{n}_{\mathbf{p}'\sigma'}(-\mathbf{q}),$$

then in real space this interaction is equivalent to an anisotropic contact interaction

$$H_I = \frac{1}{2} \sum_{\mathbf{p}\sigma \mathbf{p}\sigma'} \int d^3x d^3x' V_{\mathbf{p}\sigma, \mathbf{p}'\sigma'}(\mathbf{x} - \mathbf{x}') n_{\mathbf{p}\sigma}(\mathbf{x}) n_{\mathbf{p}'\sigma'}(\mathbf{x}'),$$

where

$$V_{\mathbf{p}\sigma, \mathbf{p}'\sigma'}(\mathbf{x}) = f_{\mathbf{p}\sigma, \mathbf{p}'\sigma'} \delta^3(\mathbf{x}) \tag{6.187}$$

is a contact interaction and

$$n_{\mathbf{p}\sigma}(x) = \int \frac{d^3q}{(2\pi)^3} c^\dagger_{\mathbf{p}-\mathbf{q}/2\sigma} c_{\mathbf{p}+\mathbf{q}/2\sigma} e^{i\mathbf{q}\cdot\mathbf{x}}$$

is the momentum and spin-resolved quasiparticle density at position $\mathbf{x}$.

(b) In practice, the contact interaction behaves more like a *hard-core* interaction. What is the range of the hard-core interaction?

(c) How must we modify (6.187) in the presence of a long-range Coulomb interaction?

Exercise 6.3 Consider how the non-linear Schrödinger equation on a one-dimensional ring,

$$i\partial_t \psi = [\frac{1}{2}(-i\partial_x)^2 - g|\psi|^2]\psi, \tag{6.188}$$

transforms between Galilean reference frames. Suppose that $\psi(x,t) = \psi(x+L,t)$ satisfies periodic boundary conditions, where L is the circumference of the ring. Let the lab frame be described by co-ordinates (x,t) and the moving frame by co-ordinates (x',t'), where $x' = x - vt$, $t' = t$.

(a) Rewriting the wavefunction in the moving frame, $\tilde{\psi}(x',t') = \psi(x,t)$, show that this wavefunction satisfies

$$i\partial_{t'}\tilde{\psi} = \left[\frac{1}{2}(-i\partial_x - v)^2 - \frac{v^2}{2} - g|\tilde{\psi}|^2\right]\tilde{\psi}. \tag{6.189}$$

(Note that, although $t = t'$ in a Galilean transformation, the derivative $\partial_{t'}$ is taken at constant x', while the derivative ∂_t is taken at constant x, so that $\partial_t\psi \equiv \left.\frac{\partial\psi}{\partial t}\right|_x = \left.\frac{\partial\tilde{\psi}}{\partial t'}\right|_{x'} + \left(\frac{\partial x'}{\partial t'}\right)_x \left.\frac{\partial\tilde{\psi}}{\partial x'}\right|_{t'} = \partial_{t'}\tilde{\psi} - v\partial_{x'}\tilde{\psi}.$)

(b) By making the gauge transformation $\tilde{\psi}(x',t') = e^{iv^2t'/2}\psi_M(x',t')$ (which still satisfies the requirement that $|\psi_M(x',t')|^2 dx' = |\psi(x,t)|^2 dx$), show that $\psi_M(x',t')$ satisfies the Schrödinger equation

$$i\partial_{t'}\psi_M = [\frac{1}{2}(-i\partial_{x'} - v)^2 - g|\psi_M|^2]\psi_M \tag{6.190}$$

corresponding to the application of a fictitious vector potential $A_N = v$. (Notice that the spatial boundary conditions have not changed.)

(c) Finally, show that, by changing the spatial boundary conditions, writing $\psi_M(x',t') = e^{ivx'}\psi_Y(x',t')$ so that if $\psi_M(x',t') = \psi_M(x'+L,t')$ then $\psi_Y(x') = e^{-ivL}\psi_Y(x')$, corresponding to twisted boundary conditions, one can absorb the fictitious vector potential,

$$i\partial_{t'}\psi_Y = [\frac{1}{2}(-i\partial_{x'})^2 - g|\psi_Y|^2]\psi_Y. \tag{6.191}$$

Show that Galilean invariance is quantized, i.e. only present at fixed boundary conditions for *quantized velocities* given by $v = n(2\pi/L)$ or, restoring m and $\hbar$, by $mv(L/2\pi) = n\hbar$, corresponding to the addition or subtraction of quantized units of angular momentum.

Exercise 6.4 Test your understanding of Landau's mass renormalization formula by generalizing it to the case when both charge and spin current are conserved. Suppose we introduce a second vector potential into (6.86) that couples to the spin current, writing

$$H[\mathbf{A}_N, \mathbf{W}] = \sum_\sigma \int d^3x \frac{1}{2m}\psi_\sigma^\dagger(x)\left[(-i\hbar\nabla - \mathbf{A}_N - \sigma\mathbf{W})^2\right]\psi_\sigma(x) + \hat{V}.$$

Whereas $\mathbf{A}$ couples to the current of particles, $\mathbf{W}$ couples to the (z component of the) spin current. Assume that V conserves spin current.

(a) By comparing the bare shift of the energies

$$\delta\epsilon_{\mathbf{p}\sigma}^{(0)} = -\frac{\mathbf{p}}{m}\cdot(\mathbf{A}_N + \sigma\mathbf{W})$$

with the shift that result from interaction feedback,

$$\delta\epsilon_{\mathbf{p}\sigma} = -\frac{\mathbf{p}}{m^*} \cdot \mathbf{A}_N - \sigma \frac{\mathbf{p}}{m_s^*} \cdot \mathbf{W},$$

show that there are two different mass renormalizations,

$$\frac{m}{m^*} = \frac{1}{1 + F_1^s}$$

$$\frac{m}{m_s^*} = \frac{1}{1 + F_1^a}. \tag{6.192}$$

(b) Show that consistency of these results requires that the dipole coupling between the up and down Fermi liquids vanishes, so that the up and down spin currents are decoupled.

Exercise 6.5 Using (6.93) and (6.107), show that the velocity of first sound can also be written in the form

$$u_1^2 = \frac{1}{3}\left(\frac{p_F}{m}\right)^2\left[1 + \frac{mp_F}{\pi^2}\int\frac{d\Omega}{4\pi}f^s(\cos\theta)(1 - \cos\theta)(1 - \cos\theta)\right].$$

Exercise 6.6 Generalize the collisionless Boltzmann equation for a charged Fermi liquid. Using your result, calculate the long-wavelength form of the plasma frequency.

Exercise 6.7 Derive the forward scattering sum rule for the t-matrix.

References

[1] D. Pines and D. Bohm, A collective description of electron interactions: II. collective vs individual particle aspects of the interactions, *Phys. Rev.*, vol. 85, p. 338, 1952.

[2] L. D. Landau, The theory of a Fermi liquid, *J. Exp. Theor. Phys.*, vol. 3, p. 920, 1957.

[3] A. Abrikosov and I. Khalatnikov, The theory of a Fermi liquid: the properties of liquid ^{3}He at low temperatures, *Rep. Prog. Phys.*, vol. 22, p. 329, 1959.

[4] P. Nozières and D. Pines, *The Theory of Quantum Liquids*, W. A. Benjamin, 1966.

[5] G. Baym and C. Pethick, *Landau Fermi-Liquid Theory: Concepts and Applications*, John Wiley & Sons, 1991.

[6] W. B. Ard, G. K. Walters, and W. M. Fairbank, Fermi–Dirac degeneracy in liquid ^{3}He below 1K, *Phys. Rev.*, vol. 95, p. 566, 1954.

[7] L. D. Landau, Oscillations in a Fermi liquid, *J. Exp. Theor. Phys.*, vol. 5, p. 101, 1957.

[8] G. G. Low and T. M. Holden, *Proc. Phys. Soc., London*, vol. 89, p. 119, 1966.

[9] A. Casey, H. Patel, J. Nyéki, B.P. Cowan, and J. Saunders, Evidence for a Mott–Hubbard transition in a two-dimensional ^{3}He fluid monolayer, *Phys. Rev. Lett.*, vol. 90, p. 115301, 2003.

[10] V. P. Silin, Theory of a degenerate electron liquid, *J. Exp. Theor. Phys.*, vol. 6, p. 387, 1957.

[11] V. P. Silin, Theory of the anomalous skin effect in metals, *J. Exp. Theor. Phys.*, vol. 6, p. 985, 1957.

[12] P. Morel and P. Nozières, Lifetime effects in condensed helium-3, *Phys. Rev.*, vol. 126, p. 1909, 1962.

[13] G.D. Mahan, *Many-Particle Physics*, Plenum, 3rd edn., 2000.

[14] H. Tsujii, H. Kontani, and K. Yoshimora, Universality in heavy fermion systems with general degeneracy, *Phys. Rev. Lett.*, vol. 94, p. 057201, 2005.

[15] M. Rice, Electron–electron scattering in transition metals, *Phys. Rev. Lett.*, vol. 20, no. 25, p. 1439, 1968.

[16] K. Kadowaki and S. B. Woods, Universal relationship of the resistivity and specific heat in heavy- fermion compounds, *Solid State Commun.*, vol. 58, p. 507, 1986.

[17] L. D. Landau, On the theory of the Fermi liquid, *J. Exp. Theor. Phys.*, vol. 8, p. 70, 1959.

[18] V. M. Galitskii, The energy spectrum of a non-ideal Fermi gas, *J. Exp. Theor. Phys.*, vol. 7, p. 104, 1958.

[19] P. Nozières and J. Luttinger, Derivation of the Landau theory of Fermi liquids. I: formal preliminaries, *Phys. Rev.*, vol. 127, p. 1423, 1962.

[20] J. M. Luttinger and P. Nozières, Derivation of the Landau theory of Fermi liquids. II: equilibrium properties and transport equation, *Phys. Rev.*, vol. 127, p. 1431, 1962.

[21] F. D. M Haldane, General relation of correlation exponents and spectral properties of one-dimensional Fermi systems: application to the anisotropic $s = \frac{1}{2}$ Heisenberg chain, *Phys. Rev. Lett.*, vol. 45, no. 16, p. 1358, 1980.

[22] G. Benfatto and G. Gallavotti, Renormalization-group approach to the theory of the Fermi surface, *Phys. Rev. B*, vol. 42, no. 16, p. 9967, 1990.

[23] R. Shankar, Renormalization-group approach to interacting fermions, *Rev. Mod. Phys.*, vol. 66, no. 1, p. 129, 1994.

7 Zero-temperature Feynman diagrams

Chapter 5 discussed adiabaticity, and we learned how Green's functions of an interacting system can be written in terms of Green's functions of the non-interacting system, weighted by the S-matrix, e.g.

$$\langle\phi|T\psi(1)\psi^\dagger(2)|\phi\rangle = \frac{\langle\phi_0|T\hat{S}\psi(1)\psi^\dagger(2)|\phi_0\rangle}{\langle\phi_0|\hat{S}|\phi_0\rangle}$$

$$\hat{S} = T\exp\left[-i\int_{-\infty}^{\infty} V(t')dt'\right], \tag{7.1}$$

where $|\phi_0\rangle$ is the ground state of H_0. In chapter 6 we showed how the concept of adiabaticity was used to establish Landau Fermi-liquid theory. Now we move on to learn how to expand the fermion Green's function and other related quantities order by order in the strength of the interaction. The Feynman diagram approach, originally developed by Richard Feynman to describe the many-body physics of quantum electrodynamics [1] and later cast into a rigorous mathematical framework by Freeman Dyson [2], provides a succinct visual rendition of this expansion, a kind of "mathematical impressionism" which is physically intuitive, without losing mathematical detail.

From the Feynman rules, we learn how to evaluate

- the ground state S-matrix

$$S = \langle\phi_0|\hat{S}|\phi_0\rangle = \sum\{\text{unlinked Feynman diagrams}\} \tag{7.2}$$

- the logarithm of the S-matrix, which is directly related to the shift in the ground state energy due to interactions

$$E - E_0 = \lim_{\tau\to\infty}\frac{\partial}{\partial\tau}\ln\langle\phi_0|S[\tau/2, -\tau/2]|\phi_0\rangle = i\sum\{\text{linked Feynman diagrams}\}, \tag{7.3}$$

where each linked Feynman diagram describes a different sequence of virtual excitations
- Green's functions,

$$G(1 - 2) = \sum\{\text{two-leg Feynman diagrams}\} \tag{7.4}$$

- response functions, a different type of Green's function, of the form

$$\mathcal{R}(1 - 2) = -i\langle\phi|[A(1), B(2)]|\phi\rangle\theta(t_1 - t_2). \tag{7.5}$$

7.1 Heuristic derivation

Feynman initially derived his diagrammatic expansion as a mnemonic device for calculating scattering amplitudes. His approach was heuristic: each diagram has a physical meaning in terms of a specific scattering process. Feynman derived a set of rules that explained how to convert the diagrams into concrete scattering amplitudes. These rules were fine-tuned and tested in the simple cases where they could be checked by other means; later, he applied his method to cases where the direct algebraic approach was impossibly cumbersome. Later, Dyson gave his diagrammatic expansion a systematic mathematical framework.

Learning Feynman diagrams is a little like learning a language. You can learn the rules, and work by the book, but to really understand them you have to work with them, gaining experience in practical situations, learning them not just as theoretical constructs, but as living tools to communicate ideas. One can be a beginner or an expert, but to make them work for you, like a language or a culture, you will have to fall in love with them!

Formally, a perturbation theory for the fully interacting S-matrix is obtained by expanding the S-matrix as a power series, then using Wick's theorem to write the resulting correlation functions as a sum of contractions:

$$(7.6)$$

$$\langle \phi_0 | \hat{S} | \phi_0 \rangle = \sum_{n=0}^{\infty} \frac{(-i)^n}{n!} \int_{-\infty}^{\infty} dt_1 \cdots dt_n \sum_{\text{contractions}} \langle \phi_0 | T \, V(t_1) \, V(t_2) \cdots V(t_n) \, | \phi_0 \rangle.$$

$$(7.7)$$

The Feynman rules tell us how to organize these contractions as a sum of diagrams, where each diagram provides a precise graphical representation of a scattering amplitude that contributes to the complete S-matrix.

Let us see how we might heuristically develop a Feynman diagram expansion for simple potential scattering, for which

$$V(1) \equiv \int d^3x_1 U(\vec{x}_1) \psi^\dagger(\vec{x}_1, t_1) \psi(\vec{x}_1, t_1), \qquad (7.8)$$

where we've suppressed spin indices into the background. When we start to make contractions we will break up each product $V(1)V(2) \cdots V(r)$ into pairs of creation and annihilation operators, replacing each pair as follows:

$$\psi(2) \cdots \psi^\dagger(1) \rightarrow (\sqrt{i})^2 \times G(2-1), \qquad (7.9)$$

where we have divided up the prefactor i into two factors of $\sqrt{i}$, which we will transfer onto the scattering amplitudes where the particles are created and annihilated. This contraction is denoted by

$$G(2 - 1) = 2 \longleftarrow 1,$$

$$(7.10)$$

representing the propagation of a particle from 1 to 2. Pure potential scattering gives us one incoming and one outgoing propagator, so we denote a single potential scattering event by the diagram

$$-iU(x) = (\sqrt{i})^2 \times -iU(x) \equiv U(x).$$

$$(7.11)$$

Here, the "$-i$" has been combined with the two factors of $\sqrt{i}$ taken from the incoming and outgoing propagators to produce a purely *real* scattering amplitude $(\sqrt{i})^2 \times -iU(x) = U(x)$.

The Feynman rules for pure potential scattering tell us that the S-matrix for potential scattering can be organized as the exponential of a sum of connected *vacuum diagrams*:

$$S = \exp\left[\bigcirc + \bigcirc + \bigcirc + \cdots \right].$$

$$(7.12)$$

The vacuum diagrams appearing in the exponential do not have any incoming or outgoing propagators because fermions are conserved at each scattering event. These diagrams describe the various processes by which electron–hole pairs can bubble out of the vacuum. Let us examine the first- and second-order contractions for potential scattering. To first order,

$$-i\langle\phi_0| V (t_1) |\phi_0\rangle = -i \sum_\sigma \int d^3x U(x) \langle\phi_0|T \psi_\sigma^\dagger (x, t_1^+) \psi_\sigma(x, t_1^-) |\phi_0\rangle.$$

$$(7.13)$$

This contraction describes a single scattering event at $(\vec{x}, t_1)$. Note that the creation operator occurs to the *left* of the annihilation operator, and, to preserve this ordering inside the time-ordered exponential, we say that the particle propagates "backwards in time" from $t = t_1^+$ to $t = t_1^-$. When we replace this term by a propagator, the backwards time propagation introduces a factor of $\zeta = -1$ for fermions, so that

$$\langle\phi_0|T \psi_\sigma^\dagger(x,t_1^+)\psi_\sigma(x,t_1^-)|\phi_0\rangle = i\zeta G(\vec{x} - \vec{x}, t_1^- - t_1^+) = i\zeta G(\vec{0}, 0^-).$$

$$(7.14)$$

We carry along the factor $U(\vec{x})$ as the amplitude for this scattering event. The result of this contraction procedure is then

$$-i \int_{-\infty}^{\infty} dt_1 \langle\phi_0| V (t_1) |\phi_0\rangle = -i(2S + 1) \int dt_1 \times \int d^3x U(x) \times i\zeta G(\vec{0}, 0^-)$$

$$= \bigcirc ,$$

$$(7.15)$$

Table 7.1 Real-space Feynman rules ($T = 0$).	
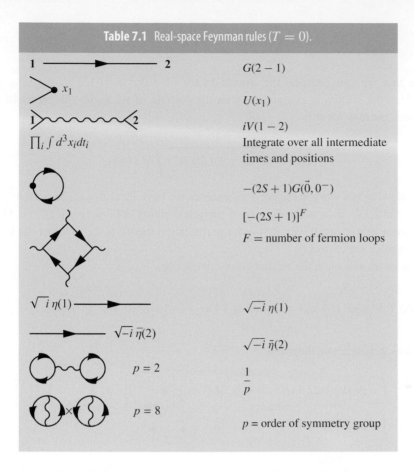	$G(2 - 1)$
	$U(x_1)$
	$iV(1 - 2)$
	Integrate over all intermediate times and positions
	$-(2S + 1)G(\vec{0}, 0^-)$
	$[-(2S + 1)]^F$
	F = number of fermion loops
	$\sqrt{-i}\,\eta(1)$
	$\sqrt{-i}\,\bar{\eta}(2)$
	$\dfrac{1}{p}$
	p = order of symmetry group

where we have translated the scattering amplitude into a single diagram. You can think of it as the spontaneous creation and re-annihilation of a single particle. Here we may tentatively infer a number of important *Feynman rules* listed in Table 7.1, associating each scattering event with an amplitude $U(x)$, connected by propagators that describe the amplitude for electron motion between scattering events. The overall amplitude involves an integration over the space–time co-ordinates of the scattering events, and, apparently, when a particle loop appears, we need to introduce the factor $\zeta(2S + 1)$ (where $\zeta = -1$ for fermions) into the scattering amplitude to account for the presence of an odd number of backwards-time propagators and the $2S + 1$ spin components of the particle field. These rules are summarized in Table 7.1

Physically, the vacuum diagram we have drawn here can be associated with the small first-order shift in the energy ΔE_1 of the particle due to the potential scattering. This in turn produces a phase shift in the scattering S-matrix,

$$S \sim \exp\left[-i\Delta E_1 \int dt\right] \sim 1 - i\Delta E_1 \int dt, \tag{7.16}$$

where the exponential has been audaciously expanded to linear order in the strength of the scattering potential. If we compare this result with our leading Feynman diagram expansion of the S-matrix,

$$\langle \phi_o | \hat{S} | \phi_o \rangle = 1 + \bullet\!\bigcirc\!,$$

we see that we can interpret the overall factor of $\int dt_1$ in (7.15) as the time period over which the scattering potential acts on the particle. If we factor this term out of the expression we may identify

$$\Delta E_1 = i\zeta(2S + 1)\overbrace{G(\vec{0}, 0^-)}^{\rho} \int d^3x U(x). \tag{7.17}$$

Here, following our work in the previous chapter, we have identified $i\zeta(2S + 1)G(\vec{0}, 0^-) = \sum_\sigma \langle \psi_\sigma^\dagger(x)\psi_\sigma(x) \rangle = \rho$ as the density of particles, giving $\Delta E_1 = \rho \int d^3x U(x)$. The correspondence of our result with first-order perturbation theory is a check that the tentative Feynman rules are correct.

Let us go on to look at the second order-contractions,

$$\langle \phi_0 | \, T \, V(t_1) \, V(t_2) \, | \phi_0 \rangle = \langle \phi_0 | T \, \overset{\sqcap}{V}(t_1) \, \overset{\sqcap}{V}(t_2) \, | \phi_0 \rangle + \langle \phi_0 | T \, \overset{\sqsupset}{V}(t_1) \, \overset{\sqsupset}{V}(t_2) | \phi_0 \rangle, \tag{7.18}$$

which now generate two diagrams:

$$\frac{1}{2!}(-i)^2 \int_{-\infty}^{\infty} dt_1 dt_2 \langle \phi_0 | T \, \overset{\sqcap}{V}(t_1) \quad \overset{\sqcap}{V}(t_2) \, | \phi_0 \rangle = \frac{1}{2}\left[\bullet\!\bigcirc\right]^2 = \left[\bigcirc \quad \bullet\!\bigcirc\right]$$

$$\frac{1}{2!}(-i)^2 \int_{-\infty}^{\infty} dt_1 dt_2 \langle \phi_0 | T \, \overset{\sqsupset}{V}(t_1) \quad \overset{\sqsupset}{V}(t_2) \, | \phi_0 \rangle = \bullet\!\bigcirc\!\bullet. \tag{7.19}$$

The first term is simply a product of two first-order terms – the beginning of an exponential combination of such terms. Notice how the square of one diagram is the original diagram, repeated twice. The factor of $\frac{1}{2}$ that occurs in the expression on the left-hand side is absorbed into this double diagram as a so-called *symmetry factor*. We shall return to this issue shortly, but, briefly, this diagram has a permutation symmetry described by a group of dimension $d = 2$; according to the Feynman rules, this generates a prefactor $1/d = \frac{1}{2}$. The second term derives from the second-order shift in the particle energies due to scattering, and, like the first-order shift, it produces a phase shift in the S-matrix. This diagram has a cyclic group symmetry of dimension $d = 2$, and once again there is a symmetry factor of $1/d = \frac{1}{2}$. This connected, second-order diagram gives rise to the scattering amplitude

$$\bullet\!\bigcirc\!\bullet = \frac{1}{2}\zeta(2S + 1) \int d1 d2 U(1)U(2)G(1 - 2)G(2 - 1), \tag{7.20}$$

where $1 \equiv (\vec{x}_1, t_1)$, so that

$$\int d1 \equiv \int dt_1 d^3x_1$$

$$G(2 - 1) \equiv G(\vec{x}_2 - \vec{x}_1, t_2 - t_1). \tag{7.21}$$

Once again, the particle loop gives a factor $\zeta(2S+1)$, and the amplitude involves an integral over all possible space–time co-ordinates of the two scattering events. You may interpret this diagram in various ways – as the creation of a particle–hole pair at $(\vec{x}_1, t_1)$ and its subsequent reannihilation at $(\vec{x}_2, t_2)$, or vice-versa. Alternatively, we can adopt an idea that Feynman developed as a graduate student with John Wheeler: the idea that an antiparticle (or hole) is a particle propagating backwards in time. From this perspective, this second order diagram represents a *single* particle that propagates around a loop in space–time. Equation (7.20) can be simplified by first making the change of variables $t = t_1 - t_2$, $T = (t_1 + t_2)/2$, so that $\int dt_1 dt_2 = \int dT \int dt$. Next, if we Fourier transform the scattering potential and Green's functions, we obtain

$$
\bigcirc = \int dT \times \frac{1}{2}\zeta(2S+1) \int dt d^3q\, d^3k |U(\vec{q}_1)|^2 G(\vec{k}+\vec{q}, t) G(\vec{k}, -t). \tag{7.22}
$$

Once again, a time integral factors out of the overall expression, and we can identify the remaining term as the *second-order* shift in the energy:

$$
\Delta E_2 = \frac{i}{2}\zeta(2S+1) \int dt \frac{d^3k}{(2\pi)^3} \frac{d^3q}{(2\pi)^3} |U(\vec{q}_1)|^2 G(\vec{k}+\vec{q}, t) G(\vec{k}, -t). \tag{7.23}
$$

To check that this result is correct, let us consider the case of fermions, where

$$
G(\mathbf{k}, t) = -i[(1 - n_{\mathbf{k}})\theta(t) - n_{\mathbf{k}}\theta(-t)]e^{-i\epsilon_{\mathbf{k}}t}, \tag{7.24}
$$

which enables us to do the integral

$$
i \int dt\, e^{-\delta|t|} G(\vec{k}+\vec{q}, t) G(\vec{k}, -t) = \frac{(1 - n_{\mathbf{k+q}})n_{\mathbf{k}}}{\epsilon_{\mathbf{k+q}} - \epsilon_{\mathbf{k}}} + (\mathbf{k} \leftrightarrow \mathbf{k+q}). \tag{7.25}
$$

We recognize the first process as the virtual creation of an electron of momentum $\vec{k}+\vec{q}$, leaving behind a hole in the state with momentum $\vec{k}$. The second term is simply a duplicate of the first with the momenta interchanged, and the sum of the two terms cancels the factor of $\frac{1}{2}$ in front of the integral. The final result,

$$
\Delta E_2 = -(2S+1) \int \frac{d^3k}{(2\pi)^3} \frac{d^3q}{(2\pi)^3} |U(\vec{q})|^2 \frac{(1 - n_{\mathbf{k+q}})n_{\mathbf{k}}}{\epsilon_{\mathbf{k+q}} - \epsilon_{\mathbf{k}}},
$$

is recognized as the second order correction to the energy derived from these virtual processes. Of course, we could have derived these results directly, but the important point is that we have established a tentative link between the diagrammatic expansion of the contractions and the perturbation expansion for the ground state energy. Moreover, we begin to see that our diagrams have a direct interpretation in terms of the virtual excitation processes that are generated by the scattering events.

To second order, our results do indeed correspond to the leading-order terms in the exponential

$$
S = 1 + \left[\bigcirc + \bigcirc \cdots \right] + \frac{1}{2!}\left[\bigcirc + \cdots \right]^2 + \cdots = \exp\left[\bigcirc + \bigcirc + \cdots \right].
$$

Before we go on to complete this connection more formally in the next section, we need to briefly discuss *source terms*, which couple directly to the creation and annihilation operators. The source terms let us examine how the S-matrix responds to incoming currents of particles. Source terms add directly to the scattering potential, so that

$$V(1) \rightarrow V(1) + \bar{\eta}(1)\psi(1) + \psi^\dagger(1)\eta(1).$$

The source-term $\eta(1)$ couples to the creation of a particle, giving rise to the process

$$\underrightarrow{\hspace{2cm}} \sqrt{-i}\,\eta(1) \equiv \sqrt{-i} \int d1 \cdots \times \eta(1), \tag{7.26}$$

where the "$-i$" from the time-ordered exponential has been combined with the "$\sqrt{i}$" associated with the propagator to produce the $\sqrt{-i} = -i\sqrt{i}$ prefactor. Similarly, the source-term $\bar{\eta}(2)$ couples to the annihilation of a particle, giving rise to the process

$$\sqrt{-i}\,\bar{\eta}(2)\underrightarrow{\hspace{2cm}} \equiv \sqrt{-i} \int d2\bar{\eta}(2) \times \cdots. \tag{7.27}$$

These terms generate diagrams which involve incoming and outgoing electrons. The simplest contraction with these terms generates the bare propagator

$$\frac{(-i)^2}{2!} \int d2d1 \langle 0| \left[V(2) + \bar{\eta}(2)\psi(2) + \psi^\dagger(2)\eta(2) \right] \left[V(1) + \bar{\eta}(1)\psi(1) + \psi^\dagger(1)\eta(1) \right] |0\rangle$$
$$= \int d1d2 \left(\sqrt{-i}\bar{\eta}(2)G(2-1)\sqrt{-i}\eta(1) \right)$$
$$= -i\bar{\eta}\underrightarrow{\hspace{2cm}}\eta.$$

$$\tag{7.28}$$

If we now include the contraction with the first scattering term, we produce the first scattering correction to the propagator:

$$\frac{(-i)^3}{3!} \int d2dX\, d1 \langle 0| \left\{ [\cdots + \bar{\eta}(2)\psi(2) + \cdots] \left[U(X)\psi^\dagger(X)\psi(X) + \cdots \right] \left[\cdots + \psi^\dagger(1)\eta(1) \right] + \text{perms} \right\} |0\rangle$$
$$= \int d1d2 \left(\sqrt{-i}\bar{\eta}(2) \int dX G(2-X)V(X)G(X-1) \sqrt{-i}\eta(1) \right)$$
$$= -i\bar{\eta}\underrightarrow{\hspace{1cm}}\bullet\underrightarrow{\hspace{1cm}}\eta,$$

$$\tag{7.29}$$

where we have only shown one of *six* equivalent contractions on the first line. This diagram is simply interpreted as a particle, created at 1, scattering at position X before propagating onwards to position 2. Notice how we must integrate over the the space–time coordinate of the intermediate scattering event at X, to obtain the total first-order scattering amplitude. Higher-order corrections will merely generate multiple insertions into the propagator and we will have to integrate over the space–time coordinate of each of these scattering events. Diagrammatically, the sum over all such diagrams generates the *renormalized propagator*, denoted by

$$G^*(2-1) = 2 \longleftarrow 1$$

$$= 2 \longleftarrow 1 \;+\; \; + \; \; + \cdots. \tag{7.30}$$

Indeed, to second order in the scattering potential, we can see that all the allowed contractions are consistent with the following exponential form for the generating functional:

$$S = exp\left[\; \bigcirc + \bigcirc + \cdots - i\bar{\eta}\,{\longleftarrow}\,\eta \right]. \tag{7.31}$$

To prove this result formally requires a more detailed proof that we discuss in the next section. However, we have illustrated the key result, that the sum of S-matrix contractions is represented by a series of diagrams which concisely represent the scattering amplitudes for all possible virtual excitations of the vacuum.

7.2 Developing the Feynman diagram expansion

A neat way to organize the Feynman diagram expansion is the source-term approach we encountered in Chapter 5. There we found we could evaluate the response of a non-interacting system to a source term which injected and removed particles. We start with the source-term S-matrix,

$$\hat{S}[\bar{\eta}, \eta] = T \exp\left(-i \int d1 [\psi^\dagger(1)\eta(1) + \bar{\eta}(1)\psi(1)] \right). \tag{7.32}$$

Here, for convenience, we shall hide details of the spin away with the space–time coordinate, so that $1 \equiv (\mathbf{x}_1, t_1, \sigma_1)$, $\psi(1) \equiv \psi_\sigma(\mathbf{x}, t)$. You can think of the quantities $\eta(1)$ and $\bar{\eta}(1)$ as "control knobs" which tune the rate at which we are adding or subtracting particles to the system. For fermions, these numbers must be anticommuting Grassman numbers which also anticommute with the Fermi field operators. In Section 5.3.2 we found that the vacuum expectation value of the S-matrix is then given by

$$S[\bar{\eta}, \eta] = \langle \phi | \hat{S}[\bar{\eta}, \eta] | \phi \rangle = \exp\left[-i \int d1 d2 \bar{\eta}(2) G(2-1)\eta(1) \right], \tag{7.33}$$

where $G(2-1) \equiv G_{\sigma_2\sigma_1}(\mathbf{x}_2 - \mathbf{x}_1, t_2 - t_1)$ is the propagator from 1 to 2. In preparation for our diagrammatic approach, we shall denote

$$\int d1 d2 \bar{\eta}(2) G(2-1)\eta(1) = \bar{\eta} \,\longleftarrow\, \eta, \tag{7.34}$$

where an integral over the space–time variables $(\mathbf{x}_1, t_1)$ and $(\mathbf{x}_2, t_2)$ and a sum over spin variables σ_1, σ_2 are implied by the diagram. The S-matrix equation can then be written

$$S[\bar{\eta}, \eta] = \exp\left[-i\bar{\eta} \overset{\blacktriangleleft}{\text{————}} \eta\right].$$

(7.35)

This is a *generating functional*: by differentiating it with respect to the source terms, we can compute the many-particle Green's functions. Note that Grassman differentials also anticommute, both with each other and with fermionic field operators.[1] Each time we differentiate the S-matrix with respect to $\bar{\eta}(1)$, we pull down a field operator inside the time-ordered product:

$$i\frac{\delta}{\delta\bar{\eta}(1)} \to \psi(1)$$

$$i\frac{\delta}{\delta\bar{\eta}(1)} \langle\phi|T\{\hat{S}\cdots\}|\phi\rangle = \langle\phi|T\{\hat{S}\cdots\psi(1)\cdots\}|\phi\rangle.$$

(7.36)

For example, the field operator has an expectation value

$$\langle\psi(1)\rangle = \frac{\langle\phi|T\{\hat{S}\psi(1)\}|\phi\rangle}{\langle\phi|\hat{S}|\phi\rangle} = \frac{1}{S[\bar{\eta}, \eta]} i\frac{\delta S[\bar{\eta}, \eta]}{\delta\bar{\eta}(1)} = i\frac{\delta \ln S[\bar{\eta}, \eta]}{\delta\bar{\eta}(1)}$$

$$= \int G(1-2)\eta(2)d2$$

$$\equiv [1 \overset{\blacktriangleleft}{\text{————}} \eta].$$

(7.37)

Notice how the differential operator $i\frac{\delta}{\delta\bar{\eta}(1)}$ "grabs hold" of the end of a propagator and connects it up to space–time coordinate 1. Likewise, each time we differentiate the S-matrix with respect to $\eta(1)$, we pull down a field creation operator inside the time-ordered product:

$$i\zeta\frac{\delta}{\delta\eta(1)} \to \psi^\dagger(1).$$

(7.38)

The appearance of an additional "ζ" in (7.38) is needed for bookkeeping: it arises because the source term anticommutes with the field operators, $\psi^\dagger(1)\eta(1) = -\eta(1)\psi^\dagger(1)$, so that

$$\frac{\delta}{\delta\eta(1)}\int dX\psi^\dagger(X)\eta(X) = \zeta\frac{\delta}{\delta\eta(1)}\int dX\eta(X)\psi^\dagger(X) = \zeta\psi^\dagger(1)$$

(7.39)

and the expectation value of the creation operator has the value

$$\langle\psi^\dagger(2)\rangle = \frac{\langle\phi|\hat{S}[\bar{\eta}, \eta]\psi^\dagger(2)|\phi\rangle}{\langle\phi|\hat{S}[\bar{\eta}, \eta]|\phi\rangle} = i\zeta\frac{\delta \ln S[\bar{\eta}, \eta]}{\delta\eta(2)}$$

$$= \int d1\bar{\eta}(1)G(1-2)$$

$$\equiv [\bar{\eta} \overset{\blacktriangleleft}{\text{————}} 2].$$

(7.40)

---

[1] For example, if $F[\bar{\eta}, \eta] = \bar{A}\eta + \bar{\eta}A + B\bar{\eta}\eta$, where $A, \bar{A}, \eta$, and $\bar{\eta}$ are Grassman numbers, while B is a commuting number, then $\frac{\partial F}{\partial\bar{\eta}} = A + B\eta$, but $\frac{\partial F}{\partial\eta} = -\bar{A} - B\bar{\eta}$ because the differential operator anticommutes with $\bar{A}$ and $\bar{\eta}$. The second derivative $\frac{\partial^2 F}{\partial\eta\partial\bar{\eta}} = -\frac{\partial^2 F}{\partial\bar{\eta}\partial\eta} = B$, illustrating that the differential operators of Grassman numbers anticommute.

If we differentiate either (7.37) with respect to $\eta(2)$, or (7.40) with respect to $\bar{\eta}(1)$ we obtain

$$\frac{\delta}{\delta\eta(2)}\langle\psi(1)\rangle\bigg|_{\eta=\bar{\eta}=0} = \frac{\delta}{\delta\bar{\eta}(1)}\langle\psi^\dagger(2)\rangle\bigg|_{\eta=\bar{\eta}=0} = -i\langle\phi|T\psi(1)\psi^\dagger(2)|\phi\rangle = G(1-2),$$
(7.41)

as expected.

In general, we can calculate arbitrary functions of the field operators by acting on the S-matrix with the appropriate function of derivative operators:

$$\langle\phi|T\hat{S}[\bar{\eta},\eta]F[\psi^\dagger,\psi]|\phi\rangle = F\left[i\zeta\frac{\delta}{\delta\eta}, i\frac{\delta}{\delta\bar{\eta}}\right]\exp\left[-i\bar{\eta}\;\overleftarrow{}\;\eta\right].$$
(7.42)

If we now set $F[\psi^\dagger,\psi] = Te^{-i\int V[\psi^\dagger,\psi]dt}$, then the *interacting* generator

$$S_I[\bar{\eta},\eta] = \langle\phi|Te^{-i\int_{-\infty}^{\infty}dt(V(\psi^\dagger,\psi)+\text{sourceterms})}|\phi\rangle$$
(7.43)

can be written completely algebraically, in the form

$$S_I[\bar{\eta},\eta] = e^{-i\int_{-\infty}^{\infty}V(i\zeta\frac{\delta}{\delta\eta},i\frac{\delta}{\delta\bar{\eta}})dt}\exp\left[-i\bar{\eta}\;\overleftarrow{}\;\eta\right].$$
(7.44)

The action of the exponentiated differential operator on the source terms generates the contractions. It is convenient to recast this expression in a form that regroups the imaginary factors. To do this, we write $\alpha = \eta, \bar{\alpha} = -i\bar{\eta}$, enabling us to rewrite the expression as

$$S_I[\bar{\alpha},\alpha] = e^{(i)^{n-1}\int_{-\infty}^{\infty}V(\zeta\frac{\delta}{\delta\alpha},\frac{\delta}{\delta\bar{\alpha}})dt}\exp\left[\bar{\alpha}\;\overleftarrow{}\;\alpha\right].$$

where we have written

$$V(i\zeta\frac{\delta}{\delta\eta}, i\frac{\delta}{\delta\bar{\eta}}) = i^n V\left(\zeta\frac{\delta}{\delta\alpha}, \frac{\delta}{\delta\bar{\alpha}}\right)$$
(7.45)

for an n-body interaction. This equation provides the basis for Feynman diagram expansions.

To develop the Feynman expansion, we need to recast our expression in a graphical form. To see how this works, let us first consider a one-particle scattering potential ($n = 1$). In this case, we write

$$i^{n-1}\int dtV\left(\zeta\frac{\delta}{\delta\alpha}, \frac{\delta}{\delta\bar{\alpha}}\right) = \int d1U(1)\left(\zeta\frac{\delta^2}{\delta\alpha(1)\delta\bar{\alpha}(1)}\right),$$
(7.46)

which we denote as

$$\int d1U(1)\left(\zeta\frac{\delta^2}{\delta\alpha(1)\delta\bar{\alpha}(1)}\right) = \;\bullet\!\!\!\overset{\displaystyle\zeta\frac{\delta}{\delta\alpha}}{\underset{\displaystyle\frac{\delta}{\delta\bar{\alpha}}}{\Big\langle}}\quad.$$
(7.47)

Notice that the basic scattering amplitude for scattering at point x is simply $U(x)$ (or $U(x)/\hbar$ if we reinstate Planck's constant). Schematically, then, our Feynman diagram expansion can be written as

$$S_I[\bar{\alpha}, \alpha] = \exp\left[\begin{array}{c} \zeta \frac{\delta}{\delta \alpha} \\ \bullet \\ \frac{\delta}{\delta \bar{\alpha}} \end{array}\right] \exp\left[\bar{\alpha} \longleftarrow \alpha\right].$$

The differential operators acting on the bare S-matrix "glue" the scattering vertices to the ends of the propagators, and thereby generate a sum of all possible Feynman diagrams. Formally, we must expand the exponentials on both sides, e.g.

$$S_I[\bar{\alpha}, \alpha] = \sum_{n,m} \frac{1}{n! m!} \left[\begin{array}{c} \zeta \frac{\delta}{\delta \alpha} \\ \bullet \\ \frac{\delta}{\delta \bar{\alpha}} \end{array}\right]^n \left[\bar{\alpha} \longleftarrow \alpha\right]^m.$$

(7.48)

The action of the differential operator on the left-hand side is to glue the m propagators together with the n vertices to make a series of Feynman diagrams. Now at first sight this sounds pretty frightening – we will have a profusion of diagrams. Let us just look at a few: do not at this stage worry about the details, just try to get a feeling for the general structure. The simplest ($n = 1, m = 1$) term takes the form

$$\left[\begin{array}{c} \zeta \frac{\delta}{\delta \alpha} \\ \bullet \\ \frac{\delta}{\delta \bar{\alpha}} \end{array}\right]\left[\bar{\alpha} \longleftarrow \alpha\right] = \zeta \int d1 V(1) \frac{\delta^2}{\delta \alpha(1) \delta \bar{\alpha}(1)} \int dX dY \bar{\alpha}(X) G(X - Y) \alpha(Y)$$

$$= \zeta \int d1 V(1) G(1^- - 1) = \bullet\!\bigcirc.$$

(7.49)

This is the simplest example of a *linked-cluster diagram*, and it results from a single contraction of the scattering potential. The sign $\zeta = -1$ occurs for fermions, because the Fermi operators need to be interchanged to write the expression as a time-ordered propagator. One can say that the expectation value involves the fermion propagating backwards in time from time t to an infinitesimally earlier time $t^- = t - \epsilon$. The term for $n = 1, m = 2$ gives rise to two sets of diagrams, as follows:

$$\frac{1}{2}\left[\begin{array}{c} \zeta \frac{\delta}{\delta \alpha} \\ \bullet \\ \frac{\delta}{\delta \bar{\alpha}} \end{array}\right]\left[\bar{\alpha} \longleftarrow \alpha\right]^2 = \bar{\alpha} \longleftarrow \bullet \longleftarrow \alpha + \left[\bullet\!\bigcirc\right] \times \bar{\alpha} \longleftarrow \alpha.$$

(7.50)

The first term corresponds to the first scattering correction to the propagator, written out algebraically as

$$\bar{\alpha} \longleftarrow \bullet \longleftarrow \alpha = \int d1 d2 \bar{\alpha}(1) \alpha(2) \int dX G(1 - X) V(X) G(X - 2),$$

whereas the second term is an unlinked diagram involving a product of the bare propagator and the first linked-cluster diagram. The Feynman rules enable us to write each possible term in the expansion of the S-matrix as a sum of unlinked diagrams. Fortunately, we are able to systematically combine these diagrams into an exponential of linked diagrams:

$$S_I(\bar{\alpha}, \alpha) = \exp\left[\sum \text{linked diagrams}\right]$$

$$= \exp\left[\,\bigcirc + \bigcirc\!\bigcirc + \cdots + \bar{\alpha}\!=\!\!\blacktriangleleft\!\!=\!\alpha\right]. \tag{7.51}$$

This compact reformulation of the S-matrix is a result of the linked-cluster theorem, a powerful result we are shortly to encounter. The Feynman rules tell us how to convert these diagrams into mathematical expressions (see Table 7.1).

Let us now look at how the same procedure works for a two-particle interaction. Heuristically, we expect a two-body interaction to involve two incoming and two outgoing propagators. We shall denote a two-body scattering amplitude by the following diagram:

$$1\,\langle\!\!\!\sim\!\!\!\sim\!\!\!\rangle\,2 = (\sqrt{i})^4 \times -iV(1-2) \equiv iV(1-2). \tag{7.52}$$

Notice how, in contrast to the one-body scattering amplitude, we pick up four factors of $\sqrt{i}$ from the external legs, so that the net scattering amplitude involves an awkward factor of i. If we now proceed using the generating function approach, we set $n = 2$ and then write

$$i^{n-1}V(\zeta\frac{\delta}{\delta\alpha}, \frac{\delta}{\delta\bar{\alpha}}) = i\frac{1}{2}\int d^3x d^3x' V(x - x')\frac{\delta}{\delta\alpha(x)}\frac{\delta}{\delta\alpha(x')}\frac{\delta}{\delta\bar{\alpha}(x')}\frac{\delta}{\delta\bar{\alpha}(x)}. \tag{7.53}$$

The two-particle scattering amplitude is simply $iV(x - x')$. We can formally denote the scattering vertex as

$$\frac{1}{2}\,\overset{\frac{\delta}{\delta\alpha(2)}}{\underset{\frac{\delta}{\delta\bar{\alpha}(2)}}{}}\,\rangle\!\!\!\sim\!\!\!\langle\,\overset{\frac{\delta}{\delta\alpha(1)}}{\underset{\frac{\delta}{\delta\bar{\alpha}(1)}}{}}\,. \tag{7.54}$$

This gives rise to the following expression for the generating functional:

$$\mathcal{S}_I[\bar{\alpha}, \alpha] = \exp\left[\frac{1}{2}\,\overset{\frac{\delta}{\delta\alpha(2)}}{\underset{\frac{\delta}{\delta\bar{\alpha}(2)}}{}}\,\rangle\!\!\!\sim\!\!\!\langle\,\overset{\frac{\delta}{\delta\alpha(1)}}{\underset{\frac{\delta}{\delta\bar{\alpha}(1)}}{}}\,\right]\exp\left[\bar{\alpha}\!\longrightarrow\!\!\blacktriangleleft\!\longrightarrow\!\alpha\right],$$

for the S-matrix of interacting particles.

As in our earlier example, the differential operators attach the scattering vertices to the propagators, generating a sum of all Feynman diagrams. Once again, we are supposed to formally expand the exponentials on both sides, e.g.

$$S_I[\bar{\alpha}, \alpha] = \sum_{n,m} \frac{1}{n!m!} \left[\frac{1}{2} \underset{\frac{\delta}{\delta\alpha(2)}}{\overset{\frac{\delta}{\delta\alpha(2)}}{}} \underset{\frac{\delta}{\delta\alpha(1)}}{\overset{\frac{\delta}{\delta\alpha(1)}}{}} \right]^n \left[\bar{\alpha} \longleftarrow \alpha \right]^m.$$

(7.55)

Let us again look at some of the leading diagrams that appear in this process. For instance,

$$\frac{1}{2!}\left[\frac{1}{2}\ \underset{\frac{\delta}{\delta\alpha(2)}}{\overset{\frac{\delta}{\delta\alpha(2)}}{}}\ \underset{\frac{\delta}{\delta\alpha(1)}}{\overset{\frac{\delta}{\delta\alpha(1)}}{}}\right]\left[\bar{\alpha}\longleftarrow\alpha\right]^2 = \bigcirc\!\!\sim\!\!\bigcirc + \bigcirc\!\!\!\!\bigcirc.$$

(7.56)

We shall see later that these are the Hartree–Fock contributions to the ground-state energy. At each of the vertices in these diagrams, we must integrate over the space–time co-ordinates and sum over the spins. Since spin is conserved along each propagator, this means that each loop has a factor of $(2S + 1)$ associated with the spin sum. Once again, for fermions we have to be careful about the minus signs. For each particle loop, there is always an odd number of fermion propagators propagating backwards in time, and this gives rise to a factor

$$\zeta(2S + 1) = -(2S + 1)$$

(7.57)

per fermion loop. The algebraic rendition of these Feynman diagrams is then

$$\frac{1}{2} \int d1d2V(1-2)\left[(2S+1)^2 G(0,0^-)^2 + \zeta(2S+1)G(1-2)G(2-1) \right].$$

(7.58)

The prefactor of $\frac{1}{2}$ is a symmetry factor, arising because each of the diagrams is invariant under a $p = 2$ dimensional group of vertex permutations, which reduces the number of independent contractions by a factor of p (see Section 7.2.1). Notice, finally, that the first Hartree diagram contains a propagator which "bites its own tail." This comes from a contraction of the density operator,

$$-i\sum_{\sigma}\langle \cdots \psi_{\sigma}^{\dagger}(\mathbf{x}, t)\psi_{\sigma}(\mathbf{x}, t) \cdots \rangle = \zeta(2S+1)G(\mathbf{x}, 0^-),$$

(7.59)

and since the creation operator lies to the left of the destruction operator, we pick up a minus sign for fermions. As a second example, consider

$$\frac{1}{3!}\left[\frac{1}{2}\ \underset{\substack{\frac{\delta}{\delta\alpha(2)}}}{\overset{\substack{\frac{\delta}{\delta\alpha(2)}}}{\times}}\ \overset{\substack{\frac{\delta}{\delta\alpha(1)}}}{\underset{\substack{\frac{\delta}{\delta\alpha(1)}}}{\Bigg]}}\left[\bar\alpha\ \underset{\longleftarrow}{\rule{3cm}{0pt}}\ \alpha\right]^3 = \bar\alpha\left[\ \ +\ \ \right]\alpha\ ,$$

corresponding to the Hartree–Fock corrections to the propagator. Notice how a similar minus sign is associated with the single fermion loop in the Hartree self-energy. By convention, the numerical prefactors are implicitly absorbed into the Feynman diagrams by introducing two more rules: one states that each fermion loop gives a factor of ζ, and the other relates the numerical prefactor to the symmetry of the Feynman diagram. When we add all of these terms, the S-matrix becomes

$$\mathcal{S}_I(\bar\alpha,\alpha) = 1 + \left[\ \ +\ \ +\ \ +\cdots\right]$$

$$+\ \bar\alpha\left[\ \ +\ \ +\ \ +\cdots\right]\alpha$$

$$+\cdots$$

$$+\ \frac{1}{2}\left[\ \times\ \right]+\left[\ \times\ \ +\cdots\right].$$

(7.60)

The diagrams on the first line are linked-cluster diagrams: they describe the creation of virtual particle–hole pairs in the vacuum. The second line of diagrams are the one-leg diagrams, which describe the one-particle propagators. There are also higher-order diagrams (not shown) with $2n$ legs, coupled to the source terms, corresponding to the n-particle Green's functions. The diagrams on the third line are *unlinked* diagrams. We shall shortly see that we can remove these diagrams by taking the logarithm of the S-matrix.

7.2.1 Symmetry factors

Remarkably, in making the contractions of the S-matrix, the prefactors in terms like (7.55) are almost completely absorbed by the combinatorics. Let us examine the number of ways of making the contractions between the two terms in (7.55). Our procedure for constructing a diagram is illustrated in Figure 7.1.

1. We label each propagator on the Feynman diagram 1 through m and label each vertex on the Feynman diagram (1) through (n).
2. The process of making a contraction corresponds to identifying each vertex and each propagator in (7.55) with each vertex and propagator in the Feynman diagram under construction. Thus the P'_rth propagator is placed at position r on the Feynman diagram,

(a)

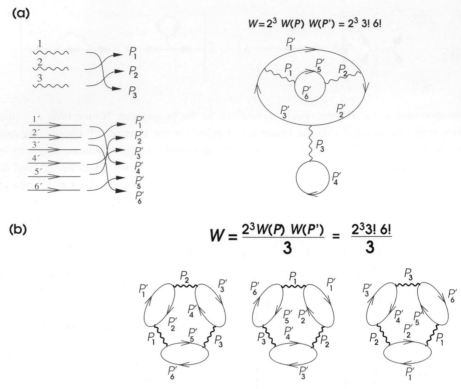

$W = 2^3 \, W(P) \, W(P') = 2^3 \, 3! \, 6!$

(b)

$$W = \frac{2^3 W(P) \, W(P')}{3} = \frac{2^3 3! \, 6!}{3}$$

Fig. 7.1 (a) Showing how six propagators and three interaction lines can be arranged on a Feynman diagram of low symmetry ($p = 1$). (b) In a Feynman diagram of high symmetry, each possible assignment of propagators and interaction lines to the diagram belongs to a p-tuplet of topologically equivalent assignments, where p is the order of the symmetry group of permutations under which the topology of the diagram is unchanged. In the example shown above, $p = 3$ is the order of the symmetry group. In this case, we need to divide the number of assignments W by a factor of p.

    and the P_kth interaction line is placed at position k on the Feynman diagram, where P is a permutation of $(1, \ldots n)$ and P' a permutation of $(1, \ldots m)$.

3. Since each interaction line can be arranged two ways at each location in the diagram, there are $2^n W(P) = 2^n n!$ ways of putting down the interaction vertices and $W(P') = m!$ ways of putting down the propagators on the Feynman diagram, giving a total of $W = 2^n n! \, m!$ ways.

4. The most subtle point is that if the connectivity of the Feynman graph is invariant under certain permutations of the vertices, then the above procedure overcounts the number of independent contractions by a *symmetry factor* p, where p is the dimension of the group of permutations under which the connectivity of the diagram is unchanged. The point is that each of the $2^n n! \, m!$ choices made in (2) actually belongs to a p-tuplet of different choices which combine the propagators and vertices in exactly the same configuration. To adjust for this overcounting, we need to divide the number of choices by the symmetry factor p, so that the number of ways of making the same Feynman graph is

$$W = \frac{2^n n! \, m!}{p}. \tag{7.61}$$

As an example, consider the simplest diagram,

$$\tag{7.62}$$

This diagram is topologically invariant under the group of permutations

$$\mathcal{G} = \{(12), (21)\}, \tag{7.63}$$

so $p = 2$. In a second example,

$$\tag{7.64}$$

the invariance group is

$$\mathcal{G} = \{(1234), (3412)\}, \tag{7.65}$$

so once again $p = 2$. By contrast, for the diagram

$$\tag{7.66}$$

the invariance group is

$$\mathcal{G} = \{(1234), (3412), (2143), (4321)\}, \tag{7.67}$$

so that $p = 4$.

7.2.2 Linked-cluster theorem

One of the major simplifications in developing a Feynman diagram expansion arises because of the *linked-cluster theorem*. Ultimately, we are more interested in calculating the logarithm of the S-matrix, $\ln S(\bar{\eta}, \eta)$. This quantity determines the energy shift due to interactions, but it also provides the n-particle (connected) Green's functions. In the Feynman diagram expansion of the S-matrix, we saw that there are two types of diagram: linked-cluster diagrams and unlinked diagrams, which are actually products of linked-cluster diagrams. The linked-cluster theorem states that the logarithm of the S-matrix involves just the sum of the linked-cluster diagrams:

$$\ln S_I[\bar{\alpha}, \alpha] = \sum \{\text{linked-cluster diagrams}\}. \tag{7.68}$$

To show this result, we shall employ a trick called the *replica trick*, invented by the British physicist Sam Edwards [3], which takes advantage of the relation

$$\ln S = \lim_{n \to 0} \left[\frac{S^n - 1}{n} \right]. \tag{7.69}$$

In other words, if we expand $S^n = e^{n \ln S} = 1 + n \ln S + O(n^2)$ as a power series in n, then the linear coefficient in the expansion will give us the logarithm of S. It proves much easier to evaluate S^n diagrammatically. To do this, we introduce n identical but independent replicas of the original system, each labeled by $\lambda = (1, n)$. The Hamiltonian of the replicated system is just $H = \sum_{\lambda=1,n}$ and, since the operators of each replica live in a completely independent Hilbert space, they commute. This permits us to write

$$(S_I[\bar{\alpha}, \alpha])^n = \langle \phi | T \exp \left[-i \int_{-\infty}^{\infty} dt \sum_{\lambda=1,n} \left(V(\psi_\lambda^\dagger, \psi_\lambda) + \text{source terms} \right) \right] | \phi \rangle. \tag{7.70}$$

When we expand this, we will generate exactly the same Feynman diagrams as in S, except that now, for each linked Feynman diagram, we will have to multiply the amplitude by n. The diagram expansion for interacting fermions will look like

$[S_I(\bar{\alpha}, \alpha)]^n = 1$

$$\tag{7.71}$$

We see that the coefficient of n in the replica expansion of S^n is equal to the sum of the linked-cluster diagrams, so that

$$\ln S_I(\bar{\alpha}, \alpha) = \lim_{n \to 0} \left[\frac{S[\bar{\alpha}, \alpha]^n - 1}{n} \right]$$

By differentiating the log of the S-matrix with respect to the source terms, we extract the one-particle Green's functions as the sum of all two-leg diagrams:

$$G(2-1) = \frac{\delta^2 \ln S_I(\bar{\alpha}, \alpha)}{\delta \alpha(1) \delta \bar{\alpha}(2)} \bigg|_{\bar{\alpha}, \alpha=0} = \sum \{\text{two-leg diagrams}\}$$

$$\tag{7.72}$$

This is a quite non-trivial result. Were we to have attempted a head-on Feynman diagram expansion of the Green's function using the Gell-Mann–Low theorem,

$$G(1-2) = -i\frac{\langle\phi|TS\psi(1)\psi^\dagger(2)|\phi\rangle}{\langle\phi|S|\phi\rangle}, \tag{7.73}$$

we would have to consider the quotient of two sets of Feynman diagrams, coming from the contractions of the denominator and numerator. Remarkably, the unlinked diagrams of the S-matrix in the numerator cancel the unlinked diagrams appearing in the Wick expansion of the denominator, leaving us with this elegant expansion in terms of two-leg diagrams.

Just as the terms which are first order in α and $\bar\alpha$ determine the one-body Green's functions, the higher-order derivatives with respect to α and $\bar\alpha$ correspond to the connected n-body Green's functions. For example, working to fourth order,

$$\ln S_I(\bar\alpha, \alpha) = \ln S[0] + \int d1d2\,\bar\alpha(1)G(1-2)\alpha(2)$$

$$+ \frac{1}{(2!)^2}\int d1d2d3d4\,\bar\alpha(1)\bar\alpha(2)G(1,2;3,4)\alpha(3)\alpha(4) + O(\bar\alpha^3,\alpha^3), \tag{7.74}$$

where the factors of $1/2!$ are introduced as part of the Taylor series expansion. Diagrammatically,

$$\tag{7.75}$$

where

$$\tag{7.76}$$

is the sum of all four-leg connected diagrams. This is the connected part of the two-particle Green's function which describes the scattering of fermions off one another. For interacting fermions, the first few diagrams for $G(1,2;3,4)$ are given by

$$\tag{7.77}$$

where the minus signs are associated with the odd parity of fermions under particle exchange.

Example 7.1

(a) By differentiating the S-matrix, show that

$$\frac{1}{S[\bar{\alpha},\alpha]}\frac{\delta^4 S}{\delta\bar{\alpha}(1)\delta\bar{\alpha}(2)\delta\alpha(3)\delta\alpha(4)} = -\langle\phi|T[\psi(1)\psi(2)\psi^\dagger(3)\psi^\dagger(4)]|\phi\rangle. \tag{7.78}$$

(b) By exponentiating the linked-cluster expansion of $\ln S[\bar{\alpha},\alpha]$, show that the two-particle Green's function is given by

$$\frac{1}{S[\bar{\alpha},\alpha]}\frac{\delta^4 S}{\delta\bar{\alpha}(1)\delta\bar{\alpha}(2)\delta\alpha(3)\delta\alpha(4)} = -\langle\phi|T[\psi(1)\psi(2)\psi^\dagger(3)\psi^\dagger(4)]|\phi\rangle$$

$$\tag{7.79}$$

Solution

(a) In terms of $\alpha \equiv \eta$ and $\bar{\alpha} \equiv -i\bar{\eta}$, the generating function given by the coupling of the source terms to the fields leads to the following S-matrix:

$$S[\bar{\alpha},\alpha] = \langle\phi_0|T\exp\left[\int dt\left(-iV(t) + \bar{\alpha}(1)\psi(1) - i\psi^\dagger(1)\alpha(1)\right)\right]|\phi_0\rangle, \tag{7.80}$$

where $|\phi_0\rangle$ is the non-interacting ground state, so that, inside the time-ordered product, derivatives with respect to α and $\bar{\alpha}$ generate field operators as follows:

$$\frac{\delta}{\delta\alpha(1)} \rightarrow -i\zeta\psi^\dagger(1), \qquad \frac{\delta}{\delta\bar{\alpha}(1)} \rightarrow \psi(1). \tag{7.81}$$

When we differentiate the S-matrix four times, we obtain

$$\frac{\delta^2 S[\bar{\alpha},\alpha]}{\delta\alpha(4)} = -i\zeta\langle\phi_0|T[\hat{S}\psi^\dagger(4)|\psi_0\rangle$$

$$\frac{\delta^4 S[\bar{\alpha},\alpha]}{\delta\bar{\alpha}(1)\delta\bar{\alpha}(2)\delta\alpha(3)\delta\alpha(4)} = (-i\zeta)^2\langle\phi_0|T[\hat{S}\psi(1)\psi(2)\psi^\dagger(3)\psi^\dagger(4)|\psi_0\rangle$$

$$= -\langle\phi_0|T[\hat{S}\psi(1)\psi(2)\psi^\dagger(3)\psi^\dagger(4)|\psi_0\rangle, \tag{7.82}$$

and hence

$$\frac{1}{S[\bar{\alpha},\alpha]}\frac{\delta^4 S[\bar{\alpha},\alpha]}{\delta\bar{\alpha}(1)\delta\bar{\alpha}(2)\delta\alpha(3)\delta\alpha(4)} = -\frac{\langle\phi_0|T[\hat{S}\psi(2)\psi^\dagger(3)\psi^\dagger(4)]|\psi_0\rangle}{\langle\phi_0|\hat{S}|\phi_0\rangle}$$

$$= -\langle\phi|T[\psi(2)\psi^\dagger(3)\psi^\dagger(4)]|\psi\rangle \tag{7.83}$$

by the Gell-Mann–Low formula, where $|\psi\rangle$ is the fully interacting ground state.

(b) If we exponentiate the linked-cluster expansion for the S-matrix

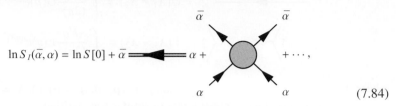

$$\ln S_I(\bar{\alpha}, \alpha) = \ln S[0] + \bar{\alpha} \blacktriangleleft \alpha + \qquad + \cdots,$$

(7.84)

we obtain

$$S_I[\bar{\alpha}, \alpha] = S[0]\left[1 + \bar{\alpha} \blacktriangleleft \alpha + \frac{1}{2!}\left(\bar{\alpha} \blacktriangleleft \alpha \right)^2 + \qquad + \cdots \right].$$

(7.85)

When we differentiate the third term with respect to the source terms, we obtain

$$\frac{\delta^4}{\delta\bar{\alpha}(1)\delta\bar{\alpha}(2)\delta\alpha(3)\delta\alpha(4)}\left[\frac{1}{2!}\left(\bar{\alpha} \blacktriangleleft \alpha \right)^2 \right]$$

$$= \frac{\delta^4}{\delta\bar{\alpha}(1)\delta\bar{\alpha}(2)\delta\alpha(3)\delta\alpha(4)}\left[\frac{1}{2!}\int d1'd4'\left(\bar{\alpha}(1')G(1'-4')\alpha(4') \right)\int d2'd3'\left(\bar{\alpha}(2')G(2'-3')\alpha(3') \right) \right]$$

$$= G(1-4)G(2-3) - G(1-3)G(2-4) \quad =$$

(7.86)

where the G are the full two-particle propagators. Putting it all together, we obtain

$$\frac{1}{S}\frac{\delta^4 S[\bar{\alpha}, \alpha]}{\delta\bar{\alpha}(1)\delta\bar{\alpha}(2)\delta\alpha(3)\delta\alpha(4)}\bigg|_{\bar{\alpha},\alpha=0} =$$

(7.87)

7.3 Feynman rules in momentum space

Though it is easiest to motivate the Feynman rules in real space, practical computations are much more readily effected in momentum space. We can easily transform to momentum space by expanding each interaction line and Green's function in terms of their Fourier components:

$$1 \longrightarrow 2 = G(X_1 - X_2) = \int \frac{d^d p}{(2\pi)^d} G(p) e^{ip(X_1 - X_2)}$$

$$1 \sim\!\!\!\sim\!\!\!\sim\!\!\!\sim 2 = V(X_1 - X_2) = \int \frac{d^d q}{(2\pi)^d} V(q) e^{iq(X_1 - X_2)},$$

$$(7.88)$$

where we have used the shorthand notation $p = (\mathbf{p}, \omega)$, $q = (\mathbf{q}, \nu)$, $X = (\mathbf{x}, t)$, and $pX = \mathbf{p} \cdot \mathbf{x} - \omega t$. We can deal with source terms in similar way, writing

$$\alpha(X) = \int \frac{d^d p}{(2\pi)^d} e^{ipX} \alpha(p). \qquad (7.89)$$

Having made these transformations, we see that the space–time co-ordinates associated with each vertex, now only appear in the phase factors. At each vertex, we can now carry out the integral over all space–time co-ordinates, which then imposes the conservation of frequency and momentum at each vertex:

$$\sim\!\!\!\sim\!\!\!\sim\!\!\!\sim_{q}^{p_2} X = \int d^d X e^{i(p_1 - p_2 - q)X} = (2\pi)^d \delta^{(d)}(p_1 - p_2 - q) .$$

$$(7.90)$$

Since momentum and energy are conserved at each vertex, this means that there is one independent energy and momentum per loop in the Feynman diagram. Thus the transformation from real-space to momentum-space Feynman rules is effected by replacing the sum over all space–time co-ordinates by the integral over all loop momenta and frequency. (Table 7.2). The convergence factor

$$e^{i\omega 0^+} \qquad (7.91)$$

is included in the loop integral. This term is only really needed when the loop contains a single propagator, propagating back to the point from which it emanated. In this case, the convergence factor builds in the information that the corresponding contraction of field operators is normal-ordered.

Actually, since all propagators and interaction variables depend only on the difference of position, the integral over all n space–time co-ordinates can be split up into an integral over the center-of-mass coordinate

$$X_{cm} = \frac{X_1 + X_2 + \cdots + X_n}{n} \qquad (7.92)$$

and the relative co-ordinates

$$\tilde{X}_r = X_r - X_1 \qquad (r > 1), \qquad (7.93)$$

as follows:

$$\prod_{r=1,n} d^d X_r = d^d X_{cm} \prod_{r=2,n} d^d \tilde{X}_r. \qquad (7.94)$$

Table 7.2 Momentum-space Feynman rules ($T = 0$).

$(\mathbf{k}, \omega)$	$G_0(\mathbf{k}, \omega)$	Fermion propagator
	$iV(q)$	Interaction
$(\mathbf{q}, \nu)$	$ig_q^2 D_0(q)$	Exchange boson
q	$U(\mathbf{q})$	Scattering potential
	$[-(2S+1)]^F$	F= number of fermion loops
$(\mathbf{q}, \nu)$	$\displaystyle\int \frac{d^d q \, d\nu}{(2\pi)^{d+1}} e^{i\nu 0^+}$	Integrate over internal loop momenta and frequency
$p=2$ $p=8$	$\dfrac{1}{p}$	p = order of symmetry group

The integral over the $\tilde{X}_r$ imposes momentum and frequency conservation, while the integral over X_{cm} can be factored out of the diagram to give an overall factor of

$$\int d^d X_{cm} = (2\pi)^d \delta^{(d)}(0) \equiv VT, \qquad (7.95)$$

where V is the volume of the system and T the time over which the interaction is turned on. This means that the proper expression for the logarithm of the S-matrix is

$$\ln S = VT \sum \{\text{linked-cluster diagrams in momentum space}\}. \qquad (7.96)$$

In other words, the phase factor associated with the S-matrix grows extensively with the volume and the time over which the interactions act.

7.3.1 Relationship between energy and the S-matrix

One of the most useful relationships of perturbation theory is the link between the S-matrix and the ground-state energy, originally derived by Jeffrey Goldstone [4]. The basic idea is very simple. When we turn on the interaction, the ground-state energy changes, which causes the phase of the S-matrix to evolve. If we turn on the interaction for a time T, then we expect that, for sufficiently long times, the phase of the S-matrix will be given by $-i\Delta ET$:

$$S[T] = \langle -\infty | \hat{U}(T/2)U^\dagger(-T/2) | -\infty \rangle \propto e^{-i\Delta ET}, \tag{7.97}$$

where $\Delta E = E_g - E_0$ is the shift in the ground state energy as a result of interactions. This means that, at long times,

$$\ln(S[T]) = -i\Delta ET + \text{constant}. \tag{7.98}$$

But from the linked-cluster theorem we know that

$$\ln S = VT \sum \{\text{linked-cluster diagrams in momentum space}\}, \tag{7.99}$$

which then means that the change in the ground-state energy due to interactions is given by

$$\Delta E = iV \sum \{\text{linked-cluster diagrams in momentum space}\}. \tag{7.100}$$

To show this result, let us turn on the interaction for a period of time T, writing the ground-state S-matrix as

$$S[T] = \langle -\infty | \hat{U}(T/2)U^\dagger(-T/2) | -\infty \rangle. \tag{7.101}$$

If we insert a complete set of energy eigenstates $\underline{1} = \sum_\lambda |\lambda\rangle\langle\lambda|$ into this expression for the S-matrix, we obtain

$$S[T] = \sum_\lambda \langle -\infty | \hat{U}(T/2) | \lambda \rangle \langle \lambda | U^\dagger(-T/2) | -\infty \rangle. \tag{7.102}$$

In the limit $T \to \infty$, the only state with an overlap with the time-evolved state $U^\dagger(-T/2)| -\infty\rangle$ will be the true ground state $|\psi_g\rangle$ of the interacting system, so we can write

$$S(T) \to \mathcal{U}(T)\mathcal{U}^\dagger(-T), \tag{7.103}$$

where $\mathcal{U}(T) = \langle -\infty | \hat{U}(T/2) | \psi_g \rangle$. Differentiating the first term in this product, we obtain

$$\frac{\partial}{\partial T}\mathcal{U}(T) = \frac{\partial}{\partial T}\langle -\infty | e^{iH_0T/2} e^{-iHT/2} | \psi_g\rangle$$

$$= \frac{i}{2}\langle -\infty | \{H_0 U(T/2) - U(T/2)H\} | \psi_g\rangle$$

$$= -\frac{i\Delta E}{2}\mathcal{U}(T). \tag{7.104}$$

Similarly, $\frac{\partial}{\partial T}\mathcal{U}^\dagger(-T) = -\frac{i\Delta E}{2}\mathcal{U}^\dagger(-T)$, so that

$$\frac{\partial S(T)}{\partial T} = -i\Delta E S(T), \tag{7.105}$$

which proves the original claim.

7.4 Examples

7.4.1 Hartree–Fock energy

As a first example of the application of Feynman diagrams, we use the linked-cluster theorem to expand the ground-state energy of an interacting electron gas to first order. To leading order in the interaction strength, the shift in the ground-state energy is given by

$$
E_g = E_0 + iV \left[\bigcirc\!\!\!\sim\!\!\!\bigcirc + \bigcirc\!\!\!\!\bigcirc \right],
\tag{7.106}
$$

corresponding to the Hartree–Fock contributions to the ground-state energy. Writing out this expression explicitly, noting that the symmetry factor associated with each diagram is $p = 2$, we obtain

$$
\Delta E_{HF} = \frac{iV}{2} \int \frac{d^3k\, d^3k'}{(2\pi)^6} \frac{d\omega\, d\omega'}{(2\pi)^2} e^{i(\omega+\omega')\delta}
$$
$$
\times \left[(-[2S+1])^2 (iV_{\mathbf{q}=0}) + (-[2S+1])(iV_{\mathbf{k}-\mathbf{k}'}) \right] G(k)G(k').
$$

In Chapter 5, equation (5.80), we obtained the result

$$
\langle c_{\mathbf{k}\sigma}^\dagger c_{\mathbf{k}\sigma} \rangle = -i \int \frac{d\omega}{2\pi} G(\mathbf{k},\omega) e^{i\omega\delta} = f_{\mathbf{k}} = \theta(k_F - |\mathbf{k}|),
\tag{7.107}
$$

so the shift in the ground state energy is given by

$$
\Delta E_{HF} = \frac{V}{2} \int \frac{d^3k\, d^3k'}{(2\pi)^6} \left[(2S+1)^2 (V_{\mathbf{q}=0}) - (2S+1)(V_{\mathbf{k}-\mathbf{k}'}) \right] f_{\mathbf{k}} f_{\mathbf{k}'}.
\tag{7.108}
$$

In the first term, we can identify $\rho = (2S+1)\sum f_{\mathbf{k}}$ as the density, so this term corresponds to the classical interaction energy of the Fermi gas. The second term is the exchange energy. This term is present because the spatial wavefunction of parallel-spin electrons is antisymmetric, which keeps them apart, producing a kind of "correlation hole" between parallel-spin electrons.

Before we end this section, let us examine the Hartree–Fock energy for the Coulomb gas. Formally, with the Coulomb interaction the Hartree interaction becomes infinite, but in practice we need not worry because, to stabilize the charged Fermi gas, we need to compensate the charge of the Fermi gas with a uniformly charged background. Since the resulting plasma of electrons plus background is neutral, the classical Coulomb energy of the combined system is identically zero, so we may drop the Hartree component of the energy. The leading-order expression for the ground-state energy of the compensated Coulomb gas of fermions is then

$$
\frac{E_g}{V} = (2S+1) \int_{\mathbf{k}} \frac{\hbar^2 k^2}{2m} f_{\mathbf{k}} - \frac{(2S+1)}{2} \int_{\mathbf{k},\mathbf{k}'} f_{\mathbf{k}} f_{\mathbf{k}'} \frac{e^2}{\epsilon_0(\mathbf{k}-\mathbf{k}')^2}.
\tag{7.109}
$$

A careful evaluation of the above integrals (see Exercise 7.1) gives

$$\frac{E_g}{V} = \rho \left[\frac{3}{5}\epsilon_F - \frac{3}{4\pi}\frac{e^2 k_F}{4\pi\epsilon_0} \right], \tag{7.110}$$

where $\rho = (2S + 1)k_F^3/(6\pi^2)$ is the density of particles. In other words, the energy per electron is given by

$$\frac{E_g}{\rho V} = \frac{3}{5}\epsilon_F - \frac{3}{4\pi}\frac{e^2 k_F}{4\pi\epsilon_0}. \tag{7.111}$$

An important parameter for the electron gas is the dimensionless separation of the electrons. The separation of electrons, R_e, in a Fermi gas is defined by

$$\frac{4\pi R_e^3}{3} = \rho^{-1},$$

where ρ is the density of electrons. The dimensionless separation r_s is defined as $r_s = R_e/a_B$, where $a_B = \frac{\hbar^2 4\pi\epsilon_0}{me^2}$ is the Bohr radius, so that

$$r_s = \frac{1}{\alpha k_F a_B}, \tag{7.112}$$

where $\alpha = \left(\frac{2}{9\pi}(2S + 1) \right)^{\frac{1}{3}} \approx 0.521$ for $S = \frac{1}{2}$. Using r_s and taking $S = \frac{1}{2}$, we can rewrite the energy of the electron gas as

$$\frac{E}{\rho V} = \frac{3}{5}\frac{R_Y}{\alpha^2 r_s^2} - \frac{3}{2\pi}\frac{R_Y}{\alpha r_s} = \frac{3}{5}\frac{R_Y}{\alpha^2 r_s^2}\left[1 - \frac{5}{2\pi}\alpha r_s \right]$$

$$= \left(\frac{2.21}{r_s^2} - \frac{0.916}{r_s} \right) R_Y, \tag{7.113}$$

where $R_Y = \frac{\hbar^2}{2ma_B^2} = 13.6\,\text{eV}$ is the Rydberg energy; see Figure 7.2 From this, we see that

- the quantity $\frac{5}{2\pi}(\alpha r_s) \sim \alpha r_s$ plays the role of an expansion paramater
- the most strongly correlated limit of the electron gas is the *dilute limit* (large r_s).

7.4.2 Exchange correlation

Let us examine the exchange correlation term in more detail. To this end, it is useful to consider the equal-time density correlation function,

$$C_{\sigma\sigma'}(\vec{x} - \vec{x}') = \langle \phi_0 | : \rho_\sigma(x)\rho_{\sigma'}(x') : |\phi_0\rangle.$$

In real space, the Hartree–Fock energy is given by

$$\langle \phi_0|\hat{V}|\phi_0\rangle = \frac{1}{2}\sum_{\sigma,\sigma'}\int d^3x\, d^3y\, V(\vec{x} - \vec{y})\langle \phi_0 | : \hat{\rho}_\sigma(\vec{x})\rho_{\sigma'}(\vec{y}) : |\phi_0\rangle$$

$$= \frac{1}{2}\sum_{\sigma,\sigma'}\int d^3x\, d^3y\, V(\vec{x} - \vec{y})C_{\sigma\sigma'}(\vec{x} - \vec{y}), \tag{7.114}$$

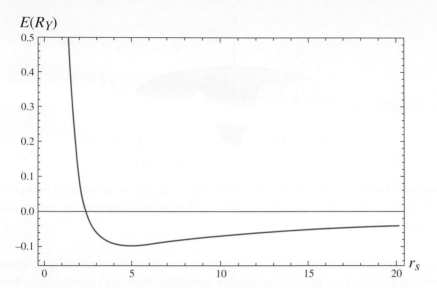

Showing the energy per electron as a function of the reduced separation r_s between electrons, after (7.113).

Fig. 7.2

so it is an integral of the interaction potential with the correlation function. Now if we look at the real-space Feynman diagrams for this energy,

$$\Delta E = i \left[\text{(diagram)} + \text{(diagram)} \right]$$

$$= -\frac{1}{2} \sum_{\sigma\sigma'} \int_{\mathbf{x},\mathbf{x}'} V(\mathbf{x}-\mathbf{x}') \left[\left(\overset{\sigma}{\text{(diagram)}}_\mathbf{x} \; \overset{\sigma'}{\text{(diagram)}}_{\mathbf{x}'} \right) + \mathbf{x} \overset{\sigma}{\text{(diagram)}} \mathbf{x}' \delta_{\sigma\sigma'} \right],$$

(7.115)

since each interaction line contributes $iV(\mathbf{x}-\mathbf{x}')$ to the total energy. The delta function in the second term derives from the connectivity of the diagram, which forces the spins σ and σ' at both density vertices to be the same. By comparing the diagrammatic expansion (7.115) with (7.114), we deduce that the Feynman diagram for the equal-time density correlation functions is

$$C_{\sigma\sigma'}(\mathbf{x}-\mathbf{x}') = - \left[\left(\overset{\sigma}{\text{(diagram)}}_\mathbf{x} \; \overset{\sigma'}{\text{(diagram)}}_{\mathbf{x}'} \right) + \left(\mathbf{x} \overset{\sigma}{\text{(diagram)}} \mathbf{x}' \right) \delta_{\sigma\sigma'} \right].$$

(7.116)

The first term is independent of the separation of $\mathbf{x}$ and $\mathbf{x}'$ and describes the uncorrelated background densities. The second term depends on $\mathbf{x} - \mathbf{x}'$ and describes the exchange correlation between the densities of parallel-spin fermions.

Written out explicitly,

$$C_{\sigma\sigma'}(\vec{x} - \vec{y}) = - \left[(\overbrace{-G(\vec{0}, 0^-)}^{-i\rho_0})^2) - \delta_{\sigma\sigma'} G(\vec{x} - \vec{y}, 0^-) G(\vec{y} - \vec{x}, 0^-) \right]$$

$$= \rho_0^2 + \delta_{\sigma\sigma'} G(\vec{x} - \vec{y}, 0^-) G(\vec{y} - \vec{x}, 0^-),$$

(7.117)

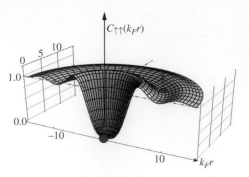

Fig. 7.3 Showing the exchange hole in the equal-time correlation function $C_{\uparrow\uparrow}(k_F r)$ for the non-interacting Fermi gas. Notice how this function vanishes at the origin, corresponding to a vanishing probability to find two "up" electrons at the same location in space.

where we have identified $G(\vec{0}, 0^-) = i\rho_0$ with the density of electrons per spin. From this we see that $C_{\uparrow\downarrow}(\vec{x} - \vec{y}) = \rho_0^2$ is independent of separation: there are no correlations between the up- and down-spin densities in the non-interacting electron ground state. However, the correlation function between parallel-spin electrons contains an additional term. We can calculate this term from the equal-time electron propagator, which in real space is given by

$$G(\vec{x}, 0^-) = \int_{\mathbf{k}} G(\mathbf{k}, 0^-) e^{i\mathbf{k}\cdot\mathbf{x}} = i \int_{\mathbf{k},} f_{\mathbf{k}} e^{i\mathbf{k}\cdot\mathbf{x}}$$

$$= i \int_{k<k_F} \frac{k^2 dk}{2\pi^2} \overbrace{\int \frac{d\cos\theta}{2} e^{ikr\cos\theta}}^{\frac{\sin kr}{kr}}$$

$$= \frac{i}{2\pi^2 r^3} [\sin(k_F r) - k_F r \cos(k_F r)] = i\rho_0 P(k_F x), \qquad (7.118)$$

where $\rho_0 = \frac{k_F^3}{6\pi^2}$ is the density per spin, while

$$P(x) = 3\left(\frac{\sin x - x\cos x}{x^3}\right) = \frac{3}{x} j_1(x) \qquad (7.119)$$

and $j_1(x)$ is the $l = 1$ spherical Bessel function. The density correlation function for parallel-spin fermions then takes the form

$$C_{\uparrow\uparrow}(r) = \rho_0^2 \left(1 - [P(k_F r)]^2\right).$$

This function is shown in Figure 7.3: at $r = 0$ it goes to zero, since the probability to find two "up" electrons in the same place actually vanishes. It is this *exchange hole* in the correlation function that gives the interacting electron fluid a predisposition towards the development of ferromagnetism and triplet paired superfluids.

7.4.3 Electron in a scattering potential

As an illustration of the utility of the Feynman diagram approach, we now consider an electron scattering off an attractive central scattering potential. Here, by re-summing the

Feynman diagrams, it is easy to show how, in dimensions $d \leq 2$, an arbitrarily weak attractive potential gives rise to bound states.

The Hamiltonian is given by

$$H = \sum_{\mathbf{k}} \epsilon_{\mathbf{k}} c_{\mathbf{k}}^{\dagger} c_{\mathbf{k}} + H_{sc}, \tag{7.120}$$

where $\epsilon_{\mathbf{k}} = k^2/2m - \mu$ and the scattering potential H_{sc} is given by

$$H_{sc} = \int d^3x U(x) \psi^{\dagger}(x) \psi(x). \tag{7.121}$$

If we Fourier transform the scattering potential, writing

$$U(x) = \int_{\mathbf{q}} U(\mathbf{q}) e^{i\mathbf{q}\cdot\mathbf{x}}, \tag{7.122}$$

then the scattering potential becomes

$$H_{sc} = \int_{\mathbf{k},\mathbf{k}'} \underbrace{U_{\mathbf{k}-\mathbf{k}'}}_{\text{amplitude to transfer momentum } \mathbf{k} - \mathbf{k}'} c_{\mathbf{k}}^{\dagger} c_{\mathbf{k}'}. \tag{7.123}$$

The Feynman diagrams for the one-electron Green's function are then

$$\tag{7.124}$$

where

$$= G^0(\mathbf{k}, \omega) = \frac{1}{\omega - \epsilon_{\mathbf{k}} - i\delta_{\mathbf{k}}} \tag{7.125}$$

denotes the propagator in the absence of potential scattering and

$$= U_{\mathbf{k}-\mathbf{k}'} \tag{7.126}$$

is the basic scattering vertex. The first diagram represents the amplitude to be transmitted without scattering; subsequent diagrams represent multiple scattering processes involving one, two, three, and more scattering events. We shall lump all scattering processes into a single amplitude, called the t-matrix, represented by

$$\tag{7.127}$$

With this shorthand notation, the diagrams for the electron propagator become

$$\tag{7.128}$$

Written out as an equation, this is

$$G(\mathbf{k}, \mathbf{k}', \omega) = \delta_{\mathbf{k}, \mathbf{k}'} G^0(\mathbf{k}, \omega) + G^0(\mathbf{k}, \omega) t_{\mathbf{k}, \mathbf{k}'}(\omega) G^0(\mathbf{k}', \omega). \tag{7.129}$$

If we look at the second, third, and higher scattering terms in the t-matrix, we see that they are a combination of the t-matrix plus the bare scattering amplitude. This enables us to rewrite the t-matrix as the following self-consistent set of Feynman diagrams:

$$\tag{7.130}$$

Written out explicitly, this is

$$t_{\mathbf{k}, \mathbf{k}'}(\omega) = U_{\mathbf{k}-\mathbf{k}'} + \sum_{\mathbf{k}''} U_{\mathbf{k}-\mathbf{k}''} G^0(\mathbf{k}'', \omega) t_{\mathbf{k}'', \mathbf{k}'}(\omega). \tag{7.131}$$

Equations (7.129) and (7.131) fully describe the scattering off the impurity.

As a simplified example of the application of these equations, let us look at the case of s-wave scattering off a point-like scattering center:

$$U(\mathbf{x}) = U\delta^{(d)}(\mathbf{x}). \tag{7.132}$$

In this case, $U(\mathbf{q}) = U$ is independent of momentum transfer. By observation, this means that the t-matrix will also be independent of momentum, i.e. $t_{\mathbf{k}, \mathbf{k}'}(\omega) = t(\omega)$. The equation for the t-matrix then becomes

$$t(\omega) = U + U \sum_{\mathbf{k}''} G^0(\mathbf{k}'', \omega) t(\omega) \tag{7.133}$$

or

$$t(\omega) = \frac{U}{1 - UF(\omega)}, \tag{7.134}$$

where

$$F(\omega) = \int \frac{d^d p}{(2\pi)^d} \frac{1}{\omega - \epsilon_{\mathbf{k}} + i\delta_{\mathbf{k}}} = \int_{-\mu}^{\Lambda} d\epsilon N(\epsilon) \frac{1}{\omega - \epsilon + i\delta \operatorname{sgn}(\epsilon)}. \tag{7.135}$$

Here $N(\epsilon)$ is the density of states and $\epsilon = -\mu$ is the bottom of the conduction sea. A high-energy cut-off has been introduced to guarantee the convergence of the integral. Physically, such a cut-off corresponds to the energy scale beyond which the scattering potential no longer behaves as a point potential. At low energies, $F(\omega) < 0$, so that, if the potential is attractive, $U < 0$, there is the possibility of poles in the t-matrix, corresponding to bound states.

It is instructive to calculate $F(\omega)$ in two dimensions, where the density of states is constant, $N(\epsilon) = N(0)$. In this case,

$$\begin{aligned} F(z) &= N(0) \int_{-\mu}^{\Lambda} d\epsilon \frac{1}{z - \epsilon + i\delta \operatorname{sgn}\epsilon} \\ &= -N(0) \ln \left[\frac{z - \Lambda}{z + \mu} \right] \approx -N(0) \ln \left[\frac{-\Lambda}{z + \mu} \right] \qquad (|z| << \Lambda). \end{aligned} \tag{7.136}$$

Here we have taken the liberty of moving into the complex plane replacing $\omega \to z$, which permits us to remove the $i\delta$ from the propagator. We have also simplified the final answer, assuming that $|z| << \Lambda$. The final answer is then

$$t(z) = \frac{U}{1 + UN(0)\ln\left[\frac{\Lambda}{-(z+\mu)}\right]}. \qquad (7.137)$$

Remarks

- For an attractive potential, $U = -|U|$,

$$t(z) = -\frac{|U|}{1 - |U|N(0)\ln\frac{\Lambda}{-(z+\mu)}} = \frac{1}{N(0)\left[-\frac{1}{UN(0)} + \ln\frac{\Lambda}{-(z+\mu)}\right]}$$

$$= \frac{1}{N(0)\ln\left(\frac{\omega_0}{-(z+\mu)}\right)}, \qquad (7.138)$$

where $\omega_0 = \Lambda e^{-\frac{1}{|U|N(0)}}$. This has a pole at the energy

$$\omega = -\omega_0 - \mu = -\Lambda e^{-\frac{1}{|U|N(0)}} - \mu,$$

corresponding to a bound state split off below the bottom of the electron sea. This energy scale ω_0 cannot be written as a power series in U, and, as such, is an elementary example of a *non-perturbative* result. The bound state appears because an infinite class of Feynman diagrams have been re-summed (see Figure 7.4). This is a special property of two dimensions. In higher dimensions the potential must exceed a threshold in order to produce a bound state. (The appearance of a bound state for electrons scattering off an arbitrarily weak attractive potential is similar to the Cooper instability.)

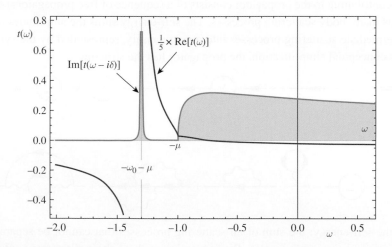

Showing real and imaginary parts of the t-matrix function for scattering off an attractive delta function potential in two dimensions. The bound state at $\omega = -\omega_0 - \mu$ develops for arbitrarily small attractive interaction in two dimensions.

Fig. 7.4

- The function $F(z)$ contains a branch cut along the real axis for $z = \omega > -\mu$, so that

$$\ln\left(\frac{-\Lambda}{\omega + \mu + i\delta}\right) = \ln\left(-\left|\frac{\Lambda}{\omega + \mu}\right|\right) + i\pi \qquad (\omega > -\mu) \tag{7.139}$$

and hence

$$t(\omega + i\delta) = \frac{U}{1 + UN(0)\ln\left[\frac{\Lambda}{\omega + \mu}\right] - i\pi UN(0)} \qquad (\omega + \mu > 0). \tag{7.140}$$

The complex value of this expression reflects the appearance of a phase shift in the scattering t-matrix. Indeed, we can write the t-matrix in standard scattering form in terms of a phase shift η, as

$$t(\omega + i\delta) = \frac{e^{i\eta}\sin\eta}{\pi N(0)}, \tag{7.141}$$

where

$$\eta = \tan^{-1}\frac{\pi UN(0)}{1 + UN(0)\ln\left[\left|\frac{\Lambda}{\omega + \mu}\right|\right]}$$

is the scattering phase shift.

7.5 The self-energy

The concept of self-energy enables us to understand the feedback of the interacting environment on a propagating particle. This is one of the most important examples of the power of Feynman diagram re-summation.

Let us consider the Green's function of a fermion in an interacting environment. Every diagram contributing to the propagator consists of a sequence of free propagators separated by various many-body scattering processes. The self-energy sums the amplitudes for all of these intermediate scattering processes into a single entity, represented by the symbol Σ. With this conceptual simplification, the propagator has the structure

$$\tag{7.142}$$

where

$$\tag{7.143}$$

denotes the self-energy: the sum of all scattering processes that cannot be separated into two by cutting a single propagator. By convention each of these diagrams contains two small stubs (without arrows) that denote the points where the diagram connects with incoming and outgoing propagators. We do not associate any propagator with these stubs. In a rather macabre terminology, the external legs of the self-energy are sometimes said to have been "amputated."

The one-particle propagator can then be expanded as a geometric series involving the self-energy, as follows:

$$
\begin{aligned}
G(\mathbf{k}, \omega) &= \quad\longrightarrow\quad + \quad\longrightarrow\!\!\!\bigcirc\!\!\!\Sigma\!\!\!\longrightarrow\quad + \quad\longrightarrow\!\!\!\bigcirc\!\!\!\Sigma\!\!\!\longrightarrow\!\!\!\bigcirc\!\!\!\Sigma\!\!\!\longrightarrow\quad + \cdots \\[2mm]
&= \quad G^0 \quad + \quad G^0 \Sigma G^0 \quad + \quad G^0 (\Sigma G^0)^2 \quad + \cdots \\[2mm]
&= \frac{G_0}{1 - \Sigma G^0} = \frac{1}{(G^0(\mathbf{k}, \omega))^{-1} - \Sigma(\mathbf{k}, \omega)} ,
\end{aligned} \tag{7.144}
$$

so that

$$
G(\mathbf{k}, \omega) = \frac{1}{\omega - \epsilon_{\mathbf{k}} - \Sigma(\mathbf{k}, \omega)}. \tag{7.145}
$$

Feynman propagator

This heuristic derivation involves the summation of a geometric series which in general will be outside its radius of convergence, but we may argue the result is true by analytic continuation. Another way to derive the same result is to notice that the second and subsequent terms in the series (7.144) can be rewritten in terms of the original Green's function, as follows:

$$
\Longrightarrow\!\!\!= \quad = \quad \longrightarrow\quad + \quad\longrightarrow\!\!\!\bigcirc\!\!\!\Sigma\!\!\!\Longrightarrow
$$

$$
G(\mathbf{k}, \omega) = G^0(\mathbf{k}, \omega) + G^0(\mathbf{k}, \omega) \Sigma(\mathbf{k}, \omega) G(\mathbf{k}, \omega). \tag{7.146}
$$

Dyson equation

This equation is called a *Dyson equation* [2]. Using it to solve for $G(\mathbf{k}, \omega)$, we also obtain (7.145).

Physically, the self-energy describes the cloud of particle–hole excitations which accompanies the propagating electron, "dressing" it into a quasiparticle. In general, the self-energy has both a real and an imaginary component:

$$
\Sigma(\mathbf{k}, \omega - i\delta) = \Sigma'(\mathbf{k}, \omega) + i\Gamma(\mathbf{k}, \omega). \tag{7.147}
$$

The imaginary component of the self-energy describes the rate of decay of the bare fermion, through the emission of particle–hole pairs.

If we use this expression to evaluate the one-particle spectral function, we obtain

$$A(\mathbf{k}, \omega) = \frac{1}{\pi} \text{Im} \, G(\mathbf{k}, \omega - i\delta) = \frac{1}{\pi} \frac{\Gamma(\mathbf{k}, \omega)}{[\omega - \epsilon_\mathbf{k} - \Sigma'(\mathbf{k}, \omega)]^2 + \Gamma(\mathbf{k}, \omega)^2}. \tag{7.148}$$

If the self-energy is small, we see that this corresponds to a Lorentzian of width Γ centered around a renormalized energy $\epsilon_\mathbf{k}^* = \epsilon_\mathbf{k} + \Sigma'(\mathbf{k}, \epsilon_\mathbf{k}^*)$. If we expand the Lorentzian around this point, we must be careful to write $\omega - \epsilon_\mathbf{k} - \Sigma'(\mathbf{k}, \omega) = (\omega - \epsilon_\mathbf{k}^*)Z_\mathbf{k}$, where $Z_\mathbf{k}^{-1} = \left(1 - \partial_\omega \Sigma'(\mathbf{k}, \omega)\right)\big|_{\omega = \epsilon_\mathbf{k}^*}$. Near the renormalized energy $\omega \sim \epsilon_\mathbf{k}^*$,

$$G(\mathbf{k}, \omega - i\delta) = \frac{Z_\mathbf{k}}{\omega - \epsilon_\mathbf{k}^* - i\Gamma_\mathbf{k}^*}, \tag{7.149}$$

where, provided $\Gamma_\mathbf{k}^*$ is small,

$$
\begin{aligned}
\epsilon_\mathbf{k}^* &= \epsilon_\mathbf{k} + \Sigma'(\mathbf{k}, \epsilon_\mathbf{k}^*) &&\text{renormalized energy} \\
\Gamma_\mathbf{k}^* &= Z_\mathbf{k} \Gamma(\mathbf{k}, \epsilon_\mathbf{k}^*) &&\text{lifetime}
\end{aligned}
\tag{7.150}
$$

can be interpreted as a quasiparticle with energy ϵ_{bk}^* and lifetime $\Gamma_\mathbf{k}^*$ (see section 6.8). Now this quasiparticle peak is not the only component to the spectral function, because it only contains a weight $Z_\mathbf{k}$, while the total weight of the spectral function is unity. The full Green's function is better represented in the form

$$G(\mathbf{k}, \omega - i\delta) = \frac{Z_\mathbf{k}}{\omega - \epsilon_\mathbf{k}^* - i\Gamma_\mathbf{k}^*} + G_{\text{inc}}(\mathbf{k}, \omega), \tag{7.151}$$

where G_{inc} represents the incoherent particle–hole continuum contribution to the Green's function. This is precisely the form of spectral function expected in a Fermi liquid (6.8), with a sharp quasiparticle pole co-existing with an incoherent background $A_{inc}(\mathbf{k}, \omega)$. From the spectral decomposition (5.123), we can relate $Z_\mathbf{k}$ to the overlap between the bare particle and the dressed quasiparticle:

$$Z_\mathbf{k} = |\langle \text{quasiparticle } \mathbf{k}\sigma | c_{\mathbf{k}\sigma}^\dagger | \phi \rangle|^2. \qquad\qquad \text{quasiparticle weight} \tag{7.152}$$

7.5.1 Hartree–Fock self-energy

The simplest example of self-energy is the Hartree–Fock self-energy, given by the two diagrams

$$\Sigma_{HF}(\mathbf{p}, \omega) = \quad\text{}\quad + \quad\text{}$$

$$= i \int_{\mathbf{p}'} \left\{ -(2S + 1)V_{\mathbf{q}=0} + V_{\mathbf{p}-\mathbf{p}'} \right\} \int \frac{d\omega}{2\pi} G^0(p') e^{i\omega 0^+}. \tag{7.153}$$

Here we see a case where we must include a convergence factor, associated with the normal ordering of the operators inside the interaction. Identifying $\int d\omega G^0(k)e^{i\omega 0^+} = 2\pi i f_{\mathbf{p}'}$, we obtain

$$\Sigma_{HF}(\mathbf{p}) = \int \frac{d^3p'}{(2\pi)^3}\left[(2S+1)V_{\mathbf{q}=0} - V_{\mathbf{p}-\mathbf{p}'}\right]f(\epsilon_{\mathbf{p}'}). \tag{7.154}$$

The first term describes a simple shift in the energy due to the interaction with the uniform density of particles. The second term describes the effect of the exchange hole (Figure 7.3) which repels fermions of the same spin polarization, lowering the density of fermions around the propagating particle. In the Hartree–Fock approximation, the fermion acquires a renormalized energy

$$\epsilon_{\mathbf{p}}^* = \epsilon_{\mathbf{p}} + \Sigma_{HF}(\mathbf{p}), \tag{7.155}$$

but since the Hartree–Fock self-energy is completely static, in this approximation the quasiparticle has an infinite lifetime and the renormalized propagator is

$$G(p) = \frac{1}{\omega - \epsilon_{\mathbf{p}}^*}.$$

The dispersion and the quasiparticle mass are renormalized by the interaction. Now in general, the effect of the Hartree–Fock self-energy will also shift the chemical potential, changing the Fermi momentum to a new value p_F^*. We can improve the Hartree–Fock solution by self-consistently feeding the renormalized Green's function back into the Hartree–Fock self-energy, as follows:

$$\Sigma_{HF}(\mathbf{p}) = \qquad\qquad + \qquad\qquad . \tag{7.156}$$

The use of this kind of self-consistent approximation is common in many-body physics. If we expand the double lines in the self-energies, we see that we are, in effect, re-summing an entire class of nested self-energy diagrams, for example,

$$\tag{7.157}$$

In Hartree–Fock theory, the effect of this change is simply to renormalize the Fermi functions used in evaluating the self-energy, so that now $f_{\mathbf{p}} = f(\epsilon_{\mathbf{p}}^*)$ reflects the quasiparticle Fermi momentum p_F^*, so that

$$\Sigma_{HF}(\mathbf{p}) = \int \frac{d^3p'}{(2\pi)^3}\left[(2S+1)V_{\mathbf{q}=0} - V_{\mathbf{p}-\mathbf{p}'}\right]f(\epsilon_{\mathbf{p}'}^*). \tag{7.158}$$

We can now relate the quasiparticle mass to the interaction. Suppose we write

$$\frac{\mathbf{p}}{m^*} = \nabla_{\mathbf{p}}\epsilon_{\mathbf{p}}^* = \left[\frac{\mathbf{p}}{m} + \nabla_{\mathbf{p}}\Sigma_{HF}(\mathbf{p})\right]. \tag{7.159}$$

Then, integrating by parts,

$$\nabla_{\mathbf{p}}\Sigma_{HF}(\mathbf{p}) = -\int_{\mathbf{p}'}\nabla_{\mathbf{p}}V_{\mathbf{p}-\mathbf{p}'}f_{\mathbf{p}'} = +\int_{\mathbf{p}'}\nabla_{\mathbf{p}'}V_{\mathbf{p}-\mathbf{p}'}f_{\mathbf{p}'} = -\int_{\mathbf{p}'}V_{\mathbf{p}-\mathbf{p}'}\nabla_{\mathbf{p}'}f_{\mathbf{p}'}. \qquad (7.160)$$

Now, since $f_{\mathbf{p}} = f(\epsilon_{\mathbf{p}}^*)$, $\nabla_{\mathbf{p}}f_{\mathbf{p}} = \nabla_{\mathbf{p}}\epsilon_{\mathbf{p}}^*\,(\partial f/\partial\epsilon^*) = -\frac{\mathbf{p}}{m^*}\delta(\epsilon_{\mathbf{p}}^*)$, we obtain

$$\begin{aligned}
\nabla_{\mathbf{p}}\Sigma_{HF}(\mathbf{p}) &= \int_{\mathbf{p}'}V_{\mathbf{p}-\mathbf{p}'}\left(\frac{\mathbf{p}'}{m^*}\right)\delta(\epsilon_{\mathbf{p}'}^*) \\
&= \frac{\mathbf{p}_F}{m^*}\int_{\mathbf{p}'}V_{\mathbf{p}-\mathbf{p}'}(\hat{\mathbf{p}}'\cdot\hat{\mathbf{p}})\delta(\epsilon_{\mathbf{p}'}^*) = \left(\frac{\mathbf{p}_F}{m^*}\right)\frac{N^*(0)}{2}\int\frac{d\Omega_{\mathbf{p}'}}{4\pi}V_{\mathbf{p}-\mathbf{p}'}\cos\theta_{\mathbf{p},\mathbf{p}'},
\end{aligned}$$

$$(7.161)$$

where $N^*(0) = m^*p_F^*/(\pi^2\hbar^3)$ is the renormalized (quasiparticle) density of states. To make contact with Landau Fermi-liquid theory, we write

$$\nabla_{\mathbf{p}}\Sigma_{HF}(\mathbf{p}) = -\frac{\mathbf{p}_F}{m^*}F_1^s,$$

where

$$F_1^s = N^*(0)\int\frac{d\Omega_{\hat{\mathbf{p}}'}}{4\pi}\left(-\frac{V_{\mathbf{p}-\mathbf{p}'}}{2}\right)\cos(\theta_{\mathbf{p},\mathbf{p}'}). \qquad (7.162)$$

This is the dipole ($l = 1$) Landau parameter expected in Hartree–Fock theory, where the quasiparticle interaction is given by $f_{\mathbf{p}\sigma,\mathbf{p}\sigma'} = V_{\mathbf{q}=0} - V_{\mathbf{p}-\mathbf{p}'}\delta_{\sigma\sigma'}$, so that $f_{\mathbf{p},\mathbf{p}'}^s = V_{\mathbf{q}=0} - \frac{1}{2}V_{\mathbf{p}-\mathbf{p}'}$ (see (6.47)). Combining (7.159) and (7.161), we then obtain

$$\frac{\mathbf{p}}{m^*}(1 + F_1^s) = \frac{\mathbf{p}}{m}, \qquad (7.163)$$

so that the renormalized mass is given by

$$\frac{m^*}{m} = 1 + F_1^s. \qquad (7.164)$$

Formally, this result is the same as that derived in Landau Fermi-liquid theory (Section 6.4.2), using the Hartree–Fock approximation to the quasiparticle interaction (6.47),

$$f_{\mathbf{p}\mathbf{p}'}^s = V_{\mathbf{q}=0} - V_{\mathbf{p}-\mathbf{p}'}. \qquad (7.165)$$

However, a more realistic theory would take into account the screening and modification of the interactions by the medium, a subject which we touch on at the end of this chapter.

7.6 Response functions

One of the most valuable applications of Feynman diagrams is to evaluate response functions. Suppose we couple the interacting system to an external source field:

$$H(t) = H_0 + H_s(t), \qquad (7.166)$$

where

$$H_s(t) = -A(t)f(t) \tag{7.167}$$

involves the coupling of an external force to a variable of the system. Examples include

$$H_s(t) = -\mu_B \int d^3 x \vec{\sigma}(x) \cdot \mathbf{B}(x, t) \qquad \text{external magnetic field}$$

$$H_s(t) = - \int d^3 x \rho(\mathbf{x}) \Phi(x, t). \qquad \text{external potential} \tag{7.168}$$

In each case, the system will respond by a change in the variable $A(t)$. To calculate this change, we use the interaction representation of $H(t)$, so that

$$A_H(t) = U^{\dagger}(t) A_I(t) U(t), \tag{7.169}$$

where, from Chapter 6,

$$U(t) = T \exp\left[-i \int_{-\infty}^{t} H_s(t') dt'\right]. \tag{7.170}$$

We shall now drop the subscript I, because $A_I(t) = A(t)$ also corresponds to the Heisenberg representation of H_0. Expanding (7.169) to linear order in H_s, we obtain

$$A_H(t) = A(t) - i \int_{-\infty}^{t} [A(t), H_s(t')] dt' + O(H_s^2). \tag{7.171}$$

Finally, taking expectation values, we obtain

$$\langle A_H(t) \rangle = \langle \phi | A(t) | \phi \rangle - i \int_{-\infty}^{t} \langle \phi | [A(t), H_s(t')] | \phi \rangle dt'. \tag{7.172}$$

But if A is zero in the absence of the applied force, i.e. $\langle \phi | A(t) | \phi \rangle = 0$, then the linear response of the system is given by

$$\langle A_H(t) \rangle = \int_{-\infty}^{\infty} dt' \chi(t - t') f(t') dt', \tag{7.173}$$

where

$$\chi(t - t') = i \langle \phi | [A(t), A(t')] | \phi \rangle \theta(t - t') \tag{7.174}$$

is called the *dynamical susceptibility* and $A(t)$ is in the Heisenberg representation of the unperturbed system.

Now in diagrammatic perturbation theory we are able to evaluate time-ordered Green's functions such as

$$\chi^T(1 - 2) = (-i)^2 \langle \phi | TA(1)A(2) | \phi \rangle. \tag{7.175}$$

Here, the prefactor $(-i)^2$ has been inserted because, almost invariably, A is a bilinear of the quantum field, so that χ^T is a two-particle Green's function. Fortunately, there is a very deep link between the dissipative response function and the fluctuations associated with a correlation function, called the *fluctuation–dissipation theorem*. The Fourier transforms of $\mathcal{R}$ and G are both governed by precisely the same many-body excitations, with precisely the same spectral functions, with one small difference: in the complex structure of $\chi(\omega)$, all the poles lie just below the real axis, guaranteeing a retarded response. By contrast, in $\chi^T(\omega)$, the positive- and negative-energy poles give rise to retarded and advanced responses, respectively. The spectral decomposition of these functions are found to be

$$
\begin{aligned}
\chi(\omega) &= \sum_\lambda \frac{2|M_\lambda|^2 \omega_\lambda}{\omega_\lambda^2 - (\omega + i\delta)^2} \\
\chi^T(\omega) &= i \sum_\lambda \frac{2|M_\lambda|^2 \omega_\lambda}{(\omega_\lambda - i\delta)^2 - \omega^2},
\end{aligned}
\tag{7.176}
$$

where $M_\lambda = \langle \lambda|A|\phi\rangle$ is the matrix element between the ground state and the excited state λ, and $\omega_\lambda = E_\lambda - E_g$ is the excitation energy. In this way, the response function can be simply related to the time-ordered response at a small imaginary frequency:

$$
\chi(\omega) = -i\chi^T(\omega + i\delta).
\tag{7.177}
$$

We can obtain the Feynman rules for the time-ordered correlation function by introducing a source term H_s and calculating the S-matrix $S[f]$. In this case,

$$
\frac{\delta^2}{\delta f(1)\delta f(2)} \ln S[f] = -\langle \phi|T[A(1)A(2)]|\phi\rangle = \chi^T(1-2) \equiv 1 \langle\!\!\!\!\!\!\!\!\bigcirc\!\!\!\!\!\!\!\!\rangle 2 \, .
\tag{7.178}
$$

Diagrammatically, the time-ordered correlation function for the quantity A is given by

$$
\chi^T(\omega) = \sum \{\text{diagrams formed by connecting two } A \text{ vertices together}\},
\tag{7.179}
$$

as summarized in Table 7.3.

Example 7.2 By introducing a chemical potential source-term

$$
V = \int d^3 x\, \delta\phi(x, t)\hat{\rho}(x)
\tag{7.180}
$$

into the original Hamiltonian, show that the change in the logarithm of the S-matrix is

$$
\ln S[\phi] = \ln S[0] + \frac{1}{2}\left[\delta\phi(1) \langle\!\!\!\!\!\!\!\!\bigcirc\!\!\!\!\!\!\!\!\rangle \delta\phi(2) \right],
\tag{7.181}
$$

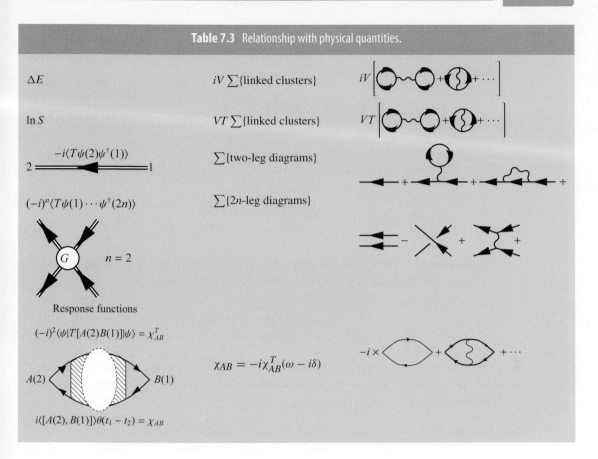

Table 7.3 Relationship with physical quantities.

where

$$(-i)^2 \langle \phi | T \delta\rho(1)\delta\rho(2) | \phi \rangle = \frac{\delta^2}{\delta\phi(1)\delta\phi(2)} \ln S[\phi] = 1 \bigtriangleup\bigtriangledown 2 .$$

(7.182)

denotes the sum of all diagrams that connect two density vertices. Use this result to show that the time-ordered density correlation function is given by

(7.183)

Solution

The added source-term creates potential scattering, denoted by the following Feynman diagram:

$$\delta\phi(1) .$$

$$(7.184)$$

When we expand the closed-loop diagrams to second order in $\delta\phi$, we obtain

$$\ln S[\phi] = \ln S[0] + \delta\phi(1) \quad \text{+} \quad \frac{1}{2}\left[\delta\phi(1) \qquad \delta\phi(2)\right].$$

$$(7.185)$$

The coefficient of the second term is identified as the density, and if this is uniform we can just factor this term out as $\int_x \delta\phi(x)\langle\rho\rangle$, where $\langle\rho\rangle$ is the density:

$$\delta\phi(1) \qquad = \rho \int d1\,\delta\phi(1) .$$

$$(7.186)$$

The third term represents the sum of all connected diagrams with two density vertices. We can then divide this sum up into *polarization bubbles*, in which the two density vertices are connected by a common fermion line,

$$(7.187)$$

and diagrams in which the fermion bubbles are connected by interaction lines.

Now each time we differentiate the generating functional with respect to ϕ, we pull down a density operator inside the time-ordered product:

$$i\frac{\delta}{\delta\phi(1)} \rightarrow \rho(1),$$

$$(7.188)$$

so that when we differentiate the logarithm of the S-matrix, we obtain

$$i\frac{\delta \ln S[\phi]}{\delta\phi(1)} = \frac{\langle\psi_0|S\rho(1)|\psi_0\rangle}{S[\phi]} = \langle\rho(1)\rangle,$$

$$(7.189)$$

where ψ_0 denotes the non-interacting ground state. When we differentiate a second time, we must be careful to differentiate both the numerator and the denominator, using

$$i\frac{\delta\langle\psi_0|S\rho(1)|\psi_0\rangle}{\delta\phi(2)} = \langle\psi_0|T\{S\rho(1)\rho(2)\}|\psi_0\rangle,$$

$$(7.190)$$

while

$$i\frac{\delta}{\delta\phi(2)}\left(\frac{1}{S[\phi]}\right) = -\frac{\langle\psi_0|T\{S\rho(2)\}|\psi_0\rangle}{S[\phi]^2},$$

$$(7.191)$$

so that

$$i^2 \frac{\delta \ln S[\phi]}{\delta \phi(2) \delta \phi(1)} = \frac{\langle \psi_0 | T\{S\rho(1)\rho(2)\} | \psi_0 \rangle}{S[\phi]} - \frac{\langle \psi_0 | S\rho(1) | \psi_0 \rangle}{S[\phi]} \frac{\langle \psi_0 | S\rho(2) | \psi_0 \rangle}{S[\phi]}$$

$$= \langle T\rho(1)\rho(2) \rangle - \langle \rho(1) \rangle \langle \rho(2) \rangle$$

$$= \langle T(\rho(1) - \langle \rho(1) \rangle)(\rho(2) - \langle \rho(2) \rangle) \rangle = \langle T\delta\rho(1)\delta\rho(2) \rangle. \qquad (7.192)$$

With this result and (7.82), we can now identify

$$\frac{\delta^2 \ln S[\phi]}{\delta \phi(2) \delta \phi(1)} = -\langle T\delta\rho(1)\delta\rho(2) \rangle = 1 \mathopen{<}\!\!\!\!\rule{0pt}{0pt}\mathclose{}\,\,\raisebox{-0.3em}{\(\bigcirc\)}\,\,\mathopen{}\!\!\!\!\rule{0pt}{0pt}\mathclose{\!>} 2 \,. \qquad (7.193)$$

7.6.1 Magnetic susceptibility of non-interacting electron gas

One of the fundamental qualities of a Fermi liquid is its non-local response to an applied field. Suppose, for example, that one introduces a localized delta-function disturbance in the magnetic field, $\delta B_z(x) = B\delta^3(x)$. Since the fermions have a characteristic wavevector of order k_F, this local disturbance will "heal" over a length scale of order $l \sim 1/k_F$. Indeed, since the maximum wavevector for low-energy particle–hole excitations is sharply cut off at $2k_F$, the response produces oscillations in the spin density with a wavelength $\lambda = 2\pi/k_F$ that decay gradually from the site of the disturbance. These oscillations are called *Friedel oscillations* (Figure 7.5). In the case of the example just cited, the change in the spin density in response to the shift in the chemical potential is given by

$$\delta M(\vec{x}) = \chi_s(\vec{x})B, \qquad (7.194)$$

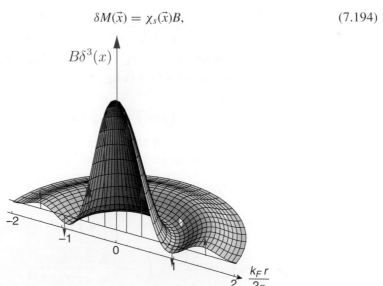

Friedel oscillations in the spin density, in response to a delta-function disturbance in the magnetic field at the origin. These oscillations may be calculated from the Fourier transform of the Lindhard function.

Fig. 7.5

where

$$\chi_s(\vec{x}) = \int_q \chi(\mathbf{q}, \omega = 0)e^{i\vec{q}\cdot\vec{x}} \tag{7.195}$$

is the Fourier transform of the dynamical spin susceptibility. We shall now calculate this quantity as an example of the application of Feynman diagrams.

From the interaction in (7.168), the magnetization is given by

$$\vec{M}(x) = \int d^4x' \underline{\chi}(x - x')\vec{B}(x'), \tag{7.196}$$

where

$$\underline{\chi}_{ab}(x) = i\langle\phi|[\sigma^a(x), \sigma^b(0)]|\phi\rangle\theta(t). \tag{7.197}$$

The electron fluid mediates this non-local response. If we Fourier transform this expression, then $\vec{M}(q) = \underline{\chi}(q)\vec{B}(q)$, where (in a relativistic shorthand)

$$\chi_{ab}(q) = i\mu_B^2 \int d^4x \langle\phi|[\sigma^a(x), \sigma^b(0)]|\phi\rangle\theta(t)e^{-iq\cdot x}. \tag{7.198}$$

We can relate $\chi_{ab}(\vec{q}, \nu) = -i\chi_{ab}^T(\vec{q}, \nu + i\delta)$, where the time-ordered Green's function is given by

$$= -\mu_B^2 \int_k \frac{d\omega}{2\pi} \mathrm{Tr}\left[\sigma^a G(k + q)\sigma^b G(k)\right] = \delta_{ab}\chi^T(q). \tag{7.199}$$

The susceptibility $\chi^T(q)$ is then

$$\chi^T(q) = -2\mu_B^2 \int_k \frac{d\omega}{2\pi}\left[\frac{1}{\omega + \nu - \tilde{\epsilon}_{k+q}}\frac{1}{\omega - \tilde{\epsilon}_k}\right], \tag{7.200}$$

where we have invoked the notation $\tilde{\epsilon}_k = \epsilon_k - i\delta\mathrm{sgn}(\epsilon_k)$. The term inside the square brackets has two poles, at $\omega = \tilde{\epsilon}_k$ and at $\omega = \tilde{\epsilon}_{k+q} - \nu$:

$$\int_\omega = \int \frac{d\omega}{2\pi}\frac{1}{(\tilde{\epsilon}_{k+q} - \tilde{\epsilon}_k) - \nu}\left[\frac{1}{\omega + \nu - \epsilon_{k+q} + i\delta_{k+q}} - \frac{1}{\omega - \epsilon_k + i\delta_k}\right].$$

We may carry out the frequency integral by completing the contour in the upper half-plane. Each Green's function gives a contribution $2\pi i \times$ Fermi function, so that

$$\chi^T(q) = -2i\mu_B^2 \int_k \frac{f_{k+q} - f_k}{(\tilde{\epsilon}_{k+q} - \tilde{\epsilon}_k) - \nu}, \tag{7.201}$$

so that the dynamical susceptibility $\chi(\mathbf{q}, \nu) = -i\chi^T(\mathbf{q}, \nu + i\delta)$ is given by

$$\chi(\mathbf{q}, \nu + i\delta) = 2\mu_B^2 \int_k \frac{f_{k+q} - f_k}{\nu - (\epsilon_{k+q} - \epsilon_k) + i\delta}. \tag{7.202}$$

dynamical spin susceptibility

There are a number of important pieces of physics encoded in the above expression that deserve special discussion:

- Spin conservation. The total spin of the system is conserved, so that the application of a strictly uniform magnetic field to the fluid cannot change the total magnetization. Indeed, in keeping with this expectation, if we take $\vec{q} \to 0$ we find $\lim_{\vec{q}\to 0} \chi(\vec{q}, \nu) = 0$.
- Static susceptibility. When we take the limit $\nu \to 0$, we obtain the magnetization response to a spatially varying magnetic field. The static susceptibility is given by

$$\chi(\mathbf{q}) = 2\mu_B^2 \int_{\mathbf{k}} \frac{f_{\mathbf{k}} - f_{\mathbf{k}+\mathbf{q}}}{(\epsilon_{\mathbf{k}+\mathbf{q}} - \epsilon_{\mathbf{k}})}. \tag{7.203}$$

This response is finite, because the spins can always redistribute themselves in response to a non-uniform field. When we take the wavelength of the applied field to infinity, i.e. $q \to 0$, we recover the Pauli susceptibililty:

$$\chi \to 2\mu_B^2 \int_{\mathbf{k}} \left(-\frac{df(\epsilon)}{d\epsilon}\right) = 2\mu_B^2 \int_{\mathbf{k}} \delta(\epsilon_{\mathbf{k}}) = 2\mu_B^2 N(0), \tag{7.204}$$

where $N(0) = \frac{mk_F}{2\pi^2}$ is the density of states per spin. The detailed momentum-dependent static susceptibility can be calculated (see Section 7.6.2), and is given by

$$\chi(\mathbf{q}) = 2\mu_B^2 N(0) F\left(\frac{q}{2k_F}\right)$$

$$F(x) = \frac{1}{4x}(1 - x^2); \ln\left|\frac{1+x}{1-x}\right| + \frac{1}{2}. \tag{7.205}$$

The function $F(x)$ is known as the *Lindhard function* [5]; see Figure 7.6. It has the property that $F(0) = 1$, while $F'(x)$ has a weak logarithmic singularity at $|x| = 1$.

- Dissipation and the imaginary part of the susceptibility. The full dynamical spin susceptibility has both a real and an imaginary part, given by

$$\chi(\mathbf{q}, \nu) = \chi'(\mathbf{q}, \nu) + i\chi''(\mathbf{q}, \nu),$$

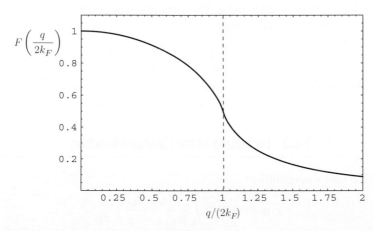

The Lindhard function. The Fourier transform of this function governs the magnetic response of a non-interacting metal to an applied field. Notice the weak singularity around $q/(2k_F) = 1$ that results from the match between the Fermi surface and the wavevector of the magnetic response.

Fig. 7.6

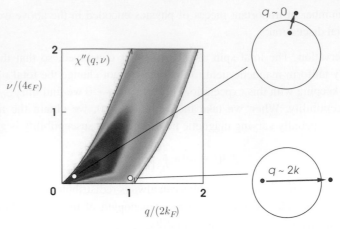

Fig. 7.7 Density plot of the imaginary part of the dynamical spin susceptibility calculated from (7.212), showing the band of width $2k_F$ that spreads up to higher energies. Excitations on the left side of the band correspond to low-momentum-transfer excitations of electrons from just beneath the Fermi surface to just above the Fermi surface. Excitations on the right-hand side of the band correspond to high-momentum-transfer processes, right across the Fermi surface.

where the imaginary part determines the dissipative part of the magnetic response. The dissipation arises because an applied magnetic field generates a cloud of electron–hole pairs which carry away the energy. If we use the Cauchy–Dirac relation $1/(x + i\delta) = P(1/x) - i\pi\,\delta(x)$ in (7.202), we obtain

$$\chi''(\mathbf{q}, \nu) = 2\mu_B^2 \int_{\mathbf{k}} \pi \delta[\nu - (\epsilon_{\mathbf{k}+\mathbf{q}} - \epsilon_{\mathbf{k}})](f_{\mathbf{k}} - f_{\mathbf{k}+\mathbf{q}}). \tag{7.206}$$

This quantity defines the density of states of particle–hole excitations. The excitation energy of a particle–hole pair is given by

$$\epsilon_{\mathbf{k}+\mathbf{q}} - \epsilon_{\mathbf{k}} = \frac{q^2}{2m} + \frac{qk}{m}\cos\theta,$$

where θ is the angle between $\mathbf{k}$ and $\mathbf{q}$. This quantity is largest when $\theta = 0$, $k = k_F$, and smallest when $\theta = \pi$, $k = k_F$, so that

$$\frac{q^2}{2m} + \frac{qk_F}{m} > \nu > \frac{q^2}{2m} - \frac{qk_F}{m}$$

defines a band of allowed wavevectors where the particle–hole density of states is finite, as shown in Figure 7.7. Outside this region, $\chi_0(\mathbf{q}, \nu)$ is purely real.

7.6.2 Derivation of the Lindhard function

The dynamical spin susceptibility

$$\chi(\mathbf{q}, \nu) = 2\mu_B^2 \int_{\mathbf{k}} \frac{f_{\mathbf{k}} - f_{\mathbf{k}+\mathbf{q}}}{(\epsilon_{\mathbf{k}+\mathbf{q}} - \epsilon_{\mathbf{k}} - \nu)} \tag{7.207}$$

can be rewritten as

$$\chi(\mathbf{q}, \nu) = 2\mu_B^2 \int_{\mathbf{k}} f_{\mathbf{k}} \left[\frac{1}{(\epsilon_{\mathbf{k}+\mathbf{q}} - \epsilon_{\mathbf{k}} - \nu)} + \frac{1}{(\epsilon_{\mathbf{k}-\mathbf{q}} - \epsilon_{\mathbf{k}} + \nu)} \right]. \tag{7.208}$$

Written out explicity, this is

$$\chi(\mathbf{q}, \nu) = 2\mu_B^2 \int_0^{k_F} \frac{k^2 dk}{2\pi^2} \int_{-1}^{1} \frac{d\cos\theta}{2} \left[\frac{1}{(\epsilon_{\mathbf{k}+\mathbf{q}} - \epsilon_{\mathbf{k}} - \nu)} + ((\nu, \mathbf{q}) \rightarrow -(\nu, \mathbf{q})) \right].$$

By replacing $\epsilon_{\mathbf{k}} \rightarrow \frac{k^2}{2m} - \mu$ and rescaling $x = k/k_F$, $\tilde{q} = q/(2k_F)$, and $\tilde{\nu} = \nu/(4\epsilon_F)$, we obtain $\chi(\mathbf{q}, \nu) = 2\mu_B^2 N(0) \mathcal{F}(\tilde{q}, \tilde{\nu})$, where

$$\mathcal{F}(\tilde{q}, \tilde{\nu}) = \frac{1}{4\tilde{q}} \int_0^1 x^2 dx \int_{-1}^{1} dc \left[\frac{1}{xc + \tilde{q} - \frac{\tilde{\nu}}{\tilde{q}}} + (\nu \rightarrow -\nu) \right] \tag{7.209}$$

is the Lindhard function. Carrying out the integral over angle, we obtain

$$\mathcal{F}(\tilde{q}, \tilde{\nu}) = \frac{1}{4\tilde{q}} \int_0^1 x \, dx \left(\ln\left[\frac{\tilde{q} - \frac{\tilde{\nu}}{\tilde{q}} + x}{\tilde{q} - \frac{\tilde{\nu}}{\tilde{q}} - x} \right] + (\tilde{\nu} \rightarrow -\tilde{\nu}) \right)$$

$$= \frac{1}{8\tilde{q}} \left(\left[1 - \left(\tilde{q} - \frac{\tilde{\nu}}{\tilde{q}} \right)^2 \right] \ln\left[\frac{\tilde{q} - \frac{\tilde{\nu}}{\tilde{q}} + 1}{\tilde{q} - \frac{\tilde{\nu}}{\tilde{q}} - 1} \right] + (\tilde{\nu} \rightarrow -\tilde{\nu}) \right) + \frac{1}{2}. \tag{7.210}$$

Its static limit, $F(\tilde{q}) = \mathcal{F}(\tilde{q}, \tilde{\nu} = 0)$,

$$F(\tilde{q}) = \frac{1}{4\tilde{q}} \left(\left[1 - \tilde{q}^2 \right] \ln\left| \frac{\tilde{q} + 1}{\tilde{q} - 1} \right| \right) + \frac{1}{2}, \tag{7.211}$$

has the properties that $F(0) = 1$ and dF/dx is singular at $x = 1$, as shown in Figure 7.6. The imaginary part of $\chi(\mathbf{q}, \nu + i\delta)$ is given by

$$\chi''(\mathbf{q}, \nu) = 2\mu_B^2 N(0) \times \frac{\pi}{8\tilde{q}} \left\{ \left(1 - \left[\tilde{q} - \frac{\tilde{\nu}}{\tilde{q}} \right]^2 \right) \theta \left[1 - \left[\tilde{q} - \frac{\tilde{\nu}}{\tilde{q}} \right]^2 \right] - (\nu \rightarrow -\nu) \right\},$$
$$\tag{7.212}$$

and is plotted in Figure 7.7.

7.7 The RPA (large-*N*) electron gas

Although the Feynman diagram approach gives us a way to generate all perturbative corrections, we still need a way to select the physically important diagrams. In general, as we have seen from the previous examples, it is important to re-sum particular classes of diagrams to obtain a physical result. What principles can be used to select classes of diagrams?

Frequently, however, there is no obvious choice of small parameter, in which case one needs an alternative strategy. For example, in the electron gas we could select diagrams according to the power of r_s entering the diagram. This would give us a high-density expansion of the properties – but what if we would like to examine a low-density electron gas in a controlled way?

One way to select Feynman diagrams in a system with no natural small parameter is to take the so-called *large-N* limit. This involves generalizing some internal degree of freedom so that it has N components. Examples include:

- the hydrogen atom in N dimensions
- an electron gas with $N = 2S + 1$ spin components
- spin systems with spin S in the limit that S becomes large
- quantum chromodynamics with N rather than three colours.

In each of these cases, the limit $N \to \infty$ corresponds to a new kind of semiclassical limit, where certain variables cease to undergo quantum fluctuations. The parameter $1/N$ plays the role of an effective $\hbar$:

$$\frac{1}{N} \sim \hbar. \tag{7.213}$$

This does not, however, mean that quantum effects have been lost, merely that their macroscopic consequences can be lumped into certain semiclassical variables.

We shall now examine the second of these examples. The idea is to take an interacting Fermi gas where each fermion has $N = 2S + 1$ possible spin components. The interacting Hamiltonian is still written

$$H = \sum_{\mathbf{k},\sigma} \epsilon_{\mathbf{k}} c^{\dagger}_{\mathbf{k}\sigma} c_{\mathbf{k}\sigma} + \frac{1}{2} \sum V_{\mathbf{q}} c^{\dagger}_{\mathbf{k}+\mathbf{q}\sigma} c^{\dagger}_{\mathbf{k}'-\mathbf{q}\sigma'} c_{\mathbf{k}'\sigma'} c_{\mathbf{k}\sigma}, \tag{7.214}$$

but now the spin summations run over $N = 2S + 1$ values rather than just two. As N is made very large, it is important that both the kinetic energy and the interaction energy scale extensively with N. For this reason, the original interaction $V_{\mathbf{q}}$ is rescaled, writing

$$V_{\mathbf{q}} = \frac{1}{N} \mathcal{V}_{\mathbf{q}}, \tag{7.215}$$

where it is understood that, as $N \to \infty$, V is to be kept fixed. The idea is to now calculate quantities as an expansion in powers of $1/N$, and at the end of the calculation to give N the value of specific interest, in our case $N = 2$. For example, if we are interested in a Coulomb gas of spin-$\frac{1}{2}$ electrons, then we study the family of problems where

$$V_{\mathbf{q}} = \frac{1}{N} \frac{\tilde{e}^2}{q^2} = \frac{\mathcal{V}_{\mathbf{q}}}{N} \tag{7.216}$$

and $\tilde{e}^2 = 2e^2/\epsilon_0$. At the end, we set $N = 2$, boldly hoping that the key features of the solution around $N = 2$ will be shared by the entire family of models. In practice, this only holds true if the density of the electron gas is large enough to avoid instabilities such as the formation of Wigner crystal. For historical reasons, the approxation that appears in the large-N limit is called the *random phase approximation* (RPA), a method developed during the 1950s. The early version of the RPA was developed by David Bohm and David Pines [6], while its reformulation in a diagrammatic language was later given by Hubbard [7].[2] The large-N treatment of the electron gas recovers the RPA electron gas in a controlled approximation.

With the above substitution, the Feynman rules are unchanged, except that now we associate a factor $1/N$ with each interaction vertex. Before we start, however, there are a few preliminaries; in particular, we need to know how to handle long-range Coulomb

---

[2] A more detailed discussion of this early history can be found in the book by Nozières and Pines [8].

interactions. We'll begin considering a general $\tilde{V}_{\mathbf{q}}$ with a finite interaction range. To be concrete, we can consider a screened Coulomb interaction

$$V_{\mathbf{q}} = \frac{\tilde{e}^2}{q^2 + \delta^2},\tag{7.217}$$

where we take $\delta \to 0$ at the end of the calculation to deal with the infinite-range interaction.

7.7.1 Jellium: introducing an inert positive background

To deal with long-range Coulomb interactions (and take $\delta \to 0$ in (7.217)), we will need to make sure that the charge of the entire system is actually neutral. The resulting medium is a radically simplified version of matter playfully referred to as "jellium" (a term first introduced by John Bardeen). In jellium, there is an inert and completely uniform background of positive charges, with charge $+|e|$ and number density $\rho_+(x) = \rho_+$ adjusted so that $\rho_+ = \rho_e$, the density of electrons. The Coulomb interaction Hamiltonian of jellium takes the form

$$H_I = \frac{1}{2} \int_{\vec{x},\vec{y}} V(x-y) : (\hat{\rho}(x) - \rho_+)(\hat{\rho}(y) - \rho_+) : = \frac{1}{2} \int_{\vec{x},\vec{y}} V(x-y) : \delta\rho(x)\delta\rho(y) :, \tag{7.218}$$

where $\hat{\rho}(x)$ is the density of electrons and $\delta\rho(x) = \hat{\rho}(x) - \rho_+$ is the fluctuation of the density. We see that the Coulomb energy of jellium is only sensitive to the fluctuations in the density. The presence of the background charge has the the effect of shifting the chemical potential of the electrons upward by an amount

$$\Delta\mu = \int V(x - x')\rho_+(x') = V_{\mathbf{q}=0}\,\rho_+.\tag{7.219}$$

This chemical potential shift can be treated as a scattering potential that is diagonal in momentum, $\Delta V_{\mathbf{k},\mathbf{k}'} = -\Delta\mu\delta_{\mathbf{k},\mathbf{k}'}$, which introduces an additional uniform potential scattering term into the electron self-energy:

$$= -\Delta\mu = -V_{\mathbf{q}=0}\,\rho_+.\tag{7.220}$$

If we compare this term with the "tadpole" diagrams in the self-energy,

$$= -i(2S + 1)V_{\mathbf{q}=0}\int_k G(k) = V_{\mathbf{q}=0}\,\rho_e,\tag{7.221}$$

where the double line indicates the use of the full fermion propagator $G(k)$, we see that, when we combine the two, provided $\rho_e = \rho_+$, they cancel one another:

$$= 0.\tag{7.222}$$

Thus by introducing a uniform positively charged background, we entirely remove the tadpole insertions.

Let us now examine how the fermions interact in this large-N Fermi gas. We can expand the effective interaction as follows:

$$(7.223)$$

The self-energy diagram for the interaction line is called a *polarization bubble*, and has the following diagrammatic expansion:

$$(7.224)$$

By summing the geometric series that appears in (7.223), we obtain

$$V_{eff} = \frac{1}{N} \frac{\mathcal{V}(q)}{1 + \mathcal{V}(q)\chi(q)}. \tag{7.225}$$

This modification of the interaction by the polarization of the medium is an example of *screening*. In the large-N limit the higher-order Feynman diagrams for $\chi(q)$ are smaller by factors of $1/N$, so in the large-N limit these terms can be neglected, giving

$$i\chi(q)N = i\chi_0(q)N + O(1) = \quad\text{<bubble diagram>}\quad + O(1). \tag{7.226}$$

The large-N approximation, where we replace $\chi(q) \rightarrow \chi_0(q)$, is the random phase approximation (RPA).

In the case of a Coulomb interaction, the screened interaction becomes

$$V_{eff}(\mathbf{q}, \nu) = \frac{1}{N} \frac{\tilde{e}^2}{q^2 \epsilon_{RPA}(\mathbf{q}, \nu)}, \tag{7.227}$$

where we have identified the quantity

$$\epsilon_{RPA}(\mathbf{q}, \omega) = 1 + \mathcal{V}(q)\chi(q) = 1 + \frac{\tilde{e}^2}{q^2} \chi_0(q) \tag{7.228}$$

as the dielectric function of the charged medium. Notice how, in the interacting medium, the interaction between the fermions has become frequency-dependent, indicating that the interactions between the particles are now *retarded*. In our discussion of the Lindhard function, we showed that $\chi_0(q) = N(0)\mathcal{F}(q/(2k_F)), \nu/(4\epsilon_F))$, where $\mathcal{F}$ is the dimensionless Lindhard function and $N(0) = \frac{mk_F}{2\pi^2\hbar^2}$ is the density of states per spin at the Fermi surface, so we may write

$$\epsilon_{RPA}(\mathbf{q}, \omega) = 1 + \lambda \left(\frac{\mathcal{F}(\tilde{q}, \tilde{\nu})}{\tilde{q}^2} \right), \tag{7.229}$$

where the dimensionless coupling constant

$$\lambda = \frac{\tilde{e}^2 N(0)}{(2k_F)^2} = \frac{1}{\pi k_F} \frac{e^2 m}{4\pi \epsilon_0 \hbar^2} = \frac{1}{\pi k_F a_B} = \left(\frac{\alpha}{\pi}\right) r_s. \tag{7.230}$$

Here a_B is the Bohr radius $\alpha = \left(\frac{4}{9\pi}\right)^{1/3} \approx 0.521$ and $r_s = (\alpha k_F a_B)^{-1}$ is the dimensionless electron separation (7.112). Notice that the accuracy of the large-N expansion places no restriction on the size of the coupling constant λ, which may take any value in the large-N limit. Summarizing,

$$\epsilon_{RPA}(\mathbf{q}, \omega) = 1 + \frac{1}{\pi k_F a_B} \left(\frac{\mathcal{F}(\tilde{q}, \tilde{\nu})}{\tilde{q}^2}\right). \tag{7.231}$$

dielectric constant of the RPA electron gas

7.7.2 Screening and plasma oscillations

At zero frequency and low momentum, $\mathcal{F} \to 1$, so the dielectric constant diverges:

$$\epsilon = \lim_{q \to 0} \epsilon(\mathbf{q}, \nu = 0) \to \infty.$$

Is this a failure of our theory?

In fact, no. The divergence of the uniform, static dielectric constant is a quintessential property of a metal. Since $\epsilon = \infty$, no static electric fields penetrate a metal. Moreover, the electron charge is completely screened. At small q, the effective interaction is

$$V_{eff}(\mathbf{q}, \nu) = \frac{1}{N} \frac{\tilde{e}^2}{q^2 + \kappa^2} \equiv \frac{e^2}{\epsilon_0(q^2 + \kappa^2)} \qquad (N = 2), \tag{7.232}$$

where

$$\kappa = \sqrt{\tilde{e}^2 N(0)} = \sqrt{2e^2 N(0)/\epsilon_0} \qquad (N = 2) \tag{7.233}$$

can be identified as an inverse screening length. κ^{-1} is the Thomas–Fermi screening length of a classical charge plasma. You can think of

$$V_{screening}(q) = \frac{e^2}{\epsilon_0(q^2 + \kappa^2)} - \frac{e^2}{\epsilon_0 q^2}$$

as the screening potential. If we Fourier transform this potential, we obtain $V_{screen}(r) = eQ(r)/(4\pi r)$ where $Q(r) = -(1 - e^{-\kappa r})$ is the screening charge. We can see that the electroni charge is fully screened at infinity, since $Q(\infty) = -1$. Note, however, that there is still a weak singularity in the susceptibility when $q \sim 2k_F$, $\chi_0(q \sim 2k_F, 0) \sim (q - 2k_F)\ln(q - 2k_F)$; Fourier transformed, this gives rise to a long-range *oscillatory* component to the interaction between the particles, of the form

$$V_{eff}(r) \propto \frac{\cos 2k_F r}{r^3} \tag{7.234}$$

(see Example 17.1). This long-range oscillatory interaction is associated with Friedel oscillations.

A second and related consequence of the screening is the emergence of collective of plasma oscillations. In the opposite limit of finite frequency but low momentum, we may approximate χ_0 by expanding it in momentum, as follows:

$$\chi_0(\mathbf{q}, \nu) = \int_k \frac{f_{\mathbf{k}+\mathbf{q}} - f_{\mathbf{k}}}{\nu - (\epsilon_{\mathbf{k}+\mathbf{q}} - \epsilon_{\mathbf{k}})} \approx \int_k \frac{(\mathbf{q} \cdot \mathbf{v_k})}{\nu - (\mathbf{q} \cdot \mathbf{v_k})} \left(\frac{df(\epsilon)}{d\epsilon} \right), \qquad (7.235)$$

where $\mathbf{v_k} = \nabla_{\mathbf{k}} \epsilon_{\mathbf{k}}$ is the group velocity. Expanding this to leading order in momentum gives

$$\chi_0(\mathbf{q}, \nu) = - \int_k \frac{(\mathbf{q} \cdot \mathbf{v_k})^2}{\nu^2} \left(-\frac{df(\epsilon)}{d\epsilon} \right) = -\frac{N(0)v_F^2}{3} \left(\frac{q^2}{\nu^2} \right) = - \left(\frac{\tilde{n}}{m} \right) \left(\frac{q^2}{\nu^2} \right), \qquad (7.236)$$

where $\tilde{n} = n/N$ is the density of electrons per spin, so that the RPA dielectric function (7.228) is given by

$$\epsilon_{RPA}(\mathbf{q}, \nu) = 1 + \frac{\tilde{e}^2}{q^2} \chi_0(\mathbf{q}, \omega) = 1 - \frac{\omega_p^2}{\nu^2}, \qquad (7.237)$$

where

$$\omega_p^2 = \frac{\tilde{e}^2 \tilde{n}}{m} = \frac{e^2 n}{\epsilon_0 m} \qquad (N = 2) \qquad (7.238)$$

is the plasma frequency. This zero in the dielectric function at $\nu = \omega_p$ indicates the presence of collective plasma oscillations in the medium at frequency ω_p. At finite q, $\omega_P(q)$ develops a collective mode.

It is instructive to examine the response of the electron gas to a time-dependent change in potential energy, $-\delta U(x, t)$ (corresponding to a change in energy $H = -\int \delta U(x, t)\rho(x)$), with Fourier transform $\delta U(q)$. In a non-interacting electron gas, the induced change in charge is

$$\delta\rho_e(q) = N\chi_0(q)\delta U(q),$$

corresponding to the diagram

$$\delta\rho_e(q) = -i \underset{\longleftarrow}{\overset{\longrightarrow}{\bigcirc}} \delta U(q).$$
$$(7.239)$$

In the RPA electron gas, the change in the electron density induced by the applied potential produces its own interaction, and the induced change in charge is given by

$$\delta\rho_e(q) = -i \left[\bigcirc + \bigcirc\!\!\sim\!\!\bigcirc + \bigcirc\!\!\sim\!\!\bigcirc\!\!\sim\!\!\bigcirc + \cdots \right] \delta U(q)$$

$$= N \left[\chi_0 + \chi_0(-V\chi_0) + \chi_0(-V\chi_0)^2 + \cdots \right] \delta U(q)$$

$$= N \left[\frac{\chi_0(q)}{1 + V_q \chi_0(q)} \right] \delta U(q). \qquad (7.240)$$

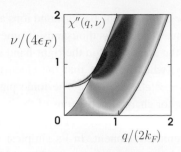

Density plot of the imaginary part of the dynamical charge susceptibility $\text{Im}[\chi_0(\mathbf{q}, \nu)/\epsilon(\mathbf{q}, \nu)]$ in the presence of the Coulomb interaction, calculated for $\frac{\alpha r_s}{\pi} = 1, (r_s \sim 6)$, using (7.231) and (7.210). Notice the split-off plasmon frequency mode, and how the charge fluctuations have moved up to frequencies above the plasma frequency.

Fig. 7.8

So we see that the dynamical charge susceptibility is renormalized by interactions

$$\chi(q) = N\frac{\chi_0(q)}{\epsilon_{RPA}(q)} = \mathcal{N}(0)\left[\frac{\mathcal{F}(\tilde{q}, \tilde{\nu})}{1 + \frac{\alpha r_s}{\pi}\mathcal{F}(\tilde{q}, \tilde{\nu})}\right] \qquad (\tilde{q} = q/2k_F, \ \tilde{\nu} = \nu/4\epsilon_F),$$

(7.241)

where $\mathcal{F}(\tilde{q}, \tilde{\nu})$ is given in (7.210) and $\mathcal{N}(0) = N \times N(0)$ is the total density of states. The imaginary part of the dynamical susceptibility $\chi(\mathbf{q}, \nu - i\delta)$ defines the spectrum of collective excitations of the RPA electron gas, shown in in Figure 7.8. Notice how the collective plasma mode is split off above the particle–hole continuum.

Remark

- The appearance of this plasma mode depends on the singular, long-range nature of the Coulomb interaction. It is rather interesting to reflect on what would have happened to the results of this section had we kept the regulating δ in the bare interaction V_q (7.217) finite. In this case the plasma frequency would be zero, while the dielectic constant would be finite. In other words, the appearance of the plasma mode and the screening of an infinite–range interaction are intimately interwined. In fact, the plasma mode in the Coulomb gas is an elementary example of a Higgs particle – a finite-mass excitation that results from the screening of a long-range (gauge) interaction. We shall discuss this topic in more depth in Section 11.6.2.

7.7.3 The Bardeen–Pines interaction

One of the most famous applications of the RPA approach is the Bardeen–Pines theory [9] for the electron–electron interaction. Whereas the treatment of jellium described so far treats the positive ionic background as a rigid medium, the Bardeen–Pines theory takes account of its finite compressibility. The ions immersed in the electron sea are thousands of times more massive than the surrounding electrons, so their motions are far more sluggish. In particular, the ionic plasma frequency is given by

$$\Omega_P^2 = \frac{(Ze)^2 n_{ion}}{\epsilon_0 M} = \frac{Ze^2 n}{\epsilon_0 M},$$

(7.242)

where $+Z|e|n_{ion}$ is the charge density of the background ions and n_{ion} is the corresponding ionic density. The ionic plasma frequency is thousands of times smaller than the electronic plasma frequency. Note that the expression on the right-hand side of (7.242) follows from the requirement of neutrality, which implies that the electron density is Z times larger than the ionic density, $|e|n = Z|e|n_{ion} = \rho_+$. The ionic plasma frequency Ω_P sets the characteristic frequency scale for charge fluctuations of the background ionic medium.

The charge polarizability of the combined electron–ion medium now contains two terms: an electron plus an ionic component. In its simplest version, the Bardeen–Pines theory treats the positive ionic background as a uniform plasma. In the RPA (large-N) approximation, the effective interaction is then

$$V_{eff} = \frac{1}{N} \frac{\mathcal{V}(q)}{1 + \mathcal{V}(q)[\chi_0(q) + \chi_{ion}(q)]} \equiv \frac{1}{N} \frac{\mathcal{V}(q)}{\epsilon(q)}, \tag{7.243}$$

where

$$i[\chi_0(q) + \chi_{ion}(q)]N = \quad \text{} \tag{7.244}$$

is the sum of the non-interacting RPA polarizabilities of the electron and ionic plasmas, where the dashed lines represent the ionic propagators. For frequencies relevant for electron–electron interactions, we can approximate the electron component of the polarizability by the low-frequency screening form:

$$\mathcal{V}(q)\chi_0(q) \sim \frac{\kappa^2}{q^2}. \tag{7.245}$$

By contrast, the large ratio of ionic to electron masses guarantees that the ionic part of the polarizability is described by its high-frequency, low-q plasma approximation (7.236), which, for the ions, is

$$\mathcal{V}(q)\chi_{ion}(q) \sim -\frac{\Omega_P^2}{v^2}. \tag{7.246}$$

With these approximations, the combined dielectric constant is then given by

$$\epsilon(q) = 1 + \frac{\kappa^2}{q^2} - \frac{\Omega_P^2}{v^2}. \tag{7.247}$$

Substituting this dielectric constant into (7.243), the effective interaction is then given by

$$V_{eff}(q) = \frac{\tilde{e}^2}{N\epsilon(q)q^2} = \frac{1}{N} \frac{\tilde{e}^2}{(q^2 + \kappa^2 - \Omega_P^2(q^2/v^2))}, \tag{7.248}$$

which we can separate into the form

$$\begin{aligned}
V_{eff}(q) &= \frac{\tilde{e}^2}{N} \left[\frac{1}{q^2 + \kappa^2} \right] \left[1 + \frac{\Omega_P^2 \frac{q^2}{v^2}}{q^2 + \kappa^2 - \Omega_P^2 \frac{q^2}{v^2}} \right] \\
&= \frac{\tilde{e}^2}{N} \left[\frac{1}{q^2 + \kappa^2} \right] \left[1 + \frac{\omega_q^2/v^2}{1 - \omega_q^2/v^2} \right],
\end{aligned} \tag{7.249}$$

where

$$\omega_q^2 = \Omega_P^2 \frac{q^2}{q^2 + \kappa^2} \tag{7.250}$$

is a renormalized plasma frequency. Replacing $\tilde{e}^2 \rightarrow (2)(e^2/\epsilon_0)$ and setting $N = 2$, we obtain

$$V_{eff}(\mathbf{q}, v) = \left[\frac{e^2}{\epsilon_0(q^2 + \kappa^2)} \right] \left(1 + \frac{\omega_q^2}{v^2 - \omega_q^2} \right). \tag{7.251}$$

Bardeen–Pines interaction

Remarks

- We see that the electron–electron interaction inside the jellium plasma has split into terms: a repulsive and instantaneous (i.e. frequency-*independent*) screened Coulomb interaction, plus a retarded (i.e. frequency-*dependent*) electron–phonon interaction:

retarded electron–phonon interaction

$$V_{eff}(\mathbf{q}, v) = \underbrace{\left[\frac{e^2}{\epsilon_0(q^2 + \kappa^2)} \right]}_{\text{screened Coulomb interaction}} + \overbrace{\left[\frac{e^2}{\epsilon_0(q^2 + \kappa^2)} \right] \frac{\omega_q^2}{v^2 - \omega_q^2}}. \tag{7.252}$$

It is the retarded attractive interaction produced by the second term that is responsible for Cooper pairing in conventional superconductors (see Exercise 7.7 and [10]).

- The plasma frequency (7.250) is renormalized by the interaction of the positive jellium with the electron sea, to form a dispersing mode with a linear dispersion $\omega_\mathbf{q} = cq$ at low frequencies, where

$$c = \frac{\Omega_P}{\kappa}. \tag{7.253}$$

Now, by (7.233),

$$\kappa^2 = \frac{e^2}{\epsilon_0} N(0) = \frac{e^2}{\epsilon_0} \left(\frac{3n}{2\epsilon_F} \right) = \left(\frac{ne^2}{\epsilon_0 m} \right) \frac{3}{v_F^2} = 3 \frac{\omega_p^2}{v_F^2}, \tag{7.254}$$

where ω_p is the electron plasma frequency, so that the sound velocity predicted by the Bardeen–Pines theory is

$$c = \frac{v_F}{\sqrt{3}} \left(\frac{\Omega_P}{\omega_p} \right) = \sqrt{\frac{Z}{3}} \left(\frac{m}{M} \right)^{\frac{1}{2}} v_F, \tag{7.255}$$

a form for the sound velocity first derived by Bohm and Staver [11]. Remarkably, this agrees within a factor of 2 with the experimental sound-velocity for a wide range of metals [9]. In this way, the Bardeen–Pines theory can account for the emergence of longitudinal phonons inside matter as a consequence of the interaction between the plasma modes of the ions and the electron sea.

- The Bardeen–Pines interaction can be used to formulate an effective Hamiltonian for the low-energy physics of jellium, known as the *Bardeen–Pines Hamiltonian*:

$$H_{BP} = \sum_{\mathbf{k}\sigma} \epsilon_{\mathbf{k}} c^{\dagger}_{\mathbf{k}\sigma} c_{\mathbf{k}\sigma} + \frac{1}{2} \sum_{\mathbf{k},\mathbf{k}'} V_{eff}(\mathbf{q}, \epsilon_{\mathbf{k}} - \epsilon_{\mathbf{k}'}) c^{\dagger}_{\mathbf{k}-\mathbf{q}\sigma} c^{\dagger}_{\mathbf{k}'+\mathbf{q}\sigma'} c_{\mathbf{k}'\sigma'} c_{\mathbf{k}\sigma}. \qquad (7.256)$$

Bardeen–Pines Hamiltonian

The Bardeen–Pines Hamiltonian is the predecessor of the Bardeen–Cooper–Schrieffer (BCS) model, and demonstrates that, while the intrinsic electron–electron interaction is repulsive, "overscreening" by the lattice causes it to develop a retarded attractive component (see Exercise 7.8).

7.7.4 Zero-point energy of the RPA electron gas

Let us now examine the linked-cluster expansion of the ground state energy. Without the tadpole insertions, the only non-zero diagrams are then

$$(7.257)$$

These diagrams are derived from the zero-point fluctuations in charge density, which modify the ground state energy $E \to E_0 + E_{zp}$. We shall select the leading contribution,

$$(7.258)$$

The nth diagram in this series has a symmetry factor $p = 2n$, and a contribution $(-\chi_0(q)\mathcal{V}(q))^n$ associated with the n polarization bubbles and interaction lines. The energy per unit volume associated with this series of diagrams is thus

$$E_{zp} = i \sum_{n=1}^{\infty} \frac{1}{2n} \int \frac{d^4q}{(2\pi)^4} (-\chi_0(q)\mathcal{V}(q))^n. \qquad (7.259)$$

By interchanging the sum and the integral, we see that we obtain a series of the form $\sum_n \frac{(-x)^n}{n} = -\ln(1 + x)$, so that the zero-point correction to the ground state energy is

$$E_{zp} = -i\frac{1}{2} \int \frac{d^4 q}{(2\pi)^4} \ln[1 + V_{\mathbf{q}} \chi_0(q)].$$

Now the logarithm has a branch cut just below the real axis for positive frequency, but just above the real axis for negative frequency. If we carry out the frequency integral by completing the contour in the lower half-plane, we can distort the contour integral around the branch cut at positive frequency, to obtain

$$
\begin{aligned}
E_{zp} &= -\frac{i}{2} \int_{\mathbf{q}} \int_0^\infty \frac{dv}{2\pi} \left[\ln[1 + \chi_0(\mathbf{q}, v + i\delta)V_{\mathbf{q}}] - \ln[1 + \chi_0(\mathbf{q}, v - i\delta)V_{\mathbf{q}}] \right] \\
&= \frac{1}{2} \int_{\mathbf{q}} \int_0^\infty \frac{dv}{\pi} \arctan\left(\frac{V_{\mathbf{q}}\chi''(\mathbf{q}, v)}{[1 + V_{\mathbf{q}}\chi'(\mathbf{q}, v)]} \right).
\end{aligned}
\tag{7.260}
$$

If we associate a "phase shift"

$$\delta(\mathbf{q}, v) = \arctan\left(\frac{V_{\mathbf{q}}\chi''(\mathbf{q}, v)}{[1 + V_{\mathbf{q}}\chi'(\mathbf{q}, v)]} \right), \tag{7.261}$$

then, by integrating by parts, we can also rewrite the zero-point fluctuation energy in the form

$$\Delta E_{zp} = - \int \frac{d^3 q}{(2\pi)^3} \int_0^\infty dv \Lambda(v) \left[\frac{v}{2} \right], \tag{7.262}$$

where

$$\Lambda(v) = \frac{1}{\pi} \frac{\partial \delta(\mathbf{q}, v)}{\partial v}. \tag{7.263}$$

We can interpret $\Lambda(\omega)$ as the "density of states" of charge fluctuations at an energy v. When the interactions are turned on, each charge fluctuation mode in the continuum experiences a scattering phase shift $\delta(\vec{q}, v)$ which has the effect of changing the density of states of charge fluctuations. The zero-point energy describes the change in the energy of the continuum due to these scattering effects.

Exercises

Exercise 7.1 Section 7.4.1 argued that, in an electron plasma with a neutralizing positive charge background (jellium), the Hartree contribution to the ground-state energy is eliminated, so that the leading-order expression for the ground-state energy is

$$\frac{E_g}{V} = (2S + 1) \int_{\mathbf{k}} \frac{\hbar^2 k^2}{2m} f_{\mathbf{k}} - \frac{(2S + 1)}{2} \int_{\mathbf{k}, \mathbf{k}'} f_{\mathbf{k}} f_{\mathbf{k}'} \frac{e^2}{\epsilon_0 (\mathbf{k} - \mathbf{k}')^2}. \tag{7.264}$$

By rewriting the momenta $k = xk_F$, $k' = yk_F$ and the direction cosine $\mathbf{k} \cdot \mathbf{k}' = xyk_F^2 \cos\theta$ as multiples of the Fermi momentum, show that the total energy per particle can be written in the form

$$\frac{E_g}{\rho V} = \frac{3}{5}\epsilon_F - \frac{3}{2\pi}\frac{e^2 k_F}{4\pi\epsilon_0}I, \tag{7.265}$$

where

$$I = \int_0^1 dxdy \int_{-1}^1 d\cos\theta \, \frac{(xy)^2}{(x^2 + y^2 - 2xy\cos\theta)}.$$

Using the result

$$\int_0^1 dx \int_0^1 dy\, xy \ln\left|\frac{x+y}{x-y}\right| = \frac{1}{2}, \tag{7.266}$$

show that the resulting energy per particle is given by

$$\frac{E_g}{\rho V} = \left[\frac{3}{5}\epsilon_F - \frac{3}{4\pi}\frac{e^2 k_F}{4\pi\epsilon_0}\right]. \tag{7.267}$$

Exercise 7.2 The separation R_e of electrons in a Fermi gas is defined by

$$\frac{4\pi R_e^3}{3} = \rho^{-1},$$

where ρ is the density of electrons. The dimensionless separation r_s is defined as $r_s = R_e/a_B$, where $a_B = \frac{4\pi\epsilon_0\hbar^2}{me^2}$ is the Bohr radius.

(a) Taking $S = \frac{1}{2}$, show that the Fermi wavevector is given by

$$k_F = \frac{1}{\alpha r_s a_B},$$

where $\alpha = \left(\frac{4}{9\pi}\right)^{\frac{1}{3}} \approx 0.521$.

(b) By rescaling the result of Exercise 7.1 in terms of the Bohr radius $a_B = \frac{\hbar^2 4\pi\epsilon_0}{me^2}$ and the Rydberg energy $R_Y = \frac{\hbar^2}{2ma_B^2} = 13.6\text{eV}$, show that the ground-state energy to leading order in the strength of the Coulomb interaction is given by

$$\frac{E}{\rho V} = \frac{3}{5}\frac{R_Y}{\alpha^2 r_s^2} - \frac{3}{2\pi}\frac{R_Y}{\alpha r_s}$$

$$= \left(\frac{2.21}{r_s^2} - \frac{0.916}{r_s}\right)R_Y. \tag{7.268}$$

(c) In what limiting case can the interaction effects in a Coulomb gas be ignored relative to the kinetic energy?

Exercise 7.3 Consider a gas of particles with interaction

$$\hat{V} = \frac{1}{2}\sum_{\vec{k}\vec{k}'\vec{q}\sigma\sigma'} V_q c^\dagger_{\vec{k}-\vec{q}\sigma} c^\dagger_{\vec{k}'+\vec{q}\sigma'} c_{\vec{k}'\sigma'} c_{\vec{k}\sigma}.$$

(a) Let $|\phi\rangle$ represent a filled Fermi sea, i.e. the ground state of the non-interacting problem. Use Wick's theorem to evaluate an expression for the expectation value of the interaction energy $\langle\phi|\hat{V}|\phi\rangle$ in the non-interacting ground state. Give a physical interpretation of the two terms that arise and draw the corresponding Feynman diagrams.

(b) Suppose $|\tilde{\phi}\rangle$ is the full ground state of the interacting system. If we add the interaction energy $\langle\tilde{\phi}|\hat{V}|\tilde{\phi}\rangle$ to the non-interacting ground-state energy, do we obtain the full ground-state energy? Please explain your answer.

(c) Draw the Feynman diagrams corresponding to the second-order corrections to the ground-state energy. Without calculation, write out each diagram in terms of the electron propagators and interaction V_q, being careful about minus signs and overall prefactors.

Exercise 7.4 Consider a d-dimensional system of fermions with spin degeneracy $N = 2S+1$, mass m, and total density $N\rho$, where ρ is the density per spin component. The fermions attract one another via the two-body potential

$$V(\mathbf{r}_i - \mathbf{r}_j) = -\alpha\delta^{(d)}(\mathbf{r}_i - \mathbf{r}_j) \qquad (\alpha > 0). \tag{7.269}$$

(a) Calculate the *total* energy *per particle*, $\epsilon_s(N,\rho)$, to first order in α.

(b) Beyond some critical value α_c, the attraction between the particles becomes so great that the gas becomes unstable, and may collapse. Calculate the dependence of α_c on the density per spin ρ. To what extent do you expect the gas to collapse in $d = 1,2,3$ when α_c is exceeded?

(c) In addition to the above two-body interaction, nucleons are also thought to interact via a repulsive three-body interaction. Write the three-body potential $V(\mathbf{r}_i, \mathbf{r}_j, \mathbf{r}_k) = \beta\delta^{(d)}(\mathbf{r}_i - \mathbf{r}_j)\delta^{(d)}(\mathbf{r}_j - \mathbf{r}_k)$ in second-quantized form.

(d) Use Feynman diagrams to calculate the ground-state energy *per particle*, $\epsilon_s(N,\rho)$, to leading order in both β and α. How does your result compare with that obtained in (a) when $N = 2$?

(e) If we neglect Coulomb interactions, why is the case $N = 4$ relevant to nuclear matter?

Exercise 7.5 (a) Consider a system of fermions interacting via a momentum-dependent interaction $V(\mathbf{q}) = \frac{1}{N}U(\mathbf{q})$, where $N = 2S + 1$ is the spin degeneracy. When N is large, the interactions in this fluid can be treated exactly. Draw the Feynman diagram expansion for the ground-state energy, identifying the leading and sub-leading terms in the $1/N$ expansion.

(b) Certain classes of Feynman diagrams in the linked-cluster expansion of the ground-state energy identically vanish. Which ones, and why?

(c) If $N\chi^{(0)}(q) = \langle\delta\rho(q)\delta\rho(-q)\rangle_0$ is the susceptibility of the non-interacting Fermi gas, i.e.

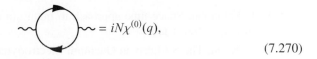

$$= iN\chi^{(0)}(q), \tag{7.270}$$

where $q = (\mathbf{q}, \nu)$, what is the effective interaction between the fermions in the large-N limit? Supposing that, in real space, $U(r) = e^2/r$ is a long-range Coulomb interaction, explain in detail what happens to the effective interaction at long distances.

Exercise 7.6 Compute the RMS quantum fluctuation $\Delta\rho = \sqrt{\langle(\rho - \rho_0)^2\rangle}$ in the charge density of the electron gas about its average density, ρ_0, in the large-N limit. Show that $\Delta\rho/\rho_0 \sim O(1/N)$, so that the density behaves as a semiclassical variable in the large-N limit.

Exercise 7.7 Show that the dynamical charge susceptibility of an interacting electron gas in the large-N limit, defined by

$$\chi(\mathbf{q}, \nu + i\delta) = \int d^3x \int_0^\infty i\langle\phi|[\rho(\mathbf{x}, t), \rho(0, 0)]|\phi\rangle e^{-i(\mathbf{q}\cdot\mathbf{x} - \omega t)}, \qquad (7.271)$$

contains a pole at frequencies

$$\omega_q = \omega_p(1 + \frac{3}{10}qv_F), \qquad (7.272)$$

where $\omega_p = \sqrt{4\pi\tilde{e}^2\tilde{n}/m}$ is the plasma frequency and $v_F = p_F/m$ is the Fermi velocity.

Exercise 7.8 Show that the Bardeen–Pines interaction (7.251) can be reformulated in terms of a screened Coulomb interaction and an electron–phonon interaction given by (see [10])

$$H_I = \frac{1}{2} \sum_{\mathbf{k},\mathbf{k}',\mathbf{q}\sigma\sigma'} V_{eff}(q)c^\dagger_{\mathbf{k}+\mathbf{q}\sigma}c^\dagger_{\mathbf{k}'-\mathbf{q}\sigma'}c_{\mathbf{k}'\sigma'}c_{\mathbf{k}\sigma} + \sum_{\mathbf{k},\mathbf{q},\sigma} g_\mathbf{q}(b^\dagger_\mathbf{q} + b_{-\mathbf{q}})c^\dagger_{\mathbf{k}-\mathbf{q}\sigma}c_{\mathbf{k}\sigma}$$

$$g_\mathbf{q} = \left(\frac{n_{ion}}{M}\right)^{\frac{1}{2}} \frac{ZqV_{eff}(q)}{[2\omega_q]^{\frac{1}{2}}}, \qquad (7.273)$$

where

$$V_{eff}(q) = \frac{e^2}{\epsilon_0(q^2 + \kappa^2)} \qquad (7.274)$$

is the screened Coulomb interaction and

$$\omega_q = \frac{q\Omega_P}{[q^2 + \kappa^2]^{\frac{1}{2}}} = \frac{q}{[q^2 + \kappa^2]^{\frac{1}{2}}} \left(\frac{(Ze)^2 n_{ion}}{\epsilon_0 M}\right)^{\frac{1}{2}} \qquad (7.275)$$

is the phonon frequency.

References

[1] R. P. Feynman, Space-time approach to quantum electrodynamics, *Phys. Rev.*, vol. 76, no. 6, p. 769, 1949.

[2] F. J. Dyson, The S Matrix in Quantum Electrodynamics, *Phys. Rev.*, vol. 75, no. 11, p. 1736, 1949.

[3] S. F. Edwards and P. W. Anderson, Theory of spin glasses, *J. Phys. F: Met. Phys.*, vol. 6, p. 965, 1975.

[4] J. Goldstone, Derivation of Brueckner many-body theory, *Proc. R. Soc. A*, vol. A239, p. 267, 1957.

[5] J. Lindhard, On the properties of a gas of charged particles, *K. Dan. Vidensk. Selsk., Mat.-Fys. Medd.*, vol. 28, p. 1, 1954.

[6] D. Bohm and D. Pines, A collective description of the electron interations: III. Coulomb interactions in a degenerate electron gas, *Phys. Rev.*, vol. 92, p. 609, 1953.

[7] J. Hubbard, The description of collective motions in terms of many-body perturbation theory, *Proc. R. Soc. A*, vol. A240, p. 539, 1957.

[8] P. Nozières and D. Pines, *The Theory of Quantum Liquids*, Perseus Books, 1999.

[9] J. Bardeen and D. Pines, Electron–phonon interaction in metals, *Phys. Rev.*, vol. 99, p. 1140, 1955.

[10] P. Morel and P. W. Anderson, Calculation of the superconducting state parameters with retarded electron–phonon interaction, *Phys. Rev.*, p. 1263, 1962.

[11] D. Bohm and T. Staver, Application of collective treatment of electron and ion vibrations to theories of conductivity and superconductivity, *Phys. Rev.*, vol. 84, p. 836, 1952.

For most purposes in many-body theory, we need to know how to include the effects of temperature. At first sight, this might be thought to lead to undue extra complexity in the mathematics, for now we need to average the quantum effects over an ensemble of states, weighted with the Boltzmann average (Figure 8.1),

$$p_\lambda = \frac{e^{-\beta E_\lambda}}{Z}. \tag{8.1}$$

It is here that some of the the most profound aspects of many-body physics come to our aid.

Remarkably, finite-temperature many-body physics is no more difficult than its zero-temperature partner, and in many ways the formulation is easier to handle. The essential step that makes this possible is due to the Japanese physicist Ryogo Kubo, who noticed[1] in the early 1950s that the quantum mechanical partition function can be regarded as a time-evolution operator in *imaginary time*:

$$\hat{\rho} \propto e^{-\beta \hat{H}} = U(-i\hbar\beta),$$

where $U(t) = e^{-i\frac{tH}{\hbar}}$ is the time-evolution operator, and by convention we write $H = H_0 - \mu N$ to take account of the chemical potential. Kubo's observation led him to realize that finite-temperature many-body physics can be compactly reformulated using imaginary rather than real time to time-evolve all states:

$$\frac{it}{\hbar} \longrightarrow \tau.$$

Kubo's observation was picked up by Takeo Matsubara, who wrote down the first imaginary-time formulation of finite-temperature many-body physics [1]. In the imaginary-time approach, the partition function of a quantum system is simply the trace of the time-evolution operator, evaluated at imaginary time $t = -i\hbar\beta$:

$$Z = \text{Tr}\, e^{-\beta H} = \text{Tr}\, U(-i\hbar\beta),$$

while the expectation value of a quantity A in thermal equilibrium is given by

$$\langle A \rangle = \frac{\text{Tr}\,[U(-i\hbar\beta)A]}{\text{Tr}\,[U(-i\hbar\beta)]},$$

an expression reminiscent of the Gell-Mann–Low formula, except that now the S-matrix is replaced by time evolution over the *finite* interval $t \in [0, -i\hbar\beta]$: the imaginary-time universe is of finite extent in the time direction! We will see that physical quantities turn

---

[1] Author's note: Philip W. Anderson recalled to me learning early ideas about the concept of imaginary time from Ryogo Kubo during his sabbatical stay in Japan, 1953–1954.

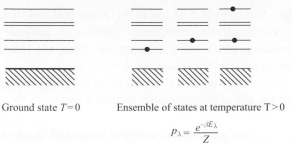

Ground state $T=0$ Ensemble of states at temperature $T>0$

$$p_\lambda = \frac{e^{-\beta E_\lambda}}{Z}$$

At zero temperature, the properties of a system are determined by the ground state. At finite temperature, we must average the properties of the system over an ensemble which includes the ground state and excited states, averaged with the Boltzmann probability weight $\frac{e^{-\beta E_\lambda}}{Z}$.

Fig. 8.1

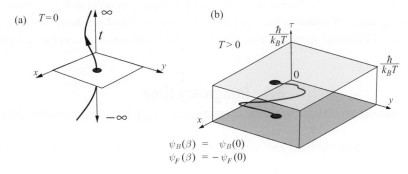

$$\psi_B(\beta) = \psi_B(0)$$
$$\psi_F(\beta) = -\psi_F(0)$$

(a) Zero-temperature field theory is carried out in a space that extends infinitely from $t = -\infty$ to $t = \infty$. (b) Finite-temperature field theory is carried out in a space that extends over a finite time, from $\tau = 0$ to $\tau = \hbar\beta$. Bosonic fields (ψ_B) are periodic over this interval, whereas fermionic fields (ψ_F) are antiperiodic over this interval.

Fig. 8.2

out to be periodic in imaginary time, over this finite interval $\tau \in [0, \hbar\beta]$. This can loosely understood as a consequence of the incoherence induced by thermal fluctuations: thermal fluctuations lead to an uncertainty $k_B T$ in energies, so

$$\tau_T = \frac{\hbar}{k_B T}$$

represents the characteristic time of a thermal fluctuation. Processes of duration longer than τ_T loose their phase coherence, so coherent quantum processes are limited within a world of finite temporal extent $\hbar\beta$ (see Figure 8.2).

One of the most valuable aspects of finite-temperature quantum mechanics, first explored by Kubo, concerns the intimate relationship between response functions and correlation functions in both real and imaginary time, which are mathematically quantified via the *fluctuation–dissipation theorem*:

Quantum/thermal fluctuations $\leftrightarrow$ dynamical response.

fluctuation–dissipation

These relationships, first exploited in detail by Kubo and now known as the *Kubo formalism*, enable us to calculate correlation functions in imaginary time, and then, by analytically continuing the Fourier spectrum, to obtain the real-time response and correlation functions at a finite temperature.

Most theoretical many-body physics is conducted in the imaginary-time formalism, and theorists rarely give the use of this wonderful method a moment's thought. It is probably fair to say that we do not understand the deep reasons why the imaginary-time formalism works. Feynman admitted in his book on statistical mechanics that he had sought but not found a reason why imaginary time and thermal equilibrium are so intimately intertwined. In relativity, it turns out that thermal density matrices are always generated in the presence of an event horizon, which excludes any transmission of information between the halves of the universe on different sides of the horizon. It would seem that a complete understanding of imaginary time may be bound up with a more complete understanding of information theory and quantum mechanics than we currently possess. Whatever the reason, it is a very pragmatic and beautiful approach, and it is this which motivates us to explore it further!

8.1 Imaginary time

The key step in the jump from zero-temperature to finite-temperatures many-body physics is the replacement

$$\frac{it}{\hbar} \to \tau \tag{8.2}$$

(where, from now on, we will work in units where $\hbar = 1$). With this single replacement we can generalize almost everything we have done at zero temperature. In zero-temperature quantum mechanics, we introduced the Schrödinger, Heisenberg, and interaction representations, going on to introduce the concept of the Green's function and develop a Feynman diagram expansion of the S-matrix. We shall now repeat this exact procedure in imaginary time, reinterpreting the various entities which appear in terms of finite-temperature statistical mechanics. Table 8.1 summarizes the key analogies between real-time zero-temperature and imaginary-time finite-temperature many-body physics.

8.1.1 Representations

The imaginary-time generalization of the Heisenberg and interaction representations precisely parallels the development in real time, but there are some minor differences that require us to go through the details here. After making the substitution $it \to \tau\hbar$, the real-time Schrödinger equation,

$$H|\psi_S\rangle = i\hbar \frac{\partial}{\partial t}|\psi_S\rangle, \tag{8.3}$$

Table 8.1	The link between real- and imaginary-time formalisms.	
Time	$t \in [-\infty, \infty]$	$it \rightarrow \tau \in [0, \beta]$
Schrödinger equation	$\lvert \psi_S(t) \rangle = e^{-itH} \lvert \psi_S(0) \rangle$	$\lvert \psi_S(\tau) \rangle = e^{-\tau H} \lvert \psi_S(0) \rangle$
Heisenberg representation	$A_H(t) = e^{itH} A_S e^{-itH}$	$A_H(\tau) = e^{\tau H} A_S e^{-\tau H}$
Interaction representation	$\lvert \psi_I(t) \rangle = e^{itH_0} \lvert \psi_S(t) \rangle$	$\lvert \psi_I(\tau) \rangle = e^{\tau H_0} \lvert \psi_S(\tau) \rangle$
Time evolution in interaction representation	$U(t) = e^{iH_0 t} e^{-iHt}$ $= T \exp\left[-i \int_0^t V_I(t') dt' \right]$	$U(\tau) = e^{H_0 \tau} e^{-H\tau}$ $= T \exp\left[-\int_0^\tau V_I(\tau') d\tau' \right]$
Perturbation expansion	$S = \langle -\infty \lvert T \exp\left[-i \int_{-\infty}^{\infty} V_I(t) dt \right] \rvert -\infty \rangle$	$\frac{Z}{Z_0} = \mathrm{Tr}\left[T e^{-\int_0^\beta V d\tau} \right]$
Wick's theorem (non-interacting particles)	$\overline{\psi(1)\psi^\dagger(2)} = \langle \phi \lvert T \psi(1) \psi^\dagger(2) \rvert \phi \rangle$	$\overline{\psi(1)\psi^\dagger(2)} = \langle T \psi(1) \psi^\dagger(2) \rangle$
Green's function	$G_{\lambda\lambda'}(t) = -i\langle \phi \lvert T \psi_\lambda(\tau) \psi_{\lambda'}^\dagger(0) \rvert \phi \rangle$	$\mathcal{G}_{\lambda\lambda'}(\tau) = -\langle T \psi_\lambda(\tau) \psi_{\lambda'}^\dagger(0) \rangle$
Feynman diagrams	$\ln S = TV \sum \{\text{linked clusters}\} = -iT\Delta E$	$\ln \frac{Z}{Z_o} = \beta V \sum \{\text{linked clusters}\}$ $= -\beta \Delta F$

becomes

$$H\lvert \psi_S \rangle = -\frac{\partial}{\partial \tau} \lvert \psi_S \rangle, \tag{8.4}$$

so the time-evolved wavefunction is given by

$$\lvert \psi_S(\tau) \rangle = e^{-H\tau} \lvert \psi_S(0) \rangle. \tag{8.5}$$

The Heisenberg representation removes all time dependence from the wavefunction, so that $\lvert \psi_H \rangle = \lvert \psi_S(0) \rangle$, and all time evolution is transfered to the operators:

$$A_H(\tau) = e^{iH(-i\tau)} A_S e^{-iH(-i\tau)} = e^{H\tau} A_S e^{-H\tau}, \tag{8.6}$$

so that the Heisenberg equation of motion becomes

$$\frac{\partial A_H}{\partial \tau} = [H, A_H].$$

If we apply this to the free-particle Hamiltonian,

$$H = \sum \epsilon_k c_k^\dagger c_k,$$

we obtain

$$\frac{\partial c_k}{\partial \tau} = [H, c_k] = -\epsilon_k c_k$$

$$\frac{\partial c_k^\dagger}{\partial \tau} = [H, c_k^\dagger] = \epsilon_k c_k^\dagger, \tag{8.7}$$

so that

$$
\begin{aligned}
c_k(\tau) &= e^{-\epsilon_k \tau} c_k \\
c_k^\dagger(\tau) &= e^{\epsilon_k \tau} c_k^\dagger
\end{aligned}. \tag{8.8}
$$

Notice a key difference from the real-time formalism: in the imaginary-time Heisenberg representation, creation and annihilation operators are no longer Hermitian conjugates $(c_k^\dagger(\tau) = (c_k(-\tau))^\dagger \neq (c_k(\tau))^\dagger)$.

We go on next to develop the interaction representation, which freezes time evolution from the non-interacting part of the Hamiltonian H_0, so that

$$
|\psi_I(\tau)\rangle = e^{H_0 \tau} |\psi_S(\tau)\rangle = e^{H_0 \tau} e^{-H\tau} |\psi_H\rangle = U(\tau)|\psi_H\rangle,
$$

where $U(\tau) = e^{H_0 \tau} e^{-H\tau}$ is the time-evolution operator. The relationship between the Heisenberg and interaction representations of operators is given by

$$
A_H(\tau) = e^{H\tau} A_S e^{-H\tau} = U^{-1}(\tau) A_I(\tau) U(\tau).
$$

In the interaction representation, states can be evolved between two times as follows:

$$
|\psi_I(\tau_1)\rangle = U(\tau_1) U^{-1}(\tau_2) |\psi_I(\tau_2)\rangle = S(\tau_1, \tau_2)|\psi_I(\tau_2)\rangle.
$$

The equation of motion for $U(\tau)$ is given by

$$
\begin{aligned}
-\frac{\partial}{\partial \tau} U(\tau) &= -\frac{\partial}{\partial \tau}\left[e^{H_0 \tau} e^{-H\tau} \right] \\
&= e^{H_0 \tau} V e^{-H\tau} \\
&= e^{H_0 \tau} V e^{-H_0 \tau} U(\tau) \\
&= V_I(\tau) U(\tau),
\end{aligned} \tag{8.9}
$$

and a similar equation applies to $S(\tau_1, \tau_2)$:

$$
-\frac{\partial}{\partial \tau} S(\tau_1, \tau_2) = V_I(\tau_1) S(\tau_1, \tau_2). \tag{8.10}
$$

These equations parallel those in real time, and, following exactly analogous procedures, we deduce that the imaginary-time-evolution operator in the interaction representation is given by a time-ordered exponential, as follows:

$$
U(\tau) = T \exp\left[-\int_0^\tau V_I(\tau) d\tau \right]
$$

$$
S(\tau_1, \tau_2) = T \exp\left[-\int_{\tau_1}^{\tau_2} V_I(\tau) d\tau \right]. \tag{8.11}
$$

One of the immediate applications of these results is to provide a perturbation expansion for the partition function. We can relate the partition function to the time-evolution operator in the interaction representation as follows:

$$Z = \mathrm{Tr}\left[e^{-\beta H}\right] = \mathrm{Tr}\left[e^{-\beta H_0}U(\beta)\right]$$

$$= \overbrace{\mathrm{Tr}\left[e^{-\beta H_0}\right]}^{Z_0}\overbrace{\left(\frac{\mathrm{Tr}\left[e^{-\beta H_0}U(\beta)\right]}{\mathrm{Tr}\left[e^{-\beta H_0}\right]}\right)}^{\langle U(\beta)\rangle_0}$$

$$= Z_0\langle U(\beta)\rangle_0, \tag{8.12}$$

enabling us to write the ratio of the interacting to the non-interacting partition function as the expectation value of the time-ordered exponential in the non-interacting system:

$$\frac{Z}{Z_0} = e^{-\beta\Delta F} = \left\langle T\exp\left[-\int_0^\beta V_I(\tau)d\tau\right]\right\rangle. \tag{8.13}$$

Notice how the logarithm of this expression gives the shift in free energy resulting from interactions. The perturbative expansion of this relation in powers of V is the basis for the finite-temperature Feynman diagram approach.

8.2 Imaginary-time Green's functions

The finite-temperature Green's function is defined as

$$\mathcal{G}_{\lambda\lambda'}(\tau - \tau') = -\langle T\psi_\lambda(\tau)\psi_{\lambda'}^\dagger(\tau')\rangle = -\mathrm{Tr}\left[e^{-\beta(H-F)}\psi_\lambda(\tau)\psi_{\lambda'}^\dagger(\tau')\right], \tag{8.14}$$

where ψ_λ can be either a fermionic or a bosonic field, evaluated in the Heisenberg representation. $F = -T\ln Z$ is the free energy and the T inside the angle brackets is the time-ordering operator. Provided H is time-independent, the resulting time–translational invariance ensures that $\mathcal{G}$ is solely a function of the time difference $\tau - \tau'$. In most cases, we will refer to situations where the quantum number λ is conserved, which will permit us to write

$$\mathcal{G}_{\lambda\lambda'}(\tau) = \delta_{\lambda\lambda'}\mathcal{G}_\lambda(\tau).$$

For the case of continuous quantum numbers λ, such as momentum, it is conventional to promote the quantum number into the argument of the Green's function, writing $\mathcal{G}(\mathbf{p}, \tau)$ rather than $\mathcal{G}_{\mathbf{p}}(\tau)$.

As an example, consider a non-interacting system with Hamiltonian

$$H = \sum \epsilon_\lambda \psi_\lambda^\dagger \psi_\lambda, \tag{8.15}$$

where $\epsilon_\lambda = E_\lambda - \mu$ is the one-particle energy, shifted by the chemical potential. Here, the equal-time expectation value of the fields is

$$\langle \psi_{\lambda'}^\dagger \psi_\lambda \rangle = \delta_{\lambda\lambda'} \begin{cases} n(\epsilon_\lambda) & \text{bosons} \\ f(\epsilon_\lambda) & \text{fermions} \end{cases} \tag{8.16}$$

where

$$n(\epsilon_\lambda) = \frac{1}{e^{\beta\epsilon_\lambda} - 1}$$

$$f(\epsilon_\lambda) = \frac{1}{e^{\beta\epsilon_\lambda} + 1} \tag{8.17}$$

are the Bose and Fermi functions, respectively. Similarly,

$$\langle \psi_\lambda \psi_{\lambda'}^\dagger \rangle = \delta_{\lambda\lambda'} \pm \langle \psi_{\lambda'}^\dagger \psi_\lambda \rangle = \delta_{\lambda\lambda'} \begin{cases} (1 + n(\epsilon_\lambda)) & \text{bosons} \\ (1 - f(\epsilon_\lambda)). & \text{fermions} \end{cases} \tag{8.18}$$

Using the time evolution of the operators,

$$\psi_\lambda(\tau) = e^{-\epsilon_\lambda \tau} \psi_\lambda(0)$$

$$\psi_\lambda^\dagger(\tau) = e^{\epsilon_\lambda \tau} \psi_\lambda^\dagger(0), \tag{8.19}$$

we deduce that

$$\mathcal{G}_{\lambda\lambda'}(\tau - \tau') = -\left[\theta(\tau - \tau')\langle \psi_\lambda \psi_{\lambda'}^\dagger \rangle + \zeta\theta(\tau' - \tau)\langle \psi_{\lambda'}^\dagger \psi_\lambda \rangle\right] e^{-\epsilon_\lambda(\tau - \tau')}, \tag{8.20}$$

where we have reintroduced $\zeta = 1$ for bosons and -1 for fermions, from Chapter 7. If we now write $\mathcal{G}_{\lambda\lambda'}(\tau - \tau') = \delta_{\lambda\lambda'}\mathcal{G}_\lambda(\tau - \tau')$, then

$$\mathcal{G}_\lambda(\tau) = -e^{-\epsilon_\lambda \tau} \begin{cases} [(1 + n(\epsilon_\lambda))\theta(\tau) + n(\epsilon_\lambda)\theta(-\tau)] & \text{bosons} \\ [(1 - f(\epsilon_\lambda))\theta(\tau) - f(\epsilon_\lambda)\theta(-\tau)]. & \text{fermions} \end{cases} \tag{8.21}$$

There are several points to notice about this Green's function:

- Apart from prefactors, at zero temperature the imaginary-time Green's function $\mathcal{G}_\lambda(\tau)$ is equal to the zero-temperature Green's function $G_\lambda(t)$, evaluated at a time $t = -i\tau$: $\mathcal{G}_\lambda(\tau) = -iG_\lambda(-i\tau)$.
- If $\tau < 0$, the Green's function satisfies the relation

$$\mathcal{G}_{\lambda\lambda'}(\tau + \beta) = \zeta\mathcal{G}_{\lambda\lambda'}(\tau),$$

so the bosonic Green's function is periodic in imaginary time, while the fermionic Green's function is antiperiodic in imaginary time, with period β (see Figure 8.3).

8.2.1 Periodicity and antiperiodicity

The (anti) periodicity observed in the last example is actually a general property of finite-temperature Green's functions. To see this, take $-\beta < \tau < 0$, then expand the Green's function as follows:

$$\mathcal{G}_{\lambda\lambda'}(\tau) = \zeta\langle \psi_{\lambda'}^\dagger(0)\psi_\lambda(\tau) \rangle$$

$$= \zeta\text{Tr}\left[e^{-\beta(H-F)}\psi_{\lambda'}^\dagger e^{\tau H}\psi_\lambda e^{-\tau H}\right]. \tag{8.22}$$

Now we can use the periodicity of the trace $\text{Tr}(AB) = \text{Tr}(BA)$ to cycle the operators on the left of the trace over to the right of the trace, as follows:

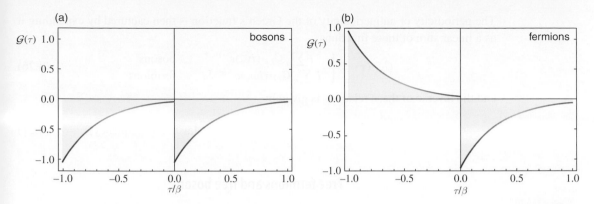

Showing (a) periodicity of the bosonic Green's function and (b) antiperiodicity of the fermionic Green's function.

Fig. 8.3

$$\mathcal{G}_{\lambda\lambda'}(\tau) = \zeta \operatorname{Tr}\left[e^{\tau H} \psi_{\lambda} e^{-\tau H} e^{-\beta(H-F)} \psi_{\lambda'}^{\dagger} \right]$$
$$= \zeta \operatorname{Tr}\left[e^{\beta F} e^{\tau H} \psi_{\lambda} e^{-(\tau+\beta)H} \psi_{\lambda'}^{\dagger} \right]$$
$$= \zeta \operatorname{Tr}\left[e^{-\beta(H-F)} e^{(\tau+\beta)H} \psi_{\lambda} e^{-(\tau+\beta)H} \psi_{\lambda'}^{\dagger} \right]$$
$$= \zeta \langle \psi_{\lambda}(\tau+\beta) \psi_{\lambda'}^{\dagger}(0) \rangle$$
$$= \zeta \mathcal{G}_{\lambda\lambda'}(\tau+\beta). \tag{8.23}$$

This periodicity or antiperiodicity was noted by Matsubara [1]. In the late 1950s, Lev Gor'kov, Alexei Abrikosov, and Igor Dzyaloshinskii [2] observed that we are in fact at liberty to extend the function $\mathcal{G}(\tau)$ outside the range $\tau \in [-\beta, \beta]$ by assuming that this periodicity, or antiperiodicity extends indefinitely along the entire imaginary-time axis. In other words, there need be no constraint on the value of τ in the periodic or antiperiodic boundary conditions

$$\mathcal{G}_{\lambda\lambda'}(\tau+\beta) = \pm\mathcal{G}_{\lambda\lambda'}(\tau).$$

With this observation, it becomes possible to carry out a Fourier expansion of the Green's function in terms of discrete frequencies. Today we use the term coined by Gor'kov, Abrikosov, and Dzyaloshinskii, calling them *Matsubara frequencies* [2].

8.2.2 Matsubara representation

The Matsubara frequencies are defined as

$$\begin{aligned} \nu_n &= 2\pi n k_B T && \text{bosons} \\ \omega_n &= \pi(2n+1)k_B T, && \text{fermions} \end{aligned} \tag{8.24}$$

where, by convention, ν_n is reserved for bosons and ω_n for fermions. These frequencies have the property that

$$\begin{aligned} e^{i\nu_n(\tau+\beta)} &= e^{i\nu_n\tau} \\ e^{i\omega_n(\tau+\beta)} &= -e^{i\omega_n\tau}. \end{aligned} \tag{8.25}$$

The periodicity or antiperiodicity of the Green's function is then captured by expanding it as a linear sum of these functions:

$$
\mathcal{G}_{\lambda\lambda'}(\tau) = \begin{cases} T\sum_n \mathcal{G}_{\lambda\lambda'}(i\nu_n)e^{-i\nu_n\tau} & \text{bosons} \\ T\sum_n \mathcal{G}_{\lambda\lambda'}(i\omega_n)e^{-i\omega_n\tau}, & \text{fermions} \end{cases}
\tag{8.26}
$$

and the inverse of these relations is given by

$$
\mathcal{G}_{\lambda\lambda'}(i\alpha_n) = \int_0^\beta d\tau \, \mathcal{G}_{\lambda\lambda'}(\tau)e^{i\alpha_n\tau} \qquad (\alpha_n = \{\text{Matsubara frequency}\}).
\tag{8.27}
$$

Free fermions and free bosons

For example, let us use (8.27) to derive the propagator for non-interacting fermions or bosons with $H = \sum \epsilon_\lambda \psi_\lambda^\dagger \psi_\lambda$. For fermions, the Matsubara frequencies are $i\omega_n = \pi(2n+1)k_BT$, so, using the real-time propagator (8.21), we obtain

$$
\mathcal{G}_\lambda(i\omega_n) = -\int_0^\beta d\tau \, e^{(i\omega_n - \epsilon_\lambda)\tau} \overbrace{(1 - f(\epsilon_\lambda))}^{[1+e^{-\beta\epsilon_\lambda}]^{-1}}
$$

$$
= -\frac{1}{i\omega_n - \epsilon_\lambda} \overbrace{\frac{(e^{(i\omega_n - \epsilon_\lambda)} - 1)}{1 + e^{-\beta\epsilon_\lambda}}}^{-1},
\tag{8.28}
$$

so that

$$
\mathcal{G}_\lambda(i\omega_n) = \frac{1}{i\omega_n - \epsilon_\lambda}. \qquad \text{free fermions}
\tag{8.29}
$$

In a similar way, for free bosons, where the Matsubara frequencies are $i\nu_n = \pi 2nk_BT$, using (8.27) and (8.21) we obtain

$$
\mathcal{G}_\lambda(i\nu_n) = -\int_0^\beta d\tau \, e^{(i\nu_n - \epsilon_\lambda)\tau} \overbrace{(1 + n(\epsilon_\lambda))}^{[1-e^{-\beta\epsilon_\lambda}]^{-1}}
$$

$$
= -\frac{1}{i\nu_n - \epsilon_\lambda} \overbrace{\frac{(e^{(i\nu_n - \epsilon_\lambda)} - 1)}{1 - e^{-\beta\epsilon_\lambda}}}^{-1},
\tag{8.30}
$$

so that

$$
\mathcal{G}_\lambda(i\nu_n) = \frac{1}{i\nu_n - \epsilon_\lambda}. \qquad \text{free bosons}
\tag{8.31}
$$

Remarks

- Notice how the finite-temperature propagators (8.29) and (8.31) are essentially identical for free fermions and free bosons. All the information about the statistics is encoded in the Matsubara frequencies.

- With the replacement $\omega \to i\omega_n$ the finite temperature propagator for free fermions (8.29) is essentially identical to the zero-temperature propagator, but notice that the inconvenient $i\delta\, \text{sgn}(\epsilon_\lambda)$ in the denominator has now disappeared.

Example 8.1 Calculate the finite-temperature Green's function,

$$D(\tau) = -\langle Tx(\tau)x(0)\rangle, \tag{8.32}$$

and its corresponding propagator,

$$D(i\nu_n) = \int_0^\beta e^{i\nu_n\tau} D(\tau), \tag{8.33}$$

for the simple harmonic oscillator

$$H = \hbar\omega \left(b^\dagger b + \frac{1}{2}\right)$$

$$x = \sqrt{\frac{\hbar}{2m\omega}}(b + b^\dagger). \tag{8.34}$$

Solution

Expanding the Green's function in terms of the creation and annihilation operators, we have

$$D(\tau) = -\frac{\hbar}{2m\omega}\langle T(b(\tau) + b^\dagger(\tau))(b(0) + b^\dagger(0))\rangle$$

$$= -\frac{\hbar}{2m\omega}\left(\langle Tb(\tau)b^\dagger(0)\rangle + \langle Tb^\dagger(\tau)b(0)\rangle\right), \tag{8.35}$$

where terms involving two creation or two annihilation operators vanish. Now, using the derivations that led to (8.21).

$$-\langle Tb(\tau)b^\dagger(0)\rangle = G(\tau) = -[(1 + n(\omega))\theta(\tau) + n(\omega)\theta(-\tau)]e^{-\omega\tau} \tag{8.36}$$

and

$$-\langle Tb^\dagger(\tau)b(0)\rangle = -[n(\omega)\theta(\tau) + (1 + n(\omega))]e^{\omega\tau}$$

$$= [(1 + n(-\omega))\theta(\tau) + n(-\omega)\theta(-\tau)]e^{\omega\tau}, \tag{8.37}$$

which corresponds to $-G(\tau)$ with the sign of ω inverted. With this observation,

$$D(\tau) = \frac{\hbar}{2m\omega}[G(\tau) - \{\omega \to -\omega\}]. \tag{8.38}$$

When we Fourier transform the first term inside the brackets, we obtain $\frac{1}{i\nu_n - \omega}$, so that

$$D(i\nu_n) = \frac{\hbar}{2m\omega}\left[\frac{1}{i\nu_n - \omega} - \frac{1}{i\nu_n + \omega}\right]$$

$$= \frac{\hbar}{2m\omega}\left[\frac{2\omega}{(i\nu_n)^2 - \omega^2}\right]. \tag{8.39}$$

This expression is identical to the corresponding zero-temperature propagator, evaluated at frequency $z = i\nu_n$.

Example 8.2 Consider a system of non-interacting fermions, described by the Hamiltonian $H = \sum_\lambda \epsilon_\lambda c_\lambda^\dagger c_\lambda$ where $\epsilon_\lambda = E_\lambda - \mu \, E_\lambda$ is the energy of a one-particle eigenstate, and μ is the chemical potential. Show that the total number of particles in equilibrium is

$$N(\mu) = T \sum \mathcal{G}_\lambda(i\omega_n) e^{i\omega_n 0^+},$$

where $\mathcal{G}_\lambda(i\omega_n) = (i\omega_n - \epsilon_\lambda)^{-1}$ is the Matsubara propagator. Using the relationship $N = -\partial F/\partial \mu$, show that free energy is given by

$$F(T, \mu) = -k_B T \sum_{\lambda, i\omega_n} \ln\left[-\mathcal{G}_\lambda(i\omega_n)^{-1}\right] e^{i\omega_n 0^+} + C(T). \tag{8.40}$$

Solution

The number of particles in state λ can be related to the equal-time Green's function, as follows:

$$N_\lambda = \langle c_\lambda^\dagger c_\lambda \rangle = -\langle T c_\lambda(0^-) c_\lambda^\dagger(0) \rangle = \mathcal{G}_\lambda(0^-).$$

Rewriting $G_\lambda(\tau) = T \sum_{i\omega_n} \mathcal{G}_\lambda e^{-i\omega_n \tau}$, we obtain

$$N(\mu) = \sum_\lambda N_\lambda = T \sum_{\lambda, i\omega_n} \mathcal{G}_\lambda(i\omega_n) e^{i\omega_n 0^+}.$$

Now since $-\partial F/\partial \mu = N(\mu)$, it follows that

$$F = -\int^\mu d\mu N(\mu) = -T \sum_{\lambda, i\omega_n} \int^\mu d\mu \frac{e^{i\omega_n 0^+}}{i\omega_n - E_\lambda + \mu}$$

$$= -T \sum_{\lambda, i\omega_n} \ln\left[\epsilon_\lambda - i\omega_n\right] e^{i\omega_n 0^+} + C(T)$$

$$= -T \sum_{\lambda, i\omega_n} \ln\left[-\mathcal{G}_\lambda(i\omega_n)^{-1}\right] e^{i\omega_n 0^+} + C(T). \tag{8.41}$$

We shall shortly see using contour integral methods that $C = 0$.

Example 8.3 Consider an electron gas where the spins are coupled to a magnetic field, so that $\epsilon_\lambda \equiv \epsilon_{\mathbf{k}} - \mu_B \sigma B$. Write down an expression for the magnetization and, by differentiating with respect to the field B, show that the temperature-dependent magnetic susceptibility is given by

$$\chi(T) = \left.\frac{\partial M}{\partial B}\right|_{B=0} = -2\mu_B^2 k_B T \sum_{\mathbf{k}, i\omega_n} G(k)^2,$$

where $\mathcal{G}(k) \equiv G(\mathbf{k}, i\omega_n)$ is the Matsubara propagator.

Solution

The magnetization is given by

$$M = \mu_B \sum_{\lambda,\sigma} \sigma \langle c^\dagger_{\mathbf{k}\sigma} c_{\mathbf{k}\sigma} \rangle = \mu_B T \sum_{\mathbf{k}\sigma, i\omega_n} \sigma \mathcal{G}_\sigma(\mathbf{k}, i\omega_n) e^{i\omega_n 0^+}.$$

Differentiating this with respect to B and then setting $B = 0$, we obtain

$$\chi = \frac{\partial M}{\partial B}\bigg|_{B=0} = -\mu_B^2 T \sum_{\mathbf{k}\sigma i\omega_n} \sigma^2 \mathcal{G}_\sigma(\mathbf{k}, i\omega_n)^2 \bigg|_{B=0}$$

$$= -2\mu_B^2 k_B T \sum_{\mathbf{k}, i\omega_n} \mathcal{G}(k)^2. \tag{8.42}$$

8.3 The contour integral method

In practice, we shall do almost all of our finite-temperature calculations in the frequency domain. To obtain practical results, we will need to be able to sum over the Matsubara frequencies, and this forces us to make an important technical digression. As an example of the kind of tasks we might want to carry out, consider how we would calculate the occupancy of a given momentum state in a Fermi gas at finite temperature, using the Matsubara propagator $G(\mathbf{p}, i\omega_n)$. This can be written in terms of the equal-time Green's function as follows:

$$\langle c^\dagger_{\mathbf{p}\sigma} c_{\mathbf{p}\sigma} \rangle = \mathcal{G}(\mathbf{p}, 0^-) = T \sum_n \frac{1}{i\omega_n - \epsilon(\mathbf{p})} e^{i\omega_n 0^+}. \tag{8.43}$$

A more involved example is the calculation of the finite-temperature dynamical spin susceptibility $\chi(q)$ of the free electron gas at wavevector and frequency $q \equiv (\mathbf{q}, i\nu_n)$. We shall see that this quantity derives from a Feynman polarization bubble diagram which gives

$$\chi(q) = -2\mu_B^2 T \sum_p \mathcal{G}(p+q)\mathcal{G}(p)$$

$$= -2\mu_B^2 \sum_{\mathbf{p}} \left(k_B T \sum_r G(\mathbf{p}+\mathbf{q}, i\omega_r + i\nu_n) G(\mathbf{p}, i\omega_r) \right), \tag{8.44}$$

where the -1 derives from the fermion loop. In both cases, we need to know how to do the sum over the discrete Matsubara frequencies, and to do this we use the method of contour integration. To make this possible, observe that the Fermi function $f(z) = 1/[e^{z\beta} + 1]$ has poles of strength $-k_B T$ at each discrete frequency $z = i\omega_n$, because

$$f(i\omega_n + \delta) = \frac{1}{e^{\beta(i\omega_n+\delta)} + 1} = -\frac{1}{\beta\delta} = -\frac{k_B T}{\delta},$$

so that, for a general function $F(i\omega_n)$, we may write

$$k_B T \sum_n F(i\omega_n) = \int_C \frac{dz}{2\pi i} F(z) f(z), \tag{8.45}$$

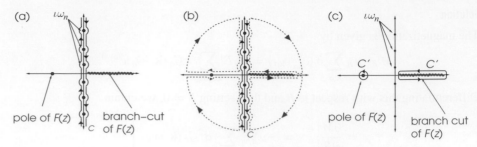

$$\textbf{Fig. 8.4}$$ (a) Contour integration around the poles in the Fermi function enables us to convert a discrete Matsubara sum $T \sum F(i\omega_n)$ to a continuous integral. (b) The integral can be distorted around the poles and branch cuts of $F(z)$. (c) Provided $F(z)$ dies away faster than $1/|z|$, we can neglect the contour at infinity.

where the contour integral C is to be taken clockwise around the poles at $z = i\omega_n$, as shown in Figure 8.4(a).

Once we have cast the sum as a contour integral, we may introduce null contours (Figure 8.4(b)) which allow us to distort the original contour C into the modified contour C' shown in Figure 8.4(c), so that now

$$k_B T \sum_n F(i\omega_n) = \int_{C'} \frac{dz}{2\pi i} F(z) f(z), \qquad (8.46)$$

where C' runs *counterclockwise* around all the poles and branch cuts in $F(z)$. Here we have used *Jordan's lemma* which guarantees that the contribution to the integral from the contour at infinity vanishes, provided the function $F(z) \times f(z)$ dies away faster than $1/|z|$ over the whole contour.

For example, in case (8.43), $F(z) = \frac{e^{z0^+}}{z - \epsilon_\mathbf{p}}$, so that $F(z)$ has a single pole at $z = \epsilon_\mathbf{p}$, and hence

$$\langle n_{\mathbf{p}\sigma} \rangle = T \sum_n \frac{1}{i\omega_n - \epsilon(\mathbf{p})} e^{i\omega_n O^+} = \int_{C'} \frac{dz}{2\pi i} \frac{1}{z - \epsilon_\mathbf{p}} e^{z0^+} f(z)$$

$$= f(\epsilon_\mathbf{p}), \qquad (8.47)$$

recovering the expected result. In this example, the convergence factor e^{z0^+} that results from the small negative time increment in the Green's function plays an important role inside the contour integral, where it gently forces the function $F(z)$ to die away faster than $1/|z|$ in the negative half-plane. Of course the original contour C integral could have been made by arbitrarily replacing $f(z)$ with $f(z) = $ constant. However, the requirement that the function die away in the positive half-plane forces us to set the constant term here to zero.

In the second example (8.44),

$$F(z) = -G(\mathbf{p} + \mathbf{q}, i\nu_n + z) G(\mathbf{p}, z) = -\frac{1}{i\nu_n + z - \epsilon_{\mathbf{p}+\mathbf{q}}} \frac{1}{z - \epsilon_\mathbf{p}},$$

which has two poles, at $z = \epsilon_\mathbf{p}$ and $z = -i\nu_n + \epsilon_{\mathbf{p}+\mathbf{q}}$. The integral for this case is then given by

$$\chi(q) = -2\mu_B^2 \sum_\mathbf{p} \int_{C'} \frac{dz}{2\pi i} G(\mathbf{p} + \mathbf{q}, z + i\nu_n) G(\mathbf{p}, z) f(z)$$

$$= -\sum_{\mathbf{p}} \left(G(\mathbf{p}, -i\nu_n + \epsilon_{\mathbf{p+q}}) f(-i\nu_n + \epsilon_{\mathbf{p+q}}) + G(\mathbf{p+q}, \epsilon_{\mathbf{p}} + i\nu_n) f(\epsilon_{\mathbf{p}}) \right). \quad (8.48)$$

The first term in the above expression deserves special attention. In this term we shall make use of the periodicity of the Fermi function to replace

$$f(-i\nu_n + \epsilon_{\mathbf{p+q}}) = f(\epsilon_{\mathbf{p+q}}).$$

This replacement may seem obvious but, later, when analytically extending $i\nu_n \to z$, we will keep this quantity *fixed*, i.e. we will *not* analytically extend $f(-i\nu_n + \epsilon_{\mathbf{p+q}}) \to f(-z + \epsilon_{\mathbf{p+q}})$. In other words, the continuation $i\nu_n \to z$ is made, keeping the location and residues of all poles in $\chi(\mathbf{q}, z)$ fixed. With this understanding, we continue and find that the resulting expression is given by

$$\chi(\mathbf{q}, i\nu_n) = 2\mu_B^2 \sum_{\mathbf{p}} \left(\frac{f_{\mathbf{p+q}} - f_{\mathbf{p}}}{i\nu_n - (\epsilon_{\mathbf{p+q}} - \epsilon_{\mathbf{p}})} \right), \quad (8.49)$$

where we have used the shorthand $f_{\mathbf{p}} \equiv f(\epsilon_{\mathbf{p}})$. The analytic extension of this quantity is then

$$\chi(\mathbf{q}, z) = 2\mu_B^2 \sum_{\mathbf{p}} \left(\frac{f_{\mathbf{p+q}} - f_{\mathbf{p}}}{z - (\epsilon_{\mathbf{p+q}} - \epsilon_{\mathbf{p}})} \right). \quad (8.50)$$

A completely parallel set of procedures can be carried for summation over Matsubara boson frequencies $i\nu_n$, by making the observation that the Bose function $n(z) = \frac{1}{e^{\beta z} - 1}$ has a string of poles at $z = i\nu_n$ of strength $k_B T$. Using a completely parallel procedure to the fermions, we obtain

$$k_B T \sum_n P(i\nu_n) = -\int_C \frac{dz}{2\pi i} P(z) n(z) = -\int_{C'} \frac{dz}{2\pi i} P(z) n(z),$$

where C is a clockwise integral around the imaginary axis and C' is a counterclockwise integral around the poles and branch cuts of $F(z)$ (see Exercise 8.1).

Example 8.4 Starting with the expression

$$F = -T \sum_{\lambda i \omega_n} \ln[(\epsilon_\lambda - i\omega_n)] e^{i\omega_n 0^+} + C(T),$$

derived in Example 8.2, use the contour integration method to show that

$$F = -T \sum_\lambda \ln\left[1 + e^{-\beta \epsilon_\lambda}\right] + C(T),$$

so that $C(T) = 0$.

Solution

Writing the free energy as a contour integral around the poles of the imaginary axis, we have

$$F = \sum_\lambda \int_P \frac{dz}{2\pi i} f(z) \ln [\epsilon_\lambda - z] \, e^{z0^+} + C(T),$$

where the path P runs counterclockwise around the imaginary axis. There is a branch cut in the function $F(z) = \ln[\epsilon_\lambda - z]$ running from $z = \epsilon_\lambda$ to $z = +\infty$. If we distort the contour P around this branch cut, we obtain

$$F = \sum_\lambda \int_{P'} \frac{dz}{2\pi i} f(z) \ln [\epsilon_\lambda - z] \, e^{z0^+} + C(T),$$

where P' runs clockwise around the branch cut, so that

$$F = \sum_\lambda \int_{\epsilon_\lambda}^\infty \frac{d\omega}{\pi} f(\omega) + C(T)$$

$$= \sum_\lambda -T \ln(1 + e^{-\beta\epsilon_\lambda}) + C(T), \tag{8.51}$$

so that $C(T) = 0$, to reproduce the standard expression for the free energy of a set of non-interacting fermions.

8.4   Generating function and Wick's theorem

The zero-temperature generating functions for free fermions and bosons, derived in Chapter 6, can be generalized to finite temperatures. Quite generally we can consider adding a source term to a free-particle Hamiltonian to form $H(\tau) = H_0 + V(\tau)$:

$$\begin{aligned} H_0 &= \sum \epsilon \psi_\lambda^\dagger \psi_\lambda \\ V(\tau) &= -\sum_\lambda [\bar\eta_\lambda(\tau)\psi_\lambda + \psi_\lambda^\dagger \eta(\tau)]. \end{aligned} \tag{8.52}$$

The corresponding finite-temperature generating functional is actually the partition function in the presence of the perturbation V. Using a simple generalization of (8.13), we have

$$Z_0[\bar\eta, \eta] = Z_0 \langle T e^{-\int_0^\beta V_I(\tau)d\tau} \rangle_0$$

$$= Z_0 \left\langle T \exp \left[\int_0^\beta d\tau \sum_\lambda \left(\bar\eta_\lambda(\tau)\psi_\lambda(\tau) + \psi_\lambda^\dagger(\tau)\eta_\lambda(\tau) \right) \right] \right\rangle_0. \tag{8.53}$$

where the driving terms are complex numbers for bosons, but anticommuting C-numbers or Grassman numbers for fermions. For free fields, the generating functional is given by

$$\frac{Z_0[\bar{\eta}, \eta]}{Z_0} = \exp\left[-\sum_\lambda \int_0^\beta d\tau_1 d\tau_2 \bar{\eta}_\lambda(1) G_\lambda(\tau_1 - \tau_2) \eta_\lambda(2)\right]$$

$$G_\lambda(\tau_1 - \tau_2) = -\langle T\psi_\lambda(\tau_1)\psi_\lambda^\dagger(\tau_2)\rangle. \tag{8.54}$$

A detailed proof of this result is given in Appendix 8A. However, a heuristic proof is obtained by appealing to the Gaussian nature of the underlying free fields. As at zero temperature, we expect the physics to be entirely Gaussian, that is, the amplitudes of fluctuation of the free fields are entirely independent of the driving terms. The usefulness of the generating function is that we can convert partial derivatives with respect to the source terms into field operators inside the expectation values:

$$\frac{\delta}{\delta\bar{\eta}(1)} \rightarrow \psi(1)$$

$$\frac{\delta}{\delta\eta(2)} \rightarrow \zeta\psi^\dagger(2), \tag{8.55}$$

where we have used the shorthand notation $\eta(1) \equiv \eta_\lambda(\tau_1)$, $\psi(1) \equiv \psi_\lambda(\tau_1)$. In particular,

$$\frac{\delta \ln Z_0[\bar{\eta}, \eta]}{\delta\bar{\eta}(1)} = \langle\psi(1)\rangle, \tag{8.56}$$

where the derivative of the logarithm of $Z_0[\bar{\eta}, \eta]$ is required to place a $Z_0[\bar{\eta}, \eta]$ in the denominator for the correctly normalized expectation value. For bosons, you can think of the source terms as an external field that induces a condensate of the field operator. At high temperatures, once the external source term is removed, the condensate disappears. However, at low temperatures in a Bose–Einstein condensate, the expectation value of the field survives even when the source terms are removed. For fermions, the idea of a genuine expectation value for the Fermi field is rather abstract, and, in this case, once the external source is removed the expectation value disappears.

We can of course take higher derivatives, and these do not vanish, even when the source terms are removed. In particular the second derivative determines the fluctuations of the quantum field, given by

$$\frac{\delta^2 \ln Z_0[\bar{\eta}, \eta]}{\delta\eta(2)\delta\bar{\eta}(1)} = \frac{\delta}{\delta\eta(2)}\left[\frac{1}{Z_0[\bar{\eta}, \eta]}\frac{\delta Z_0[\bar{\eta}, \eta]}{\delta\bar{\eta}(1)}\right]$$

$$= \frac{1}{Z_0[\bar{\eta}, \eta]}\frac{\delta^2 Z_0[\bar{\eta}, \eta]}{\delta\eta(2)\delta\bar{\eta}(1)} - \frac{1}{Z_0[\bar{\eta}, \eta]}\left[\frac{\delta Z_0[\bar{\eta}, \eta]}{\delta\eta(2)}\right]\frac{1}{Z_0[\bar{\eta}, \eta]}\left[\frac{\delta Z_0[\bar{\eta}, \eta]}{\delta\bar{\eta}(1)}\right]$$

$$= \zeta\left(\langle T\psi^\dagger(2)\psi(1)\rangle - \langle\psi^\dagger(2)\rangle\langle\psi(1)\rangle\right)$$

$$= \langle T\psi(1)\psi^\dagger(2)\rangle - \langle\psi(1)\rangle\langle\psi^\dagger(2)\rangle$$

$$= \left\langle T\Big(\psi(1) - \langle\psi(1)\rangle\Big)\Big(\psi^\dagger(2) - \langle\psi^\dagger(2)\rangle\Big)\right\rangle = \langle\delta\psi(1)\delta\psi^\dagger(2)\rangle, \tag{8.57}$$

where $\delta\psi(1) = \psi(1) - \langle\psi(1)\rangle$ represents the fluctuation of the field ψ around its mean value. If this quantity is independent of the source terms, then it follows that the fluctuations must be equal to their value in the absence of any source field, i.e.

$$\frac{\delta^2 \ln Z_0[\bar{\eta}, \eta]}{\delta \bar{\eta}_\lambda(\tau_1) \delta \eta_\lambda(\tau_2)} = \frac{\delta^2 \ln Z_0[\bar{\eta}, \eta]}{\delta \bar{\eta}_\lambda(\tau_1) \delta \eta_\lambda(\tau_2)}\bigg|_{\eta=\bar{\eta}=0} = -\mathcal{G}_\lambda(\tau_1 - \tau_2).$$

A more detailed algebraic derivation of this result is given in Appendix 8A. One of the immediate corollaries of (8.54) is that the multi-particle Green's functions can be entirely decomposed in terms of one-particle Green's functions, i.e. the imaginary-time Green's functions obey a Wick's theorem. If we decompose the original generating function (8.53) into a power series, we find that the general coefficient of the source terms is given by

$$(-1)^n \mathcal{G}(1, 2, \ldots n; 1', 2', \ldots n') = \langle T \psi(1) \cdots \psi(n) \psi^\dagger(n') \cdots \psi(1') \rangle.$$

By contrast, if we expand the right-hand side of (8.54) in the same way, we find that the same coefficient is given by

$$(-1)^n \sum_P (\zeta)^p \prod_{r=1}^n \mathcal{G}(r - P_r),$$

where p is the number of pairwise permutations required to produce the permutation P. Comparing the two results, we obtain the imaginary-time Wick's theorem

$$\mathcal{G}(1, 2, \ldots n; 1', 2', \ldots n') = \sum_P (-1)^p \prod_{r=1}^n \mathcal{G}(r - P_r).$$

Although this result is the precise analog of the zero-temperature Wick's theorem, notice that, unlike its zero-temperature counterpart, we cannot easily derive this result for simple cases by commuting the destruction operators so that they annihilate against the vacuum, since there is no finite-temperature vacuum.

Just as in the zero-temperature case, we can define a *contraction* as the process of connecting two free-field operators inside the correlation function:

$$\langle T[\cdots \psi(1) \cdots \psi^\dagger(2) \cdots] \rangle \longrightarrow \langle T[\psi(1) \psi^\dagger(2)] \rangle = -G(1 - 2)$$

$$\langle T[\cdots \psi^\dagger(2) \cdots \psi(1) \cdots] \rangle \longrightarrow \langle T[\psi^\dagger(2) \psi(1)] \rangle = -\zeta G(1 - 2),$$

so that, as before,

$$(-1)^n \langle T[\psi(1)\psi(2) \cdots \psi(n) \cdots \psi^\dagger(P_2') \cdots \psi^\dagger(P_1') \cdots \psi^\dagger(P_n')] \rangle$$
$$= \zeta^P G(1 \stackrel{\cdot}{-} P_1) G(2 \stackrel{\cdot}{-} P_2) \cdots G(n \stackrel{\cdot}{-} P_n). \tag{8.58}$$

Example 8.5 Use Wick's theorem to calculate the interaction energy of a dilute gas of spin S bosons interacting via a the interaction

$$\hat{V} = \frac{1}{2} \sum_{q,k\sigma,k',\sigma'} V(q) b^{\dagger}_{k+q\sigma} b^{\dagger}_{k',\sigma'} b_{k'+q\sigma'} b_{k\sigma}$$

at a temperature above the Bose–Einstein condensation temperature.

Solution

To leading order in the interaction strength, the interaction energy is given by

$$\langle V \rangle = \sum_{q,k,k',\sigma,\sigma'} V(q) \langle b^{\dagger}_{k+q,\sigma} b^{\dagger}_{k',\sigma'} b_{k'+q,\sigma'} b_{k\sigma} \rangle.$$

Using Wick's theorem, we evaluate

$$\langle b^{\dagger}_{k+q,\sigma} b^{\dagger}_{k',\sigma'} b_{k'+q,\sigma'} b_{k,\sigma} \rangle = \langle b^{\dagger}_{k+q,\sigma} b^{\dagger}_{k',\sigma'} b_{k'+q,\sigma'} b_{k,\sigma} \rangle + \langle b^{\dagger}_{k+q,\sigma} b^{\dagger}_{k',\sigma'} b_{k'+q,\sigma'} b_{k,\sigma} \rangle$$

$$= n_k n_{k'} \delta_{q,0} + n_k n_{k+q} \delta_{k,k'} \delta_{\sigma\sigma'} , \tag{8.59}$$

so that

$$\langle \hat{V} \rangle = \frac{1}{2} \int_{k,k'} n_k n_{k'} \left[(2S+1)^2 V_{q=0} + (2S+1) V_{k-k'} \right],$$

where $n_k = \frac{1}{e^{\beta(\epsilon_k - \mu)} - 1}$.

8.5 Feynman diagram expansion

We are now ready to generalize the Feynman approach to finite temperatures. Apart from a very small change in nomenclature, almost everything we learned for zero temperature in Chapter 7 now generalizes to finite temperature. Whereas previously we began with a Wick expansion of the S-matrix, now we must carry out a Wick expansion of the partition function

$$Z = e^{-\beta F} = Z_0 \left\langle T \exp\left[-\int_0^\beta \hat{V}(\tau) d\tau \right] \right\rangle_0 .$$

All the combinatorics of this expansion are unchanged at finite temperatures.

Now that we are at finite temperature, the free energy $F = E - ST - \mu N$ replaces the energy. The main results of this procedure can be guessed almost entirely by analogy; in particular,

- the partition function

$$Z = Z_0 \sum \{\text{unlinked Feynman diagrams}\}$$

- the change in free energy due to the perturbation V, given by

$$\Delta F = F - F_0 = -k_B T \ln \left[\frac{Z}{Z_0} \right] = -k_B T \sum \{\text{linked Feynman diagrams}\}.$$

This is the finite-temperature version of the linked-cluster theorem.

- the Matsubara one-particle Green's functions

$$\mathcal{G}(1 - 2) = \sum \{\text{two-leg Feynman diagrams}\}.$$

In fact, the Feynman rules for finite-temperature perturbation theory are almost identical to their zero-temperature counterparts. The main changes are

- replacement of $-i \rightarrow -1$ in the time-ordered exponential
- the finite range of integration in time,

$$\int_{-\infty}^{\infty} dt \longrightarrow \int_0^{\beta} d\tau,$$

which leads to the discrete Matsubara frequencies.

The effect of these changes on the real-space Feynman rules is summarized in Table 8.2.

The bookkeeping that leads to these diagrams now involves the redistribution of a -1 associated with each propagator:

$$\psi(2) \cdots \psi^{\dagger}(1) \longrightarrow (i)^2 \times \mathcal{G}(2 - 1), \tag{8.60}$$

where, as before,

$$\mathcal{G}(2 - 1) = 2 \longleftarrow 1 \tag{8.61}$$

represents the propagation of a particle from 1 to 2, but now we must redistribute an i (rather than a $\sqrt{-i}$) to each end of the propagator. When these terms are redistributed onto one-particle scattering vertices, they cancel the -1 from the time-ordered exponential:

$$-U(x) \quad = (i)^2 \times -U(x) \equiv U(x), \tag{8.62}$$

whereas, for a two-particle scattering potential $V(1 - 2)$, the four factors of i give $(i)^4 = 1$, so that the two-particle scattering amplitude is $-V(1 - 2)$:

$$1 \longrightarrow \langle 2 = (i)^4 \times -V(1 - 2) \equiv -V(1 - 2). \tag{8.63}$$

Apart from these small changes, the real-time Feynman rules are basically the same as those at zero temperature.

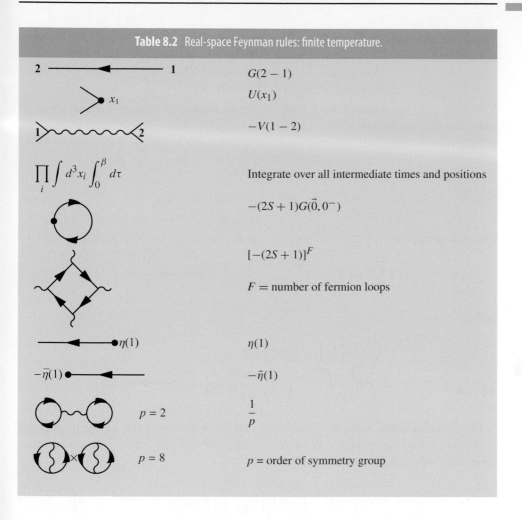

Table 8.2 Real-space Feynman rules: finite temperature.

(diagram: 2 ←— 1)	$G(2-1)$
(diagram: vertex x_1)	$U(x_1)$
(diagram: 1 ∿∿∿ 2)	$-V(1-2)$
$\prod_i \int d^3x_i \int_0^\beta d\tau$	Integrate over all intermediate times and positions
(diagram: loop)	$-(2S+1)G(\vec{0}, 0^-)$
(diagram: fermion loops)	$[-(2S+1)]^F$
	F = number of fermion loops
(diagram: —← •$\eta(1)$)	$\eta(1)$
(diagram: $-\bar\eta(1)$ •←—)	$-\bar\eta(1)$
(diagram: two loops) $p = 2$	$\dfrac{1}{p}$
(diagram: two loops) $p = 8$	p = order of symmetry group

8.5.1 Feynman rules from functional derivatives

As in Chapter 7, we can formally derive the Feynman rules from a functional derivative formulation. Using the notation

$$\int d1\,d2\,\bar\eta(1)G(1-2)\eta(2) = \bar\eta \;\underrightarrow{\hspace{2cm}}\; \eta, \tag{8.64}$$

where $d1$ and $d2$ implies integration over the space–time variables $(\vec{1}, \tau_1)$ and $(\vec{2}, \tau_2)$ and a sum over suppressed spin variables σ_1 and σ_2, we can write the non-interacting generating functional as

$$\frac{Z_0[\bar\eta, \eta]}{Z_0} = \langle \hat{S} \rangle_0 = \exp\left[-\bar\eta \;\underrightarrow{\hspace{2cm}}\; \eta\right], \tag{8.65}$$

where we have used the shorthand

$$\hat{S} = T \exp\left[\int_0^\beta d1[\bar{\eta}(1)\psi(1) + \psi^\dagger(1)\eta(1)]\right].$$

Now each time we differentiate $\hat{S}$ with respect to its source terms, we bring down an additional field operator, so that

$$\frac{\delta}{\delta\bar{\eta}(1)}\langle T \cdots \hat{S}\rangle_0 = \langle \cdots \psi(1) \cdots \hat{S}\rangle_0,$$

$$\zeta\frac{\delta}{\delta\eta(2)}\langle T \cdots \hat{S}\rangle_0 = \langle T \cdots \psi^\dagger(2) \cdots \hat{S}\rangle_0. \tag{8.66}$$

We can formally evaluate the time-ordered expectation value of any operator $F[\psi^\dagger, \psi]$ as

$$\langle T F\left[\psi^\dagger, \psi\right]\hat{S}\rangle_0 = F\left[\zeta\frac{\delta}{\delta\eta}, \frac{\delta}{\delta\bar{\eta}}\right]\exp\left[-\bar{\eta} \longleftarrow \eta\right],$$

so that

$$\frac{Z[\bar{\eta}, \eta]}{Z_0} = \left\langle T\exp\left[-\int_0^\beta \hat{V}(\tau)d\tau\right]\hat{S}\right\rangle_0$$

$$= \langle\exp\left[-\int_0^\beta d\tau V\left(\zeta\frac{\delta}{\delta\eta}, \frac{\delta}{\delta\bar{\eta}}\right)\right]\exp\left[-\bar{\eta} \longleftarrow \eta\right]\rangle.$$

The formal expansion of this functional derivative generates the Feynman diagram expansion. Changing variables to $(\alpha, \bar{\alpha}) = (\eta, -\bar{\eta})$, we can remove the minus sign associated with each propagator, to obtain

$$\frac{Z[-\bar{\alpha}, \alpha]}{Z_0} = \exp\left[(-1)^{n+1}\int_0^\beta d\tau V\left(\zeta\frac{\delta}{\delta\alpha}, \frac{\delta}{\delta\bar{\alpha}}\right)\right]\exp\left[\bar{\alpha} \longleftarrow \alpha\right] \tag{8.67}$$

for an n-body interaction. The appearance of $(-1)^{n+1}$ in the exponent indicates that we should associate a $(-1)^{n+1}$ with the corresponding scattering amplitude: the ζ inside the functional derivative keeps track of Fermi statistics, ultimately giving rise to the -1 associated with fermion loops.

As in the case of zero temperature, we may regard (8.67) as a machine for generating a series of Feynman diagrams, both linked and unlinked, so that, formally,

$$Z[\bar{\alpha}, \alpha] = Z_0 \sum\{\text{unlinked Feynman diagrams}\}.$$

8.5.2 Feynman rules in frequency–momentum space

As at zero temperature, it is generally more convenient to work in Fourier space. The transformation to Fourier transform space follows precisely parallel lines to that at zero temperature, and the Feynman rules which result are summarized in Table 8.3. We first

Table 8.3 Momentum-space Feynman rules: finite temperature.

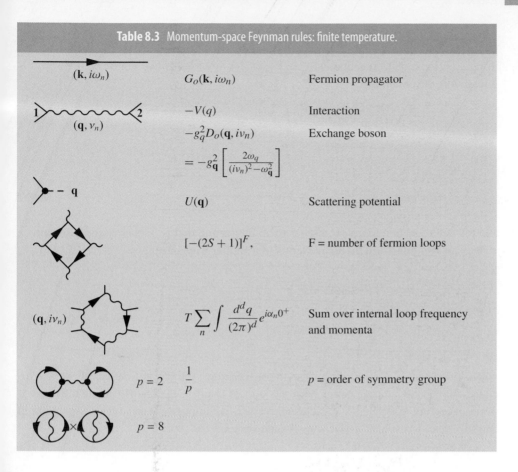

$(\mathbf{k}, i\omega_n)$	$G_o(\mathbf{k}, i\omega_n)$	Fermion propagator
$(\mathbf{q}, \nu_n)$	$-V(q)$	Interaction
	$-g_q^2 D_o(\mathbf{q}, i\nu_n)$	Exchange boson
	$= -g_{\mathbf{q}}^2 \left[\dfrac{2\omega_q}{(i\nu_n)^2 - \omega_{\mathbf{q}}^2} \right]$	
$-\mathbf{q}$	$U(\mathbf{q})$	Scattering potential
	$[-(2S+1)]^F,$	F = number of fermion loops
$(\mathbf{q}, i\nu_n)$	$T \sum_n \int \dfrac{d^d q}{(2\pi)^d} e^{i\alpha_n 0^+}$	Sum over internal loop frequency and momenta
$p = 2$	$\dfrac{1}{p}$	p = order of symmetry group
$p = 8$		

rewrite each interaction line and Green's function in a Feynman diagram in terms of their Fourier transformed variables:

$$1 \longrightarrow 2 = G(X_2 - X_1) = \sum_n \int \frac{d^{d-1}p}{(2\pi)^{d-1}} G(p) e^{ip(X_2 - X_1)}$$

$$1 \overset{q \rightarrow}{\longrightarrow} 2 = V(X_2 - X_1) = T \sum_n \int \frac{d^{d-1}q}{(2\pi)^{d-1}} V(q) e^{iq(X_2 - X_1)}, \tag{8.68}$$

where we have used the shorthand notation $p = (\mathbf{p}, i\alpha_n)$ ($\alpha_n = \omega_n$ for fermions, $\alpha_n = \nu_n$ for bosons), $q = (\mathbf{q}, i\nu_n)$, $X = (\mathbf{x}, i\tau)$, $ipX = i\mathbf{p} \cdot \mathbf{x} - i\omega_n \tau$ and $iqX = i\mathbf{q} \cdot \mathbf{x} - i\nu_r \tau$). As an example, consider a screened Coulomb interaction,

$$V(r) = \frac{e^2}{r} e^{-\kappa r}.$$

In our space–time notation, we write the interaction as

$$V(X) = V(\mathbf{x}, \tau) = \frac{e^2}{|\mathbf{x}|} e^{-\kappa |\mathbf{x}|} \times \tilde{\delta}(\tau),$$

where the delta function in time arises because the interaction is instantaneous. (Subtle point: we will in fact enforce periodic boundary conditions by taking the delta function to be a periodic delta function $\tilde{\delta}(\tau) = \sum_n \delta(\tau - n\beta)$.) When we Fourier transform this interaction, we obtain

$$
\begin{aligned}
V(Q) = V(\mathbf{q}, i\nu_r) &= \int d^4 X V(X) e^{-iQ.X} \\
&= \int d^3 x \int_0^\beta d\tau V(\mathbf{x}) \tilde{\delta}(\tau) e^{-i(\mathbf{q}\cdot\mathbf{x} - \nu_r \tau)} \\
&= V(\mathbf{q}) = \frac{4\pi e^2}{q^2 + \kappa^2}
\end{aligned}
\tag{8.69}
$$

and the delta function in time translates to an interaction that is frequency-independent.

We can also transform the source terms in a similar way, writing

$$
\eta(X) = T \sum_n \int \frac{d^3 p}{(2\pi)^3} e^{ipX} \eta(p)
$$

$$
\bar{\eta}(X) = T \sum_n \int \frac{d^3 p}{(2\pi)^3} e^{-ipX} \bar{\eta}(p),
\tag{8.70}
$$

where $ipX = i\vec{p} \cdot \vec{x} - i\alpha_n \tau$. With these transformations, the space–time coordinates associated with each scattering vertex now only appear as phase factors. By taking the integral over space–time coordinates at each such vertex, we impose conservation of momentum and (discrete) Matsubara frequencies at each vertex:

$$
\begin{array}{c}
p_2 \\
\text{\textasciitilde\textasciitilde\textasciitilde} X \\
q\to \quad p_1
\end{array}
= \int d^4 X e^{i(p_1 - p_2 - q)X} = (2\pi)^3 \beta \delta^{(3)}(\mathbf{p}_1 - \mathbf{p}_2 - \mathbf{q}) \delta_{\alpha_1 + \alpha_2 - \nu_r}.
\tag{8.71}
$$

Since momentum and frequency are conserved at each vertex, this means that there is one independent energy and frequency per loop in the Feynman diagram. To be sure that this really works, let us count the number of independent momenta that are left over after imposing a constraint at each vertex in the diagram. Consider a diagram with V vertices and P propagators. In d space–time dimensions, each propagator introduces Pd momenta. When we integrate over the space–time coordinates of the V vertices, we must be careful to split the integral into integrals over the $V - 1$ relative coordinates $\tilde{X}_j = X_{j+1} - X_j$ and over the center-of-mass coordinates:

$$
\int \prod_{j=1}^{V} d^d X_j = \int d^d X_{CM} \int \prod_{j=1}^{V-1} d^d \tilde{X}_j.
$$

This imposes $V - 1$ constraints per dimension, so the number of independent momenta is then $d[P - (V - 1)]$. Now in a general Feynman graph, the apparent number of momentum loops is the same as the number L of facets in the graph, and this is given by

$$L = \mathcal{E} + (P - V),$$

where $\mathcal{E}$ is the Euler characteristic of the object. The Euler characteristic is equal to 1 for planar diagrams, and 1 plus the number of "handles" in non-planar diagrams. For example, the diagram

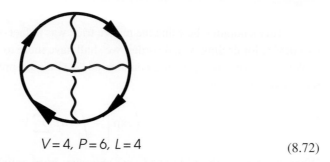

$$V = 4, \ P = 6, \ L = 4 \tag{8.72}$$

has $V = 4$ vertices, $P = 6$ propagators and it has one handle with Euler characteristic $\mathcal{E} = 2$, so that $L = 6 - 4 + 2 = 4$, as expected. Now one might have expected the number of independent momentum loops to be equal to L. However, when there are handles, this overcounts the number of independent momentum loops, for each handle added to the diagram adds only one additional momentum loop, but L increases by 2. If you look at the example above, this diagram can be embedded on a cylinder, and the interaction propagator which loops around the cylinder only counts as one momentum loop, giving a total of $4 - (2 - 1) = 3$ independent momentum loops:

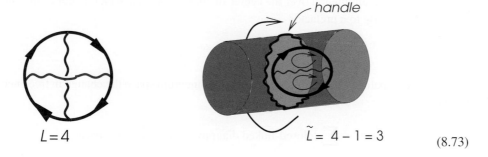

$$L = 4 \qquad\qquad \tilde{L} = 4 - 1 = 3 \tag{8.73}$$

In this way, we see that $\tilde{L} = L - (\mathcal{E} - 1) = P - V + 1$ *is* the correct number of independent momentum loops. Our momentum constraint does indeed convert the diagram from an integral over V space–time coordinates to $\tilde{L}$ independent momentum loops.

In this way, we see that the transformation from real-space to momentum-space Feynman rules is effected by replacing the sum over all internal space–time coordinates by an integral/sum over all loop momenta and frequencies. A convergence factor $e^{i\alpha_n 0^+}$ is included in the loop integral. This guarantees that if the loop contains a single propagator which propagates back to the point from which it emanated, then the corresponding contraction of field operators is normal-ordered.

8.5.3 Linked-cluster theorem

The linked-cluster theorem for imaginary time follows from the *replica trick*, as at zero temperature. In this case, we wish to compute the logarithm of the partition function

$$\ln\left(\frac{Z}{Z_0}\right) = \lim_{n\to 0}\frac{1}{n}\left[\left(\frac{Z}{Z_0}\right)^n - 1\right].$$

It is worth mentioning here that the replica trick was in fact originally invented by Edwards as a device for dealing with disorder; we shall have more to say about this in Chapter 10.

We now write the term that contains $(Z/Z_0)^n$ as the product of contributions from n replica systems, so that

$$\left(\frac{Z}{Z_0}\right)^n = \left\langle \exp\left[-\int_0^\beta d\tau \sum_{\lambda=1}^n V^{(\lambda)}(\tau)\right]\right\rangle_0.$$

When we expand the right-hand side as a sum over unlinked Feynman diagrams, each separate Feynman diagram has a replica index that must be summed over, so that a single linked diagram is of order $O(n)$, whereas a group of k unlinked diagrams is of order $O(n^k)$. In this way, as $n \to 0$ only the unlinked diagrams survive. The upshot is that the shift in the free energy ΔF produced by the perturbation $\hat{V}$ is given by

$$-\beta\Delta F = \ln(Z/Z_0) = \sum\{\text{closed-link diagrams in real space}\}.$$

Notice that, unlike the zero-temperature proof, here we do not have to appeal to adiabaticity to extract the shift in free energy from the closed-loop diagrams.

When we convert to momentum space, Fourier transforming each propagator and interaction line, an overall integral over the center-of-mass coordinates factors out of the entire diagram, giving rise to a prefactor

$$\int d^d X_{cm} = \beta(2\pi)^{d-1}\delta^{(d-1)}(0) \equiv V\beta,$$

where V is the spatial volume. Consequently, in momentum space the change in free energy is given by

$$\frac{\Delta F}{V} = -\sum\{\text{closed-linked diagrams in momentum space}\}.$$

Finally, let us say a few words about Green's functions. Since the nth-order coefficients of α and $\bar{\alpha}$ are the irreducible n-point Green's functions,

$$\ln Z[\bar{\alpha},\alpha] = -\beta\Delta F + \int d1d2\bar{\alpha}(1)\mathcal{G}(1-2)\alpha(2)$$

$$+ \frac{1}{(2!)^2}\int d1d2d3d4\bar{\alpha}(1)\bar{\alpha}(2)\alpha(3)\alpha(4)\mathcal{G}_{irr}(1,2;3,4) + \cdots . \qquad (8.74)$$

n-particle irreducible Green's functions are simply the n-particle Green's functions in which all contributions from $(n-1)$-particle Green's functions have been subtracted. Now since the nth-order coefficients in the Feynman diagram expansion of $\ln Z[\bar{\alpha},\alpha]$ are the

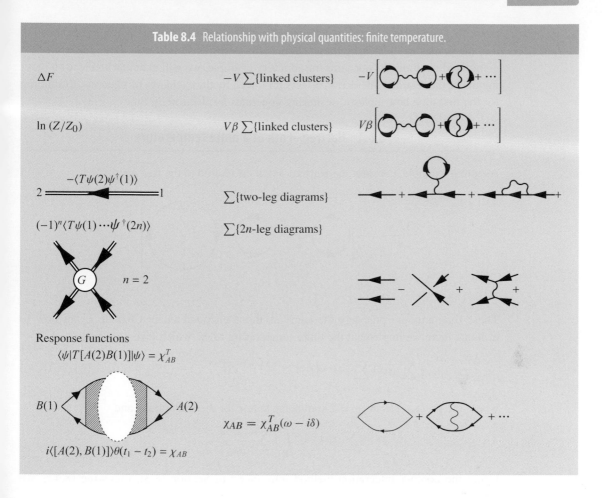

Table 8.4 Relationship with physical quantities: finite temperature.

connected $2n$-leg diagrams, it follows that the n-particle irreducible Green's functions are given by the sum of all $2n$-leg diagrams:

$$\mathcal{G}_{irr}(1, 2, \ldots n; 1', 2', \ldots n') = \sum \{\text{connected } 2n\text{-leg diagrams}\}.$$

The main links between finite-temperature Feynman diagrams and physical quantities are given in Table 8.4.

8.6 Examples of the application of the Matsubara technique

To illustrate the Matsubara technique, we shall examine three examples. In the first, we will see briefly how the Hartree–Fock approximation is modified at finite temperatures. This will give some familiarity with the techniques. In the second example, we shall examine the effect of disorder on the electron propagator. Surprisingly, the spatial fluctuations in the electron potential that arise in a disordered medium behave like a highly retarded potential, and the scattering created by these fluctuations is responsible for the Drude lifetime in a disordered medium. As our third introductory example, we will examine an electron moving under the retarded interaction effects produced by the exchange of phonons, examining for the first time how inelastic scattering generates an electron lifetime.

8.6.1   Hartree–Fock at a finite temperature

As a first example, consider the Hartree–Fock correction to the free energy,

$$\tag{8.75}$$

These diagrams are precisely the same as those encountered in Chapter 7, but now to evaluate them we implement the finite-temperature rules, which give

$$\frac{\Delta F_{HF}}{V} = \frac{1}{2} \sum_k \mathcal{G}(k) \sum_{k'} \mathcal{G}(k') \left\{ [-(2S+1)]^2 \, V(k-k') - (2S+1)V(q=0) \right\}, \quad (8.76)$$

where the prefactor is the $p = 2$ symmetry factor for these diagrams and

$$\sum_k \mathcal{G}(k) \equiv \int_{\mathbf{k}} T \sum \frac{1}{i\omega_n - \epsilon_{\mathbf{k}}} e^{i\omega_n 0^+}.$$

Using the contour integration method introduced in Section (8.3), following (8.47), we have

$$T \sum \frac{1}{i\omega_n - \epsilon_{\mathbf{k}}} e^{i\omega_n 0^+} = \int_C \frac{dz}{2\pi i} \frac{1}{z - \epsilon_{\mathbf{k}}} e^{z0^+} f(z) = f(\epsilon_{\mathbf{k}}),$$

where the contour C runs counterclockwise around the pole at $z = \epsilon_{\mathbf{k}}$, so that the first-order shift in the free energy is

$$\Delta F_{HF} = \frac{1}{2} \int_{\mathbf{k},\mathbf{k'}} \left[(2S+1)^2 (V_{\mathbf{q}=0}) - (2S+1)(V_{\mathbf{k}-\mathbf{k'}}) \right] f_{\mathbf{k}} f_{\mathbf{k'}}.$$

This is formally exactly the same as at zero temperature, except that now $f_{\mathbf{k}}$ refers to the finite-temperature Fermi–Dirac function. Notice that we could have applied exactly the same method to bosons, the main result being a change in sign of the second Fock term.

8.6.2   Electron in a disordered potential

As a second example of the application of finite-temperature methods, we shall consider the propagator for an electron in a disordered potential. This will introduce the concept of an *impurity average*.

Our interest in this problem is driven ultimately by a desire to understand the bulk properties of a disordered metal. The problem of electron transport is almost as old as our knowledge of the electron itself. The term "electron" was coined to describe the fundamental unit of charge (already measured from electrolysis) by the Irish physicist George Johnstone Stoney in 1891 [3]. Heinrich Lorentz derived his famous force law for charged ions in 1895 [4], but did not use the term "electron" until 1899. In 1897 J. J. Thomson [5] made the crucial discovery of the electron by correctly interpreting his measurement of the m/e ratio for cathode rays in terms of a new state of particulate matter "from which all chemical elements are built up." Within three years of this discovery, Paul Drude [6] had synthesized these ideas and had argued, based on the idea of a classical gas of charged electrons, that electrons would exhibit a mean free path $l = v_{electron}\tau$, where τ is the scattering rate and l the average distance between scattering events. In Drude's theory, electrons were envisioned as diffusing through the metal, and he was able to derive his famous formula for the conductivity σ:

$$\sigma = \frac{ne^2\tau}{m}.$$

Missing from Drude's pioneering picture was any notion of the Fermi–Dirac statistics of the electron fluid. He had, for example, no notion that the characteristic velocity of the electrons was given by the Fermi velocity, $v_{electron} \sim v_F$, a vastly greater velocity at low temperatures than could ever be expected on the grounds of a Maxwell–Boltzmann fluid of particles. This raises the question: how, in a fully quantum mechanical picture of the electron fluid, can we rederive Drude's basic model?

A real metal contains both disorder and electron–electron interactions; in this book we shall only touch on the simpler problem of disorder in an otherwise free electron gas. We shall actually return to this problem in earnest in the next chapter; our task here is to examine the electron propagator in a disordered medium of elastically scattering impurities. We shall consider an electron in a disordered potential:

$$H = \sum_{\mathbf{k}} \epsilon_{\mathbf{k}} c_{\mathbf{k}}^{\dagger} c_{\mathbf{k}} + V_{disorder}$$

$$V_{disorder} = \int d^3x\, U(\vec{x}) \psi^{\dagger}(\mathbf{x}) \psi^{\dagger}(\mathbf{x}), \tag{8.77}$$

where $U(\mathbf{x})$ represents the scattering potential generated by a random array of N_i impurities located at positions $\mathbf{R}_j$, each with atomic potential $\mathcal{U}(\mathbf{x} - \mathbf{R}_j)$,

$$U(\mathbf{x}) = \sum_{j} \mathcal{U}(\mathbf{x} - \mathbf{R}_j).$$

An important aspect of this Hamiltonian is that it contains no interactions between electrons, and as such the energy of each individual electron is conserved: all interactions are elastic.

We shall not be interested in calculating the value of a physical quantity for a *specific* location of impurities, but rather the value of that quantity after we have averaged over the locations of the impurities, i.e.

$$\overline{\langle A \rangle} = \int \prod_j \frac{1}{V} \, d^3 R_j \langle \hat{A}[\{\mathbf{R}_j\}] \rangle.$$

This is an elementary example of a *quenched average*, in which the *impurity average* takes place *after* the thermodynamic average. Here we'll calculate the impurity-averaged Green's function. To do this we need to know something about the fluctuations of the impurity scattering potential about its average. It is these fluctuations that scatter the electrons.

Electrons will in general scatter off the fluctuations in the potential. The average impurity potential $\overline{U(\mathbf{x})}$ plays the roll of a kind of shifted chemical potential. Indeed, if we shift the chemical potential by an amount $\Delta\mu$, the scattering potential becomes $\tilde{U}(\mathbf{x}) = U(\mathbf{x}) - \Delta\mu$, and we can always choose $\Delta\mu = \overline{U(\mathbf{x})}$ so that $\overline{\tilde{U}(\mathbf{x})} = 0$. The residual potential describes the fluctuations in the scattering potential: $\delta U(\mathbf{x}) = U(\mathbf{x}) - \overline{U(\mathbf{x})}$. We shall now drop the tilde. The fluctuations in the impurity potential are spatially correlated, and we shall shortly show that

$$\overline{\delta U(\mathbf{x}) \delta U(\mathbf{x}')} = \int_{\mathbf{q}} e^{i\mathbf{q} \cdot (\mathbf{x} - \mathbf{x}')} n_i \, |u(\mathbf{q})|^2 \,, \qquad (8.78)$$

where $u(\mathbf{q}) = \int d^3 x \mathcal{U}(\mathbf{x}) e^{-i\mathbf{q} \cdot \mathbf{x}}$ is the Fourier transform of the scattering potential and $n_i = N_i/V$ is the concentration of impurities. It is these fluctuations that scatter the electrons, and when we come to consider the impurity-averaged Feynman diagrams, we'll see that the spatial correlations in the potential fluctuations induce a sort of attractive interaction, denoted by the diagram

$$\int_{\mathbf{q}} n_i |u(\mathbf{q})|^2 e^{i\mathbf{q} \cdot (\mathbf{x} - \mathbf{x}')} = -V_{eff}(\mathbf{x} - \mathbf{x}').$$

$$(8.79)$$

Although in principle we should keep all higher moments of the impurity scattering potential in practice the leading-order moments are enough to extract a lot of the basic physics in weakly disordered metals. Notice that the fluctuations in the scattering potential are short-range – they only extend over the range of the scattering potential. Indeed, if we neglect the momentum dependence of $u(\mathbf{q})$, assuming that the impurity scattering is dominated by low-energy s-wave scattering, then we can write $u(\mathbf{q}) = u_0$. In this situation, the fluctuations in the impurity scattering potential are entirely local:

$$\overline{\delta U(\mathbf{x}) \delta U(\mathbf{x}')} = n_i u_0^2 \delta(\mathbf{x} - \mathbf{x}'). \qquad \qquad \text{white-noise potential}$$

In our discussion here, we will neglect the higher-order moments of the scattering potential, effectively assuming that it is purely Gaussian.

To prove (8.78), we first Fourier transform the potential

$$U(\mathbf{q}) = \sum_j e^{-i\mathbf{q} \cdot \mathbf{R}_j} \int d^3 x \, \mathcal{U}(\mathbf{x} - \mathbf{R}_j) e^{-i\mathbf{q} \cdot (\mathbf{x} - \mathbf{R}_j)} = u(\mathbf{q}) \sum_j e^{-i\mathbf{q} \cdot \mathbf{R}_j}, \qquad (8.80)$$

so that the locations of the impurities are encoded in the phase shifts which multiply $u(\mathbf{q})$. If we now carry out the average,

$$\overline{\delta U(\mathbf{x})\delta U(\mathbf{x}')} = \int_{\mathbf{q},\mathbf{q}'} e^{i(\mathbf{q}\cdot\mathbf{x}-\mathbf{q}\cdot\mathbf{x}')} \left(\overline{U(\mathbf{q})U(-\mathbf{q}')} - \overline{U(\mathbf{q})}\ \overline{U(-\mathbf{q}')} \right)$$

$$= \int_{\mathbf{q},\mathbf{q}'} e^{i(\mathbf{q}\cdot\mathbf{x}-\mathbf{q}\cdot\mathbf{x}')} u(\mathbf{q})u(-\mathbf{q}') \sum_{i,j} \left(\overline{e^{-i\mathbf{q}\cdot\mathbf{R}_i} e^{i\mathbf{q}'\cdot\mathbf{R}_j}} - \overline{e^{-i\mathbf{q}\cdot\mathbf{R}_i}}\ \overline{e^{i\mathbf{q}'\cdot\mathbf{R}_j}} \right). \quad (8.81)$$

Now since the phase terms are independent at different sites, the variance of the random phase term in the above expression vanishes unless $i = j$, so

$$\sum_{i,j} \left(\overline{e^{-i\mathbf{q}\cdot\mathbf{R}_i} e^{i\mathbf{q}'\cdot\mathbf{R}_j}} - \overline{e^{-i\mathbf{q}\cdot\mathbf{R}_i}}\ \overline{e^{i\mathbf{q}'\cdot\mathbf{R}_j}} \right) = N_i \int \frac{1}{V} d^3 R_j e^{-i(\mathbf{q}-\mathbf{q}')\cdot\mathbf{R}_j}$$

$$= n_i (2\pi)^3 \delta^{(3)}(\mathbf{q} - \mathbf{q}'), \quad (8.82)$$

from which

$$\overline{U(\mathbf{q})U(-\mathbf{q}')} - \overline{U(\mathbf{q})}\ \overline{U(-\mathbf{q}')} = n_i |u(\mathbf{q})|^2 (2\pi)^3 \delta^{(3)}(\mathbf{q} - \mathbf{q}'),$$

and (8.78) follows.

Now let us examine how electrons scatter off these fluctuations. If we substitute $\psi^\dagger(\mathbf{x}) = \int_{\mathbf{k}} c_k^\dagger e^{-i\mathbf{k}\cdot\mathbf{x}}$ into $\hat{V}_{disorder}$, we obtain

$$\hat{V} = \int_{\mathbf{k},\mathbf{k}'} c_k^\dagger c_{\mathbf{k}'} \delta U(\mathbf{k} - \mathbf{k}').$$

We shall represent the scattering amplitude for scattering once by

$$\delta U(\mathbf{k} - \mathbf{k}') = \left(u(\mathbf{k} - \mathbf{k}') \sum_j e^{i(\mathbf{k}-\mathbf{k}')\cdot\mathbf{R}_j} \right) - \Delta\mu\delta_{\mathbf{k}-\mathbf{k}'}, \quad (8.83)$$

where we have subtracted the scattering off the average potential. The potential transfers momentum, but does not impart any energy to the electron, and for this reason frequency is conserved along the electron propagator. Let us now write down in momentum space the Green's function of the electron:

$$\mathcal{G}(\mathbf{k}, \mathbf{k}', i\omega_n) = \quad + \quad + \quad + \quad + \cdots,$$

$$= \mathcal{G}^0(\mathbf{k}, i\omega_n)\delta_{\mathbf{k},\mathbf{k}'} + \mathcal{G}^0(\mathbf{k}, i\omega_n)\delta U(\mathbf{k} - \mathbf{k}')\mathcal{G}^0(\mathbf{k}', i\omega_n)$$

$$+ \int_{\mathbf{k}_1} \mathcal{G}^0(\mathbf{k}, i\omega_n)\delta U(\mathbf{k} - \mathbf{k}_1)\mathcal{G}^0(\mathbf{k}_1, i\omega_n)\delta U(\mathbf{k}_1 - \mathbf{k}')\mathcal{G}^0(\mathbf{k}', i\omega_n) + \cdots,$$

$$(8.84)$$

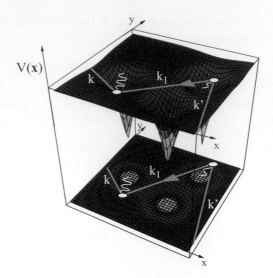

$V(\mathbf{x})$

Fig. 8.5   Double scattering event in the random impurity potential.

where the frequency $i\omega_n$ is constant along the electron line. Notice that $\mathcal{G}$ is actually a function of each impurity position! Figure 8.5 illustrates one of the scattering events contributing to the third diagram in this sum. We want to calculate the quenched average $\overline{\mathcal{G}(\mathbf{k}, \mathbf{k}', i\omega_n)}$, and to do this we need to average each Feynman diagram in the above series.

When we impurity-average the single scattering event, it vanishes:

$$\overline{\mathcal{G}^0(\mathbf{k}, i\omega_n)\delta U(\mathbf{k} - \mathbf{k}')\mathcal{G}^0(\mathbf{k}', i\omega_n)} = \mathcal{G}^0(\mathbf{k}, i\omega_n)\overbrace{\overline{\delta U(\mathbf{k} - \mathbf{k}')}}^{=0}\mathcal{G}^0(\mathbf{k}', i\omega_n),$$

but the average of a double scattering event is

$$\sum_{\mathbf{k}_1}\mathcal{G}^0(\mathbf{k}, i\omega_n)\mathcal{G}^0(\mathbf{k}_1, i\omega_n)\mathcal{G}^0(\mathbf{k}', i\omega_n) \times \overbrace{\overline{\delta U(\mathbf{k} - \mathbf{k}_1)\delta U(\mathbf{k}_1 - \mathbf{k}')}}^{n_i|u_{\mathbf{k}-\mathbf{k}'}|^2\delta_{\mathbf{k}-\mathbf{k}'}}$$

$$= \delta_{\mathbf{k}-\mathbf{k}'} \times \mathcal{G}^0(\mathbf{k}, i\omega_n)^2 n_i \sum_{\mathbf{k}_1} u(\mathbf{k} - \mathbf{k}_1)^2 \mathcal{G}^0(\mathbf{k}_1, i\omega_n)\mathcal{G}^0(\mathbf{k}, i\omega_n). \tag{8.85}$$

Notice something fascinating: after impurity averaging, momentum is now conserved. We can denote the impurity-averaged double scattering event Feynman diagram by

$$\left[\begin{array}{c} \times \quad\quad \times \\[8pt] \underleftarrow{} \end{array} \right] = \underleftarrow{\underset{\mathbf{k}-\mathbf{q}}{\overset{\overset{\mathbf{q}}{\diagup\diagdown}}{\mathbf{k}\mathbf{k}}}}$$

$$\tag{8.86}$$

where we have introduced the Feynman diagram

$$n_i|u(\mathbf{q})|^2 = -V_{eff}(\mathbf{q})$$

(8.87)

to denote the momentum transfer produced by the quenched fluctuations in the random potential. In writing the diagram this way, we bring out the notion that quenched disorder can be very loosely thought of as an interaction with an effective potential

$$V_{eff}(\mathbf{q}, i\nu_n) = \int_0^\beta d\tau e^{i\nu_n\tau} \overbrace{V_{eff}(\mathbf{q}, \tau)}^{-n_i|u(\mathbf{q})|^2} = -\beta\delta_{n0}n_i|u(\mathbf{q})|^2,$$

where the $\beta\delta_{n0} \equiv \int d\tau e^{i\nu_n\tau}$, derived from the fact that the interaction $V_{eff}(\mathbf{q}, \tau)$ does not depend on the time difference, guarantees that there is no energy transfered by the quenched scattering events. In other words, quenched disorder induces a sort of *infinitely retarded attractive potential* between electrons.[2] The notion that disorder induces interactions is an interesting one, for it motivates the idea that disorder can lead to new kinds of collective behavior.

After the impurity averaging, we notice that momentum is now conserved, so that the impurity-averaged Green's function is now diagonal in momentum space:

$$\overline{\mathcal{G}(\mathbf{k}, \mathbf{k}', i\nu_n)} = \delta_{\mathbf{k}-\mathbf{k}'}\mathcal{G}(\mathbf{k}, i\nu_n).$$

If we now carry out the impurity averaging on multiple scattering events, only repeated scattering events at the same sites will give rise to non-vanishing contributions. If we take account of all scattering events induced by the Gaussian fluctuations in the scattering potential, then we generate a series of diagrams of the form

In the Feynman diagrams, we can group all scatterings into connected self-energy diagrams, as follows:

[2] Our statement can be made formally correct in the language of replicas: this interaction takes place between electrons of the same or different replica index. In the $N \to 0$ limit, the residual interaction only acts on one electron in the same replica.

$$\Sigma(k) = \boxed{\Sigma} = \;\;\diagdown\!\!\diagdown + \;\;\diagdown\!\!\diagdown + \;\;\diagdown\!\!\diagdown + \cdots$$

$$G(k) = \Longleftarrow = \longleftarrow + \longleftarrow\!\boxed{\Sigma}\!\longleftarrow + \longleftarrow\!\boxed{\Sigma}\!\longleftarrow\!\boxed{\Sigma}\!\longleftarrow + \cdots$$

$$= [i\omega_n - \epsilon_{\mathbf{k}} - \Sigma(k)]^{-1}. \tag{8.88}$$

In the case of s-wave scattering, all momentum dependence of the scattering processes is lost, so that in this case $\Sigma(k) = \Sigma(i\omega_n)$ only depends on the frequency. In the above diagram, the double line on the electron propagator indicates that all self-energy corrections have been included. From the above, you can see that the self-energy corrections calculated from the first expression are fed into the electron propagator, which in turn is used in a self-consistent way inside the self-energy.

We shall begin by trying to calculate the above first-order diagrams for the self-energy without imposing any self-consistency. This diagram is given by

$$\Sigma(i\omega_n) = \;\;\diagdown\!\!\diagdown\;\; = n_i \sum_{\mathbf{k}'} |u(\mathbf{k} - \mathbf{k}')|^2 G(\mathbf{k}', i\omega_n)$$

$$= n_i \sum_{\mathbf{k}'} |u(\mathbf{k} - \mathbf{k}')|^2 \frac{1}{i\omega_n - \epsilon_{\mathbf{k}'}}. \tag{8.89}$$

Now we can replace the summation over momentum inside this self-energy by an integration over solid angle and energy, as follows:

$$\sum_{\mathbf{k}'} \rightarrow \int \frac{d\Omega_{\mathbf{k}'}}{4\pi} d\epsilon' N(\epsilon'),$$

where $N(\epsilon)$ is the density of states. With this replacement,

$$\Sigma(i\omega_n) = n_i u_0^2 \int d\epsilon N(\epsilon) \frac{1}{i\omega_n - \epsilon},$$

where

$$u_0^2 = \int \frac{d\Omega_{\mathbf{k}'}}{4\pi} |u(\mathbf{k} - \mathbf{k}')|^2 = \frac{1}{2} \int_{-1}^{1} d\cos\theta |u(\theta)|^2$$

is the angular average of the squared scattering amplitude. To a good approximation, this expression can be calculated by replacing the energy-dependent density of states by its value at the Fermi energy. In so doing, we neglect a small real part of the self-energy, which can in any case be absorbed by the chemical potential. This kind of approximation is extremely common in many-body physics, in cases where the key physics is dominated by electrons close to the Fermi energy. The deviations from constancy in $N(\epsilon)$ will in practice affect the real part of $\Sigma(i\omega_n)$, and these small changes can be accomodated by a shift in the chemical potential. The resulting expression for $\Sigma(i\omega_n)$ is then

$$\Sigma(i\omega_n) = n_i u_0^2 N(0) \int_{-\infty}^{\infty} d\epsilon \frac{1}{i\omega_n - \epsilon} = -i\frac{1}{2\tau} \text{sgn}(\omega_n), \tag{8.90}$$

where we have identified $\frac{1}{\tau} = 2\pi n_i u_0^2$ as the electron elastic scattering rate. We notice that this expression is entirely imaginary, and it only depends on the sign of the Matsubara frequency. Notice that in deriving this result we have extended the limits of integration to infinity, an approximation that involves neglecting terms of order $1/(\epsilon_F \tau)$.

We can now attempt to recompute $\Sigma(i\omega_n)$ with self-consistency. In this case,

$$\Sigma(i\omega_n) = \underset{\mathbf{k'}}{\rule{0pt}{0pt}} = n_i u_0^2 \sum_{\mathbf{k'}} \frac{1}{i\omega_n - \epsilon_{\mathbf{k'}} - \Sigma(i\omega_n)} . \tag{8.91}$$

If we carry out the energy integration again, we see that the imposition of self-consistency has no effect on the scattering rate:

$$\Sigma(i\omega_n) = n_i u_0^2 N(0) \int_{-\infty}^{\infty} d\epsilon \, \frac{1}{i\omega_n - \epsilon - \Sigma(i\omega_n)}$$

$$= -i \frac{1}{2\tau} \mathrm{sgn}(\omega_n). \tag{8.92}$$

Our result for the electron propagator, ignoring the *vertex corrections* to the scattering self-energy, is given by

$$G(\mathbf{k}, z) = \frac{1}{z - \epsilon_{\mathbf{k}} + i \frac{1}{2\tau} \mathrm{sgn} \, \mathrm{Im} \, z},$$

where we have boldly extended the Green's function into the complex plane. We may now make a few remarks:

- The original pole of the Green's function has been broadened. The electron *spectral function*,

$$A(\mathbf{k}, \omega) = \frac{1}{\pi} \mathrm{Im} \, G(\mathbf{k}, \omega - i\delta) = \frac{1}{\pi} \frac{(2\tau)^{-1}}{(\omega - \epsilon_{\mathbf{k}})^2 + (2\tau)^{-2}},$$

  is a Lorentzian of width $1/\tau$. The electron of momentum $\mathbf{k}$ now has a lifetime τ due to elastic scattering effects.

- Although the electron has a mean free path $l = v_F \tau$, its propagator displays no features of diffusion. The main effect of the finite scattering rate is to introduce a decay length into the electron propagation. The electron propagator does not bear any resemblance to the "diffusion propagator" $\chi = 1/(iv - Dq^2)$ – that is, the Green's function for the diffusion equation $(\partial_t - D\nabla^2)\chi = -\delta(x, t)$. The physics of diffusion and Ohm's law do not appear until we are able to examine the charge and spin response functions, and for this we have to learn how to compute the density and current fluctuations in thermal equilibrium (Chapter 9).

- The scattering rate that we have computed is often called the "classical" electron scattering rate. The neglected higher-order diagrams with vertex corrections are actually smaller than the leading-order contribution by an amount of order

$$\frac{1}{\epsilon_F \tau} = \frac{1}{k_F l}.$$

This parameter defines the size of "quantum corrections" to the Drude scattering physics, which are the origin of the physics of electron localization. To understand how this small number arises in the self-energy, consider the first vertex correction to the impurity scattering:

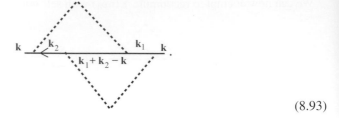

$$(8.93)$$

This diagram is given by

$$
\Sigma_2 = \overbrace{N(0) \int \frac{d\epsilon_1}{i\omega_n - \epsilon_1}}^{-i\frac{1}{2\tau}} \overbrace{N(0) \int \frac{d\epsilon_2}{i\omega_n - \epsilon_2}}^{-i\frac{1}{2\tau}} \overbrace{\int \frac{d\Omega_1 d\Omega_2}{(4\pi)^2} \frac{1}{i\omega_n - \epsilon_{\mathbf{k}_1 + \mathbf{k}_2 - \mathbf{k}}}}^{\sim \frac{-i}{k_F v_F}}
$$
$$
\sim i\frac{1}{\tau} \times \frac{1}{k_F l}, \tag{8.94}
$$

where the last term in the integral derives from the central propagator in the self-energy. In this self-energy, the momentum of the central propagator is entirely determined by the momentum of the two other internal legs, so that the energy associated with this propagator is $\epsilon_{-\mathbf{k}+\mathbf{k}_1+\mathbf{k}_2}$. This energy is only close to the Fermi energy when $\mathbf{k}_1 \sim -\mathbf{k}_2$, so that only a small fraction $1/(k_F l)$ of the possible directions of k_2 give a large contribution to the scattering processes.

8.7 Interacting electrons and phonons

The electron–phonon interaction is one of the earliest successes of many-body physics in condensed matter. In many ways, it is the condensed-matter analogue of quantum electrodynamics – and the early work on the electron–phonon problem was carried out by physicists who had their early training in the area of quantum electrodynamics.

When an electron passes through a crystal, it attracts the nearby ions, causing a local build-up of positive charge. Perhaps a better analogy is with a supersonic aircraft, for an electron moves at about Mach 100. We can confirm this with a back-of-the envelope calculation. First notice that the ratio of the sound velocity v_s to the Fermi velocity v_F is determined by the ratio of the Debye frequency to the Fermi energy:

$$
\frac{v_s}{v_F} \sim \frac{\nabla_k \omega_k}{\nabla_k \epsilon_k} \sim \frac{\omega_D/a}{\epsilon_F/a} = \frac{\omega_D}{\epsilon_F},
$$

where a is the size of the unit cell. Now an approximate estimate for the Debye frequency is given by $\omega_D^2 \sim k/M$, where M is the mass of an atomic nucleus and $k \sim \epsilon_F/a^2$ is the "spring constant" associated with atomic motions.

Thus,

$$\omega_D^2 \sim \left(\frac{\epsilon_F}{a^2}\right)\frac{1}{M}$$

and

$$\frac{\omega_D^2}{\epsilon_F^2} \sim \underbrace{\frac{1}{(\epsilon_F a^2)}}_{\sim 1/m}\frac{1}{M} \sim \frac{m}{M},$$

so that

$$\frac{v_s}{v_F} \sim \sqrt{\frac{m}{M}} \sim \frac{1}{100},$$

confirming the supersonic nature of electrons at the Fermi surface. As it moves through the crystal, an electron leaves behind a narrow wake of positive charge caused by the distortion in the crystal lattice in response to its momentary presence. This distortion attracts other electrons, long after the original disturbance has passed by. This is the origin of the weak attractive interaction induced by the exchange of virtual phonons. This attraction is highly retarded, quite unlike the repulsive Coulomb interaction which is almost instantaneous in time (the ratio of characteristic time scales being $\sim \frac{\epsilon_F}{\omega_D} \sim \sqrt{\frac{M}{m}} \sim 100$). Thus, whereas two electrons at the same place and time feel a strong mutual Coulomb repulsion, two electrons which arrive at the same place but at different times are generally subject to an attractive electron–phonon interaction. It is this attraction that is responsible for the development of superconductivity in many conventional metals.

In an electron fluid, we must take into account the quantum nature of the sound vibrations. An electron cannot continuously interact with the surrounding atomic lattice – it must do so by the emission and absorption of sound quanta or *phonons*. The basic Hamiltonian to describe the electron–phonon problem is the Fröhlich Hamiltonian, derived by Herbert Fröhlich, a German emigré to Britain, who worked in Liverpool shortly after the Second World War [7, 8]. He recognized that the electron–phonon interaction is closely analogous to the electron-photon interaction of quantum electrodynamics. Fröhlich appreciated that this interaction would give rise to an effective attraction between electrons and, together with Bardeen, was the first to identify the electron–phonon interaction as the driving force behind conventional superconductivity.

To introduce the Fröhlich Hamiltonian, we will imagine we have three phonon modes, labeled by the index $\lambda = (1, 2, 3)$, with frequencies $\omega_{\mathbf{q}\lambda}$. For the moment, we shall ignore the Coulomb interaction between electrons. The Fröhlich Hamiltonian is then

$$H_e = \sum_{\mathbf{k}\sigma} \epsilon_{\mathbf{k}} c_{\mathbf{k}\sigma}^{\dagger} c_{\mathbf{k}\sigma}$$

$$H_p = \sum_{\mathbf{q},\lambda} \omega_{\mathbf{q}\lambda}\left(a_{\mathbf{q}\lambda}^{\dagger} a_{\mathbf{q}\lambda} + \frac{1}{2}\right)$$

$$H_I = \sum_{\mathbf{k},\mathbf{q},\lambda} g_{\mathbf{q}\lambda} c_{\mathbf{k}+\mathbf{q}\sigma}^{\dagger} c_{\mathbf{k}\sigma}\left[a_{\mathbf{q}\lambda} + a_{-\mathbf{q}\lambda}^{\dagger}\right]. \qquad (8.95)$$

To understand the electron–phonon coupling, let us consider how long-wavelength fluctuations of the lattice couple to the electron energies. Let $\vec{\Phi}(\mathbf{x})$ be the displacement of the lattice at a given point $\mathbf{x}$, so that the strain tensor in the lattice is given by

$$u_{\mu\nu}(\mathbf{x}) = \frac{1}{2}\left(\nabla_\mu \Phi_\nu(\mathbf{x}) + \nabla_\nu \Phi_\mu(\mathbf{x})\right).$$

In general, we expect a small change in the strain to modify the background potential of the lattice, modifying the energies of the electrons so that, locally,

$$\epsilon(\mathbf{k}) = \epsilon_0(\mathbf{k}) + C_{\mu\nu}u_{\mu\nu}(\mathbf{x}) + \cdots.$$

Consider the following very simple model. In a free electron gas, the Fermi energy is related to the density of the electrons N/V by

$$\epsilon_F = \frac{1}{2m}\left(\frac{3\pi^2 N}{V}\right)^{\frac{2}{3}}. \tag{8.96}$$

When a portion of the lattice expands from $V \to V + dV$, the positive charge of the background lattice is unchanged, and preservation of overall charge neutrality guarantees that the number of electrons N remains constant, so the change in the Fermi energy is given by

$$\frac{\delta\epsilon_F}{\epsilon_F} = -\frac{2}{3}\frac{dV}{V} \sim -\frac{2}{3}\vec{\nabla}\cdot\vec{\Phi}.$$

On the basis of this simple model, we expect the following coupling between the displacement vector and the electron field:

$$H_I = C\int d^3x \psi_\sigma^\dagger(\mathbf{x})\psi_\sigma(\mathbf{x})\vec{\nabla}\cdot\vec{\Phi}, \qquad C = -\frac{2}{3}\epsilon_F. \tag{8.97}$$

The quantity C is often called the *deformation potential*. Now the displacement of the the phonons was studied in Chapter 3. In a general model, it is given by

$$\Phi(\mathbf{x}) = -i\sum_{\mathbf{q}\lambda} \mathbf{e}_{\mathbf{q}}^\lambda \, \Delta x_{\mathbf{q}\lambda}\left[a_{\mathbf{q}\lambda} + a_{-\mathbf{q}\lambda}^\dagger\right]e^{i\mathbf{q}\cdot\mathbf{x}},$$

where we've introduced the shorthand

$$\Delta x_{\mathbf{q}\lambda} = \left(\frac{\hbar}{2MN_s\omega_{\mathbf{q}\lambda}}\right)^{\frac{1}{2}}$$

to denote the characteristic zero-point fluctuation associated with a given mode. (N_s is the number of sites in the lattice.) The body of this expression is essentially identical to the displacement of a one-dimensional harmonic lattice (see (2.88)), dressed up with additional polarization indices. The unfamiliar quantity $\mathbf{e}_{\mathbf{q}}^\lambda$ is the *polarization vector* of the mode. For longitudinal phonons, for instance, $\mathbf{e}_{\mathbf{q}}^L = \hat{q}$. The $-i$ in front of the expression has been introduced into the definition of the phonon creation and annihilation operators so that the requirement that the Hamiltonian be Hermitian (which implies $(\mathbf{e}_{\mathbf{q}}^\lambda)^* = -(\mathbf{e}_{-\mathbf{q}}^\lambda)$)

is consistent with the convention that $\mathbf{e}$ changes sign when the momentum vector $\mathbf{q}$ is inverted.

The divergence of the phonon field is then

$$\vec{\nabla} \cdot \Phi(\mathbf{x}) = \sum_{\mathbf{q}\lambda} \mathbf{q} \cdot \mathbf{e}_{\mathbf{q}}^{\lambda} \Delta x_{\mathbf{q}\lambda} \left[a_{\mathbf{q}\lambda} + a_{-\mathbf{q}\lambda}^{\dagger} \right] e^{i\mathbf{q}\cdot\mathbf{x}}.$$

In this simple model, the electrons only couple to the longitudinal phonons, since these are the only phonons that change the density of the unit cell. When we now Fourier transform the interaction Hamiltonian, making the insertion $\psi_\sigma(\mathbf{x}) = \frac{1}{\sqrt{V}} \sum_{\mathbf{k}} c_{\mathbf{k}\sigma} e^{i\mathbf{k}\cdot\mathbf{x}}$ (8.97), we obtain

$$H_I = C \int d^3 x \, \psi_\sigma^{\dagger}(\mathbf{x}) \psi_\sigma(\mathbf{x}) \vec{\nabla} \cdot \Phi(\mathbf{x})$$

$$= \sum_{\mathbf{k},\mathbf{k}',\mathbf{q},\lambda} c_{\mathbf{k}'\sigma}^{\dagger} c_{\mathbf{k}\sigma} \left[a_{\mathbf{q}\lambda} + a_{-\mathbf{q}\lambda}^{\dagger} \right] \frac{1}{V} \overbrace{\int d^3 x \, e^{i(\mathbf{q}+\mathbf{k}-\mathbf{k}')\cdot\mathbf{x}}}^{\delta_{\mathbf{k}'-(\mathbf{k}+\mathbf{q})}} C \Delta x_{\mathbf{q}\lambda} (\mathbf{q} \cdot \mathbf{e}_{\mathbf{q}}^{\lambda})$$

$$= \sum_{\mathbf{q}\mathbf{k}\lambda} g_{\mathbf{q}\lambda} c_{\mathbf{k}+\mathbf{q}\sigma}^{\dagger} c_{\mathbf{k}\sigma} \left[a_{\mathbf{q}\lambda} + a_{-\mathbf{q}\lambda}^{\dagger} \right], \qquad (8.98)$$

where

$$g_{\mathbf{q}\lambda} = \begin{cases} Cq\Delta x_{\mathbf{q}\lambda} = Cq \left(\dfrac{\hbar}{2MN_s\omega_{\mathbf{q}\lambda}} \right)^{\frac{1}{2}} & (\lambda = \text{longitudinal}) \\ 0 & (\lambda = \text{transverse}). \end{cases}$$

Note that $N_s = V/a^3$, where a is the lattice spacing. To go over to the thermodynamic limit, we will replace our discrete momentum sums by continuous integrals, $\sum_{\mathbf{q}} \equiv V \int_{\mathbf{q}} \to \int_{\mathbf{q}}$. Rather than spending a lot of time keeping track of how the volume factor is absorbed into the integrals, it is simpler to regard $V = 1$ as a unit volume, replacing $N_s \to a^{-3}$ whenever we switch from discrete to continuous integrals. With this understanding, we will use

$$g_{\mathbf{q}} = Cq\sqrt{\hbar a^3/(2M\omega_{\mathbf{q}\lambda})} \qquad (8.99)$$

for the electron–phonon coupling to the longitudinal modes. Our simple model captures the basic aspects of the electron–phonon interaction, and it can be readily generalized. In a more sophisticated model:

- C becomes momentum-dependent and should be replaced by the Fourier transform of the atomic potential. For example, if we compute the electron–phonon potential from the change in the atomic potential V_{atomic} resulting from the displacement of atoms,

$$\delta V(\mathbf{x}) = \sum_j \delta V_{atomic}(\mathbf{x} - \mathbf{R}_j^0 - \vec{\Phi}_j) = -\sum_j \vec{\Phi}_j \cdot \vec{\nabla} V_{atomic}(\mathbf{x} - \mathbf{R}_j^0),$$

  we must replace the constant

$$C \to \frac{1}{v_{cell}} \int d^3 x \, V_{atomic}(\mathbf{x}) e^{-i\mathbf{q}\cdot\mathbf{x}} = n_{ion} V_{atomic}(\mathbf{q}) \sim Z n_{ion} V_{eff}(\mathbf{q}), \qquad (8.100)$$

where $n_{ion} = 1/a^3$ is the ionic density, Z is the atomic number, and $V_{eff}(\mathbf{q})$ is the screened Coulomb interaction. These replacements appear in the Bardeen–Pines model of the electron–phonon interaction (see 7.251 and Exercise 7.8).

- When the plane-wave functions are replaced by the detailed Bloch wavefunctions of the electron band, the electron–phonon coupling becomes dependent on both the incoming and outgoing electron momenta, so that

$$g_{\mathbf{k}'-\mathbf{k}\lambda} \rightarrow g_{\mathbf{k}',\mathbf{k}\lambda}.$$

Nevertheless, much can be learned from our simplified model. In the discussion that follows, we shall drop the polarization index, and assume that the phonon modes we refer to are exclusively longitudinal modes.

In setting up the Feynman diagrams for the Fröhlich model, we need to introduce two new elements: a diagram for the phonon propagator and a diagram to denote the vertex. If we denote $\phi_\mathbf{q} = a_\mathbf{q} + a^\dagger_{-\mathbf{q}}$, then the phonon Green's function is given by

$$D(\mathbf{q}, \tau - \tau') = -\langle T\phi_\mathbf{q}(\tau)\phi_\mathbf{q}(\tau')\rangle = T\sum_{i\nu_n} D(q)e^{-i\nu_n(\tau-\tau')}, \qquad (8.101)$$

where the propagator

$$D(q) = \frac{2\omega_\mathbf{q}}{(i\nu_n)^2 - (\omega_\mathbf{q})^2}$$

is denoted by the diagram

$$\overset{}{\underset{(\mathbf{q}, i\nu_n)}{\sim\!\!\sim\!\!\sim\!\!\sim\!\!\sim}} = D(\mathbf{q}, i\nu_n). \qquad (8.102)$$

The interaction vertex between electrons and phonon is denoted by the diagram

$$= (i)^3 \times -g_\mathbf{q} = ig_\mathbf{q}. \qquad (8.103)$$

The factor i^3 arises because we have three propagators entering the vertex, each donating a factor of i. The $-g_\mathbf{q}$ derives from the interaction Hamiltonian in the time-ordered exponential. Combining these two Feynman rules, we see that, when two electrons exchange a boson, this gives rise to the diagram

$$= (ig_\mathbf{q})^2 D(q) = -(g_\mathbf{q})^2 D(q), \qquad (8.104)$$

so that the exchange of a boson induces an effective interaction

$$V_{\text{eff}}(\mathbf{q}, z) = g_\mathbf{q}^2 \frac{2\omega_\mathbf{q}}{z^2 - \omega_\mathbf{q}^2}. \qquad (8.105)$$

Notice three things about this interaction:

- It is strongly frequency-dependent, reflecting the strongly retarded nature of the electron–phonon interaction. The characteristic phonon frequency is the Debye frequency ω_D, and the characteristic restitution time associated with the electron–phonon interaction is $\tau \sim 1/\omega_D$, whereas the corresponding time associated with the repulsive Coulomb interaction is of order $1/\epsilon_F$. The ratio $\epsilon_F/\omega_D \sim 100$ is a measure of how much more retarded the electron–phonon interaction is compared with the Coulomb potential.
- It is weakly dependent on momentum, describing an interaction that is spatially local over one or two lattice spacings.
- At frequencies below the Debye energy, $\omega \lesssim \omega_D$, the denominator in V_{eff} changes sign, and the residual low-energy interaction is actually attractive. It is this component of the interaction that is responsible for superconductivity in conventional superconductors.

We wish now to calculate the effect of the electron–phonon interaction on electron propagation. The main effect is determined by the electron–phonon self-energy. The leading-order Feynman diagram for the self-energy is given by

$$\equiv \Sigma(k) = \sum_q (ig_\mathbf{q})^2 \mathcal{G}^0(k-q) D(q),$$

(8.106)

or, written out explicitly,

$$\Sigma(\mathbf{k}, i\omega_n) = -T \sum_{\mathbf{q}, iv_n} g_\mathbf{q}^2 \left[\frac{2\omega_\mathbf{q}}{(iv_n)^2 - \omega_\mathbf{q}^2} \right] \frac{1}{i\omega_n - iv_n - \epsilon_{\mathbf{k}-\mathbf{q}}}$$

$$= -T \sum_{\mathbf{q}, iv_n} \left[\frac{1}{iv_n - \omega_\mathbf{q}} \frac{1}{i\omega_n - iv_n - \epsilon_{\mathbf{k}-\mathbf{q}}} - (\omega_\mathbf{q} \to -\omega_\mathbf{q}) \right],$$

(8.107)

where we have simplified the expression by splitting up the boson propagator into positive and negative frequency components, the latter being obtained by reversing the sign on $\omega_\mathbf{q}$. We shall carry out the Matsubara sum over the bosonic frequencies by writing it as a contour integral with the Bose function:

$$-T \sum_{iv_n} F(iv_n) = -\int_C \frac{dz}{2\pi i} n(z) F(z) = \int_{C'} \frac{dz}{2\pi i} n(z) F(z),$$

(8.108)

where C runs counterclockwise around the imaginary axis and C' runs counterclockwise around the poles in $F(z)$. In this case, we choose

$$F(z) = \frac{1}{z - \omega_\mathbf{q}} \frac{1}{i\omega_n - z - \epsilon_{\mathbf{k}-\mathbf{q}}}$$

$$= \left[\frac{1}{z - \omega_\mathbf{q}} - \frac{1}{z - (i\omega_n - \epsilon_{\mathbf{k}-\mathbf{q}})} \right] \frac{1}{i\omega_n - (\omega_\mathbf{q} + \epsilon_{\mathbf{k}-\mathbf{q}})},$$

(8.109)

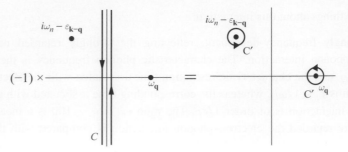

Fig. 8.6 Contours C and C' used in the evaluation of $\Sigma(\mathbf{k}, i\omega_n)$.

which has two poles, one at $z = \omega_\mathbf{q}$ and one at $z = i\omega_n - \epsilon_{\mathbf{k}-\mathbf{q}}$ (Figure 8.6). Carrying out the contour integral, we then obtain

$$
\Sigma(k) = \sum_\mathbf{q} g_\mathbf{q}^2 \left[\frac{n(\omega_\mathbf{q}) - \overbrace{n(i\omega_n - \epsilon_{\mathbf{k}-\mathbf{q}})}^{-(1-f_{\mathbf{k}-\mathbf{q}})}}{i\omega_n - (\omega_\mathbf{q} + \epsilon_{\mathbf{k}-\mathbf{q}})} - \{\omega_\mathbf{q} \to -\omega_\mathbf{q}\} \right]
$$

$$
= \sum_\mathbf{q} g_\mathbf{q}^2 \left[\frac{1 + n_\mathbf{q} - f_{\mathbf{k}-\mathbf{q}}}{i\omega_n - (\omega_\mathbf{q} + \epsilon_{\mathbf{k}-\mathbf{q}})} - \{\omega_\mathbf{q} \to -\omega_\mathbf{q}\} \right]. \tag{8.110}
$$

The second term in this expression is obtained by reversing the sign on $\omega_\mathbf{q}$ in the first term, which gives, finally,

$$
\Sigma(\mathbf{k}, z) = \sum_\mathbf{q} g_\mathbf{q}^2 \left[\frac{1 + n_\mathbf{q} - f_{\mathbf{k}-\mathbf{q}}}{z - (\epsilon_{\mathbf{k}-\mathbf{q}} + \omega_\mathbf{q})} + \frac{n_\mathbf{q} + f_{\mathbf{k}-\mathbf{q}}}{z - (\epsilon_{\mathbf{k}-\mathbf{q}} - \omega_\mathbf{q})} \right], \tag{8.111}
$$

where we have taken the liberty of analytically extending the function into the complex plane. There is a remarkable amount of physics hidden in this expression.

The terms appearing in the electron–phonon self-energy can be interpreted in terms of virtual and real phonon emission processes. Consider the zero-temperature limit, when the Bose terms $n_\mathbf{q} = 0$. If we look at the first term in $\Sigma(k)$, we see that the numerator is only finite if the intermediate electron state is empty, i.e. $|\mathbf{k}-\mathbf{q}| > k_F$. Furthermore, the poles of the first expression are located at energies $\omega_\mathbf{q} + \epsilon_{\mathbf{k}-\mathbf{q}}$, which is the energy of an electron of momentum $\mathbf{k} - \mathbf{q}$ and an emitted phonon of momentum $\omega_\mathbf{q}$, so the first process corresponds to phonon emission by an electron. If we look at the second term, then at zero temperature, the numerator is only finite if $|\mathbf{k} - \mathbf{q}| < k_F$, so the intermediate state is a hole. The pole in the second term occurs at $-z = -\epsilon_{\mathbf{k}-\mathbf{q}} + \omega_\mathbf{q}$, corresponding to a state of one hole and one phonon, so one way to interpret the second term is as the energy shift that results from the emission of virtual phonons by holes. At zero temperature, then,

$$
\Sigma(\mathbf{k}, z) = \sum_\mathbf{q} g_\mathbf{q}^2 \left[\overbrace{\frac{1 - f_{\mathbf{k}-\mathbf{q}}}{z - (\epsilon_{\mathbf{k}-\mathbf{q}} + \omega_\mathbf{q})}}^{\text{virtual phonon emission by electron}} + \overbrace{\frac{f_{\mathbf{k}-\mathbf{q}}}{z - (\epsilon_{\mathbf{k}-\mathbf{q}} - \omega_\mathbf{q})}}^{\text{virtual phonon emission by hole}} \right]. \tag{8.112}
$$

As we shall discuss in more detail in the next chapter, the analytically extended Green's function

$$G(\mathbf{k}, z) = \frac{1}{z - \epsilon_{\mathbf{k}} - \Sigma(\mathbf{k}, z)}$$

can be used to derive the real-time dynamics of the electron in thermal equilibrium. In general, $\Sigma(\mathbf{k}, \omega - i\delta) = \text{Re}\,\Sigma(\mathbf{k}, \omega - i\delta) + i\,\text{Im}\,\Sigma(\mathbf{k}, \omega - i\delta)$ will have a real and an imaginary part. The solution of the relation

$$\epsilon_{\mathbf{k}}^* = \epsilon_{\mathbf{k}} + \text{Re}\,\Sigma(\mathbf{k}, \epsilon_{\mathbf{k}}^*)$$

determines the renormalized energy of the electron due to virtual phonon emission. Let's consider the case of an electron, for which $\epsilon_{\mathbf{k}}^*$ is above the Fermi energy. The quasiparticle energy takes the form

$$\epsilon_{\mathbf{k}}^* = \epsilon_{\mathbf{k}} - \overbrace{\sum_{|\mathbf{k}-\mathbf{q}|>k_F} g_{\mathbf{q}}^2 \frac{1}{(\epsilon_{\mathbf{k}-\mathbf{q}} + \omega_{\mathbf{q}}) - \epsilon_{\mathbf{k}}^*}}^{\substack{\text{energy lowered by virtual} \\ \text{phonon emission}}} + \overbrace{\sum_{|\mathbf{k}-\mathbf{q}|<k_F} g_{\mathbf{q}}^2 \frac{1}{\epsilon_{\mathbf{k}}^* + |\epsilon_{\mathbf{k}-\mathbf{q}}| + \omega_{\mathbf{q}}}}^{\substack{\text{energy raised by blocking} \\ \text{vacuum fluctuations}}}.$$

If we approximate $\epsilon_{\mathbf{k}}^*$ on the right-hand side by its unrenormalized value $\epsilon_{\mathbf{k}}$, we obtain the second-order perturbation correction to the electron quasiparticle energy, due to virtual phonon processes. To understand these two terms, it is helpful to redraw the Feynman diagram for the self-energy so that the scattering events are explicitly time-ordered. Then we see that there are two virtual processes, depending on whether the intermediate electron line propagates forwards or backwards in time:

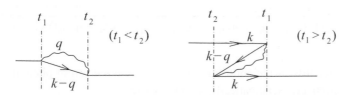

virtual phonon emission   virtual phonon and electron–hole pair

The first term is recognized as virtual scattering into an intermediate state with one photon and one electron. But what about the second term? This term involves the formation of an electron–hole pair and the subsequent reannihilation of the hole with the incoming electron. During the intermediate process, there seem to be *two* electrons (with the same spin) in the same momentum state $\mathbf{k}$. Do such virtual processes violate the exclusion principle? Fortunately, we don't have to give up on the exclusion principle, because another interpretation can be given: under close examination, unlike conventional virtual processes which lower the energy, this term actually raises the quasiparticle energy. These energy-raising processes are a "blocking effect" of the added electron on the vacuum fluctuations. In the ground state, there are virtual fluctuations

$$\text{vacuum} \rightleftharpoons \text{electron } (\mathbf{k}') + \text{hole } (-\mathbf{k}' - \mathbf{q}) + \text{phonon } (\mathbf{q}) \tag{8.113}$$

which lower the energy of the ground state. However, when a quasiparticle with momentum $\mathbf{k}$ is added, the exclusion principle prevents vacuum fluctuations with $\mathbf{k}' = \mathbf{k}$, and the elimination of these vacuum fluctuations thereby *raises* the energy of the quasiparticle. So time-ordered diagrams that appear to violate the exclusion principle actually describe the suppression of vacuum fluctuations via the exclusion principle.

At finite temperatures, both the first and the second terms in the phonon self-energy are present. The appearance of the additional Bose terms $n_\mathbf{q}$ in the emission processes is the effect of stimulated emission, whereby the occupancy of phonon states enhances the emission process (recall $\langle n + 1 | b^\dagger | n \rangle = \sqrt{n+1}$). By contrast, the new terms which vanish at zero temperature are most naturally interpreted as the absorption of thermally excited phonons, so for $\omega > 0$ (electrons)

$$\Sigma(\mathbf{k}, \omega) = \sum_\mathbf{q} g_\mathbf{q}^2 \left[\underbrace{\frac{1 - f_{\mathbf{k}-\mathbf{q}} + n_\mathbf{q}}{\omega - (\epsilon_{\mathbf{k}-\mathbf{q}} + \omega_\mathbf{q})}}_{\text{stimulated phonon emission}} + \underbrace{\frac{f_{\mathbf{k}-\mathbf{q}} + n_\mathbf{q}}{\omega - (\epsilon_{\mathbf{k}-\mathbf{q}} - \omega_\mathbf{q})}}_{\text{thermal phonon absorption}} \right]. \tag{8.114}$$

By contrast, the imaginary part of the self-energy determines the decay rate of the electron due to real phonon emission, and the decay rate of the electron is related to the quantity

$$\Gamma_\mathbf{k} = 2 \operatorname{Im} \Sigma(\mathbf{k}, \epsilon_\mathbf{k}^* - i\delta) \approx 2 \operatorname{Im} \Sigma(\mathbf{k}, \epsilon_\mathbf{k} - i\delta). \tag{8.115}$$

If we use the Cauchy–Dirac relation

$$\left[\frac{1}{x - a - i\delta} \right] = P \frac{1}{x - a} + i\pi \delta(x - a), \tag{8.116}$$

then we see that, for a weak interaction, the decay rate of the electron is given by

$$\Gamma_\mathbf{k} = 2\pi \sum_\mathbf{q} g_\mathbf{q}^2 \left[\overbrace{(1 + n_\mathbf{q} - f_{\mathbf{k}-\mathbf{q}}) \delta(\epsilon_\mathbf{k} - (\epsilon_{\mathbf{k}-\mathbf{q}} + \omega_\mathbf{q}))}^{\text{stimulated phonon emission}} + \overbrace{(n_\mathbf{q} + f_{\mathbf{k}-\mathbf{q}}) \delta(\epsilon_\mathbf{k} - (\epsilon_{\mathbf{k}-\mathbf{q}} - \omega_\mathbf{q}))}^{\text{thermal phonon absorption}} \right], \tag{8.117}$$

which we may identify as the contributions to the decay rate from phonon emission and absorption, respectively. Schematically, we may write

$$\operatorname{Im}\left[\text{diagram} \right] = \sum_\mathbf{q} \left\{ \left[\text{diagram} \right]^2 + \left[\text{diagram} \right]^2 \right\} \times 2\pi\delta(E_f - E_i), \tag{8.118}$$

so that taking the imaginary part of the self-energy "cuts" the internal lines. The relationship between the imaginary part of a forward scattering amplitude (represented by the self-energy) and real decay processes is a version of the *optical theorem* in scattering theory.

8.7.1 $\alpha^2 F$: the electron–phonon coupling function

One of the most important effects of the electron–phonon interaction is to give rise to a superconducting instability. Superconductivity is driven by the interaction of low-energy electrons very close to the Fermi surface, so the amount of energy transfered in an interaction is almost zero. For this reason, the effective interaction between the electrons is given by (8.105):

$$V_{eff}(\mathbf{q}, 0) = -\frac{2g_{\mathbf{q}}^2}{\omega_{\mathbf{q}}}. \tag{8.119}$$

Now the momentum dependence of this interaction is very weak. In our simple model, for example, $g_{\mathbf{q}}^2/2\omega_q \sim \frac{q^2}{\omega_q^2} \sim$ constant, and a weak momentum dependence implies that, to a first approximation, the effective low-energy interaction is local, extending over one unit cell and of approximate form

$$H_{eff} \approx \frac{-g}{2} \sum_{\substack{\mathbf{q},\mathbf{k},\mathbf{k}',\sigma,\sigma' \\ |\epsilon_{\mathbf{p}}| < \omega_D}} c_{\mathbf{k}+\mathbf{q}\sigma}^\dagger c_{\mathbf{k}'\sigma'}^\dagger c_{\mathbf{k}'+\mathbf{q}\sigma'} c_{\mathbf{k}\sigma}, \tag{8.120}$$

where the sum over electron momenta is restricted to a narrow band of energies within ω_D of the Fermi energy. While the interaction is formally "instantaneous," the restriction of energies within this narrow shell means that the physics is coarse-grained over time scales of order $\delta t \sim 1/\omega_D$, effectively implementing the retardation. The effective interaction strength g is the Fermi-surface average of $2g_{\mathbf{q}}^2/\omega_{\mathbf{q}}$:

$$g = \left\langle \frac{2g_{\mathbf{k}-\mathbf{k}'}^2}{\omega_{\mathbf{k}-\mathbf{k}'}} \right\rangle_{\mathbf{k},\mathbf{k}' \in FS}. \tag{8.121}$$

Bardeen and Pines were among the first to appreciate that the interaction between electrons induced by phonon exchange is highly retarded relative to the almost instantaneous Coulomb interaction, so that, for low-energy processes, the Coulomb interaction can be temporarily ignored. The attractive interaction in (8.120) was then the basis of the *Bardeen–Pines model* [9] (see Section 7.7.3) – a predecessor of the Bardeen–Cooper–Schrieffer (BCS) Hamiltonian. We can make an order-of-magnitude estimate of g by replacing

$$g \sim \frac{g_{2k_F}^2}{\omega_D} \sim \frac{1}{\omega_D} \overbrace{\left[\left(\frac{a^3}{2M\omega_D} \right) \epsilon_F^2 (2k_F)^2 \right]}^{(g_{2k_F})^2} \sim \overbrace{\left(\frac{\epsilon_F^2}{\omega_D^2} \right)}^{\frac{M}{m}} \frac{k_F^2}{2M} a^3 \sim \epsilon_F a^3, \tag{8.122}$$

where we have taken $\hbar = 1$. Thus the magnitude of the effective interaction is set by the Fermi energy. The electron–phonon coupling constant is defined as the product of the interaction strength and the electron density of states:

$$\lambda = N(0)g = N(0) \left\langle \frac{2g_{\mathbf{k}-\mathbf{k}'}^2}{\omega_{\mathbf{k}-\mathbf{k}'}} \right\rangle_{\mathbf{k},\mathbf{k}' \in FS}. \tag{8.123}$$

Since $N(0) \sim \frac{1}{\epsilon_F a^3}$, $\lambda \sim O(1)$ is a dimensionless quantity which is not reduced by the small ratio of electron to atomic mass. In typical metals, $\lambda \sim 0.1$–0.2. We'll now relate the electron–phonon self-energy to this quantity.

The electron–phonon self-energy can be simplified by the introduction of a spectral function $\alpha^2 F$ that keeps track of the frequency dependence of the electron–phonon coupling constant, where $\alpha(\omega)$ is the typical energy-dependent coupling constant and F is basically the phonon density of states. It turns out that $\alpha^2 F$ can actually be measured inside superconductors, and F can be measured by neutron scattering.

The basic idea is that the momentum dependence of the electron–phonon self-energy is far smaller than the frequency dependence, allowing it to be neglected. The dimensionless ratio between these two dependences is a small number of order ω_D/ϵ_F:

$$\left(\frac{1}{v_F}|\nabla_k \Sigma|\right) / \left(\frac{\partial \Sigma}{\partial \omega}\right) \sim \frac{\omega_D}{\epsilon_F} << 1. \tag{8.124}$$

To a good approximation, then, the electron–phonon self-energy can be averaged over the Fermi surface:

$$\Sigma(\omega) = \frac{\int dS \, \Sigma(\mathbf{k}, \omega)}{\int dS} \tag{8.125}$$

where $\int dS \equiv \int d^2k/(2\pi)^3$ is an integral over the Fermi surface. Now the sum over $\mathbf{k}'$ inside the self-energy can be replaced by a combination of an energy and a Fermi surface integral, as follows:

$$\sum_{\mathbf{k}'} \rightarrow \int dS' dk'_{perp} = \int \frac{dS'}{|d\epsilon_{k'}/dk'|} d\epsilon' = \int \frac{dS'}{v_F(S')} d\epsilon', \tag{8.126}$$

where dS' is over the surface of constant ϵ' and $v_F(S) = \mathbf{n} \cdot \nabla_{\mathbf{k}} \epsilon_{\mathbf{k}}$ is the local Fermi velocity. Making this substitution into (8.111), we obtain

$$\Sigma(\omega) = \frac{1}{\int dS} \int \frac{dS dS'}{v'_F} d\epsilon' g^2_{\mathbf{k}-\mathbf{k}'} \left[\frac{1 + n_{\mathbf{k}-\mathbf{k}'} - f(\epsilon')}{z - (\epsilon' + \omega_{\mathbf{k}-\mathbf{k}'})} + \frac{n_{\mathbf{k}-\mathbf{k}'} + f(\epsilon')}{z - (\epsilon' - \omega_{\mathbf{k}-\mathbf{k}'})}\right]. \tag{8.127}$$

If we introduce a delta function in the phonon frequency into this expression, using the identity $1 = \int d\nu \delta(\nu - \omega_{\mathbf{q}\lambda})$, then we may rewrite it as follows:

$$\Sigma(\omega) = \frac{1}{\int dS} \int d\epsilon' d\nu \int \frac{dS \, dS'}{v'_F} g^2_{\mathbf{k}-\mathbf{k}'} \delta(\nu - \omega_{\mathbf{k}-\mathbf{k}'}) \left[\frac{1 + n(\nu) - f(\epsilon')}{z - (\epsilon' + \nu)} + \frac{n(\nu) + f(\epsilon')}{z - (\epsilon' - \nu)}\right]$$

$$= \int_{-\infty}^{\infty} d\epsilon \int_0^{\infty} d\nu \alpha^2(\nu) F(\nu) \left[\frac{1 + n(\nu) - f(\epsilon')}{z - (\epsilon' + \nu)} + \frac{n(\nu) + f(\epsilon')}{z - (\epsilon' - \nu)}\right], \tag{8.128}$$

where the function

$$F(\nu) = \frac{1}{\int dS} \int \frac{dS \, dS'}{v'_F} \delta(\nu - \omega_{\mathbf{k}-\mathbf{k}'}) \tag{8.129}$$

is a measure of the phonon density of states, while

$$\alpha^2(\nu) F(\nu) = \frac{1}{\int dS} \int \frac{dS \, dS'}{v'_F} \delta(\nu - \omega_{\mathbf{k}-\mathbf{k}'}) g^2_{\mathbf{k}-\mathbf{k}'\lambda} \tag{8.130}$$

is the Fermi surface average of the phonon matrix element and density of states. With this definition, the electron–phonon coupling constant is given by

$$g = 2 \int_0^{\infty} d\nu \frac{\alpha^2(\nu) F(\nu)}{\nu}, \tag{8.131}$$

and we may rewrite the self-energy as

$$\Sigma(z) = \int_{-\infty}^{\infty} d\epsilon \int_0^{\infty} d\nu \alpha^2(\nu) F(\nu) \left[\frac{1 + n(\nu) - f(\epsilon)}{z - (\epsilon + \nu)} + \frac{n(\nu) + f(\epsilon)}{z - (\epsilon - \nu)} \right], \qquad (8.132)$$

where the energy dependence of the electron density of states has been neglected. This is a very practical form for the electron self-energy. In practice, most of the energy dependence in $\alpha^2 F$ is determined by the phonon density of states. In a conventional electron–phonon superconductor, one may infer the function $\alpha^2 F$ using the density of electron states in the superconductor, measured by tunneling in the superconducting state.

Example 8.6 Section 7.7.3 showed that the effective interaction between particles when plasma oscillations of jellium are taken into account contains a sum of a screened Coulomb interaction plus an electron–phonon interaction, given by

$$V_{eff}(\mathbf{q}, \nu) = V_{ee}(\mathbf{q}) + V_{ep}(\mathbf{q}, \nu) = \left[\frac{e^2}{\epsilon_0(q^2 + \kappa^2)} \right] \left(1 + \frac{\omega_q^2}{\nu^2 - \omega_q^2} \right), \qquad (8.133)$$

where $\omega_q^2 = \Omega_P^2 \frac{q^2}{q^2 + \kappa^2}$ is the phonon dispersion, $\Omega_P^2 = \frac{Ze^2 n}{\epsilon_0 M}$ is the ionic plasma frequency, and $\kappa^2 = \frac{2e^2}{\epsilon_0} N(0)$ is the Thomas–Fermi screening length.

(a) Use these results to show that the electron–phonon coupling associated with the retarded part of Bardeen–Pines interaction is given by

$$\frac{2g_{\mathbf{q}}^2}{\omega_{\mathbf{q}}} = \frac{1}{2N(0)} \frac{1}{1 + q^2/\kappa^2}. \qquad (8.134)$$

(b) Derive an approximate expression for λ in the Bardeen–Pines model. (For a more detailed discussion of the electron–phonon coupling constant in this model, see [10].)

Solution

(a) From the Bardeen–Pines interaction, we can identify the electron–phonon interaction as

$$V_{ep}(\mathbf{q}, \nu) = \frac{e^2}{\epsilon_0(q^2 + \kappa^2)} \left(\frac{\omega_q^2}{\nu^2 - \omega_q^2} \right) \equiv g_q^2 \frac{2\omega_q}{\nu^2 - \omega_q^2}. \qquad (8.135)$$

Comparing both sides, we see that

$$\frac{2g_q^2}{\omega_q} = \frac{1}{2N(0)} \frac{1}{(1 + q^2/\kappa^2)}. \qquad (8.136)$$

Notice that (i) this is manifestly independent of the ion mass and (ii) the more the screening (large κ), the weaker the interaction.

(b) To estimate the electron–phonon coupling constant, we need to evaluate

$$\lambda = \left\langle \frac{2g_{\mathbf{k}-\mathbf{k}'}^2 N(0)}{\omega_{\mathbf{k}-\mathbf{k}'}} \right\rangle_{FS} = \frac{1}{2} \int_{-1}^{1} \frac{d\cos\theta}{2} \frac{1}{1 + (2k_F \sin\theta/2)^2/\kappa^2}, \tag{8.137}$$

where we have replaced $q = 2k_F \sin\theta/2$ in the Fermi surface integral. Changing variables, putting $s = \sin\theta/2$, $\cos\theta = 1 - 2s^2$, $dc = -2ds^2$, we obtain

$$\lambda = \frac{1}{2} \int_{0}^{1} \frac{ds^2}{1 + ys^2} = \frac{1}{2y} \ln\left[1 + y\right], \tag{8.138}$$

where $y = (2k_F/\kappa)^2$. Now putting $\kappa^2 = 2e^2 N(0) = \frac{4}{\pi}\frac{k_F}{a_B}$, where a_B is the Bohr radius, we see that $y = \pi k_F a_B$ (see (7.230)). Using $k_F a_B = 1/(\alpha r_s)$, where $\alpha = \left(\frac{4}{9\pi}\right)^{1/3} \sim 0.5$, then $y = \frac{\pi}{\alpha r_s} \sim \frac{6}{r_s}$, so our approximate formula for λ is given by

$$\lambda \sim \frac{r_s}{12} \ln\left[1 + \frac{6}{r_s}\right]. \tag{8.139}$$

For copper, $r_s \sim 2.7$, which gives $\lambda = 0.26$, a reasonable estimate. Our result should not be taken too seriously, but it does have the interesting prediction that electron plasmas with high density, small r_s, will have smaller electron–phonon coupling constants λ.

8.7.2 Mass renormalization by the electron–phonon interaction

Our simplified expression for of the self-energy enables us to examine how electron propagation is modified by the exchange of virtual phonons. Let us expand the electron–phonon self-energy around zero frequency in the ground state. At $T = 0$, where the Bose functions vanish and the Fermi functions become step functions,

$$\Sigma(z) = \int_{-\infty}^{\infty} d\epsilon \int_{0}^{\infty} d\nu \alpha^2(\nu) F(\nu) \left[\frac{\theta(\epsilon)}{z - (\epsilon + \nu)} + \frac{\theta(-\epsilon)}{z - (\epsilon - \nu)} \right]$$
$$= \int_{0}^{\infty} d\nu \alpha^2(\nu) F(\nu) \ln\left[\frac{\nu - z}{\nu + z} \right], \tag{8.140}$$

so that, at low frequencies,

$$\Sigma(\omega) = \Sigma(0) - \lambda\omega, \tag{8.141}$$

where

$$\lambda = -\left.\frac{d\Sigma(\omega)}{d\omega}\right|_{\omega=0}$$
$$= 2 \int d\nu \frac{\alpha^2(\nu) F(\nu)}{\nu}. \tag{8.142}$$

If we look at our definition of $\alpha^2 F$, we see that this expression is the Fermi surface average of the electron–phonon coupling constant defined in (8.123).

Now at low energies we can write the electron propagator in terms of the quasiparticle energies, as follows:

$$\mathcal{G}(\mathbf{k}, \omega - i\delta) = \frac{1}{\omega - \epsilon_{\mathbf{k}} - \Sigma(\omega - i\delta)}$$

$$= \frac{1}{\omega - \underbrace{\epsilon_{\mathbf{k}} - \Sigma(\epsilon_k^* - i\delta)}_{\epsilon_{\mathbf{k}}^* - i\Gamma/2} + \lambda(\omega - \epsilon_k^*)}, \tag{8.143}$$

or

$$\mathcal{G}(\mathbf{k}, \omega - i\delta) = \frac{Z}{\omega - \epsilon_{\mathbf{k}}^* - i\Gamma^*/2}, \tag{8.144}$$

where

$$
\begin{aligned}
Z &= (1 + \lambda)^{-1} & &\text{wavefunction renormalization} \\
\epsilon_{\mathbf{k}}^* &= \epsilon_{\mathbf{k}} + \Sigma(\epsilon_k^*) & &\text{quasiparticle energy} \\
\Gamma^* &= 2Z \, \text{Im} \Sigma(\epsilon_{\mathbf{k}}^* - i\delta). & &\text{quasiparticle decay rate}
\end{aligned}
\tag{8.145}
$$

We see that, in the presence of the electron–phonon interaction, electron quasiparticles are still well defined at low temperatures. Indeed, at the Fermi surface, $\Gamma^* = 0$ in the ground state, so that electron quasiparticles are infinitely long-lived. This is an example of a Landau Fermi-liquid, discussed in Chapter 7. If we differentiate $\epsilon_{\mathbf{k}}$ with respect to $\epsilon_{\mathbf{k}}^*$, we obtain

$$\frac{d\epsilon_{\mathbf{k}}}{d\epsilon_{\mathbf{k}}^*} = (1 + \lambda) = \left(\frac{m^*}{m} \right), \tag{8.146}$$

so that the effective mass of the electron is enhanced by the cloud of virtual phonons which trails behind it. The density of states is renormalized in the same way:

$$N(0)^* = \frac{d\epsilon_{\mathbf{k}}}{d\epsilon_{\mathbf{k}}^*} N(0) = N(0)(1 + \lambda), \tag{8.147}$$

while the electron group velocity is renormalized downwards according to

$$v_F^* = \nabla_{\mathbf{k}} \epsilon_{\mathbf{k}}^* = \frac{d\epsilon_{\mathbf{k}}^*}{d\epsilon_{\mathbf{k}}} \nabla_{\mathbf{k}} \epsilon_{\mathbf{k}} = Z v_F. \tag{8.148}$$

Thus the electron–phonon interaction drives up the mass of the electron, an effect of squeezing the one-particle states more closely together and driving the electron group velocity downwards. This in turn will mean that the linear coefficient of the electronic specific heat $C_v = \gamma^* T$,

$$\gamma^* = \frac{\pi^2 k_B^2}{3} N^*(0) = \gamma_0 (1 + \lambda), \tag{8.149}$$

is enhanced.

We can give the wavefunction renormalization another interpretation. Recall that, using the method of contour integration, we can always rewrite the Matsubara representation of the Green's function,

$$G(\mathbf{k}, \tau) = T \sum_n G(\mathbf{k}, i\omega_n) e^{-i\omega_n \tau}, \tag{8.150}$$

as

$$G(\mathbf{k}, \tau) = - \int d\omega \left[(1 - f(\omega))\theta(\tau) - f(\omega)\theta(-\tau) \right] A(\mathbf{k}, \omega) e^{-\omega\tau}, \tag{8.151}$$

where $A(\mathbf{k}, \omega) = \frac{1}{\pi} \operatorname{Im} G(\mathbf{k}, \omega - i\delta)$ is the spectral function. Now, from the normalization of the fermionic commutation relation $\{c_{\mathbf{k}\sigma}, c_{\mathbf{k}\sigma}^{\dagger}\} = 1$, we deduce that the spectral function is normalized:

$$1 = \langle \{c_{\mathbf{k}\sigma}, c_{\mathbf{k}\sigma}^{\dagger}\} \rangle = \overbrace{G(\mathbf{k}, 0^-)}^{\langle c_{\mathbf{k}\sigma}^{\dagger} c_{\mathbf{k}\sigma} \rangle} - \overbrace{G(\mathbf{k}, 0^+)}^{-\langle c_{\mathbf{k}\sigma} c_{\mathbf{k}\sigma}^{\dagger} \rangle}$$

$$= \int d\omega A(\mathbf{k}, \omega). \tag{8.152}$$

The quasiparticle part of the spectral function (8.144) is a Lorentzian of width $\Gamma_{\mathbf{k}}^*$ and weight Z, and since the width $\Gamma_{\mathbf{k}}^* \to 0$ as $\epsilon_{\mathbf{k}}^*$ gets close to the Fermi energy, we deduce that for $k \sim k_F$ the quasiparticle part of the spectral function ever more closely represents a delta function of weight Z, so that

$$A(\mathbf{k}, \omega) \sim Z\delta(\omega - \epsilon_{\mathbf{k}}^*) + \text{incoherent background}, \tag{8.153}$$

where the incoherent background is required so that the total frequency integral of the spectral function is equal to unity.

Now from (8.151) we see that the ground-state occupancy of the electron momentum state $\mathbf{k}$ is given by

$$n_{\mathbf{k}\sigma} = \langle \hat{n}_{\mathbf{k}\sigma} \rangle_{T=0} = -G(\mathbf{k}, 0^-) = \int d\omega f(\omega) A(\mathbf{k}, \omega) \Big|_{T=0}$$

$$= \int_{-\infty}^{0} d\omega A(\mathbf{k}, \omega) \qquad (T = 0). \tag{8.154}$$

The presence of the quasiparticle pole in the spectral function means that, at the Fermi surface, there is a discontinuity in the occupancy, given by

$$n_{\mathbf{k}\sigma}\big|_{k=k_F^-} - n_{\mathbf{k}\sigma}\big|_{k=k_F^+} = Z = \frac{1}{1 + \lambda}, \tag{8.155}$$

as shown in Figure 8.7.

Remarks

- The survival of a sharp delta-function peak in the quasiparticle spectral function, together with this sharp precipice-like discontinuity in the momentum-space occupancy, is one of the hallmark features of the Landau Fermi-liquid. In an electron–phonon mediated superconductor, it is the coherent part of the spectral function which condenses into the pair condensate.
- One might reasonably suppose that, since the density of states $N^*(0) = (1 + \lambda)N(0)$ is enhanced, the magnetic susceptibility will follow suit. In fact the compression of the density of states produced by phonons is always located at the Fermi energy, and this means that, if the electron–phonon interaction is turned on adiabatically, it does not

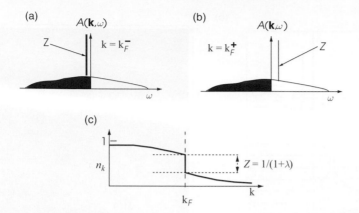

Fig. 8.7

Illustrating the relationship between the coherent, quasiparticle component in the electron spectral function, and the discontinuity in the momentum-space occupancy at the Fermi surface due to the electron–phonon interaction: (a) spectral function just below the Fermi surface: quasiparticle peak occupied; (b) spectral function just above the Fermi surface: quasiparticle peak unoccupied; (c) momentum space occupancy $n_\mathbf{k}$.

affect the Fermi momenta of either up or down electrons, so that the magnetization and hence the magnetic susceptibility are unaffected by the electron–phonon interaction.

Example 8.7 Suppose that $\alpha^2 F$ is a quadratic function of energy, given by

$$\alpha^2(\omega)F(\omega) = \begin{cases} \lambda \left(\dfrac{\omega}{\omega_D} \right)^2 & (\omega < \omega_D) \\ \\ 0 & \text{(otherwise).} \end{cases} \qquad (8.156)$$

(a) Calculate the electron–phonon self-energy.
(b) Assuming a linear dispersion, $\epsilon_k = v_F(k - k_F)$, plot the electron spectral function near the Fermi surface to illustrate the mass renormalization.

Solution

(a) Taking the result of (8.140), we have

$$\Sigma(z) = \int_0^{\omega_D} d\nu \, \alpha^2(\nu)F(\nu) \ln \left[\frac{\nu - z}{\nu + z} \right] = \lambda \int_0^{\omega_D} d\nu \left(\frac{\nu}{\omega_D} \right)^2 \ln \left[\frac{\nu - z}{\nu + z} \right]$$

$$= \lambda \omega_D \int_0^1 dx \, x^2 \ln \left[\frac{x - (z/\omega_D)}{x + (z/\omega_D)} \right] = \lambda \omega_D F \left(\frac{z}{\omega_D} \right), \qquad (8.157)$$

where

$$F(y) = \int_0^1 dx \, x^2 \ln \left[\frac{x - y}{x + y} \right] = -\frac{1}{3} y^3 \ln \left(1 - \frac{1}{y^2} \right) - \frac{y}{3} + \frac{1}{3} \ln \left(\frac{1 - y}{y + 1} \right). \qquad (8.158)$$

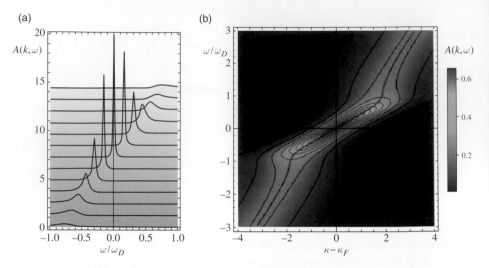

(a)

(b)

Fig. 8.8 (a) Spectral function for an electron interacting with a phonon with a quadratic $\alpha^2 F(\omega) \propto \omega^2$ (see Example 8.7). Spectral plots for successive values of the momentum are shifted upwards. (b) Contour plot of the electron spectral function, showing the renormalization of the electron velocity by the electron–phonon interaction. The red line shows the position of the maximum of the spectral function. Parameters chosen for this plot were $\lambda = 1.5$, $\omega_D = 1$, with an artificially large broadening $\delta = 0.2$ in (b).

(b) Next, we calculate

$$A(k, \omega) = \frac{1}{\pi} \text{Im} \left(\frac{1}{\omega - v_F(k - k_F) - \Sigma(\omega - i\delta)} \right). \tag{8.159}$$

Figure 8.8 shows a representative contour plot, obtained using Mathematica with parameters $v_F = 1$, $\lambda = 1.5$, $\omega_D = 1$.

8.7.3 Migdal's theorem

At first sight, one might worry about the usefulness of our leading-order self-energy correction. We have already seen that the size of the electron–phonon interaction λ is of order unity. So what permits us to ignore the vertex corrections to the self-energy?

One of the classic early results in the electron–phonon problem is Migdal's theorem [11], according to which the renormalization of the electron–phonon coupling by phonon exchange is of order $\sqrt{\frac{m}{M}}$. Migdal's theorem is a result of the huge mismatch between electron and phonon dispersion. Basically, when an electron scatters off a phonon, it moves away so fast that other phonons cannot "catch up" with the outgoing electron.

Migdal's theorem concerns the correction to the electron–phonon vertex. Diagramatically, the electron self-energy can be expanded as follows

$$(8.160)$$

which we can denote by the shorthand

$$(8.161)$$

Here, the shaded circle denotes the vertex part, given by

$$+ \cdots = ig(\mathbf{q})(1 + \Lambda(q)).$$

$$(8.162)$$

We shall discuss the leading-order vertex correction,

$$= (ig_{\mathbf{q}})\Lambda(q),$$

$$(8.163)$$

where the vertex function $\Lambda(q)$ is given by

$$\Lambda(q) = T \sum_{k' \equiv (i\omega'_n, \mathbf{k}')} (ig_{\mathbf{k}-\mathbf{k}'})^2 G(k' + q)G(k')D(k - k').$$

$$(8.164)$$

We are interested in an order-of-magnitude estimate of this quantity.

At low temperatures we can replace the summation over the Matsubara frequency by an integral:

$$T \sum_{\omega'_n} \rightarrow \int \frac{d\omega'_n}{2\pi},$$

$$(8.165)$$

so that

$$\Lambda(q) = - \int \frac{d\omega'_n}{2\pi} \int \frac{d^3 k'}{(2\pi)^3} (g_{\mathbf{k}-\mathbf{k}'})^2 G(k' + q)G(k')D(k - k').$$

$$(8.166)$$

Now the propagator

$$D(k - k') = - \frac{2\omega_{\mathbf{k}-\mathbf{k}'}}{(\omega_n - \omega'_n)^2 + \omega^2_{\mathbf{k}-\mathbf{k}'}}$$

$$(8.167)$$

vanishes as $1/(\omega'_n)^2$ in the region where $|\omega_n - \omega'_n| \gtrsim \omega_D$, so we restrict this integral, writing

$$\Lambda(q) = -\int_{-\omega_D}^{\omega_D} \frac{d\omega'_n}{2\pi} \int \frac{d^3k'}{(2\pi)^3} (g_{\mathbf{k}-\mathbf{k}'})^2 D(k-k') G(k'+q) G(k'). \tag{8.168}$$

Inside the restricted frequency integral, to obtain an estimate of this quantity we shall replace $g_{\mathbf{k}-\mathbf{k}'}^2 D(\mathbf{k}-\mathbf{k}') \sim g \times 2\omega_{\mathbf{k}-\mathbf{k}'} D(\mathbf{k}-\mathbf{k}') \sim -g$, since $2\omega_{\mathbf{k}-\mathbf{k}'} D(\mathbf{k}-\mathbf{k}') \sim -1$. To a good approximation, the frequency integral may be replaced by a single factor ω_D, so that

$$\Lambda(q) \sim \omega_D g \overbrace{\int \frac{d^3k'}{(2\pi)^3} G(k'+q) G(k')}^{\sim \frac{(k_F)^3}{\epsilon_F^2}} \Bigg|_{\omega'_n = \omega_n}. \tag{8.169}$$

Inside the momentum summation over $\mathbf{k}'$, the electron momenta are unrestricted, so the energies $\epsilon_{\mathbf{k}'}$ and $\epsilon_{\mathbf{k}'+q}$ are far from the Fermi energy and we may estimate this term as of order $\frac{(k_F)^3}{\epsilon_F^2}$. Putting these results together,

$$\Lambda \sim \frac{g}{a^3} \omega_D \frac{(k_F a)^3}{\epsilon_F^2}. \tag{8.170}$$

Since $g/a^3 \sim \lambda \epsilon_F$ and $(k_F a)^3 \sim 1$, we see that

$$\Lambda \sim \lambda \frac{\omega_D}{\epsilon_F} \sim \sqrt{\frac{m}{M}}. \tag{8.171}$$

In other words, even though the electron–phonon interaction is of order unity, the large ratio of electron to ion mass leads to a very small vertex correction.

Remarks

- Perhaps the main difficulty with the Migdal argument is that it provides a false sense of security to the theorist, giving the impression that one has "proven" that the perturbative treatment of the electron–phonon interaction is always justified. Migdal's argument is basically a dimensional analysis. The weak point of the derivation is that the dimensional analysis does not work for those scattering events where the energies of the scattered electrons are degenerate. While such scattering events may make up a small contribution to the overall phase space contributing to the self-energy, they become important because the associated scattering amplitudes can develop strong singularities that ultimately result in a catastrophic instability of the Fermi liquid. The dimensional analysis in the Migdal argument breaks down when electrons inside the loop have almost degenerate energies. For example, the Migdal calculation does not work for the case where $\mathbf{q}$ is close to a nesting vector of the Fermi surface. When $\mathbf{q}$ spans two nested Fermi surfaces, this causes $\epsilon_{\mathbf{k}'}$ and $\epsilon_{\mathbf{k}'+q}$ to become degenerate, enhancing the size of the vertex by a factor of $\epsilon_F/\omega_D \times \log(\omega_D/T)$. The singular term ultimately grows to a point where an instability to a density wave takes place, producing a charge density wave. The other parallel instability is the Cooper instability, which is a singular correction to the particle–particle scattering vertex, caused by the degeneracy of electron energies for electrons of opposite momenta.

Appendix 8A   Free fermions with a source term

In this appendix, we consider the Hamiltonian

$$
H = \overbrace{\sum_\lambda \epsilon_\lambda \psi_\lambda^\dagger \psi_\lambda}^{H_0} - \overbrace{\sum_\lambda \left[\bar\eta_\lambda(\tau)\psi_\lambda(\tau) + \psi_\lambda^\dagger(\tau)\eta_\lambda(\tau) \right]}^{-V_I} \tag{8.172}
$$

and show that the generating functional

$$
\begin{aligned}
Z_0[\bar\eta, \eta] &= Z_0 \langle T e^{-\int_0^\beta V_I(\tau)d\tau} \rangle_0 \\
&= Z_0 \left\langle T \exp\left[\int_0^\beta d\tau \sum_\lambda \left(\bar\eta_\lambda(\tau)\psi_\lambda(\tau) + \psi_\lambda^\dagger(\tau)\eta_\lambda(\tau) \right) \right] \right\rangle_0
\end{aligned} \tag{8.173}
$$

is explicitly given by

$$
\frac{Z_0[\bar\eta, \eta]}{Z_0} = \exp\left[-\sum_\lambda \int_0^\beta d\tau_1 d\tau_2 \, \bar\eta_\lambda(1) G_\lambda(\tau_1 - \tau_2) \eta_\lambda(2) \right]
$$

$$
G_\lambda(\tau_1 - \tau_2) = -\langle T\psi_\lambda(\tau_1)\psi_\lambda^\dagger(\tau_2) \rangle \tag{8.174}
$$

for both bosons and fermions.

We begin by evaluating the equation of motion of the fields in the Heisenberg representation:

$$
\frac{\partial \psi_\lambda}{\partial \tau} = [H, \psi_\lambda] = -\epsilon_\lambda \psi_\lambda(\tau) + \eta_\lambda(\tau). \tag{8.175}
$$

Multiplying this expression by the integrating factor $e^{\epsilon_\lambda \tau}$, we obtain

$$
\frac{\partial}{\partial \tau} \left[e^{\epsilon_\lambda \tau} \psi_\lambda(\tau) \right] = e^{\epsilon_\lambda \tau} \eta_\lambda(\tau), \tag{8.176}
$$

which we may integrate from $\tau' = 0$ to $\tau' = \tau$, to obtain

$$
\psi_\lambda(\tau) = e^{-\epsilon_\lambda \tau} \psi_\lambda(0) + \int_0^\tau d\tau' e^{-\epsilon_\lambda(\tau - \tau')} \eta_\lambda(\tau')d\tau'. \tag{8.177}
$$

We now take expectation values of this equation, so that

$$
\langle \psi_\lambda(\tau) \rangle = e^{-\epsilon_\lambda \tau} \langle \psi_\lambda(0) \rangle + \int_0^\tau d\tau' e^{-\epsilon_\lambda(\tau - \tau')} \eta_\lambda(\tau')d\tau'. \tag{8.178}
$$

If we impose the boundary condition $\langle \psi_\lambda(\beta) \rangle = \zeta \langle \psi_\lambda(0) \rangle$, where $\zeta = 1$ for bosons and $\zeta = -1$ for fermions, then we deduce that

$$\langle \psi_\lambda(0) \rangle = \zeta n_\lambda \int_0^\beta e^{\epsilon_\lambda \tau'} \eta_\lambda(\tau') d\tau', \tag{8.179}$$

where $n_\lambda = 1/(e^{\beta \epsilon_\lambda} - \zeta)$ is the Bose ($\zeta = 1$) or Fermi function ($\zeta = -1$). Inserting this into (8.178), we obtain

$$\langle \psi_\lambda(\tau) \rangle = \zeta n_\lambda \int_0^\beta e^{-\epsilon_\lambda(\tau - \tau')} \eta_\lambda(\tau') d\tau' + \int_0^\beta e^{-\epsilon_\lambda(\tau - \tau')} \theta(\tau - \tau') \eta_\lambda(\tau') d\tau', \tag{8.180}$$

where we have introduced a theta function in the second term in order to extend the upper limit of integration to β. Rearranging this expression, we obtain

$$\langle \psi_\lambda(\tau) \rangle = \int_0^\beta d\tau' \overbrace{e^{-\epsilon_\lambda}(\tau - \tau') \left[(1 + \zeta n_\lambda)\theta(\tau - \tau') + \zeta n_\lambda \theta(\tau' - \tau) \right]}^{-\mathcal{G}_\lambda(\tau - \tau')} \eta_\lambda(\tau')$$

$$= -\int_0^\beta d\tau' \mathcal{G}_\lambda(\tau - \tau') \eta_\lambda(\tau'), \tag{8.181}$$

so $\mathcal{G}_\lambda(\tau)$ is the imaginary-time response of the field to the source term. We may repeat the same procedure for the expectation value of the creation operator. The results of these two calculations may be summarized as

$$\langle \psi_\lambda(\tau) \rangle = \frac{\delta Z[\bar{\eta}, \eta]}{\delta \bar{\eta}(\tau)} = -\int_0^\beta d\tau' \mathcal{G}_\lambda(\tau - \tau') \eta_\lambda(\tau')$$

$$\langle \psi_\lambda^\dagger(\tau) \rangle = \frac{\delta Z[\bar{\eta}, \eta]}{\delta \eta(\tau)} = -\int_0^\beta d\tau' \bar{\eta}(\tau) \mathcal{G}_\lambda(\tau - \tau'). \tag{8.182}$$

Notice how the creation field propagates backwards in time from the source. The common integral in these two expression is

$$\ln Z[\bar{\eta}, \eta] = \ln Z_0 - \int_0^\beta d\tau d\tau' \bar{\eta}_\lambda(\tau) G_\lambda(\tau - \tau') \eta_\lambda(\tau'), \tag{8.183}$$

where the constant term $\ln Z_0$ has to be intependent of both η and $\bar{\eta}$. The exponential of this expression recovers the result (8.174).

Exercises

Exercise 8.1 Use the method of complex contour integration to carry out the Matsubara sums in the following:

(a) Derive the density of a spinless Bose gas at finite temperature from the boson propagator $D(k) \equiv D(\mathbf{k}, i\nu_n) = [i\nu_n - \omega_\mathbf{k}]^{-1}$, where $\omega_\mathbf{k} = E_\mathbf{k} - \mu$ is the energy of a boson, measured relative to the chemical potential:

$$\rho(T) = \frac{N}{V} = V^{-1} \sum_\mathbf{k} \langle T b_\mathbf{k}(0^-) b_\mathbf{k}^\dagger(0) \rangle = -(\beta V)^{-1} \sum_{i\nu_n, \mathbf{k}} D(k) e^{i\nu_n 0^+}. \tag{8.184}$$

How do you need to modify your answer to take account of Bose–Einstein condensation?

(b) Carry out the Matsubara sum in the dynamical charge susceptibility of a free Bose gas, i.e.

$$\chi_c(q, iv_n) = \quad = T \sum_{iv_n} \int \frac{d^3k}{(2\pi)^3} D(q+k)D(k).$$

$$(8.185)$$

Please analytically extend your final answer to real frequencies.

(c) Carry out the Matsubara sum in the *pair susceptibility* of a spin-$\frac{1}{2}$ free Fermi gas, i.e.

$$\chi_P(q, iv_n) = \quad = T \sum_{i\omega_r} \int \frac{d^3k}{(2\pi)^3} G(q+k)G(-k),$$

$$(8.186)$$

where $G(k) \equiv G(\mathbf{k}, i\omega_n) = [i\omega_n - \epsilon_{\mathbf{k}}]^{-1}$. (Note the direction of the arrows: why is there no minus sign for the fermion loop?) Show that the static pair susceptibility $\chi_P(0)$ is given by

$$\chi_P = \int \frac{d^3k}{(2\pi)^3} \frac{\tanh[\beta\epsilon_{\mathbf{k}}/2]}{2\epsilon_{\mathbf{k}}}.$$

$$(8.187)$$

Can you see that this quantity diverges at low temperatures? How does it diverge, and why?

Exercise 8.2 A simple model of an atom with two atomic levels coupled to a radiation field is described by the Hamiltonian

$$H = H_0 + H_I + H_{photon},$$

$$(8.188)$$

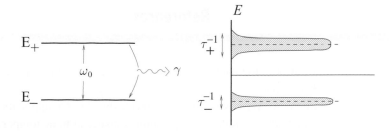

where

$$H_0 = \tilde{E}_- c_-^\dagger c_- + \tilde{E}_+ c_+^\dagger c_+$$

$$(8.189)$$

describes the atom, treating it as a *fermion*;

$$H_I = V^{-1/2} \sum_{\vec{q}} g(\omega_{\vec{q}}) \left(c_+^\dagger c_- + c_-^\dagger c_+ \right) \left[a_{\vec{q}}^\dagger + a_{-\vec{q}} \right]$$

$$(8.190)$$

describes the coupling to the radiation field (V is the volume of the box enclosing the radiation); and

$$H_{photon} = \sum_{\vec{q}} \omega_{\vec{q}} a_{\vec{q}}^{\dagger} a_{\vec{q}} \qquad (\omega_q = cq) \qquad (8.191)$$

is the Hamiltonian for the electromagnetic field. The dipole matrix element $g(\omega)$ is weak enough to be treated by second-order perturbation theory and the polarization of the photon is ignored.

(a) Calculate the self-energy $\Sigma_+(\omega)$ and $\Sigma_-(\omega)$ for an atom in the $+$ and $-$ states, respectively.

(b) Use the self-energy obtained above to calculate the lifetimes $\tau_\pm$ of the atomic states, i.e.

$$\tau_\pm^{-1} = 2\mathrm{Im}\Sigma_\pm(\tilde{E}_\pm - i\delta). \qquad (8.192)$$

If the gas of atoms is non-degenerate, i.e. the Fermi functions are all small compared with unity, $f(E_\pm) \sim 0$, show that

$$\tau_+^{-1} = 2\pi |g(\omega_0)|^2 F(\omega_0)[1 + n(\omega_0)]$$
$$\tau_-^{-1} = 2\pi |g(\omega_0)|^2 F(\omega_0) n(\omega_0), \qquad (8.193)$$

where $\omega_o = \tilde{E}_+ - \tilde{E}_-$ is the separation of the atomic levels and

$$F(\omega) = \int \frac{d^3q}{(2\pi)^3} \delta(\omega - \omega_q) = \frac{\omega^2}{2\pi c^3} \qquad (8.194)$$

is the density of state of the photons at energy ω. What do these results have to do with stimulated emission? Do your final results depend on the initial assumption that the atoms are fermions?

(c) Why is the decay rate of the upper state larger than the decay rate of the lower state by the factor $[1 + n(\omega_0)]/n(\omega_0)$?

References

[1] T. Matsubara, A new approach to quantum statistical mechanics, *Prog. Theor. Phys.*, vol. 14, p. 351, 1955.

[2] L. P. Gor'kov, A. A. Abrikosov, and I. E. Dzyaloshinskii, On application of quantum field theory methods to problems of quantum statistics at finite temperatures, *J. Exp. Theor. Phys.*, vol. 36, p. 900, 1959.

[3] G. J. Stoney, On the cause of double lines and of equidistant satellites in the spectra of gases, *Sci. Trans. R. Dublin Soc.*, vol. 4, p. 563, 1891.

[4] H. A. Lorentz, *Versuch einer Theorie der electronischen und optischen Erscheinungen in bewegten Körpen* (Search for a Theory of the Electrical and Optical Properties of Moving Bodies), E. J. Brill, 1895.

[5] J. J. Thomson, Cathode rays, *Phil. Mag.*, vol. 44, p. 293, 1897.

[6] P. Drude, Zür Elektronentheorie der Metalle (On the electron theory of metals), *Ann. Phys. (Leipzig)*, vol. 1, p. 566, 1900.

[7] H. Fröhlich, Theory of the superconducting state. I: The ground state at the absolute zero of temperature, *Phys. Rev.*, vol. 79, p. 845, 1950.

[8] H. Fröhlich, Interaction of electrons with lattice vibrations, *Proc. R. Soc. A*, vol. 215, p. 291, 1952.

[9] J. Bardeen and D. Pines, Electron–phonon interaction in metals, *Phys. Rev.*, vol. 99, p. 1140, 1955.

[10] D. Pines, Superconductivity in the Periodic System, *Phys. Rev.*, vol. 109, p. 280, 1958.

[11] A. A. Migdal, Interaction between electron and lattice vibrations in a normal metal, *J. Exp. Theor. Phys.*, vol. 7, p. 996, 1958.

9 Fluctuation–dissipation theorem and linear response theory

9.1 Introduction

In this chapter we will discuss the deep link between fluctuations about equilibrium and the response of a system to external forces. If the susceptibility of a system to external change is large, then the fluctuations about equilibrium are expected to be large. The mathematical relationship that quantifies this connection is called the *fluctuation–dissipation theorem* [1–3]. We shall discuss and derive this relationship in this chapter. It turns out that the link between fluctuations and dissipation also extends to imaginary time, enabling us to relate equilibrium correlation functions and response functions to the imaginary-time Green's function of the corresponding variables.

To describe the fluctuations and response at a finite temperature, we will introduce related types of Green's function: the correlation function $S(t)$ [4],

$$S(t - t') = \langle A(t) A(t') \rangle = \int_{-\infty}^{\infty} \frac{d\omega}{2\pi} e^{-i\omega(t-t')} S(\omega); \tag{9.1}$$

the dynamical susceptibility $\chi(t)$,

$$\chi(t - t') = i\langle [A(t), A(t')] \rangle \theta(t - t'), \tag{9.2}$$

which determines the retarded response

$$\langle A(t) \rangle = \int_{-\infty}^{\infty} dt' \chi(t - t') f(t'), \quad \langle A(\omega) \rangle = \chi(\omega) f(\omega), \tag{9.3}$$

to a force term $f(t)$ coupled to A inside the Hamiltonian $H_I = -f(t)A(t)$; and, lastly, the imaginary-time response function $\chi(\tau)$,

$$\chi(\tau - \tau') = \langle T A(\tau) A(\tau') \rangle. \tag{9.4}$$

The fluctuation–dissipation theorem [1–3] relates the Fourier transforms of these quantities, according to

$$\underbrace{S(\omega)}_{\text{fluctuations}} = 2\hbar[\ \overbrace{1}^{\text{quantum}} + \overbrace{n_B(\omega)}^{\text{thermal}}\ ]\ \underbrace{\chi''(\omega)}_{\text{dissipation}}, \tag{9.5}$$

where $\chi''(\omega) = \mathrm{Im}\, \chi(\omega)$ describes the dissipative part of the response function. In the limit $\omega \ll k_B T$, when $n(\omega) \sim k_B T/\hbar\omega$, this result reverts to the classical fluctuation–dissipation theorem,

$$S(\omega) = \frac{2k_B T}{\omega} \chi''(\omega). \tag{9.6}$$

Thus, in principle, if we know the correlation functions in thermal equilibrium, we can compute the response function of the system.

The dissipative response of the system also enters into the Kramers–Kronig expansion of the response function,

$$\chi(z) = \int \frac{d\omega}{\pi} \frac{1}{\omega - z} \chi''(\omega). \tag{9.7}$$

This allows us to interpret

$$-\frac{1}{\pi} \chi''(\omega) d\omega = \text{residue of poles between } \omega \text{ and } \omega + d\omega, \tag{9.8}$$

so that $\chi''(\omega)$ is a kind of spectral density for excitations, weighted by the matrix element of the corresponding operator. Using this expression, the dynamical susceptibility can be analytically extended into the complex plane. We will see that above the real axis $\chi(z)$ describes causal, retarded responses to an applied force, where as below the real axis, it describes a time-reversed "advanced" response to changes in the future.

In practice, the theorist takes advantage of a completely parallel fluctuation–dissipation theorem which exists in imaginary time. The imaginary-time correlation function $\chi(\tau)$ is periodic in time $\chi(\tau + \beta) = \chi(\tau)$, and has a discrete Matsubara Fourier expansion, given by

$$\chi(\tau) = \langle TA(\tau)A(0) \rangle = \frac{1}{\beta} \sum_n e^{-i\nu_n \tau} \chi_M(i\nu_n). \tag{9.9}$$

The key relation between this function and the physical response function is that

$$\chi_M(i\nu_n) = \chi(z)|_{z=i\nu_n}. \tag{9.10}$$

This relation permits us to compute the physical response function by analytically continuing the Fourier components of the imaginary-time correlation functions onto the real axis.

To understand these relations, we first need to understand the nature of the quantum mechanical response functions. We shall then carry out a *spectral decomposition* of each of the above functions, deriving the fluctuation–dissipation theorem by showing that the same underlying matrix elements enter into each expression. A heuristic understanding of the relationship between fluctuations and dissipation is obtained by examining a classical example. The main difference between the classical and quantum fluctuation–dissipation theorems is that in classical mechanics we are obliged to explicitly include the external sources of noise, whereas in the quantum case the noise is intrinsic and we can analyze the fluctuations without any specific reference to external sources of noise. Nevertheless, the classical case is highly pedagogical, and it is this limit that we shall consider first.

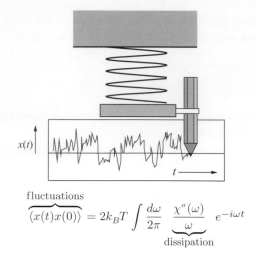

$$\underbrace{\langle x(t)x(0)\rangle}_{\text{fluctuations}} = 2k_BT \int \frac{d\omega}{2\pi} \underbrace{\frac{\chi''(\omega)}{\omega}}_{\text{dissipation}} e^{-i\omega t}$$

Fig. 9.1 Fluctuations in a classical harmonic oscillator are directly related to the dissipative response function via the fluctuation–dissipation theorem.

9.2 Fluctuation–dissipation theorem for a classical harmonic oscillator

In a classical system, to examine correlation functions we need to include an explicit source of external noise. To illustrate the procedure, consider a harmonic oscillator in thermal equilibrium inside a viscous medium. Suppose that thermal fluctuations give rise to a random force, acting on the oscillator according to the equation of motion

$$m(\ddot{x} + \omega_0^2 x) + \eta \dot{x} = f(t). \tag{9.11}$$

If we Fourier transform this relationship, we obtain

$$x(\omega) = \chi(\omega)f(\omega)$$
$$\chi(\omega) = [m(\omega_0^2 - \omega^2) - i\omega\eta]^{-1}. \tag{9.12}$$

Here $\chi(\omega)$ is the response function or susceptibility to the external force. The imaginary part of the susceptibility governs the dissipation and is given by

$$\chi''(\omega) = \frac{\omega\eta}{m(\omega_0^2 - \omega^2) + \omega^2\eta^2} = |\chi(\omega)|^2\omega\eta. \tag{9.13}$$

Now let us consider the fluctuations in thermal equilibrium. Over long time periods, we expect the two-point correlation function to be purely a function of the time difference:

$$\langle x(t)x(t')\rangle = \langle x(t-t')x(0)\rangle. \tag{9.14}$$

The power spectrum of fluctuations is defined as

$$\langle |x(\omega)|^2 \rangle = \int dt \langle x(t)x(0) \rangle e^{i\omega t} \tag{9.15}$$

and the inverse relation gives

$$\langle x(t)x(t') \rangle = \int \frac{d\omega}{2\pi} e^{-i\omega(t-t')} \langle |x(\omega)|^2 \rangle. \tag{9.16}$$

Now in thermal equilibrium the equipartition theorem tells us that

$$\frac{m\omega_0^2}{2} \langle x^2 \rangle = \frac{k_B T}{2} \tag{9.17}$$

or

$$\langle x^2 \rangle = \int \frac{d\omega}{2\pi} \langle |x(\omega)|^2 \rangle = \int \frac{d\omega}{2\pi} |\chi(\omega)|^2 \langle |f(\omega)|^2 \rangle = \frac{k_B T}{m\omega_0^2}. \tag{9.18}$$

Since the integrand is very sharply peaked around $|\omega| = \omega_0$, we replace $\langle |f(\omega)|^2 \rangle \rightarrow \langle |f(\omega_0)|^2 \rangle$ in the above expression. Replacing $|\chi(\omega)|^2 \rightarrow \frac{1}{\omega\eta} \chi''(\omega)$, we then obtain

$$\frac{k_B T}{m\omega_0^2} = \frac{\langle |f(\omega_0)|^2 \rangle}{2\eta} \int \frac{d\omega}{\pi} \frac{\chi''(\omega)}{\omega} = \frac{|f(\omega_0)|^2}{2\eta m\omega_0^2}, \tag{9.19}$$

so that the spectrum of force fluctuations is determined by the viscosity η:

$$\langle |f(\omega_0)|^2 \rangle = 2\eta k_B T. \tag{9.20}$$

Now if we assume that the noise spectrum depends only on the properties of the viscous medium in which the oscillator is embedded, and not on the properties of the oscillator, then we expect this expression to hold for any frequency ω_0, and since it is *independent* of the frequency, we conclude that the power spectrum of the force is a flat function of frequency, enabling us to replace $\omega_0 \rightarrow \omega$ in the above expression. This implies that, in thermal equilibrium, the force coupling the system to the environment is a source of white noise of an amplitude which depends on the viscosity of the medium:

$$\langle f(t)f(t') \rangle = \int \frac{d\omega}{2\pi} e^{-i\omega(t-t')} \overbrace{\langle |f(\omega)|^2 \rangle}^{2\eta k_B T} = 2\eta k_B T \delta(t - t'). \tag{9.21}$$

We can now compute the noise spectrum of fluctuations, which is given by

$$S(\omega) = \langle |x(\omega)|^2 \rangle = |\chi(\omega)|^2 \langle |f(\omega)|^2 \rangle = \langle |f(\omega)|^2 \rangle \frac{\chi''(\omega)}{\omega\eta} = \frac{2k_B T}{\omega} \chi''(\omega). \tag{9.22}$$

This expression relates the thermal fluctuations of a classical system to the dissipation, as described by the imaginary part of the response function, $\chi''(\omega)$.

9.3　Quantum mechanical response functions

Suppose we couple a force f to variable A. For later generality, it suits our needs to consider a force in both real and imaginary time, with Hamiltonian

$$H = H_0 - f(t)A \tag{9.23}$$

$$H = H_0 - f(\tau)A. \tag{9.24}$$

We shall now show that the response to these forces is given by

$$\langle A(t) \rangle = \langle A \rangle + \int_{-\infty}^{\infty} \chi(t - t')f(t')dt'$$

$$\langle A(\tau) \rangle = \langle A \rangle + \int_{0}^{\beta} \tilde{\chi}(\tau - \tau')f(\tau')d\tau' \tag{9.25}$$

$$\chi(t - t') = i\langle [A(t), A(t')] \rangle \theta(t - t')$$

$$\tilde{\chi}(\tau - \tau') = \langle T A(\tau)A(\tau') \rangle - \langle A \rangle^2, \tag{9.26}$$

where $\langle A \rangle$ is the value of A in thermal equilibrium.

Let us begin in real time. Suppose we want to look at the future response to a force $f(t)$ applied in the past. Using the interaction representation, we know that

$$A_H(t) = U^\dagger(t)A_I(t)\,U(t), \tag{9.27}$$

where

$$U(t) = T \exp\left[i \int_{-\infty}^{t} dt' A_I(t')f(t')\right]. \tag{9.28}$$

Remembering that the interaction representation corresponds to the Heisenberg representation for H_0, we can drop the subscript on $A_I(t) \equiv A(t)$, so that, to linear order in $f(t)$,

$$U(t) = 1 + i \int_{-\infty}^{t} dt' A(t')f(t')$$

$$U^\dagger(t) = 1 - i \int_{-\infty}^{t} dt' A(t')f(t'), \tag{9.29}$$

so that

$$A_H(t) = A(t) + i \int_{-\infty}^{t} dt' [A(t), A(t')]f(t'). \tag{9.30}$$

In thermal equilibrium $\langle A(t) \rangle = \langle A \rangle$, so the response to the applied force is given by

$$\langle A_H(t) \rangle = \langle A \rangle + \int_{-\infty}^{+\infty} dt'\, \chi(t - t')f(t'), \tag{9.31}$$

where

$$\chi_R(t - t') = i\langle[A(t), A(t')]\rangle\theta(t - t') \qquad (9.32)$$

retarded response function

is the *retarded response function*. The above equation is particularly interesting, for it relates a response to a quantum mechanical correlation function. For completeness, we note that the corresponding backwards-time quantity

$$\chi_A(t - t') = -i\langle[A(t), A(t')]\rangle\theta(t' - t) \qquad (9.33)$$

advanced response function

is known as the *advanced response function*, corresponding to a fictional time-reversed world in which we propagate the effects of a disturbance backwards in time!

Let us now switch to consider imaginary time. In this case, the partition function in the presence of a perturbation is

$$Z = Z_0\langle T\exp\int_0^\beta d\tau f(\tau)A_I(\tau)\rangle_0. \qquad (9.34)$$

The expectation value of $A(\tau)$ is then given by

$$\langle A(\tau)\rangle = \frac{\delta\ln Z}{\delta f(\tau)} = \frac{\langle TA(\tau)\exp\int_0^\beta d\tau' f(\tau')A_I(\tau')\rangle}{\langle T\exp\int_0^\beta d\tau' f(\tau')A_I(\tau')\rangle}$$

$$= \langle A\rangle + \int_0^\beta d\tau'\overbrace{\left[\langle TA(\tau)A(\tau')\rangle - \langle A\rangle^2\right]}^{\tilde\chi(\tau-\tau')}f(\tau') + O(f^2), \qquad (9.35)$$

so that the thermal response to the applied force is given by a Green's function that is time-ordered in imaginary time:

$$\tilde\chi(\tau - \tau') = \langle TA(\tau)A(\tau')\rangle - \langle A\rangle^2$$
$$= \langle T(A(\tau) - \langle A\rangle)(A(\tau') - \langle A\rangle)\rangle, \qquad (9.36)$$

where expectation values are to be taken at thermal equilibrium for H_0. $\chi(\tau - \tau')$ describes the thermal and quantum fluctuations of the quantity $\hat A$ in imaginary time.

9.4 Fluctuations and dissipation in a quantum world

The quantum Boltzmann formulation of many-body physics is naturally tailored to a discussion of the statistics of fluctuations and dissipation. Quantum systems are naturally noisy, and there is no need for us to add any additional noise source to examine the deep link between fluctuations and dissipation in a quantum many-body system. Indeed, the quantum fluctuation–dissipation theorem can be derived in rather mechanistic fashion by carrying out a spectral decomposition of the various response and correlation functions. The procedure is formally more direct than its classical analogue, but the algebra tends to hide the fact that the underlying physics holds precisely the same link between fluctuations – now both thermal and quantum in character – and dissipation.

To derive the quantum fluctuation–dissipation theorem, we must first spectrally decompose the correlation function $S(t - t')$ and the retarded response function $\chi_R(t - t')$.

9.4.1 Spectral decomposition I: the correlation function $S(t - t')$

This is the easiest decomposition of the three to carry out. We begin by expanding the response function in terms of a complete set of energy eigenstates which satisfy

$$H \left| \lambda \right\rangle = E_\lambda \left| \lambda \right\rangle$$

$$\sum_\lambda \left| \lambda \right\rangle \left\langle \lambda \right| = 1 \tag{9.37}$$

$$\left\langle \lambda \left| A(t) \right| \zeta \right\rangle = \left\langle \lambda \left| e^{iHt} A \, e^{-iHt} \right| \zeta \right\rangle = e^{-i(E_\zeta - E_\lambda)(t - t')} \left\langle \lambda \left| A \right| \zeta \right\rangle.$$

Using these key results, we make the expansion as follows:

$$\begin{aligned}
S(t - t') &= \left\langle A(t) A(t') \right\rangle \\
&= \sum_{\lambda, \zeta} e^{-\beta(E_\lambda - F)} \left\langle \lambda \left| A(t) \right| \zeta \right\rangle \left\langle \zeta \left| A(t') \right| \lambda \right\rangle \\
&= \sum_{\lambda, \zeta} e^{-\beta(E_\lambda - F)} \left| \left\langle \zeta \left| A \right| \lambda \right\rangle \right|^2 e^{-i(E_\zeta - E_\lambda)(t - t')}.
\end{aligned} \tag{9.38}$$

If we now Fourier transform this expression, the frequency-dependent correlation function can be written

$$\begin{aligned}
S(\omega) &= \int_{-\infty}^{\infty} dt \, e^{i\omega t} S(t) \\
&= \sum_{\lambda, \zeta} e^{-\beta(E_\lambda - F)} \left| \left\langle \zeta \left| A \right| \lambda \right\rangle \right|^2 2\pi \delta(E_\zeta - E_\lambda - \omega).
\end{aligned} \tag{9.39}$$

This is the frequency spectrum of the correlations.

9.4.2 Spectral decomposition II: the retarded response function $\chi_R(t - t')$

We now use the same spectral decomposition approach for the retarded response function. In this case, we need to take care of two operator orderings inside the commutator, which yield

$$\begin{aligned}
\chi_R(t - t') &= i \left\langle [A(t), A(t')] \right\rangle \theta(t - t') \\
&= i \sum_{\lambda, \zeta} e^{-\beta(E_\lambda - F)} \left\{ \left\langle \lambda \left| A(t) \right| \zeta \right\rangle \left\langle \zeta \left| A(t') \right| \lambda \right\rangle - \left\langle \lambda \left| A(t') \right| \zeta \right\rangle \left\langle \zeta \left| A(t) \right| \lambda \right\rangle \right\} \theta(t - t') \\
&= i \sum_{\lambda, \zeta} e^{\beta F} (e^{-\beta E_\lambda} - e^{-\beta E_\zeta}) \left| \left\langle \zeta \left| A \right| \lambda \right\rangle \right|^2 e^{-i(E_\zeta - E_\lambda)(t - t')} \theta(t - t').
\end{aligned} \tag{9.40}$$

By introducing the spectral function

$$\chi''(\omega) = \pi(1 - e^{-\beta\omega}) \sum_{\lambda,\zeta} p_\lambda \, |\langle \zeta \, |A| \, \lambda \rangle|^2 \, \delta[\omega - (E_\zeta - E_\lambda)], \tag{9.41}$$

where $p_\lambda = e^{-\beta(E_\lambda - F)}$ is the probability of being in the initial state $|\lambda\rangle$, we see that the retarded response function can be written

$$\chi_R(t) = i \int \frac{d\omega}{\pi} e^{-i\omega t} \theta(t) \, \chi''(\omega). \tag{9.42}$$

Fourier transforming this result, using

$$i \int_0^\infty dt \, e^{i(\omega - \omega' + i\delta) t} = \frac{1}{\omega' - \omega - i\delta}, \tag{9.43}$$

we can read off the Fourier transform of the retarded response function as

$$\chi_R(\omega) = \int \frac{d\omega'}{\pi} \frac{1}{\omega' - \omega - i\delta} \chi''(\omega'). \tag{9.44}$$

This is known as a *Kramers–Kronig relation*. The Kramers–Kronig relation can be used to extend the response function into the complex plane by writing

$$\chi(z) = \int \frac{d\omega'}{\pi} \frac{1}{\omega' - z} \chi''(\omega'). \tag{9.45}$$

dynamical susceptibility

This is the *dynamical susceptibility*. When we evaluate $\chi(z)$ just above the real axis, we get the retarded response function $\chi_R(\omega) = \chi(\omega + i\delta)$. The upper half-plane is thus the analytic extension of $\chi_R(\omega)$. But what about the lower half-plane? Remarkably, this gives the advanced response function, such that $\chi_A(\omega) = \chi(\omega - i\delta)$. From the definition of $\chi(z)$, we see that its poles are located exclusively along the real axis at $z = \omega'$, so that $\chi(z)$ is analytic everywhere *except* the real axis. Substituting the Cauchy–Dirac relation

$$\frac{1}{\omega' - \omega \mp i\delta} = P\frac{1}{\omega' - \omega} \pm i\pi \delta(\omega' - \omega), \tag{9.46}$$

where P denotes the principal part, into (9.45), we see that above and below the real axis at $z = \omega \pm i\delta$ the dynamical susceptibility is given by

$$\chi(\omega \pm i\delta) = \int \frac{d\omega'}{\pi} P\left(\frac{1}{\omega' - \omega}\right) \chi''(\omega') \pm i\chi''(\omega) = \chi'(\omega) \pm i\chi''(\omega), \tag{9.47}$$

so that the real part of $\chi(z)$ is continuous across the real axis, but the dissipative imaginary part has a *discontinuity* given by

$$\chi''(\omega) = \text{Im} \, \chi(\omega + i\delta) = \frac{\chi(\omega + i\delta) - \chi(\omega - i\delta)}{2i}. \tag{9.48}$$

This branch cut along the imaginary axis is a universal property of the dynamical response function.

Example 9.1   (a) By carrying out a spectral decomposition of the advanced response function $\chi_R(t-t') = -i\langle[A(t), A(t')]\rangle\theta(t'-t)$, show that its Fourier transform is determined by the same spectral function as the retarded reponse, namely

$$\chi_A(t) = -i\theta(t)\int \frac{d\omega}{\pi} e^{-i\omega t}\chi''(\omega). \tag{9.49}$$

(b) By Fourier transforming this expression, show that

$$\chi_A(\omega) = \int \frac{d\omega'}{\pi}\frac{1}{\omega'-\omega+i\delta}\chi''(\omega') = \chi(\omega-i\delta) = [\chi_R(\omega)]^*. \tag{9.50}$$

Solution

(a) To prove the first part, we carry out a spectral decomposition as follows:

$$\chi_A(t-t') = -i\langle[A(t), A(t')]\rangle\theta(t'-t)$$

$$= -i\sum_{\lambda,\zeta} e^{-\beta(E_\lambda-F)}\left\{\langle\lambda\,|A(t)|\,\zeta\rangle\langle\zeta\,|A(t')|\,\lambda\rangle - \langle\lambda\,|A(t')|\,\zeta\rangle\langle\zeta\,|A(t)|\,\lambda\rangle\right\}\theta(t'-t)$$

$$= -i\sum_{\lambda,\zeta} e^{\beta F}(e^{-\beta E_\lambda} - e^{-\beta E_\zeta})\,|\langle\zeta\,|A|\,\lambda\rangle|^2\,e^{-i(E_\zeta-E_\lambda)(t-t')}\theta(t'-t)$$

$$= -i\theta(t'-t)\int d\omega\,e^{-i\omega(t-t')}\overbrace{\sum_{\lambda,\zeta} p_\lambda(1-e^{-\beta\omega})\,|\langle\zeta\,|A|\,\lambda\rangle|^2\,\delta(\omega-(E_\zeta-E_\lambda))}^{\chi''(\omega)/\pi}$$

$$= -i\theta(t'-t)\int \frac{d\omega}{\pi}e^{-i\omega t}\chi''(\omega). \tag{9.51}$$

(b) Next, we Fourier transform the result of the last part, to obtain

$$\chi_A(\omega) = \int_{-\infty}^{0} dt\,e^{i\omega t+\delta t}\chi_A(t)$$

$$= -i\int_{-\infty}^{0} dt\,e^{i\omega t+\delta t}\int_{-\infty}^{\infty}\frac{d\omega'}{\pi}\chi''(\omega')e^{-i\omega't}. \tag{9.52}$$

Inverting the order of integration gives

$$\chi_A(\omega) = \int_{-\infty}^{\infty}\frac{d\omega'}{\pi}\chi''(\omega')\left(-i\int_{-\infty}^{0} dt e^{i(\omega-\omega'-i\delta)t}\right)$$

$$= \int_{-\infty}^{\infty}\frac{d\omega'}{\pi}\chi''(\omega')\left(\frac{1}{\omega'-\omega+i\delta}\right). \tag{9.53}$$

9.4.3   Quantum fluctuation–dissipation theorem

If we compare the relations (9.41) and (9.39), we see that

$$S(\omega) = \frac{2}{1-e^{-\beta\omega}}\chi''(\omega). \tag{9.54}$$

If we restore $\hbar$, this becomes

$$S(\omega) = \frac{2\hbar}{1 - e^{-\beta\hbar\omega}} \chi''(\omega) = 2\hbar \left[1 + n_B(\hbar\omega)\right] \chi''(\omega). \qquad (9.55)$$

quantum fluctuation–dissipation theorem

Thus, by carrying out a spectral analysis we have been able to directly link the correlation function $S(\omega)$ with the dissipative part of the retarded response function $\chi(\omega)$.

9.4.4 Spectral decomposition III: fluctuations in imaginary time

For the final of our three decompositions, we move to imaginary time and write, for $\tau - \tau' > 0$,

$$\chi(\tau - \tau') = \sum_{\lambda,\zeta} e^{-\beta(E_\lambda - F)} \left\{ \langle \lambda \left| A(\tau) \right| \zeta \rangle \langle \zeta \left| A(\tau') \right| \lambda \rangle \right\}$$

$$= \sum_{\lambda,\zeta} p_\lambda e^{-(E_\lambda - E_\zeta)(\tau - \tau')} \left| \langle \zeta \left| A \right| \lambda \rangle \right|^2, \qquad (9.56)$$

where, as before, $p_\lambda = e^{-\beta(E_\lambda - F)}$ is the Boltzmann probability of being in state $|\lambda\rangle$. Now

$$\int_0^\beta d\tau \, e^{i\nu_n\tau} e^{-(E_\lambda - E_\zeta)\tau} = \frac{1}{(E_\zeta - E_\lambda - i\nu_n)}(1 - e^{-(E_\lambda - E_\zeta)\beta}), \qquad (9.57)$$

so

$$\chi(i\nu_n) = \int_0^\beta d\tau \, e^{i\nu_n\tau} \chi(\tau)$$

$$= \sum_{\lambda,\zeta} p_\lambda (1 - e^{-\beta(E_\zeta - E_\lambda)}) \left| \langle \zeta \left| A \right| \lambda \rangle \right|^2 \frac{1}{(E_\zeta - E_\lambda - i\nu_n)}. \qquad (9.58)$$

Using (9.41), we can write this as

$$\chi(i\nu_n) = \int \frac{d\omega}{\pi} \frac{1}{\omega - i\nu_n} \chi''(\omega). \qquad (9.59)$$

But this is nothing more than the dynamical susceptibility $\chi(z)$, evaluated at $z = i\nu_n$. In other words, $\chi(i\nu_n)$ is the unique analytic extension of the dynamical susceptibility $\chi(\omega)$ into the complex plane. Our procedure to calculate response functions will be to write $\chi(i\nu_n)$ in the form (9.29), and to use this to then read off the spectral function $\chi''(\omega)$, which in turn determines the dynamical response function.

9.5 Calculation of response functions

Having made the link between the imaginary-time and real-time response functions, we are ready to discuss how we can calculate response functions from Feynman diagrams. Our procedure is to compute the imaginary-time response function and then analytically

Table 9.1 Selected operators and corresponding response functions.

Quantity	Operator $\hat{A}^a$	$A(\mathbf{k})$	Response function
Density	$\hat{\rho}(x) = \psi^\dagger(x)\psi(x)$	$\rho_{\alpha\beta} = \delta_{\alpha\beta}$	Charge susceptibility
Spin density	$\vec{S}(x) = \psi_\alpha^\dagger(x)\left(\frac{\vec{\sigma}}{2}\right)_{\alpha\beta}\psi_\beta(x)$	$\vec{M}_{\alpha\beta} = \mu_B\vec{\sigma}_{\alpha\beta}$	Spin susceptibility
Current density^a	$\frac{e}{m}\psi^\dagger(x)\left(-i\hbar\overset{\leftrightarrow}{\nabla} - e\vec{A}\right)\psi(x)$	$\vec{j} = e\vec{v}_\mathbf{k} = e\vec{\nabla}\epsilon_\mathbf{k}$	Conductivity
Thermal current^a	$\frac{\hbar^2}{2m}\psi^\dagger(x)\overset{\leftrightarrow}{\nabla}\overset{\leftrightarrow}{\partial}_t\psi(x)$	$\vec{j}_T = i\omega_n\vec{v}_\mathbf{k} = i\omega_n\vec{\nabla}\epsilon_\mathbf{k}$	Thermal conductivity

^{a} $\overset{\leftrightarrow}{\nabla} \equiv \frac{1}{2}\left(\overset{\rightarrow}{\nabla} - \overset{\leftarrow}{\nabla}\right)$, $\overset{\leftrightarrow}{\partial}_t \equiv \frac{1}{2}\left(\overset{\rightarrow}{\partial}_t - \overset{\leftarrow}{\partial}_t\right)$.

continue to real frequencies. Suppose we are interested in the response function for $\hat{A}$, where

$$\hat{A}(x) = \psi_\alpha^\dagger(x)A_{\alpha\beta}\psi_\beta(x) \tag{9.60}$$

(see table 9.1). The corresponding operator generates the vertex

$$X \;= A_{\alpha\beta}. \tag{9.61}$$

where the spin variables $\alpha\beta$ are to be contracted with the internal spin variables of the Feynman diagram. This inevitably means that the variable $A_{\alpha\beta}$ becomes part of an internal trace over spin variables. If we expand the corresponding response function $\chi(x) = \langle\hat{A}(x)\hat{A}(0)\rangle$ using Feynman diagrams, we obtain

$$\chi(\tau) = \langle\hat{A}(x)\hat{A}(0)\rangle = \sum\{\text{closed linked two-vertex diagrams}\} \tag{9.62}$$

$$= X \qquad\qquad 0. \tag{9.63}$$

For example, in a non-interacting electron system, the imaginary-time spin response function involves $A(x) = \mu_B\psi_\alpha^\dagger(x)\sigma_{\alpha\beta}\psi_\beta(x)$, so the corresponding response function is

$$\chi^{ab}(x-x') = \mu_B^2 \times \; \sigma^a_{\alpha\beta}$$

Trace over
spin variables

Trace over
spin variables

$$= -\text{Tr}\left[\sigma^a\mathcal{G}(x-x')\sigma^b\mathcal{G}(x'-x)\right]$$
$$= -\delta^{ab}2\mu_B^2\mathcal{G}(x-x')\mathcal{G}(x'-x). \tag{9.64}$$

Now to analytically continue to real frequencies, we need to transform to Fourier space, writing

$$\chi(q) = \int d^4x \, e^{-iqx} \chi(x), \qquad (9.65)$$

where the integral over time τ runs from 0 to β. This procedure converts the Feynman diagram from a real-space to a momentum-space Feynman diagram. At the measurement vertex at position x, the incoming and outgoing momenta of the fermion line give the following integral:

$$\int d^4x \, e^{-iqx} e^{i(k_{in} - k_{out})x} = \beta V \delta^4(k_{out} - k_{in} + q). \qquad (9.66)$$

As in the case of the free energy, the βV term cancels with the $1/(\beta V) \sum_k$ terms associated with each propagator, leaving behind one factor of $1/(\beta V) = T/V$ per internal momentum loop. Schematically, the effect of the Fourier transform on the measurement vertex at position x is then

$$(9.67)$$

For example, the momentum-dependent spin response function of the free electron gas is given by

$$= -\frac{1}{\beta V} \sum_k \mathrm{Tr} \left[\sigma^a \mathcal{G}(k+q) \sigma^b \mathcal{G}(k) \right] = \delta^{ab} \chi(q) \qquad (9.68)$$

where

$$\chi(\mathbf{q}, i\nu_r) = -2\mu_B^2 \int_{\mathbf{k}} T \sum_{i\omega_n} \mathcal{G}(\mathbf{k}+\mathbf{q}, i\omega_n + i\nu_r) \mathcal{G}(\mathbf{k}, i\omega_n). \qquad (9.69)$$

When we carry out the Matsubara summation in the above expression by a contour integral (see Chapter 8), we obtain

$$-T \sum_{i\omega_n} \mathcal{G}(\mathbf{k}+\mathbf{q}, i\omega_n + i\nu_r) \mathcal{G}(\mathbf{k}, i\omega_n) = -\int_{C'} \frac{dz}{2\pi i} f(z) \mathcal{G}(\mathbf{k}+\mathbf{q}, z + i\nu_r) \mathcal{G}(\mathbf{k}, z)$$

$$= \left(\frac{f_{\mathbf{k}} - f_{\mathbf{k}+\mathbf{q}}}{(\epsilon_{\mathbf{k}+\mathbf{q}} - \epsilon_{\mathbf{k}}) - i\nu_r} \right), \qquad (9.70)$$

where C' encloses the poles of the Green's functions. Inserting this into (9.69), we obtain $\chi(\mathbf{q}, i\nu_r) = \chi(\mathbf{q}, z)|_{z=i\nu_r}$, where

$$\chi(\mathbf{q}, z) = 2\mu_B^2 \int_{\mathbf{k}} \left(\frac{f_{\mathbf{k}} - f_{\mathbf{k}+\mathbf{q}}}{(\epsilon_{\mathbf{k}+\mathbf{q}} - \epsilon_{\mathbf{k}}) - i\nu_r} \right). \qquad (9.71)$$

From this we can also read off the power spectrum of spin fluctuations:

$$\chi''(\mathbf{q}, \omega) = \text{Im}\, \chi(\mathbf{q}, \omega + i\delta) = 2\mu_B^2 \int_{\mathbf{q}} \pi \delta(\epsilon_{\mathbf{q}+\mathbf{k}} - \epsilon_{\mathbf{k}} - \omega) \left[f_{\mathbf{k}} - f_{\mathbf{k}+\mathbf{q}} \right]. \tag{9.72}$$

When we come to consider conductivities, which involve the response function of current operators, we need to know how to deal with an operator that involves spatial or temporal derivatives. To do this, it is convenient to examine the Fourier transform of the operator $A(x)$:

$$\int d^4x\, e^{-iqx} \psi^\dagger(x) A \psi(x) = \sum_{k} \psi^\dagger(k - q/2) A \psi(k + q/2). \tag{9.73}$$

In current operators, A is a function of gradient terms such as $\overset{\leftrightarrow}{\nabla}$ and $\overset{\leftrightarrow}{\partial}_t$. In this case, the use of the symmetrized gradient terms ensures that, when we Fourier transform, the derivative terms are replaced by the midpoint momentum and frequency of the incoming or outgoing electron:

$$\int d^4x\, e^{-iqx} \psi^\dagger(x) A[-i\overset{\leftrightarrow}{\nabla}, i\overset{\leftrightarrow}{\partial}_t] \psi(x) = \sum_{k} \psi^\dagger(k - q/2) A(\mathbf{k}, i\omega_n) \psi(k + q/2). \tag{9.74}$$

For example, the current operator $\vec{J}(x) = \frac{e\hbar}{m}\left(-i\overset{\leftrightarrow}{\nabla}\right)$ becomes

$$J(q) = \sum_{k} e\vec{v}_{\mathbf{k}} \psi^\dagger(k - q/2) \psi(k + q/2), \tag{9.75}$$

where $\vec{v}_{\mathbf{k}} = \frac{\hbar \vec{k}}{m}$ is the electron velocity. For the thermal current operator $\vec{J}_t(\vec{x}) = \frac{\hbar^2}{m}\left(\overset{\leftrightarrow}{\nabla}\overset{\leftrightarrow}{\partial}_t\right)$,

$$\vec{J}_t(q) = \sum_{k} i\omega_n \frac{\hbar^2 \vec{k}}{m} \psi^\dagger(k - q/2) \psi(k + q/2). \tag{9.76}$$

Example 9.2 Calculate the imaginary part of the dynamical susceptibility for non-interacting electrons and show that at low energies, $\omega \ll \epsilon_F$,

$$\frac{\chi''(\mathbf{q}, \omega)}{\omega} = \begin{cases} \mu_B^2 \frac{N(0)}{v_F q} & (q < 2k_F) \\ 0 & (q > 2k_F), \end{cases} \tag{9.77}$$

where $v_F = \hbar k_F/m$ is the Fermi velocity.

Solution

Starting with (9.72) in the low-energy limit, we can write

$$\lim_{\omega \to 0} \frac{\chi''(\mathbf{q}, \omega)}{\omega} = 2\mu_B^2 \int_{\mathbf{q}} \delta(\epsilon_{\mathbf{q}+\mathbf{k}} - \epsilon_{\mathbf{k}}) \frac{f_{\mathbf{k}+\mathbf{q}} - f_{\mathbf{k}}}{\epsilon_{\mathbf{k}} - \epsilon_{\mathbf{k}+\mathbf{q}}}$$

$$= 2\mu_B^2 \int_{\mathbf{q}} \delta(\epsilon_{\mathbf{q}+\mathbf{k}} - \epsilon_{\mathbf{k}}) \left(-\frac{df}{d\epsilon_{\mathbf{k}}}\right). \tag{9.78}$$

Replacing

$$\int_{\mathbf{q}} \to \int d\epsilon N(\epsilon) \int_{-1}^{1} \frac{d\cos\theta}{2}, \tag{9.79}$$

we obtain

$$\lim_{\omega \to 0} \frac{\chi''(\mathbf{q}, \omega)}{\omega} = 2\mu_B^2 N(0) \int_{-1}^{1} \frac{d\cos\theta}{2} \delta\left(\frac{q^2}{2m} + \frac{qk_F}{m}\cos\theta\right)$$

$$= 2\mu_B^2 N(0) \frac{m}{2qk_F} = \mu_B^2 \left(\frac{N(0)}{v_F q}\right) \qquad (q < 2k_F). \qquad (9.80)$$

9.6 Spectroscopy: linking measurement and correlation

The spectroscopies of condensed matter (Table 9.2) provide an essential window on the underlying excitation spectrum, the collective modes, and ultimately the ground-state correlations of the medium. Research in condensed matter depends critically on the creative new interpretations given to measurements. It is from these interpretations that new models can be built and new insights discovered, leading ultimately to quantitative theories of matter.

Understanding the link between experiment and the microscopic world is essential for theorist and experimentalist. At the start of a career, the student is often flung into a seminar room, where it is difficult to absorb the content of the talk because the true meaning of the spectroscopy or measurements is obscure to all but the expert, so it is important to get a rough idea of how and what each measurement technique probes – to know some of the pitfalls of interpretation – and to have an idea about how one begins to calculate the corresponding quantities from simple theoretical models.

Fundamentally, each measurement is related to a given correlation function. This is seen most explicity in scattering experiments. Here, one is sending in a beam of particles and measuring the flux of outgoing particles at a given energy transfer E and momentum transfer $\mathbf{q}$. The ratio of outgoing to incoming particle flux determines the differential scattering cross-section:

$$\frac{d^2\sigma}{d\Omega d\omega} = \frac{\text{outward particle flux}}{\text{inward particle flux}}. \qquad (9.81)$$

When the particles scatter, they couple to some microscopic variable $A(x)$ within the matter, such as the spin density in neutron scattering or the particle field $\psi(x)$ itself in photoemission spectroscopy. The differential scattering cross-section this gives rise to is, in essence, a measure of the autocorrelation function of $A(x)$ at the wavevector $\mathbf{q}$ and frequency $\omega = E/\hbar$ inside the material:

$$\frac{d^2\sigma}{d\Omega d\omega} \sim \int d^4x \, \langle A(\mathbf{x}, t)A(0)\rangle e^{-i(\mathbf{q}\cdot\mathbf{x} - \omega t)} = S(\mathbf{q}, \omega). \qquad (9.82)$$

Remarkably, scattering probes matter at two points in space! How can this be? To understand this, recall that the differential scattering rate is actually an (imaginary) part of the forward scattering amplitude of the incoming particle. The amplitude for the incoming

Table 9.2 Selected spectroscopies.

Type	Name	Spectrum	A	Notes and common measurement issues
ELECTRON	STM[a]	$\dfrac{dI}{dV}(\mathbf{x}) \propto A(\mathbf{x}, \omega)\big|_{\omega = eV}$	$\psi(x)$	Surface probe. $T \sim 0$ measurement. Does the surface characterize the bulk?
	ARPES[b]	$I(\mathbf{k}, \omega) \propto f(-\omega) A(\mathbf{k}, -\omega)$	$c_{\mathbf{k}\sigma}(t)$	$p_{\perp}$ unresolved. Surface probe; no magnetic field possible.
	IPES[c]	$I(\omega) \propto \displaystyle\sum_{\mathbf{k}} [1 - f(\omega)] A(\mathbf{k}, \omega)$	$c_{\mathbf{k}\sigma}^{\dagger}(t)$	**p** unresolved. Surface probe.
SPIN	Magnetic susceptibility	$\chi_{DC} = \displaystyle\int \frac{d\omega}{\pi\omega} \chi''(\mathbf{q} = 0, \omega)$	M	$\chi \sim \frac{1}{T+\theta}$, Curie law: local moments. $\chi \sim$ constant paramagnet.
	Inelastic neutron scattering	$S(\mathbf{q}, \omega) = \dfrac{1}{1 - e^{-\beta\omega}} \chi''(\mathbf{q}, \omega)$	$S(\mathbf{q}, t)$	What is the background? Quality of crystal?
	NMR[d] Knight shift	$K_{contact} \propto \chi_{local}$	$S(\mathbf{x}, t)$	How is the orbital part subtracted?
	Nuclear relaxation rate	$\dfrac{1}{T_1} = T \displaystyle\int_q F(\mathbf{q}) \dfrac{\chi''(\mathbf{q}, \omega)}{\omega}\bigg|_{\omega = \omega_N}$		How does powdering affect sample?
CHARGE	Resistivity	$\rho = \dfrac{1}{\sigma(0)}$	$\vec{j}(\omega = 0)$	How big is the resistance ratio $R(T = 300\,\mathrm{K})/R(T = 0\,\mathrm{K})$ of the sample?
	Optical conductivity	$\sigma(\omega) = \dfrac{1}{-i\omega} \big[\langle j(\omega') j(-\omega') \rangle \big]_0^{\omega}$	$\vec{j}(\omega)$	For optical reflectivity measurements: how was the Kramers–Kronig analysis done? Spectral weight transfer.

[a] Scanning tunneling spectroscopy.
[b] Angle resolved photoemission spectroscopy.
[c] Inverse photoemission spectroscopy.
[d] Nuclear magnetic resonance.

particle to scatter in a forward direction contains the Feynman process where it omits a fluctuation of the quantity A at position x', traveling for a brief period of time as a scattered particle before reabsorbing the fluctuation at x. The amplitude for the intermediate process is nothing more than

$$(9.83)$$

$$\text{amplitude} = \overbrace{\langle A(x)A(x')\rangle}^{\text{amplitude for fluctuation}} \times \underbrace{e^{i[\mathbf{q}\cdot(\mathbf{x}-\mathbf{x}')-\omega(t-t')]}}_{\substack{\text{amplitude for particle to scatter at x'} \\ \text{and reabsorb fluctuation at x}}}. \qquad (9.84)$$

(In practice, since the whole process is translationally invariant, we can replace x by $x - x'$ and set $x' = 0$.)

The relationship between the correlation function and the scattering rate is really a natural consequence of Fermi's golden rule, according to which

$$\frac{d^2\sigma}{d\Omega d\omega} \sim \Gamma_{i\to f} = \frac{2\pi}{\hbar} \sum_f p_i |\langle f|V|i\rangle|^2 \delta(E_f - E_i), \qquad (9.85)$$

where p_i is the probability of being in the initial state $|i\rangle$. Typically, an incoming particle (photon, electron, neutron) with momentum $\mathbf{k}$ scatters into an outgoing particle state (photon, electron, neutron) with momentum $\mathbf{k}' = \mathbf{k} - \mathbf{q}$, and the system undergoes a transition from a state $|\lambda\rangle$ to a final state $|\lambda'\rangle$:

$$|i\rangle = |\lambda\rangle|\mathbf{k}\rangle, \qquad |f\rangle = |\lambda'\rangle|\mathbf{k}'\rangle. \qquad (9.86)$$

If the scattering Hamiltonian is $V \sim g \int_{\mathbf{x}} \rho(\mathbf{x})A(\mathbf{x})$, where $\rho(\mathbf{x})$ is the density of the particle beam, then the scattering matrix element is

$$\langle f|\hat{V}|i\rangle = g\int_{\mathbf{x}'} \langle \mathbf{k}'|\mathbf{x}'\rangle\langle\lambda'|A(\mathbf{x}')|\lambda\rangle\langle\mathbf{x}'|\mathbf{k}\rangle = \frac{g}{V_0}\int_{\mathbf{x}'} e^{i\mathbf{q}\cdot\mathbf{x}'}\langle\lambda'|A(\mathbf{x}')|\lambda\rangle, \qquad (9.87)$$

so the scattering rate is

$$\Gamma_{i\to f} = \frac{g^2}{V_0^2}\int_{\mathbf{x},\,\mathbf{x}'} p_\lambda\langle\lambda|A(\mathbf{x})|\lambda'\rangle\langle\lambda'|A(\mathbf{x}')|\lambda\rangle e^{-i\mathbf{q}\cdot(\mathbf{x}-\mathbf{x}')}2\pi\delta(E_{\lambda'} - E_\lambda - \omega), \qquad (9.88)$$

where $p_\lambda = e^{-\beta(E_\lambda - F)}$ is the Boltzmann probability. Now if we repeat the spectral decomposition of the correlation function made in (9.39),

$$\int dt e^{i\omega t}\langle A(\mathbf{x},t)A(\mathbf{x}',0)\rangle = 2\pi\sum_{\lambda,\lambda'} p_\lambda\langle\lambda|A(x)|\lambda'\rangle\langle\lambda'|A(\mathbf{x}')|\lambda\rangle\delta(E_{\lambda'} - E_\lambda - \omega), \qquad (9.89)$$

we see that

$$\Gamma_{i\to f} \sim \frac{g^2}{V_0^2}\int_{\mathbf{x},\mathbf{x}'} dt\, e^{i\omega t}\langle A(\mathbf{x},t)A(\mathbf{x}',0)\rangle e^{-i\mathbf{q}\cdot(\mathbf{x}-\mathbf{x}')}$$

$$= \frac{g^2}{V_0}\int d^3x\, dt\, e^{-i(\mathbf{q}\cdot\mathbf{x}-\omega t)}\langle A(\mathbf{x},t)A(0)\rangle, \qquad (9.90)$$

where the last simplification results from translational invariance. Finally, if we divide the transition rate by the incoming flux of particles $\sim 1/V_0$, we obtain the differential scattering cross-section.

For example, in an inelastic neutron scattering (INS) experiment, the neutrons couple to the electron spin density $A = S(x)$ of the material, so that

$$\frac{d^2\sigma}{d\Omega d\omega}(\mathbf{q}, \omega) \sim \int d^4x \langle S_-(\mathbf{x}, t)S_+(0)\rangle e^{-i(\mathbf{q}\cdot\mathbf{x}-\omega t)} \quad \propto \quad \frac{1}{1-e^{-\beta\omega}}\chi''(\mathbf{q}, \omega), \quad (9.91)$$

where $\chi(\mathbf{q}, \omega)$ is the dynamical spin susceptibility, which determines the magnetization $M(\mathbf{q}, \omega) = \chi(\mathbf{q}, \omega)B(\mathbf{q}, \omega)$ by a modulated magnetic field of wavevector $\mathbf{q}$, frequency ω. By contrast, in an angle-resolved photoemission (ARPES) experiment, incoming X-rays eject electrons from the material, leaving behind holes, so that $A = \psi$ is the electron annihilation operator and the intensity of emitted electrons measures the correlation function:

$$I(\mathbf{k}, \omega) \sim \int d^4x \langle \psi^\dagger(x)\psi(0)\rangle e^{-i(\mathbf{k}\cdot\mathbf{x}-\omega t)} = \overbrace{\frac{1}{1+e^{\beta\omega}}}^{f(-\omega)} A(\mathbf{k}, -\omega), \quad (9.92)$$

where the Fermi function replaces the Bose function in the fluctuation–dissipation theorem.

9.7 Electron spectroscopy

9.7.1 Formal properties of the electron Green's function

The spectral decomposition carried out for a bosonic variable A is simply generalized to a fermionic variable such as $c_{\mathbf{k}\sigma}$. The basic electron correlation functions are

$$\langle c_{\mathbf{k}\sigma}(t)c_{\mathbf{k}\sigma}^\dagger(0)\rangle = \int \frac{d\omega}{2\pi} G_>(\mathbf{k}, \omega)e^{-i\omega t}$$

$$\langle c_{\mathbf{k}\sigma}^\dagger(0)c_{\mathbf{k}\sigma}(t)\rangle = \int \frac{d\omega}{2\pi} G_<(\mathbf{k}, \omega)e^{-i\omega t}, \quad (9.93)$$

called the *greater* and *lesser Green's functions*. A spectral decomposition of these relations reveals that

$$G_>(\mathbf{k}, \omega) = \sum_{\lambda, \zeta} p_\lambda |\langle \zeta|c_{\mathbf{k}\sigma}^\dagger|\lambda\rangle|^2 2\pi \delta(E_\zeta - E_\lambda - \omega)$$

$$G_<(\mathbf{k}, \omega) = \sum_{\lambda, \zeta} p_\lambda |\langle \zeta|c_{\mathbf{k}\sigma}|\lambda\rangle|^2 2\pi \delta(E_\zeta - E_\lambda + \omega) \quad (9.94)$$

describe the positive energy distribution functions for particles ($G_>$) and the negative energy distribution function for holes ($G_<$), respectively. By relabeling $\zeta \leftrightarrow \lambda$ in (9.94) it is straightforward to show that

$$G_<(\mathbf{k}, \omega) = e^{-\beta\omega}G_>(\mathbf{k}, \omega). \quad (9.95)$$

We also need to introduce the retarded electron Green's function, given by

$$G_R(\mathbf{k}, t) = -i\langle\{c_{\mathbf{k}\sigma}(t), c_{\mathbf{k}\sigma}^\dagger(0)\}\rangle\theta(t) = \int \frac{d\omega}{2\pi} G_R(\mathbf{k}, \omega)e^{-i\omega t} \qquad (9.96)$$

(note the appearance of an anticommutator for fermions and the minus sign prefactor), which is the real-time analogue of the imaginary-time Green's function

$$\mathcal{G}(\mathbf{k}, \tau) = -\langle Tc_{\mathbf{k}\sigma}(\tau)c_{\mathbf{k}\sigma}^\dagger(0)\rangle = T\sum_n \mathcal{G}(\mathbf{k}, i\omega_n)e^{-i\omega_n\tau}. \qquad (9.97)$$

A spectral decomposition of these two functions reveals that they share the same power spectrum and Kramers–Kronig relation, and can both be related to the generalized Green's function

$$\mathcal{G}(\mathbf{k}, z) = \int d\omega \frac{1}{z - \omega} A(\mathbf{k}, \omega), \qquad (9.98)$$

where

$$G_R(\mathbf{k}, \omega) = \mathcal{G}(\mathbf{k}, \omega + i\delta) = \int d\omega' \frac{1}{\omega - \omega' + i\delta} A(\mathbf{k}, \omega)$$

$$\mathcal{G}(\mathbf{k}, i\omega_n) = \mathcal{G}(\mathbf{k}, z)|_{z=i\omega_n} = \int d\omega' \frac{1}{i\omega_n - \omega} A(\mathbf{k}, \omega'). \qquad (9.99)$$

The spectral function $A(\mathbf{k}, \omega) = \frac{1}{\pi}G(\mathbf{k}, \omega - i\delta)$ is then given by

$$A(\mathbf{k}, \omega) = \sum_{\lambda,\xi} p_\lambda \left[\overbrace{|\langle\xi|c_{\mathbf{k}\sigma}^\dagger|\lambda\rangle|^2\delta(\omega - E_\xi + E_\lambda)}^{\text{electron addition}} + \overbrace{|\langle\xi|c_{\mathbf{k}\sigma}|\lambda\rangle|^2\delta(\omega + E_\xi - E_\lambda)}^{\text{electron removal}} \right]$$

$$= \frac{1}{2\pi}[G_>(\mathbf{k}, \omega) + G_<(\mathbf{k}, \omega)], \qquad (9.100)$$

the sum of particle and hole energy distribution functions. From (9.93) and the second line of (9.100), it follows that $A(\mathbf{k}, \omega)$ is the Fourier transform of the anticommutator

$$\langle\{c_{\mathbf{k}\sigma}(t), c_{\mathbf{k}\sigma}^\dagger(0)\}\rangle = \int d\omega A(\mathbf{k}, \omega)e^{-i\omega t}. \qquad (9.101)$$

At equal times, the commutator is equal to unity, $\{c_{\mathbf{k}\sigma}, c_{\mathbf{k}\sigma}^\dagger\} = 1$, from which we deduce the normalization

$$\int d\omega\, A(\mathbf{k}, \omega) = 1. \qquad (9.102)$$

For non-interacting fermions, the spectral function is a pure delta function, but in Fermi liquids, at emergies $\omega \sim 0$ near the Fermi energy, the delta function is renormalized by a factor Z and the remainder of the spectral weight is transfered to an incoherent background (Figure 9.2):

$$A(\mathbf{k}, \omega) = Z_\mathbf{k}\delta(\omega - E_\mathbf{k}) + \text{background}. \qquad (9.103)$$

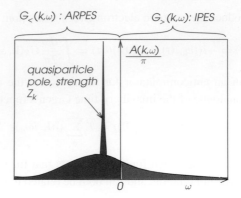

Fig. 9.2 Showing the redistribution of the quasiparticle weight into an incoherent background in a Fermi liquid.

The relations

$$G_>(\mathbf{k}, \omega) = \frac{2\pi}{1 + e^{-\beta\omega}} A(\mathbf{k}, \omega) = 2\pi(1 - f(\omega))A(\mathbf{k}, \omega) \qquad \text{particles}$$

$$G_<(\mathbf{k}, \omega) = \frac{2\pi}{1 + e^{\beta\omega}} A(\mathbf{k}, \omega) = 2\pi f(\omega)A(\mathbf{k}, \omega) \qquad \text{holes} \qquad (9.104)$$

are the fermion analogue of the fluctuation–dissipation theorem.

9.7.2 Tunneling spectroscopy

Tunneling spectroscopy is one of the most direct ways of probing the electron spectral function. The basic idea behind tunneling spectroscopy is that a tunneling probe is close enough to the surface that electrons can tunnel through the forbidden region between the probe and surface material. Traditionally, tunneling was carried out using point-contact spectroscopy, whereby a sharp probe is brought into contact with the surface, and tunneling takes place through the oxide layer separating probe and surface. The invention of the scanning tunneling microscope by Gerd Binnig and Heinrich Rohrer [5, 6] in the 1980s has revolutionized the field. In recent times this tool has developed into a practical workhorse that, under ideal surface conditions, permits the spectral function of electrons to be mapped out with sub-angstrom and sub-millivolt resolution across the surface of a conductor [7, 8] (see Figure 9.3).

In the WKB approximation [9, 10], the amplitude for an electron to tunnel between probe and surface is

$$t(x_1, x_2) \sim \exp\left[-\frac{1}{\hbar} \int_{x_1}^{x_2} \sqrt{2m[U(x) - E]}ds\right], \qquad (9.105)$$

where the integral is evaluated along the saddle-point path between probe and surface. The exponential dependence of this quantity on distance means that tunneling is dominated by the extremal path from a single atom at the end of a scanning probe, giving rise to angstrom-level spatial resolution.

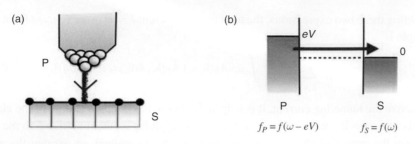

(a) Schematic diagram showing tunneling from probe (P) to sample (S). (b) Energy level occupancy of probe and sample. In this figure, we have assumed $e\phi = -|e|\phi > 0$. The red arrow shows the electron tunnel current.

Fig. 9.3

The tunneling Hamiltonian governing the flow of electrons between the probe and the sample can be written

$$\hat{V} = \sum_{\mathbf{k},\mathbf{k}'} t_{\mathbf{k},\mathbf{k}'} \left[c^\dagger_{\mathbf{k}\sigma} p_{\mathbf{k}'\sigma} + \text{H.c.} \right], \tag{9.106}$$

where $t_{\mathbf{k},\mathbf{k}'}$ is the tunneling matrix element between the probe and substrate, and $c^\dagger_{\mathbf{k}\sigma}$ and $p^\dagger_{\mathbf{k}\sigma}$ create electrons in the sample and the probe, respectively. The particle current of electrons from probe to sample is given by

$$i_{P\to S} = \frac{2\pi}{\hbar} \sum_{\mathbf{k},\mathbf{k}',\zeta,\zeta',\lambda,\lambda',\delta} p_\lambda p_{\lambda'} |t_{\mathbf{k},\mathbf{k}'}|^2 |\langle \zeta,\zeta'|c^\dagger_{\mathbf{k}\sigma} p_{\mathbf{k}'\sigma}|\lambda,\lambda'\rangle|^2 \delta(E_\zeta + E_{\zeta'} - E_\lambda - E_{\lambda'}), \tag{9.107}$$

where $|\lambda,\lambda'\rangle \equiv |\lambda\rangle|\lambda'\rangle$ and $|\zeta,\zeta'\rangle \equiv |\zeta\rangle|\zeta'\rangle$ refer to the joint many-body states of the sample (unprimed) and probe (primed), and p_λ and $p_{\lambda'}$ are the occupation probabilities of the initial states of the sample and probe. This term describes the rate of creation of electrons in the sample.

Now if we rewrite this expression in terms of the spectral functions of the probe and sample, after a little work we obtain

$$i_{P\to S} = \frac{4\pi}{\hbar} \sum_{\mathbf{k},\mathbf{k}'} |t_{\mathbf{k},\mathbf{k}'}|^2 \int d\omega A_S(\mathbf{k},\omega) \tilde{A}_P(\mathbf{k}',\omega)(1 - f(\omega)) f_P(\omega), \tag{9.108}$$

where $\tilde{A}_P(\mathbf{k},\omega)$ and $f_P(\omega)$ are the spectral function and distribution function of the voltage-biased probe, and the factor of 4π includes the two-fold spin degeneracy. You can check the validity of these expressions by expanding the spectral functions using (9.100), but the expression is simply recognized as a product of matrix element, density of states, and Fermi–Dirac electron–hole factors.

Similarly, the reverse particle current of electrons from sample to probe is

$$i_{S\to P} = \frac{2\pi}{\hbar} \sum_{\mathbf{k},\mathbf{k}',\zeta,\zeta',\lambda,\lambda',\sigma} p_\lambda p_{\lambda'} |t_{\mathbf{k},\mathbf{k}'}|^2 |\langle \zeta,\zeta'|p^\dagger_{\mathbf{k}'\sigma} c_{\mathbf{k}\sigma}|\lambda,\lambda'\rangle|^2 \delta(E_\zeta + E_{\zeta'} - E_\lambda - E_{\lambda'})$$

$$= \frac{4\pi}{\hbar} \sum_{\mathbf{k},\mathbf{k}'} |t_{\mathbf{k},\mathbf{k}'}|^2 \int d\omega A_S(\mathbf{k},\omega) \tilde{A}_P(\mathbf{k}',\omega)[1 - f_P(\omega)] f(\omega). \tag{9.109}$$

Subtracting these two expressions, the total electric current $I = e(i_{P \to S} - i_{S \to P})$ from probe to sample is

$$I = e \frac{4\pi}{\hbar} \sum_{\mathbf{k}, \mathbf{k}'} |t_{\mathbf{k}, \mathbf{k}'}|^2 \int d\omega A_S(\mathbf{k}, \omega) \, \tilde{A}_P(\mathbf{k}', \omega) [f_P(\omega) - f(\omega)]. \qquad (9.110)$$

To derive the tunneling current, it is helpful to ignore the negative charge of the electron and to continue as if e were positive. The effect of applying a potential bias ϕ to the probe is to raise the energy of the electrons in the probe by an amount $e\phi$, so that the energy distribution functions $f_P(\omega)$ and $\tilde{A}_P(\mathbf{k}, \omega)$ in the probe are shifted upwards in energy by an amount $e\phi$ with respect to their unbiased values. This means that the probe Fermi energy is at $\omega = e\phi$. In other words, $f_P(\omega) = f(\omega - e\phi)$ $(e = -|e|)$ and $\tilde{A}_P(\mathbf{k}', \omega) = A_P(\mathbf{k}', \omega - e\phi)$ (see Figure 9.3), so that

$$I = \frac{4\pi e}{\hbar} \sum_{\mathbf{k}, \mathbf{k}'} |t_{\mathbf{k}, \mathbf{k}'}|^2 \int d\omega A_S(\mathbf{k}, \omega) \, A_P(\mathbf{k}', \omega - e\phi) [f(\omega - e\phi) - f(\omega)] \qquad (e = -|e|).$$
$$(9.111)$$

We shall ignore the momentum dependence of the tunneling matrix elements, writing $|t|^2 = \overline{|t_{\mathbf{k}, \mathbf{k}'}|^2}$ and $\sum_{\mathbf{k}'} A(\mathbf{k}', \omega) = N(0)$, the density of states in the probe, to obtain

$$I(\phi, \mathbf{x}) = \frac{2e}{\hbar} \overbrace{2\pi |t|^2 N(0)}^{\Gamma} \int d\omega A_S(\omega, \mathbf{x}) [f(\omega - e\phi) - f(\omega)]$$
$$= \frac{2e\Gamma}{\hbar} \int_0^{e\phi} d\omega A_S(\omega, \mathbf{x}) [f(\omega - e\phi) - f(\omega)], \qquad (9.112)$$

where

$$A_S(\omega, \mathbf{x}) = \sum_{\mathbf{k}} A_S(\mathbf{k}, \omega; \mathbf{x}) \qquad (9.113)$$

is the *local* spectral function at position $\mathbf{x}$ on the surface of the sample. Typically, the probe is a metal with a featureless density of states, and this justifies the replacement $N(\omega) \sim N(0)$ in the above expression. The quantity $2\pi t^2 N(0) = \Gamma$ is the characteristic resonance broadening width created by tunneling out of the probe. If we now differentiate the current with respect to the applied potential, we obtain the differential conductivity,

$$G(\phi, \mathbf{x}) = \frac{dI}{d\phi} = \left(\frac{2e^2}{\hbar} \right) \Gamma \int d\omega A_S(\omega, \mathbf{x}) \overbrace{\left(-\frac{df(\omega - e\phi)}{d\omega} \right)}^{\delta(\omega - e\phi)}. \qquad (9.114)$$

At low temperatures, the derivative of the Fermi function gives a delta function in energy, so that

$$G(\phi, \mathbf{x}) = \left(\frac{2e^2 \Gamma}{\hbar} \right) A_S(\omega, \mathbf{x}) \Big|_{\omega = e\phi}. \qquad (9.115)$$

Thus by mapping out the differential conductance as a function of position, it is possible to obtain a complete spatial map of the local density of states (DOS) on the surface of the sample.

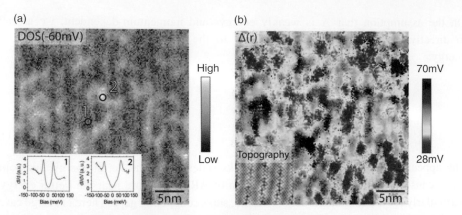

(a) Local density of states (DOS) measured by STM at -60 meV on optimally doped $Bi_2Sr_2CaCu_2O_{8+x}$ in a 30 nm $\times$ 30 nm field of view. The experiment was conducted at 5K. Inset: two spectra are shown, for locations 1 and 2. (b) Map of the gap as a function of position (called a "gap map") for the same field of view as (a). Inset: surface topography of the same sample. Figures reprinted with permission from Yang He and Jennifer Hoffman, Harvard University.

Fig. 9.4

Figure 9.4 shows two STM spectra taken on the surface of the high-temperature super-conductor $Bi_2Sr_2CaCu_2O_{8+x}$ ("BISCO") over a 30 nm square region, illustrating how STM measurements can image and quantify the atomic-scale inhomogeneities in the superconducting gap function and density of states of these materials.

9.7.3 ARPES, AIPES, and inverse PES

ARPES (angle resolved photoemission spectroscopy), AIPES (angle integrated photoemission spectroscopy), and inverse PES (inverse photoemission spectrosopy) are alternative ways of probing the hole and electron spectra in matter. The first two involve "photon in, electron out," the second "electron in, photon out." The coupling of radiation to light involves the dipole coupling term

$$H_I = -\int d^3x \vec{j}(x) \cdot \vec{A}(x), \qquad (9.116)$$

where $\vec{j}(x) = i\frac{e\hbar}{2m}\psi_\sigma^\dagger(x)\vec{\nabla}\psi_\sigma(x)$ is the paramagnetic electron current operator. Unlike STM or neutron scattering, this is a strongly coupled interaction, and the assumption that we can use Fermi's golden rule to relate the absorption to a correlation function is on much shakier ground. ARPES spectroscopy involves the absorption of a photon and the emission of a photoelectron from the material. The interpretation of ARPES spectra is based on the *sudden approximation*, namely that the dipole matrix element between the intial and final states has a slow dependence on the incoming photon energy and momentum, so that the matrix element is

$$\langle \zeta, \mathbf{k} + \mathbf{q} | -\vec{j} \cdot \vec{A} | \lambda, \mathbf{q} \rangle \sim \Lambda(\mathbf{q}, \hat{e}_\lambda)\langle \zeta | c_{\mathbf{k}\sigma} | \lambda \rangle. \qquad (9.117)$$

On the assumption that Λ is weakly energy- and momentum-dependent, we are able to directly relate the absorption intensity to the spectral density beneath the Fermi energy:

$$I_{ARPES}(\mathbf{k}, \omega) \propto f(-\omega)A(\mathbf{k}, -\omega) \qquad (9.118)$$

The appearance of the Fermi function masks states above the Fermi energy, and sometimes causes problems for the interpretation of ARPES spectra near the Fermi energy – particularly for the estimation of anisotropic, superconducting gaps. There is a large caveat to go with this equation: when photoelectrons escape from a surface, the component of their momentum perpendicular to the surface is modified by interactions with the surface. Consequently, ARPES spectroscopy cannot resolve the momenta of the spectral function perpendicular to the surface. The other consideration about ARPES is that it is essentially a surface probe: X-ray radiation has only the smallest ability to penetrate samples, so the information obtained by these methods provides strictly a surface probe of the system.

In recent years, tremendous strides have taken place in the resolution of ARPES, in large part because of the interest in probing the electron spectrum of quasi-two-dimensional cuprate superconductors. These methods have, for example, played an important role in exhibiting the anisotropic d-wave gap of these materials.

Inverse photoemission spectroscopy probes the spectral function above the Fermi energy. At present, angle resolved IPES is not as well developed, and most IPES work involves unresolved momenta:

$$I_{IPES}(\omega) \propto \sum_{\mathbf{k}}[1 - f(\omega)]A(\mathbf{k}, \omega) \qquad (9.119)$$

In certain materials, both PES and IPES spectra are available. A classic example is the spectroscopy of mixed-valent cerium compounds. In these materials, the Ce atoms have a singly occupied f-level, in the $4f^1$ configuration. PES spectroscopy can resolve the energy for the hole excitation

$$4f^1 \rightarrow 4f^0 + e^-, \qquad \Delta E_I = -E_f, \qquad (9.120)$$

where E_f is the energy of a singly occupied $4f$ level. By contrast, inverse PES reveals the energy to add an electron to the $4f^1$ state,

$$e^- + 4f^1 \rightarrow 4f^2, \qquad \Delta E_{II} = E_f + U, \qquad (9.121)$$

where U is the size of the Coulomb interaction between two electrons in an f-state. By comparing these two absorption energies, it is possible to determine the size of the Coulomb interaction energy (see Section 16.6.1).

9.8 Spin spectroscopy

9.8.1 DC magnetic susceptibility

If one measures the static DC magnetization of a medium, one is measuring the magnetic response at zero wavevector, $\mathbf{q} = 0$, and zero frequency, $\omega = 0$. By the Kramers–Kronig relation encountered in (9.44), we know that

$$\chi_{DC} = \int \frac{d\omega}{\pi} \frac{\chi''(\mathbf{q} = 0, \omega)}{\omega}. \qquad (9.122)$$

So the static magnetic susceptibility is an economy-class measurement of the magnetic fluctuation power spectrum at zero wavevector. Indeed, this link between the two measurements sometimes provides an important consistency check for neutron scattering experiments.

In static susceptibility measurements, there are two important limiting classes of behavior: Pauli paramagnetism, in which the susceptibility is derived from the polarization of a Fermi surface and is weakly temperature-dependent:

$$\chi \sim \frac{\mu_B^2}{\epsilon_F} \sim \text{constant}, \qquad \text{Pauli paramagnetism} \qquad (9.123)$$

and Curie paramagnetism, produced by unpaired electrons localized inside atoms, commonly known as *local moment*, where the magnetic susceptibilty is inversely proportional to the temperature, or, more generally,

$$\chi(T) \sim n_i \overbrace{\left(\frac{g^2 \mu_B^2 j(j+1)}{3} \right)}^{M_{\textit{eff}}^2} \times \frac{1}{T + T^*}, \qquad \text{local-moment paramagnetism} \quad (9.124)$$

where n_i is the concentration of local moments and $M_{\textit{eff}}^2$ is the effective moment produced by a moment of total angular momentum j with gyromagnetic ratio g. T^* is a measure of the interaction between local moments. For ferromagnets, $T^* = -T_c < 0$, and ferromagnetic magnetic order sets in at $T = T_c$, where the uniform magnetic susceptibility diverges. For antiferromagnets, $T^* > 0$ gives a measure of the strength of interaction between the local moments.

9.8.2 Neutron scattering

Neutrons interact weakly with matter, so that, unlike electrons or photons, they provide an ideal probe of the bulk properties of matter. Neutrons interact with atomic nuclei via an interaction of the form

$$\hat{H}_I = \alpha \int d^3 x \psi_N^\dagger(x) \psi_N(x) \rho(x), \qquad (9.125)$$

where $\rho(x)$ is the density of nuclei and $\psi_N(x)$ is the field of the neutrons. This interaction produces unpolarized scattering of the neutrons, with an inelastic scattering cross-section of the form (see Example 9.3)

$$\frac{d^2\tilde{\sigma}}{d\Omega dE} = \frac{k_f}{k_i}\left(\frac{\alpha m_N}{2\pi\hbar^2}\right)^2 S(\mathbf{q}, E),\tag{9.126}$$

where $S(\mathbf{q}, E)$ is the autocorrelation function of nuclear density fluctuations in the medium. Where do these come from? They are of course produced by phonons in the crystal. The neutrons transfer energy to the nuclei by exciting phonons, and we expect that

$$S(\mathbf{q}, E) \sim (1 + n_B(E))\delta(E - \hbar\omega_{\mathbf{q}}),\tag{9.127}$$

where $\omega_{\mathbf{q}}$ is the phonon dispersion spectrum inside the medium.

The second important interaction between neutrons and matter is that between the nuclear moment and the magnetic fields inside the material. The magnetic moment of the neutron is given by

$$\vec{M} = \gamma\mu_N\frac{\vec{\sigma}}{2},\tag{9.128}$$

where $\gamma = -1.91$ is the gyromagnetic ratio of the neutron and $\mu_N = \frac{|e|\hbar}{2m_N}$ is the neutron Bohr magneton. The interaction with the fields inside the material is then given by

$$\hat{H}_I = -\frac{\gamma\mu_N}{2}\int d^3x\,\psi_N^\dagger(x)\vec{\sigma}\,\psi_N(x)\cdot\vec{B}(x).\tag{9.129}$$

The magnetic field inside matter has two sources: the dipole field generated by the electron spins and the orbital field produced by the motion of electrons. We will only discuss the spin component here. The dipole magnetic field produced by spins is given by

$$\vec{B}(\mathbf{x}) = \int d^3x'\underline{V}(\mathbf{x} - \mathbf{x}')\cdot\vec{M}(\mathbf{x}'),\tag{9.130}$$

where $\vec{M}(\mathbf{x}) = \mu_B\psi^\dagger(\mathbf{x})\vec{\sigma}\,\psi(\mathbf{x})$ is the electron spin density, and the linear operator

$$\underline{V}(\mathbf{x} - \mathbf{x}') = -(\tilde{\nabla}\times\tilde{\nabla}\times)\frac{\mu_0}{4\pi|\mathbf{x} - \mathbf{x}'|}.\tag{9.131}$$

We can readily Fourier transform this expression by making the replacements

$$\tilde{\nabla}\to i\vec{q},\qquad\frac{1}{(4\pi|x|)}\to\frac{1}{q^2},\tag{9.132}$$

so that, in Fourier space,

$$\left[\underline{V}(\mathbf{q})\right]_{ab} = \mu_0\left[\vec{q}\times\vec{q}\times\left(\frac{1}{q^2}\right)\right]_{ab} = \mu_0\left[\hat{q}\times\hat{q}\times\right]_{ab}$$

$$= \mu_0\overbrace{\left[\delta_{ab} - \hat{q}_a\hat{q}_b\right]}^{\mathcal{P}_{ab}(\hat{q})}.\tag{9.133}$$

The only effect of the complicated dipole interaction is to remove the component of the spin parallel to the q-vector. The interaction between the neutron and electron spin density is simply written

$$H_I = g\int_{\mathbf{q}}\sigma_N(-\mathbf{q})\underline{P}(\hat{q})\cdot\vec{S}_e(\mathbf{q})\qquad(g = \mu_0\gamma\mu_N\mu_B).\tag{9.134}$$

Apart from the projector term, this is essentially a point interaction between the neutron and electron spin densities. Using this result, we can easily generalize our earlier expression for the nuclear differential scattering to the case of unpolarized neutron scattering by replacing $\alpha \to g$ and identifying

$$S_\perp(\mathbf{q}, E) = \mathcal{P}_{ab}(\hat{q})S^{ab}(\mathbf{q}, E) \tag{9.135}$$

as the projection of the spin–spin correlation function perpendicular to the q-vector. For unpolarized neutrons, the differential scattering cross-section is then

$$\frac{d^2\tilde{\sigma}}{d\Omega dE} = \frac{k_f}{k_i} r_o^2 S_\perp(\mathbf{q}, E), \tag{9.136}$$

where

$$r_0 = \left(\frac{gm_N}{2\pi\hbar^2}\right) = \frac{\gamma}{2} \overbrace{\left(\frac{\mu_0}{4\pi}\right)}^{\frac{1}{4\pi\epsilon_0 c^2}} \frac{e^2}{m}$$

$$= \left(\frac{\gamma}{2}\right) \frac{e_{cgs}^2}{mc^2} \tag{9.137}$$

is, apart from the prefactor, the classical radius of the electron.

Example 9.3 Calculate, in the imaginary-time formalism, the self-energy of a neutron interacting with matter, and use this to compute the differential scattering cross-section. Assume the interaction between the neutron and matter is given by

$$\hat{H}_I = \alpha \int d^3x \psi_N^\dagger(x)\psi_N(x)\rho(x), \tag{9.138}$$

where $\psi_N(x)$ is the neutron field and $\rho(x)$ is the density of nuclear matter.

Solution

We begin by noting that the the real-space self-energy of the neutron is given by

$$\Sigma(x - x') = \alpha^2 \langle \delta\rho(x)\delta\rho(x')\rangle \mathcal{G}(x - x'), \tag{9.139}$$

where $\langle \delta\rho(x)\delta\rho(x')\rangle = \chi(x - x')$ is the imaginary-time density response function of the nuclear matter. (Note that the minus sign in $-\alpha^2$ associated with the vertices is absent because the propagator used here, $\langle \delta\rho(x)\delta\rho(0)\rangle$, contains no minus sign prefactor.) If we Fourier transform this expression, we obtain

$$\Sigma(k) = \frac{\alpha^2}{\beta V}\sum_q \mathcal{G}(k - q)\chi(q)$$

$$= \alpha^2 \int_q T\sum_{i\nu_n} \mathcal{G}(k - q)\chi(q). \tag{9.140}$$

Carrying out the Matsubara summation, we obtain

$$\Sigma(\mathbf{k}, z) = \alpha^2 \int_q \frac{dE'}{\pi} \frac{1 + n(E') - f_{\mathbf{k}-\mathbf{q}}}{z - (E_{\mathbf{k}-\mathbf{q}} + E')} \chi''(\mathbf{q}, E'), \tag{9.141}$$

where $E_{\mathbf{k}}$ is the kinetic energy of the neutron and the Fermi function $f_{\mathbf{k}}$ of the neutron can ultimately be set to zero (there is no Fermi sea of neutrons), $f_{\mathbf{k}} \to 0$, so that

$$\Sigma(\mathbf{k}, z) = \alpha^2 \int_{\mathbf{q}} \frac{dE'}{\pi} \frac{1}{z - (E_{\mathbf{k-q}} + E')} \overbrace{(1 + n(E'))\chi''(\mathbf{q}, E')}^{S(\mathbf{q},E)}. \qquad (9.142)$$

From the imaginary part of the self-energy we deduce that the lifetime τ of the neutron is given by

$$\frac{1}{\tau} = \frac{2}{\hbar} \operatorname{Im} \Sigma(\mathbf{k}, E_{\mathbf{k}} - i\delta) = \frac{2\alpha^2}{\hbar} \int_{\mathbf{k}'} S(\mathbf{k} - \mathbf{k}', E_{\mathbf{k}} - E_{\mathbf{k}'}), \qquad (9.143)$$

where we have changed the momentum integration variable from $\mathbf{q}$ to $\mathbf{k}' = \mathbf{k} - \mathbf{q}$. Splitting the momentum integration into an integral over solid angle and an integral over energy, we have

$$\int_{\mathbf{k}'} = \int \left(\frac{m_N k_f}{8\pi^2 \hbar^2} \right) dE' d\Omega', \qquad (9.144)$$

from which we deduce that the mean free path l of the neutron is given by

$$\frac{1}{l} = \frac{1}{v_N \tau} = \frac{1}{v_N} 2 \operatorname{Im} \Sigma(\mathbf{k}, E_{\mathbf{k}} - i\delta) = \int d\Omega_{\mathbf{k}'} dE_{\mathbf{k}'} \times \left[\frac{k_f}{k_i} \left(\frac{\alpha m_N}{2\pi \hbar^2} \right)^2 S(\mathbf{q}, E) \right], \qquad (9.145)$$

where $\mathbf{q} = \mathbf{k} - \mathbf{k}'$ and $E = E_{\mathbf{k}} - E_{\mathbf{k}'}$, and $v_N = \hbar k_i / m_N$ is the incoming neutron velocity.

Normally we write $l = 1/(n_i \sigma)$, where σ is the cross-section of each scatterer and n_i is the concentration of scattering centers. Suppose $\tilde{\sigma} = n_i \sigma$ is the scattering cross-section per unit volume. Then $\tilde{\sigma} = 1/l$, so it follows that

$$\tilde{\sigma} = \frac{1}{v_N} 2 \operatorname{Im} \Sigma(\mathbf{k}, E_{\mathbf{k}} - i\delta) = \int d\Omega_{\mathbf{k}'} dE_{\mathbf{k}'} \times \left[\frac{k_f}{k_i} \left(\frac{\alpha m_N}{2\pi \hbar^2} \right)^2 S(\mathbf{q}, E) \right], \qquad (9.146)$$

from which we may identify the differential scattering cross-section as

$$\frac{d^2 \tilde{\sigma}}{d\Omega dE} = \frac{k_f}{k_i} \left(\frac{\alpha m_N}{2\pi \hbar^2} \right)^2 S(\mathbf{q}, E). \qquad (9.147)$$

9.8.3 Nuclear magnetic resonance

Knight shift, K

Nuclear magnetic resonance (NMR) or magnetic resonance imaging (MRI), as it is more commonly referred to in medical usage, is the use of nuclear magnetic absorption lines to probe the local spin environment in a material. The basic idea is that the Zeeman interaction of a nuclear spin in a magnetic field gives rise to a resonant absorption line in the microwave domain. The interaction of the nucleus with surrounding spins and orbital moments produces a *Knight shift* of this line, and it also broadens the line, giving it a width that is associated with the nuclear spin relaxation rate, $1/T_1$.

The basic Hamiltonian describing a nuclear spin is

$$H = -\mu_n \vec{I} \cdot \vec{B} + H_{hf}, \qquad (9.148)$$

where $\vec{I}$ is the nuclear spin and μ_n is the nuclear magnetic moment. The term H_{hf} describes the *hyperfine* interaction between the nuclear spin and surrounding spin degrees of freedom. The hyperfine interaction between a nucleus at site i and the nearby spins can be written

$$H_{hf} = -\vec{I}_i \cdot \vec{B}_{hf}(i)$$

$$\vec{B}_{hf}(i) = \underline{A}_{contact} \cdot \vec{S}_i + \underline{A}_{orbital} \cdot \vec{L}_i + \sum_j \underline{A}_{trans}(i - j) \cdot S_j, \qquad (9.149)$$

where $B_{hf}(i)$ is an effective field induced by the hyperfine couplings. The three terms in this Hamiltonian are derived from a local contact interaction with s-electrons at the same site, an orbital interaction, and, lastly, a transfered hyperfine interaction with spins at neighboring sites. The various tensors $\underline{A}$ are not generally isotropic, but for pedagogical purposes let us ignore the anisotropy.

The Knight shift – the shift in the magnetic resonance line – is basically the expectation value of the hyperfine field B_{hf}. In a magnetic field, the electronic spins inside the material become polarized, with $\langle S_j \rangle \sim \chi B$, where χ is the magnetic susceptibility, so in the simplest situation the Knight shift is a measure of the local magnetic susceptibility of the medium which is determined by the thermal average of the density of states $\langle N(\epsilon) \rangle$ around the Fermi energy, so

$$K \sim B_{hf} \sim \chi B \sim \langle N(\epsilon) \rangle B. \qquad (9.150)$$

One of the classic indications of the development of a gap in the electron excitation spectrum of an electronic system is the sudden reduction in the Knight shift. In more complex systems, where there are different spin sites, the temperature dependence of the Knight shift can depart from that of the global spin susceptibility.

Another application of the Knight shift is as a method to detect magnetic or antiferromagnetic order. If the electrons inside a metal develop magnetic order, then this produces a large, field-independent Knight shift that can be directly related to the size of the ordered magnetic moment:

$$K \sim \langle S_{local} \rangle. \qquad (9.151)$$

Unlike neutron scattering, NMR can distinguish between homogeneous and inhomogeneous magnetic order.

Relaxation rate, $1/T_1$

The second aspect to NMR is the broadening of the nuclear resonance. If we ignore all but the contact interaction, then the spin-flip decay rate of the local spin is determined by Fermi's golden rule:

$$\frac{1}{T_1} = \frac{2\pi}{\hbar} I^2 A_{contact}^2 S_{+-}(\omega) \bigg|_{\omega = \omega_N}, \qquad (9.152)$$

where ω_N is the nuclear resonance frequency and

$$S_{+-}(\omega) = \int_{\mathbf{q}} [1 + n_B(\omega)] \, \chi''_{+-}(\mathbf{q}, \omega)$$

$$\sim T \int \frac{d^3q}{(2\pi)^3} \frac{1}{\omega} \chi''_{+-}(\mathbf{q}, \omega) \qquad (9.153)$$

at frequencies $\omega \sim \omega_N$, so for a contact interaction the net nuclear relaxation rate is then

$$\frac{1}{T_1} = \frac{2\pi}{\hbar}I^2 A_{contact}^2 \times T \int \frac{d^3q}{(2\pi)^3}\frac{1}{\omega}\chi''_{+-}(\mathbf{q}, \omega)\Big|_{\omega=\omega_N}. \tag{9.154}$$

In a classical metal, $\chi''(\omega)/\omega \sim N(0)^2$ is determined by the square of the density of states. This leads to an NMR relaxation rate

$$\frac{1}{T_1} \propto TN(0)^2 \sim \frac{k_B T}{\epsilon_F^2}. \qquad \text{Korringa relaxation} \tag{9.155}$$

This linear dependence of the nuclear relaxation rate on temperature is referred to as a *Korringa relaxation* law, after the Dutch theorist Jan Korringa [11], who first discovered it. Korringa relaxation occurs because the Pauli principle allows only a fraction $TN(0) \sim T/\epsilon_F$ of the electrons to relax the nuclear moment. In a more general Fermi system, the NMR relaxation rate is determined by the thermally averaged square density of states:

$$\frac{1}{T_1} \sim T \int \left(-\frac{df(\omega)}{d\omega}\right) N(\omega)^2 \sim T \times [N(\omega \sim k_B T)]^2. \tag{9.156}$$

In a wide class of anisotropic superconductors with lines of nodes along the Fermi surface, the density of states is a linear function of energy. One of the classic signatures of these line nodes across the Fermi surface is then a cubic dependence of $1/T_1$ on the temperature:

$$\text{line nodes in gap} \Rightarrow N(\epsilon) \propto \epsilon \qquad \Rightarrow \frac{1}{T_1} \propto T^3. \tag{9.157}$$

In cases where the transfered hyperfine couplings are important, the non-locality introduces a momentum dependence into the hyperfine coupling, so that the coupling constant. $A_{contact}$ is now replaced by $A(\mathbf{k}) = \sum_{\vec{R}} A(\vec{R}_j)e^{-i\mathbf{k}\cdot\vec{R}_j}$:

$$\frac{1}{T_1} = \frac{2\pi}{\hbar}I^2 \times T \int \frac{d^3q}{(2\pi)^3}A(\mathbf{q})^2 \frac{1}{\omega}\chi''_{+-}(\mathbf{q}, \omega)\Big|_{\omega=\omega_N}. \tag{9.158}$$

These momentum dependences can lead to radically different temperature dependences in the relaxation rate at different sites. One of the classic examples of this behavior occurs in the normal state of high-temperature superconductors. The active physics of these materials takes place in quasi-two-dimensional layers of copper oxide, and the NMR relaxation rate can be measured at both the oxygen (O^{17}) and copper sites:

$$\left(\frac{1}{T_1}\right)_{Cu} \sim \text{constant}, \qquad \left(\frac{1}{T_1}\right)_{O} \sim T. \tag{9.159}$$

The appearance of two qualitatively different relaxation rates is surprising, because the physics of the copper oxide layers is thought to be described by a single-band model, with a single Fermi surface that can be seen in ARPES measurements. Why then are there two relaxation rates?

One explanation for this behavior has been advanced by Shastry [12] and Mila and Rice [13], who argue that there is indeed a single spin fluid, located at the copper sites. These authors noticed that, whereas the copper relaxation involves spins at the same site so that

$$A_{Cu}(\mathbf{q}) \sim \text{constant}, \tag{9.160}$$

the spin relaxation rate on the oxygen sites involves a transfered hyperfine coupling between the oxygen p_x or p_y orbitals and the neighboring copper spins. The odd parity of a p_x or p_y orbital means that the corresponding form factors have the form

$$A_{p_x}(\mathbf{q}) \sim \sin(q_x a/2). \tag{9.161}$$

Now high-temperature superconductors are doped insulators. In the insulating state, cuprate superconductors are *Mott insulators*, in which the spins on the copper sites are antiferromagnetically ordered. In the doped metallic state, the spin fluctuations on the copper sites still contain strong antiferromagnetic correlations, and they are strongly peaked around $\vec{Q}_0 \sim (\pi/a, \pi/a)$, where a is the unit cell size. But this is precisely the point in momentum space where the transfered hyperfine couplings for the oxygen sites vanish. The absence of the Korringa relaxation at the copper sites is then taken as a sign that the copper relaxation rate is driven by strong antiferromagnetic spin fluctuations which do not couple to oxygen nuclei.

9.9 Electron transport spectroscopy

9.9.1 Resistivity and the transport relaxation rate

A remarkable thing about electron transport is that one of the simplest possible measurements – the measurement of electrical resistivity – requires quite a sophisticated understanding of the interaction between matter and radiation. We shall cover this relationship in more detail in the next chapter. However, at a basic level, DC electrical resistivity can be interpreted in terms of the basic Drude formula,

$$\sigma = \frac{ne^2}{m} \tau_{tr}, \tag{9.162}$$

where $1/\tau_{tr}$ is the transport relaxation rate. In Drude theory, the electron scattering rate τ_{tr} is related to the electron mean free path l via the relation

$$l = v_F \tau, \tag{9.163}$$

where v_F is the Fermi velocity. We need to sharpen this understanding, for $1/\tau_{tr}$ is not the actual electron scattering rate, it is the rate at which currents decay in the material. For example, if we consider impurity scattering of electrons with a scattering amplitude $u(\theta)$ which depends on the scattering angle θ, the electron scattering rate is

$$\frac{1}{\tau} = 2\pi n_i N(0) \overline{|u(\theta)|^2}, \tag{9.164}$$

where

$$\overline{|u(\theta)|^2} = \int_{-1}^{1} \frac{d\cos\theta}{2} |u(\theta)|^2 \tag{9.165}$$

denotes the angular average of the scattering rate. However, as we shall see shortly, the transport scattering rate which governs the decay of electric current contains an extra weighting factor:

$$\frac{1}{\tau_{tr}} = 2\pi n_i N(0) \overline{|u(\theta)|^2 (1 - \cos\theta)}$$

$$\overline{|u(\theta)|^2 (1 - \cos\theta)} = \int_{-1}^{1} \frac{d\cos\theta}{2} |u(\theta)|^2 (1 - \cos\theta). \tag{9.166}$$

The angular weighting factor $(1 - \cos\theta)$ derives from the fact that the change in the current carried by an electron upon scattering through an angle θ is $ev_F(1 - \cos\theta)$. In other words, only large-angle scattering causes current decay. For impurity scattering this distinction is not very important, but in systems where the scattering is concentrated near $q = 0$, such as scattering off ferromagnetic spin fluctuations, the $(1 - \cos\theta)$ term substantially reduces the effectiveness of scattering as a source of resistance.

At zero temperature, the electron scattering is purely elastic, and the zero-temperature resistance R_0 is then a measure of the elastic scattering rate off impurities. At finite temperatures, electrons also experience inelastic scattering, which can be strongly temperature-dependent. One of the most important diagnostic quantities to characterize the quality of a metal is the resistance ratio – the ratio of resistance at room temperature to the resistance at absolute zero:

$$\text{resistance ratio} = \frac{R(T = 300\,\text{K})}{R(T = 0\,\text{K})}. \tag{9.167}$$

The higher this ratio, the lower the amount of impurities and the higher the quality of sample. Hardware-quality copper piping already has a resistance ratio of the order of 1000! A high resistance ratio is vital for the observation of properties which depend on the coherent ballistic motion of Bloch waves, such as de Haas–van Alphen oscillations or the development of anisotropic superconductivity, which is ultra-sensitive to impurity scattering.

With the small caveat of a distinction between transport and scattering relaxation rates, the temperature-dependent resistivity is an excellent diagnostic tool for understanding the inelastic scattering rates of electrons:

$$\rho(T) = \frac{m}{ne^2} \left(\frac{1}{\tau_{tr}(T)} \right). \tag{9.168}$$

There are three classic dependences to be familiar with:

- Electron–phonon scattering above the Debye temperature:

$$\frac{1}{\tau_{tr}} = 2\pi \lambda k_B T. \tag{9.169}$$

Linear resistivity is produced by electron–phonon scattering at temperatures above the Debye temperature, where the coefficient λ is the electron–phonon coupling constant defined in the previous chapter. In practice, this type of scattering always tends to saturate once the electron mean free path becomes comparable with the electron wavelength. It is this type of scattering that is responsible for the weak linear temperature dependence of resistivity in many metals. A note of caution: linear resistivity does not necessarily imply electron–phonon scattering! The most well-known example of linear resitivity occurs in the normal state of the cuprate superconductors, but here the resistance does not saturate at high temperatures, and the scattering mechanism is almost certainly a consequence of electron–electron scattering.

- Electron–electron or Baber scattering:

$$\frac{1}{\tau_{tr}} = \frac{\pi}{\hbar} \overline{|UN(0)|^2} N(0)(\pi k_B T)^2,$$ (9.170)

where

$$\overline{|UN(0)|^2} = N(0)^2 \int \frac{d\Omega_{\hat{k}'}}{4\pi} |U(\mathbf{k} - \mathbf{k}')|^2 (1 - \cos(\theta_{\mathbf{k},\mathbf{k}'}))$$ (9.171)

is the weighted average of the electron–electron interaction $U(\mathbf{q})$. This quadratic temperature dependence of the inelastic scattering rate can be derived from the Fermi's golden rule scattering rate,

$$\frac{1}{\tau_{tr}} = \frac{4\pi}{\hbar} \sum_{\mathbf{k}',\mathbf{k}''} |U(\mathbf{k} - \mathbf{k}')|^2 (1 - \cos\theta_{\mathbf{k},\mathbf{k}'})(1 - f_{\mathbf{k}'})(1 - f_{\mathbf{k}''})f_{\mathbf{k}'+\mathbf{k}''-\mathbf{k}}\delta(\epsilon_{\mathbf{k}'} + \epsilon_{\mathbf{k}''} - \epsilon_{\mathbf{k}'''}),$$ (9.172)

where the $4\pi = 2 \times 2\pi$ prefactor is derived from the sum over internal spin indices. If we neglect the momentum dependence of the scattering amplitude, then this quantity is determined entirely by the three-particle phase space:

$$\frac{1}{\tau_{tr}} \propto \int d\epsilon' d\epsilon'' (1 - f(\epsilon'))(1 - f(\epsilon''))f(-\epsilon' - \epsilon'')$$

$$= T^2 \int dxdy \left(\frac{1}{1 - e^{-x}}\right)\left(\frac{1}{1 - e^{-y}}\right)\left(\frac{1}{1 - e^{-(x+y)}}\right) = \frac{\pi^2}{4}T^2.$$ (9.173)

In practice, this type of resistivity is only easily observed in strongly interacting electron materials, where it is generally seen at low temperatures when a Landau Fermi-liquid develops. The T^2 resistivity is a classic hallmark of Fermi liquid behavior.

- Kondo spin-flip scattering:

In metals containing a dilute concentration of magnetic impurities, the spin-flip scattering generated by the impurities gives rise to a temperature-dependent scattering rate of the form

$$\frac{1}{\tau_{tr}} \sim n_i \frac{1}{\ln^2 \left(\frac{T}{T_K}\right)},$$ (9.174)

where T_K is the *Kondo temperature*, which characterizes the characteristic spin fluctuation rate of the magnetic impurity. This scattering is unusual because it becomes stronger at lower temperatures, giving rise to a *resistance minimum* in the resistivity.

In heavy-electron materials, Kondo spin-flip scattering is seen at high temperatures, but once a coherent Fermi liquid is formed, the resistivity drops down again at low temperatures, ultimately following a T^2 behavior.

9.9.2 Optical conductivity

Probing the electrical properties of matter at finite frequencies requires the use of optical spectroscopy. In principle, optical spectroscopy provides a direct probe of the frequency-dependent conductivity inside a conductor. The frequency-dependent conductivity is defined by the relation

$$\vec{j}(\omega) = \sigma(\omega)\vec{E}(\omega), \tag{9.175}$$

where for the moment we assume the system is isotropic, so that the conductivity tensor is a diagonal matrix.

Modern optical conductivity measurements can be made from frequencies in the infrared, of order $\omega \sim 10\,\mathrm{cm}^{-1} \sim 1\mathrm{meV}$ up to optical frequencies, of order $50000\,\mathrm{cm}^{-1} \sim 5\,\mathrm{eV}$. The most direct way of obtaining the optical conductivity is from the reflectivity, which is given by

$$r(\omega) = \frac{1 - n(\omega)}{1 + n(\omega)} = \frac{1 - \sqrt{\epsilon(\omega)}}{1 + \sqrt{\epsilon(\omega)}}, \tag{9.176}$$

where $n(\omega) = \sqrt{\epsilon(\omega)}$ is the diffractive index and $\epsilon(\omega)$ is the frequency-dependent dielectric constant. Now $\epsilon(\omega) = 1 + \chi(\omega)$, where $\chi(\omega)$ is the frequency-dependent dielectric susceptibility. Since the polarization $P(\omega) = \chi(\omega)E(\omega)$, and since the current is given by $j = \partial_t P$, it follows that $j(\omega) = -i\omega P(\omega) = -i\omega\chi(\omega)E(\omega)$, so that $\chi(\omega) = \sigma(\omega)/(-i\omega)$ and hence

$$\epsilon(\omega) = 1 + \frac{\sigma(\omega)}{-i\omega}. \tag{9.177}$$

Thus, in principle, knowledge of the complex reflectivity determines the optical conductivity.

In the simplest situations, it is only possible to measure the intensity of reflected radiation, giving $|r(\omega)|^2$. More sophisticated *elipsometry* techniques, which measure the transmission as a function of angle and polarization, can provide both the amplitude and phase of the reflectivity, but here we shall discuss the case where only the amplitude $|r(\omega)|$ is available. In this situation, experimentalists use the Kramers–Kronig relationship, which determines the imaginary part $\sigma_2(\omega)$ of the optical conductivity in terms of the real part $\sigma_1(\omega)$ (Appendix 9A):

$$\sigma_2(\omega) = \omega \int_0^\infty \frac{d\omega'}{\pi} \frac{\sigma_1(\omega')}{\omega^2 - \omega'^2}. \tag{9.178}$$

This is a very general relationship that relies on the retarded nature of the optical response. In principle, this uniquely determines the dielectric function and reflectivity. However, since the range of measurement is limited below about 5 eV, an assumption has to be made about the high-frequency behavior of the optical conductivity, where normally a Lorentzian form is assumed.

With these provisos, it becomes possible to invert the frequency-dependent reflectivity in terms of the frequency-dependent conductivity. We shall return in the next chapter to a consideration of the detailed relationship between the optical conductivity and the microscopic correlation functions. One of the curious things about conductivity is that the real rather than the imaginary part is connected with dissipation. This is because the microscopic current couples to the vector potential $\vec{A}$, rather than to the electric field. Conventionally, we write

$$\vec{j}(\omega) = -Q(\omega)\vec{A}(\omega), \tag{9.179}$$

where $Q(\omega)$ is known as the *London response kernel*. We will see shortly that the interaction of an electromagnetic field with matter involves the transverse vector potential, which couples to the currents in the material without changing the charge density. As we shall see in Section (10.2), the London response kernel is related to the following response function:

$$Q(\omega) = \left[\frac{ne^2}{m} - \langle j(\omega)j(-\omega)\rangle\right]. \tag{9.180}$$

This expression contains two parts: a leading *diamagnetic* part, which describes the high-frequency, short-time response of the medium to the vector potential, and a second, *paramagnetic* part, which describes the slow recovery of the current towards zero. We have used the shorthand

$$\langle j(\omega)j(-\omega)\rangle = i\int_0^\infty dt d^3x \langle [j(x,t), j(0)]\rangle e^{i\omega t} \tag{9.181}$$

to denote the retarded paramagnetic response of the the electron current density $j(x) = -i\frac{\hbar}{m}\psi^\dagger \vec{\nabla}\psi(x)$.

Now the London kernel $Q(z)$ is a bona fide analytic response function, with branch–cuts along the real axis, across which $Q(\omega + i\delta) = Q(\omega - i\delta)^*$. Because $\vec{E}(\omega) = -\partial_t \vec{A} \Rightarrow \vec{E}(\omega) = i\omega \vec{A}(\omega)$, it follows that $\vec{A}(\omega) = \vec{E}(\omega)/i\omega$, so that

$$\vec{j}(\omega) = -\frac{1}{i\omega}Q(\omega)\vec{E}(\omega), \tag{9.182}$$

implying that

$$\sigma(\omega) = \frac{1}{-i\omega}Q(\omega) = \frac{1}{-i\omega}\left[\frac{ne^2}{m} - \langle j(\omega)j(-\omega)\rangle\right]. \tag{9.183}$$

In this way, we see that the real part of $\sigma(\omega)$ is determined by the imaginary part of $Q(\omega)$, accounting for its dissipative properties. Another important consequence is that it is the real rather than the imaginary part of $\sigma(\omega)$ that reverses across the real axis:

$$\sigma(\omega + i\delta) = (-i(\omega + i\delta))Q(\omega + i\delta) = -[(i\omega - i\delta)Q(\omega - i\delta)]^* = -\sigma(\omega - i\delta)^*, \tag{9.184}$$

so that

$$\sigma(\omega \pm i\delta) = \pm\sigma_1(\omega) + i\sigma_2(\omega) \tag{9.185}$$

and

$$\frac{1}{2}\left[\sigma(\omega + i\delta) - \sigma(\omega - i\delta)\right] = \sigma_1(\omega). \tag{9.186}$$

9.9.3 The f-sum rule

One of the most valuable relations for the analysis of optical conductivity data is the so-called *f-sum rule*, according to which the total integrated weight under the conductivity spectrum is constrained to equal the plasma frequency of the medium:

$$\frac{2}{\pi} \int_0^\infty d\omega \sigma(\omega) = \frac{ne^2}{m} = \omega_P^2 \epsilon_0, \tag{9.187}$$

where n is the density of electronic charge and ω_P is the Plasma frequency. To understand this relation, suppose we apply a sudden pulse of electric field to a conductor:

$$E(t) = E_0 \delta(t). \tag{9.188}$$

Immediately after the pulse, the net drift velocity of the electrons is changed to $v = eE_0/m$, so the instantaneous charge current after the field pulse is

$$j(0^+) = nev = \frac{ne^2}{m} E_0, \tag{9.189}$$

where n is the density of carriers. After the current pulse, the electric current will decay. For example, in the Drude theory there is a single current relaxation time rate τ_{tr}, so that

$$j(t) = \frac{ne^2}{m} e^{-t/\tau_{tr}} E_0 \tag{9.190}$$

and thus

$$\sigma(t - t') = \frac{ne^2}{m} e^{-(t-t')/\tau_{tr}} \theta(t - t'), \tag{9.191}$$

and by Fourier transforming we deduce that

$$\sigma(\omega) = \int_0^\infty dt e^{i\omega t} \sigma(t) = \frac{ne^2}{m} \frac{1}{\tau_{tr}^{-1} - i\omega}. \tag{9.192}$$

If we integrate the real part of the conductivity over frequency, we then find

$$\int_0^\infty \sigma_1(\omega) = \frac{ne^2}{m} \int_0^\infty d\omega \frac{\tau_{tr}^{-1}}{\omega^2 + \tau_{tr}^{-2}} = \frac{\pi}{2} \left(\frac{ne^2}{m} \right), \tag{9.193}$$

recovering the f-sum rule.

   Actually, the f-sum rule does not depend on the detailed form of the current relaxation. Using the instantaneous response in (9.189) we obtain

$$J(t = 0^+) = E_o \sigma(t = \delta) = E_0 \int_{-\infty}^\infty \frac{d\omega}{2\pi} e^{-i\delta\omega} \sigma(\omega) = \frac{ne^2}{m} E_0, \tag{9.194}$$

so that

$$\int_{-\infty}^\infty \frac{d\omega}{2\pi} \sigma(\omega + i\delta) e^{-i\delta\omega} = \frac{ne^2}{m} \tag{9.195}$$

is a consequence of Newton's law! To be precise, we need to be careful here: despite temptation, we should not succumb to the urge to just cross out the slow oscillatory infinitesimal, which actually doubles the size of the integral. To see this, let us use the oscillatory term as a convergence factor that permits us to complete the integral along the real axis by closing it in the negative half-plane. The resulting integral wraps counterclockwise around the branch cut along the real axis, and can be deformed until it touches it as follows:

Writing this out explicitly using (9.186) and (9.195) leads to

$$\int_{-\infty}^{\infty} \frac{d\omega}{2\pi} \sigma(\omega + i\delta) e^{-i\delta\omega} = \oint_C \frac{dz}{2\pi} \sigma(z) = \int_{-\infty}^{\infty} \frac{d\omega \, \sigma(\omega + i\delta) - \sigma(\omega - i\delta)}{\pi \qquad 2} = \int_{-\infty}^{\infty} \frac{d\omega}{\pi} \sigma_1(\omega).$$
(9.196)

Since $\sigma_1(\omega)$ is a symmetric function of frequency, it follows that

$$\frac{2}{\pi} \int_0^{\infty} d\omega \sigma_1(\omega) = \frac{ne^2}{m} = \epsilon_0 \omega_p^2,$$
(9.197)

f-sum rule

where we have identified $\epsilon_0 \omega_p^2 = \frac{ne^2}{m}$ with the plasma frequency ω_p of the gas. This is the f-sum rule. It is important because it holds independently of the details of how the current decays. Thus in principle the total weight under the optical spectrum is always a constant, measuring the electron density, providing one integrates up to a high enough energy. Of course, the energy required to recover the entire electron density is immense. Experimentally, it is often useful to define an effective frequency-dependent electron count, given by

$$n_{eff}(\omega) = \frac{m}{e^2} \int_0^{\omega} \frac{d\omega'}{\pi} \sigma(\omega').$$
(9.198)

A plot of this quantity for metallic aluminum ($Z = 13$) is shown in Figure 9.5, showing how the integral for low energies reflects the three electrons in the conduction band, while the higher-energy integral recovers the additional ten electrons in the core states.

In practice, one is most interested in redistributions of spectral weight that accompany changes in the electron correlations. When the temperature is raised, the spectral weight in the optical conductivity tends to redistribute to higher energies. In a simple metal, the optical conductivity forms a simple *Drude peak*: a Lorentzian of width $1/\tau_{tr}$ around zero frequency. In a semiconductor, the weight inside this peak decays as $e^{-\Delta/T}$, where Δ is the semiconducting gap. In a simple insulator, the balance of spectral weight must then reappear at energies above the direct gap energy Δ_g. By contrast, in a superconductor the formation of a superconducting condensate causes the spectral weight in the optical conductivity to collapse into a delta-function peak. See Figure 9.6.

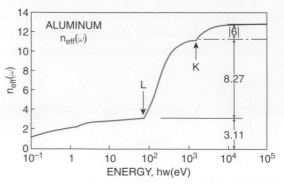

Fig. 9.5 The effective electron count $n_{eff}(\omega)$ of aluminum (see (9.91)), determined from the longitudinal optical conductivity using a combination of experiments ranging from infrared to X-ray. The high-frequency data converge to the total electron count $Z = 13$ of aluminum. Up to 100 eV, only the conduction band is important, and the effective total electron density is $n = 3$. At higher energies, the two sudden cusps in the data correspond to the L edge (excitation of the $n = 2$, L-shell electrons) and the K edge (excitation of the $n = 1$, K-shell electrons) [14]. Reprinted with permission from E. Shiles, *et al.*, *Phys. Rev. B*, vol. 22, p. 1612, 1980. Copyright 1980 by the American Physical Society.

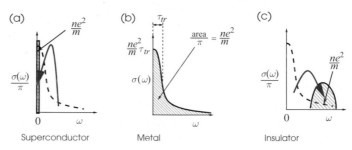

Fig. 9.6 The f-sum, rule illustrating (a) the spectral weight transfer down to the condensate in a superconductor; (b) the Drude weight in a simple metal; (c) the spectral weight transfer up to the conduction band in an insulator.

Appendix 9A   Kramers–Kronig relation

The Kramers–Kronig relation applies to any retarded linear response function, but we shall derive it here in special reference to the conductivity. In time, the current and electric field are related by the retarded response function,

$$j(t) = \int_{-\infty}^{t} dt' \sigma(t - t') E(t'),\tag{9.199}$$

which becomes $j(\omega) = \sigma(\omega)E(\omega)$ in Fourier space, where $\sigma(\omega)$ is the Fourier transform of the real-time response function $\sigma(t - t')$:

$$\sigma(\omega) = \int_{0}^{\infty} dt e^{i\omega t} \sigma(t).\tag{9.200}$$

This function can be analytically extended into the upper-half complex plain:

$$\sigma(z) = \sigma(x + iy) = \int_0^\infty dt\, e^{izt} \sigma(t) = \int_0^\infty dt\, e^{ixt - yt} \sigma(t). \tag{9.201}$$

So long as z lies above the real axis, the real part $-yt$ of the exponent is negative, guaranteeing that the integral $\sigma(z)$ is both convergent and analytic. Provided $\mathrm{Im}\, z_0 > 0$, the conductivity can be written down using Cauchy's theorem:

$$\sigma(z_0) = \int_{C'} \frac{dz}{2\pi i} \frac{\sigma(z)}{z - z_0}, \tag{9.202}$$

where C' runs counterclockwise around the point z_0. By distorting the contour onto the real axis and neglecting the contour at infinity, it follows that

$$\sigma(z_0) = \int_{-\infty}^\infty \frac{d\omega'}{2\pi i} \frac{\sigma(\omega')}{\omega' - z_0}. \tag{9.203}$$

Taking $z_0 = \omega + i\delta$ and writing $\sigma(\omega + i\delta) = \sigma_1(\omega) + i\sigma_2(\omega)$ on the real axis, we arrive at the Kramers–Kronig relations:

$$\sigma_2(\omega) = -\int_{-\infty}^\infty \frac{d\omega'}{2\pi} \frac{\sigma_1(\omega')}{\omega' - \omega} = \omega \int_0^\infty \frac{d\omega'}{\pi} \frac{\sigma_1(\omega')}{\omega^2 - \omega'^2} \tag{9.204}$$

$$\sigma_1(\omega) = \int_{-\infty}^\infty \frac{d\omega'}{2\pi} \frac{\sigma_2(\omega')}{\omega' - \omega} = \int_0^\infty \frac{d\omega'}{\pi} \frac{\omega' \sigma_2(\omega')}{\omega^2 - \omega'^2}. \tag{9.205}$$

Exercises

Exercise 9.1 Spectral decomposition. The dynamical spin susceptibility of a magnetic system is defined as

$$\chi(\mathbf{q}, t_1 - t_2) = i\langle [S^-(\mathbf{q}, t_1), S^+(-\mathbf{q}, t_2)] > \theta(t_1 - t_2), \tag{9.206}$$

where $S^\pm(\mathbf{q}) = S_x(\mathbf{q}) \pm iS_y(\mathbf{q})$ are the spin-raising and spin-lowering operators at wavevector $\mathbf{q}$, i.e.

$$S^\pm(\mathbf{q}) = \int d^3x\, e^{-i\mathbf{q}\cdot\mathbf{x}} S^\pm(\mathbf{x}), \tag{9.207}$$

so that $S^-(\mathbf{q}) = [S^+(-\mathbf{q})]^\dagger$. The dynamical spin susceptibility determines the response of the magnetization at wavevector $\mathbf{q}$ in response to an applied magnetic field at this wavevector:

$$M(\mathbf{q}, t) = (g\mu_B)^2 \int dt'\, \chi(\mathbf{q}, t - t') B(t'). \tag{9.208}$$

(a) Make a spectral decomposition and show that

$$\chi(\mathbf{q}, t) = i\theta(t) \int \frac{d\omega}{\pi} \chi''(\mathbf{q}, \omega) e^{i\omega t}, \tag{9.209}$$

where $\chi''(\mathbf{q}, \omega)$ (often called the *power spectrum* of spin fluctuations) is given by

$$\chi''(\mathbf{q}, \omega) = (1 - e^{-\beta\omega}) \sum_{\lambda, \zeta} e^{-\beta(E_\lambda - F)} |\langle \zeta | S^+(-\mathbf{q}) | \lambda \rangle|^2 \pi \delta[\omega - (E_\zeta - E_\lambda)]$$

$$(9.210)$$

and F is the free energy.

(b) Fourier transform the above result to obtain a simple integral transform which relates $\chi(\mathbf{q}, \omega)$ and $\chi''(\mathbf{q}, \omega)$. The correct result is a Kramers–Kronig transformation.

(c) In neutron scattering experiments, the inelastic scattering cross-section is directly proportional to a spectral function called $S(\mathbf{q}, \omega)$:

$$\frac{d^2\sigma}{d\Omega d\omega} \propto S(\mathbf{q}, \omega),$$

$$(9.211)$$

where $S(\mathbf{q}, \omega)$ is the Fourier transform of a correlation function:

$$S(\mathbf{q}, \omega) = \int_{-\infty}^{\infty} dt e^{i\omega t} \langle S^-(\mathbf{q}, t) S^+(-\mathbf{q}, 0) \rangle.$$

$$(9.212)$$

By carrying out a spectral decomposition, show that

$$S(\mathbf{q}, \omega) = (1 + n(\omega)) \chi''(\mathbf{q}, \omega).$$

$$(9.213)$$

This relationship, plus the one you derived in part (a), can be used to completely measure the dynamical spin susceptibility via inelastic neutron scattering.

Exercise 9.2 Compressibility or charge susceptibility sum rule. From (9.111) and its generalization to finite wavevector, we see that, in a non-superconducting state, the finiteness of the conductivity at $\omega = 0$ requires that the zero-frequency limit of the current–current correlation function is

$$\chi_{\alpha\beta}^{jj}(\mathbf{q}, \omega)|_{\omega \to 0} = \langle j_\alpha(q) j_\beta(-q) \rangle_{\omega \to 0} = \frac{ne^2}{m} \delta_{\alpha\beta}.$$

$$(9.214)$$

(a) Use the Kramers–Kronig relation to show that this leads to a sum rule on the current–current spectral function,

$$\int_0^\infty \frac{d\omega}{\pi} \frac{\operatorname{Im} \chi_{\alpha\beta}^{jj}(\mathbf{q}, \omega + i\delta)}{\omega} = \frac{ne^2}{m} \delta_{\alpha\beta}.$$

$$(9.215)$$

(b) Use the equation of continuity, $(\nabla \cdot \vec{j} = -\dot{\rho} \Leftrightarrow i\vec{q} \cdot \vec{j}(q) = -(-i\omega)\rho(q))$ to show that the charge and current correlation functions are related by

$$q^\alpha q^\beta \chi_{\alpha\beta}^{jj}(\mathbf{q}, \omega) = \omega^2 \chi_{\rho\rho}(\mathbf{q}, \omega),$$

$$(9.216)$$

where

$$\chi_{\rho\rho}(\mathbf{q}, \omega) = i \int d^3x dt \langle [\rho(x, t), \rho(0)] \rangle e^{i(\omega t - \mathbf{q} \cdot \mathbf{x})}$$

$$(9.217)$$

is the dynamical response function for charge.

(c) Use the results of the last two parts to derive the charge or *compression* sum rule,

$$\int_0^\infty \frac{d\omega}{\pi} \omega \chi''_{\rho\rho}(\mathbf{q}, \omega) = \frac{ne^2}{m} q^2. \qquad (9.218)$$

(d) Show that, if the charge susceptibility contains a single plasma pole, the long-wavelength (small-q) limit of the charge susceptibility must then have the form

$$\chi''(\mathbf{q}, \omega) = \pi \frac{ne^2}{m} \frac{q^2}{\omega_P} \delta(\omega - \omega_P) = \pi \epsilon_0 \omega_P q^2 \delta(\omega - \omega_P), \qquad (9.219)$$

where $ne^2/m = \epsilon_0 \omega_P^2$ determines the plasma frequency.

References

[1] R. Kubo, The fluctuation-dissipation theorem, *Rep. Prog. Phys.*, vol. 29, p. 255, 1966.

[2] R. Kubo, Statistical mechanical theory of irreversible processes. I: general theory and simple applications to magnetic and conduction problems, *J. Phys. Soc. Jpn.*, vol. 12, p. 570, 1957.

[3] D. Forster, *Hydrodynamic Fluctuations, Broken Symmetry and Correlation Functions*, Advanced Books Classics, Perseus Books, 1995.

[4] L. Van Hove, Correlations in space and time and Born approximation scattering in systems of interacting particles, *Phys. Rev.*, vol. 95, no. 1, p. 249, 1954.

[5] G. Binnig, H. Rohrer, C. Gerber, and E. Weibel, Tunneling through a controllable vacuum gap, *Appl. Phys. Lett.*, vol. 40, p. 178, 1982.

[6] G. Binnig, H. Rohrer, C. Gerber, and E. Weibel, Surface studies by scanning tunneling microscopy, *Phys. Rev. Lett.*, vol. 49, no. 1, p. 57, 1982.

[7] K. M. Lang, V. Madhavan, J. E. Hoffman, E. W. Hudson, H. Eisaki, S. Uchida, and J. C. Davis, Imaging the granular structure of high-T_c superconductivity in underdoped $Bi_2Sr_2CaCu_2O_8 + \delta$, *Nature*, vol. 415, no. 6870, p. 412, 2002.

[8] B. B. Zhou, S. Misra, E. H. da Silva Neto, P. Aynajian, R. E. Baumbach, J. D. Thompson, E. D. Bauer, and A. Yazdani, Visualizing nodal heavy fermion superconductivity in $CeCoIn_5$, *Nat. Phys.*, vol. 9, no. 8, p. 474, 2013.

[9] J. J. Sakurai, *Modern Quantum Mechanics*, chapter 2.4, Addison-Wesley, 1993.

[10] http://en.wikipedia.org/wiki/WKB approximation.

[11] J. Korringa, Nuclear magnetic relaxation and resonant line shift in metals, *Physica*, vol. 16, p. 601, 1950.

[12] B. S. Shastry, t–J model and nuclear magnetic relaxation in high-T_c materials, *Phys. Rev. Lett.*, vol. 63, p. 1288, 1989.

[13] F. Mila and T. M. Rice, Spin dynamics of $YBa_2Cu_3O_{6+x}$ as revealed by NMR, *Phys. Rev. B*, vol. 40, p. 11382, 1989.

[14] E. Shiles, T. Sasaki, M. Inokuti, and D. Y. Smith, Self-consistency and sum-rule tests in the Kramers–Kronig analysis of optical data: applications to aluminum, *Phys. Rev. B*, vol. 22, p. 1612, 1980.

10 Electron transport theory

10.1 Introduction

Resistivity is one of the most basic properties of conductors. Surprisingly, Ohm's law,

$$V = IR, \tag{10.1}$$

requires quite a sophisticated understanding of quantum many-body physics. In a classical electron gas, the electron current density

$$\vec{j}(x) = ne\vec{v}(x) \tag{10.2}$$

is a simple c-number related to the average drift velocity $\vec{v}(x)$ of the negatively charged electron fluid. This is the basis of the Drude model of electricity, which Paul Drude introduced shortly after the discovery of the electron [1]. Fortunately, many of the key concepts evolved in the Drude model extend to the a quantum description of electrons, where $\vec{j}(x)$ is an operator. To derive the current operator, we may appeal to the continuity equation, or alternatively we can take the derivative of the Hamiltonian with respect to the vector potential:

$$\vec{j}(x) = -\frac{\delta H}{\delta \vec{A}(x)}, \tag{10.3}$$

where

$$H = \int d^3x \left[\frac{1}{2m} \psi^\dagger(x) \left(-i\hbar \vec{\nabla} - e\vec{A}(x) \right)^2 \psi(x) - e\phi(x)\psi^\dagger(x)\psi(x) \right] + V_{INT}, \tag{10.4}$$

where the Hamiltonian is written out for electrons of charge $q = e = -|e|$. Now only the kinetic energy term depends on $\vec{A}$, so that

$$\vec{j}(x) = -\frac{ie\hbar}{m} \psi^\dagger(x) \stackrel{\leftrightarrow}{\nabla} \psi(x) - \left(\frac{e^2}{m} \right) \vec{A}(x)\rho(x), \tag{10.5}$$

where $\stackrel{\leftrightarrow}{\nabla} = \frac{1}{2} \left(\stackrel{\rightarrow}{\nabla} - \stackrel{\leftarrow}{\nabla} \right)$ is the symmetrized derivative.

   The discussion we shall follow dates back to pioneering work by Fritz London [2, 3]. London noticed, in connection with his research on superconductivity, that the current operator splits up into components, which he identified with the paramagnetic and diamagnetic responses of the electron fluid:

$$\vec{j}(x) = \vec{j}_P(x) + \vec{j}_D(x), \tag{10.6}$$

where

$$\vec{j}_P(x) = -\frac{ie\hbar}{m}\psi^\dagger(x)\overset{\leftrightarrow}{\nabla}\psi(x) \tag{10.7}$$

and

$$\vec{j}_D(x) = -\left(\frac{e^2}{m}\right)\vec{A}(x)\rho(x). \tag{10.8}$$

Although the complete expression for the current density is invariant under gauge transformations, $\psi(x) \to e^{i\phi(x)}\psi(x)$, $\vec{A}(x) \to \vec{A} - \frac{\hbar}{e}\vec{\nabla}\phi(x)$, the separate parts are not. However, in a *specific* gauge such as the London or Coulomb gauge, where $\vec{\nabla} \cdot A = 0$, they do have physical meaning. We shall identify this last term as the term responsible for the diamagnetic response of a conductor, while the first term, the paramagnetic current, is responsible for the decay of the current in a metal.

In a non-interacting system, the current operator commutes with the kinetic energy operator H_0 and is formally a constant of the motion. In a periodic crystal, electron momentum is replaced by the lattice momentum **k**, which is, in the absence of lattice vibrations, a constant of the motion, with the result that the electron current still does not decay. What is the origin of electrical resistance?

There are then two basic sources of current decay inside a conductor:

- disorder, which destroys the translational invariance of the crystal
- interactions, between the electrons and phonons and between the electrons themselves, which cause the electron momenta and currents to decay.

The key response function which determines electron current is the conductivity, relating the Fourier component of current density at frequency ω to the corresponding frequency-dependent electric field:

$$\vec{j}(\omega) = \sigma(\omega)\vec{E}(\omega). \tag{10.9}$$

We would like to understand how to calculate this response function in terms of microscopic correlation functions.

The classical picture of electron conductivity was developed by Paul Drude in 1900, while working at the University of Leipzig [1]. Although his model was introduced before the advent of quantum mechanics, many of his basic concepts carry over to the quantum theory of conductivity. Drude introduced the the concept of the *electron mean free path l*, the mean distance between scattering events. The characteristic time scale between scattering events is called the *transport scattering time* τ_{tr}. (We use the "tr" subscript to distinguish this quantity from the quasiparticle scattering time τ, because not all scattering events cause the electric current to decay.) In a Fermi gas, the characteristic velocity of electrons is the Fermi velocity and the mean free path and transport scattering time are related by the simple relation

$$l = v_F\tau_{tr}. \tag{10.10}$$

The ratio of the mean free path to the electron wavelength is determined by the product of the Fermi wavevector and the mean free path. This quantity is the same order of magnitude as the ratio of the scattering time to the characteristic time scale $\hbar/\epsilon_F$ associated with the Fermi energy, so that

$$\frac{l}{\lambda_F} = \frac{k_F l}{2\pi} \sim \frac{\tau_{tr}}{\hbar/\epsilon_F} = \frac{\epsilon \tau_{tr}}{\hbar}. \tag{10.11}$$

In very pure metals, the mean free path l of Bloch wave electrons can be tens or even hundreds of microns, $l \sim 10^{-6}$ m, so this ratio can become as large as 10^4 or even 10^6. From this perspective, the rate at which current decays in a good metal is very slow on atomic time scales.

There are two important aspects to the Drude model (see Figure 10.1):

- the diffusive nature of density fluctuations
- the Lorentzian lineshape of the optical conductivity,

$$\sigma(\omega) = \frac{ne^2}{m} \frac{1}{\tau_{tr}^{-1} - i\omega}. \tag{10.12}$$

Drude recognized that, on length scales much larger than the mean free path, multiple scattering events induce diffusion in the electron motion. On large length scales, the current and density will be related by he diffusion equation,

$$\vec{j}(x) = -D\vec{\nabla}\rho(x), \tag{10.13}$$

where $D = \frac{1}{3}\frac{l^2}{\tau_{tr}} = \frac{1}{3}v_F^2 \tau_{tr}$, which together with the continuity equation

$$\vec{\nabla} \cdot \vec{j} = -\frac{\partial \rho}{\partial t} \tag{10.14}$$

gives rise to the diffusion equation,

$$\left[-\frac{\partial}{\partial t} + D\nabla^2\right]\rho = 0. \tag{10.15}$$

The response function $\chi(q, \nu)$ of the density to small changes in potential must be the Green's function for this equation, so that in Fourier space

$$[i\nu - Dq^2]\chi(q, \nu) = 1, \tag{10.16}$$

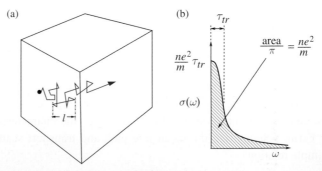

Fig. 10.1 Illustrating (a) the diffusion of electrons on length scales large compared with the mean free path l; (b) The Drude frequency-dependent conductivity. The short-time behavior of the current is determined by Newton's law, which constrains the area under the curve to equal $\int d\omega\sigma(\omega) = \pi\frac{ne^2}{m}$, a relation known as the f-sum rule.

from which we expect the response function and density–density correlation functions to contain a diffusive pole,

$$\langle \delta\rho(q, \nu)\delta\rho(-q, -\nu)\rangle \sim \frac{1}{i\nu - Dq^2}. \tag{10.17}$$

The second aspect of the Drude theory concerns the slow decay of current on the typical time scale τ_{tr}, so that, in response to an electric field pulse $E = E_0\delta(t)$, the current decays as

$$j(t) = e^{-\frac{t}{\tau_{tr}}}. \tag{10.18}$$

In the previous chapter, we discussed how, from a quantum perspective, this current is made up of two components, a diamagnetic component,

$$j_{DIA} = -\frac{ne^2}{m}A = \frac{ne^2}{m}E_0, \qquad (t > 0) \tag{10.19}$$

and a paramagnetic part associated with the relaxation of the electron wavefunction,

$$j_{PARA} = \frac{ne^2}{m}E_0(e^{-t/\tau_{tr}} - 1), \qquad (t > 0) \tag{10.20}$$

which grows to cancel this component. We would now like to see how each of these heuristic features emerges from a microscopic treatment of the conductivity and charge response functions. To do this, we need to relate the conductivity to a response function – and this brings us to the Kubo formula.

10.2 The Kubo formula

Let's now look again at the form of the current density operator. According to (10.5), it can divided into two parts:

$$\vec{j}(x) = \vec{j}_P + \vec{j}_D, \tag{10.21}$$

where

$$\vec{j}_P = -\frac{i\hbar}{m}\psi^\dagger(x) \overset{\leftrightarrow}{\nabla} \psi(x) \qquad \text{paramagnetic current}$$

$$\vec{j}_D = -\frac{e^2}{m}\int d^3x\, \rho(x)\vec{A}(x) \qquad \text{diamagnetic current} \tag{10.22}$$

are the *paramagnetic* and *diamagnetic* parts of the current. The total current operator is invariant under gauge transformations, $\psi(x) \to e^{i\phi(x)}\psi(x)$, $\vec{A}(x) \to \vec{A} + \frac{\hbar}{e}\vec{\nabla}\phi(x)$, and, strictly speaking, the two terms in this expression for the current can't be separated in a gauge-invariant fashion. However, we can separate these two terms if we work in a specific gauge. We shall choose to work in the London gauge:

$$\vec{\nabla} \cdot \vec{A} = 0. \qquad \text{London gauge} \tag{10.23}$$

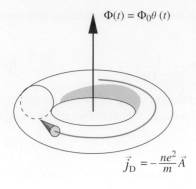
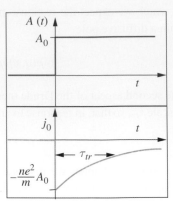

Fig. 10.2 Schematic illustration of a diamagnetic current pulse produced by a sudden change of flux through the conducting loop.

In this gauge, the vector potential is completely transverse, $\vec{q} \cdot \vec{A}(\vec{q}) = 0$. The equations of the electromagnetic field in the London gauge are

$$\left(\frac{1}{c^2} \partial_t^2 - \nabla^2 \right) \vec{A}(x) = \mu_0 \vec{j}(x)$$

$$-\nabla^2 \phi(x) = \frac{\rho(x)}{\epsilon_0}, \tag{10.24}$$

so that the potential field $\rho(x)$ is entirely determined by the distribution of charges inside the material, and the only independent external dynamical field coupling to the material is the vector potential. We shall then regard the vector potential as the only external field coupling to the material.

We shall now follow London's argument for the interpretation of these two terms. Let us carry out a thought experiment in which we imagine a toroidal piece of metal, as in Figure 10.2, in which a magnetic flux passing up through the conducting ring is turned on at $t = 0$, creating a vector potential around the ring given by $A = A_0 \theta(t) = \frac{\phi_0}{2\pi r} \theta(t)$, where r is the radius of the ring. The electric field is related to the external vector potential via the relation

$$\vec{E} = -\frac{\partial \vec{A}}{\partial t} = -A_0 \delta(t), \tag{10.25}$$

so $\vec{E} = -\vec{A}_o \delta(t)$ is a sudden inductively induced electrical pulse.

Suppose the system is described in the Schrödinger representation by the wavefunction $|\psi(t)\rangle$. Then the current flowing after time t is given by

$$\langle \vec{j}(t) \rangle = \langle \psi(t) | \vec{j}_P | \psi(t) \rangle - \frac{ne^2}{m} A_o \theta(t), \tag{10.26}$$

where we have assumed that $\langle \rho(x) \rangle = n$ is the equilibrium density of electrons in the material. We see that the second, diamagnetic term switches on immediately after the pulse. This is nothing more than the diamagnetic response – the effect of the field induced by Faraday's effect. What is so interesting is that this component of the current remains *indefinitely* after the initial step in the flux through the toroid. But the current must decay! How?

The answer is that the initial paramagnetic contribution to the current starts to develop after the flux is turned on. Once the vector potential is present, the wavefunction $|\psi(t)\rangle$ starts to evolve, producing a paramagnetic current that rises and, in a regular conductor, ultimately *exactly cancels* the time-independent diamagnetic current. From this point of view, the only difference between an insulator and a metal is the time scale required for the paramagnetic current to cancel the diamagnetic component. In an insulator this time scale is of the order of the inverse (direct) gap Δ_g, $\tau \sim \hbar/\Delta_g$, whereas in a metal it is the transport relaxation time, $\tau \sim \tau_{tr}$.

These arguments were first advanced by London. He noticed that if for some unknown reason the wavefunction of the material could become "rigid," so that it would not respond to the applied vector potential, in this special case the paramagnetic current would never build up, and one would then have a perfect diamagnet – a *superconductor*. Let's now look at this in more detail. We need to compute

$$\vec{j}(\vec{x}, t) = \langle \vec{j}_P(x, t)\rangle - \frac{ne^2}{m}\vec{A}(x, t). \tag{10.27}$$

If we are to compute the response of the current to the applied field, we need to compute the build-up of the paramagnetic part of the current. Here we can use linear response theory. The coupling of the vector potential to the paramagnetic current is simply $-\int d^3x \vec{j}(x) \cdot \vec{A}(x)$, so the response of this current is given by

$$\langle j_P^\alpha(t)\rangle = \int_{t'<t} d^3x' dt' \, i\langle [\, j_P^\alpha(x), j_P^\beta(x')]\rangle A^\beta(x'). \tag{10.28}$$

In other words, we may write

$$\vec{j}(1) = -\int d2 \underline{Q}(1-2)\vec{A}(2)$$

$$Q^{\alpha\beta}(1-2) = \frac{ne^2}{m}\delta^{\alpha\beta}\delta(1-2) - i\langle [\, j_P^\alpha(1), j_P^\beta(2)]\rangle\theta(t_1 - t_2). \tag{10.29}$$

The quantity $\underline{Q}(1-2)$ is the *London response kernel*. In the most general case, this response is non-local in both space and time. In a metal, this response is non-local over a distance given by the electron mean free path $l = v_F\tau_{tr}$. In a superconductor the response to the vector potential is non-local over the *Pippard coherence length* $\xi = v_F/\Delta$, where Δ is the superconducting gap. We can write the above result in Fourier space as

$$\vec{j}(q) = -\underline{Q}(q)\vec{A}(q), \tag{10.30}$$

where

$$Q^{\alpha\beta}(q) = \frac{ne^2}{m}\delta^{\alpha\beta} - i\langle [\, j^\alpha(q), j^\beta(-q)]\rangle. \tag{10.31}$$

We have used the cavalier notation

$$\langle [\, j^\alpha(q), j^\beta(-q)]\rangle = \int d^3x \int_0^\infty dt \langle [\, j^\alpha(x, t), j^\beta(0)]\rangle e^{-i(\vec{q}\cdot\vec{x} - vt)}. \tag{10.32}$$

Finally, if we write $\vec{E} = -\dfrac{\partial A}{\partial t}$, or $A(q) = \dfrac{1}{i\nu}E(q)$, we deduce that

$$\vec{j}(q) = \underline{\sigma}(q)\vec{E}(q)$$

$$\sigma^{\alpha\beta}(q) = -\frac{1}{i\nu}Q^{\alpha\beta}(q) = \frac{1}{-i\nu}\left\{\frac{ne^2}{m}\delta^{\alpha\beta} - i\langle[j^\alpha(q), j^\beta(-q)]\rangle\right\}. \qquad (10.33)$$

Kubo formula

This is the famous *Kubo formula* [4] that allows us to relate current fluctuations to the conductivity. In practice, the high velocity of light means that $q = \nu/c \ll k_F$ is much shorter than an electronic wavevector, so that in electronic condensed matter physics we may consider the limit $\vec{q} = 0$, writing $\sigma(\nu) = \sigma(\vec{q} = 0, \nu)$. This is the quantity that is measured in optical conductivity measurements. The DC conductivity is given by the zero-frequency limit of the uniform conductivity, i.e. $\sigma_{DC} = \lim_{\nu\to 0}\sigma(\nu)$.

In a metal, σ_{DC} is finite, which implies that $Q(\nu = 0) = 0$, so that

$$i\langle[j^\alpha(q), j^\beta(-q)]\rangle|_{q=0} = \frac{ne^2}{m}\delta^{\alpha\beta}. \qquad (10.34)$$

We shall see that this identity breaks down in a system with broken gauge invariance – and this is the origin of superconductivity. In a normal fluid, however, we can use this identity to rewrite the expression for the conductivity as

$$\sigma^{\alpha\beta}(\nu) = \frac{1}{-i\nu}\left[-i\langle[j^\alpha(\nu'), j^\beta(-\nu')]\rangle\right]_{\nu'=0}^{\nu'=\nu}. \qquad (10.35)$$

A practical calculation of conductivity depends on our ability to extract this quantity from the imaginary-time response function. We can quickly generalize expression (10.29) to imaginary time by replacing $i\langle[A(1), B(2)]\rangle \to \langle TA(1)B(2)\rangle$, so that, in imaginary time,

$$\vec{j}(1) = -\int d2\,\underline{Q}(1-2)\vec{A}(2) \qquad (1 \equiv (\vec{x}_1, \tau_1))$$

$$Q^{\alpha\beta}(1-2) = \frac{ne^2}{m}\delta^{\alpha\beta}\delta(1-2) - \langle Tj_P^\alpha(1)\,j_P^\beta(2)\rangle, \qquad (10.36)$$

so that in Fourier space our expression for the optical conductivity is given by

$$\sigma^{\alpha\beta}(i\nu_n) = -\frac{1}{\nu_n}\left[\langle j^\alpha(\nu')j^\beta(-\nu')\rangle\right]_{\nu'=0}^{\nu'=i\nu_n}, \qquad (10.37)$$

where we have used the shorthand notation

$$\langle j^\alpha(i\nu_n)j^\beta(-i\nu_n)\rangle = \int_0^\beta d\tau\, e^{i\nu_n\tau}\,\langle Tj^\alpha(\tau)j^\beta(0)\rangle. \qquad (10.38)$$

10.3 Drude conductivity: diagrammatic derivation

In the previous section we showed how the fluctuations of the electric current can be related to the optical conductivity. Let us now see how these fluctuations can be computed using

Feynman diagrams, in a disordered electron gas with dispersion $\epsilon_\mathbf{k} = \frac{k^2}{2m}$. First, let us review the Feynman rules. We shall assume that we have taken the leading-order effects of disorder into account in the electron propagator, denoted by

$$\longrightarrow = G(k) = \frac{1}{i\omega_n - \epsilon_\mathbf{k} + i \, \text{sgn} \, \omega_n \frac{1}{2\tau}} . \tag{10.39}$$

The current operator is $j^\alpha(q) = \sum e\frac{k^\alpha}{m} \psi^\dagger_{k-q/2\sigma} \psi_{k+q/2\sigma}$, which we denote by the vertex

$$\rangle \alpha \equiv e\frac{k^\alpha}{m} . \tag{10.40}$$

The set of diagrams that represents the current fluctuations can then be written

$$\tag{10.41}$$

In the above expansion, we have identified three classes of diagrams. The first diagram denotes the simplest contribution to the current fluctuation: we shall see shortly that this is already sufficient to capture the Drude conductivity. The second set of diagrams represents the leading impurity corrections to the current vertex: these terms take account of the fact that low-angle scattering does not affect the electric current, and it is these terms that are responsible for the replacement of the electron scattering rate τ by the transport relaxation rate τ_{tr}. We shall see that these terms vanish for isotropically scattering impurities, justifying our neglect of these contributions in our warm-up calculation of the conductivity.

The last set of diagrams involves crossed impurity scattering lines. We have already encountered these types of diagrams in passing, and the momentum restrictions associated with crossed diagrams lead to a reduction factor of order $O(\frac{1}{k_F l}) \sim \frac{\lambda}{l}$, or the ratio of the electron wavelength to the mean free path. These are the *quantum corrections* to the conductivity. These maximally crossed diagrams were first investigated by Langer and Neal in 1966 [5], during the early years of research into electron transport, but it was not until the late 1970s that they became associated with the physics of electron localization. More on this later.

Using the Feynman rules, the first contribution to the current fluctuations is given by

$$\beta = \langle \mathbf{j}^\alpha(i\nu_n)\mathbf{j}^\beta(-i\nu_n)\rangle$$

$$= -2e^2 T \sum_{\mathbf{k}, i\omega_r} \frac{k^\alpha k^\beta}{m^2} G(\mathbf{k}, i\omega_r + i\nu_n)G(\mathbf{k}, i\omega_r). \tag{10.42}$$

where the minus sign derives from the fermion loop and the factor of 2 derives from the sum over spin components. The difference between the fluctuations at finite and zero frequencies is then

$$\left[\langle j^\alpha(\nu)j^\beta(-\nu)\rangle\right]_0^{i\nu_n} = -2e^2 T \sum_{\mathbf{k}, i\omega_r} \frac{k^\alpha k^\beta}{m^2} \left[G(\mathbf{k}, i\omega_r + i\nu_n)G(\mathbf{k}, i\omega_r) - \{i\nu_n \to 0\}\right]. \tag{10.43}$$

Now the amplitude of current fluctuations at any one frequency involves electron states far from the Fermi surface. However, the *difference* between the current fluctuations at two low frequencies cancels out most of these contributions, and the only important remaining contributions involve electrons near the Fermi surface. This means that we can replace the momentum summation in (10.43) by an energy integral in which the density of states is approximated by a constant and the limits are extended to infinity, as follows:

$$\sum_{\mathbf{k}} \frac{k^\alpha k^\beta}{m^2} \left[\cdots\right] \to \int \frac{4\pi k^2 dk}{(2\pi)^3} \int \frac{d\Omega_{\hat{\mathbf{k}}}}{4\pi} \frac{k^\alpha k^\beta}{m^2} \left[\cdots\right]$$

$$\to \delta^{\alpha\beta} \frac{v_F^2 N(0)}{3} \int_{-\infty}^{\infty} d\epsilon \left[\cdots\right]. \tag{10.44}$$

The London kernel then becomes

$$Q^{\alpha\beta}(i\nu_n) = 2\delta^{\alpha\beta} \frac{e^2 v_F^2}{3} T$$

$$2N(0) \times \sum_{\omega_r} \int_{-\infty}^{\infty} d\epsilon \left\{ \overbrace{\left(\frac{1}{i\omega_r^+ - \epsilon + i\,\text{sgn}\,\omega_r^+/(2\tau)}\right)\left(\frac{1}{i\omega_r - \epsilon + i\,\text{sgn}\,\omega_r/(2\tau)}\right)}^{\text{Poles on opposite side if } \omega_r^+ > \omega_r} - \overbrace{\left(i\nu_n \to 0\right)}^{\text{Poles on same side}} \right\}.$$

We can now carry out the energy integral by contour methods. We shall assume that $\nu_n > 0$. Now, provided that $i\omega_r^+ > 0$ and $i\omega_r < 0$, the first term inside this summation has poles on opposite sides of the real axis, at $\epsilon = i\omega_r + i/2\tau$ and $\epsilon = i\omega_r - 1/(2\tau)$, whereas the second term has poles on the same side of the real axis. Thus, when we complete the energy integral we only pick up contributions from the first term. (It doesn't matter on which side of the real axis we complete the contour, but if we choose the contour to lie on the side where there are no poles in the second term, we can immediately see that this term gives no contribution.) The result of the integrals is then

$$Q^{\alpha\beta}(i\nu_n) = \delta^{\alpha\beta} \overbrace{\frac{2e^2 v_F^2 N(0)}{3}}^{\frac{ne^2}{m}} T \sum_{0>\omega_r>-\nu_n} \frac{2\pi i}{i\nu_n + i\tau^{-1}}$$

$$= \delta^{\alpha\beta} \frac{ne^2}{m} \frac{\nu_n}{\tau^{-1} + \nu_n}. \tag{10.45}$$

Converting the London kernel into the optical conductivity,

$$\sigma^{\alpha\beta}(i\nu_n) = \frac{1}{\nu_n} Q^{\alpha\beta}(i\nu_n) = \delta^{\alpha\beta} \frac{ne^2}{m} \frac{1}{\tau^{-1} - i(i\nu_n)}. \tag{10.46}$$

Finally, analytically continuing onto the real axis, we obtain

$$\sigma^{\alpha\beta}(\nu + i\delta) = \frac{ne^2}{m} \frac{1}{\tau^{-1} - i\nu}. \qquad \text{transverse conductivity} \tag{10.47}$$

There are a number of important points to make about this result:

- It ignores the effects of anisotropic scattering. To obtain these we need to include the ladder vertex corrections, which, we will shortly see, replace

$$\frac{1}{\tau} \to \frac{1}{\tau_{tr}} = 2\pi n_i N(0)\overline{(1 - \cos\theta)|u(\theta)|^2}, \tag{10.48}$$

where the $(1 - \cos\theta)$ term takes into account that small-angle scattering does not relax the electric current.
- It ignores localization effects that become important when $\frac{1}{k_F l} \sim 1$. In one or two dimensions, the effects of these scattering events accumulate at long distances, ultimately localizing electrons, no matter *how* weak the impurity scattering.
- Transverse current fluctuations are not diffusive. This is not surprising, since transverse current fluctuations do not involve any fluctuation in the charge density.

To improve our calculation, let us now examine the vertex corrections that we have so far neglected. Let us now reintroduce the ladder vertex corrections shown in (10.41). We shall write the current–current correlator as

$$\langle j^\alpha(q)j^\beta(-q)\rangle = \alpha \quad \begin{matrix} k \\ \bullet \blacktriangleright \\ k+q \end{matrix} \quad \beta .$$

$$\tag{10.49}$$

where the vertex correction is approximated by a sum of ladder diagrams, as follows:

$$\beta = \beta + \beta + \beta + \cdots = \Lambda e v_F^\beta. \tag{10.50}$$

We shall rewrite the vertex part as a self-consistent Dyson equation, as follows:

$$e\Lambda v_F^\beta = \beta + \beta \quad , \tag{10.51}$$

where $q = (0, i\nu_n)$ and $p' = (\vec{p}\,', i\omega_r)$. The equation for the vertex part is then

$$e v_F^\beta \Lambda(\omega_r, \nu_n) = e v_F^\beta + n_i \sum_{\vec{p}\,'} |u(\vec{p} - \vec{p}\,')|^2 G(\vec{p}\,', i\omega_r^+) G(\vec{p}\,', i\omega_r) \Lambda(\omega_r, \nu_n) e v_F'^\beta. \tag{10.52}$$

Assuming that the vertex part only depends on frequencies and has no momentum dependence, we may then write

$$\Lambda = 1 + \Lambda n_i \int \frac{d\cos\theta}{2} |u(\theta)|^2 \cos\theta \int \frac{d^3 p'}{(2\pi)^3} G(\vec{p}\,', i\omega_r^+) G(\vec{p}\,', i\omega_r). \tag{10.53}$$

We can now carry out the integral over $\vec{p}\,'$ as an energy integral, writing

$$N(0) \int d\epsilon\, G(\epsilon, i\omega_r^+) G(\epsilon, i\omega_r) = N(0) \int d\epsilon\, \frac{1}{i\tilde{\omega}_n^+ - \epsilon} \frac{1}{i\tilde{\omega}_n - \epsilon}, \tag{10.54}$$

where we use the shorthand

$$\tilde{\omega}_n = \omega_n + \text{sgn}\,\omega_n \left(\frac{1}{2\tau}\right). \tag{10.55}$$

Carrying out this integral, we obtain

$$N(0) \int d\epsilon\, G(\epsilon, i\omega_r^+) G(\epsilon, i\omega_r) = \begin{cases} \pi N(0) \frac{1}{\nu_n + \tau^{-1}} & (-\nu_n < \omega_r < 0) \\ 0 & (\text{otherwise}), \end{cases} \tag{10.56}$$

so that

$$\Lambda = 1 + \left(\frac{\tilde{\tau}^{-1}}{\nu_n + \tau^{-1}}\right) \Lambda \theta_{\nu_n, \omega_r}, \tag{10.57}$$

where $\tilde{\tau}^{-1} = 2\pi n_i N(0) \overline{\cos\theta |u(\theta)|^2}$ and $\theta_{\nu_n, \omega_r} = 1$ if $-\nu_n < \omega_r < 0$ and zero otherwise, so that

$$\Lambda = \begin{cases} \frac{\nu_n + \tau^{-1}}{\nu_n + \tau_{tr}^{-1}} & (-\nu_n < \omega_r < 0) \\ 1 & (\text{otherwise}), \end{cases} \tag{10.58}$$

where

$$\tau_{tr}^{-1} = \tau^{-1} - \tilde{\tau}^{-1} = 2\pi n_i N(0) \overline{(1 - \cos\theta) |u(\theta)|^2}. \tag{10.59}$$

When we now repeat the calculation, we obtain

$$Q^{\alpha\beta}(i\omega_n) = \frac{ne^2}{m}\delta^{\alpha\beta}T\sum_{i\omega_r}\int_{-\infty}^{\infty}d\epsilon\left[G(\epsilon,i\omega_r^+)G(\epsilon,i\omega_r) - (i\nu_n \to 0)\right]\Lambda(i\omega_r,i\nu_n)$$

$$= \frac{ne^2}{m}\delta^{\alpha\beta}T\sum_{i\omega_r}\frac{2\pi i}{i\nu_n + i\tau^{-1}}\frac{\nu_n + \tau^{-1}}{\nu_n + \tau_{tr}^{-1}}$$

$$= \frac{ne^2}{m}\left(\frac{\nu_n}{\nu_n + \tau_{tr}^{-1}}\right)\delta^{\alpha\beta}. \tag{10.60}$$

Making the analytic continuation to real frequencies, we obtain

$$\sigma(\nu + i\delta) = \frac{ne^2}{m}\frac{1}{\tau_{tr}^{-1} - i\nu}. \tag{10.61}$$

Note the following:

- Transverse current fluctuations decay at a rate $\tau_{tr}^{-1} < \tau$. By renormalizing $\tau \to \tau_{tr}$, we take into account the fact that only backwards scattering relaxes the current. τ_{tr} and τ_{tr} are only identical in the special case of isotropic scattering. This distinction between scattering rates becomes particularly marked when the scattering is dominated by low-angle scattering, which contributes to τ^{-1} but not to the decay of current fluctuations.
- There is no diffusive pole in the transverse current fluctuations. This is not surprising, since transverse current fluctuations do not change the charge density.

10.4 Electron diffusion

To display the presence of diffusion, we need to examine the density response function. Remember that a change in density is given by

$$\langle\delta\rho(q)\rangle = i\langle[\rho(q),\rho(-q)]\rangle\overbrace{\delta\mu(q)}^{-eV(q)}, \tag{10.62}$$

where V is the change in the electrical potential and

$$i\langle[\rho(q),\rho(-q)]\rangle = \int d^3x dt i\langle[\rho(x,t),\rho(0)]\rangle e^{-i\vec{q}\cdot\vec{x}+i\omega t}. \tag{10.63}$$

We shall calculate this using the same set of ladder diagrams, but now using the charge vertex. Working with Matsubara frequencies, we have

$$\langle\rho(q,i\nu_n)\rho(-q,-i\nu_n)\rangle = $$

$$= $$

$$\tag{10.64}$$

where the current vertex

$$= -e\Lambda_c(k, q). \tag{10.65}$$

Let us now rewrite (10.64) and (10.65) as equations. From (10.64), the density–density response function is given by

$$\langle \rho(q, i\nu_n)\rho(-q, -i\nu_n)\rangle = -2T\sum_k G(k + q)G(k)\Lambda_c(k, q). \tag{10.66}$$

From (10.65), the Dyson equation for the vertex is

$$\Lambda_c(k, q) = 1 + n_i \sum_{k'} |u(\mathbf{k} - \mathbf{k}')|^2 G(k' + q)G(k')\Lambda_c(k', q). \tag{10.67}$$

For convenience we will assume point scattering, so that $u = u_0$ is momentum-independent. Thus $\Lambda_c(k, q)$ only depends on k through its frequency component $i\omega_r$, so $\Lambda(k, q) = \Lambda(i\omega_r, q)$:

$$\Lambda_c(i\omega_r, q) = 1 + n_i u_0^2 \sum_{k'} G(k' + q)G(k')\Lambda_c(i\omega_r, q)$$

$$= 1 + \Pi(i\omega_r, q)\Lambda_c(i\omega_r, q) \tag{10.68}$$

or

$$\Lambda_c(i\omega_r, q) = \frac{1}{1 - \Pi(i\omega_r, q)}, \tag{10.69}$$

where the polarization bubble is given by

$$\Pi(i\omega_r, q) = n_i u_0^2 \sum_{p'} G(k' + q)G(k')$$

$$= n_i u_0^2 N(0) \int \frac{d\Omega}{4\pi} \int d\epsilon \frac{1}{i\tilde{\omega}_r^+ - (\epsilon + \vec{q} \cdot \vec{v}_F)} \frac{1}{i\tilde{\omega}_r - \epsilon}. \tag{10.70}$$

(Note the use of the tilde frequencies, as defined in (10.55).) Now if $i\nu_n > 0$, then the energy integral in $\pi(i\omega_r, q)$ will only give a finite result if $-\nu_n < \omega_r < 0$. Outside this frequency range, $\pi(i\omega_r, q) = 0$ and $\Lambda_c = 1$. Inside this frequency range, $\Pi(i\omega_r, q) = \Pi(q)$ is frequency-independent and given by

$$\Pi(q) = n_i u_0^2 N(0) \int \frac{d\Omega}{4\pi} \overbrace{\frac{2\pi i}{i\nu_n + i\tau^{-1} + \vec{q} \cdot \vec{v}_F}}^{\tau^{-1}/(2\pi)}$$

$$= \int \frac{d\Omega}{4\pi} \frac{1}{1 + \nu_n\tau - i\vec{q} \cdot \vec{v}_F\tau}. \tag{10.71}$$

Now we would like to examine the slow, very-long-wavelength charge fluctuations, which means we are interested in q small compared with the inverse mean free path, $q \ll l^{-1} = 1/(v_F\tau)$, and in frequencies that are much smaller than the inverse scattering length, $v_n \tau \ll 1$. This permits us to expand Π in powers of $\vec{q}$. We shall take the first non-zero contribution, which comes in at order q^2. With these considerations in mind, we may expand Π as follows:

$$\Pi(q) = \int \frac{d\Omega}{4\pi} \left(1 - v_n \tau + i\vec{q} \cdot \vec{v}_F \tau + i^2 (v_F \cdot \mathbf{q})^2 \tau^2 + \cdots \right)$$

$$= \left(1 - v_n \tau - \frac{v_F^2 \tau}{3} q^2 \tau + \cdots \right), \tag{10.72}$$

where we neglect terms of order $O(q^2 v_n)$. We may identify the combination $v_f^2 \tau / 3$ in the second term with the diffusion constant D. Note that, had we done this integral in d dimensions, the 3 in the denominator of the second term above would be replaced by d, but the general form for the diffusion constant in d dimensions is $D = v_f^2 \tau / d$, so that in any dimension we obtain

$$\Pi(q) = \left(1 - v_n \tau - Dq^2 \tau + \cdots \right). \tag{10.73}$$

We then obtain

$$\Lambda_c(q) = \frac{1}{1 - \Pi(q)} = \frac{\tau^{-1}}{v_n + Dq^2} \qquad (-v_n < \omega_r < 0). \tag{10.74}$$

Summarizing, the long-wavelength, low-frequency charge vertex has the form

$$\Lambda_c(i\omega_r, q) = \begin{cases} \frac{i\tau^{-1}}{v_n + Dq^2} & ((-|v_n| < \text{sgn}(v_n)\omega_r < 0)) \\ 1 & \text{(otherwise)}, \end{cases} \tag{10.75}$$

and thus the dynamical charge correlation function is given by

$$\langle \rho(q)\rho(-q) \rangle = \quad \underset{k+q}{\overset{k}{\diamondsuit}} \quad = -2N(0)T \sum_{i\omega_r} \int d\epsilon \, G(\epsilon, i\omega_r^+) G(\epsilon, i\omega_r) \Lambda_c(i\omega_r, q).$$

$$\tag{10.76}$$

Now if we evaluate this quantity at zero frequency, $v_n = 0$, where $\Lambda_c = 1$, we obtain the static susceptibility

$$\chi_0 = -2T \sum_{r,k} \frac{1}{(i\tilde{\omega}_r - \epsilon_k)^2}$$

$$= 2 \int d\epsilon N(\epsilon) \int \frac{d\omega}{2\pi i} f(\omega) \underbrace{\left\{ \frac{1}{(\omega + i/(2\tau) - \epsilon)^2} - \frac{1}{(\omega - i/(2\tau) - \epsilon)^2} \right\}}_{-2iA(\epsilon,\omega)}$$

$$= 2 \int d\epsilon N(\epsilon) \int \frac{d\omega}{2\pi i} \frac{df(\omega)}{d\omega} \underbrace{\left\{ \frac{1}{(\omega + i/(2\tau) - \epsilon)} - \frac{1}{(\omega - i/(2\tau) - \epsilon)} \right\}}_{=N(\omega)}$$

$$= 2 \int d\omega \left(-\frac{df(\omega)}{d\omega} \right) \int d\epsilon \frac{N(\epsilon)}{\pi} A(\epsilon, \omega) = 2N(0), \qquad \text{(unrenormalized)} \tag{10.77}$$

so that the static charge susceptibility is unaffected by the disorder. This enables us to write

$$\langle \rho(q)\rho(-q) \rangle = \chi_0 - 2T \sum_{i\omega_r} \int N(\epsilon)d\epsilon$$

$$\times \left[G(\epsilon, i\omega_r^+)G(\epsilon, i\omega_r)\Lambda_c(\omega_r, \nu_n) - \{\nu_n \to 0\} \right]. \quad (10.78)$$

Since this integral is dominated by contributions near the Fermi energy, we can extend the energy integral over the whole real axis, replacing

$$\int N(\epsilon)d\epsilon \to N(0) \int_{-\infty}^{\infty} d\epsilon, \quad (10.79)$$

enabling the energy integral to be carried out by contour methods, whereupon

$$\langle \rho(q)\rho(-q) \rangle = \chi_0 - 2TN(0) \sum_{i\omega_r} \int_{-\infty}^{\infty} d\epsilon \left[G(\epsilon, i\omega_r^+)G(\epsilon, i\omega_r)\Lambda_c(\omega_r, \nu_n) - \{\nu_n \to 0\} \right]$$

$$= \chi_0 - \chi_0 \overbrace{\left(\frac{\nu_n}{\nu_n + \tau^{-1}} \right)}^{\to \nu_n \tau} \left[\frac{\tau^{-1}}{\nu_n + Dq^2} \right], \quad (10.80)$$

where, again, in the last step we have assumed $|\nu_n|\tau \ll 1$. The Matsubara form for the charge susceptibility is then

$$\chi_o(\vec{q}, i\nu_n) = \chi_0 \frac{Dq^2}{|\nu_n| + Dq^2}. \quad (10.81)$$

Analytically continuing this result, we finally obtain

$$\chi(\vec{q}, \nu + i\delta) = \chi_0 \left(\frac{Dq^2}{Dq^2 - i\nu} \right). \quad (10.82)$$

Note the following:

- Density fluctuations are diffusive. Indeed, we could have anticipated the above form on heuristic grounds. The solution of the diffusion equation $D\nabla^2\rho = \frac{\partial \rho}{\partial t}$ is, in Fourier space,

$$\rho(\vec{q}, \nu) = \frac{1}{Dq^2 - i\nu}\rho(q), \quad (10.83)$$

where $\rho(q)$ is the Fourier transform of the initial charge distribution. If we require $\rho(\vec{q}, \nu = 0) = \chi_0 U(\vec{q})$, where $U(\vec{q})$ is the Fourier transform of the applied potential, then this implies (10.82).
- The order of limits is important, for, whereas

$$\lim_{q \to 0} \lim_{\nu \to 0} \chi(q, \nu) = \chi_0, \quad (10.84)$$

which is the response to a static potential of large but finite wavelength,

$$\lim_{\nu \to 0} \lim_{q \to 0} \chi(q, \nu) = 0, \tag{10.85}$$

which states that the response to a uniform potential of vanishingly small frequency is zero. The difference in these two response functions is due to the conservation of charge – if one wants to change the charge density in one place, it can only be done by redistributing the charge. If one applies a static uniform potential, the charge density does not change.

- We can use these results to deduce the longitudinal conductivity – the current response to a longitudinal electric field for which $\vec{q} \cdot \vec{E} \neq 0$. Let $\phi(q)$ be the electric potential. Then $\delta\rho(q) = \chi(q)e\phi(q)$, so that

$$\delta\rho(q) = \chi_0 \frac{Dq^2}{Dq^2 - i\nu} e\phi(q) = -\chi_0 \frac{Di\vec{q} \cdot \overbrace{(i\vec{q}\phi(q))}^{\vec{\nabla}\phi = -\vec{E}(q)}}{Dq^2 - i\nu}$$

$$= \chi_0 \left(\frac{Di\vec{q}}{Dq^2 - i\nu} \right) \cdot \vec{E}(q). \tag{10.86}$$

Now since $\dfrac{\partial \rho}{\partial t} \equiv -i\nu\rho(q)$, it follows that

$$\dot{\rho}(q) = e\chi_0 \left(\frac{D\nu\vec{q}}{Dq^2 - i\nu} \right) \cdot \vec{E}(q). \tag{10.87}$$

By continuity, $e\dfrac{\partial \rho}{\partial t} = -\vec{\nabla} \cdot \vec{j}(q) = -i\vec{q} \cdot \vec{j}(q)$, where $\vec{j}$ is the charge current, so by comparing with (10.87) we deduce that the longitudinal current is

$$j_L(q) = e^2 \chi_0 D \left(\frac{i\nu}{i\nu - Dq^2} \right) \vec{E}(q). \tag{10.88}$$

So the longitudinal conductivity contains a diffusive pole:

$$\sigma_{LONG}(q) = e^2 \chi_0 D \left(\frac{i\nu}{i\nu - Dq^2} \right). \tag{10.89}$$

Note also that at $q = 0$, $\sigma = e^2 \chi_0 D$. This can be written as the *Einstein relation*:

$$\sigma = e^2 \chi_0 D = \frac{ne^2}{m} \tau. \qquad \text{Einstein relation} \tag{10.90}$$

10.5 Anderson localization

We would like to finish our brief introduction to electron transport by touching on the concept of electron localization. The disorder that has been considered in this chapter is weak and the electron states we have considered are delocalized. We have remarked on a few occasions that disorder is like a kind of attractive but infinitely retarded inter-action, and, like other attractive interactions, it has the capacity to induce new kinds of collective behavior among the electrons. In fact, disorder actually gives rise to collective interference effects within the electron gas, which ultimately lead to the localization of the electron wavefunction. This idea was proposed in 1958 by Philip W. Anderson at Bell Laboratories in New Jersey [6], and subsequently named after him. However, it took two further decades for the idea to gain universal of acceptance. Our modern understanding of electron localization was greatly aided by a conceptual breakthrough on this problem made by Don Liciardello and David Thouless [7], working at Birm-ingham University, who proposed that the resistance of a material, or rather the inverse resistance, the conductance $G = 1/R$, is a function of scale. Thouless' idea, initially pro-posed for one dimension, was taken up by the "Gang of Four," Elihu Abrahams, Philip W. Anderson, Don Licciardello, and Tirupattur Ramakrishnan [8], working at Princeton and Rutgers University, and extended to higher dimensions, leading to the modern *scal-ing theory* of localization [9]. One of the ideas that emerged from this breakthrough is that electron localization results from the coherent interference between electron waves, which at long distances ultimately builds up to produce a disorder-driven metal–insulator transition – a kind of phase transition in which the order parameter is the conductance. Like all phase transitions, localization is sensitive to the dimensionality. Whereas in three dimensions electron localization requires that the disorder exceed a critical value, in two dimensions and one dimension an arbitrarily small amount of disorder is suffi-cient to localize electrons, and the leading-order effects of localization can already be seen in weakly disordered materials. These ideas can all be developed for weakly dis-ordered conductors by a simple extention of the Feynman diagram methods we have been using.

To develop a rudimentary conceptual understanding of electron localization, we shall follow a heuristic argument by Boris Altshuler, Arkady Aronov, Anatoly Larkin, and David Khmelnitskii [10] (see also Bergmann [11]), who pointed out that weak localization results from the constructive interference between electrons passing along time-reversed paths. Consider the amplitude for an electron to return to its starting point. In general, it can do this by passing around a sequence of scattering sites labeled 1 through n, as shown in Figure 10.3, where we identify $n \equiv 1$ as the same scattering site. The amplitude for scattering around this loop is

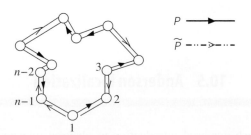

Fig. 10.3   Scattering of an electron around two time-reversed paths.

$$A_P = G_R(n, n-1)G_R(n-1, n-2)\cdots G_R(2, 1), \tag{10.91}$$

where

$$G_R(\vec{x}_1, \vec{x}_2) = \int \frac{d^d k}{(2\pi)^d} \frac{1}{\omega - \epsilon_{\mathbf{k}} + i\delta} e^{i\vec{k}\cdot(\vec{x}_1 - \vec{x}_2)} \tag{10.92}$$

is the retarded propagator describing the amplitude for an electron of frequency ω to propagate between two sites. For each path P there is a corresponding time-reversed path $\tilde{P}$. The amplitude for the same electron to follow $\tilde{P}$ starting at $1 \equiv n$ is

$$A_{\tilde{P}} = G_R(1, 2)G_R(2, 3)\cdots G_R(n-1, n). \tag{10.93}$$

The total probability associated with passage along both paths is given by

$$P = |A_P + A_{\tilde{P}}|^2 = |A_P|^2 + |A_{\tilde{P}}|^2 + 2\,\text{Re}\,[A_{\tilde{P}}^* A_P]. \tag{10.94}$$

Now if $A_P = \sqrt{p_1}e^{i\phi_1}$ and $A_{\tilde{P}} = \sqrt{p_2}e^{i\phi_2}$ then the total probability to scatter back to the starting point via the two paths,

$$p_{TOT} = p_1 + p_2 + 2\sqrt{p_1 p_2}\cos(\phi_2 - \phi_1), \tag{10.95}$$

contains an interference term $2\sqrt{p_1 p_2}\cos(\phi_2 - \phi_1)$. If the two paths were unrelated, then the impurity-averaged interference term would be zero, and we would expect $\overline{P} = p_1 + p_2$. However! The two paths *are* related by time reversal, so that $A_{\tilde{P}} = A_P$, with precisely the same magnitude and phase, and so the two processes always *constructively interfere*,

$$p_{TOT} = 4p_1, \tag{10.96}$$

without the interference term $p_{TOT} = 2p_1$, so we see that constructive interference between time-reversed paths doubles the return probabilty.

This means that an electron that enters into a random medium has a quantum mechanically *enhanced* probability of returning to its starting point – quantum electrons "bounce back" twice as often as classical electrons in a random medium! The same phenomenon causes the light from a car's headlamps to reflect backwards in a fog. We shall see that the return probability is enhanced in lower dimensions, and in one or two dimensions these effects inevitably lead to the localization of electrons for arbitrarily small amounts of disorder.

Let us now make a diagrammatic identification of these interference terms. The complex conjugate of the retarded propagator is the advanced propagator

$$G_R(2-1, \omega)^* = G(2-1, \omega + i\delta)^* = G(2-1, \omega - i\delta) == G_A(2-1, \omega). \tag{10.97}$$

The interference term

$$A_{\tilde{P}}^* A_P = \prod_{j=1}^{n-1} G_R(j+1, j; \omega)G_A(j+1, j; \omega), \tag{10.98}$$

is represented by a ladder diagram for repeated scattering of electron pairs. The sum of all such diagrams is called a *Cooperon*, because of its similarity to the pair susceptibility in superconductivity (Figure 10.4). Notice that the lower electron line involves the advanced

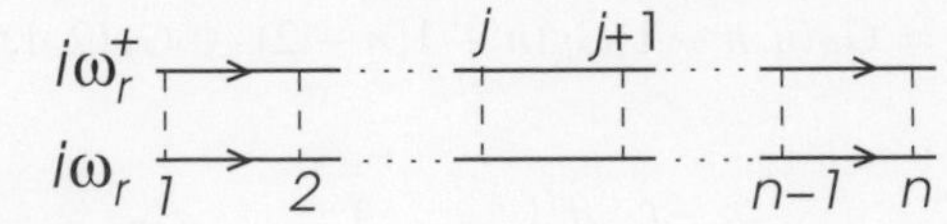

Fig. 10.4 nth-order contribution to the Cooperon.

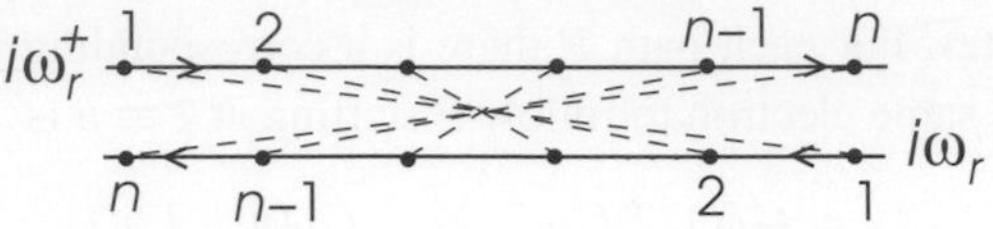

Fig. 10.5 A twisted Cooper diagram forms a maximally crossed or Langer–Neal diagram.

propagator G_A, whereas the upper line involves the retarded propagator G_R. In the Matsubara approach, the distinction between these two propagators is enforced by running a frequency $i\omega_r^+ \equiv i\omega_r + i\nu_n$ along the top line and a frequency $i\omega_r$ along the bottom. When ν_n is analytically continued and ultimately set to zero, this enforces the distinction between the two propagators. Now if we twist the Cooperon around, we see that it is equivalent to a maximally crossed or *Langer–Neal* diagram [5] (Figure 10.5).

Let us now compute the amplitudes associated with these localization corrections to the conductivity. We begin by denoting the Cooperon by a sum of ladder diagrams:

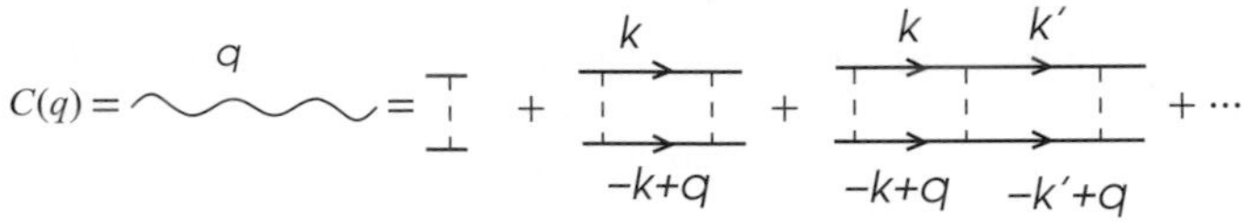

$$= \frac{n_i u_0^2}{1 - \tilde{\Pi}(q)} \tag{10.99}$$

where

$$\tilde{\Pi}(q) = n_i u_0^2 \sum_{\mathbf{k}} G_R(k) G_A(-k + q), \tag{10.100}$$

where we have denoted $G_R(k) \equiv G(\mathbf{k}, i\omega_r^+)$ and $G_A(k) \equiv G(\mathbf{k}, i\omega_r)$, implicitly assuming that ω_r^+ and ω_r are of opposite sign. Now if we look carefully at $\tilde{\Pi}$, we see that it is identical to the particle–hole bubble Π that we encountered when computing diffusive charge fluctuations in (10.70), except that the hole line has been replaced by a particle line; in so doing, we replace $\mathbf{k} + \mathbf{q} \to -\mathbf{k} + \mathbf{q}$ in the momentum of the propagator. However, thanks to time-reversal symmetry, this does not change the value of the polarization bubble, and we conclude that

$$\tilde{\Pi}(q) = \left(1 - \nu_n \tau - Dq^2 \tau + \cdots\right) \tag{10.101}$$

and thus

$$C(q) = n_i u_0^2 \frac{\tau^{-1}}{Dq^2 + |\nu_n|} = \frac{1}{2\pi N(0)\tau^2} \frac{1}{Dq^2 + |\nu_n|}. \tag{10.102}$$

We shall redraw the maximally crossed contributions to the conductivity as follows:

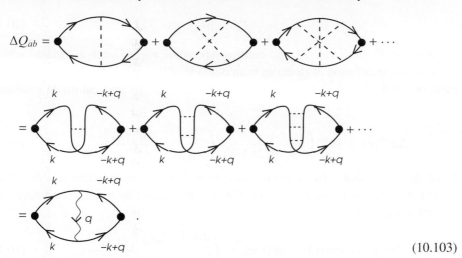

$$(10.103)$$

Written out explicitly, this gives

$$\Delta \sigma^{ab}(i\nu_n) = \frac{\Delta Q^{ab}}{\nu_n}$$

$$= \frac{2e^2 T}{\nu_n} \sum_{\substack{k=(\mathbf{k},i\omega_r) \\ q}} v^a_{\mathbf{k}} v^b_{-\mathbf{k}+\mathbf{q}} \left[C(q)G^+(k)G^-(k)G^+(-k+q)G^-(-k+q) - \{i\nu_n \to 0\} \right].$$

At this point we can simplify the diagram by observing that, to extract the most singular, long-distance effects of localization, we can ignore the smooth q dependence of the conduction electron lines. By setting $q = 0$ along the conduction lines, we decouple $\Delta\sigma$ into a product of two terms:

$$\Delta \sigma^{ab}(i\nu_n) = \frac{2e^2 T}{\nu_n} \sum_{\mathbf{q}} C(q) \overbrace{\sum_{k} v^a_{\mathbf{k}} v^b_{-\mathbf{k}}}^{-\frac{\nu_n}{2\pi T} \frac{ne^2}{m} \delta^{ab} \int d\epsilon} \left[(G^+(k))^2 (G^-(k))^2 - \{i\nu_n \to 0\} \right]$$

$$= -\frac{ne^2}{m} \delta^{ab} \frac{1}{2\pi N(0)\tau^2} \int \frac{d^d q}{(2\pi)^d} \frac{1}{Dq^2 + |\nu_n|} \int \frac{d\epsilon}{2\pi} G^2_R(\epsilon)G^2_A(\epsilon). \quad (10.104)$$

The energy integral in the second term yields

$$\int \frac{d\epsilon}{2\pi} G^2_R(\epsilon)G^2_A(\epsilon) = 2\tau^3. \quad (10.105)$$

We need to consider the upper and lower bounds to the momentum integral. The upper bound is set by the condition that $Dq^2 = \tau^{-1}$, the elastic scattering rate. The lower bound is set either by the size of the system L, in which case $q = L^{-1}$, or by the *inelastic* scattering rate τ_i^{-1}. We may define

$$\tau_0^{-1} = \max \left(\frac{D}{L^2}, \tau_i^{-1} \right) \quad (10.106)$$

as the inverse time-scale associated with the lower cut off. The quantity

$$E_{th} = \hbar \frac{D}{L^2} \qquad (10.107)$$

is called the *Thouless energy*, and corresponds to the energy scale associated with the phase-coherent diffusion of electrons from one side of the sample to the other. In an ultra-pure, or small, system, it is this scale that provides the infrared cut-off to localization effects. We may then write

$$\Delta \sigma^{ab}(\nu) = -\delta^{ab} \left(\frac{ne^2 \tau}{m} \right) \frac{1}{2\pi N(0)} \int_{(D\tau_o)^{-1/2}}^{(D\tau)^{-1/2}} \frac{d^d q}{(2\pi)^d} \frac{1}{Dq^2 - i\nu}. \qquad (10.108)$$

If we apply a sudden pulse of electric field $E = E_0 \delta(t)$, corresponding to a field with a white-noise field spectrum $E(\nu) = E_0$, the current induced by localization effects has a frequency spectrum

$$j(\nu) = \Delta \sigma(\nu) E(\nu) = \Delta \sigma(\nu) E_0 \propto \int_{(D\tau_0)^{-1/2}}^{(D\tau)^{-1/2}} \frac{d^d q}{(2\pi)^d} \frac{1}{Dq^2 - i\nu}. \qquad (10.109)$$

In highly phase-coherent systems, the characteristic time scale of the localization back-scattering response in the current pulse is given by $t \sim D/L^2$, which we recongnize as the time for electrons to diffuse across the entire sample. This is a kind of backscatter-ing "echo" produced by the phase-coherent diffusion of electrons along time-reversed paths that cross the entire sample. The momentum integral in $\Delta \sigma$ is strongly dependent on dimensionality. In three and higher dimensions this term is finite, so that the weak-localization effects are a perturbation to the Drude conductivity. However, if the dimension $d \le 2$, this integral becomes divergent, and in a non-interacting system it is cut off only by the frequency or the finite size L of the system. In two dimensions,

$$\int_{(D\tau_o)^{-1/2}}^{(D\tau)^{-1/2}} \frac{d^d q}{(2\pi)^d} \frac{1}{Dq^2 - i\nu} = \frac{1}{4\pi D} \ln \left(\frac{\tau}{\tau_0} \right), \qquad (10.110)$$

giving rise to a localization correction to the static conductivity:

$$\Delta \sigma = - \left(\frac{ne^2 \tau}{m} \right) \frac{1}{8\pi^2 N(0) D} \ln \left(\frac{\tau_0}{\tau} \right). \qquad (10.111)$$

Replacing $n\tau/m \to 2N(0)D$, we obtain

$$\Delta \sigma = - \left(\frac{e^2}{2\pi^2} \right) \ln(\frac{\tau}{\tau_0}) \to - \frac{1}{2\pi^2} \left(\frac{e^2}{\hbar} \right) \ln \left(\frac{\tau_0}{\tau} \right), \qquad (10.112)$$

where we have restored $\hbar$ to the expression. The quantity $g_0 = \frac{e^2}{\hbar} \sim \frac{1}{10} (k\Omega)^{-1}$ is known as the *universal conductance*.

There are a number of interesting consequences of these results:

- By replacing $2\pi N(0)D = \frac{1}{2} k_F l$, the total conductivity can be written

$$\sigma = \sigma_0 \left[1 - \frac{1}{2\pi k_F l} \ln \left(\frac{\tau_0}{\tau} \right) \right]. \qquad (10.113)$$

We see that the quantum interference correction to the conductivity is of order $O(1/(k_F l))$, justifying their neglect in our earlier calculations.

- If we consider the case where inelastic scattering is negligible, the localization correction to the conductivity in two dimensions is

$$
\sigma = \sigma_0 \left[1 - \frac{1}{2\pi k_F l} \ln(\frac{1}{E_{Th}\tau}) \right]
$$

$$
\sim \sigma_0 \left[1 - \frac{1}{\pi k_F l} \ln(\frac{L}{l}) \right] \tag{10.114}
$$

so that the conductivity drops gradually to zero as the size of the sample increases. The conductivity becomes of order $\frac{e^2}{\hbar}$ at the *localization length*

$$
L_c \sim l e^{k_F l}, \tag{10.115}
$$

independently of the strength of the interaction. In two dimensions, resistivity and resistance have the same dimension, so we expect that when the size of the system is equal to the localization length, the resistivity is always of order $10\text{k}\Omega$! At longer length scales, the material evolves into an insulator.

- The weak localization corrections are not divergent for dimensions greater than two, but become much stronger in dimensions below $d = 2$. It was this observation that led Abrahams, Anderson, Licciardello, and Ramakrishnan to propose the scaling theory for localization, in which $d_c = 2$ is the critical dimensionality.

We shall end this section with a brief remark about the scaling theory of localization. Stimulated by the results in two dimensions, and earlier work on one-dimensional wires by Licciardello and Thouless [7], Abrahams *et al.* [8] were led to propose that, in any dimension, conductance or inverse resistance $G = 1/R$ could always be normalized to form a dimensionless parameter

$$
g(L) = \frac{G(L)}{\frac{e^2}{\hbar}} \tag{10.116}
$$

which satisfies a one-parameter scaling equation:

$$
\frac{d \ln g(L)}{d \ln L} = \beta(g). \tag{10.117}
$$

When this quantity is large we may use the Drude model, so that $g(L) = \frac{ne^2\tau}{m} L^{d-2}$ and

$$
\beta(g) = (d - 2) \qquad (g \to \infty) \tag{10.118}
$$

is independent of g. When the conductance was small $g \to 0$ on scales longer than the localization length L_c, they argued that $g(L)$ would decay exponentially, $g(L) \sim e^{-L/L_c}$, so that, for small conductance,

$$
\beta(g) \sim -\ln g \qquad (g \to 0). \tag{10.119}
$$

By connecting these two asymptotic limits, Abrahams *et al.* [8] reasoned that the beta function for conductance would take the form shown in Figure 10.6. In dimensions $d \leq 2$, $\beta(g)$ is always negative, so the conductance always scales to zero and electrons are

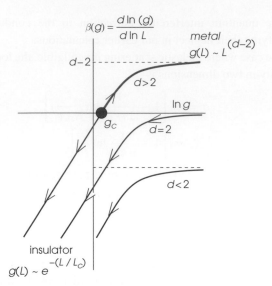

Fig. 10.6 The scaling function $\beta(g)$ deduced by Abrahams *et al.* [8] for a non-interacting metal. For $d > 2$ there is critical conductance g_c which gives rise to a disorder-driven metal–insulator transition. For $d \leq 2$ disorder always gives rise to localization and the formation of an insulator.

always localized. However, in dimensions $d > 2$ there is a disorder-driven metal–insulator transition at the critical conductance $g = g_c$. As the amount of disorder is increased when the short-distance conductance g passes below g_c, the material becomes an insulator in the thermodynamic limit. These heuristic arguments stimulated the development of a whole new field of research into the collective effects of disorder on conductors, and the basic results of the scaling theory of localization are well established in metals, where the effects of interactions between electrons are negligible. Interest in this field continues actively today, with the surprise discovery in the late 1990s that two-dimensional electron gases formed within heterojunctions appear to exhibit a metal–insulator transition – a result that confounds the one-parameter scaling theory and is thought in some circles to result from electron–electron interaction effects [12].

Exercises

Exercise 10.1 Alternative derivation of the electrical conductivity.

In our treatment of the electrical conductivity, we derived

$$\sigma^{ab}(i\nu_n) = e^2 \frac{T}{\nu_n} \sum_{\mathbf{k}, i\omega_r} v^a_{\mathbf{k}} v^b_{\mathbf{k}} \left[G(\mathbf{k}, i\omega_r + i\nu_n) G(\mathbf{k}, i\omega_r) - G(\mathbf{k}, i\omega_r)^2 \right].$$

This integral was carried out by first integrating over momentum, then integrating over frequency. This techique is hard to generalize, and it is often more convenient to integrate the expression in the opposite order. This is the topic of this question. Consider the case where

$$G(\mathbf{k}, i\omega_r) = \frac{1}{i\omega_r - \epsilon_\mathbf{k} - \Sigma(i\omega_r)}$$

and $\Sigma(i\omega_r)$ is any momentum-independent self-energy.

(a) By rewriting the momentum integral as an integral over kinetic energy ϵ and angle, show that the conductivity can be rewritten as $\sigma^{ab}(i\nu_n) = \delta^{ab}\sigma(i\nu_n)$, where

$$\sigma(i\omega_n) = \frac{ne^2}{m}\frac{1}{\nu_n}\int_{-\infty}^{\infty} d\epsilon \; T\sum_{i\omega_r}\left[G(\epsilon, i\omega_r + i\nu_n)G(\epsilon, i\omega_r) - G(\epsilon, i\omega_r)^2\right]$$

and

$$G(\epsilon, z) \equiv \frac{1}{z - \epsilon - \Sigma(z)}.$$

(b) Carry out the Matsubara sum in the above expression, to obtain

$$\sigma(i\omega_n) = \frac{ne^2}{m}\frac{1}{\nu_n}\int_{-\infty}^{\infty}\frac{d\omega}{\pi}\int_{-\infty}^{\infty} d\epsilon f(\omega)\left[G(\epsilon, \omega + i\nu_n) + G(\epsilon, \omega - i\nu_n)\right]A(\epsilon, \omega),$$

where $A(\epsilon, \omega) = \mathrm{Im}\, G(\epsilon, \omega - i\delta)$. (*Hint*: replace $T\sum_n \rightarrow -\int\frac{dz}{2\pi i}f(z)$, and notice that while $G(\epsilon, z)$ has a branch cut along $z = \omega$ with discontinuity given by $G(\epsilon, \omega - i\delta) - G(\epsilon, \omega + i\delta) = 2iA(\epsilon, \omega)$, while $G(\epsilon, z + i\nu_n)$ has a similar branch cut along $z = \omega - i\nu_n$. Wrap the contour around these branch cuts and evaluate the result.)

(c) Carry out the energy integral in the above expression, to obtain

$$\sigma(i\omega_n) = \frac{ne^2}{m}\frac{1}{\nu_n}\int_{-\infty}^{\infty}\frac{d\omega}{\pi}f(\omega)\left[\frac{1}{i\nu_n - (\Sigma(\omega + i\nu_n) - \Sigma(\omega - i\delta))}\right.$$
$$\left. - \frac{1}{i\nu_n - (\Sigma(\omega + i\delta) - \Sigma(\omega - i\nu_n))}\right].$$
$$\text{(10.120)}$$

(d) Carry out the analytic continuation in the above expression, to finally obtain

$$\sigma(\nu + i\delta) = \frac{ne^2}{m}\int_{-\infty}^{\infty} d\omega \left[\frac{f(\omega - \nu/2) - f(\omega + \nu/2)}{\nu}\right]$$
$$\times \frac{1}{-i\nu + i(\Sigma(\omega + \nu/2 + i\delta) - \Sigma(\omega - \nu/2 - i\delta))}. \quad \text{(10.121)}$$

(e) Show that your expression for the optical conductivity can be rewritten in the form

$$\sigma(\nu + i\delta) = \frac{ne^2}{m}\int_{-\infty}^{\infty} d\omega \left[\frac{f(\omega - \nu/2) - f(\omega + \nu/2)}{\nu}\right]$$
$$\times \frac{1}{\tau^{-1}(\omega, \nu) - i\nu Z(\omega, \nu)}, \quad \text{(10.122)}$$

where

$$\tau^{-1}(\omega, \nu) = \mathrm{Im}\left[\Sigma(\omega - nu/2 - i\delta) + \Sigma(\omega + \nu/2 - i\delta)\right] \quad \text{(10.123)}$$

is the average of the scattering rate at frequencies $\omega \pm \nu/2$ and

$$Z^{-1}(\omega, \nu) - 1 = -\frac{1}{\nu}\text{Re}\left[\Sigma(\omega - \nu/2) - \Sigma(\omega + \nu/2)\right]$$

is a kind of "wavefunction renormalization."

(f) Show that, if the ω dependence of Z and τ^{-1} can be neglected, one arrives at the phenomenological form

$$\sigma(\nu) = \frac{ne^2}{m}\left[\frac{1}{\tau^{-1}(\nu) - i\nu Z^{-1}(\nu)}\right].$$

This form is often used to analyze optical spectra.

(g) Show that the zero-temperature conductivity is given by the thermal average

$$\sigma(\nu + i\delta) = \frac{ne^2\tau}{m}, \tag{10.124}$$

where $\tau^{-1} = 2\,\text{Im}\Sigma(0 - i\delta)$.

References

[1] P. Drude, Zür Elektronentheorie der Metalle (On the electron theory of metals), *Ann. Phys. (Leipzig)*, vol. 1, p. 566, 1900.

[2] F. London, New Conception of Supraconductivity, *Nature*, vol. 140, p. 793, 1937.

[3] F. London, *Superfluids*, vols. 1–2, Dover Publications, 1961–1964.

[4] R. Kubo, Statistical mechanical theory of irreversible processes. I: general theory and simple applications to magnetic and conduction problems, *J. Phys. Soc. Jpn.*, vol. 12, p. 570, 1957.

[5] J. S. Langer and T. Neal, Breakdown of the concentration expansion for the impurity resistivity of metals, *Phys. Rev. Lett.*, vol. 16, p. 984, 1966.

[6] P. W. Anderson, Absence of diffusion in certain random lattices, *Phys. Rev.*, vol. 109, p. 1492, 1958.

[7] D. C. Licciardello and D. J. Thouless, Constancy of minimum metallic conductivity in two dimensions, *J. Phys. C: Solid State Phys.*, vol. 8, p. 4157, 1975.

[8] E. Abrahams, P. W. Anderson, D. C. Licciardello, and T. V. Ramakrishnan, Scaling theory of localization: absence of quantum diffusion in two dimensions, *Phys. Rev. Lett.*, vol. 42, p. 673, 1979.

[9] P. A. Lee and T. V. Ramakrishnan, Disordered electronic systems, *Rev. Mod. Phys.*, vol. 57, p. 287, 1985.

[10] B. L. Altshuler, A. G. Aronov, A. I. Larkin, and D. E. Khmelnitskii, Anomalous magnetoresistance in semiconductors, *J. Exp. Theor. Phys.*, vol. 54, p. 411, 1981.

[11] G. Bergmann, Quantitative analysis of weak localization in thin Mg films by magnetoresistance measurements, *Phys. Rev. B*, vol. 25, p. 2937, 1982.

[12] S. V. Kravchenko and M. P. Sarachik, Metal–insulator transition in two-dimensional electron systems, *Rep. Prog. Phys.*, vol. 67, p. 1, 2004.

11.1 Order parameter concept

The idea that phase transitions involve the development of an *order parameter* which lowers or breaks the symmetry is one of the most beautiful ideas of many-body physics. In this chapter, we introduce this new concept, which plays a central role in our understanding of the way complex systems transform themselves into new states of matter at low temperatures.

Lev Landau introduced the order parameter concept in 1937 [1] as a means to quantify the dramatic transformation of matter at a phase transition. Examples of such transformations abound: a snowflake forms when water freezes (Figure 11.1); iron becomes magnetic when electron spins align in a single direction; superfluidity and superconductivity develop when quantum fluids are cooled and bosons or pairs of fermions condense into a single quantum state with a well-defined phase. Phase transitions can even take place in the very fabric of space, and there is very good evidence that we are living in a broken, symmetry universe, which underwent one or more phase transitions which broke the degeneracy between the fundamental forces [2] shortly after the Big Bang. Indeed, when the sun shines on our faces we are experiencing the consequences of this broken symmetry. Remarkably, while the microscopic physics of each case is different, they are unified by a single concept.

Landau's theory associates each phase transition with the development of an order parameter ψ once the temperature drops below the transition temperature T_c:

$$
|\psi| = \begin{cases} 0 & (T > T_c) \\ |\psi_0| > 0 & (T < T_c). \end{cases}
$$

The order parameter can be a real or complex number, a vector, or a spinor that can, in general, be related to an n-component real vector $\psi(x) = (\psi_1, \psi_2, \ldots \psi_n)$ (see Table 11.1). Microscopically, each order parameter is directly related to the expectation value of a quantum operator. Thus, in an Ising ferromagnet $m = \langle \sigma_z(x) \rangle$ is the expectation value of the spin density along a particular anisotropic axis, while in a Heisenberg ferromagnet the magnetization can point in any direction, so that the order parameter is a vector pointing in the direction of the spin density $\vec{m} = \langle \vec{\sigma}(x) \rangle$. In a superconductor or superfluid, the order parameter is a complex number related to the expectation value of a bosonic field in the condensate (see Table 11.1).

The emergence of an order parameter often has dramatic macroscopic consequences in a material. In zero gravity, water droplets are perfectly spherical, yet if cooled through their freezing point they form crystals of ice with the classic six-fold symmetry of a snowflake.

Table 11.1 Examples of Order parameters.

Order parameter	Realization	Microscopic origin
$m = \psi_1$	Ising ferromagnet	$\langle \hat{\sigma}_z \rangle$
$\psi = \psi_1 + i\psi_2$	Superfluid, superconductor	$\langle \hat{\psi}_B \rangle,\ \langle \hat{\psi}_\uparrow \hat{\psi}_\downarrow \rangle$
$\vec{M} = (\psi_1, \psi_2, \psi_3)$	Heisenberg ferromagnet	$\langle \vec{\sigma} \rangle$
$\Phi = \begin{pmatrix} \psi_1 + i\psi_2 \\ \psi_3 + i\psi_4 \end{pmatrix}$	Higgs field	$\begin{pmatrix} \langle \hat{\phi}_+ \rangle \\ \langle \hat{\phi}_- \rangle \end{pmatrix}$

$\psi = 0$ $\psi \neq 0$

Fig. 11.1 Broken symmetry. The development of crystalline order within a spherical water droplet leads to the formation of a snowflake, reducing the symmetry from spherical to six-fold. Snowflake figure reproduced with permission from Kenneth G. Libbrecht.

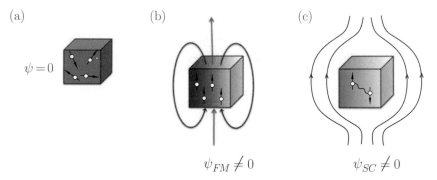

(a) (b) (c)

$\psi = 0$

$\psi_{FM} \neq 0$ $\psi_{SC} \neq 0$

Fig. 11.2 (a) In a normal metal, there is no long-range order. (b) Below the Curie temperature T_c of a ferromagnet, electron spins align to develop a ferromagnetic order parameter. The resulting metal has a finite magnetic moment. (c) Below the transitition temperature of a superconductor, electrons pair together to develop a superconducting order parameter. The resulting metal exhibits the Meissner effect, excluding magnetic fields from its interior.

We say that the symmetry of the water has "broken the symmetry" because the symmetry of the ice crystal no longer enjoys the continuous rotational symmetry of the original water droplet. Equally dramatic effects occur within quantum fluids. Thus, when a metal develops a ferromagnetic order parameter, it spontaneously develops an internal magnetic field. By contrast, when a metal develops superconducting order, it behaves as a perfect diamagnet, and will spontaneously expel magnetic fields from its interior even when cooled in a magnetic field, giving rise to what is called the *Meissner effect* (see Figure 11.2).

Part of the beauty of Landau theory is that the precise microscopic expression for the order parameter is not required in developing a theory of the macroscopic consequences of broken symmetry. The Ginzburg–Landau theory of superconductivity pre-dated the microscopic theory by seven years. Landau theory provides a coarse-grained description of the properties of matter. In general, the order parameter description is good on length scales larger than

$$\xi_0 = \text{coherence length.} \tag{11.1}$$

On length scales larger than the coherence length, the internal structure of the order parameter is irrelevant and it behaves as a smoothly varying function that has forgotten about its microscopic origins. However, physics on scales smaller than ξ_0 requires a microscopic description. For example, in a superconductor the coherence length is a measure of the size of a Cooper pair–a number that can be hundred or thousands of atom spacings–while in superfluid ^{4}He, the coherence length is basically an atom spacing.

11.2 Landau theory

11.2.1 Field-cooling and the development of order

The basic idea of Landau theory is to write the free energy as a function $F[\psi]$ of the order parameter. To keep things simple, we will begin our discussion with the simplest case, when ψ is a one-component Ising order parameter representing, for example, the magnetization of an Ising ferromagnet. We begin by considering the meaning of an order parameter and the relationship of the free energy to the microscopic physics.

We can always induce the order parameter to develop by cooling in the presence of an external field h that couples to the order parameter. In general the inverse dependence of the field on the order parameter, $h[\psi]$, will be highly non-linear, but once we know it we can convert the dependence of the energy on h to a function of ψ. Broken symmetry develops if ψ remains finite once the external field is removed.

Mathematically, an external field introduces a source term into the microscopic Hamiltonian:

$$H \rightarrow H - h \int d^3x \, \hat{\psi}(x).$$

The field h that couples linearly to the order parameter is called the *conjugate field*. For a magnet, where $\psi \equiv M$ is the magnetization, $h \equiv B$ is the external magnetic field. For a ferroelectric material, where $\psi \equiv P$ is the electric polarization, the conjugate field $h \equiv E$ is the external electric field. For many classes of order parameter, such as the pair density of a superconductor or the staggered magnetization of an antiferromagnet, although there is no naturally occurring external field that couples linearly to the order parameter the idea of a conjugate field is still a very useful concept.

The free energy of the system in the presence of an external field is a Gibbs free energy which takes account of the coupling to the field, $G[h] = F[\psi] - V\psi h$. $G[h]$ is given by

$$G[h] = -k_B T \ln\big(Z[h]\big) = -k_B T \ln\left(\text{Tr}\left[e^{-\beta(\hat{H} - h \int \hat{\psi} d^3 x)}\right]\right), \qquad (11.2)$$

where the partition function $Z[h]$ involves the trace over the many-body system. If we differentiate (11.2) with respect to h, we recover the expectation value of the induced order parameter $\psi[h] = \langle \hat{\psi} \rangle$:

$$\psi(h, V) = \frac{1}{Z[h]} \text{Tr}\left[e^{-\beta(H - h \int \psi d^3 x)} \hat{\psi}(x)\right] = -\frac{1}{V} \frac{\partial G[h]}{\partial h}. \qquad (11.3)$$

It follows that $-\delta G = \psi V \delta h$.

In a finite system, the order parameter will generally disappear once we remove the finite field. For example, if we take a molecular spin cluster and field-cool it below its bulk Curie temperature, it will develop a finite magnetization. However, once we remove the external field, thermal fluctuations will generate domains with reversed order. Each time a domain wall crosses the system, the magnetization reverses, so that on long enough time scales the magnetization will average to zero. But as the size of the system grows beyond the nano-scale, two things will happen. First, infinitesimal fields will prevent the thermal excitation of macroscopic domains. Second, even in a truly zero field, the probability to form these large domains becomes astronomically small (see Example 11.1). In this way, broken symmetry "freezes into" the system and becomes stable in the thermodynamic limit.

From this line of reasoning, it becomes clear that the development of a thermally stable order parameter requires that we take the thermodynamic limit $V \to \infty$ before we remove the external field. When we field-cool an infinitely large system below a second-order phase transition, the order parameter remains after the external field is removed. The equilibrium order parameter is then defined as

$$\psi = \lim_{h \to 0} \lim_{V \to \infty} \psi(h, V).$$

To obtain the Landau function $F[\psi]$, we must write $G[h]$ in terms of ψ; then,

$$F[\psi] = G[h] + V h \psi = G[h] - h \frac{\partial G[h]}{\partial h}.$$

This expression for $F[\psi]$ is a Legendre transformation of $G[h]$. Since $\delta G = -V \psi \delta h$, $\delta F = \delta G + V \delta(h \psi) = V h \delta \psi$, so the inverse transformation is $h = V^{-1} \frac{\partial F}{\partial \psi}$. If $h = 0$, then

$$hV = \frac{\partial F}{\partial \psi} = 0,$$

which states the intuitively obvious fact that when $h = 0$ the equilibrium value of ψ is determined by a stationary point of $F[\psi]$.

Example 11.1 Consider a cubic nanomagnet of $N = L^3$ Ising spins interacting via a nearest-neighbor ferromagnetic interaction of strength J. Suppose the dynamics can be approximated by Monte Carlo dynamics, in which each spin is "updated" after a time τ_0. At $T = 2J$ (the bulk $T_c = 4.52J$), estimate the time, in units of τ_0, required to form a domain that will cross the entire sample. If $\tau_0 = 1$ ns, estimate the minimum size L for the decay time of the total magnetization to become comparable with the time span of a PhD degree.

Solution

To form a domain wall of area $A \sim L^2$ costs free energy $\Delta F \sim 2JL^2$, occurring with probability $p \sim e^{-(\Delta F/T)}$. The time required for formation may be estimated to be

$$\tau \sim \tau_0 p^{-1} \sim \tau_0 e^{2JL^2/T},$$

where the most important aspect of the estimate is that the exponent grows with L^2. Our naive estimate does not take into account the configurational entropy (the number of ways of arranging a domain wall), but it will give a rough idea of the required size. Putting $\tau_0 \sim 10^{-9}$s and $\tau = 5\,\mathrm{y} \sim 10^8$ s for a typical PhD, this requires $\tau/\tau_0 = 10^{19} \sim e^{40}$, thus $L \sim \sqrt{40} \sim 6$. Already by about $L^3 = 40^{3/2} \sim 250$ spins, the time for the magnetization to decay is of the order of years. By $N \sim 500$, this same time scale has stretched to the age of the universe.

11.2.2 The Landau free energy

Landau theory concentrates on the region of small ψ, audaciously expanding the free energy of the many-body system as a simple polynomial:

$$f_L[\psi] = \frac{1}{V}F[\psi] = \frac{r}{2}\psi^2 + \frac{u}{4}\psi^4. \tag{11.4}$$

- The Landau free energy describes the leading dependence of the total free energy on ψ. The full free energy is given by $f_{tot} = f_n(T) + f[\psi] + O[\psi^4]$, where f_n is the energy of the "normal" state without long-range order.
- For an Ising order parameter, both the Hamiltonian and the free energy are an even function of ψ: $H[\psi] = H[-\psi]$. We say that the system possesses a *global Z_2 symmetry*, because the Hamiltonian is invariant under transformations of the Z_2 group that takes $\psi \to \pm\psi$.

Provided r and u are greater than zero, the minimum of $f_L[\psi]$ lies at $\psi = 0$. Landau theory assumes that the phase transition temperuture r changes sign, so that

$$r = a(T - T_c),$$

as illustrated in Figure 11.3(a). The minimum of the free energy occurs when

$$\frac{df}{d\psi} = 0 = r\psi + u\psi^3 \Rightarrow \psi = \begin{cases} 0 & (T > T_c) \\ \pm\sqrt{\frac{a(T_c - T)}{u}} & (T < T_c), \end{cases} \tag{11.5}$$

so that for $T < T_c$ there are two minima of the free energy function.

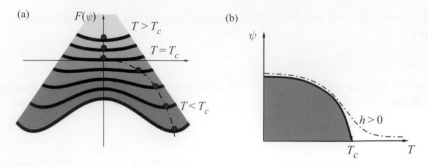

Fig. 11.3   (a) The Landau free energy $F(\psi)$ as a function of temperature for an Ising order parameter. Curves are displaced vertically for clarity. (b) Order parameter ψ as a function of temperature for a finite field $h > 0$ and an infinitesimal field $h = 0^+$.

Note the following:

- If we cool the system in a tiny external field, the sign of the order parameter reflects the sign of the field (Figure 11.3(b)):

$$\psi = \operatorname{sgn} h \sqrt{\frac{a(T_c - T)}{u}} \qquad (T < T_c). \qquad (11.6)$$

This branch cut along the temperature axis of the phase diagram is an example of a first-order phase boundary. The point $T = T_c$, $h = 0$ where the line ends is a *critical point*.

- If $u < 0$, the free energy becomes unbounded below. To cure this problem, the Landau free energy must be expanded to sixth order in ψ:

$$f[\psi] = \frac{1}{V}F[\psi] = \frac{r}{2}\psi^2 + \frac{u}{4}\psi^4 + \frac{u_6}{6}\psi^6.$$

When $u < 0$, the free energy curve develops three minima and the phase transition becomes first-order; the special point at $r = h = u = 0$ is a convergence of three critical points, called a *tri-critical point* (see Exercise 11.3).

11.2.3   Singularities at the critical point

At a second-order phase transition, the second derivatives of the free energy develop singularities. If we plug (11.6) back into the free energy $f_L[\psi]$ (11.4), we find that

$$f_L = \begin{cases} 0 & (T > T_c) \\ -\frac{a^2}{4u}(T_c - T)^2 & (T < T_c). \end{cases}$$

In this way, the free energy and the entropy $S = -\frac{\partial F}{\partial T}$ are continuous at the phase transition, but the specific heat is

$$C_V = -T\frac{\partial^2 F}{\partial T^2} = C_0(T) + \begin{cases} 0 & (T > T_c) \\ \frac{a^2 T}{2u} & (T < T_c), \end{cases} \qquad (11.7)$$

where C_0 is the background component of the specific heat not associated with the ordering process. We see that C_V jumps by an amount

$$\Delta C_V = \frac{a^2 T_c}{2u}$$

below the transition. The jump size ΔC_V has the dimensions of entropy per unit volume, and sets a characteristic size of the entropy lost per unit volume once long-range order sets in.

At a second-order transition, matter also becomes infinitely susceptible to the applied field h, as signaled by a divergence in susceptibility $\chi = \frac{\partial \psi}{\partial h}$. To see this in Landau theory, let us introduce a field by replacing

$$f(\psi) \rightarrow f(\psi) - h\psi = \frac{r}{2}\psi^2 + \frac{u}{4}\psi^4 - h\psi. \tag{11.8}$$

A finite field $h > 0$ has the effect of "tipping" the free-energy contour to the right, preferentially lowering the energy of the right-hand minimum, as illustrated in Figure 11.4. For $h \neq 0$, equilibrium requires $\partial f / \partial \psi = r\psi + u\psi^3 - h = 0$, which we can solve for $r = \frac{h}{\psi} - 4u\psi^2$. Above and below T_c, we can solve for ψ by linearizing $\psi[h] = \delta\psi + \psi_0$ around the $h = 0$ value given in (11.6), to obtain $\delta\psi = \chi(T)h + O(h^3)$ (see Figure 11.3(b)), where

$$\chi(T) = \frac{d\psi}{dh} = \frac{1}{a|T - T_c|} \times \begin{cases} 1 & (T > T_c) \\ \frac{1}{2} & (T < T_c) \end{cases} \tag{11.9}$$

describes the divergence of the susceptibility at the critical point. When we are actually at the critical point ($r = 0$), the induced order parameter is a non-linear function of field:

$$\psi = \left(\frac{h}{u}\right)^{1/3} \qquad (T = T_c). \tag{11.10}$$

The divergence of the susceptibility at the critical point means that if we cool through the critical point in the absence of a field, the tiniest stray field will produce a huge effect,

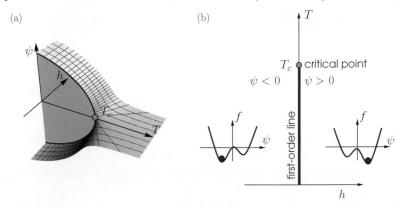

(a) (b)

Phase diagram in an applied field. A first-order line stretches along the zero-field axis, $h = 0$, up to the critical point. The equilibrium order parameter changes sign when this phase boundary is crossed. (a) Three-dimensional plot showing discontinuity in order parameter as a function of field ψ. (b) Two-dimensional phase boundary showing first-order line.

Fig. 11.4

tipping the system into either an up or down state. Once this happens, we say that the system has *spontaneously broken the Z_2 inversion symmetry* of the original Hamiltonian.

The singular power-law dependences of the order parameter, specific heat, and susceptibility near a second-order transition described by Landau theory are preserved at real second-order phase transitions, but the critical exponents are changed by the effects of spatial fluctuations of the order parameter. In general, we write

$$
\begin{aligned}
C_V &\propto (|T - T_c|)^{-\alpha} & & \text{specific heat} \\
\psi &\propto \begin{cases} (T_c - T)^{\beta} \\ h^{\frac{1}{\delta}} \end{cases} & & \text{order parameter} \\
\chi &\propto (T - T_c)^{-\gamma}, & & \text{susceptibility}
\end{aligned}
\tag{11.11}
$$

where Landau theory estimates $\alpha = 0$, $\beta = \frac{1}{2}$, $\delta = 3$, and $\gamma = 1$. Remarkably, this simple prediction of Landau theory continues to hold once the full-fledged effects of order parameter fluctuations are included. Still more remarkably, the exponents that emerge are found to be universal for each class of phase transition, independently of the microscopic physics [3].

11.2.4 Broken continuous symmetries: the "Mexican hat" potential

We now take the leap from a one- to an n-component order parameter. We shall be particularly interested in an important class of multi-component order in which the underlying physics involves a continuous symmetry that is broken by the phase transition. In this case, the *n-component* order parameter $\vec{\psi} = (\psi_1 \cdots \psi_n)$ acquires both magnitude and direction, and the discrete Z_2 inversion symmetry of the Ising model is now replaced by a continuous $O(N)$ rotational symmetry. At a phase transition, the breaking of such continuous symmetries has remarkable consequences.

The $O(N)$ symmetric Landau theory is simply constructed by replacing $\psi^2 \rightarrow |\psi|^2 = (\psi_1^2 + \cdots + \psi_n^2) = \vec{\psi} \cdot \vec{\psi}$, taking the form

$$
f_L[\vec{\psi}] = \frac{r}{2}(\vec{\psi} \cdot \vec{\psi}) + \frac{u}{4}[(\vec{\psi} \cdot \vec{\psi})]^2, \qquad O(N) \text{ invariant Landau theory}
$$

where, as before, $r = a(T - T_c)$. This Landau function is invariant under $O(N)$ rotations $\vec{\psi} \rightarrow R\vec{\psi}$ that preserve the magnitude of the order parameter. Such symmetries do not occur by accident, but owe their origin to conservation laws which protect them in both the microscopic Hamiltonian and the macroscopic Landau theory. For example, in a Heisenberg magnet the corresponding Landau theory has $O(3)$ symmetry associated with the underlying conservation of the total spin magnetization.

Once $T < T_c$, the order parameter acquires a definite magnitude and direction given by

$$
\vec{\psi} = \sqrt{\frac{|r|}{u}}\hat{n},
$$

where $\hat{n}$ is a unit (n-component) vector. By acquiring a definite direction, the order parameter breaks the $O(N)$ symmetry. In a magnet, this would correspond to the spontaneous

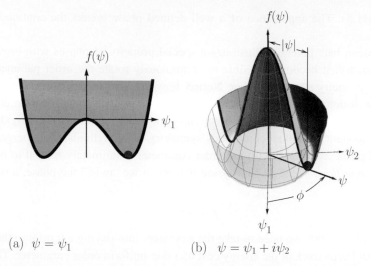

(a) $\psi = \psi_1$ (b) $\psi = \psi_1 + i\psi_2$

Dependence of free energy on order parameter for (a) an Ising order parameter $\psi = \psi_1$, showing two degenerate minima, and (b) a complex order parameter $\psi = \psi_1 + i\psi_2 = |\psi|e^{i\phi}$, where the the Landau free energy forms a Mexican hat potential in which the free energy minimum forms a rim of degenerate states with energy that is independent of the phase ϕ of the uniform order parameter.

Fig. 11.5

development of a uniform magnetization. In a superconductor or superfluid, it corresponds to the development of a macroscopic phase.

A particularly important example of a broken continuous symmetry occurs in superfluids and superconductors, where the order parameter is a single complex order parameter composed from two real order parameters, $\psi = \psi_1 + i\psi_2 = |\psi|e^{i\phi}$. In this case, the Landau free energy takes the form[1]

$$f[\psi] = r(\psi^*\psi) + \frac{u}{2}(\psi^*\psi)^2$$

$$\psi \equiv \psi_1 + i\psi_2 \equiv |\psi|e^{i\phi}. \qquad U(1) \text{ invariant Landau theory} \qquad (11.12)$$

Figure 11.5 shows the Landau free energy as a function of ψ, where the magnitude of the order parameter $|\psi|$ is represented in polar coordinates. The free-energy surface displays a striking rotational invariance, associated with the fact that the free energy is independent of the global phase of the order parameter:

$$f[\psi] = f[e^{i\alpha}\psi]. \qquad U(1) \text{ gauge invariance}$$

This is a direct consequence of the global $U(1)$ invariance of the particle fields that have condensed to develop the complex order parameter. For $T < T_c$, the negative curvature of the free-energy surface at $\psi = 0$ causes it to develop the profile of a "Mexican hat," with a continuous rim of equivalent minima where

$$\psi = \sqrt{\frac{|r|}{u}}e^{i\phi}.$$

---

[1] For complex fields, it is more convenient to work without the factor of $\frac{1}{2}$ in front of the quadratic terms. To keep the numerology simple, the interaction term is also multiplied by two.

(see Figure 11.5). The appearance of a well-defined phase breaks the continuous $U(1)$ symmetry.

The "Mexican hat" potential illustrates a special property of phases with broken continuous symmetry: it becomes possible to continuously rotate the order parameter from one broken-symmetry state to another. Notice, however, that if the order parameter is to maintain a well-defined phase or direction then it is clear that there must be an energy cost for deforming or "twisting" the direction of the order parameter. This rigidity is an essential component of broken continuous symmetry. In superfluids, the emergence of a well-defined phase associated with the order parameter is intimately related to persistent currents, or *superflow*. We shall shortly see that when we "twist" the phase, a superflow develops:

$$\vec{j} \propto \vec{\nabla}\phi.$$

To describe this rigidity, we need to take the next step, introducing a term into the energy functional that keeps track of the energy cost of a non-uniform order parameter. This leads us on to Ginzburg–Landau theory.

11.3 Ginzburg–Landau theory I: Ising order

Landau theory describes the energy cost of a uniform order parameter: a more general theory needs to account for inhomogeneous order parameters in which the amplitude varies or the direction of the order parameter is twisted. This development of Landau theory is called *Ginzburg–Landau theory*,[2] after Vitaly Ginzburg and Lev Landau [5], who developed this formalism as part of their macroscopic theory of superconductivity. We will begin our discussion of this theory with the simplest case, a one-component "Ising" order parameter.

Ginzburg–Landau theory introduces an additional energy cost $\delta f \propto |\nabla \psi|^2$ associated with gradients in the order parameter: $f_{GL}[\psi, \nabla \psi] = \frac{s}{2}|\nabla \psi|^2 + f_L[\psi(x)]$. For a single Ising order parameter, the free energy (in d dimensions) is given by

$$F_{GL}[\psi] = \int d^d x f_{GL}[\psi(x), \nabla \psi(x), h(x)]$$

$$f_{GL}[\psi, \nabla \psi, h] = \frac{s}{2}(\nabla \psi)^2 + \frac{r}{2}\psi^2 + \frac{u}{4}\psi^4 - h\psi. \tag{11.13}$$

Ginzburg–Landau free energy: one-component order

There are two points to be made here:

- Ginzburg–Landau theory is only valid near the critical point, where the order parameter is small enough to permit a leading-order expansion.

[2] The idea of using a gradient expansion of the free energy first appears in print in the work of Ginzburg and Landau. However, germs of this theory are contained in the work of Ornstein and Zernicke [4], who in 1914 developed a theory to describe critical opalescence.

- Dimensional analysis shows that $[c]/[r] = L^2$ has dimensions of length squared. The new length scale introduced by the gradient term, called the *correlation length*

$$\xi(T) = \sqrt{\frac{s}{|r(T)|}} = \xi_0 \left| 1 - \frac{T}{T_c} \right|^{-\frac{1}{2}}, \qquad \text{correlation length} \qquad (11.14)$$

sets the characteristic length scale of order-parameter fluctuations, where

$$\xi_0 = \xi(T = 0) = \sqrt{\frac{s}{aT_c}} \qquad \text{coherence length}$$

is a measure of the microscopic coherence length. Near the transition $\xi(T)$ diverges, but far from the transition it becomes comparable with the coherence length.

The traditional use of Ginzburg–Landau theory is as a *variational principle*, using the condition of stationarity, $\delta F/\delta \psi = 0$, to determine non-equilibrium configurations of the order parameter. Ginzburg–Landau theory is also the starting point for a more general analysis of thermal fluctuations around the mean-field theory. We shall return to this at the end of the chapter.

11.3.1 Non-uniform solutions of Ginzburg–Landau theory

There are two kinds of non-uniform solutions we will consider:

- linear but non-local response to a small external field
- *Soliton* or domain-wall solutions, in which the order parameter changes sign, passing through the maximum in the free energy at $\psi = 0$ (such domain walls are particular to Ising order).

To obtain the equation governing non-uniform solutions, we write

$$\delta F_{GL} = \int d^d x \, \delta \psi(x) \left[-s\nabla^2 \psi(x) + \frac{\partial f_L[\psi]}{\partial \psi(x)} \right]. \qquad (11.15)$$

Since the Ginzburg–Landau free energy must be stationary with respect to small variations in the field,

$$\frac{\delta F_{GL}}{\delta \psi(x)} = -s\nabla^2 \psi + \frac{\partial f_L[\psi]}{\partial \psi} = 0, \qquad (11.16)$$

or, more explicitly,

$$\left[(-s\nabla^2 + r) + u\psi^2 \right] \psi(x) - h(x) = 0. \qquad (11.17)$$

Susceptibility and linear response

The simplest application of Ginzburg–Landau theory is to calculate the linear response to a non-uniform applied field. For $T > T_c$, for a small linear response we can neglect the cubic term, so that $(-s\nabla^2 + r)\psi(x) = h(x)$. If we Fourier transform this equation, we obtain

$$(sq^2 + r)\psi_{\mathbf{q}} = h_{\mathbf{q}} \tag{11.18}$$

or $\psi_q = \chi_{\mathbf{q}} h_{\mathbf{q}}$, where

$$\chi_{\mathbf{q}} = \frac{1}{sq^2 + r} = \frac{1}{s(q^2 + \xi^{-2})} \tag{11.19}$$

is the momentum-dependent susceptibility and $\xi = \sqrt{s/r}$ is the correlation length defined in (11.14). Notice that $\chi_{\mathbf{q}=0} = 1/[a(T - T_c)] = r^{-1}$ is the uniform susceptibility obtained in (11.9). For large $q >> \xi^{-1}$, $\chi(q) \sim 1/q^2$ becomes strongly momentum-dependent: in other words, the response to an applied field is non-local up to the correlation length.

Example 11.2

(a) Show that in $d = 3$ dimensions, for $T > T_c$, the response of the order parameter field to an applied field is non-local and given by

$$\psi(x) = \int d^3x' \, \chi(x - x') h(x')$$

$$\chi(x - x') = \frac{\chi}{4\pi \xi^2} \frac{e^{-|x-x'|/\xi}}{|x - x'|}. \tag{11.20}$$

(b) Show that, provided $h(x)$ is slowly varying on scales of order ξ, the linear response can be approximated by

$$\psi(x) = \chi h(x).$$

Solution

(a) If we carry out the inverse Fourier transform of the response $\psi(q) = \chi(q)h(q)$, we obtain

$$\psi(x) = \int_{x'} \chi(x - x') h(x').$$

In Example (3.5) we showed that, under a Fourier transform,

$$\frac{e^{-\lambda|x|}}{|x|} \xrightarrow{\text{FT}} \frac{4\pi}{q^2 + \lambda^2},$$

so the (inverse) Fourier transform of the non-local susceptibility is

$$\chi(q) = \frac{s^{-1}}{q^2 + \xi^{-2}} \xrightarrow{\text{FT}^{-1}} \frac{1}{4\pi s} \frac{e^{-|x|/\xi}}{|x|} = \frac{\chi}{4\pi\xi^2} \frac{e^{-|x|/\xi}}{|x|}.$$

(b) At small q, we may replace $\chi(q) \approx \chi$, so that for slowly varying h in real space we can replace $\chi(x - x') \to \chi \delta^{(d)}(x - x')$. Provided h is slowly varying over lengths longer than the correlation length, $\psi(x) = \chi h(x)$.

Domain walls

Once $T < T_c$, it is energetically costly for the order parameter to deviate seriously from the equilibrium values ψ_0. Major deviations from these stable vacua can, however, take

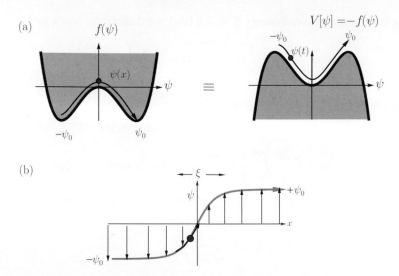

(a) $f(\psi)$ $\psi(x)$ $-\psi_0$ ψ_0 $\equiv$ $V[\psi] = -f(\psi)$ $-\psi_0$ $\psi(t)$ ψ_0

(b) ξ ψ $+\psi_0$ x $-\psi_0$

Soliton solution of Ginzburg–Landau equations. (a) The evolution of ψ in one dimension is equivalent to a particle at position ψ moving in an inverted potential $V[\psi] = -f_L[\psi]$. A soliton is equivalent to a "bounce" between maxima at $\psi = \pm\psi_0$ of $V[\psi]$. (b) The "path" that the particle traces out in time "t" $\equiv x$ defines the spatial dependence of the order parameter $\psi[x]$.

Fig. 11.6

place at *domain walls* or *solitons*, which are narrow walls of space which separate the two stable vacua of opposite sign, where $\psi = \pm\psi_0$. To change sign, an Ising order parameter must pass through zero at the center of the domain wall, passing over the hump in the free energy.

We now solve for the soliton in one dimension, where the Ginzburg–Landau equation becomes

$$s\psi'' = \frac{df_L[\psi]}{d\psi}. \tag{11.21}$$

This formula has an intriguing interpretation as Newton's law of motion for a particle of mass c moving in an inverted potential $V[\psi] = -f_L[\psi]$. This observation permits an analogy between a soliton and motion in one dimension, which enables us to to quickly develop a solution for the soliton. In this analogy, ψ plays the role of displacement while x plays the role of time. It follows that $\frac{s}{2}(\psi')^2$ is an effective "kinetic energy"[3] and the effective "energy"

$$\mathcal{E} = \frac{s}{2}(\psi')^2 - f_L[\psi]$$

is conserved and independent of x. With our simple analogy, we can map a soliton onto the problem of a particle rolling off one maximum of the inverted potential $V[\psi] = -f_L[\psi]$, "bouncing" through $\psi = 0$ out to the other maximum (Figure 11.6).

---

[3] This can be derived by multiplying (11.21) by the integrating factor ψ'. Then

$$c(\psi'\psi'') - \psi'\frac{df_L[\psi]}{d\psi} = \frac{d}{dx}\left[\frac{s}{2}(\psi')^2 - f_L[\psi]\right] = 0.$$

Fixing the conserved initial energy $\mathcal{E} = -f_L[\psi_0]$, we deduce the "velocity"

$$\psi' = \frac{d\psi}{dx} = \sqrt{\frac{2}{s}(\mathcal{E} + f_L[\psi])} = \frac{\psi_0}{\sqrt{2}\xi}\left(1 - \frac{\psi^2}{\psi_0^2}\right).$$

To make the last step, we have replaced $\psi_0^2 = \frac{|r|}{u}$ and $\xi = \sqrt{\frac{s}{|r|}}$. Solving for $dx = (\sqrt{2}\xi/\psi_0)[1 - (\tilde{\psi}/\psi_0)^2]^{-\frac{1}{2}}d\psi$ and integrating both sides yields

$$x - x_0 = \frac{\sqrt{2}\xi}{\psi_0}\int_0^\psi \frac{d\tilde{\psi}}{1 - (\tilde{\psi}/\psi_0)^2} = \sqrt{2}\xi\,\tanh^{-1}(\psi/\psi_0),$$

where $x = x_0$ is the point where the order parameter passes through zero, so that

$$\psi(x) = \psi_0\,\tanh\left(\frac{x - x_0}{\sqrt{2}\xi}\right). \qquad\qquad \text{soliton}$$

This describes a soliton solution to the Ginzburg–Landau equation, located at $x = x_0$.

Example 11.3   Show that the Ginzburg–Landau free energy of a domain wall can be written

$$\Delta F = A\frac{u}{4}\int dx[\psi_0^4 - \psi^4(x)],$$

where $A = L^{d-1}$ is the area of the domain wall. Using this result, show that surface tension $\sigma = \Delta F/A$ is given by

$$\sigma = \frac{\sqrt{8}}{3}\xi u\psi_0^4.$$

Solution

First, let us integrate by parts to write the total energy of the domain in the form

$$F = A\int dx\left[-\frac{s}{2}\psi\psi'' + f_L[\psi]\right], \tag{11.22}$$

where, for $r < 0$, $f_L[\psi] = -\frac{|r|}{2}\psi^4 + \frac{u}{4}\psi^4$. Using the Ginzburg–Landau equation (11.21),

$$s\psi'' = \frac{df_L}{d\psi} = -|r|\psi + u\psi^3.$$

Substituting into (11.22), we obtain

$$F = -A\int dx\left[-\frac{1}{2}\psi\left(-|r|\psi + u\psi^3\right) - \frac{|r|}{2}\psi^2 + \frac{u}{4}\psi^4\right]$$

$$= -uA\int dx\psi^4(x). \tag{11.23}$$

Subtracting off the energy of the uniform configuration, we then obtain

$$\Delta F = A\frac{u}{4} \int dx(\psi_0^4 - \psi^4(x)).$$

To calculate the surface tension, substitute $\psi(x) = \psi_0 \tanh[x/(\sqrt{2}\xi)]$, which gives

$$\sigma = \frac{\Delta F}{A} = \frac{u}{4}\psi_0^4 \int_{-\infty}^{\infty} dx(1 - \tanh[x/(\sqrt{2}\xi)^4]$$

$$= \frac{\xi u}{\sqrt{8}}\psi_0^4 \overbrace{\int_{-\infty}^{\infty} du(1 - \tanh[u]^4)}^{8/3} = \frac{\sqrt{8}}{3}\xi u\psi_0^4. \tag{11.24}$$

Example 11.4 Consider a two-component Dirac electron moving in one dimension through a domain wall, described by the wave equation

$$(-i\sigma_1 \nabla_x - m(x)\sigma_3)\,\psi = E\psi, \tag{11.25}$$

where the mass field forms a domain wall, changing sign at the origin according to

$$m(x) = m_0 \tanh\left(\frac{x}{\sqrt{2}\xi}\right). \tag{11.26}$$

Asymptotically, the energy of the excitations is gapped, with an excitation spectrum $E(k) = \sqrt{k^2 + m_0^2}$. Show that the domain wall gives rise to a zero-energy bound state and derive the form of its wavefunction (see Figure 11.7).

Such domain-wall zero modes are of great importance in condensed matter physics, and in higher dimensions give rise to gapless edge or surface states present in the quantum Hall effect and topological insulators.

Solution

A zero-energy bound state must satisfy

$$(-i\nabla_x \sigma_1 - m(x)\sigma_3)\,\psi = 0. \tag{11.27}$$

Consider a wavefunction of the form

$$\psi(x) = \exp\left[\sigma_2 \int_0^x \alpha(y)dy\right]\psi_0, \tag{11.28}$$

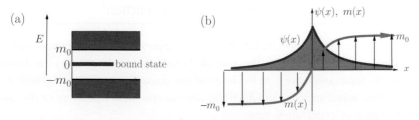

Zero-energy fermion bound state formed at domain wall. (a) Energy spectrum showing single domain-wall bound state at zero energy. (b) Form of the domain-wall bound state. See Example 11.4.

Fig. 11.7

where $\alpha(x)$ is a function to be determined. If we differentiate this with respect to x, we find $\nabla_x \psi(x) = \alpha(x)\sigma_2 \psi(x)$, so that

$$- i\sigma_1 \nabla_x \psi(x) = \overbrace{(-i\sigma_1 \sigma_2)}^{\sigma_3} \alpha(x)\psi(x) = \sigma_3 \alpha(x)\psi(x) \tag{11.29}$$

and hence

$$[-i\nabla_x \sigma_1 - m(x)\sigma_3]\,\psi = [\alpha(x) - m(x)]\sigma_3 \psi = 0 \tag{11.30}$$

if $\alpha(x) = m(x)$, so that

$$\psi(x) = \exp\left[\sigma_2 \int_0^x m(y)dy\right]\psi_0 \tag{11.31}$$

is a zero-energy state. Now in order that this be a true zero-energy bound state, it must be normalizable and decay at both positive and negative infinity. First note that, since there is a domain wall, where $m(x) = -m(-x)$ is an odd function of x, the integrand in the exponent becomes an even function of x, so we can write

$$\psi(x) = \exp\left[\sigma_2 \int_0^{|x|} m(y)dy\right]\psi_0. \tag{11.32}$$

Now if we choose

$$\psi_0 = \begin{pmatrix} 1 \\ -i \end{pmatrix} \tag{11.33}$$

to be an eigenstate of σ_2 with $\sigma_2 = -1$, $\sigma_2 \psi_0 = -\psi_0$, then the zero-energy state becomes

$$\psi(x) = \exp\left[-\int_0^{|x|} m(y)dy\right]\begin{pmatrix} 1 \\ -i \end{pmatrix}, \tag{11.34}$$

showing that the state decays at both positive and negative infinity, forming a true zero-energy bound state.

11.4 Ginzburg–Landau II: complex order and superflow

11.4.1 A "macroscopic wavefunction"

We now turn to the Ginzburg–Landau theory of complex or two-component order parameters. Here we shall focus on the use of Ginzburg–Landau theory to understand superfluids and superconductors. At the heart of our discussion is the emergence of a kind of "macroscopic wavefunction" in which the microscopic field operators of the quantum fluid $\hat{\psi}(x)$ acquire an expectation value

$$\langle \hat{\psi}(x) \rangle \equiv \psi(x) = |\psi(x)|e^{i\phi(x)}, \qquad \text{``macroscopic wavefunction''}$$

complete with phase. The magnitude of this order parameter determines the density of particles in the superfluid:

$$|\psi(x)|^2 = n_s(x),$$

while the twist or gradient of the phase determines the superfluid velocity:

$$\mathbf{v}_s(x) = \frac{\hbar}{m}\nabla\phi(x).$$

The idea that the wavefunction can acquire a kind of Newtonian reality in a superfluid or superconductor goes deeply against our training in quantum physics: at first sight, it appears to defy the Copenhagen interpretation of quantum mechanics, in which $\psi(x)$ is an unobservable variable. The bold idea suggested by Ginzburg and Landau is that $\psi(x)$ is a macroscopic manifestation of quintillions of particles–bosons–all condensed into precisely the same quantum state. Even the great figures of the field – including Landau himself – found this hard to absorb, and debate continues today. Yet on his issue history and discovery appear to have sided consistently with the bold, if perhaps naive, interpretation of the super-conducting and superfluid order parameter as an essentially real, observable property of quantum fluids.[4] It is the classic example of an *emergent phenomenon* – one of the many collective properties of matter that we are still discovering today which are not a priori self-evident from the microscopic physics.

Vitalii Ginzburg and Lev Landau introduced their theory in 1950 as a phenomenological theory of superconductivity, in which $\psi(x)$ played the role of a macroscopic wavefunction whose microscopic origin was, at the time, unknown. We shall begin by illustrating the application of this method to superfluids. For a superfluid, the Ginzburg–Landau free energy density is

$$f_{GL}[\psi, \nabla\psi] = \frac{\hbar^2}{2m}|\nabla\psi|^2 + r|\psi|^2 + \frac{u}{2}|\psi|^4. \tag{11.35}$$

Ginzburg–Landau free energy: superfluid

Before continuing, let us make a few heuristic remarks about the Ginzburg–Landau free energy:

- The Ginzburg–Landau free energy is to be interpreted as the energy density of a condensate of bosons in which the field operator behaves as a complex order parameter. This leads us to identify the coefficient of the gradient term,

$$s|\nabla\psi|^2 \equiv \frac{\hbar^2}{2m}\left\langle\nabla\hat{\psi}^\dagger\nabla\hat{\psi}\right\rangle, \tag{11.36}$$

as the kinetic energy, so that $s = \frac{\hbar^2}{2m}$.

---

[4] On more than one occasion, senior physicists have advised their students and younger colleagues against such a brash interpretation. One such story took place in Moscow in 1953. Shortly after Ginzburg–Landau theory was introduced, a young student of Landau, Alexei Abrikosov, showed that a naive classical interpretation of the order parameter field led naturally to the predication of quantized vortices and superconducting vortex lattices. Landau himself could not bring himself to make this leap and persuaded his student to shelve the theory. It was only after Feynman published a theory of vortices in superfluid helium that Landau accepted the idea, clearing the way for Abrikosov to finally publish his paper [6].

- As in the case of Ising order, the correlation length or *Ginzburg–Landau coherence length* governing the characteristic range of amplitude fluctuations of the order parameter is given by

$$\xi = \sqrt{\frac{s}{|r|}} = \sqrt{\frac{\hbar^2}{2M|r|}} = \xi_0 \left(1 - \frac{T}{T_c}\right)^{-1/2}, \tag{11.37}$$

where $\xi_0 = \xi(T = 0) = \sqrt{\frac{\hbar^2}{2maT_c}}$ is the coherence length. Beyond this length scale, only phase fluctuations survive.

- If we freeze out fluctuations in amplitude, writing $\psi(x) = \sqrt{n_s}e^{i\phi(x)}$, then $\nabla\psi = i\nabla\phi\,\psi$ and $|\nabla\psi|^2 = n_s(\nabla\phi)^2$, and the residual dependence of the kinetic energy on the twist in the phase is

$$\frac{\hbar^2 n_s}{2m}(\nabla\phi)^2 = \frac{mn_s}{2}\overbrace{\left(\frac{\hbar}{m}\nabla\phi\right)^2}^{\mathbf{v}_s^2}.$$

Since mn_s is the mass density, we see that a twist of the phase results in an increase in the kinetic energy that we may associate with a "superfluid velocity"

$$\mathbf{v_s} = \frac{\hbar}{m}\nabla\phi.$$

11.4.2 Off-diagonal long-range order and coherent states

What then is the meaning of the complex order parameter ψ? It is tempting to associate it with the expectation value of the field operator:

$$\langle\hat{\psi}(x, t)\rangle = \psi(x, t).$$

Yet, paradoxically, a field operator links states with different particle numbers, so such an expectation value can never develop in a state with a definite number of particles. One way to avoid this problem, proposed by Oliver Penrose and Lars Onsager [7, 8], is to define the order parameter in terms of correlation functions. The authors noted that, even in a state with a definite particle number, broken symmetry manifests itself as a long-distance factorization [9] of the correlation function $\langle\psi^\dagger(x)\psi(x)\rangle$:

$$\langle\psi^\dagger(x')\psi(x)\rangle \xrightarrow{|x'-x|\gg\xi} \psi^*(x')\,\psi(x) + \text{small terms} \tag{11.38}$$

off-diagonal long-range order

in terms of the order parameter. This property is called *off-diagonal long-range order* (ODLRO) [10].

However, a more modern view is that in macroscopic systems we don't need to restrict our attention to states of definite particle number; indeed, once we bring a system into contact with a bath of particles, quantum states of indefinite particle number do arise. This issue also arises in a ferromagnet, where the analogue of particle number is the conserved magnetization S_z along the z-axis. A ferromagnet of N spins polarized in the z direction has wavefunction

$$|Z\rangle = \prod_{\substack{\otimes \\ i=1,N}} |\uparrow\rangle_i.$$

However, if we cool the magnet in a field aligned along the x-axis, coupled via the Hamiltonian $H = -2BS_x = -B(S^+ + S^-)$, then once we remove the field at low temperatures the magnet remains polarized in the x direction:

$$|X\rangle = \prod_{\substack{\otimes \\ i=1,N}} |\rightarrow\rangle_i = \prod_{\substack{\otimes \\ i=1,N}} \left(\frac{|\uparrow\rangle + |\downarrow\rangle}{\sqrt{2}}\right)_i.$$

Thus the coherent exchange of spin with the environment leads to a state that contains an admixture of states of different S_z. In a similar way, we may consider cooling a quantum fluid in a field that couples to the superfluid order parameter. Such a field is created by a *proximity effect* induced by the exchange of particles with a pre-cooled superfluid in close vicinity, giving rise to a field term in the Hamiltonian such as

$$H' = -\Delta \int d^d x [\psi^\dagger(x) + \psi(x)].$$

When we cool below the superfluid transition temperature T_c in the presence of this pairing field, removing the proximity field at low temperatures, then, like a magnet, the resulting state acquires an order parameter forming a stable state of indefinite particle number.[5] To describe such states requires the many-body equivalent of wavepackets: a type of state called a *coherent state*.

Coherent states are eigenstates of the field operator

$$\hat{\psi}(x)|\psi\rangle = \psi(x)|\psi\rangle. \tag{11.39}$$

These states form an invaluable basis for describing superfluid states of matter. A coherent state can be simply written as

$$|\psi\rangle \sim e^{\sqrt{N_s}b^\dagger}|0\rangle, \qquad \text{coherent state} \tag{11.40}$$

where

$$b^\dagger = \frac{1}{\sqrt{N_s}} \int d^d x \, \psi(x)\hat{\psi}^\dagger(x)$$

coherently adds a boson to a condensate with wavefunction $\psi(x)$. Here, $N_s = \int d^d x |\psi(x)|^2$ is the *average* number of bosons in the superfluid, and the normalization is chosen so that $[b, b^\dagger] = 1$ (see Example 12.4 and Exercise 12.12.6).

Similarly, the conjugate state $\langle\psi| = \langle 0|e^{\sqrt{N_s}b}$ diagonalizes the creation operator:

$$\langle\psi|\hat{\psi}^\dagger(x) = \psi^*(x)\langle\psi|. \tag{11.41}$$

---

[5] One might well object to this line of reasoning: clearly, creating a state with a definite phase requires we have another pre-cooled superfluid prepared in a state of definite phase. But what happens if we have none to start with? It turns out that what we really can do, is to control the relative phase of two superfluids by field-cooling, and it is the relative phase that we can actually measure.

However, it is not possible to simultaneously diagonalize both creation and annihilation operators because they don't commute. Thus $|\psi\rangle$ only diagonalizes the destruction operator and $\langle\psi^*|$ only diagonalizes the creation operator.

Coherent states are really the many-body analogue of wavepackets, with the roles of momentum and position replaced by N and ϕ, respectively. Just as $\hat{p}$ generates spatial translations $e^{-iPa/\hbar}|x\rangle = |x + a\rangle$, $\hat{N}$ translates the phase (see Exercise 11.1), so that $e^{i\alpha\hat{N}}|\phi\rangle = |\phi + \alpha\rangle$ (notice the difference in the sign in the exponent). For an infinitesimal phase translation, $\langle\phi + \delta\phi| = \langle\phi|(1 - i\delta\phi\hat{N})$, so $i\frac{d}{d\phi}\langle\phi| = \langle\phi|\hat{N}$, implying that

$$\hat{N} = i\frac{d}{d\phi}.$$

This is the many-body analogue of the identity $\hat{p} \equiv -i\hbar\frac{d}{dx}$. Just as periodic boundary conditions in space give rise to discrete quantized values of momentum, the periodic nature of phase gives rise to a quantized particle number. It follows that

$$[\hat{N}, \hat{\phi}] = i,$$

implying that phase and particle number are conjugate variables which obey an uncertainty relation [6]

$$\Delta\phi\Delta N \gtrsim 1.$$

A coherent state trades in a small fractional uncertainty in particle number to gain a high degree of precision in its phase. For small quantum systems where the uncertainty in particle number is small, phase becomes ill-defined. If we write the uncertainty principle in terms of the relative error $\Delta\epsilon = \Delta N/N$, then $\Delta\phi\Delta\epsilon \gtrsim 1/N$, and we see that once $N \sim 10^{23}$, so the fractional uncertainty in particle number and the phase can be known to an accuracy of order 10^{-11}. In the thermodynamic limit this means we can localize and measure both the phase and the particle density with Newtonian precision.

Example 11.5　The coherent state (11.40) is not normalized. Show that the properly normalized coherent state

$$|\psi\rangle = e^{-N_s/2}e^{\sqrt{N_s}\hat{b}^\dagger}|0\rangle$$

$$b^\dagger = \frac{1}{\sqrt{N_s}}\int_x \psi(x)\hat{\psi}^\dagger(x) \tag{11.42}$$

is an eigenstate of the annihilation operator $\hat{\psi}(x)$ with eigenvalue $\psi(x)$, where $N_s = \int d^dx|\psi(x)|^2$.

---

[6]　The strict relation is $\Delta\phi\Delta N \geq \frac{1}{2}|[\hat{\phi}, \hat{N}]| = \frac{1}{2}$. As in the case of wavepackets, in heuristic discussions we drop the factor of $\frac{1}{2}$.

Solution

First, since $[\hat{\psi}(x), \hat{\psi}^{\dagger}(x')] = \delta^{(d)}(x - x')$, we note that

$$[b, b^{\dagger}] = \frac{1}{N_s} \int_{x,x'} \psi(x) \psi^*(x') \overbrace{[\hat{\psi}(x), \hat{\psi}^{\dagger}(x')]}^{\delta^{(d)}(x-x')} = \frac{1}{N_s} \int_x |\psi(x)|^2 = 1,$$

so that b and $b^{\dagger}$ are canonical bosons.

To obtain the normalization of a coherent state, let us expand the exponential in $|z\rangle = e^{zb^{\dagger}}|0\rangle$ in terms of eigenstates of the boson number operator $\hat{n} = b^{\dagger}b$, $|n\rangle$, as follows:

$$|z\rangle = \sum_{n=0}^{\infty} \frac{(zb^{\dagger})^n}{n!} |0\rangle = \sum_{n=0}^{\infty} \frac{z^n}{\sqrt{n!}} \overbrace{\frac{(b^{\dagger})^n}{\sqrt{n!}} |0\rangle}^{|n\rangle} = \sum_{n=0}^{\infty} \frac{z^n}{\sqrt{n!}} |n\rangle.$$

Since $\langle n'|n\rangle = \delta_{n,n'}$, taking the norm, we obtain

$$\langle z|z\rangle = \sum_n \frac{|z|^n}{n!} = e^{|z|^2}.$$

Putting $z = \sqrt{N_s}$, it follows that the normalized coherent state is $|\psi\rangle = e^{-N_s/2} e^{\sqrt{N_s} b^{\dagger}} |0\rangle$.

Since $\hat{\psi}(x)|0\rangle = 0$, the action of the field operator on the coherent state is

$$\hat{\psi}(x)|\psi\rangle = e^{-N_s/2} [\hat{\psi}(x), e^{\sqrt{N_s} b^{\dagger}}] |0\rangle. \qquad (11.43)$$

To simplify notation, let us denote $\alpha^{\dagger} = \sqrt{N_s} b^{\dagger}$. The commutator

$$[\hat{\psi}(x), \alpha^{\dagger}] = \int_{x'} \psi(x') \overbrace{[\hat{\psi}(x), \hat{\psi}^{\dagger}(x')]}^{\delta^{(d)}(x-x')} = \psi(x),$$

which in turn implies that $[\hat{\psi}(x), (\alpha^{\dagger})^r] = r\psi(x)(\alpha^{\dagger})^{r-1}$. Now expanding

$$e^{\alpha^{\dagger}} = \sum_r \frac{1}{r!} (\alpha^{\dagger})^r$$

we find that

$$[\hat{\psi}(x), e^{\hat{\alpha}^{\dagger}}] = \sum_{r=0}^{\infty} \frac{1}{r!} [\hat{\psi}(x), (\hat{\alpha}^{\dagger})^r] = \psi(x) \sum_{r=1}^{\infty} \frac{(\hat{\alpha}^{\dagger})^{r-1}}{(r-1)!} = \psi(x) e^{\hat{\alpha}^{\dagger}},$$

so that, finally,

$$\hat{\psi}(x)|\psi\rangle = e^{-N_s/2} [\hat{\psi}(x), e^{\sqrt{N_s} b^{\dagger}}] |0\rangle = \psi(x) e^{-N_s/2} e^{\sqrt{N_s} b^{\dagger}} |0\rangle = \psi(x)|\psi\rangle. \qquad (11.44)$$

Ginzburg–Landau energy for a coherent state

We shall now link the one-particle wavefunction of the condensate to the order parameter of Ginzburg–Landau theory. While coherent states are not perfect energy eigenstates, at high density they provide an increasingly accurate description of the ground-state wavefunction of a condensate. To take the expectation value of normal-ordered operators between coherent states, one simply replaces the fields by the order parameter, so that if

$$\hat{\mathcal{H}} = \frac{\hbar^2}{2m} \nabla \hat{\psi}^\dagger(x) \nabla \hat{\psi}(x) + (U(x) - \mu) \hat{\psi}^\dagger(x) \hat{\psi}(x) + \frac{u}{2} : (\hat{\psi}^\dagger(x) \hat{\psi}(x))^2 : \tag{11.45}$$

is the energy density of the microscopic fields, where $U(x)$ is the one-particle potential, then the energy density of the condensate is

$$\langle \psi | \mathcal{H}[\hat{\psi}^\dagger, \hat{\psi}] | \psi \rangle = \mathcal{H}[\psi^*, \psi] = \frac{\hbar^2}{2m} |\nabla \psi(x)|^2 + (U(x) - \mu) |\psi(x)|^2 + \frac{u}{2} |\psi(x)|^4,$$

which we recognize as a Ginzburg–Landau energy density with

$$s = \frac{\hbar^2}{2m}, \qquad r(x) = U(x) - \mu.$$

At a finite temperature, this analysis needs modification. For instance, μ will acquire a temperature dependence that permits $r(T)$ to vanish at T_c, while the relevant functional becomes the free energy, $F = E - TS$. Finally note that, at a finite temperature, $n_s(T)$ only defines the superfluid component of the total particle density n, which contains both a normal and a superfluid component, $n = n_s(T) + n_n(T)$.

11.4.3 Phase rigidity and superflow

In Ginzburg–Landau theory the energy is sensitive to a twist of the phase. If we substitute $\psi = |\psi| e^{i\phi}$ into the Ginzburg–Landau free energy, the gradient term becomes $\nabla \psi = (\nabla |\psi| + i \nabla \phi |\psi|) e^{i\phi}$, so that

$$f_{GL} = \overbrace{\frac{\hbar^2}{2m} |\psi|^2 (\nabla \phi)^2}^{\text{KE: phase rigidity}} + \overbrace{\left[\frac{\hbar^2}{2m} (\nabla |\psi|)^2 + r |\psi|^2 + \frac{u}{2} |\psi|^4 \right]}^{\text{amplitude fluctuations}}. \tag{11.46}$$

The second term resembles the Ginzburg–Landau functional for an Ising order parameter, and describes the energy cost of variations in the magnitude of the order parameter. The first term is new. This term describes the *phase rigidity*. As we learned in the previous section, amplitude fluctuations of the order parameter are confined to scales shorter than the correlation length ξ. On longer length scales the physics is entirely controlled by the phase degrees of freedom, so that

$$f_{GL} = \frac{\rho_\phi}{2} (\nabla \phi)^2 + \text{constant}. \tag{11.47}$$

The quantity $\rho_\phi = \frac{\hbar^2}{m} n_s$ is often called the *superfluid phase stiffness*.

From a microscopic point of view the phase rigidity term is simply the kinetic energy of particles in the condensate, but from a macroscopic view it is an elastic energy associated with the twisted phase. The only way to reconcile these two viewpoints is if a twist of the condensate wavefunction results in a coherent flow of particles.

To see this explicitly, let us calculate the current in a coherent state. Microscopically, the particle current operator is

$$\mathbf{J} = -i\frac{\hbar}{2m}\left(\hat{\psi}^\dagger\vec{\nabla}\hat{\psi} - \vec{\nabla}\hat{\psi}^\dagger\hat{\psi}\right),$$

so, in a coherent state,

$$\langle\psi|\mathbf{J}|\psi\rangle = -i\frac{\hbar}{2m}\left(\psi^*\vec{\nabla}\psi - \vec{\nabla}\psi^*\psi\right). \qquad (11.48)$$

If we substitute $\psi(x) = \sqrt{n_s(x)}e^{i\phi(x)}$ into this expression, we find that

$$\mathbf{J}_s = n_s\frac{\hbar}{m}\nabla\phi, \qquad (11.49)$$

so that constant twist of the phase generates a flow of matter. Writing $\mathbf{J}_s = n_s\mathbf{v}_s$, we can identify

$$\mathbf{v}_s = \frac{\hbar}{m}\nabla\phi$$

as the *superfluid velocity* generated by the twisted phase of the condensate. Conventional particle flow is achieved by the addition of excitations above the ground state, but superflow occurs through a deformation of the ground-state phase, and every single particle moves in perfect synchrony.

Example 11.6

(a) Show that, in a condensate, the quantum equations of motion for the phase and particle number can be replaced by Hamiltonian dynamics [9]:

$$\hbar\frac{dN}{dt} = i[N,H] = \frac{\partial H}{\partial\phi}$$

$$\hbar\frac{d\phi}{dt} = i[\phi,H] = -\frac{\partial H}{\partial N}, \qquad (11.50)$$

which are the analogues of $\dot{q} = \frac{\partial H}{\partial p}$ and $\dot{p} = -\frac{\partial H}{\partial q}$.

(b) Use the second of the above equations to show that, in a superfluid at chemical potential μ, the equilibrium order parameter will precess with time, according to

$$\psi(x,t) = \psi(x,0)e^{-i\mu t/\hbar}.$$

(c) If two superfluids with the same superfluid density but at different chemical potentials μ_1 and μ_2 are connected by a tube of length L, show that the superfluid velocity from $1 \to 2$ will "accelerate" according to the equation

$$\frac{dv_s}{dt} = -\frac{\hbar}{m}\frac{\mu_2 - \mu_1}{L}.$$

Solution

(a) Since $[\phi, \hat{N}] = i$, there are two alternative representations of the operators:

$$\hat{N} = -i\frac{d}{d\phi}, \qquad \hat{\phi} = \phi \qquad\qquad (11.51)$$

or, in the case that N is large enough to be considered a continuous variable,

$$\hat{\phi} = i\frac{d}{dN}, \qquad \hat{N} = N. \qquad\qquad (11.52)$$

Using (11.51), the Heisenberg equation of motion for $N(t)$ is given by

$$\frac{dN}{dt} = \frac{i}{\hbar}[N, H] = \frac{i}{\hbar}\left[-i\frac{d}{d\phi}, H(N, \phi)\right] = \frac{1}{\hbar}\frac{\partial H}{\partial \phi}, \qquad\qquad (11.53)$$

while, using (11.52), the Heisenberg equation of motion for $\phi(t)$ is given by

$$\frac{d\phi}{dt} = \frac{i}{\hbar}[\phi, H] = \frac{i}{\hbar}\left[i\frac{d}{dN}, H\right] = -\frac{1}{\hbar}\frac{\partial H}{\partial N}. \qquad\qquad (11.54)$$

(b) In a bulk superfluid, $\frac{\partial H}{\partial N} = \mu$ so, using (11.54), $\dot{\phi} = \mu/\hbar$, and hence $\phi(t) = -\frac{\mu t}{\hbar} + \phi_0$, or

$$\psi(x, t) = \psi(x, 0)e^{-i\mu t/\hbar}.$$

(c) Assuming a constant gradient of phase along the tube connecting the two superfluids, the superfluid velocity is given by

$$v_s = \frac{\hbar}{m}\nabla\phi(t) = \frac{\hbar}{m}(\phi_2(t) - \phi_1(t))/L.$$

But $\phi(2) - \phi(1) = -(\mu_2 - \mu_1)t + \text{constant}$, hence

$$\frac{dv_s}{dt} = -\frac{\hbar}{m}\frac{\mu_2 - \mu_1}{L}.$$

Vortices and topological stability of superflow

Superflow is stable because of the underlying topology of a twisted order parameter. If we wrap the system around on itself, then the single-valued nature of the order parameter implies that the change in phase around the sample must be an integer multiple of 2π:

$$\Delta\phi = \oint d\mathbf{x} \cdot \nabla\phi = 2\pi \times n_\phi,$$

corresponding to n_ϕ twists of the order parameter. But since $v_s = \frac{\hbar}{m}\nabla\phi$, this implies that the line integral or *circulation* of the superflow around the sample is quantized:

$$\omega = \oint d\mathbf{x} \cdot \mathbf{v}_s = \frac{h}{m} \times n_\phi \qquad\qquad \text{quantization of circulation}$$

(note h without a slash). Assuming translational symmetry, this implies

$$v_s = \frac{h}{mL}n_\phi, \qquad\qquad \text{quantization of velocity}$$

a phenomenon first predicted by Onsager and Feynman [11, 12]. The number of twists n_ϕ of the order parameter is a *topological invariant* of the superfluid condensate, since it cannot be changed by any continuous deformation of the phase. The only way for the superflow to decay is to create high-energy domain walls: a process that is exponentially suppressed in the thermodynamic limit. Thus the topological stability of a twisted order parameter sustains a persistent superflow.

Another topologically stable configuration of a superfluid is a *vortex*. A superfluid vortex is a singular line in the superfluid, around which the phase of the order parameter precesses by an integer multiple of 2π. If we take a circular path of radius r around the vortex, then the quantization of circulation implies

$$\omega = n_\phi \left(\frac{h}{m}\right) = \oint d\mathbf{x} \cdot \mathbf{v}_s(x) = 2\pi r v_s$$

or

$$v_s = n_\phi \times \left(\frac{\hbar}{m}\right) \frac{1}{r} \qquad (r \gtrsim \xi).$$

This formula, where the superfluid velocity appears to diverge at short distances, is no longer reliable for $r \lesssim \xi$, where amplitude variations in the order parameter become important.

Let us now calculate the energy of a superfluid vortex. Suppose the vortex is centered in the middle of a large cylinder of radius R. Then the energy per unit length is

$$\frac{F}{L} = \frac{\rho_\phi}{2} \int d^2x (\nabla \phi)^2 = \frac{\rho_\phi}{2} \int_\xi^R 2\pi r dr \left(\frac{2\pi n_\phi}{2\pi r}\right)^2 = \pi \rho_\phi \ln\left(\frac{R}{\xi}\right) \times n_\phi^2.$$

In this way we see that the energy of n_ϕ isolated vortices with unit circulation is n_ϕ times smaller than one vortex with n_ϕ-fold circulation. For this reason, vortices occur with single quanta of circulation, and their interaction is repulsive.

11.5 Ginzburg–Landau III: charged fields

11.5.1 Gauge invariance

In a neutral superfluid, the emergence of a macrosopic wavefunction with a phase leads to superfluidity. When the corresponding fluid is charged, the superflow creates a persistent electric current, forming a superconductor. One of the key properties of superconductors is their ability to actively exclude magnetic fields from their interior, a phenomenon called the Meissner effect. Ginzburg– Landau theory provides a beautiful account of this effect.

The introduction of charge into a field theory brings with it the notion of gauge invariance. From the one-body Schrödinger equation,

$$i\hbar \frac{\partial \psi}{\partial t} = \left[-\frac{\hbar^2}{2m}\left(\nabla - i\frac{e}{\hbar}\mathbf{A}\right)^2 + e\varphi(x)\right]\psi,$$

where φ is the scalar electric potential, we learn that we can change the phase of a particle wavefunction by an arbitrary amount at each point in space and time, $\psi(x, t) \rightarrow e^{i\alpha(t)}\psi(x, t)$, without altering the equation of motion, so long as the change is compensated by a corresponding gauge transformation of the electromagnetic field:

$$\mathbf{A} \rightarrow \mathbf{A} + \frac{\hbar}{e}\nabla\alpha, \qquad \varphi \rightarrow \varphi - \frac{\hbar}{e}\frac{\partial\alpha}{\partial t}. \qquad (11.55)$$

This intimate link between changes in the phase of the wavefunction and gauge transformations of the electromagnetic field threads through all of many-body physics and field theory. Once we second-quantize quantum mechanics, the same rules of gauge invariance apply to the fields that create charged particles, and when these fields or combinations of them condense, the corresponding charged order parameter also obeys the rules of gauge invariance, with the proviso that the charge e^* is the charge of the condensate field. These kinds of arguments imply that, in the Ginzburg–Landau theory of a charged quantum fluid, normal derivatives of the field are replaced by gauge-invariant derivatives:

$$\nabla \rightarrow \mathbf{D} = \nabla - \frac{ie^*}{\hbar}\mathbf{A},$$

where e^* is the charge of the condensing field. Thus the simple replacement

$$f_{GL}[\psi, \nabla\psi] \rightarrow f_{GL}[\psi, \mathbf{D}\psi]$$

incorporates the coupling of the superfluid to the electromagnetic field. To this we must add the energy density of the magnetic field $\mathbf{B}^2/(2\mu_0)$, to obtain

$$F[\psi, \mathbf{A}] = \int d^d x \left[\overbrace{\frac{\hbar^2}{2M}\left|(\nabla - \frac{ie^*}{\hbar}\mathbf{A})\psi\right|^2 + r|\psi|^2 + \frac{u}{2}|\psi|^4}^{f_\psi} + \underbrace{\frac{(\nabla \times \mathbf{A})^2}{2\mu_0}}_{f_{EM}} \right] \qquad (11.56)$$

Ginzburg–Landau free energy: charged superfluid

where M is the mass of the condensed field and $\nabla \times \mathbf{A} = \mathbf{B}$ is the magnetic field.

Note the following:

- So long as we are considering superconductors, where the condensing boson is a Cooper pair of electrons, $e^* = 2e$. Although there are cases of charged bosonic superfluids, such as a fluid of deuterium nuclei, in which $e^* = e$, for the rest of this book we shall adopt

$$e^* \equiv 2e \qquad (11.57)$$

as an equivalence.

- Under the gauge transformation

$$\psi(x) \rightarrow \psi(x)e^{i\alpha(x)}, \qquad \mathbf{A} \rightarrow \mathbf{A} + \frac{\hbar}{e^*}\nabla\alpha,$$

$\mathbf{D}\psi \rightarrow e^{i\alpha(x)}\mathbf{D}\psi$, so that $|\mathbf{D}\psi|^2$ is unchanged and the Ginzburg–Landau free energy is gauge-invariant.

- $F[\psi, A]$ really contains *two* intertwined Ginzburg–Landau theories for ψ and A, respectively, with two corresponding length scales: the coherence length $\xi = \sqrt{\frac{\hbar^2}{2M|r|}}$ governing amplitude fluctuations of ψ and and the *London penetration depth* λ_L, which sets the distance a magnetic field penetrates into the superconductor. In a uniform condensate, $\psi = \sqrt{n_s}$, the free energy dependence on the vector potential is given by

$$f[\mathbf{A}] \sim c_A \frac{(\nabla \times \mathbf{A})^2}{2} + \frac{r_A}{2}\mathbf{A}^2, \qquad (11.58)$$

where $c_A = \frac{1}{\mu_0}$ and $r_A = \frac{e^{*2}n_s}{M}$. This is a Ginzburg–Landau functional for the vector potential with a characteristic London penetration depth

$$\lambda_L = \sqrt{\frac{c_\mathbf{A}}{r_\mathbf{A}}} = \sqrt{\frac{M}{n_s e^{*2} \mu_0}}. \qquad (11.59)$$

11.5.2 The Ginzburg–Landau equations

To obtain the equations of motion, we need to take variations of the free energy with respect to the vector potential and the order parameter ψ. Variations in the vector potential recover Ampère's equation, while variations in the order parameter lead to a generalization of the non-linear Schrödinger equation obtained previously for non-uniform Ising fields. Each of these equations is of great importance: non-uniform solutions determine the physics of the domain walls between normal and superconducting regions of a type II superconductor, while the Ginzburg– Landau formulation of Ampère's equation provides an understanding of the Meissner effect.

If we vary the vector potential, then $\delta F = \delta F_\psi + \delta F_{EM}$, where

$$\delta F_\psi = -\int_x \delta\mathbf{A}(x) \cdot \overbrace{\left[-\frac{ie^*\hbar}{2M}\left(\psi^*\vec{\nabla}\psi - \vec{\nabla}\psi\,\psi\right) - \frac{e^{*2}}{M}|\psi|^2 \right]}^{\mathbf{j}(x)}$$

is the variation in the condensate energy and [7]

$$\delta F_{EM} = \frac{1}{\mu_0}\int \nabla \times \delta\mathbf{A} \cdot \mathbf{B} = \frac{1}{\mu_0}\int_x \overbrace{\nabla \cdot (\delta\mathbf{A} \times \mathbf{B})}^{=0} + \frac{1}{\mu_0}\int_x \delta\mathbf{A}(x) \cdot (\nabla \times \mathbf{B})$$

---

[7] The variation of F_{EM} is tricky. We can carry it out using index notation to integrate δF_{EM} by parts, as follows:

$$\delta F_{EM} = \frac{1}{\mu_0}\int_x \epsilon_{abc}(\nabla_b \delta A_c)B_a = \frac{1}{\mu_0}\int_x \overbrace{\epsilon_{abc}}^{=-\epsilon_{cba}}\left[\underbrace{\nabla_b(\delta A_c B_a)}_{0} - \delta A_c \nabla_b B_a\right]$$

$$= \frac{1}{\mu_0}\int_x \delta A_c(x)\epsilon_{cba}\nabla_b B_a = \frac{1}{\mu_0}\int_x \delta\mathbf{A}(x) \cdot (\nabla \times \mathbf{B}), \qquad (11.60)$$

where we have set total derivative terms to zero.

is the variation in the magnetic field energy. Setting the total variation to zero, we obtain

$$\frac{\delta F}{\delta \mathbf{A}(x)} = -\mathbf{j}(x) + \frac{\nabla \times \mathbf{B}}{\mu_0} = 0, \tag{11.61}$$

where

$$\mathbf{j}(x) = -\frac{ie^*\hbar}{2M}\left(\psi^*\vec{\nabla}\psi - \vec{\nabla}\psi^*\,\psi\right) - \frac{e^{*2}}{M}|\psi|^2\mathbf{A} \tag{11.62}$$

is the supercurrent density. In this way, we have rederived Ampère's equation, where the current density takes the well-known form of a probability current in the Schrödinger equation. However, $\psi(x)$ now assumes a macroscopic, physical significance: it is, literally, the macroscopic wavefunction of the superconducting condensate. We will shortly see how (11.61) leads to the Meissner effect.

To take variations with respect to ψ, it is useful to first integrate by parts, writing

$$F_\psi = \int_x \left[\frac{\hbar^2}{2M}\psi^*\left(-i\nabla - \frac{e^*}{\hbar}\mathbf{A}\right)^2\psi + r\psi^*\psi + \frac{u}{2}(\psi^*\psi)^2\right]. \tag{11.63}$$

If we now take variations with respect to ψ^* and ψ, we obtain

$$\delta F = \int d^d x\left(\delta\psi^*(x)\left[\frac{\hbar^2}{2M}\left(-i\nabla - \frac{e^*}{\hbar}\mathbf{A}\right)^2\psi(x) + r\psi(x) + u|\psi(x)|^2\psi(x)\right] + \text{H.c.}\right),$$

implying that

$$-\frac{\hbar^2}{2M}(\nabla - i\frac{e^*}{\hbar}\mathbf{A})^2\psi(x) + r\psi(x) + u|\psi(x)|^2\psi(x) = 0. \tag{11.64}$$

This non-linear Schrödinger equation is almost identical to (11.17) obtained for an Ising order parameter, but here $\nabla^2 \to (\nabla - i\frac{q}{\hbar}\mathbf{A})^2$ to incorporate the gauge invariance and $\psi^3 \to |\psi|^2\psi$ takes account of the complex order parameter. We will shortly see how this equation can be used to determine the surface tension σ_{sn} of a drop of superconducting fluid.

11.5.3 The Meissner effect

We now examine how a superconductor behaves in the presence of a magnetic field. It is useful to write the supercurrent (11.62)

$$\mathbf{j}(x) = -\frac{ie^*\hbar}{2M}\left(\psi^*\vec{\nabla}\psi - \text{H.c.}\right) - \frac{e^{*2}}{M}|\psi|^2\mathbf{A}$$

in terms of the amplitude and phase of the order parameter $\psi = |\psi|e^{i\phi}$ (cf. (11.46)). The derivative term $\psi^*\nabla\psi$ can be rewritten

$$\psi^*\nabla\psi = |\psi|e^{-i\phi}\vec{\nabla}(|\psi|e^{i\phi}) = i|\psi|^2\vec{\nabla}\phi + |\psi|\vec{\nabla}|\psi|,$$

so that the term $\psi^*\nabla\psi - \text{H.c.} = 2i|\psi|^2\nabla\phi$ and hence

$$\mathbf{j}(x) = \frac{e^*\hbar}{M}|\psi|^2\nabla\phi - \frac{e^{*2}}{M}|\psi|^2\mathbf{A}$$

$$= e^*n_s \overbrace{\frac{\hbar}{M}\left(\vec{\nabla}\phi - \frac{e^*}{\hbar}\mathbf{A}\right)}^{\mathbf{v}_s} = e^*n_s\mathbf{v}_s, \qquad (11.65)$$

where we have replaced $|\psi|^2 = n_s$ and identified

$$\mathbf{v}_s = \frac{\hbar}{M}\left(\nabla\phi - \frac{e^*}{\hbar}\mathbf{A}\right) \qquad (11.66)$$

as the superfluid velocity. Note that, in contrast with (11.49), either a twist in the phase or an external vector potential can promote a superflow. Under a gauge transformation, $\phi \to \phi + \alpha$, $\mathbf{A} \to \mathbf{A} + \frac{\hbar}{e^*}\nabla\alpha$, this combination is gauge-invariant. Written out explicitly, Ampère's equation then becomes

$$\nabla \times \mathbf{B} = -\mu_0\frac{n_s e^{*2}}{M}\left(\mathbf{A} - \frac{\hbar}{e^*}\nabla\phi\right). \qquad (11.67)$$

If we take the curl of this expression (assuming n_s is constant), we obtain

$$\nabla \times (\nabla \times \mathbf{B}) = \mu_0\nabla \times \mathbf{j} = -\frac{\mu_0 n_s e^{*2}}{M}\mathbf{B}, \qquad (11.68)$$

where we have used the identity $\nabla \times \nabla\phi = 0$ to eliminate the phase gradient. But $\nabla \times (\nabla \times \mathbf{B}) = \nabla(\nabla \cdot \mathbf{B}) - \nabla^2\mathbf{B} = -\nabla^2\mathbf{B}$, since $\nabla \cdot \mathbf{B} = 0$, so that

$$\nabla^2\mathbf{B} = \frac{1}{\lambda_L^2}\mathbf{B},$$

$$\frac{1}{\lambda_L^2} = \frac{\mu_0 n_s e^{*2}}{M}. \qquad (11.69)$$

Meissner effect

This equation, first derived by Fritz London on phenomenological grounds [13], expresses the astonishing property that magnetic fields are actively expelled from superconductors. The only uniform solutions that are possible are

$$\mathbf{B} = 0, n_s > 0 \qquad \text{superconductor}$$

$$\mathbf{B} \neq 0, n_s = 0. \qquad \text{normal state} \qquad (11.70)$$

One-dimensional solutions to the London equation $\nabla^2 B = B/\lambda_L^2$ take the form $B \sim B_0 e^{-\frac{x}{\lambda_L}}$, showing that, near the surface of a superconductor, magnetic fields only penetrate a distance λ_L into the condensate. The penetration depth λ_L can also be regarded as a kind of *healing length* for changes in the magnetic field inside the superconductor. The persistent supercurrents that screen the field out of the superconductor lie within this thin shell on the surface.

As we shall see, however, in the class of type II superconductors, where the coherence length is small compared with the penetration depth ($\xi < \lambda_L/\sqrt{2}$), magnetic fields can penetrate the superconductor in a non-uniform way as vortices.

Lastly, note that in a superconductor, where $M = 2m_e$ and $e^* = 2e$ are the mass and charge of the Cooper pair, respectively, while $n_s = \frac{1}{2}n_e$ is half the concentration of electrons in the condensate,

$$\frac{n_s e^{*2}}{M} = \frac{\frac{1}{2}n_e 4e^2}{2m_e} = \frac{n_e e^2}{m},$$

so the expression for the penetration depth has the same form when written in terms of the charge and mass of the electron:

$$\frac{1}{\lambda_L^2} = \mu_0 \frac{n_e e^2}{m}.$$

The critical field H_c

In a medium that is immersed in an external field, we can divide the magnetic field into an external magnetizing field **H** and the magnetization **M**. In SI units,

$$\mathbf{B} = \mu_0(\mathbf{H} + \mathbf{M}),$$

where $\mathbf{j}_{ext} = \nabla \times \mathbf{H}$ is the current density in the external coils and $\mathbf{j}_{int} = \nabla \times \mathbf{M}$ are the internal currents of the material: in a superconductor, these are the supercurrents. Now the ratio $\chi = M/H$ is the magnetic susceptibility. Since the magnetic field $\mathbf{B} = \mu_0(\mathbf{M} + \mathbf{H})$ vanishes inside a superconductor, this implies $\mathbf{M} = -\mathbf{H}$, so that[8]

$$\chi_{SC} = -1. \qquad \text{perfect diamagnet}$$

In other words, superconductors are *perfect diamagnets*, in which shielding supercurrents $\mathbf{j}_{int} = \nabla \times \mathbf{M}$ provide a perfect Faraday cage to screen out the magnetic field from the interior of the superconductor. However, the external field **H** cannot be increased without limit, and beyond a certain critical field $|\mathbf{H}| > \mathbf{H}_c$ the uniform Meissner effect can no longer be sustained.

To calculate the critical field, we need to compare the energies of the normal and superconducting states. To this end, we separate the free energy into a condensate and a field component, $F = F_\psi + F_{EM}$, where $\delta F_\psi/\delta \mathbf{B}(x) = -\mathbf{M}(x)$ is the magnetization induced by the supercurrents while $\delta F_{EM}/\delta \mathbf{B}(x) = \mu_0^{-1}\mathbf{B}(x)$ is the magnetic field. Adding these terms together,

$$\frac{\delta F}{\delta \mathbf{B}(x)} = -\mathbf{M}(x) + \frac{1}{\mu_0}\mathbf{B}(x) = \mathbf{H}.$$

Now the magnetizing field **H** is determined by the external coils, and can be taken to be constant over the scale of the coherence and penetration depth. Since it is the external field **H** that is fixed, it is more convenient to use the Gibbs free energy,

$$G[\mathbf{H}, \psi] = F[\mathbf{B}, \psi] - \int d^3x \mathbf{B}(x) \cdot \mathbf{H},$$

---

[8] Most older texts use Gaussian units, for which $\chi_{SC} = -\frac{1}{4\pi}$ in a superconductor. In Gaussian units, $\mathbf{B} = \mathbf{H} + 4\pi\mathbf{M} = (1 + 4\pi\chi)\mathbf{H}$. If $\mathbf{B} = 0$, this implies that $\chi^{SC} = -\frac{1}{4\pi}$ in Gaussian units.

which is a functional of the external field $\mathbf{H}$ and independent of the B-field ($\delta G/\delta \mathbf{B} = 0$). The second term describes the work done by the coils in producing the constant external field. This is analogous to setting $G[P] = F[V] + PV$ to include the work PV done by a piston to maintain a fluid at constant pressure. In a uniform superconductor,

$$g = \frac{G}{V} = r|\psi|^2 + \frac{u}{2}|\psi|^4 + \frac{B^2}{2\mu_0} - BH.$$

In the normal state, $\psi = 0$, $B = \mu_0 H$, so that

$$g_n = -\frac{\mu_0}{2}H^2,$$

whereas, in the superconducting state, $B = 0$ and $|\psi| = \psi_0 = \sqrt{-r/u}$, so that

$$g_{sc} = r\psi_0^2 + \frac{u}{2}\psi_0^4 = -\frac{r^2}{2u}.$$

Clearly, if $g_{sc} < g_n$, i.e. if

$$H < H_c = \sqrt{\frac{r^2}{\mu_0\, u}}, \qquad \text{critical field} \qquad (11.71)$$

the superconductor is thermodynamically stable. The free energy density of the superconductor can then be written

$$g_{sc} = -\frac{r^2}{2u} = -\frac{\mu_0}{2}H_c^2.$$

Surface energy of a superconductor

When the external field $H = H_c$, the free energy density of the normal state and the superconductor are identical, and so the two phases can coexist. The interface between the degenerate superconductor and the normal phase is a domain wall, where the Gibbs free energy per unit energy defines the surface energy:

$$\Delta G/A = \sigma_{sn},$$

where A is the area of the interface. At the interface the superconducting order parameter and the magnetic field decay away to zero over length scales of the order of the coherence length ξ and penetration depth λ_L, respectively, as illustrated in Figure 11.8.

    The surface tension σ_{ns} (surface energy) of the domain wall between the superconductor and the normal phase has a profound influence on the macroscopic behavior of the

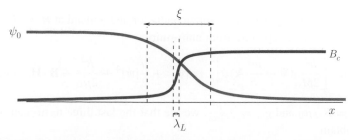

Schematic illustrating a superconductor–normal metal domain wall in a type I superconductor, where $\xi \gg \lambda_L$.   Fig. 11.8

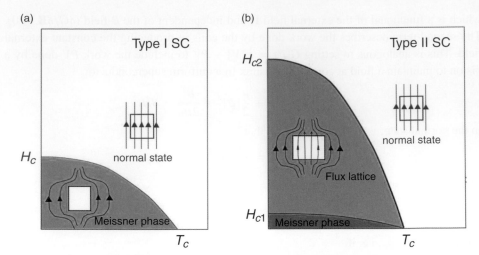

Fig. 11.9 Contrasting the phase diagrams of type I and type II superconductors. (a) In type I superconductors, application of a high field converts the Meissner phase directly into the normal state. (b) In type II superconductors, application of a modest field ($H > H_{c1}$) results in the partial penetration of the field into the superconductor to form a superconducting flux lattice, which survives up to a much higher field H_{c2}.

superconductor. The key parameter which controls the surface tension is the ratio of the magnetic penetration to the coherence length,

$$\kappa = \frac{\lambda_L}{\xi}. \qquad \text{Ginzburg–Landau parameter}$$

There are two types of superconductor (see Figure 11.9):

- **Type I** ($\kappa < \frac{1}{\sqrt{2}}$), with a positive domain-wall energy. In type I superconductors, magnetic fields are vigorously excluded from the material by a thin surface layer of screening currents (Figure 11.10(a)). At $H = H_c$ there is a first-order transition to the normal state.
- **Type II** ($\kappa > \frac{1}{\sqrt{2}}$), with a negative surface tension ($\sigma_{sn} < 0$). In type II superconductors, the surface layer of screening currents is smeared out on the scale of the coherence length, and the magnetic field penetrates much further into the superonductor (Figure 11.10(b)). There are now two critical fields, an upper critical field $H_{c2} > H_c$ and a lower critical field $H_{c1} < H_c$. Between these two fields, $H_{c1} < H < H_{c2}$, the magnetic field penetrates the bulk, forming vortices in which the high energy of the normal core is offset by the negative surface energy of the layer of screening currents.

The domain-wall energy between a superconductor and a metal at $H = H_c$ is the excess energy associated with a departure from uniformity:

$$\sigma_{sn} = \frac{1}{A}\int d^3x \left[\frac{\hbar^2}{2M}\left|\left(\nabla - \frac{ie^*}{\hbar}\mathbf{A}\right)\psi\right|^2 + r|\psi|^2 + \frac{u}{2}|\psi|^4 + \frac{B^2}{2\mu_0} - \mathbf{B}\cdot\mathbf{H}_c - g_{sc}\right]. \quad (11.72)$$

Inserting $H_c = B_c/\mu_0$ and $g_{sc} = -\frac{B_c^2}{2\mu_0}$, we see that the last three terms can be combined into one, to obtain

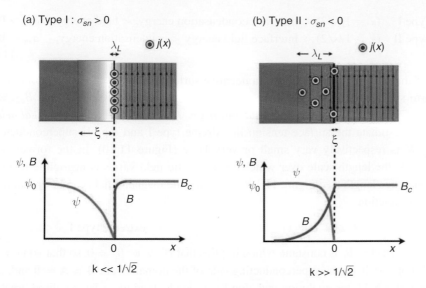

(a) Type I : $\sigma_{sn} > 0$

(b) Type II : $\sigma_{sn} < 0$

k << $1/\sqrt{2}$

k >> $1/\sqrt{2}$

Superconductor–normal domain wall in type I and type II superconductors. (a) For $\kappa = \frac{\lambda_L}{\xi} < \frac{1}{\sqrt{2}}$, the superconductor is a type I superconductor. In the limit $\kappa \rightarrow 0$ illustrated here, the magnetic field drops precipitously to zero at $x = 0$. (b) In the extreme type I limit, $\kappa >> 1/\sqrt{2}$, the magnetic field and the screening currents extend a distance $\lambda_L >> \xi$ into the superconductor.

Fig. 11.10

$$\sigma_{sn} = \frac{1}{A}\int d^3x \left[\frac{\hbar^2}{2M} \left|(\nabla - \frac{ie^*}{\hbar}\mathbf{A})\psi\right|^2 + r|\psi|^2 + \frac{u}{2}|\psi|^4 + \frac{(B - B_c)^2}{2\mu_0}\right]. \qquad (11.73)$$

By imposing the condition of stationarity, it is straightforward to show (see Example 11.6) that the domain-wall energy of a domain in the y-z plane can be cast into the compact form

$$\sigma_{sn} = \frac{B_c^2}{2\mu_0}\int_{-\infty}^{\infty} dx \left[\left(\frac{B(x)}{B_c} - 1\right)^2 - \left(\frac{\psi(x)}{\psi_0}\right)^4\right]. \qquad (11.74)$$

This compact form for the surface tension of a superconductor can be loosely interpreted as the difference of field and condensation energy:

$$\sigma_{sn} = \int_{-\infty}^{\infty} dx \left[\text{field energy} - \text{condensation energy}\right].$$

In the superconductor at the critical field, these two terms terms directly cancel one another, whereas in the normal metal, both terms are zero. It is the imperfect balance of these two energy terms at the interface that creates a non-zero surface tension. In a type I superconductor the *healing length* ξ for the order parameter is large, so the condensation energy fails to compensate for the field energy, generating a positive surface tension. By contrast, in a type II superconductor the healing length for the magnetic field λ_L is large, so the field energy fails to compensate for the condensation energy, leading to a negative surface tension. In fact, within Ginzburg–Landau theory, the surface tension vanishes at $\kappa = 1/\sqrt{2}$ (see Example 11.7), so $\kappa = 1/\sqrt{2}$ is the dividing line between the two classes of superconductor:

Type I $(\kappa < 1/\sqrt{2})$:   interface condensation energy, $<$ field energy, $\sigma_{sn} > 0$
Type II $(\kappa > 1/\sqrt{2})$:   interface field energy $<$ condensation energy, $\sigma_{sn} < 0$.

$$(11.75)$$

One of the most dramatic effects of a negative surface tension is the stabilization of non-uniform superconducting states over a wide range of fields between B_{c1} and B_{c2}, where $B_{c2} = \sqrt{2}\kappa B_c$ is the *upper critical field* and $B_{c1} \sim B_c/(\sqrt{2}\kappa)$ is the *lower critical field*.

Let us estimate the surface tension in extreme type I and type II superconductors in which K is respectively very small or very large (Figure 11.10). In the former, where $\lambda_L << \xi$, the length scale over which the magnetic field varies is negligible relative to the coherence length (see Figure 11.10(a)), so that the magnetic field can be approximated by a step function:

$$B(x) = B_c \theta(x). \qquad\qquad \text{extreme type I}$$

For $x > 0$, $B(x) = B_c$ is constant, which implies that $B'' \propto \psi^2 B_c = 0$, so that $\psi(x) = 0$ for $x \geq 0$. For $x < 0$, on the superconducting side of the domain wall, $\mathbf{B} = \mathbf{A} = \mathbf{0}$ and, in the absence of a field, the evolution equation for ψ is identical to an Ising soliton, treated in Section 11.3.1, for which the solution is $\psi/\psi_0 = \tanh(x/(\sqrt{2}\xi))$. Substituting into (11.74), the surface tension is then

$$\sigma_{sn}^{I} = \frac{B_c^2}{2\mu_0} \int_{-\infty}^{o} dx \left[1 - \tanh(x/(\sqrt{2}\xi))^4 \right] = \frac{B_c^2}{2\mu_0} \times 1.89\xi. \qquad (11.76)$$

For an extreme type II superconductor, the situation is reversed: now the longest length scale is the penetration depth. Unfortunately, since the vector potential modifies the equilibrium magnitude of the order parameter, λ_L sets the decay length of *both* the field and the order parameter. Let us nevertheless estimate the surface tension by treating the order parameter as a step function, $\psi(x) \sim \psi_0 \theta(-x)$. In this case, $A'' = \frac{1}{\lambda_L^2}(\psi/\psi_0)^2 A$, so that

$$B(x) = B_c \times \begin{cases} e^{x/\lambda_L} & (x < 0) \\ 1 & (x > 0). \end{cases} \qquad (11.77)$$

Substituting into (11.74), this then gives

$$\sigma_{sn}^{II} \approx \frac{B_c^2}{2\mu_0} \int_{-\infty}^{0} dx[(e^{x/\lambda_L} - 1)^2 - 1] = -\frac{B_c^2}{2\mu_0} \times \frac{3}{2}\lambda_L, \qquad (11.78)$$

showing that, at large κ, the surface tension becomes negative. A more detailed calculation (Example 11.8) replaces the factor $\frac{3}{2}$ by $(8/3)(\sqrt{2} - 1) = 1.1045$ [14].

Summarizing the results of detailed Ginzburg–Landau calculation:

$$\sigma_{sn} = \frac{B_c^2}{2\mu_0} \times \begin{cases} 1.89\xi & \text{extreme type I} \\ -1.10\lambda_L. & \text{extreme type II} \end{cases}$$

Example 11.7   Calculate the domain-wall energy per unit area σ_{sn} of a superconducting–normal interface lying in the $y - z$ plane, and show that it can be written

$$\sigma_{sn} = \frac{B_c^2}{2\mu_0} \int_{-\infty}^{\infty} dx \left[\left(\frac{B(x)}{B_c} - 1 \right)^2 - \left(\frac{\psi(x)}{\psi_0} \right)^4 \right]. \qquad (11.79)$$

Solution

Consider a domain wall in the y-z plane separating a superconductor at $x < 0$ from a metal at $x > 0$, immersed in a magnetic field along the z-axis. Let us take

$$\mathbf{A}(x) = (0, A(x), 0), \qquad \mathbf{B}(x) = (0, 0, A'(x)),$$

seeking a domain-wall solution in which $\psi(x)$ is real. Our boundary conditions are then

$$(\psi(x), A(x)) = \begin{cases} (\psi_0, 0) & (x \to -\infty) \\ (0, xB_c) & (x \to +\infty). \end{cases} \tag{11.80}$$

The domain-wall energy is then

$$\sigma_{sn} = \frac{G}{A} = \int dx \left[\frac{\hbar^2}{2M} \left\{ \left(\frac{d\psi}{dx} \right)^2 + \frac{e^{*2}A^2}{\hbar^2} \psi^2 \right\} + r\psi^2 + \frac{u}{2}\psi^4 + \frac{(B - B_c)^2}{2\mu_0} \right]. \tag{11.81}$$

Notice that there are no terms linear in $d\psi/dx$, because the vector potential and the gradient of the order parameter are orthogonal ($\nabla\psi \cdot \mathbf{A} = 0$). Let us rescale the x coordinate in units of the penetration length, the order parameter in units of ψ_0, and the magnetic field in units of the critical field, as follows:

$$\tilde{x} = \frac{x}{\lambda_L}, \qquad \tilde{\psi} = \frac{\psi}{\psi_0}, \qquad \tilde{A} = \frac{A}{B_c \lambda_L}, \qquad \tilde{B} = \frac{B}{B_c} = \frac{d\tilde{A}}{d\tilde{x}} \equiv \tilde{A}'.$$

In these rescaled variables, the domain-wall energy becomes

$$\sigma_{sn} = \frac{B_c^2 \lambda_L}{2\mu_0} \int dx \left[\frac{2\psi'^2}{\kappa^2} + A^2\psi^2 + \left((\psi^2 - 1)^2 - 1 \right) + (A' - 1)^2 \right], \tag{11.82}$$

where for clarity we have now dropped the tildes. The rescaled boundary conditions are $(\psi, A) \to (1, 0)$ in the superconductor at $x \ll 0$, and $(\psi, A) \to (0, x)$ deep inside the metal at $x \gg 0$.

Taking variations with respect to ψ gives

$$-\frac{\psi''}{\kappa^2} + \frac{1}{2}A^2\psi + (\psi^2 - 1)\psi = 0, \tag{11.83}$$

while taking variations with respect to A gives the dimensionless London equation,

$$A\psi^2 - A'' = 0. \tag{11.84}$$

Integrating by parts to replace $(\psi')^2 \to -\psi\psi''$ in (11.82), we obtain

$$\sigma_{sn} = \frac{B_c^2 \lambda_L}{2\mu_0} \int dx \left[-\overbrace{\frac{2\psi\psi''}{\kappa^2}}^{-A^2\psi^2 - 2(\psi^2-1)\psi^2} + A^2\psi^2 + \left((\psi^2 - 1)^2 - 1 \right) + (A' - 1)^2 \right], \tag{11.85}$$

where we have used (11.83) to eliminate ψ''. Canceling the $A^2\psi^2$ and ψ^2 terms in (11.85), we can then write the surface tension in the compact form

$$\sigma_{sn} = \frac{B_c^2 \lambda_L}{2\mu_0} \int_{-\infty}^{\infty} dx \left[(A'(x) - 1)^2 - \psi(x)^4 \right]. \tag{11.86}$$

Restoring $x \to \frac{x}{\lambda_L}$, $A'(x) \to \frac{B(x)}{B_c}$, and $\psi(x) \to \frac{\psi(x)}{\psi_0}$, we obtain (11.74).

Example 11.8   Show that the domain-wall energy changes sign at $\kappa = 1/\sqrt{2}$.

Solution

Using (11.86), we see that, in the special case where the surface tension $\sigma_{sn} = 0$, is zero, it follows that

$$A'(x) = 1 \mp \psi(x)^2,$$

where we select the upper sign to give a physical solution where the field is reduced inside the superconductor ($A' < 1$). Taking the second derivative gives $A'' = -2\psi\psi'$. But since $A'' = \psi^2 A$, it follows that $\psi' = -\frac{1}{2}A\psi$. Now we can derive an alternative expression for ψ' by integrating the second-order equation (11.84). By multiplying (11.83) by $4\tilde{\psi}'$, using (11.84) we can rewrite (11.83) as a total derivative,

$$\frac{d}{dx}\left[-\frac{2}{\kappa^2}(\psi')^2 + A^2\psi^2 + (\psi^2 - 1)^2 - A'^2 \right] = 0,$$

from which we deduce that

$$-\frac{2}{\kappa^2}(\psi')^2 + A^2\psi^2 + (\psi^2 - 1)^2 - A'^2 = \text{constant} = 0 \qquad (11.87)$$

is constant across the domain, where the value of the constant is obtained by placing $\psi = 1$, $A = A' = 0$ on the superconducting side of the domain. Substituting $A' = (1 - \psi^2)$, the last two terms cancel. Finally, putting $(\psi')^2 = \frac{1}{4}(A\psi)^2$, we obtain

$$\left(1 - \frac{1}{2\kappa^2}\right)(A\psi)^2 = 0, \qquad (11.88)$$

showing that $\kappa_c = 1/\sqrt{2}$ is the critical value where the surface tension drops to zero.

Example 11.9   Using the results of Example 11.7, show that within Ginzburg–Landau theory the surface tension of an extreme type II superconductor is [14]

$$\sigma_{sn} = -\frac{B_c^2}{2\mu_0} \times \frac{8}{3}(\sqrt{2} - 1)\lambda_L \approx -\frac{B_c^2}{2\mu_0} \times 1.10\lambda_L.$$

Solution

We start with equations (11.83) and (11.84):

$$\frac{\psi''}{\kappa^2} + \frac{1}{2}A^2\psi + (\psi^2 - 1)\psi = 0 \qquad (11.89)$$

$$A\psi^2 - A'' = 0. \qquad (11.90)$$

For an extreme type II superconductor, $\kappa \gg 1$, allowing us to neglect the derivative term in the first equation. There are then two solutions:

$$\begin{aligned} \psi^2 &= 1 - \tfrac{1}{2}A^2 & (x < 0) \\ \psi &= 0, \quad A = x + \sqrt{2} & (x > 0). \end{aligned} \qquad (11.91)$$

For $x < 0$, substituting into (11.90) we obtain

$$A(1 - A^2/2) = A''.$$
(11.92)

Multiplying both sides by the integrating factor $2A'$, we obtain

$$\frac{d}{dx}\left(A^2(1 - A^2/4)\right) = \frac{d}{dx}(A')^2$$

or $A^2(1 - A^2/4) = (A')^2 + $ constant, where the integration constant vanishes because A and A' both go to zero as $x \to -\infty$, so that

$$A' = A\sqrt{1 - A^2/4} \qquad (x < 0).$$
(11.93)

Using (11.91) in (11.86), the surface tension is

$$\sigma_{sn} = \frac{B_c^2 \lambda_L}{2\mu_0} \times I$$

$$I = \int_{-\infty}^{0} \left[(A' - 1)^2 - (1 - A^2/2)^2\right] dx.$$
(11.94)

Substituting for A' using (11.93) then gives

$$I = \int_{-\infty}^{0} \left[\left(A\sqrt{1 - A^2/4} - 1\right)^2 - (1 - A^2/2)^2\right] dx$$

$$= \int_{-\infty}^{0} \left[2A^2(1 - A^2/4) - 2A\sqrt{1 - A^2/4}\right] dx$$

$$= \int_{-\infty}^{0} \left[2(A' - 1)\right] A' dx$$

$$= \int_{0}^{\sqrt{2}} 2\left[\left(A\sqrt{1 - A^2/4} - 1\right)\right] dA = -\frac{8}{3}\left(\sqrt{2} - 1\right) \approx -1.1045,$$
(11.95)

where we have used the fact that $\psi = 0$, $A = \sqrt{2}$ at $x = 0$. It follows that, in the extreme type II superconductor,

$$\sigma_{sn} = -\frac{B_c^2}{2\mu_0} \times (1.10\lambda_L).$$

11.5.4 Vortices, flux quanta and type II superconductors

Once $H > H_{c1}$, type II superconductors support the formation of superconducing vortices. In a neutral superfluid, a superconducting vortex is a line defect around which the phase of the order parameter precesses by 2π or a multiple of 2π. In Section 11.4.3 we saw that this gave rise to a quantization of circulation. In a superconducting vortex, the rotating electric currents give rise to a trapped magnetic flux, quantized in units of the superconducting flux quantum:

$$\Phi_0 = \frac{h}{e^*} \equiv \frac{h}{2e}.$$

This quantization of magnetic flux was predicted by London and Onsager [13, 15].

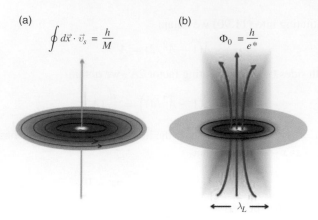

Fig. 11.11 Contrasting (a) a vortex in a neutral superfluid with (b) a vortex in a superconductor, where each unit of quantized circulation binds one quantum of magnetic flux.

To understand flux quantization, it is instructive to contrast a vortex in a neutral superfluid with a vortex in a superconducting (see Figure 11.11). In a neutral superfluid, the superfluid velocity is uniquely dictated by the gradient of the phase, $\mathbf{v}_s = \frac{\hbar}{M} \vec{\nabla}\phi$, so that around a vortex the superfluid velocity decays as $1/r$ ($v_s = n\frac{\hbar}{Mr}$). Around a vortex in a superconductor, the superfluid velocity contains an additional contribution from the vector potential:

$$\mathbf{v}_s = \frac{\hbar}{M} \vec{\nabla}\phi - \frac{e^*}{M}\mathbf{A}.$$

In the presence of a magnetic field, this term compensates for the phase gradient, lowering the supercurrent velocity and reducing the overall kinetic energy of the vortex. At distances larger than the penetration depth λ_L, the vector potential and the phase gradient almost completely cancel one another, leading to a supercurrent that decays exponentially with radius, $v_{sc} \propto e^{-r/\lambda_L}$.

If we integrate the circulation around a vortex, we find

$$\omega = \oint d\mathbf{x} \cdot \mathbf{v}_s = \frac{\hbar}{M} \overbrace{\oint d\mathbf{x} \cdot \vec{\nabla}\phi}^{\Delta\phi=2\pi n} - \frac{e^*}{M} \overbrace{\oint d\mathbf{x} \cdot \mathbf{A}}^{\Phi}, \tag{11.96}$$

where we have identified $\oint d\mathbf{x} \cdot \vec{\nabla}\phi = 2\pi n$ as the total change in phase around the vortex, while $\oint d\mathbf{x} \cdot \mathbf{A} = \int \mathbf{B} \cdot d\mathbf{S} = \Phi$ is the magnetic flux contained within the loop, so that

$$\omega = n\frac{h}{M} - \frac{e^*\Phi}{M}.$$

In this way, we see that the presence of bound magnetic flux reduces the total circulation. At large distances, energetics favor a reduction of the circulation to zero, $\lim_{R\to\infty}\omega = 0$, so that around a large loop

$$0 = n\frac{h}{M} - \frac{e^*\Phi}{M}$$

or

$$\Phi = n\left(\frac{h}{e^*}\right) = n\Phi_0, \tag{11.97}$$

where $\Phi_0 = \frac{h}{e^*}$ is the quantum of flux. In this way, each quantum of circulation generates a bound quantum of magnetic flux. The lowest-energy vortex contains a single flux, as illustrated in Figure 11.11. A simple realization of this situation occurs in a hollow superconducting cylinder (Figure 11.12). In its lowest energy state, where no supercurrent flows around the cylinder, the magnetic flux trapped inside the cylinder is quantized. If an external magnetic field is applied to the cylinder and then later removed, the cylinder is found to trap flux in units of the flux quantum $\Phi_0 = \frac{h}{2e}$ [16, 17], providing a direct confirmation of the charge of the Cooper pair.

In thermodynamic equilibrium, vortices penetrate a type II superconductor provided the applied field H lies between the upper and lower critical fields H_{c2} and H_{c1}, respectively. In an extreme type II superconductor, H_{c2} and H_{c1} differ from H_c by a factor of $\kappa = \frac{\lambda_L}{\xi}$:

$$H_{c1} \sim \frac{H_c \ln \kappa}{\sqrt{2}\kappa} \qquad (\kappa \gg 1) \tag{11.98}$$

$$H_{c2} = \sqrt{2}\kappa H_c. \tag{11.99}$$

Below H_{c1} or above H_{c2}, the system is uniformly superconducting or normal, respectively. In between, fluxoids self-organize themselves into an ordered triangular lattice, called the *Abrikosov flux lattice*. Thus H_{c1} is the first field at which it becomes energetically

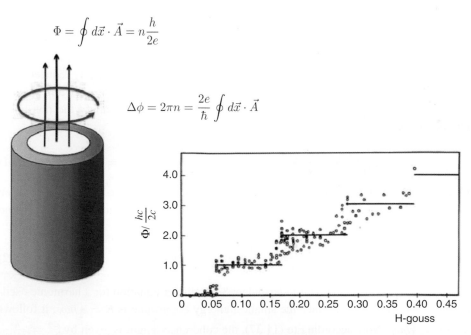

$$\Phi = \oint d\vec{x} \cdot \vec{A} = n\frac{h}{2e}$$

$$\Delta\phi = 2\pi n = \frac{2e}{\hbar} \oint d\vec{x} \cdot \vec{A}$$

Flux quantization inside a cylinder. In the lowest-energy configuration, with no supercurrent in the cylinder walls, the $\Delta\phi = 2\pi n$ twist in the phase of the order parameter around the cylinder is compensated by a quantized circulation of the vector potential, giving rise to a quantized flux. The inset shows quantized flux measured experimentally [16]. Reprinted with permission from B. S. Deaver and W. M. Fairbank, *Phys. Rev. Lett.*, vol. 7, p. 43, 1961. Copyright 1961 by the American Physical Society.

Fig. 11.12

advantageous to add a vortex to the uniform superconductor, whereas H_{c2} is the largest field at which a non-uniform superconducting solution is still stable.

For an extreme type II superconductor, H_{c1} can be determined by calculating the field at which the Gibbs free energy of a vortex

$$\Delta G_V = \epsilon_V L - \mathbf{H} \cdot \int d^3x \mathbf{B}(x)$$

$$= \epsilon_V L - H\Phi_0 L \tag{11.100}$$

becomes negative. Here L is the length of the vortex and ϵ_V is the vortex energy per unit length. For an extreme type II superconductor, this energy is roughly equal to the lost condensation energy of the core. Assuming the core to have a radius ξ, this is

$$\epsilon_V \sim \frac{r^2}{2u}\pi\xi^2 = \frac{B_c^2}{2\mu_0}\pi\xi^2.$$

Vortices will start to enter the condensate when $\Delta G_V < 0$, i.e. when

$$H_{c1}\Phi_0 \sim \frac{B_c^2}{2\mu_0}\pi\xi^2.$$

Putting $H_{c1} = B_{c1}/\mu_0$ and estimating the area over which the magnetic field is spread to be $\pi\lambda_L^2$, so that the total flux $\Phi_0 = B_{c1} \times \pi\lambda_L^2$, we obtain

$$\frac{H_{c1}}{H_c} \sim \frac{1}{\kappa},$$

so that $H_{c1} << H_c$ for an extreme type II superconductor. A more detailed calculation gives the answer quoted in (11.98).

To calculate H_{c2}, consider a metal in which the applied field is gradually reduced from a high field. H_{c2} will be the field at which the first non-uniform superconducting solution becomes possible. Non-uniform solutions of the order parameter satisfy the non-linear Schrödinger equation (11.64),

$$\frac{\hbar^2}{2M}(-i\nabla - \frac{e^*}{\hbar}\mathbf{A})^2\psi(x) + r\psi(x) + u|\psi(x)|^2\psi(x) = 0. \tag{11.101}$$

Since the developing superconducting instability will have a very small amplitude, we can ignore the cubic term. Choosing $\mathbf{A} = (0, 0, Bx)$, let us now seek solutions of ψ that depend only on x, so that

$$-\frac{\hbar^2}{2M}\psi'' + \frac{1}{2}m\omega_c^2\psi = -r\psi(x), \tag{11.102}$$

where $\omega_c = \frac{e^*B}{M}$. This is the time-independent Schrödinger equation for a harmonic oscillator with energy $E = -r$. Since the smallest energy eigenvalue is $E = \frac{1}{2}\hbar\omega_c$, it follows that $-r = \frac{1}{2}\hbar\omega_c$. Now according to (11.37), the coherence length is given by $\xi^2 = \frac{\hbar^2}{2M|r|}$, so that $|r| = \frac{\hbar^2}{2M\xi^2} = \hbar\frac{e^*B_{c2}}{M}$, so that

$$2\pi B_{c2}\xi^2 = \frac{h}{e^*} = \Phi_0, \tag{11.103}$$

where $\Phi_0 = \frac{h}{e^*}$ is the superconducting flux quantum. At the upper critical field, a tube of radius ξ contains half a flux quantum, $\Phi_0/2$.

Using (11.103), the upper critical field is given by

$$B_{c2} = \mu_0 H_{c2} = \frac{\hbar}{e^*\xi^2} = \frac{1}{e^*\xi}\sqrt{2M|r|}.$$

By contrast, using (11.71) and (11.59) the critical field B_c is given by

$$\mu_0 H_c = \sqrt{\mu_0 \frac{r^2}{u}} = \frac{1}{e^*\lambda_L}\sqrt{M|r|},$$

so that

$$\frac{H_{c2}}{H_c} = \sqrt{2}\frac{\lambda_L}{\xi} = \sqrt{2}\kappa.$$

Thus, provided $\kappa > \frac{1}{\sqrt{2}}$, the condition for type II superconductivity, the upper critical field H_{c2} exceeds the thermodynamic critical field, $H_{c2} > H_c$ (see Figure 11.9).

11.6 Dynamical effects of broken symmetry: the Anderson–Higgs mechanism

One of the most dramatic effects of broken symmetry lies in its influence on gauge fields that couple to the condensate. This effect, called the *Anderson–Higgs mechanism*, not only lies behind the remarkable Meissner effect, but is responsible for the short-range character of the weak nuclear force. When a gauge field couples to the long-wavelength phase modes of a charged order parameter, it absorbs the phase modes to become a massive gauge field that mediates a short-range (screened) force:

$$\text{gauge field + phase} \longrightarrow \text{massive gauge field.}$$

Superconductivity is the simplest, and historically the first, working model of this mechanism, which today bears the names of Philip Anderson, who first recognized its more general significance for relativistic Yang–Mills theories [18], and Peter Higgs, who later formulated these ideas in a relativistic action formulation [19]. In this section, we provide an introduction to the Anderson–Higgs mechanism, using a simple time-dependent extension of Ginzburg–Landau theory that, in essence, applies the method used by Higgs [19] to the simpler case of a $U(1)$ gauge field.

11.6.1 Goldstone mode in neutral superfluids

In the ground state, Ginzburg–Landau theory can be thought of as describing the potential energy $V[\psi] \equiv F_{GL}[\psi]|_{T=0}$ associated with a static and slowly varying configuration of the order parameter. At scales much longer than the coherent length, amplitude fluctuations of the order parameter can be neglected, and all the physics is contained in the phase of the order parameter. For a neutral superfluid, $V = \frac{1}{2}\rho_s(\nabla\phi)^2$, where ρ_s is the superfluid stiffness, given in Ginzburg–Landau theory by $\rho_s = \frac{\hbar^2 n_s}{2M}$. But to determine the dynamics, we need the Lagrangian $L = T - V$ associated with slowly varying configurations of the order parameter, where T is the kinetic energy associated with a time-dependent field configurations. The kinetic energy can also be expanded to leading order in the time-derivatives of the phase (see Exercise 11.8), so that the action governing the slow phase dynamics is

$$S = \frac{\rho_s}{2}\int dt d^3x\left[\overbrace{(\dot\phi/c^*)^2 - (\nabla\phi)^2}^{-\nabla_\mu\phi\nabla^\mu\phi}\right]. \tag{11.104}$$

In relativistic field theory, $c^* = c$ is the speed of light, and Lorentz invariance permits the action to be simplified using a 4-vector notation, $-\nabla_\mu\phi\nabla^\mu\phi$, as shown in the brackets above. The relativistic action and the Ginzburg–Landau free energy can be viewed as Minkowski and Euclidean versions of the same energy functional:

$$\overbrace{S = -\frac{\rho_s}{2}\int d^4x(\nabla_\mu\phi)^2}^{\text{Minkowski}} \longleftrightarrow \overbrace{F = \frac{\rho_s}{2}\int d^3x(\nabla\phi)^2}^{\text{Euclidean}} \tag{11.105}$$

However, in a non-relativistic superfluid, c^* is a characteristic velocity of the condensate. For example, in a paired fermionic superfluid, such as superfluid ^{3}He, $c^* = \sqrt{3}v_F$, where v_F is the Fermi velocity of the the underlying Fermi liquid. If we take variations with respect to ϕ (integrating by parts in space–time so that $\nabla\delta\phi\nabla\phi =\to -\delta\phi\nabla^2\phi$, and $\delta\dot\phi\dot\phi \to -\delta\phi\ddot\phi$), we see that ϕ satisfies the wave equation

$$\nabla^2\phi - \frac{1}{c^{*2}}\frac{\partial^2\phi}{\partial t^2} = 0 \qquad \text{(Bogoliubov phase mode, } \omega = c^*q\text{)},$$

corresponding to a phase mode that propagates at a speed c^*. This mode, often called a *Bogoliubov mode* is actually a special example of a *Goldstone mode*. The infinite-wavelength limit of this mode corresponds to a simple uniform rotation of the phase, and is an example of a naturally gapless mode that appears when a continuous symmetry is broken in a system governed by short-range forces.

Example 11.10   If density fluctuations $\delta n_s(x) = n_s(x) - n_s$ are included in the Hamiltonian of a superfluid, the ground-state energy is given by

$$H = \int d^3x\left[\frac{(n_s(x) - n_s)^2}{2\chi} + \frac{\rho_s}{2}(\nabla\phi)^2\right],$$

where $\chi = \partial N/\partial\mu$ is the charge susceptibility. From (11.55) we learned that density and phase are conjugate variables, which in the continuum satisfy Hamilton's equation that $\delta H/\delta n_s(x) = \mu(x) = -\hbar\dot\phi(x)$. Using this result, show that the Lagrangian $L = \int d^3x\frac{\delta H}{\delta n_s(x)}\delta n_s(x) - H$ can be written in the form

$$L = \frac{\rho_s}{2}\int d^3x\left[(\dot\phi/c^*)^2 - (\nabla\phi)^2\right],$$

where $(c^*)^2 = \rho_s/(\chi\hbar^2)$.

Solution

By varying the Hamiltonian with respect to the local density, we obtain the local chemical potential of the condensate:

$$\mu(x) = \frac{\delta H}{\delta n_s(x)} = \chi^{-1}\delta n_s(x). \tag{11.106}$$

By writing the condensate order parameter as $\psi(x,t) = \psi e^{i\phi(x,t)} = \psi e^{-i\frac{\mu(x)}{\hbar}t}$, we may identify $\frac{\mu(x)}{\hbar} = -\dot{\phi}$ as the rate of change of phase; thus from (11.106) we obtain

$$\hbar\dot{\phi} = -\chi^{-1}\delta n_s(x),$$

so that $(\delta n_s)^2/(2\chi) = \frac{\chi}{2}(\dot{\phi})^2$ and the Lagrangian takes the form

$$L = \int d^3x(-\hbar\dot{\phi}\delta n_s) - H = \frac{1}{2}\int d^3x\left[\chi(\hbar\dot{\phi}/c^*)^2 - \rho_s(\nabla\phi)^2\right].$$

Replacing $\hbar^2\chi = \rho_s/c^{*2}$, we obtain the result.

11.6.2 The Anderson–Higgs mechanism

The situation is subtly different when we consider a charged superfluid. In this case, changes in phase of the order parameter become coupled by the long-range electromagnetic forces, and this has the effect of turning them into gapped *plasmon* modes of the superflow and condensate charge density.

From Ginzburg–Landau theory we have already learned that, in a charge field, physical quantities such as the supercurrent and the Ginzburg–Landau free energy, depend on the the the gauge-invariant gradient of the phase, $\nabla\phi - \frac{e^*}{\hbar}\mathbf{A}$. Since the action involves time-dependent phase configurations, it must be invariant under both space- and time-dependent gauge transformations (11.55):

$$\phi \to \phi + \alpha(x,t), \qquad \mathbf{A} \to \mathbf{A} + \frac{\hbar}{e^*}\nabla\alpha, \qquad \varphi \to \varphi - \frac{\hbar}{e^*}\dot{\alpha}, \tag{11.107}$$

which means that time derivatives of the phase must occur in the gauge-invariant combination $\dot{\phi} + \frac{e^*}{\hbar}\varphi$, where φ is the electric potential. The action of a charged superluid now involves two terms:

$$S = S_\psi + S_{EM},$$

where

$$S_\psi = \int dt d^3x \frac{\rho_s}{2}\left[\frac{1}{c^{*2}}\left(\dot{\phi} + \frac{e^*}{\hbar}\varphi\right)^2 - (\nabla\phi - \frac{e^*}{\hbar}\mathbf{A})^2\right] \tag{11.108}$$

is the gauged condensate contribution to the action, and

$$S_{EM} = \frac{1}{2\mu_0}\int dt d^3x\left[\left(\frac{E}{c}\right)^2 - B^2\right] \tag{11.109}$$

is the electromagnetic Lagrangian, where $\mathbf{E} = -\frac{\partial \mathbf{A}}{\partial t} - \nabla\phi$ and $\mathbf{B} = \nabla \times \mathbf{A}$ are the electric and magnetic fields, respectively.

The remarkable thing is that, since the scalar and vector potential always occur in the same gauge-invariant combination with the phase gradients, we can redefine the electromagnetic fields to completely absorb the phase gradients, as follows:

$$\mathbf{A}' = \mathbf{A} - \frac{\hbar}{e^*}\nabla\phi, \quad \varphi' = \varphi + \frac{\hbar}{e^*}\dot{\phi}, \qquad \left(A^\mu \rightarrow \frac{\hbar}{e^*}\nabla^\mu\varphi\right).$$

Notice that in (11.108) the vector potential, which we associate with transverse electromagnetic waves, becomes coupled to gradients of the phase, which are longitudinal in character. The sum of the phase gradient and the vector potential creates a field with both longitudinal and transverse character. In terms of the new fields, the action becomes

$$S = \int dt d^3x \left\{ \overbrace{\frac{1}{2\mu_0\lambda_L^2}\left[\left(\frac{\varphi}{c^*}\right)^2 - \mathbf{A}^2\right]}^{L_\psi} + \overbrace{\frac{1}{2\mu_0}\left[\left(\frac{E}{c}\right)^2 - B^2\right]}^{L_{EM}} \right\}, \tag{11.110}$$

where $1/(\mu_0\lambda_L^2) = (\rho_s e^{*2})/(\hbar^2) = n_s e^{*2}/M$ defines the London penetration depth, and we have dropped the primes on φ and $\mathbf{A}$ in this and subsequent equations.

Amazingly, by absorbing the phase of the order parameter we arrive at a purely electromagnetic action, but one in which the phase stiffness of the condensate L_ψ imparts a new quadratic term in the action of the electromagnetic field: a *mass term*. Like a python that has swallowed its prey whole, the new gauge field is transformed into a much more sluggish object: it is heavy and weak. To see this in detail, let us re-examine Maxwell's equations in the presence of the mass term. Taking variations with respect to the fields, we obtain

$$\delta S_\psi = \int dt d^3x \left(\delta\mathbf{A}(x) \cdot \mathbf{j}(x) - \delta\varphi(x)\rho(x)\right), \tag{11.111}$$

where

$$\mathbf{j} = -\frac{1}{\mu_0\lambda_L^2}\mathbf{A}, \qquad \rho = -\frac{1}{\mu_0 c^{*2}\lambda_L^2}\varphi \tag{11.112}$$

denote the supercurrent and the voltage-induced change in charge density, while

$$\delta S_{EM} = \frac{1}{\mu_0}\int dt d^3x \left[\delta\mathbf{A} \cdot \left(\frac{1}{c^2}\dot{\mathbf{E}} - \nabla \times \mathbf{B}\right) + \delta\varphi\frac{1}{c^2}\nabla \cdot \mathbf{E}\right]. \tag{11.113}$$

Setting $\delta S = \delta S_\psi + \delta S_{EM} = 0$, the vanishing of the coefficient of $\delta\varphi$ gives Gauss' equation,

$$\frac{\delta S}{\delta\varphi} = \epsilon_0\nabla \cdot \mathbf{E} - \rho = 0, \tag{11.114}$$

while the vanishing of the coefficient of $\delta\mathbf{A}$ gives us Ampère's equation,

$$\frac{\delta S}{\delta\mathbf{A}} = \frac{1}{\mu_0}\left(\frac{1}{c^2}\dot{\mathbf{E}} - \nabla \times \mathbf{B}\right) + \mathbf{j} = 0. \tag{11.115}$$

Since $\nabla \cdot (\nabla \times \mathbf{B}) = 0$, taking the divergence of (11.115) and using (11.114) to replace $\nabla \cdot \mathbf{E} = \rho/\epsilon$ leads to a continuity equation for the supercurrent:

$$\nabla \cdot \mathbf{j} + \frac{\partial \rho}{\partial t} = -\frac{1}{\mu_0 \lambda_L^2} \left(\nabla \cdot \mathbf{A} + \frac{1}{c^{*2}} \frac{\partial \varphi}{\partial t} \right) = 0, \qquad (11.116)$$

except that now continuity also implies a gauge condition that ties ϕ to the longitudinal part of $\mathbf{A}$. For the relativistic case ($c^* = c$) this is the well-known Lorenz[9] gauge condition ($\nabla_\mu A^\mu = 0$).

If we now expand Ampère's equation in terms of $\mathbf{A}$, we obtain

$$\nabla \times \mathbf{B} = \nabla(\nabla \cdot \mathbf{A}) - \nabla^2 \mathbf{A} = -\frac{1}{\lambda_L^2} \mathbf{A} + \frac{1}{c^2} \frac{\partial}{\partial t} \left(-\frac{\partial \mathbf{A}}{\partial t} - \nabla \varphi \right), \qquad (11.117)$$

and using the continuity equation (11.116) to eliminate the potential term, we obtain

$$\left[\Box^2 - \frac{1}{\lambda_L^2} \right] \mathbf{A} = \left[1 - \left(\frac{c^*}{c} \right)^2 \right] \nabla(\nabla \cdot \mathbf{A}), \qquad (11.118)$$

where $\Box^2 = \nabla^2 - \frac{1}{c^2} \frac{\partial^2}{\partial t^2}$. In a superconductor, where $c^* \neq c$, the right-hand side of (11.118) becomes active for longitudinal modes, where $\nabla \cdot \mathbf{A} \neq 0$. If we substitute $\mathbf{A} = A_o e^{i(\mathbf{p} \cdot \mathbf{x} - E_\mathbf{p} t)/\hbar} \hat{\mathbf{e}}$ into (11.118), we find that the dispersion $E(\mathbf{p})$ of the transverse and longitudinal photons are given by

$$E(\mathbf{p}) = \begin{cases} [(m_A c^2)^2 + (pc^*)^2]^{1/2} & (\hat{\mathbf{e}} \perp \mathbf{p}, \quad \text{longitudinal}) \\[2mm] [(m_A c^2)^2 + (pc)^2]^{1/2} & (\hat{\mathbf{e}} \parallel \mathbf{p}, \quad \text{transverse}). \end{cases} \qquad (11.119)$$

Remarks

- Both photons share the same mass gap but they have widely differing velocities [18, 20]. The slower longitudinal mode of the electromagnetic field couples to density fluctuations: this is the mode associated with the exclusion of electric fields from within the superconductor, and it continues to survive in the normal metal above T_c as a consequence of electric screening.
- The rapidly moving transverse mode couples to currents: this is the new excitation of the superconductor that gives rise to the Meissner screening of magnetic fields.
- For a relativistic case, the right-hand side of (11.118) vanishes and the longitudinal and transverse photons merge into a single massive photon [19], described by a *Klein–Gordon* equation,

$$\left[\Box^2 - \left(\frac{m_A c}{\hbar} \right)^2 \right] \mathbf{A} = 0, \qquad (11.120)$$

for a vector field of mass $m_A = \hbar/(\lambda_L c)$. The generation of a finite mass in a gauge field through the absorption of the phase degrees of freedom of an order parameter into a gauge field is the essence of the Anderson–Higgs mechanism.

---

[9] Note the spelling of Lorenz, named after the Danish physicist Ludvig Lorenz, not after the more famous Dutch physicist, Henrich Lorentz!

11.6.3 Electroweak theory

The standard model for electroweak theory, developed by Sheldon Glashow, Paul Weinberg, and Abdus Salam [2, 21, 22], is a beautiful example of how the idea of broken symmetry, developed for physics in the laboratory, also provides insight into the physics of the cosmos itself. This is not abstract physics, for the sunshine we feel on our face is driven by the fusion of protons inside the Sun. The rate-limiting process is the conversion of two protons to a deuteron according to the reaction

$$p + p \rightarrow (pn) + e^+ + \nu_e,$$

where the ν_e is a neutrino. This process occurs very slowly because of the Coulomb repulsion between protons and the weakness of the weak decay process that converts a proton into a neutron. Were it not for the weakness of the weak force, fusion would burn too rapidly and the sun would have burnt out long before life could have formed on our planet. It is remarkable that the physics that makes this possible is the very same physics that gives rise to the levitation of superconductors.

Electroweak theory posits that the electromagnetic and weak forces derive from a common unified origin, in which part of the field is screened out of our universe through the development of a broken symmetry, associated with a two-component complex order parameter or *Higgs field*,

$$\Psi = \begin{pmatrix} \psi_0 \\ \psi_1 \end{pmatrix},$$

that condenses in the early universe. The coupling of its phase gradients to gauge degrees of freedom generates the massive vector bosons of the weak nuclear force via the Anderson–Higgs effect, miraculously leaving behind one decoupled gapless mode that is the photon. Fluctuations in the amplitude of the Higgs condensate give rise to a massive Higgs particle.

The basic physics of the standard model can be derived using the techniques of Ginzburg–Landau theory, by examining the interaction of the Higgs condensate with gauge fields. In its simplest version, first written down by Weinberg [2], this is given by (see Example 11.9)

$$S_\Psi = -\int d^4x \left[\frac{1}{2} |\left(\nabla_\mu - i\mathcal{A}_\mu\right)\Psi|^2 + \frac{u}{2}\left(\Psi^\dagger \Psi - 1\right)^2 \right], \tag{11.121}$$

where relativistic notation, $|\nabla_\mu \Psi|^2 \equiv |\nabla\Psi|^2 - |\dot{\Psi}|^2$, is used in the gradient term. The gauge field $\mathcal{A}_\mu$ acting on a two-component order parameter is a two-dimensional matrix made up of a $U(1)$ gauge field B_a that couples to the charge of the Higgs field, and an $SU(2)$ gauge field $\vec{A}_\mu$:

$$\mathcal{A}_\mu = g\vec{A}_\mu \cdot \vec{\tau} + g'B_\mu,$$

where $\vec{\tau}$ are the Pauli matrices and $\vec{A}_\mu = (A_\mu^1, A_\mu^2, A_\mu^3)$ is a triplet of three gauge fields that couple to the isospin of the condensate. When the Anderson–Higgs effect is taken into account, three components of the gauge fields acquire a mass, giving rise to two charged

$W^{\pm}$ bosons with mass M_W and one neutral Z boson of mass M_Z that couples to neutral currents of leptons and quarks:

$$\mathcal{A}_{\mu} \longrightarrow \begin{cases} Z,\ W^{\pm} & \text{neutral/charged vector bosons} \\ A. & \text{photon} \end{cases}$$

When S_{Ψ} is split into amplitude and phase modes of the order parameter, it divides into two parts (see Example 11.11), $S = S_H + S_W$, where

$$S_H = -\frac{1}{2} \int d^4x \left[(\nabla_{\mu}\phi_H)^2 + m_H^2 \phi_H^2 \right] \tag{11.122}$$

describes the amplitude fluctuations of the order parameter associated with the Higgs boson, where $m_H^2 = 4u$ defines its mass, while

$$S_W = -\frac{1}{2} \int d^4x \left[M_W^2 (W_{\mu}^{\dagger} W^{\mu}) + M_Z^2 (Z_{\mu} Z^{\mu}) \right] \tag{11.123}$$

determines the masses of the vector bosons.

The ratio of masses determines the weak-mixing angle θ_W:

$$\cos\theta_W = \frac{M_W}{M_Z}. \tag{11.124}$$

Experimentally, $M_Z = 91.19$ GeV/c^2 and $M_W = 80.40$ GeV/c^2, corresponding to a Weinberg angle of $\theta_W \approx 28°$. The Higgs particle was directly detected in 2012 by the CMS and Atlas collaborations at CERN, with a mass of approximately 126 GeV/c^2.

From the perspective of superconductivity, M_W and m_H define two length scales: a *penetration depth* for the screened weak fields,

$$\lambda_W = \frac{\hbar}{M_W c} \sim 2 \times 10^{-18}\,\text{m}, \tag{11.125}$$

Table 11.2 Superconductivity and electroweak physics.

	Superconductivity	Electroweak
Order parameter	ψ	$\begin{pmatrix} \psi_0 \\ \psi_1 \end{pmatrix}$
	Pair condensate	Higgs condensate
Gauge field/symmetry	$(\phi, \mathbf{A})$	$\mathcal{A}_{\mu} = g' B_{\mu} + g(\vec{A}_{\mu} \cdot \vec{\tau})$
	$U(1)$	$U(1) \times SU(2)$
Penetration depth	$\lambda_L \sim 10^{-7}$ m	$\lambda_W \sim 10^{-18}$ m
Coherence length	$\xi = \frac{v_F}{\Delta} \sim 10^{-9}$–$10^{-7}$ m	$\xi_{EW} \sim 10^{-18}$ m
Condensation mechanism	Pairing	Unknown
Screened field	$\vec{B}$	$W^{\pm}, Z$
Massless gauge field	None	Electromagnetism A_{μ}

which defines the range of the weak force, and the *coherence length* of electroweak theory [23]

$$\xi_W = \frac{\hbar}{m_{HC}} \sim 1.5 \times 10^{-18} \text{ m}. \tag{11.126}$$

Since $\lambda_W/\xi_W \sim 1$, the electroweak condensate appears to be weakly type II in character, allowing for the possibility of cosmic strings [24]. The microscopic physics that develops below the coherence length ξ_W in the vacuum is at present an open mystery that is the subject of ongoing measurements at the Large Hadron Collider.

Example 11.11

(a) Suppose the Higgs condensate is written $\Psi(x) = (1 + \phi_H(x))U(x)\Psi_0$, where ϕ_H is a real field, describing small-amplitude fluctuations of the condensate, $U(x)$ is a matrix describing the slow variations in orientation of the order parameter, and $\Psi_0 = \begin{pmatrix} 1 \\ 0 \end{pmatrix}$ is just a unit spinor. Show that the action splits into two terms, $S = S_H + S_W$, where

$$S_H = -\frac{1}{2} \int d^4x \left[(\nabla_\mu \phi_H)^2 + m_H^2 \phi_H^2 \right] \tag{11.127}$$

describes the amplitude fluctuations of the order parameter associated with the Higgs boson, where $m_H^2 = 4u$ defines its mass, while

$$S_W = -\frac{1}{2} \int d^4x |A'_\mu \Psi_0|^2 \tag{11.128}$$

determines the masses of the vector bosons.

(b) By expanding the quadratic term in (11.128), show that it is diagonalized in terms of two gauge fields:

$$S_W = -\frac{1}{2} \int d^4x \left[M_W^2 (W_\mu^\dagger W^\mu) + M_Z^2 (Z_\mu Z^\mu) \right],$$

and give the form of the fields and their corresponding masses in terms of the original fields and coupling constants.

Solution

(a) Let us substitute

$$\Psi(x) = (1 + \phi_H(x))U(x)\Psi_0,$$

where $\Psi_0 = \begin{pmatrix} 1 \\ 0 \end{pmatrix}$, into (11.121). Since $\Psi^\dagger \Psi = (1 + \phi_H)^2 \Psi_0^\dagger U^\dagger U \Psi_0 = (1 + \phi_H)^2$, to quadratic order the potential part of S_Ψ can be written as

$$\frac{u}{2}(\Psi^\dagger \Psi - 1)^2 = \frac{u}{2}(2\phi_H + \phi_H^2)^2 = \frac{m_H}{2}\phi_H^2 + O(\phi_H^3) \qquad (m_H^2 = 4u).$$

The derivatives in the gradient term can be expanded as

$$(\nabla_\mu - iA_\mu)\Psi(x) = (\nabla_\mu - iA_\mu)U\Psi_0 + \nabla_\mu\phi_H(U\Psi_0).$$

Since the derivative of a unit spinor is orthogonal to itself, the two terms in the above expression are orthogonal, so that when we take the modulus squared of the above expression we obtain

$$|(\nabla_\mu - iA_\mu)\Psi|^2 = |(\nabla_\mu - iA_\mu)U\Psi_0|^2 + (\nabla_\mu\phi_H)^2 \overbrace{|U\Psi_0|^2}^{=|\Psi_0|^2=1}$$

$$= |U^\dagger(A_\mu + i\nabla_\mu)U\Psi_0|^2 + (\nabla_\mu\phi_H)^2. \qquad (11.129)$$

Here we have introduced a prefactor $iU^\dagger$ into the first term, which does not change its magnitude. Now the combination

$$A'_\mu = U^\dagger(A_\mu + i\nabla_\mu)U$$

is a gauge transformation of A_μ which leaves the physical fields ($G_{\mu\nu} = \nabla_\mu A_\nu - \nabla_\nu A_\mu - i[A_\mu, A_\nu]$) and the action associated with the gauge fields invariant. In terms of this transformed field, the gradient terms of S_Ψ can be written simply as

$$|(\nabla_\mu - iA_\mu)\Psi|^2 = |A'_\mu\Psi_0|^2 + (\nabla_\mu\phi_H)^2,$$

so that the sum of the gradient and potential terms yields

$$L = -\frac{1}{2}|\left(\nabla_\mu - iA_\mu\right)\Psi|^2 + \frac{u}{2}\left(\Psi^\dagger\Psi - 1\right)^2$$

$$= \underbrace{-\frac{1}{2}|A'_\mu\Psi_0|^2}_{L_W} \underbrace{-\frac{1}{2}\left[(\nabla_\mu\phi_H)^2 + m_H^2\phi_H^2\right]}_{L_H}, \qquad (11.130)$$

which when integrated over space–time gives the results (11.127) and (11.128).

(b) Written out explicitly, the gradient appearing in the gauge theory mass term is

$$A'_\mu\Psi_0 = \left[g'B_\mu + g\vec{A}_\mu \cdot \vec{\tau}\right] \cdot \Psi_0$$

$$= \left[g'\begin{pmatrix} B_\mu & \\ & B_\mu \end{pmatrix} + g\begin{pmatrix} A_\mu^3 & A_\mu^1 - iA_\mu^2 \\ A_\mu^1 + iA_\mu^2 & -A_\mu^3 \end{pmatrix}\right]\begin{pmatrix} 1 \\ 0 \end{pmatrix}$$

$$= \begin{pmatrix} g'B_\mu + gA_\mu^3 \\ g(A_\mu^1 + iA_\mu^2) \end{pmatrix}, \qquad (11.131)$$

so that the mass term of the gauge fields can be written

$$L_W = -\frac{1}{2}|A_\mu\Psi|^2 = -\frac{1}{2}\left[(gA_\mu^3 + g'B_\mu)^2 + g^2|A_\mu^1 + iA_\mu^2|^2\right]$$

$$= -\frac{M_Z^2}{2}Z_\mu^2 - \frac{M_W^2}{2}|W_\mu|^2, \qquad (11.132)$$

where

$$W_\mu = A_\mu^1 + iA_\mu^2$$

$$Z_\mu = \frac{1}{\sqrt{g^2 + (g')^2}}\left(g'A_\mu^{(3)} + gB_\mu\right) \qquad (11.133)$$

are, respectively, the charged W and neutral Z bosons which mediate the weak force, $M_Z = \sqrt{g^2 + g'^2}$ and $M_W = g = M_Z cos[\theta_W]$, where θ_W is the Weinberg angle determined by

$$\cos \theta_W = \frac{g}{\sqrt{g^2 + g'^2}}.$$

11.7 The concept of generalized rigidity

The *phase rigidity* responsible for superflow, the Meissner effect, and its electroweak counterpart are each consequences of a general property of broken continuous symmetries. In any broken continuous symmetry, the order parameter can assume any one of a continuous number of directions, each with precisely the same energy. By contrast, it always costs an energy to slowly "bend" the direction of the order parameter away from a state of uniform order. This property is termed *generalized rigidity* [25]. In a superconductor or superfluid, it costs a phase-bending energy

$$U(x) \sim \frac{1}{2}\rho_s (\nabla\phi(x))^2 \tag{11.134}$$

to create a gradient of the phase. The differential of U with respect to the phase gradient $\delta U/(\hbar\delta\nabla\phi)$, which defines the superflow of particles, is directly proportional to the amount of phase bending or the gradient of the phase:

$$j_s = \frac{\delta U}{\hbar\delta\nabla\phi} = \frac{\rho_s}{\hbar}\nabla\phi. \tag{11.135}$$

This relationship holds because density and phase are conjugate variables. Anderson noted that we can generalize this concept to a wide variety of broken symmetries, each with its corresponding phase and conjugate conserved quantity. In each case, a gradient of the order parameter gives rise to a superflow of the quantity that translates the phase (see Table 11.3).

For example, broken translation symmetry leads to the superflow of momentum (sheer stress); broken spin symmetry leads to the superflow of spin. There are undoubtedly new classes of broken symmetry yet to be discovered – one of which might be broken time–translational invariance or "time crystals" [26] (see Table 11.3).

11.8 Thermal fluctuations and criticality

At temperatures that are far below or far above the critical point, the behavior of the order parameter resembles a tranquil ocean with no significant amount of thermal noise in its fluctuations. But fluctuations become increasingly important near the critical point as the correlation length diverges. At the second-order phase transition, infinitely long-range *critical fluctuations* develop in the order parameter. The study of these fluctuations requires

Table 11.3 Order parameters, broken symmetry, and rigidity.

Name	Broken symmetry	Rigidity/supercurrent
Crystal	Translation symmetry	Momentum superflow (sheer stress)
Superfluid	Gauge symmetry	Matter superflow
Superconductivity	Electromagnetic gauge symmetry	Charge superflow
Antiferromagnetism	Spin rotation symmetry	Spin superflow (x-y magnets only)
"Time crystal" [26]?	Time–translational Symmetry	Energy superflow?

that we go beyond mean-field theory. Instead of using the Ginzburg–Landau functional as a variational free energy, now we use it to determine the Boltzmann probability distribution of the thermally fluctuating order parameter, as follows:

$$p[\psi] = Z^{-1} e^{-\beta F_{GL}[\psi]} = \frac{1}{Z} \exp\left[-\beta \int d^d x \left(\frac{1}{2}\left[s(\nabla \psi)^2 + r|\psi(x)|^2\right] + u|\psi(x)|^4\right)\right],$$

where $Z = \sum_\psi e^{-\beta F_{GL}[\psi]}$ is the normalizing partition function. This is the famous "ϕ^4 field theory" of statistical mechanics (where we use ψ in place of ϕ).

The variational approach can be derived from the probability distribution function $p[\{\psi\}]$ by observing that the probability of a given configuration is sharply peaked around the mean-field solution, $\psi = \psi_0$. If we make a Taylor expansion around a nominal mean-field configuration, writing $\psi(x) = \psi_0 + \delta\psi(x)$, then

$$F_{GL}[\{\psi\}] = F_{mf} + \int_x \delta\psi(x) \overbrace{\frac{\delta F_{GL}}{\delta\psi(x)}}^{=0} + \frac{1}{2} \int_{x,x'} \delta\psi(x)\delta\psi(x') \frac{\delta^2 F_{GL}}{\delta\psi(x)\delta\psi(x')} + \cdots,$$

where the first derivative is zero because the free energy is stationary for the mean-field solution $\delta F/\delta\psi = 0$, which implies

$$F_{GL}[\{\psi\}] = F_{mf}[\psi_0] + \frac{1}{2} \int_{x,x'} \delta\psi(x)\delta\psi(x') \frac{\delta^2 F_{GL}}{\delta\psi(x)\delta\psi(x')} + \cdots.$$

The first non-vanishing terms in the free energy are second-order terms, describing a Gaussian distribution of the fluctuations of the order parameter about its average:

$$\delta\psi(x) = \psi(x) - \psi_0.$$

The amplitude of the fluctuations at long wavelengths becomes particularly intense near a critical point. This point was first appreciated by Ornstein and Zernicke [4], who observed in 1914 that light scatters strongly off the long-wavelength density fluctuations of a gas near the critical point of the liquid–gas phase transition. We now follow Ornstein and Zernicke's original treatment, and study the behavior of order parameter fluctuations above the phase transition.

To treat the fluctuations, we Fourier transform the order parameter:

$$\psi(x) = \frac{1}{\sqrt{V}} \sum_{\mathbf{q}} \psi_{\mathbf{q}} e^{i\mathbf{q}\cdot\mathbf{x}}, \qquad \psi_{\mathbf{q}} = \frac{1}{\sqrt{V}} \int d^d x \, \psi(x) e^{-i\mathbf{q}\cdot\mathbf{x}}. \qquad (11.136)$$

Here we use periodic boundary conditions in a finite box of volume $V = L^d$, with discrete wavevectors $\mathbf{q} = \frac{2\pi}{L}(l_1, l_2, \ldots l_d)$. Note that $\psi_{-\mathbf{q}} = \psi_{\mathbf{q}}^*$, since ψ (or each of its n components) is real. Substituting (11.136) into (11.15), noting that $(-s\nabla^2 + r) \to (sq^2 + r)$ inside the Fourier transform, we obtain

$$F = \frac{1}{2} \sum_{\mathbf{q}} |\psi_{\mathbf{q}}|^2 \left(sq^2 + r \right) + u \int d^d x |\psi(x)|^4, \qquad (11.137)$$

so that the quadratic term is diagonal in the momentum–space representation. Notice how we can rewrite the Ginzburg–Landau energy in terms of the (bare) susceptibility $\chi_{\mathbf{q}} = (sq^2 + r)^{-1}$ encountered in (11.19), as

$$F = \frac{1}{2} \sum_{\mathbf{q}} |\psi_{\mathbf{q}}|^2 \chi_{\mathbf{q}}^{-1} + u \int d^d x |\psi(x)|^4, \qquad (11.138)$$

so the quadratic coefficient of the Ginzburg–Landau free energy is the inverse susceptibilty.

Suppose $r > 0$ and the deviations from equilibrium, $\psi = 0$, are small enough to ignore the interaction, permitting us to temporarily set $u = 0$. In this case, F is a simple quadratic function of $\psi_{\mathbf{q}}$, and the probability distribution function is a simple Gaussian:

$$p[\psi] = Z^{-1} \exp\left[-\frac{\beta}{2} \sum_{\mathbf{q}} |\psi_{\mathbf{q}}|^2 \left(sq^2 + r \right) \right] \equiv Z^{-1} \exp\left[-\sum_{\mathbf{q}} \frac{|\psi_{\mathbf{q}}|^2}{2S_{\mathbf{q}}} \right],$$

where

$$S_{\mathbf{q}} = \langle |\psi_{\mathbf{q}}|^2 \rangle = \frac{k_B T}{sq^2 + r} = \frac{k_B T/c}{q^2 + \xi^{-2}} \qquad (11.139)$$

is the variance of the fluctuations at wavevector $\mathbf{q}$, and $\xi = \sqrt{s/r}$ is the correlation length. This distribution function is known as the *Ornstein–Zernicke* form for the Gaussian variance of the order parameter. This quantity is the direct analogue of the Green's function in many-body physics.

Note the following:

- For $q \gg \xi^{-1}$, $S_q \propto 1/q^2$ is singular or *critical*.
- Using (11.19) we see that the fluctuations of the order parameter are directly related to its static susceptibility, $S_{\mathbf{q}} = k_B T \chi_{\mathbf{q}}$. This is a consequence of the fluctuation–dissipation theorem in the classical limit.
- S_q resembles a Yukawa interaction associated with the virtual exchange of massive particles, $V(q) = 1/(q^2 + m^2)$. Indeed, short-range nuclear interactions are a result of quantum fluctuations in a pion field with correlation length $\xi \sim m^{-1}$.

Next, let us Fourier transform this result to calculate the spatial correlations:

$$S(\mathbf{x} - \mathbf{x}') = \langle \delta\psi(x)\delta\psi(x') \rangle = \frac{1}{V} \sum_{\mathbf{q},\mathbf{q}'} \overbrace{\langle \psi_{-\mathbf{q}}\psi_{\mathbf{q}'} \rangle}^{S_{\mathbf{q}}\delta_{\mathbf{q}-\mathbf{q}'}} e^{i(\mathbf{q}'\cdot\mathbf{x}'-\mathbf{q}\cdot\mathbf{x})}$$

$$= \int \frac{d^d q}{(2\pi)^d} \frac{k_B T/c}{q^2 + \xi^{-2}} e^{i\mathbf{q}\cdot(\mathbf{x}'-\mathbf{x})}, \qquad (11.140)$$

where we have taken the thermodynamic limit $V \to \infty$. This is a Fourier transform that we have encountered in conjunction with the screened Coulomb interaction, and in three dimensions we obtain

$$S(\mathbf{x} - \mathbf{x}') = \frac{k_B T}{4\pi s} \frac{e^{-|\mathbf{x}-\mathbf{x}'|/\xi}}{|\mathbf{x} - \mathbf{x}'|} \qquad (d = 3).$$

Note the following:

- The generalization of this result to d dimensions gives

$$S(\mathbf{x}) \sim \frac{e^{-x/\xi}}{x^{d-2+\eta}},$$

where Ginzburg–Landau theory predicts $\eta = 0$.
- $S(\mathbf{x})$ illustrates a very general property. On length scales below the correlation length, the fluctuations are critical, with power-law correlations, but on longer length scales, correlations are exponentially suppressed (see Figure 11.13).

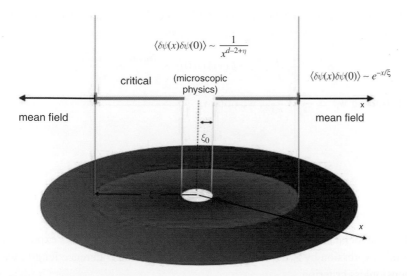

$$\langle \delta\psi(x)\delta\psi(0) \rangle \sim \frac{1}{x^{d-2+\eta}}$$

critical (microscopic physics)

$$\langle \delta\psi(x)\delta\psi(0) \rangle \sim e^{-x/\xi}$$

mean field   mean field

ξ_0

Length scales near a critical point. On length scales $\xi \gg x \gg \xi_0$, fluctuations are critical, with universal power-law correlations. On length scales larger than the correlation length ξ, fluctuations are exponentially correlated. On length scales shorter than the coherence length ξ_0, the order parameter description must be replaced by a microscopic description of the physics.

Fig. 11.13

- Ginzburg–Landau theory predicts that the correlation length diverges as

$$\xi \propto (T - T_c)^{-\nu},$$

where $\nu = \frac{1}{2}$. Remarkably, even though Ginzburg–Landau theory neglects the non-linear interactions of critical modes, these results are qualitatively correct. More precise treatments of critical phenomena show that the exponents depart from Gaussian theory in dimensions $d < 4$.

11.8.1 Limits of mean-field theory: the Ginzburg criterion

What are the limits of mean-field theory? We studied the fluctuations at temperatures $T > T_c$ by assuming that the non-linear interaction term can be ignored. This is only true provided the amplitude of fluctuations is sufficiently small. The precise formulation of this criterion was first proposed by Levanyuk [27] and Ginzburg [28]. The key observation here is that mean-field theory is only affected by fluctuations on length scales longer than the correlation length, $x \gg \xi$. Fluctuations on wavelengths shorter than the correlation length are absorbed into renormalized Landau parameters and do not produce departures from mean-field theory. To filter out the irrelevant short-wavelength fluctuations, we need to consider a coarse-grained average $\bar{\psi}$ of the order parameter over a correlation volume ξ^d. The Ginzburg criterion simply states that the variance of the averaged order parameter must be small compared with the equilibrium value, i.e.

$$\delta\bar{\psi}^2 = \frac{1}{\xi^d} \int_{|x|<\xi} d^dx \langle \delta\psi(x)\delta\psi(0)\rangle << \psi_0^2. \tag{11.141}$$

Since correlations decay exponentially on length scales longer than ξ, to get an estimate of this average, we can remove the constraint $|x| < \xi$ on the volume integral, to obtain

$$\delta\bar{\psi}^2 \sim \frac{1}{\xi^d} \int d^dx \langle \delta\psi(x)\delta\psi(0)\rangle \sim \frac{S_{\mathbf{q}=0}}{\xi^d} = \frac{k_B T_c}{s\,\xi^{d-2}}.$$

Now, substituting $\psi_0^2 = \frac{|r|}{4u} \sim \frac{s}{u}\frac{1}{\xi^2}$, we obtain

$$\frac{\delta\bar{\psi}^2}{\psi_0^2} \sim \frac{k_B T_c}{\xi^{d-4}} \frac{u}{s^2} << 1$$

or

$$\xi^{4-d} << \frac{c^2}{k_B T_c}.$$

Let us try to understand the meaning of the length scale defined by this expression. Multiplying this expression by ξ_0^{d-4}, where $\xi_0 = \sqrt{s/(aT_c)}$ is the coherence length, we obtain the dimensionless criterion

$$\left(\frac{\xi}{\xi_0}\right)^{4-d} << \xi_0^d \frac{\overbrace{s^2\xi_0^{-4}}^{(aT_c)^2}}{uk_B T_c} = \xi_0^d \frac{a^2 T_c}{uk_B}.$$

From (11.11) we recognize the combination $\frac{a^2 T_c}{u} = 8\Delta C_V$ as the jump in the specific heat, so the Ginzburg criterion can be written in the form

$$\left(\frac{\xi}{\xi_0}\right)^{4-d} << \frac{S_G}{k_B}, \qquad S_G = \Delta C_V \xi_0^d, \qquad \text{Ginzburg criterion} \qquad (11.142)$$

where we have dropped the factor of 8. The quantity $S_G = \Delta C_V \xi_0^d$ has the dimensions of entropy, and can be loosely interpreted as the entropy reduction per coherence volume ξ_0^d associated with the development of order, so that $S_G/k_B = \ln W$ is a logarithmic measure of the number of degrees of freedom W associated with the fully developed order parameter.

For models with $d > 4$, the Ginzburg criterion implies that large correlation lengths are good and, in this situation, as the correlation length diverges close to the critical point, mean-field theory becomes essentially exact. The dimension $d_U = 4$ is called the *upper critical dimension*. In a realistic situation, where $d < 4$, ξ^{4-d} diverges as the critical point is approached, so for $d < d_U = 4$ the Ginzburg criterion sets an upper bound on the correlation length and a lower bound on the distance from the phase transition. If we set $\xi/\xi_0 = |\Delta T/T_c|^{-1/2}$, the temperature deviation from T_c, ΔT must satisfy the requirement

$$\frac{|\Delta T|}{T_c} >> (S_G/k_B)^{-(4-d)2} \qquad (11.143)$$

for mean-field theory to be reliable.

From the above discussion, it is clear that systems with a large coherence length will deviate from mean-field theory only over a very narrow temperature window. Examples of systems with large coherence lengths are superconductors, superfluid ^{3}He, and spin density waves, where the ratio between the transition temperature and the Fermi temperature of the fluid, $k_B T_c/\epsilon_F << 1$. For example, in a superconductor, the entropy of condensation per unit cell is of order $k_B(\epsilon/\Delta)$, where $\Delta \sim 3.5 k_B T_c$ is the gap, while the coherence length is of order $v_F/\Delta \sim a(\epsilon_F/\Delta)$, where $v_F \sim \epsilon_F a$ is the Fermi velocity, so that the entropy of condensation per coherence length is of order

$$\Delta S_G/k_B \sim (\Delta/\epsilon_F) \times (\epsilon_F/\Delta)^3 \sim (\epsilon_F/\Delta)^2$$

and the Ginzburg criterion is

$$\frac{|\Delta T|}{T_c} >> (\Delta/\epsilon_F)^4$$

in three dimensions. Similar arguments may be applied to charge and spin density wave materials. For a typical superconductor with $T_c \sim 10\,\text{K}$, $\Delta \sim 30\,\text{K}$, $\epsilon_F \sim 10^5\,\text{K}$, this gives $\frac{|\Delta T|}{T_c} \sim (10^{-5})^4 \sim 10^{-20}$, far beyond the realm of observation. By contrast, in an insulating magnet the coherence length is of the order of the lattice spacing a, and the "Ginzburg entropy" is of order unity, so $\Delta T/T_c \sim 1$. These discussions are in accord with observations. Superconductors and charge density wave systems display perfect mean-field transitions, yet insulating magnets and superfluid ^{3}He display the classic λ-shaped specific heat curves that are a hallmark of a non-trivial specific-heat exponent α.

Exercises

Exercise 11.1 Show that the action of $U(\phi)e^{i\phi\hat{N}}$ on a coherent state $|\phi\rangle = U^{\dagger}(\phi)|\psi\rangle$ uniformly shifts the phase of the order parameter by ϕ, i.e.

$$\hat{\psi}(x)|\phi\rangle = \psi(x)e^{i\phi}|\phi\rangle,$$

so that

$$-i\frac{d}{d\phi}|\phi\rangle = \hat{N}|\phi\rangle.$$

Exercise 11.2 Consider the most general form of a two-component Landau theory:

$$f[\psi] = \frac{r}{2}(\psi_1^2 + \psi_2^2) + \frac{s}{2}(\psi_1^2 - \psi_2^2) + u(\psi_1^2 + \psi_2^2)^2 + u_2(\psi_1^4 - \psi_2^4) + u_3\psi_1^2\psi_2^2.$$

(a) Rewrite the free energy in terms of the amplitude and phase of the order parameter to demonstrate that, if s, u_2, or u_3 are finite, the free energy is no longer gauge-invariant.

(b) Rewrite the free energy as a function of ψ and ψ^*.

(c) If $s > 0$, what symmmetry is broken when $r < 0$?

(d) Write down the mean field equations for $s = 0$, $r < 0$.

(e) Sketch the phase diagram in the (u_2, u_3) plane.

Exercise 11.3 Consider the more general class of Landau theory where the interaction u can be negative:

$$f[\psi] = \frac{1}{V}F[\psi] = \frac{r}{2}\psi^2 + u_4\psi^4 + u_6\psi^6 - h\psi.$$

(a) Show that, for $h = 0$, $u < 0$, $r > 0$, the free energy contains three local minima, one at $\psi = 0$ and two others at $\psi = \pm\psi_0$, where

$$\psi_0^2 = -\frac{u}{3u_6} \pm \sqrt{\left(\frac{u}{3u_6}\right)^2 - \frac{r}{6u_6}}.$$

(b) Show that, for $r < r_c$, the solution at $\psi = 0$ becomes metastable, giving rise to a first-order phase transition at

$$r_c = -\frac{u^2}{2u_6}.$$

(*Hint*: calculate the critical value of r by imposing the second condition $f[\psi_{II}] = 0$. Solve the equation $f[\psi] = 0$ simultaneously with $f'[\psi_0] = 0$ from (a).)

(c) Sketch the (T, u) phase diagram for $h = 0$.

(d) For $r = 0$ but $h \neq 0$, show that there are three lines of critical points where $f'[\psi] = f''[\psi] = 0$, converging at the single point $r = u = h = 0$. This point is said to be a *tricritical point*.

(e) Sketch the (h, u) phase diagram for $r = 0$.

Exercise 11.4 We can construct a state of bosons in which the bosonic field operator has a definite expectation value using a coherent state as follows:

$$|\psi\rangle = \exp\left[\int d^3x\,\psi\,\hat{\psi}^\dagger(x)\right]|0\rangle.$$

The Hermitian conjugate of this state is $\langle\bar{\psi}| = \langle 0|e^{\int d^3x\hat{\psi}(x)\psi^*}$.

(a) Show that this coherent state is an eigenstate of the field destruction operator:
$\hat{\psi}(x)|\psi\rangle = \psi|\psi\rangle$.

(b) Show that overlap of the coherent state with itself is given by $\langle\bar{\psi}|\psi\rangle = e^N$, where $N = V|\psi|^2$ is the number of particles in the condensate.

(c) If

$$H = \int d^3x\left[\hat{\psi}^\dagger(x)\left(-\frac{\hbar^2}{2m}\nabla^2 - \mu\right)\psi(x) + U : (\psi^\dagger(x)\psi(x))^2 :\right]$$

is the (normal-ordered) energy density, show that the energy density $f = \frac{1}{V}\langle H\rangle$, where

$$\langle H\rangle = \frac{\langle\bar{\psi}|H|\psi\rangle}{\langle\bar{\psi}|\psi\rangle},$$

is given by

$$f = -\mu|\psi|^2 + U|\psi|^4,$$

providing a direct realization of the Landau free energy functional.

Exercise 11.5 Systematic derivation of the Ginzburg criterion.

(a) Show that the Ginzburg Landau free energy (11.138) can be written in the form

$$F = \frac{1}{2}\int d^dx'd^dx\,\psi(x')\chi_0^{-1}(x'-x)\psi(x') + u\int d^dx\,\psi(x)^4, \qquad (11.144)$$

where

$$\chi_0^{-1}(x'-x) = \delta^d(x-x')\left[-s\nabla^2 + r\right]$$

is the inverse of the susceptibility. The subscript 0 has been added to χ^{-1}, denoting that it is the "bare" susceptibilty, calculated for $u = 0$.

(b) By identifying the renormalized susceptibility with the second derivative of the free energy, show that, when interactions are taken into account,

$$\chi_q^{-1}(x'-x) \approx \left\langle\frac{\delta^2F}{\delta\psi(x)\delta\psi(x')}\right\rangle = \delta^d(x'-x)\left[-s\nabla^2 + r + 12u\langle\psi^2\rangle\right],$$

so that in momentum space

$$\chi_q = sq^2 + r + 12u\langle\psi^2\rangle_T,$$

where $\langle\psi^2\rangle_T = S(x-x')|_{x=x'}$ is the variance of the order parameter at a single point in space, evaluated at temperature T. (*Hint:* differentiate (11.17) with respect to $\psi(x)$ and take the expectation value of the resulting expression.)

(c) Show that the effects of fluctuations suppress T_c, and that, at the new suppressed transition temperature T_c^*,

$$r = r_0 = a(T_c^* - T_c) = -12u\langle\psi^2\rangle_{T_c^*} = -12u\int\frac{d^dq}{(2\pi)^d}\frac{k_BT_c^*/c}{q^2},$$

so that

$$\chi_{\mathbf{q}}^{-1} = sq^2 + (r - r_0) + 12u\left[\langle\psi^2\rangle - \langle\psi^2\rangle_{T_c^*}\right].$$

Notice how subtraction of the fluctuations at $T = T_c^*$ renormalizes $r \to r - r_0 = a(T - T_c^*)$. What is the renormalized correlation length?

(d) Finally, calculate the Ginzburg criterion by requiring that $|r - r_0| > 12u\left[\langle\psi^2\rangle - \langle\psi^2\rangle_{T_c^*}\right]$, to obtain

$$\frac{|r - r_0|}{4u} < 3\int\frac{d^dq}{(2\pi)^d}\frac{k_BT_c^*}{q^2}\left[\frac{\xi^{-2}}{q^2 + \xi^{-2}}\right]. \qquad (11.145)$$

The term inside the square brackets on the right-hand side results from the renormalization of $r \to r - r_0$. Notice how this term only involves fluctuations with $q \lesssim \xi^{-1}$, i.e. the long-wavelength fluctuations of wavelength greater than ξ. What has happened to the short-wavelength fluctuations?

(e) By approximately evaluating the integral on the right-hand side of (11.145), recast the Ginzburg criterion in the form

$$\frac{|r - r_0|}{u} << \frac{k_BT_c^*}{s}\frac{1}{\xi^{d-2}}.$$

References

[1] L. D. Landau, Theory of phase transformations, *Phys. Z. Sowjetunion*, vol. 11, no. 26, p. 545, 1937.

[2] S. Weinberg, A model of leptons. *Phys. Rev. Lett.*, vol.19, no. 21, p. 1264, 1967.

[3] N. Goldenfeld, *Lectures on Phase Transitions and the Renormalization Group*, Perseus Publishing, 1992.

[4] L. S. Ornstein and F. Zernike, Accidental deviations of density and opalescence at the critical point of a single substance, *Proc. Sect. Sci. K. Akad. Wet. Amsterdam*, vol. 17, p. 793, 1914.

[5] V. L. Ginzburg and L. D. Landau, On the theory of superconductivity. *Zh. Eksp. Teor. Fiz*, vol. 20, p. 1064, 1950.

[6] A. A. Abrikosov, Type II superconductors and the vortex lattice, *Les Prix Nobel, Nobel Foundation*, p. 59, 2003.

[7] O. Penrose, On the quantum mechanics of helium II, *Phil. Mag.*, vol. 42, p. 1373, 1951.

[8] O. Penrose and L. Onsager, Bose–Einstein condensation and liquid helium, *Phys. Rev.*, vol. 104, p. 576, 1956.

[9] P. W. Anderson, Considerations on the flow of superfluid helium, *Rev. Mod. Phys.*, vol. 38, p. 298, 1966.

[10] C. N. Yang, Concept of off-diagonal long-range order and the quantum phases of liquid he and of superconductors. *Rev. Mod. Phys.*, vol. 34, p. 694, 1962.

[11] L. Onsager, Statistical hydrodynamics, *Nuovo Cimento*, Suppl. vol. 6, p. 279, 1949.

[12] R. P. Feynman, *Progress in Low Temperature Physics*, vol. 1 North Holland, 1955.

[13] F. London, *Superfluids*, Dover Publications, vols. 1–2, 1961–1964.

[14] D. Saint-James and G. Sarma, *Type II Superconductivity*, Pergamon Press, 1969.

[15] L. Onsager, *Proceedings of the International Conference on Theoretical Physics, Kyoto and Tokyo, September 1953, Science Council of Japan*, p. 935, 1954.

[16] B. S. Deaver and W. M. Fairbank, Experimental evidence for quantized flux in superconducting cylinders. *Phys. Rev. Lett.*, vol. 7, p. 43, 1961.

[17] R. Doll and M. Näbauer, Experimental proof of magnetic flux quantization in a superconducting ring, *Phys. Rev. Lett.*, vol. 7, p. 51, 1961.

[18] P. W. Anderson, Plasmons, gauge invariance, and mass, *Phys. Rev.*, vol. 130, no. 1, p. 439, 1963.

[19] P. W. Higgs, Broken symmetries and the masses of gauge bosons, *Phys. Rev. Lett.*, vol. 13, no. 16, p. 508, 1964.

[20] P. W. Anderson, Random-phase approximation in the theory of superconductivity, *Phys. Rev.*, vol. 112, no. 6, p. 1900, 1958.

[21] S. Glashow, Partial symmetries of weak interactions, *Nucl. Phys.*, vol. 22, no. 4, p. 579, 1961.

[22] A. Salam and J. C. Ward, Electromagnetic and weak interactions, *Phys. Lett.*, vol. 13, p. 168, 1964.

[23] M. S. Carena and H. E. Haber, Higgs boson theory and phenomenology, *Prog. Part. Nucl. Phys.*, vol. 50, p. 63, 2003.

[24] M. B. Hindmarsh and T. W. B. Kibble, Cosmic strings, *Re. Prog. Phys.*, vol. 58, p. 477, 1985.

[25] P. W. Anderson, *Basic Notions of Condensed Matter Physics*, Benjamin Cummings, 1984.

[26] F. Wilczek, Quantum time crystals, *Phys. Rev. Lett.*, vol. 109, p. 160401, 2012.

[27] A. P. Levanyuk, Contribution to the theory of light scattering near the second-order phase transition points, *J. Exp. Theor. Phys.* vol. 36, p. 571, 1959.

[28] V. L. Ginzburg, Some remarks on phase transitions of the second kind and the microscopic theory of ferroelectrics, *J. Exp. Theor. Phys.*, vol. 2, p. 1824, 1960.

Path integrals

12.1 Coherent states and path integrals

In this chapter, we link the order parameter concept with microscopic many body physics by introducing the path integral formulation of quantum many-body theory. The emergence of a macroscopic order parameter in a quantum system is analogous to the emergence of classical mechanics in macroscopic quantum systems. The emergence of classical mechanics from quantum mechanics is most naturally described using wave-packets and the Feynman path integral. We shall see that a similar approach is useful for many-body systems, where the many body wavepacket states are coherent states: eigenstates of the quantum fields.

Chapter 11 introduced Landau's concept of broken symmetry, embracing the idea of an order parameter $\Psi(x)$. The beauty of the Landau approach is that it is a macroscopic description of matter: at length scales beyond the microscopic coherence length ξ_0, the emergence of an order parameter does not depend on the detailed microscopic microscopic physics that gives rise to it. In this chapter we go beneath the coherence length to examine the connection between the order parameter and the microscopic physics of a many-body system.

The basic idea of Feynman's path integral [1–3] is to reformulate the quantum mechanical amplitude as a sum of contributions from all possible paths, in which the classical action plays the role of the phase $\phi = S_{path}/\hbar$ associated with the path. The amplitude for a particle in a box to go from state $|i\rangle$ to state $|f\rangle$ is given by

$$\langle f|e^{-i\frac{Ht}{\hbar}}|i\rangle = \sum_{\text{paths } i \longrightarrow f} \exp\left[i\frac{S_{path}}{\hbar}\right]$$

where

$$S_{path} = \int_0^t dt'\left(p\dot{q} - H[p,q]\right). \tag{12.1}$$

The Feynman formulation is a precise reformulation of operator quantum mechanics. In the classical limit $\hbar \to 0$, the path integral is dominated by the paths of stationary phase, which correspond to the classical path which minimizes the action.

Feynman's idea can be extended to encompass statistical mechanics by treating the Boltzmann density matrix as a time-evolution operator in imaginary time. The trace over the density matrix is then the sum of amplitudes of paths that return to the initial configuration after an imaginary time $t = i\hbar\beta$:

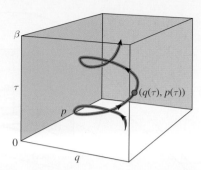

Illustrating a periodic path in imaginary time that contributes to the partition function of a single particle.

Fig. 12.1

$$Z = \text{Tr}\left[e^{-\beta\hat{H}}\right] = \sum_{\lambda} \langle\lambda|e^{-i\frac{Ht}{\hbar}}|\lambda\rangle\bigg|_{t=-i\hbar\beta}. \tag{12.2}$$

By changing variables $it/\hbar \to \tau$, so that $idt/\hbar \to d\tau$ and $p\dot{q}dt \to p\dot{q}d\tau$, we see that we can write this quantity as

$$Z = \sum_{\text{periodic paths}} \exp[-S_E],$$

(see Figure 12.1) where

$$S_E = \int_0^\beta d\tau \left(-\frac{i}{\hbar}p\partial_\tau q + H[p,q]\right). \tag{12.3}$$

We will now discuss a sophisticated extension of this idea to many-body systems, in which the path integral sums over the configurations of the particle fields rather than the trajectories of the particles themselves. The key innovation that makes this possible is the use of coherent states, which are literally eigenstates of the quantum field. In quantum optics such states, sometimes called *Glauber states*, are used to describe minimum-uncertainty wavepackets of photon fields [4]. For a single boson field, a coherent state is given by

$$|b\rangle = e^{\hat{b}^\dagger b}|0\rangle, \tag{12.4}$$

where, in this chapter, we use the roman $\hat{b}$ and $\hat{b}^\dagger$ to denote boson operators, reserving the italic b and $\bar{b}$ for the corresponding eigenvalues. Now $|0\rangle$ is a harmonic oscillator ground state defined by $\hat{b}|0\rangle = 0$, and it forms a minimum-uncertainty wavepacket centered around the origin of phase space. By contrast, the state $|b\rangle$ is the result of translating $|0\rangle$ so that it is centered around the point (q,p) in phase space, where $b = (q + ip)/\sqrt{2\hbar}$ incorporates both variables into a single complex variable (see Exercise 12.1). Paradoxically, though the state *is* an eigenstate of $\hat{b} = (\hat{q} + i\hat{p})/\sqrt{2\hbar}$, it is not an eigenstate of either $\hat{q}$ or $\hat{p}$. In a many-body problem the fields $\hat{\psi}(x)$ are defined at each point in space and in the corresponding coherent state $|\phi\rangle$:

$$\hat{\psi}(x)|\phi\rangle = \phi(x)|\phi\rangle. \tag{12.5}$$

We can still use the definition (12.4) for a coherent state, but now

$$\hat{b}^\dagger = \int d^d x\, \hat{\psi}^\dagger(x)\phi(x), \tag{12.6}$$

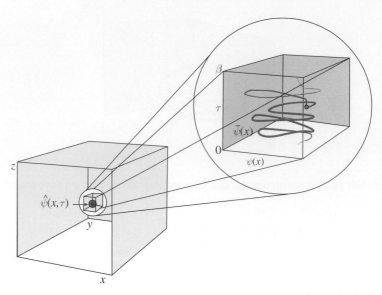

Fig. 12.2   Illustrating how the operator field at each point in space is represented by a trajectory inside the path integral.

coherently adds a boson to a condensate with wavefunction $\phi(x)$ (see Example 11.5 and Example 12.1). These states are the *wavepackets* of many-body physics. With care, we can use them as a basis set in which the matrix elements of the Hamiltonian are obtained simply by replacing the field operators with their expectation values in space–time, $\hat{\psi}(x, t) \rightarrow \phi(x, t)$. Using this procedure, the partition function can be rewritten as a path integral in which $\phi(x, t)$ defines a "history" or path over which the field at point x evolves, and Figure 12.2,

$$Z = \sum_{\text{periodic paths}} e^{-S_E[\bar{\phi}, \phi]}. \tag{12.7}$$

By convention, we denote the complex conjugate of $\phi(x)$ by $\bar{\phi}(x)$. In Chapter 2, we motivated quantum fields as the quantization of the single-particle wavefunction, identifying $\phi(x) \sim q$, $i\hbar\bar{\phi}^{\dagger}(x) \sim p$ as the corresponding canonical co-ordinates. Using this analogy, the many-body analogue of the kinetic term in (12.3) is

$$-\frac{i}{\hbar}p\partial_{\tau}q \sim \bar{\phi}(x)\partial_{\tau}\phi(x), \tag{12.8}$$

so the many-body analogue of (12.3) is expected to take the form

$$S_E = \int_0^{\beta} d\tau d^3x \left[\bar{\phi}(x, \tau)\partial_{\tau}\phi(x, \tau) + H[\bar{\phi}, \phi]\right], \tag{12.9}$$

where H is the many-body Hamiltonian with field operators replaced by the c-numbers ϕ and $\bar{\phi}$. In fact, as we'll see, this is precisely the form that is obtained when the partition function is expanded in terms of coherent states $|\phi(x, \tau)\rangle$ [5–7]. Furthermore, time-ordered Green's functions can also be rewritten as an average under the path integral, so that

$$\langle T\hat{\psi}(1)\hat{\psi}^{\dagger}(2)\rangle = \frac{1}{Z}\sum_{\text{path}}\exp\left[-S_{path}\right]\phi(1)\bar{\phi}(2),$$

where $\bar{\psi}(2)$ is the complex conjugate of $\psi(2)$. In this way, the quantum mechanics of the many-body system is transformed from an operator formalism, into a *statistical* description, with each space–time configuration of the fields weighted by the action.

Remarkably, this approach can be extended to include fermions, using an idea of Julian Schwinger [8] that generalizes the concept of c-numbers to include anticommuting Grassman numbers. For fermions, the numbers $\psi(x)$ appearing in the coherent states must *anticommute* with each other. They are thus a new kind of number, which requires some new algebraic tricks. Moreover, we'll see that we can evaluate the corresponding path integral for *all* non-interacting problems. This is already a major achievement.

A final aspect of path integrals is that interacting problems can be transformed, by the Hubbard–Stratonovich method [9, 10], into a problem of "free" particles moving in a fluctuating effective field. This technique provides an important tool for the study of broken-symmetry phase transitions:

$$Z_{\text{interacting}} \rightarrow \sum_{\{\Delta\}}\left[\text{path integral of fermions moving in field }\Delta\right], \tag{12.10}$$

where $\{\Delta\}$ denotes a given configuration of the symmetry-breaking field Δ.

12.2 Coherent states for bosons

To demonstrate the path integral approach, we will start with the bosonic version. As a warm-up, we need to establish a few key properties of the boson coherent state. The coherent state of a single boson operator $\hat{b}^{\dagger}$ is given by

$$|b\rangle = e^{\hat{b}^{\dagger}b}|0\rangle, \tag{12.11}$$

where b is a complex number. This state is an eigenstate of the annihilation operator

$$\hat{b}|b\rangle = b|b\rangle. \tag{12.12}$$

We can also form the conjugate state

$$\langle\bar{b}| = \langle 0|e^{\bar{b}\hat{b}}, \tag{12.13}$$

which is the eigenstate of the creation operator

$$\langle\bar{b}|\hat{b}^{\dagger} = \langle\bar{b}|\bar{b},$$

where $\bar{b}$ is the complex conjugate of b. Although b and $\bar{b}$ are complex conjugates of one another, they are derived from two independent real variables, and when we integrate over them we need a double integral in which we treat b and $\bar{b}$ as independent variables. The "bar" notation is adopted by convention to emphasize this linear independence.

A coherent state describes a condensate with an indefinite particle number. If we decompose it into eigenstates of particle number n by expanding in powers of b, we obtain

$$|b\rangle = \sum_n \frac{b^n}{n!}(\hat{b}^\dagger)^n|0\rangle = \sum_n |n\rangle \frac{b^n}{\sqrt{n!}}, \tag{12.14}$$

where $|n\rangle = \frac{(\hat{b}^\dagger)^n}{\sqrt{n!}}|0\rangle$ is the eigenstate of the number operator $\hat{n} = \hat{b}^\dagger\hat{b}$. In this way we see that the amplitude for a coherent state to be in a state with n particles is

$$\phi_n(b) = \langle n|b\rangle = \frac{b^n}{\sqrt{n!}}. \tag{12.15}$$

Similarly,

$$\langle \bar{b}| = \sum_m \frac{\bar{b}^m}{\sqrt{m!}}\langle m| \tag{12.16}$$

and

$$\langle \bar{b}|m\rangle = \frac{\bar{b}^m}{\sqrt{m!}}. \tag{12.17}$$

From (12.14) and (12.16), the overlap between the states $\langle \bar{b}_1|$ and $|b_2\rangle$ is given by

$$\langle \bar{b}_1|b_2\rangle = \sum_{m,n} \frac{\bar{b}_1^m}{\sqrt{m!}} \overbrace{\langle m|n\rangle}^{\delta_{mn}} \frac{b_2^n}{\sqrt{n!}} = \sum_n \frac{(\bar{b}_1 b_2)^n}{n!} = e^{\bar{b}_1 b_2}. \tag{12.18}$$

12.2.1 Matrix elements and the completeness relation

Remarkably, even though coherent states are not orthogonal, they can be used to great effectiveness as an overcomplete basis, in which the field operators are diagonal. There are two important properties that we shall repeatedly use to great advantage (see Table 12.1):

- **Matrix elements** Matrix elements of normal-ordered operators $O[\hat{b}^\dagger, \hat{b}]$ between two coherent states are obtained by simply replacing the operators $\hat{b}$ and $\hat{b}^\dagger$ with the c-numbers b and $\bar{b}$, respectively:

$$\langle \bar{b}_1|\hat{O}[\hat{b}^\dagger, \hat{b}]|b_2\rangle = O[\bar{b}_1, b_2] \times \langle \bar{b}_1|b_2\rangle = O[\bar{b}_1, b_2] \times e^{\bar{b}_1 b_2}. \tag{12.19}$$

- **Completeness** The unit operator can be decomposed in terms of coherent states as follows:

$$\hat{1} = \sum_{\bar{b},b} |b\rangle\langle \bar{b}|, \tag{12.20}$$

Table 12.1 Boson calculus.		
Completeness	$\langle b\|b\rangle = e^{\bar{b}b}$	Overcomplete basis
	$\displaystyle\int \frac{d\bar{b}db}{2\pi i} e^{-\bar{b}b} \|b\rangle\lambda\bar{b}\| = \underline{1}$	Completeness relation
	$\displaystyle\mathrm{Tr}[\hat{A}] = \int \frac{d\bar{b}db}{2\pi i} e^{-\bar{b}b} \langle\bar{b}\|\hat{A}\|b\rangle$	Trace formula
Gaussian integrals	$\displaystyle\int \prod_j \frac{d\bar{b}_j db_j}{2\pi i} e^{-\left[\bar{b}\cdot A\cdot \bar{b} - \bar{j}\cdot b - \bar{b}\cdot j\right]} = \frac{e^{\left[\bar{j}\cdot A^{-1}\cdot j\right]}}{\det A}$	

where[1]

$$\sum_{\bar{b},b} \equiv \int \frac{d\bar{b}db}{2\pi i} e^{-\bar{b}b} \tag{12.24}$$

is the normalized measure for summing over coherent states.

We present a detailed derivation of these two results in Appendix 12A, continuing now to use them to derive a path integral.

Example 12.1 Prove that in a coherent state $|b\rangle$ the probability $p(n)$ to be in a state with n particles is a Poisson distribution with average particle number $n_0 = \langle\hat{n}\rangle = \bar{b}b$ and variance $\langle\delta n^2\rangle = n_0$ (see Figure 12.3), where

$$p(n) = \frac{1}{n!}(\bar{b}b)^n e^{-\bar{b}b}. \tag{12.25}$$

Solution

To calculate the normalized probability for a coherent state $|b\rangle$ to be in a state $|n\rangle$, we calculate

$$p(n) = \frac{|\langle n|b\rangle|^2}{\langle\bar{b}|b\rangle}. \tag{12.26}$$

---

[1] Note: in quantum optics, one often encounters the normalized coherent or Glauber state,

$$|b,\bar{b}\rangle_N = \frac{1}{\sqrt{2\pi i}} e^{-\bar{b}\hat{b}+\hat{b}^\dagger b}|0\rangle = \frac{1}{\sqrt{2\pi i}} e^{-\bar{b}b/2}|b\rangle. \tag{12.21}$$

This affords the advantage of a simpler completeness relation,

$$\underline{1} = \int d\bar{b}db \, |b,\bar{b}\rangle_N \langle b,\bar{b}|_N, \tag{12.22}$$

but, unfortunately, the matrix elements of normal-ordered operators now assume a more complex form:

$$\langle b_1,\bar{b}_1|\hat{O}(\hat{b}^\dagger,\hat{b})|b_2,b_2\rangle_N = e^{\bar{b}_1 b_2 - \bar{b}_1 b_1/2 - \bar{b}_2 b_2/2} O(\bar{b},b). \tag{12.23}$$

Although the awkward prefactor in this expression vanishes if $b_1 = b_2$, it can't be neglected in a general path integral, for which b_2 and b_1 are completely independent. For this reason, this book chooses to use coherent states without the normalizing prefactor.

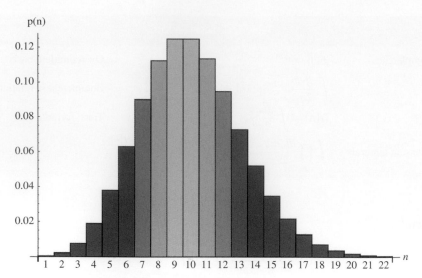

Fig. 12.3 Probability distribution function for a coherent state with $\bar{b}b = n_0 = 10$.

Now using (12.15), we have

$$\langle n|b\rangle = \frac{b^n}{\sqrt{n!}}, \qquad \langle \bar{b}|n\rangle = \frac{\bar{b}^n}{\sqrt{n!}}, \tag{12.27}$$

so that

$$|\langle n|b\rangle|^2 = \frac{(\bar{b}b)^n}{n!}. \tag{12.28}$$

Since $\langle \bar{b}|b\rangle = e^{\bar{b}b}$, it follows that

$$p(n) = \frac{|\langle n|b\rangle|^2}{\langle \bar{b}|b\rangle} = \frac{1}{n!}(\bar{b}b)^n e^{-\bar{b}b}, \tag{12.29}$$

corresponding to a Poisson distribution with average and variance $n_0 = \bar{b}b$.

To confirm the average and variance, we can calculate the average particle number, as follows:

$$n_0 = \sum_{n=1,\infty} np(n) = \sum_{n=1,\infty} \frac{1}{n-1!}(\bar{b}b)^n e^{-\bar{b}b} = \bar{b}be^{-\bar{b}b} \sum_{n=0,\infty} \frac{1}{n!}(\bar{b}b)^n = \bar{b}b. \tag{12.30}$$

To calculate the variance, we first note that

$$\langle \hat{n}^2 - \hat{n} \rangle = \sum_{n=1,\infty} n(n-1)p(n) = (\bar{b}b)^2 \sum_{n=2,\infty} \frac{1}{n-2!}(\bar{b}b)^{n-2} e^{-\bar{b}b} = (\bar{b}b)^2 = n_0^2, \tag{12.31}$$

so that $\langle \hat{n}^2 \rangle = n_0(n_0 + 1)$ and hence $\langle \delta n^2 \rangle = \langle \hat{n}^2 \rangle - n_0^2 = n_0$. Notice that $\langle \delta n^2 \rangle / \langle n \rangle^2 = 1/n_0$. When n_0 is large, the distribution function becomes Gaussian and resembles a delta function in the thermodynamic limit.

Example 12.2 Using the completeness relation, prove that, if $f(\alpha) = \langle f|\alpha\rangle$ is the overlap of coherent state $|\alpha\rangle$ with state $|f\rangle$, then

$$f(\alpha) = \int \frac{d\bar{b}db}{2\pi i} f(b) e^{\bar{b}(\alpha - b)}. \tag{12.32}$$

Solution

Write the function $f(\alpha)$ as the overlap of state $\langle f|$ with state $|\alpha\rangle$, $f(\alpha) = \langle f|\alpha\rangle$. Now insert the completeness relation into this expression to obtain

$$\langle f|\alpha\rangle = \langle f|\hat{1}|\alpha\rangle = \int \frac{d\bar{b}db}{2\pi i} \langle f|b\rangle \langle \bar{b}|\alpha\rangle e^{-\bar{b}b}$$

$$= \int \frac{d\bar{b}db}{2\pi i} f(b) e^{\bar{b}(\alpha - b)}. \tag{12.33}$$

Note the useful identity

$$\delta(\alpha - b) = \int \frac{d\bar{b}}{2\pi i} e^{\bar{b}(\alpha - b)}. \tag{12.34}$$

Example 12.3 Using the completeness relation, prove that the trace of any operator $\hat{A}[\hat{b}^\dagger, \hat{b}]$ (not necessarily normal-ordered) is given by

$$\text{Tr}[A] = \sum_{\bar{b},b} \langle \bar{b}|A|b\rangle = \int \frac{d\bar{b}db}{2\pi i} e^{-\bar{b}b} \langle \bar{b}|A|b\rangle. \tag{12.35}$$

Solution

In the particle-number basis, the trace over $\hat{A}$ is given by

$$\text{Tr}[A] = \sum_n \langle n|A|n\rangle = \sum_{n,m} \langle m|A|n\rangle \delta_{nm}. \tag{12.36}$$

From completeness,

$$\delta_{nm} = \sum_{\bar{b},b} \langle n|b\rangle \langle \bar{b}|m\rangle,$$

so that

$$\text{Tr}[A] = \sum_{\bar{b},b,n,m} \langle n|b\rangle \langle \bar{b}|m\rangle \langle m|A|n\rangle$$

$$= \sum_{\bar{b},b,m,n} \langle \bar{b} \overbrace{|m\rangle \langle m|}^{=1} \hat{A} \overbrace{|n\rangle \langle n|}^{=1} b\rangle$$

$$= \sum_{\bar{b},b} \langle \bar{b}|A|b\rangle \equiv \int \frac{d\bar{b}db}{2\pi i} e^{-\bar{b}b} \langle \bar{b}|A|b\rangle. \tag{12.37}$$

12.3 Path integral for the partition function: bosons

We now develop the path integral expression for the partition function of a single boson field, with a normal-ordered Hamiltonian $\hat{H}[\hat{b}^\dagger, \hat{b}]$. The key result to be derived is

$$
Z = \int \mathcal{D}[\bar{b}, b] e^{-S}
$$

$$
S = \int_0^\beta d\tau \left(\bar{b} \partial_\tau b + H[\bar{b}, b] \right). \tag{12.38}
$$

path integral for the partition function

All of our results can be simply generalized to include many different bosons. We begin by writing the trace required for the partition function in a coherent state basis, as

$$
Z = \text{Tr}[e^{-\beta H}] = \int \frac{d\bar{b}db}{2\pi i} e^{-\bar{b}b} \langle \bar{b} | e^{-\beta H} | b \rangle. \tag{12.39}
$$

Unfortunately, $e^{-\beta H[\hat{b}^\dagger, \hat{b}]}$ is not a normal-ordered operator, so we can't just replace the boson operators by their c-number equivalents. To achieve such a replacement, we divide the Boltzmann factor $e^{-\beta H} = U(\beta)$ (Figure 12.4) into a large number N of tiny time-slices of duration $\Delta\tau = \beta/N$:

$$
e^{-\beta H} = \left(e^{-\Delta\tau H} \right)^N. \tag{12.40}
$$

Since H is normal-ordered, $e^{-\Delta\tau H} = 1 - \Delta\tau : H : + O(\Delta\tau^2)$, so that $e^{-\Delta\tau H}$ and $: e^{-\Delta\tau H} :$ differ only at second order in $\Delta\tau$. Thus, to an accuracy $O(\Delta\tau^2) = O(1/N^2)$ per time slice, we can replace the boson operators by c-numbers in each time-slice:

$$
\langle \bar{b}_j | e^{-\Delta\tau H[\hat{b}^\dagger, \hat{b}]} | b_{j-1} \rangle = \exp \left[\bar{b}_j b_{j-1} - \Delta\tau H[\bar{b}_j, b_{j-1}] \right] + (\Delta\tau^2). \tag{12.41}
$$

This is a huge step forward, which transforms the time-slice into a purely *algebraic* expression.

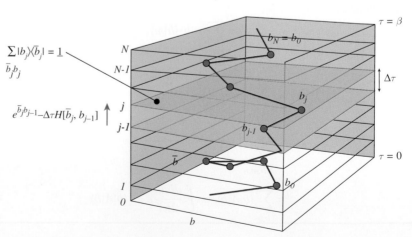

Fig. 12.4   Illustrating the division of the trajectory into N time-slices.

Let us now put this all together. The time-sliced partition function (12.39) is first written

$$Z = \int \frac{d\bar{b}_N db_0}{2\pi i} \langle \bar{b}_N | \left(e^{-\Delta\tau H} \right)^N | b_0 \rangle e^{-\bar{b}_N b_0}, \tag{12.42}$$

where we have relabeled $\bar{b} \rightarrow b_N$, $b \rightarrow b_0$ in (12.39). Next, between each time-slice we now introduce the completeness relation,

$$\hat{1} = \int \frac{d\bar{b}_j db_j}{2\pi i} | b_j \rangle e^{-\bar{b}_j b_j} \langle \bar{b}_j | \equiv \mathbb{1}_j, \tag{12.43}$$

so that the partition function becomes

$$Z = \int \frac{d\bar{b}_N db_0}{2\pi i} \langle \bar{b}_N | e^{-\Delta\tau H} \times \cdots \times \mathbb{1}_j \times e^{-\Delta\tau H} \times \mathbb{1}_{j-1} \times \cdots \times \mathbb{1}_1 \times e^{-\Delta\tau H} | b_0 \rangle e^{-\bar{b}_N b_0}$$

$$= \int \mathcal{D}_N[\bar{b}, b] \prod_{j=1}^{N} e^{-\bar{b}_j b_j} \langle \bar{b}_j | e^{-\Delta\tau H} | b_{j-1} \rangle. \tag{12.44}$$

Notice that we have identified $b_N \equiv b_0$ and $\bar{b}_N \equiv \bar{b}_0$. We have also introduced the shorthand notation

$$\mathcal{D}_N[\bar{b}, b] = \prod_{j=1}^{N} \frac{d\bar{b}_j db_j}{2\pi i} \tag{12.45}$$

for the measure.

Inserting expression (12.41) into (12.44), we then obtain

$$Z = \int \mathcal{D}_N[\bar{b}, b] \exp\left[-\sum_{j=1}^{N} \left(\bar{b}_j(b_j - b_{j-1}) + \Delta\tau H[\bar{b}_j, b_{j-1}] \right) \right] + O(N\Delta\tau^2), \tag{12.46}$$

where we have grouped the errors from all N time-slices into a final term of order $O(N\Delta\tau^2) = O(1/N)$. Since this error vanishes in the limit $N \rightarrow \infty$, we may thus write

$$Z = \lim_{N \rightarrow \infty} \int \mathcal{D}_N[\bar{b}, b] \exp[-S_N]$$

$$S_N = \sum_{j=1}^{N} \Delta\tau \left(\bar{b}_j \frac{(b_j - b_{j-1})}{\Delta\tau} + H[\bar{b}_j, b_{j-1}] \right). \tag{12.47}$$

This is the path integral representation of the partition function for a single boson field. Let us pause to reflect on this result. The integral represents a sum over all possible "histories" of the field,

$$b(\tau_j) \equiv (b_1, b_2, \ldots b_N)$$
$$\bar{b}(\tau_j) \equiv (\bar{b}_1, \bar{b}_2, \ldots \bar{b}_N). \tag{12.48}$$

This kind of integral is also called a *functional integral*, because it involves integrating over all values of the functions $b(\tau)$. When we take the thickness of each time-slice to zero, the discrete functions $b(\tau_j) \equiv b_j$ become functions of continuous time. Our identification of

$b_0 \equiv b_N$ and hence $\bar{b}_0 \equiv \bar{b}_N$ implies that the set of complete functions that we sum over is periodic in time:

$$b(\tau) = b(\tau + \beta), \qquad\qquad \bar{b}(\tau) = \bar{b}(\tau + \beta). \qquad (12.49)$$

This is a new type of integral calculus just as in conventional integral calculus, we reserve a special notation for the continuum limit:

$$\mathcal{D}_N[\bar{b}, b] \to \mathcal{D}[\bar{b}, b].$$

Assuming that the continuum limit is indeed a well-defined limit, we now replace

$$\sum_{j=1}^{N} \Delta\tau \to \int_0^\beta d\tau,$$

$$b_j \to b(\tau), \qquad \bar{b}_j \frac{(b_j - b_{j-1})}{\Delta\tau} \to \bar{b}\partial_\tau b,$$

$$H[\bar{b}_j, b_{j-1}] \to H[\bar{b}, b]. \qquad (12.50)$$

These brash replacements hide a mountain of subtlety. Unlike a conventional integral, there is no sense of "continuity" associated with the field $b(\tau)$: inside the functional integral the paths we sum over are jagged, noisy objects. However, if we look at their typical noise spectra, they have a characteristic frequency. For a harmonic oscillator, this is just the frequency of oscillation ω, but if we include interactions there will typically be a spectrum of such frequencies with some maximum frequency ω_0. The continuum limit will develop provided $\omega_0 \Delta\tau \ll 1$.

The limiting value of the path integral is then written

$$Z = \int \mathcal{D}[\bar{b}, b] e^{-S}$$

$$S = \int_0^\beta d\tau \left(\bar{b}\partial_\tau b + H[\bar{b}, b] \right). \qquad (12.51)$$

The simplest example of such a path integral is the non-interacting harmonic oscillator, in which $H = \epsilon \hat{b}^\dagger \hat{b}$. For this case,

$$Z = \int \mathcal{D}[\bar{b}, b] \exp\left[-\int_0^\beta d\tau \bar{b}\left(\partial_\tau + \epsilon \right) b \right]. \qquad (12.52)$$

This is an example of a Gaussian path integral, because the action is just a quadratic function of the fields, and we'll shortly see that we can evaluate all such path integrals in a closed form. It should be clear that this derivation does not depend on whether there are interaction terms in the Hamiltonian. We could equally well consider the case of the anharmonic oscillator, written in normal-ordered form as

$$H = \epsilon \hat{b}^\dagger \hat{b} + g : (\hat{b} + \hat{b}^\dagger)^4.$$

The partition function for this case is now

$$Z = \int \mathcal{D}[\bar{b}, b] \exp\left[-\int_0^\beta d\tau \left(\bar{b}(\partial_\tau + \epsilon) b + g(b + \bar{b})^4 \right) \right].$$

This is probably the simplest example of an interacting path integral.

12.3.1 Multiple bosons

The derivation of the previous section is easily generalized to include multiple bosons, with a Hamiltonian $H[\hat{b}_\lambda^\dagger, \hat{b}_\lambda]$, by using a multi-variable coherent state,

$$|b\rangle = \exp\left[\sum_\lambda \hat{b}_\lambda^\dagger b_\lambda\right].$$

Since this is just a product of coherent states, we can simply extend the completeness relationship as a product of the measures for each individual boson:

$$\hat{1} = \sum_{\bar{b},b} |b\rangle\langle b|, \tag{12.53}$$

where now

$$\sum_{\bar{b},b} = \int \prod_\lambda \frac{d\bar{b}_\lambda \, db_\lambda}{2\pi i} e^{-\bar{b}_\lambda b_\lambda}.$$

The procedure for developing the path integral is exactly the same: we subdivide the interval into N time-slices, approximating $e^{-\Delta\tau H}$ by its normal-ordered form. The resulting path integral is formally very similar:

$$Z = \int \mathcal{D}[\bar{b}, b] e^{-S}$$

$$S = \int_0^\beta d\tau \left(\sum_\lambda \bar{b}_\lambda \partial_\tau b_\lambda + H[\bar{b}_\lambda, b_\lambda]\right), \tag{12.54}$$

path integral for the partition function: multiple bosons

where the measure is now a product of the measure for each boson field:

$$\mathcal{D}[\bar{b}, b] = \prod_\lambda \mathcal{D}[\bar{b}_\lambda, b_\lambda].$$

For example, the path integral for a gas of free bosons with Hamiltonian $H = \sum_{\mathbf{k}} \omega_{\mathbf{k}} \hat{b}_{\mathbf{k}}^\dagger \hat{b}_{\mathbf{k}}$ has the action

$$S = \int_0^\beta \left[\sum_{\mathbf{k}} \bar{b}_{\mathbf{k}}(\partial_\tau + \omega_{\mathbf{k}}) b_{\mathbf{k}}\right].$$

12.3.2 Time-ordered expectation values

In addition to providing equilibrium thermodynamics, the path integral can also be used to calculate time-ordered expectation values. The division of time into N time-slices using coherent states can also be carried out for the evaluation of arbitrary time-ordered products of fields – and when we do so, we discover that the time-ordered product of fields maps onto a path integral over the corresponding c-number product of fields. Thus, for the two-point Green's function,

$$G(2-1) = -\langle T\hat{b}(2)\hat{b}^\dagger(1)\rangle = -\frac{\int \mathcal{D}[\bar{b},b]e^{-S}b(2)\bar{b}(1)}{\int \mathcal{D}[\bar{b},b]e^{-S}}, \qquad (12.55)$$

where we have used the notation $1 \equiv (\tau_1, X_1, \{\lambda_1\})$ to denote the continuous and discrete variables associated with the boson field. In this way, time-ordered products of operators become weighted averages of c-numbers inside the path integral. The operator form for the Green's function is written in terms of the Heisenberg fields, and to convert it into a path integral we need to rewrite the Heisenberg field operators in terms of the Schrödinger fields:

$$\hat{b}_H(2) = e^{H\tau_2}\hat{b}_S(2)e^{-H\tau_2},$$

where the time argument τ_2 of the Schrödinger field $\hat{b}_S(2)$ is a dummy variable. Now with this device the Green's function can be transformed to the Schrödinger representation, as follows:

$$\begin{aligned}
G(2-1) &= -\frac{1}{Z}\text{Tr}\left[e^{-\beta H}T\left\{\hat{b}_H(2)\hat{b}_H^\dagger(1)\right\}\right] \\
&= -\frac{1}{Z}\text{Tr}\left[e^{-\beta H}T\left\{e^{\tau_2 H}\hat{b}_S(2)e^{-\tau_2 H}e^{\tau_1 H}\hat{b}_S^\dagger(1)e^{-\tau_1 H}\right\}\right] \\
&= -\frac{1}{Z}\text{Tr}\left[T\left\{U(\beta-\tau_2)\hat{b}_S(2)U(\tau_2-\tau_1)\hat{b}_S^\dagger(1)U(\tau_1)\right\}\right] \\
&= \text{Tr}\left[T\left\{U(\beta)\hat{b}_S(2)\hat{b}_S^\dagger(1)\right\}\right], \qquad (12.56)
\end{aligned}$$

where $U(\tau) = e^{-H\tau}$ is the time-evolution operator. To write the Green's function as a path integral, we now expand the time-ordered trace in terms of N time-slices, introducing the Schrödinger operators at the time-slices τ_j and τ_k which corresponding to τ_1 and τ_2, respectively. Here's where coherent states work their marvelous magic, for we can rewrite the destruction operator as

$$\hat{b}_S(\tau_k) = \hat{b}_S(\tau_k) \times \underline{1}_k = \int \frac{d\bar{b}_k b_k}{2\pi i}e^{-\bar{b}_k b_k}|b_k\rangle \, b_k\langle\bar{b}_k|, \qquad (12.57)$$

and similarly

$$\hat{b}_S^\dagger(\tau_j) = \underline{1}_j \times \hat{b}_S^\dagger(\tau_j) = \int \frac{d\bar{b}_j b_j}{2\pi i}e^{-\bar{b}_j b_j}|b_j\rangle \, \bar{b}_j \, \langle\bar{b}_j|, \qquad (12.58)$$

so that, inside the path integral, $\hat{b}_S(2)\hat{b}_S^\dagger(1) \to b(2)\bar{b}(1)$ and

$$\text{Tr}[T\left\{U(\beta)\hat{b}_S(2)\hat{b}_S^\dagger(1)\right\}] = \int \mathcal{D}[\bar{b},b]e^{-S}b(2)\bar{b}(1),$$

from which the path integral expression (12.55) for the Green's function follows. We can easily extend these results to all higher moments, quite generally, mapping time-ordered Green's functions onto the corresponding moments under the path integral:

$$\langle T\hat{b}(1)\hat{b}(2)\cdots\hat{b}^\dagger(2')\hat{b}^\dagger(1')\rangle = \frac{\int \mathcal{D}[\bar{b},b]e^{-S}b(1)b(2)\cdots\bar{b}(2')\bar{b}(1')}{\int \mathcal{D}[\bar{b},b]e^{-S}}. \tag{12.59}$$

In this way, the path integral maps a system of interacting particles onto a statistical mechanics problem, with distribution function e^{-S}.

12.3.3 Gaussian path integrals

An important class of path integrals are the *Gaussian path integrals*, in which the action is a quadratic functional of the fields. For example, for the free-boson Hamiltonian $\hat{H} = \hat{b}_\alpha^\dagger h_{\alpha\beta}\hat{b}_\beta$ the action is

$$S_E = \int_0^\beta d\tau \bar{b}_\alpha(\partial_\tau + h_{\alpha\beta})b_\beta \equiv \int_0^\beta d\tau \bar{b}(\partial_\tau + \underline{h})b. \tag{12.60}$$

Remarkably, all Gaussian path integrals can be evaluated in a closed form, and the key result is

$$Z_G = \int \mathcal{D}[\bar{b},b]\exp\left[-\int_0^\beta d\tau \bar{b}(\partial_\tau + \underline{h})b\right] = [\det(\partial_\tau + \underline{h})]^{-1}. \tag{12.61}$$

bosonic Gaussian integral

To understand this result, it is helpful to think of the function $b_\alpha(\tau) \equiv b_{\tilde{\alpha}}$ as a huge vector labeled by the indices $\tilde{\alpha} \equiv (\alpha, \tau)$. From this perspective, a Gaussian action is a vast matrix bilinear,

$$S_E = \sum_{(\alpha,\tau),\,(\beta,\tau')} \bar{b}_\alpha(\tau)M_{\alpha\beta}(\tau,\tau')b_\beta(\tau') \equiv \bar{b} \cdot M \cdot b, \tag{12.62}$$

where

$$M_{\alpha\beta}(\tau,\tau') = \delta(\tau - \tau')(\partial_{\tau'} + h_{\alpha\beta}). \tag{12.63}$$

You may be worried about the notion of treating time integration as a summation. To assuage your doubts, it is useful to rewrite S_E in the frequency domain, where summations over time are replaced by discrete frequency summations. Since $b(\tau) = b(\tau + \beta)$, the Bose field can always be represented in terms of a discrete set of Fourier components,

$$b_\alpha(\tau) = \frac{1}{\sqrt{\beta}}\sum_n b_\alpha(i\nu_n)e^{i\nu_n\tau}. \tag{12.64}$$

In this basis

$$[\underline{M}(\tau - \tau')]_{\alpha\beta} = \delta(\tau - \tau')(\partial_{\tau'} + h_{\alpha\beta}) \longrightarrow (-i\nu_n\delta_{\alpha\beta} + h_{\alpha\beta}) = \underline{M}(i\nu_n), \tag{12.65}$$

so the action becomes a discrete summation over Matsubara frequencies:

$$S_E = \sum_{i\nu_n} \bar{b}_\alpha(i\nu_n)\left(-i\nu_n\delta_{\alpha\beta} + h_{\alpha\beta}\right)b_\beta(i\nu_n) \equiv \bar{b} \cdot M \cdot b. \qquad (12.66)$$

To integrate a Gaussian path integral, we employ the general result for a multi-dimensional Gaussian integral,

$$\int \prod_\alpha \frac{d\bar{b}_\alpha db_\alpha}{2\pi i} e^{-\bar{b}_\alpha M_{\alpha\beta} b_\beta} = \frac{1}{\det[M]}, \qquad (12.67)$$

where M is a matrix with non-zero eigenvalues. To prove this result, we transform to a basis where M is explicitly diagonal. Let $b = U \cdot a$ and $\bar{b} = \bar{a} \cdot U^\dagger$, where $U \equiv U_{\alpha\lambda}$ is the unitary matrix that diagonalizes M. Then, since $U^\dagger_{\lambda\alpha} M_{\alpha\beta} U_{\beta\lambda'} = m_\lambda \delta_{\lambda\lambda'}$, where the m_λ are the eigenvalues of M,

$$\bar{b}_\alpha M_{\alpha\beta} b_\beta = \bar{a}_\lambda m_\lambda a_\lambda$$

is explicitly diagonal. Furthermore, under a unitary transformation the measure remains unchanged. To see this, we write the transformed measure using a Jacobian,

$$\prod_\alpha d\bar{b}_\alpha db_\alpha = \prod_\alpha d\bar{a}_\alpha da_\alpha \times \frac{\delta[\bar{b}, b]}{\delta[\bar{a}, a]} = \prod_\alpha d\bar{a}_\alpha da_\alpha \left\| \begin{matrix} U^\dagger & 0 \\ 0 & U \end{matrix} \right\| = \prod_\alpha d\bar{a}_\alpha da_\alpha,$$

where unitarity guarantees that the Jacobian is unity:

$$\left\| \begin{matrix} U^\dagger & 0 \\ 0 & U \end{matrix} \right\| = \det[U^\dagger U] = 1.$$

Under these transformations, the Gaussian integral becomes diagonal and can be explicitly evaluated:

$$\int \prod_\alpha \frac{d\bar{b}_\alpha db_\alpha}{2\pi i} e^{-\bar{b}_\alpha M_{\alpha\beta} b_\beta} = \prod_\lambda \int \frac{d\bar{a}_\lambda da_\lambda}{2\pi i} e^{-m_\lambda \bar{a}_\lambda a_\lambda} = \left(\prod_\lambda m_\lambda\right)^{-1} = \frac{1}{\det[M]}, \quad (12.68)$$

where, in the last step, we have identified the determinant of M with the product of its eigenvalues, $\det[M] = \prod_\lambda m_\lambda$. Finally, if we now replace $M \to \partial_\tau + h$ we obtain the general relationship given in (12.61):

$$Z_G = \frac{1}{\det[\partial_\tau + \underline{h}]}. \qquad (12.69)$$

We can equally well write this in the frequency domain, where the determinant can be explicitly evaluated:

$$Z_G = \int \mathcal{D}[\bar{b}, b] \exp\left[-\sum_n \bar{b}(i\nu_n)(-i\nu_n + \underline{h})b(i\nu_n)\right]$$

$$= \frac{1}{\prod_n \det[-i\nu_n + \underline{h}]} = \frac{1}{\prod_{n,\lambda}(-i\nu_n + \epsilon_\lambda)}, \qquad (12.70)$$

where the ϵ_λ are the energy eigenvalues of $\underline{h}$. This expression is most usefully rewritten as an expression for the free energy,

$$F_G = -T \ln Z_G = T \sum_n \mathrm{Tr} \ln(\underline{h} - i\nu_n)e^{i\nu_n 0^+} = T \sum_{n,\lambda} \ln(\epsilon_\lambda - i\nu_n)e^{i\nu_n 0^+},$$

where we have used the identity $\ln \det A = \mathrm{Tr} \ln A$ and have introduced the convergence term $e^{i\nu_n 0^+}$. This term is motivated by the observation that derivatives of the partition function represent equal time expectation values, which are the expectation values of time-ordered operators at an infinitesimally negative time.

In ending this section, we make one last identification. For a diagonalized non-interacting Hamiltonian, the bosonic Green's function is given by

$$G_{\lambda\lambda'}(i\nu_n) = \delta_{\lambda\lambda'}(i\nu_n - \epsilon_\lambda)^{-1}, \tag{12.71}$$

so we can identify $(-i\nu_n + \epsilon_\lambda) = -G^{-1}(i\nu_n) = -G^{-1}$ as the inverse Green's function. Since this identity holds in any basis, we can identify

$$(\partial_\tau + \underline{h}) \equiv \langle b(i\nu_n)\bar{b}(i\nu_n)\rangle = -G^{-1} \tag{12.72}$$

in the time domain. An alternative expression for the Gaussian integral is then

$$Z_G = \int \mathcal{D}[\bar{b}, b] \exp\left[-\int_0^\beta \bar{b}(-G^{-1})b\right] = \frac{1}{\det[-G^{-1}]}. \tag{12.73}$$

If we take logarithms of both sides, we may write down the free energy in terms of the one-particle Green's function:

$$F = T \ln \det[-G^{-1}] = T \mathrm{Tr} \ln[-G^{-1}]. \tag{12.74}$$

This expression enables us to relate the Green's function and free energy without having to first diagonalize the Hamiltonian G^{-1}.

Example 12.4   Use the equation of motion $\partial_\tau \hat{b}(1) = [\hat{H}, \hat{b}(1)]$ to confirm that, for a system of free bosons, where $\hat{H} = \hat{b}^\dagger \underline{h} \hat{b} \equiv \hat{b}^\dagger_\alpha h_{\alpha\beta} \hat{b}_\beta$, the Green's function is given by $G = -(\partial_\tau + \underline{h})^{-1}$.

Solution

The boson Green's function is given by

$$G(1-2) = -\langle T\hat{b}(1)\hat{b}^\dagger(2)\rangle. \tag{12.75}$$

The time dependence of the Green's function has two components: a smoothly varying term derived from the time evolution of the Bose field and a discontinuous term derived from the derivatives of the time-ordering operator. To see this, let us first expand the time-ordering operator in terms of θ functions:

$$G(1-2) = -\langle \hat{b}(1)\hat{b}^\dagger(2)\rangle\theta(\tau_1 - \tau_2) - \langle \hat{b}^\dagger(2)\hat{b}(1)\rangle\theta(\tau_2 - \tau_1). \tag{12.76}$$

If we now take the derivative with respect to time, we must take account of the discontinuity in the theta functions. Using $\partial_{\tau_1}\theta(\tau_1 - \tau_2) = \delta(\tau_1 - \tau_2)$ and $\partial_{\tau_1}\theta(\tau_2 - \tau_1) = -\delta(\tau_1 - \tau_2)$, we obtain

$$\partial_{\tau_1}G(1-2) = \left(-\langle\hat{b}(1)\hat{b}^\dagger(2)\rangle\delta(\tau_1 - \tau_2) + \langle\hat{b}^\dagger(2)\hat{b}(1)\rangle\delta(\tau_1 - \tau_2)\right) - \langle T\partial_\tau\hat{b}(1)\hat{b}^\dagger(2)\rangle$$

$$= -\underbrace{\langle[\hat{b}(1),\hat{b}^\dagger(2)]\rangle\delta(\tau_1 - \tau_2)}_{\delta(1-2)} - \langle T\underbrace{\partial_\tau\hat{b}(1)}_{[H,\hat{b}(1)]}\hat{b}^\dagger(2)\rangle$$

$$= -\delta(1-2) - \langle T[H,\hat{b}(1)]\hat{b}^\dagger(2)\rangle, \tag{12.77}$$

where we have simplified the first term using the canonical commutation relations

$$\left([\hat{b}(1),\hat{b}^\dagger(2)]\delta(\tau_1 - \tau_2)\right)_{\alpha\beta} \equiv [\hat{b}_\alpha,\hat{b}^\dagger_\beta]\delta(\tau_1 - \tau_2) = \delta_{\alpha\beta}\delta(\tau_1 - \tau_2) \equiv \delta(1-2)_{\alpha\beta}$$

and used the equation of motion, $\partial_\tau b(1) = [H, b(1)]$. The commutator between the Hamiltonian and the boson field is

$$[H,\hat{b}(1)]_\alpha \equiv [H,\hat{b}_\alpha] = -[\hat{b}_\alpha,\hat{b}^\dagger_\lambda h_{\lambda\beta}\hat{b}_\beta] = -\overbrace{[\hat{b}_\alpha,\hat{b}^\dagger_\lambda]}^{\delta_{\alpha\lambda}} h_{\lambda\beta}\hat{b}_\beta = -h_{\alpha\beta}\hat{b}_\beta \equiv -[\underline{h}\cdot\hat{b}(1)]_\alpha.$$

Putting this all together, we have

$$\partial_{\tau_1}G(1-2) = -\delta(1-2) - \underline{h}\cdot G(1-2) \tag{12.78}$$

or

$$(\partial_{\tau_1} + \underline{h})G(1-2) = -\delta(1-2). \tag{12.79}$$

If we write this expression succinctly as

$$(\partial_\tau + \underline{h})G = -1, \tag{12.80}$$

we see that

$$G = -(\partial_\tau + \underline{h})^{-1}. \tag{12.81}$$

If you are uncomfortable with treating integrals over the time domain as a matrix multiplication, you can Fourier transform (12.78), writing

$$G(\tau - \tau') = T\sum_n G(i\nu_n)e^{-i\nu_n(\tau-\tau')}, \tag{12.82}$$

so that $\partial_\tau \to -i\nu_n$. Then (12.80) becomes

$$(i\nu_n - \underline{h})\cdot G(i\nu_n) = \underline{1} \tag{12.83}$$

and hence

$$G(i\nu_n) = (i\nu_n - \underline{h})^{-1}, \tag{12.84}$$

which is the Fourier transform of (12.81).

Example 12.5   Calculate the free energy of a free bosonic gas, where $\hat{H} = \sum_{\mathbf{k}} \epsilon_{\mathbf{k}} \hat{b}_{\mathbf{k}}^{\dagger} \hat{b}_{\mathbf{k}}$ using the path integral method.

Solution

We begin by writing the action in the frequency domain as

$$S_E = - \sum_{\mathbf{k}, i\nu_n} b(\mathbf{k}, i\nu_n) G(\mathbf{k}, i\nu_n)^{-1} b(\mathbf{k}, i\nu_n)$$

$$G(\mathbf{k}, i\nu_n)^{-1} = (i\nu_n - \epsilon_{\mathbf{k}}). \tag{12.85}$$

The partition function is given by

$$e^{-\beta F} = \frac{1}{\det[-G^{-1}]}, \tag{12.86}$$

so that

$$F = T \ln \det[-G^{-1}] = T \operatorname{Tr} \ln[-G^{-1}] = T \sum_{\mathbf{k}, i\nu_n} \ln(\epsilon_{\mathbf{k}} - i\nu_n) e^{i\nu_n 0^+}, \tag{12.87}$$

where we have introduced the convergence factor $e^{i\nu_n 0^+}$ and used the identity $\ln \det A = \operatorname{Tr} \ln A$.

Carrying out the frequency summation using complex contour methods, we have

$$F = - \sum_{\mathbf{k}} \oint \frac{dz}{2\pi i} n(z) \ln(\epsilon_{\mathbf{k}} - z), \tag{12.88}$$

where the integral is counterclockwise around the branch cut on the real axis. This branch cut runs out from $\omega = \epsilon_{\mathbf{k}}$ to positive infinity, with a discontinuity of 2π across the branch cut. Rewriting the integral along this discontinuity, we have

$$F = - \sum_{\mathbf{k}} \int_{-\infty}^{\infty} \frac{d\omega}{2\pi i} n(\omega) \overbrace{\left[\ln(\epsilon_{\mathbf{k}} - \omega + i\delta) - \ln(\epsilon_{\mathbf{k}} - \omega - i\delta) \right]}^{2\pi i\, \theta(\omega - \epsilon_{\mathbf{k}})} = - \sum_{\mathbf{k}} \int_{\epsilon_{\mathbf{k}}}^{\infty} d\omega\, n(\omega)$$

$$= -T \sum_{\mathbf{k}} \left[\ln(1 - e^{-\beta\omega}) \right]_{\epsilon_{\mathbf{k}}}^{\infty} = +T \sum_{\mathbf{k}} [\ln(1 - e^{-\beta\epsilon_{\mathbf{k}}})]. \tag{12.89}$$

12.3.4   Source terms in Gaussian integrals

Source terms provide a means of probing the correlations and fluctuations described by a path integral. For Gaussian path integrals, the result of introducing source terms can be evaluated to obtain

$$Z_G[\bar{j}, j] = \int \mathcal{D}[\bar{b}, b] \exp\left\{-\int_0^\beta d1\left[\bar{b}\left(\partial_\tau + \underline{h}\right)b - \bar{j}(1)\right)\cdot b(1) - \bar{b}(1)\cdot j(1)\right]\right\}$$

$$= \frac{\exp\left[-\int_0^\beta d1d2\bar{j}(1)G(1-2)j(2)\right]}{\det[\partial_\tau + \underline{h}]}, \qquad (12.90)$$

bosonic Gaussian path integral with source terms

where we have used the schematic notation $1 \equiv (\tau_1, X_1, \{\lambda_1\})$, $2 \equiv (\tau_2, X_2, \{\lambda_1\})$ to denote the time, position, and all other relevant indices of the boson field and $\int_0^\beta d1 = \sum_{\lambda_1}\int_0^\beta d\tau_1 \int d^d X_1$ to denote the corresponding integration over continuous variables and summation over discrete quantum numbers. The expansion of the left- and right-hand sides of this expression as a power series provides the Wick expansion of multi-particle Green's functions of the boson field. Differentiating first the left- and then the right-hand side with respect to $\bar{j}(1)$, we obtain

$$\langle \hat{b}(1) \rangle \equiv \frac{\int \mathcal{D}[\bar{b}, b]e^{-S}b(1)}{\int \mathcal{D}[\bar{b}, b]e^{-S}} = \frac{1}{Z_G[\bar{j}, j]}\frac{\delta Z_G[\bar{j}, j]}{\delta\bar{j}(1)} = -\int_0^\beta d2G(1-2)j(2). \qquad (12.91)$$

Taking second derivatives and setting the source terms to zero, we obtain

$$\langle T\hat{b}(1)\hat{b}^\dagger(2)\rangle_{\bar{j}, j=0} \equiv \frac{\int \mathcal{D}[\bar{b}, b]e^{-S}b(1)\bar{b}(2)}{\int \mathcal{D}[\bar{b}, b]e^{-S}} = \frac{1}{Z_G[\bar{j}, j]}\frac{\delta^2 Z_G[\bar{j}, j]}{\delta\bar{j}(2)\delta\bar{j}(1)}\Big|_{\bar{j}, j=0}$$

$$= -G(1-2), \qquad (12.92)$$

while higher-order differentials give us the Wick expansion:

$$\frac{1}{Z_G[\bar{j}, j]}\frac{\delta^{2n}Z_G[\bar{j}, j]}{\delta j(1')\cdots\delta\bar{j}(1)}\Big|_{\bar{j}, j=0} =$$

$$(-1)^n \sum_P G(1-P_1')G(2-P_2')\cdots G(n-P_n') = \frac{\int \mathcal{D}[\bar{b}, b]e^{-S}b(1)b(2)\cdots\bar{b}(2')\bar{b}(1')}{\int \mathcal{D}[\bar{b}, b]e^{-S}}$$

$$\equiv \langle T\hat{b}(1)\hat{b}(2)\cdots\hat{b}^\dagger(2')\hat{b}^\dagger(1')\rangle. \qquad (12.93)$$

In this remarkable fashion, the correlation functions of non-interacting bosons in imaginary time are identified with the classic properties of Gaussian-distributed random variables.

To prove (12.91), we take (12.67) and shift the integration variables inside the integral:

$$b \to b - M^{-1}j, \qquad \bar{b} \to \bar{b} - \bar{j}M^{-1}. \qquad (12.94)$$

Under this simple shift, the measure remains unchanged while the action term $\bar{b} \cdot M \cdot b$ becomes

$$\bar{b} \cdot M \cdot b \to (\bar{b} - \bar{j}M^{-1}) \cdot M \cdot (b - M^{-1}j) = \bar{b} \cdot M \cdot b - (\bar{j} \cdot b + \bar{b} \cdot j) + \bar{j} \cdot M^{-1} \cdot j. \qquad (12.95)$$

Since the integral is unchanged under this change of variables, it follows that

$$e^{-\bar{j}\cdot Mj}\int \prod_\alpha \frac{d\bar{b}_\alpha db_\alpha}{2\pi i}e^{-(\bar{b}_\alpha M_{\alpha\beta}b_\beta - \bar{j}_\alpha b_\alpha - \bar{b}_\alpha j_\alpha)} = \frac{1}{\det[M]}. \qquad (12.96)$$

In other words,

$$\int \prod_\alpha \frac{d\bar{b}_\alpha db_\alpha}{2\pi i} e^{-(\bar{b}_\alpha M_{\alpha\beta} b_\beta - \bar{j}_\alpha b_\alpha - \bar{b}_\alpha j_\alpha)} = \frac{e^{\bar{j}\cdot M^{-1}j}}{\det[M]} \qquad (12.97)$$

(see Table 12.1). If we rewrite this expression by replacing $M \rightarrow -G^{-1} = (\partial_\tau + h)$, we obtain the key result (12.91). As usual, if you are uncomfortable with the change from discrete to continuous variables, this procedure can first be carried out using the discrete variables in Fourier space, followed by an inverse Fourier transformation back into real space.

12.4 Fermions: coherent states and Grassman mathematics

We now generalize the results of the previous section to fermions, using Grassman numbers to set up a completely parallel derivation of the fermionic path integral in terms of coherent states.

Feynman's original derivation of path integrals applied purely to bosonic fields, and its extension to fermions was begun in the 1950s. The idea of using anticommuting numbers, both as eigenvalues of fermion fields and as fermionic source terms, was proposed in a seminal paper by Julian Schwinger in 1953 [8]. Early proposals for path integrals for fermions were made by Paul Matthews and Abdus Salam in 1955 [11] and by David Candlin in 1956 [5]. The first explicit formulation of the fermionic action in terms of Grassman numbers, with a derivation using fermion coherent states, was made by by J. L. Martin in 1959 [6]. The mathematical foundations of fermionic path integrals were extensively developed in the 1960s by Felix Berezin [12], and the extension of the fermionic path integral to imaginary time and finite temperature was later provided by David Sherrington and Sam Edwards [7, 13]. However it is only in the last few decades that the method has become a commonly used tool in quantum many-body physics.

To illustrate the basic approach, we shall consider a single fermionic field $\hat{c}^\dagger$. The coherent state for this field is

$$|c\rangle = e^{\hat{c}^\dagger c}|0\rangle$$

and its conjugate is

$$\langle \bar{c}| = \langle 0| e^{\bar{c}\hat{c}}.$$

In this text we've reserved roman symbols $\hat{c}^\dagger$ and $\hat{c}$ for the creation and annihilation operators, to distinguish them from their expectation values $\bar{c}$ and c. Here c and $\bar{c}$ are anticommuting *Grassman numbers*. Note that in common usage the notation $c^\dagger$ is often used interchangeably to describe both the operator and its Grassman counterpart $\bar{c}$.

There are a number of caveats you need to remember about Grassman numbers. On the one hand, the quantities c and $\bar{c}$ are numbers which *commute* with all observables $\hat{O}$: $c\hat{O} = \hat{O}c$. On the other hand, to correctly represent the anticommuting algebra of the original Fermi fields, Grassman numbers *anticommute* among themselves *and* with other Fermi operators, so that

$$c\bar{c} + \bar{c}c = 0, \qquad c\hat{\psi} + \hat{\psi}c = 0. \tag{12.98}$$

But c must also anticommute with itself, which means that

$$c^2 = \bar{c}^2 = 0. \tag{12.99}$$

But how can we possibly deal with numbers which when squared give zero? Though this seems absurd, we'll see that anticommuting or Grassman numbers do form a non-trivial calculus and that, ultimately, the leap to this new type of number is no worse and no more remarkable than the jump from real to complex numbers.

The main effect of the anticommuting properties of Grassman numbers is to drastically reduce the set of possible functions and the set of possible linear operations one can carry out on such functions. For example, the Taylor series expansion of Grassman functions has to truncate at first order in any particular variable. Thus a function of two variables

$$f(\bar{c}, c) = f_o + \bar{c}f_1 + \tilde{f}_1 c + f_{12}\bar{c}c$$

only has four terms! The coherent state also truncates, so that

$$|c\rangle = |0\rangle + \hat{c}^\dagger c|0\rangle$$
$$= |0\rangle + |1\rangle c, \tag{12.100}$$

so that the overlap between the "n" fermion state ($n = 0, 1$) and the coherent state is given by

$$\langle n|c\rangle = c^n \qquad (n = 0, 1).$$

To develop a path integral representation for fermions, one needs to know how to carry out Grassman calculus. The key properties of Grassman calculus are summarized in Table 12.2. In particular, you will notice that the *only* formal differences from bosons are that the measure contains a different normalization:

$$\sum_{\bar{b},b} = \int \frac{d\bar{b}db}{2\pi i} e^{-\bar{b}b} \rightarrow \sum_{\bar{c},\,c} = \int d\bar{c}dc e^{-\bar{c}c}, \tag{12.101}$$

the trace formula contains an additional minus sign:

$$\mathrm{Tr}[A]_B = \sum_{\bar{b},b} \langle \bar{b}|A|b\rangle \rightarrow \mathrm{Tr}[A]_F = \sum_{\bar{c},c} \langle -\bar{c}|A|c\rangle, \tag{12.102}$$

and both the Jacobian and the Gaussian integrals are the *inverses* of their bosonic counterparts.

12.4.1　Completeness and matrix elements

Coherent states are overcomplete, for

$$\langle \bar{c}|c\rangle = \langle 0|(1 + \bar{c}\hat{c})(1 + \hat{c}^\dagger c)|0\rangle = 1 + \bar{c}c = e^{\bar{c}c}. \tag{12.103}$$

	Table 12.2 Grassman calculus.	
Algebra	$c_1 c_2 = -c_2 c_1$	Anticommute with fermions and other Grassman numbers
	$c\hat{b} = \hat{b}c, \quad c\hat{\psi} = -\hat{\psi}c$	Commute with bosons, anticommute with fermi operators
Functions	$f[\bar{c}, c] = f_o + \bar{c}f_1 + \tilde{f}_1 c + f_{12}\bar{c}c$	Since $c^2 = 0$, truncate at linear order in each variable
Calculus	$\partial f = -\tilde{f}_1 - f_{12}\bar{c}$	Differentiation
	$\bar{\partial} f = f_1 + f_{12}c$	
	$\int dc \equiv \partial_c$	$\int dc 1 = \partial_c 1 = 0$
		$\int dc\, c = \partial_c c = 1$
Completeness	$\langle c|c \rangle = e^{\bar{c}c}$	Overcomplete basis
	$\int d\bar{c}dc\, e^{-\bar{c}c} |c\rangle\langle\bar{c}| = \underline{1}$	Completeness relation
	$\text{Tr}[\hat{A}] = \int d\bar{c}dc\, e^{-\bar{c}c} \langle -\bar{c}|\hat{A}|c \rangle$	Trace formula
Change of variable	$J\left(\dfrac{c_1 \dots c_r}{\xi_1 \dots \xi_r}\right) = \left|\dfrac{\partial(c_1, \dots c_r)}{\partial(\xi_1, \dots \xi_r)}\right|^{-1}$	Jacobian (inverse of bosonic Jacobian)
Gaussian integrals	$\int \prod_j d\bar{c}_j dc_j e^{-\left[\bar{c}\cdot A\cdot\bar{c} - \bar{j}\cdot c - \bar{c}\cdot j\right]} = \det A \times e^{\left[\bar{j}\cdot A^{-1}\cdot j\right]}$	

Notice the formal parallel with the overlap of bosonic coherent states. To derive the completeness relation, we start with the identity

$$\int d\bar{c}dc\, e^{-\bar{c}c} c^n \bar{c}^m = \delta_{nm} \qquad (n, m = 0, 1). \qquad (12.104)$$

Then, by writing $c^n - \langle n|c \rangle$, $\bar{c}^m = \langle \bar{c}|m \rangle$, we see that the overlap between the eigenstates $|n\rangle$ of definite particle number is given by

$$\delta_{nm} = \langle n|m \rangle = \int d\bar{c}dc\, e^{-\bar{c}c} \langle n|c \rangle\langle\bar{c}|m \rangle = \langle n| \int d\bar{c}dc\, e^{-\bar{c}c} |c\rangle\langle\bar{c}| |m \rangle, \qquad (12.105)$$

from which it follows that

$$\int d\bar{c}dc |c\rangle\langle\bar{c}| e^{-\bar{c}c} = |0\rangle\langle 0| + |1\rangle\langle 1| \equiv \underline{1}. \qquad (12.106)$$

completeness relation

Alternatively, we may write

$$\sum_{\bar{c},c} |c\rangle\langle\bar{c}| = \underline{1},$$

where

$$\sum_{\bar{c},\,c} \equiv \int d\bar{c}dce^{-\bar{c}c} \qquad (12.107)$$

is the measure for fermionic coherent states. The exponential factor $e^{-\bar{c}c} = 1/\langle\bar{c}|c\rangle$ provides the normalizing factor to take account of the overcompleteness.

Matrix elements between coherent states are easy to evaluate. If an operator $A[\hat{c}^{\dagger}, \hat{c}]$ is normal ordered, then, since the coherent states are eigenvectors of the quantum fields, it follows that

$$\langle\bar{c}|\hat{A}|c\rangle = \langle\bar{c}|c\rangle A[\bar{c}, c] = e^{\bar{c}c}A[\bar{c}, c]. \qquad (12.108)$$

That is,

$$\langle\bar{c}|\hat{A}|c\rangle = e^{\bar{c}c} \times \text{ c-number formed by replacing } A[\hat{c}^{\dagger}, \hat{c}] \to A[\bar{c}, c]. \qquad (12.109)$$

This wonderful feature of coherent states enables us, at a swoop, to convert normal-ordered operators into c-numbers.

The last result we need is the trace of A. We might guess that the appropriate expression is

$$\text{Tr}\,\hat{A} = \sum_{\bar{c},c}\langle\bar{c}|\hat{A}|c\rangle.$$

This is almost right, but in fact it turns out that the anticommuting properties of the Grassmans force us to introduce a minus sign into this expression:

$$\text{Tr}\,\hat{A} = \sum_{\bar{c},c}\langle-\bar{c}|\hat{A}|c\rangle = \int d\bar{c}dce^{-\bar{c}c}\langle-\bar{c}|\hat{A}|c\rangle, \qquad (12.110)$$

Grassman trace formula

which, we shall shortly see, gives rise to the antisymmetric boundary conditions of fermionic fields. To prove the above result, we rewrite (12.105) as

$$\delta_{nm} = \langle n|m\rangle = \int d\bar{c}dce^{-\bar{c}c}\langle-\bar{c}|m\rangle\langle n|c\rangle, \qquad (12.111)$$

where the minus sign arises from anticommuting c and $\bar{c}$. We can now rewrite the trace as

$$\text{Tr}\,A = \sum_{n,m}\langle m|A|n\rangle\delta_{nm}$$

$$= \sum_{n,m}\int d\bar{c}dce^{-\bar{c}c}\langle-\bar{c}|m\rangle\langle m|A|n\rangle\langle n|c\rangle$$

$$= \int d\bar{c}dce^{-\bar{c}c}\langle-\bar{c}|\hat{A}|c\rangle. \qquad (12.112)$$

We shall make extensive use of the completeness and trace formulae (12.106) and (12.110) in developing the path integral. Both expressions are simply generalized to many fields c_j

by making the appropriate change in the measure and replacing $\bar{c}c$ in the exponent by the dot product:

$$d\bar{c}dc \rightarrow \prod_j d\bar{c}_j dc_j, \tag{12.113}$$

$$\bar{c}c \rightarrow \sum_j \bar{c}_j c_j.$$

12.4.2 Path integral for the partition function: fermions

This section very closely parallels the derivation of the bosonic path integral in Section 12.3, but for completeness we include all relevant steps. To begin with, we consider a single fermion, with Hamiltonian

$$H = \epsilon \hat{c}^\dagger \hat{c}. \tag{12.114}$$

Using the trace formula (12.110), the partition function

$$Z = \text{Tr}\, e^{-\beta H} \tag{12.115}$$

can be rewritten in terms of coherent states as

$$Z = -\int d\bar{c}_N dc_0 e^{\bar{c}_N c_0} \langle \bar{c}_N | e^{-\beta H} | c_0 \rangle, \tag{12.116}$$

where the labeling anticipates the next step. Now we expand the exponential into a sequence of time-slices (see Figure 12.5):

$$e^{-\beta H} = \left(e^{-\Delta\tau H}\right)^N, \qquad \Delta\tau = \beta/N. \tag{12.117}$$

Between each time-slice we introduce the completeness relation

$$\int d\bar{c}_j dc_j |c_j\rangle \langle \bar{c}_j | e^{-\bar{c}_j c_j} = 1, \tag{12.118}$$

so that

$$Z = -\int d\bar{c}_N dc_0 e^{\bar{c}_N c_0} \prod_{j=1}^{N-1} d\bar{c}_j dc_j e^{-\bar{c}_j c_j} \prod_{j=1}^{N} \langle \bar{c}_j | e^{-H\Delta\tau} | c_{j-1} \rangle, \tag{12.119}$$

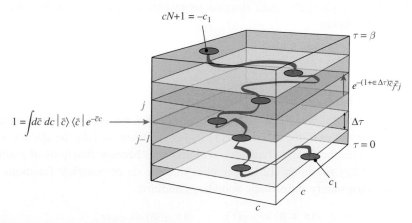

$cN+1 = -c_1$

$\tau = \beta$

$e^{-(1+\epsilon\Delta\tau)\bar{c}_j \bar{c}_j}$

$1 = \int d\bar{c}\, dc\, |\bar{c}\rangle \langle \bar{c} | e^{-\bar{c}c}$

$\Delta\tau$

$\tau = 0$

c_1

c

c

Division of Grassmanian time evolution into time-slices.

Fig. 12.5

where the first integral is associated with the trace and the subsequent integrals with the $N - 1$ completeness relations. Now if we define

$$c_N = -c_0, \tag{12.120}$$

we are able to identify the Nth time-slice with the zero-th time-slice. In this way, the integral associated with the trace

$$- \int d\bar{c}_N dc_0 e^{\bar{c}_N c_0} \langle \bar{c}_N | \cdots | c_0 \rangle = \int d\bar{c}_N dc_N e^{-\bar{c}_N c_N} \langle \bar{c}_N | \cdots | c_0 \rangle \tag{12.121}$$

can be absorbed into the other $N - 1$ integrals. Furthermore, we notice that the fields entering into the discrete path integral are *antiperiodic*.

With this observation,

$$Z = \int \prod_{j=1}^{N} d\bar{c}_j dc_j e^{-\bar{c}_j c_j} \langle \bar{c}_j | e^{-H\Delta\tau} | c_{j-1} \rangle. \tag{12.122}$$

Provided each time-slice is of sufficiently brief duration, we can replace $e^{-\Delta\tau H}$ by its normal-ordered form, so that

$$\langle \bar{c}_j | e^{-H\Delta\tau} | c_{j-1} \rangle = e^{\bar{c}_j c_{j-1}} e^{-H[\bar{c}_j c_{j-1}]\Delta\tau} + O(\Delta\tau^2), \tag{12.123}$$

where $H[\bar{c}, c] = \epsilon \bar{c} c$ is the normal-ordered Hamiltonian, with Grassman numbers replacing operators.

Combining (12.116) and (12.119) we can write

$$Z = \mathrm{Lt}_{N \to \infty} Z_N$$

$$Z_N = \int \prod_{j=1}^{N} d\bar{c}_j dc_j \exp\left[-S\right]$$

$$S = \sum_{j=1}^{N} \left[\bar{c}_j (c_j - c_{j-1}) / \Delta\tau + \epsilon \bar{c}_j c_{j-1} \right] \Delta\tau, \tag{12.124}$$

As in the bosonic case, this path integral represents a sum over all possible values "histories" of the fields:

$$c(\tau_j) \equiv \{c_1, c_2, \ldots c_N\}, \tag{12.125}$$

$$\bar{c}(\tau_j) \equiv \{\bar{c}_1, \bar{c}_2, \ldots \bar{c}_N\} \tag{12.126}$$

as illustrated in Figure 12.5.

This kind of integral is also called a *functional integral*, because it involves integrating over all possible values of the functions $c(\tau)$ and $\bar{c}(\tau)$. When we take the thickness of the time-slices to zero, the discrete functions $c(\tau)$ and $\bar{c}(\tau)$ become functions of continuous time. The boundary condition (12.120) implies that the set of complete functions which we sum over must satisfy antiperiodic boundary conditions

$$c(\tau + \beta) = -c(\tau), \qquad \bar{c}(\tau + \beta) = -\bar{c}(\tau).$$

In the continuum limit $N \to \infty$, we now replace

$$\bar{c}_j(c_j - c_{j-1})/\Delta\tau \to \bar{c}\partial_\tau c,$$

$$\sum_j \Delta\tau \to \int_0^\beta d\tau. \tag{12.127}$$

The sense in which c_j becomes "close" to c_{j+1} needs to be carefully understood. Suppose we rewrite the antiperiodic c_j in terms of their frequency components as

$$c_j = \frac{1}{\sqrt{\beta}} \sum_{|n| \leq N/2} c(i\omega_n) e^{-i\omega_n \tau_j}.$$

Then, in this new basis,

$$\sum_j \bar{c}_j(c_j - c_{j-1}) = \sum_{|n| \leq N/2} \bar{c}(i\omega_n) \left[\frac{1 - e^{-i\omega_n \Delta\tau}}{\Delta\tau} \right] c(i\omega_n).$$

In practice, the path integral is dominated by functions c_j with a maximum characteristic temporal frequency $\max(|\omega_n|) \sim \epsilon$, so that as $\Delta\tau \to 0$ we can replace

$$\left[\frac{1 - e^{-i\omega_n \Delta\tau}}{\Delta\tau} \right] \to -i\omega_n,$$

which is the Fourier transform of ∂_τ.

With these provisos, the continuum limit of the action and path integral are then

$$S = \int_0^\infty d\tau \left[\bar{c}(\partial_\tau + \epsilon)c \right]$$

$$Z = \int \mathcal{D}[\bar{c}, c] \exp\left[-S \right], \tag{12.128}$$

where we use the notation

$$\mathcal{D}[\bar{c}, c] = \prod_{\tau_l} d\bar{c}(\tau_l) dc(\tau_l).$$

At first sight, it might seem a horrendous task to carry out the integral over all possible functions $c(\tau)$. How can we possibly do this in a controlled fashion? The clue to this problem lies in the observation that the set of functions $c(\tau)$ and its conjugate $\bar{c}(\tau)$ are spanned by a *discrete* but complete set of antiperiodic functions, as follows:

$$c(\tau) = \frac{1}{\sqrt{\beta}} \sum_n c_n e^{-i\omega_n \tau}.$$

We can integrate over all possible functions $c(\tau)$ by integrating over all possible values of the coefficients c_n, and since the transformation which links these two bases is unitary, the Jacobian which links the two bases is unity, i.e.

$$\mathcal{D}[\bar{c}, c] \equiv \prod_n d\bar{c}_n dc_n.$$

It is much easier to visualize and work with a discrete basis. We can transform to this basis by replacing $\partial_\tau \rightarrow -i\omega_n$ in the action, rewriting it as

$$S = \sum_n \bar{c}_n(-i\omega_n + \epsilon)c_n.$$

Now the path integral is just a discrete Gaussian integral,

$$Z = \int \prod_n d\bar{c}_n dc_n \exp\left[-\sum_n \bar{c}_n(-i\omega_n + \epsilon)c_n\right] = \prod_n(-i\omega_n + \epsilon),$$

so that the free energy is given by

$$F = -T\ln Z = -T\sum_n \ln(\epsilon - i\omega_n)e^{i\omega_n 0^+}.$$

Here we have added a small convergence factor $e^{i\omega_n 0^+}$ because the time evolution from $\tau = 0$ to $\tau = \beta$ is equivalent to time evolution from $\tau = 0$ to $\tau = 0^-$.

We can show that this reverts to the standard expression for one-particle free energy by replacing the Matsubara sum with a contour integral:

$$F = T - \oint \frac{dz}{2\pi i} f(z)\ln[\epsilon_\lambda - z]e^{z0^+}, \qquad (12.129)$$

where the contour integral passes counterclockwise around the poles of the Fermi function at $z = i\omega_n$, and the choice of $f(z)$ is dictated by the convergence factor. We take the logarithm to have a branch cut which extends from $z = \epsilon_\lambda$ to infinity. By deforming the integral around this branch cut, we obtain

$$F = -\int_\epsilon^\infty \frac{d\omega}{2\pi i} f(\omega)\left[\ln(\epsilon - \omega - i\delta) - (\text{c.c.})\right]$$

$$= -\int_\epsilon^\infty d\omega f(\omega)$$

$$= -T\ln[1 + e^{-\beta\epsilon}], \qquad (12.130)$$

which is the well-known free energy of a single fermion.

Of course, here we have used a sledgehammer to crack a walnut, but the virtue of the method is the ease with which it can be generalized to more complex problems. Three important points need to be made about this result:

• This result can easily be generalized to an arbitrary number of Fermi fields. In this case,

$$S = \int_0^\infty d\tau\left[\sum_\lambda \bar{c}_\lambda \partial_\tau c_\lambda + H[\bar{c}, c]\right],$$

and the measure for the path integral becomes

$$\mathcal{D}[\bar{c}, c] = \prod_{\tau_l, r} d\bar{c}_\lambda(\tau_l)dc_\lambda(\tau_l).$$

- The derivation did not depend on any details of H, and can thus be simply generalized to interacting Hamiltonians. In both cases, the conversion of the normal-order Hamiltonian occurs by simply replacing operators with the appropriate Grassman variables:

$$: H[\hat{c}^\dagger, \hat{c}] :\longrightarrow H[\bar{c}, c].$$

- The amplitude associated with a particular path can be split into the product of two terms, as follows:

$$e^{-S_{PATH}} = \exp\left[i\gamma - \int_0^\beta H[\bar{c}, c]d\tau\right], \qquad (12.131)$$

where the quantity

$$\gamma = i\sum_\lambda \int_0^\beta d\tau\, \bar{c}_\lambda \partial_\tau c_\lambda \qquad (12.132)$$

is known as a *Berry phase*, named after the British physicist Michael Berry. The Berry phase can be identified as the phase picked up by the coherent state $|\psi\rangle = \exp\left[\sum_\lambda \hat{c}_\lambda^\dagger c_\lambda\right]|0\rangle$ as it is adiabatically evolved along the path defined by the functions $c_\lambda(\tau)$. For this reason, the time-derivative term in a path integral is often refered to as the *Berry phase term*.

- Because the Jacobian for a unitary transformation is unity, we can change basis inside the path integral. For example, if we start with the action for a gas of fermions,

$$S = \int_0^\beta d\tau \sum_{\mathbf{k}} \bar{c}_{\mathbf{k}}(\partial_\tau + \epsilon_{\mathbf{k}})c_{\mathbf{k}},$$

where $\epsilon_{\mathbf{k}} = (k^2/2m) - \mu$, we can transform to a completely discrete basis by Fourier transforming in time:

$$c_{\mathbf{k}} = \frac{1}{\sqrt{\beta}} \sum_n c_{\mathbf{k}n} e^{i\omega_n \tau},$$

$$\partial_\tau \rightarrow -i\omega_n$$

$$\mathcal{D}[\bar{c}, c] \rightarrow \prod_{\mathbf{k},n} d\bar{c}_{\mathbf{k}n} dc_{\mathbf{k}n}. \qquad (12.133)$$

In this discrete basis, the action becomes

$$S = \sum_{\mathbf{k},n}(\epsilon_{\mathbf{k}} - i\omega_n)\bar{c}_{\mathbf{k}n} c_{\mathbf{k}n}.$$

This basis usually proves very useful for practical calculations.

- We can also transform to a continuum real-space basis, as follows:

$$c_{\mathbf{k}} = \frac{1}{\sqrt{V}} \int d^3x\, \psi(\mathbf{x}) e^{-i\mathbf{k}\cdot\mathbf{x}}$$

$$\epsilon_{\mathbf{k}} \rightarrow -\frac{\nabla^2}{2m} - \mu$$

$$\mathcal{D}[\bar{c}, c] \rightarrow \mathcal{D}[\bar{\psi}, \psi]. \qquad (12.134)$$

In the new basis, the action becomes

$$S = \int_0^\beta d\tau \int d^3x \bar{\psi}(\mathbf{x}) \left[\partial_\tau - \frac{\nabla^2}{2m} - \mu \right] \psi(\mathbf{x}).$$

The discrete and continuous measures, (12.133) and (12.134), respectively, are equivalent:

$$\prod_{\mathbf{k},n} d\bar{c}_{\mathbf{k}n} dc_{\mathbf{k}n} \equiv \mathcal{D}[\bar{\psi}, \psi]$$

because the space of continuous functions $\psi(x)$ is spanned by a complete but discrete set of basis functions:

$$\psi(\mathbf{x}, \tau) = \frac{1}{\sqrt{\beta V}} \sum_{\mathbf{k},n} c_{\mathbf{k}n} e^{i(\mathbf{k}\cdot\mathbf{x} - \omega_n \tau)}.$$

We can integrate over all possible functions $\psi(\mathbf{x}, \tau)$ by integrating over all values of the discrete vector $c_{\mathbf{k}n}$.

12.4.3 Gaussian path integral for fermions

For non-interacting fermions the action only involves bilinears of the Fermi fields, so the path integral is of Gaussian form and can always be evaluated. To discuss the most general case, we shall include source terms in the original Hamiltonian, writing

$$H(\tau) = \sum_\lambda \left[\epsilon_\lambda \hat{c}_\lambda^\dagger \hat{c}_\lambda - \bar{j}_\lambda(\tau) \hat{c}_\lambda - \hat{c}_\lambda^\dagger j_\lambda(\tau) \right],$$

where $\hat{c}_\lambda^\dagger$ is Schrödinger field that creates a fermion in the eigenstate with energy ϵ_λ. With source terms, the partition function becomes a *generating functional*,

$$Z[\bar{j}, j] = \mathrm{Tr} \left[T \exp\left\{ -\int_0^\beta d\tau H(\tau) \right\} \right].$$

Derivatives of the generating functional generate the irreducible Green's functions of the fermions; for instance,

$$\frac{\delta \ln Z[\bar{j}, j]}{\delta \bar{j}(1)} = \langle c(1) \rangle \tag{12.135}$$

$$\frac{\delta^2 \ln Z[\bar{j}, j]}{\delta j(2) \delta \bar{j}(1)} = \langle T[c(1) c^\dagger(2)] \rangle - \langle c(2) \rangle \langle c^\dagger(1) \rangle, \tag{12.136}$$

where

$$\langle \ldots \rangle = \frac{1}{Z[\bar{j}, j]} \mathrm{Tr} \left[T \exp\left\{ -\int_0^\beta d\tau H(\tau) \right\} \ldots \right].$$

Transforming to a path integral representation,

$$Z[\bar{j}, j] = \int \mathcal{D}[\bar{c}, c] e^{-S} \tag{12.137}$$

$$S = \int d\tau \left[\bar{c}(\tau)(\partial_\tau + \underline{h}) c(\tau) - \bar{j}(\tau) c(\tau) - \bar{c}(\tau) j(\tau) \right], \tag{12.138}$$

where $\underline{h}_{\alpha\beta} = \epsilon_\alpha \delta_{\alpha\beta}$ is the one-particle Hamiltonian. One can carry out functional derivatives on this integral without actually evaluating it. For example, we find that

$$\langle c(1) \rangle = \frac{1}{Z[\bar{j},j]} \int \mathcal{D}[\bar{c},c]c(1)e^{-S} \tag{12.139}$$

$$\langle T[c(1)c^\dagger(2)] \rangle = \frac{1}{Z[\bar{j},j]} \int \mathcal{D}[\bar{c},c]c(1)\bar{c}(2)e^{-S}. \tag{12.140}$$

Notice how the path integral automatically furnishes us with time-ordered expectation values.

Fortunately, the path integral is Gaussian, allowing us to use the general result obtained in Appendix 12D,

$$\int \prod_j d\bar{\xi}_j d\xi_j \exp[-\bar{\xi} \cdot A \cdot \xi + \bar{j} \cdot \xi + \bar{\xi} \cdot j] = \det A \exp[\bar{j} \cdot A^{-1} \cdot j]. \tag{12.141}$$

In the case considered here, $A = \partial_\tau + \underline{h}$, so we can do the integral to obtain

$$Z[\bar{j},j] = \int \mathcal{D}[\bar{c},c] \exp\left[-\int d\tau \left[\bar{c}(\tau)(\partial_\tau + \underline{h})c(\tau) - \bar{j}(\tau)c(\tau) - \bar{c}(\tau)j(\tau) \right] \right]$$

$$= \det[\partial_\tau + \underline{h}] \exp\left[-\int d\tau d\tau' \bar{j}(\tau)\underline{G}[\tau - \tau']j(\tau') \right], \tag{12.142}$$

where

$$\underline{G}[\tau - \tau'] = -(\partial_\tau + \underline{h})^{-1}. \tag{12.143}$$

By differentiating (12.142) with respect to j and $\bar{j}$, we are able to identify

$$\frac{\delta^2 \ln Z}{\delta j(\tau')\delta \bar{j}(\tau)}\bigg|_{\bar{j},j=0} = (\partial_\tau + \underline{h})^{-1} = \langle c(\tau)c^\dagger(\tau') \rangle = -\underline{G}[\tau - \tau'], \tag{12.144}$$

so the inverse of the Gaussian coefficient in the action $-[\partial_\tau + \underline{h}]^{-1}$ directly determines the imaginary-time Green's function of these non-interacting fermions. Higher-order moments of the generating functional provide a derivation of Wick's theorem.

From the partition function in (12.142), the free energy is then given by

$$F = -T\ln Z = -T\ln \det[\partial_\tau + \underline{h}] = -T\mathrm{Tr}\ln[\partial_\tau + \underline{h}] = T\mathrm{Tr}\ln[-G^{-1}],$$

where we have used the result $\ln \det[A] = \mathrm{Tr}\ln[\partial_\tau + \underline{h}]$.

To explicitly compute the free energy, it is useful to transform to Fourier components:

$$c_\lambda(\tau) = \frac{1}{\sqrt{\beta}} \sum_n c_{\lambda n} e^{-i\omega_n \tau}$$

$$j_\lambda(\tau) = \frac{1}{\sqrt{\beta}} \sum_n j_{\lambda n} e^{-i\omega_n \tau}. \tag{12.145}$$

In this basis,

$$(\partial_\tau + \epsilon_\lambda) \longrightarrow (-i\omega_n + \epsilon_\lambda)$$

$$\underline{G} = -(\partial_\tau + \epsilon_\lambda)^{-1} \longrightarrow (i\omega_n - \epsilon_\lambda)^{-1}, \tag{12.146}$$

so that

$$S = \sum_{\lambda,n}\left[[-i\omega_n + \epsilon_\lambda]\bar{c}_{\lambda n}c_{\lambda n} - \bar{j}_{\lambda n}c_{\lambda n} - \bar{c}_{\lambda n}j_{\lambda n}\right], \tag{12.147}$$

whereupon

$$\det[\partial_\tau + \underline{h}] = = \prod_{\lambda,n}(-i\omega_n + \epsilon_\lambda)$$

$$Z[\bar{j},j] = \prod_{\lambda,n}(-i\omega_n + \epsilon_\lambda)\exp\left[\sum_{\lambda,n}(-i\omega_n + \epsilon_\lambda)^{-1}\bar{j}_{\lambda n}j_{\lambda n}\right]. \tag{12.148}$$

If we set $j = 0$ in Z, we obtain the free energy in terms of the fermionic Green's function:

$$F = -T\sum_{\lambda,n}\ln[-i\omega_n + \epsilon_\lambda].$$

As in the case of a single field, by replacing the Matsubara sum with a contour integral we obtain

$$F = \sum_\lambda \oint \frac{dz}{2\pi i}f(z)\ln[\epsilon_\lambda - z] \tag{12.149}$$

$$= -T\sum_\lambda \ln[1 + e^{-\beta\epsilon_\lambda}]. \tag{12.150}$$

If we differentiate Z with respect to its source terms, we obtain the Green's function:

$$-\frac{\delta^2\ln Z}{\delta\bar{j}_{\lambda n}\delta j_{\lambda'n'}} = [\underline{G}]_{\lambda n,\lambda'n'} = \delta_{\lambda\lambda'}\delta_{nn'}\frac{1}{i\omega_n - \epsilon_\lambda}.$$

Example 12.6 Consider the Grassman integral

$$I = \det\begin{pmatrix} A & B \\ C & D \end{pmatrix} = \int \prod_{j=1,N}d\bar{\alpha}_j d\alpha_j \prod_{k=1,M}d\bar{\beta}_k d\beta_k \exp\left[(\bar{\alpha},\bar{\beta})\begin{pmatrix} A & B \\ C & D \end{pmatrix}\begin{pmatrix} \alpha \\ \beta \end{pmatrix}\right], \tag{12.151}$$

where A and B are square matrices of dimension N and M, respectively, and α and β are column Grassman vectors of length N and M, respectively. By integrating out the β variables first, prove the identity

$$\det\begin{pmatrix} A & B \\ C & D \end{pmatrix} = \det[A - BD^{-1}C]\det D. \tag{12.152}$$

Solution

By expanding the argument of the exponential in I, we can rewrite the integral in the form

$$I = \int \int \prod_{j=1,N} d\bar{\alpha}_j d\alpha_j \exp\left[-\bar{\alpha}A\alpha\right]Y[\bar{\alpha},\alpha], \tag{12.153}$$

where the inner integral is an integral over the β variables:

$$Y[\bar{\alpha},\alpha] = \int \prod_{k=1,M} d\bar{\beta}_k d\beta_k \exp\left[-\left(\bar{\beta}D\beta + (\bar{\alpha}B)\beta + \bar{\beta}(C\alpha)\right)\right]. \qquad (12.154)$$

The inner integral is of the form (12.141) with $\bar{j} = \bar{\alpha}B$ and $j = C\alpha$, so it can be evaluated as

$$Y[\bar{\alpha},\alpha] = \det[D] \exp\left[\bar{\alpha}BD^{-1}C\alpha\right].$$

Putting this back into I, we then obtain

$$I = \det D \int \int \prod_{j=1,N} d\bar{\alpha}_j d\alpha_j \exp\left[-\bar{\alpha}\left(A - BD^{-1}C\right)\alpha\right] = \det D \det\left(A - BD^{-1}C\right),$$
$$(12.155)$$

thus proving the result.

12.5 The Hubbard–Stratonovich transformation

12.5.1 Heuristic derivation

The *Hubbard–Stratonovich transformation*, named after John Hubbard and Ruslan Stratonovich [9, 10, 14], provides a means of representing the interactions between fermions in terms of an exchange boson. It is in essence a way of replacing an instantaneous interaction by a force-carrying boson that describes the fluctuations of an emergent order parameter. Using this method it becomes possible to formally "integrate out" the microscopic fermions, rewriting the problem as an effective field theory describing the thermal and quantum fluctuations of the order parameter as a path integral with a new *effective action*. The method also provides an important formal basis for the order-parameter and mean-field description of broken-symmetry states.

To motivate this approach, we begin with a heuristic derivation. Consider a simple attractive point interaction between particles $V(\mathbf{x} - \mathbf{x}') = -g\delta(\mathbf{x} - \mathbf{x}')$, given by the interaction Hamiltonian

$$H_I = -\frac{g}{2} \int_{\mathbf{x}} \rho(\mathbf{x})^2. \qquad (12.156)$$

We can write the partition function as a path integral,

$$Z = \int \mathcal{D}[\psi] \exp\left[-\int_{\mathbf{x},\tau} \left(\bar{\psi}(x)(\partial_\tau + \underline{h})\psi(x) - \frac{g}{2}\rho(x)^2\right)\right]. \qquad (12.157)$$

If we expand the logarithm of the partition function diagrammatically, then we get a series of linked-cluster diagrams,

$$\ln(Z/Z_0) = \quad + \quad + \quad + \quad + \cdots,$$

(12.158)

where the point interaction is represented by the Feynman diagram

$$1 \quad\text{———}\quad 2 = g\delta(1-2).$$

(12.159)

Rather that thinking of an instantaneous contact interaction, we can regard this diagram as the exchange of a force-carrying boson, writing the diagram as

$$1 \quad\text{———}\quad 2 = \underbrace{(i)^2}_{\text{vertices}} \times \overbrace{-g\delta(1-2)}^{-\langle T\phi(1)\phi(2)\rangle},$$

(12.160)

where the vertices $(-i)$ derive from an interaction $S'_I = \int_{\mathbf{x},\tau} \rho(x)\phi(x)$, between the fermions and the boson with imaginary-time Green's function

$$G(1-2) = -\langle T\phi(1)\phi(2)\rangle = -g\delta(1-2).$$

(12.161)

But this implies that the exchange boson has a white-noise correlation function $\langle T\phi(1)\phi(2)\rangle = \delta(1-2)$: this kind of white-noise correlation is exactly what we expect for a field governed by a simple Gaussian path integral, where

$$\frac{\int D[\phi]\phi(1)\phi(2)e^{-S_\phi}}{\int D[\phi]e^{-S_\phi}} = g\delta(1-2),$$

(12.162)

with the Gaussian action

$$S_\phi = \int_{\mathbf{x}} \int_0^\beta d\tau\, \frac{\phi(x)^2}{2g}.$$

(12.163)

By adding $S_\phi + S'_I$ to the free fermion action, we can thus represent original point interaction by a fluctuating white-noise potential,

$$-\frac{g}{2}\rho(x)^2 \rightarrow \rho(x)\phi(x) + \frac{\phi(x)^2}{2g}.$$

(12.164)

If we now insert this transformed interaction into the action, the transformed path integral expression of the partition function becomes

$$Z = \int \mathcal{D}[\psi,\phi]\exp\left[-\int_{\mathbf{x},\tau}\left(\bar{\psi}(x)[\partial_\tau + \underline{h} + \phi(x)]\psi(x) + \frac{1}{2g}\phi(x)^2\right)\right].$$

(12.165)

Note the following:

- Although our derivation is heuristic, we shall shortly see that the Hubbard–Stratonovich transformation is *exact* so long as we allow $\phi(x) = \phi(\mathbf{x}, \tau)$ to describe a fluctuating quantum variable inside the path integral.
- If we replace $\phi(\mathbf{x}, \tau)$ by its average value, $\phi(\mathbf{x}, \tau) \rightarrow \langle \phi(\mathbf{x}, \tau) \rangle = \phi(\mathbf{x})$, we obtain a *mean-field theory*. Suppose, instead of carrying out the Hubbard–Stratonovich transformation, we choose to expand the density in powers of its fluctuations $\delta\rho(x)$ about its average value $\langle \rho(\mathbf{x}) \rangle$, writing $\rho(x) = \langle \rho(\mathbf{x}) \rangle + \delta\rho(x)$. The interaction can then be written

$$
\begin{aligned}
H_I &= -\frac{g}{2} \int_{\mathbf{x}} (\langle \rho(\mathbf{x}) \rangle + \delta\rho(x))^2 \\
&= -\frac{g}{2} \int_{\mathbf{x}} \left[\langle \rho(\mathbf{x}) \rangle^2 + 2\langle \rho(\mathbf{x}) \rangle \delta\rho(x) \right] + O(\delta\rho(x)^2).
\end{aligned}
\tag{12.166}
$$

If we neglect the term which is second-order in the fluctuations, then resubstitute $\delta\rho(x) = \rho(x) - \langle \rho(\mathbf{x}) \rangle$, we obtain

$$
H_I \approx -\frac{g}{2} \int_{\mathbf{x}} \left[2\langle \rho(\mathbf{x}) \rangle \rho(x) - \langle \rho(\mathbf{x}) \rangle^2 \right] = \int_{\mathbf{x}} \left[\rho(x)\phi(\mathbf{x}) + \frac{\phi(\mathbf{x})^2}{2g} \right],
\tag{12.167}
$$

where we have replaced $-g\langle \rho(\mathbf{x}) \rangle = \phi(\mathbf{x})$. This approximate mean-field Hamiltonian (12.167) resembles the result of the Hubbard–Stratonovich transformation (12.164).

With care, this kind of reasoning can be extended to a whole host of interactions between various kinds of charge, spin, and current densities, including both non-local interactions and repulsive interactions. For example, in the Hubbard and Anderson models, the interaction can be written as an attractive interaction in the magnetic channel of the form that is factorized as follows:

$$
-\frac{U}{2}(n_\uparrow - n_\downarrow)^2 \rightarrow (n_\uparrow - n_\downarrow)M + \frac{M^2}{2U},
\tag{12.168}
$$

corresponding to electrons exchanging fluctuations of the magnetic Weiss field M. The coupling between the field M and the electrons can sometimes stabilize a broken–symmetry state where M develops an expectation value – leading to a magnet. The Hubbard–Stratonovich transformation can also be applied to complex fields, permitting the following factorization:

$$
H_I = -gA^\dagger A \rightarrow \bar{A}\Delta + \bar{\Delta}A + \frac{\bar{\Delta}\Delta}{g},
\tag{12.169}
$$

where Δ is a complex field. Notice how we have switched $A^\dagger \rightarrow \bar{A}$ to emphasize that the replacement is only exact *under the path integral* (or alternatively, if you wish to switch to operators, under the time-ordering operator). This kind of interaction occurs in a BCS superconductor, where the pairing interaction

$$
H_I = -g \sum_{\mathbf{k},\mathbf{k'}} c^\dagger_{\mathbf{k}\uparrow} c^\dagger_{-\mathbf{k}\downarrow} c_{-\mathbf{k'}\downarrow} c_{\mathbf{k'}\uparrow} = -g \overbrace{\sum_{\mathbf{k}} c^\dagger_{\mathbf{k}\uparrow} c^\dagger_{-\mathbf{k}\downarrow}}^{A^\dagger} \overbrace{\sum_{\mathbf{k}} c_{-\mathbf{k}\downarrow} c_{\mathbf{k}\uparrow}}^{A}.
$$

In this case, *under the path integral* the interaction can be rewritten in terms of electrons moving in a fluctuating pair field:

$$H_I \rightarrow \bar{\Delta} \sum_{\mathbf{k}} c_{-\mathbf{k}\downarrow} c_{\mathbf{k}\uparrow} + \sum_{\mathbf{k}} \bar{c}_{\mathbf{k}\uparrow} \bar{c}_{-\mathbf{k}\downarrow} \Delta + \frac{\bar{\Delta}\Delta}{g}.$$

Once superconductivity develops, Δ develops an expectation value, playing the role of an order parameter.

12.5.2 Detailed derivation

Let us examine the above procedure in detail. To be concrete, consider an attractive inter-action of the form $H_I = -g \sum_j A_j^\dagger A_j$, where A_j represents an electron bilinear (such as the pair density or spin density of an x-y spin). Consider a fermion path integral on a lattice with interactions $H_I = -g \sum_j A_j^\dagger A_j$,

$$Z = \int \mathcal{D}[\bar{c}, c] \exp\left[-\int_0^\beta d\tau \left(\bar{c} (\partial_\tau + \underline{h}) c - g \sum_j \bar{A}_j A_j \right) \right], \tag{12.170}$$

where, inside the path integral, we have replaced $A^\dagger \rightarrow \bar{A}$. The next step is to introduce a *white-noise variable* α_j, described by the path integral

$$Z_\alpha = \int \mathcal{D}[\bar{\alpha}, \alpha] \exp\left[-\sum_j \int_0^\beta d\tau \frac{\bar{\alpha}_j \alpha_j}{g} \right]. \tag{12.171}$$

The weight function

$$\exp\left[-\sum_j \int_0^\beta d\tau \frac{\bar{\alpha}_j \alpha_j}{g} \right]$$

is a Gaussian distribution function for a white-noise field with correlation function [2]

$$\langle \bar{\alpha}_i(\tau) \alpha_j(\tau') \rangle = g \delta_{ij} \delta(\tau - \tau'). \tag{12.172}$$

---

[2] To show this, it is helpful to consider the generating functional

$$\Lambda[\bar{j}, j] = \int \mathcal{D}[\bar{\alpha}, \alpha] \exp\left[-\sum_r \int_0^\beta d\tau \left(\frac{\bar{\alpha}_r \alpha_r}{g} - \bar{j}_r \alpha_r - \bar{\alpha}_r j_r \right) \right].$$

By changing variables, $\alpha_r \rightarrow \alpha_r + g j_r$ we can absorb the terms linear in j to obtain

$$\Lambda[\bar{j}, j] = \exp\left[g \sum_r \int_0^\beta d\tau (\bar{j}_r(\tau) j_r(\tau)) \right].$$

Differentiating this with respect to $j_r(\tau)$, we find that

$$\frac{\partial^2 \ln \Lambda[j, \bar{j}]}{\partial \bar{j}_r(\tau) \partial j_{r'}(\tau')} \bigg|_{j,\bar{j}=0} = \langle \alpha_r(\tau) \bar{\alpha}_{r'}(\tau') \rangle = g \delta_{rr'} \delta(\tau - \tau').$$

Now the product of these two path integrals,

$$Z \times Z_\alpha = \int \mathcal{D}[\bar{c}, c] \int \mathcal{D}[\bar{\alpha}, \alpha] \exp\left[-\int_0^\beta d\tau \left(\bar{c}(\partial_\tau + \underline{h})c - \sum_j \overbrace{\left(-g\bar{A}_j A_j + \frac{\bar{\alpha}_j \alpha_j}{g}\right)}^{H_I'(j)}\right)\right],$$

$$(12.173)$$

describes two independent systems. As written, the "α" integrals are on the inside of the path so that, for all configurations of the $\alpha_j(\tau)$ field explored in the inner α integral, the space–time configuration of the $A_j(\tau)$ set by the outer integral are frozen and can hence be regarded as "constants," fixed at each point in space–time. This permits us to define a new variable,

$$\Delta_j(\tau) = \alpha_j(\tau) - gA_j(\tau),$$

and its corresponding conjugate, $\bar{\Delta}_j = \bar{\alpha}_j - g\bar{A}_j$. Formally this is just a shift in the integration variable, so the measure is unchanged and we can write $\mathcal{D}[\bar{\Delta}, \Delta] = \mathcal{D}[\bar{\alpha}, \alpha]$. The transformed interaction becomes

$$H_I' = \sum_j \left\{-g\bar{A}_j A_j + \frac{(\bar{\Delta}_j + g\bar{A}_j)(\Delta_j + gA_j)}{g}\right\}$$

$$= \sum_j \left\{\bar{A}_j \Delta_j + \bar{\Delta}_j A_j + \frac{\bar{\Delta}_j \Delta_j}{g}\right\}. \qquad (12.174)$$

In this way, we arrive at a transformed interaction in which the new variable Δ_j is linearly coupled to the electron operator A_j. If we now re-invert the order of integration inside the path integral (12.173), we obtain

$$Z = \int \mathcal{D}[\bar{\Delta}, \Delta] \exp\left[-\sum_j \int_0^\beta d\tau \frac{\bar{\Delta}_j \Delta_j}{g}\right] \int \mathcal{D}[\bar{c}, c] e^{-\tilde{S}}$$

$$\tilde{S} = \int_0^\beta d\tau \left(\bar{c}\partial_\tau c + H_E[\bar{\Delta}, \Delta]\right), \qquad (12.175)$$

where

$$H_E[\bar{\Delta}, \Delta] = \bar{c}\underline{h}c + \sum_j \left\{\bar{A}_j \Delta_j + \bar{\Delta}_j A_j\right\} \qquad (12.176)$$

represents the action for electrons moving in the fluctuating field Δ_j. Notice that, since A and $\bar{A}$ represent fermion bilinear terms, H_E is *itself a bilinear Hamiltonian*.

These noisy fluctuations mediate the interaction between the fermions, much as an exchange boson mediates interactions in the vacuum. More schematically,

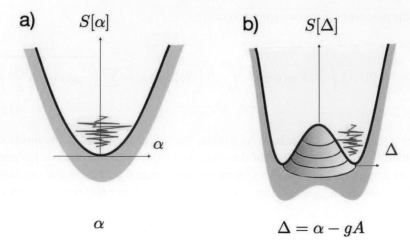

Fig. 12.6 (a) Action for initial white-noise variable α. (b) Action for shifted variable Δ is shifted off-center when the related quantity A has a predisposition towards developing an expectation value.

$$Z = \sum_{\{\Delta\}} \exp\left(-\sum_j \int d\tau \frac{|\Delta_j|^2}{g} \right) \times \left[\text{ path integral of fermions moving in field } \Delta \right],$$

$$(12.177)$$

where the summation represents a sum over all possible configurations $\{\Delta\}$ of the auxiliary field Δ. The transformed field

$$\Delta_j = \alpha_j - gA_j$$

is a combination of a white-noise field α_j and the physical field $-gA_j$, so its fluctuations now acquire the correlations associated with the electron fluid. Indeed, when the associated variable A is prone to the development of a broken-symmetry expectation value, the distribution function for Δ becomes concentrated around a non-zero value (Figure 12.6). We call Δ_j a *Weiss field* after Weiss, who first introduced such a field in the context of magnetism.

12.5.3 Effective action

Since the fermionic action inside the path integral is actually Gaussian, we can formally integrate out the fermions as follows:

$$e^{-S_\psi[\bar{\Delta},\Delta]} = \int \mathcal{D}[\bar{c}, c] e^{-\tilde{S}} = \det[\partial_\tau + \underline{h}_E[\bar{\Delta}, \Delta]], \qquad (12.178)$$

where $\underline{h}_E$ is the matrix representation of H_E. The full path integral may thus be written

$$Z = \int \mathcal{D}[\bar{\Delta}, \Delta] e^{-S_E[\bar{\Delta},\Delta]},$$

where

$$S_E[\bar{\Delta}, \Delta] = \sum_j \int d\tau \frac{\bar{\Delta}_j \Delta_j}{g} - \ln \det[\partial_\tau + \underline{h}_E[\bar{\Delta}, \Delta]]$$

$$= \sum_j \int d\tau \frac{\bar{\Delta}_j \Delta_j}{g} - \text{Tr} \ln[\partial_\tau + \underline{h}_E[\bar{\Delta}, \Delta]]. \qquad (12.179)$$

effective action

Here we have made the replacement $\ln \det \rightarrow \text{Tr} \ln$. This quantity is called the *effective action* of the field Δ. The additional fermionic contribution to this action can profoundly change the distribution of the field Δ. For example, if S_E develops a minima around $\Delta = \Delta_o \neq 0$, then $\Delta = -A/g$ will acquire a vacuum expectation value. This makes the Hubbard–Stratonovich transformation an invaluable tool for studying the development of broken symmetry in interacting Fermi systems.

12.5.4 Generalizations to real variables and repulsive interactions

The method outlined in the previous section can also be applied to real fields. If we have an interaction between real fields, we can introduce a real white-noise field as follows:

$$H_I = -\frac{g}{2} \sum_j A_j^2 \rightarrow \sum_j \left\{ -\frac{g}{2} A_j^2 + \frac{q_j^2}{2g} \right\}. \qquad (12.180)$$

Then, by redefining $q_j = Q_j + gA_j$, one obtains

$$-\frac{g}{2} \sum_j A_j^2 \rightarrow \sum_j \left\{ Q_j A_j + \frac{Q_j^2}{g} \right\}. \qquad (12.181)$$

For example, we can use the Hubbard–Stratonovich transformation to replace an attractive interaction between fermions by a white-noise potential with variance g:

$$H_I = -\frac{g}{2} \sum_j (n_j)^2 \rightarrow \sum_{j\sigma} V_j n_j + \frac{V_j^2}{2g},$$

where $n_j = n_{j\uparrow} + n_{j\downarrow}$.

But what about repulsive interactions? These require a little more care, because we can't just change the sign of g in (12.181), for the integral over the white-noise fields will no longer be convergent. Instead, after introducing the dummy white-noise fields as before,

$$H_I = \frac{g}{2} A_j^2 \rightarrow \sum_j \left\{ \frac{g}{2} A_j^2 + \frac{q_j^2}{2g} \right\}, \qquad (12.182)$$

we shift each variable in the path integral $q_j(\tau)$ by an imaginary amount, $q_j(\tau) = Q_j(\tau) + igA_j(\tau)$, to obtain [3]

$$\frac{g}{2}\sum_j A_j^2 \rightarrow \sum_j \left\{ iQ_j A_j + \frac{Q_j^2}{2g} \right\}. \qquad (12.183)$$

Note finally that, if one replaces $Q_j = -i\tilde{Q}_j$, this takes the form

$$\frac{g}{2}\sum_j A_j^2 \rightarrow \sum_j \left\{ \tilde{Q}_j A_j - \frac{\tilde{Q}_j^2}{2g} \right\}. \qquad (12.184)$$

At first sight, this looks like the generalization of (12.181) to negative g, except that now the integrals over each $Q_j(\tau)$ traverse the imaginary rather than the real axis.

Example 12.7   Using the Hubbard–Stratonovich transformation, show that the Coulomb interaction can be decoupled in terms of a fluctuating potential as follows:

$$H_I = \frac{1}{2}\int_{\mathbf{x},\,\mathbf{x}'} \rho(\mathbf{x})\rho(\mathbf{x}')\frac{e^2}{4\pi\epsilon_0|\mathbf{x}-\mathbf{x}'|} \rightarrow \int_{\mathbf{x}}\left[e\rho(\mathbf{x})\phi(\mathbf{x}) - \epsilon_0\frac{(\nabla\phi)^2}{2} \right]. \qquad (12.185)$$

What is the interpretation of the new term, quadratic in the potential field (and why is the sign negative)?

Solution

Because of the non-local nature of the Coulomb interaction, it is more transparent to make this transformation in momentum space. Writing

$$\rho(x) = \int_{\mathbf{q}} \rho_{\mathbf{q}} e^{i\mathbf{q}\cdot\mathbf{x}}, \qquad \frac{1}{4\pi\epsilon_0|\mathbf{x}-\mathbf{x}'|} = \int_{\mathbf{q}}\frac{1}{\epsilon_0 q^2}e^{i\mathbf{q}\cdot(\mathbf{x}-\mathbf{x}')}, \qquad (12.186)$$

where $\int_{\mathbf{q}} \equiv \int \frac{d^3q}{(2\pi)^3}$, the interaction becomes

$$H_I = \frac{1}{2}\int_{\mathbf{q}}\frac{(e\rho_{\mathbf{q}})(e\rho_{-\mathbf{q}})}{\epsilon_0 q^2}.$$

We now add in a dummy white-noise term:

$$H_I \rightarrow H_I' = \frac{1}{2}\int_{\mathbf{q}}\left[\frac{(e\rho_{\mathbf{q}})(e\rho_{-\mathbf{q}})}{\epsilon_0 q^2} - \epsilon_0 q^2 \phi_{\mathbf{q}}\phi_{-\mathbf{q}} \right],$$

---

[3] One might be worried about the legitimacy of shifting a real field by an imaginary quantity. However, just as the integral

$$\int_{-\infty}^{\infty} dQ e^{-Q^2/2} = \int_{-\infty+iA}^{\infty+iA} dQ e^{-Q^2/2}$$

is unaffected by a constant shift of the variable Q by an imaginary amount, $Q \rightarrow Q + iA$, a multi-variable path integral

$$\int D[Q]e^{-\int d\tau Q(\tau)^2/2}$$

is similarly unaffected by shifting the integration variable $Q(\tau)$ by an amount $iA(\tau)$, $Q(\tau) \rightarrow Q(\tau) + iA(\tau)$.

with the understanding that, in the path integral, the $\phi_{\mathbf{q}}$ field is to be integrated along the imaginary axis $\phi_{\mathbf{q}} = i\tilde{\phi}_{\mathbf{q}}$. Now if we shift $\phi_{\mathbf{q}} \rightarrow \phi_{\mathbf{q}} - \frac{e\rho_{\mathbf{q}}}{\epsilon_0 q^2}$, we obtain

$$H_I' = \int_{\mathbf{q}} \left[(e\rho_{\mathbf{q}})\phi_{-\mathbf{q}} - \frac{\epsilon_0}{2} q^2 \phi_{\mathbf{q}}\phi_{-\mathbf{q}} \right].$$

Finally, Fourier transforming back into real space ($q^2 \rightarrow -\nabla^2$), we obtain

$$H_I' = \int_{\mathbf{x}} \left[e\rho(x)\phi(x) + \frac{\epsilon_0}{2} \phi\nabla^2\phi \right]. \tag{12.187}$$

Integrating the last term by parts gives

$$H_I' = \int_{\mathbf{x}} \left[e\rho(x)\phi(x) - \overbrace{\frac{\epsilon_0}{2}(\nabla\phi)^2}^{-\epsilon_0 E^2/2} \right]. \tag{12.188}$$

We can identify the last term in this expression as $-\epsilon_0 E^2/2$, which is the electrostatic contribution to the action. The minus sign can be traced back to the fact that, inside the electromagnetic (Maxwell) action

$$S_{EM} = \int d^3x d\tau \left[\frac{B^2}{2\mu_0} - \frac{\epsilon_0 E^2}{2} \right], \tag{12.189}$$

the electrostatic contribution to the action enters with the opposite sign to the magnetic part. The complete path integral for interacting electrons in this representation is then

$$Z = \int \mathcal{D}[\bar{\psi}, \psi, \phi] \exp\left[-\int_0^\beta d\tau \int d^3x \left(\bar{\psi}\left(-\frac{1}{2m}\nabla^2 + e\phi(x) - \mu \right)\psi - \frac{\epsilon_0}{2}(\nabla\phi)^2 \right) \right].$$

Thus, by carrying out a Hubbard–Stratonovich transformation, the action becomes local. This formulation is ideal for the development of RPA approximations to the electron gas, while mean-field solutions of this path integral can be used to explore the formation of Wigner crystals.

Appendix 12A   Derivation of key properties of bosonic coherent states

Here we derive the matrix elements and the completeness properties of bosonic coherent states.

Matrix elements

Matrix elements of normal-ordered operators $O[\hat{b}^\dagger, \hat{b}]$ between two coherent states are obtained simply by replacing the operators $\hat{b}$ and $\hat{b}^\dagger$ by the c-numbers b and $\bar{b}$, respectively:

$$\langle \bar{b}_1 | \hat{O}[\hat{b}^\dagger, \hat{b}] | b_2 \rangle = O[\bar{b}_1, b_2] \times \langle \bar{b}_1 | b_2 \rangle = O[\bar{b}_1, b_2] \times e^{\bar{b}_1 | b_2}. \tag{12.190}$$

To derive the matrix elements of coherent states, we first note that the properties of coherent states guarantee that

$$\langle \bar{b}|(\hat{b}^{\dagger})^n \hat{b}^m|b\rangle = (\bar{b}^{\dagger})^n b^m \langle \bar{b}|b\rangle = (\bar{b})^n b^m e^{\bar{b}b}. \tag{12.191}$$

Thus, if $\hat{O}[\hat{b}^{\dagger}, \hat{b}] = \sum_{m,n} O_{mn}(\hat{b}^{\dagger})^m b^n$ is a *normal-ordered* operator (all annihilation operators on the right), it follows that

$$\langle \bar{b}|\hat{O}[\hat{b}^{\dagger}, \hat{b}]|b\rangle = \sum_{m,n} O_{mn} \bar{b}^m b^n \times \langle \bar{b}|b\rangle = O[\bar{b}, b] \times e^{\bar{b}b}$$

or

$$\hat{O}[b^{\dagger}, b] \xrightarrow{\text{coherent states}} O[\bar{b}, b] \times \langle \bar{b}|b\rangle.$$

Note that, if one has an operator that is not normal-ordered, then one has to normal-order the operator prior to applying this theorem. For example, if $O = (\hat{b} + \hat{b}^{\dagger})^2$, then $O = :O:$ $+1$, and $\langle \bar{b}|O|b\rangle = [(b + \bar{b})^2 + 1]e^{\bar{b}b}$.

Completeness

The unit operator can be decomposed in terms of coherent states as follows:

$$\hat{1} = \sum_{\bar{b},b} |b\rangle\langle\bar{b}|, \tag{12.192}$$

where

$$\sum_{\bar{b},b} \equiv \int \frac{d\bar{b}db}{2\pi i} e^{-\bar{b}b} \tag{12.193}$$

is the normalized measure for summing over coherent states. To demonstrate the completeness relation, we will first derive the orthogonality relation between the wavefunctions $\phi_n(b) = \langle n|b\rangle$ of the coherent states:

$$I_{nm} = \int \frac{d\bar{b}db}{2\pi i} e^{-\bar{b}b} \langle n|b\rangle\langle\bar{b}|m\rangle = \delta_{nm}. \tag{12.194}$$

To prove this, let us substitute $b = re^{i\phi}$ and $\bar{b} = re^{-i\phi}$. Although $\bar{b}$ and b are complex conjugates of each other, they are derived from two independent real variables, and so the measure for integrating over them is two-dimensional. We can transform the measure into polar co-ordinates by introducing a Jacobian, as follows:

$$d\bar{b}db = \overbrace{\left\| \begin{array}{cc} \frac{\delta\bar{b}}{\delta r} & \frac{\delta\bar{b}}{\delta\phi} \\ \frac{\delta b}{\delta r} & \frac{\delta b}{\delta\phi} \end{array} \right\|}^{\delta[\bar{b},b]/\delta[r,\phi]} drd\phi = \left\| \begin{array}{cc} e^{-i\phi} & -ire^{-i\phi} \\ e^{i\phi} & ire^{i\phi} \end{array} \right\| drd\phi = 2ir\, drd\phi,$$

so that (12.194) factorizes into a radial and an angular integral:

$$
I_{nm} = \frac{1}{\sqrt{n!\,m!}} \int \frac{d\bar{b}db}{2\pi i} \bar{b}^n \bar{b}^m e^{-\bar{b}b} = \frac{1}{\sqrt{n!\,m!}} \int_0^\infty 2r dr\, r^{n+m} e^{-r^2} \times \overbrace{\int_0^{2\pi} \frac{d\phi}{2\pi} e^{i\phi(n-m)}}^{\delta mn},
$$
(12.195)

where we have substituted $\langle n|b \rangle = \frac{1}{\sqrt{n!}} b^n$ and $\langle \bar{b}|m \rangle = \frac{1}{\sqrt{m!}} \bar{b}^m$. The angular integral vanishes unless $n = m$. Changing variables $r^2 \to x$, $2r dr = dx$, in the first integral, we then obtain

$$
I_{nm} = \frac{\delta_{nm}}{n!} \int_0^\infty dx\, x^n e^{-x} = \delta_{nm},
$$
(12.196)

proving the orthogonality relation. Now since $\delta_{nm} = \langle n|m \rangle$, we can write the orthogonality relation (12.194) as

$$
\langle n|m \rangle = \int \frac{d\bar{b}db}{2\pi i} e^{-\bar{b}b} \langle n|b \rangle \langle \bar{b}|m \rangle = \langle n| \left(\int \frac{d\bar{b}db}{2\pi i} e^{-\bar{b}b} |b \rangle \langle \bar{b}| \right) |m \rangle.
$$

Since this holds for all states $|n \rangle$ and $|m \rangle$, it follows that the quantity in brackets is the unit operator:

$$
\hat{1} = \int \frac{d\bar{b}db}{2\pi i} e^{-\bar{b}b} |b \rangle \langle \bar{b}| = \int \frac{d\bar{b}db}{2\pi i} \frac{|b \rangle \langle \bar{b}|}{\langle \bar{b}|b \rangle} \equiv \sum_{\bar{b},b} |b \rangle \langle \bar{b}|. \quad \text{Completeness relation} \quad (12.197)
$$

Appendix 12B   Grassman differentiation and integration

Differentiation is defined to have the normal linear properties of the differential operator. We denote

$$
\partial_c \equiv \frac{\partial}{\partial c}, \quad \partial_{\bar{c}} \equiv \frac{\partial}{\partial \bar{c}},
$$
(12.198)

so that

$$
\partial_c c = \partial_{\bar{c}} \bar{c} = 1.
$$
(12.199)

If we have a function

$$
f(\bar{c},c) = f_0 + \bar{f}_1 c + \bar{c} f_1 + f_{12} \bar{c} c,
$$
(12.200)

then differentiation from the left-hand side gives

$$
\partial_c f = \bar{f}_1 - f_{12} \bar{c}
$$
$$
\partial_{\bar{c}} f = f_1 + f_{12} c,
$$
(12.201)

where the minus sign in the first expression occurs because the $\bar{\partial}$ operator must anticommute with c. But how do we define integration? This proves to be much easier for Grassman variables than for regular c-numbers. The great sparseness of the space of functions dramatically restricts the number of linear operations we can apply to functions, forcing differentiation and integration to become the *same* operation:

$$\int dc \equiv \partial_c, \qquad \int d\bar{c} \equiv \partial_{\bar{c}}. \tag{12.202}$$

In other words,

$$\int d\bar{c}\bar{c} = 1, \qquad \int dc c = 1, \qquad \int d\bar{c} = \int dc = 0. \tag{12.203}$$

Appendix 12C   Grassman calculus: change of variables

Suppose we change variables, writing

$$\begin{pmatrix} c_1 \\ \vdots \\ c_r \end{pmatrix} = A \begin{pmatrix} \xi_1 \\ \vdots \\ \xi_r \end{pmatrix}, \tag{12.204}$$

where A is a c-number matrix. Then we would like to know how to evaluate the Jacobian for this transformation, which is defined so that

$$\int dc_1 \cdots dc_r [\ldots] = \int J \left(\frac{c_1 \cdots c}{\xi_1 \cdots \xi_r} \right) d\xi_1 \cdots d\xi_r [\ldots]. \tag{12.205}$$

Now since integration and differentiation are identical for Grassman variables, we can evaluate the fermionic Jacobian using the chain rule for differentiation, as follows:

$$\int dc_1 \cdots dc_r [\ldots] = \frac{\partial^r}{\partial c_1 \cdots \partial c_r} [\ldots]$$

$$= \sum_P \left(\frac{\partial \xi_{P_1}}{\partial c_1} \cdots \frac{\partial \xi_{P_r}}{\partial c_r} \right) \frac{\partial^r}{\partial \xi_{P_1} \cdots \partial \xi_{P_r}} [\ldots], \tag{12.206}$$

where $P = \begin{pmatrix} 1 & \cdots & r \\ P_1 & \cdots & P_r \end{pmatrix}$ is a permutation of the sequence $(1 \cdots r)$. But we can order the differentiation in the second term, picking up a factor $(-1)^P$, where P is the signature of the permutation, to obtain

$$\int dc_1 \cdots dc_r [\ldots] = \sum_P (-1)^P \left(\frac{\partial \xi_{P_1}}{\partial c_1} \cdots \frac{\partial \xi_{P_r}}{\partial c_r} \right) \frac{\partial^r}{\partial \xi_1 \cdots \partial \xi_r} [\ldots]$$

$$= \det[A^{-1}] \frac{\partial^r}{\partial \xi_1 \cdots \partial \xi_r} [\ldots]$$

$$= \int \det[A^{-1}] d\xi_1 \cdots d\xi_r [\ldots], \tag{12.207}$$

where we have recognized the prefactor as the determinant of the inverse transformation $\xi = \underline{A}^{-1}c$. From this result, we can read off the Jacobian of the transformation as

$$J\left(\frac{c_1 \dots c_r}{\xi_1 \dots \xi_r}\right) = \det[A]^{-1} = \left|\frac{\partial(c_1, \dots c_r)}{\partial(\xi_1, \dots \xi_r)}\right|^{-1}, \tag{12.208}$$

which is precisely the inverse of the bosonic Jacobian. This has important implications for supersymmetric field theories, where the Jacobians of the bosons and fermions precisely cancel. For our purposes, however, the most important point is that, for a unitary transformation, the Jacobian is unity.

Appendix 12D   Grassman calculus: Gaussian integrals

The basic Gaussian integral is simply

$$\int d\bar{c}dc e^{-a\bar{c}c} = \int d\bar{c}dc(1 - a\bar{c}c) = a. \tag{12.209}$$

If we now introduce a set of N variables, then

$$\int \prod_j d\bar{c}_j dc_j \exp - \left[\sum_j a_j \bar{c}_j c_j\right] = \prod_j a_j. \tag{12.210}$$

Suppose we now carry out a unitary transformation, for which the Jacobian is unity. Then, since

$$c = U\xi, \qquad \bar{c} = \bar{\xi}U^\dagger,$$

the integral becomes

$$\int \prod_j d\bar{\xi}_j d\xi_j \exp[-\bar{\xi} \cdot A \cdot \xi] = \prod_j a_j,$$

where $A_{ij} = \sum_l U^\dagger_{il} a_l U_{lj}$ is the matrix with eigenvalues a_l. It follows that

$$\int \prod_j d\bar{\xi}_j d\xi_j \exp[-\bar{\xi} \cdot A \cdot \xi] = \det A. \tag{12.211}$$

Finally, by shifting the variables $\xi \to \xi + A^{-1}j$, where j is an arbitrary vector, we find that

$$Z[j] = \int \prod_j d\bar{\xi}_j d\xi_j \exp[-(\bar{\xi} \cdot A \cdot \xi + \bar{j} \cdot \xi + \bar{\xi} \cdot j)] = \det A \exp[\bar{j} \cdot A^{-1} \cdot j]. \tag{12.212}$$

This is the basic Gaussian integral for Grassman variables. Notice that, using the result $\ln \det A = \mathrm{Tr}\, \ln A$, it is possible to take the logarithm of both sides to obtain

$$S[j] = -\ln Z[j] = -\mathrm{Tr}\, \ln A - \bar{j} \cdot A^{-1} \cdot j. \tag{12.213}$$

The main use of this integral is for evaluating the path integral for free-field theories. In this case, the matrix $A \to -G^{-1}$ becomes the inverse propagator for the fermions, and $\xi_n \to \psi(i\omega_n)$ is the Fourier component of the Fermi field at Matsubara frequency $i\omega_n$.

Exercises

Exercise 12.1 In this problem, consider $\hbar = 1$. Suppose $|0\rangle$ is the ground state of a harmonic oscillator, where $b|0\rangle = 0$. Consider the state formed by simultaneously translating this state in momentum and position space as follows:

$$|p, x\rangle = \exp\left[-i(x\hat{p} - p\hat{x})\right]|0\rangle.$$

By rewriting $\hat{b} = (\hat{x} + i\hat{p})/\sqrt{2}$, $z = (x + ip)/\sqrt{2}$, show that this state can be rewritten as

$$|p, x\rangle = e^{b^\dagger z - \bar{z}b}|0\rangle.$$

Using the relation $e^{A+B} = e^A e^B e^{\frac{1}{2}[A,B]}$, provided $[A, [A, B]] = [B, [A, B]] = 0$, show that $|p, x\rangle$ is equal to a normalized coherent state

$$|p, x\rangle \equiv |z\rangle e^{-\bar{z}z/2} = e^{b^\dagger z}|0\rangle e^{-\frac{1}{2}\bar{z}z},$$

showing that the coherent state $|z\rangle$ represents a minimum-uncertainty wavepacket centered at (q, p) in phase space.

Exercise 12.2 Repeat the calculation of Section 12.3 without taking the continuum limit. Show that the path integral for a single boson with Hamiltonian $H = \epsilon b^\dagger b$ with a large but finite number of time-slices is given by

$$\ln Z_N = \sum_{n=1}^{N} \ln\left(\epsilon - i\nu_n F(i\nu_n \Delta\tau/2)\right),$$

where $F(x) = (1 - e^{-x})/x$. If you approximate each term in the sum by its value at $\Delta\tau = 0$ and then take $N \to \infty$, the result obviously converges to the continuum limit. But the error contribution from N such terms appears to be of order $O(N \times \Delta\tau) = O(1)$. Use contour integration to show that this is fortunately an overestimate, and that the actual error is $O(\Delta\tau) = O(1/N)$.

Exercise 12.3 Using path integrals, calculate the partition function for a single Zeeman-split electronic level described by the action

$$S = \int d\tau \bar{f}_\alpha \left(\delta_{\alpha\beta}\partial_\tau + \vec{\sigma}_{\alpha\beta} \cdot \vec{B}\right) f_\beta.$$

Why is your answer not the same as the partition function of a spin $S = \frac{1}{2}$ in a magnetic field?

Exercise 12.4 Suppose

$$\mathcal{M} = e^{\frac{1}{2}\sum_{ij} A_{ij} c^\dagger_i c^\dagger_j},$$

where A_{ij} is an $N \times N$ antisymmetric matrix and the $c^{\dagger}_j$ are a set of N canonical Fermi creation operators. Using coherent states, calculate

$$\text{Tr}[\mathcal{M}\mathcal{M}^{\dagger}],$$

where the trace is over the 2^N-dimensional Hilbert space of fermions. (*Hint*: notice that $\mathcal{M}\mathcal{M}^{\dagger}$ is already normal ordered, so that by using the trace formula you can rewrite this in terms of a simple Grassman integral.)

Exercise 12.5 Suppose $H = \epsilon c^{\dagger}c$ represents a single fermion state. Consider the approximation to the partition function obtained in Section 12.4.2 by first dividing up the period $\tau \in [0, \beta]$ into N equal time-slices:

$$Z_N = \text{Tr}[(e^{-\Delta\tau H})^N], \tag{12.214}$$

was given as

$$Z_N = \int \prod_{j=1}^{N} d\bar{c}_j dc_j \exp[-S_N] \tag{12.215}$$

$$S_N = \sum_{j=1}^{N} \left[\bar{c}_j(c_j - c_{j-1})/\Delta\tau + \epsilon\bar{c}_j c_{j-1} \right].$$

(a) Show that Z_3 can be written as a *toy functional integral*,

$$Z_3 = \int d\bar{c}_3 dc_3 d\bar{c}_2 dc_2 d\bar{c}_1 dc_1 \exp\left\{ -(\bar{c}_3, \bar{c}_2, \bar{c}_1) \begin{bmatrix} 1 & -\alpha & 0 \\ 0 & 1 & -\alpha \\ \alpha & 0 & 1 \end{bmatrix} \begin{pmatrix} c_3 \\ c_2 \\ c_1 \end{pmatrix} \right\}, \tag{12.216}$$

where $\alpha = 1 - \Delta\tau\epsilon$. In this formula, the discrete time-line is labeled as follows:

$$
\begin{array}{cccc}
\bar{c}_3 & c_2 \;\; \bar{c}_2 & c_1 \;\; \bar{c}_1 & c_0 = -\bar{c}_3 \\
\mid & \mid & \mid & \mid \\
\beta = \tau_3 & \tau_2 & \tau_1 & 0
\end{array}
\tag{12.217}
$$

where $\bar{c}_j$, c_j are the conjugate Grassman variables at each discrete time $\tau_j = j\Delta\tau$.

(b) Evaluate Z_3.

(c) Generalize the result to N time-slices and obtain an expression for Z_N. What is the limiting value of your result as $N \to \infty$?

Exercise 12.6 The one dimensional electron gas is prone to the development of charge-density wave instabilities. The treatment of these instabilities bears a close resemblance to the BCS theory of superconductivity. Suppose we have a one-dimensional conductor, described by the Hamiltonian

$$H - \mu N = H_0 + H_I$$

$$H_0 = -t \sum_{j, \sigma} \left(\psi^{\dagger}_{(j+1)\,\sigma} \psi_{j\sigma} + \psi^{\dagger}_{j\sigma} \psi_{(j+1)\,\sigma} \right)$$

$$H_I = -g \sum_{j} n_{j\uparrow} n_{j\downarrow}, \tag{12.218}$$

where $g > 0$ and $\psi^\dagger_{j\sigma}$ creates an electron with spin $\sigma = \pm\frac{1}{2}$ at site j. The separation between sites is taken to be unity and the chemical potential has been chosen to be *zero*, giving a half-filled band.

(a) Show that H_0 can be diagonalized in the form

$$H_0 = -\sum_{k\,\sigma}(2t\cos k)c^\dagger_{k\sigma}c_{k\sigma}, \tag{12.219}$$

where $c_{k\sigma} = \frac{1}{\sqrt{N}}\sum_j \psi_{j\sigma}e^{-ikj}$, $k = \frac{2\pi}{N}(0, 1, \ldots N - 1)$. Note that the band is exactly half-filled, so that the Fermi surfaces are separated by a distance π in momentum space and the average electron density is 1 per site.

(b) Suppose a staggered potential $V_j = -(-1)^j\Phi$ is applied to the conductor. This will induce a staggered charge density to the sample:

$$\langle n_{j\sigma}\rangle = \frac{1}{2} + (-1)^j\Delta_j/g. \tag{12.220}$$

At low temperatures, the staggered order will remain even after the applied potential is removed. Why? If the RMS fluctuations in the staggered charge density can be ignored, show that the interaction Hamiltonian can be recast in the form

$$H_I \to \sum_j\left((-1)^j\Delta_j\hat{n}_j + \frac{\Delta_j^2}{g}\right) + O(\delta\hat{n}_j^2). \tag{12.221}$$

(c) How can the above transformation be elevated to the status of an exact result using a path integral? (Note that the order parameter is no longer complex; does this change your discussion?)

(d) Calculate the excitation spectrum in the presence of the uniformly staggered order parameter $\Delta_j = \Delta$. (*Hint*: write the mean field Hamiltonian in momentum space and treat the terms that scatter from one side of the Fermi surface in an analogous fashion to the pairing terms in superconductivity. You may find it useful to work with the spinor $\Psi_{k\sigma} = \begin{pmatrix} c_{k\sigma} \\ c_{k+\pi\sigma} \end{pmatrix}$.)

(e) Calculate the free energy $F[\Delta]$ and sketch your result as a function of temperature. Write down the gap equation for the value of $\Delta(T)$ that develops spontaneously at low temperatures.

References

[1] R. P. Feynman, Space-Time Approach to Non-Relativistic Quantum Mechanics, *Rev. Mod. Phys.*, vol. 20, no. 2, p. 367, 1948.

[2] R. P. Feynman and A. R. Hibbs, Quantum Mechanics and Path Integrals, McGraw-Hill, 1965.

[3] J. Z. Justin, *Path Integrals in Quantum Mechanics*, Oxford University Press, 2004.

[4] R. J. Glauber, Coherent and incoherent states of the radiation field, *Phys. Rev.*, vol. 131, no. 6, p. 2766, 1963.

[5] D. J. Candlin, On sums over trajectories for systems with Fermi statistics, *Nuovo Cimento*, vol. 4, p. 231, 1956.

[6] J. L. Martin, The Feynman principle for a Fermi system, *Proc. R. Soc. A*, vol. 251, p. 543, 1959.

[7] D. Sherrington, A new method of expansion in the quantum many-body problem: III. The density field, *Proc. Phys. Soc.*, vol. 91, 1967.

[8] J. Schwinger, The theory of quantized fields. IV, *Phys. Rev.*, vol. 92, no. 5, p. 1283, 1953.

[9] R. L. Stratonovich, On a method of calculating quantum distribution functions, *Sov. Phys.–Doklady*, vol. 2, 1958.

[10] J. Hubbard, Calculation of partition functions, *Phys. Rev. Lett.*, vol. 3, no. 2, p. 77, 1959.

[11] P. T. Matthews and A. Salam, Propagators of quantized field, *Nuovo Cimento*, vol. 2, p. 367, 1955.

[12] F. A. Berezin, *The Method of Second Quantization*, Academic Press, 1966.

[13] S. F. Edwards and D. Sherrington, A new method of expansion in the quantum many-body problem, *Proc. Phys. Soc.*, vol. 90, p. 3, 1967.

[14] P. T. Matthews and A. Salam, The Green's functions of quantised fields, *Nuovo Cimento*, vol. 12, 1954.

13 Path integrals and itinerant magnetism

To illustrate the Hubbard–Stratonovich transformation, we now examine its application to the treatment of itinerant magnetism in the class of Hubbard models. Without spin, all matter would be magnetically inert (neither diamagnetic nor paramagnetic). Quantum mechanics provides an explanation of magnetism as a consequence of the orientational ordering of electron spins. This connection between magnetism and spin is one of the huge accomplishments of quantum mechanics.

13.1 Development of the theory of itinerant magnetism

Let us begin with a few remarks about the development of the theory of magnetism [1, 2]. A century ago, the ferromagnetism of simple metals such as iron, cobalt, or nickel was an unsolved mystery. In 1906 the French physicist Pierre Weiss, working at ETH, Zurich, discovered that if you look at a ferromagnet on a small enough scale, it consists of magnetic domains. This led him to propose the first *mean-field theory*, introducing the concept of an emergent *molecular* contribution to the effective internal magnetic field [3, 4]:

$$\mathbf{H}_E = \mathbf{H} + \overbrace{I\mathbf{M}}^{\text{molecular Weiss field}}.$$ (13.1)

But the origin of the Weiss field was unknown. Worse, it quickly became clear that magnetism can't be understood using classical mechanics, for according to the *Bohr–van Leeuwen theorem* independently proven by Neils Bohr and Hendrika van Leeuwen [5, 6], a fluid of (spinless) classical electrons *in thermal equilibrium* has zero magnetization,[1] even in a field [7].

---

[1] The Bohr–van Leeuwen theorem follows from the field-independence of the partition function of a classical fluid of interacting particles. The classical partition function is written

$$Z[\mathbf{A}] = \int \prod_{i=1,N} d^3 p_i d^3 x_i e^{-\beta H[\mathbf{A}]},$$ (13.2)

where

$$H[\mathbf{p}, \mathbf{x}] = \sum_i \frac{(\mathbf{p}_i - e\mathbf{A}(\mathbf{x}_i))^2}{2m} + \sum_{i<j} U(\mathbf{x}_i - \mathbf{x}_j) + e\phi(\mathbf{x}_i).$$ (13.3)

All the magnetic field dependence of $Z[\mathbf{A}]$ resides in the vector potential term, given by $\mathbf{A} = \frac{1}{2}\mathbf{B} \times \mathbf{x}$ in the Landau gauge. However, one can always make a change of variable $\mathbf{p}' = \mathbf{p} + e\mathbf{A}(\mathbf{x})$, $\mathbf{x}' = \mathbf{x}$, for which the Jacobian is unity, completely absorbing all dependence on the external magnetic field, so that $Z[\mathbf{A}] = Z[\mathbf{A} = 0]$. The equilibrium magnetization $\mathbf{M} = -T\delta \ln Z/\delta \mathbf{B}(x)$ is therefore zero. This also implies that the isothermal magnetic susceptibility of a classical plasma is zero. Note, however, that a classical electron gas does have a diamagnetic response when a field is applied adiabatically rather than isothermally.

This mystery was ultimately resolved by quantum mechanics, by the discovery of *spin* and later through the theory of superconductors, which are, as we shall discuss shortly, perfect quantum diamagnets. In 1928, Werner Heisenberg, working at Leipzig, made the critical link between magnetization and electron spin polarization; he also identified the Coulomb exchange interaction as the driving force for ferromagnetism [8] and the origin of the mysterious "*I*" in Weiss' theory. In the 1930s, Edmund Stoner at Leeds University and John Slater at Harvard University developed the basis for an itinerant theory of ferromagnetism in metals [9–11]. A key idea here is that strong interactions drive a metal to become unstable towards the development of a spontaneous spin polarization. In the simplest case a ferromagnet develops, but later Albert Overhauser, working at Ford Labs in the early 1960s, showed that the instability can also occur at a finite wavevector $\mathbf{Q}$ to form a *spin density wave* [12], as in the case of metallic chromium. This instability occurs when the product of the electron interaction I and the "bare" magnetic susceptibility of the non-interacting electron gas at this wavevector $\chi_0(\mathbf{Q})$ reaches unity:

$$I_{\mathbf{Q}}\chi_0(\mathbf{Q}) = 1. \qquad \text{Stoner criterion}$$

Later in the 1960s, Junjiro Kanamori [13] at Osaka University and John Hubbard [14] in Harwell, England reformulated the theory of magnetism using the model we now call the Hubbard model. Sebastian Doniach and Stanley Engelsberg [15] at Imperial College London, and Norman Berk and Robert Schrieffer [16] at the University of Pennsylvania, refined this work, demonstrating that quantum fluctuations of the magnetization play a crucial role: these fluctuations act to suppress the magnetization and become particularly strong near the point of zero-temperature instability or *quantum critical point* (QCP), a point brought out through the landmark work in the mid-1970s of Peter Young, working at Oxford University, and John Hertz, working at the University of Chicago [17, 18], later extended to embrace itinerant antiferromagnetism by Andrew Millis, working at Bell Laboratories [19]. Young and Hertz were among the first to recognize that generalizing statistical mechanics to include quantum fluctuations of magnetization required a Lagrangian path formulation in space–time. In the 1980s, Gilbert Lonzarich, working with his students Nicholas Bernhoeft, Thorsteinn Sigfusson, and Louis Taillefer at the Cavendish Laboratory in Cambridge, started an experimental program of study of magnetic fluctuations, adapting the quantum fluctuation approach into a more physical language [20, 21] and pioneering the experimental measurement of quantum criticality in weak itinerant ferromagnets such as Ni_3Al, $ZrZn_2$, and $MnSi$ using neutron scattering and de Haas–van Alphen techniques.

Itinerant magnetism is only one part of the story of magnetism, for in magnetic materials where the electrons are localized, the magnetization derives from *localized magnetic moments*. High-performance neodynium–iron alloy magnets derive their strength from localized moments–at the neodynium sites. Many of the most fascinating systems of current study, such as the high-temperature cuprate and iron-based superconductors, appear to lie in a murky region between "intineracy" and "localization," where electrons are on the brink of localization. This is a topic we shall return to in Chapter 16.

13.2   Path integral formulation of the Hubbard model

We encountered the Hubbard model in Chapter 4. It consists of a single band of electrons moving on a tight-binding lattice, with a localized interaction of strength U described by the Hamiltonian

$$H = \sum_{\mathbf{k},\sigma} \epsilon_{\mathbf{k}} c^{\dagger}_{\mathbf{k}\sigma} c_{\mathbf{k}\sigma} + U \sum_{j} n_{j\uparrow} n_{j\downarrow}, \tag{13.4}$$

where

$$c_{\mathbf{k}\sigma} = \frac{1}{\sqrt{N_s}} \sum c^{\dagger}_{j\sigma} e^{i\mathbf{k}\cdot\mathbf{r}_j}$$

creates an electron of wavevector $\mathbf{k}$ with energy $\epsilon_{\mathbf{k}}$. To explore magnetism in this model we shall use a path integral approach to itinerant magnetism using a Hubbard–Stratonovich decoupling of the interaction [22, 23]. As a first step, we rewrite the interaction in terms of the spin operators [22, 23]:

$$U n_{j\uparrow} n_{j\downarrow} = -\frac{U}{2}(n_{j\uparrow} - n_{j\downarrow})^2 + \frac{U}{2}(n_{j\uparrow} + n_{j\downarrow}), \tag{13.5}$$

where we have used the fact that $n_{j\uparrow}^2 = n_{j\uparrow}$. As written, the above decoupling emphasizes the magnetic fluctuations along the z-axis. Indeed, we might have made the decoupling around any spin quantization axis, and, since we are interested in keeping track of magnetic fluctuations along all axes, it makes sense to average over all three directions, writing the decoupling as

$$U n_{j\uparrow} n_{j\downarrow} = -\frac{U}{6}\left(\boldsymbol{\sigma}_j\right)^2 + \frac{U}{2}(n_{j\uparrow} + n_{j\downarrow}), \tag{13.6}$$

where we have introduced the notation $\boldsymbol{\sigma}_j = (c^{\dagger}_{j\alpha} \boldsymbol{\sigma}_{\alpha\beta} c_{j\beta})$ for the magnetization at site j. The second term in this expression can be absorbed into a redefinition of the chemical potential, by writing $\mu = \mu' + U/2$. The minus sign in front of the magnetic interaction exhibits the exchange effect of the Coulomb interaction, whereby a repulsion between charges induces a *ferromagnetic attraction between spins*.

We now formulate the problem as a path integral:

$$Z = \int \mathcal{D}[c] e^{-S}$$

$$S = \int_0^{\beta} d\tau \left[\sum_{\mathbf{k},\sigma} \bar{c}_{\mathbf{k}\sigma}(\partial_\tau + \epsilon_{\mathbf{k}}) c_{\mathbf{k}\sigma} - \frac{I}{2} \sum_j (\boldsymbol{\sigma}_j)^2 \right] \quad (I = U/3), \tag{13.7}$$

where we have introduced the coupling constant $I = U/3$. At this point we carry out a Hubbard–Stratonovich transformation. Adding a white-noise field m_j into the action, so that

$$-\frac{I}{2} \sum_j (\vec{\sigma}_j)^2 \rightarrow -\frac{I}{2} \sum_j (\vec{\sigma}_j)^2 + \sum_j \frac{m_j^2}{2I}, \tag{13.8}$$

and then shifting $\mathbf{m}_j = \mathbf{M}_j - I\sigma_j$, we obtain

$$-\frac{I}{2}(\vec{\sigma}_j)^2 \rightarrow -\mathbf{M}_j(\tau) \cdot \vec{\sigma}_j + \frac{\mathbf{M}_j(\tau)^2}{2I}, \tag{13.9}$$

where $\mathbf{M}_j(\tau)$ is a fluctuating Weiss field. We have chosen the sign of the first term to reflect the role of the Weiss field as an "effective magnetic field." The transformed partition function

$$Z = \int \mathcal{D}[\mathbf{M}, \bar{c}, c] e^{-S[\bar{c},c,\mathbf{M}]},$$

$$S[\bar{c}, c, \mathbf{M}] = \int_0^\beta d\tau \left(\sum_{\mathbf{k},\sigma} \bar{c}_{\mathbf{k}\sigma}(\partial_\tau + \epsilon_{\mathbf{k}})c_{\mathbf{k}\sigma} + \sum_j \left[-\mathbf{M}_j \cdot \vec{\sigma}_j + \frac{\mathbf{M}_j^2}{2I} \right] \right) \tag{13.10}$$

describes electrons moving through a lattice of fluctuating magnetization. We can emphasize this interpretation by moving the magnetization integral to the outside, writing

$$Z = \int \mathcal{D}[\mathbf{M}] e^{-S_E[\mathbf{M}]}, \tag{13.11}$$

where the effective action,

$$e^{-S_E[\mathbf{M}]} = \int \mathcal{D}[\bar{c}, c] e^{-S[\bar{c},c,\mathbf{M}]}, \tag{13.12}$$

describes the action associated with a particular space–time configuration $\{\mathbf{M}_j(\tau)\}$ of the magnetization. Since the exponential $S[\bar{c}, c, \mathbf{M}]$ in (13.12) is a quadratic function of fermion fields, the integral is Gaussian and can be evaluated in closed form. To carry out the integral, it is convenient to Fourier transform the fields, writing $c_{j\sigma} = \frac{1}{\sqrt{N_s}} \sum_{\mathbf{k}} c_{\mathbf{k}\sigma} e^{i\mathbf{k}\cdot\mathbf{x}_j}$, so that

$$\sum_{j\sigma} \mathbf{M}_j \cdot \vec{\sigma}_j = \sum_{j\sigma} \mathbf{M}_j \cdot (\bar{c}_{j\alpha}\vec{\sigma}_{\alpha\beta}c_{j\beta}) = \sum_{\mathbf{k},\mathbf{k}',\sigma} \bar{c}_{\mathbf{k}'\alpha}(\mathbf{M}_{\mathbf{k}'-\mathbf{k}} \cdot \vec{\sigma}_{\alpha\beta})c_{\mathbf{k}\beta}, \tag{13.13}$$

where $\mathbf{M}_{\mathbf{q}} = \frac{1}{N_s} \sum_j \mathbf{M}_j e^{-i\mathbf{q}\cdot\mathbf{R}_j}$ is the Fourier transform of the magnetization. The effective action can be written in the compact form

$$e^{-S_E[\mathbf{M}]} = \int \mathcal{D}[\bar{c}, c] \exp\left[-\int_0^\beta d\tau \left(\bar{c}(\partial_\tau + \underline{h}_E[\mathbf{M}])c + \sum_j \frac{\mathbf{M}_j^2}{2I} \right) \right], \tag{13.14}$$

where

$$[\underline{h}_E]_{\mathbf{k}',\mathbf{k}} = \epsilon_{\mathbf{k}}\delta_{\mathbf{k},\mathbf{k}'} - \mathbf{M}_{\mathbf{k}'-\mathbf{k}}(\tau) \cdot \vec{\sigma} \tag{13.15}$$

describes the effective Hamiltonian for the electrons moving in the (time-dependent) magnetization field. Carrying out the Gaussian integral over $\bar{c}$ and c using (12.142) then gives

$$e^{-S_E[\mathbf{M}]} = \det\left[\partial_\tau + h_E[\mathbf{M}] \right] \exp\left[-\sum_j \int_0^\beta d\tau \frac{\mathbf{M}_j^2}{2I} \right] \tag{13.16}$$

or, more explicitly,

$$S_E[\mathbf{M}] = \overbrace{-\mathrm{Tr}\ln\left[(\partial_\tau + \epsilon_{\mathbf{k}})\delta_{\mathbf{k}',\mathbf{k}} - \mathbf{M}_{\mathbf{k}'-\mathbf{k}} \cdot \vec{\sigma}\right]}^{-\ln\mathrm{Det}[\partial_\tau + h_E]} + \sum_j \int_0^\beta d\tau \frac{\mathbf{M}_j^2}{2I}. \tag{13.17}$$

Note the following:

- In general, we can only evaluate S_E analytically for simple static configurations of $\mathbf{M}_j(\tau) = \mathbf{M}_j$. These provide the basis for mean-field theories.
- The factor $e^{-S_E[M]}$ in (13.16) resembles a Boltzmann distribution in classical statistical mechanics. However, in striking distinction from its classical counterpart, in certain non-uniform configurations of the magnetization the weight function $e^{-S_E[M]}$ acquires *negative* values. These configurations are in many ways the most interesting configurations of the path integral, and when they proliferate, standard Metropolis Monte Carlo approaches become exceedingly inaccurate. This is "the minus sign problem" of many-body physics – one of the major unsolved problems of numerical many body physics.
- The single-orbital Hubbard model can be readily generalized to multi-orbital models. In this situation, the interaction between channels is a Hund's interaction (see Example 13.8).

It is also useful to cast the effective action in terms of Feynman diagrams. To do this, we first rewrite the magnetization in terms of its Matsubara Fourier modes:

$$\mathbf{M}_q \equiv \mathbf{M}_{\mathbf{q}}(i\nu_n) = \frac{1}{\beta}\int_0^\beta d\tau\, \mathbf{M}_{\mathbf{q}}(\tau)e^{i\nu_n\tau} = \frac{1}{\beta N_s}\sum_j \int_0^\beta d\tau\, \mathbf{M}_j(\tau)e^{i(\nu_n\tau - \mathbf{q}\cdot\mathbf{R}_j)}. \tag{13.18}$$

In Fourier space, we replace $\partial_\tau \to -i\omega_n$ in the fermionic determinant of (13.17) to obtain

$$S_E[\mathbf{M}] = -\mathrm{Tr}\ln\left[(-i\omega_n + \epsilon_{\mathbf{k}})\delta_{k,k'} - \mathbf{M}_{k-k'} \cdot \vec{\sigma}\right] + N_s\beta\sum_q \frac{|\mathbf{M}_q|^2}{2I}. \tag{13.19}$$

We can factor out $(-i\omega + \epsilon_{\mathbf{k}})$ inside the logarithm, which permits us to split it into two terms:

$$S_E[\mathbf{M}] = -\mathrm{Tr}\ln\left[(-i\omega_n + \epsilon_{\mathbf{k}})(1 + (i\omega_n - \epsilon_{\mathbf{k}})^{-1}\mathbf{M}_{k-k'} \cdot \vec{\sigma})\right] + N_s\beta\sum_q \frac{|\mathbf{M}_q|^2}{2I}$$

$$= -\mathrm{Tr}\ln\left[(-i\omega_n + \epsilon_{\mathbf{k}})\right] - \overbrace{\mathrm{Tr}\ln\left[1 - G_0(k)V_{k,k'}\right]}^{\mathrm{Tr}\ln(1 - G_0 V)} + N_s\beta\sum_q \frac{|\mathbf{M}_q|^2}{2I}, \tag{13.20}$$

where

$$G_0(k) = (i\omega_n - \epsilon_{\mathbf{k}})^{-1}, \qquad V_{k,k'} = -\mathbf{M}_{k-k'} \cdot \vec{\sigma}. \tag{13.21}$$

Here we have used the identity $\mathrm{Tr}\,\ln(AB) = \mathrm{Tr}\,\ln A + \mathrm{Tr}\,\ln B$ to separate the terms inside the logarithm. The first term in (13.20) can be normalized with respect to the volume of space–time to give the free energy density of the non-interacting system,

$$\mathcal{F}_0 = \frac{S_0}{N_s\beta} = -\frac{1}{N_s\beta}\mathrm{Tr}\,\ln\left[(-i\omega_n + \epsilon_{\mathbf{k}})\right].$$

The second term in (13.20) is the change in the free energy of the fermions due to the magnetization field: the overbrace shows how we can rewrite it in terms of the bare propagator $G_0 = (i\omega_n - \epsilon_{\mathbf{k}})^{-1}$ and the scattering potential $V_{k',k} = -\mathbf{M}_{k'-k}\cdot\vec{\sigma}$. This term can be reinterpreted as an infinite sum of Feynman diagrams, describing repeated scattering off the exchange field:

$$\mathrm{Tr}\,\ln(1 - G_0 V) = \mathrm{Tr}\left[-G_0 V - \frac{1}{2}(G_0 V)^2 - \frac{1}{3}(G_0 V)^3 + \cdots\right]$$

$$(13.22)$$

where the prefactor $N_s\beta$ results from the integration over center-of-mass time and position. Here the scattering vertex

$$(13.23)$$

denotes scattering off the fluctuating Weiss field. The prefactor $N_s\beta$, the volume of space–time, is included because we are working in Fourier space, with the convention that all internal momentum and frequency sums are normalized with a measure $\frac{1}{N_s\beta}\sum_{\mathbf{k},i\omega_n}$. The effective free energy (per site) $\mathcal{F}_E[\mathbf{M}] = S_E/(N_s\beta)$ can then be written diagrammatically as

$$(13.24)$$

13.3 Saddle points and the mean-field theory of magnetism

To explore broken-symmetry solutions, we now make a saddle-point approximation, approximating the partition function by its value at the saddle point $\mathbf{M} = \mathbf{M}_0$:

$$Z = \int \mathcal{D}[M] e^{-S_E[\mathbf{M}]} \approx e^{-S_E[\mathbf{M}_0]}, \qquad (13.25)$$

where

$$\left. \frac{\delta S_E[\mathbf{M}]}{\delta \mathbf{M}} \right|_{\mathbf{M}=\mathbf{M}^{(0)}} = 0. \qquad (13.26)$$

Equations (13.25) and (13.26) contain the essence of mean-field theory and deserve some discussion. We discussed in Chapter 11 how a system develops a spontaneously broken symmetry when the Landau functional $F[M]$ develops a minimum at a non-zero value of the order parameter. A full-fledged calculation of this functional would involve calculating the full path integral $Z[h]$ with a symmetry-breaking field h in place, using a Legendre transformation to calculate $S[M] = S[h] - h\delta S/\delta h$, ultimately taking h to zero at the end of the calculation. The mean-field approach approximates $S[\mathbf{M}] \approx S_E[\mathbf{M}]$. Such saddle-point or mean-field solutions serve as the staging point to compute the fluctuations around the broken-symmetry state. The ultimate consistency of any mean-field approximation depends on the fluctuations being small enough that they do not wash out the broken-symmetry solution.

If we differentiate $S_E[\mathbf{M}]$ in (13.12), we see that the saddle-point condition (13.26) implies

$$\frac{\delta S_E}{\delta \mathbf{M}_j} = \frac{1}{e^{-S_E}} \int \mathcal{D}[\bar{c}, c] \overbrace{\left(\frac{\mathbf{M}_j}{I} - \bar{c}_j \vec{\sigma} c_j \right)}^{\frac{\delta S[\bar{c}, c, \mathbf{M}]}{\delta \mathbf{M}_j}} e^{-S[\bar{c}, c, \mathbf{M}]} = \frac{\mathbf{M}_j}{I} - \langle c_j^{\dagger} \vec{\sigma} c_j \rangle \Big|_{h_E}, \qquad (13.27)$$

where we have used (13.10) to calculate $\delta S[\bar{c}, c, \mathbf{M}]/\delta \mathbf{M}_j$. In this way the saddle-point condition (13.26) automatically satisfies the mean-field relation

$$\left. \frac{\delta S[\mathbf{M}]}{\delta \mathbf{M}_j} \right|_{\mathbf{M}=\mathbf{M}_0} = 0 \iff \mathbf{M}_j^{(0)} = I \langle c_j^{\dagger} \vec{\sigma} c_j \rangle \Big|_{h_E[\mathbf{M}^{(0)}]}.$$

$$\text{saddle-point condition} \qquad \text{mean-field theory} \qquad (13.28)$$

This makes life a lot easier: instead of laboring to impose the self-consistency condition on the right-hand side, we can simply generate mean-field solutions by minimizing the effective action. Generally, we're interested in a static saddle point, where $\mathbf{M}_j(\tau) = \mathbf{M}_j^{(0)}$. In this situation, the effective action is directly related to the mean-field partition function

$$e^{-S_E[\mathbf{M}^{(0)}]} = \text{Tr} \left[e^{-\beta \hat{H}_{MF}} \right], \qquad (13.29)$$

where

$$\hat{H}_{MF} = c^{\dagger} \underline{h}_E[\mathbf{M}^{(0)}] \, c + \sum_j \frac{(\mathbf{M}_j^{(0)})^2}{2I} \qquad (13.30)$$

is read off from the action in the path integral (13.14).

In a ferromagnet, the magnetization is uniform: for convenience we choose the spin polarization along the z-axis, writing

$$\mathbf{M}_j^{(0)} = M\hat{\mathbf{z}} , \qquad (13.31)$$

or, in Fourier space, $\mathbf{M_q} = M\delta_{\mathbf{q}}\,\hat{\mathbf{z}}$. In this case, the mean-field Hamiltonian is diagonal:

$$H_{MF} = \sum_{\mathbf{k}\sigma} c_{\mathbf{k}\sigma}^\dagger (\epsilon_{\mathbf{k}} - \sigma M) c_{\mathbf{k}\sigma} + N_s \frac{M^2}{2I} \qquad (13.32)$$

since $M_q = M\delta_{q0}$. We see that, when M is finite, the up and down Fermi surfaces are now exchange-split by an amount $\Delta = 2M$. By carrying out the Gaussian integral over the Fermi fields, or substituting into (13.19), we can immediately write down the effective action as

$$S_E[M] = -\sum_{\mathbf{k},i\omega_n} \mathrm{Tr}\ln\left[\epsilon_{\mathbf{k}} - M\sigma_z - i\omega_n\right] + N_s\beta \frac{M^2}{2I}. \qquad (13.33)$$

The result of carrying out the Matsubara sum on this expression gives the well known form

$$\mathcal{F}_E[M] = -\frac{1}{N_s\beta} \sum_{\mathbf{k},\sigma} \ln\left[1 + e^{-\beta(\epsilon_{\mathbf{k}} - \sigma M)}\right] + \frac{M^2}{2I}$$

$$= -\frac{T}{2} \int d\epsilon N(\epsilon) \sum_\sigma \ln\left[1 + e^{-\beta(\epsilon - \sigma M)}\right] + \frac{M^2}{2I}, \qquad (13.34)$$

where $\mathcal{F}_E = S_E/(\beta N_s)$ is the free energy per unit volume, and we have rewritten the momentum summation as an integral over the density of states per site $N(\epsilon)$.

To find the stationary point of the action, we differentiate it with respect to M to get

$$-\frac{\partial \mathcal{F}_E[M]}{\partial M} = 0 = \frac{M}{I} - \frac{1}{2}\sum_{\sigma=\pm 1}\overbrace{\int d\epsilon N(\epsilon)f(\epsilon - \sigma M)\sigma}^{\langle\sigma^z\rangle} \qquad (13.35)$$

or

$$M = \frac{I}{2}\sum_{\sigma=\pm 1}\int d\epsilon N(\epsilon)f(\epsilon - \sigma M)\sigma, \qquad (13.36)$$

which expresses the mean-field condition $M = I\langle\sigma^z\rangle$. We can obtain the second-order phase transition temperature T_c by letting $M \to 0^+$. Replacing $f(\epsilon - \sigma M) \to f(\epsilon) - \sigma M f'(\epsilon)$ gives

$$1 = I\overbrace{\int d\epsilon N(\epsilon)\left(-\frac{df}{d\epsilon}\right)\bigg|_{T=T_c}}^{\chi_0(T_c)} = I\chi_0(T_c), \qquad \text{Stoner criterion}$$

where we have identified the bracketed term as the spin susceptibility of the non-interacting gas at T_c. At a finite temperature the Stoner criterion defines the Curie temperature T_c of

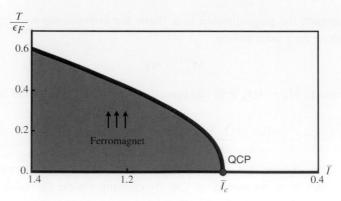

Fig. 13.1   Phase diagram for the 3D Stoner model, computed using using (13.39) and (13.40). The horizontal axis is the coupling constant $\bar{I} = IN(0)$. The transition temperature goes to zero at the quantum critical point (QCP), located at the critical coupling constant $\bar{I}_c = 1$.

the electron gas. In the ground state at absolute zero, we can replace the derivative of the Fermi function by a delta function, $-df/d\epsilon \rightarrow \delta(\epsilon)$, so the Stoner criterion becomes

$$I_c N(0) = 1, \qquad \text{Stoner criterion for } T = 0$$

where $I = I_c$ is the critical value of the interaction I, beyond which the paramagnetic *ground state* becomes unstable to magnetism, as shown in Figure (13.1). This is a *quantum phase transition*, driven not by thermal but by quantum fluctuations.

Example 13.1   Map out the critical temperature $T_c(I)$ as a function of interaction strength I for the 3D continuum Stoner model where the density of states $N(\epsilon) = N(0)\sqrt{\frac{\epsilon + \mu}{\epsilon_F}}$, where ϵ_F is the Fermi temperature and $N(0)$ the density of states at the Fermi surface.

Solution

In three dimensions, the Stoner criterion can be written

$$1 = IN(0) \int_0^\infty dE \sqrt{\frac{E}{\epsilon_F}} \frac{f(E - \mu)[1 - f(E - \mu)]}{T_c}$$

$$= I\sqrt{\frac{T_c}{2\epsilon_F}} \int_0^\infty dx \sqrt{x}\, \text{sech}^2[x - \mu\beta_c/2]. \qquad (13.37)$$

If we were interested in the problem at constant chemical potential, we could stop here. However, if we wish to take account of the drift of the chemical potential at finite temperature, we need to impose the condition of constant particle density n_0,

$$n_0 = N(0) \int_0^\infty dE \sqrt{\frac{E}{\epsilon_F}} f(E - \mu)$$

$$= N(0)\epsilon_F \left(\frac{T_c}{\epsilon_F}\right)^{\frac{3}{2}} \int_0^\infty dx \sqrt{x} \frac{1}{e^{x - \mu\beta_c} + 1}. \qquad (13.38)$$

At zero temperature, this gives $n_0 = \frac{2}{3}N(0)\epsilon_F$, so that

$$\frac{2}{3}N(0)\epsilon_F = N(0)\epsilon_F \left(\frac{T_c}{\epsilon_F}\right)^{\frac{3}{2}} \int_0^\infty dx\sqrt{x}\frac{1}{e^{x-\mu\beta_c}+1},$$

enabling us to write T_c as a parametric function of $y = \mu\beta_c$:

$$T_c(y) = \epsilon_F \left[\frac{3}{2}\int_0^\infty dx\sqrt{x}\frac{1}{e^{x-y}+1}\right]^{-2/3}. \tag{13.39}$$

Inserting (13.39) into (13.37), we can also write $\bar{I} = IN(0)$ as a parametric function of $y = \mu\beta_c$:

$$\bar{I}(y) = \frac{\left[\frac{3}{2}\int_0^\infty dx\sqrt{x}\frac{1}{e^{x-y}+1}\right]^{1/3}}{\frac{1}{\sqrt{2}}\int_0^\infty dx\sqrt{x}\,\text{sech}^2[x-y/2]}. \tag{13.40}$$

Figure 13.1 shows the phase diagram computed using (13.39) and (13.40).

To finish this section, let us calculate the Landau expansion of the free energy. If we make a binomial expansion of the logarithm in $S_E[M]$ in powers of M, we obtain

$$-\frac{T}{2}\sum_\sigma \ln[1+e^{-\beta(\epsilon-\sigma M)}] = -T\ln[1+e^{-\beta\epsilon}] + \sum_{r=1}^\infty \frac{M^{2r}}{(2r)!}\frac{d^{2r-1}f(\epsilon)}{d\epsilon^{2r-1}}, \tag{13.41}$$

where odd powers of M vanish and $f(\epsilon)$ is the Fermi function. Thus

$$\mathcal{F}[M] = \mathcal{F}_0 + \sum_{r=1}^\infty \frac{M^{2r}}{(2r)!}\int d\epsilon N(\epsilon)\frac{d^{2r-1}f(\epsilon)}{d\epsilon^{2r-1}} + \frac{M^2}{2I}. \tag{13.42}$$

If we integrate in (13.42) by parts, we obtain

$$\mathcal{F}[M] = \mathcal{F}_0 - \sum_r \frac{M^{2r}}{(2r)!}\overbrace{\int d\epsilon\left(-\frac{df}{d\epsilon}\right)N^{(2r-2)}(\epsilon)}^{\overline{N^{(2r-2)}(0)}} + \frac{M^2}{2I}, \tag{13.43}$$

where $N^{(r)} = d^rN(\epsilon)/d\epsilon^r$ is the rth derivative of the density of states and $\overline{N^{(r)}(0)}$ is its corresponding thermal average around the Fermi surface. If we take terms up to M^4, we obtain

$$\mathcal{F} = \mathcal{F}_0 + \frac{1}{2}M^2\left(\frac{1}{I} - \chi_0(T)\right) + \frac{M^4}{4!}\overline{(-N''(0))} + O(M^6), \tag{13.44}$$

where $\overline{(-N''(0))}$ denotes the thermal average of the second derivative of the density of states around the Fermi energy. This is the Landau energy function predicted by the Stoner theory of itinerant ferromagnet.

Note the following:

- The quartic coefficient in the free energy is positive only if $\overline{N''(0)} < 0$ is negative, i.e. if the density of states has a downward curvature. If this requirement is not met, the ferromagnetic phase transition becomes first-order. Most transition metal ferromagnets, such as iron and cobalt, involve narrow bands in three dimensions with a large negative curvature of the density of states, and the transition is second-order. However, in quasi-two-dimensional systems where the density of states has mostly positive curvature, the ferromagnetic phase transition is expected to be first-order.

- The mean-field parameters in the above action are likely to be modified by fluctuations. In our mean-field theory, an isotropic decoupling gave $I = U/3$, but had we chosen an Ising decoupling just in the z direction, we would have obtained $I = U$, which is most likely an overestimate of I. Mean-field theories cannot in general give a very reliable indication of the absolute size of such parameters.

- There is a formal *large-N limit* in which the above mean-field theory does become exact (see Example 13.2). We may take the limit $N \to \infty$ in the family of multi-band (N-band) models described by the action

$$S = \int_0^\beta d\tau \left[\sum_{\mathbf{k}, \lambda \sigma} \bar{c}_{\mathbf{k}\lambda\sigma} (\partial_\tau + \epsilon_\mathbf{k}) c_{\mathbf{k}\lambda\sigma} - \frac{I}{2N} \sum_j \left(\sum_\lambda \sigma_\lambda(j) \right)^2 \right], \qquad (13.45)$$

where the band index $\lambda \in [1, N]$. Here the interaction I can be regarded as a Hund's interaction between the different bands. For large-N, the action of this model grows extensively with N, and in this situation the path integral becomes saturated by the saddle-point solution, so the mean-field theory becomes exact.

Example 13.2 Consider a multi-orbital Hubbard model involving N orbitals interacting purely via a Hund's interaction, described by

$$H = \sum_{\mathbf{k}, \lambda, \sigma} \epsilon_\mathbf{k} c^\dagger_{\mathbf{k}\lambda\sigma} c_{\mathbf{k}\lambda\sigma} - \frac{I}{2N} \sum_j \left(\sum_\lambda \sigma_\lambda(j) \right)^2, \qquad (13.46)$$

where $\lambda \in [1, N]$ is the orbital index and we use the shorthand notation

$$\sigma_\lambda(j) = c^\dagger_{j\lambda\alpha} \sigma_{\alpha\beta} c_{j\lambda\beta} \qquad (13.47)$$

to denote the spin density at site j.

(a) Formulate the partition function as a path integral.
(b) Show that the effective action grows extensively with the number of channels N.
(c) What do you expect to happen in this model as the number of orbital channels becomes large?

Solution

(a) Inside a path integral, we can carry out a Hubbard–Stratonovich transformation on the Hund's interaction, replacing

$$-\frac{I}{2N} (\vec{\sigma}_j)^2 \to \mathbf{M}_j \cdot \sigma_j + N \frac{\mathbf{M}_j \cdot \mathbf{M}_j}{2I}$$

$$= \sum_\lambda \left(\mathbf{M}_j \cdot \sigma_{j\lambda} + \frac{M_j^2}{2I} \right), \qquad (13.48)$$

where we have denoted $\sigma_{j\lambda} = c^\dagger_{j\lambda\alpha}\vec\sigma_{\alpha\beta}c_{j\lambda\beta}$ for the spin density at site j in orbital channel λ. Notice that, in the factorized interaction, each orbital at site j interacts with a common Weiss field M_j, so we are able to write the interaction as a sum over N orbital channels.

We may now write the partition function as

$$Z = \int \mathcal{D}[\mathbf{M}]e^{-NS_E[\mathbf{M}]}, \tag{13.49}$$

where

$$e^{-NS_E[\mathbf{M}]} = \int \mathcal{D}[\bar{c}, c] \exp\left[-\sum_\lambda \int_0^\beta d\tau \left(\bar{c}_\lambda(\partial_\tau + \underline{h}_E[\mathbf{M}])c_\lambda + \sum_j \frac{\mathbf{M}_j^2}{2I} \right) \right]. \tag{13.50}$$

Here we have used the same shorthand notation for the channel-diagonal one-particle Hamiltonian,

$$[\underline{h}_E]_{\mathbf{k}',\mathbf{k}} = \epsilon_\mathbf{k}\delta_{\mathbf{k},\mathbf{k}'} - \mathbf{M}_{\mathbf{k}'-\mathbf{k}}(\tau) \cdot \vec\sigma, \tag{13.51}$$

that was used in the single-channel case. On the right-hand side of (13.50) the factor of N before the $\sum_j \frac{\mathbf{M}_j^2}{2I}$ term has been absorbed into the sum over channels λ. Also, in anticipation that the action grows with N, we have written it as NS_E. To prove this is the case, we need to show that S_E is independent of N.

(b) We can write the expression for the action as the product of N identical replicas of a single orbital problem, since

$$e^{-NS_E[\mathbf{M}]} = \prod_\lambda \int \mathcal{D}[\bar{c}_\lambda, c_\lambda] \exp\left[-\int_0^\beta d\tau \left(\bar{c}_\lambda(\partial_\tau + \underline{h}_E[\mathbf{M}])c_\lambda + \sum_j \frac{\mathbf{M}_j^2}{2I} \right) \right]$$
$$= Z_\lambda[\mathbf{M}]^N, \tag{13.52}$$

where $Z_\lambda = e^{-S_E[\mathbf{M}]}$ is the partition function for a single channel. By carrying out the formal path integral, the action for a single channel is

$$S_E[\mathbf{M}] = -\mathrm{Tr}\ln\left[\partial_\tau + h_E[\mathbf{M}]\right] + \sum_j \int_0^\beta d\tau \frac{M_j^2(\tau)}{2I}, \tag{13.53}$$

which is independent of the channel index λ and the number of channels. When we sum over the N orbitals, the total action NS_E is then extensive in N.

(c) We can rewrite the partition function in the suggestive form

$$Z = \int \mathcal{D}[M] \exp\left[-\frac{S_E}{\hbar_E} \right], \tag{13.54}$$

where

$$\hbar_E = \frac{1}{N} \tag{13.55}$$

is a kind of "synthetic Planck's constant." From this we see that, as $N \to \infty$, the fluctuations in $\mathbf{M}_j$ will scale as

$$\langle \delta \mathbf{M}_j^2 \rangle \sim O\left(\frac{1}{N}\right), \tag{13.56}$$

so that the magnetization will behave as a *classical* variable. In this way, the large-N limit of this model will be well described by the mean-field theory.

Example 13.3 (a) Show from the Landau energy (13.44) that, near the quantum critical point at $T = 0, I = I_c = 1/N(0)$, the magnetic moment is given by

$$M = \sqrt{\left(\frac{I - I_c}{II_c}\right)\frac{6}{-N''(0)}} \sim \sqrt{I - I_c}. \tag{13.57}$$

(b) By expanding the density of states in a power series about the Fermi energy, show that the transition temperature predicted by (13.44) is

$$T_c = \sqrt{\frac{6}{\pi^2}\left(\frac{1}{I} - N(0)\right)\frac{1}{(-N''(0))}}.$$

Solution

(a) We begin by writing the Landau free energy as

$$\mathcal{F} = \frac{rM^2}{2} + \frac{uM^4}{4},$$

where $r = I^{-1} - \overline{N(0)}$, $u = \overline{-N''(0)}/6$. At zero temperature,

$$r = \left(\frac{1}{I} - \frac{1}{I_c}\right), \qquad u = \frac{-N''(0)}{6},$$

where $I_c = 1/N(0)$. Setting $\partial F/\partial M^2 = 0$, we obtain $rM + uM^3 = 0$, or

$$M = \sqrt{\frac{r}{u}} = \sqrt{\left(\frac{I - I_c}{II_c}\right)\frac{6}{-N''(0)}} \sim \sqrt{I - I_c}.$$

(b) Carrying out a Taylor expansion of the density of states,

$$\overline{N(0)} = \int d\epsilon \left(-\frac{df}{d\epsilon}\right)\left[N(0) + \epsilon N'(0) + \frac{\epsilon^2}{2}N''(0)\right] = N(0) + \frac{\pi^2 T^2}{6}N''(0),$$

it follows that, at a small finite temperature,

$$r(T) = \left(\frac{1}{I} - \frac{1}{I_c} + \frac{\pi^2 T^2}{6}(-N''(0))\right).$$

Setting $r(T_c) = 0$, it follows that

$$T_c = \sqrt{\frac{6}{\pi^2}\left(\frac{1}{I} - N(0)\right)\frac{1}{(-N''(0))}}.$$

13.4 Quantum fluctuations in the magnetization

The beauty of the saddle-point approach is that it allows one to go beyond the mean-field theory to examine the fluctuations in the order parameter. The basic idea is to expand the magnetization in fluctuations around the saddle point (Figure 13.2), writing

$$\mathbf{M}_j(\tau) = \mathbf{M}^{(0)} + \delta\mathbf{M}_j(\tau) \tag{13.58}$$

or, in Fourier space,

$$\mathbf{M}_q = \mathbf{M}^{(0)}\delta_{q=0} + \delta\mathbf{M}_q \qquad (q \equiv (\mathbf{q}, i\nu_n)). \tag{13.59}$$

Because the effective action is stationary with respect to variations in $\mathbf{M}$ at the saddle point, the leading-order corrections to the effective action are quadratic in the fluctuations:

$$S_E[\mathbf{M}] = S_E[\mathbf{M}^{(0)}] + \frac{1}{2}\sum_q \frac{\delta^2 S}{\delta M_q^a \delta M_{-q}^b}\delta M_q^a \delta M_{-q}^b + O(\delta\mathbf{M}^3).$$

Notice that all linear terms in the fluctuations vanish by virtue of the fact that the mean-field action is stationary with respect to fluctuations. Provided the fluctuations are small compared to the order parameter, one can use the quadratic approximation to the effective action to examine the leading fluctuations of the magenization in the ferromagnetic state.

In a magnet these fluctuations take place against a *broken-symmetry* background. The electrons scattering off the fluctuations are partially spin-polarized and governed by the renormalized propagator, denoted by the double line:

$$\Longrightarrow_{k} = \underline{G}(k) = (i\omega_n - \epsilon_\mathbf{k} - \sigma_z M)^{-1},$$

where we have underlined $\underline{G}(k)$ to emphasize that it is a two-dimensional, albeit diagonal, matrix.

Let us now expand the effective action $S_E[\mathbf{M}]$ in (13.19) in the fluctuations by substituting $M_{k-k'} = M\delta_{k-k'} + \delta M_{k-k'}$ to obtain

$$\mathcal{F}_E[\mathbf{M}] = -\frac{1}{N_s\beta}\mathrm{Tr}\ln\left[-\underline{G}(k)^{-1}\delta_{k,k'} - \delta\mathbf{M}_{k-k'}\cdot\vec{\sigma}\right] + \sum_q \frac{|M\hat{\mathbf{z}}\delta_q + \delta\mathbf{M}_q|^2}{2I}. \tag{13.60}$$

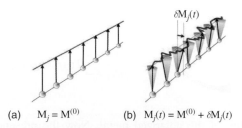

$\delta M_j(t)$

(a) $M_j = M^{(0)}$ (b) $M_j(t) = M^{(0)} + \delta M_j(t)$

Illustrating (a) mean-field theory (b) fluctuations about mean-field theory.

Fig. 13.2

If we now expand this expression in powers of $\delta \mathbf{M}_q$, we get a Feynman diagram expansion in terms of the renormalized propagators, as follows

$$\mathcal{F}_E = -\frac{1}{N_s \beta} \mathrm{Tr} \ln[-\underline{G}(k)^{-1}] - \left[\text{◯} + \text{◯} + \text{▷} + \text{◯} + \cdots \right]$$
$$+ \sum_q \frac{|M\hat{\mathbf{z}}\delta_q + \delta \mathbf{M}_q|^2}{2I}, \tag{13.61}$$

where the wavy line denotes scattering off the order-parameter fluctuations. The $N_s \beta$ term in the denominator of the trace is absorbed into the momentum/frequency loop summation $\frac{1}{N_s \beta} \sum_q$ that is implicit inside the Feynman diagrams. Now since the action is stationary with respect to fluctuations, all terms linear in $\delta \mathbf{M}_q$ must cancel, which leads to

$$\Delta \mathcal{F}_E[\mathbf{M}] = -\left[\text{◯} + \text{▷} + \text{◯} + \cdots \right] + \sum_q \frac{|\delta \mathbf{M}_q|^2}{2I},$$

$$\tag{13.62}$$

where $\Delta \mathcal{F}_E[\mathbf{M}] = \mathcal{F}_E[\mathbf{M}] - \mathcal{F}_E[\mathbf{M}^{(0)}]$. Only the first diagram and the final term in this expression are quadratic in $\delta \mathbf{M}_q$. Combining them and dropping the higher-order terms, we obtain the *Gaussian action* for the magnetization fluctuations:

$$\Delta \mathcal{F}_G[\mathbf{M}] = \frac{1}{2} \sum_q \delta M_{-q}^a \left[\frac{\delta_{ab}}{I} - \sigma^a \underset{k}{\overset{k+q}{\bigcirc}} \sigma^b \right] \delta M_q^b$$

$$= \frac{1}{2} \sum_q \delta M_{-q}^a \left[\frac{\delta_{ab}}{I} - \chi_{ab}^{(0)}(q) \right] \delta M_{-q}^b, \tag{13.63}$$

Gaussian action of fluctuations

where

$$\chi_{ab}^{(0)}(q) = \sigma^a \underset{k}{\overset{k+q}{\bigcirc}} \sigma_b = -\frac{1}{\beta N_s} \sum_k \mathrm{Tr}\left[\sigma_a G(k+q) \sigma_b G(k) \right] \tag{13.64}$$

is the bare susceptibility of the polarized metal. Now the presence of a magnetization means that the off-diagonal terms $\chi_{xy}^{(0)}(q) = \chi_{yx}^{(0)}(-q)$ are non-zero. To diagonalize the magnetic fluctuations, it is convenient to work in terms of the raising and lowering components of the transverse spin, $\sigma_\pm = \frac{1}{2}(\sigma_x \pm i\sigma_y)$, and the corresponding components of the magnetization,

$$M_q^\pm = M_q^x \pm iM_q^y.$$

The non-zero components of the transverse susceptibility are then

$$\chi_{+-}^{(0)}(q) = -\frac{1}{\beta N_s} \sum_k \text{Tr}\left[\sigma_+ G(k+q)\sigma_- G(k)\right]$$

$$\chi_{-+}^{(0)}(q) = -\frac{1}{\beta N_s} \sum_k \text{Tr}\left[\sigma_- G(k+q)\sigma_+ G(k)\right] = \chi_{+-}^{(0)}(-q), \qquad (13.65)$$

where the identity $\chi_{-+}^{(0)}(q) = \chi_{+-}^{(0)}(-q)$ follows by changing variables $k \to k - q$ inside the sum.

Rewriting $\mathbf{M} \cdot \boldsymbol{\sigma} = M_q^z \sigma_z + M^+ \sigma_- + M^- \sigma_+$, the Gaussian effective action then becomes

$$\Delta \mathcal{F}_G[\mathbf{M}] = \frac{1}{2} \sum_q \left[\delta M_{-q}^z \left(\frac{1}{I} - \chi_{zz}^{(0)}(q)\right) \delta M_q^z + \delta M_{-q}^- \left(\frac{1}{2I} - \chi_{+-}^{(0)}(q)\right) \delta M_q^+ \right.$$

$$\left. + \delta M_{-q}^+ \left(\frac{1}{2I} - \chi_{-+}^{(0)}(q)\right) \delta M_q^- \right]. \qquad (13.66)$$

Now since the magnetization is a real variable, it follows that $\overline{\delta M^\pm}_q = \delta M^\mp_{-q}$ (where we use a bar to denote the complex conjugate), so we can rewrite this expression in the form

$$\Delta \mathcal{F}_G[\mathbf{M}] = \frac{1}{2} \sum_q \left[\delta M_{-q}^z \left(\frac{1}{I} - \chi_{zz}^{(0)}(q)\right) \delta M_q^z + \overline{\delta M^+}_q \left(\frac{1}{2I} - \chi_{+-}^{(0)}(q)\right) \delta M_q^+ \right.$$

$$\left. + \delta M_{-q}^+ \left(\frac{1}{2I} - \chi_{-+}^{(0)}(q)\right) \overline{\delta M^+}_{-q} \right]. \qquad (13.67)$$

It is this quadratic functional that provides the argument for the Gaussian distribution function of the magnetic fluctuations $p[M_q] = Z^{-1} e^{-\Delta S[\mathbf{M}]} = e^{-\beta N_s \Delta \mathcal{F}_G[\mathbf{M}]}$. Now by (13.65), $\chi_{-+}^{(0)}(q) = \chi_{+-}^{(0)}(q)$, so we can combine the last terms into one. The final results, describing the distribution function for the Gaussian magnetic fluctuations about the Stoner mean-field theory for an itinerant ferromagnet, are

$$p[M_q] \propto e^{-\Delta S[\mathbf{M}]} = e^{-\beta N_s \Delta \mathcal{F}_G[\mathbf{M}]} \qquad (13.68)$$

$$\Delta \mathcal{F}_G[\mathbf{M}] = \sum_q \left[\frac{1}{2} \delta M_{-q}^z \left(\frac{1}{I} - \chi_{zz}^{(0)}(q)\right) \delta M_q^z + \overline{\delta M^+}_q \left(\frac{1}{2I} - \chi_{+-}^{(0)}(q)\right) \delta M_q^+ \right].$$

$$(13.69)$$

From the Gaussian form of this distribution, we can immediately read off the fluctuations in magnetization. Denoting

$$\langle \delta M_q^\alpha \delta M_{-q'}^\beta \rangle = \frac{1}{\beta N_s} \delta_{q,q'} \times \langle \delta M_q^\alpha \delta M_{-q}^\beta \rangle, \tag{13.70}$$

the fluctuations in magnetization are given by

$$\langle \delta M_q^z \delta M_{-q}^z \rangle = \frac{1}{\frac{1}{I} - \chi_{zz}^{(0)}(q)}$$

$$\langle \delta M_q^+ \delta M^+{}_{-q} \rangle = \frac{1}{\frac{1}{2I} - \chi_{+-}^{(0)}(q)}. \tag{13.71}$$

Let us now convert these results into spin correlation functions. If we go back to the original Hubbard–Stratonovich transformation (13.9), we recall that, to decouple the interaction, we had to introduce a dummy white-noise variable; let us call it $\mathbf{m}_j(\tau)$, with distribution function $\langle m_j^a(\tau) m_j^b(\tau') \rangle = I \delta^{ab} \delta(\tau - \tau')$, or $\langle m_q^a m_{-q}^b \rangle = I \delta^{ab}$. To carry out the Hubbard–Stratonovich transformation, we redefined this variable, writing $\mathbf{m}_j(\tau) \to \mathbf{M}_j - I \vec{\sigma}_j$. It follows that the variable we are working with is related to the original white-noise variable by $\mathbf{M}_j(\tau) = \mathbf{m}_j(\tau) + I \vec{\sigma}_j(\tau)$. Consequently, the Gaussian fluctuations in the magnetization are given by

$$\langle \sigma_q^a \sigma_{-q}^b \rangle = \frac{1}{I^2} \left[\langle \delta M_q^a \delta M_{-q}^b \rangle - \overbrace{\langle \delta m_q^a \delta m_{-q}^b \rangle}^{I \delta^{ab}} \right].$$

It follows that

$$\langle \sigma_q^z \sigma_{-q}^z \rangle = \chi_{zz}(q) = \frac{1}{I^2} \left[\frac{1}{\frac{1}{I} - \chi_{zz}^{(0)}(q)} - I \right] = \frac{\chi_{zz}^{(0)}(q)}{1 - I \chi_{zz}^{(0)}(q)} \qquad \text{longitudinal}$$

$$\langle \sigma_q^+ \sigma_{-q}^- \rangle = \chi_{+-}(q) = \frac{1}{I^2} \left[\frac{1}{\frac{1}{2I} - \chi_{+-}^{(0)}(q)} - 2I \right] = \frac{\chi_{+-}^{(0)}(q)}{1 - 2I \chi_{+-}^{(0)}(q)}. \qquad \text{transverse} \qquad (13.72)$$

RPA spin fluctuations

These are the celebrated *RPA spin fluctuations* of an itinerant ferromagnet.

It is particularly interesting to examine the transverse spin fluctuations in (13.72). A uniform transverse spin fluctuation corresponds to a rotation of the magnetization, which costs no energy due to the rotational invariance of the system. If we carry out a slow twist of the magnetization, this costs an energy that goes to zero as the pitch of the twist goes to infinity. The corresponding normal mode is the *Goldstone mode* of the magnet.

One can analytically calculate the transverse spin fluctuations of a ferromagnet with a quadratic dispersion $\epsilon_{\mathbf{k}} = \frac{k^2}{2m} - \mu$, because the bare susceptibilities $\chi^{(0)}(q)$ can be calculated as Lindhard functions. The transverse bare susceptibility (per unit cell) is given by

$$\chi^{(0)}_{+-}(q) = -\frac{1}{N_s\beta} \sum_{\mathbf{k},i\omega_n} [\sigma_+ G(k+q)\sigma_- G(k)] = \sigma^+ \quad \overset{k+q\downarrow}{\underset{k\uparrow}{\bigcirc}} \quad \sigma^-$$

$$= -\frac{1}{N_s\beta} \sum_{\mathbf{k},i\omega_n} \left[G_\downarrow(k+q)G_\uparrow(k) \right]$$

$$= a^3 \int_{\mathbf{k}} \frac{f_{\mathbf{k}\uparrow} - f_{\mathbf{k}+\mathbf{q}\downarrow}}{(\epsilon_{\mathbf{k}+\mathbf{q}\downarrow} - \epsilon_{\mathbf{k}\uparrow}) - i\nu_n} , \tag{13.73}$$

where $\epsilon_{\mathbf{k}\sigma} = \epsilon_{\mathbf{k}} - \sigma M$ ($\sigma = \uparrow, \downarrow$). This sort of expression is a type of Lindhard function already encountered in Chapter 7. Following the same lines as Section 7.62, we analytically continue to real frequencies, and rewrite the integrals as follows:

$$\chi^{(0)}_{+-}(\mathbf{q}, \nu) = a^3 \int_{\mathbf{k}} \left(\frac{f_{\mathbf{k}\uparrow}}{(\epsilon_{\mathbf{k}+\mathbf{q}} - \epsilon_{\mathbf{k}}) - (\nu - 2M)} + \frac{f_{\mathbf{k}\downarrow}}{(\epsilon_{\mathbf{k}-\mathbf{q}} - \epsilon_{\mathbf{k}}) + (\nu - 2M)} \right)$$

$$= \sum_{\sigma=\pm} a^3 \int_0^{k_{F\sigma}} \frac{k^2 dk}{2\pi^2} \int \frac{d\cos\theta}{2} \left[\frac{1}{(\epsilon_{\mathbf{k}+\mathbf{q}} - \epsilon_{\mathbf{k}}) - \sigma(\nu - 2M)} \right]$$

$$= \frac{1}{2} \sum_{\sigma} \left(\frac{mk_{F\sigma}}{\pi^2} \right) \mathcal{F} \left(\frac{q}{2k_{F\sigma}}, \sigma \frac{\nu - 2M}{4\epsilon_F} \right), \tag{13.74}$$

where $\epsilon_{F\sigma} = \epsilon_F + \sigma M$ and $k_{F\sigma} = k_F(1 + \frac{M}{\epsilon_F})^{\frac{1}{2}}$ are the Fermi energy and momenta of the spin $\sigma = (\uparrow, \downarrow)$ Fermi surfaces, and

$$\mathcal{F}[\tilde{q}, \tilde{\nu}] = \frac{1}{8\tilde{q}} \left[(1 - A^2) \ln \left(\frac{A+1}{A-1} \right) + 2A \right], A = \tilde{q} - \frac{\tilde{\nu}}{\tilde{q}} \tag{13.75}$$

is the Lindhard function.

Figure 13.3 shows a density plot of the transverse dynamical spin susceptibility $\chi''_{+-}(\mathbf{q}, \nu) = \operatorname{Im} \chi_{+-}(\mathbf{q}, \nu - i\delta)$ predicted by the Gaussian (RPA) theory. The spectrum of magnetic fluctuations about the mean-field theory is determined by the energies at which one can excite a particle–hole pair by flipping a spin. Unlike a non-magnetic metal, the energy to flip a spin at $\mathbf{q} = 0$ is twice the Weiss field, $\epsilon_{\mathbf{k}\downarrow} - \epsilon_{\mathbf{k}\uparrow} = 2M$. The continuum of spin-flip particle–hole excitations is thus lifted up at low momenta, forming what is known as the *Stoner continuum*. The threshold energy for a spin-flip excitation finally drops to zero at the wavevector $q = k_{F\uparrow} - k_{F\downarrow}$. Below the Stoner continuum is a sharp Goldstone mode, labeled by the dotted line in Figure 13.3, corresponding to a low-energy pole in the dynamical susceptibility, located at frequencies $\omega_{\mathbf{q}}$ determined by the condition

$$2I\chi_{+-}(\mathbf{q}, \omega_{\mathbf{q}}) = 1.$$

A careful evaluation of this condition shows that

$$\omega_{\mathbf{q}} = Z(M/\epsilon_F)\frac{q^2}{2m}, \tag{13.76}$$

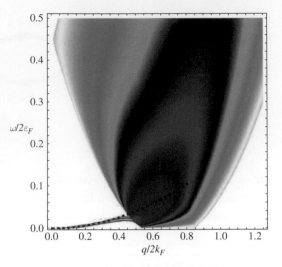

Fig. 13.3 The energy spectrum of quantum magnetic fluctuations in an itinerant ferromagnet. This spectrum was computed for a magnetization $M = 0.9\epsilon_F$, corresponding to an almost fully polarized Fermi sea.

where

$$Z(x) = \frac{4}{5x} \left[\frac{(1+x)^{5/2} - (1-x)^{5/2} - 5}{(1+x)^{3/2} - (1-x)^{3/2}} \right]. \tag{13.77}$$

This is the relation used to determine the dotted curve in Figure 13.3.

Exercises

Exercise 13.1 Mean-field theory for antiferromagnetic spin density wave.

Develop the mean-field theory for a three dimensional tight-binding cubic lattice with commensurate antiferromagnetic order parameter

$$\mathbf{M}_j = \mathbf{M}e^{i\mathbf{Q}\cdot\mathbf{R}_j}, \tag{13.78}$$

where $\mathbf{Q} = (\pi, \pi, \pi)$.

(a) Show that the mean-field free energy can be written in the form

$$H_{MF} = \sum_{\mathbf{k}\in\frac{1}{2}BZ} \psi_{\mathbf{k}}^{\dagger} \begin{pmatrix} \epsilon_{\mathbf{k}} - \mu & \mathbf{M}\cdot\boldsymbol{\sigma} \\ \mathbf{M}\cdot\boldsymbol{\sigma} & \epsilon_{\mathbf{k}+\mathbf{Q}} - \mu \end{pmatrix} \psi_{\mathbf{k}} + \mathcal{N}_s \frac{M^2}{2I}, \tag{13.79}$$

where $M = |\mathbf{M}|$ is the magnitude of the staggered magnetization, $\psi_{\mathbf{k}}$ denotes the four-component spinor

$$\psi_{\mathbf{k}} = \begin{pmatrix} c_{\mathbf{k}\uparrow} \\ c_{\mathbf{k}\downarrow} \\ c_{\mathbf{k}+\mathbf{Q}\uparrow} \\ c_{\mathbf{k}+\mathbf{Q}\downarrow} \end{pmatrix}, \tag{13.80}$$

$\epsilon_{\mathbf{k}} = -2t(c_x + c_y + c_z)$, $(c_l \equiv \cos k_l, \ l = x, y, z)$ is the kinetic part of the energy, and the summation is restricted to the magnetic Brillouin zone (one-half the original Brillouin zone).

(b) On a tight-binding lattice the kinetic energy has the nesting property that $\epsilon_{\mathbf{k}+\mathbf{Q}} = -\epsilon_{\mathbf{k}}$. Show that the energy eigenvalues of the mean-field Hamiltonian have a BCS form,

$$E_{\mathbf{k}\pm} = \pm\sqrt{\epsilon_{\mathbf{k}}^2 + M^2} - \mu, \tag{13.81}$$

corresponding to an excitation spectrum with gap M. Notice that the gap is offset by an amount μ.

(c) Show that the mean-field free energy takes the form

$$F = \sum_{\mathbf{k},p=\pm 1} -T \ln\left[2\cosh\left(\frac{\beta E_{\mathbf{k}p}}{2}\right)\right] + N_s\left(\frac{M^2}{2I} - 2\mu\right). \tag{13.82}$$

(d) By minimizing the free energy with respect to M, show that the gap equation for M is given by

$$\frac{1}{2}\sum_{\mathbf{k},p=\pm 1}\tanh\left(\frac{\sqrt{\epsilon_{\mathbf{k}}^2 + M^2} - \mu p}{2T}\right)\frac{1}{\sqrt{\epsilon_{\mathbf{k}}^2 + M^2}} = \frac{1}{I}. \tag{13.83}$$

(e) Show that, at half-filling, the nesting guarantees that a transition to a spin-density wave will occur for arbitrarily small interaction strength I. What do you think will happen at a finite doping ($\mu \neq 0$)?

(f) Calculate the phase diagram, assuming that the order remains commensurate at finite doping.

Exercise 13.2 Generalize the calculation of the Landau energy for a paramagnetic metal near a ferromagnetic quantum critical point (i.e. $IN(0) = 1 - \epsilon$) carried out in Example 13.3, to include dynamical fluctuations.

(a) Show, using the Lindhard function for the dynamical spin susceptibility, that the leading terms in the action can be written in the form

$$S = \frac{1}{2}\sum_{\mathbf{q},i\nu_n}\mathbf{M}(-\mathbf{q}, -i\nu_n)\left[r + (cq)^2 + a\frac{|\nu_n|}{q}\right]\mathbf{M}(\mathbf{q}, i\nu_n) + \frac{u}{4}\int d^4x M^4(x). \tag{13.84}$$

Explicitly calculate a and c.

(b) Why is there no static cubic term in the action?

(c) Show that the characteristic decay rate of a magnetic fluctuation is given by

$$\Gamma(\mathbf{q}) = \frac{|\mathbf{q}|}{a}(r + (cq)^2). \tag{13.85}$$

Why does the decay rate vanish as $\mathbf{q} \to 0$?

Exercise 13.3 Using the leading Gaussian fluctuations about the mean-field theory, calculate the Landau parameters for a paramagnetic metal on the verge of ferromagnetic instability.

References

[1] J. H. Van Vleck, *Quantum Mechanics: The Key to Understanding Magnetism* (Nobel Lecture, December 8, 1977), World Scientific, 1992.

[2] L. Hoddeson, G. Baym, and M. Eckert, The development of the quantum mechanical electron theory of metals: 1928–1933, *Rev. Mod. Phys.*, vol. 59, no. 1, p. 287, 1987.

[3] P. Weiss, La variation du ferromagnetisme du temperature, *Comptes Rendus*, vol. 143, p. 1136, 1906.

[4] P. Weiss, La constante du champ moléculaire: equation d'état magnétique et calorimétrie, *J. Phys. Radium*, vol. 1, p. 163, 1930.

[5] N. Bohr, *Early work 1905–1911.* vol. 1 of Neils Bohr collected work, ed. L. Rosenfield and J. Rud Nielsen Elsevier, 1972.

[6] H. J. van Leeuwen, Problèmes de la théorie électronique du magnetisme, *J. Phys. Radium*, vol. 2, p. 361, 1921.

[7] J. H. Van Vleck, *The theory of electric and magnetic susceptibilities.*, Clarendon Press, 1932.

[8] W. Heisenberg, Zur Theorie des Ferromagnetismus, *Z. Phys. A*, vol. 49, p. 619, 1928.

[9] E. C. Stoner, Free electrons and ferromagnetism, *Proc. Leeds Phil. Lit. Soc.*, vol. 2, p. 50, 1930.

[10] J. C. Slater, Cohesion in monovalent metals, *Phys. Rev.*, vol. 35, no. 5, p. 509, 1930.

[11] E. C. Stoner and E. P. Wohlfarth, A mechanism of magnetic hysteresis in heterogeneous alloys, *Philos. Trans. R. Soc. A*, vol. 240, p. 599, 1948.

[12] A. W. Overhauser, Spin density waves in an electron gas, *Phys. Rev.*, vol. 128, no. 3, p. 1437, 1962.

[13] J. Kanamori, Electron correlation and ferromagnetism of transition metals, *Prog. Theor. Phys.*, vol. 30, no. 3, p. 275, 1963. The Hubbard model, with the modern notation "U" for the interaction, was independently introduced by Kanamori in equation (1) of this paper.

[14] J. Hubbard, Electron correlations in narrow energy bands, *Proc. R. Soc. A*, vol. 276, p. 238, 1963.

[15] S. Doniach and S. Engelsberg, Low-temperature properties of nearly ferromagnetic Fermi liquids, *Phys. Rev. Lett.*, vol. 17, no. 14, p. 750, 1966.

[16] N. F. Berk and J. R. Schrieffer, Effect of ferromagnetic spin correlations on superconductivity, *Phys. Rev. Lett.*, vol. 17, no. 8, p. 433, 1966.

[17] A. P. Young, Quantum effects in the renormalization group approach to phase transitions, *J. Phys. C: Solid State Phys.)*, vol. 8, p. L309, 1975.

[18] J. A. Hertz, Quantum critical phenomena, *Phys. Rev. B*, vol. 14, p. 1165, 1976.

[19] A. J. Millis, Effect of a nonzero temperature on quantum critical points in itinerant fermion systems, *Phys. Rev. B*, vol. 48, p. 7183, 1993.

[20] T. I. Sigfusson and N. R. Bernhoeft, The de Haas-van Alphen effect, exchange splitting and Curie temperature in the weak itinerant ferromagnetic Ni_3Al, *J. Phys. F: Met. Phys.*, vol. 14, p. 2141, 1984.

[21] G. G. Lonzarich and L. Taillefer, Effect of spin fluctuations on the magnetic equation of state of ferromagnetic or nearly ferromagnetic metals, *J. Phys. C: Solid State*, vol. 18, p. 4339, 1985.

[22] S. Q. Wang, W. E. Evenson, and J. R. Schrieffer, Theory of itinerant ferromagnets exhibiting localized-moment behavior above the Curie point, *Phys. Rev. Lett.*, vol. 23, p. 92, 1969.

[23] J. Hubbard, The magnetism of iron, *Phys. Rev. B*, vol. 19, p. 2626, 1979.

Superconductivity and BCS theory

14.1 Introduction: early history

Superconductivity, the phenomenon whereby the resistance of a metal spontaneously drops to zero upon cooling below its critical temperature, was discovered over a hundred years ago by Heike Kamerlingh Onnes in 1911. However, it took another 46 years for the development of the conceptual framework required to understand this collective phenomenon as a condensation of electron pairs. During this time, many great physicists, including Bohr, Einstein, Heisenberg, Bardeen, and Feynman, had tried to develop a microscopic theory of the phenomenon. Today, superconductivity has been observed in a wide variety of materials (see Table 14.1), with transition temperatures reaching up as high as high as 134 K.

The development of the theory of superconductivity leading to BCS theory really had two parts – one phenomenological, the second microscopic. Let me mention some highlights:

- The discovery of the Meissner effect in 1933 by Walther Meissner and Robert Ochsenfeld [1]. When a metal is cooled in a small magnetic field, the flux is spontaneously excluded as the metal becomes superconducting (see Figure 14.1). The Meissner effect demonstrates that a superconductor is, in essence, a perfect diamagnet.
- Rigidity of the wavefunction. In 1937 Fritz London, working at Oxford [2, 3], proposed that a persistent supercurrent is a property of the *ground state* associated with its *rigidity* against the application of a field. London's idea applies to the full many-body wavefunction, but he initially developed it using a phenomenological one-particle wavefunction $\psi(x)$ that today we call the *superconducting order parameter*. He noted that the quantum mechanical current contains a *paramagnetic* and a *diamagnetic* component, writing

$$ j = \frac{\hbar e}{2im}(\psi^* \vec{\nabla} \psi - \psi \vec{\nabla} \psi^*) - \left(\frac{e^2}{m}\right)\psi^* \psi \vec{A}. \tag{14.1} $$

In the ground state in the absence of a field ($\vec{A} = 0$), the current vanishes, so the ground-state wavefunction ψ_0 must be uniform. Normally, the wavefunction is highly sensitive to an external magnetic field, but London reasoned that, if the wavefunction is somehow *rigid* and hence unchanged to linear order in the magnetic field, $\psi(x) = \psi_0(x) + O(B^2)$, where ψ_0 is the ground-state wavefunction, then, to leading order in a field, the current carried by the uniform quantum state is

Symmetry	Superfluid/superconductor	T_c	Mechanism
Table 14.1 Selected superfluids/superconductors.			
s	Hg	4.2 K	Phonon-mediated
	Pb	7.2 K	
	NbGe$_3$	23 K	
	MgB$_2$	39 K	
p	^{3}He	2.5 mK	Magnetic interactions
	UPt$_3$	0.51 K	
	Sr$_2$RuO$_4$	0.93 K	
d	CeCu$_2$Si$_2$	0.65 K	
	PuCoGa$_7$	18.5 K	
	HgBa$_2$Ca$_2$Cu$_3$O$_8$	134 K	
$s^{\pm}$	Sr$_{0.5}$Sm$_{0.5}$FeAsF	56 K	

(a) Metal, $T > T_c$ (b) Super conductor, $T < T_c$

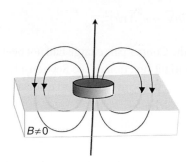
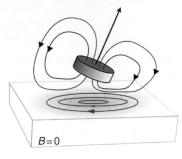

(a) A magnet rests on top of a normal metal, with its field lines penetrating the metal. (b) Once cooled below T_c, the superconductor spontaneously excludes magnetic fields, generating persistent supercurrents at its surface, causing the magnet to levitate.

Fig. 14.1

$$\vec{j} = -\frac{e^2}{m}|\psi_0|^2\vec{A} + \cdots. \tag{14.2}$$

In London's equation we see a remarkable convergence of the classical and the quantum: it is certainly a classical equation of motion in that it involves purely macroscopic variables, yet on the other hand it contains a naked vector potential $\vec{A}$ rather than the magnetic field $\vec{B} = \nabla \times \vec{A}$, a feature which reflects the broken gauge symmetry of the quantum ground state.

London's equation provides a natural explanation of the Meissner effect. To see this, we use Ampère's relation $\vec{j} = \mu_0^{-1}\nabla \times \vec{B}$ to rewrite the current in London's terms of the magnetic field:

$$\nabla \times \vec{B} = -\frac{1}{\lambda_L^2}\vec{A} \qquad \left(\frac{1}{\lambda_L^2} = \mu_0\frac{e^2}{m}|\psi_0|^2\right), \tag{14.3}$$

where the quantity λ_L defined above is the *London penetration depth*. Taking the curl of (14.3), we eliminate the vector potential to obtain

$$\overbrace{\nabla \times \nabla \times \vec{B}}^{-\nabla^2 \vec{B}} = -\frac{1}{\lambda_L^2} \overbrace{\nabla \times \vec{A}}^{\vec{B}} \tag{14.4}$$

or

$$\nabla^2 B = \frac{1}{\lambda_L^2} \vec{B}, \tag{14.5}$$

where we have substituted $\nabla \times (\nabla \times \vec{B}) = \vec{\nabla}(\nabla \cdot \vec{B}) - \nabla^2 \vec{B} = -\nabla^2 \vec{B}$, using the divergence-free nature of the magnetic field. The solutions of this equation describe magnetic fields $B(x) \sim B_0 e^{\pm x/\lambda_L}$ which decay inside the superconductor over a London penetration depth. This exclusion of magnetic fields inside superconductors is precisely the Meissner effect.

- Ginzburg–Landau theory [4]. In 1950, Lev Landau and Vitaly Ginzburg in Moscow reinterpreted London's phenomenological wavefunction $\psi(x)$ as a *complex-order parameter*. Using arguments of gauge invariance, they reasoned that the free energy must contain a gradient term that instills the rigidity of the order parameter:

$$f = \int d^3x \frac{1}{2m^*} |(-i\hbar\vec{\nabla} - e^*\vec{A})\psi|^2. \tag{14.6}$$

(At this stage, the identification of $e^* = 2e$ as the Cooper pair charge had not been made.) The vitally important aspect of this gauge-invariant functional (see Section 11.5) is that, once $\psi \neq 0$, the electromagnetic field develops a mass, giving rise to a super-current

$$\vec{j}(x) = -\delta f/\delta \vec{A}(x) = -\frac{(e^*)^2}{m^*} |\psi_0|^2 \vec{A}(x) \tag{14.7}$$

for a uniform $\psi = \psi_0$.

Following the Second World War, physicists set to work to try to develop a microscopic theory of superconductivity. The development of quantum field theory and new experimental techniques, such as microwaves – a biproduct of radar – and the availability of isotopes after the Manhattan Project, meant that a new intellectual offensive could begin. The landmark events included:

- Theory of the electron–phonon interaction. In 1949–1950, Herbert Fröhlich at Purdue and Liverpool universities [5] formulated the electron–phonon interaction as a direct analogue of photon exchange in electromagnetism. He showed that it gives rise to a low-energy attractive interaction,

$$V_{eff}(\mathbf{k}, \mathbf{k}') = -g_{\mathbf{k}-\mathbf{k}'}^2 \frac{2\omega_{\mathbf{k}-\mathbf{k}'}}{\omega_{\mathbf{k}-\mathbf{k}'}^2 - (\epsilon_{\mathbf{k}} - \epsilon_{\mathbf{k}'})^2}, \tag{14.8}$$

where $\epsilon_{\mathbf{k}}$ and $\epsilon_{\mathbf{k}'}$ are the energies of incoming and outgoing electrons, while $\omega_{\mathbf{q}}$ is the phonon frequency. $V_{eff}(\mathbf{k}, \mathbf{k}')$ is attractive for low-energy transfer, $|\epsilon_{\mathbf{k}} - \epsilon_{\mathbf{k}'}| << \omega_{\mathbf{k}-\mathbf{k}'}$.

- Discovery of the isotope effect. In 1950, Emanuel Maxwell at the National Bureau of Standards [7] and the group of Bernard Serin at Rutgers University [8] observed a reduction in the superconducting transition temperature with the isotopic mass in mercury.

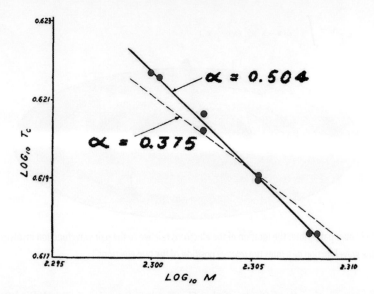

Fig. 14.2

Superconducting transition temperature as a function of isotopic mass for mercury, showing the $-\frac{1}{2}$ exponent, implying phonon-driven superconductivity. Reprinted with permission from B. Serin, *et al.*, *Phys. Rev.*, vol. 80, p. 761, 1950. Copyright 1950 by the American Physical Society.

It now became clear that the electron–phonon interaction provided the key to superconductivity. Indeed, in *any* theory in which the transition temperature is proportional to the Debye temperature, the expected dependence on isotopic mass M is given by [9]

$$T_c \propto \omega_D \sim \frac{1}{\sqrt{M}} \Rightarrow \frac{d \ln T_c}{d \ln M} = -\frac{1}{2}. \tag{14.9}$$

Careful analysis showed agreement with the $-\frac{1}{2}$ exponent [6] (see Figure 14.2), but what was the mechanism?

- Discovery of the coherence length. In 1953 Brian Pippard at the Cavendish Laboratory in Cambridge [10, 11] proposed, based on his thesis work on the anomalous skin depth in dirty superconductors, that the character of superconductivity changes at short distances, below a scale he named the *coherence length* ξ. Pippard showed that, at these short distances, the local London relation between current and vector potential is replaced by a non-local relationship. Pippard's result means that Ginzburg–Landau theory is inadequate at distances shorter than the coherence length ξ, demanding a microscopic theory.

- Gap hypothesis. In 1955 John Bardeen, who had recently resigned from Bell Laboratories to pursue his research into the theory of superconductivity at the University of Illinois Urbana-Champaign, proposed that if a gap Δ developed in the electron spectrum this would account for the wavefunction rigidity proposed by London and would also give rise to Pippard's coherence length $\xi \sim v_F/\Delta$, where v_F is the Fermi velocity [12]. What was now needed was a model and mechanism to create the gap.

- Bardeen–Pines Hamiltonian. In 1955 John Bardeen and David Pines at the University of Illinois Urbana-Champaign [13] rederived the Fröhlich interaction as a second-quantized

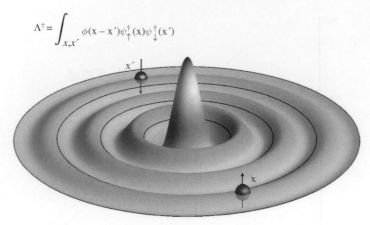

$$\Lambda^\dagger = \int_{x,x'} \phi(x-x')\psi_\uparrow^\dagger(x)\psi_\downarrow^\dagger(x')$$

Fig. 14.3 Illustration of a Cooper pair. (Note: the location of the electrons relative to the pair wavefunction involves artistic license since the wavefunction describes the *relative* position of the two electrons.)

model, incorporating the effects of the Coulomb interaction in a "Jellium model" in which the ions form a smeared positive background (see Section 7.7.3). The Bardeen–Pines effective interaction takes the form

$$V_{BP}(\mathbf{q},\nu) = \frac{e^2}{\epsilon_0(q^2+\kappa^2)}\left[1+\frac{\omega_\mathbf{q}^2}{\nu^2-\omega_\mathbf{q}^2}\right], \tag{14.10}$$

where κ^{-1} is the Thomas–Fermi screening length and the phonon frequency $\omega_\mathbf{q}$ is related to the plasma frequency of the ions $\Omega_p^2 = (Ze)^2 n_{ion}/(\epsilon_0 M)$ via the relation $\omega_\mathbf{q} = (q/[q^2+\kappa^2]^{1/2})\Omega_p$. The Bardeen–Pines interaction is seen to contain two terms: a frequency-independent Coulomb interaction, and a strongly frequency-dependent electron–phonon interaction. In the time domain, the former corresponds to an instantaneous Coulomb repulsion, while the latter is a highly retarded attractive interaction. This interaction became the basis for BCS theory.

The stage was set for Bardeen–Cooper–Schrieffer (BCS) theory.

14.2 The Cooper instability

In the fall of 1956, Bardeen's postdoc Leon Cooper, at the University of Illinois Urbana-Champaign, solved one of the most famous "warm-up" problems of all time. Considering two electrons moving above the Fermi surface of a metal, Cooper found that an arbitrarily weak electron–electron attraction induces a two-particle bound state that will destabilize the Fermi surface [14].

Cooper imagined adding a pair of electrons above the Fermi surface in a state with no net momentum, described by the wavefunction

$$|\Psi\rangle = \Lambda^\dagger|FS\rangle, \tag{14.11}$$

where

$$\Lambda^\dagger = \int d^3x d^3x' \phi(\mathbf{x} - \mathbf{x}') \psi_\downarrow^\dagger(\mathbf{x}) \psi_\uparrow^\dagger(\mathbf{x}') \tag{14.12}$$

creates a pair of electrons, while $|FS\rangle = \prod_{k<k_F} c_{\mathbf{k}\uparrow}^\dagger c_{-\mathbf{k}\downarrow}^\dagger |0\rangle$ defines the filled sea. If we Fourier transform the fields, writing $\psi_\sigma^\dagger(\mathbf{x}) = \frac{1}{\sqrt{V}} \sum_{\mathbf{k}} c_{\mathbf{k}\sigma}^\dagger e^{-i\mathbf{k}\cdot\mathbf{x}}$, then the pair creation operator can be recast as a sum over pairs in momentum space:

$$\Lambda^\dagger = \sum_{\mathbf{k}} \phi_{\mathbf{k}} c_{\mathbf{k}\downarrow}^\dagger c_{-\mathbf{k}\uparrow}^\dagger, \tag{14.13}$$

Cooper pair creation operator

where

$$\phi_{\mathbf{k}} = \int d^3x e^{-i\mathbf{k}\cdot\mathbf{x}} \phi(\mathbf{x}) \tag{14.14}$$

is the Fourier transform of the spatial pair wavefunction. This result tells us that a real-space pair of fermions can be decomposed into a sum of momentum-space pairs, weighted by the amplitude $\phi_{\mathbf{k}}$. The properties of the pair (and the superconductor it will give rise to) are encoded in the pair wavefunction $\phi_{\mathbf{k}}$. In the phonon-mediated superconductors considered by BCS, $\phi_{\mathbf{k}} \sim f(k)$ is an isotropic s-wave function, but in a rapidly growing class of *anisotropically paired superfluids* of great current interest, including superfluid ^{3}He, heavy-fermion, and iron- and copper-based high-temperature superconductors $\phi_{\mathbf{k}}$ is anisotropic changing sign *somewhere* in momentum space to lower the repulsive interaction energy, giving rise to a *nodal pair wavefunction*.

When an electron pair is created, electrons can only be added above the Fermi surface, so that

$$|\Psi\rangle = \Lambda^\dagger |FS\rangle = \sum_{|\mathbf{k}|>k_F} \phi_{\mathbf{k}} |\mathbf{k}_P\rangle, \tag{14.15}$$

where $|\mathbf{k}_P\rangle \equiv |\mathbf{k}\uparrow, -\mathbf{k}\downarrow\rangle = c_{\mathbf{k}\uparrow}^\dagger c_{-\mathbf{k}\downarrow}^\dagger |FS\rangle$. Now suppose that the Hamiltonian has the form

$$H = \sum_{\mathbf{k}} \epsilon_{\mathbf{k}} c_{\mathbf{k}\sigma}^\dagger c_{\mathbf{k}\sigma} + \hat{V}, \tag{14.16}$$

where $\hat{V}$ contains the details of the electron–electron interaction; if $|\Psi\rangle$ is an eigenstate with energy E, then

$$H|\Psi\rangle = \sum_{|\mathbf{k}|>k_F} 2\epsilon_{\mathbf{k}} \phi_{\mathbf{k}} |\mathbf{k}_P\rangle + \sum_{|\mathbf{k}|, |\mathbf{k}'|>k_F} |\mathbf{k}_P\rangle \langle \mathbf{k}_P|\hat{V}|\mathbf{k}'_P\rangle \phi_{\mathbf{k}'}. \tag{14.17}$$

Identifying this with $E|\Psi\rangle = E\sum_{\mathbf{k}} \phi_{\mathbf{k}}|\mathbf{k}_P\rangle$, so comparing the amplitudes to be in the state $|\mathbf{k}_P\rangle$,

$$E\phi_{\mathbf{k}} = 2\epsilon_{\mathbf{k}} \phi_{\mathbf{k}} + \sum_{|\mathbf{k}'|>k_F} \langle \mathbf{k}_P|\hat{V}|\mathbf{k}'_P\rangle \phi_{\mathbf{k}'}. \tag{14.18}$$

Fig. 14.4   Virtual phonon exchange process responsible for the BCS interaction. The process $|\mathbf{k}\uparrow, -\mathbf{k}\downarrow\rangle \rightarrow |\mathbf{k}'\uparrow, -\mathbf{k}'\downarrow\rangle$ can be thought of as the consequence of Bragg diffraction of a virtual standing wave: one electron in the pair $|\mathbf{k}\uparrow, -\mathbf{k}\downarrow\rangle$ diffracts from $\mathbf{k} \rightarrow \mathbf{k}'$, creating a virtual standing wave (phonon) of momentum $\mathbf{k} - \mathbf{k}'$. Later, the second diffracts from $-\mathbf{k} \rightarrow -\mathbf{k}'$, reabsorbing the virtual phonon.

The beauty of this equation is that the details of the electron interactions are entirely contained in the pair scattering matrix element $V_{\mathbf{k},\mathbf{k}'} = \langle \mathbf{k}_P | \hat{V} | \mathbf{k}'_P \rangle$. Microscopically, this scattering is produced by the exchange of virtual phonons (in conventional superconductors), and the scattering matrix element is determined by the electron–phonon propagator

$$V_{\mathbf{k},\mathbf{k}'} = g^2_{\mathbf{k}-\mathbf{k}'} D(\mathbf{k}' - \mathbf{k}, \epsilon_{\mathbf{k}} - \epsilon_{\mathbf{k}'}), \tag{14.19}$$

as illustrated in Figure 14.4. Cooper noted that this matrix element is not strongly momentum-dependent, only becoming attractive within an energy ω_D of the Fermi surface, and this motivated a simplified model interaction in which

$$V_{\mathbf{k},\mathbf{k}'} = \begin{cases} -g_0/V & (|\epsilon_{\mathbf{k}}|, \, |\epsilon_{\mathbf{k}'}| < \omega_D) \\ 0 & (\text{otherwise}). \end{cases} \tag{14.20}$$

This is a piece of pure physics *haiku*, a brilliant simplification that makes BCS theory analytically tractable. Much more is to come, but for the moment it enables us to simplify (14.18):

$$(E - 2\epsilon_{\mathbf{k}})\phi_{\mathbf{k}} = -\frac{g_0}{V} \sum_{0 < \epsilon_{\mathbf{k}'} < \omega_D} \phi_{\mathbf{k}'}, \tag{14.21}$$

so that by solving for $\phi_{\mathbf{k}}$,

$$\phi_{\mathbf{k}} = -\frac{g_0/V}{E - 2\epsilon_{\mathbf{k}}} \sum_{0 < \epsilon_{\mathbf{k}'} < \omega_D} \phi_{\mathbf{k}'}, \tag{14.22}$$

then summing both sides over $\mathbf{k}$ and factoring out $\sum_{\mathbf{k}} \phi_{\mathbf{k}}$, we obtain the self-consistent equation

$$1 = -\frac{1}{V} \sum_{0 < \epsilon_{\mathbf{k}} < \omega_D} \frac{g_0}{E - 2\epsilon_{\mathbf{k}}}. \tag{14.23}$$

Replacing the summation by an integral over energy, $\frac{1}{V}\sum_{0<\epsilon_{\mathbf{k}}<\omega_D} \to N(0)\int_0^{\omega_D}$, where $N(0)$ is the density of states per spin per unit volume at the Fermi energy, the resulting equation gives

$$1 = g_0 N(0)\int_0^{\omega_D}\frac{d\epsilon}{2\epsilon - E} = -\frac{1}{2}g_0 N(0)\ln\left[\frac{2\omega_D - E}{-E}\right] \approx -\frac{1}{2}g_0 N(0)\ln\left[\frac{2\omega_D}{-E}\right],$$

(14.24)

where, anticipating the smallness of $|E| << \omega_D$, we have approximated $2\omega_D - E \approx 2\omega_D$. In other words, the energy of the Cooper pair is given by

$$E = -2\omega_D e^{-\frac{2}{g_0 N(0)}}.$$

(14.25)

Remarks

- The Cooper pair is a bound state beneath the particle–hole continuum (see Figure 14.5).
- In his seminal paper, Cooper notes that the Cooper pair is a boson, an operator governed by a bosonic (commutator) algebra. (We will see shortly that it can be regarded as the transverse component of a very large isospin.) This changes everything, for, as pairs, electrons can *condense macroscopically*.
- A generalization of the above calculation to finite momentum (see Example 14.1) shows that the Cooper pair has a *linear* dispersion $E_{\mathbf{p}} - E = v_F p$ (see Figure 14.5), reminiscent of a collective mode.

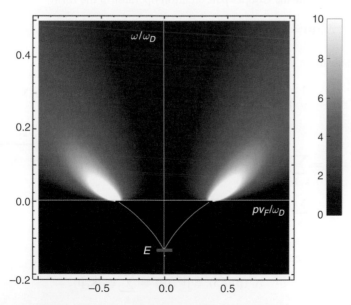

Formation of a Cooper pair beneath the two-particle continuum. This density plot shows the density of states of pair excitations obtained from the imaginary part of the pair susceptibility $\chi''(E, \mathbf{p})$ (see Example 14.1). At a finite momentum, the Cooper pair energy defines a collective bosonic mode beneath the quasiparticle continuum with dispersion $E_{\mathbf{p}} \approx E(0) + v_F|p|$.

Fig. 14.5

Example 14.1 Generalize Cooper's calculation to a pair with finite momentum. In particular:

(a) Show that the operator that creates a Cooper pair at a finite momentum $\mathbf{p}$,

$$\Lambda^{\dagger}(\mathbf{p}) = \int d^3x\, d^3x'\, \phi(\mathbf{x} - \mathbf{x}')\psi_{\uparrow}^{\dagger}(\mathbf{x})\psi_{\downarrow}^{\dagger}(\mathbf{x}')e^{i\mathbf{p}\cdot(\mathbf{x}+\mathbf{x}')/2}, \tag{14.26}$$

can be rewritten in the form

$$\Lambda^{\dagger}(\mathbf{p}) = \sum_{\mathbf{k}} \phi(\mathbf{k}) c_{\mathbf{k}+\mathbf{p}/2\uparrow}^{\dagger} c_{-\mathbf{k}+\mathbf{p}/2\downarrow}^{\dagger}. \tag{14.27}$$

(b) Show that the energy $E_{\mathbf{p}}$ of the pair state $\Lambda^{\dagger}(\mathbf{p})|FS\rangle$ is given by the roots $z = E_{\mathbf{p}}$ of the equation

$$1 + \frac{g_0}{V} \sum_{0 < \epsilon_{\mathbf{k}\pm\mathbf{p}/2} < \omega_D} \frac{1}{z - (\epsilon_{\mathbf{k}+\mathbf{p}/2} + \epsilon_{\mathbf{k}-\mathbf{p}/2})} = 0. \tag{14.28}$$

Demonstrate that this equation predicts a linear dispersion given by

$$E_{\mathbf{p}} = -2\omega_D e^{-\frac{2}{g_0 N(0)}} + v_F|p|. \tag{14.29}$$

Solution

(a) Introducing center-of-mass variables $\mathbf{X} = (\mathbf{x} + \mathbf{x}')/2$ and $\mathbf{r} = \mathbf{x} - \mathbf{x}'$, using $d^3x\, d^3x' = d^3X\, d^3r$, we rewrite the Cooper pair creation operator in the form

$$\Lambda^{\dagger}(\mathbf{p}) = \int d^3r\, d^3X\, e^{i\mathbf{p}\cdot\mathbf{X}} \phi(\mathbf{r})\, \psi_{\uparrow}^{\dagger}(\mathbf{X} + \mathbf{r}/2)\psi_{\downarrow}^{\dagger}(\mathbf{X} - \mathbf{r}/2). \tag{14.30}$$

If we substitute $\psi_{\sigma}^{\dagger}(x) = \frac{1}{\sqrt{V}} \sum_{\mathbf{k}} c_{\mathbf{k}\sigma}^{\dagger} e^{-i\mathbf{k}\cdot\mathbf{x}}$, we then obtain

$$\Lambda^{\dagger}(p) = \frac{1}{V} \int d^3r\, d^3X e^{i\mathbf{p}\cdot\mathbf{X}} \phi(\mathbf{r}) \sum_{\mathbf{k}_1,\mathbf{k}_2} c_{\mathbf{k}_1\uparrow}^{\dagger} c_{\mathbf{k}_2\downarrow}^{\dagger} e^{-i\mathbf{k}_1\cdot(\mathbf{X}+\mathbf{r}/2)} e^{i\mathbf{k}_2\cdot(\mathbf{X}-\mathbf{r}/2)}$$

$$= \sum_{\mathbf{k}_1,\mathbf{k}_2} c_{\mathbf{k}_1\uparrow}^{\dagger} c_{-\mathbf{k}_2\downarrow}^{\dagger} \overbrace{\int d^3r\, \phi(\mathbf{r})e^{i\mathbf{r}\cdot(\mathbf{k}_1+\mathbf{k}_2)/2}}^{\phi((\mathbf{k}_1+\mathbf{k}_2)/2)} \overbrace{\frac{1}{V} \int d^3R\, e^{i[\mathbf{p}-(\mathbf{k}_1-\mathbf{k}_2)]\cdot\mathbf{X}}}^{\delta_{\mathbf{p}-(\mathbf{k}_1-\mathbf{k}_2)}}$$

$$= \sum_{\mathbf{k}} \phi(\mathbf{k})\, c_{\mathbf{k}+\mathbf{p}/2\uparrow}^{\dagger} c_{-\mathbf{k}+\mathbf{p}/2\downarrow}^{\dagger}, \tag{14.31}$$

where we have replaced $(\mathbf{k}_1 + \mathbf{k}_2)/2 \to \mathbf{k}$ in the last step.

(b) Denote a Cooper pair with momentum $\mathbf{p}$ by

$$\Lambda^{\dagger}(\mathbf{p})|FS\rangle \equiv |\psi(\mathbf{p})\rangle = \sum_{\mathbf{k}} \phi_{\mathbf{k}}|\mathbf{k}, \mathbf{p}\rangle, \tag{14.32}$$

where $|\mathbf{k}, \mathbf{p}\rangle = c_{\mathbf{k}+\mathbf{p}/2\uparrow}^{\dagger} c_{-\mathbf{k}+\mathbf{p}/2\downarrow}^{\dagger}|FS\rangle$. Applying $H|\Psi(\mathbf{p})\rangle = E_{\mathbf{p}}|\Psi(\mathbf{p})\rangle$, using (14.16),

$$E_{\mathbf{p}} \sum_{\mathbf{k}} \phi_{\mathbf{k}}|\mathbf{k}, \mathbf{p}\rangle = \sum_{|\mathbf{k}\pm\frac{\mathbf{p}}{2}|>k_F} (\epsilon_{\mathbf{k}+\mathbf{p}/2} + \epsilon_{\mathbf{k}-\mathbf{p}/2})\, \phi_{\mathbf{k}}|\mathbf{k}, \mathbf{p}\rangle + \sum_{|\mathbf{k}|,\, |\mathbf{k}'|>k_F} |\mathbf{k}, \mathbf{p}\rangle\langle\mathbf{k}, \mathbf{p}|\hat{V}|\mathbf{k}', \mathbf{p}\rangle\phi_{\mathbf{k}'}.$$

Assume that $\langle \mathbf{k}, \mathbf{p} | \hat{V} | \mathbf{k}', \mathbf{p} \rangle \phi_{\mathbf{k}'} = -g_0/V$ is independent of $\mathbf{p}$. Comparing coefficients of $|\mathbf{k}, \mathbf{p}\rangle$,

$$E_{\mathbf{p}} \phi_{\mathbf{k}} = (\epsilon_{\mathbf{k}+\mathbf{p}/2} - \epsilon_{\mathbf{k}-\mathbf{p}/2}) \phi_{\mathbf{k}} - \frac{g_0}{V} \sum_{0 < \epsilon_{\mathbf{k}' \pm \mathbf{p}/2} < \omega_D} \phi_{\mathbf{k}'}. \tag{14.33}$$

Solving for $\phi_{\mathbf{k}}$,

$$\phi_{\mathbf{k}} = \frac{g_0/V}{\epsilon_{\mathbf{k}+\mathbf{p}/2} + \epsilon_{\mathbf{k}-\mathbf{p}/2} - E_{\mathbf{p}}} \sum_{0 < \epsilon_{\mathbf{k}' \pm \mathbf{p}/2} < \omega_D} \phi_{\mathbf{k}'}. \tag{14.34}$$

Summing both sides over momentum $\mathbf{k}$ and removing the common factor $\sum_{\mathbf{k}} \phi_{\mathbf{k}}$, we then obtain

$$1 - \frac{g_0}{V} \sum_{0 < \epsilon_{\mathbf{k} \pm \mathbf{p}/2} < \omega_D} \frac{1}{\epsilon_{\mathbf{k}+\mathbf{p}/2} + \epsilon_{\mathbf{k}-\mathbf{p}/2} - E_{\mathbf{p}}} = 0. \tag{14.35}$$

It is convenient to cast this as the zero of the function $\mathcal{G}^{-1}[E_{\mathbf{p}}, \mathbf{p}] = 0$, where

$$\mathcal{G}^{-1}[z, \mathbf{p}] = 1 - g_0 \chi_0(z, \mathbf{p}), \tag{14.36}$$

and

$$\chi_0(z, \mathbf{p}) = \frac{1}{V} \sum_{0 < \epsilon_{\mathbf{k} \pm \mathbf{p}/2} < \omega_D} \frac{1}{\epsilon_{\mathbf{k}+\mathbf{p}/2} + \epsilon_{\mathbf{k}-\mathbf{p}/2} - z} \tag{14.37}$$

can be interpreted as the bare pair susceptibility of the conduction sea. Now, taking $\epsilon_{\mathbf{k}} = k^2/2m - \mu$ in the momentum summation, we must impose the condition

$$\epsilon_{\mathbf{k} \pm \mathbf{p}/2} = \epsilon_{\mathbf{k}} \pm \frac{\mathbf{p} \cdot \mathbf{v}_F}{2} + \frac{p^2}{8m} > 0, \tag{14.38}$$

or $\epsilon_{\mathbf{k}} > \frac{p v_F}{2} |\cos \theta| - \frac{p^2}{8m}$. Replacing the momentum summation by an integral over energy and angles,

$$\chi_0[z, p] = \frac{N(0)}{2} \int_{-1}^{1} \frac{d \cos \theta}{2} \int_{\frac{p v_F}{2} |\cos \theta| - p^2/8m}^{\omega_D} \frac{d\epsilon}{2\epsilon + p^2/4m - z}$$

$$= \frac{N(0)}{2} \int_{0}^{1} d \cos \theta \ln \left[\frac{2\omega_D}{p v_F \cos \theta - z} \right]. \tag{14.39}$$

Finally, carrying out the integral over θ, one obtains

$$\chi_0(z, p) = \frac{N(0)}{2} \tilde{\chi}_0 \left[\frac{z}{2\omega_D}, \frac{p v_F}{2\omega_D} \right], \tag{14.40}$$

where

$$\tilde{\chi}_0[\tilde{z}, \tilde{p}] = \ln \left(\frac{1}{\tilde{p} - \tilde{z}} \right) + \left[1 + \frac{\tilde{z}}{\tilde{p}} \ln \left(1 - \frac{\tilde{p}}{\tilde{z}} \right) \right]. \tag{14.41}$$

Thus for small $v_F p \ll |E|$, using (14.36),

$$\mathcal{G}^{-1}[E, p] = 1 - \frac{g_0 N(0)}{2} \ln \left[\frac{2\omega_D}{v_F p - E} \right], \tag{14.42}$$

so the bound-state pole occurs at $\mathcal{G}^{-1}(E_{\mathbf{p}}, \mathbf{p}) = 0$ or

$$E_{\mathbf{p}} = -2\omega_D \exp\left[-\frac{2}{g_0 N(0)}\right] + v_F p. \tag{14.43}$$

The linear spectrum is a signature of a collective bosonic mode. Incidentally, the quantity

$$\chi''(E, \mathbf{p}) = \text{Im}[\chi_0(z, \mathbf{p})/(1 - g_0 \chi_0(z, \mathbf{p}))]|_{z=E-i\delta} \tag{14.44}$$

can be interpreted as a spectral function giving the density of Cooper pairs above and below the particle–particle continuum. It is this quantity that is plotted in Figure 14.5.

14.3 The BCS Hamiltonian

After Cooper's discovery, it took a further six months of intense exploration of candidate wavefunctions before Bardeen, Cooper and Schrieffer succeeded in formulating the theory of superconductivity in terms of a *pair condensate* [15]. It was the graduate student in the team, J. Robert Schrieffer, who took the next leap.[1] Schrieffer's insight was to identify the superconducting ground state as a coherent state of the Cooper pair operator:

$$|\psi_{BCS}\rangle = \exp[\Lambda^\dagger]|0\rangle, \tag{14.45}$$

where $|0\rangle$ is the electron vacuum and $\Lambda^\dagger = \sum_{\mathbf{k}} \phi_{\mathbf{k}} c_{\mathbf{k}\uparrow}^\dagger c_{-\mathbf{k}\downarrow}^\dagger$ is the Cooper pair creation operator (14.13). If we expand the exponential as a product in momentum space,

$$|\psi_{BCS}\rangle = \prod_{\mathbf{k}} \exp[\phi_{\mathbf{k}} c_{\mathbf{k}\uparrow}^\dagger c_{-\mathbf{k}\downarrow}^\dagger]|0\rangle = \prod_{\mathbf{k}}(1 + \phi_{\mathbf{k}} c_{\mathbf{k}\uparrow}^\dagger c_{-\mathbf{k}\downarrow}^\dagger)|0\rangle. \tag{14.46}$$

BCS wavefunction

In the second step, we have truncated the exponential to linear order because all higher powers of the pair operator vanish: $(c_{\mathbf{k}\uparrow}^\dagger c_{-\mathbf{k}\downarrow}^\dagger)^n = 0$ $(n > 1)$. This remarkable coherent state mixes states of different particle number, giving rise to a state of off-diagonal long-range order in which

$$\langle\psi_{BCS}|c_{-\mathbf{k}\downarrow} c_{\mathbf{k}\uparrow}|\psi_{BCS}\rangle \propto \phi_{\mathbf{k}}. \tag{14.47}$$

---

[1] Following a conference at the Stevens Institute of Technology on the many–body problem, inspired by a wavefunction that Tomonaga had derived, Schrieffer wrote down a candidate wavefunction for the ground-state superconductivity. He recalls the event in his own words [16]:

> So I guess it was on the subway, I scribbled down the wave function and I calculated the beginning of that expectation value and I realized that the algebra was very simple. I think it was somehow in the afternoon and that night at this friend's house I worked on it. And the next morning, as I recall, I did the variational calculation to get the gap equation and I solved the gap equation for the cutoff potential.

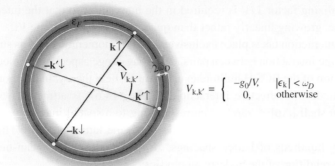

$$V_{k,k'} = \begin{cases} -g_0/V, & |\epsilon_k| < \omega_D \\ 0, & \text{otherwise} \end{cases}$$

In the BCS Hamiltonian, the matrix $V_{\mathbf{k},\mathbf{k}'}$ acts attractively on pairs of electrons within ω_D of the Fermi surface. Provided the repulsive interaction at higher energies is not too large, a superconducting instability results.

Fig. 14.6

But what Hamiltonian explicitly gives rise to pairing? A clue came from the Cooper instability, which depends on the scattering amplitude $V_{\mathbf{k},\mathbf{k}'} = \langle \mathbf{k}_P | \hat{V} | \mathbf{k}'_P \rangle$ between *zero-momentum pairs*. BCS [15] incorporated this feature into a model Hamiltonian:

$$H = \sum_{\mathbf{k}\sigma} \epsilon_{\mathbf{k}\sigma} c^\dagger_{\mathbf{k}\sigma} c_{\mathbf{k}\sigma} + \sum_{\mathbf{k},\mathbf{k}'} V_{\mathbf{k},\mathbf{k}'} c^\dagger_{\mathbf{k}\uparrow} c^\dagger_{-\mathbf{k}\downarrow} c_{-\mathbf{k}'\downarrow} c_{\mathbf{k}'\uparrow}. \tag{14.48}$$

BCS Hamiltonian

In the universe of possible superconductors and superfluids, the interaction $V_{\mathbf{k},\mathbf{k}'}$ can take a wide variety of symmetries, but in its s-wave manifestation it is simply an isotropic attraction that develops within a narrow energy shell of electrons within a Debye energy of the Fermi surface, ω_D (Figure 14.6):

$$V_{\mathbf{k},\mathbf{k}'} = \begin{cases} -g_0/V & (|\epsilon_{\mathbf{k}}| < \omega_D) \\ 0 & (\text{otherwise}). \end{cases} \tag{14.49}$$

The s-wave BCS Hamiltonian then takes the form

$$H = \sum_{|\epsilon_{\mathbf{k}}|<\omega_D,\, \sigma} \epsilon_{\mathbf{k}} c^\dagger_{\mathbf{k}\sigma} c_{\mathbf{k}\sigma} - \frac{g_0}{V} A^\dagger A.$$

$$A^\dagger = \sum_{|\epsilon_{\mathbf{k}}|<\omega_D} c^\dagger_{\mathbf{k}\uparrow} c^\dagger_{-\mathbf{k}\downarrow}, \qquad A = \sum_{|\epsilon_{\mathbf{k}'}|<\omega_D} c_{-\mathbf{k}'\downarrow} c_{\mathbf{k}'\uparrow}. \tag{14.50}$$

s-wave BCS Hamiltonian

Remarks

• The BCS Hamiltonian is a *model* Hamiltonian capturing the low-energy pairing physics.

- The normalizing factor $1/V$ is required in the interaction so that the interaction energy is extensive, growing linearly rather than quadratically with volume V.
- The BCS interaction takes place exclusively at zero momentum, and as such involves an infinite-range interaction between pairs. This long-range aspect of the model permits the exact solution of the BCS Hamiltonian using mean-field theory. In the more microscopic Fröhlich model the effective interaction (Figure (14.6)) is attractive within a narrow momentum shell $|\Delta\mathbf{p}| \sim \omega_D/v_F$, corresponding to a spatial interaction range of order $1/|\Delta\mathbf{p}| \sim v_F/\omega_D \sim O(\epsilon_F/\omega_D) \times a$, where a is the lattice spacing. This length scale is typically hundreds of lattice spacings, so the infinite-range mean-field theory is a reasonable rendition of the underlying physics.

14.3.1 Mean-field description of the condensate

The key consequence of the BCS model is the development of a state with off-diagonal long-range order (see Section 11.4.2). The pair operator $\hat{A}$ is extensive, and in a superconducting state its expectation value is proportional to the volume of the system $\langle\hat{A}\rangle \propto V$. The pair density

$$\Delta = |\Delta|e^{i\phi} = -\frac{g_0}{V}\langle\hat{A}\rangle = -g_0 \int_{|\epsilon_{\mathbf{k}}|<\omega_D} \frac{d^3k}{(2\pi)^3}\langle c_{-\mathbf{k}\downarrow}c_{\mathbf{k}\uparrow}\rangle \tag{14.51}$$

is an intensive, macroscopic property of superconductors that has both an amplitude $|\Delta|$ *and* a phase ϕ. This is the order parameter. It sets the size of the gap in the excitation spectrum and gives rise to the emergent phase variable whose rigidity supports superconductivity.

Like the pressure in a gas, the order parameter Δ is an emergent many-body property. Just as fluctuations in pressure $\langle\delta P^2\rangle \sim O(1/V)$ become negligible in the thermodynamic limit, fluctuations in Δ can be similarly ignored. Of course, the reasoning needs to be refined to encompass a quantum variable, formally requiring a path-integral approach. The important point is that the change in action $\delta S[\delta\Delta] = S[\Delta + \delta\Delta_0] - S[\Delta]$ associated with a small variation in Δ about a stationary point scales extensively in volume: $\delta S[\delta\Delta] \sim V \times \delta\Delta^2$, so that the corresponding distribution function can be expanded as a Gaussian,

$$\mathcal{P}[\Delta] \propto e^{-S[\delta\Delta]} \sim \exp\left[-\frac{\delta\Delta^2}{O(1/V)}\right], \tag{14.52}$$

which is exquisitely peaked about $\Delta = \Delta_0$, with variance $\langle\delta\Delta^2\rangle \propto 1/V$, justifying a mean-field treatment.

Let us now expand the BCS interaction in powers of the fluctuation operator $\delta\hat{A} = \hat{A} - \langle\hat{A}\rangle$:

$$-\frac{g_0}{V}A^\dagger A = \overbrace{\bar{\Delta}A + A^\dagger\Delta + V\frac{\bar{\Delta}\Delta}{g_0}}^{O(V)} - \overbrace{\frac{g_0}{V}\delta A^\dagger\delta A}^{O(1)}. \tag{14.53}$$

Now the first three terms are extensive in volume, but since $\langle \delta A^\dagger \delta A \rangle \sim O(V)$ the last term is intensive $O(1)$, and can be neglected in the thermodynamic limit. We shall shortly see how this same decoupling is accomplished in a path integral using a Hubbard–Stratonovich transformation. The resulting mean-field Hamiltonian for BCS theory is then

$$H_{MFT} = \sum_{\mathbf{k}\sigma} \epsilon_{\mathbf{k}} c^\dagger_{\mathbf{k}\sigma} c_{\mathbf{k}\sigma} + \sum_{\mathbf{k}} \left[\bar{\Delta} c_{-\mathbf{k}\downarrow} c_{\mathbf{k}\uparrow} + c^\dagger_{\mathbf{k}\uparrow} c^\dagger_{-\mathbf{k}\downarrow} \Delta \right] + \frac{V}{g_0} \bar{\Delta} \Delta, \qquad (14.54)$$

BCS theory: mean-field Hamiltonian

in which Δ needs to be determined self-consistently by minimizing the free energy.

14.4 Physical picture of BCS theory: pairs as spins

Let us discuss the physical meaning of the pairing terms in the BCS mean-field Hamiltonian (14.54)

$$H_P(\mathbf{k}) = \left(\bar{\Delta} c_{-\mathbf{k}\downarrow} c_{\mathbf{k}\uparrow} + c^\dagger_{\mathbf{k}\uparrow} c^\dagger_{-\mathbf{k}\downarrow} \Delta \right). \qquad (14.55)$$

On the one hand, the term $\bar{\Delta} c_{-\mathbf{k}\downarrow} c_{\mathbf{k}\uparrow}$ converts two particles into the condensate:

$$\text{Pair creation}: \qquad e^- + e^- \rightleftharpoons \text{pair}^{2-}. \qquad (14.56)$$

Alternatively, by writing $c_{-\mathbf{k}\downarrow} = h^\dagger_{\mathbf{k}\downarrow}$ as a hole creation operator, we see that $H_P(\mathbf{k}) \equiv (h^\dagger_{\mathbf{k}\uparrow} \bar{\Delta}) c_{\mathbf{k}\uparrow} + \text{H.c.}$ describes the scattering of a single electron into a condensed pair (represented by $\bar{\Delta}$) and a hole, a process called *Andreev reflection*, named after its discoverer, Alexander Andreev:

$$\text{Andreev reflection}: \qquad e^- \rightleftharpoons \text{pair}^{2-} + h^+. \qquad (14.57)$$

While the first process builds the condensate, the second coherently mixes particle and holes. We will denote the Andreev scattering process by a Feynman diagram:

Andreev reflection differs from conventional reflection in that

- it elastically scatters electrons into holes, reversing *all* components of the velocity[2]

---

[2] Andreev noticed that, although the momentum of the hole is the same as the incoming electron, its group velocity $\nabla_{\mathbf{k}}(-\epsilon_{-\mathbf{k}}) = \nabla_{\mathbf{k}}(-\epsilon_{\mathbf{k}}) = -\nabla_{\mathbf{k}}\epsilon_{\mathbf{k}}$ is reversed. This led him to predict that such scattering at the interface of a superconductor leads to non-specular reflection of electrons, which scatter back as holes moving in the opposite direction to the incoming electrons.

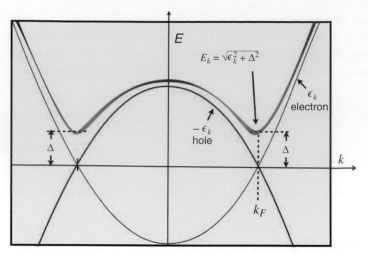

Fig. 14.7 Illustrating the excitation spectrum of a superconductor. Andreev scattering mixes the electron excitation spectrum (blue) with the hole excitation spectrum (red), producing the gap Δ in the quasiparticle excitation spectrum. The quasiparticles at the Fermi momentum are linear combinations of electrons and holes, with an indefinite charge.

- it *conserves* spin, momentum, *and* current, for a hole in the state $(-\mathbf{k}, \downarrow)$ has spin up, momentum $+\mathbf{k}$, and carries a current $I = (-e) \times (-\nabla\epsilon_\mathbf{k}) = e\nabla_\mathbf{k}\epsilon_\mathbf{k}$.

Now the particle and hole dispersions are given by

$$
\begin{aligned}
\text{particle:} \quad & \epsilon_\mathbf{k} \\
\text{hole:} \quad & -\epsilon_{-\mathbf{k}},
\end{aligned}
\tag{14.58}
$$

as denoted by the blue and red lines, respectively, in Figure 14.7. These lines intersect at the Fermi surface, so that the Andreev mixing between electrons and holes in a superconductor opens up a gap that eliminates the Fermi surface, giving rise to a dispersion which, we will shortly show, takes the form

$$
E_\mathbf{k} = \sqrt{\epsilon_\mathbf{k}^2 + |\Delta|^2},
\tag{14.59}
$$

as illustrated in Figure 14.7. The quasiparticle operators now become linear combinations of electron and hole states with corresponding quasiparticle operators

$$
a_{\mathbf{k}\sigma}^\dagger = u_\mathbf{k} c_{\mathbf{k}\sigma}^\dagger + \text{sgn}(\sigma) v_\mathbf{k} c_{-\mathbf{k}-\sigma}.
\tag{14.60}
$$

14.4.1 Nambu spinors

We now introduce Nambu's spinor formulation of BCS theory, which we'll employ to expose the beautiful magnetic analogy between pairs and spins, discovered by Yoichiro Nambu [17] working at the University of Chicago and Philip W. Anderson [18] at AT&T Bell Laboratories. The analogue of a superconductor is an antiferromagnet, for both superconductivity and antiferromagnetism involve an order parameter which (unlike ferromagnetism), does *not* commute with the Hamiltonian. Superconductivity involves an analogous quantity to spin, which we will call *isospin*, which describes orientations in charge space. The pairing field Δ can be regarded as a transverse field in isospin space.

To bring out this physics, it is convenient to introduce the charge analogue of the electron spinor, the *Nambu spinor*, defined as

$$\psi_{\mathbf{k}} = \begin{pmatrix} c_{\mathbf{k}\uparrow} \\ c_{-\mathbf{k},\downarrow}^{\dagger} \end{pmatrix} \qquad \begin{array}{l} \text{electron} \\ \text{hole} \end{array} \qquad (14.61)$$

with the corresponding Hermitian conjugate

$$\psi_{\mathbf{k}}^{\dagger} = \left(c_{\mathbf{k}\uparrow}^{\dagger}, \ c_{-\mathbf{k}\downarrow} \right). \qquad (14.62)$$

Nambu spinors behave like conventional electron fields, with an algebra

$$\{\psi_{\mathbf{k}\alpha}, \ \psi_{\mathbf{k}'\beta}^{\dagger}\} = \delta_{\alpha\beta} \delta_{\mathbf{k},\mathbf{k}'}, \qquad (14.63)$$

but instead of up and down electrons, they describe electrons and holes. These spinors enable us to unify the kinetic and pairing energy terms into a single *vector* field, analogous to a magnetic field, that acts in isospin space.

The kinetic energy can be written as

$$\sum_{\mathbf{k}} \epsilon_{\mathbf{k}}(c_{\mathbf{k}\uparrow}^{\dagger} c_{\mathbf{k}\uparrow} - c_{-\mathbf{k}\downarrow} c_{-\mathbf{k}\downarrow}^{\dagger} + 1) = \left(c_{\mathbf{k}\uparrow}^{\dagger}, \ c_{-\mathbf{k}\downarrow} \right) \begin{bmatrix} \epsilon_{\mathbf{k}} & 0 \\ 0 & -\epsilon_{\mathbf{k}} \end{bmatrix} \begin{pmatrix} c_{\mathbf{k}\uparrow} \\ c_{-\mathbf{k}\downarrow}^{\dagger} \end{pmatrix} + \sum_{\mathbf{k}} \epsilon_{\mathbf{k}}, \qquad (14.64)$$

where the sign reversal in the lower component derives from anticommuting the down-spin electron operators. The energy $-\epsilon_{\mathbf{k}}$ is the energy to create a hole. We will drop the constant remainder term $\sum_{\mathbf{k}} \epsilon_{\mathbf{k}}$. We can now combine the kinetic and pairing terms into a single matrix:

$$\epsilon_{\mathbf{k}} \sum_{\sigma} c_{\mathbf{k}\sigma}^{\dagger} c_{\mathbf{k}\sigma} + \left[\bar{\Delta} c_{-\mathbf{k}\downarrow} c_{\mathbf{k}\uparrow} + c_{\mathbf{k}\uparrow}^{\dagger} c_{-\mathbf{k}\downarrow}^{\dagger} \Delta \right] = \left(c_{\mathbf{k}\uparrow}^{\dagger}, \ c_{-\mathbf{k}\downarrow} \right) \begin{bmatrix} \epsilon_{\mathbf{k}} & \Delta \\ \bar{\Delta} & -\epsilon_{\mathbf{k}} \end{bmatrix} \begin{pmatrix} c_{\mathbf{k}\uparrow} \\ c_{-\mathbf{k}\downarrow}^{\dagger} \end{pmatrix}$$

$$= \psi_{\mathbf{k}}^{\dagger} \begin{bmatrix} \epsilon_{\mathbf{k}} & \Delta_1 - i\Delta_2 \\ \Delta_1 + i\Delta_2 & -\epsilon_{\mathbf{k}} \end{bmatrix} \psi_{\mathbf{k}}$$

$$= \psi_{\mathbf{k}}^{\dagger} [\epsilon_{\mathbf{k}} \tau_3 + \Delta_1 \tau_1 + \Delta_2 \tau_2] \psi_{\mathbf{k}}, \qquad (14.65)$$

where we denote $\Delta = \Delta_1 - i\Delta_2$, $\bar{\Delta} = \Delta_1 + i\Delta_2$ and we have introduced the *isospin matrices*

$$\vec{\tau} = (\tau_1, \tau_2, \tau_3) = \left(\begin{bmatrix} 0 & 1 \\ 1 & 0 \end{bmatrix}, \begin{bmatrix} 0 & -i \\ i & 0 \end{bmatrix}, \begin{bmatrix} 1 & 0 \\ 0 & -1 \end{bmatrix} \right). \qquad (14.66)$$

By convention the symbol $\vec{\tau}$ is used to distinguish a Pauli matrix in charge space from a spin σ acting in spin space. Putting this all together, the mean-field Hamiltonian can now be rewritten

$$H = \sum_{\mathbf{k}} \psi_{\mathbf{k}}^{\dagger} (\vec{h}_{\mathbf{k}} \cdot \vec{\tau}) \psi_{\mathbf{k}} + V \frac{\bar{\Delta}\Delta}{g_0}, \qquad (14.67)$$

where

$$\vec{h}_{\mathbf{k}} = (\Delta_1, \Delta_2, \epsilon_{\mathbf{k}}) \qquad (14.68)$$

plays the role of a Zeeman field acting in isospin space.

14.4.2 Anderson's domain-wall interpretation of BCS theory

Anderson noted that the isospin operators $\psi_{\mathbf{k}}^{\dagger}\vec{\tau}\psi_{\mathbf{k}}$ have the properties of spin-$\frac{1}{2}$ operators acting in charge space. The z component of the isospin is

$$\tau_{3\mathbf{k}} = \psi_{\mathbf{k}}^{\dagger}\tau_3\psi_{\mathbf{k}} = (c_{\mathbf{k}\uparrow}^{\dagger}c_{\mathbf{k}\uparrow} - c_{-\mathbf{k}\downarrow}c_{-\mathbf{k}\downarrow}^{\dagger}) = (n_{\mathbf{k}\uparrow} + n_{-\mathbf{k}\downarrow} - 1), \qquad (14.69)$$

so the up and down states correspond to the doubly occupied and empty pair state, respectively:

$$\begin{aligned}
\tau_{3\mathbf{k}} = +1: &\quad |\Uparrow_{\mathbf{k}}\rangle \;\equiv\; |2\rangle = c_{\mathbf{k}\uparrow}^{\dagger}c_{-\mathbf{k}\downarrow}^{\dagger}|0\rangle \\
\tau_{3\mathbf{k}} = -1: &\quad |\Downarrow_{\mathbf{k}}\rangle \;\equiv\; |0\rangle.
\end{aligned} \qquad (14.70)$$

By contrast, the transverse components of the isospin describe pair creation and annihilation:

$$\begin{aligned}
\hat{\tau}_{1\mathbf{k}} = \psi_{\mathbf{k}}^{\dagger}\tau_1\psi_{\mathbf{k}} = &\quad c_{\mathbf{k}\uparrow}^{\dagger}c_{-\mathbf{k}\downarrow}^{\dagger} + c_{-\mathbf{k}\downarrow}c_{\mathbf{k}\uparrow} \\
\hat{\tau}_{2\mathbf{k}} = \psi_{\mathbf{k}}^{\dagger}\tau_2\psi_{\mathbf{k}} = &-i(c_{\mathbf{k}\uparrow}^{\dagger}c_{-\mathbf{k}\downarrow}^{\dagger} - c_{-\mathbf{k}\downarrow}c_{\mathbf{k}\uparrow}).
\end{aligned} \qquad (14.71)$$

In a normal metal, the isospin points "up" in the occupied states below the Fermi surface, and "down" in the empty states above the Fermi surface (Figure 14.8(a)). Now since the Hamiltonian is $H = \sum_{\mathbf{k}} \psi_{\mathbf{k}}^{\dagger}(\vec{h}_{\mathbf{k}} \cdot \vec{\tau})\psi_{\mathbf{k}}$, the quantity

$$\vec{B}_{\mathbf{k}} = -\vec{h}_{\mathbf{k}} = -(\Delta_1, \Delta_2, \epsilon_{\mathbf{k}}) \qquad (14.72)$$

is thus a momentum-dependent Weiss field, setting a natural quantization axis for the electrons at momentum $\mathbf{k}$: in the ground state, the fermion isospins line up with this field. In the normal state, the natural isospin quantization axis is the charge or "z-axis," but in the

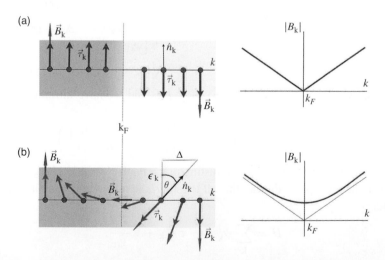

Fig. 14.8 Showing the domain-wall configuration of the isospin $\vec{\tau}_{\mathbf{k}}$ and direction of pairing field $\hat{n}_{\mathbf{k}}$ near the Fermi momentum: (a) a normal metal, in which the Weiss field $B_{\mathbf{k}}$ vanishes linearly at the Fermi energy, and (b) a superconductor in which the Weiss field remains finite at the Fermi energy, giving rise to a gap in the excitation spectrum.

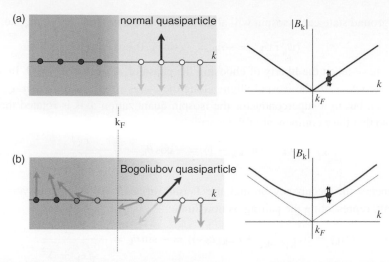

Illustrating how the excitation of quasiparticle pairs corresponds to an "isospin flip," which forms a pair of up and down quasiparticles with energy $2|B_k|$: (a) quasiparticle pair formation in the normal state where the quasiparticle spectrum is gapless; (b) formation of a Bogoliubov quasiparticle pair in the superconducting state where the excitation spectrum is gapped.

Fig. 14.9

superconductor, the presence of a pairing condensate tips the quantization axis, mixing particle and hole states (Figure 14.8(b)).

With this analogy one can identify the reversal of an isospin out of its ground-state configuration as the creation of a pair of quasiparticles "above" the condensate. Since this costs an energy $2|\vec{B}_k|$, the magnitude of the Weiss field

$$E_k \equiv |\vec{B}_k| = \sqrt{\epsilon_k^2 + |\Delta|^2} = \text{quasiparticle energy} \qquad (14.73)$$

must correspond to the energy of a single quasiparticle. In a metal ($\Delta = 0$), the Weiss field vanishes at the Fermi surface so it costs no energy to create a quasiparticle there (Figure 14.9(a)), but in a superconductor the Weiss field has magnitude $|\Delta|$ so the quasiparticle spectrum is now gapped (Figure 14.9(b)).

Let us write $\vec{B}_k = -E_k \hat{n}_k$, where the unit vector

$$\hat{n}_k = \left(\frac{\Delta_1}{E_k}, \frac{\Delta_2}{E_k}, \frac{\epsilon_k}{E_k} \right) \qquad (14.74)$$

points upwards far above the Fermi surface, and downwards far beneath it. In a normal metal, $\hat{n}_k$ (see Figure 14.8) reverses at the Fermi surface forming a sharp "Ising-like" domain wall, but in a superconductor the $\hat{n}$ vector is aligned at an angle θ to the $\hat{z}$ axis, where

$$\cos \theta_k = \frac{\epsilon_k}{E_k}. \qquad (14.75)$$

This angle rotates continuously as one passes through the Fermi energy, so the domain wall is now spread out over an energy range of order Δ, forming a kind of Bloch domain wall in isospin space, as shown in Figure 14.8.

In the ground state each isospin will align parallel to the field $\vec{B}_{\mathbf{k}} = -E_{\mathbf{k}}\hat{n}_{\mathbf{k}}$, i.e.

$$\langle \psi_{\mathbf{k}}^{\dagger} \vec{\tau} \psi_{\mathbf{k}} \rangle = -\hat{n}_{\mathbf{k}} = -(\sin\theta_{\mathbf{k}}, 0, \cos\theta_{\mathbf{k}}), \tag{14.76}$$

where we have taken the liberty of choosing the phase of Δ so that $\Delta_2 = 0$. In a normal ground state ($\Delta = 0$) the isospin aligns along the z-axis, $\langle \tau_{3\mathbf{k}} \rangle = \langle n_{\mathbf{k}\uparrow} + n_{-\mathbf{k}\downarrow} - 1 \rangle = \text{sgn}(k_F - k)$, but in a superconductor the isospin quantization axis is rotated through an angle $\theta_{\mathbf{k}}$ so that the z component of the isospin is

$$\langle \tau_{3\mathbf{k}} \rangle = \langle n_{\mathbf{k}\uparrow} + n_{-\mathbf{k}\downarrow} - 1 \rangle = -\cos\theta_{\mathbf{k}} = -\frac{\epsilon_{\mathbf{k}}}{\sqrt{\epsilon_{\mathbf{k}}^2 + \Delta^2}}, \tag{14.77}$$

which smears the occupancy around the Fermi surface, while the transverse isospin component, representing the pairing, is now finite:

$$\langle \tau_{1\mathbf{k}} \rangle = \langle (c_{\mathbf{k}\uparrow}^{\dagger} c_{-\mathbf{k}\downarrow}^{\dagger} + c_{-\mathbf{k}\downarrow} c_{\mathbf{k}\uparrow}) \rangle = -\sin\theta_{\mathbf{k}} = -\frac{\Delta}{\sqrt{\epsilon_{\mathbf{k}}^2 + \Delta^2}}. \tag{14.78}$$

Now since we have chosen $\Delta_2 = 0$, $\langle \tau_{2\mathbf{k}} \rangle = -i\langle (c_{\mathbf{k}\uparrow}^{\dagger} c_{-\mathbf{k}\downarrow}^{\dagger} - c_{-\mathbf{k}\downarrow} c_{\mathbf{k}\uparrow}) \rangle = 0$, it follows that $\langle c_{-\mathbf{k}\downarrow} c_{\mathbf{k}\uparrow} \rangle = -\frac{1}{2} \sin\theta_{\mathbf{k}}$. Imposing the self-consistency condition $\Delta = -\frac{g_0}{V} \sum_{\mathbf{k}} \langle c_{-\mathbf{k}\downarrow} c_{\mathbf{k}\uparrow} \rangle$ (14.51), one then obtains the *BCS gap equation*:

$$\Delta = \frac{g_0}{V} \sum_{\mathbf{k}} \frac{1}{2} \sin\theta_{\mathbf{k}} = g_0 \int_{|\epsilon_{\mathbf{k}}| < \omega_D} \frac{d^3 k}{(2\pi)^3} \frac{\Delta}{2\sqrt{\epsilon_{\mathbf{k}}^2 + \Delta^2}}. \tag{14.79}$$

BCS gap equation ($T = 0$)

Since the momentum sum is restricted to a narrow region of the Fermi surface, one can replace the momentum sum by an energy integral, to obtain

$$1 = g_0 N(0) \int_{-\omega_D}^{\omega_D} d\epsilon \frac{1}{2\sqrt{\epsilon^2 + \Delta^2}} = g_0 N(0) \sinh^{-1}\left(\frac{\omega_D}{\Delta}\right) \approx g_0 N(0) \ln\left[\frac{2\omega_D}{\Delta}\right], \tag{14.80}$$

so, in the superconducting ground state, the BCS gap is given by

$$\Delta = 2\omega_D e^{-\frac{1}{g_0 N(0)}}. \tag{14.81}$$

Remarks

- Note the disappearance of the factor of 2 in the exponent that appeared in Cooper's original calculation (14.25).
- The magnetic analogy has many intriguing consequences. One can immediately see that, like a magnet, there must be collective pair excitations, in which the isospins fluctuate about their ground-state orientations. Like magnons, these excitations form quantized collective modes. In a neutral superconductor, this leads to a gapless "sound" (Bogoliubov or Goldstone) mode, but in a charged superconductor the condensate phase mixes with the electromagnetic vector potential via the Anderson–Higgs mechanism (see Section 11.6) to produce the massive photon responsible for the Meissner effect.

14.4.3 The BCS ground state

In the vacuum $|0\rangle$, electron isospin operators all point "down," $\tau_{3\mathbf{k}} = -1$. To construct the ground state in which the isospins are aligned with the Weiss field, we need to construct a state in which each isospin is rotated relative to the vacuum. This is done by rotating the isospin at each momentum $\mathbf{k}$ through an angle $\theta_{\mathbf{k}}$ about the y-axis, as follows:

$$|\theta_{\mathbf{k}}\rangle = \exp\left[-i\frac{\theta_{\mathbf{k}}}{2}\psi_{\mathbf{k}}^{\dagger}\tau_y\psi_{\mathbf{k}}\right]|\Downarrow_{\mathbf{k}}\rangle = \left(\cos\frac{\theta_{\mathbf{k}}}{2} - i\sin\frac{\theta_{\mathbf{k}}}{2}\psi_{\mathbf{k}}^{\dagger}\tau_y\psi_{\mathbf{k}}\right)|\Downarrow_{\mathbf{k}}\rangle$$

$$= \cos\frac{\theta_{\mathbf{k}}}{2}|\Downarrow_{\mathbf{k}}\rangle - \sin\frac{\theta_{\mathbf{k}}}{2}|\Uparrow_{\mathbf{k}}\rangle. \tag{14.82}$$

The ground state is a product of these isospin states:

$$|BCS\rangle = \prod_{\mathbf{k}}|\theta_{\mathbf{k}}\rangle = \prod_{\mathbf{k}}\left(\cos\frac{\theta_{\mathbf{k}}}{2} + \sin\frac{\theta_{\mathbf{k}}}{2}c_{-\mathbf{k}\downarrow}^{\dagger}c_{\mathbf{k}\uparrow}^{\dagger}\right)|0\rangle, \tag{14.83}$$

where we have absorbed the minus sign by anticommuting the two electron operators. Following BCS, the coefficients $\cos\left(\frac{\theta_{\mathbf{k}}}{2}\right)$ and $\sin\left(\frac{\theta_{\mathbf{k}}}{2}\right)$ are labeled $u_{\mathbf{k}}$ and $v_{\mathbf{k}}$, respectively, writing

$$|BCS\rangle = \prod_{\mathbf{k}}|\theta_{\mathbf{k}}\rangle = \prod_{\mathbf{k}}\left(u_{\mathbf{k}} + v_{\mathbf{k}}c_{-\mathbf{k}\downarrow}^{\dagger}c_{\mathbf{k}\uparrow}^{\dagger}\right)|0\rangle, \tag{14.84}$$

where

$$u_{\mathbf{k}} \equiv \cos\left(\frac{\theta_{\mathbf{k}}}{2}\right) = \sqrt{\frac{1}{2}\left[1 + \underbrace{\cos\theta_{\mathbf{k}}}_{\epsilon_{\mathbf{k}}/E_{\mathbf{k}}}\right]} = \sqrt{\frac{1}{2}\left[1 + \frac{\epsilon_{\mathbf{k}}}{E_{\mathbf{k}}}\right]}$$

$$v_{\mathbf{k}} = \sin\left(\frac{\theta_{\mathbf{k}}}{2}\right) = \sqrt{\frac{1}{2}\left[1 - \cos\theta_{\mathbf{k}}\right]} = \sqrt{\frac{1}{2}\left[1 - \frac{\epsilon_{\mathbf{k}}}{E_{\mathbf{k}}}\right]}. \tag{14.85}$$

Remarks

- Dropping the normalization, the BCS wavefunction can be rewritten as a coherent state (14.45),

$$|BCS\rangle = \prod_{\mathbf{k}}\left(1 + \phi_{\mathbf{k}}c_{\mathbf{k}\uparrow}^{\dagger}c_{-\mathbf{k}\downarrow}^{\dagger}\right)|0\rangle) = \exp\left[\sum_{\mathbf{k}}\phi_{\mathbf{k}}c_{\mathbf{k}\uparrow}^{\dagger}c_{-\mathbf{k}\downarrow}^{\dagger}\right]|0\rangle = \exp\left[\Lambda^{\dagger}\right]|0\rangle, \tag{14.86}$$

where $\phi_{\mathbf{k}} = -\frac{v_{\mathbf{k}}}{u_{\mathbf{k}}}$ determines the Cooper pair wavefunction.
- We can thus expand the exponential in (14.86) as a coherent sum of pair-states:

$$|BCS\rangle = \sum_{n}\frac{1}{n!}(\Lambda^{\dagger})^{n}|0\rangle = \sum_{n}\frac{1}{\sqrt{n!}}|n\rangle, \tag{14.87}$$

where $|n\rangle = \frac{1}{\sqrt{n!}}(\Lambda^{\dagger})^{n}|0\rangle$ is a state containing n pairs.

The BCS wavefunction breaks gauge invariance, because it is not invariant under gauge transformations $c_{\mathbf{k}\sigma}^{\dagger} \to e^{i\alpha} c_{\mathbf{k}\sigma}^{\dagger}$ of the electron operators:

$$|BCS\rangle \to |\alpha\rangle = \prod_{\mathbf{k}}(1 + e^{2i\alpha}\phi_{\mathbf{k}} c_{\mathbf{k}\uparrow}^{\dagger} c_{-\mathbf{k}\downarrow}^{\dagger})|0\rangle = \sum \frac{e^{i2n\alpha}}{\sqrt{n!}}|n\rangle. \tag{14.88}$$

Under this transformation, the order parameter $\Delta = -g_0/V \sum_{\mathbf{k}}\langle\alpha|c_{-\mathbf{k}\downarrow}c_{\mathbf{k}\uparrow}|\alpha\rangle$ acquires a phase $\Delta \to e^{2i\alpha}|\Delta|$. On the other hand, the energy of the BCS state is unchanged by a gauge transformation, so the states $|\alpha\rangle$ must form a family of degenerate broken-symmetry states.

The action of the number operator $\hat{N}$ on this state may be represented as a differential with respect to phase:

$$\hat{N}|\alpha\rangle = \sum \frac{1}{\sqrt{n!}} 2n e^{i2n\alpha}|n\rangle = -i\frac{d}{d\alpha}|\alpha\rangle, \tag{14.89}$$

so that

$$\hat{N} \equiv -i\frac{d}{d\alpha}. \tag{14.90}$$

In this way, we see that the particle number is the generator of gauge transformations. Moreover, the phase of the order parameter is conjugate to the number operator, $[\alpha, N] = i$, and like position and momentum, or energy and time, the two variables therefore obey an uncertainty principle,

$$\Delta\alpha\,\Delta N \gtrsim 1. \tag{14.91}$$

Just as a macroscopic object with a precise position has an ill-defined momentum, a pair condensate with a sharply defined phase (relative to other condensates) is a physical state of matter – a macroscopic Schrödinger cat state – with an *ill-defined particle number*.

For the moment, we're ignoring the charge of the electron, but once we restore it, we will have to keep track of the vector potential, which also changes under gauge transformations.

14.5 Quasiparticle excitations in BCS theory

Let us now construct the quasiparticles of the BCS Hamiltonian. Recall that, for any one-particle Hamiltonian $H = \psi_{\alpha}^{\dagger} h_{\alpha\beta} \psi_{\beta}$, we can transform to an energy basis where the operators $a_k^{\dagger} = \psi_{\beta}^{\dagger}\langle\beta|k\rangle$ diagonalize $H = \sum_k E_k a_k^{\dagger} a_k$. Now the $\langle\beta|k\rangle$ are the eigenvectors of $h_{\alpha\beta}$, since $\langle\alpha|\hat{H}|k\rangle = E_k\langle\alpha|k\rangle = h_{\alpha\beta}\langle\beta|k\rangle$, so to construct quasiparticle operators we must project the particle operators onto the eigenvectors of $h_{\alpha\beta}$, $a_k^{\dagger} = \psi_{\beta}^{\dagger}\langle\beta|k\rangle$.

We now seek to diagonalize the BCS Hamiltonian, written in Nambu form:

$$H = \sum_{\mathbf{k}} \psi_{\mathbf{k}}^{\dagger} (\vec{h}_{\mathbf{k}} \cdot \vec{\tau})\psi_{\mathbf{k}} + \frac{V}{g_0}\bar{\Delta}\Delta.$$

The two-dimensional Nambu matrix

$$\underline{h}_{\mathbf{k}} = \epsilon_{\mathbf{k}}\tau_3 + \Delta_1\tau_1 + \Delta_2\tau_2 \equiv E_{\mathbf{k}}\hat{n}_{\mathbf{k}} \cdot \vec{\tau} \tag{14.92}$$

has two eigenvectors with isospin quantized parallel and antiparallel to $\hat{n}_{\mathbf{k}}$, [3]

$$\hat{n}_{\mathbf{k}} \cdot \vec{\tau} \begin{pmatrix} u_{\mathbf{k}} \\ v_{\mathbf{k}} \end{pmatrix} = + \begin{pmatrix} u_{\mathbf{k}} \\ v_{\mathbf{k}} \end{pmatrix}, \qquad \hat{n}_{\mathbf{k}} \cdot \vec{\tau} \begin{pmatrix} -v_{\mathbf{k}}^* \\ u_{\mathbf{k}}^* \end{pmatrix} = - \begin{pmatrix} -v_{\mathbf{k}}^* \\ u_{\mathbf{k}}^* \end{pmatrix}, \qquad (14.93)$$

and corresponding energies $\pm E_{\mathbf{k}} = \pm\sqrt{\epsilon_{\mathbf{k}}^2 + |\Delta_{\mathbf{k}}|^2}$. We can combine (14.93) into a single equation,

$$(\hat{n}_{\mathbf{k}} \cdot \vec{\tau})U_{\mathbf{k}} = U_{\mathbf{k}}\tau_3, \qquad (14.94)$$

where

$$U_{\mathbf{k}} = \begin{pmatrix} u_{\mathbf{k}} & -v_{\mathbf{k}}^* \\ v_{\mathbf{k}} & u_{\mathbf{k}}^* \end{pmatrix} \qquad (14.95)$$

is the unitary matrix formed from the eigenvectors of $\underline{h}_{\mathbf{k}}$. If we now project $\psi_{\mathbf{k}}^{\dagger}$ onto the eigenvectors of $h_{\mathbf{k}}$, we obtain the quasiparticle operators for the BCS Hamiltonian:

$$a_{\mathbf{k}\uparrow}^{\dagger} = \psi_{\mathbf{k}}^{\dagger} \cdot \begin{pmatrix} u_{\mathbf{k}} \\ v_{\mathbf{k}} \end{pmatrix} = c_{\mathbf{k}\uparrow}^{\dagger} u_{\mathbf{k}} + c_{-\mathbf{k}\downarrow} v_{\mathbf{k}}$$

$$a_{-\mathbf{k}\downarrow} = \psi_{\mathbf{k}}^{\dagger} \cdot \begin{pmatrix} -v_{\mathbf{k}}^* \\ u_{\mathbf{k}}^* \end{pmatrix} = c_{-\mathbf{k}\downarrow} u_{\mathbf{k}}^* - c_{\mathbf{k}\uparrow}^{\dagger} v_{\mathbf{k}}^*. \qquad (14.96)$$

Bogoliubov transformation

This transformation, mixing particles and holes, is named after its inventor, Nikolai Bogoliubov. If one takes the complex conjugate of the quasihole operator and reverses the momentum, one obtains $a_{\mathbf{k}\downarrow}^{\dagger} = c_{\mathbf{k}\downarrow}^{\dagger} u_{\mathbf{k}} - c_{-\mathbf{k}\uparrow} v_{\mathbf{k}}$, which defines the spin-down quasiparticle. The general expression for the spin-up and spin-down quasiparticles can be written

$$a_{\mathbf{k}\sigma}^{\dagger} = c_{\mathbf{k}\sigma}^{\dagger} u_{\mathbf{k}} + \text{sgn}(\sigma)c_{-\mathbf{k}-\sigma} v_{\mathbf{k}}. \qquad (14.97)$$

Let us combine the two expressions (14.96) into a single Nambu spinor $a_{\mathbf{k}}^{\dagger}$:

$$a_{\mathbf{k}}^{\dagger} = (a_{\mathbf{k}\uparrow}^{\dagger}, a_{-\mathbf{k}\downarrow}) = \psi_{\mathbf{k}}^{\dagger} \overbrace{\begin{pmatrix} u_{\mathbf{k}} & -v_{\mathbf{k}}^* \\ v_{\mathbf{k}} & u_{\mathbf{k}}^* \end{pmatrix}}^{U_{\mathbf{k}}} = \psi_{\mathbf{k}}^{\dagger} U_{\mathbf{k}}. \qquad (14.98)$$

Taking the Hermitian conjugate $a_{\mathbf{k}} = U_{\mathbf{k}}^{\dagger}\psi_{\mathbf{k}}$, then $\psi_{\mathbf{k}} = U_{\mathbf{k}}a_{\mathbf{k}}$, since $UU^{\dagger} = 1$. Using (14.94),

$$\psi_{\mathbf{k}}^{\dagger}h_{\mathbf{k}}\psi_{\mathbf{k}} = a_{\mathbf{k}}^{\dagger}\overbrace{U_{\mathbf{k}}^{\dagger}h_{\mathbf{k}}U_{\mathbf{k}}}^{U_{\mathbf{k}}E_{\mathbf{k}}\tau_3} a_{\mathbf{k}} = a_{\mathbf{k}}^{\dagger}E_{\mathbf{k}}\tau_3 a_{\mathbf{k}}, \qquad (14.99)$$

so that, as expected,

$$H = \sum_{\mathbf{k}} a_{\mathbf{k}}^{\dagger} E_{\mathbf{k}}\tau_3 a_{\mathbf{k}} + V\frac{\bar{\Delta}\Delta}{g_0} \qquad (14.100)$$

---

[3] Here complex conjugation is required to ensure that the complex eigenvectors are orthogonal when the gap is complex.

is diagonal in the quasiparticle basis. Written out explicitly,

$$H = \sum_{\mathbf{k}} E_{\mathbf{k}} \left(a_{\mathbf{k}\uparrow}^{\dagger} a_{\mathbf{k}\uparrow} - a_{-\mathbf{k}\downarrow} a_{-\mathbf{k}\downarrow}^{\dagger} \right) + V \frac{\bar{\Delta}\Delta}{g_0}. \tag{14.101}$$

If we rewrite the Hamiltonian in the form

$$H = \sum_{\mathbf{k}\sigma} E_{\mathbf{k}} \left(a_{\mathbf{k}\sigma}^{\dagger} a_{\mathbf{k}\sigma} - \frac{1}{2} \right) + V \frac{\bar{\Delta}\Delta}{g_0}, \tag{14.102}$$

we can interpret the excitation spectrum in terms of quasiparticles of energy $E_{\mathbf{k}} = \sqrt{\epsilon_{\mathbf{k}}^2 + |\Delta|^2}$ and a ground-state energy[4]

$$E_g = -\sum_{\mathbf{k}} E_{\mathbf{k}} + V \frac{\bar{\Delta}\Delta}{g_0}. \tag{14.104}$$

Now if the density of Bogoliubov quasiparticles per spin is $N_s(E)$, then, since the number of quasiparticle states is conserved, $N_s(E)dE = N_n(0)d|\epsilon|$ (where $N_n(0) = 2N(0)$ is the quasiparticle density of states in the normal state). It follows that

$$N_s^*(E) = N_n(0) \frac{d|\epsilon_{\mathbf{k}}|}{dE_{\mathbf{k}}} = N_n(0) \left(\frac{E}{\sqrt{E^2 - |\Delta|^2}} \right) \theta(E - |\Delta|), \tag{14.105}$$

where we have written $\epsilon_{\mathbf{k}} = \sqrt{E_{\mathbf{k}}^2 - |\Delta|^2}$ to obtain $d\epsilon_{\mathbf{k}}/dE_{\mathbf{k}} = E_{\mathbf{k}}/\sqrt{E_{\mathbf{k}}^2 - |\Delta|^2}$. The theta function describes the absence of states in the gap (see Figure 14.10(a)). Notice how the Andreev scattering causes states to pile up in a square-root singularity above the gap; this feature is called a *coherence peak*.

One of the most direct vindications of BCS theory derives from tunneling measurements of the excitation spectrum, in which the differential tunneling conductance is proportional to the quasiparticle density of states:

$$\frac{dI}{dV} \propto N_s(eV) = N_n(0) \frac{eV}{\sqrt{eV^2 - \Delta^2}} \theta(eV - |\Delta|). \tag{14.106}$$

The observation of such tunneling spectra in superconducting aluminum in 1960 by Ivar Giaever [19] provided the first direct confirmation of the energy gap predicted by BCS theory (see Figure 14.10(b)).

---

[4] Note that, if we were to restore the constant term $\sum_{\mathbf{k}} \epsilon_k$ dropped in (14.64), the ground-state energy becomes

$$E_g = \sum_{\mathbf{k}} (\epsilon_{\mathbf{k}} - E_{\mathbf{k}}) + V \frac{\bar{\Delta}\Delta}{g_0}. \tag{14.103}$$

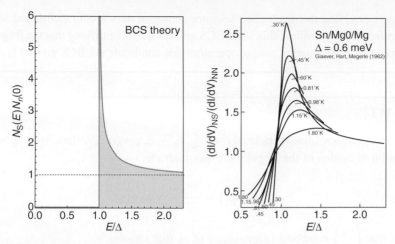

Contrasting (a) quasiparticle density of states with (b) measured tunneling density of states in Sn-MgO-Mg superconducting-normal tunnel junctions. In practice, finite temperature, disorder, variations in gap size around the Fermi surface, and strong-coupling corrections to BCS theory lead to small deviations from the ideal ground-state BCS density of states. Reprinted with permission from I. Giaever, *et al.*, *Phys. Rev.*, vol. 126, p. 941, 1962. Copyright 1962 by the American Physical Society.

Fig. 14.10

Example 14.2 Show that the BCS ground state is the vacuum for the Bogoliubov quasiparticles, i.e. that the destruction operators $a_{\mathbf{k}\sigma}$ annihilate the BCS ground state.

Solution

One way to confirm this is to directly construct the quasiparticle vacuum $|\psi\rangle$, by repeatedly applying the pair destruction operators to the electron vacuum, so that

$$|\psi\rangle = \prod_{\mathbf{k}} a_{-\mathbf{k}\downarrow} a_{\mathbf{k}\uparrow} |0\rangle$$

$$\Rightarrow a_{\mathbf{k}\sigma}|\psi\rangle = 0 \qquad (14.107)$$

for all $\mathbf{k}$, since the square of a destruction operator is zero, so $|\psi\rangle$ is the quasiparticle vacuum. Using the form (14.97),

$$a_{\mathbf{k}\uparrow} = u_{\mathbf{k}} c_{\mathbf{k}\uparrow} + v_{\mathbf{k}} c^{\dagger}_{-\mathbf{k}\downarrow}$$

$$a_{-\mathbf{k}\downarrow} = u_{\mathbf{k}} c_{-\mathbf{k}\downarrow} - v_{\mathbf{k}} c^{\dagger}_{\mathbf{k}\uparrow}, \qquad (14.108)$$

where for convenience we assume that $u_{\mathbf{k}}$ and $v_{\mathbf{k}}$ are real, we find

$$\prod_{\mathbf{k}} a_{-\mathbf{k}\downarrow} a_{\mathbf{k}\uparrow}|0\rangle = \prod_{\mathbf{k}} (u_{\mathbf{k}} c_{-\mathbf{k}\downarrow} - v_{\mathbf{k}} c^{\dagger}_{\mathbf{k}\uparrow})(u_{\mathbf{k}} c_{\mathbf{k}\uparrow} + v_{\mathbf{k}} c^{\dagger}_{-\mathbf{k}\downarrow})|0\rangle$$

$$= \prod_{\mathbf{k}} (u_{\mathbf{k}} v_{\mathbf{k}} c_{-\mathbf{k}\downarrow} c^{\dagger}_{-\mathbf{k}\downarrow} - (v_{\mathbf{k}})^2 c^{\dagger}_{\mathbf{k}\uparrow} c^{\dagger}_{-\mathbf{k}\downarrow})|0\rangle$$

$$= \prod_{\mathbf{k}} v_{\mathbf{k}} \times \prod_{\mathbf{k}} (u_{\mathbf{k}} + v_{\mathbf{k}} c^{\dagger}_{-\mathbf{k}\downarrow} c^{\dagger}_{\mathbf{k}\uparrow})|0\rangle \propto |BCS\rangle, \qquad (14.109)$$

where terms involving the destruction operator acting on the vacuum vanish and are omitted. Apart from normalization, this is the BCS ground state, confirming that the Bogoliubov quasiparticle operators are the unique operators that annihilate the BCS ground state.

Example 14.3

(a) If the Bogoliubov quasiparticle $\alpha_{\mathbf{k}\uparrow}^{\dagger} = c_{\mathbf{k}\uparrow}^{\dagger} u_{\mathbf{k}} + c_{-\mathbf{k}\downarrow} v_{\mathbf{k}} \alpha_{-\mathbf{k}\downarrow}$, then, starting with the equation of motion of the Bogoliubov quasiparticle,

$$[H, \alpha_{\mathbf{k}\uparrow}^{\dagger}] = \frac{\partial \alpha_{\mathbf{k}\uparrow}^{\dagger}}{\partial \tau} = E_{\mathbf{k}} \alpha_{\mathbf{k}\uparrow}^{\dagger}, \tag{14.110}$$

show that $\begin{pmatrix} u_{\mathbf{k}} \\ v_{\mathbf{k}} \end{pmatrix}$ must be an eigenvector of $h_{\mathbf{k}}$ that satisfies

$$\underline{h}_{\mathbf{k}} \begin{pmatrix} u_{\mathbf{k}} \\ v_{\mathbf{k}} \end{pmatrix} = \begin{pmatrix} \epsilon_{\mathbf{k}} & \Delta \\ \Delta & -\epsilon_{\mathbf{k}} \end{pmatrix} \begin{pmatrix} u_{\mathbf{k}} \\ v_{\mathbf{k}} \end{pmatrix} = E_{\mathbf{k}} \begin{pmatrix} u_{\mathbf{k}} \\ v_{\mathbf{k}} \end{pmatrix}. \tag{14.111}$$

(b) By solving the eigenvalue problem assuming the gap is real, show that

$$u_{\mathbf{k}}^2 = \frac{1}{2} \left[1 + \frac{\epsilon_{\mathbf{k}}}{\sqrt{\epsilon_{\mathbf{k}}^2 + \Delta^2}} \right]$$

$$v_{\mathbf{k}}^2 = \frac{1}{2} \left[1 - \frac{\epsilon_{\mathbf{k}}}{\sqrt{\epsilon_{\mathbf{k}}^2 + \Delta^2}} \right]. \tag{14.112}$$

Solution

(a) We begin by writing

$$\alpha_{\mathbf{k}\uparrow}^{\dagger} = \psi_{\mathbf{k}}^{\dagger} \cdot \begin{pmatrix} u_{\mathbf{k}} \\ v_{\mathbf{k}} \end{pmatrix} \tag{14.113}$$

where $\psi_{\mathbf{k}}^{\dagger} = (c_{\mathbf{k}\uparrow}^{\dagger}, c_{-\mathbf{k}\downarrow})$ is the Nambu spinor. Since $[H, \psi_{\mathbf{k}}^{\dagger}] = \psi_{\mathbf{k}}^{\dagger} \underline{h}_{\mathbf{k}}$, it follows that

$$[H, \alpha_{\mathbf{k}\uparrow}^{\dagger}] = \psi_{\mathbf{k}}^{\dagger} \underline{h}_{\mathbf{k}} \begin{pmatrix} u_{\mathbf{k}} \\ v_{\mathbf{k}} \end{pmatrix}. \tag{14.114}$$

Comparing (14.110) and (14.114), we see that the spinor $\begin{pmatrix} u_{\mathbf{k}} \\ v_{\mathbf{k}} \end{pmatrix}$ is an eigenvector of $h_{\mathbf{k}}$:

$$\underline{h}_{\mathbf{k}} \begin{pmatrix} u_{\mathbf{k}} \\ v_{\mathbf{k}} \end{pmatrix} = \begin{pmatrix} \epsilon_{\mathbf{k}} & \Delta \\ \Delta & -\epsilon_{\mathbf{k}} \end{pmatrix} \begin{pmatrix} u_{\mathbf{k}} \\ v_{\mathbf{k}} \end{pmatrix} = E_{\mathbf{k}} \begin{pmatrix} u_{\mathbf{k}} \\ v_{\mathbf{k}} \end{pmatrix}. \tag{14.115}$$

(b) Taking the determinant of the eigenvalue equation, $\det[\underline{h}_{\mathbf{k}} - E_{\mathbf{k}}\underline{1}] = E_{\mathbf{k}}^2 - \epsilon_{\mathbf{k}}^2 - \Delta^2 = 0$, and imposing the condition that $E_{\mathbf{k}} > 0$, we obtain $E_{\mathbf{k}} = \sqrt{\epsilon_{\mathbf{k}}^2 + \Delta^2}$.

Expanding the eigenvalue equation (14.115),

$$(E_{\mathbf{k}} - \epsilon_{\mathbf{k}})u_{\mathbf{k}} = \Delta v_{\mathbf{k}}$$

$$\Delta u_{\mathbf{k}} = (E_{\mathbf{k}} + \epsilon_{\mathbf{k}})v_{\mathbf{k}}. \qquad (14.116)$$

Multiplying these two equations, we obtain $(E_{\mathbf{k}} - \epsilon_{\mathbf{k}})u_{\mathbf{k}}^2 = (E_{\mathbf{k}} + \epsilon_{\mathbf{k}})v_{\mathbf{k}}^2$, or $\epsilon_{\mathbf{k}}(u_{\mathbf{k}}^2 + v_{\mathbf{k}}^2) = \epsilon_{\mathbf{k}} = E_{\mathbf{k}}(u_{\mathbf{k}}^2 - v_{\mathbf{k}}^2)$, since $u_{\mathbf{k}}^2 + v_{\mathbf{k}}^2 = 1$. It follows that $u_{\mathbf{k}}^2 - v_{\mathbf{k}}^2 = \epsilon_{\mathbf{k}}/E_{\mathbf{k}}$. Combining this with $u_{\mathbf{k}}^2 + v_{\mathbf{k}}^2 = 1$, we obtain the results given in (14.112).

14.6 Path integral formulation

Following our discussion of the physics, let us return to the math to examine how the BCS mean-field theory is succinctly formulated using path integrals. The appearance of single pairing fields A and $A^\dagger$ in the BCS Hamiltonian makes it particularly easy to apply path-integral methods. We begin by writing the problem as a path integral:

$$Z = \int \mathcal{D}[\bar{c}, c] e^{-S}, \qquad (14.117)$$

where

$$S = \int_0^\beta \sum_{\mathbf{k}\sigma} \bar{c}_{\mathbf{k}\sigma}(\partial_\tau + \epsilon_{\mathbf{k}})c_{\mathbf{k}\sigma} - \frac{g_0}{V}\bar{A}A. \qquad (14.118)$$

Here the condition $|\epsilon_{\mathbf{k}}| < \omega_D$ is implicit in all momentum sums. Next, we carry out the Hubbard–Stratonovich transformation (see Chapter 13):

$$- g\bar{A}A \rightarrow \bar{\Delta}A + A\bar{\Delta} + \frac{V}{g_0}\bar{\Delta}\Delta, \qquad (14.119)$$

where $\bar{\Delta}(\tau)$ and $\Delta(\tau)$ are fluctuating complex fields. Inside the path integral this substitution is formally exact, but its real value lies in the static mean-field solution it furnishes for superconductivity. We then obtain

$$Z = \int \mathcal{D}[\bar{\Delta}, \Delta, \bar{c}, c] e^{-S}$$

$$S = \int_0^\beta d\tau \left\{ \sum_{\mathbf{k}\sigma} \bar{c}_{\mathbf{k}\sigma}(\partial_\tau + \epsilon_{\mathbf{k}})c_{\mathbf{k}\sigma} + \bar{\Delta}A + A\bar{\Delta} + \frac{V}{g_0}\bar{\Delta}\Delta \right\}. \qquad (14.120)$$

The Hamiltonian part of this expression can be compactly reformulated in terms of Nambu spinors, following precisely the same steps used for the operator Hamiltonian. To transform the Berry phase term (see (12.132)), we note that, since the Nambu spinors satisfy a conventional anticommutation algebra, they must have precisely the same Berry phase term as conventional fermions, i.e. $\int d\tau \, \bar{c}_{\mathbf{k}\sigma} \partial_\tau c_{\mathbf{k}\sigma} = \int d\tau \, \bar{\psi}_{\mathbf{k}} \partial_\tau \psi_{\mathbf{k}}$.[5]

Putting this all together, the partition function and the action can now be rewritten:

$$Z = \int \mathcal{D}[\bar{\Delta}, \Delta, \bar{\psi}, \psi] e^{-S}$$

$$S = \int_0^\beta d\tau \left\{ \sum_{\mathbf{k}} \bar{\psi}_{\mathbf{k}} (\partial_\tau + \underline{h}_{\mathbf{k}}) \psi_{\mathbf{k}} + \frac{V}{g_0} \bar{\Delta} \Delta \right\}, \qquad (14.122)$$

where $\underline{h}_{\mathbf{k}} = \epsilon_{\mathbf{k}} \tau_3 + \Delta_1 \tau_1 + \Delta_2 \tau_2$, with $\Delta = \Delta_1 - i\Delta_2$, $\bar{\Delta} = \Delta_1 + i\Delta_2$. Since the action is explicitly quadratic in the Fermi fields, we can carry out the Gaussian integral of the Fermi fields to obtain

$$Z = \int \mathcal{D}[\bar{\Delta}, \Delta] e^{-S_E[\bar{\Delta}, \Delta]}$$

$$e^{-S_E[\bar{\Delta}, \Delta]} = \prod_{\mathbf{k}} \det[\partial_\tau + \underline{h}_{\mathbf{k}}(\tau)] e^{-V \int_0^\beta d\tau \frac{\bar{\Delta}\Delta}{g_0}} \qquad (14.123)$$

for the effective action, where we have separated the fermionic determinant into a product over each decoupled momentum. Thus

$$S_E[\bar{\Delta}, \Delta] = V \int_0^\beta d\tau \frac{\bar{\Delta}\Delta}{g_0} + \sum_{\mathbf{k}} \mathrm{Tr} \ln(\partial_\tau + \underline{h}_{\mathbf{k}}), \qquad (14.124)$$

where we have replaced $\ln \det \to \mathrm{Tr} \ln$. This is the action of electrons moving in a *time-dependent* pairing field $\Delta(\tau)$.

14.6.1 Mean-field theory as a saddle point of the path integral

Although we can only explicitly calculate S_E in static configurations of the pair field, in BCS theory it is *precisely* these configurations that saturate the path integral in the thermodynamic limit ($V \to \infty$). To see this, consider the path integral

$$Z = \int \mathcal{D}[\bar{\Delta}, \Delta] e^{-S_E[\bar{\Delta}, \Delta]}. \qquad (14.125)$$

Every term in the effective action is extensive in the volume V, so if we find a static configuration of $\Delta = \Delta_0$ which minimizes $S_E = V S_0$, so that $\delta S_E / \delta \Delta = 0$, fluctuations $\delta \Delta$

---

[5] We can confirm this result by anticommuting the down-spin Grassmans in the Berry phase, then integrating by parts:

$$S_B = \sum_{\mathbf{k}} \int_0^\beta d\tau \left[\bar{c}_{\mathbf{k}\uparrow} \partial_\tau c_{\mathbf{k}\uparrow} - (\partial_\tau c_{-\mathbf{k}\downarrow}) \bar{c}_{-\mathbf{k}\downarrow} \right] = \sum_{\mathbf{k}} \int_0^\beta d\tau \left[\bar{c}_{\mathbf{k}\uparrow} \partial_\tau c_{\mathbf{k}\uparrow} + c_{-\mathbf{k}\downarrow} \partial_\tau \bar{c}_{-\mathbf{k}\downarrow} - \overbrace{\partial_\tau (c_{-\mathbf{k}\downarrow} \bar{c}_{-\mathbf{k}\downarrow})}^{\to 0} \right]$$

$$= \sum_{\mathbf{k}} \int_0^\beta d\tau \left[\bar{\psi}_{\mathbf{k}} \partial_\tau \psi_{\mathbf{k}} \right]. \qquad (14.121)$$

The antiperiodicity of the Grassman fields in imaginary time causes the total derivative to vanish.

around this configuration will cost a free energy that is of order $O(V)$, i.e. the amplitude for a small fluctuation is given by

$$e^{-S} = e^{-VS_0 + O(V \times |\delta\Delta|^2)}. \tag{14.126}$$

The appearance of V in the coefficient of this Gaussian distribution implies the variance of small fluctuations around the minimum will be of order $\langle \delta\Delta^2 \rangle \sim O(1/V)$ so that, to a good approximation,

$$Z \approx Z_{BCS} = e^{-S_E[\bar{\Delta}_0, \Delta_0]}. \tag{14.127}$$

This is why the mean-field approximation to the path integral is essentially exact for the BCS model. Note that we can also expand the effective action as a Gaussian path integral:

$$Z_{BCS} = \int \mathcal{D}[\bar\psi, \psi] e^{-S_{MFT}}$$

$$S_{MFT} = \int_0^\beta d\tau \left\{ \sum_{\mathbf{k}} \bar\psi_{\mathbf{k}} (\partial_\tau + \overbrace{\epsilon_{\mathbf{k}}\tau_3 + \Delta_1\tau_1 + \Delta_2\tau_2}^{h_{\mathbf{k}}}) \psi_{\mathbf{k}} + \frac{V}{g_0} \bar\Delta\Delta \right\}, \tag{14.128}$$

in which the saddle-point solution $\Delta^{(0)}(\tau) \equiv \Delta = \Delta_1 - i\Delta_2$ is assumed to be static. Since this is a Gaussian integral, we can immediately carry out the the integral to obtain

$$Z_{BCS} = \prod_{\mathbf{k}} \det(\partial_\tau + h_{\mathbf{k}}) \exp\left[-\frac{V\beta}{g_0} \bar\Delta\Delta \right].$$

It is far easier to work in Fourier space, writing the Nambu fields in terms of their Fourier components:

$$\psi_{\mathbf{k}}(\tau) = \frac{1}{\sqrt{\beta}} \sum_n \psi_{\mathbf{k}n} e^{-i\omega_n\tau}. \tag{14.129}$$

In this basis,

$$\partial_\tau + h \to [-i\omega_n + \underline{h}_{\mathbf{k}}], \tag{14.130}$$

and the path integral is now diagonal in momentum and frequency:

$$Z_{BCS} = \int \prod_{\mathbf{k}n} d\bar\psi_{\mathbf{k}n} d\psi_{\mathbf{k}n} e^{-S_{MFT}[\bar\psi_{\mathbf{k}n}, \psi_{\mathbf{k}n}]}$$

$$S_{MFT}[\bar\psi_{\mathbf{k}n}, \psi_{\mathbf{k}n}] = \sum_{\mathbf{k}\,n} \bar\psi_{\mathbf{k}n} (-i\omega_n + h_{\mathbf{k}}) \psi_{\mathbf{k}n} + \beta V \frac{\bar\Delta\Delta}{g_0}. \tag{14.131}$$

Remarks

- The distribution function $P[\psi_{\mathbf{k}}]$ for the fermion fields is Gaussian:

$$P[\psi_{\mathbf{k}n}] \sim e^{-S_{MFT}} \propto \exp[-\bar\psi_{\mathbf{k}n}(-i\omega_n + h_{\mathbf{k}})\psi_{\mathbf{k}n}], \tag{14.132}$$

so that the amplitude of fluctuations (see 12.144) is given by

$$\langle \psi_{\mathbf{k}n} \bar{\psi}_{\mathbf{k}n} \rangle = -\mathcal{G}(\mathbf{k}, i\omega_n) = [-i\omega_n + h_{\mathbf{k}}]^{-1}, \qquad (14.133)$$

which is the electron Green's function in the superconductor. We shall study this in the next section.

- We can now evaluate the determinant

$$\det[\partial_\tau + \underline{h}_{\mathbf{k}}] = \prod_n \det[-i\omega_n + \underline{h}_{\mathbf{k}}] = \prod_n [\omega_n^2 + \epsilon_{\mathbf{k}}^2 + |\Delta|^2]. \qquad (14.134)$$

With these results, we can fully evaluate the partition function

$$Z_{BCS} = \prod_n [\omega_n^2 + \epsilon_{\mathbf{k}}^2 + |\Delta|^2] \times e^{-\frac{\beta V |\Delta|^2}{g_0}} = e^{-S_E}, \qquad (14.135)$$

and the effective action is then

$$\mathcal{F}[\Delta, T] = \frac{S_E}{\beta} = -T \sum_{\mathbf{k}n} \ln[\omega_n^2 + \epsilon_{\mathbf{k}}^2 + |\Delta|^2] + V \frac{|\Delta|^2}{g_0}. \qquad (14.136)$$

free energy: BCS pair condensate

This is the mean-field free-energy for the BCS model.

Remarks

- This quantity provides a microscopic realization of the Landau free energy of a super-conductor, discussed in Chapter 11. Notice how $\mathcal{F}$ is invariant under changes in the phase of the gap function so that $\mathcal{F}[\Delta, T] = \mathcal{F}[\Delta e^{i\phi}, T]$, which follows from particle conservation. (The number operator, which commutes with H, is the generator of phase translations.)
- Following our discussion in Chapter 11, we expect that below T_c the free energy $\mathcal{F}[\Delta, T]$ develops a minimum at finite $|\Delta|$, forming a "Mexican hat" potential (Figure 14.11).
- Notice the appearance of the quasiparticle energy $E_{\mathbf{k}} = \sqrt{\epsilon_{\mathbf{k}}^2 + |\Delta|^2}$ inside the logarithm.

To identify the equilibrium gap Δ, we minimize $\mathcal{F}$ with respect to $\bar{\Delta}$, which leads to the BCS gap equation,

$$\frac{\partial \mathcal{F}}{\partial \bar{\Delta}} = -\sum_{\mathbf{k}n} \frac{\Delta}{\omega_n^2 + E_{\mathbf{k}}^2} + V \frac{\Delta}{g_0} = 0 \qquad (14.137)$$

or

$$\frac{1}{g_0} = \frac{1}{\beta V} \sum_{\mathbf{k}n} \frac{1}{\omega_n^2 + E_{\mathbf{k}}^2}.$$

BCS gap equation

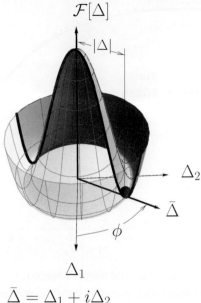

$$\bar{\Delta} = \Delta_1 + i\Delta_2$$

Showing the form of $\mathcal{F}[\Delta]$ for $T < T_c$. The free energy is a minimum at a finite value of $|\Psi|$. The free energy is invariant under changes in phase of the gap, which are generated by the number operator $\hat{N} \propto -i\frac{d}{d\phi}$. See Exercise 14.4.

Fig. 14.11

If we now convert the Matsubara sum to a contour integral, we obtain

$$\frac{1}{\beta}\sum_n \frac{1}{\omega_n^2 + E_{\mathbf{k}}^2} = -\oint \frac{dz}{2\pi i} f(z) \frac{1}{z^2 - E_{\mathbf{k}}^2} = -\oint \frac{dz}{2\pi i} f(z) \frac{1}{2E_{\mathbf{k}}} \left[\frac{1}{z - E_{\mathbf{k}}} - \frac{1}{z + E_{\mathbf{k}}} \right]$$

$$= -\sum_{\mathbf{k}} \overbrace{(f(E_{\mathbf{k}}) - f(-E_{\mathbf{k}}))}^{=2f(E_{\mathbf{k}})-1} \frac{1}{2E_{\mathbf{k}}} = \frac{\tanh(\beta E_{\mathbf{k}}/2)}{2E_{\mathbf{k}}}, \tag{14.138}$$

where the integral runs counterclockwise around the poles at $z = \pm E_{\mathbf{k}}$. Thus the gap equation can be rewritten as

$$\frac{1}{g_0} = \int_{|\epsilon_{\mathbf{k}}|<\omega_D} \frac{d^3 k}{(2\pi)^3} \left[\frac{\tanh(\beta E_{\mathbf{k}}/2)}{2E_{\mathbf{k}}} \right], \qquad \text{BCS gap equation II} \qquad (14.139)$$

where we have reinstated the implicit energy shell restriction $|\epsilon_{\mathbf{k}}| < \omega_D$. If we approximate the density of states by a constant $N(0)$ per spin over the narrow shell of states around the Fermi surface, we may replace the momentum sum by an energy integral, so that

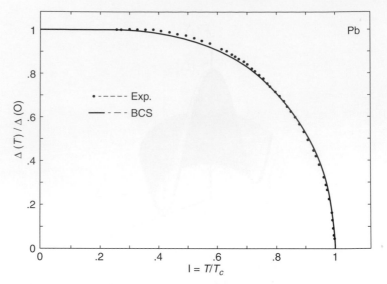

Fig. 14.12 Comparison between the dependence of the gap on the reduced temperature T/T_C and the gap measured by tunneling in superconducting lead. Reprinted with permission from R. F. Gasparovic, *et al.*, *Solid State Commun.*, vol. 4, p. 59, 1966. Copyright 1966 Elsevier.

$$\frac{1}{g_0 N(0)} = \int_0^{\omega_D} d\epsilon \left[\frac{\tanh(\beta\sqrt{\epsilon^2 + \Delta^2}/2)}{\sqrt{\epsilon^2 + \Delta^2}} \right]. \tag{14.140}$$

At absolute zero, the hyperbolic tangent becomes unity. If we subtract this equation from its zero-temperature value, it becomes

$$\int_0^\infty d\epsilon \left[\frac{\tanh(\beta\sqrt{\epsilon^2 + \Delta^2}/2)}{\sqrt{\epsilon^2 + \Delta^2}} - \frac{1}{\sqrt{\epsilon^2 + \Delta_0^2}} \right] = 0, \tag{14.141}$$

where $\Delta_0 = \Delta(T = 0)$ is the zero-temperature gap. Since the argument of the integrand now rapidly converges to zero at high energies, we can set the upper limit of integration to zero. This is a useful form for the numerical evaluation of the temperature dependence of the gap. Figure 14.12 contrasts the BCS prediction of the temperature-dependent gap obtained from (14.141), with the gap measured from tunneling in lead.

Example 14.4 Carry out the Matsubara sum in (14.136) to derive a an explicit form for the free energy of the superconducting condensate in terms of the quasiparticle excitation energies:

$$\mathcal{F} = -2TV \int_{|\epsilon_\mathbf{k}| < \omega_D} \frac{d^3 k}{(2\pi)^3} \left[\ln[2\cosh(\beta E_\mathbf{k}/2)] \right] + V\frac{|\Delta|^2}{g_0}. \tag{14.142}$$

Solution

Using the contour integration method, we can rewrite (14.136) as

$$\mathcal{F} = -\sum_{\mathbf{k}} \oint \frac{dz}{2\pi i} f(z)\ln[z^2 - E_{\mathbf{k}}^2] + V\frac{|\Delta|^2}{g_0}, \qquad (14.143)$$

where the integral runs counterclockwise around the poles of the Fermi function. The logarithm inside the integral can be split up into two terms,

$$\ln[z^2 - E_{\mathbf{k}}^2] \to \ln[E_{\mathbf{k}} - z] + \ln[-E_{\mathbf{k}} - z], \qquad (14.144)$$

which we immediately recognize as the contributions from fermions with energies $\pm E_{\mathbf{k}}$, so that the result of carrying out the contour integral is

$$\mathcal{F} = -TV \int \frac{d^3k}{(2\pi)^3} \left[\ln[1 + e^{-\beta E_{\mathbf{k}}}] + \ln[1 + e^{\beta E_{\mathbf{k}}}] \right] + V\frac{|\Delta|^2}{g_0}$$

$$= -2TV \int_{|\epsilon_{\mathbf{k}}|<\omega_D} \frac{d^3k}{(2\pi)^3} \left[\ln[2\cosh(\beta E_{\mathbf{k}}/2)] \right] + V\frac{|\Delta|^2}{g_0}. \qquad (14.145)$$

14.6.2 Computing Δ and T_c

To compute T_c we shall take the Matsubara form of the gap equation (14.136), which we rewrite by replacing the sum over momenta by an integral near the Fermi energy, $\frac{1}{V}\sum_{\mathbf{k}} \to N(0)\int d\epsilon$, to get

$$\frac{1}{g_0} = TN(0) \sum_{n} \int_{-\infty}^{\infty} d\epsilon \frac{1}{\omega_n^2 + \epsilon_{\mathbf{k}}^2 + \Delta^2} = \pi TN(0) \sum_{|\omega_n|<\omega_D} \frac{1}{\sqrt{\omega_n^2 + \Delta^2}}, \qquad (14.146)$$

where we have extended the limits of integration over energy to infinity. By carrying out the integral over energy first, we are forced to impose the cut-off on the Matsubara frequencies.

If we now take $T \to 0$ in this expression, we may replace

$$T\sum_{n} = T\sum \frac{\Delta\omega_n}{2\pi T} \to \int \frac{d\omega}{2\pi}, \qquad (14.147)$$

so that at zero temperature (setting $T = 0$) we obtain

$$1 = gN(0) \int_0^{\omega_D} \frac{d\epsilon}{\sqrt{\epsilon^2 + \Delta^2}} = gN(0)\left[\sinh^{-1}\left(\frac{\omega_D}{\Delta}\right)\right] \approx gN(0)\ln\left(\frac{2\omega_D}{\Delta}\right), \quad (14.148)$$

where we have assumed $gN(0)$ is small, so that $\omega_D/\Delta \gg 1$. We may now solve for the zero-temperature gap, to obtain

$$\Delta = 2\omega_D e^{-\frac{1}{gN(0)}}. \qquad (14.149)$$

This recovers the form of the gap first derived in Section 14.4.2.

To calculate the transition temperature T_c, we note that, just below the transition temperature, the gap becomes infinitesimally small, so that $\Delta(T_c^-) = 0$. Substituting this into (14.147), we obtain

$$\frac{1}{gN(0)} = \pi T_c \sum_{|\omega_n| < \omega_D} \frac{1}{|\omega_n|} = 2\pi T_c \sum_{n=0}^{\infty} \left(\frac{1}{\omega_n} - \frac{1}{\omega_n + \omega_D} \right), \tag{14.150}$$

where we have imposed the limit on ω_n by subtracting an identical term, with $\omega_n \to \omega_n + \omega_D$. Simplifying this expression gives

$$\frac{1}{gN(0)} = \sum_{n=0}^{\infty} \left(\frac{1}{n + \frac{1}{2}} - \frac{1}{\omega_n + \frac{1}{2} + \frac{\omega_D}{2\pi T_c}} \right). \tag{14.151}$$

At this point we can use an extremely useful identity of the digamma function $\psi(z) = \frac{d}{dz} \ln \Gamma(z)$,

$$\psi(z) = -\zeta - \sum_{n=0}^{\infty} \left(\frac{1}{z+n} - \frac{1}{1+n} \right), \tag{14.152}$$

where $\zeta = 0.577216 = -\psi(1)$ is the Euler constant, so that

$$\frac{1}{gN(0)} = \overbrace{\psi(\frac{1}{2} + \frac{\omega_D}{2\pi T_c})}^{\approx \ln(\omega_D/(2\pi T_c))} - \psi(\frac{1}{2}) = \ln\left(\frac{\omega_D e^{-\psi(\frac{1}{2})}}{2\pi T_c} \right). \tag{14.153}$$

We have approximated $\psi(z) \approx \ln z$ for large $|z|$. Thus,

$$T_c = \overbrace{\left(\frac{e^{-\psi(1/2)}}{2\pi} \right)}^{\approx 1.13} \omega_D e^{-\frac{1}{g_0 N(0)}}. \tag{14.154}$$

Notice that the details of the way we introduced the cut-off into the sums affects both the gap Δ in (14.149) and the transition temperature in (14.154). However, the ratio of twice the gap to T_C,

$$\frac{2\Delta}{T_c} = 8\pi e^{\psi(\frac{1}{2})} \approx 3.53 \tag{14.155}$$

is *universal* for BCS superconductors, because the details of the cut-off cancel out of this ratio. Experiments confirm that this ratio of gap to transition is indeed observed in phonon-mediated superconductors.

14.7 The Nambu–Gor'kov Green's function

To describe the propagation of electrons and the Andreev scattering between electron and hole requires a matrix Green's function, formed from two Nambu spinors. This object, written

$$\mathcal{G}_{\alpha\beta}(\mathbf{k}, \tau) = -\langle T \psi_{\mathbf{k}\alpha}(\tau) \psi_{\mathbf{k}\beta}^\dagger(0) \rangle, \tag{14.156}$$

is called the *Nambu–Gor'kov Green's function*. Written out more explicitly, it takes the form

$$\mathcal{G}(\mathbf{k}, \tau) = -\left\langle T \begin{pmatrix} c_{\mathbf{k}\uparrow}(\tau) \\ \bar{c}^{\dagger}_{-\mathbf{k}\downarrow}(\tau) \end{pmatrix} \otimes (c^{\dagger}_{\mathbf{k}\uparrow}(0), c_{-\mathbf{k}\downarrow}(0)) \right\rangle$$

$$= -\begin{bmatrix} \langle Tc_{\mathbf{k}\uparrow}(\tau)c^{\dagger}_{\mathbf{k}\uparrow}(0)\rangle & \langle Tc_{\mathbf{k}\uparrow}(\tau)c_{-\mathbf{k}\downarrow}(0)\rangle \\ \langle Tc^{\dagger}_{-\mathbf{k}\downarrow}(\tau)c^{\dagger}_{\mathbf{k}\uparrow}(0)\rangle & \langle Tc^{\dagger}_{-\mathbf{k}\downarrow}(\tau)c_{-\mathbf{k}\downarrow}(0)\rangle \end{bmatrix}. \tag{14.157}$$

The unusual off-diagonal components

$$F(\mathbf{k}, \tau) = -\langle Tc_{\mathbf{k}\uparrow}(\tau)c_{-\mathbf{k}\downarrow}(0)\rangle, \qquad \bar{F}(\mathbf{k}, \tau) = -\langle Tc^{\dagger}_{-\mathbf{k}\downarrow}(\tau)c^{\dagger}_{\mathbf{k}\uparrow}(0)\rangle \tag{14.158}$$

in $\mathcal{G}(\mathbf{k}, \tau)$ describe the amplitude for an electron to convert to a hole as it Andreev scatters off the condensate.

Now from (12.142) and (14.131) the Green's function is given by the inverse of the Gaussian action, $\mathcal{G} = -(\partial_{\tau} - \mathcal{H})^{-1}$, or, in Matsubara space,

$$\mathcal{G}(\mathbf{k}, i\omega_n) = [i\omega_n - h_{\mathbf{k}}]^{-1} \equiv \frac{1}{(i\omega_n - h_{\mathbf{k}})}, \tag{14.159}$$

where we use the notation $\frac{1}{M} \equiv M^{-1}$ to denote the inverse of the matrix M. Now since $h_{\mathbf{k}} = \epsilon_{\mathbf{k}}\tau_3 + \Delta_1\tau_1 + \Delta_2\tau_2$ (14.92) is a sum of Pauli matrices, its square is diagonal: $h_{\mathbf{k}}^2 = \epsilon_{\mathbf{k}}^2 + \Delta_1^2 + \Delta_2^2 = E_{\mathbf{k}}^2$ and thus $(i\omega_n - h_{\mathbf{k}})(i\omega_n + h_{\mathbf{k}}) = (i\omega_n)^2 - E_{\mathbf{k}}^2$. Using the matrix identity $\frac{1}{B} = A\frac{1}{BA}$, we may then write

$$\underline{\mathcal{G}}(k) = (i\omega_n + h_{\mathbf{k}})\frac{1}{(i\omega_n - h_{\mathbf{k}})(i\omega_n + h_{\mathbf{k}})} = \frac{(i\omega_n + h_{\mathbf{k}})}{[(i\omega_n)^2 - E_{\mathbf{k}}^2]}. \tag{14.160}$$

Written out explicitly, this is

$$\underline{\mathcal{G}}(\mathbf{k}, i\omega_n) = \frac{1}{(i\omega_n)^2 - E_{\mathbf{k}}^2} \begin{bmatrix} i\omega_n + \epsilon_{\mathbf{k}} & \Delta \\ \bar{\Delta} & i\omega_n - \epsilon_{\mathbf{k}} \end{bmatrix}, \tag{14.161}$$

where $E_{\mathbf{k}} = \sqrt{\epsilon_{\mathbf{k}}^2 + \Delta^2}$ is the quasiparticle energy.

To gain insight, let us obtain the same results diagrammatically. Andreev scattering converts a particle into a hole, which we denote by the Feynman scattering vertices

$$\bar{\Delta}c_{-\mathbf{k}\downarrow}c_{\mathbf{k}\uparrow} \equiv \overset{k}{\longrightarrow} \overset{\bar{\Delta}}{\times} \overset{-k}{\longleftarrow} \bar{\Delta}$$

$$\Delta c^{\dagger}_{\mathbf{k}\uparrow}c^{\dagger}_{-\mathbf{k}\downarrow} \equiv \overset{-k}{\longleftarrow} \overset{\Delta}{\times} \overset{k}{\longrightarrow} \Delta.$$

$$\tag{14.162}$$

The bare propagators for the electron and hole are the diagonal components of the bare Nambu propagator:

$$\underline{\mathcal{G}}_0(k) = \frac{1}{i\omega_n - \epsilon_{\mathbf{k}}\tau_3} = \begin{bmatrix} \frac{1}{i\omega_n - \epsilon_{\mathbf{k}}} & \\ & \frac{1}{i\omega_n + \epsilon_{\mathbf{k}}} \end{bmatrix}. \tag{14.163}$$

We denote these two components by the diagrams

$$
\begin{aligned}
\xrightarrow{\ \ k\ \ } \quad &\equiv G_0(k) \quad = \frac{1}{i\omega_n - \epsilon_{\mathbf{k}}} \\
\xleftarrow{\ -k\ } \quad &\equiv -G_0(-k) = \frac{1}{i\omega_n + \epsilon_{\mathbf{k}}}\, .
\end{aligned}
$$

$$(14.164)$$

(There is a minus sign in the second term because we have commuted creation and annihilation operators to construct the hole propagator.) The Feynman diagrams for the conventional propagator are given by

$$(14.165)$$

involving an even number of Andreev reflections. This enables us to identify a self-energy term that describes the Andreev scattering off a hole state:

$$
\Sigma(k) = \frac{|\Delta|^2}{i\omega_n + \epsilon_{\mathbf{k}}}\, .
$$

$$(14.166)$$

We may then redraw the propagator as

$$
\begin{aligned}
G(k) &= \longrightarrow \;+\; \longrightarrow\!\boxed{\Sigma}\!\longrightarrow \;+\; \longrightarrow\!\boxed{\Sigma}\!\longrightarrow\!\boxed{\Sigma}\!\longrightarrow \;+\;\cdots \\
&= \frac{1}{i\omega_n - \epsilon_{\mathbf{k}} - \Sigma(i\omega_n)} = \frac{1}{i\omega_n - \epsilon_{\mathbf{k}} - \frac{|\Delta|^2}{i\omega_n+\epsilon_{\mathbf{k}}}} = \frac{i\omega_n + \epsilon_{\mathbf{k}}}{(i\omega_n)^2 - E_{\mathbf{k}}^2}\, .
\end{aligned}
$$

$$(14.167)$$

In a similar way, the anomalous propagator is given by

$$(14.168)$$

so that

$$
F(k) = \frac{\Delta}{i\omega_n + \epsilon_{\mathbf{k}}}\,\frac{1}{i\omega_n - \epsilon_{\mathbf{k}} - \frac{|\Delta|^2}{i\omega_n+\epsilon_{\mathbf{k}}}} = \frac{\Delta}{(i\omega_n)^2 - E_{\mathbf{k}}^2}\, .
$$

$$(14.169)$$

Example 14.5   Decompose the Nambu–Gor'kov Green's function in terms of its quasiparticle poles, and show that the diagonal part can be written

$$
G(k) = \frac{u_{\mathbf{k}}^2}{i\omega_n - E_{\mathbf{k}}} + \frac{v_{\mathbf{k}}^2}{i\omega_n + E_{\mathbf{k}}}\, .
$$

$$(14.170)$$

Solution

To carry out this decomposition, it is convenient to introduce the projection operators

$$P_+(\mathbf{k}) = \frac{1}{2}(\underline{1} + \hat{n} \cdot \vec{\tau}), \qquad P_-(\mathbf{k}) = \frac{1}{2}(\underline{1} - \hat{n} \cdot \vec{\tau}), \tag{14.171}$$

which satisfy $P_+^2 = P_+$, $P_-^2 = P_-$, and $P_+ + P_- = 1$, and furthermore,

$$P_+(\mathbf{k})(\hat{n}_\mathbf{k} \cdot \vec{\tau}) = P_+(\mathbf{k}), \qquad P_-(\mathbf{k})(\hat{n}_\mathbf{k} \cdot \vec{\tau}) = -P_-(\mathbf{k}), \tag{14.172}$$

so that these operators conveniently project the isospin onto the directions $\pm n_\mathbf{k}$.

We can use the projectors $P_\pm(\mathbf{k})$ to project the Nambu propagator as follows:

$$\begin{aligned}
\underline{\mathcal{G}} &= (P_+ + P_-)\frac{1}{i\omega_n - E_\mathbf{k}\hat{n} \cdot \vec{\tau}} \\
&= P_+\frac{1}{i\omega_n - E_\mathbf{k}} + P_-\frac{1}{i\omega_n + E_\mathbf{k}}.
\end{aligned} \tag{14.173}$$

We can interpret these two terms as the quasiparticle and quasihole parts of the Nambu propagator. If we explicitly expand this expression, using

$$\hat{n} = \left(\frac{\epsilon}{E_\mathbf{k}}, \frac{\Delta_1}{E_\mathbf{k}}, \frac{\Delta_2}{E_\mathbf{k}}\right), \tag{14.174}$$

then

$$P_\pm = \frac{1}{2}\underline{1} \pm \begin{bmatrix} \frac{\epsilon_\mathbf{k}}{E_\mathbf{k}} & \frac{\Delta}{2E_\mathbf{k}} \\ \frac{\Delta}{2E_\mathbf{k}} & -\frac{\epsilon_\mathbf{k}}{2E_\mathbf{k}} \end{bmatrix}, \tag{14.175}$$

where $\Delta = \Delta_1 - i\Delta_2$, and we find that the diagonal part of the Green's function is given by

$$\begin{aligned}
G(k) &= \frac{1}{2}\left(1 + \frac{\epsilon_\mathbf{k}}{E_\mathbf{k}}\right)\frac{1}{i\omega_n - E_\mathbf{k}} + \frac{1}{2}\left(1 - \frac{\epsilon_\mathbf{k}}{E_\mathbf{k}}\right)\frac{1}{i\omega_n + E_\mathbf{k}} \\
&= \frac{u_\mathbf{k}^2}{i\omega_n - E_\mathbf{k}} + \frac{v_\mathbf{k}^2}{i\omega_n + E_\mathbf{k}},
\end{aligned} \tag{14.176}$$

confirming that $u_\mathbf{k}$ and $v_\mathbf{k}$ determine the overlap between the electron and the quasiparticle and quasihole, respectively.

Example 14.6 The semiconductor analogy

One useful way to regard superconductors is via the *semiconductor analogy*, in which the quasiparticles are treated like the positive and negative energy excitations of a semiconductor.

(a) Divide the Brillouin zone up into two equal halves and redefine a set of positive and negative energy quasiparticle operators according to

$$\left.\begin{aligned}
\alpha_{\mathbf{k}\sigma+}^\dagger &= a_{\mathbf{k}\sigma}^\dagger \\
\alpha_{\mathbf{k}\sigma-}^\dagger &= \mathrm{sgn}(\sigma)a_{-\mathbf{k}-\sigma}
\end{aligned}\right\} \quad (\mathbf{k} \in \tfrac{1}{2}\mathrm{BZ}). \tag{14.177}$$

Rewrite the BCS Hamiltonian in terms of these new operators, and show that the excitation spectrum can be interpreted in terms of an empty band of positive energy excitations and a filled band of negative energy excitations.

(b) Show that the BCS ground-state wavefunction can be regarded as a filled sea of negative energy quasiparticle states and an empty sea of positive energy quasiparticle states.

Solution

(a) Dividing the Brillouin zone into two halves, the BCS Hamiltonian can be rewritten

$$
\begin{aligned}
H &= \sum_{\mathbf{k}\in\frac{1}{2}BZ} E_\mathbf{k}(a^\dagger_{\mathbf{k}\uparrow} a_{\mathbf{k}\uparrow} - a_{-\mathbf{k}\downarrow} a^\dagger_{-\mathbf{k}\downarrow}) + \sum_{\mathbf{k}\in\frac{1}{2}BZ} E_\mathbf{k}(a^\dagger_{-\mathbf{k}\uparrow} a_{-\mathbf{k}\uparrow} - a_{\mathbf{k}\downarrow} a^\dagger_{\mathbf{k}\downarrow}) \\
&= \sum_{\mathbf{k}\in\frac{1}{2}BZ,\sigma} E_\mathbf{k}(a^\dagger_{\mathbf{k}\sigma} a_{\mathbf{k}\sigma} - a_{-\mathbf{k}\sigma} a^\dagger_{-\mathbf{k}\sigma}) \\
&= \sum_{\mathbf{k}\in\frac{1}{2}BZ,\sigma} E_\mathbf{k}(\alpha^\dagger_{\mathbf{k}\sigma+} \alpha_{\mathbf{k}\sigma+} - \alpha^\dagger_{\mathbf{k}-} \alpha_{\mathbf{k}\sigma-}),
\end{aligned}
\tag{14.178}
$$

corresponding to two bands of positive and negative energy quasiparticles.

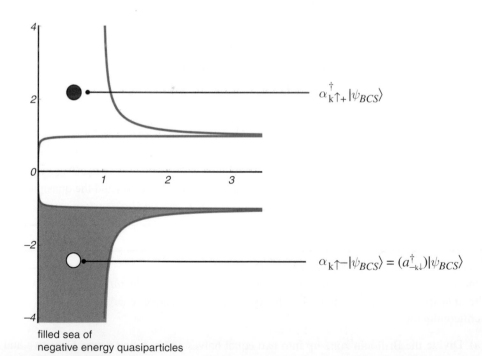

filled sea of
negative energy quasiparticles

Fig. 14.13 Semiconductor analogy for BCS theory (see Example 14.6). The BCS ground state can be regarded as a filled sea of negative energy quasiparticles. Positive energy excitations are created by adding positive quasiparticles, $\alpha^\dagger_{\mathbf{k}\sigma+}|\psi_{BCS}\rangle$, or removing negative energy quasiparticles, $\alpha_{\mathbf{k}\sigma-}|\psi_{BCS}\rangle$.

(b) Following Example 14.2, the BCS ground state can be written (up to a normalization) as

$$|\psi_{BCS}\rangle = \sum_{\mathbf{k}} a_{-\mathbf{k}\downarrow} a_{\mathbf{k}\uparrow} |0\rangle. \tag{14.179}$$

Factoring the product into the two halves of the Brillouin zone, we may rewrite this as

$$|\psi_{BCS}\rangle = \prod_{\mathbf{k}\in\frac{1}{2}BZ} (a_{\mathbf{k}\downarrow} a_{\mathbf{k}\uparrow})(a_{-\mathbf{k}\downarrow} a_{-\mathbf{k}\uparrow})|0\rangle$$

$$= \overbrace{\prod_{\mathbf{k}\in\frac{1}{2}BZ,\sigma} \alpha_{\mathbf{k}\sigma+}}^{\text{empty sea of positive energy quasiparticles}} \underbrace{\prod_{\mathbf{k}\in\frac{1}{2}BZ,\sigma} \alpha^{\dagger}_{\mathbf{k}\sigma-}}|0\rangle, \tag{14.180}$$

$$\underbrace{\phantom{\prod_{\mathbf{k}\in\frac{1}{2}BZ,\sigma} \alpha^{\dagger}_{\mathbf{k}\sigma-}}}_{\text{filled sea of negative energy quasiparticles}}$$

corresponding to an empty sea of positive energy quasiparticles and a filled sea of negative energy quasiparticles (see Figure 14.13).

14.7.1 Tunneling density of states and coherence factors

In a superconductor, the particle–hole mixing transforms the character of the quasiparticle, changing the matrix elements for scattering, introducing terms we call *coherence factors* into the physical response functions. These effects produce dramatic features in the various spectroscopies of the superconducting condensate.

Let us begin by calculating the tunneling density of states, which probes the spectrum to add and remove particles from the condensate. In a tunneling experiment the differential conductance is directly proportional to the local spectral function:

$$\frac{dI}{dV} \propto A(\omega)|_{\omega=eV}, \tag{14.181}$$

where

$$A(\omega) = \frac{1}{\pi} \mathrm{Im} \sum_{\mathbf{k}} G(\mathbf{k}, \omega - i\delta). \tag{14.182}$$

The mixed particle–hole character of the quasiparticle $a^{\dagger}_{\mathbf{k}\uparrow} = u_{\mathbf{k}} c^{\dagger}_{\mathbf{k}\uparrow} + v_{\mathbf{k}} c_{-\mathbf{k}\downarrow}$ means that quasiparticles can be created by adding or removing electrons from the condensate. Taking the decomposition of the Green's function in terms of its poles (14.176),

$$G(\mathbf{k}, z) = \frac{\omega + \epsilon_{\mathbf{k}}}{z^2 - E_{\mathbf{k}}^2} = \frac{1}{2}\left(1 + \frac{\epsilon_{\mathbf{k}}}{E_{\mathbf{k}}}\right)\frac{1}{z - E_{\mathbf{k}}} + \frac{1}{2}\left(1 - \frac{\epsilon_{\mathbf{k}}}{E_{\mathbf{k}}}\right)\frac{1}{z + E_{\mathbf{k}}}$$

$$= \frac{u_{\mathbf{k}}^2}{z - E_{\mathbf{k}}} + \frac{v_{\mathbf{k}}^2}{z + E_{\mathbf{k}}}, \tag{14.183}$$

it follows that

$$A(\mathbf{k}, \omega) = \frac{1}{\pi} \mathrm{Im}\, G(\mathbf{k}, \omega - i\delta) = u_{\mathbf{k}}^2 \delta(\omega - E_{\mathbf{k}}) + v_{\mathbf{k}}^2 \delta(\omega + E_{\mathbf{k}}). \tag{14.184}$$

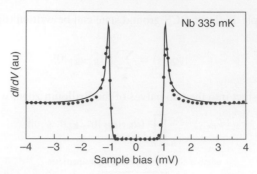

Fig. 14.14 Comparison of the experimental tunneling spectrum and the BCS spectrum in superconducting Nb at $T = 335$ mK [22]. Reprinted with permission from S. H. Pan, *et al.*, *Appl. Phys. Lett.*, vol. 73, p. 2992, 1998. Copyright 1998 by the American Insitute of Physics.

The positive energy part of this expression corresponds to the process of creating a quasi-particle by adding an electron, while the negative energy part corresponds to the creation of a quasiparticle by adding a hole. The amplitudes

$$|u_{\mathbf{k}}|^2 = |\langle \mathrm{qp} : \mathbf{k}\sigma |c^{\dagger}_{\mathbf{k}\sigma}|\psi_{BCS}\rangle|^2$$
$$|v_{\mathbf{k}}|^2 = |\langle \mathrm{qp} : \mathbf{k}\sigma |c_{-\mathbf{k}-\sigma}|\psi_{BCS}\rangle|^2 \qquad (14.185)$$

describe the probability to create a quasiparticle through the addition or removal of an electron, respectively. In this way, the tunneling density of states contains both negative and positive energy components.

Now we can sum over the momenta in (14.182), replacing the momentum sum by an integral over energy. In this case,

$$A(\omega) = \frac{N(0)}{\pi} \mathrm{Im} \int_{-\infty}^{\infty} d\epsilon \frac{\omega + \epsilon}{(\omega - i\delta)^2 - \epsilon^2 - |\Delta|^2} = -N(0)\mathrm{Im}\frac{\omega}{\sqrt{\Delta^2 - (\omega - i\delta)^2}}$$

$$= N(0)\,\mathrm{Re}\left[\frac{|\omega|}{\sqrt{(\omega - i\delta)^2 - \Delta^2}}\right] = N(0)\frac{|\omega|}{\sqrt{\omega^2 - \Delta^2}}\theta(|\omega| - \Delta), \qquad (14.186)$$

where we have used $\mathrm{Im}[\sqrt{\Delta^2 - (\omega - i\delta)^2}] = \sqrt{\omega^2 - \Delta^2}\,\mathrm{sgn}(\omega)\,\theta(|\omega| - \Delta)$. Curiously, this result is identical (up to a factor of ½ derived from the energy average of the coherence factors) to the quasiparticle density of states, except that there is both a positive and a negative energy component to the spectrum. In weakly coupled phonon-paired super-conductors such as niobium, experimental tunneling spectra are in good accord with BCS theory (see Figure 14.14). In more strongly coupled electron–phonon superconductors, wiggles develop in the spectrum related to the detailed phonon spectrum.

Other forms of spectroscopy probe the condensate by scattering electrons. In general a one-particle observable $\hat{A}$, such as spin or charge density, can be written as

$$\hat{A} = \sum_{\mathbf{k}\alpha, \mathbf{k}'\beta} A_{\alpha\beta}(\mathbf{k}, \mathbf{k}')c^{\dagger}_{\mathbf{k}\alpha}c_{\mathbf{k}'\beta}, \qquad (14.187)$$

where $A_{\alpha\beta}(\mathbf{k}, \mathbf{k}') = \langle \mathbf{k}\alpha|\hat{A}|\mathbf{k}'\beta\rangle$ are the electron matrix elements of the operator $\hat{A}$. For example, for the charge operator $\hat{\rho}_{\mathbf{q}} = e\sum_{\mathbf{k}\sigma} c^{\dagger}_{\mathbf{k}+\mathbf{q}\sigma}c_{\mathbf{k}\sigma}$, $A_{\alpha\beta}(\mathbf{k}, \mathbf{k}') = e\delta_{\alpha\beta}\delta_{\mathbf{k}-(\mathbf{k}'+\mathbf{q})}$

Table 14.2 Coherence factors.

Name	$\hat{A}$	$A_{\alpha\beta}(\mathbf{k}, \mathbf{k}')$	θ	Coherence factor
Density	$\hat{\rho}_{\mathbf{q}}$	$\delta_{\alpha\beta}\delta_{\mathbf{k}-(\mathbf{k}'+\mathbf{q})}$	$+1$	$uu' - vv'$
Magnetization	$\vec{M}_{\mathbf{q}}$	$\left(\frac{g\mu_B}{2}\right)\vec{\sigma}_{\alpha\beta}\delta_{\mathbf{k}-(\mathbf{k}'+\mathbf{q})}$	-1	$uu' + vv'$
Current	$\vec{J}_{\mathbf{q}}$	$\delta_{\alpha\beta}[(\mathbf{k}' + \mathbf{q}/2) - e\vec{A}]\delta_{\mathbf{k}-(\mathbf{k}'+\mathbf{q})}$	-1	$uu' + vv'$

(see Table 14.2). Let us now rewrite this expression in terms of Bogoliubov quasiparticle operators, substituting $c^\dagger_{\mathbf{k}\alpha} = u_{\mathbf{k}}a_{\mathbf{k}\alpha} - \text{sgn}(\alpha)v_{\mathbf{k}}a^\dagger_{-\mathbf{k}-\alpha}$ (where we have taken the gap, $u_{\mathbf{k}}$, and $v_{\mathbf{k}}$ to be real), so that the operator expands into the long expression

$$\hat{A} = \sum_{\mathbf{k}\alpha\mathbf{k}'\beta} A_{\alpha\beta}(\mathbf{k}, \mathbf{k}')\left[(uu'a^\dagger_{\mathbf{k}\alpha}a_{\mathbf{k}'\beta} - vv'\tilde{\alpha}\tilde{\beta}a_{-\mathbf{k}-\alpha}a^\dagger_{-\mathbf{k}'-\beta})\right.$$

$$\left. - (uv'\tilde{\beta}a^\dagger_{\mathbf{k}\alpha}a^\dagger_{-\mathbf{k}'-\beta} + \text{H.c.})\right]. \tag{14.188}$$

We have used the shorthand $\tilde{\alpha} = \text{sgn}(\alpha)$, $\tilde{\beta} = \text{sgn}(\beta)$, and $u \equiv u_{\mathbf{k}}$, $u' \equiv u_{\mathbf{k}'}$ and so on. This expression can be simplified by taking account of the time-reversal properties of $\hat{A}$. Under time reversal, $A \to -i\sigma_2 A^T i\sigma_2 = \theta A$, where $\theta = \pm 1$ is the parity of the operator under time reversal. In longhand,[6]

$$A_{\alpha\beta}(\mathbf{k}, \mathbf{k}') \to \tilde{\alpha}\tilde{\beta}A_{-\beta\,-\alpha}(-\mathbf{k}', -\mathbf{k}) = \theta A_{\alpha\beta}(\mathbf{k}, \mathbf{k}'). \tag{14.189}$$

Using this property, we can rewrite $\hat{A}$ as

$$\hat{A} = \sum_{\mathbf{k}\alpha,\mathbf{k}'\beta} A(\mathbf{k}, \mathbf{k}')_{\alpha\beta}\left[(uu' - \theta vv')a^\dagger_{\mathbf{k}\alpha}a_{\mathbf{k}'\beta}\right.$$

$$\left. + \frac{1}{2}\left((uv' - \theta vu')a^\dagger_{\mathbf{k}\alpha}a^\dagger_{-\mathbf{k}-\beta}\tilde{\beta} + \text{H.c.}\right)\right]. \tag{14.190}$$

We see that, in the pair condensate, the matrix element for quasiparticle scattering is renormalized by the *coherence factor*

$$A_{\alpha\beta}(\mathbf{k}, \mathbf{k}') \to A_{\alpha\beta}(\mathbf{k}, \mathbf{k}') \times (u_{\mathbf{k}}u_{\mathbf{k}'} - \theta v_{\mathbf{k}}v_{\mathbf{k}'}), \tag{14.191}$$

while the matrix element for creating a pair of quasiparticles has been modified by the factor

$$A_{\alpha\beta}(\mathbf{k}, \mathbf{k}') \to A_{\alpha\beta}(\mathbf{k}, \mathbf{k}') \times (u_{\mathbf{k}}v_{\mathbf{k}'} - \theta v_{\mathbf{k}}u_{\mathbf{k}'}). \tag{14.192}$$

Remarks

- At the Fermi energy, $|u_{\mathbf{k}}| = |v_{\mathbf{k}}| = \frac{1}{\sqrt{2}}$, so that for time-reversed even operators ($\theta = 1$) the coherence factors vanish on the Fermi surface.

[6] For example, for the magnetization density at wavevector $\mathbf{q}$, where $\vec{A}(\mathbf{k}, \mathbf{k}') = \vec{\sigma}\delta_{\mathbf{k}-(\mathbf{k}'+\mathbf{q})}$, using the result $\vec{\sigma}^T = i\sigma_2\vec{\sigma}i\sigma_2$, we obtain $-i\sigma_2\vec{A}^T(-\mathbf{k}', -\mathbf{k})i\sigma_2 = -i\sigma_2\vec{\sigma}i\sigma_2\delta_{-\mathbf{k}'-(-\mathbf{k}+\mathbf{q})} = -\vec{\sigma}\delta_{\mathbf{k}-(\mathbf{k}'+\mathbf{q})}$, corresponding to an odd time-reversal parity, $\theta = -1$.

- If we square the quasiparticle scattering coherence factor, we obtain

$$
(uu' - \theta vv')^2 = u^2(u')^2 + v^2(v')^2 - 2\theta(uv)(u'v')
$$

$$
= \frac{1}{4}\left(1 + \frac{\epsilon}{E}\right)\left(1 + \frac{\epsilon'}{E'}\right) + \frac{1}{4}\left(1 - \frac{\epsilon}{E}\right)\left(1 - \frac{\epsilon'}{E'}\right) - 2\theta\left(\frac{\Delta^2}{4EE'}\right)
$$

$$
= \frac{1}{2}\left(1 + \frac{\epsilon\epsilon'}{EE'} - \theta\frac{\Delta^2}{EE'}\right), \tag{14.193}
$$

with the notation $\epsilon = \epsilon_{\mathbf{k}}$, $\epsilon' = \epsilon_{\mathbf{k}'}$, $E = E_{\mathbf{k}}$, and $E' = E_{\mathbf{k}'}$.

- If we employ the semiconductor analogy, using positive ($\lambda = +$) and negative energy ($\lambda = -$) quasiparticles (see Example 14.6), with energies $E_{\mathbf{k}\lambda} = \mathrm{sgn}\,(\lambda)E_{\mathbf{k}}$ ($\lambda = \pm$) and modified Bogoliubov coefficients,

$$
u_{\mathbf{k}\lambda} = \sqrt{\frac{1}{2}\left(1 + \frac{\epsilon_{\mathbf{k}}}{E_{\mathbf{k}\lambda}}\right)}, \qquad v_{\mathbf{k}\lambda} = \sqrt{\frac{1}{2}\left(1 - \frac{\epsilon_{\mathbf{k}}}{E_{\mathbf{k}\lambda}}\right)}. \tag{14.194}
$$

Then

$$
(u_{\mathbf{k}}v_{\mathbf{k}'} - \theta v_{\mathbf{k}}u_{\mathbf{k}'})a^{\dagger}_{\mathbf{k}\sigma}a^{\dagger}_{-\mathbf{k}'} = (u_{\mathbf{k}+}u_{\mathbf{k}'-} - \theta v_{\mathbf{k}+}v_{\mathbf{k}'-})\alpha^{\dagger}_{\mathbf{k}\sigma+}a_{\mathbf{k}'\sigma'-}, \tag{14.195}
$$

so that the creation of a pair of quasiparticles can be regarded as an interband scattering of a valence negative energy quasiparticle into a conduction positive energy quasiparticle state. This has the advantage that all processes can be regarded as quasiparticle scattering, with a single coherent factor for all processes:

$$
\hat{A} = \frac{1}{2}\sum_{\mathbf{k}\sigma\lambda,\,\mathbf{k}'\sigma'\lambda'} A_{\sigma\sigma'}(\mathbf{k},\mathbf{k}')(uu' - \theta vv') \times \alpha^{\dagger}_{\mathbf{k}\sigma\lambda}\alpha_{\mathbf{k}'\sigma'\lambda'}. \tag{14.196}
$$

Once the condensate forms, the coherence factors renormalize the charge, spin, and current matrix elements of a superconductor. For example, in a metal the NMR relaxation rate is determined by the thermal average of the density of states:

$$
\frac{1}{T_1 T} \propto \int\left(-\frac{df}{dE}\right)N(E)^2|\langle E\uparrow|S^+|E\downarrow\rangle|^2 = \int\left(-\frac{df}{dE}\right)N(E)^2 = N(0)^2 \tag{14.197}
$$

at temperatures much smaller than the Fermi energy. However, in a superconductor we need to take account of the strongly energy-dependent quasiparticle density of states

$$
N(E) \to N(0)\frac{|E|}{\sqrt{E^2 - \Delta^2}}, \tag{14.198}
$$

while in this case the matrix elements

$$
|\langle E\uparrow|S^+|E\downarrow\rangle|^2 \to |\langle E\uparrow|S^+|E\downarrow\rangle|^2(u(E)^2 + v(E)^2) = 1
$$

are unrenormalized, so that the NMR relaxation rate becomes

$$
\left(\frac{1}{T_1 T}\right)_s \Big/ \left(\frac{1}{T_1 T}\right)_n = \int dE\left(-\frac{df}{dE}\right)\frac{E^2}{E^2 - \Delta^2}\theta(|E| - \Delta)
$$

$$
= \frac{1}{2}\int_{\Delta}^{\infty} dE\left(-\frac{df}{dE}\right)\frac{E^2}{E^2 - \Delta^2}. \tag{14.199}
$$

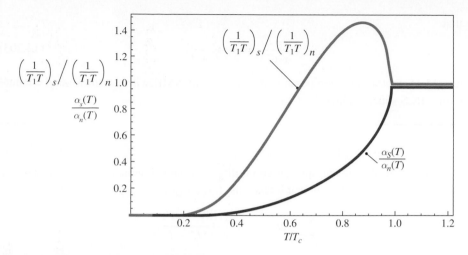

Showing the effect of coherence factors NMR and ultrasonic attenuation in a superconductor, calculated in BCS theory. **Fig. 14.15**
The orange line displays the NMR relaxation rate, showing the Hebel–Slichter peak. The blue line shows the
ultrasound attenuation. The integrals entering the NMR relaxation rate are formally divergent for $T < T_c$ and were
regulated by introducing a small imaginary damping rate $i\delta$ to the frequency where $\delta/\Delta = 0.005$.

The NMR relaxation rate is thus sensitive to the coherence peak in the density of states,
which leads to a sharp peak in the NMR relaxation rate just below the transition temperature, known as the *Hebel–Slichter peak* (Figure 14.15).[7] By contrast, the absorption
coefficient for ultrasound is proportional to the imaginary part of the charge susceptibility
at $\mathbf{q} = 0$, which in a normal metal is given by

$$\alpha_n(T) \propto \int dE \left(-\frac{df}{dE} \right) N(E) \overbrace{|\langle E|\rho_{\mathbf{q}=0}|E\rangle|^2}^{=1} \sim N(0), \qquad (14.200)$$

but in the superconductor this becomes

$$\alpha_s(T) \propto \int dE \left(-\frac{df}{dE} \right) N_s(E) |\langle E|\rho_{\mathbf{q}=0}|E\rangle|^2 \times (u(E)^2 - v(E)^2). \qquad (14.201)$$

However, in this case the renormalization of the matrix elements identitically cancels the
renormalization of the density of states:

$$N_s(E)(u^2 - v^2) = N(0)\theta(|E| - \Delta).$$

So there is no net coherence factor effect and

$$\alpha_s(T) \propto N(0) \int_{-\infty}^{\infty} dE \left(-\frac{df}{dE} \right) \theta(|E| - \Delta) = N(0)2f(\Delta), \qquad (14.202)$$

---

[7] Equation (14.199) contains a logarithmic divergence from the coherence peak. In practice, this is cut off by
the quasiparticle scattering. To obtain a finite result, one can replace $E \to E - i/(2\tau)$ and use the expression
$N(E) = \mathrm{Im}(E/\sqrt{\Delta^2 - (E - i/(2\tau))^2})$ to regulate the logarithmic divergence.

so that

$$\frac{\alpha_s(T)}{\alpha_n(T)} = \frac{2}{e^{\Delta/T} + 1}.$$ (14.203)

Figure 14.15 contrasts the temperature dependence of NMR with the ultrasound attenuation for a BCS superconductor.

Example 14.7

(a) Calculate the dynamical spin susceptibility of a superconductor using the Nambu Green's function, and show that it takes the form $\chi_{ab}(q) = \delta_{ab}\chi(q)$, where

$$\chi(q) = 2 \sum_{\mathbf{k},\eta,\eta'} (uu' + vv')^2 \frac{f(E') - f(E)}{\nu - (E' - E)}$$

$$= 2 \sum_{\mathbf{k},\eta,\eta'} \left(\frac{1}{2}\left(1 + \frac{\epsilon\epsilon' + \Delta^2}{EE'}\right) \right)^2 \frac{f(E') - f(E)}{\nu - (E' - E)},$$ (14.204)

where $\eta = \pm$, $\eta' = \pm$ and we have employed the (semiconductor analogy) notation $u \equiv u_{\mathbf{k}\eta}$, $u' \equiv u_{\mathbf{k}+\mathbf{q}\eta'}$, $E \equiv E_{\mathbf{k}}\mathrm{sgn}\,(\eta)$, $E' \equiv E_{\mathbf{k}+\mathbf{q}}\mathrm{sgn}\,(\eta')$, and so on.

(b) Assuming that the NMR relaxation rate is given by the expression

$$\frac{1}{T_1 T} \propto \sum_{\mathbf{q}} \left. \frac{\chi''(\mathbf{q}, \nu - i\delta)}{\nu} \right|_{\nu \to 0},$$ (14.205)

show that

$$\frac{1}{T_1 T} \propto \int \left(-\frac{df}{dE}\right) N(E)^2.$$ (14.206)

Solution

(a) The dynamical susceptibility in imaginary time is given by

$$\chi_{ab}(\mathbf{q}, i\nu_n) = \langle M_a(q)M_b(-q)\rangle = \int_0^\beta d\tau \langle TM_a(\mathbf{q}, \tau)M_b(-\mathbf{q}, 0)\rangle e^{i\nu_n \tau}.$$ (14.207)

Since the system is spin isotropic, we can write $\chi_{ab}(q) = \delta_{ab}\chi(q)$, using the z component of the magnetic susceptibility to calculate $\chi(q) = \langle M_z(q)M_z(-q)\rangle$. In Nambu notation,

$$M_z(-\mathbf{q}) = \sum_{\mathbf{k}}(c^\dagger_{\mathbf{k}+\mathbf{q}\uparrow}c_{\mathbf{k},\uparrow} - c^\dagger_{\mathbf{k}+\mathbf{q}\downarrow}c_{\mathbf{k},\downarrow}) = \sum_{\mathbf{k}}(c^\dagger_{\mathbf{k}+\mathbf{q}\uparrow}c_{\mathbf{k}\uparrow} + c_{\mathbf{k}\downarrow}c^\dagger_{\mathbf{k}+\mathbf{q}\downarrow})$$

$$= \sum_{\mathbf{k}}(c^\dagger_{\mathbf{k}+\mathbf{q}\uparrow}c_{\mathbf{k}\uparrow} + c_{-\mathbf{k}-\mathbf{q}\downarrow}c^\dagger_{-\mathbf{k}\downarrow})$$

$$= \sum_{\mathbf{k}} \psi^\dagger_{\mathbf{k}+\mathbf{q}} \cdot \psi_{\mathbf{k}},$$ (14.208)

where we have anticommuted the down fermion operators and relabeled $\mathbf{k} \to -\mathbf{k} + \mathbf{q}$. Thus the z component of the magnetization is a unit matrix in Nambu space. The vertex for the magnetization is thus

$$
\begin{array}{c}
\mathbf{k+q} \\[4pt]
\raisebox{-0.5em}{\includegraphics{}} \quad = M_z(-q),
\end{array}
$$

$$= M_z(-q), \tag{14.209}$$

and we can guess that the Feynman diagram for the susceptibility is

$$\langle M_z(q)M_z(-q)\rangle = \quad = -\frac{1}{\beta}\sum_k \mathrm{Tr}\left[\mathcal{G}(k+q)\mathcal{G}(k)\right],$$

where the fermion lines represent the Nambu propagator.

Let us confirm this result. The dynamical susceptibility is written

$$\chi(\mathbf{q},\tau) = \sum_{\mathbf{k},\mathbf{k}'}\langle T\psi^{\dagger}_{\mathbf{k}'-\mathbf{q}}(\tau)\cdot\psi_{\mathbf{k}'}(\tau)\,\psi^{\dagger}_{\mathbf{k}+\mathbf{q}}(0)\cdot\psi_{\mathbf{k}}(0)\rangle. \tag{14.210}$$

Since the mean field theory describes a non-interacting system, we can evaluate this expression using Wick's theorem:

$$
\begin{aligned}
\chi(\mathbf{q},\tau) &= \sum_{\mathbf{k},\mathbf{k}'}\langle T\psi^{\dagger}_{\mathbf{k}'-\mathbf{q}\alpha}(\tau)\psi_{\mathbf{k}'\alpha}(\tau)\psi^{\dagger}_{\mathbf{k}+\mathbf{q}\beta}(0)\psi_{\mathbf{k}\beta}(0)\rangle \\[6pt]
&= -\sum_{\mathbf{k}}\mathcal{G}_{\alpha\beta}(\mathbf{k}+\mathbf{q},\tau)\mathcal{G}_{\beta\alpha}(\mathbf{k},-\tau) \\[6pt]
&= -\sum_{\mathbf{k}}\mathrm{Tr}[\mathcal{G}(\mathbf{k}+\mathbf{q},\tau)\mathcal{G}(\mathbf{k},-\tau)].
\end{aligned}
\tag{14.211}
$$

Notice that the anomalous contractions of the Nambu spinors, such as $\langle T\psi_{\mathbf{k}\alpha}(\tau)\psi_{\mathbf{k}'\beta}(0)\rangle$, equal 0 because these terms describe triplet correlations that vanish in a singlet superconductor. For example, $\langle T\psi_{\mathbf{k}1}(\tau)\psi_{\mathbf{k}'2}(0)\rangle = \langle Tc_{\mathbf{k}\uparrow}(\tau)c^{\dagger}_{\mathbf{k}'\downarrow}(0)\rangle = 0$.

If we Fourier analyze this, $\chi(q) \equiv \chi(\mathbf{q},i\nu_r) = \int_0^{\beta}\chi(\mathbf{q},\tau)e^{i\nu_r\tau}$, we obtain

$$
\begin{aligned}
\chi(\mathbf{q},i\nu_r) &= -T^2\sum_{\mathbf{k},n,m}\int_0^{\beta}d\tau\,\mathrm{Tr}\left[\mathcal{G}(\mathbf{k}+\mathbf{q},i\omega_m)\mathcal{G}(\mathbf{k},i\omega_n)\right]e^{i(\nu_r-\omega_m+\omega_n)\tau} \\[6pt]
&= -T\sum_{\mathbf{k},i\omega_n}\mathrm{Tr}\left[\mathcal{G}(\mathbf{k}+\mathbf{q},i\omega_n+i\nu_r)\mathcal{G}(\mathbf{k},i\omega_n)\right] \\[6pt]
&= -T\sum_k\mathrm{Tr}\left[\mathcal{G}(k+q)\mathcal{G}(k)\right].
\end{aligned}
\tag{14.212}
$$

Now if we choose a real gap,

$$\mathcal{G}(\mathbf{k}, z) = \frac{z + \epsilon_{\mathbf{k}}\tau_3 + \Delta\tau_1}{z^2 - E_{\mathbf{k}}^2}, \tag{14.213}$$

we deduce that

$$\mathrm{Tr}\left[\mathcal{G}(k')\mathcal{G}(k)\right] = \mathrm{Tr}\left[\frac{z' + \epsilon_{\mathbf{k}'}\tau_3 + \Delta\tau_1}{z'^2 - E_{\mathbf{k}}^2}\frac{z + \epsilon_{\mathbf{k}}\tau_3 + \Delta\tau_1}{z^2 - E_{\mathbf{k}}^2}\right]$$

$$= 2\left[\frac{zz' + \epsilon_{\mathbf{k}}\epsilon_{\mathbf{k}'} + \Delta^2}{(z^2 - E_{\mathbf{k}}^2)(z^2 - E_{\mathbf{k}'}^2)}\right]. \tag{14.214}$$

If we first carry out the Matsubara summation in the expression of the susceptibility, then by converting the summation to a contour integral we obtain

$$\chi(q) = -2\sum_{\mathbf{k}}\oint\frac{dz}{2\pi i}f(z)\left[\frac{z(z + i\nu_r) + \epsilon_{\mathbf{k}}\epsilon_{\mathbf{k}+\mathbf{q}} + \Delta^2}{(z^2 - E_{\mathbf{k}}^2)((z + i\nu_r)^2 - E_{\mathbf{k}+\mathbf{q}}^2)}\right], \tag{14.215}$$

where the contour passes clockwise around the poles in the Green's functions.

To do this integral, it is useful to rewrite the denominators of the Green's functions using the relation

$$\frac{1}{z^2 - E_{\mathbf{k}}^2} = \frac{1}{2E_{\mathbf{k}}}\frac{1}{z - E_{\mathbf{k}}} - \frac{1}{2E_{\mathbf{k}}}\frac{1}{z + E_{\mathbf{k}}}$$

$$= \sum_{\lambda=\pm1}\frac{1}{z - E_{\mathbf{k}\lambda}}\frac{1}{2E_{\mathbf{k}\lambda}}, \tag{14.216}$$

where we have introduced (cf. semiconductor analogy, Example 14.6) $E_{\mathbf{k}\lambda} = \mathrm{sgn}(\lambda)E_{\mathbf{k}}$. Similarly,

$$\frac{z}{z^2 - E_{\mathbf{k}}^2} = \sum_{\lambda=\pm}\frac{1}{2(z - E_{\mathbf{k}\lambda})}.$$

With this device, the integral becomes

$$\chi(q) = -2\sum_{\mathbf{k},\lambda=\pm,\lambda'=\pm}\oint\frac{dz}{2\pi i}f(z)\left[\frac{1}{4} + \frac{\epsilon_{\mathbf{k}}\epsilon_{\mathbf{k}+\mathbf{q}} + \Delta^2}{(4E_{\mathbf{k}\lambda}E_{\mathbf{k}+\mathbf{q}\lambda'})}\right]\frac{1}{(z - E_{\mathbf{k}\lambda})(z + i\nu_r - E_{\mathbf{k}+\mathbf{q}\lambda'})}$$

$$= \sum_{\mathbf{k},\lambda=\pm,\lambda'=\pm}\overbrace{\left[\frac{1}{2} + \frac{\epsilon_{\mathbf{k}}\epsilon_{\mathbf{k}+\mathbf{q}} + \Delta^2}{2E_{\mathbf{k}\lambda}E_{\mathbf{k}+\mathbf{q}\lambda'}}\right]}^{(uu'+vv')^2}\frac{f(E_{\mathbf{k}+\mathbf{q}\lambda'}) - f(E_{\mathbf{k}\lambda})}{i\nu_r - (E_{\mathbf{k}+\mathbf{q}\lambda'} - E_{\mathbf{k}\lambda})}$$

$$= \sum_{\mathbf{k},\lambda=\pm,\lambda'=\pm}(u_{\mathbf{k}\lambda}u_{\mathbf{k}+\mathbf{q}\lambda'} + v_{\mathbf{k}\lambda}v_{\mathbf{k}+\mathbf{q}\lambda'})^2\frac{f(E_{\mathbf{k}+\mathbf{q}\lambda'}) - f(E_{\mathbf{k}\lambda})}{i\nu_r - (E_{\mathbf{k}+\mathbf{q}\lambda'} - E_{\mathbf{k}\lambda})}, \tag{14.217}$$

thereby proving (14.204).

(b) If we analytically continue the susceptibility onto the real axis, then

$$\chi(\mathbf{q}, \nu - i\delta) = \sum_{\mathbf{k},\lambda=\pm,\lambda'=\pm}(u_{\mathbf{k}\lambda}u_{\mathbf{k}+\mathbf{q}\lambda'} + v_{\mathbf{k}\lambda}v_{\mathbf{k}+\mathbf{q}\lambda'})^2\frac{f(E_{\mathbf{k}+\mathbf{q}\lambda'}) - f(E_{\mathbf{k}\lambda})}{\nu - i\delta - (E_{\mathbf{k}+\mathbf{q}\lambda'} - E_{\mathbf{k}\lambda})}. \tag{14.218}$$

Taking the imaginary part,

$$\frac{\chi''(\mathbf{q}, \nu - i\delta)}{\nu}$$

$$= \pi \sum_{\mathbf{k},\lambda=\pm,\lambda'=\pm} (uu' + vv')^2 \frac{f(E_{\mathbf{k}\lambda} + \nu) - f(E_{\mathbf{k}\lambda})}{\nu} \delta(E_{\mathbf{k}+\mathbf{q}\lambda'} - E_{\mathbf{k}\lambda}) \qquad (14.219)$$

so that

$$\frac{\chi''(\mathbf{q}, \nu - i\delta)}{\nu}\Bigg|_{\nu\to 0} = \pi \sum_{\mathbf{k},\lambda=\pm,\lambda'=\pm} \left(-\frac{df(E_{\mathbf{k}\lambda})}{dE_{\mathbf{k}\lambda}}\right) \delta(E_{\mathbf{k}+\mathbf{q}\lambda'} - E_{\mathbf{k}\lambda}). \qquad (14.220)$$

Summing over momentum,

$$\frac{1}{T_1 T} \propto \sum_{\mathbf{q}} \frac{\chi''(\mathbf{q}, \nu - i\delta)}{\nu}\Bigg|_{\nu\to 0}$$

$$= \pi \sum_{\mathbf{k},\lambda=\pm} \sum_{\mathbf{k'},\ \lambda=\pm'} \left(-\frac{df(E_{\mathbf{k}\lambda})}{dE_{\mathbf{k}\lambda}}\right) \delta(E_{\mathbf{k}+\mathbf{q}\lambda'} - E_{\mathbf{k}\lambda})$$

$$= \pi N(0)^2 \int dE \left(\frac{|E|}{\sqrt{E^2 - \Delta^2}}\right)^2 \left(-\frac{df(E)}{dE}\right), \qquad (14.221)$$

where we have replaced the summation over momentum and semiconductor index λ by an integral over the quasiparticle and quasihole density of states:

$$\sum_{\mathbf{k},\lambda=\pm} \rightarrow \int dE N_s(|E|) = N(0) \int dE \left(\frac{|E|}{\sqrt{E^2 - \Delta^2}}\right). \qquad (14.222)$$

14.8 Twisting the phase: the superfluid stiffness

One of the key features in a superconductor is the emergence of a complex order parameter, with a phase. It is the rigidity of this phase that endows the superconductor with its ability to sustain a superflow, a feature held in common between superfluids and superconductors. However, superconductors stand apart from their neutral counterparts because the phase of the condensate is directly coupled to the electromagnetic field. The important point, as we saw in Chapter 11, is that the phase of the order parameter and the vector potential are linked by gauge invariance, so that a twisted phase and a uniform vector are gauge-equivalent. This feature implies that, once a gauge stiffness develops, the electromagnetic field acquires a mass. We shall now derive these features from the microscopic perspective of BCS theory.

To explore a twisted phase, we need to consider an order parameter with position dependence, so that now the interaction that gives rise to superconductivity cannot be infinitely long-range. For this purpose we use Gor'kov's coarse-grained continuum version of BCS theory, where

$$H = \int d^3x \left[\psi_\sigma^\dagger \left(\frac{1}{2m} (-i\hbar\nabla - e\vec{A})^2 - \mu \right) \psi_\sigma - g(\psi_\uparrow^\dagger \psi_\downarrow^\dagger \psi_\downarrow \psi_\uparrow) \right]. \qquad (14.223)$$

For compactness, the position arguments of the fields are no longer shown explicitly, $\psi_\sigma(x) \equiv \psi_\sigma$. This is a coarse-grained version of the microscopic Hamiltonian, in which the delta-function interaction represents the effective interaction on scales larger than v_F/ω_D.

Under the Hubbard–Stratonovich transformation, the interaction becomes

$$- g(\psi_\uparrow^\dagger \psi_\downarrow^\dagger \psi_\downarrow \psi_\uparrow) \rightarrow \bar{\Delta}\psi_\downarrow\psi_\uparrow + \psi_\uparrow^\dagger \psi_\downarrow^\dagger \Delta + \frac{\bar{\Delta}\Delta}{g}, \qquad (14.224)$$

where the gap function $\Delta(x)$ can acquire spatial dependence. The transformed Hamiltonian is then

$$H = \int d^3x \left[\psi_\sigma^\dagger \left(\frac{1}{2m} (-i\hbar\nabla - e\vec{A})^2 - \mu \right) \psi_\sigma + \bar{\Delta}\psi_\downarrow\psi_\uparrow + \psi_\uparrow^\dagger \psi_\downarrow^\dagger \Delta + \frac{\bar{\Delta}\Delta}{g} \right], \qquad (14.225)$$

where, at the mean-field saddle point, $\Delta(x) = -g\langle \psi_\downarrow(x)\psi_\uparrow(x)\rangle$. The curious thing is that, once the interaction is factorized in this way, we must take account of the transformation of the charged condensate field under the gauge transformation.

14.8.1 Implications of gauge invariance

The kinetic energy part of the Hamiltonian is invariant under the gauge transformations:

$$\psi_\sigma(x) \rightarrow e^{i\alpha(x)}\psi_\sigma(x)$$
$$\vec{A}(x) \rightarrow \vec{A}(x) + \frac{\hbar}{e}\vec{\nabla}\alpha(x). \qquad (14.226)$$

However, in order that the pairing terms remain invariant under a gauge transformation, we must also transform

$$\Delta(x) \rightarrow e^{2i\alpha(x)}\Delta(x), \qquad (14.227)$$

reflecting the fact that the pair condensate carries charge $2e$. The free energy of the condensate must therefore be invariant under the combined transformations (14.226) and (14.227). If we write the gap as an amplitude and phase term, $\Delta(x) = |\Delta(x)|e^{i\phi(x)}$, we see that under a gauge transformation the phase of the gap picks up twice the shift of a single electron field:

$$\phi(x) \rightarrow \phi(x) + 2\alpha(x). \qquad (14.228)$$

Now if the phase becomes *rigid* beneath T_c, so that there is an energetic cost to bending the phase, then the free energy must contain a phase-stiffness term

$$\mathcal{F} \sim \frac{\rho_s}{2} \int_x (\nabla\phi)^2. \qquad (14.229)$$

We've seen such terms in the Ginzburg–Landau theory of a neutral superfluid, but now they must appear when we expand the total energy in powers of the gradient of the order parameter. However, in a charged superfluid such a coupling term is not gauge-invariant under the combined transformation $\phi \rightarrow \phi + 2\alpha$, $\vec{A} \rightarrow \vec{A} + \frac{\hbar}{e}\vec{\nabla}\alpha(x)$. Indeed, gauge invariance of the free energy under these two transformations requires that the gradient of the phase and the vector potential can only appear as the gauge-invariant combination $\vec{\nabla}\phi - \frac{2e}{\hbar}\vec{A}$, so the phase stiffness term must take the form

$$\mathcal{F} = \frac{\rho_s}{2}\int_x \left(\vec{\nabla}\phi(x) - \frac{2e}{\hbar}\vec{A}(x)\right)^2 = \frac{Q}{2}\int_x \left(\vec{A}(x) - \frac{\hbar}{2e}\vec{\nabla}\phi(x)\right)^2, \qquad (14.230)$$

where we have substituted[8]

$$Q = \frac{(2e)^2}{\hbar^2}\rho_s. \qquad (14.233)$$

If we now look back at (14.230), we see that the electric current carried by the condensate is

$$\vec{j}(x) = -\frac{\delta\mathcal{F}}{\delta\vec{A}(x)} = -Q\left(\vec{A}(x) - \frac{\hbar}{2e}\vec{\nabla}\phi(x)\right), \qquad (14.234)$$

so we can identify Q with the *London kernel* in Chapter 10 in the study of electron transport, except that in a superconductor Q is finite in the DC limit.

Imagine a superconductor of length L in which the phase of the order parameter is twisted, so that $\Delta(L) = e^{i\Delta\phi}\Delta(0)$. Let us consider a uniform twist, so that

$$\Delta(x) = e^{i\vec{a}\cdot\vec{x}}\Delta_0, \qquad (14.235)$$

where $\vec{a} = \frac{\Delta\phi}{L}\hat{x}$. Now this twist of the order parameter can be removed by a gauge transformation

$$\Delta(x) \rightarrow e^{-i\vec{a}\cdot\vec{x}}\Delta(x) = \Delta_0$$
$$\vec{A} \rightarrow \vec{A} - \frac{\hbar}{2e}\vec{a}, \qquad (14.236)$$

so a twist in the order parameter is gauge-equivalent to a uniform vector potential $\vec{A} \equiv -\frac{\hbar}{2e}\vec{a} = -\frac{\hbar}{2e}\vec{\nabla}\phi$. We might have guessed this by noting that the combination $\vec{A} - \frac{\hbar}{2e}\vec{\nabla}\phi$ in the supercurrent formula (14.234) has to be the same in all gauges because it represents a physical quantity: it is gauge-invariant. This means that the effective (gauge-invariant) twist between the two ends of a superconductor is given by

$$\text{effective twist} = \overbrace{\Delta\phi}^{\text{phase twist}} - \overbrace{\frac{2e}{\hbar}\int_0^L \vec{A}\cdot\vec{dl}}^{\text{electromagnetic twist}}. \qquad (14.237)$$

---

[8] Notice the sheer power of this argument: by using gauge invariance, we have been able to deduce that a stiffness of the phase in a charged condensate gives rise to an electromagnetic mass term. As we discussed in Section 11.6.2, since $\mathcal{F}_{EM}$ is invariant under gauge transformations, it becomes possible to redefine the vector potential to absorb the phase of the order parameter, forming a massive field with both longitudinal and transverse components:

$$\vec{A}_H(x) = \vec{A}(x) - \frac{\hbar}{2e}\vec{\nabla}\phi(x). \qquad (14.231)$$

Once the phase of the order parameter is absorbed into the electromagnetic field,

$$\mathcal{F} \sim \frac{Q}{2}\int_x \vec{A}_H(x)^2 + \mathcal{F}_{EM}[A] \qquad (14.232)$$

and the vector potential has acquired a mass. This is the Anderson–Higgs mechanism, whereby a gauge field "eats" the phase of a condensate, losing manifest gauge invariance by acquiring a mass [18, 23, 24].

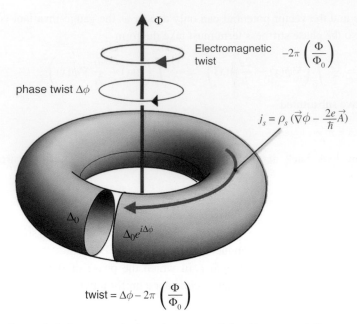

Electromagnetic twist $-2\pi\left(\dfrac{\Phi}{\Phi_0}\right)$

phase twist $\Delta\phi$

$j_s = \rho_s\left(\vec{\nabla}\phi - \dfrac{2e}{\hbar}\vec{A}\right)$

Δ_0

$\Delta_0 e^{i\Delta\phi}$

$$\text{twist} = \Delta\phi - 2\pi\left(\frac{\Phi}{\Phi_0}\right)$$

Fig. 14.16   Illustrating the phase-twist in the superconducting order parameter induced by a magnetic flux.

Each of the terms on the right is gauge-dependent, but their sum is a physical quantity. From a computational point of view, it means we can evaluate the phase stiffness without actually changing the phase of the order parameter, by calculating the change in the condensate energy due to an external field of magnitude $\vec{A} = -\frac{\hbar}{2e}\vec{\nabla}\phi$.

This reasoning has interesting consequences when we connect up the two ends of a superconductor to form a torus. Now we can induce an electromagnetic twist by passing a magnetic flux Φ through the torus (see Figure 14.16), inducing a circulating vector potential around the torus such that $\oint \vec{A} \cdot \vec{dl} = \Phi$. The supercurrent and the energy of the condensate will depend on the effective twist,

$$\text{effective twist} = \Delta\phi - \frac{2e}{\hbar}\oint \vec{A} \cdot \vec{dl} = \Delta\phi - \frac{2e}{\hbar}\Phi, \qquad (14.238)$$

where Φ is the magnetic flux threading the torus. Whereas the phase change $\Delta\phi$ along a superconducting strip is not gauge-invariant, the phase change around a torus is a topological invariant which must be a multiple of 2π, $\Delta\phi = 2\pi n$, and it is gauge-invariant. The supercurrent around the torus and the total energy of the condensate thus depend on the quantity

$$\Delta\phi - \frac{2e}{\hbar}\Phi = 2\pi\left(n - \frac{\Phi}{\Phi_0}\right), \qquad (14.239)$$

where

$$\Phi_0 = \frac{h}{2e} \equiv \frac{2\pi\hbar}{2e} \qquad (14.240)$$

is the superconducting flux quantum. In this situation, the supercurrent and the energy are minimized when the flux is quantized as a multiple of Φ_0, $\Phi = n\Phi_0$.

14.8.2 Calculating the phase stiffness

Let us now continue to calculate the phase stiffness or *superfluid density* of a BCS superconductor using the reasoning of the previous section, by applying an equivalent vector potential $\vec{A} = -\frac{\hbar}{2e}\vec{\nabla}\phi$. Such a field changes the dispersion according to $\epsilon_{\vec{k}} \to \epsilon_{\vec{k}-e\vec{A}}$, so, inside $h_{\mathbf{k}}$,

$$
\epsilon_{\vec{k}}\tau_3 = \begin{pmatrix} \epsilon_{\vec{k}} & \\ & -\epsilon_{-\vec{k}} \end{pmatrix} \to \begin{pmatrix} \epsilon_{\vec{k}-e\vec{A}} & \\ & -\epsilon_{-\vec{k}-e\vec{A}} \end{pmatrix}
$$

$$
= \begin{pmatrix} \epsilon_{\vec{k}-e\vec{A}} & \\ & -\epsilon_{\vec{k}+e\vec{A}} \end{pmatrix} \equiv \epsilon_{\vec{k}-e\vec{A}\tau_3}\tau_3, \tag{14.241}
$$

i.e.

$$
h_{\vec{k}} \to h_{\vec{k}-e\vec{A}\tau_3} = \epsilon_{\vec{k}-e\vec{A}\tau_3}\tau_3 + \Delta\tau_1. \tag{14.242}
$$

The free energy in a field is then

$$
F = -T \sum_{\mathbf{k},i\omega_n} \operatorname{Tr}\ln[\epsilon_{\vec{k}-e\vec{A}\tau_3}\tau_3 + \Delta\tau_1 - i\omega_n] + \frac{\Delta^2}{g}. \tag{14.243}
$$

We need to calculate

$$
Q_{ab} = -\frac{1}{V}\frac{\partial^2 F}{\partial A_a \partial A_b}. \tag{14.244}
$$

Taking the first derivative with respect to the vector potential gives us the steady-state diamagnetic current:

$$
\langle J_a \rangle = \frac{1}{V}\frac{\partial F}{\partial A_a} = -\frac{1}{\beta V}\sum_{k\equiv(\mathbf{k},i\omega_n)} \operatorname{Tr}\left[e\nabla_a\epsilon_{\vec{k}-e\vec{A}\tau_3} G(k - eA\tau_3) \right], \tag{14.245}
$$

where $G(k - eA\tau_3) = [i\omega_n - h_{\vec{k}-e\vec{A}\tau_3}]^{-1} = [i\omega_n - \epsilon_{\vec{k}-e\vec{A}\tau_3}\tau_3 - \Delta\tau_1]^{-1}$.

Taking one more derivative,

$$
Q_{ab} = \frac{1}{V}\frac{\partial^2 F}{\partial A_a \partial A_b}\bigg|_{A=0} = \frac{e^2}{\beta V}\sum_k \left(\overbrace{\left(\nabla^2_{ab}\epsilon_{\vec{k}}\right)\operatorname{Tr}\left[\tau_3 G(k)\right]}^{\text{diamagnetic part}} + \overbrace{\left(\nabla_a\epsilon_{\vec{k}}\nabla_b\epsilon_{\vec{k}}\right)\operatorname{Tr}\left[G(k)G(k)\right]}^{\text{paramagnetic part}} \right). \tag{14.246}
$$

Here we have used the fact that $\delta(GG^{-1}) = \delta G G^{-1} + G\delta G^{-1} = 0$ to derive $\delta G = -G\delta G^{-1}G$, which then led to the result $\frac{\partial}{\partial A_b}G(k - eA\tau_3) = -G(k - eA\tau_3)e\nabla_b\epsilon_{\vec{k}-e\vec{A}\tau_3}G(k - eA\tau_3)$, in which we then set $A = 0$. We may identify the above expression as a sum of the diamagnetic and paramagnetic parts of the superfluid stiffness. The first is associated with the instantaneous diamagnetic response of the wavefunction; the second is the retarded paramagnetic correction to the current that occurs as a result of the relaxation of the wavefunction. The diamagnetic part of the response can be integrated by parts, to give

$$\frac{e^2}{\beta V} \sum_{k,n} \left(\nabla^2_{ab}\epsilon_{\vec{k}}\right) \mathrm{Tr}\left[\tau_3 G(k)\right] = -\frac{e^2}{\beta V} \sum_{k,n} \nabla_a\epsilon_{\vec{k}} \mathrm{Tr}\left[\tau_3 \nabla_b G(k)\right]$$

$$= -\frac{e^2}{\beta V} \sum_{k,n} \left(\nabla_a\epsilon_{\vec{k}}\nabla_b\epsilon_{\vec{k}}\right) \mathrm{Tr}\left[\tau_3 G(k)\tau_3 G(k)\right], \quad (14.247)$$

where we have used $\nabla_b G = -G\nabla_b G^{-1}G = G\nabla_b\epsilon_k \tau_3 G$ to derive the last line. Notice how this term is identical to the paramagnetic term, apart from the τ_3 insertions. We now add these two terms, to obtain

$$Q_{ab} = -\frac{e^2}{\beta V} \sum_{k} \nabla_a\epsilon_{\vec{k}}\nabla_b\epsilon_{\vec{k}} \left(\overbrace{\mathrm{Tr}\left[\tau_3 G(k)\tau_3 G(k)\right]}^{\text{diamagnetic part}} - \overbrace{\mathrm{Tr}\left[G(k)G(k)\right]}^{\text{paramagnetic part}} \right). \quad (14.248)$$

Notice that, when pairing is absent, the τ_3 commute with $\mathcal{G}(k)$, and the diamagnetic and paramagnetic contributions exactly cancel. We can make this explicit by writing

$$Q_{ab} = -\frac{e^2}{2\beta V} \sum_{k} \nabla_a\epsilon_{\vec{k}}\nabla_b\epsilon_{\vec{k}} \mathrm{Tr}\left[[\tau_3, G(k)]^2\right]. \quad (14.249)$$

Now

$$[\tau_3, G(k)] = 2i\frac{\Delta\tau_2}{(i\omega_n)^2 - E^2_{\mathbf{k}}}, \quad (14.250)$$

so

$$-\mathrm{Tr}\left[[\tau_3, G(k)]^2\right] = 8\frac{\Delta^2}{[\omega^2_n + \epsilon^2_{\mathbf{k}} + \Delta^2]^2}, \quad (14.251)$$

so that

$$Q_{ab} = \frac{4e^2}{\beta V} \sum_{k} \nabla_a\epsilon_{\vec{k}}\nabla_b\epsilon_{\vec{k}} \frac{\Delta^2}{[(\omega_n)^2 + \epsilon^2_{\mathbf{k}} + \Delta^2]^2}. \quad (14.252)$$

Remarkably, although the diamagnetic and paramagnetic parts of the superfluid stiffness involve electrons far away from the Fermi surface, the difference between the two is dominated by terms where $\omega^2_n + \epsilon^2_{\mathbf{k}} \sim \Delta^2$, i.e. by electrons near the Fermi surface. This enables us to replace the summation over $\mathbf{k}$ by an integral over energy:

$$\frac{4}{V} \sum_{\mathbf{k}} \nabla_a\epsilon_{\vec{k}}\nabla_b\epsilon_{\vec{k}} \{\ldots\} = 2N(0) \int_{-\infty}^{\infty} d\epsilon \int \overbrace{\frac{d\Omega_{\mathbf{k}}}{4\pi} v_a v_b}^{\frac{1}{3}v^2_F \delta_{ab}} \{\ldots\} = \frac{2\delta_{ab}}{3}N(0)v^2_F \int_{-\infty}^{\infty} d\epsilon \{\ldots\}.$$
$$(14.253)$$

Note that a factor of 2 is absorbed into the total density of states of up and down electrons. We have taken advantage of the rapid convergence of the integrand to extend the limits of the integral over energy to infinity. Replacing $\frac{1}{3}N(0)v^2_F = \frac{n}{m}$, we can now write $Q_{ab} = Q\delta_{ab}$, where

$$Q(T) = \frac{ne^2}{m}T \sum_{n} \int_{-\infty}^{\infty} d\epsilon \frac{2\Delta^2}{(\epsilon^2 + \omega^2_n + \Delta^2)^2} = \left(\frac{ne^2}{m}\right)\pi T \sum_{n} \frac{\Delta^2}{(\omega^2_n + \Delta^2)^{\frac{3}{2}}}. \quad (14.254)$$

To evaluate this expression, it is useful to note that the argument of the summation is a total derivative, so that

$$Q(T) = \left(\frac{ne^2}{m}\right) \pi T \sum_n \frac{\partial}{\partial \omega_n} \left(\frac{\omega_n}{(\omega_n^2 + \Delta^2)^{1/2}}\right). \qquad (14.255)$$

Now at absolute zero we can replace $T \sum_n \to \int \frac{d\omega}{2\pi}$, so that

$$Q(0) \equiv Q_0 = \left(\frac{ne^2}{m}\right) \overbrace{\int_{-\infty}^{\infty} \frac{d\omega}{2} \frac{d}{d\omega} \left(\frac{\omega}{(\omega^2 + \Delta^2)^{1/2}}\right)}^{=1} = \left(\frac{ne^2}{m}\right). \qquad (14.256)$$

In other words, *all* of the electrons have condensed to form a perfect diamagnet. This is a rather remarkable result, for the pairing only extends within a narrow shell around the Fermi surface and one might have thought that only a tiny fraction T_c/ϵ_F of the Fermi sea would contribute to the stiffness, i.e. that $Q \sim O(T_c/\epsilon_F) \times ne^2/m \ll ne^2/m$, but this is *not* the case. The fact that all the electrons contribute to the superfluid stiffness means the wavefunction is completely rigid, so that no paramagnetic current develops at absolute zero in response to an applied vector potential.

At a finite temperature this is no longer the case, due to the presence of excited quasiparticles. To evaluate the stiffness at a finite temperature, we rewrite the Matsubara sum as a clockwise contour integral around the poles of the Fermi function:

$$Q(T) = \pi Q_0 \oint_{\text{Im axis}} \frac{dz}{2\pi i} f(z) \frac{d}{dz} \left(\frac{z}{\sqrt{\Delta^2 - z^2}}\right). \qquad (14.257)$$

By deforming the integral to run counterclockwise around the branch cuts along the real axis and then integrating by parts, we obtain

$$\begin{aligned}
Q(T) &= Q_0 \pi \oint_{\text{real axis}} \frac{dz}{2\pi i} f(z) \frac{d}{dz} \left(\frac{z}{\sqrt{\Delta^2 - z^2}}\right) \\
&= Q_0 \int_{-\infty}^{\infty} d\omega f(\omega) \frac{d}{d\omega} \text{Im} \left(\frac{z}{\sqrt{\Delta^2 - z^2}}\right)_{z=\omega-i\delta} \\
&= Q_0 \left[f(\omega) \text{Im} \left(\frac{z}{\sqrt{\Delta^2 - z^2}}\right)_{z=\omega-i\delta} \right]_{-\infty}^{\infty} \\
&\quad + Q_0 \int_{-\infty}^{\infty} d\omega \left(-\frac{df(\omega)}{d\omega}\right) \text{Im} \left(\frac{z}{\sqrt{\Delta^2 - z^2}}\right)_{z=\omega-i\delta}.
\end{aligned} \qquad (14.258)$$

Now a careful calculation of the imaginary part of the integrand gives

$$\begin{aligned}
\text{Im} \left(\frac{\omega}{\sqrt{\Delta^2 - (\omega - i\delta)^2}}\right) &= \text{Im} \left(\frac{\omega}{\sqrt{-(\omega^2 - \Delta^2) + i\delta \, \text{sgn}(\omega)}}\right) \\
&= \left(-\frac{|\omega|}{\sqrt{\omega^2 - \Delta^2}}\right) \theta(\omega^2 - \Delta^2),
\end{aligned} \qquad (14.259)$$

so the finite-temperature stiffness can then be written

$$Q(T) = Q_0 \left[1 - 2 \int_{\Delta(T)}^{\infty} d\omega \left(-\frac{df(\omega)}{d\omega} \right) \left(\frac{\omega}{\sqrt{\omega^2 - \Delta^2}} \right) \right], \qquad (14.260)$$

where the factor of 2 derives from folding over the contribution from the negative region of the integral. The second term in this expression is nothing more than the thermal average of the quasiparticle density of states $N_{qp}(E) = N(0) \frac{E}{\sqrt{E^2 - \Delta^2}}$. This term, with its factor of 2, can thus be interpreted as the reduction in the condensate fraction by a thermal depopulation of the condensate into quasiparticles. We can alternatively rewrite this expression as a formula for the temperature-dependent penetration depth:

$$\frac{1}{\lambda_L^2(T)} = \frac{1}{\lambda_L^2(0)} \left[1 - \overline{\left(\frac{A(E)}{N(0)} \right)} \right], \qquad (14.261)$$

where $1/\lambda_L^2(0) = \mu_0 n e^2 / m$ and $A(E)$ is the tunneling density of states given in (14.186), thermally averaged over both positive and negative energies.

Exercises

Exercise 14.1 Show, using the Cooper wavefunction, that the mean-squared radius of a Cooper pair is given by

$$\xi^2 = \frac{\int d^3 r \, r^2 |\phi(\mathbf{r})|^2}{\int d^3 r \, |\phi(\mathbf{r})|^2} = \frac{4}{3} \left(\frac{v_F}{E} \right)^3.$$

Exercise 14.2 Generalize the Cooper pair calculation to higher angular momenta. Consider an interaction that has an attractive component in a higher angular momentum channel, such as

$$N(0)V_{\mathbf{k},\mathbf{k}'} = \begin{cases} -g_l(2l+1)P_l(\hat{\mathbf{k}} \cdot \hat{\mathbf{k}}') & (|\epsilon_{\mathbf{k}}|, |\epsilon_{\mathbf{k}}'| < \omega_0) \\ 0 & (\text{otherwise}), \end{cases} \qquad (14.262)$$

where you may assume l is even.

(a) By decomposing the Legendre polynomial in terms of spherical harmonics, $(2l+1)P_l(\hat{\mathbf{k}}, \hat{\mathbf{k}}') = 4\pi \sum_m Y_{lm}(\mathbf{k})Y_{lm}^*(\hat{\mathbf{k}})$, show that this interaction gives rise to bound Cooper pairs with a finite angular momentum given by

$$|\psi_P\rangle = \sum_{\mathbf{k}} \phi_{\mathbf{k}m} Y_{lm}(\hat{\mathbf{k}}) c_{\mathbf{k}\uparrow}^\dagger c_{-\mathbf{k}\downarrow}^\dagger |0\rangle,$$

with a bound-state energy given by

$$E = -2\omega_0 \exp\left[-\frac{2}{g_l N(0)} \right].$$

(b) A general interaction will have several harmonics,

$$V_{\mathbf{k},\mathbf{k}'} = \frac{1}{V} \sum_l g_l (2l+1) P_l(\hat{\mathbf{k}} \cdot \hat{\mathbf{k}}'),$$

not all of them attractive. In which channel(s) will the pairs tend to condense?

(c) Why can't you use this derivation for the case when l is odd?

Exercise 14.3 Generalize the BCS solution to the case where the gap has a finite phase $\Delta = |\Delta| e^{i\phi}$. Show that, in this case, the eigenvectors of the BCS mean-field Hamiltonian are

$$u_{\mathbf{k}} = e^{i\phi/2} \left(\frac{1}{2} + \frac{\epsilon_{\mathbf{k}}}{2E_{\mathbf{k}}} \right)^{\frac{1}{2}}$$

$$v_{\mathbf{k}} = e^{-i\phi/2} \left(\frac{1}{2} - \frac{\epsilon_{\mathbf{k}}}{2E_{\mathbf{k}}} \right)^{\frac{1}{2}}, \qquad (14.263)$$

while the BCS ground state is given by

$$|BCS(\phi)\rangle = \prod_{\mathbf{k}} (u_{\mathbf{k}}^* + v_{\mathbf{k}}^* c_{-\mathbf{k}\downarrow}^{\dagger} c_{\mathbf{k}\uparrow}^{\dagger})|0\rangle. \qquad (14.264)$$

Exercise 14.4 Explicit calculation of the free energy.

(a) Assuming that the Debye frequency is a small fraction of the bandwidth, show that the difference between the superconducting and normal-state free energies can be written as the integral

$$\mathcal{F}_S - \mathcal{F}_N = -2TN(0) \int_{-\omega_D}^{\omega_D} d\epsilon \ln \left[\frac{\cosh\left(\frac{\sqrt{\epsilon^2 + |\Delta|^2}}{2T} \right)}{\cosh\left(\frac{\epsilon}{2T} \right)} \right] + V \frac{|\Delta|^2}{g_0}.$$

Why is this free energy invariant under changes in the phase of the gap parameter, $\Delta \to \Delta e^{i\phi}$?

(b) By differentiating the above expression with respect to Δ, confirm the zero-temperature gap equation,

$$\frac{V}{gN(0)} = \int_0^{\omega_D} \frac{d\epsilon}{\sqrt{\epsilon^2 + \Delta_0^2}},$$

where $\Delta_0 = \Delta(T=0)$ is the zero-temperature gap, and use this result to eliminate g_0, to show that the free energy can be written

$$\mathcal{F}_S - \mathcal{F}_N = N(0)\Delta_0^2 \, \Phi\left[\frac{\Delta}{\Delta_0}, \frac{T}{\Delta_0} \right],$$

where the dimensionless function

$$\Phi(\delta, t) = \int_0^{\infty} dx \left\{ -4t \ln \left[\frac{\cosh\left(\frac{\sqrt{x^2 + \delta^2}}{2t} \right)}{\cosh\left(\frac{x}{2t} \right)} \right] + \frac{\delta^2}{\sqrt{x^2 + 1}} \right\}.$$

Here, the limit of integration has been moved to infinity. Why can we do this without loss of accuracy?

(c) Use Mathematica or Maple to plot the free energy obtained from the above result, confirming that the minimum is at $\Delta/\Delta_0 = 1$ and the transition occurs at $T_c = 2\Delta_0/3.53$.

References

[1] W. Meissner and R. Ochsenfeld, Ein neuer Effekt bei Eintritt der Supraleitfähigkeit, *Naturwissenschaften*, vol. 21, no. 44, p. 787, 1933.

[2] F. London, New conception of supraconductivity, *Nature*, vol. 140, p. 793, 1937.

[3] F. London, *Superfluids*, vols. 1–2, Dover Publications, 1961–1964.

[4] V. L. Ginzburg and L. D. Landau, On the theory of superconductivity, *Zh. Eksp. Teor. Fiz*, vol. 20, p. 1064, 1950.

[5] H. Fröhlich, Theory of the superconducting state. I The ground state at the absolute zero of temperature, *Phys. Rev.*, vol. 79, p. 845, 1950.

[6] B. Serin, C. A. Reynolds, and L. B. Nesbitt, Mass dependence of the superconducting transition temperature of mercury, *Phys. Rev.*, vol. 80, no. 4, p. 761, 1950.

[7] E. Maxwell, Isotope effect in the superconductivity of mercury, *Phys. Rev.*, vol. 78, no. 4, p. 477, 1950.

[8] C. A. Reynolds, B. Serin, W. H. Wright, and L. B. Nesbitt, Superconductivity of isotopes of mercury, *Phys. Rev.*, vol. 78, no. 4, p. 487, 1950.

[9] H. Fröhlich, Isotope effect in superconductivity, *Proc. Phys. Soc., London*, vol. 63, p. 778, 1950.

[10] *Oral History Transcript: Sir A. Brian Pippard*, Interview with Sir A. Brian Pippard on the history of superconductivity, September 14, 1982, Center for History of Physics, American Institute of Physics.

[11] A. B. Pippard, An experimental and theoretical study of the relation between magnetic field and current in a superconductor, *Proc. R. Soc. A*, vol. A216, p. 547, 1953.

[12] J. Bardeen, Theory of the Meissner effect in superconductors, *Phys. Rev.*, vol. 97, no. 6, p. 1724, 1955.

[13] J. Bardeen and D. Pines, Electron–phonon interaction in metals, *Phys. Rev.*, vol. 99, p. 1140, 1955.

[14] L. N. Cooper, Bound electron pairs in a degenerate Fermi gas, *Phys. Rev.*, vol. 104, no. 4, p. 1189, 1956.

[15] J. Bardeen, L. N. Cooper, and J. R. Schrieffer, Theory of superconductivity, *Phys. Rev.*, vol. 108, no. 5, p. 1175, 1957.

[16] *Schrieffer's Story: How We Got an Explanation of Superconductivity*, excerpts from an interview with J. Robert Schrieffer, Moments of Discovery, American Institute of Physics.

[17] Y. Nambu, Quasi-particles and gauge invariance in the theory of superconductivity, *Phys. Rev.*, vol. 117, p. 648, 1960.

[18] P. W. Anderson, Random-phase approximation in the theory of superconductivity, *Phys. Rev.*, vol. 112, no. 6, p. 1900, 1958.

[19] I. Giaever, Energy gap in superconductors measured by electron tunneling, *Phys. Rev. Lett.*, vol. 5, p. 147, 1960.

[20] I. Giaever, H. R. Hart, and K. Megerle, Tunneling into superconductors at temperatures below 1K, *Phys. Rev.*, vol. 126, p. 941, 1962.

[21] R. F. Gasparovic, B. N. Taylor, and R. E. Eck, Temperature dependence of the superconducting energy gap of Pb, *Solid State Commun.*, vol. 4, p. 59, 1966.

[22] S. H. Pan, E. W. Hudson, and J. C. Davis, Vacuum tunneling of superconducting quasiparticles from atomically sharp scanning tunneling microscope tips, *Appl. Phys. Lett.*, vol. 73, p. 2992, 1998.

[23] P. W. Anderson, Plasmons, gauge invariance, and mass, *Phys. Rev.*, vol. 130, no. 1, p. 439, 1963.

[24] P. W. Higgs, Broken symmetries and the masses of gauge bosons, *Phys. Rev. Lett.*, vol. 13, no. 16, p. 508, 1964.

15 Retardation and anisotropic pairing

This chapter continues our discussion of superconductivity, considering the effects of repulsive interactions and the physics of anisotropic Cooper pairing. According to an apocryphal story, Landau is reputed to have said that "nobody has yet repealed Coulomb's law" [1]. In the BCS theory of superconductors, there is no explicit appearance of the the repulsive Coulomb interaction between paired electrons. How then do real-world superconductors produce electron pairs, despite the presence of the strong interaction between them?

This chapter we will examine two routes by which Nature is able to satisfy the Coulomb interaction. In conventional superconductors, the attraction between electrons develops because the positive screening charge created by the ionic lattice around an electron remains in place long after the electron has moved away. This process that gives rise to a short-time repulsion between electrons is followed by a *retarded attraction* which drives s-wave pairing. However, since the 1980s physicists have been increasingly fascinated by *anisotropic superconductors*. In these systems, it is the repulsive interaction between the fermions that drives the pairing. The mechanism by which this takes place is through the development of nodes in the pair wavefunction – often by forming a higher angular momentum Cooper pair. The two classic examples of this physics are the p-wave pairs of superfluid ^{3}He and the d-wave pairs of cuprate high-temperature superconductors.

In truth, the physics community is still trying to understand the full interplay of superconductivity and the Coulomb force. The discovery of room-temperature superconductivity will surely involve finding a quantum material where strong correlations within the electron fluid lead to a large reduction in the sum total of kinetic and Coulomb energy.[1]

15.1 BCS theory with momentum-dependent coupling

We now illustrate these two different ways in which superconductors "overcome" the Coulomb interaction, by returning to the more generalized version of BCS theory with a momentum-dependent interaction:

---

[1] In weakly interacting systems we are trying to reduce the Coulomb energy in the face of a large kinetic energy, but in strongly interacting systems we are more often trying to reduce the kinetic energy in the face of large Coulomb interactions.

$$H = \sum_{k\sigma} \epsilon_k c_{k\sigma}^\dagger c_{k\sigma} + \overbrace{\sum_{k,k'} V_{k,k'}(c_{k\uparrow}^\dagger c_{-k\downarrow}^\dagger)(c_{-k'\downarrow} c_{k'\uparrow})}^{H_I}. \tag{15.1}$$

Notice how we have deliberately included a $+$ sign in front of the interaction H_I, to emphasize its predominantly repulsive character. As before, we carry out a Hubbard–Stratonovich decoupling of the interaction:

$$H_I \to \sum_k [\bar\Delta_k c_{-k\downarrow} c_{k\uparrow} + \text{H.c.}] - \sum_{k,k'} \bar\Delta_k V_{k,k'}^{-1} \Delta_{k'}, \tag{15.2}$$

where $V_{k,k'}^{-1}$ is the inverse of the matrix $V_{k,k'}$. While this is formally exact inside a path integral, following s-wave BCS theory we seek a mean-field theory in which the Δ_k are static. The only place where $V_{k,k'}$ appears is in the last term, so we can immediately diagonalize the resulting BCS theory to obtain a quasiparticle dispersion $E_k = \sqrt{\epsilon_k^2 + |\Delta_k|^2}$ in which the function Δ_k is obtained self-consistently by minimizing the free energy.

We can immediately generalize the mean-field free energy obtained for the momentum-independent interaction (14.145),

$$\mathcal{F} = -2T \sum_k \left[\ln[2\cosh(\beta E_k/2)] \right] - \sum_{k,k'} \bar\Delta_k V_{k,k'}^{-1} \Delta_{k'}, \tag{15.3}$$

and if we differentiate with respect to $\bar\Delta_k$, we obtain

$$\frac{\delta F}{\delta \bar\Delta_k} = -\tanh(\beta E_k/2) \frac{\Delta_k}{2E_k} - \sum_{k'} V_{k,k'}^{-1} \Delta_{k'} = 0. \tag{15.4}$$

Inverting this equation by multiplying by $V_{k,k'}$, we obtain the BCS gap equation:

$$\Delta_k = -\sum_{k'} V_{k,k'} \left(\frac{\Delta_{k'}}{2E_{k'}} \tanh\left(\frac{\beta E_{k'}}{2} \right) \right). \tag{15.5}$$

BCS gap equation: momentum-dependent coupling

The zero-temperature limit of this equation takes the simpler form

$$\Delta_k = -\sum_{k'} V_{k,k'} \left(\frac{\Delta_{k'}}{2E_{k'}} \right). \tag{15.6}$$

Note the minus sign in front of this equation! If the interaction is uniformly attractive, so that $V_{k,k'} < 0$ is negative, then this equation is satisfied by a uniformly positive gap function. However, in general the interaction $V_{k,k'}$ will contain repulsive (i.e. positive) terms, so a uniformly positive gap function cannot satisfy the gap equation, giving rise to gap *nodes* where the gap changes sign. The most satisfying kind of solution occurs if the sign of the gap function can satisfy

$$\text{sgn}(\Delta_k) = -\text{sgn}(V_{k,k'})\,\text{sgn}(\Delta_{k'}), \tag{15.7}$$

so that *regions of phase space that are linked by a repulsive interaction will have opposite gap signs*, whereas regions linked by an attractive interaction will have the same sign. This is the situation that leads to the largest gap and the largest mean-field transition temperature. We shall see that this can occur in two ways:

- In electron–phonon superconductors, where the interaction is repulsive at high energies, the gap function $\Delta(\epsilon)$ is largely isotropic in momentum space, but is energy-dependent and changes sign at an energy comparable with the Debye frequency.
- In anisotropic superconductors, the gap function $\Delta_{\mathbf{k}}$ becomes strongly momentum-dependent and acquires nodes in momentum space.

This last mechanism appears to be at work in all electronically mediated superconductors: organic, heavy-fermion, high-temperature cuprate and iron-based superconductors. We shall now illustrate this physics by using the BCS gap equation.

Example 15.1   The simplest anisotropic pair potential takes a factorizable form $V_{\mathbf{k},\mathbf{k}'} = -\frac{g_0}{V}\gamma_{\mathbf{k}}\gamma_{\mathbf{k}'}$, where $\gamma_{\mathbf{k}}$ is real and normalized, $\sum_{\mathbf{k}}(\gamma_{\mathbf{k}})^2 = 1$. In this case,

$$H_I = -\frac{g_0}{V}A^{\dagger}A, \tag{15.8}$$

but now the pairs acquire a spatial form factor

$$A = \sum_{\mathbf{k}}\left(\gamma_{\mathbf{k}}c_{-\mathbf{k}\downarrow}c_{\mathbf{k}\uparrow}\right), \qquad A^{\dagger} = \sum_{\mathbf{k}}\left(\gamma_{\mathbf{k}}c_{\mathbf{k}\uparrow}^{\dagger}c_{-\mathbf{k}\downarrow}^{\dagger}\right). \tag{15.9}$$

For example, in a simple model of d-wave pairing in a square two-dimensional lattice of side length a, $\gamma_{\mathbf{k}} = \cos(k_x a) - \cos(k_y)$.

(a) Show that the action for this case is identical to that of s-wave pairing, except that the gap $\Delta \rightarrow \Delta_{\mathbf{k}} = \Delta\gamma_{\mathbf{k}}$ now acquires a form-factor $\gamma_{\mathbf{k}}$. Write the action for the path integral.
(b) Derive the gap equation for the factorizable interaction above.

Solution

(a) We carry out a Hubbard–Stratonovich decoupling of the interaction that is formally the same as for s-wave pairing:

$$H_I = -\frac{g_0}{V}\bar{A}A \rightarrow \bar{\Delta}A + \bar{A}\Delta + \frac{V}{g_0}\bar{\Delta}\Delta. \tag{15.10}$$

Now we substitute $\bar{A} = \sum_{\mathbf{k}}\gamma_{\mathbf{k}}\bar{c}_{\mathbf{k}\uparrow}\bar{c}_{-\mathbf{k}\downarrow}$ and $A = \sum_{\mathbf{k}}\gamma_{\mathbf{k}}c_{-\mathbf{k}\downarrow}c_{\mathbf{k}\uparrow}$ to obtain

$$H_I = \sum_{\mathbf{k}}\gamma_{\mathbf{k}}\left(\bar{\Delta}c_{-\mathbf{k}\downarrow}c_{\mathbf{k}\uparrow} + \text{H.c.}\right) + \frac{V}{g_0}\bar{\Delta}\Delta.$$

Following the approach of Section 13.6, written in a Nambu notation the action for the path integral is then

$$S = \int_0^\beta d\tau \sum_{\mathbf{k}} \bar{\psi}_{\mathbf{k}}(\partial_\tau + \epsilon_{\mathbf{k}}\tau_3)\psi_{\mathbf{k}} + H_I$$

$$= \int_0^\beta d\tau \left\{ \sum_{\mathbf{k}} \bar{\psi}_{\mathbf{k}}(\partial_\tau + \underline{h}_{\mathbf{k}})\psi_{\mathbf{k}} + \frac{V}{g_0}\bar{\Delta}\Delta \right\}, \tag{15.11}$$

where $h_{\mathbf{k}} = \epsilon_{\mathbf{k}}\tau_3 + \gamma_{\mathbf{k}}(\bar{\Delta}\tau_- + \Delta\tau_+)$ and $\tau_\pm = \frac{1}{2}(\tau_1 \pm i\tau_2)$.

(b) Approximating the path integral by a mean-field saddle-point approximation, where $\bar{\Delta}(\tau) = \bar{\Delta}$ is a real constant, the mean-field free energy is then given by

$$F = -T \sum_{\mathbf{k},i\omega_n} \mathrm{Tr}\ln\left(-i\omega_n + h_{\mathbf{k}}\right) + \frac{V}{g_0}|\Delta|^2$$

$$= -T \sum_{\mathbf{k}} \ln\left[2\cosh\left(\frac{\beta E_{\mathbf{k}}}{2}\right)\right] + \frac{V}{g_0}|\Delta|^2, \tag{15.12}$$

where $E_{\mathbf{k}} = \sqrt{\epsilon_{\mathbf{k}}^2 + \gamma_{\mathbf{k}}^2|\Delta|^2}$. Finally, differentiating with respect to $\bar{\Delta}$, we obtain

$$\frac{\partial F}{\partial \bar{\Delta}} = 0 = -\sum_{\mathbf{k}} \frac{\gamma_{\mathbf{k}}^2 \Delta}{2E_{\mathbf{k}}} \tanh\left(\frac{\beta E_{\mathbf{k}}}{2}\right) + \frac{V}{g_0}\Delta, \tag{15.13}$$

from which we have the gap equation

$$\frac{V}{g_0} = \sum_{\mathbf{k}} \frac{\gamma_{\mathbf{k}}^2}{2E_{\mathbf{k}}} \tanh\left(\frac{\beta E_{\mathbf{k}}}{2}\right). \tag{15.14}$$

In the continuum limit, this becomes

$$\frac{1}{g_0} = \int_{\mathbf{k}} \frac{\gamma_{\mathbf{k}}^2}{2E_{\mathbf{k}}} \tanh\left(\frac{\beta E_{\mathbf{k}}}{2}\right). \tag{15.15}$$

15.2 Retardation and the Coulomb pseudopotential

In Chapter 7, we encountered the Bardeen–Pines model interaction,

$$V_{eff}(\mathbf{q}, \omega) = \left[\frac{e^2}{\epsilon_0(q^2 + \kappa^2)}\right]\left(1 + \frac{\omega_q^2}{\omega^2 - \omega_q^2}\right), \tag{15.16}$$

where the first term describes the instantaneous Coulomb interaction and the second describes the retarded attractive component due to phonons. Notice that the Bardeen–Pines interaction, taken in its entirety, is *always* repulsive, but is less so at low energies.

A simple BCS model that captures the character of this interaction has the form $V_{\mathbf{k},\mathbf{k}'} = V_{\textit{eff}}(\omega)|_{\omega=\epsilon_{\mathbf{k}}-\epsilon_{\mathbf{k}'}}$, where

$$V_{\textit{eff}}(\omega) = N(0)^{-1} \times \begin{cases} \mu - g & (|\omega| < \omega_D) \\ \mu & \text{(otherwise)}, \end{cases} \tag{15.17}$$

corresponding to an attractive electron–phonon interaction of strength $-g/N(0)$ operating at energy scales lower than the Debye frequency ω_D, superimposed on an instantaneous Coulomb repulsive interaction of strength $+\mu/N(0)$ (Figure 15.1), where μ is a dimensionless coupling constant representing the Fermi surface average of the Coulomb interaction.[2]

If we Fourier transform this interaction to the time domain, we obtain

$$N(0)V_{\textit{eff}}(t) = N(0) \int \frac{d\omega}{2\pi} V_{\textit{eff}}(\omega)e^{-i\omega t} = \mu \int_{-\infty}^{\infty} \frac{d\omega}{2\pi} e^{-i\omega t} - g \int_{-\omega_D}^{\omega_D} \frac{d\omega}{2\pi} e^{-i\omega t}$$

$$= \underbrace{\mu\delta(t)}_{\text{Instantaneous repulsion}} - \underbrace{\frac{g\omega_D}{\pi}\left(\frac{\sin \omega_D t}{\omega_D t}\right)}_{\text{retarded attraction}}, \tag{15.18}$$

where t is the time between "emission" and "absorption" of the exchange boson. We see that the interaction contains an instantaneous delta-function repulsion and a retarded attraction with an oscillatory tail. It is the second term that drives the pairing.

We now show that the retardation has the effect of renormalizing the effective Coulomb interaction down to a much weaker value,

$$\mu^* = \frac{\mu}{1 + \mu \ln(D/\omega_D)}, \tag{15.19}$$

where D is the half-bandwidth and ω_D is the Debye energy. Typically, the ratio $D/\omega_D \sim 10^5\,\text{K}/500\,\text{K} \sim 10^2$, so that $\ln(D/\omega_D) \sim 5$ and, even if the bare Coulomb coupling constant is of order unity, the renormalized Coulomb coupling constant $\mu^* \sim 1/6$. Provided $g - \mu^* > 0$, the renormalized s-wave pairing interacton is attractive and superconductivity develops.

We shall slightly modify interaction (15.17), and write the BCS interaction in the form

$$V_{\textit{eff}}(\mathbf{k}, \mathbf{k}') = N(0)^{-1} \times \begin{cases} \mu - g & (|\epsilon_{\mathbf{k}}|, |\epsilon_{\mathbf{k}'}| < \omega_D) \\ \mu & \text{(otherwise)}. \end{cases} \tag{15.20}$$

Let us assume a constant density of states $N(\epsilon) = N(0)$, replacing the momentum sum by an energy integral, $\sum_{\mathbf{k}'} \to N(0) \int d\epsilon$, and denoting $\Delta(\epsilon_{\mathbf{k}}) = \Delta_{\mathbf{k}}$. Then the gap equation (15.6) becomes

$$\Delta(\epsilon) = -N(0) \int_{-D}^{D} d\epsilon' V(\epsilon, \epsilon') \frac{\Delta(\epsilon')}{2E(\epsilon')}, \tag{15.21}$$

---

[2] Caution: by convention we adopt the "μ" notation for the repulsive interaction, in the full knowledge that it clashes with our notation for the chemical potential.

where D is the half-bandwidth. Since the interaction has a stepwise dependence on energy, we seek a corresponding solution for the gap function:

$$\Delta(\epsilon) = \begin{cases} \Delta_1 & (|\epsilon| < \omega_D) \\ \Delta_2 & (\omega_D < |\epsilon| < D). \end{cases} \tag{15.22}$$

Evaluating the pairing equation in these two energy regimes, we obtain

$$\Delta_1 = (g - \mu) \int_0^{\omega_D} d\epsilon \frac{\Delta_1}{\sqrt{\epsilon^2 + \Delta_1^2}} - \mu \int_{\omega_D}^{D} d\epsilon \frac{\Delta_2}{\sqrt{\epsilon^2 + \Delta_2^2}}$$

$$\Delta_2 = -\mu \int_0^{\omega_D} d\epsilon \frac{\Delta_1}{\sqrt{\epsilon^2 + \Delta_1^2}} - \mu \int_{\omega_D}^{D} d\epsilon \frac{\Delta_2}{\sqrt{\epsilon^2 + \Delta_2^2}}. \tag{15.23}$$

Carrying out the integrals, assuming Δ_1 and Δ_2 are small in magnitude compared with ω_D,

$$\Delta_1 = (g - \mu)\Delta_1 \ln\left(\frac{2\omega_D}{\Delta_1}\right) - \mu\Delta_2 \ln\left(\frac{D}{\omega_D}\right) \tag{15.24}$$

$$\Delta_2 = -\mu\Delta_1 \ln\left(\frac{2\omega_D}{\Delta_1}\right) - \mu\Delta_2 \ln\left(\frac{D}{\omega_D}\right). \tag{15.25}$$

Solving (15.25) for Δ_2, we obtain

$$\Delta_2 = -\mu^* \ln\left(\frac{2\omega_D}{\Delta_1}\right)\Delta_1, \tag{15.26}$$

where

$$\mu^* = \frac{\mu}{1 + \mu \ln\left(\frac{D}{\omega_D}\right)} \tag{15.27}$$

is the screened *Coulomb pseudopotential*. Substituting (15.26) into (15.24) then gives

$$\Delta_1 = (g - \mu)\Delta_1 \ln\left(\frac{2\omega_D}{\Delta_1}\right) + \frac{\mu^2}{1 + \mu \ln\left(\frac{D}{\omega_D}\right)}\Delta_1 \ln\left(\frac{D}{\omega_D}\right) \ln\left(\frac{2\omega_D}{\Delta_1}\right)$$

$$= \Delta_1(g - \mu^*) \ln\left(\frac{2\omega_D}{\Delta_1}\right). \tag{15.28}$$

In other words,

$$\Delta_1 = 2\omega_D \exp\left[-\frac{1}{(g - \mu^*)}\right],$$

$$\Delta_2 = -\frac{\mu^*}{g - \mu^*}\Delta_1. \tag{15.29}$$

This solution is illustrated in Figure 15.1.

Note the following:

- A superconducting solution can be obtained even when $g - \mu < 0$ is repulsive, provided $g - \mu^* > 0$. In certain metals, such as the alkali metals, μ^* is still too great for s-wave pairing to develop.
- The gap $\Delta(\omega)$ changes sign to become negative in the region where the pairing interaction is negative. By including a node in frequency space, it changes sign exactly where the Coulomb interaction becomes repulsive. In the time domain the

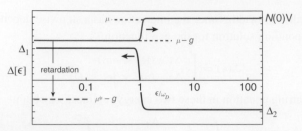

Fig. 15.1 Showing model interaction (orange line) and energy-dependent gap function that it gives rise to. The gap function changes sign around $\epsilon = \omega_D$ to minimize the effect of the Coulomb interaction. In this example $\mu - g > 0$, but because of renormalization the effective interaction at low energies becomes $\mu^* - g < 0$.

gap function contains an essentially instantaneous negative component and a retarded positive component, of the form

$$\Delta(t) = -|\Delta_2|\delta(t) + \Delta_1 \frac{\omega_D}{\pi} \left(\frac{\sin \omega_D t}{\omega_D t} \right).$$

It is thus adapted to the interaction so as to minimze the energy.

- The renormalization of the Coulomb interaction can be understood as the screening effect of high-frequency virtual pair fluctuations on energy scales between ω_D and the bandwidth D. Whereas pair fluctuations enhance an attractive interaction, they screen a repulsive interaction, driving it logarithmically towards weak coupling at low energies. The renormalized Coulomb interaction can be written self-consistently as follows:

$$-\frac{\mu^*}{N(0)} \qquad\qquad -\frac{\mu}{N(0)} \qquad\qquad +\left(-\frac{\mu}{N(0)}\right)\left(\sum_{|\epsilon_{k''}|>\omega_D} \frac{1}{2|\epsilon_{k''}|}\right)\left(-\frac{\mu^*}{N(0)}\right).$$

Carrying out the integral over the intermediate virtual pairs, we obtain

$$\mu^* = \mu - \mu\mu^* \left[\int_{-D}^{-\omega_D} + \int_{\omega_D}^{D} \right] \frac{d\epsilon}{2|\epsilon|} = \mu - \mu\mu^* \ln\left(\frac{D}{\omega_D}\right), \qquad (15.30)$$

from which we obtain the result $\mu^* = \mu/[1 + \mu \ln(D/\omega_D)]$.

15.3   Anisotropic pairing

At the turn of the 1960s, physicists on both sides of the Iron Curtain independently predicted that fermions interacting via repulsive interactions can develop pair condensates with pairs which carry finite orbital and spin angular momentum, giving rise to gap

Table 15.1 Pair potentials.

Interaction	Singlet $V^S_{\mathbf{k},\mathbf{k}'}$	Triplet $V^T_{\mathbf{k},\mathbf{k}'}$
$\frac{1}{2}V_{\mathbf{q}}:\rho_{-\mathbf{q}}\rho_{\mathbf{q}}:$	$\frac{1}{2}(V_{\mathbf{k}-\mathbf{k}'}+V_{\mathbf{k}+\mathbf{k}'})$	$\frac{1}{2}(V_{\mathbf{k}-\mathbf{k}'}-V_{\mathbf{k}+\mathbf{k}'})$
e.g. $V(\mathbf{q})=-g$	$-g$	0
$\frac{1}{2}J_{\mathbf{q}}:\vec{S}_{-\mathbf{q}}\cdot\vec{S}_{\mathbf{q}}:$	$-\frac{3}{4}\left(\frac{J_{\mathbf{k}-\mathbf{k}'}+J_{\mathbf{k}+\mathbf{k}'}}{2}\right)$	$\frac{1}{4}\left(\frac{J_{\mathbf{k}-\mathbf{k}'}-J_{\mathbf{k}+\mathbf{k}'}}{2}\right)$
e.g. $J_{\mathbf{q}}=2J(\cos q_x+\cos q_y)$	$-\frac{3}{2}J\left(\cos k_x\cos k'_x+\cos k_y\cos k'_y\right)$	$\frac{1}{2}J\left(\sin k_x\sin k'_x+\sin k_y\sin k'_y\right)$

functions that are anisotropic in momentum space. This idea was to have a wide applicability, accounting for superfluidity in ^{3}He and in certain kinds of nuclear matter; it also holds the key to high-temperature superconductivity.

Let us now explore the relationship between interactions and the pair potential $V_{\mathbf{k},\mathbf{k}'}$. We will examine two important examples, a repulsive potential,

$$V = \frac{1}{2}\sum V_{\mathbf{q}}\left[:\rho_{-\mathbf{q}}\rho_{\mathbf{q}}:\right] = \frac{1}{2}\sum_{\mathbf{k}_1,\mathbf{k}_2,\mathbf{q}} V_{\mathbf{q}}c^\dagger_{\mathbf{k}_1+\mathbf{q}\sigma}c^\dagger_{\mathbf{k}_2-\mathbf{q}\sigma'}c_{\mathbf{k}_2\sigma'}c_{\mathbf{k}_1\sigma}, \tag{15.31}$$

and a "magnetic" interaction,

$$V_{mag} = \frac{1}{2}\sum_{\mathbf{q}} J_{\mathbf{q}}\left[\vec{S}_{-\mathbf{q}}\cdot\vec{S}_{\mathbf{q}}\right], \tag{15.32}$$

where $J_{\mathbf{q}}$ is an effective interaction between quasiparticles on the Fermi surface. For example, in the *spin fluctuation* model discussed in Section 13.4, this interaction is given by $J_{\mathbf{q}}\sim -I/(1-I\chi_0(\mathbf{q}))$, where $\chi_0(q)$ is the momentum-dependent spin susceptibility and $I=U/3$ is the local repulsive Hubbard interaction. Such magnetic interactions are enhanced in the vicinity of a magnetic instability or *quantum critical point*. A summary of the pair potentials associated with these interactions, which we shall now derive, is given in Table 15.1.

Let's first consider the potential interaction (15.31). The pairing potential $V_{\mathbf{k},\mathbf{k}'}$ it gives rise to is determined by the influence of the interaction on Cooper pairs, which have zero total momentum. The BCS interaction is thus a projection of those terms in the interaction in which the incoming and outgoing Cooper pairs have zero momentum, so that $\mathbf{k}_1 = -\mathbf{k}_2 = \mathbf{k}'$ and $\mathbf{k}_1+\mathbf{q} = -(\mathbf{k}_2-\mathbf{q}) = \mathbf{k}$, hence $\mathbf{q} = \mathbf{k}-\mathbf{k}'$. The resulting interaction is

$$V_{BCS} = \frac{1}{2}\sum_{\mathbf{k},\mathbf{k}',\sigma,\sigma'} V_{\mathbf{k}-\mathbf{k}'}c^\dagger_{\mathbf{k}\sigma}c^\dagger_{-\mathbf{k}\sigma'}c_{-\mathbf{k}'\sigma'}c_{\mathbf{k}'\sigma} = V^{\uparrow\uparrow}_{BCS} + V^{\downarrow\downarrow}_{BCS} + V^{\uparrow\downarrow}_{BCS}, \tag{15.33}$$

which we have split up into terms according to the spin of the fermions in the pair. Let us first focus on the opposite-spin pairing term $V^{\uparrow\downarrow}_{BCS}$,

$$V^{\uparrow\downarrow}_{BCS} = \sum_{\mathbf{k},\mathbf{k}'} V_{\mathbf{k}-\mathbf{k}'}(c^\dagger_{\mathbf{k}\uparrow}c^\dagger_{-\mathbf{k}\downarrow})(c_{-\mathbf{k}'\downarrow}c_{\mathbf{k}'\uparrow}) = \sum_{\mathbf{k},\mathbf{k}'} V_{\mathbf{k}-\mathbf{k}'}\Psi^\dagger_{\mathbf{k}}\Psi_{\mathbf{k}'}, \tag{15.34}$$

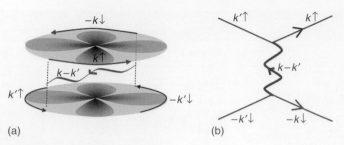

Fig. 15.2 (a) Illustrating transfer of momentum between Cooper pairs in condensate. (b) Feynman diagram representation of the transfer of momentum $\mathbf{q} = \mathbf{k} - \mathbf{k}'$ between Cooper pairs.

as shown in Figure 15.2, where we have lumped $V^{\uparrow\downarrow}$ and $V^{\downarrow\uparrow}$ together, absorbing the factor of $\frac{1}{2}$. Now despite appearances, an opposite-spin pair is not a well-defined singlet *or triplet*, since this requires appropriately symmetrized wavefunctions. If $F(\mathbf{k})_{\alpha\beta} = \langle \mathbf{k}\alpha, -\mathbf{k}\beta | \mathbf{k}_P \rangle$ is the pair wavefunction, then we can define a spatial parity P by $F(-\mathbf{k})_{\alpha\beta} = PF(\mathbf{k})_{\alpha\beta}$ and a spin exchange parity X by $F(\mathbf{k})_{\beta\alpha} = XF(\mathbf{k})_{\alpha\beta}$. Since this interaction is inversion-symmetric, $V_{\mathbf{q}} = V_{-\mathbf{q}}$, it preserves the parity of the Cooper pairs. The spin-exchange parity discriminates between spin singlets with $X = -1$ and spin triplets with $X = +1$. The joint process of spin and momentum exchange exchanges two fermions and must give an exchange eigenvalue of -1, $F(-\mathbf{k})_{\beta\alpha} = XPF(\mathbf{k})_{\alpha\beta} = -F(\mathbf{k})_{\alpha\beta}$, so that $XP = -1$. Even-parity pairs are thus singlets, $(P, X) = (+, -)$, whereas odd-parity pairs are triplets, $(P, X) = (-, +)$. To display the singlet and triplet pair scattering, we divide the interaction into symmetric and antisymmetric parts:

$$
V^{\uparrow\downarrow}_{BCS} = \sum_{\mathbf{k},\mathbf{k}'} \left[\overbrace{\left(\frac{V_{\mathbf{k}-\mathbf{k}'} + V_{\mathbf{k}+\mathbf{k}'}}{2} \right)}^{V^{S}_{\mathbf{k},\mathbf{k}'}} + \overbrace{\left(\frac{V_{\mathbf{k}-\mathbf{k}'} - V_{\mathbf{k}+\mathbf{k}'}}{2} \right)}^{V^{T}_{\mathbf{k},\mathbf{k}'}} \right] \Psi^{\dagger}_{\mathbf{k}} \Psi_{\mathbf{k}'}, \tag{15.35}
$$

where

$$
V^{S,T}_{\mathbf{k},\mathbf{k}'} = \frac{1}{2} \left(V_{\mathbf{k}-\mathbf{k}'} \pm V_{\mathbf{k}+\mathbf{k}'} \right) \tag{15.36}
$$

are the BCS pairing interactions in the singlet and triplet channels, respectively. Now because the first term is even in $\mathbf{k}$ and $\mathbf{k}'$ while the second is odd, the summations over momenta will project out pairs of definite parity: the first term scatters even-parity singlets while the second scatters odd-parity triplets, represented as

$$
\begin{aligned}
\Psi^{S\dagger}_{\mathbf{k}} &= (c^{\dagger}_{\mathbf{k}\uparrow} c^{\dagger}_{-\mathbf{k}\downarrow} + c^{\dagger}_{-\mathbf{k}\uparrow} c^{\dagger}_{\mathbf{k}\downarrow}), & \Psi^{S\dagger}_{\mathbf{k}} &= +\Psi^{S\dagger}_{-\mathbf{k}} \\
\Psi^{T\dagger}_{\mathbf{k}} &= (c^{\dagger}_{\mathbf{k}\uparrow} c^{\dagger}_{-\mathbf{k}\downarrow} - c^{\dagger}_{-\mathbf{k}\uparrow} c^{\dagger}_{\mathbf{k}\downarrow}), & \Psi^{T\dagger}_{\mathbf{k}} &= -\Psi^{T\dagger}_{-\mathbf{k}}.
\end{aligned} \tag{15.37}
$$

In terms of these operators, the unequal spin pairing interaction takes the form

$$
\begin{aligned}
V^{\uparrow\downarrow}_{BCS} &= \frac{1}{4} \sum_{\mathbf{k},\mathbf{k}'} \left[V^{S}_{\mathbf{k},\mathbf{k}'} \Psi^{S\dagger}_{\mathbf{k}} \Psi^{S}_{\mathbf{k}'} + V^{T}_{\mathbf{k},\mathbf{k}'} \Psi^{T\dagger}_{\mathbf{k}} \Psi^{T}_{\mathbf{k}'} \right] \\
&= \sum_{\mathbf{k},\mathbf{k}' \in \frac{1}{2}BZ} \left[V^{S}_{\mathbf{k},\mathbf{k}'} \Psi^{S\dagger}_{\mathbf{k}} \Psi^{S}_{\mathbf{k}'} + V^{T}_{\mathbf{k},\mathbf{k}'} \Psi^{T\dagger}_{\mathbf{k}} \Psi^{T}_{\mathbf{k}'} \right],
\end{aligned} \tag{15.38}
$$

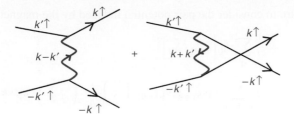

Direct and exchange scattering of a triplet pair.

Fig. 15.3

where we have restricted the sum over momentum space to one-half of momentum space, reflecting the fact that singlet and triplet pairs are only independently defined in half of momentum space ($\mathbf{k} \in \frac{1}{2} BZ$). Now the equal spin pairing terms also involve triplet pairs, which also interact via the triplet interaction $V^T_{\mathbf{k},\mathbf{k}'}$. We find

$$V^{\uparrow\uparrow}_{BCS} + V^{\downarrow\downarrow}_{BCS} = \sum_{\mathbf{k},\mathbf{k}' \in \frac{1}{2} BZ} (V_{\mathbf{k}-\mathbf{k}'} - V_{\mathbf{k}+\mathbf{k}'}) \left[(c^\dagger_{\mathbf{k}\uparrow} c^\dagger_{-\mathbf{k}\uparrow})(c_{-\mathbf{k}'\uparrow} c_{\mathbf{k}'\uparrow}) + (c^\dagger_{\mathbf{k}\downarrow} c^\dagger_{-\mathbf{k}\downarrow})(c_{-\mathbf{k}'\downarrow} c_{\mathbf{k}'\downarrow}) \right].$$

$$(15.39)$$

The appearance of scattering of amplitude $V_{\mathbf{k}-\mathbf{k}'}$ and amplitude $V_{\mathbf{k}+\mathbf{k}'}$ can be understood as a result of the exchange scattering term shown in Figure 15.3. A compact way to represent both parallel and unequal spin pair operators is to use the vector of $S = 1$ triplet pair operators:

$$\vec{\Psi}^{T\dagger}_{\mathbf{k}} = c^\dagger_{\mathbf{k}\alpha} \left(\vec{\sigma} i\sigma_2 \right)_{\alpha\beta} c^\dagger_{-\mathbf{k}\beta} = \begin{cases} c^\dagger_{\mathbf{k}\downarrow} c^\dagger_{-\mathbf{k}\downarrow} - c^\dagger_{\mathbf{k}\uparrow} c^\dagger_{-\mathbf{k}\uparrow}, & x \\ i(c^\dagger_{\mathbf{k}\downarrow} c^\dagger_{-\mathbf{k}\downarrow} + c^\dagger_{\mathbf{k}\uparrow} c^\dagger_{-\mathbf{k}\uparrow}), & y \\ c^\dagger_{\mathbf{k}\uparrow} c^\dagger_{-\mathbf{k}\downarrow} + c^\dagger_{\mathbf{k}\uparrow} c^\dagger_{-\mathbf{k}\downarrow}, & z \end{cases} \qquad (15.40)$$

refering to the x, y, and z components of the pair operator. The z component describes unequal spin pairing, while the x and y components describe linear combinations of equal spin pairing. Under a rotation, the triplet creation operator $\vec{\Psi}^{T\dagger}_{\mathbf{k}}$ transforms as a vector. In this notation, the BCS interaction is written

$$\hat{V}_{BCS} = \sum_{\mathbf{k},\mathbf{k}' \in \frac{1}{2} BZ} \left(V^S_{\mathbf{k},\mathbf{k}'} \Psi^{S\dagger}_{\mathbf{k}} \Psi^S_{\mathbf{k}'} + V^T_{\mathbf{k},\mathbf{k}'} \vec{\Psi}^{T\dagger}_{\mathbf{k}} \cdot \vec{\Psi}^T_{\mathbf{k}'} \right). \qquad (15.41)$$

Note the following:

- If one is only interested in singlet pairing, one can drop the triplet pairing terms and consider the interaction

$$V_{BCS} = \sum V^S_{\mathbf{k},\mathbf{k}'} (c^\dagger_{\mathbf{k}\uparrow} c^\dagger_{-\mathbf{k}\downarrow})(c_{-\mathbf{k}'\downarrow} c_{\mathbf{k}'\uparrow}). \qquad (15.42)$$

- One can decompose pairs into their orbital angular momentum components. Since the parity of a pair is related to its orbital angular momentum quantum number L by $P = (-1)^L$, even-parity superconductors involve even L (s, d, ... wave), while odd-parity triplet pairs involve L odd (p, f, ... wave).

Let us now return to consider the pair potential induced by the magnetic interaction

$$
\begin{aligned}
V_{mag} &= \frac{1}{2} \sum_{\mathbf{q}} J_{\mathbf{q}} \left[\vec{S}_{-\mathbf{q}} \cdot \vec{S}_{\mathbf{q}} \right] \\
&= \frac{1}{2} \sum_{\mathbf{k}_1, \mathbf{k}_2, \mathbf{q}} J_{\mathbf{q}} c^{\dagger}_{\mathbf{k}_1 + \mathbf{q}\alpha} c^{\dagger}_{\mathbf{k}_2 - \mathbf{q}\gamma} \left(\frac{\vec{\sigma}}{2} \right)_{\alpha\beta} \cdot \left(\frac{\vec{\sigma}}{2} \right)_{\gamma\delta} c_{\mathbf{k}_2\delta} c_{\mathbf{k}_1\beta},
\end{aligned}
\tag{15.43}
$$

where the $J_{\mathbf{q}}$ is an effective renormalized interaction between the quasiparticles. For example, in the cuprate superconductors, nearest-neighbor antiferromagnetic interactions derive from the vicinity to a Mott transition; these give rise to an antiferromagnetic interaction of the form $J_{\mathbf{q}} = 2J(\cos q_x a + \cos q_y a)$, where a is the separation of Cu atoms in a two-dimensional square lattice.

Now we need to consider the spin dependence of the interaction, determined by the matrices $\left(\frac{\vec{\sigma}}{2} \right)_{\alpha\beta} \cdot \left(\frac{\vec{\sigma}}{2} \right)_{\gamma\delta} \equiv \vec{S}_1 \cdot \vec{S}_2$. We note that the eigenvalue of $\vec{S}_1 \cdot \vec{S}_2$ is different for singlet and triplet states:

$$
\vec{S}_1 \cdot \vec{S}_2 = \begin{cases} +\frac{1}{4} & \text{(triplet)} \\ -\frac{3}{4} & \text{(singlet)}. \end{cases}
\tag{15.44}
$$

Since the symmetric and antisymmetric parts of the interaction filter out the singlet and triplet pairs, respectively, these eigenvalues must now enter as prefactors into the pairing potentials, giving

$$
\begin{aligned}
V^S_{\mathbf{k},\mathbf{k}'} &= -\frac{3}{4} \left(\frac{J_{\mathbf{k}-\mathbf{k}'} + J_{\mathbf{k}+\mathbf{k}'}}{2} \right) \\
V^T_{\mathbf{k},\mathbf{k}'} &= \frac{1}{4} \left(\frac{J_{\mathbf{k}-\mathbf{k}'} - J_{\mathbf{k}+\mathbf{k}'}}{2} \right).
\end{aligned}
\tag{15.45}
$$

Remarks

- Antiferromagnetic interactions ($J_{\mathbf{k}-\mathbf{k}'} > 0 \Rightarrow V^S_{\mathbf{k},\mathbf{k}'} < 0$) attract in the anisotropic singlet channel, whereas ferromagnetic interactions ($J_{\mathbf{k}-\mathbf{k}'} < 0 \Rightarrow V^T_{\mathbf{k},\mathbf{k}'} < 0$) attract in the triplet channel:

> antiferromagnetic interaction $\leftrightarrow$ singlet (mainly d-wave) anisotropic pairing
> ferromagnetic interaction $\leftrightarrow$ triplet (mainly p-wave) anisotropic pairing.

- The idea that ferromagnetic interactions could drive triplet pairing in nearly ferromagnetic metals, such as palladium, was first proposed by Layzer and Fay in the early 1970s [2]. In 1986 the discovery of antiferromagnetic spin fluctuations in the heavy-fermion superconductor UPt$_3$ led three separate groups (Zazie Béal Monod, Claude Bourbonnais, and Victor Emery at Orsay, Sherbrooke, and Brookhaven National Laboratory [3]; Kazu Miyake, Stefan Schmitt-Rink, and Chandra Varma at Bell Laboratories [4]; and Douglas Scalapino, Eugene Loh, and Jorge Hirsh at the University of California, Santa Barbara [5]), to propose that antiferromagnetic fluctuations drive d-wave superconductivity.

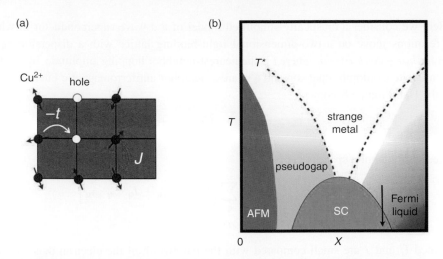

Fig. 15.4

(a) Superconductivity in cuprate superconductors involves the two-dimensional motion of electrons on a square lattice. The undoped material contains a square lattice of Cu^{2+} ions, each carrying a localized $S = \frac{1}{2}$ moment. When holes (or electrons) are introduced into the lattice via doping, the spins become mobile and the residual antiferromagnetic interactions drive d-wave pairing. A simplified model treats this as a single band of electrons of concentration $1 - x$, moving on a square lattice with hopping strength $-t$ and nearest-neighbor antiferromagnetic interaction J. (b) Schematic phase diagram of cuprate superconductors where x is the degree of hole doping. A commensurate antiferromagnetic insulator (pink) forms at small x, while at higher doping a superconducting dome develops. The normal state contains a pseudogap at low doping, forming a strange metal at optimal doping, with a linear resistivity. Fermi-liquid-like properties only develop at high doping, and it is only in this regime that the superconducting instability can be treated as a bona-fide Cooper pair instability of a Fermi liquid.

The basic connection between anisotropic singlet superconductivity and antiferromagnetic interactions is relevant to a wide variety of superconductors.

• If one is only interested in anisotropic singlet pairing, it is sufficient to work with an interaction of the form

$$V_{BCS} = -\frac{3}{4} \sum_{\mathbf{k},\mathbf{k'}} \left(\frac{J_{\mathbf{k-k'}} + J_{\mathbf{k+k'}}}{2} \right) (c_{\mathbf{k}\uparrow}^{\dagger} c_{-\mathbf{k}\downarrow}^{\dagger})(c_{-\mathbf{k'}\downarrow} c_{\mathbf{k'}\uparrow}).$$

15.4 d-wave pairing in two-dimensions

One of the most dramatic examples of anisotropic pairing is provided by the d-wave pairing in the copper oxide layers of the cuprate superconductors. These materials form antiferromagnetic Mott insulators, but when electrons or holes are introduced into the layers by doping, the magnetism is destroyed and the doped Mott insulator develops d-wave superconductivity. The normal state of these materials is not well understood, and for most of the phase diagram it cannot be treated as a Fermi liquid. For instance, at optimal doping these materials exhibit a linear resistivity $\rho(T) = AT + \rho_0$ due to electron–electron scattering that cannot be understood within the Fermi liquid framework. However, in the over-doped materials Fermi liquid behavior appears to recover and a BCS treatment is thought to be applicable.

Here we consider a drastically simplified model of a d-wave superconductor in which the fermions move on a two-dimensional tight-binding lattice with a dispersion $\epsilon_{\mathbf{k}} = -2t(\cos k_x a + \cos k_y a) - \mu$, where t is the nearest-neighbor hopping amplitude, interacting via an onsite Coulomb repulsion and a nearest-neighbor antiferromagnetic interaction, so that the Hamiltonian becomes

$$H = \sum_{\mathbf{k}} \epsilon_{\mathbf{k}} c_{\mathbf{k}\sigma}^{\dagger} c_{\mathbf{k}\sigma} + \sum_{j} U n_{j\uparrow} n_{j\downarrow} + J \sum_{(i,j)} \vec{S}_i \cdot \vec{S}_j, \tag{15.46}$$

while, in momentum space,

$$H = \sum_{\mathbf{k}} \epsilon_{\mathbf{k}} c_{\mathbf{k}\sigma}^{\dagger} c_{\mathbf{k}\sigma} + \frac{1}{2} \sum_{\mathbf{q}} \left[U\rho_{-\mathbf{q}} \cdot \rho_{\mathbf{q}} + J_{\mathbf{q}} \vec{S}_{-\mathbf{q}} \cdot \vec{S}_{\mathbf{q}} \right]$$

$$J_{\mathbf{q}} = 2J(\cos q_x a + \cos q_y a). \tag{15.47}$$

Provided U and J are small compared with the bandwidth of the electron band, we can treat this as a Fermi liquid with a BCS interaction in the singlet channel given by

$$V_{\mathbf{q}}^{singlet} = U - \frac{3J}{2}(\cos q_x a + \cos q_y a).$$

Here, following the previous section, we have multiplied the spin-dependent interaction by $-3/4$ to take care of the expectation value of $\vec{S}_1 \cdot \vec{S}_2 = -3/4$ in the singlet channel. When we replace $\mathbf{q} \to \mathbf{k} - \mathbf{k}'$ and symmetrize on momenta to obtain the singlet interaction, we obtain

$$V_{\mathbf{k},\mathbf{k}'} = \frac{1}{2} \left[V^{singlet}(\mathbf{k} - \mathbf{k}') + V^{singlet}(\mathbf{k} + \mathbf{k}') \right] = U - \frac{3J}{2}(c_x c_{x'} + c_y c_{y'}), \tag{15.48}$$

where we have used the notation $c_x \equiv \cos k_x$, $c_y \equiv \cos k_y$, and so on. The mean-field BCS Hamiltonian is then

$$H_{BCS} = \sum_{\mathbf{k}\sigma} \epsilon_{\mathbf{k}} c_{\mathbf{k}\sigma}^{\dagger} c_{\mathbf{k}\sigma} + \sum_{\mathbf{k},\mathbf{k}'} \left(U - \frac{3J}{2}(c_x c_{x'} + c_y c_{y'}) \right) c_{\mathbf{k}\uparrow}^{\dagger} c_{-\mathbf{k}\downarrow}^{\dagger} c_{-\mathbf{k}'\downarrow} c_{\mathbf{k}'\uparrow}.$$

Let us immediately jump forward to look at the gap equation,

$$\Delta_{\mathbf{k}} = -\int \frac{d^2 k'}{(2\pi)^2} V_{\mathbf{k},\mathbf{k}'} \frac{\Delta_{\mathbf{k}'}}{2E_{\mathbf{k}'}} \tanh\left(\frac{\beta E_{\mathbf{k}'}}{2} \right). \tag{15.49}$$

In the gap equation, the interaction will preserve the symmetries of the pair. If we divide the interaction into an s-wave and a d-wave term, $V_{\mathbf{k},\mathbf{k}'} = V_{\mathbf{k},\mathbf{k}'}^S + V_{\mathbf{k},\mathbf{k}'}^D$, as follows,

$$V_{\mathbf{k},\mathbf{k}'}^S = \overbrace{U}^{\text{s-wave}} \overbrace{- \frac{3}{4}J(c_x + c_y)(c_{x'} + c_{y'})}^{\text{extended s-wave}} \qquad \text{(s-wave)}$$

$$V_{\mathbf{k},\mathbf{k}'}^D = -\frac{3}{4}J(c_x - c_y)(c_{x'} - c_{y'}) \qquad \text{(d-wave)}, \tag{15.50}$$

then we see that the s-wave term is invariant under 90° rotations of $\mathbf{k}$ or $\mathbf{k}'$, whereas the d-wave term changes sign:

$$V_{\mathbf{k},\mathbf{k}'}^S = +V_{\mathbf{k},R\mathbf{k}'}^S, \qquad V_{\mathbf{k},\mathbf{k}'}^D = -V_{\mathbf{k},R\mathbf{k}'}^D,$$

where $R\mathbf{k} = (-k_y, k_x)$. Notice how the onsite Coulomb interaction is absent from the d-channel. A condensate with d-symmetry,

$$\Delta_{\mathbf{k}}^D = \Delta_D(c_x - c_y),$$

such that $\Delta_{R\mathbf{k}}^D = -\Delta_{\mathbf{k}}^D$, will couple to Cooper pairs via the d-wave interaction, because its integral with s-wave functions must change sign under $\pi/2$ rotations and is hence zero, $\sum_{\mathbf{k}'} V_{\mathbf{k},\mathbf{k}'}^S \Delta_{\mathbf{k}'}^D (\ldots) = 0$. By contrast, a condensate with *extended s-wave symmetry*, with the form

$$\Delta_{\mathbf{k}}^S = \Delta_1 + \Delta_2(c_x + c_y),$$

for which $\Delta_{R\mathbf{k}}^S = +\Delta_{\mathbf{k}}^S$, will vanish when integrated with the d-wave part of the interaction, $\sum_{\mathbf{k}'} V_{\mathbf{k},\mathbf{k}'}^D \Delta_{\mathbf{k}'}^S (\ldots) = 0$. In this case the two types of pairing are symmetry decoupled; moreoever, *the symmetry of the d-wave pair condensate orthogonalizes against the local Coulomb pseudopotential*.

Let us now look more carefully at the d-wave condensate, where the gap function $\Delta_{\mathbf{k}}^D = \Delta_D(c_x - c_y)$ vanishes along *nodes* along the diagonals $k_x = \pm k_y$. The corresponding quasiparticle energy

$$E_{\mathbf{k}} = \sqrt{\epsilon_{\mathbf{k}}^2 + \Delta_D^2(c_x - c_y)^2} \qquad (15.51)$$

must therefore vanish at the intersection of the nodes (where $\Delta_{\mathbf{k}} = 0$) and the Fermi surface (where $\epsilon_{\mathbf{k}} = 0$), as illustrated in Figure 15.5. At the nodal points the dispersion can be linearized in momentum, so that

$$E \sim \sqrt{(v_F \delta k_\perp^2) + (v_\Delta \delta k_\parallel)^2}$$

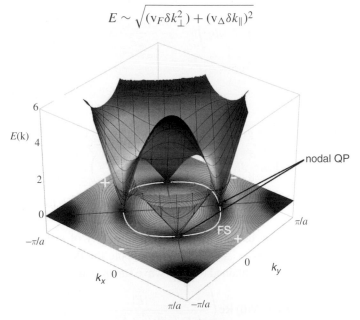

Energy dispersion for a two-dimensional d-wave superconductor with a $d_{x^2-y^2}$ gap function. The upper part of the plot shows a cut-away three-dimensional plot of the dispersion, showing the banana-shaped quasiparticle cones of quasiparticle excitations around the nodes. In the lower contour plot, the position of the nodal excitations is seen to occur at the intersections of the Fermi surface (white line) and the nodal lines of the gap function (red line).

Fig. 15.5

where $v_F = \partial E / \partial k_\perp$ is the Fermi velocity at the node and $v_\Delta = \partial E / \partial k_\parallel = \sqrt{2}\Delta_D a \sin\left(\frac{k_F a}{\sqrt{2}}\right)$ is the group velocity parallel to the Fermi surface created by the pairing. These excitations form a "Dirac cone" of excitations.

Let us now write out the gap equation for the d-wave solution in full:

$$\Delta_D(c_x - c_y) = -\int \frac{d^2 k'}{(2\pi)^2} \overbrace{\left(-\frac{3}{4}J(c_x - c_y)(c_{x'} - c_{y'})\right)}^{V^D_{\mathbf{k},\mathbf{k'}}} \frac{\Delta_D(c_{x'} - c_{y'})}{2E_{\mathbf{k'}}} \tanh\left(\frac{\beta E_{\mathbf{k'}}}{2}\right). \tag{15.52}$$

Fortunately, the d-wave form factor $c_x - c_y$ drops out of both sides, to give

$$1 = \frac{3}{4}J \int \frac{d^2 k}{(2\pi)^2} \frac{(c_x - c_y)^2}{2E_{\mathbf{k}}} \tanh\left(\frac{\beta E_{\mathbf{k}}}{2}\right). \tag{15.53}$$

Though it is straightforward to evaluate this kind of integral numerically, to get a feel of the physics let us suppose that the interaction only extends by an energy ω_{SF} around the Fermi energy, and that, furthermore, the band-filling around the Γ ($\mathbf{k} = 0$) point is small enough to use a quadratic approximation, $\epsilon_{\mathbf{k}} = -4t - \mu + tk^2$. In this case, the 2D density of states per spin $N(0) = \frac{1}{4\pi t}$ is a constant, while the gap function is

$$\Delta_D(c_y - c_x) = \Delta_D(k_x^2 - k_y^2) = \Delta_0 \cos 2\theta, \tag{15.54}$$

where $\Delta_0 = \Delta_D(k_F a)^2/2$ and a is the lattice spacing. Notice the characteristic $\Delta(\theta) \propto \cos 2\theta$ form, characteristic of an $l = 2$, d-wave Cooper pair. Now the gap equation becomes

$$1 = \frac{3}{4}JN(0) \int_{-\omega_{SF}}^{\omega_{SF}} d\epsilon \int_0^{2\pi} \frac{d\theta}{2\pi} \frac{\cos^2 2\theta}{2E} \tanh\left(\frac{\beta E}{2}\right)$$

$$E = \sqrt{\epsilon^2 + (\Delta_0 \cos 2\theta)^2}. \tag{15.55}$$

BCS gap equation: d-wave pairing

At T_c the average over angle gives $\frac{1}{2}$, so the equation for T_c is

$$1 = \frac{3}{8}JN(0) \int_0^{\omega_{SF}} d\epsilon \frac{1}{\epsilon} \tanh\left(\frac{\epsilon}{2T_c}\right). \tag{15.56}$$

This is identical to the BCS gap equation, but with $g = \frac{3}{8}JN(0)$, with the same formal form for $T_c = 1.13\omega_{SF}e^{-1/g}$.

It is particularly interesting to compute the d-wave density of states. Let us continue to use our approximation $\Delta(\theta) = \Delta_0 \cos 2\theta$. To compute the density of states, we must average the density of states we obtained for an s-wave superconductor (14.186) over angle:

$$N_D^*(E) = N(0) \operatorname{Re}\left\langle \frac{|E|}{\sqrt{(E - i\delta)^2 - \Delta^2 \cos^2 2\theta}} \right\rangle_\theta, \tag{15.57}$$

where $\langle \ldots \rangle_\theta \equiv \int \frac{d\theta}{2\pi}(\ldots)$ and the real part cleverly builds in the fact that the density of states vanishes when $|E| < |\Delta(\theta)|$. We can recast this expression as a standard elliptic integral by making the change of variable $2\theta \to \phi - \pi/2$. The resulting integral over ϕ is then

$$\frac{N_D^*(E)}{N(0)} = \mathrm{Re}\left[\int_0^\pi \frac{d\phi}{\pi} \frac{|E|}{\sqrt{(E - i\delta)^2 - \Delta^2 \sin^2 \phi}}\right] = \Phi\left[\frac{E - i\delta}{\Delta}\right], \quad (15.58)$$

where

$$\Phi[x] = \frac{2}{\pi}\mathrm{Re}\left[K\left(\frac{1}{x^2}\right)\right] \quad (15.59)$$

is expressed in terms of the elliptic function

$$K(x) = \int_0^{\pi/2} \frac{d\phi}{\sqrt{1 - x \sin^2 \phi}}, \quad (15.60)$$

known from the study of the pendulum.[3] This function is plotted in Figure 15.6. The clean gap of the s-wave superconductor is now replaced by a V-shaped structure, with a low-lying linear density of states derived from the Dirac cones in the excitation spectrum, and a sharp coherence peak in the density of states around $E \sim \pm\Delta$. We can understand the linear density of states at low energies by remembering that, for a relativistic spectrum $E = ck$, the density of states is $\frac{1}{2\pi}k\frac{dk}{dE} = \frac{|E|}{(2\pi c^2)}$. For these anisotropic Dirac cones, we must replace $c^2 \rightarrow v_F v_\Delta$; taking into account the four nodal cones and remembering the tricky factor of $\frac{1}{2}$ that enters because of the energy average of the coherence factors in the tunneling density of states (14.184), we obtain

$$N^*(E) = \frac{1}{2} \times 4 \times \frac{1}{2\pi} \frac{|E|}{v_F v_\Delta} = \overbrace{\frac{k_F}{2\pi v_F}}^{N(0)} \frac{|E|}{\Delta} = N(0)\frac{|E|}{\Delta}, \quad (15.61)$$

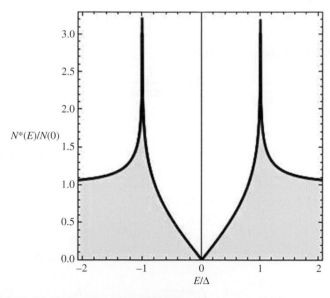

Density of states $N^*(E)/N(0))$ for a d-wave superconductor.

Fig. 15.6

[3] Note: here we use the notation used by Mathematica, with x multiplying $\sin^2 \phi$.

where we have put $v_\Delta = \partial E/\partial k_\parallel = (k_F)^{-1}\partial\Delta(\theta)/\partial\theta = 2\Delta/k_F$, identifying $N(0) = \frac{m}{2\pi} = \frac{k_F}{2\pi v_F}$.

Lastly, let us take a brief look at the alternative s-wave solution, where $\Delta_\mathbf{k} = \Delta_1 + \Delta_2(c_x + c_y)$. The first, momentum-independent term is entirely local, whereas the second term describes s-waves pairing with nearest neighbors. The gap equation

$$\Delta_\mathbf{k}^S = -\int \frac{d^2k'}{(2\pi)^2} \overbrace{\left(U - \frac{3}{4}J(c_x + c_y)(c_{x'} + c_{y'}) \right)}^{V_{\mathbf{k},\mathbf{k'}}^S} \overbrace{\frac{\Delta_\mathbf{k}^S}{2E_{\mathbf{k'}}}}^{\Delta_1 + \Delta_2(c_{x'} + c_{y'})} \tanh\left(\frac{\beta E_{\mathbf{k'}}}{2}\right) \quad (15.62)$$

is more complicated because there is cross-talk between the local and extended s-wave terms. To simplify our discussion, suppose we confine the interaction to within an energy ω_{SF} of the Fermi surface and assume that the filling of the Fermi surface is small enough that we can take $\mathbf{k} = \mathbf{k'} \sim 0$ in the pair potential. Then the effective s-wave coupling constant will be

$$V_{\mathbf{k},\mathbf{k'}} = U - \frac{3}{4}J \overbrace{(c_x + c_y)}^{\sim 2} \overbrace{(c_{x'} + c_{y'})}^{\sim 2} \to U - 3J, \quad (15.63)$$

which is only attractive providing $J > U/3$. We see that, for a single Fermi surface, the attraction in the extended-s-wave channel is suppressed by the Coulomb interaction, entirely vanishing if $J < J_c = U/3$. In fact, extended s-wave solutions are possible, and are believed to occur in the iron-based superconductors, but they require compensating Fermi surfaces in regions where $c_x + c_y$ have opposite signs, so that the Fermi surface average of the gap function vanishes, permitting a decoupling of the pairing from the repulsive Coulomb interaction.

Example 15.2 For a single Dirac cone of excitations with dispersion

$$E_\mathbf{k} = \sqrt{(v_x k_x)^2 + (v_y k_y)^2}, \quad (15.64)$$

show that the density of states is given by

$$N(E) = \frac{E}{2\pi v_x v_y}.$$

Solution

We write the density of states as

$$N(E) = \sum_\mathbf{k} \delta(E - E_\mathbf{k}) = \int \frac{dk_x dk_y}{(2\pi)^2} \delta\left(E - \sqrt{(v_x k_x)^2 + (v_y k_y)^2} \right). \quad (15.65)$$

Changing variables, $x = v_x k_x$, $y = v_y k_y$, then

$$N(E) = \int \frac{dx dy}{(2\pi)^2} \delta\left(E - \sqrt{(x^2 + y^2)} \right). \quad (15.66)$$

Changing $x = r\cos\theta$, $y = r\sin\theta$, then the measure becomes $dx dy \to r dr d\theta$ and the integral is

$$N(E) = \frac{1}{v_x v_y} \int \frac{d\theta\, r dr}{(2\pi)^2} \delta(E - r) = \frac{E}{2\pi v_x v_y}.$$

Example 15.3

(a) Carry out a Hubbard–Stratonovich decoupling of the BCS Hamiltonian on a two-dimensional lattice, where the pair potential is

$$V_{\mathbf{k},\mathbf{k}'} = \mathcal{N}_s^{-1}\left(U - \frac{3}{2}J(c_x c_{x'} + c_y c_{y'})\right), \qquad (15.67)$$

($c_x \equiv \cos k_x a$, $c_y \equiv \cos k_y a$), and show that the mean-field action takes the form

$$S_{MFT} = \int_0^\beta \left\{ \sum_{\mathbf{k}} \bar{\psi}_{\mathbf{k}}(\partial_\tau + \overbrace{\epsilon_{\mathbf{k}} \tau_3 + \bar{\Delta}_{\mathbf{k}} \tau_- + \Delta_{\mathbf{k}} \tau_+}^{h_{\mathbf{k}}})\psi_{\mathbf{k}} \right.$$
$$\left. + \mathcal{N}_s\left[\frac{4}{3J}(\bar{\Delta}_{2S}\Delta_{2S} + \bar{\Delta}_D \Delta_D) - \frac{\bar{\Delta}_{1S}\Delta_{1S}}{U}\right] \right\}, \qquad (15.68)$$

where

$$\Delta_{\mathbf{k}} = \Delta_{1S} + \Delta_{2S}(c_x + c_y) + \Delta_D(c_x - c_y) \qquad (15.69)$$

is the momentum-dependent gap function.
(b) Write down the mean-field free energy.
(c) Assuming a d-wave solution (i.e. $\Delta_D \neq 0, \Delta_1 = \Delta_2 = 0$), rederive the gap equation for this problem.
(d) For a single Fermi surface, why will a d-wave condensate have a higher T_c than an extended s-wave condensate?

Solution

(a) Let us factorize the interaction into s- and d-wave component, as follows:

$$V_{\mathbf{k},\mathbf{k}'} = \frac{U}{\mathcal{N}_s}\gamma_{1S}(\mathbf{k})\gamma_{1S}(\mathbf{k}') - \frac{3J}{4\mathcal{N}_s}\left[\gamma_{2S}(\mathbf{k})\gamma_{2S}(\mathbf{k}') + \gamma_D(\mathbf{k})\gamma_D(\mathbf{k}')\right], \qquad (15.70)$$

where $\gamma_{1S}(\mathbf{k}) = 1$, $\gamma_{2S}(\mathbf{k}) = c_x + c_y$, $\gamma_D(\mathbf{k}) = c_x - c_y$ are a set of normalized s-, extended s-, and d-wave form factors, respectively. We can then write the interaction Hamiltonian as

$$H_I = \frac{U}{\mathcal{N}_s}A_{1S}^\dagger A_{1S} - \frac{3J}{4\mathcal{N}_s}\left[A_{2S}^\dagger A_{2S} + A_D^\dagger A_D\right], \qquad (15.71)$$

where

$$A_\Gamma = \sum_{\mathbf{k}} \phi_\Gamma(\mathbf{k})c_{-\mathbf{k}\downarrow}c_{\mathbf{k}\uparrow} \qquad (\Gamma \in \{1S, 2S, D\}) \qquad (15.72)$$

create s-, extended s-, and d-wave pairs, respectively. If we carry out a Hubbard–Stratonovich decoupling of each of the product terms in this interaction, we then obtain

$$H_I \rightarrow \sum_{\Gamma \in \{1S,2S,D\}} (\bar{\Delta}_\Gamma A_\Gamma + \text{H.c.}) + \frac{4\mathcal{N}_s}{3J}(\bar{\Delta}_{2S}\Delta_{2S} + \bar{\Delta}_D \Delta_D) - \frac{\mathcal{N}_s}{U}\bar{\Delta}_{1S}\Delta_{1S}$$
$$= \sum_{\mathbf{k}} (\bar{\Delta}_{\mathbf{k}}c_{-\mathbf{k}\downarrow}c_{\mathbf{k}\uparrow} + \bar{c}_{\mathbf{k}\uparrow}\bar{c}_{-\mathbf{k}\downarrow}\Delta_{\mathbf{k}}) + \frac{4\mathcal{N}_s}{3J}(\bar{\Delta}_{2S}\Delta_{2S} + \bar{\Delta}_D \Delta_D)$$
$$- \frac{\mathcal{N}_s}{U}\bar{\Delta}_{1S}\Delta_{1S}, \qquad (15.73)$$

where $\Delta_\mathbf{k} = \sum_\Gamma \gamma_\Gamma(\mathbf{k})\Delta_\Gamma = \Delta_{2S}(c_x + c_y) + \Delta_D(c_x - c_y)$. Then the complete transformed Hamiltonian takes the form

$$
\begin{aligned}
H &= \sum_\mathbf{k} \epsilon_\mathbf{k} c^\dagger_{\mathbf{k}\sigma} c_{\mathbf{k}\sigma} + \sum_\mathbf{k}(\Delta_\mathbf{k} c^\dagger_{\mathbf{k}\uparrow} c^\dagger_{-\mathbf{k}\downarrow} + \text{H.c.}) \\
&\quad + \mathcal{N}_s\left(\frac{4}{3J}(\bar{\Delta}_D\Delta_D + \bar{\Delta}_{2S}\Delta_{2S}) - \frac{1}{U}\bar{\Delta}_{1S}\Delta_{1S}\right) \\
&= \sum_\mathbf{k} \psi^\dagger_\mathbf{k}\left(\epsilon_\mathbf{k}\tau_3 + \Delta_\mathbf{k}\tau^+ + \bar{\Delta}_\mathbf{k}\tau_-\right)\psi_\mathbf{k} \\
&\quad + \mathcal{N}_s\left(\frac{4}{3J}(\bar{\Delta}_D\Delta_D + \bar{\Delta}_{2S}\Delta_{2S}) - \frac{1}{U}\bar{\Delta}_{1S}\Delta_{1S}\right),
\end{aligned}
\tag{15.74}
$$

where we've dropped the constant remainder $\sum_\mathbf{k} \epsilon_\mathbf{k}$. The corresponding action is given by

$$
S = \int_0^\beta \left\{ \sum_\mathbf{k} \bar{\psi}_\mathbf{k}(\partial_\tau + \overbrace{\epsilon_\mathbf{k}\tau_3 + \bar{\Delta}_\mathbf{k}\tau_- + \Delta_\mathbf{k}\tau_+}^{h_\mathbf{k}})\psi_\mathbf{k} \right.
$$
$$
\left. + \mathcal{N}_s\left[\frac{4}{3J}(\bar{\Delta}_{2S}\Delta_{2S} + \bar{\Delta}_D\Delta_D) - \frac{\Delta_{1S}\Delta_{1S}}{U}\right] \right\}.
$$

(b) Carrying out the Gaussian path integral over the Fermi fields for constant gap functions, we obtain

$$
Z_{MF} = e^{-\beta F_{MF}} = \int \mathcal{D}[\bar{\psi}, \psi] e^{-S},
$$

where

$$
\begin{aligned}
F_{MF} &= -T\ln Z_{MF} = -T\sum_{\mathbf{k},i\omega_n} \ln\det[-i\omega_n + h_\mathbf{k}] \\
&\quad + \mathcal{N}_s\left[\frac{4}{3J}(\bar{\Delta}_{2S}\Delta_{2S} + \bar{\Delta}_D\Delta_D) - \frac{\Delta_{1S}\Delta_{1S}}{U}\right] \\
&= -T\sum_{\mathbf{k},i\omega_n} 2\ln[(i\omega_n)^2 - \epsilon_\mathbf{k}^2 - \bar{\Delta}_\mathbf{k}\Delta_\mathbf{k}] \\
&\quad + \mathcal{N}_s\left[\frac{4}{3J}(\bar{\Delta}_{2S}\Delta_{2S} + \bar{\Delta}_D\Delta_D) - \frac{\Delta_{1S}\Delta_{1S}}{U}\right] \\
&= -T\sum_\mathbf{k} \ln\left[2\cos\left(\frac{\beta E_\mathbf{k}}{2}\right)\right] \\
&\quad + \mathcal{N}_s\left[\frac{4}{3J}(\bar{\Delta}_{2S}\Delta_{2S} + \bar{\Delta}_D\Delta_D) - \frac{\Delta_{1S}\Delta_{1S}}{U}\right],
\end{aligned}
\tag{15.75}
$$

where the last line follows from carrying out the Matsubara sum and $E_\mathbf{k} = \sqrt{\epsilon_\mathbf{k}^2 + |\Delta_\mathbf{k}|^2}$.

(c) Suppose Δ_D is the only non-zero component of the gap function. Then

$$F_{MF} = -T \sum_\mathbf{k} \ln \left[2 \cos \left(\frac{\beta E_\mathbf{k}}{2} \right) \right] + \mathcal{N}_s \frac{4}{3J} (\bar{\Delta}_{2S} \Delta_{2S}), \qquad (15.76)$$

where $E_\mathbf{k} = \sqrt{\epsilon_\mathbf{k}^2 + \gamma_D(\mathbf{k})^2 \bar{\Delta}_D \Delta_D}$.

Taking the derivative of F_{MF} with respect to $\bar{\Delta}_D$, we obtain

$$\frac{\delta F_{MF}}{\delta \bar{\Delta}_D} = 0 = - \sum_\mathbf{k} \tanh \left(\frac{\beta E_\mathbf{k}}{2} \right) \frac{\gamma_D(\mathbf{k})^2 \Delta_D}{2E_\mathbf{k}} + \mathcal{N}_s \frac{4\Delta_D}{3J}, \qquad (15.77)$$

giving us the gap equation,

$$\frac{4}{3J} = \int_\mathbf{k} \tanh \left(\frac{\beta E_\mathbf{k}}{2} \right) \frac{\gamma_D(\mathbf{k})^2}{2E_\mathbf{k}}. \qquad (15.78)$$

(d) Whereas the d-wave condensate is completely decoupled from the repulsive U, so that $\partial^2 F_{MF}/\partial \Delta_{1S} \Delta_D = 0$, the extended s-wave component always mixes with the local s-wave component, which leads to a reduction of the effective coupling constant, so the d-wave Cooper instability will typically occur at a higher temperature. If we set the differentials of the free energy with respect to Δ_{1S} and Δ_{2S} to zero, we obtain two coupled gap equations, which, written in shorthand, are

$$\frac{4\Delta_{2S}}{3J} = \Delta_{2S} \langle \gamma_{1S}^2 \rangle + \Delta_{1S} \langle \gamma_{1S} \gamma_{2S} \rangle$$

$$-\frac{\Delta_{1S}}{U} = \langle \gamma_{1S}^2 \rangle \Delta_{1S} + \langle \gamma_{1S} \gamma_{2S} \rangle \Delta_{2S}, \qquad (15.79)$$

where we have used the shorthand $\langle \ldots \rangle = \sum_\mathbf{k} \frac{1}{2E_\mathbf{k}} \tanh \left(\frac{\beta E_\mathbf{k}}{2} \right) (\ldots)_\mathbf{k}$ (although $\gamma_{1S} = 1$, we have kept it in its symbolic form to show the symmetry of the equations). The two equations are coupled, because in general $\langle \gamma_{1S} \gamma_{2S} \rangle \neq 0$ for two s-wave form factors. We can eliminate Δ_{1S} from the second equation, to obtain

$$\Delta_{1S} = - \frac{\langle \gamma_{1S} \gamma_{2S} \rangle}{\langle \gamma_{1S}^2 \rangle + \frac{1}{U}} \Delta_{2S}. \qquad (15.80)$$

In other words, providing $\langle \gamma_{1S} \gamma_{2S} \rangle \neq 0$, the extended s-wave solution will always induce a finite onsite s-wave pairing, which costs a lot of Coulomb repulsion energy. Substituting this into the first of the mean-field equations (15.79), we obtain

$$\frac{4}{3J_{eff}} = \left(\frac{4}{3J} + \frac{\langle \gamma_{1S} \gamma_{2S} \rangle^2}{\frac{1}{U} + \langle \gamma_{1S}^2 \rangle} \right) = \langle \gamma_{2S}(\mathbf{k})^2 \rangle = \int_\mathbf{k} \tanh \left(\frac{E_\mathbf{k}}{2T_c} \right) \gamma_{2S}(\mathbf{k})^2. \qquad (15.81)$$

Since $1/J_{eff}$ is increased, we see that the effective coupling constant J_{eff} is reduced by the cross-talk between the extended s-wave channel and the onsite Coulomb interaction, suppressing the extended s-wave T_c. When the higher T_c d-wave condensate develops, this opens up a gap in the spectrum, pre-empting any lower-temperature s-wave instability. This is presumably why d-wave pairing predominates in the cuprate superconductors.

An important exception to this case occurs when there are multiple Fermi surface sheets which live in sectors of the extended s-wave form factor which have opposite sign. In this case, the average $\langle \gamma_{1s} \gamma_{2s} \rangle \sim 0$ and the larger average gap of the s-wave solution then favors extended s-wave over d-wave.

Example 15.4

(a) Show that the Nambu Green's function for a singlet superconductor with a momentum-dependent gap is
$$\mathcal{G}(\mathbf{k}, i\omega_n) = [i\omega_n - \epsilon_{\mathbf{k}}\tau_3 - \Delta_{\mathbf{k}}\tau_1]^{-1}, \tag{15.82}$$
where the gap function $\Delta_{\mathbf{k}} = \Delta_{-\mathbf{k}}$ assumed to be real.

(b) Using the Nambu Green's function, compute the tunneling density of states for a *three-dimensional* d-wave superconductor with gap $\Delta_{\mathbf{k}} = \Delta \cos 2\phi$.

Solution

(a) The Nambu Hamiltonian for a singlet superconductor with a momentum-dependent gap $\Delta_{\mathbf{k}} = \Delta(\phi) = \Delta \cos 2\phi$ is given by
$$H = \sum_{\mathbf{k}} \psi_{\mathbf{k}}^{\dagger} \underline{h}_{\mathbf{k}} \psi_{\mathbf{k}}$$
$$\underline{h}_{\mathbf{k}} = \epsilon_{\mathbf{k}}\tau_3 + \Delta \cos 2\theta \tau_1, \tag{15.83}$$
where we taken the gap to be real. The Nambu Green's function is then
$$\mathcal{G}(\mathbf{k}, \omega) = \frac{1}{\omega - \underline{h}_{\mathbf{k}}} = \frac{\omega + \underline{h}_{\mathbf{k}}}{\omega^2 - (\epsilon_{\mathbf{k}}^2 + \Delta^2 \cos^2 2\phi)}.$$

(b) The diagonal part of the Nambu Green's function is given by
$$[\mathcal{G}(k)]_{11} = \frac{\omega + \epsilon_{\mathbf{k}}}{\omega^2 - (\epsilon_{\mathbf{k}}^2 + \Delta^2 \cos^2 2\phi)}$$
and the tunneling density of states is given by
$$N(\omega) = \frac{1}{\pi} \sum_{\mathbf{k}} \text{Im} \left(\frac{\omega + \epsilon_{\mathbf{k}}}{(\omega - i\delta)^2 - E_{\mathbf{k}}^2} \right)$$
$$= \frac{1}{\pi} N(0) \int \frac{d\phi}{2\pi} \int d\epsilon \, \text{Im} \left(\frac{\omega + \epsilon}{(\omega - i\delta)^2 - \epsilon^2 + \Delta(\phi)^2} \right)$$
$$= -N(0) \int \frac{d\phi}{2\pi} \text{Im} \left(\frac{\omega}{\sqrt{\Delta^2 \cos^2 2\phi - (\omega - i\delta)^2}} \right)$$
$$= N(0) \int_0^{\pi/2} \frac{d\phi}{\pi/2} \text{Re} \left(\frac{|\omega|}{\sqrt{(\omega - i\delta)^2 - \Delta^2 \sin^2 \phi}} \right)$$
$$= \frac{2N(0)}{\pi} \text{Re} \, K \left(\frac{\Delta}{\omega - i\delta} \right), \tag{15.84}$$

where

$$K(x) = \int_0^{\pi/2} \frac{dx}{\sqrt{1 - x^2 \sin^2 \phi}}.$$

The last few stages of this calculation are the same as those in the derivation of the s-wave density of states in (14.185). We see that the form of the mean-field density of states of a three-dimensional d-wave system is the same as the density of states of a two-dimensional one.

Example 15.5

(a) By generalizing the approach taken in Section 13.8 for an s-wave superconductor, compute the London stiffness of a d-wave superconductor with gap $\Delta(\phi) = \Delta \cos \phi$, showing that it takes the form

$$Q_D(T) = Q_0 \left[1 - \int_{-\infty}^{\infty} d\omega \int \frac{d\phi}{2\pi} \left(-\frac{df(\omega)}{d\omega} \right) \mathrm{Re} \left(\frac{\omega}{\sqrt{\omega^2 - \Delta(\phi)^2}} \right) \right]$$

$$\Delta(\theta) = \Delta \cos(2\phi). \tag{15.85}$$

(b) Contrast the temperature dependence of the penetration depth in an s-wave and a clean d-wave superconductor.

Solution

This question is a little subtle at the beginning, because the d-wave gap has momentum dependence $\Delta_{\mathbf{k}}$, and it is not immediately clear whether, when a vector potential is included, we should make the Peierls replacement $\Delta_{\mathbf{k}} \to \Delta_{\mathbf{k}-e\mathbf{A}}$ or not.

One way to rationalize this is to notice that, in Nambu notation, the correct gauge-invariant Peierls replacement is $\mathbf{k} \to \mathbf{k} - e\mathbf{A}\tau_3$, so that in pairing terms of the form $\Delta_{\mathbf{k}}\tau_1$ we must replace

$$\Delta_{\mathbf{k}}\tau_1 \to \frac{1}{2}\{\Delta_{\mathbf{k}-e\mathbf{A}\tau_3}, \tau_1\} = \Delta_{\mathbf{k}} - \frac{e}{2}\nabla_{\mathbf{k}}\Delta_{\mathbf{k}} \overbrace{\{\tau_3, \tau_1\}}^{=0} = \Delta_{\mathbf{k}} + O(A^2), \tag{15.86}$$

so there is no correction to the current operator derived from the pairing, and the only important dependence of the BCS Hamiltonian on the vector potential comes from the kinetic energy $\epsilon_{\mathbf{k}-e\mathbf{A}\tau_3}\tau_3$ (14.241).

An alternative and more convincing way to argue the above is to explicitly introduce the vector potential into the pairing interaction using a Peierls substitution in real space. Consider the local pairing interaction, $-g \int_{\mathbf{x}} \Psi_D^\dagger(x)\Psi_D(x)$, where

$$\Psi_D^\dagger = \int_{\mathbf{R}} \gamma_D(\mathbf{R})\psi_\uparrow^\dagger(x + \mathbf{R}/2)\psi_\downarrow^\dagger(x - \mathbf{R}/2) \tag{15.87}$$

creates a d-wave pair with spatial form factor $\gamma_D(\mathbf{R})$ centered at $\mathbf{x}$. If we write the interaction out in full, it takes the form

$$H_I = -g \int_{\mathbf{x},\mathbf{R},\mathbf{R}'} \gamma_D(\mathbf{R})\gamma_D(\mathbf{R}') \left(\psi_\uparrow^\dagger(\mathbf{x}+\mathbf{R}/2)\psi_\downarrow^\dagger(\mathbf{x}-\mathbf{R}/2) \right) \left(\psi_\downarrow(\mathbf{x}-\mathbf{R}'/2)\psi_\uparrow(\mathbf{x}+\mathbf{R}'/2) \right)$$

$$= -g \int_{\mathbf{x},\mathbf{R},\mathbf{R}'} \gamma_D(\mathbf{R})\gamma_D(\mathbf{R}') : \left(\psi_\uparrow^\dagger(\mathbf{x}+\mathbf{R}/2)\psi_\uparrow(\mathbf{x}+\mathbf{R}'/2) \right)$$

$$\times \left(\psi_\downarrow^\dagger(\mathbf{x}-\mathbf{R}/2)\psi_\downarrow(\mathbf{x}-\mathbf{R}'/2) \right) : , \tag{15.88}$$

which involves the normal-ordered product of two hopping terms. To make this gauge-invariant, we need to make a Peierls substitution on each hopping term, replacing

$$\psi_\uparrow^\dagger(\mathbf{x}+\mathbf{R}/2)\psi_\uparrow(\mathbf{x}+\mathbf{R}'/2) \rightarrow \psi_\uparrow^\dagger(\mathbf{x}+\mathbf{R}/2)\psi_\uparrow(\mathbf{x}+\mathbf{R}'/2)\, \overbrace{e^{-i\int_{\mathbf{x}+\mathbf{R}/2}^{\mathbf{x}+\mathbf{R}'/2}\mathbf{A}\cdot d\mathbf{l}}}^{e^{-i\mathbf{A}(\mathbf{x})\cdot(\mathbf{R}-\mathbf{R}')/2}}$$

$$\psi_\downarrow^\dagger(\mathbf{x}-\mathbf{R}/2)\psi_\downarrow(\mathbf{x}-\mathbf{R}'/2) \rightarrow \psi_\downarrow^\dagger(\mathbf{x}-\mathbf{R}/2)\psi_\downarrow(\mathbf{x}-\mathbf{R}'/2)\, \overbrace{e^{-i\int_{\mathbf{x}-\mathbf{R}/2}^{\mathbf{x}-\mathbf{R}'/2}\mathbf{A}\cdot d\mathbf{l}}}^{e^{i\mathbf{A}(\mathbf{x})\cdot(\mathbf{R}-\mathbf{R}')/2}}, \tag{15.89}$$

where the Peierls factors have been evaluated ignoring gradients in the vector potential. We notice that the two Peierls factors cancel, so there is no dependence of the pairing term on the external vector potential.

(a) We can now follow the methodology of Section 13.8, including the momentum dependence of the gap, throughout the calculation. We obtain

$$Q_{ab} = \frac{4e^2}{\beta V} \sum_k \nabla_a \epsilon_\mathbf{k} \nabla_b \epsilon_\mathbf{k} \frac{\Delta_\mathbf{k}^2}{[(\omega_n)^2 + \epsilon_\mathbf{k}^2 + \Delta_\mathbf{k}^2]^2}. \tag{15.90}$$

Carrying out the integral over energy for each direction, and the summation over the Matsubara frequencies following the method of Section 14.8, then gives an angular-averaged version of (14.260):

$$Q(T) = Q_0 \left[1 - \int_{-\infty}^{\infty} d\omega \left(-\frac{df(\omega)}{d\omega} \right) \int \frac{d\phi}{2\pi} \right.$$

$$\left. \text{Re}\left(\frac{|\omega|}{\sqrt{(\omega-i\delta)^2 - \Delta^2 \cos^2 2\phi}} \right) \right], \tag{15.91}$$

where we have taken the real part of the integrand to eliminate terms where $|\omega| < |\Delta(\phi)|$.

We recognize the last term as the thermal average of the density of states, so that

$$Q(T) = Q_0 \left[1 - \overline{\left(\frac{A(\omega)}{N(0)} \right)} \right],$$

where (see (15.58))

$$A(\omega) = \frac{2N(0)}{\pi} \text{Re}\, K\left(\frac{\Delta}{\omega - i\delta} \right)$$

and $K(x)$ is the elliptic integral (15.60).

(b) At low temperatures, the density of states is given by $A(\omega)/N(0) = (|\omega|/\Delta)$, so that the thermally averaged density of states

$$\overline{\left(\frac{A(\omega)}{N(0)}\right)} = \frac{k_B T}{\Delta} 2 \int_0^\infty \frac{x}{(e^x + 1)(e^{-x} + 1)} = \frac{k_B T}{\Delta} \ln 4 \qquad (15.92)$$

grows linearly with temperature. Thus in a d-wave superconductor the inverse penetration depth $\frac{1}{\lambda_L^2} \propto Q(T)$ will exhibit a linear dependence on temperature at low temperatures, rather than the exponential dependence expected from a fully gapped s-wave superconductor:

$$1 - \frac{\lambda_L^2(0)}{\lambda_L^2(T)} \sim \frac{k_B T}{\Delta} \qquad (k_B T << \Delta).$$

(Note that in a dirty d-wave superconductor the density of states is constant at low temperatures, which leads to a quadratic temperature dependence of the inverse penetration depth at the lowest temperatures.)

15.5 Superfluid ^{3}He

15.5.1 Early history: theorists predict a new superfluid

As our second example of anisotropic pairing, we discuss the remarkable case of superfluid ^{3}He. As the 1950s came to an end and the wider significance of the BCS pairing instability was appreciated, the condensed-matter community began to realize that ^{3}He might form a BCS superfluid condensate, avoiding the mutual repulsion of the atoms by pairing in a higher angular momentum channel. Four independent groups (Lev Pitaevksii [6] at the Kapitza institute in Moscow; David Thouless at the Lawrence Radiation Laboratory, University of California, Berkeley [7]; Victor Emery and Andrew Sessler at the University of California, Berkeley [8]; and the Gang of Four, Keith Brueckner and Toshio Soda at the University of California, La Jolla, and Philip W. Anderson with Pierre Morel at Bell Laboratories, New Jersey [4] [9, 10]) came up with the idea of anisotropic pairing. Although these early papers examined both p- and d-wave pairs, each of them used *bare* nuclear interaction parameters as input to the BCS theory, and on the basis of these calculations came to the conclusion that the leading attractive channel was the $l = 2$, d-wave channel, predicting a d-wave superfluid condensate would develop in ^{3}He around $T_c = 50–150$ mK. The theory community would later be vindicated in their prediction of anistropic superfluidity in ^{3}He, but at a much lower temperature and with a p-wave rather than a d-wave symmetry.

During the 1960s the theory of anisotropic superfluidity developed rapidly, providing the framework for p-wave pairing that would ultimately be used to understand ^{3}He. In 1961 Morel and Anderson [10] introduced the ground state of what would later be identified

---

[4] Pierre Morel was officially a scientific attache at the French Embassy in New York City.

as the "A" phase, while in 1963 Roger Balian at the Centre d'Etude Nucléaires, Saclay, and Richard Werthamer at Bell Laboratories [11] discovered, an isotropic triplet paired ground state that would later be identified as the "B" phase. Gradually, towards the end of the 1960s, it became clear that the use of a bare interaction parameter as an input to BCS theory needed to be corrected for many-body effects, particularly with ladder diagram corrections to the pair scattering amplitude [12]. In a pioneering work, Walter Kohn at the University of California, San Diego, and Joaquin Luttinger at Columbia University, New York, [13] showed that, when many-body corrections to the Cooper channel interaction are considered, the sharpness of the Fermi surface guarantees that Fermi liquids are inevitably unstable to anisotropic pairing in some higher angular momentum channel. Using an input delta-function potential, Kohn and Luttinger derived an approximate asymptotic formula for T_c as a function of angular momentum l in ^{3}He, given by

$$T_c(l) \sim \epsilon_F \exp\left\{-\frac{\pi^2}{(k_F a)^2} l^4\right\}, \tag{15.93}$$

where l is the angular momentum of the pair, ϵ_F and k_F are the Fermi energy and momentum, respectively, and a is the diameter of the ^{3}He atom. Curiously, Kohn and Luttinger chose to illustrate this equation for $l = 2$, d-wave pairing, which for $k_F a \sim 2$ gives $T_c \sim 10^{-17} \epsilon_F$. Had they made the bold but uncontrolled insertion of $l = 1$, they would have obtained $T_c \sim 0.05 \epsilon_F \sim 50$ mK, surely an indication that p-wave pairing is a stronger candidate than d-wave! Then in 1967 D. Fay and A. Layzer, working at the Stevens Institute of Technology, New Jersey, made the critical observation [14] that in dilute neutral fluids many-body effects, which tend to ferromagnetically enhance interactions, will also generally lead to p-wave pairing.

It was not until 1972 that Douglas Osheroff, Robert Richardson, and David Lee at Cornell University finally discovered superfluidity in ^{3}He, developing at 2.65 mK [15] (see Figure 15.7). From the anomalies in the NMR response, this team was able to identify two phases: a high-temperature A phase and a low-temperature B phase in which most of the magnetic response disappeared. By carefully analyzing the detailed NMR measurements

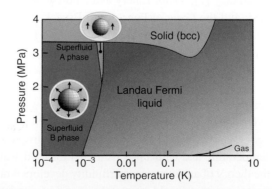

Fig. 15.7 Phase diagram of ^{3}He, showing the superfluid A and B phases, [22] with icons representing the gap anisotropy. Note that 0.1MPa = 1 atm. Adapted with permission from D. Vollhardt and P. Wölfle, *The Superfluid Phases of Helium 3*, Dover, 2013. Copyright 2013 by Dieter Vollhardt and Peter Wölfle.

carried out on these phases, Anthony Leggett, working at Sussex University [16], was able to show [17–19] that the pair symmetry of the A phase is triplet and probably corresponds to the Anderson–Morel state (now called the Anderson–Brinkman–Morel state). The pair symmetry of the B phase was later identified with the isotropic and fully gapped Balian–Werthamer state [20].

Curiously, although the early ^{3}He theorists predicted the wrong pair symmetry for ^{3}He, their efforts were not in vain, for d-wave pairing was realized seven years later in superconductors, with the discovery of the first anisotropic superconductor, $CeCuSi_2$, by Frank Steglich at Cologne University [21]. We now know many examples of d-wave superconductors, including the high-temperature cuprate superconductors.

15.5.2 Formulation of a model

The beauty of ^{3}He is that its isotropy provides us with a model system. The Fermi surface is perfectly spherical and in this case the pairing interaction between the quasiparticles depends only on the relative angle between the initial and final pair momenta $\mathbf{k}$ and $\mathbf{k}'$, i.e. $V_{\mathbf{k},\mathbf{k}'} = V(\cos\theta_{\mathbf{k},\mathbf{k}'})$. This implies that the pairing interaction can be decomposed as a multipole expansion involving Legendre polynomials:

$$V_{\mathbf{k},\mathbf{k}'} = \sum_l (2l+1) V_l P_l(\hat{\mathbf{k}} \cdot \hat{\mathbf{k}}'). \tag{15.94}$$

This is reminiscent of the multipole expansion of Fermi liquid interactions (6.38). Using the orthogonality relation $\int \frac{dc}{2} P_l(c) P_{l'}(c) = \delta_{l,l'}/(2l+1)$, the parameters V_l are given by

$$V_l = \int_{-1}^{1} \frac{d\cos\theta}{2} P_l(\cos\theta) V(\cos\theta) \qquad l \in \begin{cases} \text{even} & \text{(singlet)} \\ \text{odd} & \text{(triplet).} \end{cases} \tag{15.95}$$

These are the higher angular momentum analogues of the BCS s-wave interaction parameter. Now the parity of the Legendre polynomials alternates with l, $P_l = (-1)^l$ ($P_l(-x) = (-1)^l P_l(x)$), so the even l define singlet pair potentials while the odd l define triplet ($S=1$) pair potentials.

Using the relationship $(2l+1) P_l(\hat{\mathbf{k}} \cdot \hat{\mathbf{k}}') = 4\pi \sum_{m=-l}^{l} Y_{lm}^*(\hat{\mathbf{k}}) Y_{lm}(\hat{\mathbf{k}}')$, we can factorize the anisotropic BCS interaction in the form

$$V_{\mathbf{k},\mathbf{k}'} = \sum_{l,m} V_l \, y_{lm}^*(\hat{\mathbf{k}}) y_{lm}(\hat{\mathbf{k}}'), \tag{15.96}$$

where we have used the notation $y_{lm} = \sqrt{4\pi}\, Y_{lm}$ to denote spherical harmonics normalized to give unit norm when averaged over the sphere $\int \frac{d\Omega}{4\pi} y_{lm}^* y_{lm} = \delta_{l,l'} \delta_{m,m'}$. This is the same kind of factorized interaction encountered in the previous section, and we can treat it in the same way. For ^{3}He, the hard-core repulsion between the atoms rules out an s-wave instability[5] and it is the p-wave ($l=1$) triplet ($S=1$) channel that takes over. Approximating $V_1 = -g/V$ and ignoring all other channels, then

[5] Curiously, in optical atom traps in which the atomic interactions among highly dilute fermions can be tuned through a Feshbach resonance, it is possible to produce an attractive s-wave interaction, so a conventional BCS instability does occur.

$$V_{\mathbf{k},\mathbf{k}'} = -\frac{g}{V}3\cos(\mathbf{k}\cdot\mathbf{k}') = -\frac{3g}{V}(\hat{k}_a\hat{k}_a), \qquad (15.97)$$

where $\hat{k}_a = k_a/k_F$ and the sum over the repeated index $a = 1, 2, 3$ is implied. The BCS Hamiltonian for a triplet superfluid is then [11]

$$H_{BCS} = \sum_{\mathbf{k}\sigma}\epsilon_{\mathbf{k}}c^\dagger_{\mathbf{k}\sigma}c_{\mathbf{k}\sigma} - \frac{3g_0}{V}\sum_{\mathbf{k},\mathbf{k}'\in\frac{1}{2}BZ}(\vec{\Psi}^\dagger_{\mathbf{k}}\hat{k}_\ell)\cdot(\hat{k}'_\ell\vec{\Psi}_{\mathbf{k}'})$$

$$\vec{\Psi}_{\mathbf{k}} = c_{-\mathbf{k}\alpha}\left(-i\sigma_2\vec{\sigma}\right)_{\alpha\beta}c_{\mathbf{k}\beta}$$

$$\vec{\Psi}^\dagger_{\mathbf{k}} = c^\dagger_{\mathbf{k}\alpha}\left(\vec{\sigma}i\sigma_2\right)_{\alpha\beta}c^\dagger_{-\mathbf{k}\beta}. \qquad (15.98)$$

Notice that there are now three *triplet* channels ($\vec{\Psi}_{\mathbf{k}} \equiv \Psi^a_{\mathbf{k}}$, $a = 1, 2, 3$) and three *orbital* channels ($\hat{k}_l$, $l = x, y, z$) in which the pairing takes place. The summation over momentum in the interaction takes place over one-half the Brillouin zone.

15.5.3 Gap equation

If we carry out a Hubbard–Stratonovich transformation, we get

$$H_{MFT} = \sum_{\mathbf{k}\sigma}\epsilon_{\mathbf{k}}c^\dagger_{\mathbf{k}\sigma}c_{\mathbf{k}\sigma} + \sum_{\mathbf{k}\in\frac{1}{2}BZ}\left[\vec{\Psi}^\dagger_{\mathbf{k}}\cdot(\vec{\Delta}_l)\,\hat{k}_l + \text{H.c.}\right] + \frac{V}{3g_0}(\vec{\Delta}^*_l\cdot\vec{\Delta}_l). \qquad (15.99)$$

The three vectors $\vec{\Delta}_l$ ($l = x, y, z$) define a three-dimensional matrix $\Delta^a_l \equiv (\vec{\Delta}_l)^a$ which links the spin and orbital degrees of freedom. If we denote $\vec{\Delta}_{\mathbf{k}} = \sum_{l=x,y,z}\vec{\Delta}_l\hat{k}_l$, then, since $\int\frac{d\Omega_{\hat{k}}}{4\pi}\hat{k}_l\hat{k}_m = \frac{1}{3}\delta_{lm}$, it follows that

$$\Delta^a_l = (\vec{\Delta}_l)^a = 3\int\frac{d\Omega_{\hat{k}}}{4\pi}(\vec{\Delta}_{\mathbf{k}})^a\hat{k}_l. \qquad (15.100)$$

Thus we can write

$$\frac{V}{3g_0}(\vec{\Delta}^*_l\cdot\vec{\Delta}_l) = \frac{3V}{g_0}\int\frac{d\Omega_{\hat{k}}}{4\pi}\frac{d\Omega_{\hat{k}'}}{4\pi}\vec{\Delta}_{\mathbf{k}}\cdot\vec{\Delta}_{\mathbf{k}'}(\hat{k}\cdot\hat{k}') \equiv -\int_{\hat{k},\hat{k}'}\vec{\Delta}^*_{\mathbf{k}}V^{-1}_{\mathbf{k},\mathbf{k}'}\vec{\Delta}_{\mathbf{k}'}, \qquad (15.101)$$

where we have identified $V^{-1}_{\mathbf{k},\mathbf{k}'} \equiv -\frac{V}{g_0}(3\hat{k}\cdot\hat{k}')$ and denoted $\int_{\hat{k}} = \int\frac{d\Omega_{\hat{k}}}{4\pi}$. The mean-field Hamiltonian is then

$$H_{MFT} = \sum_{\mathbf{k}\sigma}\epsilon_{\mathbf{k}}c^\dagger_{\mathbf{k}\sigma}c_{\mathbf{k}\sigma} + \sum_{\mathbf{k}\in\frac{1}{2}BZ}\left(\vec{\Psi}^\dagger_{\mathbf{k}}\cdot\vec{\Delta}_{\mathbf{k}} + \text{H.c.}\right) + \frac{3V}{g_0}\int\frac{d\Omega_{\hat{k}}}{4\pi}\frac{d\Omega_{\hat{k}'}}{4\pi}\vec{\Delta}_{\mathbf{k}}\cdot\vec{\Delta}_{\mathbf{k}'}(\hat{k}\cdot\hat{k}').$$

$$(15.102)$$

Now, to diagonalize this mean-field theory we need to cast it into spinors. Triplet pairing mixes up and down electrons, which obliges us to use a four-component spinor called a *Balian–Werthamer spinor* [11] after its inventors:

$$\psi_{\mathbf{k}} \equiv \begin{pmatrix} c_{\mathbf{k}} \\ i\sigma_2 c^{\dagger}_{-\mathbf{k}} \end{pmatrix} \equiv \begin{pmatrix} c_{\mathbf{k}\uparrow} \\ c_{\mathbf{k}\downarrow} \\ c^{\dagger}_{-\mathbf{k}\downarrow} \\ -c^{\dagger}_{-\mathbf{k}\uparrow} \end{pmatrix}. \qquad \text{Balian–Werthamer spinor} \qquad (15.103)$$

The upper two entries are the destruction operators for particles of momentum $\mathbf{k}$, while the lower two,

$$\begin{pmatrix} a_{\mathbf{k}\uparrow} \\ a_{\mathbf{k}\downarrow} \end{pmatrix} \equiv \begin{pmatrix} c^{\dagger}_{-\mathbf{k}\downarrow} \\ -c^{\dagger}_{-\mathbf{k}\uparrow} \end{pmatrix}, \qquad (15.104)$$

are the destruction operators for *holes* of momentum $\mathbf{k}$. Hole-destruction operators are the time reversal (denoted by the operator θ) of the corresponding particle-creation operators, and the minus sign in the lower entry appears on time reversal of a down-spin state, $a^{\dagger}_{\mathbf{k}\downarrow} = \theta c_{\mathbf{k}\downarrow}\theta^{-1} = -c_{-\mathbf{k}\uparrow}$.[6] Notice how the $i\sigma_2$ that appears in the triplet pair operators is now neatly absorbed into the spinor. Moreoover, the BW spinor obeys canonical anticommutation rules:

$$\{\psi_{\mathbf{k}\alpha}, \psi^{\dagger}_{\mathbf{k}'\beta}\} = \delta_{\mathbf{k},\mathbf{k}'}\delta_{\alpha\beta}.$$

Of course, we have doubled the number of components in the spinor, so we must now restrict the momentum to one-half of momentum space, $\mathbf{k} \in \frac{1}{2}BZ$. The payoff is that we now have a rotationally invariant representation in which the spin operator is defined in terms of block-diagonal Pauli matrices:

$$\vec{\sigma}_4 \equiv \underline{1} \otimes \vec{\sigma} = \left(\begin{array}{c|c} \vec{\sigma} & \\ \hline & \vec{\sigma} \end{array} \right), \qquad (15.105)$$

while the Nambu matrices are now block matrices:

$$\vec{\tau}_4 \equiv \vec{\tau} \otimes \underline{1} = \left\{ \left(\begin{array}{c|c} & \underline{1} \\ \hline \underline{1} & \end{array} \right), \left(\begin{array}{c|c} & -i\underline{1} \\ \hline i\underline{1} & \end{array} \right), \left(\begin{array}{c|c} \underline{1} & \\ \hline & -\underline{1} \end{array} \right) \right\}. \qquad (15.106)$$

In this notation, the BCS Hamiltonian can be succinctly rewritten as

$$H_{MFT} = \sum_{\mathbf{k} \in \frac{1}{2}BZ} \psi^{\dagger}_{\mathbf{k}} h_{\mathbf{k}} \psi_{\mathbf{k}} + \frac{3V}{g_0} \int \frac{d\Omega_{\hat{k}}}{4\pi} \frac{d\Omega_{\hat{k}'}}{4\pi} \vec{\Delta}_{\mathbf{k}} \cdot \vec{\Delta}_{\mathbf{k}'}(\hat{k} \cdot \hat{k}')$$

$$h_{\mathbf{k}} = \left(\begin{array}{c|c} \epsilon_{\mathbf{k}} & \vec{\Delta}_{\mathbf{k}} \cdot \vec{\sigma} \\ \hline \vec{\Delta}^{*}_{\mathbf{k}} \cdot \vec{\sigma} & -\epsilon_{\mathbf{k}} \end{array} \right) \equiv \epsilon_{\mathbf{k}}\tau_3 + (\vec{\Delta}_{\mathbf{k}} \cdot \vec{\sigma})\tau_+ + (\vec{\Delta}^{*}_{\mathbf{k}} \cdot \vec{\sigma})\tau_-, \quad (15.107)$$

where $\tau_{\pm} = \frac{1}{2}(\tau_1 \pm i\tau_2)$. It is common to denote the direction of the gap function in spin space by the complex *d-vector* $\vec{d}_{\mathbf{k}}$,

$$\vec{\Delta}_{\mathbf{k}} = \Delta \vec{d}_{\mathbf{k}}, \qquad (15.108)$$

which is normalized so that its angular average over the Fermi surface is unity:

$$\int \frac{d\Omega_{\mathbf{k}}}{4\pi} |\vec{d}_{\mathbf{k}}|^2 = 1. \qquad (15.109)$$

---

[6] You can also verify that the diagonal and off-diagonal matrix elements of the spin operator are the same for particles and for holes, so that $h^{\dagger}_{\mathbf{k}}\vec{\sigma}h_{\mathbf{k}} = c_{-\mathbf{k}}(i\sigma_2)\vec{\sigma}(-i\sigma_2)c_{-\mathbf{k}} = c^{\dagger}_{-\mathbf{k}}\vec{\sigma}c_{-\mathbf{k}}$, where the last step follows because $\vec{\sigma}^T = -\sigma_2\vec{\sigma}\sigma_2$.

The d-vector is an emergent property of the Fermi surface, and the textures it gives rise to in momentum space define the state of the condensate.

If we take the determinant of $\omega - h_{\mathbf{k}}$ by multiplying out its two-dimensional block diagonals, we find

$$
\begin{aligned}
\det(\omega - h_{\mathbf{k}}) &= \det\left[(\omega^2 - \epsilon_{\mathbf{k}}^2)\underline{1} - (\vec{\Delta}_{\mathbf{k}}^* \cdot \vec{\sigma})(\vec{\Delta}_{\mathbf{k}} \cdot \vec{\sigma})\right] \\
&= \det\left[(\omega^2 - \epsilon_{\mathbf{k}}^2)\underline{1} - \Delta^2(|\vec{d}_{\mathbf{k}}|^2 + i\vec{d}_{\mathbf{k}}^* \times \vec{d}_{\mathbf{k}} \cdot \vec{\sigma})\right] \\
&= \det\left[(\omega^2 - \epsilon_{\mathbf{k}}^2)\underline{1} - \Delta^2(|\vec{d}_{\mathbf{k}}|^2 + 2\vec{d}_{1\mathbf{k}} \times \vec{d}_{2\mathbf{k}} \cdot \vec{\sigma})\right],
\end{aligned} \tag{15.110}
$$

where we have used the identity $\sigma^a \sigma^b = \delta^{ab} + i\epsilon^{abc}\sigma^c$ on the second line and decomposed $\vec{d}_{\mathbf{k}} = \vec{d}_{1\mathbf{k}} - i\vec{d}_{2\mathbf{k}}$ into its real and imaginary parts on the last line. The quasiparticle energies determined by pairing matrix $h_{\mathbf{k}}$ are then

$$
E_{\mathbf{k}\pm} = \sqrt{\epsilon_{\mathbf{k}}^2 + \Delta^2(|\vec{d}_{\mathbf{k}}|^2 \pm 2|\vec{d}_{1\mathbf{k}} \times \vec{d}_{2\mathbf{k}}|)}.
$$

There are in fact two superfluid phases of ^{3}He, and, in both, $\vec{d}_{1\mathbf{k}}$ and $\vec{d}_{2\mathbf{k}}$ are parallel, and the gap functions take the form

$$
\vec{\Delta}_{\mathbf{k}} = \Delta \times \begin{cases} \hat{k}_x\hat{\mathbf{x}} + \hat{k}_y\hat{\mathbf{y}} + \hat{k}_z\hat{\mathbf{z}} & \text{BW or B phase} \\ \sqrt{\tfrac{3}{2}}(\hat{k}_x + i\hat{k}_y)\hat{\mathbf{z}}. & \text{ABM or A phase} \end{cases} \tag{15.111}
$$

The BW or B phase is named after Balian and Werhammer. In this phase the d-vector points radially outwards from the Fermi sea, forming a topological "hedgehog" configuration (see Figure 15.8(a)) with a uniform gap and quasiparticle energy given simply by

$$
E_{\mathbf{k}} = \sqrt{\epsilon_{\mathbf{k}}^2 + \Delta^2}. \qquad \text{B phase}
$$

The B phase, with a full gap, dominates the phase diagram. The ABM or A-phase, named after its discoverers, Anderson, Brinkman, and Morel, develops in a small sliver of the phase diagram under pressures of about 2 MPa (see Figure 15.7(b)). This phase involves pairing in a single triplet orbital channel with a uniform ("z") direction of the d-vector; now the magnitude of the gap is momentum-dependent:

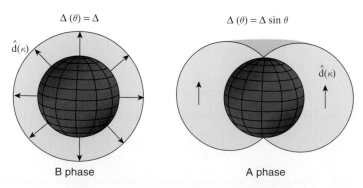

Showing the gap structure and d-vector orientation for the B and A phases of superfluid ^{3}He.

$$\vec{\Delta}_{\mathbf{k}} = \sqrt{\frac{3}{2}} \Delta \sin\theta e^{i\phi} \hat{\mathbf{z}}. \qquad \text{A phase}$$

This function vanishes at the poles, giving rise to a quasiparticle excitation spectrum

$$E_{\mathbf{k}} = \sqrt{\epsilon_{\mathbf{k}}^2 + \frac{3}{2}\Delta^2 \sin^2\theta}. \qquad \text{A phase}$$

The derivation of the mean-field equations for these two solutions is simplified by the observation that, for both of them, the potential energy term is

$$\frac{3V}{g_0} \int \frac{d\Omega_{\hat{k}}}{4\pi} \frac{d\Omega_{\hat{k}'}}{4\pi} \vec{\Delta}_{\mathbf{k}} \cdot \vec{\Delta}_{\mathbf{k}'}(\hat{k} \cdot \hat{k}') = \frac{V}{g_0}\Delta^2.$$

The free energy of the mean-field theory then takes precisely the same form as in BCS theory:

$$F_{MFT} = -2T \sum_{\mathbf{k}} \ln\left(2\cosh\frac{\beta E_{\mathbf{k}}}{2}\right) + \frac{V}{g_0}\Delta^2.$$

If we differentiate with respect to Δ^2 we obtain the gap equation:

$$\frac{1}{g_0 N(0)} = \int_{-1}^{1} \frac{d\cos\theta}{2} \int_{-\omega_D}^{\omega_D} d\epsilon \frac{\Delta(\theta)^2/\Delta^2}{\sqrt{\epsilon^2 + \Delta(\theta)^2}} \tanh\left[\frac{\sqrt{\epsilon^2 + \Delta(\theta)^2}}{2}\right].$$

According to this analysis, the A and B phases have identical mean-field transition temperatures. However, at lower temperatures the B phase wins out because its fully gapped Fermi surface gives rise to a lower free energy.

Example 15.6 Consider a single triplet Cooper pair described by the state

$$|\Psi\rangle = \frac{1}{\sqrt{2}}\left(\hat{d} \cdot \vec{\Psi}_{\mathbf{k}}^{\dagger}\right)|0\rangle = \frac{1}{\sqrt{2}}\hat{d} \cdot \left(c_{\mathbf{k}}^{\dagger}\vec{\sigma} i\sigma_2 c_{-\mathbf{k}}^{\dagger}\right)|0\rangle,$$

where $\hat{d}$ is a real unit vector.

(a) Show that

$$\vec{S}|\Psi\rangle = \frac{i}{\sqrt{2}}(\hat{d} \times \vec{\Psi}_{\mathbf{k}}^{\dagger})|0\rangle$$

and use this to prove that the spin of the state is $S = 1$, i.e.

$$S^2|\Psi\rangle = 2|\Psi\rangle, \qquad (15.112)$$

while the component of the spin in the direction of the d-vector vanishes:

$$(\hat{\mathbf{d}} \cdot \vec{S})|\Psi\rangle = 0 \qquad (15.113)$$

and the expectation value of the magnetic moment is zero, i.e. $\langle\Psi|\vec{S}|\Psi\rangle = 0$.

(b) Show that the expectation value is

$$\langle \Psi | S^a S^b | \Psi \rangle = \delta^{ab} - \hat{d}^a \hat{d}^b, \tag{15.114}$$

so that $\langle S^2 \rangle = S(S+1) = 2$, corresponding to a spin-quadrupole with a fluctuating moment in the plane perpendicular to the d-vector.

Solution

(a) The effective spin operator for this state only involves momenta $\pm \mathbf{k}$, so we may use $\vec{S} = \frac{1}{2}[c_{\mathbf{k}}^{\dagger} \vec{\sigma} c_{\mathbf{k}} + c_{-\mathbf{k}}^{\dagger} \vec{\sigma} c_{-\mathbf{k}}]$. To determine the action of the spin operator on the triplet pair, we need to commute it past the triplet pair operator onto the vacuum. The commutator is

$$\left[S^a, (\vec{\Psi}_{\mathbf{k}}^{\dagger})^b \right] = \left[\left(c_{\mathbf{k}}^{\dagger} \sigma^a c_{\mathbf{k}} + c_{-\mathbf{k}}^{\dagger} \sigma^a c_{-\mathbf{k}} \right), \left(c_{\mathbf{k}}^{\dagger} \sigma^b i\sigma_2 c_{-\mathbf{k}}^{\dagger} \right) \right]$$

$$= c_{\mathbf{k}}^{\dagger} \left(\sigma^a \sigma^b i\sigma_2 + \sigma^b i\sigma_2 (\sigma^a)^T \right) c_{-\mathbf{k}}^{\dagger},$$

where the first and second terms derive from the positive and negative momentum components of the spin operator. Using $\sigma_2 (\sigma^a)^T = -\sigma^a \sigma_2$ we obtain

$$\left[S^a, (\vec{\Psi}_{\mathbf{k}}^{\dagger})^b \right] = \frac{1}{2} c_{\mathbf{k}}^{\dagger} \left[\sigma^a, \sigma^b \right] i\sigma_2 c_{-\mathbf{k}}^{\dagger} = i\epsilon_{abc} c_{\mathbf{k}}^{\dagger} \sigma^c i\sigma_2 c_{-\mathbf{k}}^{\dagger} \tag{15.115}$$

or

$$\left[S^a, \hat{d} \cdot \left(c_{\mathbf{k}}^{\dagger} \vec{\sigma} i\sigma_2 c_{-\mathbf{k}}^{\dagger} \right) \right] = i\epsilon_{abc} d^b \left(c_{\mathbf{k}}^{\dagger} \sigma^c i\sigma_2 c_{-\mathbf{k}}^{\dagger} \right) = i(\hat{d} \times \vec{\Psi}_{\mathbf{k}}^{\dagger})^a, \tag{15.116}$$

and hence

$$\vec{S} | \Psi \rangle = \frac{1}{\sqrt{2}} [\vec{S}, (\hat{d} \cdot \vec{\Psi}_{\mathbf{k}}^{\dagger})] | 0 \rangle = \frac{i}{\sqrt{2}} \left(\hat{d} \times \vec{\Psi}_{\mathbf{k}}^{\dagger} \right) | 0 \rangle. \tag{15.117}$$

Using (15.115), we have

$$S^a (\vec{\Psi}_{\mathbf{k}}^{\dagger})^b | 0 \rangle = i\epsilon_{abc} (\vec{\Psi}_{\mathbf{k}}^{\dagger})^c | 0 \rangle, \tag{15.118}$$

so that

$$S^2 (\vec{\Psi}_{\mathbf{k}}^{\dagger})^b | 0 \rangle = S^a S^a (\vec{\Psi}_{\mathbf{k}}^{\dagger})^b | 0 \rangle$$

$$= i\epsilon_{abc} S_a (\vec{\Psi}_{\mathbf{k}}^{\dagger})^c | 0 \rangle$$

$$= \overbrace{i\epsilon_{abc} i\epsilon_{acd}}^{2\delta_{bd}} (\vec{\Psi}_{\mathbf{k}}^{\dagger})^d | 0 \rangle = 2(\vec{\Psi}_{\mathbf{k}}^{\dagger})^b | 0 \rangle, \tag{15.119}$$

so, writing this in vector notation,

$$S^2 \vec{\Psi}_{\mathbf{k}}^{\dagger} | 0 \rangle = 2 \vec{\Psi}_{\mathbf{k}}^{\dagger} | 0 \rangle. \tag{15.120}$$

Hence $S^2 | \Psi \rangle = S^2 (\hat{d} \cdot \vec{\Psi}_{\mathbf{k}}^{\dagger}) | 0 \rangle = 2 | \Psi \rangle$, corresponding to a spin of 1.

If we evaluate the expectation value of the moment, we get

$$\langle \Psi | \vec{S} | \Psi \rangle = \frac{1}{2} \langle 0 | (\hat{d}^* \cdot \vec{\Psi}_{\mathbf{k}})(\hat{d} \times \vec{\Psi}_{\mathbf{k}}^{\dagger}) | 0 \rangle = i\hat{d} \times \hat{d}^*. \tag{15.121}$$

In our case $\hat{d}$ is real, so that $\langle\vec{S}\rangle = 0$. Note, however, that if $\hat{d} = \hat{d}_1 + i\hat{d}_2$ is complex, then $\langle\vec{S}\rangle = 2\hat{d}_1 \times \hat{d}_2$, so that if $\hat{d}_1$ and $\hat{d}_2$ are not parallel, the Cooper pair state carries a net magnetic moment.

(b) Taking the result (15.117), we have

$$\langle\Psi|S^a S^b|\Psi\rangle = \frac{1}{2}\epsilon_{apq}\epsilon_{brs}d^p d^r \overbrace{\langle 0|\left(\bar{\Psi}_{\mathbf{k}}\right)^q \left(\bar{\Psi}_{\mathbf{k}}^\dagger\right)^s|0\rangle}^{2\delta^{qs}} = \epsilon_{apq}\epsilon_{brq}$$

$$= \left(\delta^{ab}\delta^{pr} - \delta^{ar}\delta^{pb}\right)d^p d^r$$

$$= \delta^{ab} - d^a d^b, \tag{15.122}$$

so the moment fluctuations of the pair lie in the plane perpendicular to the d-vector.

Example 15.7 Derive the BCS pair wavefunction for the B phase of ^{3}He.

Solution

By analogy with the case of singlet pairing, we expect the ground state to be a coherent state of a triplet pair,

$$|\Psi\rangle = \exp\left[\Lambda_T^\dagger\right]|0\rangle, \tag{15.123}$$

where

$$\Lambda_T^\dagger = \frac{1}{2}\sum_{\mathbf{k}}\phi_{\mathbf{k}}(\hat{k}\cdot\vec{\Psi}_{\mathbf{k}}^\dagger) \tag{15.124}$$

creates a triplet pair and $\phi_{\mathbf{k}} = \phi_{-\mathbf{k}}$ is an even function of momentum. The factor of $\frac{1}{2}$ is included as a normalization that takes account of the fact that $\Psi_{\mathbf{k}}^\dagger$ is only independent in one-half of momentum space.

Now the ground state is annihilated by the quasiparticle destruction operators. For the triplet B phase we write the quasiparticle-creation operators as

$$a_{\mathbf{k}}^\dagger = \psi_{\mathbf{k}}^\dagger \cdot \begin{pmatrix} u_{\mathbf{k}} \\ v_{\mathbf{k}} \end{pmatrix} = c_{\mathbf{k}\sigma}^\dagger u_{\mathbf{k}\sigma} + \tilde{c}_{\mathbf{k}\sigma}v_{\mathbf{k}\sigma}, \tag{15.125}$$

where $\tilde{c}_{\mathbf{k}\alpha} = c_{-\mathbf{k}\beta}[-i\sigma_2]_{\beta\alpha}$ and the $u_{\mathbf{k}\sigma}$ and $v_{\mathbf{k}\sigma}$ are two-component spinors. For the B phase, we can take the mean-field Hamilonian to be

$$H_{MFT} = \sum_{\mathbf{k}\in\frac{1}{2}BZ}\psi_{\mathbf{k}}^\dagger h_{\mathbf{k}}\psi_{\mathbf{k}}, \qquad h_{\mathbf{k}} = \begin{pmatrix} \epsilon_{\mathbf{k}} & \Delta(\hat{k}\cdot\vec{\sigma}) \\ \Delta(\hat{k}\cdot\vec{\sigma}) & -\epsilon_{\mathbf{k}} \end{pmatrix}. \tag{15.126}$$

Now since $[H, a_{\mathbf{k}}] = E_{\mathbf{k}}a_{\mathbf{k}}$, it follows that

$$\begin{pmatrix} \epsilon_{\mathbf{k}} & \Delta(\hat{k}\cdot\vec{\sigma}) \\ \Delta(\hat{k}\cdot\vec{\sigma}) & -\epsilon_{\mathbf{k}} \end{pmatrix}\begin{pmatrix} u_{\mathbf{k}} \\ v_{\mathbf{k}} \end{pmatrix} = E_{\mathbf{k}}\begin{pmatrix} u_{\mathbf{k}} \\ v_{\mathbf{k}} \end{pmatrix}. \tag{15.127}$$

(Notice that, if we choose a spin quantization axis parallel to $\hat{k}$, then this eigenvalue equation is identical to singlet pairing.)

Now we must find the condensate that is annihilated by the quasiparticle operators:

$$a_{\mathbf{k}} = (u_{\mathbf{k}}^{\dagger}, v_{\mathbf{k}}^{\dagger}) \cdot \psi_{\mathbf{k}} = u_{\mathbf{k}\sigma}^{\dagger} c_{\mathbf{k}\sigma} + v_{\mathbf{k}\sigma}^{\dagger} \tilde{c}_{\mathbf{k}\sigma}^{\dagger}. \tag{15.128}$$

To commute the quasiparticle operator with the pair creation operator, we note that

$$[a_{\mathbf{k}}, c_{\mathbf{k}'\sigma}^{\dagger}] = u_{\mathbf{k}\sigma}^{\dagger} \delta_{\mathbf{k},\mathbf{k}'}, \tag{15.129}$$

so that

$$[a_{\mathbf{k}}, \Lambda_T^{\dagger}] = \frac{1}{2}\left[a_{\mathbf{k}}, \sum_{\mathbf{k}'} \phi_{\mathbf{k}'} c_{\mathbf{k}'}^{\dagger} (\hat{k}' \cdot \vec{\sigma}) i\sigma_2 c_{-\mathbf{k}'}^{\dagger}\right]$$

$$= \frac{1}{2}\phi_{\mathbf{k}}\left[u_{\mathbf{k}}^{\dagger}(\hat{k} \cdot \vec{\sigma}) i\sigma_2 c_{-\mathbf{k}}^{\dagger} + c_{-\mathbf{k}}^{\dagger}(\hat{k} \cdot \vec{\sigma}) i\sigma_2 (u_{\mathbf{k}}^{\dagger})^T\right]$$

$$= \frac{1}{2}\phi_{\mathbf{k}}\left[u_{\mathbf{k}}^{\dagger}(\hat{k} \cdot \vec{\sigma}) i\sigma_2 c_{-\mathbf{k}}^{\dagger} + u_{\mathbf{k}}^{\dagger} \overbrace{-i\sigma_2(\hat{k} \cdot \vec{\sigma}^T)}^{(\hat{k}\cdot\vec{\sigma})i\sigma_2} c_{-\mathbf{k}}^{\dagger}\right]$$

$$= \phi_{\mathbf{k}}\left[u_{\mathbf{k}}^{\dagger}(\hat{k} \cdot \vec{\sigma}) \tilde{c}_{-\mathbf{k}}^{\dagger}\right], \tag{15.130}$$

where we have used $\sigma_2 \vec{\sigma}^T = -\vec{\sigma}\sigma_2$ and the fact that $\phi_{\mathbf{k}} = \phi_{-\mathbf{k}}$. Now by (15.127),

$$u_{\mathbf{k}}^{\dagger}(\hat{k} \cdot \vec{\sigma}) = \frac{(E_{\mathbf{k}} + \epsilon_{\mathbf{k}})}{\Delta} v_{\mathbf{k}}^{\dagger}, \tag{15.131}$$

so that

$$\frac{|u_{\mathbf{k}}|}{|v_{\mathbf{k}}|} = \frac{(E_{\mathbf{k}} + \epsilon_{\mathbf{k}})}{\Delta}, \tag{15.132}$$

enabling the commutator of the quasiparticle operator with the pair creation operator to be written in the compact form

$$[\alpha_{\mathbf{k}}, \Lambda_T^{\dagger}] = \frac{|u_{\mathbf{k}}|}{|v_{\mathbf{k}}|} \phi_{\mathbf{k}} (v_{\mathbf{k}}^{\dagger} \tilde{c}_{-\mathbf{k}}^{\dagger}). \tag{15.133}$$

As in the case of singlet pairing, if we choose

$$\phi_{\mathbf{k}} = -\frac{|v_{\mathbf{k}}|}{|u_{\mathbf{k}}|} \tag{15.134}$$

then

$$[\alpha_{\mathbf{k}}, \Lambda_T^{\dagger}] = -v_{\mathbf{k}}^{\dagger} \cdot \tilde{c}_{-\mathbf{k}}^{\dagger}, \tag{15.135}$$

and since $\tilde{c}_{\mathbf{k}}$ commutes with $\Lambda_T^{\dagger}$, it follows that

$$[\alpha_{\mathbf{k}}, (\Lambda_T^{\dagger})^n] = -n(\Lambda_T^{\dagger})^{n-1} v_{\mathbf{k}}^{\dagger} \tilde{c}_{\mathbf{k}}, \tag{15.136}$$

so that

$$[\alpha_{\mathbf{k}}, \exp[\Lambda_T^{\dagger}]] = -\exp[\Lambda_T^{\dagger}] v_{\mathbf{k}}^{\dagger} \tilde{c}_{\mathbf{k}}. \tag{15.137}$$

This means that

$$\alpha_{\mathbf{k}} \exp[\Lambda_T^{\dagger}] = \exp[\Lambda_T^{\dagger}]\alpha_{\mathbf{k}} - \exp[\Lambda_T^{\dagger}] v_{\mathbf{k}}^{\dagger} \tilde{c}_{\mathbf{k}} = \exp[\Lambda_T^{\dagger}] u_{\mathbf{k}}^{\dagger} \cdot c_{\mathbf{k}}, \tag{15.138}$$

so that $\alpha_{\mathbf{k}}$ annihilates the coherent state:

$$\alpha_{\mathbf{k}} \exp[\Lambda_T^{\dagger}]|0\rangle = \exp[\Lambda_T^{\dagger}]u_{\mathbf{k}}^{\dagger} \cdot c_{\mathbf{k}}|0\rangle = 0, \tag{15.139}$$

proving that

$$|\Psi\rangle = \exp\left[-\frac{1}{2}\sum \frac{|v_{\mathbf{k}}|}{|u_{\mathbf{k}}|}(\hat{k} \cdot \vec{\Psi}_{\mathbf{k}}^{\dagger})\right]|0\rangle \tag{15.140}$$

is the ground state.

Note that we have to be careful in reducing this to the usual multiplicative BCS form, for the square of the triplet pair operator is *not* zero. If one splits the sum over momentum space into two parts, $k_z > 0$ and $k_z < 0$, then the Cooper pair operator can be written as

$$\Lambda_T^{\dagger} = \sum_{k_z>0} \phi_{\mathbf{k}} c_{\mathbf{k}}^{\dagger}\left(\frac{\hat{k} \cdot \vec{\sigma} + 1}{2}\right)\tilde{c}_{-\mathbf{k}} + \sum_{k_z<0} \phi_{\mathbf{k}} c_{\mathbf{k}}^{\dagger}\left(\frac{\hat{k} \cdot \vec{\sigma} - 1}{2}\right)\tilde{c}_{-\mathbf{k}}$$

$$= \sum_{\mathbf{k}} \phi_{\mathbf{k}} c_{\mathbf{k}}^{\dagger}\left(\frac{\hat{k} \cdot \vec{\sigma} + \mathrm{sgn}(k_z)}{2}\right)\tilde{c}_{-\mathbf{k}}. \tag{15.141}$$

The additional singlet term that has been added and subtracted from the upper and lower halves of momentum space cancel with each other. Now the terms inside the pair operators are projection operators, and the squares of these operators do vanish. We can now expand the coherent triplet paired state as a BCS product, as follows:

$$|\Psi\rangle = \prod_{\mathbf{k}}\left(|u_{\mathbf{k}}| - |v_{\mathbf{k}}|c_{\mathbf{k}}^{\dagger}\left(\frac{\hat{k} \cdot \vec{\sigma} + \mathrm{sgn}(k_z)}{2}\right)\tilde{c}_{-\mathbf{k}}\right)|0\rangle. \tag{15.142}$$

Example 15.8

(a) Show that the Nambu Green's function for ^{3}HeB is given by

$$\mathcal{G}(k) = [i\omega_n - \epsilon_{\mathbf{k}}\tau_3 - (\vec{\Delta}_{\mathbf{k}} \cdot \vec{\sigma})\tau_1]^{-1} = \frac{i\omega_n + \epsilon_{\mathbf{k}}\tau_3 + (\vec{\Delta}_{\mathbf{k}} \cdot \vec{\sigma})\tau_1}{(i\omega_n)^2 - E_{\mathbf{k}}^2}.$$

(b) Calculate the magnetic susceptibility of the B phase of ^{3}He. Show that the ground-state condensate has a finite Pauli susceptibility equal to 2/3 of the normal state.

Solution

(a) As in the case of singlet pairing, we can write the propagator as $G(\mathbf{k}) = -\frac{1}{\partial_\tau + h_{\mathbf{k}}}$. Let us start with the imaginary-time propagator, which we will write

$$\mathcal{G}(\mathbf{k}, \tau) = -\langle T\psi_{\mathbf{k}}(\tau)\psi_{\mathbf{k}}^{\dagger}(0)\rangle \tag{15.143}$$

or, written out explicitly, $\mathcal{G}_{\alpha\beta}(\mathbf{k}, \tau) = -\langle T\psi_{\mathbf{k}\alpha}(\tau)\psi_{\mathbf{k}\beta}^{\dagger}(0)\rangle$, where $\psi_{\mathbf{k}\alpha}$ is a Balian–Werthamer spinor. The expectation values are to be evaluated with the mean-field Hamiltonian $H = \sum_{\mathbf{k}\in\frac{1}{2}BZ} \psi_{\mathbf{k}}^{\dagger} h_{\mathbf{k}} \psi_{\mathbf{k}}$, where

$$h_{\mathbf{k}} = \epsilon_{\mathbf{k}}\tau_3 + (\vec{\Delta}_{\mathbf{k}} \cdot \vec{\sigma})\tau_1. \tag{15.144}$$

When we take account of the time-ordering, the equation of motion for $\mathcal{G}$ is

$$
\begin{aligned}
\partial_\tau \mathcal{G}(\mathbf{k}, \tau) &= -\delta(\tau)\langle\{\psi_{\mathbf{k}}, \psi_{\mathbf{k}}^\dagger\}\rangle - \langle T(\partial_\tau \psi_{\mathbf{k}}(\tau))\psi_{\mathbf{k}}^\dagger(0)\rangle \\
&= -\delta(\tau)\underline{1} - \langle T[H, \psi_{\mathbf{k}}(\tau)]\psi_{\mathbf{k}}^\dagger(0)\rangle \\
&= -\delta(\tau)\underline{1} - h_{\mathbf{k}}\mathcal{G}(\mathbf{k}, \tau),
\end{aligned}
\tag{15.145}
$$

where we have used $\psi_{\mathbf{k}}(\tau) = e^{H\tau}\psi_{\mathbf{k}}e^{-H\tau}$ and $\partial_\tau \psi_{\mathbf{k}}(\tau) = [H, \psi_{\mathbf{k}}(\tau)] = -h_{\mathbf{k}}\psi_{\mathbf{k}}$. It follows that

$$(\partial_\tau + h_{\mathbf{k}})\mathcal{G}(\mathbf{k}, \tau) = -\delta(\tau)\underline{1} \tag{15.146}$$

or $\mathcal{G}(\mathbf{k}, \tau) = -1/[\partial_\tau + h_{\mathbf{k}}]$. Fourier transforming this expression in time ($\mathcal{G}(\mathbf{k}, \tau) \to \mathcal{G}(\mathbf{k}, i\omega_n)$, $\partial_\tau \to -i\omega_n$), it follows that $(-i\omega_n + h_{\mathbf{k}})\mathcal{G}(k) = -1$, or

$$G(\mathbf{k}, i\omega_n) = \frac{1}{i\omega_n - \epsilon_{\mathbf{k}}\tau_3 - (\vec{\Delta}_{\mathbf{k}} \cdot \vec{\sigma})\tau_1} = \frac{i\omega_n + \epsilon_{\mathbf{k}}\tau_3 + (\vec{\Delta}_{\mathbf{k}} \cdot \vec{\sigma})\tau_1}{(i\omega_n)^2 - E_{\mathbf{k}}^2}, \tag{15.147}$$

where, for ^{3}He-B, we can take $\Delta_{\mathbf{k}} = \Delta\hat{k}$, so that $(\vec{\Delta}_{\mathbf{k}} \cdot \vec{\sigma})^2 = \Delta^2$.

(b) In a magnetic field, the free energy becomes

$$F = -\frac{T}{2}\sum_k \operatorname{Tr}\ln[-\mathcal{G}^{-1}(k) - \mu_N\vec{\sigma} \cdot \vec{B}] + \text{field-independent terms}, \tag{15.148}$$

where the factor of $\frac{1}{2}$ derives from expanding the summation over one-half the Brillouin zone to the entire momentum space and μ_N is the nuclear moment of the ^{3}He-atom. We can either differentiate this twice with respect to the field or write the spin susceptibility as a mean-field polarization bubble, to obtain

$$
\chi^{ab} = -\frac{\partial^2 F}{\partial B_a \partial B_b} = a \qquad\qquad b \quad = -\frac{T\mu_N^2}{2}\sum_k \operatorname{Tr}\left[\sigma^a \mathcal{G}(k)\sigma^b \mathcal{G}(k)\right].
$$

$$\tag{15.149}$$

Inserting (15.147), we obtain

$$\chi^{ab} = -\frac{T\mu_N^2}{2}\sum_k \operatorname{Tr}\left[\sigma^a \frac{i\omega_n + \epsilon_{\mathbf{k}}\tau_3 + (\vec{\Delta}_{\mathbf{k}} \cdot \vec{\sigma})\tau_1}{(i\omega_n)^2 - E_{\mathbf{k}}^2}\sigma^b \frac{i\omega_n + \epsilon_{\mathbf{k}}\tau_3 + (\vec{\Delta}_{\mathbf{k}} \cdot \vec{\sigma})\tau_1}{(i\omega_n)^2 - E_{\mathbf{k}}^2}\right]. \tag{15.150}$$

Now we can carry out the traces over the Nambu and Pauli matrices separately. Carrying out the trace over the Nambu components, we obtain

$$\chi^{ab} = -T\mu_N^2\sum_k \frac{1}{[(i\omega_n)^2 - E_{\mathbf{k}}^2]^2}\left([(i\omega_n)^2 + \epsilon_{\mathbf{k}}^2]\operatorname{Tr}[\sigma^a\sigma^b] + \left[\sigma^a(\vec{\Delta}_{\mathbf{k}} \cdot \vec{\sigma})\sigma^b(\vec{\Delta}_{\mathbf{k}} \cdot \vec{\sigma})\right]\right).$$

Now $\text{Tr}[\sigma^a \sigma^b] = 2\delta^{ab}$. To calculate $\text{Tr}[\sigma^a \sigma^b \sigma^c \sigma^d]$, one can cyclically anticommute σ^a around the trace (using $\sigma^a \sigma^b = 2\delta^{ab} - \sigma^b \sigma^a$), picking up the remainders, to obtain

$$\text{Tr}[\sigma^a \sigma^b \sigma^c \sigma^d] = 2 \left(\delta^{ab}\delta^{cd} - \delta^{ac}\delta^{bd} + \delta^{ad}\delta^{bc} \right),$$

so that

$$\text{Tr}[\sigma^a (\vec{\Delta}_\mathbf{k} \cdot \vec{\sigma}) \sigma^b (\vec{\Delta}_\mathbf{k} \cdot \vec{\sigma})] = 2[2\Delta_\mathbf{k}^a \Delta_\mathbf{k}^b - \delta^{ab} \Delta_\mathbf{k} \cdot \Delta_\mathbf{k}] = 2\Delta^2 [2\hat{k}^a \hat{k}^b - \delta^{ab}], \quad (15.151)$$

so the susceptibility can be rewritten

$$\chi^{ab} = -2T\mu_N^2 \sum_k \frac{1}{[(i\omega_n)^2 - E_\mathbf{k}^2]^2} \left([(i\omega_n)^2 + \epsilon_\mathbf{k}^2]\delta^{ab} + \Delta^2[2\hat{k}^a\hat{k}^b - \delta^{ab}] \right)$$

$$= -2T\mu_N^2 \sum_k \frac{1}{[(i\omega_n)^2 - E_\mathbf{k}^2]^2} \left([(i\omega_n)^2 + \epsilon_\mathbf{k}^2 + \Delta^2]\delta^{ab} + 2\Delta^2[\hat{k}^a\hat{k}^b - \delta^{ab}] \right).$$

$$(15.152)$$

After the momentum sums, $\hat{k}^a\hat{k}^b \to \frac{1}{3}\delta^{ab}$ so the susceptibility is isotropic, $\chi^{ab} = \chi(T)\delta^{ab}$, where

$$\chi = -2T\mu_N^2 \sum_k \frac{1}{[(i\omega_n)^2 - E_\mathbf{k}^2]^2} \left([(i\omega_n)^2 + E_\mathbf{k}^2] - \frac{4}{3}\Delta^2 \right). \quad (15.153)$$

The first term is recognized as the Pauli susceptibility of a singlet BCS superconductor, which drops exponentially to zero as $T \to 0$, while the second term must be interpreted as an additional contribution derived from the polarizability of the triplet condensate. The evaluation of the Matsubara sums follows the same lines as for a singlet superconductor. We obtain

$$\chi = -2\mu_N^2 \sum_\mathbf{k} \oint_{z=\pm E_\mathbf{k}} \frac{dz}{2\pi i} f(z) \frac{1}{(z - E_\mathbf{k})^2 (z + E_\mathbf{k})^2} \left(z^2 + E_\mathbf{k}^2 - \frac{4}{3}\Delta^2 \right)$$

$$= -2\mu_N^2 \sum_\mathbf{k} \left\{ \frac{\partial}{\partial z} \left[f(z) \frac{1}{(z + E_\mathbf{k})^2} \left(z^2 + E_\mathbf{k}^2 - \frac{4}{3}\Delta^2 \right) \right]_{z=E_\mathbf{k}} + (E_\mathbf{k} \to -E_\mathbf{k}) \right\}$$

$$= 2\mu_N^2 \sum_\mathbf{k} \left\{ -f'(E_\mathbf{k}) \left(1 - \frac{2\Delta^2}{3E_\mathbf{k}^2} \right) + (1 - 2f(E_\mathbf{k})) \frac{\Delta^2}{3E_\mathbf{k}^3} \right\}. \quad (15.154)$$

At zero temperature, the first term vanishes. The second term becomes

$$\chi(T=0) = 2\mu_N^2 N(0) \int_{-\infty}^{\infty} d\epsilon \left(\frac{\Delta^2}{3[\epsilon^2 + \Delta^2]^{3/2}} \right)$$

$$= 2\mu_N^2 N(0) \left[\frac{\epsilon}{3\sqrt{\epsilon^2 + \Delta^2}} \right]_{-\infty}^{\infty} = \frac{2}{3} \times 2\mu_N^2 N(0), \quad (15.155)$$

so the zero-temperature susceptibility is 2/3 of the normal-state Pauli susceptibility. This intrinsic susceptibility of the condensate is present because the triplet pairs become slightly spin-polarized in a magnetic field.

We can actually do a little better than this, however, by noticing that, at a finite temperature (denoting $E = \sqrt{\epsilon^2 + \Delta^2}$),

$$\frac{2}{3} = \frac{1}{3} \int_{-\infty}^{\infty} d\epsilon \frac{d}{d\epsilon} \left[\frac{\epsilon}{\sqrt{\epsilon^2 + \Delta^2}} [1 - 2f(E)] \right]$$

$$= \frac{1}{3} \int_{-\infty}^{\infty} d\epsilon \frac{d}{d\epsilon} \left[\frac{\Delta^3}{E^2} [1 - 2f(E)] - 2f'(E) \left(1 - \frac{\Delta^2}{E^2} \right) \right], \qquad (15.156)$$

which we recognize as the argument of the second part of the integral in (15.154). We can thus rewrite the susceptibility as

$$\chi(T) = \frac{1}{3} \chi_S(T) + \frac{2}{3} \chi_P,$$

where $\chi_P = 2\mu_N^2 N(0)$ is the Pauli susceptibility of the normal state and

$$\chi_S(T) = 2\mu_N^2 N(0) \int_{-\infty}^{\infty} d\epsilon [-f'(\sqrt{\epsilon^2 + \Delta^2})] = 2\mu_N^2 N(0) Y \left[\frac{\Delta}{2T} \right], \qquad (15.157)$$

where

$$Y[x] = \frac{1}{2} \int_{-\infty}^{\infty} \frac{du}{\cosh^2[\sqrt{u^2 + x^2}]} \qquad (15.158)$$

is called the *Yoshida function*, after its inventor, Kei Yoshida. The final expression for the susceptibility of the B phase is then

$$\chi_B(T) = \chi_P \left[\frac{2}{3} + \frac{1}{3} Y[\Delta/2T] \right]. \qquad (15.159)$$

Exercises

Exercise 15.1 The standard two-component Nambu spinor approach does not allow a rotationally invariant treatment of the electron spin and the Zeeman coupling of fermions to a magnetic field. This drawback can be overcome by switching to a four-component Balian–Werthamer spinor, denoted by

$$\psi_{\mathbf{k}} = \begin{pmatrix} c_{\mathbf{k}}^{\dagger} \\ -i\sigma_2 (c_{\mathbf{k}}^{\dagger})^T \end{pmatrix} = \begin{pmatrix} c_{\mathbf{k}\uparrow} \\ -c_{\mathbf{k}\downarrow} \\ c_{-\mathbf{k}\downarrow}^{\dagger} \\ c_{-\mathbf{k}\uparrow}^{\dagger} \end{pmatrix}. \qquad (15.160)$$

(a)   Show, using this notation, that the total electron spin can be written

$$\vec{S} = \frac{1}{4} \sum_{\mathbf{k}} \psi_{\mathbf{k}}^{\dagger} \vec{\sigma}_{(4)} \psi_{\mathbf{k}}, \qquad (15.161)$$

where

$$\vec{\sigma}_4 = \begin{pmatrix} \vec{\sigma} & 0 \\ 0 & \vec{\sigma} \end{pmatrix} \qquad (15.162)$$

is the four-component Pauli matrix. (You may find it useful to use the relationship $\vec{\sigma}^T = i\sigma_2 \vec{\sigma} i\sigma_2$.) In practical usage, the subscript 4 is normally dropped.

(b) Show that, in a Zeeman field, the BCS Hamiltonian

$$H_{MFT} = \sum_{k\sigma} c^{\dagger}_{k\alpha} [\epsilon_k \delta_{\alpha\beta} - \vec{\sigma}_{\alpha\beta} \cdot \vec{B}] c_{k\beta} + \sum_k \left[\bar{\Delta} c_{-k\downarrow} c_{k\uparrow} + c^{\dagger}_{k\uparrow} c^{\dagger}_{-k\downarrow} \Delta \right] + \frac{V}{g_0} \bar{\Delta}\Delta$$

(15.163)

can be rewritten using Balian–Werthamer spinors in the compact form

$$H_{MFT} = \frac{1}{2} \sum_k \psi^{\dagger}_k \left[\underline{h}_k - \vec{\sigma}_4 \cdot \vec{B} \right] \psi_k + \frac{V}{g_0} \bar{\Delta}\Delta,$$

(15.164)

where $\underline{h}_k = \epsilon_k \tau_1 + \Delta_1 \tau_1 + \Delta_2 \tau_2$ as before, but the $\vec{\tau}$ now refer to the four-dimensional Nambu matrices

$$\vec{\tau} = \left(\begin{bmatrix} 0 & \underline{1} \\ \underline{1} & 0 \end{bmatrix}, \begin{bmatrix} 0 & -i\underline{1} \\ i\underline{1} & 0 \end{bmatrix}, \begin{bmatrix} \underline{1} & 0 \\ 0 & -\underline{1} \end{bmatrix} \right).$$

(15.165)

(c) Show that the quasiparticle energies in a field are given by $\pm E_k - \sigma B$.

Exercise 15.2 Pauli limited type II superconductors.

The BCS Hamiltonian introduced in describes a *Pauli limited superconductor*, in which the Zeeman coupling of the paired electrons with the magnetic field dominates over the orbital coupling to the magnetic field. In the flux lattice of a Pauli limited type II superconductor, the magnetic field penetrates the condensate and can be considered to be approximately uniform.

(a) Assuming that the orbital coupling of the electron to the magnetic field is negligible, use the Balian–Werthamer approach developed in the previous problem to formulate BCS theory in a uniform Zeeman field, as a path integral. Show that the free energy can be written

$$F = -\frac{T}{2} \sum_k \text{Tr} \ln[\partial_\tau + \underline{h}_k - \vec{\sigma}_4 \cdot \vec{B}] + \frac{V}{g_0} \bar{\Delta}\Delta$$

$$= -\frac{T}{2} \sum_{k,i\omega_n,\sigma} \ln\left[E_k^2 - (i\omega_n - \sigma B)^2 \right] + \frac{V}{g_0} \bar{\Delta}\Delta$$

$$= -T \sum_{k,\sigma} \ln\left[2\cosh \frac{\beta(E_k - \sigma B)}{2} \right] + \frac{V}{g_0} \bar{\Delta}\Delta.$$

(15.166)

(b) Show that the gap equation for a Pauli limited superconductor becomes

$$\frac{1}{g_0} = \frac{1}{2} \sum_{k,\sigma} \tanh\left(\frac{\beta(E_k - \sigma B)}{2} \right) \frac{1}{2E_k}.$$

Use this expression to show that the upper critical field is given by $g\mu_B B_{c2}/2 = \Delta/2$, where Δ is the zero-temperature value of the gap.

(c) Pauli limited superconductors usually undergo a first-order transition to the flux state at a higher field than the one just estimated. Why is this?

References

[1] V. L. Ginzburg, Landau's attitude toward physics and physicists, *Physics Today*, vol. 42, p. 54, no. 5 1989.

[2] A. Layzer and D. Fay, Superconducting pairing tendancy in nearly ferromagnetic systems, *Int. J. Magn*, vol. 1, no. 2, p. 135, 1971.

[3] M. T. Béal Monod, C. Bourbonnais, and V. J. Emery, Possible superconductivity in nearly antiferromagnetic itinerant fermion systems, *Phys. Rev. B*, vol. 34, p. 7716, 1986.

[4] K. Miyake, S. Schmitt-Rink, and C. M. Varma, Spin-fluctuation-mediated even-parity pairing in heavy-fermion superconductors, *Phys. Rev. B*, vol. 34, p. 6554, 1986.

[5] D. J. Scalapino, E. Loh, and J. E. Hirsch, d-wave pairing near a spin-density-wave instability, *Phys. Rev. B*, vol. 34, p. 8190, 1986.

[6] L. P. Pitaevskii, On the Superfluidity of liquid ^{3}He, *J. Exp. Theor. Phys.*, Vol. 10, p. 1267, 1960.

[7] D. J. Thouless, Perturbation theory in statistical mechanics and the theory of superconductivity, *Ann. Phys.*, vol. 10, p. 553, 1960.

[8] V. J. Emery and A. M. Sessler, Possible phase transition in liquid ^{3}He, *Phys. Rev.*, vol. 119, p. 43, 1960.

[9] K. A. Brueckner, T. Soda, P. W. Anderson, and P. Morel, Level structure of nuclear matter and liquid ^{3}He, *Phys. Rev.*, vol. 118, p. 1442, 1960.

[10] P. W. Anderson and P. Morel, Generalized Bardeen-Cooper-Schrieffer states and the proposed low-temperature phase of liquid ^{3}He, *Phys. Rev.*, vol. 123, p. 1911, 1961.

[11] R. Balian and N. Werthamer, Superconductivity with pairs in a relative p-wave, *Phys. Rev.*, vol. 131, p. 1, 1963.

[12] V. J. Emery, Theories of liquid helium three, *Ann. Phys.*, vol. 28, no. 1, p. 1, 1964.

[13] W. Kohn and J. M. Luttinger, New mechanism for superconductivity, *Phys. Rev. Lett.*, vol. 15, p. 524, 1965.

[14] D. Fay and A. Layzer, Superfluidity of low density fermion systems, *Phys. Rev. Lett.*, vol. 20, no. 5, p. 187, 1968.

[15] D. D. Osheroff, R. C. Richardson, and D. M. Lee, Evidence for a new phase of solid ^{3}He, *Phys. Rev. Lett.*, vol. 28, p. 885, 1972.

[16] D. D. Osheroff, W. J. Gully, R. C. Richardson, and D. M. Lee, New magnetic phenomena in liquid He$_3$ below 3 mK, *Phys. Rev. Lett.*, vol. 29, p. 920, 1972.

[17] A. J. Leggett, Interpretation of recent results on He 3 below 3 mK: a new liquid phase?, *Phys. Rev. Lett.*, 1972.

[18] A. J. Leggett, Microscopic theory of NMR in an anisotropic superfluid (^{3}He A), *Phys. Rev. Lett.*, vol. 31, p. 352, 1973.

[19] A. J. Leggett, NMR lineshifts and spontaneously broken spin–orbit symmetry. I general concepts, *J. Phys. C*, vol. 6, p. 3187, 1973.

[20] W. F. Brinkman, J. W. Serene, and P. W. Anderson, Spin-fluctuation stabilization of anisotropic superfluid states, *Phys. Rev.* A: At., Mol., Opt. Phys. vol. 10, no. 6, p. 2386, 1974.

[21] F. Steglich, J. Aarts, C. D. Bredl, W. Leike, D. E. Meshida, W. Franz, and H. Schäfer, Superconductivity in the presence of strong Pauli paramagnetism: $CeCu_2 Si_2$, *Phys. Rev. Lett.*, vol. 43, p. 1892, 1979.

[22] D. Vollhardt and P. Wölfle, *Superfluid Phases of Helium 3*, Taylor and Francis, 1990.

Local moments and the Kondo effect

16.1 Strongly correlated electrons

One of the fascinating growth areas in condensed matter physics concerns *strongly correlated systems*: states of matter in which the many-body interaction energies dominate the kinetic energies, becoming large enough to qualitatively transform the macroscopic properties of the medium. The growing list of strongly correlated systems includes the following:

- **Cuprate superconductors**, where interactions among electrons in localized $3d$-shells form an antiferromagnetic *Mott insulator*, which develops high-temperature superconductivity when doped
- **Heavy-electron compounds**, in which localized magnetic moments immersed within the metal give rise to electron quasiparticles with effective masses in excess of a thousand bare-electron masses
- **Fractional quantum Hall systems**, where strong interactions in the lowest Landau level of a two-dimensional electron fluid generate an incompressible state with quasiparticles of fractional charge and statistics
- **Quantum dots**, which are tiny pools of electrons in semiconductors that act as artificial atoms. As the gate voltage is changed, the electron repulsion in the dot causes a *Coulomb blockade*, whereby electrons can only be added one-by-one to the quantum dot
- **Cold atomic gases**, in which the interactions between neutral atoms governed by two-body resonances can be magnetically tuned to create a whole new world of strongly correlated quantum fluids.

In each case, the electron system has been tuned – by electronic or nuclear chemistry, by geometry or nanofabrication, to give rise to a quantum state with novel collective properties, in which the interactions between the particles are large compared with their kinetic energies. The next three chapters will introduce one corner of this field: the physics of local moments and heavy-fermion compounds. A large class of strongly correlated materials have atoms with partially filled d- or f-orbitals. Heavy-electron materials are an extreme example, in which one component of the electron fluid is highly localized, usually inside f-orbitals, giving rise to the formation of magnetic moments. The interaction of localized magnetic moments with the conduction sea provides the driving force for the strongly correlated electron physics in these materials.

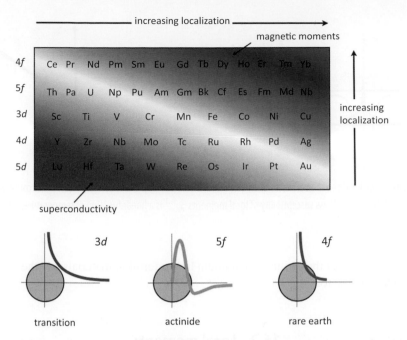

Fig. 16.1 Kmetko–Smith diagram [68], showing the broad trends towards increasing electron localization in d- and f-electron compounds.

Within the Periodic Table, there are broad trends that govern strongly correlated electron behavior. The most strongly interacting electrons tend to reside in partially filled orbitals that are well localized around the nucleus. The weak overlap between orbitals of neighboring atoms promotes the formation of narrow electron bands, while the interactions between electrons in the highly localized orbitals are strong.

In order of increasing degree of localization, the unfilled electron orbitals of the central rows of the Periodic Table may be ordered

$$5d < 4d < 3d < 5f < 4f.$$

There are two trends operating here: first, orbitals with higher principle quantum numbers contain more radial nodes and tend to be more delocalized, so that $5d < 4d < 3d$ and $5f < 4f$. Second, as we move from d- to f-orbitals, or along a row of the Periodic Table, the increased nuclear charge pulls the orbitals towards the nucleus. These trends are summarized in the *Kmetko–Smith diagram* in Figure 16.1, in which the central rows of the Periodic Table are stacked in order of increasing localization. Moving up and to the right in this diagram leads to increasingly localized atoms. In metals lying on the bottom left-hand side of this diagram, the d-orbitals are highly itinerant, giving rise to conventional superconductivity at low temperatures. By contrast, in metals towards the top right-hand side of the diagram, the electrons in the rare-earth or actinide ions are localized, forming magnets or, more typically, antiferromagnets.

The materials that lie in the crossover between these two regions are particularly interesting, for these materials are "on the brink of magnetism." With some exceptions,

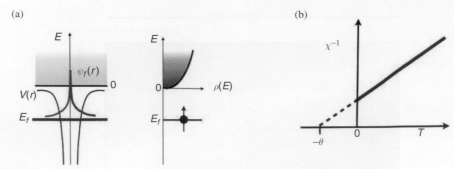

Fig. 16.2 (a) In isolation, the localized atomic states of an atom form a stable, sharp excitation lying below the continuum. (b) The inverse of the Curie–Weiss susceptibility of local moments χ^{-1} is a linear function of temperature, intersecting zero at $T = -\theta$.

it is in this region that the cerium and uranium heavy-fermion materials and the iron-based superconductors are found.

16.2 Local moments

To understand heavy-electron materials, we need to understand how electrons localize to form magnetic moments, and how these local moments interact with the conduction sea of electrons. The simplest example of a localized moment is an unpaired electron bound in an isolated atom or ion (Figure 16.2(a)). At temperatures far below the ionization energy $|E_f|$, the only remaining degree of freedom of this localized electron is its magnetic moment, described by the operator

$$\vec{M} = \mu_B \vec{\sigma},$$

where $\vec{\sigma}$ denotes the Pauli matrices and $\mu_B = \frac{e\hbar}{2m}$ is the Bohr magneton. In a magnetic field, the Hamiltonian describing low-energy physics is simply $H = -\vec{M} \cdot \vec{B} = -\mu_B \vec{\sigma} \cdot \vec{B}$, giving rise to a *Curie susceptibility*,

$$\chi(T) = \frac{\partial M}{\partial B} = -\frac{\partial^2 F}{\partial B^2} = \frac{\mu_B^2}{T}.$$

The classic signature of local moments is the appearance of Curie paramagnetism, with a high-temperature magnetic susceptibility of the Curie–Weiss form (see Figure 16.2(b))

$$\chi \approx n_i \frac{M^2}{3(T + \theta)}, \qquad M^2 = g^2 \mu_B^2 J(J + 1), \qquad (16.1)$$

where n_i is the concentration of magnetic moments and M is the magnetic moment with total angular momentum quantum number J and gyromagnetic ratio (g-factor) g. Here θ is the *Curie–Weiss temperature*, a phenomenological scale which takes account of interactions between spins.[1] For a pure spin, $J = S$ is the total spin and $g = 2$, but for rare-earth and actinide ions, the orbital and spin angular momenta combine into a single entity with angular momentum $\vec{J} = \vec{L} + \vec{S}$, for which g lies between 1 and 2. For example, a Ce^{3+} ion contains a single unpaired $4f$-electron in the state $4f^1$, with $l = 3$ and $s = \frac{1}{2}$. Spin-orbit coupling gives rise to a low-lying multiplet with $j = 3 - \frac{1}{2} = \frac{5}{2}$, consisting of $2j + 1 = 6$ degenerate orbitals $|4f^1 : Jm\rangle$, $(m_J \in [-\frac{5}{2}, \frac{5}{2}])$ with an associated magnetic moment $M = 2.64\mu_B$.

16.3 Asymptotic freedom in a cryostat: a brief history of the theory of magnetic moments

Though the concept of localized moments was employed in the earliest applications of quantum theory to condensed matter,[2] a full theoretical understanding of the *mechanism* of moment formation did not develop until the late 1950s, when experimentalists began to systematically study impurities in metals. In 1933 three Dutch physicists, W. J. de Haas, J. H. de Boer, and G. J. van den Berg, working at the Kamerlingh Onnes Laboratory at Leiden University [3, 4], discovered a curious "resistance minimum" in the resistivity of gold, copper, and silver. Whereas resistivity normally falls as the temperature is lowered (Figure 16.3), they found that, below a certain temperature, it starts to rise again as the temperature is further lowered. It took 30 more years before materials physicists learned to control the concentration of magnetic impurities in the parts per million range required for the study of individual impurities. The control of purity evolved during the 1950s, with the development of new techniques needed for semiconductor physics, such as zone refining.

In the early 1960s, Al Clogston, Bernd Matthias, Myriam Sarachik, and collaborators [2, 5], working at Bell Laboratories, showed that when small concentrations n_i of magnetic ions such as iron are added to a metallic host, they do not always form magnetic moments. For example, iron impurities in pure niobium do not develop a local moment, but they do so in the niobium–molybdenum alloy $Nb_{1-x}Mo_x$ once the concentration of molybdeneum exceeds 40% ($x > 0.4$). One of their remarkable observations was that the presence of magnetic moments triggered the development of a resistivity minimum [2].

These developments stimulated the theorists to make new progress on the theory of magnetic moments. Working at the Laboratoire de Physique des Solides, in Orsay, Jacques Friedel and his graduate student André Blandin invented the concept of a *virtual bound-state resonance* to describe the interaction of electrons with the d-orbitals of magnetic ions [6]. Building on these ideas, in 1961 Philip W. Anderson, working at Bell Laboratories, formulated a second-quantized model for the formation of magnetic moments

[1] A positive θ indicates an antiferromagnetic interaction between spins, while a negative θ is associated with ferromagnetic interactions. giving rise to a divergence of the susceptibility at the Curie temperature, $T_c = -\theta$.

[2] The concept of a local moment appears in Heisenberg's original paper on ferromagnetism [1]. Landau and Néel invoked the notion of the localized moment in their 1932 papers on antiferromagnetism, and in 1933 Kramers used this idea again in his theory of magnetic superexchange.

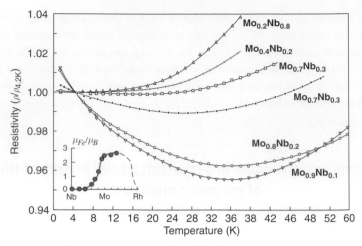

Fig. 16.3   Resistance minimum in Mo_xNb_{1-x} [2]. Once the Mo concentration exceeds 40%, the dissolved iron atoms in the alloy become magnetic and a resistance minimum develops. Insert: growth of the iron moment with Mo and Re alloying [5]. Main figure reprinted with permission from M. Sarachik, *et al.*, *Phys. Rev. A*, vol. 135, p. 1041, 1964. Copyright 1964 by the American Physical Society. Inset figure reprinted with permission from A. M. Clogston, *et al.*, *Phys. Rev.*, vol. 125, p. 541, 1962. Copyright 1962 by the American Physical Society.

in metals [7] which we now call the *Anderson model*. In his paper Anderson deduced that magnetic moments would develop an antiferromagnetic coupling [7–11] with the spin density of the surrounding fluid, described by the interaction

$$H_I = J\vec{\sigma}(0) \cdot \vec{S}, \qquad (16.2)$$

where $\vec{S}$ is the spin of the local moment and $\vec{\sigma}(0)$ is the spin density of the electron fluid.

Working at the Electro-Technical Laboratory in Tokyo, the physicist Jun Kondo confirmed Anderson's prediction of an antiferromagnetic interaction [8], and set out to examine its consequences for electron scattering. Kondo found that, when he calculated the magnetic scattering rate $\frac{1}{\tau}$ of electrons to one order higher than the Born approximation, the cubic term contained a logarithmic temperature dependence:

$$\frac{1}{\tau} \propto \left[J\rho + 2(J\rho)^2 \ln \frac{D}{T} \right], \qquad (16.3)$$

where ρ is the density of states of electrons in the conduction sea and D is the half-width of the electron band, an interaction that forms the basis of what we now call the *Kondo model*. Kondo noted that as the temperature is lowered the coupling constant, the scattering rate, and the resistivity start to rise. When the rise in magnetic scattering overcame the phonon scattering, the resistance would develop a resistance minimum. In this way, Kondo solved the 30-year-old mystery of the resistance minimum, by linking it with the antiferromagnetic interaction between spins and their surroundings.

A deeper understanding of the logarithmic term in this scattering rate required the renormalization group concept [12–18]. Here, the key new concept is that the physics of a spin in a metal depends on the energy scale at which it is probed. The *Kondo effect* is an

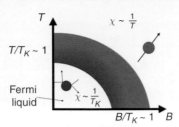

Free local moment

Fig. 16.4

Schematic diagram showing the crossover from the high-temperature, high-field regime where local moments are asymptotically free, to the low-temperature, low-field regime where they are screened by the Kondo effect to produce a resonant scattering center.

example of the phenomenon of *asymptotic freedom* that also governs quark physics. Like the quark, at high energies the local moments inside metals are asymptotically free, but at temperatures and energies below a characteristic scale, the *Kondo temperature*,

$$T_K \sim De^{-1/(2J\rho)}, \qquad (16.4)$$

the local moment interacts so strongly with the surrounding electrons that it becomes screened into a singlet state, or confined at low energies, ultimately forming a Landau Fermi-liquid [17, 18] (Figure 16.4).

The next three chapters will look at the development of these ideas. In this chapter we introduce the Anderson model, examining its connection with the Kondo model and then going on to study the renormalization associated with the Kondo effect.

16.4 Anderson's model of local moment formation

Anderson's model for moment formation, proposed in 1961 [7], combines two essential ideas [19]:

- The localizing influence of Coulomb interactions. Rudolph Peierls and Neville Mott [20, 21] had reasoned in the 1940s that a sufficiently strong Coulomb repulsion between electrons in an atomic state would blockade the passage of electrons, converting a metal into what is now called a *Mott insulator*. These ideas were independently explored by J. H. Van Vleck and Henry Hurwitz in an early attempt to understand magnetic ions in metals [22, 23].
- The formation of an electronic resonance. In the 1950s Friedel and Blandin [6, 23, 24] proposed that electrons scattering off transition metal atoms form virtual bound-state resonances. Anderson reinterpreted the resonance formation idea as a physical tunneling process between localized *d*- or *f*-orbitals and the conduction sea: an electronic version of nuclear alpha decay.

Anderson unified these ideas in a second-quantized Hamiltonian,

$$
H = \overbrace{\sum_{\mathbf{k},\sigma} \epsilon_{\mathbf{k}} n_{\mathbf{k}\sigma} + \sum_{\mathbf{k},\sigma} \left[V(\mathbf{k}) c_{\mathbf{k}\sigma}^{\dagger} f_{\sigma} + V^*(\mathbf{k}) f_{\sigma}^{\dagger} c_{\mathbf{k}\sigma} \right]}^{H_{resonance}} + \underbrace{E_f n_f + U n_{f\uparrow} n_{f\downarrow}}_{H_{atomic}}, \qquad (16.5)
$$

Anderson model

where H_{atomic} describes the atomic limit of an isolated magnetic ion involving a Kramers doublet of energy E_f. Although Anderson's model applies to any localized orbital, with rare-earth ions in mind to prepare outselves for the link with heavy electrons in rare-earth materials, we shall refer here to f-electrons.

The engine of magnetism in the Anderson model is the Coulomb interaction

$$
U = \frac{e^2}{4\pi\epsilon_0} \int_{\mathbf{r},\mathbf{r}'} \frac{1}{|\mathbf{r} - \mathbf{r}'|} \rho_f(\mathbf{r}) \rho_f(\mathbf{r}')
$$

of a doubly occupied f-state, where $\rho_f(\mathbf{r}) = |\Psi_f(\mathbf{r})|^2$ is the electron density in a single atomic orbital $\psi_f(\mathbf{r})$. The operator $c_{\mathbf{k}\sigma}^{\dagger}$ creates a conduction electron of momentum $\mathbf{k}$, spin σ, and energy $\epsilon_{\mathbf{k}} = E_{\mathbf{k}} - \mu$, while

$$
f_{\sigma}^{\dagger} = \int_{\mathbf{r}} \Psi_f(\mathbf{r}) \hat{\psi}_{\sigma}^{\dagger}(r), \qquad (16.6)
$$

creates an f-electron in the atomic f-state. Unlike the electron continuum in a vacuum, a conduction band in a metal has a finite energy width, so in the model the energies are taken to lie in the range $\epsilon_{\mathbf{k}} \in [-D, D]$. $H_{resonance}$ describes the hybridization with the Bloch waves of the conduction sea that develops when the ion is immersed in a metal. The quantity

$$
V(\mathbf{k}) = \langle \mathbf{k} | V_{ion} | f \rangle = \int d^3 r \, e^{-i\mathbf{k}\cdot\mathbf{r}} V_{ion}(r) \Psi_f(\vec{r}) \qquad (16.7)
$$

is the hybridization between the ionic potential and a plane wave. This term is the result of applying first-order perturbation theory to the degenerate states of the conduction sea and the atomic f-orbital.

To understand the formation of local moments, we shall examine two limiting types of behavior:

- the atomic limit, where the hybridization vanishes
- virtual bound state formation, described by the limit where the interaction vanishes.

16.4.1 The atomic limit

The atomic physics of an isolated ion, described by

$$
H_{atomic} = E_f n_f + U n_{f\uparrow} n_{f\downarrow}, \qquad (16.8)
$$

is the engine at the heart of the Anderson model that drives moment formation. The four atomic quantum states are

$$\left.\begin{array}{ll} |f^2\rangle & E(f^2) = 2E_f + U \\ |f^0\rangle & E(f^0) = 0 \end{array}\right\} \quad \text{non-magnetic}$$

$$|f^1\uparrow\rangle, \quad |f^1\downarrow\rangle \qquad E(f^1) = E_f. \qquad \text{magnetic}$$

(16.9)

The cost of adding or removing an electron from the magnetic f^1 state is given by

$$\left.\begin{array}{ll} \text{adding:} & E(f^2) - E(f^1) = U + E_f \\ \text{removing:} & E(f^0) - E(f^1) = -E_f \end{array}\right\} \Rightarrow \Delta E = \frac{U}{2} \pm \left(E_f + \frac{U}{2}\right). \quad (16.10)$$

In other words, provided

$$U/2 > |E_f + U/2|, \tag{16.11}$$

the ground state of the atom is a two-fold degenerate magnetic doublet (Figure 16.5). Indeed, provided it is probed at energies below the smallest charge excitation energy, $\Delta E_{min} = U/2 - |E_f + U/2|$, only the spin degrees of freedom remain, and the system behaves as a local moment – a *quantum top*. The interaction between such a local moment and the conduction sea gives rise to the Kondo effect that will be the main topic of this chapter.

Although we shall be mainly interested in positive, repulsive U, we note that in the attractive region of the phase diagram ($U < 0$) the atomic ground state can form a degenerate *charge doublet* ($|f^0\rangle, |f^2\rangle$) or *isospin*. For $U < 0$, when $E_f + U/2 = 0$ the doubly occupied state $|f^2\rangle$ and the empty state $|f^0\rangle$ become degenerate. This is the charge analogue of the magnetic doublet that exists for $U > 0$, and when coupled to the sea of electrons it gives rise to an effect known as the *charge Kondo effect*. Such

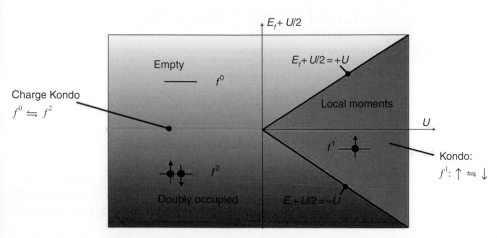

Phase diagram for the Anderson impurity model in the atomic limit. For $U > |E_f + U/2|$, the ground state is a magnetic doublet. When $U < 0$, the ground state is a degenerate charge doublet, provided $E_f + U/2 = 0$.

Fig. 16.5

charge doublets are thought to be important in certain *negative U* materials, such as Tl doped PbTe.

Example 16.1 Derivation of the non-interacting Anderson model.

Consider an isolated ion, where the f-state is a solution of the one-particle Schrödinger equation

$$\left[-\nabla^2 + \hat{V}_{ion}\right]|f\rangle = E_f^{ion}|f\rangle. \tag{16.12}$$

Here $V_{ion}(r)$ is the ionic potential and $E_f^{ion} < 0$ is the energy of the atomic f-level. In a metal, the positive ionic background draws the continuum downwards to become degenerate with the f-level, as shown in Figure 16.6. A convenient way to model this situation is to use a "muffin-tin" potential,

$$V(r) = (V_{ion}(r) + W)\,\theta(R_0 - r), \tag{16.13}$$

equal to the ionic potential, shifted upwards by an amount W inside the muffin-tin radius R_0. The f-state is now an approximate eigenstate of $\mathcal{H} = -\nabla^2 + \hat{V}$ that is degenerate with the continuum.

Derive the non-interacting component of the Anderson model using degenerate perturbation theory, evaluating the matrix elements of $\mathcal{H}$ between the conduction states $|\mathbf{k}\rangle$ and the local f-state $|f\rangle$. You may assume that the muffin-tin R_0 is much smaller than the Fermi wavelength, so that the conduction electron matrix elements $V_{\mathbf{k},\mathbf{k}'} = \langle \mathbf{k}|V|\mathbf{k}'\rangle$ are negligible.

Solution

To carry out degenerate perturbation theory we must first orthogonalize the f-state to the continuum, $|\tilde{f}\rangle = |f\rangle - \sum_{\epsilon_{\mathbf{k}} \in [-D,D]} |\mathbf{k}\rangle\langle\mathbf{k}|f\rangle$, where D is the conduction electron bandwidth. Now we need to evaluate the matrix elements of $\mathcal{H} = -\nabla^2 + V$. If we set

$$V_{\mathbf{k},\mathbf{k}'} = \int_{r<R_0} d^3 r\, e^{i(\mathbf{k}'-\mathbf{k})\cdot\mathbf{r}}(V_{ion}(r) + W), \tag{16.14}$$

then the conduction electron matrix elements are

$$\langle\mathbf{k}|\mathcal{H}|\mathbf{k}'\rangle = E_{\mathbf{k}}\delta_{\mathbf{k},\mathbf{k}'} + V_{\mathbf{k},\mathbf{k}'} \approx E_{\mathbf{k}}\delta_{\mathbf{k},\mathbf{k}'} \tag{16.15}$$

while $\langle\tilde{f}|\mathcal{H}|\tilde{f}\rangle \approx E_f^{ion}$ is the f-level energy.

The hybridization is given by the off-diagonal matrix element,

$$V(\mathbf{k}) = \langle\mathbf{k}|\mathcal{H}|\tilde{f}\rangle = \langle\mathbf{k}| -\nabla^2 + \hat{V}|\tilde{f}\rangle = E_{\mathbf{k}}\langle\mathbf{k}|\tilde{f}\rangle + \langle\mathbf{k}|\hat{V}|\tilde{f}\rangle = \langle\mathbf{k}|\hat{V}|\tilde{f}\rangle, \tag{16.16}$$

where we have used the orthogonality $\langle\mathbf{k}|\tilde{f}\rangle = 0$ to eliminate the kinetic energy. In fact, since the f-state is highly localized, its overlap with the conduction electron states is small, $\langle\mathbf{k}|f\rangle \approx 0$, so we can now drop the tilde, approximating $\langle\mathbf{k}|\hat{V}|\tilde{f}\rangle \approx \langle\mathbf{k}|\hat{V}_{ion} + W|f\rangle \approx \langle\mathbf{k}|\hat{V}_{ion}|f\rangle$, so that

$$V(\mathbf{k}) \approx \langle\mathbf{k}|V_{ion}|f\rangle = \int d^3 r\, e^{-i\mathbf{k}\cdot\mathbf{r}} V_{ion}(r)\psi_f(\mathbf{r}). \tag{16.17}$$

In this way, the only surviving term contributing to the hybridization is the atomic potential – only this term has the high-momentum Fourier components to create a significant overlap between the low-momentum conduction electrons and the localized f-state. Putting these results together, the non-interacting Anderson model can then be written

$$\hat{H}_{resonance} = \sum_{k} \overbrace{(E_k + W - \mu)}^{\epsilon_k} c^{\dagger}_{k\sigma} c_{k\sigma} + \sum_{k\sigma} (V(k) c^{\dagger}_{k\sigma} f_{\sigma} + \text{H.c.}) + \overbrace{(E_f^{ion} - \mu)}^{E_f} n_f.$$

16.4.2 Virtual bound state formation: the non-interacting resonance

When the magnetic ion is immersed in a sea of electrons, the f-electrons within the core of the atom can tunnel out, hybridizing with the Bloch states of the surrounding electron sea [6], as shown in Figure 16.6.

In the absence of interactions, this physics is described by

$$H_{resonance} = \sum_{k,\sigma} \epsilon_k n_{k\sigma} + \sum_{k\sigma} \left[V(k) c^{\dagger}_{k\sigma} f_{\sigma} + \text{H.c.} \right] + E_f n_f, \tag{16.18}$$

where $c^{\dagger}_{k\sigma}$ creates an electron of momentum $\mathbf{k}$, spin σ, and energy $\epsilon_k = E_k - \mu$ in the conduction band. The hybridization broadens the localized f-state, and in the absence of interaction gives rise to a resonance of width Δ given by Fermi's golden rule:

$$\Delta = \pi \sum_{\vec{k}} |V(\mathbf{k})|^2 \delta(\epsilon_k - E_f). \tag{16.19}$$

This is really an average of the density of states $\rho(\epsilon) = \sum_k \delta(\omega - \epsilon_k)$ with the hybridization $|V(\mathbf{k})|^2$. For future reference, we shall define

$$\Delta(\epsilon) = \pi \sum_{\vec{k}} |V(\mathbf{k})|^2 \delta(\epsilon_k - \epsilon) = \pi \overline{\rho(\epsilon) V^2(\epsilon)} \tag{16.20}$$

as the *hybridization function*.

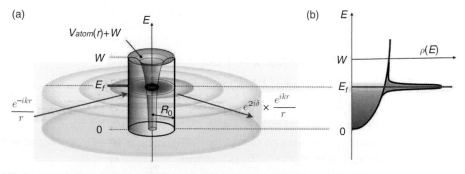

(a) The immersion of an atomic f-state in a conduction sea leads to hybridization between the localized f-state and the degenerate conduction electron continuum, forming (b) a resonance in the density of states.

Fig. 16.6

Let us now examine the resonant scattering off a non-interacting f-level, using Feynman diagrams. We'll denote the propagator of the bare f-electron by a full line, and that of the conduction electron by a dashed line, as follows:

$$G_f^{(0)}(\omega) = \frac{1}{\omega - E_f}$$

$$G^{(0)}(\mathbf{k}, \omega) = \frac{1}{\omega - \epsilon_\mathbf{k}}.$$

$$(16.21)$$

For simplicity, we will ignore the momentum dependence of the hybridization, taking $V(\mathbf{k}) = V(\mathbf{k})^* \equiv V$. The hybridization is a kind of off-diagonal potential scattering, which we denote by a filled dot, as follows:

$$(16.22)$$

Now the hybridization permits the f-electron to tunnel back and forth into the continuum, a process we can associate with the self-energy diagram:

$$= \Sigma_c(\omega) = \sum_\mathbf{k} \frac{V^2}{\omega - \epsilon_\mathbf{k}}.$$

$$(16.23)$$

We can view this term as an effective scattering potential for the f-electrons, one that is frequency-dependent and hence retarded in time, reflecting the fact that an f-electron can spend large amounts of time out in the conduction band. The Feynman diagrams describing the multiple scattering of the f-electron off this potential are then

$$(16.24)$$

Each time the electron tunnels into the conduction band, it does so with a different momentum, so the momenta of the conduction electrons are independently summed over the intermediate states. As in previous chapters, we can sum these terms as a geometric series to obtain a familiar-looking self-energy correction to the f-propagator:

$$G_f(\omega) = G_f^{(0)} \left[1 + \Sigma_c G_f^{(0)} + \left(\Sigma_c G_f^{(0)} \right)^2 + \cdots \right] = [\omega - E_f - \Sigma_c(\omega)]^{-1}. \quad (16.25)$$

Now for a broad conduction band there is a very useful approximation for Σ_c.
To derive it, we rewrite the momentum sum in the self-energy as an energy integral with the density of states, replacing $\sum_\mathbf{k} \rightarrow \int d\epsilon \rho(\epsilon)$, so that

$$\Sigma_c(\omega) = \int \frac{d\epsilon}{\pi} \rho(\epsilon) \frac{\pi V^2}{\omega - \epsilon} = \int \frac{d\epsilon}{\pi} \frac{\Delta(\epsilon)}{\omega - \epsilon}, \quad (16.26)$$

where $\Delta(\epsilon) = \pi\rho(\epsilon)V^2$. In the complex plane, $\Sigma_c(\omega)$ has a branch cut along the real axis, with a discontinuity in its imaginary part proportional to the hybridization:

$$\mathrm{Im}\,\Sigma_c(\omega \pm i\delta) = \int \frac{d\epsilon}{\pi}\Delta(\epsilon)\,\mathrm{Im}\overbrace{\frac{1}{\omega - \epsilon \pm i\delta}}^{\mp i\pi\delta(\omega-\epsilon)} = \mp\Delta(\omega). \qquad (16.27)$$

Consider the particular case where $\Delta(\epsilon) = \Delta$ is constant for $\epsilon \in [-D, D]$, so that

$$\Sigma(\omega \pm i\delta) = \frac{\Delta}{\pi}\int_{-D}^{D}\frac{d\epsilon}{\omega - \epsilon \pm i\delta} = \frac{\Delta}{\pi}\ln\left[\frac{\omega \pm i\delta + D}{\omega \pm i\delta - D}\right]$$

$$= \frac{\Delta}{\pi}\ln\overbrace{\left|\frac{\omega + D}{\omega - D}\right|}^{O(\omega/D)} \mp i\Delta\theta(D - |\omega|), \qquad (16.28)$$

which is a function with a branch cut stretching from $\omega = -D$ to $\omega = +D$. The frequency-dependent part of $\mathrm{Re}\,\Sigma_c = O(\omega/D)$ is negligible in a broad band. We can extend this observation to more general functions $\Delta(\omega)$ that vary slowly over the width of the resonance (lumping any constant part of Σ_c into a shift of E_f). With this observation, for a broad band we drop the real part of Σ_c, writing it in the form

$$\Sigma_c(\omega + i\omega') = -i\Delta\,\mathrm{sgn}(\omega'), \qquad (16.29)$$

where ω' is the imaginary part of the frequency (at the Matsubara frequencies, $\Sigma_c(i\omega_n) = -i\Delta\,\mathrm{sgn}\omega_n$). On the real axis, the f-propagator, takes a particularly simple form,

$$G_f(\omega - i\delta) = \frac{1}{(\omega - E_f - i\Delta)}, \qquad (16.30)$$

that describes a resonance with a width Δ, centered around energy E_f, with a Lorentzian density of states

$$\rho_f(\omega) = \frac{1}{\pi}\,\mathrm{Im}\,G_f(\omega - i\delta) = \frac{\Delta}{(\omega - E_f)^2 + \Delta^2}.$$

Now let us see how the conduction electrons scatter off this resonance. Consider the repeated scattering of the conduction electrons, represented by the dashed line, off the f-level, as follows:

Using (16.24) we see that the third and higher terms can be concisely absorbed into the second term by replacing the bare f-propagator by the full (broadened) f-propagator, as follows:

$$(16.31)$$

$$G(\mathbf{k}', \mathbf{k}, \omega) = \delta_{\mathbf{k}',\mathbf{k}}G^{(0)}(\mathbf{k}, \omega) + G^{(0)}(\mathbf{k}, \omega)V^2 G_f(\omega)G^{(0)}(\mathbf{k}', \omega).$$

We can identify

$$t(\omega) = V^2 G_f(\omega) \tag{16.32}$$

as the scattering t-matrix of the resonance. In fact, this relationship holds quite generally, even when interactions are present, because the only way conduction electrons can scatter is by passing through the localized f-state. The full conduction electron propagator can then be written

$$G(\mathbf{k}', \mathbf{k}, \omega) = \delta_{\mathbf{k}',\mathbf{k}} G^{(0)}(\mathbf{k}, \omega) + G^{(0)}(\mathbf{k}, \omega) t(\omega) G^{(0)}(\mathbf{k}', \omega). \tag{16.33}$$

Scattering theory tells us that the t-matrix is related to the S-matrix $S(\omega) = e^{2i\delta(\omega)}$, where $\delta(\omega)$ is the scattering phase shift, by the relation $S = 1 - 2\pi i \rho t(\omega + i\eta)$ (here we use η as the infinitesimal to avoid confusion with the notation for the phase shift), or

$$t(\omega + i\eta) = \frac{1}{-2\pi i \rho}(S(\omega) - 1) = -\frac{1}{\pi\rho} \times \frac{1}{cot\delta(\omega) - i}. \tag{16.34}$$

Substituting our explicit form of the f-Green's function,

$$t(\omega + i\delta) = V^2 G_f(\omega + i\eta) = \frac{1}{\pi\rho} \times \frac{\overbrace{\pi\rho V^2}^{\Delta}}{\omega - E_f + i\Delta} = -\frac{1}{\pi\rho} \times \frac{1}{(\frac{E_f - \omega}{\Delta}) - i}. \tag{16.35}$$

Comparing (16.34) and (16.35), we see that the scattering phase shift is given by

$$\delta_f(\omega) = \cot^{-1}\left(\frac{E_f - \omega}{\Delta}\right) = \tan^{-1}\left(\frac{\Delta}{E_f - \omega}\right). \tag{16.36}$$

$\delta_f(\omega)$ is a monotonically increasing function, rising from $\delta_f = 0$ at $\omega \ll 0$ to $\delta_f = \pi$ at high energies. On resonance, $\delta(E_f) = \pi/2$, corresponding to the strongest kind of *unitary scattering*.

16.4.3 The Friedel sum rule

Remarkably, the phase shift $\delta_f \equiv \delta_f(0)$ at the Fermi surface sets the amount of charge bound inside the resonance. We can see this by using the f-spectral function to calculate the ground-state occupancy:

$$n_f = 2\int_{-\infty}^0 d\omega \rho_f(\omega) = 2\int_{-\infty}^0 \frac{d\omega}{\pi} \frac{\Delta}{(\omega - E_f)^2 + \Delta^2} = \frac{2}{\pi}\cot^{-1}\left(\frac{E_f}{\Delta}\right) \equiv 2 \times \frac{\delta_f}{\pi}. \tag{16.37}$$

Note that when $\delta(0) = \pi/2$, $n_f = 1$. This is a particular example of the *Friedel sum rule*, a very general relation between the number Δn of particles bound in a potential well and the sum of the scattering phase shifts at the Fermi surface:

$$\Delta n = \sum_\lambda \frac{\delta_\lambda}{\pi}, \tag{16.38}$$

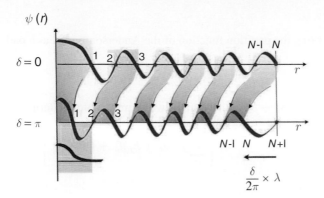

Fig. 16.7

Illustrating the Friedel sum rule. As the scattering phase shift grows, the nodes of the eigenstates at the Fermi surface are drawn into the potential well. Each time the phase shift passes through π, one more node passes into the well, leading to one more bound electron. In this illustration, one bound state has formed as one node passes through the boundary, increasing the phase shift by π.

where δ_λ denotes the scattering phase shift in the partial-wave state labeled by the orbital quantum numbers λ (Figure 16.7).[3]

We can understand the Friedel sum rule by looking at the scattering wavefunction far from the impurity. The asymptotic radial wavefunctions of the incoming and the phase-shifted outgoing electrons on the Fermi surface take the form

$$\psi(r) \sim \left[\frac{e^{-ik_F r}}{r} + e^{2i\delta_f} \frac{e^{ik_F r}}{r} \right] \sim \frac{e^{i\delta} \sin(k_F r + \delta_f)}{r},$$

which corresponds to a radial wave in which the wavefunction of the electrons is shifted by an amount

$$\Delta r = -\frac{\delta_f}{k_F} = -\frac{\lambda_F}{2} \times \frac{\delta_f}{\pi}.$$

Thus, for a positive phase shift, electrons are *drawn inwards* by the scattering process. Each time δ_f passes through π, one more node of the wavefunction passes through the boundary at infinity, corresponding to an additional bound electron. Anderson has called Friedel's sum rule a "node-counting theorem."

Example 16.2   Anderson model as a path integral.
Formulate the Anderson model as a path integral and show that the conduction electrons can be *integrated out*, giving rise to an action of the following form [26]:

$$S_F = \sum_{\sigma, i\omega_n} \bar{f}_{\sigma n} \left\{ -i\omega_n + E_f - i\Delta \, \text{sgn} \, \omega_n \right\} f_{\sigma n} + \int_0^\beta d\tau \, U n_\uparrow n_\downarrow, \qquad (16.39)$$

where $f_{\sigma n} \equiv \beta^{-\frac{1}{2}} \int_0^\beta d\tau \, e^{i\omega_n \tau} f_\sigma(\tau)$ is the Fourier transform of the f-electron field.

---

[3] For a spherical atom without spin–orbit coupling, $\lambda = (l, m, \sigma)$, where l, m, and σ are the angular momentum and spin quantum numbers. With spin–orbit coupling, $\lambda = (j, m)$ denote the quantum numbers of total angular momentum j.

Solution

We begin by writing the partition function of the Anderson model as a path integral:

$$Z = \int \mathcal{D}[f, c] e^{-S}, \tag{16.40}$$

where the action $S = S_A + S_B$ is the sum of two terms, an atomic term

$$S_A = \int_0^\beta d\tau \left[\sum_\sigma \bar{f}_\sigma (\partial_\tau + E_f) f_\sigma + U n_{f\uparrow} n_{f\downarrow} \right]$$

and a bath term

$$S_B = \int_0^\beta d\tau \left\{ \sum_{\mathbf{k}\sigma} \bar{c}_{\mathbf{k}\sigma} (\partial_\tau + \epsilon_{\mathbf{k}\sigma}) c_{\mathbf{k}\sigma} + V \left[\bar{f}_\sigma c_{\mathbf{k}\sigma} + \bar{c}_{\mathbf{k}\sigma} f_\sigma \right] \right\} \tag{16.41}$$

describing the hybridization with the surrounding sea of conduction electrons.

We can rearrange the path integral so that the conduction electron integral is carried out first:

$$Z = \int \mathcal{D}[f] e^{-S_F} \overbrace{\int \mathcal{D}[c] e^{-S_B}}^{Z_B[\{f\}]}, \tag{16.42}$$

where $Z_B[\{f\}]$ contains the change to the f-electron induced by integrating out the conduction electrons. The bath action is free of interactions and can be written schematically as a quadratic form,

$$S_B = \bar{c} \cdot A \cdot c + \bar{c} \cdot j + \bar{j} \cdot c, \tag{16.43}$$

where $A \equiv (\partial_\tau + \epsilon_{\mathbf{k}})\delta(\tau - \tau')$ is the matrix sandwiched between the fields $c \equiv c_{\mathbf{k}\sigma}(\tau)$ and $\bar{c} = \bar{c}_{\mathbf{k}\sigma}(\tau)$, while $j(\tau) = V f_\sigma(\tau)$ and $\bar{j} = \bar{f}_\sigma(\tau) V$ are source terms. You may find it reassuring to recast S_B in Fourier space, where $A = (-i\omega_n + \epsilon_{\mathbf{k}})$ is explicitly diagonal.

Using the standard result for Gaussian fermion integrals,

$$Z_B = \int \mathcal{D}[c] e^{-\bar{c}Ac - \bar{j}c + \bar{c}j} = \det A \times \exp[\bar{j} \cdot A^{-1} \cdot j],$$

or, explicitly,

$$Z_B[\{f\}] = \overbrace{\det[\partial_\tau + \epsilon_{\mathbf{k}}]}^{Z_C = e^{-\beta F_C}} \exp \left[\int_0^\beta d\tau \bar{f}_\sigma \left(\sum_{\mathbf{k}} \frac{V^2}{\partial_\tau + \epsilon_{\mathbf{k}}} \right) f_\sigma \right]. \tag{16.44}$$

The first term is the partition function Z_C of the conduction sea in the absence of the magnetic ion. Substituting $Z_B[\{f\}]$ back into the full path integral (16.42) and combining the quadratic terms then gives

$$Z = Z_C \times \int \mathcal{D}[f] \exp \left[-\int d\tau \left\{ \bar{f}_\sigma \left(\partial_\tau + E_f - \sum_{\mathbf{k}} \frac{V^2}{\partial_\tau + \epsilon_{\mathbf{k}}} \right) f_\sigma + U n_\uparrow n_\downarrow \right\} \right].$$

If we transform the first term into Fourier space, substituting $f_\sigma(\tau) = \beta^{-1/2} \sum_n f_{\sigma n} e^{-i\omega_n \tau}$, $\bar{f}_\sigma(\tau) = \beta^{-1/2} \sum_n \bar{f}_{\sigma n} e^{i\omega_n \tau}$, so that $\partial_\tau \to -i\omega_n$, the action can be written

$$
S_F = \sum_{\sigma, i\omega_n} \bar{f}_{\sigma n} \underbrace{\left\{ -i\omega_n + E_f + \overbrace{\sum_{\mathbf{k}} \frac{V^2}{i\omega_n - \epsilon_{\mathbf{k}}}}^{-i\Delta \,\mathrm{sgn}\, \omega_n} \right\}}_{-G_f^{-1}(i\omega_n)} f_{\sigma n} + \int_0^\beta d\tau\, U n_\uparrow n_\downarrow. \tag{16.45}
$$

The quadratic coefficient of the f-electrons is the inverse f-electron propagator of the non-interacting resonance. We immediately recognize the self-energy term $\Sigma_c(i\omega_n) = -i\Delta \,\mathrm{sgn}\, \omega_n$ introduced in (16.23). From this path integral derivation, we can see that this term accounts for the effect of the conduction bath electrons, even in the presence of interactions. If we now use the large-bandwidth approximation $\Sigma(i\omega_n) = -i\Delta \,\mathrm{sgn}\, \omega_n$ introduced in (16.29), the action can be compactly written

$$
S_F = \sum_{\sigma, i\omega_n} \bar{f}_{\sigma n} \left\{ -i\omega_n + E_f - i\Delta \mathrm{sgn}\, \omega_n \right\} f_{\sigma n} + \int_0^\beta d\tau\, U n_\uparrow n_\downarrow. \tag{16.46}
$$

16.4.4 Mean-field theory

In the Anderson model, the Coulomb interaction and hybridization compete with one another. Crudely speaking, we expect that when the Coulomb interaction exceeds the hybridization, local moments will develop. To gain an insight into the effect of hybridization on local moment formation, Anderson [7] carried out a Hartree mean-field treatment of the repulsive U interaction, decoupling

$$
U n_\uparrow n_\downarrow \to U n_\uparrow \langle n_\downarrow \rangle + U \langle n_\uparrow \rangle n_\downarrow - U \langle n_\uparrow \rangle \langle n_\downarrow \rangle + O(\delta n^2). \tag{16.47}
$$

We can understand this factorization as the result of a saddle-point description of the path integral, treated in more detail in Excercise 16.3. Using this mean-field approximation, Anderson concluded that, for the symmetric Anderson model, local moments would develop, provided

$$
U \gtrsim U_c = \pi \Delta. \tag{16.48}
$$

Let us now rederive this mean-field result. From (16.47), the mean-field effect of the interactions is to shift the f-level position:

$$
E_f \to E_{f\sigma} = E_f + U \langle n_{f-\sigma} \rangle. \tag{16.49}
$$

Using (16.37), this implies that the scattering phase shifts for the up and down channels,

$$
\delta_{f\sigma} = \cot^{-1} \left(\frac{E_{f\sigma}}{\Delta} \right), \tag{16.50}
$$

are no longer equal. Using the Friedel sum rule (16.37), we then obtain the mean-field equations:

$$
\langle n_{f\sigma} \rangle = \frac{\delta_{f\sigma}}{\pi} = \frac{1}{\pi} \cot^{-1} \left(\frac{E_f + U \langle n_{f-\sigma} \rangle}{\Delta} \right). \tag{16.51}
$$

It is convenient to introduce an occupancy $n_f = \sum_\sigma \langle n_{f\sigma}\rangle$ and magnetization $M = \langle n_{f\uparrow}\rangle - \langle n_{f\downarrow}\rangle$, so that $\langle n_{f\sigma}\rangle = \frac{1}{2}(n_f + \sigma M)$ ($\sigma = \pm 1$). The mean-field equations for the occupancy and magnetization are then

$$n_f = \frac{1}{\pi}\sum_{\sigma=\pm 1}\cot^{-1}\left(\frac{E_f + U/2(n_f - \sigma M)}{\Delta}\right) \tag{16.52}$$

$$M = \frac{1}{\pi}\sum_{\sigma=\pm 1}\sigma\cot^{-1}\left(\frac{E_f + U/2(n_f - \sigma M)}{\Delta}\right). \tag{16.53}$$

To find the critical size of the interaction strength where a local moment develops, we set $M \to 0^+$ in (16.52) to obtain $\frac{E_f + U_c n_f/2}{\Delta} = \cot\left(\frac{\pi n_f}{2}\right)$. Linearing (16.53) in M, we obtain

$$1 = \frac{U_c}{\pi\Delta}\frac{1}{1 + \left(\frac{E_f + U n_f/2}{\Delta}\right)^2} = \frac{U_c}{\pi\Delta}\sin^2\left(\frac{\pi n_f}{2}\right), \tag{16.54}$$

so that, for $n_f = 1$,

$$U_c = \pi\Delta. \tag{16.55}$$

For larger values of $U > U_c$, there are two solutions, corresponding to up and down spin polarization of the f-state. We will see that this is an oversimplified description of the local moment, but it gives us an approximate picture of the physics. The total density of states now contains two Lorentzian peaks, located at $E_f \pm UM$:

$$\rho_f(\omega) = \frac{1}{\pi}\left[\frac{\Delta}{(\omega - E_f - UM)^2 + \Delta^2} + \frac{\Delta}{(\omega - E_f + UM)^2 + \Delta^2}\right].$$

The critical curve obtained by plotting U_c and E_f as a parametric function of n_f is shown in Figure 16.8.

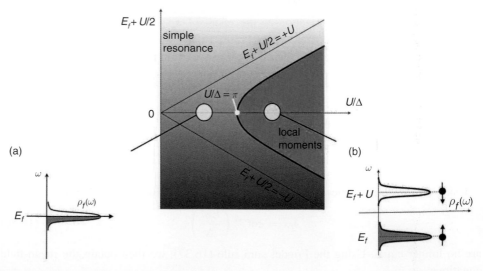

Fig. 16.8 Mean field phase diagram of the Anderson model, illustrating how the f-electron resonance splits to form a local moment. (a) $U < \pi\Delta$, single half-filled resonance; (b) $U > \pi\Delta$, up and down components of the resonance are split by an energy U.

Anderson's mean-field theory allows a qualitative understanding of the experimentally observed formation of local moments. When dilute magnetic ions are dissolved in a metal to form an alloy, the formation of a local moment is dependent on whether the ratio $U/\pi\Delta$ is larger than or smaller than 1. When iron is dissolved in pure niobium, the failure of the moment to form reflects the higher density of states and larger value of Δ in this alloy. When iron is dissolved in molybdenum, the lower density of states causes $U > U_c$, and local moments form [5].

Example 16.3 Factorizing the interaction in the Anderson model.

(a) Show that the interaction in the Anderson model can be decoupled via a Hubbard–Stratonovich decoupling to yield

$$\int_0^\beta d\tau\, U n_\uparrow n_\downarrow \rightarrow \int_0^\beta d\tau \left[\phi_\uparrow n_\uparrow + \phi_\downarrow n_\downarrow - \frac{\phi_\uparrow \phi_\downarrow}{U} \right], \qquad (16.56)$$

where $\phi_\sigma = \phi_0 + i\lambda(\tau) - \sigma h(\tau)$ is the sum of a real and an imaginary field.

(b) Derive the mean-field partition function obtained by assuming that the path integral over ϕ can be approximated by the saddle-point configuration where ϕ_σ is independent of time, given by

$$Z_{MF} = \int \mathcal{D}[f] e^{-S_{MF}[\phi_\sigma, f]}$$

$$S_{MF} = \sum_{\sigma, i\omega_n} \bar{f}_{\sigma n} \left[-G_{f\sigma}^{-1}(i\omega_n) \right] f_{\sigma n} + \frac{\beta}{U} \phi_\uparrow \phi_\downarrow, \qquad (16.57)$$

where

$$G_{f\sigma}^{-1}(i\omega_n) = i\omega_n - E_f - \phi_\sigma + i\Delta\, \text{sgn}\, \omega_n$$

is the inverse mean-field f-propagator.

(c) Carry out the Gaussian integral in (16.57) to show that the mean-field free energy is

$$F_{MF} = -k_B T \sum_{\sigma, i\omega_n} \ln \left[-G_{f\sigma}^{-1}(i\omega_n) \right] - \frac{1}{U} \phi_\uparrow \phi_\downarrow,$$

and, by setting $\partial F/\partial \phi_\sigma = 0$, derive the mean-field equations

$$\phi_{-\sigma} = U\langle n_{f\sigma} \rangle = U \int_{-\infty}^\infty \frac{d\omega}{\pi} f(\omega) \frac{\Delta}{(\omega - E_f - \phi_\sigma)^2 + \Delta^2}.$$

Solution

(a) The interaction in the Anderson model can be rewritten as a sum of two terms,

$$U n_\uparrow n_\downarrow = \overbrace{\frac{U}{4}(n_\uparrow + n_\downarrow)^2}^{\text{charge}} - \overbrace{\frac{U}{4}(n_\uparrow - n_\downarrow)^2}^{\text{spin}}.$$

that we can loosely interpret as a repulsion between charge fluctuations and an attraction between spin fluctuations. Following the results of Section 12.5, inside the path integral the attractive magnetic interaction can be decoupled in terms of a fluctuating Weiss $h(\tau)$ field, while the repulsive charge interaction can be decoupled in terms of a fluctuating potential field $\phi(\tau) = \phi_0 + i\lambda(\tau)$, as follows:

$$
\begin{aligned}
-\frac{1}{2} \times \frac{U}{2}(n_\downarrow - n_\uparrow)^2 &\rightarrow -h(n_\uparrow - n_\downarrow) + \frac{h^2}{2 \times (U/2)} \\
+\frac{1}{2} \times \frac{U}{2}(n_\uparrow + n_\downarrow)^2 &\rightarrow \phi(n_\uparrow + n_\downarrow) - \frac{\phi^2}{2 \times (U/2)},
\end{aligned}
\tag{16.58}
$$

with the understanding that, for repulsive $U > 0$, fluctuations of $\phi(\tau)$ are integrated along the imaginary axis, $\phi(\tau) = \phi_0 + i\lambda(\tau)$. Adding these terms gives

$$
\int_0^\beta d\tau\, U n_\uparrow n_\downarrow \rightarrow \int_0^\beta d\tau \left[(\phi - \sigma h)n_\sigma + \frac{h^2 - \phi^2}{U} \right] = \int_0^\beta d\tau \left[\phi_\uparrow n_\uparrow + \phi_\downarrow n_\downarrow + \frac{\phi_\uparrow \phi_\downarrow}{U} \right],
\tag{16.59}
$$

where $\phi_\sigma = \phi - \sigma h$. The decoupled path integral then takes the form

$$
Z_F = \int \mathcal{D}[\phi_\sigma] \int \mathcal{D}[f] e^{-S_F[\phi_\sigma, f]}
$$

$$
S_F = \int d\tau \left\{ \bar{f}_\sigma \left(\partial_\tau + E_f + \phi_\sigma - \sum_{\mathbf{k}} \frac{V^2}{\partial_\tau + \epsilon_{\mathbf{k}}} \right) f_\sigma - \frac{1}{U}\phi_\uparrow \phi_\downarrow \right\}.
\tag{16.60}
$$

Note how the Weiss fields ϕ_σ shift the f-level position: $E_f \rightarrow E_f + \phi_\sigma(\tau)$. In this way, the Anderson model can be regarded as a resonant level immersed in a white-noise magnetic field that modulates the splitting between the up and down spin resonances.

(b) Anderson's mean-field treatment corresponds to to a saddle-point approximation to the integral over the ϕ_σ fields. At the saddle-point, $\langle \delta S/\delta \phi_\sigma \rangle = 0$. From (16.60), we obtain

$$
\frac{\delta S_F}{\delta \phi_\sigma} = \bar{f}_\sigma f_\sigma - \frac{1}{U}\phi_{-\sigma},
$$

so the saddle-point condition $\langle \delta S_F/\delta \phi_\sigma \rangle = 0$ implies $\phi_{-\sigma} = U\langle n_{f\sigma} \rangle$, recovering the Hartree mean-field theory. We can clearly seek solutions in which $\phi_\sigma(\tau) = \phi_\sigma^{(0)}$ is a constant. With this understanding, the saddle-point approximation is

$$
Z_F \approx Z_{MF} = \int \mathcal{D}[f] e^{-S_F[\phi_\sigma^{(0)}, f]},
\tag{16.61}
$$

where

$$
S_{MF} = \int d\tau \left\{ \bar{f}_\sigma \left(\partial_\tau + E_f + \phi_\sigma^{(0)} - \sum_{\mathbf{k}} \frac{V^2}{\partial_\tau + \epsilon_{\mathbf{k}}} \right) f_\sigma - \frac{1}{U}\phi_\uparrow^{(0)} \phi_\downarrow^{(0)} \right\}.
\tag{16.62}
$$

Now since $\phi^{(0)}$ is a constant, we can Fourier transform the first term in this expression, replacing $\partial_\tau \rightarrow -i\omega_n$, to obtain

$$
S_{MF} = \sum_{\sigma,\, i\omega_n} \bar{f}_{\sigma n} \overbrace{\left(-i\omega_n + E_f + \phi_\sigma^{(0)} - \underbrace{\sum_{\mathbf{k}} \frac{V^2}{-i\omega_n + \epsilon_{\mathbf{k}}}}_{-i\,\mathrm{sgn}(\omega_n)\Delta} \right)}^{-G_{f\sigma}^{-1}(i\omega_n)} f_{\sigma n} - \frac{\beta}{U}\phi_\uparrow^{(0)} \phi_\downarrow^{(0)},
\tag{16.63}
$$

where, in the broad-bandwidth limit, we can replace

$$G_{f\sigma}^{-1}(i\omega_n) = i\omega_n - E_f - \phi_\sigma^{(0)} + i\,\text{sgn}(\omega_n)\Delta. \tag{16.64}$$

(c) Carrying out the Gaussian integral in (16.61), we obtain

$$Z_{MF} = \det[-G_{f\sigma}^{-1}(i\omega_n)]e^{\frac{\beta}{U}\phi_\uparrow\phi_\downarrow} = \prod_{\sigma,i\omega_n}[-G_{f\sigma}^{-1}(i\omega_n)]e^{\frac{\beta}{U}\phi_\uparrow\phi_\downarrow},$$

or

$$F_{MF} = -k_B T \ln Z_{MF} = -k_B T \sum_{\sigma,i\omega_n} \ln\left[-G_{f\sigma}^{-1}(i\omega_n)\right]e^{i\omega_n 0^+} - \frac{1}{U}\phi_\uparrow\phi_\downarrow, \tag{16.65}$$

where we have included the convergence factor $e^{i\omega_n 0^+}$. By (16.64), $\dfrac{\partial G_{f\sigma}^{-1}(i\omega_n)}{\partial\phi_\sigma} = -1$, so by differentiating (16.65) with respect to ϕ_σ we obtain

$$0 = k_B T \sum_{i\omega_n} G_{f\sigma}(i\omega_n)e^{i\omega_n 0^+} - \frac{1}{U}\phi_{-\sigma}, \tag{16.66}$$

or

$$\phi_{-\sigma} = U\langle n_{f\sigma}\rangle = U k_B T \sum_{i\omega_n} G_{f\sigma}(i\omega_n)e^{i\omega_n 0^+}.$$

Carrying out the sum over the Matsubara frequencies by the standard contour integral method, we obtain

$$\phi_{-\sigma} = -U\oint_{\text{Im axis}}\frac{dz}{2\pi i}f(z)G_{f\sigma}(z) = U\oint_{\text{Re axis}}\frac{dz}{2\pi i}f(z)G_{f\sigma}(z)$$

$$= U\int_{-\infty}^{\infty}\frac{d\omega}{\pi}f(\omega)\,\text{Im}\,G_{f\sigma}(\omega - i\delta)$$

$$= U\int_{-\infty}^{\infty}\frac{d\omega}{\pi}f(\omega)\frac{\Delta}{(\omega - E_f - \phi_{0\sigma})^2 + \Delta^2}. \tag{16.67}$$

16.5 The Coulomb blockade: local moments in quantum dots

A modern realization of the physics of local moments is found in *quantum dots*. Quantum dots are tiny electron pools in a doped semiconductor, small enough so that the electron states inside the dot are quantized, loosely resembling the electronic states of an atom. Quantum-dot behavior also occurs in nanotubes. Unlike a conventional atom, the separation of the electronic states in quantum dots is of the order of meV rather than electron volts. The overall position of the quantum-dot energy levels can be changed by applying a gate voltage to the dot. It is then possible to pass a small current through the dot by placing it between two leads. The differential conductance $G = dI/dV$ is directly proportional to the density of states $\rho(\omega)$ inside the dot, $G \propto \rho(0)$. Experimentally, when G is measured as a function of gate voltage V_g, the differential conductance is observed to develop a periodic structure, with a period of a few meV [27].

This phenomenon is known as a *Coulomb blockade* [28, 29], and it results from precisely the same physics that is responsible for moment formation. A simple model for a quantum dot considers it as a sequence of single-particle levels at energies ϵ_λ, interacting via a single Coulomb potential U according to the model

$$H_{dot} = \sum_\lambda (\epsilon_\lambda + eV_g)n_{\lambda\sigma} + \frac{U}{2}N(N-1), \tag{16.68}$$

where $n_{\lambda\sigma}$ is the occupancy of the spin σ state of the λ level, $N = \sum_{\lambda\sigma} n_{\lambda\sigma}$ is the total number of electrons in the dot, and V_g is the gate voltage. This is a simple generalization of the single-atom part of the Anderson model. Notice that the capacitance of the dot is $C = e^2/U$.

The energy difference between the n and $n + 1$ electron states of the dot is given by

$$E(n+1) - E(n) = nU + \epsilon_{\lambda_n} - |e|V_g,$$

where λ_n is the one-particle state into which the nth electron is being added. As the gate voltage is raised, the quantum dot fills each level sequentially, as illustrated in Figure 16.9, and when $|e|V_g = nU + \epsilon_{\lambda_n}$ the nth level becomes degenerate with the Fermi energy of each lead. At this point, electrons can pass coherently through the resonance, giving rise to a sharp peak in the conductance. At maximum conductance, the transmission and reflection of electrons is unitary, and the conductance of the quantum dot will reach a substantial fraction of the quantum of conductance, e^2/h per spin. A calculation of the zero-temperature

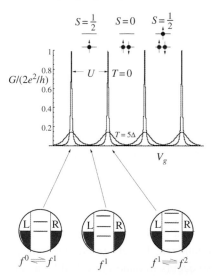

Fig. 16.9   Variation of zero-bias conductance $G = dI/dV$ with gate voltage in a quantum dot. Coulomb interactions mean that, for each additional electron in the dot, the energy to add one electron increases by U. When the charge on the dot is integral, the Coulomb interaction blocks the addition of electrons and the conductance is suppressed. When the energy to add an electron is degenerate with the Fermi energy of the leads, unitary transmission occurs and, for symmetric leads, $G = 2e^2/h$.

conductance through a single non-interacting resonance coupled symmetrically to two leads gives

$$G(V_g) = \frac{2e^2}{h} \frac{\Delta^2}{(\epsilon_\lambda - |e|V_g)^2 + \Delta^2}, \tag{16.69}$$

where the factor of 2 derives from two spin channels. This gives rise to a conductance peak when the gate voltage $|e|V_g = \epsilon_\lambda$. At a finite temperature, the Fermi distribution of the electrons in the leads is thermally broadened, and the conductance involves a thermal average about the Fermi energy:

$$G(V_g, T) = \frac{2e^2}{h} \int d\epsilon \left(-\frac{\partial f}{\partial \epsilon} \right) \frac{\Delta^2}{(\epsilon_\lambda - |e|V_g - \epsilon)^2 + \Delta^2}, \tag{16.70}$$

where $f(\epsilon) = 1/(e^{\beta\epsilon} + 1)$ is the Fermi function. When there are multiple levels, each successive level contributes to the conductance, to give

$$G(V_g, T) = \sum_{n \geq 0} \frac{2e^2}{h} \int d\epsilon \left(-\frac{\partial f}{\partial \epsilon} \right) \frac{\Delta^2}{(nU + \epsilon_{\lambda_n} - |e|V_g - \epsilon)^2 + \Delta^2},$$

where the nth level is shifted by the Coulomb blockade.

The effect of a bias voltage on these results is interesting. In this situation, the energy distribution functions of the two leads are now shifted relative to one another. A crude model for the effect of a voltage is obtained by replacing the Fermi function by an average over both leads, so that $f'(\epsilon) \rightarrow \frac{1}{2} \sum_{\pm} f'(\epsilon \pm \frac{eV_{sd}}{2})$, which has the effect of splitting the conductance peaks into two, peaked at voltages

$$|e|V_g = \epsilon_{\lambda_n} + nU \pm |e|V_{sd}/2, \tag{16.71}$$

as shown in Figure 16.10.

It is remarkable that the physics of moment formation and the Coulomb blockade operate in both artificial mesoscopic devices and naturally occurring magnetic ions.

16.6 The Kondo effect

Although Anderson's mean-field theory provides a mechanism for moment formation, it raises new questions. While the mean-field treatment of the local moment would be appropriate for an ordered magnet involving a macroscopic number of spins rigidly locked together, for a single magnetic impurity there will always be a finite quantum mechanical amplitude for the spin to tunnel between an up and down configuration:

$$e_\downarrow^- + f_\uparrow^1 \rightleftharpoons e_\uparrow^- + f_\downarrow^1.$$

This tunneling rate τ_{sf}^{-1} defines a temperature scale

$$k_B T_K = \frac{\hbar}{\tau_{sf}},$$

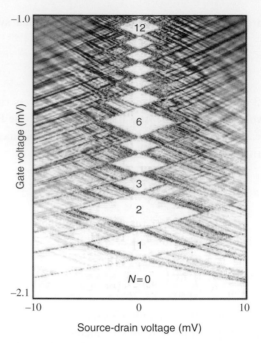

Gate voltage (mV)

Source-drain voltage (mV)

Fig. 16.10 Experimentally measured conductance for a voltage-biased quantum dot, showing the splitting of the Coulomb blockade into two components, shifted up and down by the voltage bias, $\pm eV_{sd}/2$. In the white, diamond-shaped regions, $G(V_{sd}) \approx 0$ as a result of the Coulomb blockade. The number of particles N is fixed in each of the diamond regions. The lines outside the diamonds, running parallel to the sides of he diamonds, identify excited states. Reprinted with permission from L. P. Kouwenhoven, *et al.*, *Science*, vol. 5, p. 1788, 1997. Copyright 1997 by the American Association for the Advancement of Science.

called the *Kondo temperature*, which sets the crossover between local moment behavior, where the spin is free, and the low-temperature physics, where the spin and conduction electrons are entangled to form a spin singlet. The mechanism by which the local moment is screened involves a remarkable piece of quantum physics called the *Kondo effect*, named after the Japanese physicist Jun Kondo [8].

Understanding the physics of this crossover posed a major problem for the theoretical physics community, and it took about a decade to resolve. The basic issue is that perturbation theory fails below the Kondo temperature. Indeed, the basic process by which a local moment disappears or *quenches* at low temperatures is analogous to the physics of quark confinement, and the understanding of the phenomenon required a new understanding of the renormalization group [12–14, 16, 17, 31, 32], culminating in Ken Wilson's numerical renormalization solution of the Kondo model [16].

The Kondo problem has proven to be extremely fertile ground for both experimental and theoretical physics, and at each stage it has held new surprises. After the problem was solved using Wilson's numerical renormalization methods, the theoretical community was astonished in 1980 by the independent discoveries of Natan Andrei [33, 34], then working at New York University, and of Paul Weigman, at the Landau Institute Moscow, [35], that the Kondo impurity model is integrable and that its full many-body spectrum can be exactly diagonalized by the method of *Bethe ansatz*. While these results did not change the conclusions derived from renormalization, they added a precise analytic foundation to the field. Later work by Paul Weigman and Alexei Tsvelik in Moscow [36] showed that even the original Anderson impurity model is also exactly diagonalizable. More recently, the Kondo and Anderson impurity models have enjoyed a new life as "impurity solvers" for the dynamical mean-field theory approach to the physics of strongly interacting materials, invented by Antoine Georges, Gabriel Kotliar, Walter Metzner, and Dieter Vollhardt [37, 38]. A treatment of these remarkable developments lies outside the scope of this introductory text.

The Kondo effect has many manifestations in condensed matter physics: not only does it govern the quenching of magnetic moments in a magnetic alloy or a quantum dot [27], but it is also responsible for the formation of heavy fermions in dense Kondo lattice materials (heavy-fermion compounds), where the local moments transform into composite quasi-particles with masses sometimes in excess of a thousand bare-electron masses [39]. We will see that the Kondo temperature depends exponentially on the strength of the Anderson interaction parameter U. In the symmetric Anderson model, where $E_f = -U/2$,

$$T_K = \sqrt{\frac{2U\Delta}{\pi}} \exp\left(-\frac{\pi U}{8\Delta}\right). \tag{16.72}$$

We will derive the key elements of this basic result using perturbative renormalization group reasoning [40], but it is also obtained from the exact Bethe ansatz solution of the Anderson model [41–43].

One can view the physics of local moments from two complementary perspectives (see Figure (16.11)):

- An *adiabatic picture* which starts with the non-interacting resonant ground state ($U = 0$) of the Anderson model, and then considers the effect of dialing up the interaction term U

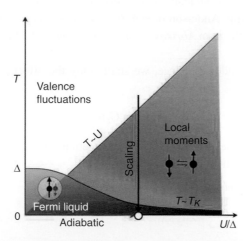

The phase diagram of the symmetric Anderson model. Below a scale $T \sim U$, local moments develop. The Kondo temperature T_K plays the role of the renormalized resonant level width. Below a temperature $T \sim T_K$, the local moments become screened by the conduction sea via the Kondo effect, to form a Fermi liquid.

Fig. 16.11

- A *scaling approach*, which starts with the interacting but isolated atom ($V(k) = 0$), and considers the effect of immersing it in an electron sea, gradually integrating out lower and lower-energy electrons.

The adiabatic approach involves "dialing up" the interaction, as shown by the horizontal arrow in Figure 16.11. From the adiabatic perspective, the ground state remains in a Fermi liquid. In principle, one might imagine the possibiity of a phase transition at some finite interaction strength U, but in a single-impurity model, with a finite number of local degrees of freedom, we don't expect any symmetry-breaking phase transitions. In the scaling approach, we follow the physics as a function of ever-decreasing energy scale, loosely equivalent to "dialing down" the temperature, as shown by the vertical arrow in Figure 16.11. The scaling approach starts from an atomic perspective: it allows us to understand the formation of local moments and, at lower temperatures, how a Fermi liquid can develop through the interaction of an isolated magnetic moment with an electron sea.

We shall first discuss one of the most basic manifestations of the Kondo effect: the appearance of a Kondo resonance in the spectral function of the localized electron. This part of our analysis will involve rather qualitative reasoning based on the ideas of adiabaticity introduced in earlier chapters. Afterwards we adopt the scaling approach, deriving the Kondo model and describing low-energy coupling between the local moments and conduction electrons by using a *Schrieffer–Wolff transformation* of the Anderson model. Finally, we discuss the concept of renormalization and apply it to the Kondo model, following the evolution of the physics from the local moment to the Fermi liquid.

16.6.1   Adiabaticity and the Kondo resonance

The adiabatic approach allows us to qualitatively understand the emergence of a remarkable resonance in the excitation spectrum of the localized f-electron, the *Kondo resonance*. This resonance is simply the adiabatic renormalization of the Friedel–Anderson resonance seen in the non-interacting Anderson model. Its existence was first inferred by Abrikosov and Suhl [44, 45], and the term *Abrikosov–Suhl resonance* is the historically correct name for the resonance.

To understand the Kondo resonance, we shall study the effects of interactions on the f-spectral function

$$A_f(\omega) = \frac{1}{\pi} \operatorname{Im} G_f(\omega + i\eta), \qquad (16.73)$$

where $G_f(\omega - i\delta) = $ is the advanced f-Green's function. From a spectral decomposition (Section 9.7.1), we know that

$$A_f(\omega) = \begin{cases} \overbrace{\displaystyle\sum_\lambda \left| \langle \lambda | f_\sigma^\dagger | \phi_0 \rangle \right|^2 \delta(\omega - [E_\lambda - E_0])}^{\text{energy distribution for adding one } f\text{-electron}} & (\omega > 0) \\[2em] \underbrace{\displaystyle\sum_\lambda |\langle \lambda | f_\sigma | \phi_0 \rangle|^2 \delta(\omega - [E_0 - E_\lambda])}_{\text{energy distribution for removing one } f\text{-electron}} & (\omega < 0), \end{cases} \qquad (16.74)$$

where E_λ and E_0 are the excited and ground-state energies, respectively. For negative energies, $\omega < 0$, this spectrum corresponds to the energy spectrum of electrons emitted in X-ray photoemission, while for positive energies, $\omega > 0$, the spectral function can be measured from inverse X-ray photoemission [46, 47]. The weight beneath the Fermi energy determines the f-charge of the ion:

$$\langle n_f \rangle = 2 \int_{-\infty}^{0} d\omega A_f(\omega). \tag{16.75}$$

In a magnetic ion such as a cerium atom in a $4f^1$ state, this quantity is just a little below unity.

Figure (16.12.) illustrates the effect of the interaction on the f-spectral function. In the non-interacting limit ($U = 0$), the f-spectral function is a Lorentzian of width Δ. If we turn on the interaction U, being careful to shift the f-level position beneath the Fermi energy to maintain a constant occupancy, the resonance splits into three peaks: two at energies $\omega = E_f$ and $\omega = E_f + U$ corresponding to the energies for a valence fluctuation, plus an additional central Kondo resonance associated with the spin fluctuations of the local moment.

When the interaction is much larger than the hybridization width, $U \gg \Delta$, one might expect no spectral weight left at low energies. But it turns out that the spectral

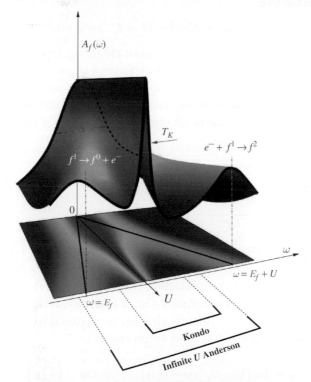

Schematic illustration of the formation of a Kondo resonance in the f-spectral function $A_f(\omega)$ as interaction strength U is turned on. Here, the interaction is turned on while maintaining a constant f-occupancy, by shifting the bare f-level position beneath the Fermi energy. The lower part of diagram is the density plot of the f-spectral function, showing how the non-interacting resonance at $U = 0$ splits into upper and lower atomic peaks at $\omega = E_f$ and $\omega = E_f + U$.

Fig. 16.12

function at the Fermi energy is an adiabatic invariant determined by the scattering phase shift δ_f:

$$A_f(\omega = 0) = \frac{\sin^2 \delta_f}{\pi \Delta}. \tag{16.76}$$

This result, due to Langreth [48, 49], guarantees that a Kondo resonance is always present at the Fermi energy. Now the total spectral weight $\int_{-\infty}^{\infty} d\omega A_f(\omega) = 1$ is conserved, so if $|E_f|$ and U are both large compared with Δ, most of this weight will be lie far from the Fermi energy, leaving a small residue $Z << 1$ in the Kondo resonance. If the area under the Kondo resonance is Z, since the height of Kondo resonance is fixed at $\sim 1/\Delta$, the renormalized hybridization width Δ^* must be of order $Z\Delta$. This scale is set by the Kondo temperature, so that $Z\Delta \sim T_K$.

The Langreth relation (16.76) follows from the analytic form of the f-Green's function near the Fermi energy. For a single magnetic ion, we expect that the interactions between electrons can be increased continuously without any risk of instabilities, so that the excitations of the strongly interacting case remain in one-to-one correspondence with the excitations of the non-interacting case $U = 0$, forming a *local Fermi liquid*. In this local Fermi liquid, the interactions give rise to an f-electron self-energy, which at zero temperature takes the form

$$\Sigma_I(\omega - i\eta) = \Sigma_I(0) + (1 - Z^{-1})\omega + iA\omega^2, \tag{16.77}$$

at low energies. As discussed in Chapter 7, the quadratic energy dependence of $\Sigma_I(\omega) \sim \omega^2$ follows from the Pauli exclusion principle, which forces a quadratic energy dependence of the phase space for the emission of a particle–hole pair. The *wavefunction renormalization* Z, representing the overlap with the state containing one additional f-quasiparticle, is less than unity, $Z < 1$. Using the result (16.77), the low-energy form of the f-electron propagator is

$$G_f^{-1}(\omega - i\eta) = \omega - E_f - i\Delta - \Sigma_I(\omega) = Z^{-1}\Big[\omega - \overbrace{Z(E_f + \Sigma_I(0))}^{E_f^*} - i\overbrace{Z\Delta}^{\Delta^*} - iO(\omega^2)\Big]$$

$$G_f(\omega - i\eta) = \frac{Z}{\omega - E_f^* - i\Delta^* - iO(\omega^2)}. \tag{16.78}$$

This corresponds to a renormalized resonance of reduced weight $Z < 1$, located at postion E_f^* with renormalized width $\Delta^* = Z\Delta$. Now by (16.32) and (16.34), the f-Green's function determines the t-matrix of the conduction electrons, $t(\omega + i\eta) = V^2 G_f(\omega + i\eta) = -(\pi\rho)^{-1} e^{i\delta(\omega)} \sin \delta(\omega)$, so the phase of the f-Green's function at the Fermi energy determines the scattering phase shift δ_f, hence $G_f(0 + i\eta) = (G_f(0 - i\eta))^* = -|G_f(0)|e^{i\delta_f}$. This implies that the scattering phase shift at the Fermi energy is

$$\delta_f = \text{Im}\left(\ln[-G_f^{-1}(\omega - i\eta)]\right)\Big|_{\omega=0} = \tan^{-1}\left(\frac{\Delta^*}{E_f^*}\right). \tag{16.79}$$

Eliminating $E_f^* = \Delta^* \cot \delta_f$ from (16.78), we obtain

$$G_f(0 + i\eta) = -\frac{Z}{\Delta^*} e^{-i\delta_f} \sin \delta_f = -\frac{1}{\Delta} e^{-i\delta_f} \sin \delta_f, \tag{16.80}$$

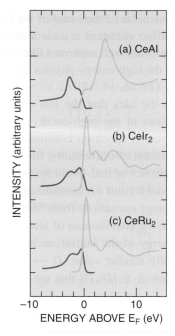

Spectral functions for three different cerium f-electron materials, measured using X-ray photoemission (below the Fermi energy) and inverse X-ray photoemission (above the Fermi energy). CeAl is an antiferromagnet and does not display a Kondo resonance. Reprinted with permission from J. W. Allen, *et al.*, *Phys. Rev.*, vol. 28, p. 5347, 1983. Copyright 1983 by the American Physical Society.

Fig. 16.13

so that

$$A_f(0) = \frac{1}{\pi} \operatorname{Im} G_f(0 - i\eta) = \frac{\sin^2 \delta_f}{\pi \Delta} \tag{16.81}$$

is an adiabatic invariant.

Photoemission studies do reveal the three-peaked structure characteristic of the Anderson model in many Ce systems, such as $CeIr_2$ and $CeRu_2$ [47] (see Figure 16.13). Materials in which the Kondo resonance is wide enough to be resolved are the more mixed-valent materials in which the f-valence departs significantly from unity. Three-peaked structures have also been observed in certain U $5f$ materials such as UPt_3 and UAl_2 materials [50], but it has not yet been resolved in UBe_{13}. A three-peaked structure has recently been observed in $4f$ Yb materials, such as $YbPd_3$, where the $4f^{13}$ configuration contains a single f-hole, so that the positions of the three peaks are reversed relative to Ce [51].

16.7 Renormalization concept

The Anderson model illustrates a central theme of condensed matter physics: the existence of physics on several widely spaced energy scales. In particular, the scale at which local

- Fixed point. If the cut-off energy scale drops below the lowest energy scale in the problem, then there are no further changes to occur in the Hamiltonian, which will now remain invariant under the scaling procedure (so that the β function of all remaining parameters in the Hamiltonian must vanish). This fixed-point Hamiltonian describes the essence of the low-energy physics.

Local-moment physics involves a sequence of such cross-overs (Figure 16.14.). The highest energy scales in the Anderson model are associated with valence fluctuations into the empty and doubly occupied states:

$$f^1 \rightleftharpoons f^2 \qquad \Delta E_I = U + E_f > 0$$
$$f^1 \rightleftharpoons f^0 \qquad \Delta E_{II} = -E_f > 0. \tag{16.85}$$

The successive elimination of these processes leads to two cross-overs. Suppose ΔE_I is the largest scale; then, once $D < \Delta E_I$, charge fluctuations into the doubly occupied state are eliminated and the remaining low-energy Hilbert space of the atom is

$$D < E_f + U: \qquad |f^0\rangle, \qquad |f^1, \sigma\rangle \qquad \left(\sigma = \pm\frac{1}{2}\right). \tag{16.86}$$

The operators that span this space are called *Hubbard operators* [52], and they are denoted as follows:

$$X_{\sigma 0} = |f^1, \sigma\rangle\langle f^0| = P f_\sigma^\dagger, \qquad X_{0\sigma} = |f^0\rangle\langle f^1, \sigma| = f_\sigma^\dagger P,$$
$$X_{\sigma\sigma'} = |f^1, \sigma\rangle\langle f^1, \sigma'|, \tag{16.87}$$

where $P = (1 - n_{f\uparrow} n_{f\downarrow})$ projects out doubly occupied states. (Note that the Hubbard operators $X_{\sigma 0} = P f_\sigma^\dagger$ cannot be treated as simple creation operators, for they do not satisfy the canonical anticommutation algebra.) The corresponding renormalized Hamiltonian is the *Infinite U Anderson model*.

$$H = \sum_{\mathbf{k},\sigma} \epsilon_{\mathbf{k}} n_{\mathbf{k}\sigma} + \left[V(\mathbf{k}) c_{\mathbf{k}\sigma}^\dagger X_{0\sigma} + V(\mathbf{k})^* X_{\sigma 0} c_{\mathbf{k}\sigma} \right] + E_f \sum_\sigma X_{\sigma\sigma}. \tag{16.88}$$

Infinite U Anderson model

In this model, all the interactions are hidden inside the Hubbard operators.

Finally, once $D < \Delta E_{II}$, the low-energy Hilbert space no longer involves the f^2 or f^0 states. The object left behind is a *quantum top* – a quantum mechanical object with purely spin degrees of freedom and a two-dimensional[5] Hilbert space:

$$|f^1, \sigma\rangle \qquad \left(\sigma = \pm\frac{1}{2}\right).$$

---

[5] In the simplest version of the Anderson model, the local moment is $S = \frac{1}{2}$, but in more realistic atoms much larger moments can be produced. For example, an electron in a Ce^{3+} ion atom lives in a $4f^1$ state. Here spin–orbit coupling combines orbital and spin angular momentum into a total angular moment $j = l - \frac{1}{2} = \frac{5}{2}$. The cerium ion that forms thus has a spin $j = \frac{5}{2}$ with a spin degeneracy of $2j + 1 = 6$. In multi-electron atoms, the situation can become still more complex, involving Hund's coupling between atoms.

Now the residual spin degrees of freedom still interact with the surrounding conduction sea, for virtual charge fluctuations, in which an electron temporarily migrates off of or onto the ion, lead to spin exchange between the local moment and the conduction sea. There are two such virtual processes:

$$
\begin{aligned}
e_\uparrow + f_\downarrow^1 &\leftrightarrow f^2 \leftrightarrow e_\downarrow + f_\uparrow^1 & \Delta E_I &\sim U + E_f \\
e_\uparrow + f_\downarrow^1 &\leftrightarrow e_\uparrow + e_\downarrow \leftrightarrow e_\downarrow + f_\uparrow^1 & \Delta E_{II} &\sim -E_f.
\end{aligned}
\tag{16.89}
$$

In both cases, spin exchange only takes place in the singlet channel, $S = 0$, state. From second-order perturbation theory, we know that these virtual charge fluctuations will selectively lower the energy of the singlet configurations by an amount of order $\Delta E = -J$, where

$$
J \sim V^2 \left[\frac{1}{\Delta E_1} + \frac{1}{\Delta E_2} \right] = V^2 \left[\frac{1}{-E_f} + \frac{1}{E_f + U} \right].
\tag{16.90}
$$

Here V is the size of the hybridization matrix element near the Fermi surface. The selective reduction in the energy of the singlet channel constitutes an effective antiferromagnetic interaction between the conduction electrons and the local moment. If we introduce $\vec{\sigma}(0) = \sum_{k,k'} c_{k\alpha}^\dagger \vec{\sigma}_{\alpha\beta} c_{k'\beta}$, measuring the electron spin at the origin, then the effective interaction that lowers the energy of singlet combinations of conduction and f-electrons will have the form $H_{eff} \sim J\vec{\sigma}(0) \cdot \vec{S}_f$. The resulting low-energy Hamiltonian that describes the interaction of a spin with a conduction sea is the deceptively simple Kondo model:

$$
H = \sum_{k\sigma} \epsilon_k c_{k\sigma}^\dagger c_{k\sigma} + \overbrace{J\psi^\dagger(0)\vec{\sigma}\,\psi(0) \cdot \vec{S}_f}^{\Delta H}.
\tag{16.91}
$$

Kondo model

This heuristic argument was ventured in 1961 in Anderson's paper [7] on local moment formation. At the time, the antiferromagnetic sign in this interaction was entirely unexpected, for it had long been assumed that exchange forces always induce a ferromagnetic interaction between the conduction sea and local moments. The innocuous-looking sign difference has deep consequences for the physics of local moments at low temperatures, giving rise to an interaction that grows as the temperature is lowered, ultimately leading to a final crossover into a low-energy Fermi liquid fixed point. The remaining sections of this chapter are devoted to following this process in detail.

16.8 Schrieffer–Wolff transformation

We now carry out the transformation that links the Anderson and Kondo models via a canonical transformation, first introduced by Schrieffer and Wolff [10, 11]. This transformation is a kind of one-step renormalization process in which the valence fluctuations are integrated out of the Anderson model. When a local moment forms, hybridization with the conduction sea induces virtual charge fluctuations. It's useful to consider dividing the Hamiltonian into two terms,

$$H = H_1 + \lambda \mathcal{V},$$

where λ is an expansion parameter. Here,

$$H_1 = H_{band} + H_{atomic} = \left[\begin{array}{c|c} H_L & 0 \\ \hline 0 & H_H \end{array} \right]$$

is diagonal in the low-energy f^1 (H_L) and the high-energy f^2 or f^0 (H_H) subspaces, whereas the hybridization term

$$\mathcal{V} = H_{mix} = \sum_{\mathbf{k}\sigma} [V_{\vec{k}} c^\dagger_{\mathbf{k}\sigma} f_\sigma + \text{H.c.}] = \left[\begin{array}{c|c} 0 & V^\dagger \\ \hline V & 0 \end{array} \right]$$

provides the off-diagonal matrix elements between these two subspaces. The idea of the Schrieffer–Wolff transformation is to carry out a canonical transformation that returns the Hamiltonian to block-diagonal form:

$$\mathcal{U} \left[\begin{array}{c|c} H_L & \lambda V^\dagger \\ \hline \lambda V & H_H \end{array} \right] \mathcal{U}^\dagger = \left[\begin{array}{c|c} H^* & 0 \\ \hline 0 & H' \end{array} \right]. \tag{16.92}$$

This is a *renormalized* Hamiltonian, and the block-diagonal part of this matrix, $H^* = P_L H' P_L$, in the low-energy subspace provides an *effective* Hamiltonian for the low-energy physics. We now set $\mathcal{U} = e^S$, referring to S as the *action operator*. Now since $\mathcal{U}^\dagger = \mathcal{U}^{-1} = e^{-S}$, this implies that the action operator $S^\dagger = -S$ is anti-Hermitian. Writing S as a power series in λ,

$$S = \lambda S_1 + \lambda^2 S_2 + \cdots,$$

and using the identity $e^A B e^{-A} = B + [A, B] + \frac{1}{2!}[A, [A, B]] + \cdots$, (16.92) can be expanded in powers of λ as follows:

$$e^S(H_1 + \lambda \mathcal{V})e^{-S} = H_1 + \lambda \left(\mathcal{V} + [S_1, H_1] \right) + \lambda^2 \left(\frac{1}{2}[S_1, [S_1, H]] + [S_1, \mathcal{V}] + [S_2, H_1] \right) + \cdots.$$

Since $\mathcal{V}$ is not diagonal, by requiring

$$[S_1, H_1] = -\mathcal{V}, \tag{16.93}$$

we can eliminate all off-diagonal components to leading order in λ. To second order,

$$e^S(H_1 + \lambda \mathcal{V})e^{-S} = H_1 + \lambda^2 \left(\frac{1}{2}[S_1, \mathcal{V}] + [S_2, H_1] \right) + \cdots. \tag{16.94}$$

Since $[S_1, \mathcal{V}]$ is block-diagonal, we can satisfy (16.92) to second order by requiring $S_2 = 0$, so that, to this order, the renormalized Hamiltonian has the form

$$H^* = H_L + \lambda^2 \Delta H, \tag{16.95}$$

where

$$\Delta H = \frac{1}{2} P_L [S_1, \mathcal{V}] P_L + \cdots \tag{16.96}$$

is an interaction term induced by virtual fluctuations into the high-energy manifold. Writing the action operator in matrix form,

$$S = \left[\begin{array}{c|c} 0 & -s^\dagger \\ \hline s & 0 \end{array}\right], \tag{16.97}$$

and substituting into (16.93), we obtain $V = -sH_L + H_H s$. Now since $(H_L)_{ab} = E_a^L \delta_{ab}$ and $(H_H)_{ab} = E_a^H \delta_{ab}$ are diagonal, it follows that

$$s_{ab} = \frac{V_{ab}}{E_a^H - E_b^L}, \qquad -s_{ab}^\dagger = \frac{V_{ab}^\dagger}{E_a^L - E_b^H}, \tag{16.98}$$

or, more schematically,

$$S = \sum_{H,L} \left(|H\rangle \frac{\langle H|V|L\rangle}{E_H - E_L} \langle L| - \text{H.c.} \right) + O(V^3). \tag{16.99}$$

From (16.98) we obtain

$$\Delta H_{LL'} = -\frac{1}{2}(V^\dagger s + s^\dagger V)_{LL'} = -\frac{1}{2}\sum_H (V_{LH}^\dagger V_{HL'}) \left[\frac{1}{E_H - E_L} + \frac{1}{E_H - E_{L'}} \right]. \tag{16.100}$$

Some important points about this result:

- We recognize (16.100) as a simple generalization of second-order perturbation theory, including off-diagonal matrix elements by averaging over initial- and final-state energy denominators.
- ΔH can also be written

$$\Delta H_{LL'} = \frac{1}{2}[T(E_L) + T(E_{L'})],$$

where

$$\hat{T}(E) = P_L V \frac{P_H}{E - H_1} V P_L$$

$$T_{LL'}(E) = \sum_{|H\rangle} \left[\frac{V_{LH}^\dagger V_{HL'}}{E - E_H} \right] \tag{16.101}$$

is the leading-order expression for the many-body scattering t-matrix induced by scattering off V. We can thus relate ΔH to a scattering amplitude, and schematically represent it by a Feynman diagram, illustrated in Figure 16.15.
- If the separation of the low- and high-energy subspaces is large, we can take $E_L \sim E_{L'}$, so that

$$\Delta H = T(E_L) = -\frac{1}{\Delta E_{HL}}(V P_H V), \tag{16.102}$$

where $\Delta E_{HL} = E_H - E_L$ is the energy of excitation into the high-energy subspace and $P_H = \sum_{|H\rangle} |H\rangle\langle H|$.

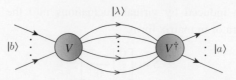

Fig. 16.15   A t-matrix representation of the interaction induced between states $|b\rangle$ and $|a\rangle$ by integrating out the virtual fluctuations into the high-energy states $|\lambda\rangle$.

We now apply this method to the Anderson model for which the atomic ground state is a local-moment f^1 configuration. In this case, there are two high-energy intermediate states, corresponding to f^0 and f^2 configurations. When a conduction electron or hole is excited into the localized f-state to create these excited-state configurations, the corresponding excitation energies are $\Delta E(f^1 \to f^0) = -E_f$ and $\Delta E(f^1 \to f^2) = E_f + U$. The hybridization $\mathcal{V} = \sum_{\mathbf{k}\sigma} \left[V(\mathbf{k}) c^{\dagger}_{\mathbf{k}\sigma} f_\sigma + \text{H.c.} \right]$ generates virtual fluctuations into these excited states. Using (16.102), the interaction induced by these fluctuations is given by

$$\Delta H = -\frac{VP[f^2]V}{E_f + U} - \frac{VP[f^0]V}{-E_f}$$

$$= -\sum_{k\alpha, k'\beta} V^*_{k'} V_k \left[\frac{\overbrace{(c^{\dagger}_{k\alpha} f_\alpha)(f^{\dagger}_\beta c_{k'\beta})}^{f^1 + e^- \leftrightarrow f^2}}{E_f + U} + \frac{\overbrace{(f^{\dagger}_\beta c_{k'\beta})(c^{\dagger}_{k\alpha} f_\alpha)}^{f^1 \leftrightarrow f^0 + e^-}}{-E_f} \right] P_{n_f = 1}, \qquad (16.103)$$

where $P_{n_f = 1} = (n_{f\uparrow} - n_{f\downarrow})^2$ projects into the subspace of unit occupancy. Using the Fierz identity[6] $2\delta_{\alpha\gamma}\delta_{\eta\beta} = \delta_{\alpha\beta}\delta_{\eta\gamma} + \vec{\sigma}_{\alpha\beta} \cdot \vec{\sigma}_{\eta\gamma}$, we may recast the spin exchange terms in terms of Pauli matrices, as follows:

$$(c^{\dagger}_{k\alpha} f_\alpha)(f^{\dagger}_\beta c_{k'\beta}) = (c^{\dagger}_{k\alpha} f_\gamma)(f^{\dagger}_\eta c_{k'\beta}) \times \overbrace{(\delta_{\alpha\gamma}\delta_{\eta\beta})}^{\frac{1}{2}(\delta_{\alpha\beta}\delta_{\eta\gamma} + \vec{\sigma}_{\alpha\beta} \cdot \vec{\sigma}_{\eta\gamma})}$$

$$= \frac{1}{2} c^{\dagger}_{k\alpha} c_{k'\alpha} - (c^{\dagger}_{k\alpha} \vec{\sigma}_{\alpha\beta} c_{k'\beta}) \cdot \vec{S}_f, \qquad (16.104)$$

and similarly

$$(f^{\dagger}_\beta c_{k'\beta})(c^{\dagger}_{k\alpha} f_\alpha) = -\frac{1}{2} c^{\dagger}_{k\alpha} c_{k'\alpha} - (c^{\dagger}_{k\alpha} \vec{\sigma}_{\alpha\beta} c_{k'\beta}) \cdot \vec{S}_f \qquad (16.105)$$

(where we have replaced $n_f = 1$ and dropped residual constants in both cases). The operator

$$\vec{S}_f \equiv f^{\dagger}_\sigma \left(\frac{\vec{\sigma}_{\alpha\beta}}{2} \right) f_\beta, \qquad (n_f = 1) \qquad (16.106)$$

describes the spin of the f-electron. The induced interaction is then

$$\Delta H = \sum_{k\alpha, k'\beta} J_{k,k'} c^{\dagger}_{k\alpha} \vec{\sigma}_{\alpha\beta} c_{k'\beta} \cdot \vec{S}_f + H', \qquad (16.107)$$

---

[6] This identity is obtained by expanding an arbitrary two-dimensional matrix A in terms of Pauli matrices. If we write $A_{\alpha\beta} = \frac{1}{2}\text{Tr}[A\underline{1}]\delta_{\alpha\beta} + \frac{1}{2}\text{Tr}[A\vec{\sigma}] \cdot \vec{\sigma}_{\alpha\beta}$ and read off the coefficients of A inside the traces, we obtain the inequality.

where

$$
J_{k,k'} = V_{k'}^* V_k \left[\overbrace{\frac{1}{E_f + U}}^{f^1 + e^- \leftrightarrow f^2} + \overbrace{\frac{1}{-E_f}}^{f^1 \leftrightarrow f^0 + e^-} \right] \tag{16.108}
$$

is the Kondo coupling constant.

Notice how, in the low-energy subspace, the occupancy of the f-state is constrained to $n_f = 1$. This fermionic representation (16.106) of the spin operator proves to be very useful. Apart from a constant, the second term,

$$
H' = -\frac{1}{2} \sum_{k,k'\sigma} V_{k'}^* V_k \left[\frac{1}{E_f + U} + \frac{1}{E_f} \right] c_{k\sigma}^\dagger c_{k'\sigma},
$$

is a residual potential scattering term. This term vanishes for the particle–hole symmetric case $E_f = -(E_f + U)$, and will be dropped since it does not involve the internal dynamics of the local moment. Summarizing, the effect of the high-frequency valence fluctuations is to induce an antiferromagnetic coupling between the local spin density of the conduction electrons and the local moment:

$$
H = \sum_{k\sigma} \epsilon_k c_{k\sigma}^\dagger c_{k\sigma} + \sum_{k,k'} J_{k,k'} c_{k\alpha}^\dagger \vec{\sigma} c_{k'\beta} \cdot \vec{S}_f. \tag{16.109}
$$

This is the famous Kondo model. For many purposes, the k-dependence of the coupling constant can be dropped, so that the model takes the simpler form shown in (16.91):

$$
H = \sum_{k\sigma} \epsilon_k c_{k\sigma}^\dagger c_{k\sigma} + \overbrace{J \vec{\sigma}(0) \cdot \vec{S}_f}^{\Delta H}, \tag{16.110}
$$

Kondo model

where $\psi_\alpha(0) = \sum_k c_{k\alpha}$ is the electron operator at the origin and $\vec{\sigma}(0) = \psi^\dagger(0)\vec{\sigma}\psi(0)$ is the spin density at the origin. In other words, there is a simple point interaction between the spin density of the metal at the origin and the local moment.

Example 16.4 Details of the Schrieffer–Wolff transformation.
Show that the action operator S for the canonical transformation $H \rightarrow H^* = e^S H e^{-S}$ that effects the Schrieffer–Wolff transformation from the Anderson model (16.5) to the Kondo model (16.110) is given by [10, 11]

$$
S = \sum_{k,\sigma} \left[V_k c_{k\sigma}^\dagger f_\sigma \left(\frac{1 - n_{f-\sigma}}{\epsilon_k - E_f} + \frac{n_{f\sigma}}{\epsilon_k - (E_f + U)} \right) - \text{H.c.} \right] + O(V^3). \tag{16.111}
$$

Solution

Using (16.98), we may write the action operator S as

$$S = \sum_{H,L} \left[|H\rangle \frac{\langle H|V|L\rangle}{E_H - E_L} \langle L| - \text{H.c.} \right] + O(V^3), \tag{16.112}$$

where, L and H denote the low- and high-energy subspaces, respectively. For the Anderson model, $\hat{V} = \sum_{\mathbf{k},\sigma} (V_{\mathbf{k}} c^\dagger_{\mathbf{k}\sigma} f_\sigma + \text{H.c.})$ is the hybridization, while the low-energy Hilbert space is the states with $n_f = 1$. The projector into the low-energy subspace H is $P_L = (n_{f\uparrow} - n_{f\downarrow})^2$, so we may write $\langle H|V|L\rangle = \langle H|VP_L|L\rangle$, so that

$$S = \sum_{H,L,\mathbf{k}\sigma} \left(|H\rangle \langle H| \frac{(V_{\mathbf{k}} c^\dagger_{\mathbf{k}\sigma} f_\sigma + V^*_{\mathbf{k}} f^\dagger_\sigma c_{\mathbf{k}\sigma})(n_{f\uparrow} - n_{f\downarrow})^2}{E_H - E_L} |L\rangle \langle L| - \text{H.c.} \right) + O(V^3). \tag{16.113}$$

Now the initial state has energy E_f while the excited state is either a state with one conduction electron and no f-electrons, with energy $\epsilon_{\mathbf{k}}$, or a state with two f-electrons and one conduction hole, with energy $2E_f + U - \epsilon_{\mathbf{k}}$, so we may write

$$S = \sum_{\mathbf{k}\sigma} \left[V_{\mathbf{k}} \frac{c^\dagger_{\mathbf{k}\sigma} f_\sigma (1 - n_{f\,-\sigma})}{\epsilon_{\mathbf{k}} - E_f} + V^*_{\mathbf{k}} \frac{f^\dagger_\sigma c_{\mathbf{k}\sigma} n_{f\,-\sigma}}{E_f + U - \epsilon_{\mathbf{k}}} - \text{H.c.} \right] + O(V^3), \tag{16.114}$$

where we have replaced $f_\sigma(n_{f\,\uparrow} - n_{f\downarrow})^2 = f_\sigma(1 - n_{f\,-\sigma})^2 = f_\sigma(1 - n_{f\,-\sigma})$ and $f^\dagger_\sigma(n_{f\,\uparrow} - n_{f\downarrow})^2 = f^\dagger_\sigma n^2_{f\,-\sigma} = f^\dagger_\sigma n_{f\,-\sigma}$. Rearranging this a little, we obtain

$$S = \sum_{\mathbf{k},\sigma} \left[V_{\mathbf{k}} \left((1 - n_{f\,-\sigma}) \frac{c^\dagger_{\mathbf{k}\sigma} f_\sigma}{\epsilon_{\mathbf{k}} - E_f} + n_{f\,-\sigma} \frac{c^\dagger_{\mathbf{k}\sigma} f_\sigma}{\epsilon_{\mathbf{k}} - (E_f + U)} \right) - \text{H.c.} \right] + O(V^3). \tag{16.115}$$

Example 16.5   Composite nature of the f-electron.
The Kondo model only involves the spin of the f-electron, and the f-creation and annihilation operators have apparently completely disappeared. To find out what has happened to them, consider adding a source term for the f-electrons,

$$H_S = \sum_\sigma (f^\dagger_\sigma \eta_\sigma + \bar{\eta}_\sigma f)_\sigma, \tag{16.116}$$

into the Anderson impurity model, so that now $H \to H[\bar{\eta}, \eta] = H + H_S$, so that the functional derivatives of the partition function

$$Z[\eta_\sigma, \eta_\sigma] = Z_0 \left\langle T \exp\left[-\int_0^\beta d\tau f^\dagger_\sigma(\tau) \eta_\sigma(\tau) + \bar{\eta}_\sigma(\tau) f_\sigma(\tau) \right] \right\rangle \tag{16.117}$$

generate correlation functions of the fermion operators:

$$\frac{\delta}{\delta \bar{\eta}_\sigma} \to f_\sigma(\tau), \qquad -\frac{\delta}{\delta \eta_\sigma} \to f^\dagger_\sigma(\tau). \tag{16.118}$$

(a) Repeat the Schrieffer–Wolff transformation for the case of constant hybridization $V_{\mathbf{k}} = V$ and particle–hole symmetry to show that the Kondo model with source terms now becomes

$$H_K[\bar{\eta}, \eta] = \sum_{\mathbf{k}\sigma} \epsilon_{\mathbf{k}} c_{\mathbf{k}\sigma}^{\dagger} c_{\sigma} + J\left(\psi^{\dagger}(0) + V^{-1}\bar{\eta}\right)\vec{\sigma}\left(\psi(0) + V^{-1}\eta\right)\cdot\vec{S}. \quad (16.119)$$

(b) By differentiating this expression with respect to $\bar{\eta}_{\sigma}$, show that in the Kondo model the original f-electron operator has now become a *composite operator* involving a combined conduction electron and spin-flip, as follows:

$$f_{\alpha} \equiv \frac{\delta H_K[\bar{\eta}, \eta]}{\delta\bar{\eta}_{\alpha}} = \frac{J}{V}\left(\sigma_{\alpha\beta}\cdot\vec{S}\right)\psi(0)_{\beta}. \quad (16.120)$$

When a Fermi liquid develops, it is this object that behaves like a resonant bound-state fermion.

Solution

(a) In the Anderson model, we can absorb the source term into the hybridization, writing it in the form

$$V = \sum \left(V\psi_{\sigma}^{\dagger}(0) + \bar{\eta}_{\sigma}\right)f_{\sigma} + \text{H.c.}, \quad (16.121)$$

so that in the hybridization we have replaced $\psi_{\sigma}(0) \rightarrow \psi_{\sigma}(0) + \frac{1}{V}\eta_{\sigma}$. If we now repeat the Schrieffer–Wolff transformation, the spin exchange term in the Kondo model takes the form

$$H_K[\bar{\eta}, \eta] = \sum_{\mathbf{k}\sigma} \epsilon_{\mathbf{k}} c_{\mathbf{k}\sigma}^{\dagger} c_{\sigma} + J\left(\psi^{\dagger}(0) + V^{-1}\bar{\eta}\right)\vec{\sigma}\left(\psi(0) + V^{-1}\eta\right)\cdot\mathbf{S}. \quad (16.122)$$

(b) If we now differentiate H_K with respect to $\bar{\eta}$, we obtain

$$f_{\sigma} \equiv \frac{\delta H_K[\bar{\eta}, \eta]}{\delta\eta_{\sigma}}\bigg|_{\eta,\bar{\eta}=0} = \frac{J}{V}\left[(\vec{\sigma}\cdot\vec{S})\psi(0)\right]_{\sigma}. \quad (16.123)$$

16.9 "Poor man's" scaling

We now apply the scaling concept to the Kondo model. This was originally carried out by Anderson and Yuval [12–14] using a method formulated in the time rather than the energy domain. The method presented here follows Anderson's "poor man's" scaling approach [31, 32], in which the evolution of the coupling constant is followed as the bandwidth of the conduction sea is reduced. The Kondo model is written

$$H = \sum_{|\epsilon_k|<D} \epsilon_k c_{k\sigma}^{\dagger} c_{k\sigma} + H^{(I)}$$

$$H^{(I)} = J(D) \sum_{|\epsilon_k|,|\epsilon_{k'}|<D} c_{k\alpha}^\dagger \vec{\sigma}_{\alpha\beta} c_{k'\beta} \cdot \vec{S}_f, \tag{16.124}$$

where the density of conduction electron states $\rho(\epsilon)$ is taken to be constant. The poor man's renormalization procedure follows the evolution of $J(D)$ that results from reducing D by progressively integrating out the electron states at the edge of the conduction band. In the poor man's procedure, the bandwidth is not rescaled to its original size after each renormalization. This avoids the need to renormalize the electron operators so that, instead of (16.84), $H(D') = \tilde{H}_L$.

To carry out the renormalization procedure, we integrate out the high-energy spin fluctuations using the t-matrix formulation for the induced interaction ΔH, derived in the previous section. Formally, the induced interaction is given by

$$\Delta H_{ab} = \frac{1}{2}[T_{ab}(E_a) + T_{ab}(E_b)],$$

where

$$T_{ab}(E) = \sum_{\lambda \in |H\rangle} \left[\frac{H_{a\lambda}^{(I)} H_{\lambda b}^{(I)}}{E - E_\lambda^H} \right],$$

where the energy of state $|\lambda\rangle$ lies in the range $[D', D]$. There are two possible intermediate states that can be produced by the action of $H^{(I)}$ on a one-electron state: either (I) the electron state is scattered directly or (II) a virtual electron–hole pair is created in the intermediate state. In process I, the t-matrix can be represented by the Feynman diagram

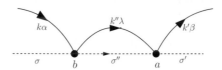

for which the t-matrix for scattering into a high-energy electron state is

$$T^{(I)}(E)_{k'\beta\sigma';k\alpha\sigma} = \sum_{\epsilon_{k''}\in[D-\delta D,D]} \left[\frac{1}{E - \epsilon_{k''}} \right] J^2(\sigma^a\sigma^b)_{\beta\alpha}(S^aS^b)_{\sigma'\sigma}$$

$$\approx J^2\rho\delta D \left[\frac{1}{E - D} \right] (\sigma^a\sigma^b)_{\beta\alpha}(S^aS^b)_{\sigma'\sigma}. \tag{16.125}$$

In process II,

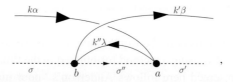

the formation of a particle–hole pair involves a conduction electron line that crosses itself, leading to a negative sign. Notice how the spin operators of the conduction sea and antiferromagnet reverse their relative order in process II, so that the t-matrix for scattering into a high-energy hole state is given by

$$T^{(II)}(E)_{k'\beta\sigma';k\alpha\sigma} = - \sum_{\epsilon_{k''}\in[-D,-D+\delta D]} \left[\frac{1}{E-(\epsilon_k+\epsilon_{k'}-\epsilon_{k''})} \right] J^2 (\sigma^b\sigma^a)_{\beta\alpha} (S^aS^b)_{\sigma'\sigma}$$

$$= -J^2\rho\delta D \left[\frac{1}{E-D} \right] (\sigma^b\sigma^a)_{\beta\alpha} (S^aS^b)_{\sigma'\sigma}, \qquad (16.126)$$

where we have assumed that the energies ϵ_k and $\epsilon_{k'}$ are negligible compared with D. Adding (16.125) and (16.126) gives

$$\delta H^{int}_{k'\beta\sigma';k\alpha\sigma} = \hat{T}^{(I)} + \hat{T}^{(II)} = -\frac{J^2\rho|\delta D|}{D} [\sigma^a,\sigma^b]_{\beta\alpha} S^aS^b$$

$$= -\frac{1}{2}\frac{J^2\rho|\delta D|}{D} \overbrace{[\sigma^a,\sigma^b]_{\beta\alpha}}^{2i\epsilon^{abc}\sigma^c} \overbrace{[S^a,S^b]}^{i\epsilon^{abd}S^d}$$

$$= \frac{J^2\rho|\delta D|}{D} \epsilon^{abc}\epsilon^{abd} \overbrace{\sigma^c_{\beta\alpha}S^d}^{2\delta_{cd}}$$

$$= 2\frac{J^2\rho|\delta D|}{D} \vec{\sigma}_{\beta\alpha}\cdot\vec{S}_{\sigma'\sigma}. \qquad (16.127)$$

In this way we see that the virtual emission of a high-energy electron and hole generates an antiferromagnetic correction to the original Kondo coupling constant:

$$J(D-|\delta D|) = J(D) + 2J^2\rho\frac{|\delta D|}{D} = J(D) - 2J^2\rho\frac{\delta D}{D}, \qquad (16.128)$$

since we have reduced the bandwidth, $\delta D = -|\delta D|$. In other words,

$$\frac{\partial J\rho}{\partial \ln D} = -2(J\rho)^2 \qquad (16.129)$$

or, in terms of the dimensionless coupling constant $g = \rho J$,

$$\frac{\partial g}{\partial \ln D} = \beta(g) = -2g^2 + O(g^3). \qquad (16.130)$$

Now since $\beta(g=0) = 0$ at $g=0$, scaling comes to halt: we say that $g=0$ is a *fixed point*. It is instructive to rewrite the scaling equation in the form

$$\frac{\partial \ln g}{\partial \ln(D_0/D)} = 2g + O(g^2), \qquad (16.131)$$

where D_0 is the initial bandwidth. From this form, we see that the direction of the scaling depends on the sign of $g = J\rho$ (see Figure 16.16). As we reduce the size D of the cut-off,

- for antiferromagnetic $g > 0$, the magnitude of g grows. We say that the fixed point is *repulsive*. In other words, spin fluctuations antiscreen the antiferrromagnetic interaction, causing it to grow at low energies.
- for ferromagnetic $g < 0$, scaling reduces the magnitude of g, driving one ever closer to the weak coupling fixed point at $g = 0$. In this case, the fixed point is attractive and spin fluctuations screen the interaction, causing J to scale logarithmically slowly to zero at low energies.

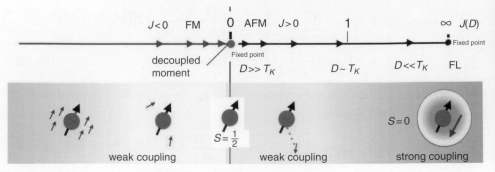

Fig. 16.16 Schematic illustration of renormalization group flow in the Kondo model. For $J < 0$ (ferromagnetic), the coupling constant scales to an attractive fixed point at $J = 0$, forming a decoupled local moment. For $J > 0$ (antiferromagnetic), scaling proceeds from a repulsive weak-coupling fixed point, via a crossover to an attractive strong-coupling fixed point in which the local moment is screened by the conduction electrons, removing its internal degrees of freedom to form a Fermi liquid. In the diagram, FM = ferromagnet, AFM = antiferromagnet, and FL = Fermi liquid.

To examine these two cases in more detail, we integrate the scaling equation between the initial bandwidth D_0 and D', writing

$$\int_{g_0}^{g(D)} \frac{dg'}{g'^2} = -2 \int_{\ln D_0}^{\ln D'} d \ln D'' \tag{16.132}$$

or

$$\left(\frac{1}{g_0} - \frac{1}{g(D')} \right) = -2 \ln(D'/D_0), \tag{16.133}$$

where $g_0 = J\rho = g(D_0)$ is the unrenormalized coupling constant at the original bandwidth D_0. In this case,

$$g(D') = \frac{g_0}{1 - 2g_0 \ln \frac{D_0}{D'}}. \tag{16.134}$$

Let us look at the ferromagnetic and antiferromagnetic cases seperately.

Ferromagnetic interaction, $g < 0$

In this case,

$$g(D') = -\frac{|g_0|}{1 + 2|g_0| \ln(D_0/D')}, \tag{16.135}$$

which corresponds to a very gradual decoupling of the local moment from the surrounding conduction sea. The interaction is said to be marginally irrelevant, because it scales logarithmically to zero, and at all scales the problem remains perturbative.

Antiferromagnetic interaction, $g > 0$

For the antiferromagnetic case ($g > 0$), the solution to the scaling equation is

$$g(D') = \frac{g_o}{1 - 2g_o \ln(D/D')} = \frac{1}{2} \frac{1}{\ln(D'/D_0) + \frac{1}{2g_0}} = \frac{1}{2} \frac{1}{\ln\left[\frac{D'}{D_0 \exp(-1/(2g_0))}\right]}, \quad (16.136)$$

where we have divided numerator and denominator by $2g_0$. It follows that

$$2g(D') = \frac{1}{\ln(D'/T_K)},$$

where we have introduced the Kondo temperature

$$T_K = D_0 \exp\left[-\frac{1}{2g_o}\right]. \quad (16.137)$$

The Kondo temperature T_K is an example of a dynamically generated scale.

Were we to take this equation literally, we would say that g diverges at the scale $D' = T_K$. This interpretation is too literal, because the scaling has only been calculated to order g^2. Nevertheless it does show that the Kondo interaction can only be treated perturbatively at energies large compared with the Kondo temperature. We also see that, once we have written the coupling constant in terms of the Kondo temperature, all reference to the original cut-off energy scale vanishes from the expression. This cut-off independence of the problem is an indication that the physics of the Kondo problem does not depend on the high-energy details of the model: there is only one relevant energy scale, the Kondo temperature.

By calculating the higher-order diagrams shown in Figure 16.17, it is straightforward, though somewhat technical (see Exercise 16.8), to show that the beta function to order g^3 is given by

$$\frac{\partial g}{\partial \ln D} = \beta(g) = -2g^2 + 2g^3 + O(g^4). \quad (16.138)$$

One can integrate this equation to obtain

$$\ln\left(\frac{D'}{D}\right) = \int_{g_o}^{g} \frac{dg'}{\beta(g')} = -\frac{1}{2} \int_{g_o}^{g} dg' \left[\frac{1}{g'^2} + \frac{1}{g'} + O(1)\right], \quad (16.139)$$

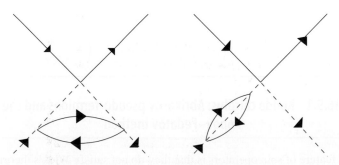

Diagrams contributing to the third-order term in the beta function. See Exercise 16.8.

Fig. 16.17

where we have expanded the numerator in $\frac{1}{\beta(g)} \approx -\frac{1}{g^2(1-g)} - \frac{1}{g^2}(1 + g) + O(1)$. A better estimate of the temperature T_K where the system scales to strong coupling is obtained by setting $D' = T_K$ and $g = 1$ in this equation, which gives

$$\ln\left(\frac{T_K}{D}\right) = -\frac{1}{2}\int_{g_0}^{1} dg' \left[\frac{1}{g'^2} + \frac{1}{g'}\right] = -\frac{1}{2g_0} + \frac{1}{2}\ln 2g_0 + O(1), \tag{16.140}$$

where we have dropped the terms of order unity on the right-hand side. Thus, up to a prefactor, the dependence of the Kondo temperature on the bare coupling constant is given by

$$T_K = D_0\sqrt{2g_0}e^{-\frac{1}{2g_0}}. \tag{16.141}$$

The square-root $\sqrt{g_0}$ dependence on the coupling constant is often dropped in qualitative discussions, but it is important for more quantitative comparisons.

Example 16.6   Consider the symmetric Anderson model. At energy scales greater than $U/2$ the impurity is mixed-valent. However, once the cut-off $D \sim U/2$, one must carry out a Schrieffer–Wolff transformation.

(a) Show that the Kondo coupling constant of the symmetric Anderson model is $g_0 = J\rho = 4\Delta/(\pi U)$, where $\Delta = \pi\rho V^2$ is the bare resonant level width of the Anderson model.

(b) Using (16.141) with a cut-off $D = U/2$, derive the following form (16.72) for the Kondo temperature of the symmetric Anderson model:

$$T_K = \sqrt{\frac{2U\Delta}{\pi}} \exp\left(-\frac{\pi U}{8\Delta}\right).$$

Solution

(a) In the symmetric Anderson model, $E_f = -U/2$. Assuming that the hybridization $V_k = V$ is constant, then from (16.108) the Kondo coupling constant is given by

$$g_0 = J\rho = V^2\rho \left[\frac{1}{E_f + U/2} + \frac{1}{-E_f}\right] = 4\frac{V^2\rho}{U} = \frac{4\Delta}{\pi U}, \tag{16.142}$$

where $\Delta = \pi\rho V^2$.

(b) Using (16.141) with $D = U/2$, we obtain

$$T_K = \frac{U}{2}\sqrt{\frac{8\Delta}{\pi U}} \exp\left(-\frac{8\Delta}{\pi U}\right) = \sqrt{\frac{2U\Delta}{\pi}} \exp\left(-\frac{\pi U}{8\Delta}\right). \tag{16.143}$$

16.9.1   Kondo calculus: Abrikosov pseudo-fermions and the Popov–Fedatov method

The awkward feature of spin operators is that they do not satisfy Wick's theorem, so that we cannot treat them directly in a Feynman diagram expansion. Kondo calculus requires that we overcome this difficulty, and a variety of methods have been developed. One tool that is particularly useful is the Abrikosov pseudo-fermion representation [44], in which the spin operator, is factorized in terms of a spin-$\frac{1}{2}$ fermion field $f_\sigma^\dagger$, as follows:

$$\vec{S} = f_\alpha^\dagger \left(\frac{\vec{\sigma}}{2}\right)_{\alpha\beta} f_\beta. \tag{16.144}$$

This has the advantage that one can now take advantage of Wick's theorem. In Abrikosov's representation of a spin-$\frac{1}{2}$ operator, the up and down states are now represented by the states

$$|\sigma\rangle = f_\sigma^\dagger |0\rangle \qquad (\sigma = \uparrow, \downarrow). \tag{16.145}$$

However, in using the f-electron one has inadvertently expanded the Hilbert space, introducing two unphysical states: the empty state $|0\rangle$ and the doubly occupied state $|\uparrow\downarrow\rangle = f_\downarrow^\dagger f_\uparrow^\dagger |0\rangle$, which need to be eliminated by requiring that

$$n_f = 1. \tag{16.146}$$

Conveniently, this constraint commutes with the spin operator, and hence is a constant of the motion, provided the f-electrons only enter as spin operators in the Hamiltonian. An ingenious way of imposing this constraint has been developed by Popov and Fedotov [53]. Their method introduces a complex chemical potential for the pseudo-fermions:

$$\mu = -i\pi \frac{T}{2}.$$

The partition function of the Hamiltonian is written as an unconstrained trace over the conduction and pseudofermion Fock spaces:

$$Z = \mathrm{Tr}\left[e^{-\beta(H + i\pi \frac{T}{2}(n_f - 1))}\right]. \tag{16.147}$$

Now since the Hamiltonian conserves n_f, we can divide this trace into contributions from the d^0, d^1, and d^2 subspaces, as follows:

$$Z = e^{i\pi/2} Z(f^0) + Z(f^1) + e^{-i\pi/2} Z(f^2).$$

But since $S_f = 0$ in the f^2 and d^0 subspaces, $Z(f^0) = Z(f^2)$, so that the contributions to the partition function from these two unwanted subspaces exactly cancel. In the Popov–Fedotov approach, the bare f-propagators have the form

$$\mathcal{G}_f(i\tilde{\omega}_n) = \frac{1}{i\omega_n + \mu} = \frac{1}{i\omega_n - i\pi T/2} = \frac{1}{i2\pi T(n + \frac{1}{4})}, \tag{16.148}$$

corresponding to a shifted Matsubara frequency $\tilde{\omega}_n = 2\pi T(n + \frac{1}{4})$. You can test this method by applying it to a free spin in a magnetic field (see Example 16.7). The method can also be extended to deal with spin-1 operators using $-\mu = i\pi T/3$.

Example 16.7   Test the Popov–Fedotov trick [53]. Consider the magnetization of a free electron with Hamiltonian

$$H = \epsilon_\sigma f_\sigma^\dagger f_\sigma. \tag{16.149}$$

Show that with $\epsilon_\sigma = -\sigma B$ you obtain the wrong field dependence of the magnetization $M = \tanh\left(\frac{B}{2T}\right)$, but that with the Popov–Fedotov form $\epsilon_\sigma = -\sigma B + \frac{i\pi T}{2}$ you recover the Brillouin formula for a free spin,

$$M = \tanh\left(\frac{B}{T}\right). \tag{16.150}$$

Solution

If we write

$$F = -T \sum_{\sigma=\pm} \ln\left[1 + e^{-\beta\epsilon_\sigma}\right], \tag{16.151}$$

then the magnetization is given by

$$M = -\frac{\partial F}{\partial B} = \sum_\sigma \sigma f(\epsilon_\sigma). \tag{16.152}$$

If we evaluate this expression with $\epsilon_\sigma = -\sigma B$, we obtain the wrong form for the magnetization:

$$M = \frac{1}{e^{-\beta B} + 1} - \frac{1}{e^{\beta B} + 1} = \frac{(e^{\beta B} - e^{-\beta B})}{2 + 2\cosh\beta B}. \tag{16.153}$$

We can see the problem: the extra contribution to the partition function from the empty and doubly occupied sites gives $Z = 2 + 2\cosh\beta B$ rather than $Z = 2\cosh\beta B$. If we simplify this expression, we obtain

$$M = \frac{\sinh\beta B}{1 + \cosh\beta B} = \frac{2\sinh(\beta B/2)\cosh(\beta B/2)}{2\cosh^2(\beta B/2)} = \tanh\left(\frac{B}{2T}\right),$$

which has the wrong field dependence. By contrast, if we use $\epsilon_o = -\sigma B + \frac{i\pi T}{2}$ we obtain

$$M = \frac{1}{ie^{-\beta B} + 1} - \frac{1}{ie^{\beta B} + 1} = \frac{(e^{\beta B} - e^{-\beta B})}{2\cosh\beta B} = \tanh\left(\frac{B}{T}\right), \tag{16.154}$$

recovering the Brillouin function.

Example 16.8   Explicitly calculate the Kondo scattering amplitude

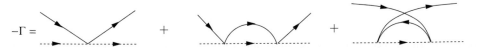

$$-\Gamma = \qquad\qquad + \qquad\qquad + $$

to second order in J, using the Popov–Fedotov scheme. By examining the scattering amplitude on the Fermi surface, show that the Kondo coupling constant is logarithmically enhanced according to the formula

$$J\rho \to J\rho + 2(J\rho)^2 \ln\left[\frac{De^{\pi/2 - \psi(\frac{1}{2})}}{2\pi T}\right], \tag{16.155}$$

where $\psi(x)$ is the digamma function.

Solution

We represent the conduction electron and f-electron propagators by the diagrams

$$\longrightarrow = G_c(i\omega_n, \mathbf{k}) = \frac{1}{i\omega_n - \epsilon_\mathbf{k}}$$

$$\dashrightarrow = G_f(i\omega_n) = \frac{1}{i\omega_n - \lambda_f},$$

$$(16.156)$$

where $\lambda_f \equiv i\frac{\pi T}{2}$ is the imaginary chemical potential that cancels the doubly occupied and empty states. The first-order Kondo scattering amplitude is given by

$$\Gamma_0 = \quad = -J\left(\vec{\sigma}_{\beta\alpha} \cdot \vec{S}_{\sigma'\sigma}\right) \qquad \left(\vec{S}_{\sigma'\sigma} \equiv \left(\frac{\vec{\sigma}}{2}\right)_{\sigma'\sigma}\right).$$

$$(16.157)$$

Here the minus sign derives from the Feynman rules for an interaction vertex, and we have used the shorthand $\vec{S} \equiv \frac{\vec{\sigma}}{2}$ for the f-electron spin matrix elements.

The second-order scattering processes are given by

$$\Gamma_I = \quad = J^2 \left(\sigma^b \sigma^a\right)_{\beta\alpha} \left(S^b S^a\right)_{\sigma'\sigma} \pi_1(i\nu_n)$$

$$(16.158)$$

$$\Gamma_{II} = \quad = -J^2 \left(\sigma^a \sigma^b\right)_{\beta\alpha} \left(S^b S^a\right)_{\sigma'\sigma} \pi_2(i\nu_n),$$

$$(16.159)$$

where $i\nu_n$ is the total energy in the particle–particle and particle–hole channels, respectively. Note the -1 prefactor in Γ_{II} and the inversion of the order of the conduction electron Pauli matrices, which both derive from the crossing of the incoming and outgoing conduction electron lines. Here,

$$= \pi_1(i\nu_n) = T \sum_{i\omega_r, \mathbf{k}} \frac{1}{i\omega_r - \epsilon_\mathbf{k}} \frac{1}{i\nu_n - i\omega_r - \lambda_f},$$

$$(16.160)$$

and

$$= \pi_2(i\nu_n) = -T \sum_{i\omega_r, \mathbf{k}} \frac{1}{i\omega_r - \epsilon_\mathbf{k}} \frac{1}{i\nu_n + i\omega_r - \lambda_f}$$

$$(16.161)$$

are the Kondo polarization bubbles in the particle–particle and particle–hole channels, respectively.

Now let us calculate the polarization bubbles (16.160) and (16.161). If we reverse the sign of $i\omega_r \to -i\omega_r$ in the internal summation in (16.160) we obtain

$$\pi_1(i\nu_n) = -T \sum_{i\omega_r,\mathbf{k}} \frac{1}{i\omega_r + \epsilon_{\mathbf{k}}} \frac{1}{i\nu_n + i\omega_r - \lambda_f}, \tag{16.162}$$

and assuming a particle–hole symmetric conduction electron density of states, we can replace $\epsilon_{\mathbf{k}} \to -\epsilon_{\mathbf{k}}$, so that

$$\pi_1(i\nu_n) = \pi_2(i\nu_n) = -T \sum_{i\omega_r,\mathbf{k}} \frac{1}{i\omega_r - \epsilon_{\mathbf{k}}} \frac{1}{i\nu_n + i\omega_r - \lambda_f}. \tag{16.163}$$

Now this is a well-known fermion bubble, and we can use our standard method of contour integration to carry out the summation over the internal Matsubara frequency $i\omega_r$, to obtain

$$\pi_2(i\nu_n) = \sum_{\mathbf{k}} \frac{f(\lambda) - f(\epsilon_{\mathbf{k}})}{i\nu_n - (\lambda_f - \epsilon_{\mathbf{k}})} = \int d\epsilon \rho(\epsilon) \frac{f(\lambda) - f(\epsilon)}{i\nu_n - (\lambda_f - \epsilon)}, \tag{16.164}$$

where $\rho(\epsilon)$ is the density of states per spin. The summation over energy in this integral is a bit tricky. If we use a flat density of states, then at zero temperature

$$\pi_2(i\nu_n) = \frac{\rho}{2} \int_{-D}^{D} \frac{\mathrm{sgn}\epsilon}{i\nu_n + \epsilon} = \rho \ln\left(\frac{D}{|\nu_n|}\right),$$

so the frequency provides the lower logarithmic cut-off. When we do the calculation at a finite temperature, we expect that if $T \gg |\nu_n|$ then the temperature becomes the cut-off, so that our back-of-the-envelope estimate of this integral is

$$\pi_2(i\nu_n) \sim \rho \ln\left(\frac{D}{\max(|\nu_n|, T)}\right).$$

To calculate the precise form of the integral takes more work, but can be done for a Lorentzian density of states $\rho(\epsilon) = \rho\Phi(\epsilon)$, where $\Phi(x) = D^2/(\epsilon^2 + D^2)$. Here we quote the result (see Appendix 16A):

$$\int d\epsilon \, \Phi(\epsilon) \left(\frac{f(\lambda_f) - f(\epsilon)}{\epsilon - \xi}\right) = \ln \frac{D}{2\pi T} - \psi\left(\frac{1}{2} + \frac{\xi\beta}{2\pi i}\right) - i\frac{\pi}{2}\tanh(\beta\lambda_f/2), \tag{16.165}$$

provided $\mathrm{Im}\,\xi > 0$ (for the opposite sign, one takes the complex conjugate of the above). Putting in $\lambda_f = i\pi T/2$, $\xi = i(\pi T/2 - \nu_n)$, we then obtain

$$\pi_2(i\nu_n) = \rho \int d\epsilon \, \Phi(\epsilon) \left[\frac{f(\lambda_f) - f(\epsilon)}{\epsilon - (\lambda_f - i\nu_n)}\right] = \rho\left[\ln \frac{D}{2\pi T} - \psi\left(\frac{1}{2} + \frac{\pi T/2 - \nu_n}{2\pi T}\right) + \frac{\pi}{2}\right]$$

$$= \rho\left[\ln \frac{De^{\pi/2}}{2\pi T} - \psi\left(\frac{1}{2} + \frac{\pi T/2 - \nu_n}{2\pi T}\right)\right] \qquad (\pi T/2 - \nu_n > 0). \tag{16.166}$$

Strictly speaking, our result only holds for $\mathrm{Im}\,\xi > 0$, i.e. when $\pi/2 - \nu_n = |A| > 0$. The other sign, where $\pi/2 - \nu_n = -|A| < 0$, is obtained by taking the complex conjugate of the result for positive $\pi/2 - \nu_n = |A|$. But since the right-hand side is real, taking the complex conjugate has no effect, so we see that the result only depends on the magnitude $|\pi T/2 - \nu_n|$, enabling us to write

$$\pi_2(i\nu_n) = \rho\left[\ln\frac{De^{\pi/2}}{2\pi T} - \psi\left(\frac{1}{2} + \frac{|\nu_n - \pi T/2|}{2\pi T}\right)\right]. \tag{16.167}$$

Notice that the analytic continuation of this expression contains a branch cut along the line $\mathrm{Im}\, z = i\pi T/2$, a consequence of using a non-Hermitian Hamiltonian (this can be fixed by using a shifted Matsubara frequency for the f-lines). It follows that

$$\pi_2(z) = \begin{cases} \rho\left[\ln\frac{De^{\pi/2}}{2\pi T} - \psi\left(\frac{1}{2} + \frac{z - i\pi T/2}{2\pi iT}\right)\right] & (\mathrm{Im}\, z > \pi T/2) \\ \rho\left[\ln\frac{De^{\pi/2}}{2\pi T} - \psi\left(\frac{1}{2} - \frac{z^* + i\pi T/2}{2\pi iT}\right)\right] & (\mathrm{Im}\, z < \pi T/2). \end{cases} \tag{16.168}$$

Adding up the second-order amplitudes, we obtain

$$-\Gamma(z_{pp}, z_{ph})_{\beta\sigma';\alpha\sigma} = -J\left(\vec{\sigma}_{\beta\alpha} \cdot \vec{S}_{\sigma'\sigma}\right) + J^2\rho[\pi_2(z_{pp})(\sigma^b\sigma^a)_{\beta\alpha} - \pi_2(z_{ph})(\sigma^b\sigma^a)_{\beta\alpha}]\left(S^b S^a\right)_{\sigma'\sigma}. \tag{16.169}$$

Notice that the logarithmically divergent parts of the particle–hole and particle–particle scattering are the same, while the low-energy parts differ by finite amounts. However, if we examine the onshell scattering on the Fermi surface, i.e. with $z_{ph} = z_{pp} = i\pi T/2$, then we obtain

$$-\Gamma = -J\left(\vec{\sigma}_{\beta\alpha} \cdot \vec{S}_{\sigma'\sigma}\right) + J^2\rho\ln\frac{De^{\pi/2 - \psi(1/2)}}{2\pi T}[\sigma^b, \sigma^a]_{\beta\alpha}(S^b S^a)_{\sigma'\sigma}$$

$$= -J\left(\vec{\sigma}_{\beta\alpha} \cdot \vec{S}_{\sigma'\sigma}\right) + J^2\rho\ln\frac{De^{\pi/2 - \psi(1/2)}}{2\pi T} \overbrace{2i\sigma^c_{\beta\alpha}}^{i S^c}\epsilon_{bac}(S^b S^a)_{\sigma'\sigma}$$

$$= -\left[J + 2J^2\rho\ln\frac{De^{\pi/2 - \psi(1/2)}}{2\pi T}\right]\left(\vec{\sigma}_{\beta\alpha} \cdot \vec{S}_{\sigma'\sigma}\right), \tag{16.170}$$

explicitly demonstrating the logarithmic renormalization of the coupling constant.

16.9.2 Universality and the resistance minimum

Provided the Kondo temperature is far smaller than the cut-off, then at low energies it is the only scale governing the physics of the Kondo effect. For this reason, we expect all physical quantities to be expressed in terms of universal functions involving the ratio of the temperature or field to the Kondo scale. For example, the susceptibility

$$\chi(T) = \frac{1}{T}F\left(\frac{T}{T_K}\right) \tag{16.171}$$

and the quasiparticle scattering rate

$$\frac{1}{\tau(T)} = \frac{1}{\tau_o}G\left(\frac{T}{T_K}\right) \tag{16.172}$$

both display universal behavior. If we change the cut-off of the model, adjusting the bare coupling constant g_0 so that T_K is fixed, the physical quantities will be unchanged. If we replace $g_0 \rightarrow g(D)$ in (16.140), then all models with $J(D)\rho = g(D)$, where

$$\ln\left(\frac{T_K}{D}\right) = -\frac{1}{2g(D)} + \frac{1}{2}\ln 2g(D), \qquad (16.173)$$

will have the same Kondo temperature and thus the same low-temperature behavior. However, we can view this another way: as the temperature is lowered, quantum processes become coherent at increasingly lower energies, and the effective cut-off for quantum processes is T. Thus, as the temperature is lowered, the coupling constant g_0 is renormalized to a new value,

$$g_0 \rightarrow g(T), \qquad (16.174)$$

where

$$\ln\left(\frac{T_K}{T}\right) = -\frac{1}{2g(T)} + \frac{1}{2}\ln 2g(T). \qquad (16.175)$$

In this way, lowering the temperature drives the system along the renormalization trajectory from weak to strong coupling.

We can check the existence of universality by examining these properties in the weak-coupling limit, where $T >> T_K$. Here, we find

$$\frac{1}{\tau(T)} = 2\pi J^2 \rho S(S+1) n_i = \frac{2\pi}{\rho} S(S+1) n_i g_0^2 \qquad \left(S = \frac{1}{2}\right) \qquad (16.176)$$

$$\chi(T) = \frac{n_i}{T}[1 - 2J\rho] = \frac{n_i}{T}\left[1 - 2g_0\right], \qquad (16.177)$$

where n_i is the density of impurities.

Now scaling, if it's correct, implies that at lower temperatures $J\rho \rightarrow J\rho + 2(J\rho)^2 \ln\frac{D}{T}$, so that to next leading order we expect

$$\frac{1}{\tau} = n_i \frac{2\pi}{\rho} S(S+1)\left[J\rho + 2(J\rho)^2 \ln\frac{D}{T}\right]^2 \qquad (16.178)$$

$$\chi(T) = \frac{n_i}{T}\left[1 - 2J\rho - 4(J\rho)^2 \ln\frac{D}{T} + O((J\rho)^3)\right]. \qquad (16.179)$$

These results are confirmed in second-order perturbation theory as the result of adding in the one-loop corrections to the scattering vertices. The first result was obtained by Jun Kondo in his pioneering study [9]. Kondo was looking for a consequence of the antiferromagnetic superexchange interaction predicted by Anderson [7], so he computed the electron scattering rate to third order in the magnetic coupling. The logarithm which appears in the electron scattering rate means that, as the temperature is lowered, the rate at which electrons scatter off magnetic impurities rises. It is this phenomenon that gives rise to the famous Kondo resistance minimum.

But we can use universality to go much further, and actually deduce the form of the universal functions $F[x]$ and $G[x]$ in (16.171) and (16.172), at least in weak coupling when the temperature is large compared with the Kondo temperature, $T/T_K >> 1$. Let us rearrange (16.175) into

$$g(T) = \frac{1}{2\ln\left(\frac{T}{T_K}\right) + \ln 2g(T)}, \tag{16.180}$$

which we may iterate to obtain

$$2g(T) = \frac{1}{\ln\left(\frac{T}{T_K}\right) + \frac{1}{2}\ln\left(\frac{1}{\ln\frac{T}{T_K}+\ln 2g}\right)} = \frac{1}{\ln\left(\frac{T}{T_K}\right)} + \frac{\ln(\ln(T/T_K))}{2\ln^2\left(\frac{T}{T_K}\right)} + \cdots, \tag{16.181}$$

where the expansion has been made assuming $\ln T/T_K >> \ln g$. At high temperature, by substituting $T_K = De^{-1/2J\rho}$ we can check that the leading-order term is simply

$$2g(T) = \frac{1}{\ln\left(\frac{T}{T_K}\right)} = \frac{1}{\frac{1}{2J\rho} + \ln T/D} = \frac{2J\rho}{1 + 2J\rho\ln T/D}$$

$$= 2\left[J\rho + 2(J\rho)^2\ln\left(\frac{D}{T}\right)\right] + O[(J\rho)^3], \tag{16.182}$$

the leading logarithmic correction to $g(T)$. By using scaling we are thus able to re-sum diagrams far beyond leading-order perturbation theory. Using this expression to make the replacement $J\rho \to g(T)$ in the leading-order perturbation theory (16.176), we obtain

$$\chi(T) = \frac{n_i}{T}\left[1 - \frac{1}{\ln(T/T_K)} - \frac{1}{2}\frac{\ln(\ln(T/T_K))}{\ln^2(T/T_K)} + \cdots\right] \tag{16.183}$$

$$\frac{1}{\tau(T)} = n_i\frac{\pi S(S+1)}{2\rho}\left[\frac{1}{\ln^2(T/T_K)} + \frac{\ln(\ln(T/T_K))}{\ln^3(T/T_K)} + \cdots\right]. \tag{16.184}$$

From the second result, we see that the electron scattering rate has the scale-invariant form in (16.172)

$$\frac{1}{\tau(T)} = \frac{1}{\tau_0}\mathcal{G}(T/T_K), \tag{16.185}$$

where $\frac{1}{\tau_0} \propto \frac{n_i}{\rho}$ represents the intrinsic scattering rate off the Kondo impurity. The quantity $1/\rho$ is essentially the Fermi energy of the electron gas, and $1/\tau_0 \sim \frac{n_i}{\rho}$ is the unitary scattering rate, the maximum possible scattering rate that is obtained when an electron experiences a resonant $\pi/2$ scattering phase shift. From this result, we see that, at absolute zero, the electron scattering rate will rise to the value $\frac{1}{\tau(T)}|_{T=0} = \frac{n_i}{\rho}\mathcal{G}(0)$, indicating that at strong coupling the scattering rate is of the same order as the unitary scattering limit. We shall now see how this same result comes naturally out of a strong-coupling analysis.

Example 16.9

(a) Use the Popov–Fedotov scheme to compute the leading correction to the impurity magnetic susceptibility, given by the diagrams

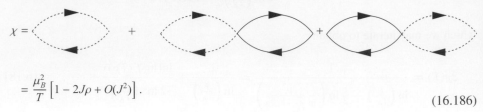

$$\chi =$$

$$= \frac{\mu_B^2}{T}\left[1 - 2J\rho + O(J^2)\right].$$

$$(16.186)$$

(b) Based on scaling arguments, what is the form of the J^2 correction to the susceptibility?
(c) What diagrams are responsible for the logarithmic correction to the susceptibility?

Solution

(a) For these calculations, let us temporarily set $\mu_B = 1$. We need to calculate the f-electron susceptibility, given by

$$\chi_f \delta^{ab} = \sigma^a \left\langle \cdots \right\rangle \sigma^b = -T\sum_{i\omega_n} \mathrm{Tr}[\sigma^a G_f(i\omega_n)\sigma^b G_f(i\omega_n)] = \chi_f \delta^{ab}.$$

$$(16.187)$$

So

$$\chi_f = -2T\sum_{i\omega_n} \frac{1}{(i\omega_n - \lambda_f)^2} = \frac{\partial}{\partial\lambda_f} \, 2T\sum_{i\omega_n} \overbrace{\frac{1}{i\omega_n - \lambda_f}}^{-f(\lambda_f)}$$

$$= 2\left(-\frac{\partial f(\lambda_f)}{\partial\lambda_f}\right) = \frac{2f_\lambda(1 - f_\lambda)}{T} = \frac{1}{T}, \qquad (16.188)$$

where the factor of 2 derives from the trace over the spin degrees of freedom and we have used $f_\lambda(1 - f_\lambda) = 1/[(i + 1)(-i + 1)] = \frac{1}{2}$. Similarly, the conduction electron susceptibility given by

$$\chi_c \delta^{ab} = \sigma^a \left\langle \cdots \right\rangle \sigma^b = -T\sum_{\mathbf{k},i\omega_n} \mathrm{Tr}\left[\sigma^a G_\mathbf{k}(i\omega_n)\sigma^b G_\mathbf{k}(i\omega_n)\right]$$

$$= -2\delta^{ab}T\sum_{\mathbf{k},i\omega_n} \frac{1}{(i\omega_n - \epsilon_\mathbf{k})^2} = 2\delta^{ab}\sum_\mathbf{k} \frac{\partial}{\partial\epsilon_\mathbf{k}} T\sum_{i\omega_n} \overbrace{\frac{1}{i\omega_n - \epsilon_\mathbf{k}}}^{-f(\epsilon_\mathbf{k})}$$

$$= 2\delta^{ab}\sum_\mathbf{k}\left(-\frac{\partial f(\epsilon_\mathbf{k})}{\partial\epsilon_\mathbf{k}}\right) = 2\rho\delta^{ab}. \qquad (16.189)$$

Now the first-order diagrams are given by

$$\chi_A \delta^{ab} = \sigma^a \left(\text{diagram} \right) \sigma^b = (\chi_f \delta^{ac}) \left(-\frac{J}{2}\right) (\chi_c \delta^{cb}) = -\frac{J\rho}{T} \tag{16.190}$$

and

$$\chi_B \delta^{ab} = \sigma^a \left(\text{diagram} \right) \sigma^b = (\chi_c \delta^{ac}) \left(-\frac{J}{2}\right) (\chi_f \delta^{cb}) = -\frac{J\rho}{T}. \tag{16.191}$$

Adding the results together, the first-order correction to the impurity susceptibility is given by

$$\chi = \frac{\mu_B^2}{T}(1 - 2J\rho) + O((J\rho)^2), \tag{16.192}$$

where we have reinstated μ_B.

(b) We expect the second-order corrections to the susceptibility to be obtained by renormalizing the coupling constant. Following the results of Example 16.7, the renormalization is given by

$$J\rho \rightarrow J\rho + 2(J\rho)^2 \ln\left[\frac{D}{T}\right]. \tag{16.193}$$

Since the renormalization group only provides leading logarithmic accuracy, we have dropped the dimensionless constants inside the logarithm. We therefore expect that the second-order corrections to the susceptibility will take the form

$$\chi(T) = \frac{\mu_B^2}{T}\left[1 - 2\left(J\rho + 2(J\rho)^2 \ln\left[\frac{D}{T}\right]\right)\right]$$

$$= \frac{\mu_B^2}{T}\left[1 - 2J\rho + 4(J\rho)^2 \ln\left[\frac{D}{T}\right]\right] + O[(J\rho)^3]. \tag{16.194}$$

(c) The logarithmic corrections to the susceptibility derive from the vertex insertions into the first-order diagrams, given by

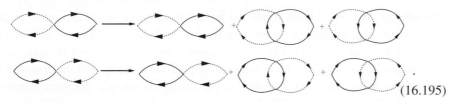

$$\tag{16.195}$$

There are other contributions to the susceptibility, such as self-energy corrections and corrections to the external magnetic vertex, but none will be logarithmically divergent corrections. Moreover, the conservation of spin will mean that many self-energy and vertex corrections will cancel one another.

16.10　Nozières Fermi-liquid theory

The weak-coupling analysis tells us that somewhere around the Kondo temperature, the running Kondo coupling constant g becomes of order one, $O(1)$. Although perturbative renormalization group methods cannot go past this point, Anderson and Yuval [12–14] argued that it is not unreasonable to suppose that the coupling constant continues scaling to infinity. This is the simplest possibility and, if true, it means that the strong-coupling limit is an attractive fixed point, a fixed point that is stable under the renormalization group. Anderson and Yuval conjectured that the Kondo singlet would be paramagnetic, with a temperature-independent magnetic susceptibility and a universal linear specific heat given by $C_V = \gamma_K \frac{T}{T_K}$ at low temperatures.

The first controlled treatment of this crossover regime was carried out by Kenneth Wilson at Cornell University in Ithaca, New York, using the numerical renormalization group method that he invented [16]. His numerical renormalization method was able to confirm the conjectured renormalization of the Kondo coupling constant to infinity. This limit is called the *strong-coupling* limit of the Kondo problem. Wilson carried out an analysis of the strong-coupling limit, and was able to show that the specific heat would be a linear function of temperature, like a Fermi liquid. He showed that the linear specific heat could be written in a universal form:

$$C_V = \gamma T$$
$$\gamma = \frac{\pi^2}{3} \frac{0.4128 \pm 0.002}{8T_K}. \tag{16.196}$$

Wilson also compared the ratio between the magnetic susceptibility and the linear specific heat with the corresponding value in a non-interacting system, computing

$$W = \frac{\chi/\chi^0}{\gamma/\gamma^0} = \frac{\chi}{\gamma}\left(\frac{\pi^2 k_B^2}{3(\mu_B)^2}\right) = 2 \tag{16.197}$$

within the accuracy of the numerical calculation.

16.10.1　Strong-coupling expansion

Remarkably, Wilson's second result can be rederived using an exceptionally elegant set of arguments due to Philippe Nozières at the Institut Laue-Langevin in Grenoble, France [17], that leads to an explicit form for the strong-coupling fixed-point Hamiltonian. Nozières began by considering a local moment coupled via a strong antiferromagnetic Kondo coupling to an electron sea. Wilson had previously shown that the local moment only couples to the s-wave scattering channel, so that the Kondo model can be mapped onto a spin coupled to the origin of a one-dimensional tight-binding chain, as illustrated in Figure 16.18.

The tight-binding representation of the one-dimensional Kondo model is

$$H_{lattice} = -t \sum_{j=0,\infty} [c_\sigma^\dagger(j+1)c_\sigma(j) + \text{H.c.}] + Jc_\alpha^\dagger(0)\vec{\sigma}_{\alpha\beta}c_\beta(0) \cdot \vec{S}_f. \tag{16.198}$$

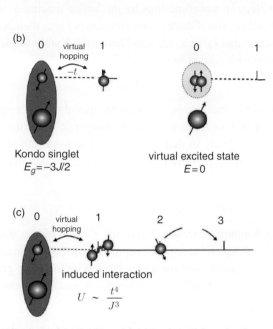

Illustrating the strong-coupling limit of the Kondo model. (a) Singlet with bound-state energy $-3J/2$ and a singlet–triplet gap of size $2J$ interacts with narrow band of width $2t$. (b) Kondo singlet formed at origin because the addition or removal of electrons costs energy. (c) Virtual fluctuations of charge in and out of the Kondo singlet induces interactions between electrons at site 1.

Fig. 16.18

Since the renormalization trajectories link the weak-coupling model, where $J/t << 1$, to the strong-coupling model, where $J/t >> 1$, Nozières reasoned that the strong-coupling physics of this model could be used to understand the low-temperature physics of the Kondo model. When $J >> t$, in the ground state, the local moment couples to an electron at the origin, forming a Kondo singlet, denoted by

$$|GS\rangle = \frac{1}{\sqrt{2}} (|\Uparrow\downarrow\rangle - |\Downarrow\uparrow\rangle), \quad (16.199)$$

where the thick arrow refers to the spin state of the local moment and the thin arrow refers to the spin state of the electron at site 0. The ground-state energy of the Kondo singlet is

$$E_g = J\left[S(S+1) - \frac{3}{2}\right] = -\frac{3}{2}J. \quad (16.200)$$

The kinetic energy of the electrons in the one-dimensional band now becomes a perturbation to the Kondo singlet. Any electron which hops from site 1 to site 0 will automatically break the Kondo singlet, raising its energy by $3J/2$. This will have the effect of excluding electrons (or holes) from the origin. The fixed-point Hamiltonian must then take the form

$$H_{lattice} = -t \sum_{j=1,\infty} [c_\sigma^\dagger(j+1)c_\sigma(j) + \text{H.c.}] + \text{weak interaction}, \quad (16.201)$$

where the second term refers to the weak interactions induced by virtual fluctuations onto site 0. If the wavefunction of electrons far from the impurity has the form $\psi(x) \sim \sin(k_F x)$, where k_F is the Fermi momentum, then the exclusion of electrons from site 1 has the effect of phase-shifting the electron wavefunctions by the lattice spacing a, so that now $\psi(x) \sim \sin(k_F x - \delta)$, where $\delta = k_F a$. But if there is one electron per site, then $2(2k_F a/(2\pi)) = 1$ by the Luttinger sum rule, so that $k_F = \pi/(2a)$ and hence the Kondo singlet acts as a spinless, elastic scattering center with scattering phase shift

$$\delta = \pi/2. \tag{16.202}$$

The appearance of $\delta = \pi/2$ can also be deduced by appeal to the Friedel sum rule, which determines the number of bound electrons at the magnetic impurity site in terms of the phase shift, $\sum_{\sigma=\uparrow,\downarrow} \frac{\delta}{\pi} = 2\delta/\pi$, so that $\delta = \pi/2$. By considering virtual fluctuations of electrons between site 1 and 0, Nozières argued that the induced interaction at site 1 must take the form

$$\Delta H = U n_{1\uparrow} n_{1\downarrow}, \qquad U \sim \frac{t^4}{J^3}, \tag{16.203}$$

because fourth-order hopping processes lower the energy of the singly occupied state relative to the doubly occupied state. This repulsive interaction among the conduction electrons is actually a marginal operator under the renormalization group, leading to the conclusion that the effective Hamiltonian describes a weakly interacting *local* Landau Fermi-liquid.

16.10.2 Phase-shift formulation of the local Fermi liquid

Nozières then used the results of the strong-coupling expansion to formulate a local Fermi-liquid description of the strong-coupling fixed point of the Kondo model. He formulated this theory in terms of an occupancy-dependent phase shift as follows. In a non-interacting impurity problem, the asymptotic wavefunctions experience a scattering phase shift, with a radial wavefunction that takes the form

$$\psi(r) \sim \frac{\sin(kr + \delta(E_k))}{r}. \tag{16.204}$$

If we put the system inside a large sphere of radius R, with the boundary condition $\psi(R) = 0$, then $kR + \delta(E_k) = n\pi$ determines the allowed momenta of the quasiparticles, given by $k_n = n\frac{\pi}{R} - \frac{\delta(E_k)}{R}$. The level spacing in the absence of scattering is $\Delta\epsilon = \frac{\partial E}{\partial k}\frac{\pi}{R}$. Now in the presence of the scattering phase shift, momenta are reduced by an amount $\Delta k = -\frac{\delta[E_k]}{R}$, so the corresponding energy levels are shifted downwards by an amount

$$E_k \rightarrow E_k - \frac{\partial E}{\partial k}\frac{\delta(E_k)}{R}$$

$$= E_k - \Delta\epsilon\frac{\delta(E_k)}{\pi}. \tag{16.205}$$

In the impurity Landau Fermi-liquid, the ground-state energy will be a functional of quasiparticle occupancies $\{n_{k\sigma}\}$. As in bulk Landau Fermi-liquid theory, the differential of the total energy with respect to occupancies defines a renormalized quasiparticle energy as follows:

$$\frac{\delta E[\{n_{k'\sigma}\}]}{\delta n_{k\sigma}} = E_k - \frac{\Delta \epsilon}{\pi} \delta_\sigma(\{n_{k'\sigma'}\}, E_k). \tag{16.206}$$

Now, interactions cause the phase shift to depend on the quasiparticle occupancies $n_{k\sigma}$. Expanding to linear order in the deviation of quasiparticle occupancies about the ground state, $\delta n_{k\sigma} = n_{k\sigma} - n_{k\sigma}^{(0)}$, then

$$\delta_\sigma(\{n_{k'\sigma'}\}, E) = \frac{\pi}{2} + \alpha(E - \epsilon_F) + \Phi \sum_{k'} \delta n_{k',-\sigma}, \tag{16.207}$$

where ϵ_F is the Fermi energy. The term with coefficient Φ is the impurity analogue of the Landau Fermi-liquid interaction $f_{k\sigma, k'\sigma'}$. This term describes the onsite interaction between opposite spin states of the Fermi liquid. Nozières argued that, when the chemical potential of the conduction sea is changed, the occupancy of the localized d-state will not change, and because the Friedel sum rule links the $\pi/2$ phase shift to the occupancy, this implies that phase shift is invariant under changes in μ, i.e. the Kondo resonance is pinned to the Fermi energy. Now under a shift $\Delta\mu$, there is a change in the occupancy of the quasiparticle states, $\sum_k \delta n_{k-\sigma} \to \rho\Delta\mu$, so that changing the chemical potential modifies the phase shift at the Fermi surface by an amount

$$\Delta \delta_\uparrow = \delta(\{n_{k'\sigma'}\}, \epsilon_F + \Delta\mu) - \delta(\{n_{k'\sigma'}^{(0)}\}, \epsilon_F)$$

$$= \alpha\Delta\mu + \Phi \overbrace{\sum_{k'}(n_{k'\downarrow} - n_{k'\downarrow}^{(0)})}^{\rho\Delta\mu}$$

$$= (\alpha + \Phi\rho)\Delta\mu = 0. \tag{16.208}$$

In other words, the interaction term must compensate for the energy dependence of the phase shift, so that the magnitude $\Phi = -\alpha/\rho$ is set by the density of states α.

We are now in a position to calculate the impurity contribution to the magnetic susceptibility and specific heat. First note that the density of quasiparticle states is given by

$$\rho = \frac{dN}{dE} = \rho_o + \frac{1}{\pi}\frac{\partial\delta}{\partial\epsilon} = \rho_o + \frac{\alpha}{\pi}, \tag{16.209}$$

so that the low-temperature specific heat is given by $C_V = (\gamma_{bulk} + \gamma)T$, where

$$\gamma = 2\left(\frac{\pi^2 k_B^2}{3}\right)\frac{\alpha}{\pi}, \tag{16.210}$$

where the prefactor 2 is derived from the spin-up and spin-down bands. It is convenient to write this in the form

$$\gamma = \left(\frac{\pi^2 k_B^2}{3}\right)\tilde{\gamma},$$

where $\tilde{\gamma} = 2\alpha/\pi$. Now in a magnetic field, the impurity magnetization is given by

$$M = \frac{\delta_\uparrow}{\pi} - \frac{\delta_\downarrow}{\pi}. \tag{16.211}$$

Since the Fermi energies of the up and down quasiparticles are shifted, $\epsilon_{F\sigma} \to \epsilon_F - \sigma B$, we have $\sum_k \delta n_{k\sigma} = \sigma\rho B$, so that the phase shift at the Fermi surface in the up and down scattering channels becomes

$$\delta_\sigma = \frac{\pi}{2} + \alpha \delta \epsilon_{F\sigma} + \Phi \left(\sum_k \delta n_{k\sigma} \right)$$

$$= \frac{\pi}{2} + \alpha \sigma B - \Phi \rho \sigma B$$

$$= \frac{\pi}{2} + 2\alpha \sigma B, \tag{16.212}$$

so that the presence of the interaction term *doubles* the size of the change in the phase shift due to a magnetic field. The impurity magnetization then becomes

$$M_i = \chi_s B = \frac{\delta_\uparrow}{\pi} - \frac{\delta_\downarrow}{\pi} = 2 \left(\frac{2\alpha}{\pi} \right) B. \tag{16.213}$$

If we reinstate the magnetic moment μ_B of the electron, this becomes

$$\chi_s = \mu_B^2 \tilde{\chi}_s = \mu_B^2 \left(\frac{4\alpha}{\pi} \right). \tag{16.214}$$

This is twice the value expected for a rigid resonance, and it means that the Wilson ratio is

$$W = \frac{\chi_s}{\gamma} \left(\frac{\pi^2 k_B^2}{3(\mu_B)^2} \right) = \frac{\tilde{\chi}_s}{\tilde{\gamma}} = 2. \tag{16.215}$$

Example 16.10   Consider what happens to the Fermi-liquid ground state of the Anderson impurity model in which the change of the phase shifts $\Delta \delta_\uparrow = \Delta \delta_\downarrow = \Delta \delta$ due to a shift $\Delta \mu$ in the chemical potential is now finite, due to the finite charge susceptibility χ_c, so that

$$\frac{\Delta(\delta_\uparrow + \delta_\downarrow)}{\pi} = \frac{2\Delta \delta}{\pi} = \chi_c \Delta \mu, \tag{16.216}$$

where $\Delta \mu$ is the change in the chemical potential.

(a) By generalizing the Nozières Fermi-liquid theory to the Anderson model, use the above result to show that (16.208) becomes

$$\Delta \delta = \frac{\pi}{2} \chi_c \Delta \mu = (\alpha + \Phi \rho) \Delta \mu \tag{16.217}$$

in the Anderson impurity model, so that the charge susceptibility is given by

$$\chi_c = \frac{2}{\pi} (\alpha + \Phi \rho).$$

(b) By generalizing the calculation of the susceptibility to the Anderson impurity model, show that the specific heat, charge, and spin susceptibility are related by the Yamada–Yoshida identity,

$$2\tilde{\gamma} = \tilde{\chi}_s + \chi_c, \tag{16.218}$$

where

$$\gamma = \left(\frac{\pi^2 k_B^2}{3} \right) \tilde{\gamma}, \qquad \chi_s = \mu_B^2 \tilde{\chi}_s. \tag{16.219}$$

(c) What happens to the charge susceptibility when U is large and negative? (This is the charge Kondo model)

Solution

(a) In the impurity Anderson model, we write

$$\delta_\sigma(\{n_{k'\sigma'}\}, \epsilon_k) = \delta_0 + \alpha(\epsilon_k - \epsilon_F) + \Phi\left(\sum_k \delta n_{k,-\sigma}\right), \tag{16.220}$$

where, by the Friedel sum rule, $2\delta_0/\pi = \Delta n_e$, the number of electrons bound by the impurity. If we add extra quasiparticles to the Fermi sea, shifting the Fermi energy up by an amount $\Delta\mu$, the change in the phase shift at the Fermi energy is given by

$$\Delta\delta_\sigma = \delta_\sigma(\{n_{k'\sigma'}\}, \Delta\mu + \epsilon_F) - \delta_\sigma(\{n_{k'\sigma'}^{(0)}\}, \epsilon_F) = \alpha\Delta\mu + \Phi\left(\sum_{k'}\delta n_{k'-\sigma}\right)$$

$$= (\alpha + \Phi\rho)\Delta\mu. \tag{16.221}$$

But the change in the impurity charge is then given by

$$\chi_c\Delta\mu = \frac{2\Delta\delta}{\pi} = \frac{2}{\pi}(\alpha + \Phi\rho)\Delta\mu,$$

so that

$$\chi_c = \frac{2}{\pi}(\alpha + \Phi\rho). \tag{16.222}$$

Now the linear specific heat is determined by the impurity density of states (16.210),

$$\gamma = \left(\frac{\pi^2 k_B^2}{3}\right)\frac{2}{\pi}\frac{\partial\delta}{\partial\epsilon} = \left(\frac{\pi^2 k_B^2}{3}\right)\overbrace{\frac{2\alpha}{\pi}}^{=\tilde{\gamma}}, \tag{16.223}$$

so we may write

$$\chi_c = \tilde{\gamma} + \left(\frac{2\Phi\rho}{\pi}\right)$$

$$\left(\frac{2\Phi\rho}{\pi}\right) = \chi_c - \tilde{\gamma}. \tag{16.224}$$

(b) Setting $\mu_B = 1$, the magnetization of the impurity is given by

$$M = \chi_s B = \left(\frac{\delta_\uparrow}{\pi} - \frac{\delta_\downarrow}{\pi}\right), \tag{16.225}$$

where, in a field, the Fermi surface splits into two, with the shift in the Fermi energy given by σB, so that the changes in the phase shifts are given by

$$\delta_\sigma = \delta_0 + \alpha\sigma B + \Phi\overbrace{\left(\sum_{k'}\delta n_{k'-\sigma}\right)}^{-\rho\sigma B}$$

$$= \delta_0 + (\alpha - \Phi\rho)\sigma B. \tag{16.226}$$

Using (16.224), it follows that

$$M = \tilde{\chi}_s = \frac{\delta_\uparrow}{\pi} - \frac{\delta_\downarrow}{\pi} = 2\frac{\alpha - \Phi\rho}{\pi} = \tilde{\gamma} - (\chi_c - \tilde{\gamma}) = 2\tilde{\gamma} - \chi_c, \qquad (16.227)$$

from which the Yamada–Yoshida identity

$$2\tilde{\gamma} = \chi_c + \chi_s \qquad (16.228)$$

follows. Notice that, if we restore μ_B, then $\chi_s \to \chi_s/\mu_B^2 = \tilde{\chi}_s$.

(c) Notice that, in the non-interacting impurity ($U = 0$), $\chi_c = \tilde{\chi}_s = \gamma$. In the limit that $U < 0$ is large and negative, the spin susceptibilty is suppressed to zero, so that the charge Wilson ratio is

$$\frac{\chi_c}{\tilde{\gamma}} = 2. \qquad (16.229)$$

This is a result of the charge Kondo effect.

16.10.3 Experimental observation of the Kondo effect

Experimentally, there is now a wealth of observations that confirm our understanding of the single-impurity Kondo effect. Here is a brief itemization of some of the most important observations (Figure 16.19).

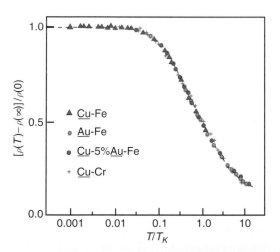

Fig. 16.19 Temperature dependence of excess resistivity associated with scattering from an impurity spin [54]. The resistivity saturates at the unitarity limit at low temperatures, due to the formation of the Kondo resonance. The notation A-M denotes dilute magnetic impurities M dissolved in host metal A. Reprinted and adapted Figure VII. 14 from R. M. White and T. Geballe, Long-range order in solids, in *Solid State Physics*, ed. H. Ehrenreich, F. Seitz, and D. Turnbull, vol. 15, p. 283, 1979. Copyright 1979 Elsevier.

- A resistance minimum appears when local moments develop in a material. For example, in $Nb_{1-x}Mo_x$ alloys, a local moment develops for $x > 0.4$, and the resistance is seen to develop a minimum beyond this point [2, 5].
- Universality is seen in the specific heat $C_V = \frac{n_i}{T} F(T/T_K)$ of metals doped with dilute concentrations of impurities. Thus the specific heat of Cu-Fe (iron impurities in copper) can be superimposed on the specific heat of Cu-Cr, with a suitable rescaling of the temperature scale [54].
- Universality is observed in the differential conductance of quantum dots [55, 56] and spin-fluctuation resistivity of metals with a dilute concentration of impurities [57]. Actually, both properties are dependent on the same thermal average of the imaginary part of the scattering t-matrix:

$$\rho_i = n_i \frac{ne^2}{m} \int d\omega \left(-\frac{\partial f}{\partial \omega} \right) 2 \, \text{Im} \, [T(\omega)]$$

$$G = \frac{2e^2}{\hbar} \int d\omega \left(-\frac{\partial f}{\partial \omega} \right) \pi \rho \, \text{Im} \, [T(\omega)]. \tag{16.230}$$

Putting $\pi \rho \int d\omega \left(-\frac{\partial f}{\partial \omega} \right) \text{Im} \, T(\omega) = t(\omega/T_K, T/T_K)$, we see that these properties have the form

$$\rho_i = n_i \frac{2ne^2}{\pi m \rho} t(T/T_K)$$

$$G = \frac{2e^2}{\hbar} t(T/T_K), \tag{16.231}$$

where $t(T/T_K)$ is a universal function. This result is born out by experiment (Figure 16.19).

16.11 Multi-channel Kondo physics

In practice, magnetic moments in real materials exhibit many different variants on the original $S = \frac{1}{2}$ Kondo model, a point first emphasized by Philippe Nozières and André Blandin [18]. Here we end with a brief discussion of two important variants of the original $SU(2)$ Kondo spin model:

- The multi-channel Kondo model, in which the spin interacts with k different screening channels.
- The spin-S Kondo model, in which the impurity has spin $S > \frac{1}{2}$. This is important in multi-electron orbitals, in which the localized electrons are coupled together to form a large spin S by the Hund's interaction.

The k-channel spin-S Kondo model (Figure 16.20) which incorporates both of these features is written

$$H = J \sum_{\lambda=1}^{k} \vec{\sigma}_\lambda(0) \cdot \vec{S} + \sum_{\substack{k,\sigma=\pm \\ \lambda=1,k}} \epsilon_k c_{k\lambda\sigma}^\dagger c_{k\lambda\sigma}. \tag{16.232}$$

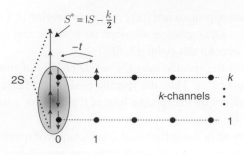

Fig. 16.20 The k-channel Kondo model, in which k-channels screen a single impurity spin $S > \frac{1}{2}$. The impurity spin S at the origin can be described as a composite of $2S$ elementary $S = \frac{1}{2}$ spins. At strong coupling, the net spin at the origin is $S^* = |S - k/2|$.

Here $\vec{S}$ denotes a spin-S local moment, with $\vec{S}^2 = S(S + 1)$, while $\vec{\sigma}_\lambda(0) \equiv c_\lambda^\dagger(0)\vec{\sigma}c_\lambda(0)$ is the spin density at the origin in channel $\lambda \in 1, f$.

Multi-channel $SU(2)$ Kondo models describe the physics of $3d$ magnetic moments in metals. For example, Mn^{2+} dissolved in copper has a $3d^5$ configuration with five d-electrons Hund's-coupled together into a single $S = 5/2$ with no orbital degrees of freedom. Microscopically, the d-states hybridize with the corresponding partial-wave states of the conduction electrons, where

$$c_{km\sigma}^\dagger = \int \frac{d\Omega_\mathbf{k}}{4\pi} Y_{2m}(\hat{\mathbf{k}})c_{\mathbf{k}\sigma}^\dagger \qquad (m \in [-2, 2]) \qquad (16.233)$$

creates a conduction electron in the $l = 2$, $L_z = m$-partial-wave state. Once the virtual valence fluctuations are eliminated by a Schrieffer–Wolff Hamiltonian, one arrives at a spin $S = 5/2$ Kondo model with $k = 5$ channels.

Like its single-channel cousin, the weak-coupling fixed point of the antiferromagnetic multi-channel Kondo model is always unstable. Indeed, the scaling equations we deduced in Section 16.7 are independent, to order g^2 in both S and k, so that, to leading order in the scaling, the multi-channel Kondo temperature is still given by $T_K \sim D_0 e^{-\frac{1}{2J\rho}}$. However, at strong coupling, the behavior of the multi-channel Kondo model depends critically on the relative sizes of S and k. To understand this behavior, as before we consider a tight-binding model, only now with k tight-binding chains interacting with a single spin S (Figure 16.20):

$$H = J\sum_{\lambda=1}^{k} \vec{\sigma}_\lambda(0) \cdot \vec{S} - t \sum_{\substack{j=0,\infty \\ \lambda=1,k}} [c_{\lambda\sigma}^\dagger(j+1)c_{\lambda\sigma}(j) + \text{H.c.}], \qquad (16.234)$$

where $\vec{\sigma}_\lambda(0) = c_{\lambda\alpha}^\dagger(0)\vec{\sigma}_{\alpha\beta}c_\lambda(0)$ is the spin density at the origin on the λth chain.

We can think of the impurity spin at the origin as being composed of $2S$ elementary spin-$\frac{1}{2}$ objects, symmetrically combined into a single spin S. When $J >> t$, the the local moment binds as many electrons as it can into an antiferromagnetic configuration. Since there are k spin-$\frac{1}{2}$ conduction electron channels, the total S total spin of the screened local moment is then, $S^* = S - k/2$. In fact, there are three cases (see Figure 16.21):

a. *Perfect screening*, $k = 2S$, in which the local moment and conduction electrons form a singlet with $S^* = 0$

b. *Underscreening*, $k < 2S$, in which the local moment spin is partially reduced to a spin $S^* = S - k/2$

c. *Overscreening*, $k > 2S$, in which the screening effect "overshoots" to produce a spin $S^* = k/2 - S$.

The fully screened case a ($k = 2S$) is closely analogous to the strong-coupling fixed point of the $S = \frac{1}{2}$ one-channel Kondo model: as in this case, the spin-S impurity forms a singlet with $k = 2S$ electrons at the origin. This site creates a strongly repulsive scattering potential at the origin, which scatters electrons in each channel at the unitarity limit, $\delta_\lambda = \pi/2$. The residual charge fluctuations between site 1 and the origin then induce a weak Fermi liquid interaction at site 1, forming a Fermi liquid with an effective Hamiltonian

$$H_{eff} = U_{\lambda\lambda'} n_\lambda(1) n_{\lambda'}(1), \qquad U_{\lambda\lambda'} \sim \frac{t^4}{J^3}. \qquad (16.235)$$

This is the situation for Mn^{2+} in copper.

Cases b and c are more interesting, for now the underscreened or overscreened impurity spin has an internal degree of freedom, so we must carefully examine its residual inter-actions with the conduction sea. In fact, virtual charge fluctuations between electrons or holes at site 1 onto site 0 will generate a new Kondo interaction of strength $O(t^2/J)$. We'll now see that the sign of this interaction is ferromagnetic for the underscreened case but antiferromagnetic for the overscreened case, with dramatic consequences for the stability of the strong-coupling fixed point.

To understand the residual interaction between the underscreened moment and site 1, let us consider first the underscreened fixed point where $k < 2S$. Suppose the original spin is up, so that the k screening electrons are predominantly down. The exclusion principle only allows electrons on site 1 to virtually hop to the origin and back if they are up, i.e. antipar-allel to the screening electrons at the origin. This implies that the conduction electron at site 1 lowers its energy via virtual fluctuations when it is parallel to the local moment $\vec{S}^*$, and this gives rise to a ferromagnetic interaction:

$$H^* = -J_{eff} \sum_{\lambda=1,k} \vec{\sigma}_\lambda(1) \cdot \vec{S}^* \qquad (J_{eff} \sim t^2/J). \qquad \text{underscreened Kondo model}$$

$$(16.236)$$

Now this ferromagnetic J_{eff} will scale logarithimically to zero, implying that the residual moment S^* asymptotically decouples from the conduction sea and the strong-coupling fixed point is stable. Since J_{eff} and the original Kondo coupling constant are inversely related, the scaling of J_{eff} to zero also implies that the Kondo coupling constant J scales to infinite coupling, as we expect if the strong-coupling fixed point is stable.

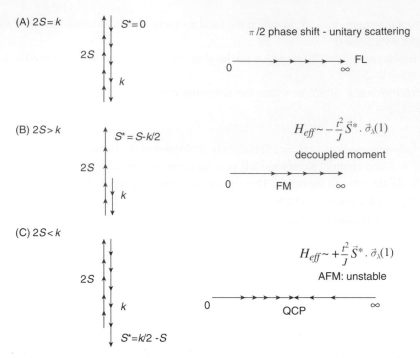

Fig. 16.21 Renormalization group flows of the antiferromagnetic multi-channel Kondo model: (a) fully screened case, giving rise to a Fermi liquid at strong coupling; (b) underscreening, giving rise to a decoupled moment at strong coupling; (c) overscreening, in which both weak- and strong-coupling fixed points are antiferromagnetic and unstable, flowing to an intermediate-coupling quantum critical point. FL = Fermi liquid; FM = Ferromagnetic; AFM = antiferromagnetic; QCP = quantum critical point.

As we approach the underscreened fixed point at low temperatures, we expect the low-temperature magnetic susceptibility to take the form

$$\chi(T) \sim \frac{S^*(S^*+1)}{3T}\left[1 + O\left(\overbrace{\frac{1}{\ln T_K/T}}^{J_{eff}\rho}\right)\right] \quad \begin{array}{l}(S^* = S - k/2), \\ \text{underscreened Kondo effect}\end{array}$$

(16.237)

where the correction term is due to the residual ferromagnetic interaction, scaling logarithimically to zero.

By contrast, in the overscreened case, the unscreened moment and the overscreened moment S^* have an antiparallel configuration, so that virtual charge fluctuations lower the energy of an electron at site 1 when its spin is antiparallel with the overscreened moment, thus inducing an *antiferromagnetic* interaction:

$$H^* = +J_{eff}\sum_{\lambda=1,k}\vec{\sigma}_\lambda(1)\cdot\vec{S}^* \qquad (J_{eff}\sim t^2/J). \qquad \text{overscreened Kondo model}$$

(16.238)

Now if we scale J_{eff} with the bandwidth of the strong-coupling fixed point, J_{eff} will grow, so the strong-coupling fixed point is unstable. Since the original coupling constant g and J_{eff} are inversely related, this implies that scaling moves us away from the overscreened fixed point, back towards weak coupling. But since both weak and strong coupling are unstable, this implies the presence of a new *intermediate-coupling* fixed point. It is this intermediate-coupling fixed point, rather than the strong-coupling fixed point, that determines the low-temperature physics of the original weak-coupling overscreened Kondo model. The presence of this fixed point was first deduced by Nozières and Blandin [18] using this reasoning. They noted that, with hindsight, the β function for the k-channel model is given by

$$\beta(g) = -2g^2 + 2kg^3 + O(g^4), \tag{16.239}$$

suggesting that indeed at large k there is an intermediate-coupling fixed point where $\beta(g^*) = 0$ at $g^* \sim \frac{1}{k}$. Since its discovery, a variety of powerful methods, including exact solution by Bethe ansatz [58, 59], bosonization [60], and conformal field theory [61] have confirmed the reasoning of Nozières and Blandin, and have added much detail to the nature of this intriguing new fixed point. These studies show that the intermediate-coupling fixed point of the overscreened Kondo model is not described by Fermi-liquid theory, but is instead an unstable *quantum critical point* with a self-similar excitation spectrum in which the spins develop anamalous scaling exponents in time. This *non-Fermi liquid* ground state also has a finite but fractional zero-point entropy, which for the case $k = 2$, $S = \frac{1}{2}$ is $S = \frac{1}{2} \ln 2$, corresponding to a decoupled Majorana fermion. The interested reader who wishes to pursue this topic further is encouraged to read some of the references at the end of the chapter [60–62]. The overscreened Kondo effect is comparatively rare, because typically the size of the local moment is precisely balanced with the number of screening channels $S = 2k$. However, it does appear to be realized in a number of very interesting situations, including integer-spin U and Pr atoms [63–66], which have non-Kramers doublets which undergo a *quadrupolar Kondo effect* [63], and also in certain specially designed quantum dots [67].

Appendix 16A   Derivation of Kondo integral

We wish to derive the result quoted in (16.165),

$$I = \int d\epsilon\, \Phi(\epsilon) \left(\frac{f(\lambda_f) - f(\epsilon)}{\epsilon - \xi} \right) = \ln \frac{D}{2\pi T} - \psi \left(\frac{1}{2} + \frac{\xi\beta}{2\pi i} \right) - i\frac{\pi}{2} \tanh(\beta\lambda_f/2), \tag{16.240}$$

where $\Phi(\epsilon) = D^2/(D^2 + \epsilon^2)$ imposes the high-energy cut-off and $\xi = \lambda + i\Delta$ $(\Delta > 0)$ lies in the upper half of the complex plane. We first write the integral as the difference $I = I_1 - I_2$, where

$$I_1 = \int d\epsilon\, \Phi(\epsilon) \frac{f(\lambda_f)}{\epsilon - \xi}$$

$$I_2 = \int d\epsilon\, \Phi(\epsilon) \frac{f(\epsilon)}{\epsilon - \xi}. \tag{16.241}$$

These integrals can be carried out using contour integration. Now $\Phi(z) = D^2/((z+iD)(z-iD))$, so in the integral for I_1 there are two poles in the upper half-plane at $z = iD$ and $z = \xi = \lambda + i\Delta$ and only one pole in the lower half-plane at $z = -iD$. If we close the contour in the lower half-plane, we only pick up the single pole at $z = -iD$, which gives

$$I_1 = -2\pi i \left(\frac{iD}{2} \right) \frac{f_\lambda}{-iD - \xi} \to i\pi f_\lambda \qquad (16.242)$$

in the large-D limit. To calculate the second integral, I_2, we begin by writing

$$\frac{D^2}{D^2 + z^2} \frac{1}{z - \xi} = \frac{1}{2} \left(\frac{-iD}{z - iD} + \{D \to -D\} \right) \frac{1}{z - \xi}$$

$$= \frac{1}{2} \left(\frac{1}{z - \xi} - \frac{1}{z - iD} \right) + \{D \to -De^{i0^+}\}, \qquad (16.243)$$

so we can calculate the answer for positive D, change the sign of D, and average over the two results. (In practice, we need to give D an infinitesimal phase $D \to De^{i0^+}$ so that the pole at $z = -iDe^{i\delta}$ is $f(-iD + D0^+) \to 0$.) If we now integrate the first term of this expression with $f(z)$, closing the contour in the lower half-plane, then the only poles we pick up are the poles in $f(z)$ of strength $-T$ at $z = -i\omega_n = -i(2n + 1)2\pi T$, leading to

$$I_2 = \oint dz \frac{1}{2} \left(\frac{1}{z - \xi} - \frac{1}{z - iD} \right) f(z) + \left\{ D \to -De^{i0^+} \right\}$$

$$= \frac{1}{2}(-2\pi i)(-T) \sum_{n=0}^{\infty} \left(-\frac{1}{i\omega_n + \xi} + \frac{1}{i\omega_n + iD} \right) + \left\{ D \to -De^{i0^+} \right\}, \quad (16.244)$$

where the $-2\pi i$ line derives from the left-handed contour in the lower half-plane and the $-T$ from the pole strength of $f(z)$ at $z = -i\omega_n$. Simplifying this result,

$$I_2 = \frac{1}{2} \sum_{n=0}^{\infty} \left(\frac{1}{n + \frac{1}{2} + \frac{D}{2\pi T}} - \frac{1}{n + \frac{1}{2} + \frac{\xi}{2\pi iT}} \right) + \left\{ D \to -De^{i0^+} \right\}. \qquad (16.245)$$

Now at this point we use the series form of the digamma function $\psi(z) = d \ln \Gamma(z)/dz$,

$$\psi(z) = \sum_{n=0}^{\infty} \left(\frac{1}{n + 1} - \frac{1}{n + z} \right) - \zeta, \qquad (16.246)$$

where $\zeta = 0.577216 = -\psi(1)$ is Euler's constant. From this relation, we deduce that

$$\psi(z) - \psi(a) = \sum_{n=0}^{\infty} \left(\frac{1}{n + a} - \frac{1}{n + z} \right), \qquad (16.247)$$

so that

$$I_2 = \frac{1}{2} \left[\psi \left(\frac{1}{2} + \frac{\xi}{2\pi iT} \right) - \psi \left(\frac{1}{2} + \frac{D}{2\pi T} \right) \right] + \left\{ D \to -De^{i0^+} \right\}. \qquad (16.248)$$

Now for large z, $\psi(z) \to \ln z$, so that, for a large cut-off,

$$
I_2 = \frac{1}{2}\left[\psi\left(\frac{1}{2} + \frac{\xi}{2\pi i T}\right) - \ln\left(\frac{D}{2\pi T}\right)\right] + \left\{D \to -De^{i0^+}\right\}
$$

$$
= \psi\left(\frac{1}{2} + \frac{\xi}{2\pi i T}\right) - \ln\left(\frac{D}{2\pi T}\right) + i\frac{\pi}{2}, \tag{16.249}
$$

where the remainder term derives from $-\frac{1}{2}\ln -e^{i0^+} = i\frac{\pi}{2}$. Subtracting I_2 (16.249) from I_1 (16.242) then gives

$$
I = I_1 - I_2 = \ln\left(\frac{D}{2\pi T}\right) - \psi\left(\frac{1}{2} + \frac{\xi}{2\pi i T}\right) - i\frac{\pi}{2}(1 - 2f_\lambda)
$$

$$
= \ln\left(\frac{D}{2\pi T}\right) - \psi\left(\frac{1}{2} + \frac{\xi}{2\pi i T}\right) - i\frac{\pi}{2}\tanh\left[\frac{\beta\lambda_f}{2T}\right]. \tag{16.250}
$$

Exercises

Exercise 16.1

(a) Using the identity $n_{f\sigma}^2 = n_{f\sigma}$, show that the atomic part of the Anderson model can be written in the form

$$
H_{atomic} = (E_f + \frac{U}{2})n_f + \frac{U}{2}\left[(n_f - 1)^2 - 1\right]. \tag{16.251}
$$

What happens when $E_f + U/2 = 0$?

(b) Using the completeness relation

$$
\underbrace{|f^0\rangle\langle f^0| + |f^2\rangle\langle f^2|}_{(n_f - 1)^2} + \underbrace{|\uparrow\rangle\langle\uparrow| + |\downarrow\rangle\langle\downarrow|}_{\dfrac{\vec{S}^2}{S(S+1)}} = 1 \qquad \left(S = \frac{1}{2}\right),
$$

show that the interaction can also be written in the form

$$
H_{atomic} = (E_f + \frac{U}{2})n_f - \frac{2U}{3}\vec{S}^2, \tag{16.252}
$$

which makes it clear that the repulsive U term induces a "magnetic attraction" that favors formation of a local moment.

(c) Derive the Hubbard–Stratonovich decoupling for {(16.252)}.

Exercise 16.2 By expanding a plane-wave state in terms of spherical harmonics,

$$
\langle\mathbf{r}|\mathbf{k}\rangle = e^{i\mathbf{k}\cdot\mathbf{r}} = 4\pi\sum_{l,m} i^l j_l(kr)Y^*_{lm}(\hat{\mathbf{k}})Y_{lm}(\hat{\mathbf{r}}),
$$

show that the overlap between a state $|\psi\rangle$, with wavefunction $\langle\vec{x}|\psi\rangle = R(r)Y_{lm}(\hat{r})$, and a plane wave is given by $V(\vec{k}) = \langle\vec{k}|V|\psi\rangle = V(k)Y_{lm}(\hat{k})$, where

$$
V(k) = 4\pi i^{-l}\int dr\, r^2 V(r)R(r)j_l(kr). \tag{16.253}
$$

Exercise 16.3

(a) Show that $\delta = \cot^{-1}\left(\frac{E_d}{\Delta}\right)$ is the scattering phase shift for scattering off a resonant level at position E_d.

(b) Show that the energy of states in the continuum is shifted by an amount $-\Delta\epsilon\,\delta(\epsilon)/\pi$, where $\Delta\epsilon$ is the separation of states in the continuum.

(c) Show that the increase in the density of states is given by $\partial\delta/\partial E = \rho_d(E)$ (see Chapter 2).

Exercise 16.4 Generalize the scaling equations to the anisotropic Kondo model with an anisotropic interaction

$$H_I = J^x \sigma^x(0)S^x + J^y \sigma^y(0)S^y + J^z \sigma^z(0)S^z, \tag{16.254}$$

where $\sigma^a(0) = \sum_{k,k'} c_k^\dagger \sigma^a c_{k'}$ is the local spin density of the conduction sea and $S^{x,y,z}$ are the three components of the localized magnetic moment. Show that the scaling equations take the form

$$\frac{\partial J_a}{\partial \ln D} = -2J_b J_c \rho + O(J^3),$$

where (a,b,c) are a cyclic permutation of (x,y,z). Show that, in the special case where $J_x = J_y = J_\perp$, the scaling equations become

$$\frac{\partial J_\perp}{\partial \ln D} = -2J_z J_\perp \rho + O(J^3)$$

$$\frac{\partial J_z}{\partial \ln D} = -2(J_z)^2 \rho + O(J^3), \tag{16.255}$$

so that $J_z^2 - J_\perp^2 = $ constant. Draw the corresponding scaling diagram.

Exercise 16.5 Consider the symmetric Anderson model, with a symmetric band structure at half-filling, given by

$$H = \sum_k \epsilon_k c_{k\sigma}^\dagger c_{k\sigma} + \sum_k (V c_{k\sigma}^\dagger f_\sigma + \text{H.c.}) + \frac{U}{2}(n_f - 1)^2, \tag{16.256}$$

where the density of states satisfies $\rho(\epsilon) = \rho(-\epsilon)$. In this model, the d^0 and d^2 states are degenerate and there is the possibility of a charged Kondo effect when the interaction U is negative. Show that, under the particle-hole transformation

$$\begin{aligned}
c_{k\uparrow} &\rightarrow c_{k\uparrow}, & d_\uparrow &\rightarrow d_\uparrow \\
c_{k\downarrow} &\rightarrow -c_{k\downarrow}^\dagger, & d_\downarrow &\rightarrow -d_\downarrow^\dagger,
\end{aligned} \tag{16.257}$$

the sign of U reverses, so that the positive U model is transformed to the negative U model. Show that, under this transformation, spin operators of the local moment are transformed into Nambu *isospin operators* which describe the charge and pair degrees of freedom of the d-state, i.e.

$$\vec{S} = f^\dagger \left(\frac{\vec{\sigma}}{2}\right) f \leftrightarrow \vec{T} = \tilde{f}^\dagger \left(\frac{\vec{\tau}}{2}\right) \tilde{f}, \tag{16.258}$$

where $\tilde{f}^{\dagger} = (f^{\dagger}{}_{\uparrow}, f_{\downarrow})$ is the Nambu spinor for the f-electrons and $\vec{\tau}$ is used to denote the Pauli (Nambu) matrices acting in particle–hole space. Use this transformation to prove that, when U is negative, a charged Kondo effect will occur at exactly half-filling, involving quantum fluctuations between the degenerate d^0 and d^2 configurations. If the spin susceptibilty is enhanced in the spin Kondo effect, what susceptibilities are enhanced in the charge Kondo effect? This negative U physics is believed to occur at Tl ions in the doped semiconductor PbTe [68]. If magnetic impurities are prone to magnetic order, what kind of order is expected for negative U impurities?

Exercise 16.6 Consider the infinite U Anderson model, in which U is taken to infinity, eliminating double occupancy. The Hamiltonian for this limit is given by

$$H = \sum_{k} \epsilon_k c^{\dagger}_{k\sigma} c_{k\sigma} + \sum_{k} V(c^{\dagger}_{k\sigma} X_{0\sigma} + X_{\sigma 0} c_{k\sigma}) + E_f \sum_{\sigma} X_{\sigma\sigma}, \qquad (16.259)$$

where $X_{0\sigma} = P_0 f_{\sigma}$ and $X_{\sigma 0} = f^{\dagger}_{\sigma} P_0$, where $P_0 = (1 - n_{\uparrow})(1 - n_{\downarrow})$ projects into the empty state, are the Hubbard operators linking the singly occupied f-state and the empty state, while $X_{\sigma\sigma'} = f^{\dagger}_{\sigma} P_0 f_{\sigma'} = |f^1 : \sigma\rangle\langle f^1 : \sigma'|$ is the generalized spin operator, acting on the singly occupied states.

(a) What happens to the Schrieffer–Wolff transformation in this infinite U limit?

(b) Generalize the Schrieffer–Wolff transformation in the infinite U limit to the case where the electrons have N spin components, i.e. in every equation the spin summation runs from $\sigma \in [1, N]$?

Exercise 16.7 The Kondo model relevant to spin-orbit coupled ions with total angular momentum j, where $N = 2j + 1$ is the number of spin components, is called the *Coqblin–Schrieffer model*, and takes the from

$$H = \sum_{k\sigma \in [-j, j]} \epsilon_k c^{\dagger}_{k\sigma} c_{k\sigma} + J \sum_{\alpha, \beta \in [-j, j]} c^{\dagger}_{\alpha} c_{\beta} S_{\beta\alpha},$$

where $S_{\beta\alpha} = |\beta\rangle\langle\alpha|$ is the spin operator linking the $J_z = \alpha$ and $J_z = \beta$ states of the local moment. This model has an $SU(N)$ symmetry, rather than the usual $SU(2)$ symmetry of the $S = \frac{1}{2}$ Kondo model. The Nozières–Fermi liquid picture is now generalized to describe the $SU(N)$ Fermi liquid that develops at low temperatures.

(a) Begin by generalizing 16.207, to N components, to obtain

$$\delta_{\sigma}(\{n_{k'\sigma'}\}, E) = \frac{\pi}{N} + \alpha(E - \mu) + \Phi \sum_{k', \sigma' \neq \sigma} \delta n_{k', \sigma'}. \qquad (16.260)$$

Why is the phase shift π/N for an $SU(N)$ Kondo model?

(b) If the charge susceptibility of the impurity is zero, show that α and Φ must satisfy the relation $\alpha + (N - 1)\Phi\rho = 0$.

(c) Show that the Wilson ratio now becomes $W = N/(N - 1)$.

Exercise 16.8 This question takes you through the calculation of the two-loop corrections to the Kondo beta function. In this calculation, work at zero temperature, replacing Matsubara summations by their continuum version, $T \sum_{i\alpha_n} F(i\alpha_n) \rightarrow \int \frac{d\alpha}{2\pi} F[i\alpha]$.

(a) As a preliminary to the calculation, calculate the f-electron self-energy and show that

$$\cdots\!-\!<\!\!-\!\!-\!\!-\!\!->\!\!-\!>\!\!-\!-\!\!-\!\!= \Sigma(z) = \frac{3}{2}\sigma(z) , \tag{16.261}$$

where (for $\text{Im}\,[z] > 0$)

$$\sigma(z) = -(J\rho)^2 z \ln\left[\frac{De}{iz}\right]. \tag{16.262}$$

Hint: you may find it useful to use the result of Example 16.7, which calculated the Kondo polarization bubble

$$\cdots\!-\!->\!-\!-\!= \Pi(z) = \rho \ln\left(\frac{D}{-iz}\right), \qquad (\text{Im}\,z > 0). \tag{16.263}$$

Using this result, one can rewrite the f-electron self-energy as

$$\Sigma(i\omega) = -\left(\frac{3}{2}\right) \int \frac{d\omega'}{2\pi} \Pi(i\omega + i\omega')(-i\pi\rho\,\text{sgn}(\omega')). \tag{16.264}$$

(b) Show that, at low energies, the f-electron propagator acquires a wavefunction renormalization $G_f(i\omega) = Z(i\omega)/i\omega$, where

$$Z(i\omega) = (1 - \frac{\partial \Sigma(i\omega)}{\partial i\omega})^{-1} = \left[1 - \frac{3}{2}(J\rho)^2 \ln\left[\frac{D}{-i\omega}\right]\right], \tag{16.265}$$

and that the leading-order corrections to the Kondo coupling coming from the f-electron self-energies are then

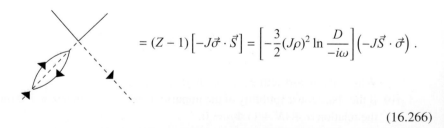

$$= (Z - 1)\left[-J\vec{\sigma} \cdot \vec{S}\right] = \left[-\frac{3}{2}(J\rho)^2 \ln\frac{D}{-i\omega}\right](-J\vec{S} \cdot \vec{\sigma}) . \tag{16.266}$$

(c) Now consider the vertex correction to the Kondo interaction, and show that this can be rewritten as a derivative of the same diagram appearing in the f-electron self-energy, so that

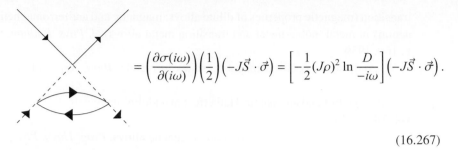

$$= \left(\frac{\partial\sigma(i\omega)}{\partial(i\omega)}\right)\left(\frac{1}{2}\right)(-J\vec{S}\cdot\vec{\sigma}) = \left[-\frac{1}{2}(J\rho)^2 \ln\frac{D}{-i\omega}\right](-J\vec{S}\cdot\vec{\sigma}).$$

(16.267)

(d) Summing the results of (b) and (c), and combining them with the results of Example 16.8, show that the leading logarithmic corrections to the Kondo interaction up to two-loop order are

$$-\Gamma = \;\cdots \;+\; \cdots \;+\; \cdots \;+\; \cdots \;+\; \cdots$$

$$= \left(-J\vec{S}\cdot\vec{\sigma}\right) \times \left(1 + 2(J\rho)\ln\frac{D}{-i\omega} - 2(J\rho)^2 \ln\frac{D}{-i\omega}\right),$$

(16.268)

so that, if we reduce the bandwidth from D to D', the renormalization of $J(D)$ required to keep the low-energy scattering amplitudes and physics fixed is given by

$$J\rho(D') = J\rho + [2(J\rho)^2 - 2(J\rho)^3]\ln\left(\frac{D}{D'}\right),$$

(16.269)

leading to the scaling equation for $g = J\rho$,

$$\frac{\partial g}{\partial \ln D'} = \beta(g) = -2g^2 + 2g^3 + \cdots.$$

(16.270)

References

[1] W. Heisenberg, Zur Theorie des Ferromagnetismus, *Z. Phys. A*, vol. 49, p. 619, 1928.

[2] M. Sarachik, E. Corenzwit, and L. D. Longinotti, Resistivity of Mo-Nb and Mo-Re alloys containing 1% Fe, *Phys. Rev. A*, vol. 135, p. 1041, 1964.

[3] W. J. de Haas, J. H. de Boer, and G. J. van den Berg, The electrical resistance of gold, copper and lead at low temperatures, *Physica*, vol. 1, p. 1115, 1933.

[4] D. K. C. MacDonald and K. Mendelssohn, Resistivity of pure metals at low temperatures I: the alkali metals, *Proc. R. Soc. London*, vol. 202, p. 523, 1950.

[5] A. M. Clogston, B. T. Matthias, M. Peter, H. J. Williams, E. Corenzwit, and R. C. Sherwood, Local magnetic moment associated with an iron atom dissolved in various transition metal alloys, *Phys. Rev.*, vol. 125, p. 541, 1962.

[6] A. Blandin and J. Friedel, Propriétés magnétiques des alliages dilués: interactions magnétiques et antiferromagnétisme dans les alliages du type métal noble-métal de

transition (magnetic properties of dilute alloys: magnetic and antiferromagnetic interactions in metal, noble-metal and transition metal alloys), *J. Phys. Radium*, vol. 19, p. 160, 1959.

[7] P. W. Anderson, Localized magnetic states in metals, *Phys. Rev.*, vol. 124, p. 41, 1961.

[8] J. Kondo, g-shift and anomalous Hall effect in gadolinium metals, *Prog. Theor. Phys.*, vol. 28, p. 846, 1962.

[9] J. Kondo, Resistance minimum in dilute magnetic alloys, *Prog. Theor. Phys.*, vol. 32, p. 37, 1964.

[10] J. R. Schrieffer and P. Wolff, Relation between the Anderson and Kondo Hamiltonians, *Phys. Rev.*, vol. 149, p. 491, 1966.

[11] B. Coqblin and J. R. Schrieffer, Exchange interaction in alloys with cerium impurities, *Phys. Rev.*, vol. 185, p. 847, 1969.

[12] P. W. Anderson and G. Yuval, Exact results in the Kondo problem: equivalence to a classical one-dimensional Coulomb gas, *Phys. Rev. Lett.*, vol. 45, p. 370, 1969.

[13] P. W. Anderson and G. Yuval, Exact results for the Kondo problem: one-body theory and extension to finite temperature, *Phys. Rev. B*, vol. 1, p. 1522, 1970.

[14] P. W. Anderson and G. Yuval, Some numerical results on the Kondo problem and the inverse square one-dimensional Ising model, *J. Phys. C*, vol. 4, p. 607, 1971.

[15] M. Fowler and A. Zawadowski, Scaling and the renormalization group in the Kondo effect, *Solid State Commun.*, vol. 9, p. 471, 1971.

[16] K. G. Wilson, The renormalization group: critical phenomena and the Kondo problem, *Rev. Mod. Phys.*, vol. 47, p. 773, 1975.

[17] P. Nozières, A "Fermi liquid" description of the Kondo problem at low temperatures, *J. Phys., Colloq.*, vol. 37, p. 1, 1976.

[18] P. Nozières and A. Blandin, Kondo effect in real metals, *J. Phys.*, vol. 41, p. 193, 1980.

[19] P. W. Anderson, Local moments and localized states, *Rev. Mod. Phys.*, vol. 50, p. 191, 1978.

[20] N. F. Mott and R. Peierls, Discussion of the paper by de Boer and Verwey, *Proc. R. Soc.*, vol. 49, p. 72, 1937.

[21] N. F. Mott, The basis of the electron theory of metals, with special reference to the transition metals, *Proc. Phys. Soc. London, Sect. A*, vol. 62, p. 416, 1949.

[22] J. H. Van Vleck, Models of exchange coupling in ferromagnetic media, *Rev. Mod. Phys.*, vol. 25, p. 220, 1953.

[23] H. Hurwitz, The Theory of Magnetism in Alloys. Unpublished PhD thesis, Harvard University, 1941.

[24] J. Friedel, On Some electrical and magnetic properties of magnetic solid solutions, *Can . J. Phys.*, vol. 34, p. 1190, 1956.

[25] J. Friedel, Lecture notes on the electronic structure of alloys, *Nuovo Cimento*, Suppl, vol. 7, p. 287, 1958.

[26] D. Morandi and D. Sherrington, Functional integral transformation of the Anderson model into the s-d exchange model, *J. Phys. F: Met. Phys.*, vol. 4, p. L51, 1974.

[27] L. Kouwenhoven and L. Glazman, The revival of the Kondo effect, *Phys. World*, vol. 14, p. 33, 2001.

[28] T. E. Kopley, P. L. McEuen, and R. G. Wheeler, Resonant tunneling through single electronic states and its suppression in a magnetic field, *Phys. Rev. Lett.*, vol. 61, no. 14, p. 1654, 1988.

[29] H. Grabert and M. H. Devoret, *Single Charge Tunneling Coulomb Blockade Phenomena in Nanostructures*, Plenum, 1992.

[30] L. P. Kouwenhoven, T. H. Oosterkamp, M. W. S. Danoesastro, M. Eto, D. G. Austing, T. Honda, and S. Tarucha, Excitation spectra of circular, few-electron quantum dot, *Science*, vol. 5, p. 1788, 1997.

[31] P. W. Anderson, Kondo effect IV: out of the wilderness, *Comments Solid State Phys.*, vol. 5, p. 73, 1973.

[32] P. W. Anderson, *Poor Man's derivaton of scaling laws for the Kondo problem*, J. *Phys. C*, vol. 3, p. 2346, 1970.

[33] N. Andrei, Diagonalization of the Kondo Hamiltonian, *Phys. Rev. Lett.*, vol. 45, p. 379, 1980.

[34] N. Andrei, K. Furuya, and J. H. Lowenstein, Solution of the Kondo problem, *Rev. Mod. Phys.*, vol. 55, p. 331, 1983.

[35] P. B. Weigman, Exact solution of s-d exchange model at $T = 0$, *JETP Lett.*, 1980.

[36] A. Tsvelik and P. Wiegman, The exact results for magnetic alloys, *Adv. Phys.*, vol. 32, p. 453, 1983.

[37] A. Georges, G. Kotliar, W. Krauth, and M. Rozenberg, Dynamical mean-field theory of strongly correlated fermion systems and the limit of infinite dimensions, *Rev. Mod. Phys.*, vol. 68, p. 13, 1996.

[38] G. Kotliar, S. Y. Savrasov, K. Haule, V. S. Oudovenko, O. Parcollet, and C. A. Marianetti, Electronic structure calculations with dynamical mean-field theory, *Rev. Mod. Phys.*, vol. 78, p. 865, 2006.

[39] G. Stewart, Heavy-fermion systems, *Rev. Mod. Phys.*, vol. 56, p. 755, 1984.

[40] F. D. and M. Haldane, Scaling theory of the asymmetric Anderson model, *Phys. Rev. Lett.*, vol. 40, p. 416, 1978.

[41] P. B. Wiegmann, Towards an exact solution of the Anderson model, *Phys. Lett. A*, vol. 80, p. 163, 1980.

[42] N. Kawakami and A. Okiji, Exact expression of the ground-state energy for the symmetric Anderson model, *Phys. Lett. A*, vol. 86, p. 483, 1981.

[43] A. Okiji and N. Kawakami, Thermodynamic properties of the Anderson model, *Phys. Rev. Lett.*, vol. 50, p. 1157, 1983.

[44] A. A. Abrikosov, Electron scattering on magnetic impurities in metals and anomalous resistivity effects, *Physics*, vol. 2, p. 5, 1965.

[45] H. Suhl, Formation of local magnetic moments in metals, *Phys. Rev. A*, vol. 38, p. 515, 1965.

[46] J. W. Allen, S. J. Oh, O. Gunnarsson, K. Schönhammer, M. B. Maple, M. S. Torikachvili, and I. Lindau, Electronic structure of cerium and light rare-earth intermetallics, *Adv. Phys.*, vol. 35, p. 275, 1986.

[47] J. W. Allen, S. J. Oh, M. B. Maple, and M. S. Torikachvili, Large Fermi-level resonance in the electron-addition spectrum of CeRu$_2$ and CeIr$_2$, *Phys. Rev.*, vol. 28, p. 5347, 1983.

[48] D. Langreth, Friedel sum rule for Anderson's model of localized impurity states, *Phys. Rev.*, vol. 150, p. 516, 1966.

[49] J. S. Langer and V. Ambegaokar, Friedel sum rule for a system of interacting electrons, *Phys. Rev.*, vol. 121, p. 1090, 1961.

[50] J. W. Allen, S. J. Oh, L. E. Cox, W. P. Ellis, S. Wire, Z. Fisk, J. L. Smith, B. B. Pate, I. Lindau, and J. Arko, Spectroscopic evidence for the 5f Coulomb interaction in UAl$_2$ and UPt$_3$, *Phys. Rev. Lett.*, vol. 2635, p. 54, 1985.

[51] L. Z. Liu, J. W. Allen, C. L. Seaman, *et al.*, Kondo resonance in $Y_{1-x}U_xPd_3$, *Phys. Rev. Lett.*, vol. 68, p. 1034, 1992.

[52] J. Hubbard, Electron correlations in narrow energy bands. II: the degenerate band case, *Proc. R. Soc. London, Ser. A.*, vol. 277, p. 237, 1964.

[53] V. N. Popov and S. A. Fedotov, The functional-integration method and diagram technique for spin systems, *J. Exp. Theor. Phys.*, vol. 67, p. 535, 1988.

[54] R. H. White and T. H. Geballe, Long-range order in solids, in *Solid State Physics*, ed. H. Ehrenreich, F. Seitz, and D. Turnbull, vol. 15, p. 283, Academic Press, 1979.

[55] S. M. Cronenwett, T. H. Oosterkamp, and L. P. Kouwenhoven, A tunable Kondo effect in quantum dots, *Science*, vol. 281, p. 540, 1998.

[56] W. G. van der Wiel, S. De Franceschi, T. Fujisawa, J. M. Elzerman, S. Tarucha, and L. P. Koewenhoven, The Kondo effect in the unitary limit, *Science*, vol. 289, p. 2105, 2000.

[57] F. T. Hedgcock and C. Rizzuto, Influence of magnetic ordering on the low-temperature electrical resistance of dilute Cd-Mn and Zn-Mn alloys, *Phys. Rev.*, vol. 517, p. 163, 1963.

[58] N. Andrei and C. Destri, Solution of the multichannel Kondo problem, *Phys. Rev. Lett.*, vol. 52, p. 364, 1984.

[59] A. M. Tsvelick and P. B. Wiegmann, Exact solution of the multichannel Kondo problem, scaling, and integrability, *J. Stat. Phys.*, vol. 38, no. 1–2, p. 125, 1985.

[60] V. Emery and S. Kivelson, Mapping of the two-channel Kondo problem to a resonant-level model, *Phys. Rev. B*, vol. 46, no. 17, p. 10812, 1992.

[61] A. W. W. Ludwig and I. Affleck, Exact, asymptotic, three-dimensional, space- and time-dependent Green's functions in the multichannel Kondo effect, *Phys. Rev. Lett.*, vol. 67, p. 3160, 1991.

[62] P. Coleman, L. B. Ioffe, and A. M. Tsvelik, Simple formulation of the two-channel Kondo model, *Phys. Rev. B*, vol. 52, p. 6611, 1995.

[63] D. L. Cox, Quadrupolar Kondo effect in uranium heavy-electron materials?, *Phys. Rev. Lett.*, vol. 59, p. 1240, 1987.

[64] D. Cox and A. Zawadowski, *Exotic Kondo Effects in Metals*, CRC Press, 1999.

[65] A. Yatskar, W. P. Beyermann, R. Movshovich, and P. C. Canfield, Possible correlated-electron behavior from quadrupolar fluctuations in PrInAg$_2$, *Phys. Rev. Lett.*, vol. 77, p. 3637, 1996.

[66] T. Kawae, T. Yamamoto, K. Yurue, N. Tateiwa, K. Takeda, and T. Kitai, Non-Fermi liquid behavior in dilute quadrupolar system $Pr_xLa_{1-x}Pb_3$ with $x \leq 0.05$, *J. Phys. Soc. Jpn.*, vol. 72, p. 2141, 2003.

[67] R. M. Potok, I. G. Rau, H. Shtrikman, Y. Oreg, and D. Goldhaber-Gordon, Observation of the two-channel Kondo effect, *Nature*, vol. 446, no. 7132, p. 167, 2007.

[68] Y. Matsushita, H. Bluhm, T. H. Geballe, and I. R. Fisher, Evidence for charge Kondo effect in superconducting Tl-doped PbTe, *Phys. Rev. Lett.*, vol. 94, p. 157002, 2005.

Heavy electrons

Although the single-impurity Kondo problem was essentially solved by the early 1970s, it took a further decade before the physics community was ready to accept the notion that the same phenomenon could occur in a dense Kondo lattice of local moments, forming quasiparticles with greatly enhanced masses that we now call *heavy electrons*. The early resistance to change was rooted in a number of misconceptions about spin physics and the Kondo effect. Some of the first heavy-electron systems to be discovered are superconductors, yet it was well known that small concentrations of magnetic ions, typically a few percent, suppress conventional superconductivity, so the appearance of superconductivity in a dense magnetic system appeared at first sight to be impossible. Indeed, the observation of superconductivity in UBe_{13} in 1973 [1] was dismissed as an artifact, and ten more years passed before it was revisited and acclaimed as a heavy-fermion superconductor, in which the Kondo effect quenches the local moments to form a new kind of *heavy-fermion metal* [2, 3]. In this chapter, we will study some of the key physics of Kondo lattices that makes this possible.

17.1 The Kondo lattice and the Doniach phase diagram

Local-moment metals normally develop antiferromagnetic order at low temperatures. A magnetic moment induces a cloud of Friedel oscillations in the spin density of a metal with a magnetization profile given by

$$\langle \vec{M}(\mathbf{x}) \rangle = -J \int d^3x' \chi(\mathbf{x} - \mathbf{x}') \langle \vec{S}(\mathbf{x}') \rangle, \tag{17.1}$$

where J is the strength of the Kondo coupling and

$$\chi(\mathbf{x}) = \int_{\vec{q}} \chi(\mathbf{q}) e^{i\mathbf{q}\cdot\mathbf{x}},$$

$$\chi(\mathbf{q}) = 2 \int_{\mathbf{k}} \frac{f(\epsilon_{\mathbf{k}}) - f(\epsilon_{\mathbf{k+q}})}{\epsilon_{\mathbf{k+q}} - \epsilon_{\mathbf{k}}} \tag{17.2}$$

is the the non-local susceptibility of the metal. If a second local moment is introduced at location $\mathbf{x}$, then it couples to $\langle M(\mathbf{x}) \rangle$, shifting the energy by an amount $J\vec{S}(\mathbf{x}) \cdot \langle \vec{M}(\mathbf{x}) \rangle$, giving rise to a long-range magnetic interaction called the *RKKY interaction* (named after Ruderman, Kittel, Kasuya, and Yosida [4]):

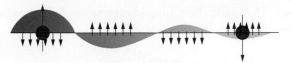

Illustrating how the polarization of spin around a magnetic impurity gives rise to Friedel oscillations, inducing an RKKY interaction between the spins.

Fig. 17.1

$$H_{RKKY} = \frac{1}{2} \sum_{\mathbf{x},\mathbf{x}'} \overbrace{-J^2 \chi(\mathbf{x} - \mathbf{x}')}^{J_{RKKY}(\mathbf{x}-\mathbf{x}')} \vec{S}(\mathbf{x}) \cdot \vec{S}(\mathbf{x}'), \qquad (17.3)$$

where the factor of $\frac{1}{2}$ arises because of the summation over $\mathbf{x}$ and $\mathbf{x}'$. The sharp discontinuity in electron occupancy at the Fermi surface manifests itself as $q = 2k_F$ Friedel oscillations in the RKKY interaction (see Example 17.1),

$$J_{RKKY}(r) \sim J^2 \rho \frac{\cos 2k_F r}{|r|^3}, \qquad (17.4)$$

where r is the distance from the impurity and ρ is the conduction electron density of states per spin (see Figure 17.1). The oscillatory nature of this magnetic interaction tends to frustrate the interaction between spins, so that, in alloys containing a dilute concentration of magnetic transition metal ions, the RKKY interaction gives rise to a frustrated, glassy magnetic state known as a spin glass, but in dense systems the RKKY interaction typically gives rise to an ordered antiferromagnet with a Néel temperature $T_N \sim J^2 \rho$.

The first heavy-electron materials to be discovered are now called *Kondo insulators* [5]. In the late 1960s, Anthony Menth, Ernest Buehler, and Ted Geballe at AT&T Bell Laboratories [6] discovered an unusual metal, SmB_6, containing magnetic Sm^{3+} ions. While apparently a magnetic metal with a Curie–Weiss susceptibility at room temperature, on cooling SmB_6 transforms continuously into a paramagnetic insulator with a tiny 10 meV gap. The subsequent discovery of similar behavior in SmS under pressure led Brian Maple and Dieter Wohlleben [7], working at the University of California, San Diego, to propose that quantum mechanically coherent valence fluctuations in rare-earth ions destabilize magnetism, allowing the f-spin to delocalize into the conduction sea. SmS and SmB_6 are special cases, where the additional heavy f-quasiparticles dope the metal to form a highly correlated insulator. More typically, however, this process gives rise to a heavy-fermion metal.

The first heavy-fermion metal, $CeAl_3$ was discovered by Klaus Andres, John Graebner, and Hans Ott in 1976 [3]. Like many other heavy-fermion metals, this metal displays:

- a Curie–Weiss susceptibility $\chi \sim (T + \theta)^{-1}$ at high temperatures
- a paramagnetic spin susceptibility $\chi \sim$ constant at low temperatures, in this case below 1 K
- a dramatically enhanced linear specific heat $C_V = \gamma T$ at low temperatures, where in $CeAl_3$ $\gamma \sim 1600$ mJ/(mol K^2) is about 1600 times larger than in copper
- a quadratic temperature dependence of the low-temperature resistivity $\rho = \rho_0 + AT^2$.

Andres, Graebner, and Ott proposed that the ground-state excitations of $CeAl_3$ were those of a *Landau Fermi-liquid*, in which the effective mass of the quasiparticles is about 1000 bare electron masses. The Landau Fermi-liquid expressions for the magnetic susceptibility χ and the linear specific heat coefficient γ are

$$\chi = (\mu_B)^2 \frac{N^*(0)}{1 + F_0^a}$$

$$\gamma = \frac{\pi^2 k_B^2}{3} N^*(0), \tag{17.5}$$

where $N^*(0) = \frac{m^*}{m} N(0)$ is the renormalized density of states and F_0^a is the spin-dependent part of the s-wave interaction between quasiparticles. What could be the origin of this huge mass renormalization? Like other cerium heavy-fermion materials, the cerium atoms in this metal are in a $Ce^{3+}(4f^1)$ configuration, and because they are spin–orbit coupled, they form huge local moments with a spin of $J = 5/2$. In their paper, Andres, Ott, and Graebner suggested that a lattice version of the Kondo effect is responsible.

Three years later, in 1979, Frank Steglich and collaborators, working at the Technical Hochshule in Darmstadt, Germany [2], discovered that the heavy-fermion metal $CeCu_2Si_2$ becomes superconducting at 0.5 K. This pioneering result was initially treated with great scepticism, but today we recognize it as the discovery of electronically mediated superconductivity, establishing not only that heavy fermions form within a Kondo lattice, but that they can also pair to form heavy-fermion superconductors.

These discoveries prompted Neville Mott [8] and Sebastian Doniach [9] to propose that heavy-electron systems should be modeled as a Kondo–lattice, where a dense array of local moments interact with the conduction sea via an antiferromagnetic interaction J. The simplest Kondo lattice Hamiltonian [10] is

$$H = \sum_{\mathbf{k}\sigma} \epsilon_{\mathbf{k}} c_{\mathbf{k}\sigma}^\dagger c_{\mathbf{k}\sigma} + J \sum_j \vec{S}_j \cdot c_{j\alpha}^\dagger \vec{\sigma}_{\alpha\beta} c_{j\beta}, \tag{17.6}$$

where

$$c_{j\alpha}^\dagger = \frac{1}{\sqrt{\mathcal{N}_s}} \sum_{\mathbf{k}} c_{\mathbf{k}\alpha}^\dagger e^{i\mathbf{k}\cdot\mathbf{R}_j} \tag{17.7}$$

creates an electron at site j. Mott and Doniach pointed out that there are two energy scales in the Kondo lattice: the Kondo temperature T_K and the RKKY scale E_{RKKY}:

$$T_K = De^{-1/(2J\rho)}$$

$$E_{RKKY} = J^2\rho. \tag{17.8}$$

For small $J\rho$, $E_{RKKY} \gg T_K$, leading to an antiferromagnetic ground state, but when $J\rho$ is large, $T_K \gg E_{RKKY}$, stabilizing a ground state in which every site in the lattice resonantly scatters electrons. Working at Stanford, Remi Jullien, John Fields, and Sebastian Doniach were later able to confirm the correctness of this argument in a simplified one-dimensional "Kondo necklace" model [11], finding that the transition between the two regimes is a continuous quantum phase transition in which the characteristic scale of the antiferromagnet and paramagnet drops to zero at the transition (see Figure 17.2). This led Doniach to conjecture [9] that the general transition between the antiferromagnet and the dense Kondo state is a continuous quantum phase transition. In the Kondo lattice ground state which ensues, Bloch's theorem ensures that the resonant elastic scattering at each site will generate a renormalized f-band, of width $\sim T_K$. In contrast with the impurity

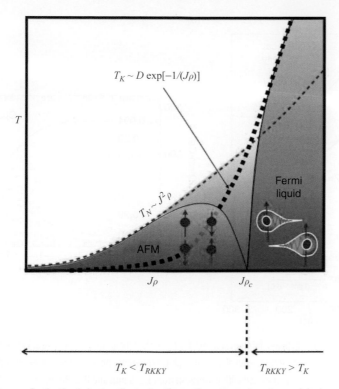

$$T_K \sim D \exp[-1/(J\rho)]$$

T

$T_N \sim J^2 \rho$

AFM

Fermi
liquid

$J\rho$ $J\rho_c$

$T_K < T_{RKKY}$ $T_{RKKY} > T_K$

Doniach phase diagram for the Kondo lattice, illustrating the antiferromagnetic regime and the heavy-fermion regime   **Fig. 17.2**
for $T_K < T_{RKKY}$ and $T_K > T_{RKKY}$, respectively. The effective Fermi temperature of the heavy Fermi liquid is
indicated by a solid line. Experimental evidence suggests that in many heavy-fermion materials this scale drops to
zero at the antiferromagnetic quantum critical point.

Kondo effect, here elastic scattering at each site acts coherently. For this reason, as the
heavy-electron metal develops at low temperatures, its resistivity drops towards zero (see
Figure 17.3(a)).

One of the fascinating aspects of the Kondo lattice concerns the Luttinger sum rule.
Richard Martin [14], working at the Xerox Palo Alto Research Center, pointed out that
the Kondo impurity and lattice models can both be regarded as the result of adiabatically
increasing the interaction strength U in a corresponding Anderson model, while preserving
the valence of the magnetic ion. During this process, the conservation of charge gives
rise to "node-counting" sum rules. In the previous chapter we saw that, for an impurity,
the scattering phase shift at the Fermi energy counts the number of localized electrons,
according to the Friedel sum rule,

$$\sum_\sigma \frac{\delta_\sigma}{\pi} = n_f = 1.$$

This sum rule survives to large U, and reappears as the constraint on the scattering phase
shift created by the Kondo. In the lattice, the corresponding sum rule is the 'Luttinger sum
rule', which states that the Fermi surface volume counts the number of electrons, which at
small U is just the number of localized ($4f$, $5f$, or $3d$) and conduction electrons. When U
becomes large, the number of localized electrons is now the number of spins, so that

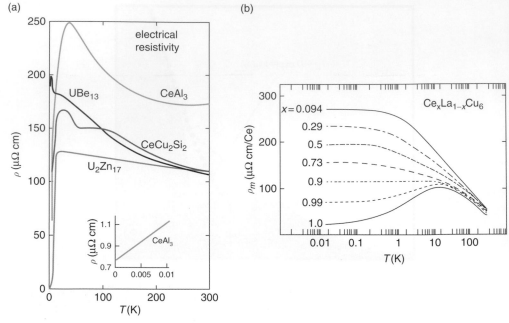

Fig. 17.3 Development of coherence in heavy-fermion systems. (a) The resistivity drops dramatically at low temperatures in heavy-fermion materials, indicating the development of phase coherence between the scattering of the lattice of screened magnetic ions [12]. Reproduced with permission from J. L. Smith and P. S. Riseborough, *J. Magn. Magn. Mater.*, vol. 47–48, p. 545, 1985. Copyright 1985 Elsevier. (b) Resistivity in $Ce_{1-x}La_xCu_6$, showing the transition from the single-impurity ($x \to 0$) form to the coherent-lattice ($x = 1$) limit [13]. Reproduced with permission from Y. Onuki and T. Komatsubara, *J. Magn. Magn. Mater.*, vol. 63–64, p. 281, 1987. Copyright 1987 Elsevier.

$$2\frac{\mathcal{V}_{FS}}{(2\pi)^3} = n_e + n_{spins}. \tag{17.9}$$

Although this sum rule is derived by Martin's adiabatic argument, it is an independent property of the Kondo lattice, and it holds independently of the origin of the localized moments. In other words, we could imagine a Kondo lattice of nuclear spins, provided the Kondo temperature were large enough to guarantee a paramagnetic state. Physically, we can regard the conduction sea as a magnetically polar fluid which magnetically solvates the localized moments, causing them to dissolve into the conduction sea, where they become mobile heavy electrons (see Figure 17.4). A more detailed derivation of Martin's original proposal has been provided by Oshikawa, at the Institute for Solid State Physics, Tokyo [15].

Experimentally, there is a great deal of support for the above picture. It is possible, for example, to follow the effect of progressively increasing the concentration of Ce in the non-magnetic host $LaCu_6$ (Figure 17.3(b)). At dilute concentrations, the resistivity rises to a maximum at low temperatures. At dense concentrations, the resistivity shows the same high-temperature behavior, but at low temperatures coherence between the sites leads to a

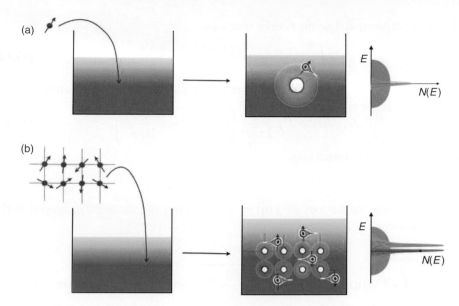

Fig. 17.4

Schematic illustration of the Kondo effect. (a) Single spin in a conduction sea "ionizes" into a Kondo singlet and a heavy fermion orbiting in the vicinity of the Kondo singlet, forming a Kondo resonance at the Fermi surface. (b) Immersion of a lattice of spins in a conduction sea injects a resonance at each site in the lattice, giving rise to a new band of delocalized heavy fermions with a hybridization gap. The density of carriers is increased in the Kondo lattice.

dramatic drop in the resistivity. The thermodynamics of the dense and dilute systems are essentially identical, but the transport properties display the effects of coherence.

The most direct evidence that the Fermi surface of f-electron systems counts the f-electrons derives from quantum oscillation (de Haas–van Alphen and Shubnikov–de Haas oscillation) measurements of the Fermi surface [16, 17]. Typically, in the heavy Fermi liquid, the measured de Haas–van Alphen orbits are consistent with band-structure calculations in which the f-electrons are assumed to be delocalized. By contrast, the measured masses of the heavy electrons often exceed the band-structure calculated masses of the narrow f-band by an order of magnitude or more. Perhaps the most remarkable discovery of recent years is the observation that the volume of the f-electron Fermi surface appears to "jump" to a much smaller value when the f-electrons antiferromagnetically order, indicating that, once the Kondo effect is interrupted by magnetism, the heavy f-electrons relocalize [18].

Example 17.1   The RKKY interaction between two moments in a Fermi liquid is given by

$$J_{RKKY}(\mathbf{x}) = -J^2 \int \frac{d^3q}{(2\pi)^3} e^{i\mathbf{q}\cdot\mathbf{x}} \chi(\mathbf{q}), \qquad (17.10)$$

where, as shown in Chapter 7, $\chi(\mathbf{q}) = 2\rho F[q/2k_F]$ is the static magnetic susceptibility and

$$F(x) = \frac{1}{2} + \frac{1-x^2}{4x} \ln\left|\frac{1+x}{1-x}\right| \qquad (17.11)$$

is the Lindhard function. Use the Fourier transform

$$\int_0^\infty dy \sin(\alpha y) \ln \left| \frac{1+y}{1-y} \right| = \pi \frac{\sin \alpha}{\alpha} \qquad (17.12)$$

to show that the real-space RKKY interaction between local moments is given by

$$J_{RKKY}(r) = J^2 \rho \frac{1}{2\pi^2 r^3} \left[\cos 2k_F r - \frac{\sin 2k_F r}{2k_F r} \right], \qquad (17.13)$$

where r is measured in lattice units.

Solution

We begin by using the isotropy of $\chi(\mathbf{q}) = \chi(q)$ to carry out the angular integral in the Fourier transform:

$$\chi(\mathbf{x}) = \int_{\mathbf{q}} e^{i\mathbf{q}\cdot\mathbf{x}} \chi(\mathbf{q}) = \int_0^\infty \frac{4\pi q^2 dq}{(2\pi)^3} \overbrace{\int \frac{d\Omega}{4\pi} e^{i\mathbf{q}\cdot\mathbf{x}}}^{\frac{\sin qr}{qr}} \chi(q) = \frac{1}{2\pi^2 r} \int_0^\infty dq\, q \sin(qr) \chi(q)$$

$$= \frac{2\rho}{2\pi^2 r} \int_0^\infty dq\, q \sin(qr) F\left[\frac{q}{2k_F} \right]. \qquad (17.14)$$

If we change variable to $y = q/(2k_F)$, so that $dq\, q = (2k_F)^2 y\, dy$, we obtain

$$\chi(r) = \frac{\rho}{\pi^2 r}(2k_F)^2 \int_0^\infty dy\, y \sin(2k_F ry) F[y] = \frac{\rho}{\pi^2 r}(2k_F)^2 G[2k_F r], \qquad (17.15)$$

where

$$G[\alpha] = \int_0^\infty dy \sin(\alpha y) \left(\frac{y}{2} + \frac{(1-y^2)}{4} \ln \left| \frac{1+y}{1-y} \right| \right). \qquad (17.16)$$

Notice that near $y \sim 1$ the singular part of the integrand goes as $dy(y-1) \ln(y-1)$, and since the singular part of the integral has dimension $[y^2] \equiv [\alpha^{-2}]$, we expect this integral to have a $1/\alpha^2 \sim 1/(k_F r)^2$ dependence. To Fourier transform the last two terms in (17.16), we use the result

$$\int_0^\infty dy \sin(\alpha y) \ln \left| \frac{1+y}{1-y} \right| = \pi \frac{\sin \alpha}{\alpha}. \qquad (17.17)$$

(This result is obtained as the inverse Fourier transform of $\sin \alpha/\alpha$.) By differentiating both sides twice with respect to α, we then obtain

$$\int_0^\infty dy \sin(\alpha y)(1-y^2) \ln \left| \frac{1+y}{1-y} \right| = \pi \left(1 + \frac{d^2}{d\alpha^2} \right) \frac{\sin \alpha}{\alpha}$$

$$= -2\pi \left(\frac{\cos \alpha}{\alpha^2} - \frac{\sin \alpha}{\alpha^3} \right), \qquad (17.18)$$

with the expected $1/\alpha^2$ dependence. To complete the job we need to Fourier transform the first term in (17.16). If we differentiate $\int_{-\infty}^\infty dx \cos \alpha y = 2\pi \delta(\alpha)$ with respect to α, we obtain

$$\int_0^\infty dy\, y \sin \alpha y = -\pi \delta'(\alpha). \qquad (17.19)$$

Combining (17.18) and (17.19), we obtain

$$G[\alpha] = \frac{\pi}{2} \left[\frac{\sin\alpha}{\alpha^3} - \frac{\cos\alpha}{\alpha^2} - \delta'(\alpha) \right]. \tag{17.20}$$

When inserted into (17.15) we finally obtain

$$J_{RKKY}(r) = -J^2 \chi(r) = \frac{J^2 \rho}{2\pi r^3} \left[\cos 2k_F r - \frac{\sin 2k_F r}{2k_F r} \right], \tag{17.21}$$

where we have dropped the $\delta'(2k_F r)$ term. Notice that at small distances $J_{RKKY}(r) < 0$ is a ferromagnetic interaction.

17.2 The Coqblin–Schrieffer model

17.2.1 Construction of the model

The stabilization of the heavy-fermion state in f-electron materials owes its origins to the strong spin–orbit coupling, which locks the spin and orbital angular momentum into a large half-integer moment that is unquenched by crystal fields. For example, in Ce^{3+} ions, the $4f^1$ electron is spin–orbit coupled into a state with $j = 3 - \frac{1}{2} = \frac{5}{2}$, giving a spin degeneracy of $N = 2j + 1 = 6$. Ytterbium heavy-fermion materials involve the $Yb:4f^{13}$ configuration, which is most readily understood as a single hole in the filled $4f^{14}$ f-shell, with one hole in the upper spin–orbit multiplet with angular momentum $j = 3 + \frac{1}{2} = \frac{7}{2}$, or $N = 8$. The large spin degeneracy $N = 2j + 1$ of the local moments has the effect of enhancing the Kondo temperature to a point where the zero-point spin fluctuations destroy magnetism.

The presence of large spin–orbit coupling requires a generalization of the Kondo model developed by Coqblin and Schrieffer [19]: they considered a spin–orbit coupled version of the infinite U Anderson model in which the z component of the electron angular momentum, $M \in [-j, j]$, runs from $-j$ to j:

$$H = \sum_{k,M} \epsilon_k c^\dagger_{kM} c_{kM} + E_f \sum_M |f^1 : M\rangle\langle f^1 : M| + \sum_{k,M} V \left[c^\dagger_{kM} |f^0\rangle\langle f^1 : M| + \text{H.c.} \right]. \tag{17.22}$$

In this model, both the f- and the conduction electrons carry spin indices M that run from $-j$ to j. This strange feature is a consequence of rotational invariance, which causes total angular momentum $\vec{J}$ to be conserved by hybridization process: this means that the hybridization is diagonal in a basis where partial-wave states of the conduction sea are written in states of definite j. In this basis, spin–orbit coupled f-states hybridize diagonally with partial wave-states of the conduction electrons in the same spin–orbit coupled j states as the f-electron. Suppose $|k\sigma\rangle$ represents a plane wave of momentum $\mathbf{k}$; then one can construct a state of definite orbital angular momentum l by integrating the plane wave with a spherical harmonic, as follows:

$$|klm\sigma\rangle = \int \frac{d\Omega}{4\pi} |\mathbf{k}\sigma\rangle Y_{lm}(\hat{\mathbf{k}}). \tag{17.23}$$

When spin–orbit interactions are strong, one must work with a partial wave of definite j, obtained by combining these states in the following linear combinations:

$$|kM\rangle = \sum_{\sigma=\pm\frac{1}{2}} |klM - \sigma, \sigma\rangle \left(lM - \sigma, \frac{1}{2}\sigma \middle| jM \right), \tag{17.24}$$

where $(lm, \frac{1}{2}\sigma | jM)$ is the Clebsch–Gordan coefficient between the spin–orbit coupled state $|jM\rangle$ and the l–s coupled state $|lm, \sigma\rangle$. These coefficients can be explicitly evaluated as

$$\left(lM - \sigma, \frac{1}{2}\sigma \middle| jM \right) = \begin{cases} \sqrt{\frac{1}{2} + \frac{2M\sigma}{(2l+1)}}, & j = l + \frac{1}{2} \\ \text{sgn}\,\sigma\sqrt{\frac{1}{2} - \frac{2M\sigma}{2l+1}}, & j = l - \frac{1}{2} \end{cases} \quad \left(M \in [-j,j], \ \sigma = \pm\frac{1}{2} \right). \tag{17.25}$$

Putting this all together, a partial-wave state of definite j, M can then be written as

$$c_{kM}^\dagger = \sum_{\sigma=\pm\frac{1}{2}} \int \frac{d\Omega}{4\pi} \, c_{\mathbf{k},\sigma}^\dagger \, \mathcal{Y}_{\sigma,M}(\hat{\mathbf{k}}), \tag{17.26}$$

where

$$\mathcal{Y}_{\sigma,M}(\mathbf{k}) = \left(l, M - \sigma ; \frac{1}{2}\sigma \middle| jM \right) Y_{l,M-\sigma}(\hat{\mathbf{k}}) \tag{17.27}$$

is a spin–orbit coupled spherical harmonic. Note that the spin–orbit coupled partial-wave states form a complete basis for an impurity model involving a single spherically symmetric magnetic site. This is no longer the case in a lattice, where the set of partial waves at different sites is overcomplete, and an electron which sets off in one partial-wave state at one site can arrive in another partial-wave state at another site.

When $E_f << 0$, the valence of the ion approaches unity ($n_f \to 1$) and one can integrate out the virtual fluctuations $f^1 \rightleftharpoons f^0 + e^-$ via a Schrieffer–Wolff transformation, to obtain the Coqblin–Schrieffer model,

$$H_{CS} = \sum_{kM} \epsilon_k c_{kM}^\dagger c_{kM} - J \sum_{k,k',M,M'} (f_M^\dagger c_{k'M})(c_{kM'}^\dagger f_{M'}) \qquad (M, M' \in [-j,j]), \tag{17.28}$$

where $J = V^2/|E_f|$ is the amplitude for the virtual process. The second term describes a virtual fluctuation in which an f-electron with $j_z = M'$ jumps out into the conduction sea, creating a state with excitation energy of order $|E_f|$, only to be subsequently replaced by an electron with $j_z = M$. Notice how the f-charge $Q = n_f$ of the impurity is *conserved*, by the spin-exchange interaction, $[H, n_f] = 0$, so that the interaction in the Coqblin–Schrieffer model only involves the spin degrees of freedom. It is sometimes useful to rewrite the Coqblin–Schrieffer model in the form

$$H_{CS} = \sum_{kM} \epsilon_k c_{kM}^\dagger c_{kM} + J \sum_{k,k',M,M'} c_{kM}^\dagger c_{k'M'} \overbrace{\left(f_M^\dagger f_M - \frac{1}{N} n_f \delta_{M,M'} \right)}^{S_{M'M}} + \hat{V} \qquad (M, M' \in [-j,j]), \tag{17.29}$$

where $S_{M'M}$ is the $SU(N)$ generalization of a traceless Pauli spin operator. This form of the model emphasizes that the interaction is a pure spin-exchange process. In writing this expression, we have omitted the elastic scattering term $\hat{V} = J\left(\frac{n_f}{N}\right)\sum_{k,k',M}(c_{kM}^{\dagger}c_{k'M} - \delta_{k,k'})$ which results from the rearrangement of the operators. This term does not renormalize, and may be absorbed into a potential scattering phase shift of the conduction electrons off the impurity, or a shift of the chemical potential (in the Kondo lattice).

Example 17.2   In a certain tetragonal crystalline environment, the low-lying ground state of a Ce^{3+} ion is a $|j = 5/2, M_J = \pm\frac{3}{2}\rangle$ state. The hybridization of this state with Bloch waves of momentum $|\mathbf{k}| = k$ is described by the Hamiltonian

$$H_{mv} = V \sum_{M=\pm\frac{3}{2}} \int \frac{k^2 dk}{2\pi^2} \left[c_{kM}^{\dagger}f_M + \text{H.c.}\right], \tag{17.30}$$

where V is the strength of hybridization near the Fermi energy and $c_{kM}^{\dagger}$ creates a conduction electron in an $l = 3, j = 5/2, M_J = \pm\frac{3}{2}$ partial-wave state of wavevector k.

(a) Recast H_{mv} using a plane-wave basis for the conduction electrons.
(b) Show that the hybridization vanishes along the z-axis of momentum space. Why does this happen?

Solution

(a) We begin by rewriting the partial-wave states as plane waves. Using (17.26), we have

$$c_{kM}^{\dagger} = \sum_{\sigma=\pm\frac{1}{2}} \int \frac{d\Omega}{4\pi} c_{\mathbf{k},\sigma}^{\dagger} \mathcal{Y}_{\sigma M}(\mathbf{k}), \tag{17.31}$$

where

$$\mathcal{Y}_{\sigma M}(\mathbf{k}) = \left(3, M - \sigma ; \frac{1}{2}\sigma \Big| \frac{5}{2}M\right) Y_{3,M-\sigma}(\hat{\mathbf{k}})$$

$$= \text{sgn}\,\sigma \sqrt{\frac{1}{2} - \frac{\text{sgn}(M\sigma)}{14}} Y_{3,M-\sigma}(\hat{\mathbf{k}}) \qquad \left(\sigma = \pm\frac{1}{2}, M = \pm\frac{3}{2}\right). \tag{17.32}$$

The hybridization Hamiltonian is then written

$$H_{mv} = V \sum_{\mathbf{k}\sigma,M} \left[c_{\mathbf{k}\sigma}^{\dagger} \mathcal{Y}_{\sigma M}(\hat{\mathbf{k}})f_M + \text{H.c.}\right]. \tag{17.33}$$

(b) Now the Clebsch–Gordan coefficients are either $\pm\sqrt{3/7}$ or $\pm\sqrt{4/7}$, so that

$$\mathcal{Y}_{\sigma M}(\mathbf{k}) = \begin{pmatrix} \mathcal{Y}_{\frac{1}{2}\,\frac{3}{2}}(\hat{\mathbf{k}}) & \mathcal{Y}_{\frac{1}{2}\,-\frac{3}{2}}(\hat{\mathbf{k}}) \\ \mathcal{Y}_{-\frac{1}{2}\,\frac{3}{2}}(\hat{\mathbf{k}}) & \mathcal{Y}_{-\frac{1}{2}\,-\frac{3}{2}}(\hat{\mathbf{k}}) \end{pmatrix} = \begin{pmatrix} \sqrt{\frac{3}{7}}Y_{3,1}(\hat{\mathbf{k}}) & \sqrt{\frac{4}{7}}Y_{3,-2}(\hat{\mathbf{k}}) \\ -\sqrt{\frac{4}{7}}Y_{3,2}(\hat{\mathbf{k}}) & -\sqrt{\frac{3}{7}}Y_{3,-1}(\hat{\mathbf{k}}) \end{pmatrix}. \tag{17.34}$$

The spherical harmonics are given by (Mathematica: SphericalHarmonicY)

$$Y_{3,\pm1}(\hat{\mathbf{k}}) = \mp\sqrt{\frac{21}{64\pi}}(\hat{k}_x \pm i\hat{k}_y)(4\hat{k}_z^2 - 1) \propto \mp(\hat{k}_x \pm i\hat{k}_y)$$

$$Y_{3,\pm2}(\hat{\mathbf{k}}) = \sqrt{\frac{105}{32\pi}}(\hat{k}_x \pm i\hat{k}_y)^2\hat{k}_z \propto \mp(\hat{k}_x \pm i\hat{k}_y)^2. \tag{17.35}$$

We see that, near the z-axis of momentum space, the off-diagonal components of $\mathcal{Y}$ vanish quadratically with k, whereas the diagonal components vanish linearly, so that

$$\mathcal{Y}(\hat{\mathbf{k}}) \sim \begin{pmatrix} \hat{k}_x + i\hat{k}_y & 0 \\ 0 & (\hat{k}_x - i\hat{k}_y) \end{pmatrix}, \tag{17.36}$$

which vanishes linearly with $(\hat{k}_x, \hat{k}_y)$ along the z-axis. This mismatch occurs because plane waves traveling along the z-axis carry $\pm\frac{1}{2}$ units of angular momentum in their direction of motion, and therefore cannot hybridize with the high-spin $M_J = \pm\frac{3}{2}$ f-states, giving rise to a vorticity in the hybridization. This phenomenon is believed to be important for the semi-metallic behavior in the compounds CeNiSn and CeRhSb [20], sometimes called *failed Kondo insulators*.

17.2.2 Enhancement of the Kondo temperature

To get an idea of how the Kondo effect is modified by the large degeneracy, consider the first-order renormalization of the interaction, which is given by the diagrams

$$J_{eff}(D') = \quad \text{(diagram)} \quad + \quad \text{(diagram)}$$

$$= J + NJ^2\rho \ln\left(\frac{D}{D'}\right), \tag{17.37}$$

where the cross on the intermediate conduction electron state indicates that all states in the energy window $|\epsilon_k| \in [D', D]$ are integrated out. The important point to notice here is that the rate of renormalization has been enhanced by a factor of N, due to the multiplicity of intermediate hole states. We can immediately see that the second term is comparable with the first at a scale $D' \equiv T_K = D \exp\left[-\frac{1}{NJ\rho}\right]$, with an N-fold enhancement of the coupling constant. More precisely, we see that the beta function $\beta(g) = \partial g(D')/\partial \ln D' = -g^2$, where $g(D') = N\rho J_{eff}(D')$. A more extensive calculation shows that the beta function to third order takes the form

$$\beta(g) = \frac{dg}{d\ln D'} = -g^2 + \frac{g^3}{N}. \tag{17.38}$$

The beta function describes a family of Kondo models with different cut-offs D' but the same low-energy physics. We can determine T_K as the temperature where the coupling constant becomes of order unity, $g \sim 1$. If we integrate out the conduction electrons with energy greater than T_K, we find

$$\int_{T_K}^{D} d\ln D' = \ln\left(\frac{D}{T_K}\right) = \int_1^{NJ\rho} \frac{dg}{-g^2 + g^3/N} \approx \int_1^{NJ\rho} dg \left[-\frac{1}{g^2} - \frac{1}{Ng}\right]$$

$$= \frac{1}{NJ\rho} - 1 - \frac{1}{N}\ln(NJ\rho), \tag{17.39}$$

which leads to the Kondo temperature

$$T_K = De(NJ\rho)^{\frac{1}{N}} \exp\left[-\frac{1}{NJ\rho}\right], \tag{17.40}$$

so that large degeneracy enhances the Kondo temperature in the exponential factor. By contrast, the RKKY interaction strength is given by $T_{RKKY} \sim J^2\rho$, and it does not involve any N-fold enhancement factors. Thus, in systems with large spin degeneracy, the enhancement of the Kondo temperature favors the formation of the heavy-fermion ground state.

In practice, rare-earth ions are exposed to the crystal fields of their host, which splits the $(N = 2j+1)$-fold degeneracy into many multiplets. Even in this case, the large degeneracy is helpful, because the crystal field splitting is small compared with the bandwidth. At energies D' large compared with the crystal field splitting T_x, $D' >> T_x$, the physics is that of an N-fold degenerate ion, whereas at energies D' small compared with the crystal field splitting, the physics is typically that of a Kramers doublet, i.e.

$$\frac{\partial g}{\partial \ln D} = \begin{cases} -g^2 & (D >> T_x) \\ -\frac{2}{N}g^2 & (D << T_x), \end{cases} \tag{17.41}$$

from which we see that, at low-energy scales, the leading-order renormalization of g is given by

$$\frac{1}{g(D')} = \frac{1}{NJ\rho} - \ln\left(\frac{D}{T_x}\right) - \frac{2}{N}\ln\left(\frac{T_x}{D'}\right),$$

where the first logarithm describes the high-energy screening with spin degeneracy N, and the second logarithm describes the low-energy screening with spin degeneracy 2. This expression is ~ 0 when $D' \sim T_K^*$, the Kondo temperature, so that

$$0 = \frac{1}{NJ\rho} - \ln\left(\frac{D}{T_x}\right) - \frac{2}{N}\ln\left(\frac{T_x}{T_K^*}\right),$$

from which we deduce that the renormalized Kondo temperature has the form [21]

$$T_K^* = D\exp\left(-\frac{1}{2J_0\rho}\right)\left(\frac{D}{T_X}\right)^{\frac{N}{2}-1}.$$

Here the first factor is the expression for the Kondo temperature of a spin factor $\frac{1}{2}$ Kondo model. The second captures the enhancement of the Kondo temperature derived from the renormalization on scales larger than the crystal field splitting. For $T_x \sim 100\,\text{K}$, $D \sim 1000\,\text{K}$ and $N = 6$, the enhancement factor is of order $10^{6/2-1} = 100$. In short, spin–orbit coupling substantially enhances the Kondo temperature even in the presence of crystal fields, and this is an important source of stabilization for the Kondo lattice in local-moment rare-earth and actinide materials. The absence of this effect in transition metal systems means that they are much more prone to the formation of spin glasses rather than heavy-fermion metals.[1]

---

[1] To obtain heavy-fermion behavior in transition metal systems, one needs magnetic frustration. A good example of such behavior is provided by the pyrochlore transition metal heavy-fermion system LiV_2O_3; see [22].

17.3   Large-N expansion for the Kondo lattice

17.3.1   Preliminaries

In the early 1980s, Anderson [23] pointed out that the large spin degeneracy $N = 2j + 1$ furnishes a small parameter $1/N$ which could be used to develop a controlled expansion about the limit $N \rightarrow \infty$. Anderson's observation opened up a new approach to the heavy-fermion problem: the *large-N expansion* [24, 25].

In 1983, two groups, Dennis Newns and Nicholas Read at Imperial College, London, working with Sebastian Doniach at Stanford University [26], and Piers Coleman, the author, working with Philip W. Anderson at Princeton [27], realized that, in the large-N limit, the *RKKY* interaction in the Kondo model could be ignored relative to the Kondo effect. The basic idea is simple: since the Kondo temperature in the N-fold degenerate Coqblin–Schrieffer model is given by (17.40),

$$ T_K = De(NJ\rho)^{\frac{1}{N}} \exp\left[-\frac{1}{NJ\rho} \right], \tag{17.42} $$

then to take the large-N limit at fixed Kondo temperature, one must keep $\tilde{J} = NJ$ fixed. However, if one takes $N \rightarrow \infty$, the rescaled RKKY interaction $J^2\rho = \frac{1}{N^2}\tilde{J}^2\rho \sim O(1/N^2)$ is of order $1/N^2$, and hence vanishes in the large-N limit (Figure 17.5(b)), so the Kondo lattice is stable against magnetism in the large-N limit.

Building on this idea, and taking advantage of Edward Witten's large-N approach to the Gross–Neveu problem [24] and earlier path-integral formulations of the Kondo problem [28, 29], Nicholas Read and Dennis Newns formulated a large-N path integral approach for the Kondo lattice [26, 30, 31], work later extended by Assa Auerbach and Kathryn Levin at the University of Chicago [32]. We shall later examine how this method can be extended to include valence fluctuations in the infinite Anderson model using *slave bosons* [30, 31, 33–35].

The basic idea is to take a limit where every term in the Hamiltonian grows extensively with N. In the path integral for the partition function, the corresponding action then grows extensively with N, so that

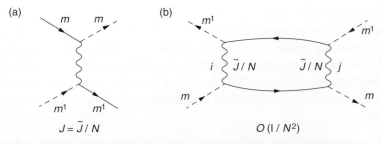

Fig. 17.5   Diagrams from [27]: (a) The Kondo exchange process is of order $O(1/N)$ so that (b) the RKKY interaction, which exchanges spin between two sites, is of order $O(1/N^2)$ and may be neglected in the large-N limit.

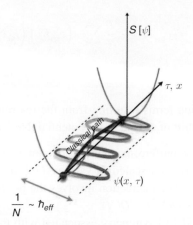

Schematic diagram illustrating the convergence of a quantum path integral about a semiclassical trajectory in the large-N limit.

Fig. 17.6

$$Z = \int \mathcal{D}[\psi] e^{-NS} = \int \mathcal{D}\psi \exp\left[-\frac{S}{1/N}\right] \equiv \int \mathcal{D}[\psi] \exp\left[-\frac{S}{\hbar_{eff}}\right]. \tag{17.43}$$

Here

$$\frac{1}{N} \sim \hbar_{eff}$$

behaves as an effective Planck's constant for the theory, focusing the path integral into a non-trivial "semiclassical" or mean-field solution as $\hbar_{eff} \to 0$. As $N \to \infty$, the quantum fluctuations of intensive variables $\hat{a}$, such as the electron density per spin, become smaller and smaller, scaling as $\langle \delta a^2 \rangle / \langle a^2 \rangle \sim 1/N$, causing the path integral to focus around a non trivial mean-field trajectory. In this way, one can obtain new results by expanding around the solvable large-N limit in powers of $\frac{1}{N}$. For the Kondo model, we are lucky, because much of the important physics of is already captured by the large-N limit (Figure 17.6).

For simplicity, we shall consider a *toy Kondo lattice*, in which all electrons have a spin degeneracy $N = 2j + 1$, interacting with the local moment at each site via a Coqblin–Schrieffer interaction,

$$H = \sum_{\mathbf{k}\alpha} \epsilon_{\mathbf{k}} c^{\dagger}_{\mathbf{k}\alpha} c_{\mathbf{k}\alpha} + \frac{J}{N} \sum_{j,\alpha\beta} c^{\dagger}_{j\beta} c_{j\alpha} S_{\alpha\beta}(j), \tag{17.44}$$

where $c^{\dagger}_{j\alpha} = \frac{1}{\sqrt{N_s}} \sum_{\mathbf{k}} c^{\dagger}_{\mathbf{k}\alpha} e^{-i\mathbf{k}\cdot\vec{R}_j}$ creates an electron localized at site j, and the spin of the local moment at position $\mathbf{R}_j$ is represented by pseudo-fermions:

$$S_{\alpha\beta}(j) = f^{\dagger}_{j\alpha} f_{j\beta} - \frac{n_f(j)}{N}\delta_{\alpha\beta}. \tag{17.45}$$

This representation requires that we set a value for the conserved f-occupancy $n_f(j)$ at each site. In preparation for a path integral approach, we rewrite the interaction in the factorized form encountered in (17.28), so that now

$$H = \sum_{\mathbf{k}\alpha} \epsilon_{\mathbf{k}} c^\dagger_{\mathbf{k}\alpha} c_{\mathbf{k}\alpha} - \frac{J}{N} \sum_{j,\alpha\beta} : \left(c^\dagger_{j\beta} f_{j\beta} \right) \left(f^\dagger_{j\alpha} c_{j\alpha} \right) :, \qquad (17.46)$$

Read–Newns model for the Kondo lattice

where the potential scattering terms resulting from the rearrangement of the f-operators have been absorbed into a shift of the chemical potential. Notice the following:

- In this factorized form, the antiferromagnetic Kondo interaction is attractive.
- The coupling constant has been scaled to vary as J/N, to ensure that the interaction grows extensively with N. The interaction involves the product of two terms that scale as $O(N)$, scaling as $J/N \times O(N^2) \sim O(N)$.
- The model has a global $SU(N)$ symmetry associated with the conservation of the total magnetization.
- This model neglects the effects of spin–orbit coupling and the non-conservation of spin that is present in a typical rare-earth or actinide Kondo lattice material.
- The Coqblin–Schrieffer model also has a *local gauge invariance*: the absence of f-charge fluctuations allows us to change the phase of the f-electrons *independently* at each site:

$$f_{j\sigma} \to e^{i\phi_j} f_{j\sigma}. \qquad (17.47)$$

The appearance of local gauge symmetries in a strongly correlated electron problem is actually a general phenomenon. Here, the incompressible nature of the f-electrons gives rise to a constraint on the Hilbert space, which manifests itself as a gauge field.

Finally, before we continue, we need to decide what value to give the conserved charge $n_f = Q$. Most times, in the physical models of interest, $n_f = 1$ at each site, so one might be inclined to explicitly maintain this condition. However, the large-N expansion requires that the action is extensive in N, and this forces us to consider more general classes of solution where $n_f = Q$ also scales with N so that the f-filling factor $q = Q/N$ is finite as $N \to \infty$. With this device, even if we only impose the constraint $\langle n_f \rangle = Q$ on the average, the RMS fluctuations $\sqrt{\langle \delta n_f^2 \rangle} \sim O(\sqrt{N})$ can be neglected relative to $Q \sim O(N)$. Thus if we're interested in a Kramers doublet Kondo model, we take the half-filled case $q = \frac{1}{2}$, $Q = N/2$, but if we want to understand a $j = 7/2$ Yb^{3+} atom without crystal fields, then in the physical system $N = 2j + 1 = 8$, and we should fix $q = Q/N = \frac{1}{8}$.

17.4 The Read–Newns path integral

To construct the path integral, we need to first take care of the constraint $n_f = Q$. We want to write the partition function as a trace:

$$Z = \mathrm{Tr}\left[e^{-\beta H} \prod_j P_Q(j) \right], \qquad (17.48)$$

where $P_Q(j)$ projects out the states with $n_f(j) = Q$ at site j. The constraints $P_Q(j)$ commute with the Hamiltonian and can be rewritten as a Fourier transform:

$$P_Q(j) = \delta_{n_{fj}, Q} = \int_0^{2\pi} \frac{d\alpha_j}{2\pi} \exp\left[-i\alpha_j(n_{fj} - Q)\right] = \int_0^{2\pi iT} \frac{d\lambda_j}{2\pi iT} \exp\left[-\beta\lambda_j(n_{fj} - Q)\right],$$

$$(17.49)$$

where $\lambda_j = i\alpha_j T$ plays the role of a local chemical potential, integrated between $\lambda_j = 0$ and $\lambda_j = 2\pi iT$ along the imaginary axis. Substituting this expression for $P_Q(j)$ in the partition function, we obtain

$$Z = \int \mathcal{D}[\lambda]\mathrm{Tr}\left[e^{-\beta H[\lambda]}\right], \qquad (17.50)$$

where now

$$H[\lambda] = \sum_{\mathbf{k}\alpha} \epsilon_{\mathbf{k}} c_{\mathbf{k}\alpha}^\dagger c_{\mathbf{k}\alpha} - \frac{J}{N} \sum_{j,\alpha\beta} :\left(c_{j\beta}^\dagger f_{j\beta}\right)\left(f_{j\alpha}^\dagger c_{j\alpha}\right): + \sum_j \lambda_j(n_{fj} - Q) \qquad (17.51)$$

and formally $\mathcal{D}[\lambda] = \prod_j \frac{d\lambda_j}{2\pi iT}$. Now, following the lines of Chapter 11, we rewrite the trace as a path integral:

$$Z = \int \mathcal{D}[\psi^\dagger, \psi, \lambda] \exp\left[-\int_0^\beta d\tau \overbrace{\left(\psi^\dagger \partial_\tau \psi + H[\bar{\psi}, \psi, \lambda]\right)}^{L[\psi^\dagger, \psi, \lambda]}\right], \qquad (17.52)$$

where $\psi^\dagger \equiv (\{c^\dagger\}, \{f^\dagger\})$ schematically represent the conduction and f-electron fields, while ψ is its conjugate. Inside the path integral we shall use $\psi^\dagger$ and ψ to represent the Grassman co-ordinates of the path integral, with the understanding that when used outside the path integral these symbols represents the corresponding field operators. Written in full, the Lagrangian is

$$L[\psi^\dagger, \psi, \lambda] = \sum_{\mathbf{k},\sigma} c_{\mathbf{k}\sigma}^\dagger (\partial_\tau + \epsilon_{\mathbf{k}}) c_{\mathbf{k}\sigma} + \sum_j f_{j\sigma}^\dagger (\partial_\tau + \lambda_j) f_{j\sigma}$$

$$- \frac{J}{N} \sum_{j,\alpha\beta} \left(c_{j\beta}^\dagger f_{j\beta}\right)\left(f_{j\alpha}^\dagger c_{j\alpha}\right) - \sum_j \lambda_j Q. \qquad (17.53)$$

The next step is to carry out a Hubbard–Stratonovich transformation on the interaction:

$$- \frac{J}{N} \sum_{\alpha\beta} \left(c_{j\beta}^\dagger f_{j\beta}\right)\left(f_{j\alpha}^\dagger c_{j\alpha}\right) \rightarrow \sum_\alpha \left[\bar{V}_j \left(c_{j\alpha}^\dagger f_{j\alpha}\right) + \left(f_{j\alpha}^\dagger c_{j\alpha}\right) V_j\right] + N\frac{\bar{V}_j V_j}{J}. \qquad (17.54)$$

In the original Kondo model, we started out with an interaction between electrons and spins. Now, by carrying out the Hubbard–Stratonovich transformation, we have formulated the interaction as the exchange of a charged boson:

$$-\frac{J}{N}\sum_{\mathbf{k},\mathbf{k'},\alpha,\beta}(c^{\dagger}_{\beta}f_{\beta})(f^{\dagger}_{\alpha}c_{\alpha}) \tag{17.55}$$

where the solid lines represent the conduction electron propagators and the dashed lines represent the f-electron operators. Notice how the bare amplitude associated with the exchange boson is frequency-independent, i.e. the interaction is instantaneous. Physically, we may interpret this exchange process as due to an intermediate valence fluctuation.

The path integral now involves an additional integration over the hybridization fields V and $\bar{V}$:

$$Z = \int \mathcal{D}[\bar{V}, V, \lambda] \int \mathcal{D}[\psi^{\dagger}, \psi] \exp\left[-\overbrace{\int_{0}^{\beta}(\psi^{\dagger}\partial_{\tau}\psi + H[\bar{V}, V, \lambda])}^{S[\bar{V},V,\lambda,\,\psi^{\dagger},\psi]}\right]$$

$$H[\bar{V}, V, \lambda] = \sum_{\mathbf{k}}\epsilon_{\mathbf{k}}c^{\dagger}_{\mathbf{k}\sigma}c_{\mathbf{k}\sigma} + \sum_{j}\left[\bar{V}_{j}\left(c^{\dagger}_{j\sigma}f_{j\sigma}\right) + \left(f^{\dagger}_{j\sigma}c_{j\sigma}\right)V_{j}\right.$$

$$\left.+\lambda_{j}(n_{fj} - Q) + N\frac{\bar{V}_{j}V_{j}}{J}\right], \tag{17.56}$$

Read–Newns path integral for the Kondo lattice

where we have suppressed summation signs for repeated spin indices (summation convention).

The importance of the Read–Newns path integral is that it allows us to develop a mean-field description of the many-body Kondo scattering processes that captures the physics and is asymptotically exact as $N \to \infty$. In this approach, the condensation of the hybridization field describes the formation of bound states between spins and electrons that cannot be dealt with in perturbation theory. Bound states induce long-range temporaral correlations in scattering and, indeed, once the hybridization condenses, the interaction lines break up into independent anomalous scattering events, denoted by

The hybridization V in the Read–Newns action carries the local $U(1)$ gauge charge of the f-electrons, giving rise to an important local gauge invariance:

$$f_{j\sigma} \rightarrow e^{i\phi_j} f_{j\sigma}, \qquad V_j \rightarrow e^{i\phi_j} V_j, \qquad \lambda_j \rightarrow \lambda_j - i\dot{\phi}_j(\tau). \qquad (17.57)$$

Read–Newns gauge transformation

It is often useful to use this invariance to choose a gauge in which V_j is real and gauge neutral. We do this by absorbing the phase of the hybridization $V_j = |V_j|e^{i\phi_j}$ into the f-electron. Let us examine how the action at site j transforms when we redefine the f-electrons to absorb this phase:

$$S_K(j) = \int_0^\beta d\tau \left[f_{j\sigma}^\dagger (\partial_\tau + \lambda_j) f_{j\sigma} + \left(\overbrace{|V_j|e^{-i\phi_j}}^{\bar{V}_j} c_{j\sigma}^\dagger f_{j\sigma} + \overbrace{|V_j|e^{i\phi_j}}^{V_j} f_{j\sigma}^\dagger c_{j\sigma} \right) + N\frac{|V_j|^2}{J_K} - \lambda_j Q \right]$$

$$\xrightarrow{f_j \rightarrow e^{i\phi_j} f_j} \int_0^\beta d\tau \left[f_{j\sigma}^\dagger (\partial_\tau + \lambda_j + i\dot{\phi}_j) f_{j\sigma} + |V_j| \left(c_{j\sigma}^\dagger f_{j\sigma} + f_{j\sigma}^\dagger c_{j\sigma} \right) \right.$$

$$\left. + N\frac{|V_j|^2}{J_K} - \lambda_j Q \right]. \qquad (17.58)$$

In our starting model, the constraint field was constant, but in this *radial gauge* it has acquired a time dependence derived from the precession of phase ϕ. If we define the dynamical variable $\lambda_j(\tau) = \lambda_j + i\dot{\phi}_j$, this becomes

$$S_K(j) = \int_0^\beta d\tau \left[f_{j\sigma}^\dagger [\partial_\tau + \lambda_j(\tau)] f_{j\sigma} + |V_j| \left(c_{j\sigma}^\dagger f_{j\sigma} + f_{j\sigma}^\dagger c_{j\sigma} \right) \right.$$

$$\left. + N\frac{|V_j|^2}{J_K} - \lambda_j(\tau) Q \right] + iQ \overbrace{\int_0^\beta d\tau \dot{\phi}_j}^{iQ\Delta\phi_j = i2\pi Qn}. \qquad (17.59)$$

The remainder term comes from making the change of variables in the constraint term $\lambda_j = \lambda_j(\tau) - i\dot{\phi}_j$. Fortunately, this term is an exact integral, and since the change in the phase of the hybridization is an integral multiple of 2π it adds an overall phase $e^{i2\pi n Q} = 1$ to the path integral, and hence can be dropped. In this radial gauge, the Read–Newns path integral becomes

$$Z = \int \mathcal{D}[|V|, \lambda] \int \mathcal{D}[\psi^\dagger, \psi] \exp\left[-\overbrace{\int_0^\beta (\psi^\dagger \partial_\tau \psi + H[|V|, \lambda])}^{S[|V|\lambda, \, \psi^\dagger, \psi]} \right]$$

$$H[|V|, \lambda] = \sum_{\mathbf{k}} \epsilon_{\mathbf{k}} c_{\mathbf{k}\sigma}^\dagger c_{\mathbf{k}\sigma} + \sum_j \left[|V_j| \left(c_{j\sigma}^\dagger f_{j\sigma} + f_{j\sigma}^\dagger c_{j\sigma} \right) \right.$$

$$\left. + \lambda_j (n_{fj} - Q) + N\frac{|V_j|^2}{J} \right]. \qquad (17.60)$$

Read–Newns path integral: radial gauge

By absorbing the phase, the constraint field becomes a dynamical potential field, integrated along the entire imaginary axis (see Example 17.2 for details). Subsequently, when we use the radial gauge we will drop the modulus sign. The interesting feature about this Hamiltonian is that, with the real hybridization, the conduction and f-electrons now transform under a single global $U(1)$ gauge transformation, i.e. the f-electrons have become *charged*. We will return to this issue in Sections 17.8 and 18.6.

17.4.1 The effective action

We now develop the large-N expansion by calculating the effective action. We'll begin without fixing the gauge. The interior fermion integral in the path integral (17.60) defines an effective action $S_E[\bar{V}, V, \lambda]$ by the relation

$$\exp\left[-NS_E[\bar{V}, V, \lambda]\right] \equiv Z_E[\bar{V}, V, \lambda] = \int \mathcal{D}[\psi^\dagger, \psi] \exp\left[-S[\bar{V}, V, \lambda, \psi^\dagger, \psi]\right], \quad (17.61)$$

where we have defined $Z_E = e^{-NS_E}$. Using (17.56),

$$S = \int_0^\beta d\tau \left[\sum_{\mathbf{k}} c_{\mathbf{k}\sigma}^\dagger \left(\partial_\tau + \epsilon_{\mathbf{k}}\right) c_{\mathbf{k}\sigma} + \sum_j \left(f_{j\sigma}^\dagger (\partial_\tau + \lambda_j) f_{j\sigma} + (\bar{V}_j c_{j\sigma}^\dagger f_{j\sigma} + V_j f_{j\sigma}^\dagger c_{j\sigma}) \right.\right.$$
$$\left.\left. + N \frac{\bar{V}_j V_j}{J} - \lambda_j Q \right) \right]. \quad (17.62)$$

The extensive growth of the effective action with N means that at large N the integration in (17.56) is dominated by its stationary points:

$$Z = \int \mathcal{D}[\lambda, \bar{V}, V] \exp\left[-NS_E[\bar{V}, V\lambda]\right] \approx \left.\exp\left[-NS_E[\bar{V}, V, \lambda]\right]\right|_{\text{saddle point}}. \quad (17.63)$$

If we identify $NS_E = -\ln Z_E$, so that $N\delta S_E = -\delta Z_E/Z_E$, then, differentiating (17.61) with respect to $\bar{V}_j$ and λ_j, we see that the saddle-point conditions impose the self-consistent relations

$$\frac{\delta NS_E}{\delta \bar{V}_j(\tau)} = \frac{1}{Z_E} \int \mathcal{D}[\psi^\dagger, \psi] \overbrace{\left(c_{j\sigma}^\dagger(\tau) f_{j\sigma}(\tau) + \frac{NV_j(\tau)}{J} \right)}^{\delta S/\delta \bar{V}_j(\tau)} e^{-S} = \langle c_{j\sigma}^\dagger f_{j\sigma}\rangle(\tau) + \frac{N}{J} V_j(\tau) = 0$$

$$\frac{\delta NS_E}{\delta \lambda_j(\tau)} = \frac{1}{Z_E} \int \mathcal{D}[\psi^\dagger, \psi] \left(n_f(j, \tau) - Q \right) e^{-S} = \langle n_f(j, \tau)\rangle - Q = 0, \quad (17.64)$$

where repeated spin indices imply summation. The second equation in (17.64) is the satisfaction of the constraint, on the average. The first relation, which can be written $V_j = -\frac{J}{N} \langle c_{j\sigma}^\dagger f_{j\sigma}\rangle$, is recognized as the mean-field self-consistency associated with the Hubbard–Stratonovich factorization. We can denote this self-consistency by the Feynman diagram

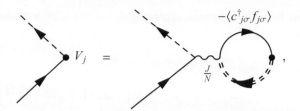

indicating that the condensation of the boson V is self-consistently induced by an anomalous hybridization. Fortunately, we will not have to solve these equations by explicitly calculating the expectation values; instead, as we found in previous chapters (see Section 13.3), they are implicity imposed by finding the stationary point of the action.

In practice, we shall seek static solutions, using the radial gauge to absorb the phase of the hybridization, so that $\bar{V}_j(\tau) = V_j(\tau) = |V_j|$, $\lambda_j(\tau) = \lambda_j$. In this case the saddle-point partition function $Z_E[V, \lambda]$ is simply the partition function of the static mean-field Hamiltonian $H_{MF} = H[V, \lambda]$, $Z_E = \text{Tr} e^{-\beta H_{MF}}$. Now we may write the action in the form

$$S = \int_0^\beta d\tau \left[\sum_\sigma \psi_\sigma^\dagger (\partial_\tau + \underline{h}) \psi_\sigma + \sum_j \left(N \frac{V_j^2}{J} - \lambda_j \underline{Q} \right) \right], \qquad (17.65)$$

where the matrix $\underline{h}[V, \lambda]$ is a mean-field Hamiltonian, read off from (17.56). For instance, in a tight-binding representation,

$$H[V, \lambda] = \sum_{i,j,\sigma} \left(c_{i\sigma}^\dagger, \ f_{i\sigma}^\dagger \right) \overbrace{\begin{bmatrix} (t_{ij} - \mu \delta_{ij}) & \bar{V}_j \delta_{ij} \\ V_j \delta_{ij} & \lambda_j \delta_{ij} \end{bmatrix}}^{\underline{h}[V, \lambda]} \begin{pmatrix} c_{j\sigma} \\ f_{j\sigma} \end{pmatrix} + \sum_j \left(N \frac{|V_j|^2}{J} - \lambda_j Q \right), \quad (17.66)$$

where the t_{ij} are the hopping matrix elements obtained by Fourier transforming $\epsilon_{\mathbf{k}} = \sum_{\mathbf{R}_{ij}} (t(\mathbf{R}_{ij}) - \mu \delta_{ij}) e^{-i\mathbf{k} \cdot \mathbf{R}_{ij}}$.

Since the action is Gaussian in the Fermi fields, the Fermi integral can be carried out using (12.142) in terms of the determinant of the action:

$$\int \mathcal{D}[\psi^\dagger, \psi] \exp\left[-\int_0^\beta d\tau \sum_\sigma \psi_\sigma^\dagger (\partial_\tau + \underline{h}) \psi_\sigma \right] = (\det[\partial_\tau + \underline{h}])^N$$

$$= \exp\left[N \ln \det[\partial_\tau + \underline{h}] \right]$$

$$= \exp\left[N \text{Tr} \ln[\partial_\tau + \underline{h}] \right], \qquad (17.67)$$

where the power N derives from the N identical integrals over each spin component of ψ_σ. In the last line, we have replaced $\ln \det \to \text{Tr} \ln$. Thus

$$N S_E[V, \lambda] = N \left[-\text{Tr} \ln\left(\partial_\tau + \underline{h} \right) + \sum_j \int_0^\beta d\tau \left(\frac{|V_j|^2}{J} - \lambda_j q \right) \right]. \qquad (17.68)$$

Since $Z_E = e^{-\beta F_{MF}} = e^{-N S_E}$, where F_{MF} is the mean-field free energy, it follows that

$$F_{MF}[V, \lambda] = \frac{1}{\beta} S_E[V, \lambda] = -\frac{N}{\beta} \text{Tr} \ln\left(\partial_\tau + \underline{h}[V, \lambda] \right) + \sum_j \left(\frac{N|V_j|^2}{J} - \lambda_j Q \right). \quad (17.69)$$

If we switch to the frequency domain, replacing $\partial_\tau \rightarrow -i\omega_n$ by a Matsubara frequency, we may also write

$$
F_{MF} = -NT \sum_{i\omega_n} \text{Tr} \ln\left[-\mathcal{G}^{-1}(i\omega_n) \right] + \sum_j \left(\frac{N|V_j|^2}{J} - \lambda_j Q \right)
$$

$$
\mathcal{G}^{-1} = (i\omega_n - \underline{h}[V, \lambda]), \tag{17.70}
$$

where we have identified $\mathcal{G}^{-1} = (i\omega_n - \underline{h}[V, \lambda])$ with the inverse Green's function. Sometimes, it's convenient to re-express S_E in terms of the eigenvalues E_ζ of the Hamiltonian. If we diagonalize the Hamiltonian, so that $\underline{h} \rightarrow E_\zeta \delta_{\zeta\zeta'}$, then $\text{Tr} \ln(-i\omega_n + \underline{h}) = \sum_\zeta \ln(E_\zeta - i\omega_n)$. We can also do the Matsubara sum, under which $-T \sum_{i\omega_n} \ln(E_\zeta - i\omega_n) \rightarrow -T \ln(1 + e^{-\beta E_\zeta})$, so that the free energy can also be written

$$
F_E[V, \lambda] = -NT \sum_\zeta \ln\left(1 + e^{-\beta E_\zeta} \right) + \sum_j \left(\frac{N|V_j|^2}{J} - \lambda_j Q \right). \tag{17.71}
$$

Equations (17.70) and (17.71) are complementary: the former reflects the path-integral approach, the latter a more conventional mean-field approach. Let us now apply them to the Kondo impurity and lattice models.

Example 17.3   This example shows in detail how to derive the measure of the Read–Newns path integral. The initial Kondo lattice path integral involves static constraint fields λ_j, integrated over a finite range of the imaginary axis: $\lambda_j \in [0, i2\pi T]$, as follows:

$$
Z = \prod_j \int_0^{2\pi iT} \frac{d\lambda_j}{2\pi iT} \int \mathcal{D}[V, \psi] \exp\left[-\int_0^\beta (\bar{\psi} \partial_\tau \psi + H[V, \lambda]) \right]. \tag{17.72}
$$

By inserting the identity $\int \mathcal{D}[g_j] = 1$ into the Kondo path integral, where $\mathcal{D}[g_j]$ denotes the integration over the entire orbit of gauge transformations $g_j(\tau) = e^{i\phi_j(\tau)}$, show that λ_j is promoted to a dynamical variable $\lambda_j'(\tau) = \lambda_j + i\dot{\phi}_j(\tau)$, integrated over the entire imaginary axis.

Solution

If we insert the identity $\prod_j \int \mathcal{D}[g_j] = 1$ into the path integral, it becomes

$$
Z = \int \mathcal{D}[\lambda, g, V, \psi] e^{-S[\lambda, V, \psi]}. \tag{17.73}
$$

At this point, g is just a dummy variable. We need to (a) carry out a gauge transformation to absorb g into the fields, and (b) rewrite the measure of integration in terms of the transformed fields.

(a) Change of variables.

The first step is to show that the action is unchanged by the gauge transformation

$$f_j(\tau) = e^{i\phi_j(\tau)}f'_j(\tau), \qquad V_j(\tau) = e^{i\phi_j(\tau)}V'_j(\tau). \tag{17.74}$$

Under this transformation, the Hamiltonian is unchanged but the f-Berry phase term (see (12.132)) acquires an additional $i\dot{\phi}_j$ term from the time dependence of $g_j(\tau) = e^{i\phi_j(\tau)}$, as follows:

$$f^\dagger(\partial_\tau + \lambda_j)f \longrightarrow f'^\dagger e^{-i\phi_j}(\partial_\tau + \lambda_j)e^{i\phi_j}f' = f'^\dagger(\partial_j + \lambda_j + i\dot{\phi}_j)f'_j. \tag{17.75}$$

To absorb this term we must also transform the λ_j field, introducing the dynamical variable $\lambda'_j(\tau) = \lambda_j + i\dot{\phi}_j(\tau)$. Subtly, under the transformation the constraint term adds a phase shift to the action:

$$S[V, \lambda, \psi] = \int_0^\beta d\tau \sum_j \left[f'^\dagger_j(\partial_\tau + \lambda'_j)f'_j - (\lambda'_j - i\dot{\phi}_j)Q \right] + \cdots$$

$$= S[V', \lambda', \psi'] + iQ \sum_j \int_0^\beta d\tau \dot{\phi}_j. \tag{17.76}$$

Now $Q\int_0^\beta d\tau \dot{\phi}_j = Q\Delta\phi_j$ is determined by the change in ϕ_j between $\tau = 0$ and $\tau = \beta$. Since $g_j = e^{i\phi_j}$ is periodic in time, $\Delta\phi_j$ is an integer multiple M_j of 2π, and since Q is an integer, the phase shift is a multiple of 2π, leaving e^{-S} invariant:

$$\exp\left(-S[V, \lambda, \psi]\right) = \exp\left(-S[V', \lambda', \psi'] - 2\pi i \sum_j (QM_j)\right) = \exp\left(-S[V', \lambda', \psi']\right). \tag{17.77}$$

(b) Change of measure.

Since the gauge transformation is unitary, the measure for the hybridization and f-electron fields is unchanged (phase factors cancel):

$$\prod_\tau d\bar{V}_j(\tau)dV_j(\tau) = \prod_\tau d\bar{V}'_j(\tau)dV'_j(\tau), \qquad \prod_\tau df_j^\dagger(\tau)df_j(\tau) = \prod_\tau df_j'^\dagger(\tau)df'_j(\tau). \tag{17.78}$$

Next, we show that the remaining measure $\mathcal{D}[\lambda, g] = \mathcal{D}[\lambda']$, with a flat measure of integration over the dynamical variable $\lambda'_j(\tau) = \lambda_j + i\dot{\phi}_j$. Since $\phi_j(\beta) = \phi_j(0) + 2\pi M_j$ is periodic up to a multiple of 2π, we may write

$$\phi_j(\tau) = 2\pi T M_j \tau + \tilde{\phi}_j(\tau), \tag{17.79}$$

which describes a path for $g_j(\tau) = e^{i\phi_j(\tau)}$ that wraps M_j times around the origin. The second term is a periodic function of τ that can be decomposed into its Matsubara

Fourier components, $\tilde{\phi}_j(\tau) = \sum_n \tilde{\phi}_n(j)e^{-i\nu_n\tau}$. The original measure for integrating over the static λ_j and g_j is

$$\mathcal{D}[\lambda_j, g_j] = \sum_{M_j} \int_0^{2\pi iT} d\lambda_j \prod_\tau d\tilde{\phi}_j(\tau)$$

$$= \sum_{M_j} \int_0^{2\pi iT} d\lambda_j \prod_n d\tilde{\phi}_n(j), \qquad (17.80)$$

where, in the last line, the measure for the integration over $\tilde{\phi}_j$ has been replaced by the integration over its Matsubara Fourier components.

Now the dynamical variable $\lambda'_j(\tau) = \lambda_j + i\dot{\tilde{\theta}}_j = \lambda_j + 2\pi iTM_j + i\dot{\tilde{\phi}}_j(\tau)$ has a Fourier series

$$\lambda'_j(\tau) = \sum_n \lambda'_n(j)e^{-i\nu_n\tau}, \qquad (17.81)$$

where $\lambda'_0(j) = \lambda_j + 2\pi iTM_j$ and $\lambda'_n(j) = i(-i\nu_n)\tilde{\theta}_n(j) = \nu_n\tilde{\phi}_n(j)$. When we integrate over λ_j, the range of the $\lambda'_0(j) = \lambda_j + 2\pi iTM_j$ runs from $2\pi iTM_j$ to $2\pi iT(M_j + 1)$ along the imaginary axis, so that when we sum over all M_j, $\lambda'_0(j)$ runs over the entire imaginary axis (see figure below).

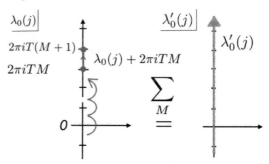

It follows that the combination

$$\sum_{M_j} \int_{2\pi iTM_j}^{2\pi iT(M_j+1)} d\lambda'_0(j) \equiv \int_{-i\infty}^{i\infty} d\lambda'_0(j) \qquad (17.82)$$

gives an unconstrained integral over the static part $\lambda'_0(j)$ of $\lambda'_j(\tau)$.

For $n \neq 0$, the Fourier coefficients $\lambda'_n(j) = \nu_n\tilde{\phi}_n(j)$ are directly proportional, so, up to a normalization, their measures are equal, so that

$$\prod_n d\tilde{\phi}_n(j) = \mathcal{N}^{-1}d\phi_0(j) \prod_{n\neq 0} d\lambda'_n(j),$$

where $\mathcal{N}$ is a normalizing factor $\prod_n = \prod_{n\neq 0} \nu_n$ that we shall drop. Thus, by integrating over all possible $\tilde{\phi}_j$, we integrate over all finite frequency Fourier components of $\lambda'_j(\tau)$.

Combining the static and dynamical parts of the measure, it follows that

$$\mathcal{D}[\lambda_j, g_j] \equiv \mathcal{D}[\lambda'_j] = \prod_j d\phi_0(j) \prod_n d\lambda'_n(j),$$

and since $\mathcal{D}[V, \psi] = \mathcal{D}[V', \psi']$,

$$\mathcal{D}[g, \lambda, V, \psi] = \mathcal{D}[\phi_0]\mathcal{D}[\lambda', V', \psi'], \tag{17.83}$$

where the measure for the dynamical field λ' is flat and $\mathcal{D}[\phi_0] = \prod_j d\phi_0(j)$ is the integral over the static phases $\phi_0(j)$. Since the action is independent of ϕ_0, we can drop the overall integral over the static phases, enabling us to replace $\mathcal{D}[\phi_0] \to 1$ and write

$$\mathcal{D}[g, \lambda, V, \psi] \equiv \mathcal{D}[\lambda', V', \psi']. \tag{17.84}$$

17.5 Mean-field theory of the Kondo impurity

17.5.1 The impurity effective action

The large-N mean-field theory of the Kondo effect maps the original Hamiltonian onto a self-consistently determined resonant level model, which we will write in the form

$$H_{MF} = \sum_{\sigma} \left(\cdots c_{\mathbf{k}\sigma}^{\dagger} \cdots , \ f_{\sigma}^{\dagger} \right) \left(\begin{array}{c|c} \epsilon_{\mathbf{k}}\delta_{\mathbf{k},\mathbf{k}'} & \bar{V} \\ \hline V & \lambda \end{array} \right) \left(\begin{array}{c} \vdots \\ c_{\mathbf{k}'\sigma} \\ \vdots \\ f_{\sigma} \end{array} \right) + \frac{NV^2}{J} - \lambda Q. \tag{17.85}$$

In Section 16.4.2 we learned that the single resonance described by this model is located at an energy λ, with a hybridization width $\Delta = \pi \rho V^2$ (see (16.19)). By minimizing the free energy of this system, we need to figure out how λ and Δ are related to the Kondo coupling constant. Let us first evaluate the free energy of the resonance. We can read off $\underline{h}$ from (17.85), so from (17.70), the mean-field free energy is given by

$$F_{MF} = -TN \sum_n \ln \det \left(\begin{array}{c|c} (\epsilon_{\mathbf{k}} - i\omega_n)\delta_{\mathbf{k},\mathbf{k}'} & \bar{V} \\ \hline V & \lambda - i\omega_n \end{array} \right) + \left(\frac{NV^2}{J} - \lambda Q \right). \tag{17.86}$$

Using the result $\det \begin{pmatrix} D & C \\ B & A \end{pmatrix} = \det D \det \left[A - BD^{-1}C \right]$, we can integrate out the conduction electrons to write

$$F_{MF} = -TN \sum_n \ln \left[-i\omega_n + \lambda + \sum_{\mathbf{k}} \frac{|V|^2}{i\omega_n - \epsilon_{\mathbf{k}}} \right] + \left(\frac{NV^2}{J} - \lambda Q \right) + F_C, \tag{17.87}$$

where $F_C = -TN \sum_{\mathbf{k},n} \ln(\epsilon_{\mathbf{k}} - i\omega_n)$ is the conduction electron free energy. Using the large-bandwidth approximation, $\sum_{\mathbf{k}} \frac{|V|^2}{i\omega_n - \epsilon_{\mathbf{k}}} = -i\Delta \text{sgn}(\omega_n) \equiv \Delta_n$ (see (16.29)), this becomes

$$F_{MF}[V, \lambda] = -\frac{N}{\beta} \sum_n \ln \left[-i\omega_n + \lambda + i\Delta_n \right] e^{i\omega_n 0^+} + N \left(\frac{|V|^2}{J_K} - \lambda Q \right) + F_C. \tag{17.88}$$

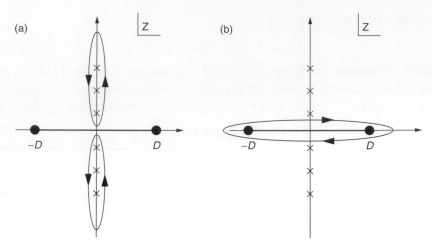

Fig. 17.7 Contour used in evaluating free energy: (a) undistorted contour; (b) contour distorted to run around branch cut in the f-electron Green's function.

We now carry out the Matsubara summation by the standard method, replacing $-T \sum_n F(i\omega_n) \rightarrow \oint \frac{dz}{2\pi i} f(z) F[z]$, where the contour runs counterclockwise around the poles in the Fermi function (Figure 17.7(a)). Now the logarithm contains a branch cut along the real axis, where $\Delta_n = \Delta \operatorname{sgn}(\operatorname{Im} z)$ jumps from $i\Delta$ below the real axis to $-i\Delta$ above it. If we introduce a finite bandwidth D, this branch cut runs from $z = -D$ to $z = +D$.

Distorting the contour to run clockwise around this branch cut (Figure 17.7(b)) we obtain

$$F_{MF}[V, \lambda] = N \int_{-D}^{D} \frac{d\omega}{2\pi i} f(\omega) \left(\ln\left[-\omega + \lambda - i\Delta\right] - \ln\left[-\omega + \lambda + i\Delta\right] \right)$$

$$+ N \left(\frac{|V|^2}{J_K} - \lambda Q \right) + F_C$$

$$= -N \int_{-D}^{D} d\omega f(\omega) \left(\overbrace{\frac{1}{\pi} \operatorname{Im} \ln\left[\lambda + i\Delta - \omega\right]}^{\delta_f(\omega)/\pi} \right) + N \left(\frac{|V|^2}{J_K} - \lambda Q \right) + F_C,$$

$$(17.89)$$

where we have made the identification $\delta(\omega) = \operatorname{Im} \ln\left[\lambda + i\Delta - \omega\right] = \tan^{-1}\left(\frac{\Delta}{\lambda - \omega}\right)$ as the scattering phase shift of the impurity (see (16.33)). We then obtain

$$F_{MF}[V, \lambda] = -N \int_{-D}^{D} d\omega \left(\frac{\delta_f(\omega)}{\pi} \right) f(\omega) + N \left(\frac{|V|^2}{J_K} - \lambda Q \right) + F_C. \qquad (17.90)$$

We can give this result a simple interpretation: the effect of the resonant phase shift changes the allowed momenta of the radial partial-wave states, which in turn causes the one-particle eigenstates of the continuum to move by a fraction δ_f/π of the energy-level spacing $\Delta\epsilon$ according to the relation (see (16.205)) $\tilde{\epsilon}_k = \epsilon_k - \frac{\delta(\epsilon_k)}{\pi} \Delta\epsilon$, where k labels the eigenstates. The corresponding change in the free energy of the continuum is then

$$\Delta F = \sum_k \frac{\partial}{\partial \epsilon_k} \left(-T \ln \left[1 + e^{-\beta \epsilon_k} \right] \right) \left[-\frac{\delta_f(\epsilon_k)}{\pi} \Delta \epsilon \right] = -\sum_k f(\epsilon_k) \frac{\delta(\epsilon_k)}{\pi} \Delta \epsilon$$

$$\equiv -\int d\epsilon \frac{\delta_f(\epsilon)}{\pi} f(\epsilon), \tag{17.91}$$

where we have replaced the discrete summation by an integral. The first term in (17.90) is precisely this shift in the continuum free energy.

Example 17.4

(a) Diagonalize the impurity resonant level Hamiltonian

$$H_{MF} = \sum_{k\sigma} \epsilon_k c_{k\sigma}^\dagger c_{k\sigma} + \sum_{k\sigma} V[c_{k\sigma}^\dagger f_\sigma + f_\sigma^\dagger c_{k\sigma}] + \lambda \sum_\sigma n_{f\sigma} \tag{17.92}$$

and compute the scattering phase shift of the resonant level.

(b) Show that injection of an f-state into the continuum induces a resonant correction to the total the density of states:

$$\rho \rightarrow \rho^*(E) = \rho + \frac{1}{\pi} \frac{\Delta}{(E - \lambda)^2 + \Delta^2}. \tag{17.93}$$

Solution

(a) To diagonalize the Hamiltonian, we write it in the form

$$H = \sum_{\gamma\sigma} E_\gamma a_{\gamma\sigma}^\dagger a_{\gamma\sigma}, \tag{17.94}$$

where the quasiparticle operators a_γ are related to the original operators via the one-particle eigenstates, $a_{\gamma\sigma}^\dagger = \sum_k c_{k\sigma}^\dagger \langle k|\gamma\rangle + f_\sigma^\dagger \langle f|\gamma\rangle \equiv \sum_k \alpha_k c_{k\sigma}^\dagger + \beta f_\sigma^\dagger$. Now if we denote the amplitudes of the one-particle eigenstates $|\gamma\rangle$ by $\langle \eta|\gamma\rangle \equiv (\cdots \langle k'|\gamma\rangle \cdots, \langle f|\gamma\rangle)$, then since $h_{\eta\eta'}\langle \eta'|\gamma\rangle = \langle \eta|H|\gamma\rangle = E_\gamma \langle \eta|\gamma\rangle$ it follows that the amplitudes $\langle \eta|\gamma\rangle$ must satisfy the eigenvalue equation

$$\underline{h} \cdot \begin{pmatrix} \vdots \\ \alpha_k \\ \vdots \\ \beta \end{pmatrix} = \left(\begin{array}{c|c} \epsilon_k \delta_{k,k'} & \bar{V} \\ \hline V & \lambda \end{array} \right) \begin{pmatrix} \vdots \\ \alpha_{k'} \\ \vdots \\ \beta \end{pmatrix} = E_\gamma \begin{pmatrix} \vdots \\ \alpha_k \\ \vdots \\ \beta \end{pmatrix} \tag{17.95}$$

or

$$\epsilon_k \alpha_k + V\beta = E_\gamma \alpha_k$$

$$V \sum_{k'} \alpha_{k'} + \lambda\beta = E_\gamma \beta. \tag{17.96}$$

(If you like, you can rederive this by expanding the quasiparticle operators on both sides of (17.94) in terms of the conduction and f-electron fields, carrying out the commutator and then comparing coefficients of $c_{k\sigma}^\dagger$ and $f_\sigma^\dagger$ (see Example 14.3).) Solving for α_k using the first equation, and substituting into the second, we obtain

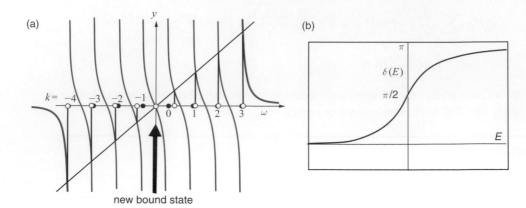

(a)

(b)

new bound state

Fig. 17.8 (a) Graphical solution of the equation $y = \lambda + \sum_k \frac{V^2}{y - \epsilon_k}$, for eight equally spaced conduction electron energies for a resonance located at $\lambda = 0$ (arrow). Notice how the injection of a bound state at $y = 0$ displaces electron band states away from the Fermi surface, increasing the number of eigenstates by one. (b) Energy dependence of the scattering phase shift.

$$E_\gamma - \lambda - \sum_k \frac{V^2}{E_\gamma - \epsilon_k} = 0. \tag{17.97}$$

We can recognize this solution as a pole of the f-Green's function, $G_f(E_\gamma)^{-1} = 0$ (see (16.23) and (16.25)).

The solutions of the eigenvalue equation (17.97) are illustrated graphically in Figure 17.8. Suppose the energies of the conduction sea are given by the $2M$ discrete values

$$\epsilon_k = \left(k + \frac{1}{2}\right) \Delta \epsilon \qquad (k \in \{-M, \ldots M - 1\}), \tag{17.98}$$

distributed symmetrically above and below the Fermi energy. Consider the particle–hole case when the f-state is exactly half-filled, i.e. when $\lambda = 0$. From the diagram, we see that one solution to the eigenvalue equation corresponds to $E_\gamma = 0$, i.e. the original $2M$ band-electron energies have been displaced to both lower and higher energies, forming a band of $2M + 1$ eigenvalues: the resonance has injected one new eigenstate into the band. Each new eigenvalue is shifted infinitesimally relative to the original conduction electron energies, according to

$$E_\gamma = \epsilon_\gamma - \Delta \epsilon \frac{\delta(E_\gamma)}{\pi}, \tag{17.99}$$

where $\delta(E_\gamma) \in [0, \pi]$ is the resonant scattering phase shift.

Let us now determine the dependence of $\delta[E]$ on the conduction electron energy. Substituting the phase shift into the eigenvalue equation (17.97), we obtain

$$E_\gamma = \lambda + \sum_{n=\gamma+1-M}^{\gamma+M} \frac{V^2}{\Delta \epsilon (n - \frac{\delta_\gamma}{\pi})} \rightarrow \lambda + \frac{\Delta}{\pi} \sum_{n=-\infty}^{\infty} \frac{1}{(n - \frac{\delta_\gamma}{\pi})}. \tag{17.100}$$

Here we have identified $\rho \equiv \frac{1}{\Delta \epsilon}$ as the conduction electron density of states, writing $\Delta = \pi V^2 / \Delta \epsilon = \pi V^2 \rho$ as the resonance level width. We have also taken the liberty of extending the bounds of the summation to infinity. Using contour integration methods, recognizing that $\cot z$ has poles at $z = \pi n$ of strength one,

$$\sum_{n=-\infty}^{\infty} \frac{\pi}{(\pi n - \delta_\gamma)} = \Sigma_n \oint_{poles\, z=\pi n} \frac{dz}{2\pi i} \frac{\pi \cot z}{z - \delta_\gamma}$$

$$= -\oint_{pole\, at\, z=\delta_\gamma} \frac{dz}{2\pi i} \frac{\pi \cot z}{z - \delta_\gamma} = -\pi \cot \delta(E_\gamma). \quad (17.101)$$

Using this result, (17.100) becomes

$$E_\gamma = \lambda - \Delta \cot \delta[E_\gamma] \Rightarrow \tan \delta[E_\gamma] = \frac{\Delta}{\lambda - E_\gamma}. \quad (17.102)$$

(b) From (17.99) we deduce that

$$\frac{d\epsilon}{dE} = 1 + \frac{\Delta\epsilon}{\pi} \frac{d\delta(E)}{dE}$$

$$= 1 + \frac{1}{\pi\rho} \frac{d\delta(E)}{dE}, \quad (17.103)$$

where $\rho = 1/\Delta\epsilon$ is the density of states in the continuum. The new density of states $\rho^*(E)$ is given by $\rho^*(E)dE = \rho d\epsilon$, so that

$$\rho^*(E) = \rho(0)\frac{d\epsilon}{dE} = \rho + \rho_i(E), \quad (17.104)$$

where

$$\rho_i(E) = \frac{1}{\pi} \frac{d\delta(E)}{dE} = \frac{1}{\pi} \frac{\Delta^2}{(E - \lambda)^2 + \Delta^2} \quad (17.105)$$

corresponds to the enhancement of the conduction electron density of the states due to injection of the resonant bound state.

17.5.2 Minimization of free energy

With the results from the previous section, let us now calculate the free energy and minimize it to self-consistently evaluate λ and Δ. The shift in the free energy due to the Kondo effect is then

$$\Delta F = -N \int_{-D}^{D} \frac{d\epsilon}{\pi} f(\epsilon) \text{Im} \ln[\xi - \epsilon] - \lambda Q + \frac{N\Delta}{\pi J\rho}, \quad (17.106)$$

where we have introduced the complex number $\xi = \lambda + i\Delta$ whose real and imaginary parts represent the position and width of the resonant level, respectively. This integral can be done at finite temperature, but for simplicity let us carry it out at $T = 0$, when the Fermi function becomes a step function, $f(x) = \theta(-x)$. This gives

$$\Delta E = \frac{N}{\pi} \text{Im} \left[(\xi - \epsilon) \ln \left[\frac{\xi - \epsilon}{e} \right] \right]_{\epsilon=-D}^{\epsilon=0} - \lambda Q + \frac{N\Delta}{\pi J\rho}$$

$$= \frac{N}{\pi} \text{Im} \left[\xi \ln \left[\frac{\xi}{eD} \right] - D \ln \left[\frac{D}{e} \right] \right] - \lambda Q + \frac{N\Delta}{\pi J\rho}, \quad (17.107)$$

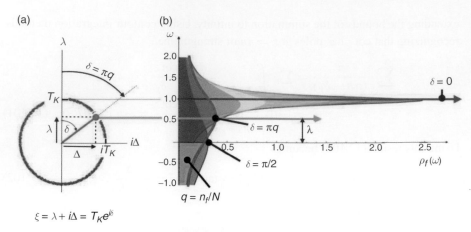

$$\xi = \lambda + i\Delta = T_K e^{i\delta}$$

Fig. 17.9   Illustrating the mean-field solution to the impurity Kondo model. (a) Showing how the real and imaginary parts of the resonant level position $\xi = \lambda + i\Delta$ lie on a circle of radius T_K, with a phase shift $\delta = \pi q = \pi n_f/N$. (b) Showing the corresponding f density of states $\rho_f(\omega)$ for a range of different occupancies.

where we have expanded $(\xi + D) \ln\left[\frac{D+\xi}{e}\right] \to D \ln\left[\frac{D}{e}\right] + \xi \ln D$ to obtain the second line. We can further simplify this expression by noting that

$$-\lambda Q + \frac{N\Delta}{\pi J \rho} = -\frac{N}{\pi} \mathrm{Im}\left[\xi \ln\left[e^{-\frac{1}{\rho J} + i\pi q}\right]\right], \qquad (17.108)$$

where $q = Q/N$, so that

$$\Delta E = \frac{N}{\pi} \mathrm{Im}\left[\xi \ln\left[\frac{\xi}{e T_K e^{i\pi q}}\right]\right], \qquad (17.109)$$

where we have dropped the constant term and introduced the Kondo temperature $T_K = D e^{-\frac{1}{J\rho}}$. The stationary point $\partial E/\partial \xi = 0$ is given by (see Figure 17.9)

$$\xi = \lambda + i\Delta = T_K e^{i\pi q} \qquad \left\{ \begin{array}{rcl} T_K & = & \sqrt{\lambda^2 + \Delta^2} \\ \tan(\pi q) & = & \frac{\Delta}{\lambda}. \end{array} \right. \qquad (17.110)$$

Notice the following:

- The phase shift $\delta = \pi q$ is the same in each spin scattering channel, reflecting the singlet nature of the ground state. The relationship between the filling of the resonance and the phase shift $Q = \sum_\sigma \frac{\delta_\sigma}{\pi} = N\frac{\delta}{\pi}$ is Friedel's sum rule.
- The energy is stationary with respect to small variations in λ and Δ. It is only a local minimum once the condition $\partial E/\partial \lambda = 0 \equiv (\langle \hat{n}_f \rangle - Q)$ is imposed, which gives $\lambda = \Delta \cot(\pi q)$ and hence

$$\Delta E = \frac{N}{\pi}\left[\Delta \ln\left[\frac{\Delta}{e T_K \sin \pi q}\right]\right]. \qquad (17.111)$$

Plotted as a function of V, this is the classic "Mexican hat" potential, with a minimum where $\partial E/\partial V = 0$ at $\Delta = \pi \rho |V|^2 = T_K \sin \pi q$ (Figure 17.9).

- According to (17.104), the enhancement of the density of states at the Fermi energy is

$$
\begin{aligned}
\rho^*(0) &= \rho + \frac{\Delta}{\pi(\Delta^2 + \lambda^2)} \\
&= \rho + \frac{\sin^2(\pi q)}{\pi T_K}
\end{aligned}
\tag{17.112}
$$

per spin channel. When the temperature is changed or a magnetic field is introduced, one can neglect changes in Δ and λ, since the free energy is stationary. This implies that, in the large-N limit, the susceptibility and linear specific heat are those of a non-interacting resonance of width Δ. The change in linear specific heat $\Delta C_V = \Delta\gamma T$ and the change in the paramagnetic susceptibility $\Delta\chi$ are given by

$$
\begin{aligned}
\Delta\gamma &= \left[\frac{N\pi^2 k_B^2}{3}\right]\rho_i(0) = \left[\frac{N\pi^2 k_B^2}{3}\right]\frac{\sin^2(\pi q)}{\pi T_K} \\
\Delta\chi &= \left[N\frac{j(j+1)(g\mu_B)^2}{3}\right]\rho_i(0) = \left[N\frac{j(j+1)(g\mu_B)^2}{3}\right]\frac{\sin^2(\pi q)}{\pi T_K}.
\end{aligned}
\tag{17.113}
$$

Notice how it is the Kondo temperature that determines the size of these two quantities. The dimensionless Wilson ratio of these two quantities is

$$
W = \left[\frac{(\pi k_B)^2}{(g\mu_B)^2 j(j+1)}\right]\frac{\Delta\chi}{\Delta\gamma} = 1.
$$

At finite N, fluctuations in the mean-field theory can no longer be ignored. These fluctuations induce *interactions* among the quasiparticles, and the Wilson ratio becomes

$$
W = \frac{1}{1 - \frac{1}{N}}.
$$

The dimensionless Wilson ratios of a large variety of heavy-electron materials lie remarkably close to this value.

17.6 Mean-field theory of the Kondo lattice

17.6.1 Diagonalization of the Hamiltonian

We can now make the jump from the single-impurity problem to the lattice. The virtue of the large-N method is that, while approximate, it can be readily scaled up to the lattice. We'll now recompute the effective action for the lattice, using equation (17.70). Let us assume that the hybridization and constraint fields at the saddle point are uniform, with $V_j = V$ and $\lambda_j = \lambda$ at every site. In fact, even if we start with a $V_j = V e^{-i\phi j}$ with a different phase at each site, we can always absorb the phase ϕ_j using the Read–Newns gauge transformation (17.57) to absorb the additional phase onto the f-electron field. We then have a translationally invariant mean-field Hamiltonian. We begin by rewriting the mean-field Hamiltonian in momentum space as follows:

$$H_{MFT} = \sum_{\mathbf{k}\sigma} \left(c^\dagger_{\mathbf{k}\sigma}, f^\dagger_{\mathbf{k}\sigma}\right) \overbrace{\begin{pmatrix} \epsilon_\mathbf{k} & V \\ \bar{V} & \lambda \end{pmatrix}}^{\underline{h}(\mathbf{k})} \begin{pmatrix} c_{\mathbf{k}\sigma} \\ f_{\mathbf{k}\sigma} \end{pmatrix} + N\mathcal{N}_s \left(\frac{|V|^2}{J} - \lambda q\right) \tag{17.114}$$

$$= \sum_{\mathbf{k}\sigma} \psi^\dagger_{\mathbf{k}\sigma}\, \underline{h}(\mathbf{k})\, \psi_{\mathbf{k}\sigma} + N\mathcal{N}_s \left(\frac{|V|^2}{J} - \lambda q\right).$$

Here, $f^\dagger_{\mathbf{k}\sigma} = \frac{1}{\sqrt{\mathcal{N}_s}} \sum_j f^\dagger_{j\sigma} e^{i\mathbf{k}\cdot\mathbf{R}_j}$ is the Fourier transform of the f-electron field and we have introduced the two-component notation

$$\psi_{\mathbf{k}\sigma} = \begin{pmatrix} c_{\mathbf{k}\sigma} \\ f_{\mathbf{k}\sigma} \end{pmatrix}, \qquad \psi^\dagger_{\mathbf{k}\sigma} = \left(c^\dagger_{\mathbf{k}\sigma}, f^\dagger_{,\mathbf{k}\sigma}\right), \qquad \underline{h}(\mathbf{k}) = \begin{pmatrix} \epsilon_\mathbf{k} & V \\ \bar{V} & \lambda \end{pmatrix}. \tag{17.115}$$

We should think of H_{MFT} as a renormalized Hamiltonian, describing the low-energy quasiparticles moving through a self-consistently determined array of resonant scattering centers. Later, we will see that the f-electron operators are composite objects, formed as bound states between spins and conduction electrons.

The mean-field Hamiltonian can be diagonalized in the form

$$H_{MFT} = \sum_{\mathbf{k}\sigma} \left(a^\dagger_{\mathbf{k}\sigma}, b^\dagger_{\mathbf{k}\sigma}\right) \begin{pmatrix} E_{\mathbf{k}+} & 0 \\ 0 & E_{\mathbf{k}-} \end{pmatrix} \begin{pmatrix} a_{\mathbf{k}\sigma} \\ b_{\mathbf{k}\sigma} \end{pmatrix} + Nn \left(\frac{\bar{V}V}{J} - \lambda q\right). \tag{17.116}$$

Here $a^\dagger_{\mathbf{k}\sigma} = u_\mathbf{k} c^\dagger_{\mathbf{k}\sigma} + v_\mathbf{k} f^\dagger_{\mathbf{k}\sigma}$ and $b^\dagger_{\mathbf{k}\sigma} = -v_\mathbf{k} c^\dagger_{\mathbf{k}\sigma} + u_\mathbf{k} f^\dagger_{\mathbf{k}\sigma}$ are linear combinations of $c^\dagger_{\mathbf{k}\sigma}$ and $f^\dagger_{\mathbf{k}\sigma}$, playing the role of quasiparticle operators with corresponding energy eigenvalues

$$\det\left[E^\pm_\mathbf{k} \underline{1} - \begin{pmatrix} \epsilon_\mathbf{k} & V \\ \bar{V} & \lambda \end{pmatrix} \right] = (E_{\mathbf{k}\pm} - \epsilon_\mathbf{k})(E_{\mathbf{k}\pm} - \lambda) - |V|^2 = 0 \tag{17.117}$$

or

$$E_{\mathbf{k}\pm} = \frac{\epsilon_\mathbf{k} + \lambda}{2} \pm \left[\left(\frac{\epsilon_\mathbf{k} - \lambda}{2}\right)^2 + |V|^2 \right]^{\frac{1}{2}}, \tag{17.118}$$

and eigenvectors taking the BCS form

$$\left\{ \begin{matrix} u_\mathbf{k} \\ v_\mathbf{k} \end{matrix} \right\} = \left[\frac{1}{2} \pm \frac{(\epsilon_\mathbf{k} - \lambda)/2}{2\sqrt{\left(\frac{\epsilon_\mathbf{k} - \lambda}{2}\right)^2 + |V|^2}} \right]^{\frac{1}{2}}. \tag{17.119}$$

The hybridized dispersion described by these energies is shown in Figure 17.10.

Note the following:

- Hybridization builds an upper and a lower band, separated by a *direct* hybridization gap of size $2V$ and a much smaller *indirect* gap. If we put $\epsilon_\mathbf{k} = \pm D$, we see that the upper and lower edges of the gap are given by

(a)
(b)

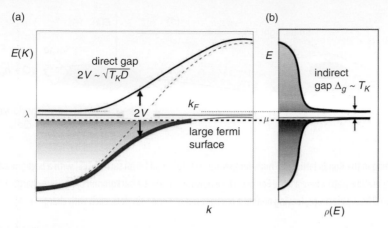

(a) Dispersion for the Kondo lattice mean-field theory. (b) Renormalized density of states, showing hybridization gap Δ_g.

Fig. 17.10

$$E^\pm = \frac{\mp D + \lambda}{2} \pm \sqrt{\left(\frac{\mp D - \lambda}{2}\right)^2 + V^2} \approx \lambda \pm \frac{V^2}{D} \qquad (D >> \lambda), \qquad (17.120)$$

so the indirect gap has a size $\Delta_g \sim 2V^2/D$, where D is the half-bandwidth. From our mean-field solution to the Kondo impurity problem, we can anticipate $V^2/D \sim V^2\rho \sim T_K$, so that $\Delta_g \sim T_K$, the single-ion Kondo temperature, which implies that $V \sim \sqrt{T_K D}$.

- In the special case when the chemical potential lies in the gap, a *Kondo* insulator is formed.
- The effective mass of the Fermi surface is *opposite* to the conduction sea, so a conduction sea of electrons is transformed into a heavy-fermion sea of holes.
- The Fermi surface volume *expands* in response to the formation of heavy electrons (see Figure 17.11). The enlarged Fermi surface volume now counts the total number of occupied quasiparticle states,

$$N_{tot} = \langle \sum_{k\lambda\sigma} n_{k\lambda\sigma} \rangle = \langle \hat{n}_f + \hat{n}_c \rangle, \qquad (17.121)$$

where $n_{k\lambda\sigma} = a^\dagger_{k\lambda\sigma} a_{k\lambda\sigma}$ is the number operator for the quasiparticles and n_c is the total number of conduction electrons. This means

$$N_{tot} = N\frac{V_{FS}a^3}{(2\pi)^3} = Q + n_c, \qquad (17.122)$$

where a^3 is the volume of the unit cell. This is rather remarkable, for the expansion of the Fermi surface implies an increased charge density in the Fermi sea. Since charge is conserved, we are forced to conclude that there is a compensating $+Q|e|$ charge density per unit cell provided by the Kondo singlets formed at each site, as illustrated in Figure 17.11.

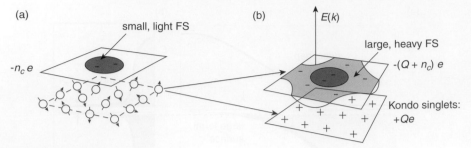

Fig. 17.11 Spin quenching in the Kondo lattice: (a) high-temperature state: small Fermi surface (FS) with a background of spins; (b) low-temperature state where a large Fermi surface develops against a background of positive charge. Each spin "ionizes" into Q heavy electrons, leaving behind a background of Kondo singlets, each with charge $+Qe$.

- We can construct the mean-field ground state from the quasiparticle operators, as follows:

$$|MF\rangle = \prod_{|\mathbf{k}|<k_{F}\sigma} b^{\dagger}_{\mathbf{k}\sigma}|0\rangle = \prod_{|\mathbf{k}|<k_{F}\sigma} (-v_{\mathbf{k}}c^{\dagger}_{\mathbf{k}\sigma} + u_{\mathbf{k}}f^{\dagger}_{\mathbf{k}\sigma})|0\rangle. \qquad (17.123)$$

However, this state only satisfies the constraint on the average. We can improve it by imposing the constraint, forming a *Gutzwiller wavefunction* [36–38],

$$|GW\rangle = P_{Q} \prod_{|\mathbf{k}|<k_{F}\sigma} (-v_{\mathbf{k}}c^{\dagger}_{\mathbf{k}\sigma} + u_{\mathbf{k}}f^{\dagger}_{\mathbf{k}\sigma})|0\rangle, \qquad (17.124)$$

where, using (17.48),

$$P_{Q} = \prod_{j} P_{Q}(j) = \int_{0}^{2\pi} \prod_{j} \frac{d\alpha_{j}}{2\pi} e^{i\sum_{j}\alpha_{j}(\hat{n}_{f}(j)-Q)}. \qquad (17.125)$$

The action of the constraint gives rise to a highly incompressible Fermi liquid, in which the compressibility is far smaller than the density of states.

17.6.2 Mean-field free energy and saddle point

Let us now use the results of the previous section to calculate the mean-field free energy F_{MFT} and determine self-consistently the parameters λ and V which set the scales of the Kondo lattice. Using (17.70) we obtain

$$F_{MF} = -NT \sum_{\mathbf{k},i\omega_{r}} \mathrm{Tr}\ln\left[\overbrace{-i\omega_{r} + \begin{pmatrix} \epsilon_{\mathbf{k}} & V \\ V & \lambda \end{pmatrix}}^{-\mathcal{G}_{\mathbf{k}}^{-1}(i\omega_{r})}\right] + \mathcal{N}_{s}\left(\frac{N|V|^{2}}{J} - \lambda Q\right), \qquad (17.126)$$

where $\mathcal{N}_{s}$ is the number of sites in the lattice. Note that translational invariance means that momentum is conserved and the Green's function is diagonal in momentum, so we can rewrite the trace over the momentum as a sum over $\mathbf{k}$. Let us remind ourselves of the steps taken between (17.70) and (17.71). We begin by re-writing the trace of the logarithm as a determinant, which we then factorize in terms of the energy eigenvalues:

$$\text{Tr}\ln\left[-i\omega_r\underline{1} + \begin{pmatrix} \epsilon_{\mathbf{k}} & V \\ V & \lambda \end{pmatrix}\right] = \ln\det\left[-z\underline{1} + \begin{pmatrix} \epsilon_{\mathbf{k}} & V \\ V & \lambda \end{pmatrix}\right] = \ln\left[\overbrace{(\epsilon_{\mathbf{k}} - i\omega_r)(\lambda - i\omega_r)}^{(E_{\mathbf{k}+}-i\omega_r)(E_{\mathbf{k}-}-i\omega_r)} - V^2\right]$$

$$= \sum_{n=\pm}\ln(E_{\mathbf{k}n} - i\omega_r). \qquad (17.127)$$

Next, by carrying out the summation over Matsubara frequencies, using the result $-T\sum_{i\omega_r}\ln(E_{\mathbf{k}n} - i\omega_r) = -T\ln(1 + e^{-\beta E_{\mathbf{k}n}})$, we obtain

$$\frac{F}{N} = -T\sum_{\mathbf{k},\pm}\ln\left[1 + e^{-\beta E_{\mathbf{k}\pm}}\right] + \mathcal{N}_s\left(\frac{V^2}{J} - \lambda q\right). \qquad (17.128)$$

Let us discuss the ground state, in which only the lower band contributes to the free energy. As $T \to 0$, we can replace $-T\ln(1 + e^{-\beta E_{\mathbf{k}}}) \to \theta(-E_{\mathbf{k}})E_{\mathbf{k}}$, so the ground-state energy $E_0 = F(T = 0)$ involves an integral over the occupied states of the lower band:

$$\frac{E_0}{N\mathcal{N}_s} = \int_{-\infty}^{0} dE\rho^*(E)E + \left(\frac{V^2}{J} - \lambda q\right), \qquad (17.129)$$

where we have introduced the density of heavy-electron states $\rho^*(E) = \sum_{\mathbf{k},\pm}\delta(E - E_{\mathbf{k}}^{(\pm)})$. Now by (17.117) the relationship between the energy E of the heavy electrons and the energy ϵ of the conduction electrons is

$$E = \epsilon + \frac{V^2}{E - \lambda}.$$

As we sum over momenta $\mathbf{k}$ within a given energy shell, there is a one-to-one correspondence between each conduction electron state and each quasiparticle state, so we can write $\rho^*(E)dE = \rho(\epsilon)d\epsilon$, where the density of heavy-electron states is

$$\rho^*(E) = \rho\frac{d\epsilon}{dE} = \rho\left(1 + \frac{V^2}{(E - \lambda)^2}\right). \qquad (17.130)$$

Here we have approximated the underlying conduction electron density of states by a constant $\rho = 1/(2D)$. The originally flat conduction electron density of states is now replaced by a hybridization gap, flanked by two sharp peaks of approximate width $\pi\rho V^2 \sim T_K$ (Figure 17.10). Note that the lower bandwidth is lowered by an amount $-V^2/D$. With this information, we can carry out the integral over the energies, to obtain

$$\frac{E_0}{N\mathcal{N}_s} = \rho\int_{-D-V^2/D}^{0} dE E\left(1 + \frac{V^2}{(E - \lambda)^2}\right) + \left(\frac{V^2}{J} - \lambda q\right), \qquad (17.131)$$

where we have assumed that the upper band is empty and the lower band is partially filled. Carrying out the integral, we obtain

$$\frac{E_0}{N\mathcal{N}_s} = -\frac{\rho}{2}\left(D + \frac{V^2}{D}\right)^2 + \frac{\Delta}{\pi}\int_{-D}^{0} dE\left(\frac{1}{E - \lambda} + \frac{\lambda}{(E - \lambda)^2}\right) + \left(\frac{V^2}{J} - \lambda q\right)$$

$$= -\frac{D^2\rho}{2} + \frac{\Delta}{\pi}\ln\left(\frac{\lambda}{D}\right) + \left(\frac{V^2}{J} - \lambda q\right), \tag{17.132}$$

where we have replaced $\Delta = \pi\rho V^2$ and have dropped terms of order $O(\Delta^2/D)$. We can rearrange this expression, absorbing the bandwidth D and Kondo coupling constant into a single Kondo temperature $T_K = De^{-\frac{1}{J\rho}}$, as follows:

$$\frac{E_0}{N\mathcal{N}_s} = -\frac{D^2\rho}{2} + \frac{\Delta}{\pi}\ln\left(\frac{\lambda}{D}\right) + \left(\frac{\pi\rho V^2}{\pi\rho J} - \lambda q\right)$$

$$= -\frac{D^2\rho}{2} + \frac{\Delta}{\pi}\ln\left(\frac{\lambda}{D}\right) + \left(\frac{\Delta}{\pi\rho J} - \lambda q\right)$$

$$= -\frac{D^2\rho}{2} + \frac{\Delta}{\pi}\ln\left(\frac{\lambda}{De^{-\frac{1}{J\rho}}}\right) - \lambda q$$

$$= -\frac{D^2\rho}{2} + \frac{\Delta}{\pi}\ln\left(\frac{\lambda}{T_K}\right) - \lambda q. \tag{17.133}$$

This describes the energy of a family of Kondo lattice models with different $J(D)$ and cutt-off D, but fixed Kondo temperature. If we impose the constraint $\frac{\partial E_0}{\partial\lambda} = \langle n_f\rangle - Q = 0$, we obtain $\frac{\Delta}{\pi\lambda} - q = 0$, so

$$\frac{E_0(V)}{N\mathcal{N}_s} = \frac{\Delta}{\pi}\ln\left(\frac{\Delta}{\pi q e T_K}\right) - \frac{D^2\rho}{2} \qquad (\Delta = \pi\rho|V|^2). \tag{17.134}$$

Let us pause for a moment to consider this energy functional qualitatively. There are two points to be made:

- The energy surface $E_0(V)$ is actually independent of the phase of $V = |V|e^{i\phi}$ (see Figure 17.12), and has the form of a "Mexican hat" (Figure 17.12) at low temperatures. The minimum of this functional will then determine a familiy of saddle-point values $V = |V_0|e^{i\phi}$, where ϕ can have any value. If we differentiate the ground-state energy with respect to Δ, we obtain

$$0 = \frac{1}{\pi}\ln\left(\frac{\Delta}{\pi q T_K}\right)$$

or

$$\Delta = \pi q T_K,$$

confirming that $\Delta \sim T_K$.

- The mean-field value of the constraint field λ is determined relative to the Fermi energy μ. Were we to introduce a slowly varying external potential field to the conduction electron sea, then the chemical potential would be locally shifted so that $\mu \to \mu + e\phi(t)$. So long as the field $\phi(t)$ is varied at a rate that is slow compared with the Kondo temperature, the constraint field will always track with the chemical potential, and, since the constraint field is pinned to the chemical potential, $\lambda \to \lambda + e\phi(t)$. In the process, the constraint term will become

$$\lambda(\hat{n}_f(j) - Q) \to \lambda(\hat{n}_f(j) - Q) + e\phi(t)(\hat{n}_f(j) - Q). \tag{17.135}$$

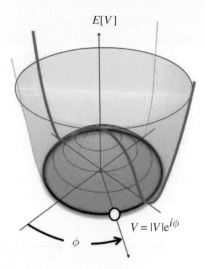

$$E[V]$$

$$V = |V|e^{i\phi}$$

$$\phi$$

"Mexican hat" potential for the Kondo Lattice, evaluated at constant $\langle n_f \rangle = Q$ as a function of a complex hybridization $V = |V|e^{i\phi}$.

Fig. 17.12

Since the f-electrons now couple to the external potential $e\phi$ we have to ascribe a physical charge $e = -|e|$ to them. By contrast, the $-Q$ term in the constraint must be interpreted as a *background positive charge* $|e|Q \equiv |e|$ per site. These lines of reasoning indicate that we should think of the Kondo effect as a *many-body ionization phenomenon*, in which the neutral local moment splits up into a negatively charged heavy electron and a stationary positive background charge that we can associate with the formation of a Kondo singlet.

17.6.3 Kondo lattice Green's function

Let's now take a look at the matrix Green's function, given by

$$\mathcal{G}_{\mathbf{k}}(\tau) = -\langle \psi_{\mathbf{k}\sigma}(\tau)\psi_{\mathbf{k}\sigma}^{\dagger}(0)\rangle \equiv \begin{bmatrix} G_c(\mathbf{k}, \tau) & G_{cf}(\mathbf{k}, \tau) \\ G_{fc}(\mathbf{k}, \tau) & G_f(\mathbf{k}, \tau) \end{bmatrix}, \qquad (17.136)$$

where $G_c(\mathbf{k}, \tau) = -\langle c_{\mathbf{k}}(\tau)c_{\mathbf{k}\sigma}^{\dagger}(0)\rangle$, $G_{cf}(\mathbf{k}, \tau) = -\langle c_{\mathbf{k}}(\tau)f_{\mathbf{k}\sigma}^{\dagger}(\tau)\rangle$, and so on. The anomalous off-diagonal members of this Green's function remind us of the Gor'kov functions in BCS theory, and develop with the coherent hybridization. Using the two-component notation (17.115), and the results of Section 12.4.3, this Green's function can be written

$$\mathcal{G}_{\mathbf{k}}(\tau) = -(\partial_{\tau} + \underline{h}_{\mathbf{k}})^{-1} \xrightarrow{\text{F.T.}} \mathcal{G}_{\mathbf{k}}(z) = (z - \underline{h}_{\mathbf{k}})^{-1} \qquad (17.137)$$

or, more explicitly,

$$\mathcal{G}_{\mathbf{k}}(z) = (z - \underline{h}_{\mathbf{k}})^{-1} = \begin{pmatrix} z - \epsilon_{\mathbf{k}} & -V \\ -V & z - \lambda \end{pmatrix}^{-1} = \begin{pmatrix} G_c(\mathbf{k}, z) & G_{cf}(\mathbf{k}, z) \\ G_{fc}(\mathbf{k}, z) & G_f(\mathbf{k}, z) \end{pmatrix}$$

$$= \frac{1}{(z - \epsilon_{\mathbf{k}})(z - \lambda) - V^2} \begin{pmatrix} z - \lambda & V \\ V & z - \epsilon_{\mathbf{k}} \end{pmatrix}, \qquad (17.138)$$

where we have taken the liberty of analytically extending $i\omega_r \to z$ into the complex plane. Now we can read off the Green's functions. In particular, the hybridized conduction electron Green's function is

$$G_c(\mathbf{k}, z) = \underrightarrow{\quad\quad} = \frac{z - \lambda}{(z - \epsilon_{\mathbf{k}})(z - \lambda) - V^2}$$

$$= \frac{1}{z - \epsilon_{\mathbf{k}} - \frac{V^2}{z - \lambda}} \equiv \frac{1}{z - \epsilon_{\mathbf{k}} - \Sigma_c(z)}, \tag{17.139}$$

which we can interpret physically as conduction electrons scattering off resonant f-states at each site, giving rise to a momentum-conserving self-energy:

$$\Sigma_c(z) = \overset{V}{\underset{O(1)}{\bullet}} \,\text{-}\,\text{-}\,\text{-}\,\blacktriangleright\,\text{-}\,\text{-}\,\text{-}\, \overset{V}{\bullet} = \frac{V^2}{z - \lambda}. \tag{17.140}$$

We can treat this process as a pole at energy $z = \lambda$ in the condution t-matrix. We shall argue later that this pole represents the formation of a composite fermion. A similar process occurs in the impurity Kondo model, but in that case the scattering is local, and connects all wavevectors, whereas in the lattice, coherence implies momentum is conserved. Notice that the denominator in each of the Green's functions involves the same quasiparticle poles, since $(z - \epsilon_{\mathbf{k}})(z - \lambda) - V^2 = (z - E_{\mathbf{k}}^+)(z - E_{\mathbf{k}}^-)$, and hence near the Fermi surface at $E_{\mathbf{k}_F} = 0$ the conduction Green's function can be written

$$G_c(z \sim E_{\mathbf{k}}) = \frac{Z_{\mathbf{k}}}{z - E_{\mathbf{k}}^-}, \tag{17.141}$$

where

$$Z_{\mathbf{k}} = (1 - \partial_z \Sigma_c(z))^{-1}\,|_{z=0} = \frac{1}{1 + \frac{V^2}{\lambda^2}} \sim \frac{T_K}{D} \sim \frac{m}{m^*} << 1, \tag{17.142}$$

where we have identified the scales $V^2/D \sim T_K$ (hence $V^2 \sim DT_K$) and $\lambda \sim T_K$ with the single-ion Kondo temperature. We see that the strength of the quasiparticle pole in the conduction electrons, related to the mass renormalization, is very small.

Similarly, the f-Green's function is

$$G_f(\mathbf{k}, z) = = = = \blacktriangleright = = \; = \frac{z - \epsilon_{\mathbf{k}}}{(z - \epsilon_{\mathbf{k}})(z - \lambda) - V^2} = \frac{1}{i\omega_r - \lambda - \frac{V^2}{z - \epsilon_{\mathbf{k}}}}. \tag{17.143}$$

Finally, the *anomalous* Green's functions are given by

$$G_{cf}(\mathbf{k}, z) = = = \blacktriangleright = \bullet\!\!-\!\!\longrightarrow = \frac{1}{z - \epsilon_{\mathbf{k}}} V G_f(\mathbf{k}, i\omega_n) = \frac{V}{(z - \epsilon_{\mathbf{k}})(z - \lambda) - V^2}, \tag{17.144}$$

which we can interpret as the result of hybridization. We will return to use these expressions to calculate the low-energy part of the tunneling spectrum.

17.7 Kondo insulators

The Kondo insulator is the simplest version of the Kondo lattice, in which the formation of Kondo singlets leads to a fully gapped, insulating state. While the term "Kondo insulator" dates back to the early 1990s [5], these are the oldest heavy-fermion materials. The first heavy-fermion or Kondo insulator, SmB_6, was discovered in 1969 by Menth, Buehler, and Geballe at AT&T Bell Laboratories [6], followed closely by SmS under pressure [7]. It was these materials that inspired Neville Mott to propose that Kondo insulators involve a kind of excitonic ordering between localized f-electrons and conduction electrons [8], driving the emergent hybridization that we have been discussing. A predecessor of the large-N path integral approach to Kondo insulators was proposed in 1979 by Claudine Lacroix and Michel Cyrot at the Laboratoire Louis Néel in Grenoble [28]. At the time of writing this book, SmB_6 has once again been thrust into the main-stream of research, with the proposal [39] that this is an example of a topological insulator – a *topological Kondo insulator* with robust conducting surfaces [40, 41]. This is a topic we will return to in Chapter 18 when we consider mixed valence.

17.7.1 Strong-coupling expansion

In many ways, the Kondo insulator is the simplest ground state of the Kondo lattice. Let us begin by returning to the $SU(2)$ Kondo lattice model:

$$H = -t \sum_{(i,j)\sigma}(c_{i\sigma}^\dagger c_{j\sigma} + \text{H.c.}) + J \sum_{j,\alpha\beta} \vec{\sigma}_j \cdot \vec{S}_j \qquad (\vec{\sigma}_j \equiv (c_{j\beta}^\dagger \vec{\sigma}_{\beta\alpha} c_{j\alpha})), \qquad (17.145)$$

corresponding to a tight-binding Kondo lattice where the electrons at each site are coupled antiferromagnetically to a local moment. We can gain a lot of insight by examining the strong-coupling limit, in which the dispersion of the conduction sea is much smaller than J, so that $t/J << 1$ is a small parameter. In this limit, the intersite hopping is a perturbation to the onsite Kondo insteraction:

$$H \overset{t/J \to 0}{\longrightarrow} J \sum_{j,\alpha\beta} \vec{\sigma}_j \cdot \vec{S}_j + O(t), \qquad (17.146)$$

and the ground state corresponds to the formation of a spin singlet at each site, denoted by the wavefunction

$$|KI\rangle = \prod_j \frac{1}{\sqrt{2}}\left(\Uparrow_j\downarrow_j - \Downarrow_j\uparrow_j\right), \qquad (17.147)$$

where the double and single arrows denote the localized moment and conduction electron, respectively, as illustrated in Figure 17.13(a).

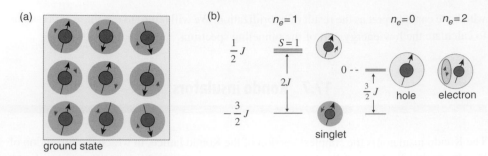

Fig. 17.13 (a) Illustrating the ground state of the Kondo insulator in the strong-coupling limit. (b) Excitations of the Kondo insulator, showing triplet excitation with spin gap $2J$ and $S = \frac{1}{2}$ hole and electron excitations with excitation energy $\frac{3}{2}J$.

Each singlet has a ground-state energy $E = -\frac{3}{2}J$ per site and a singlet–triplet spin gap of magnitude $\Delta E = 2J$. Moreover, if we remove an electron from site i, we break a Kondo singlet and create an unpaired spin with excited energy $\frac{3}{2}J$,

$$|\mathrm{qp}^+, i\uparrow\rangle = \Uparrow_i \prod_{j\neq i} \frac{1}{\sqrt{2}}\left(\Uparrow_j\downarrow_j - \Downarrow_j\uparrow_j\right) = \sqrt{2}c_{i\downarrow}|KI\rangle, \tag{17.148}$$

while if we add an electron, we create an electron quasiparticle, corresponding to an unpaired local moment and a doubly occupied conduction electron orbital,

$$|\mathrm{qp}^-, i\uparrow\rangle = \Uparrow_i \left(\uparrow_i\downarrow_i\right) \prod_{j\neq i} \frac{1}{\sqrt{2}}\left(\Uparrow_j\downarrow_j - \Downarrow_j\uparrow_j\right) = \sqrt{2}c_{j\uparrow}^\dagger|KL\rangle, \tag{17.149}$$

as illustrated in Figure 17.13(b).

If we now reintroduce the hopping $-t$ between sites, then these quasiparticle excitations become mobile, as illustrated in Figure 17.14(a) and (b). From the explicit form of the states, we see that the matrix elements to hop these quasiparticles between nearest-neighbor (n.n.) sites are given by

$$\langle \mathrm{qp}^\pm, i\sigma |H|\mathrm{qp}^\pm, j\sigma\rangle = \pm\frac{t}{2} \qquad \big((i,j) \in \text{n.n.}\big), \tag{17.150}$$

corresponding to one-half the bare hopping, giving quasiparticle energies

$$E_{\mathrm{qp}\pm}(\mathbf{k}) = \pm t(c_x + c_y + c_z) + \frac{3}{2}J. \tag{17.151}$$

To transform from the quasiparticle to the electron basis, we need to reverse the sign of the hole (qp^+) dispersion to obtain the valence band dispersion, so that the band energies predicted by the strong-coupling limit of the Kondo lattice are

$$E_{\mathbf{k}}^\pm = -t(c_x + c_y + c_z) \pm \frac{3}{2}J, \tag{17.152}$$

separated by an energy $3J$ as shown in Figure 17.14(c). Note that these are hard-core fermions that cannot occupy the same lattice site simultaneously.

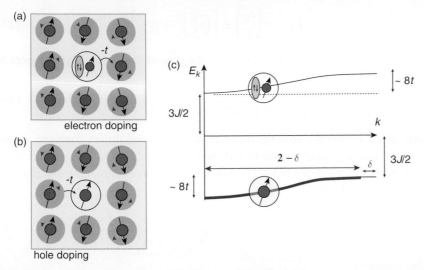

Fig. 17.14

Showing (a) electron and (b) hole doping of strong-coupling Kondo insulator. (c) Dispersion of strong-coupling Kondo insulator. A small amount of hold doping δ gives rise to a Fermi surface containing $2 - \delta = (1 - \delta) + 1$ heavy electrons, showing that the Fermi surface counts the number of electrons and spins.

In this way, the half-filled strong-coupling Kondo lattice forms an insulator with a charge gap of size $3J$ and a spin gap of size $2J$. Notice finally that if we dope the insulator with an amount δ of holes, we form a band of heavy-fermions. We can regard the resulting "hole" Fermi surface as containing $2 - \delta = (1 - \delta) + 1$ heavy electrons, which is one more than we expect based on the density of conduction electrons. In this way, we see once again that the Fermi surface of the Kondo lattice counts both the electrons and the spins.

17.7.2 Large-N treatment of the Kondo insulator

Let us now re-examine the Kondo insulator using the large-N expansion. Here, our advantage is that we are not restricted to strong coupling. The Kondo insulator is a special case of the large-N Kondo lattice, in which the chemical potential lies between the upper conduction and lower valence band. We start with the mean-field hybridization model (17.50):

$$H_{MF} = \sum_{\mathbf{k}} \epsilon_{\mathbf{k}} c_{\mathbf{k}\sigma}^{\dagger} c_{\mathbf{k}\sigma} + \sum_{j} \left[\bar{V} \left(c_{j\sigma}^{\dagger} f_{j\sigma} \right) + \left(f_{j\sigma}^{\dagger} c_{j\sigma} \right) V_j + \lambda(n_{fj} - Q) + N \mathcal{N}_s \frac{|V|^2}{J} \right].$$

(17.153)

We can simplify the problem by considering the special case of particle–hole symmetry, with $Q = N/2$ and $n_e = N/2$ per site, in which case the mean-field constraint $\langle n_f \rangle = N/2$ is satisfied with $\lambda = 0$ and we only need to optimize the value of the hybridization. Following the steps of (17.111)–(17.115), the mean-field dispersions for the Kondo lattice are

$$E_{\mathbf{k}\pm} = \left[\frac{\epsilon_{\mathbf{k}}}{2}\right] \pm \left[\left(\frac{\epsilon_{\mathbf{k}}}{2}\right)^2 + |V|^2\right], \tag{17.154}$$

where we have set $\lambda = 0$. Following (17.125), the ground-state energy is then given by

$$\frac{E_g}{N} = \sum_{\mathbf{k}} \left(\left[\frac{\epsilon_{\mathbf{k}}}{2}\right] - \left[\left(\frac{\epsilon_{\mathbf{k}}}{2}\right)^2 + |V|^2\right]\right) + \mathcal{N}_s \frac{|V|^2}{J}. \tag{17.155}$$

This expression is strongly reminiscent of BCS theory, and differs from the mean-field theory of the heavy-fermion metal, in that the energy involves an unrestricted sum over momenta in the lower valence band. If we differentiate this expression with respect to the hybridization, we obtain

$$\frac{1}{N\mathcal{N}_s}\frac{\partial E_g}{\partial |V|^2} = 0 = -\int_{\mathbf{k}} \frac{1}{\sqrt{\epsilon_{\mathbf{k}}^2 + 4V^2}} + \frac{1}{J}, \tag{17.156}$$

which is a kind of gap equation. To get an approximate treatment of the problem, let us replace the momentum integral by an energy integration. Assuming the conduction electron density of states $\rho = \frac{1}{2D}$ can be treated as constant, the Kondo insulator gap equation becomes

$$\frac{1}{J} = \rho \int_{-D}^{D} \frac{d\epsilon}{\sqrt{\epsilon^2 + 4V^2}} = 2\rho \sinh^{-1}\left(\frac{D}{2V}\right). \tag{17.157}$$

Putting $\rho = \frac{1}{2D}$, we then obtain

$$V = \frac{D}{2\sinh\left(\frac{D}{J}\right)}. \tag{17.158}$$

As we increase the half-bandwidth D from a value that is small to a value that is large compared with J, we see that V interpolates from $V = J/2$ at strong coupling to $V = D\exp\left[-\frac{1}{2J\rho}\right] = \sqrt{DT_K}$, where $T_K = De^{-\frac{1}{J\rho}}$. We can also calculate the indirect gap of the insulator, determined by the value of the dispersion when the conduction electrons are at the edge of the band, i.e.

$$\Delta_g = E_{\mathbf{k}+}|_{\epsilon_{\mathbf{k}}=-D} - E_{\mathbf{k}-}|_{\epsilon_{\mathbf{k}}=D} = \sqrt{D^2 + 4V^2} - D$$

$$= 2V\left[\sqrt{\left(\frac{D}{2V}\right)^2 + 1} - \left(\frac{D}{2V}\right)\right]$$

$$= 2V\left[\cosh\frac{D}{J} - \sinh\frac{D}{J}\right]$$

$$= 2V\exp\left[-\frac{D}{J}\right] = \frac{2D}{e^{\frac{2D}{J}} - 1}, \tag{17.159}$$

where we have used (17.158) to make the substitutions on the third and fourth lines. We see that, in the large-N limit, the gap undergoes a crossover from $\Delta_g = J$ to $\Delta_g = 2T_K$, as shown in Figure 17.15.

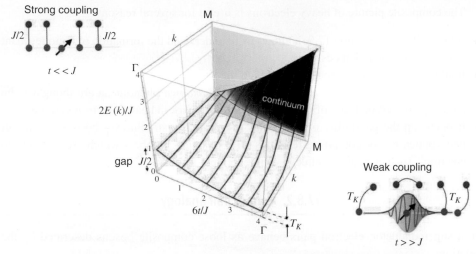

Evolution of Kondo insulator from strong to weak coupling, calculated in the large-N limit. The points $\Gamma \equiv (0, 0, 0)$ and $M \equiv (\pi, \pi, \pi)$ correspond to the origin and the zone-center of the Brillouin zone. As the bandwidth t of the conduction sea is increased, the quasiparticle excitation energy drops from $O(J)$ to $O(T_K)$. At strong coupling the excitations are immobile unpaired spins, but at weak coupling, the gapped excitations are mobile heavy-fermions. At finite N, we expect that, when T_K becomes too small, the insulator will ultimately become unstable to antiferromagnetism.

Fig. 17.15

17.8 The composite nature of the f-electron

17.8.1 A thought experiment: a Kondo lattice of nuclear spins

In electronic materials the Kondo effect involves localized f- or d-electrons. However, a Kondo effect could occur equally well with a *nuclear* spin. This might seem absurd, yet nuclear spins do couple antiferromagnetically with conduction electrons to produce RKKY interactions that drive nuclear antiferromagnetism. In practice the coupling is far too small to destabilize the nuclear magnetism and produce a nuclear Kondo effect. Nevertheless, we learn something from the thought experiment in which the nuclear spin coupling to electrons is strong enough to overcome the nuclear magnetism. In this case, resonant bound states would form with the nuclear spin lattice, giving rise to *charged* heavy electrons, presumably with an expanded Fermi surface.

From this line of argument we see that, while it's tempting to associate the heavy fermion in the Kondo effect with a physical f- or d-electron localized inside the local moment, from a renormalization group perspective the heavy electron is an emergent excitation: a fermionic bound state formed between the conduction sea and the neutral localized moments. The only memory of the underlying localized electrons is encoded in the spatial symmetry of the Kondo coupling, which of course for rare-earth systems is an f-form factor.

The composite picture of heavy electrons is useful for several reasons:

- As we will see in Section 17.9, it allows us to understand the formation of Fano resonant structures in Kondo lattices.
- It allows us to envisage processes in which the Kondo effect breaks down, leading to the loss of the large Fermi surface. Such *Kondo breakdown* phenomena are thought to be the origin of certain types of non-Fermi-liquid behavior in heavy-electron systems.
- It opens up the possibility of new kinds of composite structures – bosons that might pair-condense – or composite fermions with quantum numbers which are different to electrons, such as neutral, spinless, or integer-spin fermions.

17.8.2   Cooper pair analogy

In a superconductor, electron pairs behave as loose composite bosons described by the relation

$$\overline{\psi_\uparrow(x)\psi_\downarrow(x')} = -F(x - x').$$
(17.160)

Here $F(x - x') = -\langle T\psi_\uparrow(1)\psi_\downarrow(2)\rangle$ is the anomalous Gor'kov Green's function which determines the Cooper pair wavefunction, extended over the coherence length $\xi \sim v_F/T_c$. We can treat the pair operator as a c-number because the pairs condense.

A similar phenomenon takes place in the Kondo effect, but here the bound state develops between spins and electrons, forming a fermion, rather than a boson. For an isolated Kondo impurity, the analogue of the coherence length in the superconductor is the *Kondo screening length* $\xi_K \sim v_F/T_K$, but in a lattice the renormalization of the heavy-fermion velocity means that this screening length is of the order of a lattice spacing. In this situation, it is perhaps more useful to think in terms of a *screening time* $\tau_K \sim \hbar/T_K$, rather than a length, governing the electron spin-flip correlations. Both Cooper pairs and heavy electrons involve a binding process that spans decades of energy up to a cut-off, be it the Debye energy ω_D in superconductivity or the (much larger) bandwidth D in the Kondo effect [42, 43].

To follow this analogy in greater depth, recall that in the path integral the Kondo interaction factorizes as

$$\frac{J}{N}c_\beta^\dagger S_{\alpha\beta}c_\alpha \longrightarrow \bar{V}\left(c_\alpha^\dagger f_\alpha\right) + \left(f_\alpha^\dagger c_\alpha\right)V + N\frac{\bar{V}V}{J},$$
(17.161)

so by comparing the right- and left-hand sides, we see that the composite operators $S_{\beta\alpha}c_\beta$ and $c_\beta^\dagger S_{\alpha\beta}$ behave as a single fermion denoted by the contractions

$$\frac{1}{N}\sum_\beta \overline{S_{\beta\alpha}c_\beta} = \left(\frac{\bar{V}}{J}\right)f_\alpha, \qquad \frac{1}{N}\sum_\beta \overline{c_\beta^\dagger S_{\alpha\beta}} = \left(\frac{V}{J}\right)f_\alpha^\dagger.$$
(17.162)

composite fermion

Physically, this means that the spins bind high-energy electrons, transforming themselves into composites which then hybridize with the conduction electrons. The resulting *heavy fermions* can be thought of as moments ionized in the magnetically polar electron fluid to form mobile, negatively charged heavy electrons while leaving behind a positively charged *Kondo singlet*. Microscopically, the many-body amplitude to scatter an electron off a local moment develops a bound-state pole, which for large N we can denote by the diagrams

The leading diagram describes a kind of condensation of the hybridization field; the second and higher terms describe the smaller $O(1/N)$ fluctuations around the mean-field theory.

By analogy with superconductivity, we can associate a wavefunction with the temporal correlations between spin-flips and conduction electrons, as follows:

$$\frac{1}{N}\sum_{\beta} \overline{c_{\beta}(\tau)S_{\beta\alpha}(\tau')} = g(\tau - \tau')\hat{f}_{\alpha}(\tau'), \tag{17.163}$$

where the spin-flip correlation function $g(\tau - \tau')$ is an analogue of the Gor'kov function, extending over a coherence time $\tau_K \sim \hbar/T_K$. Notice that, in contrast to the Cooper pair, this composite object is a fermion and thus requires a distinct operator $\hat{f}_{\alpha}$ for its expression. The Fourier (Laplace) decomposition of $g(\tau)$ describes the spectral distribution of electrons and spin-flips inside the composite f-electron, which we may calculate as follows:

$$\frac{1}{N}\sum_{\beta} \overline{c_{\beta}(\tau)S_{\beta\alpha}(\tau')} = \frac{1}{N}\sum_{\beta} \overline{c_{\beta}(\tau)f_{\beta}^{\dagger}(\tau')}f_{\alpha}(\tau')$$

$$= \frac{1}{N}\sum_{\beta} \langle Tc_{\beta}(\tau)f_{\beta}^{\dagger}(\tau')\rangle f_{\alpha}(\tau')$$

$$= -G_{cf}(\tau - \tau')f_{\alpha}(\tau'). \tag{17.164}$$

In this way, we identify

$$g(\tau - \tau') = \langle Tc_{\beta}(\tau)f_{\beta}^{\dagger}(\tau')\rangle = -G_{cf}(\tau - \tau') \tag{17.165}$$

with the anomalous Green's function between the f- and conduction electrons at the same site.

A detailed calculation (see Example 17.5) shows that $g(\tau)$ is logarithimically correlated at short times, but decays as $\frac{1}{\tau}$ at times $|\tau| >> \frac{\hbar}{T_K}$:

$$g(\tau) \sim \begin{cases} \rho V \ln\left(\frac{T_K \tau}{\hbar}\right) & (\hbar/D << \tau << \hbar/T_K) \\ \frac{1}{\tau} & (\tau >> \hbar/T_K). \end{cases} \tag{17.166}$$

The short-time logarithimic correlations between the spin-flip and electron ($\tau << \hbar/T_K$) represent the weak-coupling interior of the composite fermion, whereas the long-time power law correlations reflect the development of the Fermi liquid correlations at long times.

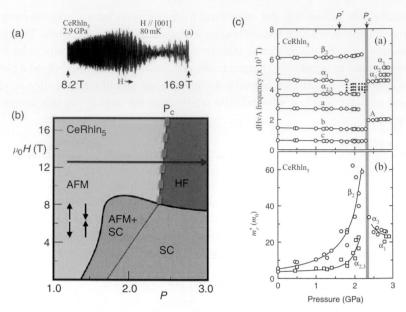

Fig. 17.16   Jump in the Fermi surface area of CeRhIn$_5$ at an antiferromagnetic phase transition, imaged from de Haas–van Alphen (dHvA) oscillations. (a) de Haas–van Alphen signal [44] at 2.9 GPa. (b) Schematic zero-temperature pressure-field phase diagram for CeRhIn$_5$, showing antiferromagnetic, superconducting, and heavy-fermion regions of the phase diagram [45]. The red arrow shows the approximate region of the phase diagram swept by the dHvA measurements. (c) Pressure dependence of dHvA frequencies [44], which are a measure of the Fermi surface area of extremal orbits on the Fermi surface. The data show the jump in Fermi surface area at the critical pressure where antiferromagnetism disappears, and a corresponding divergence in effective mass. Panels (a) and (c) reprinted with permission from H. Shishido, *et al.*, *J. Phys. Soc. Jpn.*, vol. 74, no. 4, p. 1103, 2005. Copyright 2005 by the Physical Society of Japan. Panel (b) reprinted with permission from T. Park, *et al.*, *Nature*, vol. 440, no. 7080, p. 65, 2006. Copyright 2006 Macmillan Publishers.

The internal structure of the composite fermion, spread over several decades up to the bandwidth, guarantees that the composite f-state is orthogonal to the low-energy conduction electrons, behaving as an emergent electron field, injected into the low-energy Fermi sea. The physical manifestation of this phenomenon is an expansion of the Fermi surface by the composite fermions. A particularly dramatic example of this expansion is seen in the material CeRhIn$_5$, which is an antiferromagnetic metal at ambient pressures but becomes superconducting as the f-electrons delocalize at higher pressures (Figure 17.16). De Haas–van Alphen experiments on the normal state show that the Fermi surface expands as the mobile f-electrons are formed. Similar effects are also seen in Hall-constant measurements. Most remarkably of all, in cases where the Fermi surface expands to fill the entire Brillouin zone, the resulting system becomes an insulator, a *Kondo insulator*.

Example 17.5   Calculate the internal spin-flip correlation function of the composite f-electron,

$$\frac{1}{N}\sum_{\beta} \overline{c_{\beta}(\tau) S_{\beta\alpha}(\tau')} = g(\tau - \tau')\hat{f}_{\alpha}(\tau'), \tag{17.167}$$

in the large-N expansion. Carry this out using a Fourier decomposition,

$$g(\tau) = -T \sum_{\mathbf{k}, i\omega_n} G_{cf}(\mathbf{k}, i\omega_n) e^{-i\omega_n \tau}, \tag{17.168}$$

where $G_{cf}(\mathbf{k}, \tau) = -\langle c_{\mathbf{k}\sigma}(\tau) f_{\mathbf{k}\sigma}(0) \rangle$ is the anomalous propagator between the conduction and f-state.

Solution

Transforming to Fourier space, we have

$$G_{cf}(\mathbf{k}, i\omega_n) = \equiv \equiv \blacktriangleright \equiv \bullet \longrightarrow = \left(\frac{V}{i\omega_n - \epsilon_k} \right) \frac{1}{i\omega_n - \lambda - \frac{V^2}{i\omega_n - \epsilon_k}}$$

$$= \frac{V}{(i\omega_n - \epsilon_k)(i\omega_n - \lambda) - V^2}, \tag{17.169}$$

where the double dashed line is the full f-electron propagator. We can approximate the summation over momentum in (17.168) as an integral over energy:

$$G_{cf}(z) = \sum_{\mathbf{k}} G_{cf}(\mathbf{k}, z) = \rho \int_{-D}^{D} d\epsilon \frac{V}{(z - \epsilon)(z - \lambda) - V^2}$$

$$= \frac{\rho V}{z - \lambda} \ln \left[\frac{(z + D)(z - \lambda) - V^2}{(z - D)(z - \lambda) - V^2} \right]. \tag{17.170}$$

This function contains two branch cuts along the real axis, corresponding to the upper and lower bands, which run from $E_1^{\pm} \rightarrow E_2^{\pm}$, where $(E_{1,2}^{\pm} \pm D)(E_{1,2}^{\pm} - \lambda) - V^2 = 0$. The low-energy ends of the branch cut $|E_2^+| \sim |E_1^-| \sim V^2/D \sim T_K$ are of the order of the Kondo scale, whereas the high-energy ends $|E_1^-| \sim |E_2^+| \sim D$ are set by the bandwidth:

There are thus two energy scales in this function – the bandwidth D and the Kondo temperature $T_K \sim \lambda$. The internal structure of the composite fermion is thus determined by the spectral function

$$g(\omega) = -\frac{1}{\pi} \text{Im} \, G_{cf}(\omega - i\delta) = -\frac{\rho V}{\omega - \lambda} \sum_{\pm} [\theta(\omega - E_1^{\pm}) - \theta(\omega - E_2^{\pm})], \tag{17.171}$$

as shown in Figure 17.17.

In the time domain,

$$g(\tau) = -\int_{-\infty}^{\infty} d\omega g(\omega) \left[(1 - f(\omega))\theta(\tau) - f(\omega)\theta(-\tau) \right] e^{-\omega\tau}. \tag{17.172}$$

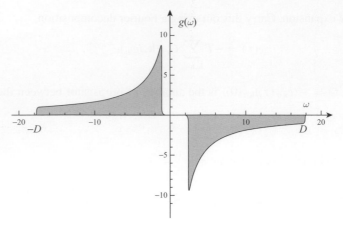

Fig. 17.17 Spectral distribution function $g(\omega)$ (17.171) describing the internal correlations of spin and electron inside a composite f-electron. See Example 17.5.

For simplicity, let's examine the case where the Fermi energy is in the lower band ($\lambda > 0$). Now by (17.167) and (17.162), the bound-state amplitude V is given by the equal-time Green's function,

$$\frac{V}{J} = g(0^-) = -V\rho \int_{-D}^{0} \frac{d\omega}{\omega - \lambda} = \rho V \ln \frac{D}{\lambda}, \qquad (17.173)$$

from which we deduce that

$$\frac{V}{J} = \rho V \ln \frac{D}{\lambda} \Rightarrow \lambda = De^{-\frac{1}{J\rho}} = T_K, \qquad (17.174)$$

as obtained earlier from the minimization of the energy. Note that the argument in the bound-state integral (17.173) depends on the inverse of the energy, right out to the band-width. If we divide the band on a logarithmic scale into n equal parts, where the ratio of the lower and upper energies is $s > 1$, we see that each decade of energy counts equally to the bound-state amplitude:

$$\frac{V}{J} = -\rho V \int_{-D}^{-\lambda} d\epsilon \frac{1}{\epsilon} = -\rho V \left\{ \int_{-D}^{-D/s} + \int_{-D/s}^{-D/s^2} + \cdots + \int_{-D/s^{n-1}}^{-\lambda} \right\} \frac{d\epsilon}{\epsilon}$$

$$= \rho V \left\{ \ln s + \ln s + \cdots + \ln \frac{Ds^{-n+1}}{\lambda} \right\}, \qquad (17.175)$$

demonstrating that the low-energy heavy-fermion bound state is formed from electron states that are spread out over decades of energy out to the bandwidth.

Finally, returning to the time dependence,

$$g(\tau) = -\int_{-D}^{0} d\omega \frac{\overbrace{V\rho}^{-g(\omega)}}{\omega - \lambda} e^{-\omega\tau} \qquad (\tau < 0), \qquad (17.176)$$

we see that there are two main frequency domains:

$$g(\omega) \sim \begin{cases} -\dfrac{1}{\omega} & (D \gg \omega \gg T_K) \\[2ex] \dfrac{1}{T_K} & (\omega \ll T_K). \end{cases} \tag{17.177}$$

By dimensional analysis ($[\int \frac{d\omega}{\omega}] \sim [\tau^0] \sim \ln \tau$, $[\int d\omega e^{-\omega\tau}] \sim \frac{1}{\tau}$) we then obtain

$$g(\tau) \sim \begin{cases} \rho V \ln\left(\dfrac{T_K \tau}{\hbar}\right) & (\hbar/D \ll \tau \ll \hbar/T_K) \\[2ex] \dfrac{1}{\tau} & (\tau \gg \hbar/T_K), \end{cases} \tag{17.178}$$

so that up to the Kondo time $\hbar/T_K$ the correlations are logarithmic in time, but beyond this time scale they decay more rapidly with the inverse of the time.

17.9 Tunneling into heavy-electron fluids

How do electrons tunnel into a Kondo lattice? Since direct tunneling into localized magnetic orbitals is prevented by the Coulomb blockade, the naive expectation is that electrons can only tunnel into the conduction sea. In fact this is not the case, for the formation of composite fermions allows a new tunneling process, that of *cotunneling*, by which tunneling electrons can directly interconvert into composite heavy-fermions.

While electrons cannot directly tunnel into f-states, quantum mechanics allows them to virtually hop on and then hop off the f-site, exchanging their spin, and this process gives rise to a spin-exchange process between the tunneling tip and the Kondo lattice of the form

$$H_{cotunnel} \sim \frac{t_{tip} V_0}{U} \sum_{\sigma,\sigma'} \left[(c^\dagger_{0\sigma} f_{0\sigma'})(f^\dagger_{0\sigma} p_{0\sigma}) + \text{H.c.} \right], \tag{17.179}$$

where V_0 and t_{tip} are the hybridization between the the f-state on the surface at position 0 and the conduction and tip electrons, respectively. Here $p_{0\sigma}$ destroys an electron in the tip and $c^\dagger_{0\sigma}$ creates an electron directly beneath it in the conduction sea of the Kondo lattice. This term is a kind of Kondo coupling between the tip and the sample.

In the mid 1960s A. F. Wyatt [46], working at Bell laboratories, New Jersey, discovered that the conductance curves for tunneling from niobium or tantalum metal through thin insulating layers into aluminum contained unusual zero-bias anomalies that could be split in a magnetic field. This led theorists Phillip W. Anderson and J. Appelbaum at Bell Laboratories to propose that magnetic ions actively participate in electron tunneling via a process we now call cotunneling [49, 50], in which the electron tunneling via the localized f-state induces a spin-flip of localized moments. Cotunneling occurs when electrons tunnel into magnetic quantum dots or isolated magnetic atoms adsorbed on surfaces [49–52]. One of its most notable consequences is the formation of a *Fano resonance* [53], created via an interference between the direct and cotunneling processes [54]. This same physics occurs when electrons tunnel into Kondo lattices. The first scanning tunneling experiments into Kondo lattices were carried out in 2010 by the Séamus Davis group working at Cornell University [55], and have since been used to visualize hybridized heavy bands in Kondo lattice systems [56].

17.9.1 The cotunneling Hamiltonian

Cotunneling can be understood as a result of quantum mechanical mixing between states in the tunneling tip and the localized orbitals of the Kondo lattice [57].

When a tip is introduced above site 0 on the surface of the lattice, tunneling between the f-state and the probe electrons modifies the hybridization as follows:

$$\mathcal{V}_0 \left(c_{0\sigma}^\dagger f_{0\sigma} + \text{H.c.} \right) \rightarrow (\mathcal{V}_0 c_{0\sigma}^\dagger + t_f p_{0\sigma}^\dagger) f_{0\sigma} + \text{H.c.}, \qquad (17.180)$$

where $\mathcal{V}_0$ is the bare hybridization, $p_{0\sigma}^\dagger$ creates an electron at the tip of the probe, and t_f is the amplitude to tunnel an f-electron to the probe.[2] In this way the orbital hybridizing with the f-state is modified as follows:

$$c_{0\sigma} \rightarrow c_{0\sigma} + \frac{t_f}{\mathcal{V}_0} p_{0\sigma}. \qquad (17.181)$$

Now to follow the effect of this mixing on the low-energy Hamiltonian, we need to carry out a Schrieffer–Wolff transformation.

To follow the effect of this mixing on the Kondo lattice model, we need to carry out a Schrieffer–Wolff transformation on an Anderson model for the lattice and tunneling tip [48, 50, 58]. When we integrate out the high-energy valence fluctuations to obtain a Kondo

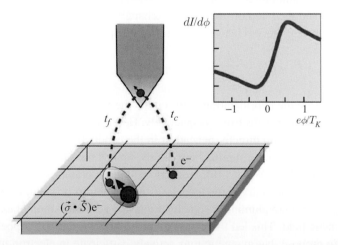

Fig. 17.18 Electron tunneling into a heavy-fermion material involves two parallel processes: direct tunneling with amplitude t_c into the conduction sea, and cotunneling with amplitude t_f into a composite combination of the conduction electron and local magnetic f-moments. These composite states are expected to develop coherence below the Kondo temperature T_K. Inset shows a typical differential conductance curve observed for tunneling into a single Kondo ion.

---

[2] Note that in this section we have adopted a caligraphic $\mathcal{V}$ to denote the hybridization to avoid confusion between hybridization and the bias voltage V.

model using a Schrieffer–Wolff transformation (see Section 16.6), this same replacement must be made in the Kondo interaction at site 0. The result of this procedure is then

$$H_K(0) \rightarrow -J_K \left[\left(c_{0\alpha}^\dagger + \frac{t_f}{\mathcal{V}_0} p_{0\alpha}^\dagger \right) f_{0\alpha} \right] \left[f_{0\beta}^\dagger \left(c_{0\beta} + \frac{t_f}{\mathcal{V}_0} p_{0\beta} \right) \right]$$

$$= H_K(0) - \tilde{t}_f \left[(p_{0\alpha}^\dagger f_{0\alpha})(f_{0\beta}^\dagger c_{0\beta}) + \text{H.c.} \right] + O(\frac{\tilde{t}_f^2}{\mathcal{V}^2})$$

$$\equiv H_K(0) + \tilde{t}_f \left[p_{0\alpha}^\dagger \left(\vec{\sigma}_{\alpha\beta} \cdot \vec{S}_f(0) - \frac{1}{2}\delta_{\alpha\beta} \right) c_{0\beta} + \text{H.c.} \right] + O\left(\frac{\tilde{t}_f^2}{\mathcal{V}_0^2} \right), \quad (17.182)$$

where $\tilde{t}_f = J t_f / \mathcal{V}$ and we have used the identity $f_\alpha f_\beta^\dagger = \frac{1}{2}\delta_{\alpha\beta} - \vec{S} \cdot \vec{\sigma}_{\alpha\beta}$,[3] and have dropped terms of order $\tilde{t}_f^2$. The spin-dependent part of the second term is the cotunneling term, describing correlated spin-flip tunneling. When we add the cotunneling and the direct tunneling together, the complete Hamiltonian becomes

$$H = H_{KL} + \sum_{\mathbf{k}} \epsilon_{\mathbf{k}} p_{\mathbf{k}\sigma}^\dagger p_{\mathbf{k}\sigma} + H_T, \quad (17.183)$$

where $p_{\mathbf{k}\sigma}^\dagger$ creates a Bloch wave of momentum $\mathbf{k}$ in the tunneling tip,

$$H_T = t_c \hat{p}_{0\alpha}^\dagger \left[c_{0\alpha} + \frac{\tilde{t}_f}{t_c} \left(\vec{\sigma}_{\alpha\beta} \cdot \vec{S}_f(0) \right) c_{0\beta} \right] + \text{H.c.} \quad (17.184)$$

describes the direct and cotunneling contributions to the tunneling Hamiltonian, and $p_{0\alpha}^\dagger = \frac{1}{\mathcal{N}_s^{-1/2}} \sum_{\mathbf{k}} p_{\mathbf{k}\alpha}^\dagger$ creates an electron at the tip of the tunneling probe. In other words, to take into account the cotunneling, formally all we have to do is to replace the field $c_{0\sigma}$ in the conduction sea by the composite $\psi_{0\sigma}$ as follows:

$$c_{0\sigma} \rightarrow \psi_{0\sigma} = \left[c_{0\sigma} + \left(\frac{\tilde{t}_f}{t_c} \right) \left(\vec{\sigma}_{\alpha\beta} \cdot \vec{S}_f(0) \right) c_{0\beta} \right]. \quad (17.185)$$

The additional composite component modifies the tunneling current, and the two components interfere with one another to produce *Fano lineshapes* in the tunneling spectra, as we now show.

17.9.2 Tunneling conductance and the "Fano lattice"

To calculate the tunneling conductance, we can adapt the formula obtained in (9.114),

$$\frac{dI}{d\phi}(\phi, \mathbf{x}) = g(e\phi, \mathbf{x}) = \left(\frac{Ne^2 \Gamma}{\hbar} \right) A_\psi(\omega, \mathbf{x}) \Big|_{\omega = e\phi}, \quad (17.186)$$

where ϕ is the applied potential and we have adapted the formula to take account of N spin channels; $\Gamma = 2\pi |t_c|^2 \rho_{tip}$ and ρ_{tip} is the density of states in the tip. The spectral function

$$A_\psi(\omega, \mathbf{x}) = \frac{1}{\pi} \operatorname{Im} G_\psi(\mathbf{x}, \omega - i\delta) = \int dt e^{i\omega t} \left\langle \left\{ \psi_\sigma(\mathbf{x}, t), \psi_\sigma^\dagger(\mathbf{x}, 0) \right\} \right\rangle \quad (17.187)$$

---

[3] When we decompose $f_\alpha f_\beta^\dagger = \hat{m}_0 \delta_{\alpha\beta} + \vec{m} \cdot \vec{\sigma}_{\alpha\beta}$, then we find $m_0 = \frac{1}{2}[f_\alpha f_\beta^\dagger \delta_{\beta\alpha}] = \frac{1}{2}(2 - n_f) = \frac{1}{2}$, while $\vec{m} = \frac{1}{2} f_\alpha f_\beta^\dagger \vec{\sigma}_{\beta\alpha} = -\frac{1}{2} f_\alpha^\dagger f_\beta \vec{\sigma}_{\beta\alpha} = -\vec{S}$.

is the local electron spectral function, which as usual we obtain by analytically extending the imaginary-time Green's function. To take into account the cotunneling, formally all we have to do is to make the replacement (17.185)

$$\psi_{0\sigma} = \left[c_{0\sigma} + \left(\frac{\tilde{t}_f}{t_c} \right) \left(\vec{\sigma}_{\alpha\beta} \cdot \vec{S}_f(0) \right) c_{0\beta} \right]. \tag{17.188}$$

Let us now use the large-N limit of the Kondo lattice to compute the tunneling conductance. As we saw in Section 17.8.2, the mean-field theory provides a representation of the composite fermion $\left(\vec{\sigma}_{\alpha\beta} \cdot \vec{S}_f(j) \right) \hat{c}_{j\beta}$ in (17.185) as a single fermionic operator,

$$\sum_{\beta} \left(\vec{\sigma}_{\alpha\beta} \cdot \vec{S}_f(j) \right) \hat{c}_{j\beta} \to \frac{V}{J} \hat{f}_{j\alpha}, \tag{17.189}$$

where the amplitude $\frac{V}{J} = -\langle \hat{f}_{j\beta}^\dagger \hat{c}_{j\beta} \rangle$. Thus, in terms of composite f-electrons, we can rewrite the direct and cotunneled electron operators in (17.185) as

$$\hat{\psi}_{0\alpha} = \hat{c}_{0\alpha} + \left(\frac{\tilde{t}_f}{t_c} \right) \hat{f}_{0\alpha}, \tag{17.190}$$

where the complex amplitude for tunneling into the composite fermion state is $\tilde{t}_f = \frac{V}{J} t_f$. The Green's function for the ψ field is then

$$G_\psi(z) = \sum_{\mathbf{k}} \left(1, \frac{\tilde{t}_f}{t_c} \right) \cdot \begin{bmatrix} G_c(\mathbf{k}, z) & G_{cf}(\mathbf{k}, z) \\ G_{fc}(\mathbf{k}, z) & G_f(\mathbf{k}, z) \end{bmatrix} \cdot \begin{pmatrix} 1 \\ \tilde{t}_f/t_c \end{pmatrix}, \tag{17.191}$$

where $G_c(z)$, $G_{cf}(z)$, $G_{fc}(z)$, and $G_f(z)$ are the propagators between the heavy f- and conduction electrons.

It is instructive to contrast the tunneling conductance expected in a Kondo lattice with that of a single Kondo impurity. In the case of a single Kondo impurity, the local propagators are

$$G_f(\omega - i\delta) = \;=\;=\;=\;\blacktriangleright\;=\;=\; = \frac{1}{\omega - \lambda - i\Delta} \tag{17.192}$$

for the local f-propagator and

$$G_{fc}(\omega - i\delta) = G_{cf}(\omega - i\delta) = \sum_{k} =\;=\;\blacktriangleright\;=\;\bullet\!\!\longrightarrow_{k} = (i\pi\rho V) G_f(\omega - i\delta) \tag{17.193}$$

for the mixed propagator, where the single solid line denotes the bare conduction electron propagator $1/(z - \epsilon_{\mathbf{k}})$. Notice here that, since we are dealing with an impurity problem, momentum is not conserved, so the conduction propagator involves a sum over all momenta:

$$\sum_{\mathbf{k}} \longrightarrow_{\mathbf{k}} = \sum_{\mathbf{k}} \frac{1}{i\omega_n - \epsilon_{\mathbf{k}}} = -i\pi\rho\,\mathrm{sgn}(\omega_n), \tag{17.194}$$

for a broad band with a constant density of states. Finally,

$$
G_c(\omega - i\delta) = \sum_{k,k'} \overset{}{\underset{k}{\Longrightarrow}} = \sum_{k'} \underset{k}{\longrightarrow} + \sum_{k,k'} \underset{k}{\longrightarrow}\!\!\bullet = \blacktriangleright = \bullet\!\!\underset{k'}{\longrightarrow}
$$

$$
= i\pi\rho + (i\pi\rho V)^2 G_f(\omega - i\delta). \tag{17.195}
$$

Substituting these expressions into (17.191), we obtain

$$
G_\psi^{imp}(\omega - i\delta) = \frac{(i\pi\rho V + \tilde{t}_f/t_c)^2}{\omega - \lambda - i\Delta} + i\pi\rho, \tag{17.196}
$$

where ρ is the density of states of the conduction electrons and $\Delta = \pi\rho V^2 \simeq T_K$ is the width of the Kondo resonance. By dividing numerator and denominator of the first term by Δ, we can rewrite this expression in terms of dimensionless quantities:

$$
G_\psi^{imp}(\omega - i\delta) = \pi\rho \left[\frac{(i+q)^2}{\omega' - i} + i \right], \tag{17.197}
$$

where $q = \tilde{t}_f/(\pi V t_c \rho)$ is the ratio of the f and c tunneling amplitudes [54] and $\omega' = (\omega - \lambda)/\Delta$.

If we take the imaginary part of this expression, we obtain

$$
\begin{aligned}
A_\psi^{imp}(\omega) &= \frac{1}{\pi}\mathrm{Im}[G_\psi^{imp}(\omega - i\delta)] = \rho\,\mathrm{Im}\left[\frac{(\omega' + i)(i+q)^2}{(\omega'^2 + 1)} + i \right] \\
&= \rho \left[\frac{-1 + q^2 + 2q\omega'}{(\omega'^2 + 1)} + 1 \right] \\
&= \rho \left[\frac{(q+\omega')^2 - (1+\omega'^2)}{(\omega'^2 + 1)} + 1 \right] = \rho\frac{(q+\omega')^2}{1+\omega'^2}.
\end{aligned} \tag{17.198}
$$

Using (17.186), the differential conductance $\frac{dI}{d\phi} \equiv g(e\phi)$ for the impurity is

$$
g_{imp}(e\phi) = N\frac{\Gamma e^2}{\hbar}\rho\left.\frac{|q+\omega'|^2}{1+\omega'^2}\right|_{\omega'=(e\phi-\lambda)/\Delta}. \tag{17.199}
$$

Figure 17.19 illustrates the characteristic *Fano lineshape* predicted by this formula.

Now we turn to the case of the Kondo lattice. The diagrams for the Green's functions are identical, except that now we have momentum conservation along the propagator. Using (17.143), (17.139), and (17.144) in (17.191), we have

$$
G_\psi(z) = \left(1, \frac{i_f}{t_c}\right) \cdot \begin{bmatrix} G_c(z) & G_{cf}(z) \\ G_{fc}(z) & G_f(z) \end{bmatrix} \cdot \begin{pmatrix} 1 \\ \tilde{t}_f/t_c \end{pmatrix}
$$

$$
= \sum_k \left(1, \frac{i_f}{t_c}\right) \cdot \left[\begin{array}{c} \Longrightarrow \quad\quad \text{-}\text{-}\text{-}\bullet\!\!\Longrightarrow \\[2mm] \Longrightarrow\!\!\bullet\text{-}\text{-}\text{-} \quad \text{-}\text{-}\!+\!\text{-}\bullet\!\!\Longrightarrow \end{array} \right] \cdot \begin{pmatrix} 1 \\ \tilde{t}_f/t_c \end{pmatrix},
$$

$$
\tag{17.200}
$$

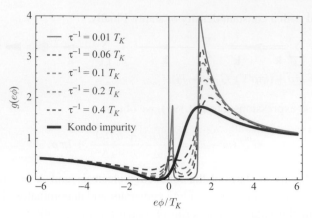

Fig. 17.19 Differential tunneling conductance $g(\phi)$ for a single Kondo impurity case (blue line) given by (17.199), and for a Kondo lattice (green line) given by (17.203). A typical Fano lineshape in the single Kondo impurity case gets replaced with a double-peaked resonance line in the Kondo lattice case. The dashed lines illustrate the effect of disorder, which destroys the coherence, closing the gap in the density of states curve. Here $\tilde{t}_f/t_c = 0.2, q = 1.2, \lambda/T_K = 0.6$, while $D_1 = 40T_K, D_2 = 80T_K$.

where, as in previous sections, the double lines denote the full conduction propagator $G_c(\mathbf{k}, z) = [(\omega - \epsilon_\mathbf{k}) - \mathcal{V}^2/(\omega - \lambda)]^{-1}$ while the single dashed lines denote the bare f-propagator $\frac{1}{z-\lambda}$. Writing this out in full, we have

$$G_\psi(z) = \sum_\mathbf{k} \left(1, \ \frac{\tilde{t}_f}{t_c}\right) \cdot \left[\begin{array}{cc} G_c(\mathbf{k}, z) & \frac{\mathcal{V}}{z-\lambda}G_c(\mathbf{k}, z) \\ \frac{\mathcal{V}}{z-\lambda}G_c(\mathbf{k}, z) & \frac{1}{z-\lambda} + \frac{\mathcal{V}}{z-\lambda}G_c(\mathbf{k}, z)\frac{\mathcal{V}}{z-\lambda} \end{array}\right] \cdot \left(\begin{array}{c} 1 \\ \tilde{t}_f/t_c \end{array}\right)$$

$$= \sum_\mathbf{k} \left(\frac{(1 + \frac{\tilde{t}_f}{t_c}\frac{\mathcal{V}}{z-\lambda})^2}{z - \epsilon_\mathbf{k} - \frac{\mathcal{V}^2}{z-\lambda}} + \frac{(\tilde{t}_f/t_c)^2}{z - \lambda}\right). \tag{17.201}$$

In this expression we have separated out the terms that are proportional to the conduction electron propagator. Notice that the poles of this expression are at the quasiparticle energies $z = E_{\mathbf{k}\pm} = \frac{\epsilon_\mathbf{k}+\lambda}{2} \pm \sqrt{\left(\frac{\epsilon_{\mathbf{k}-\lambda}}{2}\right)^2 + \mathcal{V}^2}$, and the poles at $z = \lambda$, in the first and second terms actually cancel.

The momentum summation in $G_\psi^{KL}(\omega)$ (17.200) can be carried out analytically assuming a constant conduction electron density of states ρ, as follows:

$$G_\psi^{KL}(z) = \rho \int_{-D_1}^{D_2} d\epsilon \left(\frac{(1 + \frac{\tilde{t}_f}{t_c}\frac{\mathcal{V}}{z-\lambda})^2}{z - \epsilon - \frac{\mathcal{V}^2}{z-\lambda}} + \frac{(\tilde{t}_f/t_c)^2}{z - \lambda}\right)$$

$$= \rho\left(1 + \frac{q\Delta}{z-\lambda}\right)^2 \ln\left[\frac{z + D_1 - \frac{\mathcal{V}^2}{z-\lambda}}{z - D_2 - \frac{\mathcal{V}^2}{z-\lambda}}\right] + \frac{2D\rho\tilde{t}_f^2/t_c^2}{z - \lambda}. \tag{17.202}$$

Here $-D_1$ and D_2 are the lower and upper conduction band edges, respectively, and $2D = D_1 + D_2$ is the bandwidth. Using (17.186), the final expression for the tunneling conductance is then

$$g(e\phi) = N\left(\frac{\Gamma e^2}{\hbar}\right)\rho\ \frac{1}{\pi}\text{Im}[\tilde{G}_\psi^{KL}(e\phi - i\delta)],\qquad(17.203)$$

where

$$\tilde{G}_\psi^{KL}(\omega) = \left(1 + \frac{q\Delta}{\omega - \lambda}\right)^2 \ln\!\left[\frac{\omega + D_1 - \frac{\mathcal{V}^2}{\omega - \lambda}}{\omega - D_2 - \frac{\mathcal{V}^2}{\omega - \lambda}}\right] + \frac{2D\rho \tilde{t}_f^2/t_c^2}{\omega - \lambda}.\qquad(17.204)$$

The differential tunneling conductance predicted by this formula has two pronounced peaks at $e\phi \sim \lambda$, separated by a narrow hybridization gap $\Delta_g \sim 2\mathcal{V}^2/D$ in the single-particle spectrum, as shown in Figure 17.19.

In practice, experimental tunneling results will be modified by the effects of disorder. A phenomenological quasiparticle elastic relaxation rate τ^{-1} may be introduced into the theory by replacing $\omega \to \omega - i\tau^{-1}$ in (17.204). The results of this procedure are shown in Figure 17.19. As we see, disorder removes the sharp peak structure in the tunneling conductance $g(e\phi)$ (17.203). The resulting lineshape of the tunneling conductance $dI/d\phi(e\phi)$ is an asymmetric smooth curve.

17.10 Optical conductivity of heavy electrons

17.10.1 Heuristic discussion

The optical conductivity of heavy-fermion metals deserves special discussion. According to the f-sum rule (see Section 9.9.3), the total integrated optical conductivity is determined by the plasma frequency (9.197):

$$\frac{2}{\pi}\int_0^\infty d\omega\sigma(\omega) = f_1 = \left(\frac{ne^2}{m}\right),\qquad(17.205)$$

where n is the density of electrons. In the absence of local moments, this is the total spectral weight inside the Drude peak of the optical conductivity. But what happens to the distribution of the spectral weight when the heavy-electron fluid forms? Physically, while we expect this sum rule to be preserved, a new quasiparticle Drude peak will form, corresponding to the heavy-electron Drude peak:

$$\frac{2}{\pi}\int_0^{T_K} d\omega\sigma(\omega) = f_2 = \frac{ne^2}{m^*} = f_1\frac{m}{m^*}.\qquad(17.206)$$

In other words, the spectral weight will divide into a small heavy-fermion Drude peak of total weight f_2, where

$$\sigma(\omega) = \frac{ne^2}{m^*}\frac{1}{(\tau^*)^{-1} - i\omega},\qquad(17.207)$$

separated by an energy of order $V \sim \sqrt{T_K D}$ from an interband component associated with excitations between the lower and upper Kondo bands [59, 60]. This second term carries the bulk $\sim f_1$ of the spectral weight (Figure 17.20).

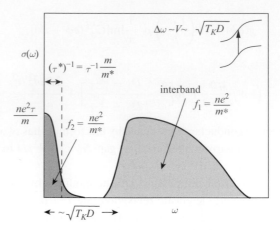

Fig. 17.20 Schematic showing separation of the optical sum rule in a heavy-fermion system into a high-energy interband component of weight $f_2 \sim ne^2/m$ and a low-energy Drude peak of weight $f_1 \sim ne^2/m^*$.

17.10.2 Calculation of the optical conductivity, including interband term

Let us now illustrate this phenomenon using the Kubo formula (10.37), which we rewrite here:

$$\sigma^{\alpha\beta}(i\nu_n) = -\frac{1}{\nu_n} \left[\langle Tj^\alpha(\nu')j^\beta(-\nu') \rangle \right]_{\nu'=0}^{\nu'=i\nu_n}. \tag{17.208}$$

If we take the model in which the bare f-states have no dispersion, then the current operator is given by

$$\vec{j} = e \sum_{\mathbf{k}\sigma} \vec{\nabla}_{\mathbf{k}} \epsilon_{\mathbf{k}} c^\dagger_{\mathbf{k}\sigma} c_{\mathbf{k}\sigma}. \tag{17.209}$$

Our calculation of the resistivity will follow the lines of Section 10.3, except that now the conduction electron propagator is modified by the effects of hybridization. Using Section 17.6.3, the hybridized conduction electron propagator is

$$G_c(\mathbf{k}, i\omega_n) = \frac{1}{i\tilde{\omega}_n - \epsilon_{\mathbf{k}} - \Sigma_c(i\tilde{\omega}_n)}, \qquad \Sigma_c(z) = \frac{V^2}{i\omega_n - \lambda}, \tag{17.210}$$

where we have introduced

$$i\tilde{\omega}_n = i\omega_n + i\,\mathrm{sgn}\,\omega_n \frac{\Gamma}{2} \tag{17.211}$$

to include an isotropic white-noise scattering rate Γ acting on the conduction electrons. The conductivity is then given by

$$\sigma(i\nu_n)\delta_{ab} = -\frac{e^2}{\nu_n}\left[a \sim\!\!\bigcirc\!\!\sim b\right]_0^{i\nu_n}$$

$$= -\frac{e^2}{\nu_n}\left[T\sum_{\kappa=(\mathbf{k},i\omega_r)}\left(\nabla_a\epsilon_{\mathbf{k}}\nabla_b\epsilon_{\mathbf{k}}G_c(\mathbf{k},i\omega_r+i\nu_n)G_c(\mathbf{k},i\omega_r)\right) - (i\nu_n \to 0)\right].$$

(17.212)

Our calculation will now follow the lines of Section 10.3, except that we now have to deal with the frequency dependence of the self-energy. On the other hand, if we assume the scattering is isotropic, we can neglect the current vertex corrections. We will assume that the hybridization is small enough compared with the bandwidth to take the bandwidth cut-off to infinity, but in order to keep the interband physics we need to be quite careful with the internal frequency integrals. Our procedure is:

- Sum over the loop momentum $\mathbf{k}$ by replacing the momentum sum with an energy integral, $\sum_{\mathbf{k}} \to \rho\int d\epsilon$, the bounds of which can be taken to infinity.
- Carry out the frequency sum at zero temperature, replacing $T\sum_{i\omega_r} \to \int \frac{dz}{2\pi i}$.

Our first step is to replace the sum over $\mathbf{k}$ by an integral over energy, $\sum_{\mathbf{k}} \nabla_a\epsilon_{\mathbf{k}}\nabla_b\epsilon_{\mathbf{k}} \to \frac{1}{3}\rho v_F^2\delta_{ab}\int d\epsilon = \frac{n}{m}\delta_{ab}\int d\epsilon$, where v_F is the conduction electron Fermi velocity and we have replaced $\rho v_F^2/3 = \frac{n}{m}$, where n is the conduction electron density and m the conduction electron mass. Since the direct hybridization gap $(2V)$ of the Kondo lattice does not depend on the bandwidth of the conduction electrons, we can extend the integration to infinity to obtain

$$\sigma(i\nu_n) = -\left(\frac{ne^2}{m}\right)\frac{1}{\nu_n}$$

$$\times\left[T\sum_{i\omega_r}\int_{-\infty}^{\infty}\frac{d\epsilon}{(i\tilde{\omega}_r^+ - \epsilon - \Sigma_c(i\omega_r^+))(i\tilde{\omega}_r^- - \epsilon - \Sigma_c(i\omega_r^-))} - (i\nu_n \to 0)\right],$$

(17.213)

where $\omega_r^{\pm} = \omega_r \pm \nu_n/2$. If we carry out the energy integral as a contour integral, then the poles are at $z^{\pm} = i\tilde{\omega}_r^{\pm} - \Sigma_c(i\omega_r^{\pm})$. Although these poles look a little complicated, please note that they lie on the same side of the real axis as $i\omega_r^{\pm}$. Now $i\omega_r^{\pm}$ and hence $z^{\pm}$ are on opposite sides of the real axis, providing $|\omega_r| < |\nu_n|/2$. This is precisely the same condition we obtained for an electron gas without hybridization, so that when we complete the integral over energy as a contour around the upper half-plane, we will only get a finite result if this condition is satisfied. The result of this reasoning is

$$\sigma(i\nu_n) = \left(\frac{ne^2}{m}\right)\frac{1}{\nu_n}\left[2\pi iT\sum_{|\omega_r|<\nu_n/2}\frac{1}{i\nu_n + i\Gamma - (\Sigma^+ - \Sigma^-)}\right] \qquad (\nu_n > 0),$$

(17.214)

where we have introduced $\Sigma^{\pm} = \Sigma_c(i\omega_n^{\pm})$. (Notice that the contribution from the term where we replace $i\nu_n \to 0$ now disappears.)

Next, we take the zero-temperature limit, so that $T \sum_{i\omega_r} \to \int_{-i\infty}^{i\infty} \frac{dz}{2\pi i}$, to obtain

$$\sigma(i\nu) = \left(\frac{ne^2}{m}\right) \frac{1}{\nu_n} \int_{-i\nu/2}^{i\nu/2} dz \left(\frac{1}{i\nu + i\Gamma - (\Sigma(z + i\nu/2) - \Sigma(z - i\nu/2))}\right) \qquad (\nu > 0).$$
(17.215)

To check that we are on the right track, let us look at the low-frequency limit of this expression. In this limit, we can replace

$$i\nu - (\Sigma(z + i\nu/2) - \Sigma(z - i\nu/2)) \to i\nu \left(1 - \frac{\partial \Sigma(z)}{\partial z}\right) = Z^{-1} i\nu,$$
(17.216)

where $Z = \left(1 + \frac{V^2}{\lambda^2}\right)^{-1}$ is recognized as the quasiparticle weight, so the low-frequency conductivity becomes

$$\sigma(\omega) = \left(\frac{ne^2}{m^*}\right) \frac{1}{\Gamma^* - i\nu},$$
(17.217)

where $m^* = m/Z$ is the renormalized mass and $\Gamma^* = Z\Gamma$ is the renormalized Drude width, so we have recovered the Drude peak with an overall weight reduced by the factor Z.

Let us now continue to see if we can capture the interband part of the conductivity. Fortunately, the argument of (17.145) simplifies into a two-pole structure as follows:

$$\frac{1}{i\nu + i\Gamma - (\Sigma(z + i\nu/2) - \Sigma(z - i\nu/2))} = \frac{1}{i\tilde{\nu}} \left[1 - \frac{V^2[i\nu]}{(z - \lambda)^2 + \left(\frac{\nu}{2}\right)^2 + V^2[i\nu]}\right]$$
$$= \frac{1}{i\tilde{\nu}} \left[1 - \frac{V^2[i\nu]}{(z - z_+)(z - z_-)}\right], \qquad (17.218)$$

where $\tilde{\nu} = \nu + \Gamma$ and

$$z_\pm = \lambda \pm \sqrt{\left(\frac{i\nu}{2}\right)^2 - V^2[i\nu]}, \qquad V^2[i\nu] = V^2 \frac{i\nu}{i\nu + i\Gamma},$$
(17.219)

so that when we do the integral over z, we obtain

$$\int_{-i\nu/2}^{i\nu/2} dz \left(\frac{1}{i\nu - (\Sigma(z + i\nu/2) - \Sigma(z - i\nu/2))}\right)$$
$$= \frac{i\nu}{i\tilde{\nu}} \left[1 - \frac{V^2}{i\tilde{\nu}(z_+ - z_-)} \left(\ln\left[\frac{\frac{i\nu}{2} - z_+}{-\frac{i\nu}{2} - z_+}\right] - \ln\left[\frac{\frac{i\nu}{2} - z_-}{-\frac{i\nu}{2} - z_-}\right]\right)\right]. \qquad (17.220)$$

Inserting this into (17.215), the optical conductivity is given by

$$\sigma(i\nu) = \left(\frac{ne^2}{m}\right) \frac{1}{\Gamma - i(i\nu)}$$
$$\times \left[1 - \frac{V^2}{i(\nu + \Gamma)(z_+ - z_-)} \left(\ln\left[\frac{\frac{i\nu}{2} - z_+}{-\frac{i\nu}{2} - z_+}\right] - \ln\left[\frac{\frac{i\nu}{2} - z_-}{-\frac{i\nu}{2} - z_-}\right]\right)\right]. \qquad (17.221)$$

$$\sigma(\omega) = \sigma_1(\omega) + i\sigma_2(\omega)$$

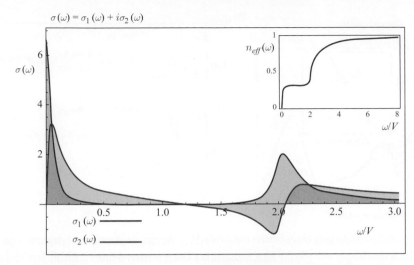

Plot of the real and imaginary parts of the optical conductivity obtained from (17.222) using mean-field conduction electron propagators. Inset shows the integrated spectral weight $n_{eff}(\omega) = \frac{2m}{e^2} \int_0^\omega \sigma_1(x)\frac{dx}{\pi}$, showing contributions from Drude and interband parts.

Fig. 17.21

Finally, analytically continuing $i\nu \to \omega + i\delta$, we obtain

$$\sigma(\omega + i\delta) = \left(\frac{ne^2}{m}\right)\frac{1}{\Gamma - i\omega}\left[1 + \frac{V^2}{(\omega + i\Gamma)(z_+ - z_-)}\left(\ln\left[\frac{z_+ + \frac{\omega}{2}}{z_+ - \frac{\omega}{2}}\right] - \ln\left[\frac{z_- + \frac{\omega}{2}}{z_- - \frac{\omega}{2}}\right]\right)\right],$$

$$z_\pm = \lambda \pm \sqrt{\left(\frac{\omega}{2}\right)^2 - V^2\frac{\omega}{\omega + i\Gamma}}. \qquad (17.222)$$

Note that although it is tempting to combine the final logarithms into a single term, unfortunately the dangerous identity $\ln f(z) + \ln g(z) = \ln[f(z)g(z)]$ fails to preserve the branch cut structure of the logarithms and can't be used here if one wants to preserve the full analytic structure of the conductivity. Figure 17.21 shows a plot of the optical conductivity obtained with this function. Notice the formation of the Drude peak and the direct gap of size $2V$.

These basic features – the formation of a narrow Drude peak, the presence of a hybridization gap $V \sim \sqrt{T_K D}$ that scales as the square root of the Kondo temperature – have been confirmed in optical measurements on heavy-electron systems [61–64]. In particular, the relationship $V \sim \sqrt{T_K D}$ implies that

$$\left(\frac{V}{T_K}\right)^2 \propto \frac{D}{T_K} \sim \frac{m^*}{m}. \qquad (17.223)$$

Experimentally, the characteristic scale T_K can be determined from specific heat measurements, while V can be directly inferred from optical conductivity measurements. Figure 17.22 shows that this relationship is approximately followed for a wide range of materials.

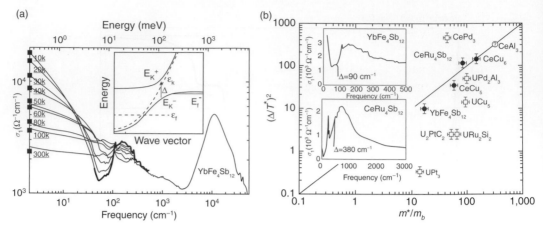

(a)

Energy (meV)

Fig. 17.22 (a) Measured optical conductivity heavy-fermion metal YbFe$_4$Sb$_{12}$, showing development of hybridization gap. (b) Scaling of hybridization gap with effective mass, measured in a variety of heavy-fermion materials. Reprinted with permission from S. V. Dordevic, *et al., Phys. Rev. Lett.*, vol. 86, p. 684, 2001. Copyright 2001 by the American Physical Society.

17.11 Summary

In this chapter we have examined Doniach's argument that the enhancement of the Kondo temperature over and above the characteristic RKKY magnetic interaction energy between spins leads to the formation of a heavy-electron ground state. This enhancement is thought to be generated by the large spin degeneracies of rare-earth or actinide ions. A simple mean-field theory of the Kondo model and Kondo lattice which ignores the RKKY interactions provides a unified picture of heavy-electron formation and the Kondo effect in terms of the formation of a composite quasiparticle between high-energy conduction band electrons and local moments. This basic physical effect is local in space, but non-local in time. Certain analogies can be struck between Cooper pair formation and the formation of the heavy-electron bound state; in particular, the charge on the f-electron can be seen as a direct consequence of the temporal phase stiffness of the Kondo bound state. This bound state hybridizes with conduction electrons, producing a single isolated resonance in a Kondo impurity and an entire renormalized Fermi surface in the Kondo lattice.

Exercises

Exercise 17.1 Directly calculate the phase stiffness $\rho_\phi = -\frac{d^2 F}{d\lambda^2}$ of the large-N Kondo impurity model, and show that, at $T = 0$,

$$\rho_\phi = \frac{N}{\pi}\left(\frac{\sin(\pi q)}{T_K}\right).$$

Exercise 17.2 Kondo effect in an insulator and an s-wave superconductor.

Use the large-N mean-field theory to contrast the effect of an insulating and superconducting gap on the Kondo resonance in the single-impurity Kondo effect. Take the large-N limit for a Kondo resonance lying at the Fermi energy, with $Q/N = \frac{1}{2}$. In a metallic environment, the mean-field f-electron (composite fermion) spectral function is a simple Lorentzian,

$$A_f = \frac{1}{\pi}\text{Im}[G_f(\omega - i\delta)] = \frac{1}{\pi}\text{Im}\left[\frac{1}{\omega - i\Delta}\right] = \frac{1}{\pi}\left(\frac{\Delta}{\omega^2 + \Delta^2}\right), \qquad (17.224)$$

where $\Delta = \pi\rho|V|^2 \sim T_K$ is the resonant level width. Once a gap develops at the Fermi surface, provided the Kondo temperature is large enough compared with the gap a Kondo effect can still take place, and the f-propagator is modified to the form

$$G_f(\omega - i\delta) = \frac{1}{\omega - \Sigma(\omega)}. \qquad (17.225)$$

(a) Assuming a constant density of states with an insulating gap Δ_g, show that, in the insulator,

$$\Sigma_I(\omega - i\delta) = i\Delta + \frac{\Delta}{\pi}\ln\left[\frac{\omega - \Delta_g - i\delta}{\omega + \Delta_g - i\delta}\right]. \qquad (17.226)$$

Plot the spectral function of the f-electron.

(b) Show that, in an s-wave superconductor with gap function Δ_g, the self-energy becomes

$$\Sigma_{sc}(\omega - i\delta) = i\Delta\frac{1}{\sqrt{(\omega - i\delta)^2 - \Delta_g^2}}(\omega\mathbf{1} + \Delta_g\tau_1), \qquad (17.227)$$

where τ_1 is the first Nambu matrix. Plot this spectral function and contrast it with the result obtained in the insulator. What is the meaning of the anomalous component of the f-spectral function?

(c) Discuss the main physical difference between the Kondo effect in an insulator, compared with a Kondo effect in an s-wave superconductor.

References

[1] E. Bucher, J. P. Maita, G. W. Hull, R. C. Fulton, and A. S. Cooper, Electronic properties of beryllides of the rare earth and some actinides, *Phys. Rev. B*, vol. 11, p. 440, 1975.

[2] F. Steglich, J. Aarts, C. D. Bredl, W. Leike, D. E. Meshida, W. Franz, and H. Schäfer, Superconductivity in the presence of strong Pauli paramagnetism: $CeCu_2Si_2$, *Phys. Rev. Lett.*, vol. 43, p. 1892, 1976.

[3] K. Andres, J. Graebner, and H. R. Ott, 4f-virtual-bound-state formation in $CeAl_3$ at low temperatures, *Phys. Rev. Lett.*, vol. 35, p. 1779, 1975.

[4] M. A. Ruderman and C. Kittel, Indirect exchange coupling of nuclear magnetic moments by conduction electrons, *Phys. Rev.*, vol. 96, p. 99, 1954.

[5] G. Aeppli and Z. Fisk, Kondo insulators, *Comments Condens. Matter Phys.*, vol. 16, p. 155, 1992.

[6] A. Menth, E. Buehler, and T. H. Geballe, Magnetic and semiconducting properties of SmB_6, *Phys. Rev. Lett.*, vol. 22, p. 295, 1969.

[7] M. B. Maple and D. Wohlleben, Nonmagnetic 4f shell in the high-pressure phase of SmS, *Phys. Rev. Lett.*, vol. 27, no. 8, p. 511, 1971.

[8] N. F. Mott, Rare-earth compounds with mixed valencies, *Philos. Mag.*, vol. 30, no. 2, p. 403, 1974.

[9] S. Doniach, Kondo lattice and weak antiferromagnetism, *Physica*, vol. 91B, p. 231, 1977.

[10] T. Kasuya, A theory of metallic ferro- and antiferromagnetism on Zener's model, *Prog. Theor. Phys.*, vol. 16, no. 1, p. 45, 1956.

[11] R. Jullien, J. Fields, and S. Doniach, Zero-temperature real-space renormalization-group method for a Kondo-lattice model Hamiltonian, *Phys. Rev. B*, vol. 16, no. 11, p. 4889, 1977.

[12] J. L. Smith and P. S. Riseborough, Actinides, the narrowest bands, *J. Magn. Magn. Mater.*, vol. 47–48, p. 545, 1985.

[13] Y. Onuki and T. Komatsubara, Heavy fermion state $CeCu_6$, *J. Magn. Magn. Mater.*, vol. 63–64, p. 281, 1987.

[14] R. M. Martin, Fermi-surface sum rule and its consequences for periodic Kondo and mixed-valence systems, *Phys. Rev. Lett.*, vol. 48, p. 362, 1982.

[15] M. Oshikawa, Topological approach to Luttinger's theorem and the Fermi surface of a Kondo lattice, *Phys. Rev. Lett.*, vol. 84, p. 3370, 2000.

[16] L. Taillefer, R. Newbury, G. G. Lonzarich, Z. Fisk, and J. L. Smith, Direct observation of heavy quasiparticles in UPt_3 via the dHvA effect, *J. Magn. Magn. Mater.*, vol. 63–64, p. 372, 1987.

[17] L. Taillefer and G. G. Lonzarich, Heavy-fermion quasiparticles in UPt_3, *Phys. Rev. Lett.*, vol. 60, p. 1570, 1988.

[18] H. Shishido, R. Settai, H. Harima, and Y. Onuki, A drastic change of the Fermi surface at a critical pressure in CeRhIn 5: dHvA study under pressure, *J. Phys. Soc. Jpn.*, vol. 74, p. 1103, 2005.

[19] B. Coqblin and J. R. Schrieffer, Exchange interaction in alloys with cerium impurities, *Phys. Rev.*, vol. 185, p. 847, 1969.

[20] H. Ikeda and K. Miyake, A theory of anisotropic semiconductor of heavy fermions, *J. Phys. Soc. Jpn.*, vol. 65, no. 6, p. 1769, 1996.

[21] M. Mekata, S. Ito, N. Sato, T. Satoh, and N. Sato, Spin fluctuations in dense Kondo alloys, *J. Magn. Magn. Mater.*, vol. 54, p. 433, 1986.

[22] C. Urano, M. Nohara, S. Kondo, F. Sakai, H. Takagi, T. Shiraki, and T. Okubo, *Phys. Rev. Lett.*, vol. 85, p. 1052, 2000.

[23] P. W. Anderson, *Valence Fluctuations in Solids*, North Holland, 1981.

[24] E. Witten, Chiral symmetry, the 1/N expansion and the SU(N) Thirring model, *Nucl. Phys. B*, vol. 145, p. 110, 1978.

[25] O. Gunnarsson and K. Schönhammer, Electron spectroscopies for Ce compounds in the impurity model, *Phys. Rev. B*, vol. 28, p. 4315, 1983.

[26] N. Read and D. M. Newns, On the solution of the Coqblin–Schrieffer Hamiltonian by the large-N expansion technique, *J. Phys. C*, vol. 16, p. 3274, 1983.

[27] P. Coleman, 1/N expansion for the Kondo lattice, *Phys. Rev.*, vol. 28, p. 5255, 1983.

[28] C. Lacroix and M. Cyrot, Phase diagram of the Kondo lattice, *Phys. Rev. B*, vol. 20, p. 1969, 1979.

[29] S. Chakravarty, abstract of unpublished preprint, *Proc. Am. Phys. Soc. (March)*, 1982.

[30] P. Coleman, New approach to the mixed-valence problem, *Phys. Rev. B*, vol. 29, p. 3035, 1984.

[31] N. Read and D. M. Newns, A new functional integral formalism for the degenerate Anderson model, *J. Phys. C*, vol. 29, p. L1055, 1983.

[32] A. Auerbach and K. Levin, Kondo bosons and the Kondo lattice: microscopic basis for the heavy Fermi liquid, *Phys. Rev. Lett.*, vol. 57, p. 877, 1986.

[33] S. E. Barnes, New method for the Anderson model, *J. Phys. F: Met. Phys.*, vol. 6, p. 1375, 1976.

[34] A. J. Millis and P. A. Lee, Large-orbital-degeneracy expansion for the lattice Anderson model, *Phys. Rev. B*, vol. 35, no. 7, p. 3394, 1987.

[35] P. Coleman, Mixed valence as an almost broken symmetry, *Phys. Rev. B*, vol. 35, p. 5072, 1987.

[36] M. C. Gutzwiller, Effect of correlation on the ferromagnetism of transition metals, *Phys. Rev. Lett.*, vol. 10, no. 5, p. 159, 1963. The Hubbard model was written down independently by Gutzwiller in equation (11) of this paper.

[37] W. F. Brinkman and T. M. Rice, Application of Gutzwiller's variational method to the metal insulator transition, *Phys. Rev. B*, vol. 2, p. 4302, 1970.

[38] B. H. Brandow, Variational theory of valence fluctuations: ground states and quasi-particle excitations of the Anderson lattice model, *Phys. Rev. B*, vol. 33, p. 215, 1986.

[39] M. Dzero, K. Sun, V. Galitski, and P. Coleman, Topological Kondo insulators, *Phys. Rev. Lett.*, vol. 104, p. 106408, 2010.

[40] S. Wolgast, Ç. Kurdak, K. Sun, J. W. Allen, D.-J. Kim, and Z. Fisk, Low-temperature surface conduction in the Kondo insulator SmB_6, *Phys. Rev. B*, vol. 88, p. 180405, 2013.

[41] D. J. S. Thomas, T. Grant, J. Botimer, Z. Fisk, and J. Xia, Surface Hall effect and non-local transport in SmB_6: evidence for surface conduction, *Sci. Rep.*, vol. 3, p. 3150, 2014.

[42] S. Burdin, A. Georges, and D. R. Grempel, Coherence scale of the Kondo lattice, *Phys. Rev. Lett.*, vol. 85, p. 1048, 2000.

[43] T. A. Costi and N. Manini, Low-energy scales and temperature-dependent photoemission of heavy fermions, *J. Low Temp. Phys.*, vol. 126, p. 835, 2002.

[44] H. Shishido, R. Settai, H. Harima, and Y. Ōnuki, A drastic change of the Fermi surface at a critical pressure in CeRhIn$_5$: dHvA study under pressure, *J. Phys. Soc. Jpn.*, vol. 74, no. 4, p. 1103, 2005.

[45] T. Park, F. Ronning, H. Q. Yuan, M. B. Salamon, R. Movshovich, J. L. Sarrao, and J. D. Thompson, Hidden magnetism and quantum criticality in the heavy-fermion superconductor CeRhIn$_5$, *Nature*, vol. 440, no. 7080, p. 65, 2006.

[46] A. F. G. Wyatt, Anomalous densities of states in normal tantalum and niobium, *Phys. Rev Lett.*, vol. 13, p. 401, 1964.

[47] J. Appelbaum, s-d exchange model of zero-bias tunneling anomalies, *Phys. Rev. Lett.*, vol. 17, p. 91, 1966.

[48] P. W. Anderson, Localized magnetic states and Fermi-surface anomalies in tunneling, *Phys. Rev. Lett.*, vol. 17, p. 95, 1966.

[49] D. V. Averin and Y. V. Nazarov, Virtual electron diffusion during quantum tunneling of the electric charge, *Phys. Rev. Lett.*, vol. 65, no. 19, p. 2446, 1990.

[50] M. Pustilnik and L. Glazman, Kondo effect in real quantum dots, *Phys. Rev. Lett.*, vol. 87, p. 216601, 2001.

[51] D. Goldhaber-Gordon, H. Shtrikman, and D. Mahalu, Kondo effect in a single-electron transistor, *Nature*, vol. 391, no. 6663, p. 156, 1998.

[52] M. I. Katsnelson, A. I. Lichtenstein, and H. Van Kempen, Real-space imaging of an orbital Kondo resonance on the Cr (001) surface, *Nature*, vol. 415, no. 6871, p. 507, 2002.

[53] U. Fano, Effects of configuration interaction on intensities and phase shifts, *Phys. Rev.*, vol. 124, no. 6, p. 1866, 1961.

[54] V. Madhavan, Tunneling into a single magnetic atom: spectroscopic evidence of the Kondo resonance, *Science*, vol. 280, no. 5363, p. 567, 1998.

[55] A. R. Schmidt, M. H. Hamidian, P. Wahl, and F. Meier, Imaging the Fano lattice to 'hidden order' transition in URu$_2$Si$_2$, *Nature*, vol. 465, p. 570, 2010.

[56] P. Aynajian, E. H. da Silva Neto, A. Gyenis, R. E. Baumbach, J. D. Thompson, Z. Fisk, E. D. Bauer, and A. Yazdani, Visualizing heavy fermions emerging in a quantum critical Kondo lattice, *Nature*, vol. 486, no. 7402, p. 201, 2012.

[57] M. Maltseva, M. Dzero, and P. Coleman, Electron cotunneling into a Kondo lattice, *Phys. Rev. Lett.*, vol. 103, no. 20, 2009.

[58] J. R. Schrieffer and P. Wolff, Relation between the Anderson and Kondo Hamiltonians, *Phys. Rev.*, vol. 149, p. 491, 1966.

[59] A. J. Millis, Effect of a nonzero temperature on quantum critical points in itinerant fermion systems, *Phys. Rev. B*, vol. 48, p. 7183, 1993.

[60] L. Degiorgi, F. Anders, and G. Gruner, Charge excitations in heavy electron metals, *Eur. Phys. J. B*, vol. 19, p. 167, 2001.

[61] S. V. Dordevic, D. N. Basov, N. R. Dilley, E. D. Bauer, and M. B. Maple, Hybridization gap in heavy fermion compounds, *Phys. Rev. Lett.*, vol. 86, p. 684, 2001.

[62] B. Bucher, Z. Schlesinger, D. Mandrus, *et al.*, Charge dynamics of Ce-based compounds: Connection between the mixed valent and Kondo-insulator states, *Phys. Rev. B*, vol. 53, p. R2948, 1996.

[63] W. P. Beyerman, G. Gruner, Y. Dlicheouch, and M. B. Maple, Frequency-dependent transport properties of UPt$_3$, *Phys. Rev. B*, vol. 37, p. 10353, 1988.

[64] S. Donovan, A. Schwartz, and G. Grüner, Observation of an Optical Pseudogap in UPt$_3$, *Phys. Rev. Lett.*, vol. 79, p. 1401, 1997.

Chapter 17 discussed the mean-field theory of the Kondo lattice. In this final chapter we return to discuss fluctuations, in their various forms.

In particular, we shall discuss how we can include valence fluctuations in our description of the Kondo lattice. Mixed-valence ions undergo coherent charge fluctuations between different valence states, such as the fluctuation of a mixed-valent cerium ion: $Ce^{3+} \leftrightharpoons Ce^{4+} + e^-$, in which electrons hop in an out of the cerium $4f$ state:

$$4f^1 \leftrightharpoons 4f^0 + e^-. \tag{18.1}$$

If we draw this process as a heuristic Feynman diagram,

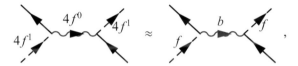

we see that the intermediate $4f^0$ state can be regarded as a kind of exchange boson. We'll see how this qualitative observation leads us to the *slave boson*: a kind of intermediate boson that mediates the valence fluctuations. There is an interesting link between the slave boson and the gauge bosons of high-energy physics. The passage from the Kondo to the mixed-valence problem is akin to the passage from Fermi's point contact theory of the weak interaction to the intermediate vector boson description of the electroweak force.

We shall also discuss the fluctuations around mean-field theory that we neglected in the previous chapter. Unlike a thermodynamic mean-field theory, where the small parameter controlling the strength of fluctuations is the macroscopic volume, here the control parameter N is finite, most likely to be two! Thus the use of a large-N mean-field theory really does require a lot of chutzpah, and it is quite important to understand the nature of the physics contained in the $1/N$ fluctuations about the mean field theory.

The strict large-N limit approximates the crossover from unscreened to fully screened local moments as a sharp phase transition. We shall see how, at finite N, phase fluctuations convert the off-diagonal long-range order of the slave boson field or hybridization to power-law correlations:

$$\langle b_j(\tau) b_j^\dagger(0) \rangle \sim t^{-\alpha/N}. \tag{18.2}$$

These fluctuations about the mean-field theory restore the gauge symmetry, converting the sharp transition of the mean-field theory into a smooth crossover (Section 18.3). However, at the same time, they preserve the infinitely long-range correlations (in time) and, for this reason, the mean-field starting point is still useful.

Finally, we'll see how, using the slave boson approach, we can describe the phenomenon of topological Kondo insulators.

18.1 The slave boson and mixed valence

To understand the Kondo lattice, we used Abrikosov's trick to factorize the spin operator as a fermion bilinear, $S_{\alpha\beta} \rightarrow f_\alpha^\dagger f_\beta$. Now we'd like to ask how we can also take account of strongly correlated charge fluctuations or *mixed valence*. Mixed valence describes the effect of coherent quantum fluctuations within a restricted set of valence configurations, such as the fluctuations between the Ce^{3+} ($4f^1$) and Ce^{4+}($4f^0$) configurations of a cerium ion in heavy-fermion materials or between the magnetic $(CuO_2)^{2-}$ and hole-doped $(CuO_2)^{1-}$ configurations of a cuprate superconductor.

When interactions are strong, the allowed configurations for valence fluctuations are restricted. For example, consider the infinite U Anderson model, involving valence fluctuations $f^0 + e^- \rightleftharpoons f^1$:

$$H = \sum_{\mathbf{k},\,\sigma \in [1,N]} \epsilon_{\mathbf{k}} c_{\mathbf{k}\sigma}^\dagger c_{\mathbf{k}\sigma} + \sum_{\mathbf{k}\sigma \in \{1,N\}} [V_{\mathbf{k}} c_{\mathbf{k}\sigma}^\dagger X_{0\sigma} + V_{\mathbf{k}}^* X_{\sigma 0} c_{\mathbf{k}\sigma}] + E_f \sum_{\sigma=1,N} X_{\sigma\sigma}. \quad (18.3)$$

infinite U Anderson impurity model

Here the $\hat{X}_{\alpha\beta} = |\alpha\rangle\langle\beta|$ are *Hubbard operators* linking the the restricted set of atomic states

$$|\alpha\rangle = \begin{cases} |f^0\rangle \\ |f^1 : \sigma\rangle & (\sigma = 1, \dots N), \end{cases} \quad (18.4)$$

which denote the empty ($\alpha = 0$) and N magnetic states ($\alpha \in [1,N]$) of an ion. These operators enforce the "no double occupancy" constraint $n_f \leq 1$. For instance, in mixed-valent cerium, the N magnetic states refer to the $N = 6$ magnetic configurations of the $4f^1$, $j = 5/2$ configuration of the Ce^{3+} ion. If we denote $P \equiv |f^0\rangle\langle f^0|$ as the projector onto the empty state, then the Hubbard operators

$$X_{\sigma 0} = |f^1 : \sigma\rangle\langle f^0| \equiv f_\sigma^\dagger P_0 \qquad X_{0\sigma} \equiv |f^0\rangle\langle f^1 : \sigma| = P_0 f_\sigma \quad (18.5)$$

are anticommuting operators which describe the addition or removal of an f-electron. By contrast,

$$X_{\sigma\sigma'} = |f^1 : \sigma\rangle\langle f^1 : \sigma'| \qquad (\sigma, \sigma' \in [1,N]) \quad (18.6)$$

are bosonic operators describing the spin.[1] These operators pose two difficulties:

---

[1] Notice that the proper traceless spin operator is given by

$$S_{\alpha\beta} = X_{\alpha\beta} - \frac{1}{N}\delta_{\alpha\beta}X_{\alpha\alpha}. \quad (18.7)$$

- They do not satisfy a canonical commutation or anticommutation algebra, so Wick's theorem is not obeyed.
- The constraint $n_f \leq 1$ is non-holonomic and not naturally encompassed by field theory.

A solution to this problem, devised by Stuart Barnes, at Trinity College, Dublin, and the author, working at Princeton University [1, 2], is to factorize the Hubbard operators in terms of canonical operators, to form slave bosons.[2] Shortly thereafter, Nicholas Read and Denis Newns at Imperial College London incorporated the slave boson into a path integral approach [3]. Andrew Millis and Patrick Lee, working at MIT, later carried out the first in-depth examination of the application of the approach to the Anderson lattice [4].

The basic idea is to introduce an auxiliary *slave boson field* $b^\dagger$ which denotes the f^0 configuration, such that

$$|f^0\rangle = b^\dagger |0\rangle,$$
$$|f^1 : \sigma\rangle = f_\sigma^\dagger |0\rangle \tag{18.8}$$

describes the restricted Hilbert space. In this language, mixed valence is the interconversion of f-electrons to slave bosons:

$$f \rightleftharpoons b + e^-, \qquad\qquad\qquad \text{'slave boson'}$$

$$\tag{18.9}$$

The corresponding Hubbard operators are then

$$X_{\sigma 0} = f_\sigma^\dagger b, \qquad X_{0\sigma} = b^\dagger f_\sigma$$
$$X_{00} = b^\dagger b, \qquad X_{\sigma\sigma'} = f_\sigma^\dagger f_{\sigma'}. \tag{18.10}$$

slave boson formulation of Hubbard operators

This representation faithfully reproduces the super-algebra of Hubbard operators:

$$[X_{\alpha\beta}, X_{\gamma\delta}]_\pm = X_{\alpha\delta}\delta_{\beta\gamma} \pm X_{\gamma\beta}\delta_{\alpha\delta}, \tag{18.11}$$

where the $+$ sign is used when both Hubbard operators are fermionic. Moreover, each operator commutes with the conserved charge

$$Q = b^\dagger b + n_f, \tag{18.12}$$

measuring the number of bosons and fermions at each site, $[X_{\alpha\beta}, Q] = 0$, so the non-holomorphic constraint $n_f \leq 1$ is now replaced by the holonomic constraint $Q = 1$. We can identify Q as a Casimir of the supergroup defined by the Hubbard operators, permitting us to generalize its representations to any $Q \leq N$. This permits us to take a large-N limit, following the same scheme as for the Kondo lattice, namely keeping the ratio $q = Q/N$

---

[2] Author's note: The term "slave boson" was proposed to me in spring 1983 by Chandra Varma, during a conversation with him and Duncan Haldane at Bell Laboratories, and I included this term in the original preprint, but dropped it in the final publication. Nevertheless, the name proved popular and stuck with the community.

fixed as $N \to \infty$, so that Q becomes an extensive variable. Finally, in a many-site lattice, the above replacements are made at each site, adding a site label j to each operator, so that $X_{\sigma 0}(j) = f_{j\sigma}^{\dagger} b_j$ thus the the infinite U Anderson lattice model becomes

$$H = \sum_{\mathbf{k}, \, \sigma \in [1,N]} \epsilon_{\mathbf{k}} c_{\mathbf{k}\sigma}^{\dagger} c_{\mathbf{k}\sigma} + \sum_{\mathbf{k}\sigma \in \{1,N\}} [V_{\mathbf{k}} c_{\mathbf{k}\sigma}^{\dagger} (b^{\dagger} f_{\sigma}) + V_{\mathbf{k}}^{*} (f_{\sigma}^{\dagger} b) c_{\mathbf{k}\sigma}] + E_f n_f + \lambda (n_f + n_b - Q).$$

(18.13)

slave boson formulation of the infinite U Anderson model

Notice the following:

- The reformulation of the infinite U constraint using slave bosons introduces a conserved quantity Q at each site, leading to a gauge theory.
- The slave boson describes the deviation of the valence from the maximum value $n_f = Q$, and as such is a field that keeps track of valence fluctuations.
- In the mean field we can explore solutions in which the slave boson condenses. In the Anderson model, this leads to the renormalization of the hybridization:

$$V_{\mathbf{k}} \to \tilde{V}_{\mathbf{k}} = V_{\mathbf{k}} \langle b_j \rangle,$$

(18.14)

giving a natural account of the narrowing of the resonance width that results from the Kondo effect.

- The constraint also has the effect of renormalizing the position of the f-level:

$$E_f \to \tilde{E}_f = E_f + \lambda.$$

(18.15)

It is this effect that leads to the formation of the Kondo resonance near the Fermi energy, even when the bare f-energy $E_f < 0$ is far below (or above) the Fermi energy.

18.2 Path integrals with slave bosons

Now that we have formulated the Hubbard operators in terms of canonical fields, using the coherent state representations developed in Chapter 14, we can immediately formulate the infinite U Anderson model as a path integral. The result of this procedure is to re-incorporate valence fluctuations into the Kondo lattice. Consider the infinite U Anderson lattice model, written as

$$H = \sum_{\mathbf{k}, \, \sigma \in [1,N]} \epsilon_{\mathbf{k}} c_{\mathbf{k}\sigma}^{\dagger} c_{\mathbf{k}\sigma} + \sum_{j\sigma} \frac{V_0}{\sqrt{N}} \left[c_{j\sigma}^{\dagger} (b_j^{\dagger} f_{j\sigma}) + \text{H.c.} \right] + \sum_j E_f n_f(j),$$

(18.16)

where for pedagogy, we assume an s-wave, momentum-independent hybridization of strength V_0. Here we have taken the liberty of rescaling the hybridization with respect to the spin degeneracy, anticipating that, as N becomes large, $\langle b \rangle \sim O(\sqrt{N})$, so that each term in the Hamiltonian is extensive in N, permitting a controlled large-N expansion. If we now represent the fields as coherent states, we can immediately write down a path integral for the partition function:

$$Z = \int \mathcal{D}[\bar{b}, b, \lambda] \int \mathcal{D}[\bar{\psi}, \psi] \exp\left[-\left(S_c + \sum_j S_A(j) \right) \right]$$

$$S_c = \int_0^\beta d\tau \sum_{\mathbf{k}\sigma} c^\dagger_{\mathbf{k}\sigma} (\partial_\tau + \epsilon_\mathbf{k}) c_{\mathbf{k}\sigma}$$

$$S_A(j) = \int_0^\beta d\tau \left[f^\dagger_{j\sigma}(\partial_\tau + \lambda_j + E_f) f_{j\sigma} + b^\dagger_j(\partial_\tau + \lambda_j) b_j + \frac{V_0}{\sqrt{N}} \left[c^\dagger_{j\sigma}(b^\dagger_j f_{j\sigma}) + \text{H.c.} \right] - \lambda_j Q \right],$$

$$(18.17)$$

infinite U Anderson lattice model: slave boson formulation

where S_c and $S_A(j)$ are the conduction and f-electron parts of the action, and we use the summation convention for the repeated spin indices σ. Here, following our earlier convention, we use the dagger notation to represent Hermitian conjugates of both operators and their c-number expectation values inside the path integral. We have also introduced the Lagrange multiplier field λ_j at each site to impose the constraint $\hat{Q} - Q = n_f(j) + n_b(j) - Q = 0$. By comparing this expression with the corresponding one for the Kondo lattice (17.56), we recognize that the slave boson field plays the same role as the hybridization field used to factorize the Coqblin–Schrieffer Hamiltonian, but with a number of differences:

- The bare f-level position E_f appears explicitly in the Lagrangian, shifted by the amount λ_j. This shift is responsible for the formation of the Kondo resonance. In the Kondo regime, we expect the renormalized f-level position $\lambda_j + E_f \sim T_K \sim 0$ to be of the order of the Kondo temperature, so that $\lambda_j \sim -E_f = |E_f|$.
- The slave boson carries gauge charge, and couples to the constraint field λ_j.
- The dynamics of mixed valence are included via the bare slave boson action $b^\dagger_j(\partial_\tau + \lambda_j)b_j$, associated with charge fluctuations at a high-frequency scale $\omega \sim \lambda_j \sim |E_f|$.

Since we expect the characteristic fluctuations of the boson field to occur at high frequencies of order $|E_f|$, one of the first things we're going to do is to shift the constraint field $\lambda_j \to \lambda_j - E_f$, so that the shifted λ field is a small quantity (typically of the order of the Kondo temperature). The f-electron part of the action now becomes

$$S_A(j) = \int_0^\beta d\tau \left[f^\dagger_{j\sigma}(\partial_\tau + \lambda_j) f_{j\sigma} + b^\dagger_j(\partial_\tau + \lambda_j - E_f) b_j \right.$$

$$\left. + \frac{V_0}{\sqrt{N}} \left(c^\dagger_{j\sigma} b^\dagger_j f_{j\sigma} + \text{H.c.} \right) - (\lambda_j - E_f) Q \right].$$

$$(18.18)$$

As in the Kondo lattice, the presence of a gauge symmetries associated with the local conservation of Q (see 18.20) allows us to choose a gauge in which the slave boson field is purely real. Let $r_j = |b_j|$ be the magnitude of the slave boson field at site j; then, under the transformations

$$b_j \rightarrow r_j e^{i\phi_j}$$
$$f_{j\sigma} \rightarrow e^{i\phi_j} f_{j\sigma}, \tag{18.19}$$

the action at site j becomes

$$S_A(j) = \int_0^\beta d\tau \left[f_{j\sigma}^\dagger (\partial_\tau + \lambda_j + i\dot\phi_j) f_{j\sigma} + r_j(\partial_\tau + (\lambda_j - E_f + i\dot\phi_j)) r_j \right.$$
$$\left. + \frac{V_0 r_j}{\sqrt{N}} \left(c_{j\sigma}^\dagger f_{j\sigma} + \text{H.c.} \right) - (\lambda_j - E_f) Q \right]. \tag{18.20}$$

If we now shift the constraint field to absorb the phase velocity terms $\dot\phi_j$, we obtain

$$S_A(j) \xrightarrow{\lambda_j \rightarrow \lambda_j - i\dot\phi_j} \int_0^\beta d\tau \left[f_{j\sigma}^\dagger (\partial_\tau + \lambda_j) f_{j\sigma} + \frac{V_0 r_j}{\sqrt{N}} \left(c_{j\sigma}^\dagger f_{j\sigma} + \text{H.c.} \right) + (\lambda_j - E_f)(r_j^2 - Q) \right]$$
$$+ \overbrace{\int_0^\beta d\tau (\frac{1}{2}\partial_\tau r_j^2 + iQ\dot\phi_j)}^{2\pi i \times \text{integer}}$$
$$\equiv \int_0^\beta d\tau \left[f_{j\sigma}^\dagger (\partial_\tau + \lambda_j) f_{j\sigma} + \frac{V_0 r_j}{\sqrt{N}} \left(c_{j\sigma}^\dagger f_{j\sigma} + \text{H.c.} \right) + (\lambda_j - E_f)(r_j^2 - Q) \right], \tag{18.21}$$

where the final remainder term in the first line is a perfect differential which vanishes (up to an integer multiple of $2\pi i$) under periodic boundary conditions. The resulting action is then

$$S = \int_0^\beta d\tau \left\{ \sum_{\mathbf{k}\sigma} c_{\mathbf{k}\sigma}^\dagger (\partial_\tau + \epsilon_\mathbf{k}) c_{\mathbf{k}\sigma} + \sum_{j\sigma} \left(f_{j\sigma}^\dagger (\partial_\tau + \lambda_j) f_{j\sigma} + \frac{V_0 r_j}{\sqrt{N}} [c_{j\sigma}^\dagger f_{j\sigma} + \text{H.c.}] \right) \right.$$
$$\left. + \sum_j (\lambda_j - E_f)(r_j^2 - Q) \right\}. \tag{18.22}$$

infinite U Anderson lattice: radial gauge

Example 18.1

(a) Show that, at stationary points of the effective action for the infinite U Anderson model,

$$S_{\textit{eff}}[b, \lambda] = -\ln \left[\int \mathcal{D}[\bar\psi, \psi] \exp[-S] \right], \tag{18.23}$$

the uniform expectation value of the slave boson field is given by

$$\langle b_j \rangle = -\frac{V_0}{\sqrt{N}(\lambda - E_f)} \sum_\sigma \langle c_{j\sigma}^\dagger f_\sigma \rangle. \tag{18.24}$$

(b) By comparing this with the corresponding saddle-point condition for a Kondo lattice model, show that one can identify the renormalized hybridization V as

$$V_j = \frac{V_0}{N} \langle b_j \rangle \tag{18.25}$$

and the Kondo coupling constant as

$$J = \frac{|V_0|^2}{\lambda - E_f}. \tag{18.26}$$

Solution

(a) If we take the functional derivative of the effective action,

$$S = \int_0^\beta d\tau \left[\sum_{\mathbf{k}\sigma} c^\dagger_{\mathbf{k}\sigma}(\partial_\tau + \epsilon_{\mathbf{k}})c_{\mathbf{k}\sigma} + \sum_{j\sigma} f^\dagger_{j\sigma}(\partial_\tau + \lambda_j)f_{j\sigma} + \sum_j b^\dagger_j(\partial_\tau + \lambda_j - E_f)b_j \right.$$
$$\left. + \sum_{j\sigma} \frac{V_0}{\sqrt{N}} \left[c^\dagger_{j\sigma}(b^\dagger_j f_{j\sigma}) + (f^\dagger_{j\sigma} b_j)c_{j\sigma} \right] - \sum_j (\lambda_j - E_f)Q \right], \tag{18.27}$$

and set it to zero at the saddle point, we obtain

$$0 = \left\langle \frac{\delta S}{\delta b^\dagger_j(\tau)} \right\rangle = \left(\partial_\tau + \lambda_j - E_f \right)\langle b \rangle + \frac{V_0}{\sqrt{N}} \sum_\sigma \langle c^\dagger_{j\sigma} f_{j\sigma} \rangle. \tag{18.28}$$

Assuming a static saddle point, $\partial_\tau \langle b \rangle = 0$, and a uniform solution $\lambda_j \equiv \lambda$, we then obtain

$$\langle b_j \rangle = -\frac{V_0}{\sqrt{N}(\lambda - E_f)} \sum_\sigma \left\langle c^\dagger_{j\sigma} f_\sigma \right\rangle, \tag{18.29}$$

which confirms the relationship between the hybridization and the expectation value of the slave boson field.

(b) Now consider the Kondo lattice action,

$$S = \int_0^\beta d\tau \left[\sum_{\mathbf{k}} c^\dagger_{\mathbf{k}\sigma}(\partial_\tau + \epsilon_{\mathbf{k}})c_{\mathbf{k}\sigma} + \sum_j \left(f^\dagger_{j\sigma}(\partial_\tau + \lambda_j)f_{j\sigma} + (\bar{V}_j c^\dagger_{j\sigma} f_{j\sigma} + V_j f^\dagger_{j\sigma} c_{j\sigma}) \right. \right.$$
$$\left. \left. + N\frac{\bar{V}_j V_j}{J} - \lambda_j Q \right) \right]. \tag{18.30}$$

Comparing this with the Anderson lattice action (18.27), we may identify the renormalized hybridization with the combination

$$V_j = \frac{V_0 b_j}{\sqrt{N}}, \tag{18.31}$$

so, at the saddle point,

$$V = \langle V_j \rangle = \frac{V_0}{\sqrt{N}} \langle b_j \rangle. \tag{18.32}$$

Using (18.29), we obtain

$$V = -\frac{V_0^2}{N(\lambda - E_f)} \sum_\sigma \left\langle c_{j\sigma}^\dagger f_\sigma \right\rangle. \tag{18.33}$$

Now from the stationary point of (18.30) we obtain

$$V_j = -\frac{J}{N} \sum_\sigma \left\langle c_{j\sigma}^\dagger f_\sigma \right\rangle. \tag{18.34}$$

Comparing these two results, we see that

$$J = \frac{|V_0|^2}{\lambda - E_f}, \tag{18.35}$$

where we have added the modulus sign to take account of cases where V_0 is complex.

18.2.1 The link between the Kondo and Anderson lattices

Once we have transformed into the radial gauge, the dynamics of the slave boson field have been absorbed into the constraint field. The resulting action is remarkably like that of the Kondo lattice, and this allows us to identify the slave boson with the emergent hybridization of the Kondo lattice:

$$\frac{r_j}{\sqrt{N}} \equiv \frac{V_j}{V_0}, \tag{18.36}$$

equivalence of slave boson and emergent hybridization

where V_j is taken to be real. The constraint term now becomes

$$(\lambda_j - E_f)(r_j^2 - Q) \rightarrow N\frac{V_j^2}{J(\lambda)} - \lambda_j Q + E_f Q, \tag{18.37}$$

where we have identified

$$\frac{1}{J(\lambda)} = \frac{-E_f + \lambda}{V_0^2} = \frac{|E_f| + \lambda}{V_0^2} = \frac{1}{J_0} + \frac{\lambda}{V_0^2} \tag{18.38}$$

as the Schrieffer–Wolff relation for the inverse Kondo coupling $1/J(\lambda)$. It is the explicit dependence on λ that captures the valence fluctuations. We can drop the constant $E_f Q$ term so that

$$S = \int_0^\beta d\tau \left\{ \sum_{\mathbf{k}\sigma} c_{\mathbf{k}\sigma}^\dagger (\partial_\tau + \epsilon_{\mathbf{k}}) c_{\mathbf{k}\sigma} + \sum_{j\sigma} \left[f_{j\sigma}^\dagger (\partial_\tau + \lambda_j) f_{j\sigma} + V_j (c_{j\sigma}^\dagger f_{j\sigma} + \text{H.c.}) \right. \right.$$

$$\left. \left. + NV_j^2 \left(\frac{1}{J_0} + \frac{\lambda_j}{V_0^2} \right) \right] - \lambda_j Q \right\}. \tag{18.39}$$

Apart from the additional cubic coupling between the hybridization and constraint, this resembles a Kondo lattice. Yet the information about the dynamics of the valence fluctuations is still there: we can follow the transformation from dynamics into the cubic coupling term as follows:

$$b^\dagger \partial_\tau b \to r^2 i\dot\phi \equiv \lambda r^2 \equiv N\lambda \left(\frac{V}{V_0}\right)^2. \tag{18.40}$$

Notice also that, if we differentiate the action S with respect to λ_j, setting $\langle \delta S/\delta \lambda_j \rangle = 0$ we obtain

$$V_j = \frac{V_0}{\sqrt{N}} \sqrt{Q - \langle n_f \rangle} = \sqrt{(q - \langle n_f/N \rangle)}, \tag{18.41}$$

showing that, as the valence approaches Q, the hybridization renormalizes to zero. In this way, the slave boson approach includes the Gutzwiller renormalization [5] of the hybridization.

Example 18.2

(a) By adding the additional coupling term $\Delta H_{MV} = N(V/V_0)^2 \lambda - E_f Q$ to the mean-field impurity energy obtained in (17.109), show that the mean-field energy for the infinite U Anderson model can be written

$$\Delta E_A = N \, \mathrm{Im}\left[\xi \ln\left[\frac{\xi}{eD}\right] + \frac{(\xi - E_{fc})^2}{2\Delta_0} \right],$$

where $E_{fc} = E_f + i\Delta_0 q$ and $\xi = \lambda + i\tilde\Delta$ with $\Delta_0 = \pi\rho V_0^2$ and $\tilde\Delta = \pi\rho |\tilde V|^2$.

(b) Show that the mean-field equations can be written

$$\xi + \frac{\Delta_0}{\pi} \ln\left(\frac{\xi}{D}\right) = E_f + i\Delta_0 q = E_{fc}. \tag{18.42}$$

(c) How can one rewrite the mean-field equations in a cut-off–invariant fashion?

Solution

(a) From (17.109), the Kondo impurity energy is given by

$$\Delta E_K = \frac{N}{\pi} \, \mathrm{Im}\left[\xi \ln\left[\frac{\xi}{eT_K e^{i\pi q}}\right] \right]$$

$$= N \, \mathrm{Im}\left[\frac{\xi}{\pi} \ln\left[\frac{\xi}{eD}\right] - \frac{(E_f + i\Delta_0 q)\xi}{\Delta_0} \right]$$

$$= N \, \mathrm{Im}\left[\frac{\xi}{\pi} \ln\left[\frac{\xi}{eD}\right] - \frac{E_{fc}\xi}{\Delta_0} \right], \tag{18.43}$$

where in the second line we have replaced $T_K = De^{-1/J\rho} \rightarrow D\exp[\pi E_f/\Delta_0]$. Now by (18.37) and (18.38), the correction required to derive the infinite U Anderson ground-state energy is

$$\Delta E_A = \Delta E_K + N\left(\frac{|\tilde{V}|^2\lambda}{V_0^2} + E_f q\right) = \Delta E_K + N\,\mathrm{Im}\left[\frac{(E_f + i\Delta_0 q)^2}{2\Delta_0} + \frac{(\lambda + i\tilde{\Delta})^2}{2\Delta_0}\right]$$

$$= \Delta E_K + N\,\mathrm{Im}\left[\frac{E_{fc}^2}{2\Delta_0} + \frac{\xi^2}{2\Delta_0}\right]. \tag{18.44}$$

Combining (18.43) and (18.44) we obtain

$$\Delta E_A = \mathrm{Im}\left[\frac{\xi}{\pi}\ln\left[\frac{\xi}{eD}\right] + \frac{(E_{fc} - \xi)^2}{2\Delta_0}\right]. \tag{18.45}$$

(b) Taking variations of ΔE_A with respect to $d\xi = d\lambda + id\tilde{\Delta}$ then gives

$$\xi + \frac{\Delta_0}{\pi}\ln\left(\frac{\xi}{D}\right) = E_f + i\Delta_0 q. \tag{18.46}$$

(c) If we introduce the invariant f-level position

$$E_f^* = E_f + \frac{\Delta_0}{\pi}\ln\left(\frac{D}{\Delta_0}\right), \tag{18.47}$$

then the mean-field equation beomes

$$\xi + \frac{\Delta_0}{\pi}\ln\left(\frac{\xi}{\Delta_0}\right) = E_f^* + i\Delta_0 q. \tag{18.48}$$

Writing the real and imaginary parts of this expression out explicilty

$$\lambda + \frac{\Delta_0}{\pi}\ln\left(\frac{\sqrt{\lambda^2 + \Delta^2}}{\Delta_0}\right) = E_f^*$$

$$\overbrace{\frac{N\Delta}{\Delta_0}}^{\langle b^\dagger b\rangle} + \overbrace{N\tan^{-1}\left(\frac{\Delta}{\lambda}\right)}^{\langle n_f\rangle = N\delta_f/\pi} = Q. \tag{18.49}$$

We can identify the second of these expressions as the mean-field constraint $\langle b^\dagger b\rangle + \langle n_f\rangle = Q$, where the first term is the number of slave bosons and the second is the Friedel sum rule expression for the total number of f-electrons.

We see that:

(i) the family of models with different band cut-offs D, but the same value of E_f^*, i.e.

$$E_f(D) = E_f^* - \frac{\Delta_0}{\pi}\ln\left(\frac{D}{\Delta}\right), \tag{18.50}$$

lie on a scaling trajectory. In other words, as the cut-off reduces, the f-level position renormalizes upwards, a scaling phenomenon first discovered by Duncan Haldane [6] while working at the Institut Laue-Langevin.

(ii) When $|E_f^*|/\Delta_0$ is large, we can write

$$\xi = \Delta \exp\left[\frac{\pi(E_f^* + i\Delta_0 q)}{\Delta_0}\right] = T_K e^{i\pi q}, \tag{18.51}$$

where $T_K = \Delta \exp\left[\frac{-\pi|E_f^*|}{\Delta_0}\right]$, corresponding to the Kondo limit of the model.

Example 18.3 Large-N phase transition into the Kondo phase.

The large-N Kondo impurity at a finite temperature has a mean-field free energy

$$\frac{F}{N} = -T\sum_{\omega_n} \ln\left[\lambda - i\Delta \mathrm{sgn}(\omega_n) - i\omega_n\right] + \left(\frac{V^2}{J} - \lambda q\right). \tag{18.52}$$

To regulate this, it is useful to subtract off the free energy of an impurity with $q = 0$ and a shifted $\xi \to \xi + D$, so that

$$\frac{F}{N} = -T\sum_{\omega_n} \left[\ln\left(\lambda - i\Delta \mathrm{sgn}(\omega_n) - i\omega_n\right) - (\lambda \to \lambda + D)\right] + \left(\frac{V^2}{J} - \lambda q\right). \tag{18.53}$$

(a) Using the relationship

$$\ln\left[\frac{\Gamma(z + a)}{\Gamma(z)}\right] = \sum_{n=0}^{\infty} \left[\ln(n + z) - \ln(n + z + a) + \frac{a}{1 + n}\right] - aC, \tag{18.54}$$

obtained by integrating (14.152) with respect to z, show that the mean-field free energy can be evaluated as

$$\frac{F}{N} = -2T\,\mathrm{Re}\ln\left[\frac{\tilde{\Gamma}[\xi + D]}{\tilde{\Gamma}(\xi)}\right] + \left(\frac{V^2}{J} - q\lambda\right), \tag{18.55}$$

where $\tilde{\Gamma}(z) = \Gamma\left(\frac{1}{2} + \frac{z}{2\pi i T}\right)$, $\xi = \lambda + i\Delta$ is the complex resonance-level energy.

(b) From the stationary point of the free energy, show that the mean-field equations at finite temperature can be written

$$\psi\left(\frac{1}{2} + \frac{\xi}{2\pi i T}\right) = \ln\frac{T_K e^{i\pi q}}{2\pi Ti}. \tag{18.56}$$

How can you interpret the real and imaginary parts of this equation?

(c) Determine the large-N transition temperature where Δ becomes finite, and show that, in the symmetric case where $Q = N/2$,

$$T_c = \frac{T_K}{2\pi} e^{-\psi(\frac{1}{2})} = 1.13 T_K. \tag{18.57}$$

(d) Suppose a magnetic field splits the f-level into N non-degenerate states distributed evenly with energies $\lambda \pm g\mu_B b$, where b is distributed evenly over the range ($b \in [-B/2, B/2]$). Show that for $q = \frac{1}{2}$ the large-N phase boundary between the screened and unscreened phases is given by

$$\text{Re}\left(\frac{\ln\Gamma\left(\frac{1}{2}+\frac{g\mu_B B}{4\pi iT}\right)}{\frac{g\mu_B B}{4\pi iT}}\right) = \ln\frac{T_K}{2\pi T}. \tag{18.58}$$

Plot out this phase boundary.

Solution

(a) Let us begin by rewriting the free energy as simple summation:

$$\frac{F}{N} = -2T\,\text{Re}\sum_n [\ln(\xi + i\omega_n) - \ln(\xi + D + i\omega_n)] + \left(\frac{V^2}{J} - \lambda q\right)$$

$$= -2T\,\text{Re}\sum_{n\geq 0}\left[\ln\left(n + \frac{1}{2} + \frac{\xi}{2\pi iT}\right) - \ln\left(n + \frac{1}{2} + \frac{\xi}{2\pi iT} - i\frac{D}{2\pi T}\right)\right]$$

$$+ \left(\frac{V^2}{J} - \lambda q\right). \tag{18.59}$$

Taking the real part of (18.54),

$$\text{Re}\ln\left[\frac{\Gamma(z + ia)}{\Gamma(z)}\right] = \text{Re}\sum_{n=0}^{\infty}[\ln(n + z) - \ln(n + z + ia)], \tag{18.60}$$

For real a, we can rewrite

$$\frac{F}{N} = -2T\,\text{Re}\left(\ln\left[\frac{\Gamma\left(\frac{\xi+D}{2\pi iT} + \frac{1}{2}\right)}{\Gamma\left(\frac{\xi}{2\pi iT} + \frac{1}{2}\right)}\right]\right) + \left(\frac{V^2}{J} - \lambda q\right). \tag{18.61}$$

(b) Since D is large, we can approximate $\ln\Gamma(x) \approx x\ln\frac{x}{e}$ and $\psi(x) \approx \ln x$, rewriting

$$\ln\Gamma\left(\frac{\xi + D}{2\pi iT} + \frac{1}{2}\right) = \frac{D}{2\pi iT}\ln\frac{D}{2\pi iT} + \frac{\xi}{2\pi iT}\ln\frac{D}{2\pi iT}, \tag{18.62}$$

where we are not interested in the terms extensive in D. We may now cast the free energy in the form

$$\frac{F}{N} = 2T\,\text{Re}\left(\ln\tilde{\Gamma}(\xi) - \frac{\xi}{2\pi iT}\ln\frac{D}{2\pi iT}\right) + \left(\frac{V^2}{J} - \lambda q\right)$$

$$= \frac{1}{\pi}\text{Im}\left(2\pi iT\ln\tilde{\Gamma}(\xi) - \xi\ln\frac{D}{2\pi iT}\right) + \frac{1}{\pi}\left(\frac{\Delta}{J\rho} - \lambda\pi q\right)$$

$$= \frac{1}{\pi}\text{Im}\left(2\pi iT\ln\tilde{\Gamma}(\xi) - \xi\ln\frac{D}{2\pi iT}\right) + \frac{1}{\pi}\text{Im}\left[\xi\left(-\frac{1}{J\rho} + i\pi q\right)\right]$$

$$= \frac{1}{\pi}\text{Im}\left(2\pi iT\ln\tilde{\Gamma}(\xi) - \xi\ln\frac{T_K e^{i\pi q}}{2\pi iT}\right). \tag{18.63}$$

Taking variations with respect to $\delta\xi = \delta\lambda + i\delta\Delta$, using $\partial\ln\Gamma(z)/dz = \psi(z)$, we then obtain

$$\tilde{\psi}(\xi) - \ln\frac{T_K e^{i\pi q}}{2\pi iT} = 0. \tag{18.64}$$

The imaginary part of this equation is

$$\frac{1}{\pi}\operatorname{Im}\tilde{\psi}(\xi) = q - \frac{1}{2}. \tag{18.65}$$

This looks like the constraint equation in disguise. To confirm this, let us write the constraint equation as

$$q = \int \frac{d\omega}{\pi} f(\omega)\operatorname{Im}\frac{1}{z - \xi} = \operatorname{Im}\frac{1}{\pi}\left[\tilde{\psi}(\xi) - \ln\left(\frac{D\beta}{2\pi i}\right)\right] = \frac{1}{\pi}\operatorname{Im}\tilde{\psi}(\xi) + \frac{1}{2}, \tag{18.66}$$

which recovers (18.65).

(c) To determine the temperature where $\Delta(T_c) = 0$, we take the real part of (18.64),

$$\operatorname{Re}\psi\left(\frac{1}{2} + \frac{\xi}{2\pi i T}\right) = \ln\frac{T_K}{2\pi T}, \tag{18.67}$$

and set $\xi = \lambda$ to be real. In the particle–hole symmetric case, $\lambda = 0$, so we obtain

$$\psi(\frac{1}{2}) = \ln\frac{T_K}{2\pi T_c} \tag{18.68}$$

or

$$T_c = \frac{T_K}{2\pi}e^{-\psi(\frac{1}{2})} = 1.13 T_K. \tag{18.69}$$

(d) If we introduce a magnetic field, we smear the f-state into N non-degenerate states located at $\lambda \pm g\mu_B b$ ($b \in [-B/2, B/2]$). The large-N mean-field equations now need to be averaged over b:

$$\frac{1}{B}\int_{-B/2}^{B/2} db\, \psi\left(\frac{1}{2} + \frac{\xi - g\mu_B b}{2\pi i T}\right)$$

$$= -\frac{1}{B}\left[\frac{2\pi i T}{g\mu_B}\ln\Gamma\left(\frac{1}{2} + \frac{\xi - g\mu_B b}{2\pi i T}\right)\right]_{-B/2}^{B/2}$$

$$= \frac{2\pi i T}{g\mu_B B}\left[\ln\Gamma\left(\frac{1}{2} + \frac{\xi + g\mu_B B}{4\pi i T}\right) - \ln\Gamma\left(\frac{1}{2} + \frac{\xi - g\mu_B B}{4\pi i T}\right)\right] = \ln\frac{T_K e^{i\pi q}}{2\pi i T}. \tag{18.70}$$

For $q = \frac{1}{2}$ we have particle–hole symmetry, so that at the mean-field phase transition $\xi = 0$, giving

$$\operatorname{Re}\left(\frac{4\pi i T \ln\Gamma\left(\frac{1}{2} + \frac{g\mu_B B}{4\pi i T}\right)}{g\mu_B B}\right) = \ln\frac{T_K}{2\pi T}. \tag{18.71}$$

It is useful to parametrically plot $t = T/T_K$ as a function of $b = g\mu_B B/T$, as follows:

$$g\mu_B B/T_K = bt, \quad t = \frac{1}{2\pi}\exp\left[-\operatorname{Re}\frac{4\pi i \ln\Gamma\left(\frac{1}{2} + \frac{b}{4\pi i}\right)}{b}\right]$$

$$= \frac{1}{2\pi}\left|\exp\left[-\frac{4\pi i \ln\Gamma\left(\frac{1}{2} + \frac{b}{4\pi i}\right)}{b}\right]\right|. \tag{18.72}$$

The corresponding Mathematica code is listed below:

```
1  Tc [b_] :=   1/(2  π)  Abs [ Exp [−  4 π I LogGamma [1/2 + b /(4 π I ) ]/   b ]];
2  ParametricPlot  [{ Tc [x]  x ,  Tc [ x ]}, {x,  0,  200},  PlotRange  −> All ,  PlotStyle  −>  Directive
   [ Thickness  [0.02]],   AspectRatio  −> 1,   AxesLabel  −>{μ_BB/T_K , T_K}, AxesStyle −> Bold]
```

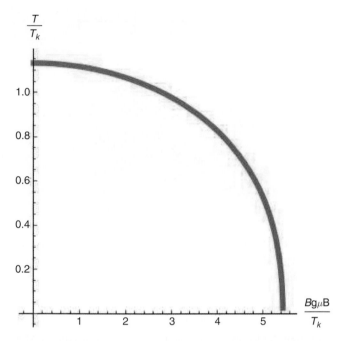

At finite N, this phase boundary is smeared into a crossover (see Section 18.3).

18.3   Fluctuations about the large-*N* limit

In all of the previous sections, we have neglected the fluctuations about the large-N limit. Of course, in a real system, N is finite, most likely to be 2 for a low-lying Kramers doublet, so in practice the use of a large-N limit cannot be justified by the smallness of $1/N$ alone, but rather by the hope that the basic physics adiabatically continues all the way to small N. In this penultimate section, we examine some of the basic effects of the fluctuations about the large-N limit. In particular:

- The loss of strict long-range order at finite N is a consequence of two related pieces of physics, the *X-ray orthogonality catastrophe* [7], in which a sudden change in a local scattering potential produces a power-law relaxation of the wavefunction, and *Elitzur's theorem* [8], according to which fields with local gauge symmetries cannot develop true long-range order (Figure 18.1).
- Fluctuations generate the residual Fermi liquid interactions between the heavy quasiparticles.

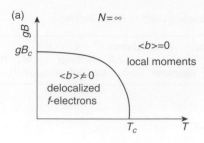

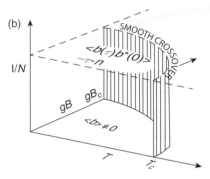

Fig. 18.1   (a) In the large-N limit, the crossover between strong and weak coupling in the Kondo model is described by a sharp phase transition. (b) At finite N, the large-N phase transition becomes a crossover [9]. Reprinted with permission from P. Coleman, *J. Magn. Magn. Mater.*, vol. 47–48, p. 323, 1985. Copyright 1985 Elsevier.

- Fluctuations in the phase of the hybridization and slave boson generate an internal gauge field. We'll see that the Kondo effect can be regarded as a kind of Meissner effect which locks the physical electromagnetic field to the internal gauge fields so that they move together at low frequency. This is the origin of the charge of the heavy electrons in a Kondo lattice.

18.3.1   Effective action and Gaussian fluctuations

We recall that, in the path integral for the Kondo lattice, we integrate out the the fermions to form an effective action S_E which is a function of the auxiliary fields. The path integral is then an integral of the exponential of the effective action over the auxiliary fields (17.63):

$$Z = \int \mathcal{D}[\lambda, \bar{V}, V] \exp\left[-N S_E[\bar{V}, V, \lambda]\right], \tag{18.73}$$

where (17.69)

$$\frac{S_E[V, \lambda]}{\beta} = -\frac{N}{\beta} \mathrm{Tr} \ln\left(\partial_\tau + \underline{h}[V, \lambda]\right) + \sum_j \left(\frac{N|V_j|^2}{J} - \lambda_j Q\right) \tag{18.74}$$

is the effective action. The first term results from the Gaussian integral over the internal fermions, moving under the influence of the Hamiltonian $h[V, \lambda]$. We can regard S_E as a kind of "machine" that gives us back the action for any given space–time configuration of the auxiliary fields.

In the large-N limit, the integral over the auxiliary fields was approximated by its saddle-point value, and $h[V, \lambda]$ was time-invariant. We now want to reintroduce small fluctuations around the saddle point, $V_j(\tau) = V_{MF} + \delta V_j(\tau)$ and $\lambda_j(\tau) = \lambda_{MF} + \delta\lambda_j(\tau)$, and integrate over them, taking advantage of the large-N limit to approximate the remaining integral by its leading Gaussian approximation. As we discussed in Chapter 13, this is a type of RPA expansion about the ordered state. In Chapter 13 we considered fluctuations about a magnetically ordered ground state

Here however, we have a new issue: because the order parameter we are dealing with is subject to a $U(1)$ gauge symmetry, the effective action is invariant under local translations in the phase:

$$V \to V_g = e^{i\phi_j(\tau)} V_j, \qquad \lambda \to \lambda_g = \lambda_j + i\dot{\phi}_j, \qquad (18.75)$$

such that $S[V_g, \lambda_g] = S(V, \lambda)$. This defines an "orbit" of degenerate space–time configurations $\{V_g, \lambda_g\}$. To control the divergent infrared fluctuations along the orbit, we choose a given point on the gauge orbit, and expand in fluctuations that are "perpendicular" to that point. Integrations along the degenerate gauge orbit factor out of the path integral. Schematically:

$$\int \mathcal{D}[\bar{V}, V, \lambda] e^{-S} = \int dg \int \mathcal{D}[|V|, \lambda] e^{-S[|V|, \lambda]}. \qquad (18.76)$$

This is the procedure of *gauge fixing*, something that, fortunately, we have already done by moving to the Read–Newns radial gauge.

With these provisos, we now expand the effective action in the radial gauge about the saddle point. We can follow the same basic development as in Section 13.4. Let us denote the fluctuations by

$$\begin{pmatrix} \delta V_j(\tau) \\ \delta\lambda_j(\tau) \end{pmatrix} = \vec{\Phi}_j(\tau) = \sum_q \vec{\Phi}_q e^{iq\cdot X}, \qquad (18.77)$$

where $X \equiv (\mathbf{R_j}, \tau)$, $q \equiv (\mathbf{q}, i\nu_n)$, and N_s is the number of sites. The two components of Φ_j describe the amplitude and phase fluctuations of the slave boson.

Now the fluctuations scatter the electrons, introducing the additional scattering term

$$S_I = \sum_{k,q,\sigma} (c^\dagger_{k+q\sigma}, f^\dagger_{k+q\sigma}) \begin{pmatrix} 0 & \delta V_q \\ \delta V_q & \delta\lambda_q \end{pmatrix} \begin{pmatrix} c_{k\sigma} \\ f_{k\sigma} \end{pmatrix} = \sum_{k,q,\sigma} \psi^\dagger_{k+q\sigma} \delta h_q \psi_{k\sigma} \qquad (18.78)$$

into the action, where

$$\delta h_q = \begin{pmatrix} 0 & \delta V_q \\ \delta V_q & \delta\lambda_q \end{pmatrix} = \delta V_q \tau_1 + \delta\lambda_q \frac{1}{2}(1 - \tau_3) \equiv \vec{\Phi}_q \cdot \vec{\gamma} \qquad (18.79)$$

is a kind of fluctuating Weiss field, acting in hybridization space. Here, we have denoted $\vec{\gamma} = (\gamma_1, \gamma_2) = (\tau_1, \frac{1}{2}(1 - \tau_3))$. Diagrammatically, the interaction vertex is

$$(18.80)$$

Following the same procedure as in Chapter 13, we expand the effective action as a loop expansion:

$$\mathcal{F}_E = \frac{S_E}{N_s\beta} = -\frac{1}{N_s\beta} \mathrm{Tr}\ln[-\underline{G}(k)^{-1}] - \left[\quad \right]$$
$$+ N \sum_j \left[\frac{V_j^2}{J(\lambda_j)} - \lambda_j Q \right], \qquad (18.81)$$

where, as discussed in Section 18.2.1, the valence fluctuation dynamics are included by using

$$\frac{1}{J(\lambda)} = \frac{1}{J_0} + \frac{\lambda}{V_0^2}. \qquad (18.82)$$

The linear terms cancel at the saddle point, which by its definition is stationary to linear order in the fluctuations. In the Gaussian approximation, we take the second term in the loop expansion and drop higher-order terms. The Gaussian correction to the saddle-point action is then

$$\Delta\mathcal{F}_E = \frac{\Delta S_E}{N_s\beta} = N \sum_q \frac{|\delta V_q|^2}{J} + \frac{2\delta V_{-q}\delta\lambda_q V}{V_0^2} - \frac{1}{2} \sum_q \Phi^a_{-q} \left[\gamma_a \quad \gamma_b \right] \Phi^b_q + O(\Phi^3)$$

$$= \frac{N}{2} \sum_q \Phi^a_{-q} \left[\overbrace{\begin{pmatrix} \frac{2}{J} & \frac{2V}{V_0^2} \\ \frac{2V}{V_0^2} & 0 \end{pmatrix}}^{-D^{-1}(q)} - \chi_{ab}(q) \right]_{ab} \Phi^b_q + O(\Phi^3) \qquad (18.83)$$

where we identify $J \equiv J(\lambda)$ and

$$N\chi_{ab}(q) = \gamma_a \underbrace{\bigcirc}_{k}^{k+q} \gamma_b$$

is the susceptibility to the fluctuations. Notice how the trace over the N spin components has been factored out of this expression. Here we have identified the inverse propagator of the fluctuations $-D^{-1}(q)$ as the Gaussian coefficient. From this we can immediately read off the Gaussian fluctuations in the auxiliary fields as

$$\langle \Phi^a_{-q} \Phi^b_q \rangle = -D_{ab}(q) = \frac{1}{N} \left[\begin{pmatrix} \frac{2}{J} & \frac{2V}{V_0^2} \\ \frac{2V}{V_0^2} & 0 \end{pmatrix} - \underline{\chi}(q) \right]_{ab}^{-1}. \tag{18.84}$$

To calculate the fluctuations in detail, we use the Feynman rules to write the susceptibility out in terms of the Green's functions of the mean-field theory:

$$\chi_{ab}(q) = -\frac{1}{N_s \beta} \sum_{\kappa} \text{Tr}\left[\gamma^a \mathcal{G}(k+q) \gamma^b \mathcal{G}(k) \right]. \tag{18.85}$$

The Feynman diagrams for these susceptibilities can be written out in terms of the f-propagator as follows:

$$N\chi_{rr}(q) = \quad N\chi_{r\lambda}(q) = \quad N\chi_{\lambda r}(q) = \quad N\chi_{\lambda\lambda}(q) = \tag{18.86}$$

where a double dashed line represents the full f-propagator $\mathcal{G}_{ff}(k)$, a single full line denotes the bare conduction propagator $G_c(k) = (i\omega_n - \epsilon_k)^{-1}$, and a hatched vertex denotes the conduction emission and absorption process

$$= V[G_c(k) + G_c(k+q)], \tag{18.87}$$

where the orange dot denotes hybridization V. Let us take a look at the single-impurity calculation, where $k \equiv i\omega_n$ is a single loop frequency; also, momentum is not conserved at the hybridization vertex, and the bare conduction propagator is then given by

$G_c(k) \rightarrow G_c(i\omega_n) = \sum_k (i\omega_n - \epsilon_k)^{-1} = -i\pi\rho \, \text{sgn}(\omega_n)$. The result of the frequency summations gives $\chi_{ab}(i\nu_n) = \chi_{ab}(i|\nu_n|)$, where, for $\nu_n \geq 0$ (see Appendix 18A),

$$\frac{2}{J} - \chi_{rr}(i\nu_n) = 2\frac{1}{g(i\nu_n)} \text{Re} \, \mathcal{L}(i\nu_n) \qquad (18.88)$$

$$\frac{2V}{V_0^2} - \chi_{r\lambda}(i\nu_n) = \frac{2V}{V_0^2} - \chi_{\lambda r}(i\nu_n) = 2\left(\frac{V}{\nu_n}\right) \text{Im} \, \mathcal{L}(i\nu_n) \qquad (18.89)$$

$$\chi_{\lambda\lambda}(i\nu_n) = 2g(i\nu_n)\left(\frac{V}{\nu_n}\right)^2 \text{Re} \, \mathcal{L}(i\nu_n). \qquad (18.90)$$

Here we have denoted

$$g(i\nu_n) = \frac{i\nu_n}{i\nu_n + 2i\Delta} \qquad (18.91)$$

and

$$\mathcal{L}(i\nu_n) = \rho\left[\tilde{\psi}(\xi + i\nu_n) - \tilde{\psi}(\xi)\right] + \frac{i\nu_n}{V_0^2}, \qquad (18.92)$$

where

$$\tilde{\psi}(z) = \psi\left(\frac{1}{2} + \frac{z}{2\pi iT}\right) \qquad (18.93)$$

is defined in terms of the digamma function $\psi(z) = \frac{\partial \ln \Gamma(z)}{\partial z}$ and $\xi = \lambda + i\Delta = T_K e^{i\pi q}$ is the complex impurity position (see (17.110)). The explicit disappearance of J from the right-hand side of (18.88) means that the physics only depends implicitly on the coupling constant, via the emergent Kondo scale $T_K = D \exp e^{-1/J\rho}$, so that all models with the same T_K have precisely the same fluctuations about the large-N saddle point.

Note the following:

- In the particle–hole symmetric case, where $Q = N/2$ and the scattering phase shift of the Kondo singlet is $\delta = \pi/2$, $\lambda = 0$ so that $\xi = i\Delta = iT_K$, $\mathcal{L}$ is entirely real, so that in this case the radial and phase fluctuations decouple.
- The analytic extension to real frequencies is accomplished by the substitutions

$$\mathcal{L}(z) = \mathcal{L}(i\nu_n)|_{i\nu_n \to z} = \rho\left[\psi\left(\frac{1}{2} + \frac{\xi + z}{2\pi iT}\right) - \psi\left(\frac{1}{2} + \frac{\xi}{2\pi iT}\right)\right] + \frac{z}{V_0^2}$$

$$\mathcal{L}^*(z) = \mathcal{L}(i\nu_n)^*|_{i\nu_n \to z} = \rho\left[\psi\left(\frac{1}{2} + \frac{z - \xi^*}{2\pi iT}\right) - \psi\left(\frac{1}{2} - \frac{\xi^*}{2\pi iT}\right)\right] - \frac{z}{V_0^2}, \quad (18.94)$$

with the understanding that "Re $\mathcal{L}(z)$"= $[\mathcal{L}^*(z) + \mathcal{L}^*(z)]/2$ and "Im $\mathcal{L}(z)$"= $[\mathcal{L}^*(z) - \mathcal{L}^*(z)]/2i$.

- in the zero-temperature limit, $\psi(z) \sim \ln z$, so that

$$\mathcal{L}(z) \rightarrow \rho \ln\left(\frac{\xi + z}{\xi}\right) + \frac{z}{V_0^2} \qquad (T \rightarrow 0). \qquad (18.95)$$

Putting these results together, the inverse propagator is given by

$$-D^{-1}(i\nu_n) = N\begin{pmatrix} 2/g(i\nu_n) \, \text{Re} \, \mathcal{L}(i\nu_n) & 2V/\nu_n \, \text{Im} \, \mathcal{L}(i\nu_n) \\ 2V/\nu_n \, \text{Im} \, \mathcal{L}(i\nu_n) & -2[g(i\nu_n)V^2/\nu_n^2] \, \text{Re} \, \mathcal{L}(i\nu_n) \end{pmatrix}. \qquad (18.96)$$

By inverting this matrix, we obtain the Gaussian fluctuations for the single-impurity Kondo model:

$$-D_{ab}(i\nu_n) = \langle \Phi_a(-i\nu_n)\Phi_b(i\nu_n) \rangle = \begin{pmatrix} \langle \delta V(-i\nu_n)\delta V(i\nu_n) \rangle & \langle \delta V(-i\nu_n)\delta\lambda(i\nu_n) \rangle \\ \langle \delta\lambda(-i\nu_n)\delta V(i\nu_n) \rangle & \langle \delta\lambda(-i\nu_n)\delta\lambda(i\nu_n) \rangle \end{pmatrix}$$

$$= \frac{1}{2N\mathcal{L}(i\nu_n)\mathcal{L}^*(i\nu_n)} \begin{pmatrix} g(i\nu_n) \operatorname{Re} \mathcal{L}(i\nu_n) & (\nu_n/V) \operatorname{Im} \mathcal{L}(i\nu_n) \\ (\nu_n/V) \operatorname{Im} \mathcal{L}(i\nu_n) & -[\nu_n^2/V^2 g(i\nu_n)] \operatorname{Re} \mathcal{L}(i\nu_n) \end{pmatrix}.$$

$$(18.97)$$

We shall now use these results to illustrate various important features of the fluctuations.

18.3.2 Fermi liquid interactions

The low-energy Fermi liquid properties of the Kondo impurity or lattice are directly related to the fluctuations we have just discussed. As the quasiparticles move through the Kondo lattice, they exchange fluctuations that generate quasiparticle interactions, as illustrated in Figure 18.2.

Formally, we can imagine integrating over the high-frequency components of the Fermi fields, keeping track of the renormalized interactions that are induced between the residual low-energy components of the electron fields. In the large-N limit, we can relate these fluctuations to the Gaussian fluctuations. In particular, since the λ fluctuations couple to the f-electrons, the induced Fermi liquid interaction between them produces a kind of "Hubbard U" interaction among the f-electrons, of the form

$$H_{FL} = \frac{1}{2} \sum_q U(q)\delta n(q)\delta n(-q), \qquad (18.98)$$

where, at low frequencies,

$$U^*(\mathbf{q}) = -\langle \delta\lambda(-q)\delta\lambda(q) \rangle \Big|_{i\nu_n=0} = D_{\lambda\lambda}(i\mathbf{q}, \nu_n = 0). \qquad (18.99)$$

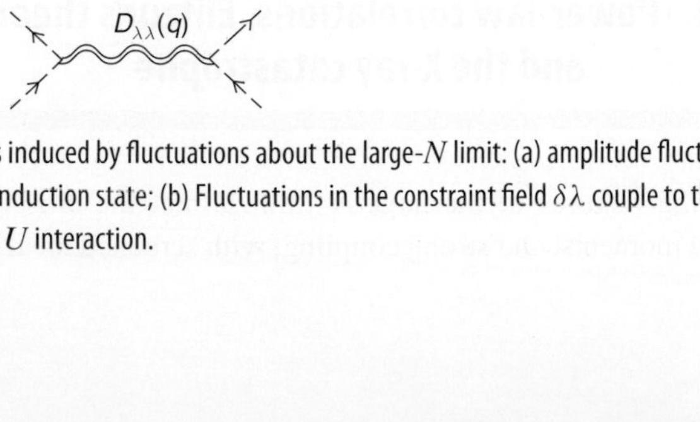

Showing the interactions induced by fluctuations about the large-N limit: (a) amplitude fluctuations in δV scatter between f-states and conduction state; (b) Fluctuations in the constraint field $\delta\lambda$ couple to the f-electrons, inducing a renormalized repulsive U interaction.

Fig. 18.2

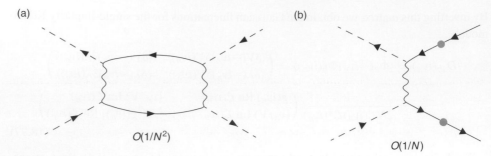

(a)   (b)

$O(1/N^2)$ $O(1/N)$

Fig. 18.3   Showing (a) bare RKKY interaction at high temperatures, of order $O(1/N^2)$; (b) renormalized RKKY interaction, of order $O(1/N)$ a driver of superconductivity.

From the fluctuations,

$$D_{\lambda\lambda}(iv_n) = -\frac{1}{4Ng(iv_n)}\left(\frac{1}{\mathcal{L}(iv_n)} + \frac{1}{\mathcal{L}^*(iv_n)}\right)\left(\frac{iv_n}{V}\right)^2. \tag{18.100}$$

If we evaluate this in the Kondo limit ($V_0 \to \infty$), we obtain

$$D_{\lambda\lambda}(iv_n) \approx -\frac{iv_n + 2\Delta}{4N\rho V^2}\left[\frac{\xi}{1 - \frac{iv_n}{2\xi}} - \frac{\xi^*}{1 + \frac{iv_n}{2\xi}}\right]$$

$$= -\frac{\pi(iv_n + 2i\Delta)}{4N\Delta}\left[(\xi - \xi^*) + iv_n\right] = \frac{\pi}{N}(\Delta + |v_n|), \tag{18.101}$$

so that, taking $iv_n \to 0$,

$$U^* = \frac{\pi\Delta}{N}. \tag{18.102}$$

Thus the phase fluctuations of the slave boson field generate the onsite repulsive U of the heavy Fermi liquid.

The amplitude fluctuations of the slave boson field also generate Fermi liquid interactions. In fact, in the lattice we can interpret the intersite interactions mediated by these fluctuations as a residue of the magnetic RKKY interaction. In the high-temperature phase, the RKKY interaction can be regarded as the exchange of two slave bosons; thus at high temperatures this interaction is an $O(1/N^2)$ effect (Figure 18.3(a)). However, in the low-temperature phase, when the slave boson condenses, one of these fluctuations is removed and the residue of the RKKY interaction is now a term of order $O(1/N)$ (Figure 18.3(b)). These magnetic interactions are important drivers of heavy-fermion superconductivity [10].

18.4   Power-law correlations, Elitzur's theorem, and the X-ray catastrophe

One of the intriguing features of the large-N limit is that the crossover between weak coupling, with local moments, and strong coupling, with screened local moments, becomes a phase transition. At finite N, the slave boson field develops a power-law correlation in time (Figure 18.1):

$$\langle b(\tau)b^\dagger(0)\rangle \sim \tau^{-\frac{\pi}{N}}, \tag{18.103}$$

and the large-N phase transition becomes a crossover. These power-law correlations link the fluctuations about the mean-field theory to two interesting pieces of physics: Elitzur's theorem and the X-ray orthogonality catastrophe, which we now discuss.

In 1975, Schmuel Elitzur, working at Tel Aviv University, showed, in what is now called *Elitzur's theorem*, that order parameters which carry gauge charge *do not* spontaneously develop an expectation value [8]. Of course, one can always polarize an order parameter by applying an external field, and when a *global* symmetry is broken, the broken symmetry remains even after the polarizing field is removed. Elitzur showed, however, that when the order parameter has a gauge symmetry, local fluctuations cause the order parameter field to explore the entire gauge orbit, vanishing once the polarizing source field J is removed (for details, see Appendix 18B), so that

$$\lim_{J\to 0}\langle\phi\rangle_J = \int dg\langle\phi_g\rangle = 0. \tag{18.104}$$

Elitzur's theorem plays an analogous role for gauge fields to the Mermin–Wagner theorem for two-dimensional superfluids. In each case the theorem removes the possibility of strict off-diagonal long-range order. However, in both cases the phase stiffness can still survive through the development of power-law correlations with an infinite correlation length or correlation time.

Elitzur's theorem is relevant to the Kondo lattice because of the $U(1)$ gauge invariance associated with the Coulomb constraint $n_f + n_b = Q$. In the radial gauge (see (17.83), we see that the phase average of the slave boson field vanishes:

$$\langle b\rangle = Z^{-1}\int \mathcal{D}[\bar{b},b]e^{-S[\bar{b},b]} = Z^{-1}\int\frac{d\phi_0}{2\pi}\int \mathcal{D}[r,\lambda]r(\tau)e^{i\phi}e^{-S[r,\lambda]} = 0. \tag{18.105}$$

But let's look at the temporal correlations. The calculation here is similar to the computation of correlations in the power-law phase of an x-y spin model, and follows an adaptation by Nicholas Read [11] of Edward Witten's treatment of the large-N Gross–Neveu model [12]. The basic point is that the long-time correlations of the slave boson field are determined by the divergent phase fluctuations:

$$\langle b(\tau)b^\dagger(\tau)\rangle \propto r^2\langle e^{i(\phi(\tau)-\phi(0))}\rangle. \tag{18.106}$$

In the Gaussian approximation, the average of the exponential is given by

$$\langle e^{i(\phi(\tau)-\phi(0))}\rangle = \exp\left[-\frac{1}{2}\langle(\phi(\tau)-\phi(0))^2\rangle\right]. \tag{18.107}$$

We can calculate the correlation function in the exponent from fluctuations in the phase, as follows:

$$\begin{aligned}
\frac{1}{2}\langle(\phi(\tau)-\phi(0))^2\rangle &= \frac{1}{2}\langle(\delta\phi(\tau)-\delta\phi(0))^2\rangle \\
&= \langle\delta\phi(0)^2\rangle - \langle\delta\phi(\tau)\delta\phi(0)\rangle \\
&= T\sum_n\langle\delta\phi(-i\nu_n)\delta\phi(i\nu_n)\rangle\left(1-e^{-i\nu_n\tau}\right).
\end{aligned} \tag{18.108}$$

Now the constraint field λ is related to the phase fluctuations, since $\lambda(\tau) = i\partial_\tau \phi + \lambda_0$. We can use this to relate fluctuations in λ and ϕ in the frequency domain, $\delta\lambda(iv_n) = i(-iv_n)\delta\phi(iv_n)$, from which we deduce that

$$\langle \delta\phi(-iv_n)\delta\phi(iv_n)\rangle = \langle \delta\lambda(-iv_n)\delta\lambda(iv_n)\rangle/(iv_n)^2 = -D_{\lambda\lambda}(iv_n)/(iv_n)^2. \qquad (18.109)$$

It follows that

$$\frac{1}{2}\langle(\phi(\tau) - \phi(0))^2\rangle = -T\sum_n \frac{D_{\lambda\lambda}(iv_n)}{(iv_n)^2}\left(1 - e^{-iv_n\tau}\right). \qquad (18.110)$$

Now at zero temperature we can replace the summation over Matsubara frequencies by a continuous integral along the imaginary axis, so that

$$\frac{1}{2}\langle(\phi(\tau) - \phi(0))^2\rangle = -\int_{-i\infty}^{i\infty} \frac{dz}{2\pi i}\frac{D_{\lambda\lambda}(z)}{z^2}\left(1 - e^{-z\tau}\right)$$
$$= \int_0^\infty \frac{d\omega}{\pi}\mathrm{Im}[D_{\lambda\lambda}(\omega - i\delta)]\frac{(1 - e^{-\omega\tau})}{\omega^2}, \qquad (18.111)$$

where we have disorted the contour of integration around the positive frequency part of the branch cut on the real axis. At long times, this Laplace transform picks up the low-frequency part of $D_{\lambda\lambda}(\omega)$. Now from (18.101), at low frequencies $|v_n| << T_K$, $D_{\lambda\lambda}(iv_n) = \frac{\pi}{N}(\Delta + |v_n|) = \frac{\pi}{N}(\Delta - i(iv_n)\mathrm{sgn}(v_n))$, so, in the complex plane near $z = 0$, $D_{\lambda\lambda}(z) = \frac{\pi}{N}(\Delta \mp iz)$, depending on whether we are in the upper ($-$) or lower ($+$) half-plane. This means that, below the real axis, $\mathrm{Im}[D_{\lambda\lambda}(\omega - i\delta)] \sim \frac{\pi\omega}{N}$, from which we deduce that, at long times,

$$\frac{1}{2}\langle(\phi(\tau) - \phi(0))^2\rangle = \frac{1}{N}\int_0^{\Lambda_K} \frac{d\omega}{\omega}(1 - e^{-\omega\tau}), \qquad (18.112)$$

where we have introduced an upper cut-off of the order of the Kondo temperature into the integral. This integral contains an infrared divergence which is cut off at frequencies $v \sim 1/\tau$, so that

$$\frac{1}{2}\langle(\phi(\tau) - \phi(0))^2\rangle = \frac{1}{N}\int_0^{\Lambda_K} \frac{d\omega}{\omega}(1 - e^{-\omega\tau}) \sim \frac{1}{N}\int_{\frac{1}{\tau}}^{\Lambda_K} \frac{d\omega}{\omega} = \frac{1}{N}\ln[bT_K\tau], \quad (18.113)$$

where b is of order one. Inserting this into the exponent of (18.107), we obtain

$$\langle b(\tau)b^\dagger(\tau)\rangle = r^2\exp\left[-\frac{1}{2}\langle(\phi(\tau) - \phi(0))^2\rangle\right] = r^2\exp\left[-\frac{1}{N}\ln(bT_K\tau)\right] \propto (\tau T_K)^{-1/N}. \qquad (18.114)$$

Note the following:

- The development of power-law correlations of the slave boson reflects the rigidity of the development of a phase stiffness in time. If we expand the action in fluctuations in $\delta\lambda$, then, at low frequencies,

$$\delta S = -\chi_{\lambda\lambda}\frac{1}{2}\int_0^\beta d\tau\,\delta\lambda(\tau)^2 = \frac{\chi_{\lambda\lambda}}{2}\int_0^\beta d\tau\,(\partial_\tau\phi)^2, \qquad (18.115)$$

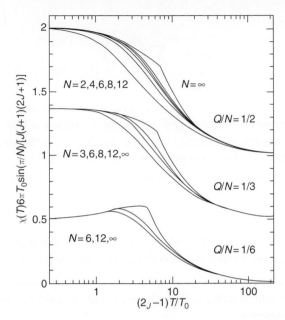

Fig. 18.4

Magnetic susceptibilities calculated in the Kondo model as a function of spin degeneracy N using the Bethe ansatz solution, contrasted with the large-N mean-field theory [13]. In the large-N limit, there is a sharp phase transition between high-temperature local moment behavior and low-temperature Fermi liquid behavior. At finite N, this becomes a crossover that becomes increasingly sharp as $N \to \infty$. Reprinted with permission from P. Coleman and N. Andrei, *J. Phys. C: Solid State Phys.*, vol. 19, no. 17, p. 3211, 1986. Copyright 1986 by the American Institute of Physics.

where $\chi_{\lambda\lambda} = \Delta/(\pi T_K^2) \sim 1/T_K$ is the static value of the susceptibility $\chi_{\lambda\lambda}(i\nu_n)$ given in (19.65), and we have replaced $\delta\lambda = i\partial_\tau\phi$. The amplitude of the phase fluctuations then has the form

$$\langle|\theta(i\nu_n)|^2\rangle \sim \frac{1}{\chi_{\lambda\lambda}\nu_n^2}\left(1 + \frac{|\nu_n|}{T_K}\right), \tag{18.116}$$

so the rapidity with which the phase fluctuations grow with time depends inversely on the phase stiffness.

- At a finite temperature, the phase fluctuations will grow linearly with time, leading to an exponential decay in the slave boson field (see Exercise 18.2). Thus true long-range correlations of the slave boson field only exist at zero temperature, and so, at any finite spin degeneracy N, there is only a crossover into the Kondo screened phase. This is borne out explicitly in the Bethe ansatz solution of the single-impurity model (see Figure 18.4).

An alternative way to understand these power law correlations is through a link with the *X-ray orthogonality catastrophe*. Philip Anderson [7] considered the change in the ground state of an electron gas due to the sudden ionization of an atomic core state. The formation of a positive ion then changes the scattering phase shift, giving rise to an orthogonality catastrophe, in which the final state has no overlap with the new ground state. In a subsequent work, Philippe Nozières and Cyrano De Dominicis [14] showed that the subsequent decay of the final state has a power-law form,

$$\langle\psi_I|e^{-iHt}|\psi_I\rangle \propto t^{-\alpha}, \tag{18.117}$$

where the exponent

$$\alpha = \sum_\lambda \left(\frac{\Delta\delta_\lambda}{\pi}\right)^2 \tag{18.118}$$

involves a sum over the change in phase shift in each scattering channel λ.

The action of a slave boson on the ground state is like an inverse photoionization, for the slave boson increases the value of $Q = n_f + n_b$ by 1, and in the Kondo limit this means that the final state has one additional f-electron, so by the Friedel sum rule the change in the scattering phase shift in each of the N scattering channels is $\Delta\delta = \pi/N$, so that $\alpha = N(1/N)^2 = 1/N$ and we recover the result of the large-N calculation:

$$\langle Q|b(\tau)b^\dagger(0)|Q\rangle \propto \tau^{-\frac{1}{N}}. \tag{18.119}$$

The virtue of the link with the orthogonality catastrophe, is that it doesn't depend on the large-N limit; moreoover, the power-law correlations will also hold in a Kondo lattice.

Example 18.4   Use orthogonality catastrophe arguments to calculate the X-ray exponent for the correlation function

$$\langle \Psi_{MV}|b^\dagger(\tau)b(0)|\Psi_{MV}\rangle \propto \tau^{-\alpha}, \tag{18.120}$$

where $|\Psi_{MV}\rangle$ is the ground state of an infinite U Anderson impurity model with average f-occupancy $\langle \hat{n}_f\rangle = n_f$.

Solution

The action of the slave boson is to reduce $Q = 1$ to $Q = 0$. By the Friedel sum rule, the initial phase shift is $\delta_f = \pi\frac{n_f}{N}$ (so that $\sum \delta_{f\sigma} = N\delta_f = \pi n_f$), whereas the final phase shift is zero. The change in scattering phase shift is thus $\Delta\delta = -\frac{\pi n_f}{N}$, so that $\alpha = \sum(\Delta\delta_\lambda/\pi)^2 = n_f/N$ and

$$\langle \Psi_{MV}|b^\dagger(\tau)b(0)|\Psi_{MV}\rangle \propto \tau^{-\frac{n_f}{N}}. \tag{18.121}$$

18.5   The spectrum of spin and valence fluctuations

In this section we calculate the spectrum of spin and charge fluctuations in a mixed-valent impurity, calculating the two spectral functions

$$\chi_s(i\nu_n) = \langle \delta S_-(-i\nu_n)\delta S_+(i\nu_n)\rangle$$
$$\chi_c(i\nu_n) = \langle \delta n_f(-i\nu_n)\delta n_f(i\nu_n)\rangle, \tag{18.122}$$

where

$$S_+ = f_\alpha^\dagger f_\beta, \qquad\qquad S_- = f_\beta^\dagger f_\alpha, (\alpha \neq \beta) \tag{18.123}$$

are the spin-raising and spin-lowering operators of the f-electrons (for arbitrary choice of $\alpha \neq \beta$).

We can calculate the spin dynamics very quickly, simply by identifying the Feynman diagram for the spin fluctuations with the polarization diagram, $\chi_s(i\nu_n) = \chi_{\lambda\lambda}(i\nu_n)$, so, using (18.90),

$$\chi_s(i\nu_n) = \frac{1}{N} \underbrace{}_{k} \overset{k+q}{} = 2\frac{i\nu_n}{i\nu_n + 2i\Delta}\left(\frac{V}{V_n}\right)^2 \operatorname{Re} \mathcal{L}(i\nu_n).$$

(18.124)

Now since there is a constraint $n_f + n_b = Q$, it follows that fluctuations in n_f and n_b are anticorrelated, $\delta n_f = -\delta n_b$, so that

$$\chi_c(i\nu_n) = \langle \delta n_b(-i\nu_n)\delta n_b(i\nu_n)\rangle.$$

(18.125)

Moreover, in the radial gauge $\delta n_b = \delta r^2 = 2r\delta r$. Now since $r = \sqrt{N}\frac{V}{V_0}$, it follows that we can compute the charge fluctuations from the fluctuations in the hybridization as follows:

$$\chi_c(i\nu_n) = \frac{N^2}{V_0^2}\langle \delta V(-i\nu_n)\delta V(i\nu_n)\rangle = -\frac{N^2}{V_0^2}D_{VV}(i\nu_n).$$

(18.126)

Now, from the fluctuations computed in (18.97),

$$-D_{VV}(i\nu_n) = \frac{g(i\nu_n)}{4N}\left(\frac{1}{\mathcal{L}(i\nu_n)} + \frac{1}{\mathcal{L}^*(i\nu_n)}\right).$$

(18.127)

Putting this all together, we have

$$\chi_s(i\nu_n) = \frac{i\nu_n}{(i\nu_n + 2i\Delta)}\left(\frac{V}{V_n}\right)^2\left[\mathcal{L}(i\nu_n) + \mathcal{L}^*(i\nu_n)\right]$$

$$\chi_c(i\nu_n) = N\frac{i\nu_n}{4V_0^2(i\nu_n + 2i\Delta)}\left(\frac{1}{\mathcal{L}(i\nu_n)} + \frac{1}{\mathcal{L}^*(i\nu_n)}\right),$$

(18.128)

where, from (18.92),

$$\mathcal{L}(i\nu_n) = \rho\left[\tilde{\psi}(\xi + i\nu_n) - \tilde{\psi}(\xi)\right] + \frac{i\nu_n}{V_0^2}.$$

(18.129)

Analytically extending this into the complex plane, we introduce

$$\mathcal{L}_{MV}(z) = V_0^2\mathcal{L}(z) = \frac{\Delta_0}{\pi}\left[\tilde{\psi}(\xi + z) - \tilde{\psi}(\xi)\right] + z$$

$$\mathcal{L}_{MV}^*(z) = V_0^2\mathcal{L}^*(z) = \frac{\Delta_0}{\pi}\left[\tilde{\psi}(z - \xi^*) - \tilde{\psi}(-\xi)\right] - z,$$

(18.130)

where we have introduced $\Delta_0 = \pi\rho V_0^2$ as the bare hybridization width. The dynamical valence fluctuations are then determined by the functions

$$\chi_s(\omega + i\delta) = \frac{\Delta}{\pi\omega(\omega + 2i\Delta)}\left[\left(\tilde{\psi}(\xi) - \tilde{\psi}(\omega + \xi)\right) + \left(\tilde{\psi}(-\xi^*) - \tilde{\psi}(\omega - \xi^*)\right)\right]$$

$$\frac{\chi_c(\omega + i\delta)}{N} = \frac{\omega}{4(\omega + 2i\Delta)}\left(\frac{1}{\mathcal{L}_{MV}(\omega)} + \frac{1}{\mathcal{L}_{MV}^*(\omega)}\right).$$

(18.131)

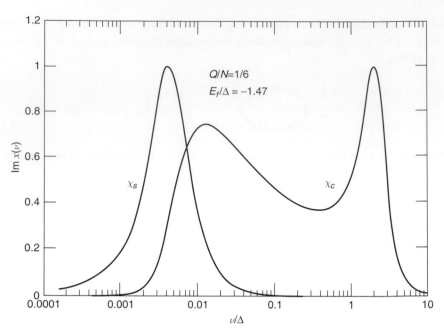

Fig. 18.5 Showing dynamical spin and charge susceptibilities for the infinite U Anderson model in the large-N limit.

At zero temperature, these expressions simplify as follows:

$$\chi_s(\omega + i\delta) = \frac{\Delta}{\omega(\omega + 2i\Delta)} \ln\left[\frac{\lambda^2 + \Delta^2}{\lambda^2 + (\Delta - i\omega)^2}\right]$$

$$\frac{\chi_c(\omega + i\delta)}{N} = \frac{\omega}{4(\omega + 2i\Delta)}\left(\frac{1}{\frac{\Delta_0}{\pi}\ln\left[\frac{\xi+\omega}{\xi}\right] + \omega} + \frac{1}{\frac{\Delta_0}{\pi}\ln\left[\frac{\xi^*-\omega}{\xi^*}\right] - \omega}\right). \quad (18.132)$$

Figure 18.5 shows that both the spin and charge fluctuations contain a low-frequency term associated with the Kondo resonance, but the charge fluctuations also contain a high-frequency valence fluctuation component, peaked around the energy $\omega_0 \gg T_K$, where

$$\text{Re } \mathcal{L}_{MV}^*(\omega_0) \approx \frac{\Delta_0}{\pi}\text{Re}\ln\left(\frac{\omega_0}{\xi}\right) - \omega_0 = 0. \quad (18.133)$$

However, from (18.48) we know that the mean-field value of ξ is determined by the equation

$$\text{Re}\left[\xi + \frac{\Delta_0}{\pi}\ln\left(\frac{\xi}{\Delta_0}\right)\right] = E_f^* = E_f + \frac{\Delta_0}{\pi}\ln\left(\frac{D}{\Delta_0}\right), \quad (18.134)$$

so we can rewrite the equation for ω_0 as

$$\frac{\Delta_0}{\pi}\text{Re}\ln\left(\frac{\omega_0}{\Delta_0}\right) - \omega_0 = E_f^* - \lambda \approx E_f^*, \quad (18.135)$$

so that the position of the maximum is located approximately at the position of the renormalized f-level, given by $\omega_0 \approx -E_f^* = |E_f^*|$.

18.6 Gauge invariance and the charge of the *f*-electron

The large-N expansion also provides a gauge theoretic interpretation of how local moments acquire charge as a kind of Anderson–Higgs mechanism [15] which locks the gauge fields of the conduction and *f*-electrons together. Here we will follow the development of ideas by Matthias Vojta, Todadri Senthil, and Subir Sachdev [16], working at Harvard, later adapted by Andrew Schofield, Brad Marston, and the author [17]. The basic idea is to regard the Kondo lattice as composed of two fluids:

- an electron fluid (c) coupled to the electromagnetic field ($\Phi, \vec{A}$)
- an incompressible, initially neutral *spinon* fluid of *f*-electrons coupled to an internal gauge field ($\lambda, \vec{\mathcal{A}}$).

When the Kondo effect develops, the coherent hybridization between the two fluids locks the internal and external gauge fields together. Specifically, the difference field develops a stiffness or mass term, described by the effective action

$$S_{eff} = \int dx^3 d\tau \left[\frac{\rho_f}{2} (\vec{\nabla}\phi_j - \mathcal{A} + e\vec{A})^2 - \frac{\chi_f}{2} (i\partial_\tau \phi + \delta\lambda - e\Phi)^2 \right], \tag{18.136}$$

where ϕ_j is the phase of the hybridization field $V_j = |V_j| e^{i\phi_j}$, while $\chi_f \sim 1/T_K$ and ρ_f are the corresponding temporal and spatial phase stiffnesses. The quantity $\delta\lambda = \lambda - \lambda_0$ is the equilibrium value of the constraint field in the absence of an external potential. This resembles the Ginzburg–Landau theory of a superconductor, except that it involves a *difference* field, so the corresponding "Meissner" effect causes the internal gauge field to lock to the electromagnetic field, as shown in Figure 18.6.

The remarkable thing about these results is that the static saddle points with respect to $\vec{\mathcal{A}}$ and λ_j of the free energy impose the constraints

$$\vec{\mathcal{A}}(x) - e\vec{A}(x) = 0, \qquad \delta\lambda_j - e\Phi_j = 0$$
$$\Rightarrow \vec{\mathcal{A}}(x) = e\vec{A}(x), \qquad \delta\lambda_j = e\Phi_j, \tag{18.137}$$

causing the internal and external fields to lock together. It is this process that causes the *f*-electrons to become charged. The system is not a superconductor because the sum of the two fields is still massless and propagates through the Kondo lattice as a physical electromagnetic field (albeit with a renormalized plasma frequency). The locking of the gauge fields means that at low frequencies the constraint field λ_j tracks with the external potential,

$$\delta\lambda_j(n_f(j) - Q) \longrightarrow e\Phi_j(n_f(j) - Q). \tag{18.138}$$

The two terms in this expression have a natural interpretation:

- The coupling $e\Phi_j n_f(j)$ signals that the composite *f*-electrons are charged.
- The term $-e\Phi_j Q$ implies the presence of a positively charged background which compensates the additional charge of the heavy electrons in the enlarged Fermi sea.

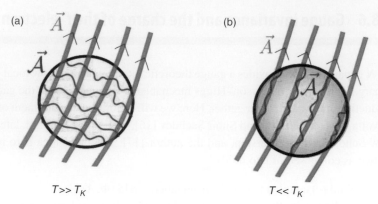

Fig. 18.6 Gauge theory picture of the Kondo lattice. (a) At high temperature $T \gg T_K$, the internal gauge field of the f-electrons is decoupled from the electromagnetic field. (b) At low temperature, $T \ll T_K$, the internal field locks together with the electromagnetic field.

In this way, we see that the local moment has "ionized" into a mobile negative electron plus a localized positive charge which we may associate with the Kondo singlet.

To see how this works in greater detail, it is helpful to consider a Kondo lattice model with an additional nearest-neighbor antiferromagnetic Heisenberg interaction, described by

$$H = H_c + H_K + H_M, \tag{18.139}$$

where

$$H_c = \sum_{ij,\sigma} \left(t_{ij} e^{-ie \int_j^i \vec{A} \cdot d\vec{x}} + (e\Phi_i - \mu)\delta_{ij} \right) c_{i\sigma}^\dagger c_{j\sigma} \tag{18.140}$$

describes the conduction band, while

$$H_K = -\frac{J}{N} \sum_{\alpha\beta} (c_{j\alpha}^\dagger f_{j\alpha})(f_{j\beta}^\dagger c_{j\beta})$$

$$H_H = -\frac{J_H}{N} \sum_{(i,j)} (f_{i\alpha}^\dagger f_{j\alpha})(f_{j\beta}^\dagger f_{i\beta}) \tag{18.141}$$

describe the onsite Kondo coupling (H_K) and the super-exchange between neighboring spin sites (H_H). A scalar and a vector potential ($\Phi, \vec{A}$) have been introduced into H_c using the *Peierls substitution* to include the vector potential as a phase factor in the hopping. For uniform vector and scalar potentials, H_c can be rewritten as $H_c = \sum(\epsilon_{\vec{k}-e\vec{A}} + e\Phi) c_{\vec{k}\sigma}^\dagger c_{\vec{k}\sigma}$, where the dispersion $\epsilon_{\vec{k}} = \sum_{\vec{R}} t(\vec{R}) e^{-i\vec{k}\cdot\vec{R}} - \mu$ is the kinetic energy of the conduction electrons.

The internal gauge fields appear when we formulate the problem as a path integral. When we factorize the Kondo interaction, we can always use the radial gauge to absorb the phase of the hybridization into the f-electron field. When we factorize the Heisenberg interaction,

$$H_H \to H_H = \sum_{(i,j)} \left(|\chi_{ij}| e^{-i \int_j^i \vec{A} \cdot d\vec{x}} f_{i\alpha}^\dagger f_{j\alpha} + \text{H.c.} \right) + \frac{N |\chi_{ij}|^2}{J_H},$$

the phase of the bond variables gives rise to a dynamical vector potential $\mathcal{A}$. This kind of decoupling was first introduced to describe spin liquids by Philip Anderson, and also by Lev Ioffe and Anatoly Larkin in the context of the RVB model of high-temperature super-conductivity [18, 19]. Without the hybridization, the bond variables describe the motion of spinons in a neutral spin fluid. If the χ_{ij} are uniform, the dispersion of the spinons is given by $j_{\mathbf{k}} = \sum_{\mathbf{R}} \chi(\mathbf{R}) e^{-i \mathbf{k} \cdot \mathbf{R}} + \lambda$, corresponding to a spinon fluid with a Fermi surface. The main difference between the internal field $(\lambda, \mathcal{A})$ and the electromagnetic field is that the internal field has no field energy $\frac{1}{2\mu_0}[B^2 + (E/c)^2]$ so the saddle-point conditions which normally generate Gauss' and Ampère's laws are simply

$$-\frac{\delta S}{\delta \lambda} = n_f - Q = 0, \qquad -\frac{\delta S}{\delta \vec{\mathcal{A}}} = j = 0, \tag{18.142}$$

enforcing the incompressible nature of the spin fluid.

When $V_j = 0$, there are two independent gauge symmetries for the electrons and for the spinons, but once the hybridization becomes finite, a single gauge invariance applies equally to both:

$$c_j \to c_j e^{-i\alpha}, \qquad f_j \to f_j e^{-i\alpha}$$
$$(e\Phi, e\vec{A}) \to (e\Phi + i\partial_\tau \alpha, e\vec{A} + \vec{\nabla}\alpha)$$
$$(\lambda, \vec{\mathcal{A}}) \to (\lambda + i\partial_\tau \alpha, \vec{\mathcal{A}} + \vec{\nabla}\alpha). \tag{18.143}$$

The emergence of a single gauge transformation for all fermions signals that the Kondo effect causes the f-spinons to acquire charge. Moreover, only the difference fields $e\vec{A} - \vec{\mathcal{A}}$ and $e\Phi - \lambda$ are invariant under the $U(1)$ gauge transformation, so the gauge-invariant long-wavelength action can thus only depend on the difference between these two fields.

Let us look at this in more detail. In the absence of the hybridization, the energy of the conduction electrons is $\epsilon_{\mathbf{k}-e\vec{A}}$. Adding an external field $\vec{A}$ simply shifts all the momenta of the quasiparticles by $-e\vec{A}$ and, since we sum over all momenta to get the total free energy or action, the action remains unchanged. However, once the hybridization is switched on, the energy of the quasiparticles becomes

$$E_{\mathbf{k}}^\pm = \frac{1}{2}(\epsilon_{\mathbf{k}-e\vec{A}} + j_{\mathbf{k}-\vec{\mathcal{A}}}) \pm \sqrt{\left(\frac{(\epsilon_{\mathbf{k}-e\vec{A}} - j_{\mathbf{k}-\vec{\mathcal{A}}})}{2}\right)^2 + V^2}. \tag{18.144}$$

Since the vector potential only couples to the conduction electrons, its introduction is not equivalent to a simple shift of momentum, so the energy now depends on $\vec{A}$ as if we have a superconductor! If we differentiate the free energy twice with respect to $\vec{A}$, we get

$$\rho_f^{(1)} \delta_{ab} = \frac{1}{e^2} \frac{\partial^2 F}{\partial A_a \partial A_b} = -\left(\text{} \right)$$

$$= T \sum_{\kappa=(\mathbf{k}, i\omega_r)} \left[\nabla_{ab}^2 \epsilon_{\mathbf{k}} G_c(\kappa) + \nabla_a \epsilon_{\mathbf{k}} \nabla_a \epsilon_{\mathbf{k}} G_c(\kappa)^2 \right], \tag{18.145}$$

where the single and double vertices reflect the first and second derivatives of the energy with respect to momentum. (The minus sign on the first line appears because the diagrammatic expansion of the free energy is $(-1)\times$ the sum of linked-cluster diagrams.) Were the spinon fluid dispersionless, so that $j_{\mathbf{k}} =$ constant, this expression would be a perfect differential $T \sum_\kappa \nabla_a(\nabla_b G_c(\kappa)) = 0$ and vanish, but otherwise, this stiffness becomes finite. At first sight this looks like a superconductor.

However, if we change the internal and external vector potentials by the same amount, i.e. $e\vec{A} = \vec{\mathcal{A}}$, then this is equivalent to uniformly shifting the momenta, and in this case the effective action is unchanged. The effective action must therefore be a function of $e\vec{A} - \vec{\mathcal{A}}$, vanishing when the two fields are equal. In other words, the Anderson–Higgs mass of the Kondo lattice is associated with the difference of the internal and external fields, with a long-wavelength effective action of the form

$$S_{eff} = \int dx^3 d\tau \left[\frac{\rho_f}{2} (\mathcal{A} - e\vec{A})^2 - \frac{\chi_f}{2}(\lambda - e\Phi)^2 \right]. \tag{18.146}$$

The above expression holds in the radial gauge. If we return to the original Cartesian gauge, restoring the complex $V_j = |V_j| e^{i\phi}$, then we must replace $\lambda_j \to \lambda_j + i\partial_\tau \phi$ and $\vec{\mathcal{A}} \to \vec{\mathcal{A}} - \nabla\phi$, giving rise to a more general expression,

$$S_{eff} = \int dx^3 d\tau \left[\frac{\rho_f}{2} (\vec{\nabla}\phi_j - \vec{\mathcal{A}} + e\vec{A})^2 - \frac{\chi_f}{2}(i\partial_\tau \phi + \delta\lambda - e\Phi)^2 \right], \tag{18.147}$$

as discussed earlier. Thus the acquisition of charge by the f-electrons can be regarded as a consequence of the stiffness that develops in the phase of the hybridization. A similar phase stiffness also occurs in the *chiral Gross–Neveu* model, a (1+1)-dimensional, relativistic version of the Kondo model [12].

From this discussion, we see that the very same stiffness is shared by both the internal and the external vector potential in particular the coupled stiffness,

$$\rho_f^{(2)} \delta_{ab} = -\frac{1}{e} \frac{\partial^2 F}{\partial A_a \partial \mathcal{A}_b} = \nabla_a j_{\mathbf{k}} \text{} \nabla_b \epsilon_{\mathbf{k}},$$

$$= -T \sum_{\kappa=(\mathbf{k}, i\omega_r)} \left[\nabla_a j_{\mathbf{k}} G_{fc}(\kappa) \nabla_b \epsilon_{\mathbf{k}} G_{cf}(\kappa) \right], \tag{18.148}$$

where the vertices reflect the current operators $\nabla_{\mathbf{k}} j_{\mathbf{k}}$ and $\nabla_{\mathbf{k}} \epsilon_{\mathbf{k}}$ of the f- and c-fields, respectively, is also equal to $\rho_f^{(1)}$.

To prove the equivalence $\rho_f^{(1)} = \rho_f^{(2)}$, we use the fact that, when we uniformly shift the momenta of all particles, physical quantities must not change. This is a restatement of gauge invariance, because a uniform shift $\delta\mathbf{k}$ in momentum is equivalent to a gauge transformation on both the c- and f-fields. Consider the invariance of the conduction electron current under such a shift, which leads to

$$\sum \delta\mathbf{k}_a \cdot \nabla_a \left(\nabla_b \epsilon_{\mathbf{k}} \, \bigcirc \right) = 0.$$

(18.149)

Each time we differentiate a propagator with respect to its momentum, we introduce a current operator. We can prove this by differentiating the matrix propagator $\nabla_a(z - h(\mathbf{k}))^{-1} = (z - h(\mathbf{k}))^{-1}\nabla_a h(\mathbf{k})(z - h(\mathbf{k}))^{-1}$. Notice, however, that $\nabla_a h(\mathbf{k})$ contains both a conduction and an f-component, so that

$$\sum_{\mathbf{k}} \nabla_a \left(\nabla_b \epsilon_{\mathbf{k}} \, \bigcirc \right) = \sum_{\mathbf{k}} \nabla_b \epsilon_{\mathbf{k}} \, \bigcirc \nabla_a \epsilon_{\mathbf{k}} + \nabla_b \epsilon_{\mathbf{k}} \, \bigcirc \nabla_a j_{\mathbf{k}}$$

$$+ \nabla_{ab}^2 \epsilon_{\mathbf{k}} \, \bigcirc = 0,$$

(18.150)

where the last term derives from differentiating the group velocity $\nabla_b \epsilon_{\mathbf{k}}$. If you like, you can check that this result also holds algebraically. This is an example of a *Ward identity*, and we note that, although we have only calculated it in the large-N limit, it actually holds to all orders in the $1/N$ expansion.

Rearranging the diagrams, we obtain the equality

$$-\left(\nabla_{ab}^2 \epsilon_{\mathbf{k}} \, \bigcirc + \nabla_a \epsilon_{\mathbf{k}} \, \bigcirc \nabla_b \epsilon_{\mathbf{k}} \right) = \nabla_a j_{\mathbf{k}} \, \bigcirc \nabla_b \epsilon_{\mathbf{k}}$$

(18.151)

or $\rho_f^{(1)} = \rho_f^{(2)}$. Notice that, were the original spinons to lose their dispersion, i.e. $\nabla_{\mathbf{k}} j_{\mathbf{k}} = 0$, then $\rho_f = 0$ and both sides of the equality would vanish.

18.7 Topological Kondo insulators

One of the amazing things about physics is that, just when a topic is thought closed, new insights come to light. One such surprise is the discovery of a completely new class of insulators: *topological insulators* [21–28]. These developments led to the realization that certain Kondo insulators discovered more than 40 years ago are *topological*, with topologically protected conducting surface states.

18.7.1 The rise of topology

In 2006, Liang Fu, Charles Kane, and Eugene Mélé [28] at the University of Pennsylvania in Philadelphia made the discovery that spin–orbit coupled insulators can acquire a topologically protected twist to their wavefunction, leading to robust conducting surface states. Although the reductionist, microscopic physics required nothing more in principle than non-interacting band theory, the emergent physics of these materials required the concept of topologically protected ground states that was unknown to an entire generation of semiconductor experts.

The idea of topological insulators grew out of the pioneering work of Robert Laughlin, David Thouless, Duncan Haldane, and collaborators on the quantum Hall insulator [22–24], in which the topological twist of the wavefunction gives rise to a perfectly quantized Hall constant. Later work of Kane and Mélé, Shoucheng Zhang, Andrei Bernevig, Taylor Hughes, Joel Moore, Leon Balents, and Rahul Roy [25–27, 29] led to the concept of the spin Hall insulator, which is basically two time-reversed copies of the quantum Hall insulator. This then paved the way for the jump to three-dimensional Z_2 topological insulators.

18.7.2 The Z_2 index

Conventional insulators are the vacuum "in miniature," for we can imagine adiabatically expanding them until we reach the true vacuum of empty space.[3] But this is not true for topological insulators, which are topologically distinct from the vacuum of empty space and cannot be adiabatically deformed into it. Now this has interesting consequences, for we can imagine smearing out the surface of an insulator so that the spatial gradients in the Hamiltonian are far smaller than typical Fermi velocities. In this situation, the path from the insulator to the vacuum of empty space becomes an adiabatic deformation of the Hamiltonian. But if the two vacua are topologically distinct, then the gap cannot remain open along this path, for this would imply an adiabatic path that links them. Consequently, the insulating gap has to collapse at the interface between two topologically distinct gapped states, to produce a gapless surface. Examples of such topological gapped states of matter are:

- superfluid ^{3}HeB, in which the $\hat{d}$ vector has a topological winding number, giving rise to Majorana surface states [30, 31]
- the integer quantum Hall effect, with chiral edge states
- the quantum spin Hall effect, in two-dimensional topological insulators
- three-dimensional Z_2 topological insulators.

Conventional surface states are incredibly volatile and almost never survive as macroscopic conducting surfaces, since they are highly sensitive to disorder, which produces Anderson localization of the electrons, and surface reconstruction, which eliminates them altogether. However, *topologically protected* surface states are robust against both

---

[3] This is a piece of imagination that requires a lot of chutzpah. We have to imagine expanding the lattice while at the same time dialing down the Coulomb interaction so that the electrons do not Mott localize in the process.

Anderson localization and surface reconstruction. The first two-dimensional Z_2 insulators were predicted in mercury/cadmium telluride (HgTe/CdTe) quantum wells by Andrei Bernevig, Taylor Hughes, and Shou-Cheng Zhang [26] at Stanford University in 2006, and were discovered in 2007 by Lauren Mölencamp and collaborators at the University of Würzberg [32]. The first experimentally realized three-dimensional topological insulator was bismuth antimonide ($Bi_{1-x}Sb_x$) [33], discovered by the groups of Robert Cava and Zahid Hasan at Princeton in 2008.

In 2007, Liang Fu and Charles Kane showed that, if an insulator has both time-reversal and inversion symmetry [34], the Z_2 index is uniquely determined by the the parities δ_{in} of the Bloch states at the high-symmetry points Γ_i of the valence band, which determine a Z_2 index,

$$Z_2 = \prod_{\Gamma_i} \delta(\Gamma_i) = \begin{cases} +1 & \text{conventional insulator} \\ -1, & \text{topological insulator} \end{cases} \qquad (18.152)$$

Fu–Kane formula for the Z_2 index of topological insulators

where $\delta(\Gamma_i) = \prod_n \delta_{in}$ is the the product of the parities of the occupied bands at the high-symmetry points in the Brillouin zone. This beautiful formula allows one to compute whether an insulator state is topological, merely by checking whether the index $Z_2 = -1$, without any detailed knowledge of the ground-state wavefunction. For a cubic insulator, with four equivalent high-symmctry points at the Γ, X, M, and R points, this formula reduces to $Z_2 = \delta_\Gamma \delta_X \delta_M \delta_R$. (The X and M points occur three times, but $\delta_{X,M}^3 = \delta_{X,M}$.)

18.7.3 Topology: solution to the mystery of SmB_6

In 2010, Maxim Dzero, Kai Sun, Victor Galitski and the author [35], working at the University of Maryland and at Rutgers Unversity, New Jersey, proposed that Kondo insulators can form strongly interacting versions of the Z_2 topological insulator, which they named *topological Kondo insulators* (TKI). The important point here is that:

- Kondo insulators are the distant adiabatic descendents of band insulators
- compared with the renormalized gap, spin–orbit coupling in a Kondo insulator is essentially infinite
- f-states are odd-parity, so that each time they cross into the valence band, the Z_2 index changes sign.

The TKI proposal provides an appealing, tentative resolution to a long-standing mystery in the Kondo insulator SmB_6, which for more than 30 years was known to exhibit a low-temperature resistivity plateau [20, 36] (see Figure 18.7), which can now be understood as a consequence of topologically protected surface states [35, 37]. In 2012, teams at the University of Michigan [38] and the University of California, Irvine [39], confirmed the existence of robust surface states in SmB_6, and in 2014 Xu *et al.* [40] detected the spin-polarized structure of the surface states in these materials that tentatively confirms their topological character.

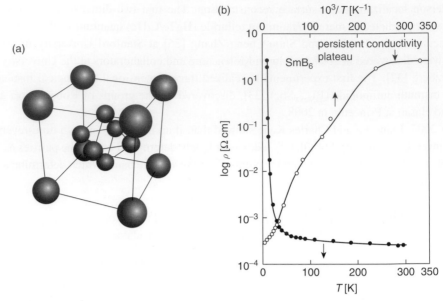

Fig. 18.7 (a) Crystal structure of SmB_6, where the gold-colored atoms are Sm and the silver-colored atoms are B. (b) Resistivity of SmB_6 as measured by Allen, Batlogg, and Wachter in 1979 [20], showing the conductivity plateau developing at low temperatures. This plateau is now known to derive from surface states, believed to be formed at the surface of a topological Kondo insulator. Reprinted with permission from J. W. Allen, *et al.*, *Phys. Rev.*, vol. 28, p. 5347, 1979. Copyright 1979 by the American Physical Society. [20]

The key ideas of the topological Kondo insulator are nicely illustrated by the spin–orbit coupled infinite U Anderson lattice model [41]:

$$H_{TKI} = H_c + H_f + H_{cf}$$
$$H_c = t_c \sum_{(i,j)} \left[c_{i\sigma}^\dagger c_{j\sigma} + \text{H.c} \right] - \mu \sum_j c_{j\sigma}^\dagger c_{j\sigma},$$
$$H_f = -t_f \sum_{(i,j)} \left[(f_{i\sigma}^\dagger b_i)(b_j^\dagger f_{j\sigma}) + \text{H.c.} \right] + \sum_j \left[\lambda_j f_{j\sigma}^\dagger f_{j\sigma} + (\lambda_j - E_f)(b_j^\dagger b_j - Q) \right]$$
$$H_{cf} = V \sum_{i,j} \left[c_{i\sigma}^\dagger \Phi_{\sigma\alpha}(\mathbf{R}_i - \mathbf{R}_j) b_j^\dagger f_{j\alpha} e^{-i\mathbf{k}\cdot\mathbf{R}_j} + \text{H.c.} \right]. \tag{18.153}$$

topological Kondo insulator model

Here,

- H_c is a tight-binding model for the conduction band
- H_f describes the nearest-neighbor hopping of f-electrons in an infinite U Hubbard model, written in a slave-boson formulation
- H_{cf} describes the hybridization with the spin–orbit coupled f-electron with form factor $\Phi_{\sigma\alpha}(\mathbf{R})$

- The constraint $n_f + n_b = Q$ is imposed at each site by the field λ_j ($Q = 1$ in a realistic model).

We'll consider a simplified model in which the simplified form factor

$$\Phi(\mathbf{R}) = \begin{cases} -i\hat{R} \cdot \frac{\vec{\sigma}}{2}. & (\mathbf{R} \in \text{n.n.}) \\ 0 & (\text{otherwise}) \end{cases} \tag{18.154}$$

describes the spin–orbit mixing between states with orbital angular momentum l differing by 1, such as f- and d-, or p- and s-orbitals (see Example 17.7). The odd parity of the form factor $\Phi(\mathbf{R}) = -\Phi(-\mathbf{R})$ derives from the odd-parity f-orbitals, while the prefactor $-i$ ensures that the hybridization is invariant under time reversal. The Fourier transform of this form factor, $\Phi(\mathbf{k}) = \sum_{\mathbf{R}} \Phi(\mathbf{R}) e^{i\mathbf{k}\cdot\mathbf{R}}$, is then

$$\Phi(\mathbf{k}) = \vec{s}_{\mathbf{k}} \cdot \vec{\sigma}, \tag{18.155}$$

where the s-vector $\vec{s}_{\mathbf{k}} = (\sin k_1, \sin k_2, \ldots \sin k_D)$ is the periodic equivalent of the unit momentum vector $\hat{\mathbf{k}}$. Notice how $\vec{s}(\Gamma_i) = 0$ vanishes at the high-symmetry points.

We'll study this model at a mean-field level in which the slave boson and constraint have a constant magnitude $\langle b_j \rangle = |b|$ and $\lambda_j = \lambda$, leading to a mean-field Hamiltonian

$$H^*_{TKI} = \sum_{\mathbf{k}\sigma,\alpha} \left(c^\dagger_{\mathbf{k}\sigma}, f^\dagger_{\mathbf{k}\sigma} \right) \overbrace{\begin{pmatrix} \epsilon_{\mathbf{k}} & \tilde{V}\vec{s}_{\mathbf{k}} \cdot \vec{\sigma} \\ \tilde{V}\vec{s}_{\mathbf{k}} \cdot \vec{\sigma} & \epsilon_{f\mathbf{k}} \end{pmatrix}}^{\underline{h}(\mathbf{k})}_{\sigma\alpha} \begin{pmatrix} c_{\mathbf{k}\alpha} \\ f_{\mathbf{k}\alpha} \end{pmatrix} + N_s\lambda \left(|b|^2 - Q \right), \tag{18.156}$$

where $\epsilon_{\mathbf{k}} = 2t_c \sum_{l=1,D} c_l - \mu$, $\tilde{V} = V|b|$ is a renormalized hybridization, and $\epsilon_{f\mathbf{k}} = -2t_f|b|^2 \sum_{l=1,D} c_l + E_f + \lambda$ is the renormalized f-dispersion.

Note the following:

- The condensed slave boson describes the Gutzwiller renormalization of the hybridization and f-hopping.
- We could have derived (18.277) starting from a Kondo–Heisenberg model (18.139).
- The f-band dispersion must be opposite to the conduction band (i.e. one hole-like, the other electron-like), to open up a gap.

This Hamiltonian is similar to the Hamiltonian of superfluid ^{3}He-B (Section 15.5): here the odd-parity triplet hybridization described by the vector $s_{\mathbf{k}}$ is analogous to the odd-parity triplet gap function of ^{3}He-B, described by the d-vector $\mathbf{d}(\mathbf{k})$.

The quasiparticle eigenvalues are given by

$$E^\pm(\mathbf{k}) = \epsilon^{(+)}_{\mathbf{k}} \pm \sqrt{(\epsilon^-_{\mathbf{k}})^2 + \tilde{V}^2 |\vec{s}_{\mathbf{k}}|^2}, \tag{18.157}$$

where $\epsilon^{(\pm)}_{\mathbf{k}} = \frac{1}{2}(\epsilon_{\mathbf{k}} \pm \epsilon_{f\mathbf{k}})$.

Example 18.5 Consider an Anderson impurity model involving a spin–orbit coupled p-orbital with $j = \frac{1}{2}$, $S = \frac{1}{2}$, and $l = 1$, hybridizing with a conduction band. By adapting the results of Section 17.2.1, the form of the hybridization between plane waves and the spin–orbit coupled p-state is

$$H_{hyb} = \sum_{\mathbf{k},(\sigma,M)\in\pm\frac{1}{2}} \left(V\Phi_{\sigma M}(\mathbf{k})c^{\dagger}_{\mathbf{k}\sigma}p_M + \text{H.c.} \right), \tag{18.158}$$

where

$$\Phi_{\sigma,M}(\mathbf{k}) = \mathcal{Y}_{\sigma,M}(\mathbf{k}) = \left(l, M - \sigma \; ; \frac{1}{2}\sigma \Big| jM \right) Y_{l,M-\sigma}(\hat{\mathbf{k}}) \Bigg|_{l=1,j=\frac{1}{2}}. \tag{18.159}$$

(a) Using these results, show that the hybridization form factor can be written

$$\Phi(\mathbf{k}) = \frac{1}{\sqrt{4\pi}}(\hat{\mathbf{k}} \cdot \vec{\sigma}). \tag{18.160}$$

(b) Suppose we now go to a cubic lattice, in which each atom has a p- and an s-state. The p-state hybridizes with s-states at positions $\pm\hat{x}$, $\pm\hat{y}$, and $\pm\hat{z}$. Extend the result of part (a) to show that a cubic Anderson lattice of such spin–orbit coupled p-electrons hybridized with neighboring s-states in the conduction band will involve a hybridization of the form

$$V(\mathbf{k}) \sim V(\sin k_x \sigma_x + \sin k_y \sigma_y + \sin k_z \sigma_z). \tag{18.161}$$

Solution

(a) We begin by noting (see (17.25)) that, for $j = l - \frac{1}{2} = \frac{1}{2}$,

$$\left(lM - \sigma, \frac{1}{2}\sigma \Big| jM \right) = \text{sgn}\,\sigma \sqrt{\frac{1}{2} - \frac{2}{3}M\sigma} \equiv \begin{pmatrix} \sqrt{\frac{1}{3}} & \sqrt{\frac{2}{3}} \\ -\sqrt{\frac{2}{3}} & -\sqrt{\frac{1}{3}} \end{pmatrix}_{\sigma M}, \tag{18.162}$$

where $(\sigma, M) \in \pm\frac{1}{2}$, and

$$Y_{1m}(\hat{\mathbf{k}}) = \begin{pmatrix} Y_{11}(\mathbf{k}) \\ Y_{10}(\mathbf{k}) \\ Y_{1-1}(\mathbf{k}) \end{pmatrix} = \sqrt{\frac{3}{4\pi}} \begin{pmatrix} -(\hat{k}_x + i\hat{k}_y)/\sqrt{2} \\ \hat{k}_z \\ (\hat{k}_x - i\hat{k}_y)/\sqrt{2} \end{pmatrix}, \tag{18.163}$$

so that

$$\mathcal{Y}_{\sigma M}(\mathbf{k}) = \begin{pmatrix} \sqrt{\frac{1}{3}}Y_{10} & \sqrt{\frac{2}{3}}Y_{1-1} \\ -\sqrt{\frac{2}{3}}Y_{11} & -\sqrt{\frac{1}{3}}Y_{10} \end{pmatrix} = \frac{1}{\sqrt{4\pi}} \begin{pmatrix} \hat{k}_z & (\hat{k}_x - i\hat{k}_y) \\ (\hat{k}_x + i\hat{k}_y) & -\hat{k}_z \end{pmatrix} \equiv \frac{1}{\sqrt{4\pi}}(\hat{\mathbf{k}} \cdot \vec{\sigma}). \tag{18.164}$$

(b) Consider the overlap between the p-state at the origin, and an s-state on an atom at position $\hat{z}$ along the z-axis. The Slater–Koster integral between the p-state and the s-state must have the form $V\sigma_z = -iV\hat{z}\cdot\vec{\sigma}$ (where the prefactor $-i$ is chosen to display the time-reversal symmetry). We can obtain the overlap matrix element for an atom at position $\mathbf{r}$ by rotating our coordinate system. Under the transformation $\hat{z} \to \hat{\mathbf{r}}$, the hybridization becomes $V(\mathbf{r}) \sim -iV\hat{\mathbf{r}}\cdot\vec{\sigma}$. When we Fourier transform this hybridization over the six nearest-neighbor atoms, we obtain

$$V(\mathbf{k}) = \sum V(\mathbf{r})e^{i\mathbf{k}\cdot\mathbf{r}} = -iV\sum(\hat{\mathbf{r}}\cdot\vec{\sigma})e^{i\mathbf{k}\cdot\mathbf{r}} = 2V(s_x\sigma_x + s_y\sigma_y + s_y\sigma_z) \qquad (s_l \equiv \sin k_l).$$
$$(18.165)$$

18.7.4 The Shockley chain

It is particularly useful to begin by looking at the one-dimensional case, where $\vec{s}_{\mathbf{k}} \cdot \vec{\sigma} \to \sin k_z \sigma_z \equiv \sin k_z$. Here, the effects of spin–orbit coupling can be absorbed by a gauge transformation, $c_{k\sigma} \to \sigma c_{k\sigma}$. This model was first proposed in 1939 by William Shockley, the future co-inventor of the transistor, working at Bell Laboratories somewhere near the current High Line in New York City. Shockley considered a chain of hybridized s- and p-orbitals [42] (see Figure 18.8(a)). In real space the Shockley chain is written

$$H_c = t_c \sum_j (c^\dagger_{j+1\sigma} c_{j\sigma} + \text{H.c.}) - \tilde{t}_f \sum_j (f^\dagger_{j+1\sigma} f_{j\sigma} + \text{H.c.})$$

$$+ \sum_j \frac{\tilde{V}}{2}[(c^\dagger_{j+1\sigma} - c^\dagger_{j-1\sigma})f_{j\sigma} + \text{H.c.}], \qquad (18.166)$$

where we have retained the symbol "f" for the odd-parity orbitals and have also absorbed the imaginary prefactors into their definition. Shockley found that end states formed as the p-states crossed into the valence band (Figure 18.8(b)). This led him to speculate that a

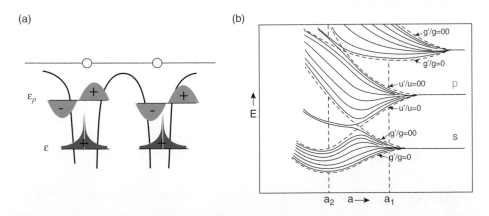

(a) One-dimensional Shockley chain, involving odd-parity hybridization between p-states and s-states. (b) Figure from the 1939 paper by William Shockley [42], showing the development of an edge state when the increased overlap between orbitals causes the p-band to cross into the valence band. Reprinted with permission from W. Shockley, *Phys. Rev.*, vol. 56, p. 317, 1939. Copyright 1939 by the American Physical Society.

Fig. 18.8

higher-dimensional version of this mechanism would lead to conducting surfaces in diamond. However, Shockley did not know about the additional role of spin–orbit coupling
and topology required for protected surface states in higher dimension, and the fruition of
his far-reaching idea had to await new insights and discoveries.

In one dimension, the Fu–Kane formula becomes $Z_2 = \delta_\Gamma \delta_X$, corresponding to the
parities of the occupied states at $k = 0$ and $k = \pi$. Let us consider what happens in a
finite chain with *periodic boundary conditions* as we lower the f-band into the conduction
band (Figure 18.9). At the high-symmetry points, the hybridization vanishes, when $\epsilon_{k=0}^-$
becomes zero, the bulk spectrum becomes gapless, and there is a band crossing between
the odd-parity f-state and the even-parity conduction state, changing the sign of the parity
$\delta_\Gamma \to -1$ so that $Z_2 \to -1$ (Figure 18.9(a)).

Now consider what happens for a finite chain with *open boundary conditions*. The ability of the odd- and even-parity states to pass through one another depends on the bulk
translational symmetry, which is lost on a chain which has ends. In a chain of length L,
states acquire an uncertainty in momentum

$$\Delta k \sim \frac{1}{L}, \tag{18.167}$$

so that the f-states at the bottom of the conduction band and the conduction states at the
top of the valence band now mix with each other in the vicinity of the Γ point, via a
matrix element of order $|V(k \sim 1/L)| \sim 1/L$. So once there are edges, the states at the
band edges now repel one another: they don't cross and now get stuck at the Fermi energy,
forming edge states separated by an energy of order $1/L$, as illustrated in Figure 18.9(b).
When we include spin, these edge states are Kramers doublets, localized at each end of
the chain.

A particularly easy way to visualize these end states [43, 44] is to adiabatically deform
the Hamiltonian into the limit where $\Delta = t_c = \tilde{t}_f = \frac{V}{2}$. If we define symmetric and
antisymmetric combinations of c- and f-electrons,

$$s_j^\dagger = \frac{1}{\sqrt{2}}(c_{j\sigma}^\dagger + f_{j\sigma}^\dagger)$$

$$a_j^\dagger = \frac{1}{\sqrt{2}}(c_{j\sigma}^\dagger - f_{j\sigma}^\dagger), \tag{18.168}$$

in this limit the Shockley chain can be simply written

$$H = \Delta \sum_j \left[c_{j+1\sigma}^\dagger c_{j\sigma} + c_{j+1\sigma}^\dagger f_{j\sigma} \right.$$
$$\left. -f_{j+1\sigma}^\dagger c_{j\sigma} - f_{j+1\sigma}^\dagger c_{j\sigma} + \text{H.c.} \right]$$
$$= \Delta \sum_j \left(a_{j+1\sigma}^\dagger s_{j\sigma} + \text{H.c.} \right), \tag{18.169}$$

so that the states are linked along the right-facing diagonals. Here we can see manifestly
the formation of decoupled edge states at either end of the chain.

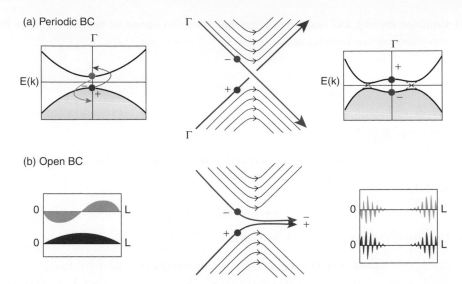

(a) Periodic BC

(b) Open BC

Fig. 18.9

Illustrating band crossing at the Γ point in the one-dimensional Shockley chain. (a) For periodic boundary conditions, the states at Γ have opposite parity and do not hybridize at the Γ point. The states at Γ invert without mixing, changing the Z_2 parity of the bulk. (b) For open boundary conditions, the uncertainty $\Delta k \sim 1/L$ in the momentum of the states at Γ causes the two states to mix, preventing the crossing at Γ and leaving two almost degenerate states stranded at the Fermi energy.

Example 18.6 Consider a particle–hole symmetric Shockley chain in which the hopping in the f- and conduction bands are equal and opposite, so that

$$h(k) = \begin{pmatrix} \epsilon_k & V \sin k \sigma_z \\ V \sin k \sigma_z & -\epsilon_k \end{pmatrix} = \epsilon_k \tau_3 + V \sin k \tau_1 \sigma_z, \qquad (18.170)$$

where $\epsilon_k = 2t \cos k + \epsilon$. This gives rise to a bulk dispersion of the form $E_k = \sqrt{(\epsilon_k^2 + \tilde{V}^2 \sin^2 k)}$, so that a bulk band crossing at Γ occurs for $\epsilon > -2t$. Consider an evanescent edge state of the form

$$\psi(z) = \exp\left[i(k + i\kappa \hat{A})z\right]\psi_0, \qquad (18.171)$$

where $\hat{A}$ is a matrix with eigenvalues $a = \pm 1$ (i.e. $A^2 = 1$), corresponding to decaying and growing modes, while ψ_0 is an eigenvector to be determined. Note that $-i\partial_z \psi = (k - i\kappa \hat{A})\Psi_0$.

(a) Derive the eigenvalue equation for the bound state.
(b) By solving the eigenvalue equation, derive the allowed values of κ and k and show that the energy of the edge state is zero.
(c) Show that an edge state forms providing

$$|\epsilon| > \sqrt{(2t)^2 + V^2}.$$

(d) Combine ingoing and outgoing evanescent waves to derive an explicit form for the wavefunction that vanishes at $z = 0$.

Solution

(a) Let us seek an $\hat{A}$ that commutes with each term in the Hamiltonian. In this case replacing the bulk momentum by $k \to -i\partial_z$, we may write

$$h[-i\partial_z]\psi(z) = \exp\left[i(k + i\kappa\hat{A})z\right]h(k + i\kappa\hat{A})\psi_0, \tag{18.172}$$

so that

$$h(k + i\kappa\hat{A})\psi_0 = E\psi_0.$$

(b) Now using the fact that $\hat{A}^2 = 1$, and $\sin i\kappa = -i\sinh\kappa$, we can expand

$$\cos(k + i\kappa\hat{A}) = \cos k \cosh\kappa + i\sin k \sinh\kappa \,\hat{A} \equiv c\tilde{c} + is\tilde{s}\,\hat{A},$$
$$\sin(k + i\kappa\hat{A}) = \sin k \cosh\kappa - i\cos k \sinh\kappa\hat{A} \equiv s\tilde{c} - ic\tilde{s}\,\hat{A}, \tag{18.173}$$

where $(c, s) \equiv (\cos k, \sin k)$ and $(\tilde{c}, \tilde{s}) \equiv (\cosh\kappa, \sinh\kappa)$. Then

$$h(k + i\kappa\hat{A}) = (2t(c\tilde{c} - 1) + \epsilon)\tau_3 + Vs\tilde{c}\tau_1\sigma_z + (2its\tilde{s}\,\tau_3 - iVc\tilde{s}\tau_1\sigma_z)\hat{A}$$
$$= c[2t\tilde{c}\tau_3 - iV\tilde{s}\tau_1\sigma_z\hat{A}] + \epsilon\tau_3 + s[V\tilde{c}\tau_1\sigma_z + 2it\tilde{s}\tau_3\hat{A}]. \tag{18.174}$$

Now we can cast the terms in the last term into the same matrix form by choosing $\tau_3\hat{A} = i\tau_1\sigma_z$, i.e. $\hat{A} = -\tau_2\sigma_z$, so that

$$h(k + i\kappa\hat{A}) = c[2t\tilde{c}\tau_3 - iV\tilde{s}\tau_1\sigma_z\hat{A}] + \epsilon\tau_3 + s\tau_1\sigma_z[V\tilde{c} - 2t\tilde{s}]$$
$$= [(2t\tilde{c} - V\tilde{s})c + \epsilon]\tau_3 + s\tau_1\sigma_z[V\tilde{c} - 2t\tilde{s}]. \tag{18.175}$$

Now the only way that $\hat{A} = -\tau_2\sigma_z$ commutes with each term in the Hamiltonian is if each term vanishes. If we put $V\tilde{c} = 2t\tilde{s}$, or

$$\tanh\kappa = \frac{V}{2t}, \tag{18.176}$$

then the last term vanishes. Writing $(\tilde{c}, \tilde{s}) = (2t, V)/\sqrt{(2t)^2 + V^2}$, we then have

$$h(k + i\kappa\hat{A}) = [\sqrt{(2t)^2 - V^2}\cos k + \epsilon]\tau_3. \tag{18.177}$$

Finally, to make this term vanish, we must choose

$$\cos k = \frac{\epsilon}{\sqrt{(2t)^2 - V^2}}$$

$$k = \pm\cos^{-1}\left(\frac{\epsilon}{\sqrt{(2t)^2 - V^2}}\right), \tag{18.178}$$

corresponding to reflected ($k > 0$) and incoming waves ($k < 0$), respectively. Notice that these values of k correspond approximately to the two wavevectors where the upper or lower bands cross.

(c) We can find solutions for k providing $|\cos k| \leq 1$, i.e. $\epsilon^2 < (2t)^2 - V^2$ or $V^2 + \epsilon^2 < (2t)^2$. In this region of parameter space, the ground state is topological and forms edge states.

(d) In order to make the wavefunction vanish at $z = 0$, we must combine an outgoing and an ingoing wave, so that

$$\psi_{\pm}(z) = \sin kz e^{-\kappa z} \psi_{0\pm}, \qquad (18.179)$$

where $\psi_{0\pm}$ corresponds to the Kramers pair of eigenstates of $\hat{A} = -\tau_2 \sigma_2$ with $a = +1$. These two states have either $\tau_2 = +1$ and $\sigma_z = +1$ or $\tau_2 = -1$ and $\sigma_z = -1$, so that

$$\psi_0 \equiv \begin{pmatrix} \psi_{0c\uparrow} \\ \psi_{0c\downarrow} \\ \psi_{0f\uparrow} \\ \psi_{0f\downarrow} \end{pmatrix}. \qquad (18.180)$$

Then the two time-reversed eigenvalues are

$$\psi_{0+} = \begin{pmatrix} 1 \\ i \end{pmatrix}_{\tau_2=+1} \otimes \begin{pmatrix} 1 \\ 0 \end{pmatrix}_{\sigma_z=-1} = \begin{pmatrix} 1 \\ 0 \\ i \\ 0 \end{pmatrix} \qquad (18.181)$$

and

$$\psi_{0-} = \begin{pmatrix} 1 \\ -i \end{pmatrix}_{\tau_2=-1} \otimes \begin{pmatrix} 0 \\ 1 \end{pmatrix}_{\sigma_z=-1} = \begin{pmatrix} 0 \\ 1 \\ 0 \\ -i \end{pmatrix}. \qquad (18.182)$$

18.7.5 Two dimensions: the spin Hall effect

With modern insight, the arguments developed for one dimension can be extended to two and three dimensions. Let's see how this works in two dimensions, where the formation of spin-polarized edge states gives rise to the phenomenon of the spin Hall effect. In this case,

$$\vec{s}_{\mathbf{k}} \cdot \vec{\sigma} = \sin k_x \sigma_x + \sin k_y \sigma_y, \qquad (18.183)$$

and the spin–orbit coupling plays a crucial role. If we define the spinors

$$\psi_{\mathbf{k}}^{\dagger} = (c_{\mathbf{k}\uparrow}^{\dagger}, c_{\mathbf{k}\downarrow}^{\dagger}, f_{\mathbf{k}\uparrow}^{\dagger}, f_{\mathbf{k}\downarrow}^{\dagger}), \qquad \psi_{\mathbf{k}} = \begin{pmatrix} c_{\mathbf{k}\uparrow} \\ c_{\mathbf{k}\downarrow} \\ f_{\mathbf{k}\uparrow} \\ f_{\mathbf{k}\downarrow} \end{pmatrix}, \qquad (18.184)$$

then the Hamiltonian can be written in the compact form

$$H_{MF} = \sum_{\mathbf{k}} \psi_{\mathbf{k}}^{\dagger} \underline{h}(\mathbf{k}) \psi_{\mathbf{k}} + \mathcal{N}_s \lambda (|b|^2 - Q)$$

$$h(\mathbf{k}) = \begin{pmatrix} \epsilon_{\mathbf{k}} & \tilde{V}\vec{s}_{\mathbf{k}} \cdot \vec{\sigma} \\ \tilde{V}\vec{s}_{\mathbf{k}} \cdot \vec{\sigma} & \epsilon_{f\mathbf{k}} \end{pmatrix} = \epsilon_{\mathbf{k}}^{(+)}\underline{1} + \epsilon_{\mathbf{k}}^{(-)}\gamma_3 + \tilde{V}(s_x\gamma_1 + s_y\gamma_2), \quad (18.185)$$

where we have introduced the gamma matrices,

$$\vec{\gamma} \equiv (\gamma_1, \gamma_2, \gamma_3) = \left[\begin{pmatrix} & \sigma_x \\ \sigma_x & \end{pmatrix}, \begin{pmatrix} & \sigma_y \\ \sigma_y & \end{pmatrix}, \begin{pmatrix} 1 & \\ & -\underline{1} \end{pmatrix} \right]. \quad (18.186)$$

To bring out the analogy with superfluid ^{3}He-B alluded to before, it is useful to introduce a d-vector,

$$\vec{d}(\mathbf{k}) = -\left(\tilde{V}s_x, \tilde{V}s_y, \epsilon_{\mathbf{k}}^{(-)} \right), \quad (18.187)$$

which defines the texture of the insulating ground state. Now

$$\underline{h}(\mathbf{k}) = \epsilon_{\mathbf{k}}^{(+)} - \vec{d}(\mathbf{k}) \cdot \vec{\gamma}. \quad (18.188)$$

and the magnitude of the d-vector $d(\mathbf{k}) = \sqrt{(\epsilon_{\mathbf{k}}^-)^2 + \tilde{V}^2|\vec{s}_{\mathbf{k}}|^2}$ determines the magnitude of the band splitting, with quasiparticle eigenvalues $E^{\pm}(\mathbf{k}) = \epsilon_{\mathbf{k}}^{(+)} \pm d_{\mathbf{k}}$. The unit d-vector $\hat{d}(\mathbf{k}) = \vec{d}(\mathbf{k})/d(\mathbf{k})$ defines the hybridized f- and conduction character of a state in the valence band, with the correspondence

$$\hat{d}(\mathbf{k}) = \begin{cases} \Uparrow & \text{conduction state} \\ \Downarrow . & f\text{-state} \end{cases} \quad (18.189)$$

At the high-symmetry points where the hybridization vanishes, the d-vector points up or down along the $\hat{z}$ axis. The texture defined by the d-vector may be thought of as a kind of "tent" in momentum space, supported at the high-symmetry points by poles which point either *up* or *down*. Reversals of the tent poles introduce topological defects into the canvas of the wavefunction, as illustrated in Figure 18.10.

Now let us repeat, for two dimensions, the thought experiment in which we gradually lowered the f-state into the conduction band (see Figure 18.10). On a square lattice the Fu–Kane formula becomes

$$Z_2 = \delta_{\Gamma}(\delta_X)^2\delta_M = \delta_{\Gamma}\delta_M, \qquad \text{square lattice} \quad (18.190)$$

because the parities of the two X points cancel. Thus, as in one dimension, the Γ and M points determine the topology. Suppose the f-level is initially empty, lying well above the occupied sea of conduction electrons. In this case, the $\hat{d}(\mathbf{k})$ vector points down throughout the valence band and $Z_2 = +1$, as shown in Fig 18.10(a). If we reduce E_f, there is a band crossing at the Γ point ($\mathbf{k} = 0$) once $\epsilon_{\mathbf{k}=0}^{(-)} = 0$. At the point of band crossing, when $\epsilon_{\mathbf{k}=0}^{(-)} = 0$, the bulk dispersion near the Γ point takes the form

$$E_{\mathbf{k}}^{\pm} = \pm\tilde{V}\sqrt{k_x^2 + k_y^2} + \epsilon_0^{(+)}, \quad (18.191)$$

so at the quantum phase transition, when $\epsilon_{\mathbf{k}=0}^- = 0$, the bulk becomes a semi-metal with a gapless Dirac cone of excitations in the bulk. Now once band crossing takes place, the gap opens up again and now the $\hat{d}(\mathbf{k})$ vector inverts at the Γ point, so that $\hat{d}(\mathbf{k})$ develops a *skyrmion* texture in which the d-vector rotates from *up* to *down* as one moves towards the

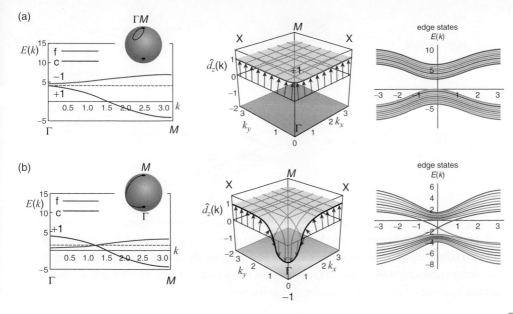

Fig. 18.10

Illustrating the development of a two-dimensional Z_2 topological insulator. (a) Prior to band crossing, the $\hat{d}(\mathbf{k})$ vector forms a uniform texture with no topological defects. In passing from Γ to M, the d-vector returns to its intial direction and there are no edge states. (b) After band crossing at the Γ point, the $\hat{d}(\mathbf{k})$ vector forms a skyrmion defect about the Γ point. The ground state becomes topologically distinct from the vacuum and gapless states form at the edges.

Γ-point as shown in Figure 18.10. The presence of the skyrmion texture in the ground state is indicated by the change in sign of the Fu–Kane formula, so that $Z_2 = -1$.

Another way of measuring the topology is through the *Chern number* of the d-vector. If we consider a small square of momentum space, then the solid angle subtended by the $\hat{d}$ vector is

$$\delta\Omega = \delta k_x \delta k_y \hat{d}(\mathbf{k}) \cdot \left(\frac{\partial \hat{d}(\mathbf{k})}{\partial k_x} \times \frac{\partial \hat{d}(\mathbf{k})}{\partial k_y} \right). \tag{18.192}$$

The Chern number n is the total number of times the d-vector wraps around the sphere:

$$n = \frac{\Omega}{4\pi} = \frac{1}{4\pi} \int d^2k \, \hat{d}(\mathbf{k}) \cdot \left(\frac{\partial \hat{d}(\mathbf{k})}{\partial k_x} \times \frac{\partial \hat{d}(\mathbf{k})}{\partial k_y} \right). \tag{18.193}$$

Such non-trivial winding of the topology of an insulator is well known in the integer quantum Hall state and, according to a result of Thouless, Kohmoto, Nightingale, and den Nijs [23], it is associated with the quantization of the Hall constant, given by $\sigma_H = -n\frac{e^2}{h}$. Here, though, it is spin–orbit coupling that provides the equal and opposite fields that act on *up* and *down* electrons, leading to two equal and opposite anomalous, quantized Hall constants for the *up* and *down* components of the insulator, given by $\sigma_{xy}^{\Uparrow} = -\sigma_{xy}^{\Downarrow} = -n\frac{e^2}{h}$. Although these two terms cancel, they give rise to an anomalous spin Hall conductance,

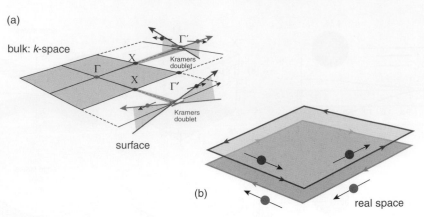

(a)

bulk: *k*-space

surface

(b)   real space

Fig. 18.11 Illustrating the formation of helical edge states in a two-dimensional Z_2 insulator. (a) The lines connecting the bulk high-symmetry points in momentum space have a single band crossing (here at the Γ point), forming a topologically protected Kramers doublet at the surface. The absence of inversion symmetry causes the Kramers doublet to split into two counter-propagating states of opposite spin polarization. (b) Spin-polarized edge states in real space, counter propagating around the edge to produce a quantized spin Hall conductance.

given by $\sigma_{xy}^{spin} = \frac{\hbar}{2e}(\sigma_{xy}^{\Uparrow} - \sigma_{xy}^{\Downarrow}) = -n(\mathrm{mod}\,2)\frac{e}{2\pi}$ (Figure 18.11).[4] Here the factor $\hbar/2$ is associated with the spin of up and down electrons, while we divide by e to factor out the electron charge included in the Hall current. Thus an electric field E_y applied in the y direction will induce a spin current $-\frac{e}{4\pi^2}E_y$ in the x direction. This topological effect, known as the *quantum spin Hall effect*, results from spin-polarized edge states runnning in opposite directions.

Let us now see how the twisted topology gives rise to edge states. In two dimensions, the edge excitations are one-dimensional with conserved momentum k, so, for an edge running along the x-axis of real space, k_x is a conserved quantity. Now the k_x or k_y axes in the bulk run between the high-symmetry Γ and X points. On the surface, these lines intersect with the surface high-symmetry point Γ'. Moreoever, for momenta running along these normals to the surface, the problem reduces to the Shockley chain (Figure 18.11(a)).

From this reasoning, we deduce that topologically protected Kramers doublets can form at $k_x = 0$ or π on the surface. Now although there is time-reversal symmetry at the edge, there is no inversion symmetry (since inversion maps one edge to the opposite one), so at a finite surface momentum ($k_x \neq 0$, the doublet splits into an *up* and a *down* band propagating in opposite directions. Along an edge that runs in the direction $\hat{\mathbf{n}}$, the edge-state Hamiltonian takes the form

$$H(k) \sim k\tilde{V}(\hat{\mathbf{n}} \cdot \vec{\sigma}), \qquad (18.194)$$

---

[4] A more sophisticated argument links Z_2 topological behavior to a topological spin-pumping effect. When a fictional half flux quantum is introduced in momentum space, this has the result of transfering either an odd or an even number of spins between surfaces. If the number of spins transfered is odd, the time-reversal properties of the surface are changed and the insulator is said to be a z_2 topological insulator. For this reason, one has to replace the integer n in the formula for the quantized spin Hall constant by $n \bmod 2$. See [28] for a detailed discussion.

giving rise to counter-dispersing *helical* edge states in which the component of the spin in the direction of motion is $+1$ ($\mathbf{n} \cdot \vec{\sigma} = \pm 1$) (Figure 18.11(b)). The formation of these helical edge states is a natural consequence of the topological twist in the Hamiltonian and the underlying valence band wavefuction. The two-dimensional version of Z_2 topological insulators is also called the spin Hall effect, because it contains two time-reversed copies of the edge states produced in the quantum Hall effect.

Example 18.7   Calculation of surface states on a strip.
One of the simplest ways to examine topological surface states is by numerically diagonalizing the Hamiltonian on a strip of width W. Consider the case of a two-dimensional model.

(a) Show that the mean-field Hamiltonian for a two-dimensional Kondo insulator with periodic boundary conditions can be block-diagonalized as follows:

$$H_{MF} = \sum_{\mathbf{k}} \left(\psi_{\mathbf{k}\uparrow}^{\dagger} \underline{h}(\mathbf{k}) \psi_{\mathbf{k}\uparrow} + \psi_{\mathbf{k}\downarrow}^{\dagger} \underline{h}^{*}(\mathbf{k}) \psi_{\mathbf{k}\downarrow} \right), \quad (18.195)$$

where $\psi_{\mathbf{k}\sigma}^{\dagger} = (c_{\mathbf{k}\sigma}^{\dagger}, f_{\mathbf{k}-\sigma}^{\dagger})$ combines a conduction and f-electron with *oppsite* spins and

$$\underline{h}(\mathbf{k}) = \begin{pmatrix} \epsilon_{\mathbf{k}} & \tilde{V}(s_x - is_y) \\ \tilde{V}(s_x + is_y) & \epsilon_{f\mathbf{k}} \end{pmatrix} \quad (18.196)$$

is a two-dimensional Hamiltonian.

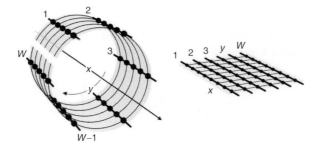

(b) Suppose we Fourier transform in the y direction, so that the Hamiltonian has a tight-binding form in the transverse direction but maintains its momentum-space structure in the x direction. Show that the one-particle Hamiltonian can be written in the form

$$\mathcal{H}_{jl}(k_x) = \frac{1}{W} \sum_{k_y = \frac{2\pi n}{L}} \exp\left[ik_y(j - l) \right] \underline{h}(k_x, k_y) = a(k)\delta_{jl} + b\delta_{j,l-1} + b^{T}\delta_{j,l+1}, \quad (18.197)$$

where

$$a(k_x) = \begin{pmatrix} -2tc_x - \mu & Vs_x \\ Vs_x & 2tc_x + \lambda \end{pmatrix}, \qquad b = \begin{pmatrix} -t & -V/2 \\ V/2 & t_f \end{pmatrix}. \quad (18.198)$$

(c) Show that, for open boundary conditions, the Hamiltonian has the block-diagonal form

$$
\mathcal{H}_{ij}(k) =
\begin{pmatrix}
a(k) & b^T & 0 & & & & & & 0 \\
b & a(k) & b^T & 0 & \cdots & & & & \\
0 & b & a(k) & b^T & 0 & & & & \\
& \ddots & \ddots & \ddots & \ddots & \ddots & & & \\
& & & 0 & b & a(k) & b^T & 0 & \\
& & \cdots & & 0 & b & a(k) & b^T & \\
0 & & & & & 0 & b & a(k)
\end{pmatrix}.
\tag{18.199}
$$

(d) Examine the band structure numerically and show that one-dimensional edge states develop once the bands cross at the Γ point.

Solution

(a) In terms of the original spinor $\psi_{\mathbf{k}} = (c^{\dagger}_{\mathbf{k}\uparrow}, c^{\dagger}_{\mathbf{k}\downarrow}, f^{\dagger}_{\mathbf{k}\uparrow}, f^{\dagger}_{\mathbf{k}\downarrow})$, the Hamiltonian is

$$
\underline{h}_0(\mathbf{k}) =
\begin{pmatrix}
\epsilon_{\mathbf{k}} & V\vec{\sigma} \cdot \vec{s} \\
V\vec{\sigma} \cdot \vec{s} & \epsilon_{f\mathbf{k}}
\end{pmatrix}.
\tag{18.200}
$$

Now for two dimensions the off-diagonal terms can be rewritten

$$
\vec{\sigma} \cdot \vec{s} = \sigma_+(s_x - is_y) + \sigma_-(s_x + is_y),
\tag{18.201}
$$

so that the up conductions are exclusively hybridized with the down f-electrons and vice-versa. Thus if we rewrite

$$
\psi_{\mathbf{k}} =
\begin{pmatrix}
\psi_{\mathbf{k}\uparrow} \\
\psi_{\mathbf{k}\downarrow}
\end{pmatrix}
=
\begin{pmatrix}
c_{\mathbf{k}\uparrow} \\
f_{\mathbf{k}\downarrow} \\
c_{\mathbf{k}\downarrow} \\
f_{\mathbf{k}\uparrow}
\end{pmatrix},
\tag{18.202}
$$

then in this basis the Hamiltonian becomes block diagonal:

$$
\underline{h}_0(\mathbf{k}) \rightarrow
\begin{pmatrix}
\epsilon_{\mathbf{k}} & \tilde{V}(s_x - is_y) & & \\
\tilde{V}(s_x + is_y) & \epsilon_{f\mathbf{k}} & & \\
& & \epsilon_{\mathbf{k}} & \tilde{V}(s_x + is_y) \\
& & \tilde{V}(s_x - is_y) & \epsilon_{f\mathbf{k}}
\end{pmatrix}
=
\begin{pmatrix}
h(\mathbf{k}) & \\
& h^*(\mathbf{k})
\end{pmatrix},
\tag{18.203}
$$

with

$$
h(\mathbf{k}) =
\begin{pmatrix}
\epsilon_{\mathbf{k}} & \tilde{V}(s_x - is_y) \\
\tilde{V}(s_x + is_y) & \epsilon_{f\mathbf{k}}
\end{pmatrix}.
\tag{18.204}
$$

(b) When we Fourier transform in the y direction,

$$\mathcal{H}_{ij}(k_x) = \sum_{k_y} \langle i|k_y\rangle h(k_x, k_y)\langle k_y|j\rangle = \frac{1}{W}\sum e^{ik_y(y_j - y_i)} h(k_x, k_y). \tag{18.205}$$

Now we may write the Hamiltonian $h(\mathbf{k})$ as

$$h(\mathbf{k}) = \begin{pmatrix} (-2tc_x - \mu) & \tilde{V}s_x \\ \tilde{V}s_x & 2t_f c_x + \lambda_f \end{pmatrix} + e^{ik_y}\begin{pmatrix} -t & -V/2 \\ V/2 & t_f \end{pmatrix} + e^{-ik_y}\begin{pmatrix} -t & V/2 \\ -V/2 & t_f \end{pmatrix}. \tag{18.206}$$

Under the Fourier transformation,

$$\frac{1}{W}\sum e^{ik_y(y_j - y_i)} \times \left\{\begin{matrix} 1 \\ e^{ik_y} \\ e^{-ik_y} \end{matrix}\right\} = \left\{\begin{matrix} \delta_{ij} \\ \delta_{i(j+1)} \\ \delta_{i(j-1)} \end{matrix}\right\}, \tag{18.207}$$

so that

$$h(\mathbf{k}) \to \mathcal{H}_{ij}(k_x)\big|_{periodic} = \begin{pmatrix} (-2tc_x - \mu) & \tilde{V}s_x \\ \tilde{V}s_x & 2t_f c_x + \lambda_f \end{pmatrix}\delta_{ij} + \begin{pmatrix} -t & -V/2 \\ V/2 & t_f \end{pmatrix}\delta_{i,j+1}$$

$$+ \begin{pmatrix} -t & V/2 \\ -V/2 & t_f \end{pmatrix}\delta_{+i,j-1}$$

$$= a(k_x) + b\delta_{i,j+1} + b^T\delta_{i,j-1}$$

$$= \begin{pmatrix} a(k) & b^T & 0 & & & & 0 & b \\ b & a(k) & b^T & 0 & \cdots & & & 0 \\ 0 & b & a(k) & b^T & 0 & & & \\ & \ddots & \ddots & \ddots & \ddots & \ddots & & \\ & & & 0 & b & a(k) & b^T & 0 \\ 0 & & \cdots & & 0 & b & a(k) & b^T \\ b^T & 0 & & & & 0 & b & a(k) \end{pmatrix}. \tag{18.208}$$

(c) To convert periodic to open boundary conditions, we remove the links between the first and the last sites, i.e. we remove the top-right and bottom-left entries in $\mathcal{H}_{ij}(k)$, so that

$$\mathcal{H}_{ij}(k_x)\big|_{open} = \begin{pmatrix} a(k) & b^T & 0 & & & & 0 \\ b & a(k) & b^T & 0 & \cdots & & \\ 0 & b & a(k) & b^T & 0 & & \\ & \ddots & \ddots & \ddots & \ddots & \ddots & \\ & & & 0 & b & a(k) & b^T & 0 \\ & & \cdots & & 0 & b & a(k) & b^T \\ 0 & & & & & 0 & b & a(k) \end{pmatrix}. \tag{18.209}$$

(d) To calculate the spectrum, we write a Mathematica code to build the above matrix, and then diagonalize it. Here is the code I wrote:

Strip calculation forTopological Insulator width W, Length L

In[1]:= (* First set up the matrix and test it for W=5 *)

```
matty[W_] := Module[{hb}, hb = ConstantArray[0, {W, W}];
   For[i = 1, i ≤ W, i++,
     For[j = 1, j ≤ W, j++,
       If[i == j, hb[[i, j]] = a];
       If[j - i == 1, hb[[i, j]] = bT];
       If[j - i == -1, hb[[i, j]] = b];]]; hb];
matty[5]
```

Out[2]=
$$\begin{pmatrix} a & bT & 0 & 0 & 0 \\ b & a & bT & 0 & 0 \\ 0 & b & a & bT & 0 \\ 0 & 0 & b & a & bT \\ 0 & 0 & 0 & b & a \end{pmatrix}$$

In[3]:=

(* Fill in the details for sub−matrices a and b *)

```
hstrip[W_] := Module[{}, aa = ( 2 t Cos[kx] - μ      V Sin[kx]                    );
                                   V Sin[kx]   2 tf (2 - Cos[kx]) + λf
```

$$hstrip[W\_] := Module\Big[\{\}, aa = \begin{pmatrix} 2\,t\,Cos[kx] - \mu & V\,Sin[kx] \\ V\,Sin[kx] & 2\,tf\,(2 - Cos[kx]) + \lambda f \end{pmatrix};$$

$$bb = \begin{pmatrix} t & -V/2 \\ V/2 & -tf \end{pmatrix};$$

```
   ArrayFlatten[matty[W] //. {a → aa, b → bb, bT → Transpose[bb]}]];
Clear[t, V, tf, μ, λf];
hstrip[3]
```

Out[3]=
$$\begin{pmatrix}
2\,t\cos(kx) - \mu & V\sin(kx) & t & \frac{V}{2} & 0 & 0 \\
V\sin(kx) & \lambda f + 2\,tf\,(2 - \cos(kx)) & -\frac{V}{2} & -tf & 0 & 0 \\
t & -\frac{V}{2} & 2\,t\cos(kx) - \mu & V\sin(kx) & t & \frac{V}{2} \\
\frac{V}{2} & -tf & V\sin(kx) & \lambda f + 2\,tf\,(2 - \cos(kx)) & -\frac{V}{2} & -tf \\
0 & 0 & t & -\frac{V}{2} & 2\,t\cos(kx) - \mu & V\sin(kx) \\
0 & 0 & \frac{V}{2} & -tf & V\sin(kx) & \lambda f + 2\,tf\,(2 - \cos(kx))
\end{pmatrix}$$

In[4]:= (* Define the Bulk Dispersion *)

```
Clear[t, V, tf, μ, λf]; enc[kx_, ky_] := 2 t (Cos[kx] + Cos[ky] - 2);
enf[kx_, ky_] := 2 tf (2 - (Cos[kx] + Cos[ky])) + λf;
```

$$En[kx\_, ky\_, P\_] := \left(\frac{enc[kx, ky] + enf[kx, ky]}{2} \right) +$$

$$P\sqrt{\left[\left(\frac{enc[kx, ky] - enf[kx, ky]}{2} \right)^2 + (V\,Sin[kx])^2 + (V\,Sin[ky])^2 \right]};$$

In[9]:=

(* Set up the plotting module *)

```
Plotty[x_, W_, L_] := Module[{EigVals, hstore, k}, λf = x;
   t = 1; V = 2.; μ = 4 t; tf = 1; hstore = hstrip[W];
EigVals = ConstantArray[0, L]; For[i = 1, i ≤ L, i++,
```

$$k = -\pi + \frac{2\,\pi}{L}\,i;$$

```
   Eig = Sort[Eigenvalues[hstore /. kx → k]];
   EigVals[[i]] = Eig;]; Show[GraphicsRow[{
Plot[{En[k, 0, +1], En[k, 0, -1]}, {k, 0, π}, AxesLabel → {"kₓ", "E⁺ₖ"},
       PlotLabel → "Bulk Dispersion"], ListLinePlot[Transpose[EigVals],
       DataRange → {-π, π}, PlotRange → All, PlotLabel → "Edge States",
       AxesLabel → {"kₓ", "E(k)"}]}, ImageSize → Large]]];
```

(* Display the plots as a function of the f–level chemical potential *)
`Manipulate[Plotty[λf, 10, 30], {λf, 4, -5, Appearance -> "Open"}]`

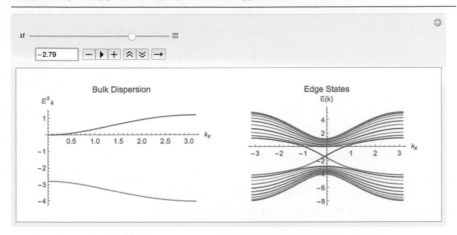

Example 18.8 Calculation of the spin Hall constant.

Consider the two-dimensional spin–orbit coupled Anderson model,

$$\underline{h}(\mathbf{k}) = \begin{pmatrix} \epsilon_{\mathbf{k}} & \tilde{V}\vec{s}_{\mathbf{k}} \cdot \vec{\sigma} \\ \tilde{V}\vec{s}_{\mathbf{k}} \cdot \vec{\sigma} & \epsilon_{f\mathbf{k}} \end{pmatrix} \equiv \epsilon_{\mathbf{k}}^{(+)} + \vec{d}(\mathbf{k}) \cdot \vec{\gamma}. \tag{18.210}$$

(a) Show that the charge current operator is given by

$$\vec{j}^{E} = \frac{e}{\hbar}\vec{\nabla}_{\mathbf{k}}h(\mathbf{k}) = \frac{e}{\hbar}\left(\vec{\nabla}_{\mathbf{k}}\epsilon_{\mathbf{k}}^{+} + \vec{\nabla}_{\mathbf{k}}\vec{d}(\mathbf{k}) \cdot \vec{\gamma}\right). \tag{18.211}$$

(b) In the two-dimensional spin–orbit coupled Anderson model, spin is not a conserved quantity. However, the matrix $S_3 = \gamma_3 \sigma_3$ commutes with the Hamiltonian, so that the corresponding *spin current* is given by the product of the charge and spin operator:

$$\vec{j}^{S} = -\frac{\hbar}{2}\frac{1}{\hbar}\vec{\nabla}_{\mathbf{k}}h(\mathbf{k})\gamma_3\sigma_3 = \frac{1}{2}\left(\vec{\nabla}_{\mathbf{k}}\epsilon_{\mathbf{k}}^{+} + \vec{\nabla}_{\mathbf{k}}\vec{d}(\mathbf{k}) \cdot \vec{\gamma}\right)\gamma_3\sigma_3. \tag{18.212}$$

Generalize the Kubo formula from the Hall current to the spin Hall current and show that the spin Hall constant is given by

$$\sigma_{SH} = \lim_{\nu_n \to 0}\frac{1}{i\nu_n}\langle j_x^S(i\nu_n)j_y^E(-i\nu_n)\rangle\Big|_0^{i\nu_n}. \tag{18.213}$$

(c) Calculate the spin Hall constant using the above formula, to show that

$$\sigma_{SH} = -\frac{e}{2}\int\frac{d^2k}{(2\pi)^2}\,\hat{d}(\mathbf{k}) \cdot \left(\frac{\partial\hat{d}(\mathbf{k})}{\partial k_x} \times \frac{\partial\hat{d}(\mathbf{k})}{\partial k_y}\right) = -\frac{ne}{2\pi}. \tag{18.214}$$

Solution

(a) For convenience, let us set $\hbar = 1$ in this calculation; we can restore it at the end. To calculate the charge current, we replace $\mathbf{k} \to \mathbf{k} - e\vec{A}$ and calculate

$$\vec{j}^{\,E} = -\nabla_{\vec{A}} h(\mathbf{k} - e\vec{A}) = e\nabla_{\mathbf{k}} h(\mathbf{k} - e\vec{A}). \tag{18.215}$$

Setting $\vec{A} = 0$, we obtain

$$\vec{j}^{\,E} = e\left(\vec{\nabla}_{\mathbf{k}}\epsilon_{\mathbf{k}}^+ + \vec{\nabla}_{\mathbf{k}}\vec{d}(\mathbf{k}) \cdot \vec{\gamma}\right). \tag{18.216}$$

Notice how the d-vector introduces an additional spin-dependent contribution to the current. This term is reminiscent of the $-e\vec{A}$ term in the current $\vec{j} = \vec{p} - e\vec{A}$ of a free particle, and we may interpret $\nabla_{\mathbf{k}}d(\mathbf{k})$ as a kind of spin–orbit induced vector potential that is equal and opposite for *up* and *down* spin electrons.

(b) From (10.38), the Kubo formula for the Hall constant is

$$\sigma_H = \sigma_{xy}(i\nu_n) = -\frac{1}{\nu_n}\left[\langle j_x^E(\nu') j_y^E(-\nu')\rangle\right]_{\nu'=0}^{\nu'=i\nu_n}. \tag{18.217}$$

We can simply generalize this expression by replacing the electric current response by the spin current response, substituting $j^E \to j^S$, i.e.

$$\sigma_{SH}(i\nu_n) = -\frac{1}{\nu_n}\left[\langle j_x^S(\nu') j_y^E(-\nu')\rangle\right]_{\nu'=0}^{\nu'=i\nu_n}. \tag{18.218}$$

(We only substitute in the first j^E, since the second term is the current coupled to the external electric field.)

(c) Expanding the Kubo formula for the spin Hall constant, we need to take the zero-frequency limit of

$$\sigma_{SH} = \frac{1}{\nu_n} T \sum_{\kappa} \text{Tr}\left[\mathcal{G}(\kappa) j_x^S \mathcal{G}(\kappa + \nu) j_y^E\right], \tag{18.219}$$

where we denote $\kappa = (\mathbf{k}, i\omega_n)$, $\kappa + \nu = (\mathbf{k}, i\omega_n + i\nu_r)$, and

$$\mathcal{G}(\kappa) = \frac{1}{i\omega_n - \epsilon_{\mathbf{k}}^+ - \vec{d}(\mathbf{k}) \cdot \vec{\gamma}} = \sum_{p=\pm} P_p \frac{1}{i\omega_n - E_{\mathbf{k}}^p} \qquad (p = \pm 1), \tag{18.220}$$

where

$$P_{\pm} = \frac{1}{2}\left(1 \pm \vec{d}(\mathbf{k}) \cdot \vec{\gamma}\right) \tag{18.221}$$

projects onto the upper and lower bands.

Substituting the Green's functions into (18.219),

$$\sigma_{SH} = \frac{1}{\nu_n} \sum_{p,q,\mathbf{k}} \text{Tr}\left[P_p(\mathbf{k}) j_x^S P_q(\mathbf{k}) j_y^E\right] T \sum_{i\omega_n} \frac{1}{i\omega_n - E_{\mathbf{k}}^p} \frac{1}{i\omega_n + i\nu_r - E_{\mathbf{k}}^q}$$

$$= \frac{1}{\nu_n} \sum_{p,q,\mathbf{k}} \text{Tr}\left[P_p(\mathbf{k}) j_x^S P_q(\mathbf{k}) j_y^E\right] \frac{f(E_{\mathbf{k}}^p) - f(E_{\mathbf{k}}^q)}{i\nu_n - (E_{\mathbf{k}}^q - E_{\mathbf{k}}^p)}, \tag{18.222}$$

where we have used standard contour techniques to carry out the Matsubara summation in the line. (See, for example, the calculation of dynamical spin susceptibility in (8.50)). Evaluating this at zero temperature, in order that the Fermi functions do not cancel we must have $q = -p$, so that

$$
\sigma_{SH} = -\frac{1}{\nu_n} \sum_{p=\pm,\mathbf{k}} \text{Tr}\left[P_p(\mathbf{k})j_x^S P_{-p}(\mathbf{k})j_y^E\right] \frac{p}{i\nu_n + 2pd(\mathbf{k})}
$$

$$
= -\frac{e}{2}\frac{1}{\nu_n} \sum_{p=\pm,\mathbf{k}} \text{Tr}\left[\gamma_3\sigma_3 P_p(\mathbf{k})\nabla_x h(\mathbf{k})P_{-p}(\mathbf{k})\nabla_y h(\mathbf{k})\right] \frac{p}{i\nu_n + 2pd(\mathbf{k})}, \quad (18.223)
$$

where we have put $E_{\mathbf{k}}^+ - E_{\mathbf{k}}^- = 2d(\mathbf{k})$ and $\nabla_l \equiv \frac{\partial}{\partial k_l}$. Now the spin matrix σ_3 does not appear explicitly anywhere in P_p or $h(\mathbf{k})$, so the only way to produce a non-vanishing spin trace is to combine the σ_3 with the σ_1 and σ_2 appearing in the $d_1(\mathbf{k})\gamma_1 = d_1(\mathbf{k})\tau_1\sigma_1$ and $d_2(\mathbf{k})\gamma_2 = d_2(\mathbf{k})\tau_1\sigma_2$ terms. Likewise, to eliminate the γ_3 appearing at the beginning of the trace, we must also pick up a $d_3(\mathbf{k})\gamma_3$ term. The only non-vanishing matrix elements thus contain three $\vec{d}(\mathbf{k})$ terms: one from $\nabla_x h$, one from $\nabla_y h$, and one from either P_p or P_{-p}. Calculating these matrix elements, we obtain

$$
\text{Tr}\left[\gamma_3\sigma_3 P_p \nabla_x h P_{-p} \nabla_y h\right] = \text{Tr}\left[\gamma_3\sigma_3 P_p(\nabla_x \vec{d}\cdot\vec{\gamma})P_{-p}(\nabla_y \vec{d}\cdot\vec{\gamma})\right]
$$

$$
= \frac{p}{4}\text{Tr}\left[\gamma_3\sigma_3 \hat{d}^a\gamma_a(\nabla_x d^b\gamma_b)(\nabla_y d^c\gamma_c)\right]
$$

$$
- \frac{p}{4}\text{Tr}\left[\gamma_3\sigma_3(\nabla_x d^b\gamma_b)\hat{d}^a\gamma_a(\nabla_y d^c\gamma_c)\right]
$$

$$
= \hat{d}^a\nabla_x d^b\nabla_y d^c\frac{p}{2}\overbrace{\text{Tr}\left[\gamma_3\sigma_3\gamma_a\gamma_b\gamma_c\right]}^{4i\epsilon^{abc}} = 2ip\left[\hat{d}\cdot(\nabla_x\vec{d}\times\nabla_y\vec{d})\right].
$$
$$
(18.224)
$$

The two terms in the second line are equal because the γ matrices anticommute, allowing us to combine them into a single term on the third line. The hat indicates $\hat{d} = \vec{d}/d$. To evaluate the trace over the γ matrices in the final line, we note that $\gamma_1\gamma_2 = i\sigma_3 = i\gamma_3(\gamma_3\sigma_3)$, so that $\text{Tr}[\gamma_1\gamma_2\gamma_3(\gamma_3\sigma_3)] = 4i$, and since the γ_as anticommute it follows that $\text{Tr}[\gamma_3\sigma_3\gamma_a\gamma_b\gamma_c] = 4i\epsilon_{abc}$.

Putting this all together,

$$
\sigma_{SH}(i\nu_n) = e\frac{1}{i\nu_n}\sum_{\mathbf{k}}\hat{d}\cdot(\nabla_x\vec{d}\times\nabla_y\vec{d})\left[\frac{1}{i\nu_n - 2d(\mathbf{k})} + \frac{1}{i\nu_n + 2d(\mathbf{k})}\right]
$$

$$
= e\frac{1}{i\nu_n}\sum_{\mathbf{k}}\hat{d}\cdot(\nabla_x\vec{d}\times\nabla_y\vec{d})\frac{2i\nu_n}{(i\nu_n)^2 - (2d(\mathbf{k}))^2}. \quad (18.225)
$$

Taking the zero-frequency limit, we obtain

$$
\sigma_{SH} = -\frac{e}{2}\int\frac{d^2k}{(2\pi)^2}\frac{\hat{d}\cdot(\nabla_x\vec{d}\times\nabla_y\vec{d})}{d(\mathbf{k})^2}
$$

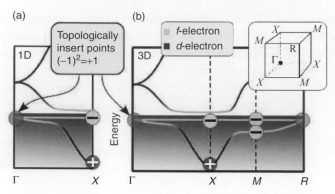

Fig. 18.12 Schematic band structure illustrating (a) a one-dimensional Kondo insulator with local cubic symmetry and (b) a three-dimensional model for SmB$_6$. The d-band crosses through the f-band, producing a band inversion at the three X points, which drives the formation of a topological Kondo insulator. Reprinted from [45].

$$= -\frac{e}{2\pi} \int \frac{d^2k}{4\pi} \, \hat{d}(\mathbf{k}) \cdot \overbrace{\left(\frac{\partial \hat{d}(\mathbf{k})}{\partial k_x} \times \frac{\partial \hat{d}(\mathbf{k})}{\partial k_y} \right)}^{n} = -\left(\frac{e}{2\pi} \right) n. \quad (18.226)$$

One last thing: we need to check how many powers of $\hbar$ need to be restored to make our final result dimensionally correct. The dimensions of $[\sigma_{SH}] = $ [spin current/voltage] or $[\sigma_{SH}] = [\hbar/t]/[E/Q] = [E]/[E/Q] = [Q]$, so that σ_{SH} has the dimensions of charge. Since our final result is already dimensionally correct, it is unchanged when we restore $\hbar$.

18.7.6 Three dimensions and SmB$_6$

Remarkably, the two-dimensional arguments can be pushed on to three dimensions. In our mean-field Hamiltonian, the four-dimensional gamma matrices

$$\vec{\gamma} \equiv (\gamma_1, \gamma_2, \gamma_3, \gamma_4) = \left[\begin{pmatrix} & \sigma_x \\ \sigma_x & \end{pmatrix}, \begin{pmatrix} & \sigma_y \\ \sigma_y & \end{pmatrix}, \begin{pmatrix} & \sigma_z \\ \sigma_y & \end{pmatrix}, \begin{pmatrix} 1 & \\ & -1 \end{pmatrix} \right] \quad (18.227)$$

form a basis for a four-dimensional $\hat{d}(\mathbf{k})$ vector that develops a topologically stable texture in three-dimensional momentum space. Each time a band inversion between even- and odd-parity states occurs at the high-symmetry points, the Z_2 index changes sign, changing the ground-state topology and producing a transition between conventional and topological insulating behavior.

In three dimensions the surface states have two conserved momenta and the spin-split states that emanate from the surface Kramers doublets are two-dimensional *Dirac cones* of spin-polarized edge states, defined by a surface Hamiltonian of the form $H = V(\vec{k} \cdot \vec{\sigma})$, where $\vec{k}$ is the surface momentum. In its trivial form, SmB$_6$ would be a band insulator, with a $4f^6$ filled $j = 5/2$ band of f-electrons.[5] However, $5d$-electrons hybridize with the f-band, causing a band crossing (see Figure 18.12) around the three X points. The band structure of the d-electrons is basically identical to that of LaB_6 (whose band structure is identical to SmB_6 but lacks the magnetic f-electrons). The crossing of the d-bands is predicted by band theory [47] and has been confirmed by ARPES spectroscopy [48, 49]. This leads to a change in sign of the Z_2 index:

$$Z_2 = \delta_\Gamma \overbrace{(\delta_X)^3}^{-1} \delta_M^3 \delta_R = -1.$$

The band crossing at the three X points then gives rise to three topologically protected Dirac cones on each surface. The interest in SmB_6 derives from the possible impact of strong interactions on its topologically protected edge states, which are not yet understood beyond the Gutzwiller renormalization provided by the mean-field Hamiltonian.

One of the unsolved mysteries at the time of going to press concerns the unexpectedly high group velocity of the surface states. The author's research group believes this is due to the breakdown of the Kondo effect at the Fermi surface [41], but this story takes us beyond the scope of this book.

18.8 Summary

This chapter has provided a more in-depth coverage of fluctuations and mixed valence in the infinite U Anderson model. By using the slave boson, we are able to capture both the low-frequency Kondo physics and the high-frequency valence fluctuation physics. The slave boson formulation enables us to treat the effect of valence fluctuations on the Kondo lattice. Fluctuations are also important for us to understand how the f-electrons acquire charge, and we saw that the formation of the heavy-fermion metal or Kondo insulator can be regarded as a kind of Anderson–Higgs effect, in which the external electromagnetic field locks together with the internal gauge fields of the Kondo lattice, giving rise to charged heavy fermions. Finally, as an application of these methods, we returned to the problem of Kondo insulators, examining the possibility of topological Kondo insulators.

Many aspects of heavy fermions are beyond the scope of this introductory volume. We have not touched on the physics of heavy-fermion superconductivity [50–53], nor have we discussed the development of antiferromagnetism and the fascinating quantum criticality that develops at the zero-temperature onset of antiferromagnetism [54–58]. These are areas of ongoing research with close connections to many of the unsolved problems in correlated-electron physics, including the interplay of magnetism and superconductivity [51], the possibility of new classes of quantum phase transition [59, 60], and the development of new technical methods to describe these challenging systems.

---

[5] Indeed, the cubic insulator samarium sulphide (SmS) is exactly of this sort, and when a pressure is applied, the d-band cuts into the f-band to produce a band crossing, forming an insulator that is thought to be isoelectronic with topological SmB_6 [46].

Here I mention four areas of particular challenge and opportunity:

- Developing a unified mean-field description of the Doniach phase diagram that extends to include magnetism and the Kondo effect. This is presumably a prerequisite if we are to understand the unusual quantum critical points that separate the heavy-fermion metal from the antiferromagnet.
- Exploring superconductivity, and finding ways to describe the interplay of the Kondo effect and superconductivity. One development of interest here is the use of the symplectic $SP(N)$ group rather than the $SU(N)$ group, which allows a large-N limit that can encompass both particle–hole and Cooper-pair singlets [61–64].
- The rapid development of computational methods, including dynamical mean-field theory [65–67] and density matrix renormalization methods [68, 69] for computing the properties of many-body fermion systems.
- Developing a description of the novel antiferromagnetic quantum critical points (zero-temperature instabilities) and of the strange metallic properties (including unusual electrical properties) [59, 60, 70, 71], Hall transport (sometimes with two relaxation times for the current and Hall current ([70, 72–74]), and thermodynamics that accompany them.

Appendix 18A   Fluctuation susceptibilities in the Kondo and infinite U Anderson models

Here we compute the susceptibilities associated with fluctuations about the mean-field theory of the single-impurity Kondo and infinite U Anderson model. The susceptibilities of interest are

$$(18.228)$$

The full lines denote the bare local conduction Green's function,

$$\xrightarrow[i\omega_n]{} = G_c(i\omega_n) = \sum_k \frac{1}{i\omega_n - \epsilon_k} = -i\pi\rho\,\mathrm{sgn}(\omega_n),$$

$$(18.229)$$

while double lines denote the full f-propagator,

$$= = = = \blacktriangleright = = = = \mathcal{G}_f(i\omega_n) = \frac{1}{i\omega_n - \lambda + i\Delta \mathrm{sgn}(\omega_n)}.$$

$$\underset{i\omega_n}{} \qquad\qquad\qquad\qquad\qquad\qquad\qquad\qquad\qquad (18.230)$$

The hatched vertex

$$= \alpha(i\omega_r, i\nu_n)$$
$$= V[G_c(i\omega_r) + G_c(i\omega_r + i\nu_r)] \qquad (18.231)$$
$$= -i\pi V \rho[\mathrm{sgn}(\omega_r) + \mathrm{sgn}(\omega_r + \nu_n)].$$

Writing out the Feynman diagrams, we obtain

$$\chi_{rr}(i\nu_n) = -T\sum_{i\omega_r} \mathcal{G}_f(i\omega_r + i\nu_n)\mathcal{G}_f(i\omega_r)[\alpha(i\omega_n, i\nu_n)]^2 + \Gamma(i\nu_n)$$

$$\chi_{r\lambda}(i\nu_n) = \chi_{\lambda r}(i\nu_n) = -T\sum_{i\omega_r} \mathcal{G}_f(i\omega_r + i\nu_n)\mathcal{G}_f(i\omega_r)\alpha(i\omega_n, i\nu_n)$$

$$\chi_{\lambda\lambda}(i\nu_r) = -T\sum_{i\omega_r} \mathcal{G}_f(i\omega_r + i\nu_n)\mathcal{G}_f(i\omega_r), \qquad (18.232)$$

where

$$\Gamma(i\nu_n) = -T\sum_{i\omega_r} \left[\mathcal{G}_f(i\omega_r)[G_c(i\omega_r + i\nu_n) + G_c(i\omega_r - i\nu_n)]\right]$$

$$= 2\,\mathrm{Re}\left[-T\sum_{i\omega_r} \mathcal{G}_f(i\omega_r + i\nu_n)G_c(i\omega_r)\right]. \qquad (18.233)$$

We now convert the summations over the Matsubara frequencies to contour integrals around the branch cuts (bc) of the arguments, using

$$-T\sum_{i\omega_n} F(i\omega_n) = -\oint_{\mathrm{bc}} \frac{dz}{2\pi i} f(z)F(z), \qquad (18.234)$$

to obtain

$$\chi_{rr}(i\nu_n) = I_2(i\nu) + \Gamma(i\nu_n)$$
$$\chi_{r\lambda}(i\nu_n) = \chi_{\lambda r}(i\nu_n) = I_1(i\nu_n)$$
$$\chi_{\lambda\lambda}(i\nu_r) = I_0(i\nu_n), \qquad (18.235)$$

where now

$$\Gamma(i\nu_n) = -2\,\mathrm{Re}\left[\oint_{\mathrm{bc}} \frac{dz}{2\pi i} f(z)\mathcal{G}_f(z + i\nu_n)G_c(z)\right] \qquad (18.236)$$

and we have defined

$$I_m(i\nu_n) = -\oint_{\mathrm{bc}} \frac{dz}{2\pi i} f(z)\mathcal{G}_f(z + i\nu_n)\mathcal{G}_f(z)[\alpha(z, i\nu_n)]^m. \qquad (18.237)$$

We will now calculate these two expressions by integrating along the branch cuts. The analytical extensions of the integrands into the complex plane that we will need are

$$\mathcal{G}_f(z) = 1/[z - \lambda + i\mathrm{sgn}(z_I)\Delta], \qquad (18.238)$$

where $z_I = \text{Im}\, z$, which has a branch cut along the real axis, and the vertex

$$
\alpha(z, i\nu_n) = \begin{cases} -2i\pi\rho V & (\text{Im}\, z > 0) \\ 0 & (\text{Im}\, z \in [-\nu_n, 0]) \\ 2i\pi\rho V & (\text{Im}\, z < 0), \end{cases} \tag{18.239}
$$

which has two branch cuts, one along $\text{Im}\, z = 0$ and one along $\text{Im}\, z = -\nu_n$. The contour of integration is then.

$$
\oint_{bc} dz \quad \begin{array}{l} z = x \\ z = x - i\nu_n \end{array}
$$

In carrying out the integrals, we will make extensive use of the result (Appendix 16A)

$$
\int_{-\infty}^{\infty} \frac{dx}{\pi} \phi(x/D) f(x) \frac{1}{x - z} = \frac{1}{\pi} \left[\tilde{\psi}(z) - \ln \frac{D\beta}{2\pi i} \right], \tag{18.240}
$$

where $\tilde{\psi}(z) = \psi(\frac{1}{2} + \frac{z\beta}{2\pi i})$ is defined in terms of the digamma function, while $\phi(x/D) = (1 + (x/D)^2)^{-1}$ defines a Lorentzian band cut-off of width D.

Let us first evaluate $\Gamma(i\nu_n)$. Shrinking the contour of integration around the branch cut in $G_c(z)$ along $\text{Im}\, z = 0$, and around the branch cut in $\mathcal{G}_f(z + i\nu_n)$ along $\text{Im}\, z = -i\nu_n$, we obtain

$$
\Gamma(i\nu_n) = -2\, \text{Re} \int \frac{dx}{2\pi i} \left[\mathcal{G}_f(x + i\nu_n)(2i\pi\rho) - \left(\mathcal{G}_f(x - i\delta) - \mathcal{G}_f(x + i\delta) \right)(i\pi\rho) \right]
$$

$$
= -2\pi\rho \int_{-\infty}^{\infty} \frac{dx}{\pi} f(x)\, \text{Re} \left[\mathcal{G}_f(x + i\nu_n) + G_f(x + i\delta) \right]
$$

$$
= -2\pi\rho \int_{-\infty}^{\infty} \frac{dx}{\pi} f(x)\, \text{Re} \left[\frac{1}{x + i\nu_n + \xi^*} + \frac{1}{x - \xi^*} \right]
$$

$$
= -2\pi\rho \int_{-\infty}^{\infty} \frac{dx}{\pi} f(x)\, \text{Re} \left[\frac{1}{x - i\nu_n - \xi} + \frac{1}{x - \xi} \right]. \tag{18.241}
$$

Using the integral (18.240), we then obtain

$$
\Gamma(i\nu_n) = -2\rho\, \text{Re} \left[\tilde{\psi}(\xi + i\nu_n) + \tilde{\psi}(\xi) - 2\ln \frac{D\beta}{2\pi i} \right]
$$

$$
= -2\, \text{Re}\, [\mathcal{L}(i\nu_n)] - 2\rho\, \text{Re} \left[\tilde{\psi}(\xi) - \ln \frac{D\beta}{2\pi i} \right], \tag{18.242}
$$

where

$$
\mathcal{L}(i\nu_n) = \rho \left[\tilde{\psi}(i\nu_n + \xi) - \tilde{\psi}(\xi) \right]. \tag{18.243}
$$

Now the mean-field saddle-point condition is given by (see (18.64))

$$
\tilde{\psi}(\xi) - \ln \frac{T_K e^{i\pi q}}{2\pi i T} = 0. \tag{18.244}
$$

Taking the real part,

$$
\text{Re}\, \tilde{\psi}(\xi) - \ln \frac{T_K e^{i\pi q}}{2\pi i T} = \text{Re}\, \tilde{\psi}(\xi) - \ln \frac{D\beta}{2\pi} + \frac{1}{J\rho} \tag{18.245}
$$

or

$$\rho \operatorname{Re}\left[\tilde{\psi}(\xi) - \ln \frac{D\beta}{2\pi i}\right] = -\frac{1}{J}, \tag{18.246}$$

we can replace the last term in $\Gamma(i\nu_n)$ by $2/J$:

$$\Gamma(i\nu_n) = -2 \operatorname{Re}\left[\mathcal{L}(i\nu_n)\right] + \frac{2}{J}. \tag{18.247}$$

Next let us calculate $I_m(i\nu_n)$. Shrinking the contour of integration around the two branch cuts, we obtain

$$I_m(i\nu_n) = -\sum_{P=\pm 1} \int_{-\infty}^{\infty} \frac{dx}{\pi} f(x) \operatorname{Im}\left[\mathcal{G}_f(x - i\delta)\mathcal{G}_f(x + i\nu_n P)[\alpha(x - i\delta, i\nu_n P)]^m\right]. \tag{18.248}$$

For $m > 0$, since $\alpha(x - i\delta, i\nu_n) = 0$ (for $\nu_n > 0$), only the contribution with $P = -1$, $\alpha = +2i\pi\rho V$ contributes, and we obtain

$$\begin{aligned}
I_m(i\nu_n) &= -\int_{-\infty}^{\infty} \frac{dx}{\pi} f(x) \operatorname{Im}\left[\frac{1}{x - \xi}\frac{1}{x - i\nu_n - \xi}(2i\pi\rho V)^m\right] \\
&= -\int_{-\infty}^{\infty} \frac{dx}{\pi} f(x) \operatorname{Im}\left[\frac{1}{i\nu_n}\left(\frac{1}{x - i\nu_n - \xi} - \frac{1}{x - \xi}\right)(2i\pi\rho V)^m\right] \\
&= -\frac{1}{\pi} \operatorname{Im}\left[\frac{1}{i\nu_n}(\tilde{\psi}(i\nu_n + \xi) - \tilde{\psi}(\xi))(2i\pi\rho V)^m\right], \tag{18.249}
\end{aligned}$$

from which we deduce

$$I_2(i\nu_n) = -\frac{4\Delta}{\nu_n}\operatorname{Re}[\mathcal{L}(i\nu_n)]$$

$$I_1(i\nu_n) = -\frac{2V}{\nu_n}\operatorname{Im}[\mathcal{L}(i\nu_n)]. \tag{18.250}$$

Carrying out the integral for the case where $m = 0$, we obtain

$$\begin{aligned}
I_0(i\nu_n) &= -\int_{-\infty}^{\infty} \frac{dx}{\pi} f(x) \operatorname{Im}\left[\frac{1}{x - \xi}\frac{1}{x - i\nu_n - \xi} + \frac{1}{x - \xi}\frac{1}{x + i\nu_n - \xi^*}\right] \\
&= -\int_{-\infty}^{\infty} \frac{dx}{\pi} f(x) \operatorname{Im}\left[\left(\frac{1}{x - \xi} - \frac{1}{x - i\nu_n - \xi}\right)\frac{1}{-i\nu_n} + \left(\frac{1}{x - \xi} - \frac{1}{x + i\nu_n - \xi^*}\right)\right. \\
&\qquad\qquad\qquad\left. \times \frac{1}{i\nu_n + 2i\Delta}\right] \\
&= -\int_{-\infty}^{\infty} \frac{dx}{\pi} f(x) \operatorname{Im}\left[\left(\frac{1}{x - \xi} - \frac{1}{x - i\nu_n - \xi}\right)\frac{1}{-i\nu_n} + \left(\frac{1}{x - \xi} - \frac{1}{x - i\nu_n - \xi}\right)\right. \\
&\qquad\qquad\qquad\left. \times \frac{1}{i\nu_n + 2i\Delta}\right], \tag{18.251}
\end{aligned}$$

where in the last term we have used the relationship $\operatorname{Im} F(z) = -\operatorname{Im} F(z)^*$ to replace $i\nu_n - \xi^* \to -i\nu_n - \xi$ in the last term. Finishing the calculation,

$$
\begin{aligned}
I_0(iv_n) &= \int_{-\infty}^{\infty} \frac{dx}{\pi} f(x) \, \mathrm{Im}\left[\left(\frac{1}{x - iv_n - \xi} - \frac{1}{x - \xi}\right)\frac{2i\Delta}{(iv_n)(iv_n + 2i\Delta)}\right] \\
&= \frac{2\Delta}{(v_n + 2\Delta)v_n}\frac{1}{\pi}\mathrm{Re}\left[\tilde{\psi}(iv_n + \xi) - \tilde{\psi}(\xi)\right] \\
&= \frac{2V^2}{(v_n + 2\Delta)v_n}\mathrm{Re}\left[\mathcal{L}(iv_n)\right] \\
&= 2g(iv_n)\left(\frac{V}{v_n}\right)^2 \mathrm{Re}\left[\mathcal{L}(iv_n)\right],
\end{aligned} \tag{18.252}
$$

where

$$
g(iv_n) = \left(\frac{iv_n}{(iv_n + 2i\Delta)}\right).
$$

Finally, inserting (18.242), (18.250), and (18.252) into (18.235), we find that

$$
\frac{2}{J} - \chi_{rr}(iv_r) = -2\frac{1}{g(iv_r)}\,\mathrm{Re}\,\mathcal{L}(iv_r) \tag{18.253}
$$

$$
\chi_{r\lambda}(iv_r) = \chi_{\lambda r}(iv_r) = -2\left(\frac{V}{v_r}\right)\mathrm{Im}\,\mathcal{L}(iv_r) \tag{18.254}
$$

$$
\chi_{\lambda\lambda}(iv_r) = 2g(iv_r)\left(\frac{V}{v_r}\right)^2 \mathrm{Re}\,\mathcal{L}(iv_r). \tag{18.255}
$$

The explicit disappearance of J from the right-hand side of (18.253) means that the physics only depends implicitly on the coupling constant, via the emergent Kondo scale $T_K = De^{-1/J\rho}$, so that all models with the same T_K have precisely the same fluctuations about the large-N saddle point.

Appendix 18B   Elitzur's theorem

This is an outline of Elitzur's theorem, following a discussion given by Drouffe and Zuber [75]. Suppose one has a field theory in which the elementary field is ϕ, such that under a group of local transformations $\phi \to \phi_g$ the action is invariant, $S(\phi) = S(\phi_g)$. If $f(\phi)$ is a function of the fields that has no component transforming according to the trivial representation of g, then, when f is averaged over a gauge orbit,

$$
\int Dg\, f(\phi_g) = 0. \tag{18.256}
$$

Examples of such quantities are matter fields and gauge fields. Now this alone does not mean that the average of f vanishes. To explore this possibility, we have to examine what happens as we go to the thermodynamic limit ($N_s \to \infty$) in the presence of source terms, which we then remove. Schematically, we must take

$$
\langle f(\phi)\rangle_{N_s,J} = \frac{1}{Z_{N_s,J}}\int \mathcal{D}[\phi]\, f(\phi)\exp\left[-S[\phi] + J \cdot \phi\right] \tag{18.257}
$$

in the thermodynamic limit $N_s \to \infty$, and then we remove the source field $J \to 0$:

$$\langle f(\phi) \rangle = \lim_{J \to 0} \lim_{N_s \to \infty} \langle f(\phi) \rangle_{N_s, J}. \tag{18.258}$$

Suppose we now make the change of variable $\phi \to \phi_g$ inside the path integral (18.257). Then, since the action and the measure are invariant under this transformation, we can write

$$\langle f(\phi) \rangle_{N_s, J} = \frac{1}{Z_{N_s, J}} \int \mathcal{D}[\phi] \mathcal{D}[g] f(\phi_g) \exp\left[-S[\phi] + J \cdot \phi_g\right], \tag{18.259}$$

where we have integrated over all such transformations g, using a normalized measure $\int \mathcal{D}[g] = 1$. Usually one factors ϕ into two subsets, $\{\phi''\}$, which is unchanged by the transformation g, $\phi'' = \phi''_g$, and its complement $\{\phi'\}$, writing $J \cdot \phi_g = J \cdot \phi'_g$. Elitzur's theorem rides on the observation that, for sufficiently small sources $||J|| < \epsilon$, one can bound the deviation of the source term from unity:

$$\left| \exp\left[J \cdot \phi'\right] - 1 \right| \leq \eta(\epsilon), \tag{18.260}$$

with $\eta(\epsilon)$ vanishing as ϵ goes to zero, and being independent of both ϕ_g and N_s. The independence or uniformity of this limit in ϕ relies on the compact nature of the group (so that ϕ is a kind of phase that has an upper bound). The uniformity of the limit with N_s relies on the locality of the gauge transformation, which requires that the number of degrees of freedom ϕ' involved in the transformation is independent of N_s. This is not the case for a global gauge invariance, where $J \cdot \phi'$ is extensive in N_s, so the theorem fails. However, in systems with local gauge invariance, the source term only involves an intensive number of ϕ' fields, and the bound works. In this case, one can write $\exp\left[J \cdot \phi'_g\right] = 1 + \left(\exp\left[J' \cdot \phi'_g\right] - 1\right)$. Now from (18.258), the first part vanishes in the integration over g. Written out explicitly, this is

$$\langle f(\phi) \rangle_{N_s, J} = \frac{1}{Z_{N_s, J}} \int \mathcal{D}[\phi] \exp\left[-S[\phi]\right] \int \mathcal{D}[g] f(\phi_g) \left(1 + (\exp[J \cdot \phi'_g] - 1)\right)$$

$$= \frac{1}{Z_{N_s, J}} \int \mathcal{D}[\phi] \exp\left[-S[\phi]\right] \int \mathcal{D}[g] f(\phi_g) \left(\exp[J \cdot \phi'_g] - 1\right). \tag{18.261}$$

The second term is bounded by the absolute value of $\exp[J \cdot \phi'_g] - 1$ and the maximum $||f||$ over ϕ of $|f(\phi)|$, so that

$$|\langle f(\phi) \rangle_{N_s, J}| \leq ||f|| \times \frac{1}{Z_{N_s, J}} \int \mathcal{D}[\phi] \exp\left[-S[\phi]\right] \int \mathcal{D}[g] \left|\exp[J \cdot \phi'_g] - 1\right|$$

$$\equiv ||f|| \times \left|\exp[J \cdot \phi'] - 1\right|$$

$$\leq \eta(\epsilon) ||f||, \tag{18.262}$$

where the ability to set the bound $\eta(\epsilon)$ on the second term depends on the locality of the gauge transformation. (This is the step that fails for global symmetries.) Finally, taking the limits $N_s \to \infty$ and then $J \to 0$,

$$\langle f(\phi) \rangle = 0. \tag{18.263}$$

Exercises

Exercise 18.1 Consider the infinite U Hubbard model:

$$= -t \sum_{(i,j),\sigma} \left(X_{\sigma 0}(i) X_{0\sigma}(j) + \text{H.c.} \right) - \mu \sum_{j,\sigma} X_{\sigma\sigma}, \tag{18.264}$$

infinite U Hubbard model

describing the hopping of electrons between sites with an infinite U constraint that maintains $n_f \leq 1$.

(a) Reformulate the model using the slave boson approach and show that in the radial gauge the action becomes

$$S = \int_0^\beta d\tau \left[\sum_j f^\dagger_{j\sigma} \partial_\tau f_{j\sigma} + H \right], \tag{18.265}$$

where

$$H = -t \sum_{(i,j),\sigma} r_i r_j \left[f^\dagger_{i\sigma} f_{j\sigma} + \text{H.c.} \right] + \sum_{j,\sigma} \left[\lambda_j \left(n_f(j) + r_j^2 - Q \right) - \mu n_f(j) \right], \tag{18.266}$$

where r_j is the amplitude of the slave boson field.

(b) Calculate the mean-field equations for the infinite U Hubbard model, assuming $\lambda_j = \lambda$ and $r_j = r$ are uniform.

(c) Show that the mean-field free energy is not a local minimum of the slave boson amplitude r unless one first imposes the constraint $\partial \langle H \rangle / \partial \lambda = 0$.

(d) Show that the effective mass of the electrons diverges as $n_f \to Q$.

Exercise 18.2

(a) By including the valence fluctuation terms (18.40)

$$\Delta H_{MV} = N \sum_j \lambda_j \left(\frac{|\tilde{V}_j|^2}{V_0^2} \right) \tag{18.267}$$

to the mean-field expression (17.133) for the Kondo lattice energy, show that the mean-field energy of the infinite U Anderson lattice can be written

$$\tilde{E}_A[\Delta, \lambda] = \frac{E_A}{N \mathcal{N}_s} = -\frac{D^2 \rho}{2} + \frac{\Delta}{\pi} \ln \left(\frac{\lambda}{D} \right) + (\lambda - E_f) \left[\frac{\Delta}{\Delta_0} - q \right]. \tag{18.268}$$

(b) Show that, by imposing the constraint $\partial \tilde{E}_A / \partial \lambda = 0$, you obtain

$$\lambda = \frac{N\Delta}{\pi \left(Q - \langle b^\dagger b \rangle \right)} = \frac{\Delta}{\pi \left(q - \frac{\Delta}{\Delta_0} \right)}, \tag{18.269}$$

so that

$$\tilde{E}_A[\Delta] = -\frac{D^2 \rho}{2} + \frac{\Delta}{\pi} \ln \left(\frac{\Delta}{\pi (q - \frac{\Delta}{\Delta_0}) D e} \right) + E_f \left[q - \frac{\Delta}{\Delta_0} \right]. \tag{18.270}$$

(c) By minimizing (18.270) with respect to Δ, show that the renormalized resonant level width for the Anderson lattice is given by a Fermi function of the f-level position with an effective temperature Δ_0 / π, as follows:

$$\Delta = \frac{q \Delta_0}{1 + \exp \left[-\frac{\pi E_f^*}{\Delta_0} \right]} = \frac{q \Delta_0}{1 + \exp \left[\frac{\pi |E_f^*|}{\Delta_0} \right]}, \tag{18.271}$$

where

$$E_f^* = E_f + \frac{\Delta_0}{\pi} \ln \left(\frac{\pi D}{\Delta_0} \right). \tag{18.272}$$

Exercise 18.3 (a) Generalize the calculation of phase fluctuations carried out in (18.273) to real time, to show that

$$\frac{1}{2} \langle (\phi(t) - \phi(0))^2 \rangle = \int_0^\infty \frac{d\omega}{\pi} \operatorname{Im} D_{\lambda\lambda}(\omega - i\delta) \frac{(1 - \cos \omega t)}{\omega^2}. \tag{18.273}$$

(b) Use the fluctuation–dissipation theorem to generalize this to finite temperatures, showing that

$$\frac{1}{2} \langle (\phi(t) - \phi(0))^2 \rangle = \int_0^\infty \frac{d\omega}{\pi} (2n(\omega) + 1) \operatorname{Im} D_{\lambda\lambda}(\omega - i\delta) \frac{(1 - \cos \omega t)}{\omega^2}. \tag{18.274}$$

(c) Assuming that $\operatorname{Im} D_{\lambda\lambda}(\omega - i\delta) \approx \frac{\pi \omega}{N}$, show that the thermal fluctuations in the phase cause the phase fluctuations to grow linearly in time, i.e. that the finite-temperature correlations of the slave boson field in the single impurity model decay exponentially with time.

Exercise 18.4 Generalize the method of Example 18.6 to derive the surface states for a particle–hole symmetric topological Kondo insulator in three dimensions, where

$$h(\mathbf{k}) = \begin{pmatrix} \epsilon_{\mathbf{k}} & V \vec{s}_{\mathbf{k}} \cdot \vec{\sigma} \\ V \vec{s}_{\mathbf{k}} \cdot \vec{\sigma} & -\epsilon_{\mathbf{k}} \end{pmatrix} = \epsilon_{\mathbf{k}} \tau_3 + V \tau_1 (\sin k_x \sigma_x + \sin k_y \sigma_z + \sin k_z \sigma_z), \tag{18.275}$$

where $\epsilon_{\mathbf{k}} = 2t(\cos k_x + \cos k_y + \cos k_z)$.

(a) Show that the surface dispersion on the 001 surface is given by

$$E(\mathbf{k}) = \pm V \sqrt{\sin^2 k_x + \sin^2 k_y}. \tag{18.276}$$

(b) Show that the penetration depth is the same as in the Shockley chain.

(c) What is the relationship between the bound state in this particle–hole symmetric topological insulator and the topological bound state at the surface of superfluid ^{3}He-B?

References

[1] P. Coleman, New approach to the mixed-valence problem, *Phys. Rev. B*, vol. 29, p. 3035, 1984.

[2] S. E. Barnes, New method for the Anderson model, *J. Phys. F*, vol. 6, p. 1375, 1976.

[3] N. Read and D. M. Newns, A new functional integral formalism for the degenerate Anderson model, *J. Phys. C*, vol. 29, p. L1055, 1983.

[4] A. J. Millis and P. A. Lee, Large-orbital-degeneracy expansion for the lattice Anderson model, *Phys. Rev. B*, vol. 35, no. 7, p. 3394, 1987.

[5] M. C. Gutzwiller, Effect of correlation on the ferromagnetism of transition metals, *Phys. Rev. Lett.*, vol. 10, no. 5, p. 159, 1963. The Hubbard model was written down independently by Gutzwiller in equation (11) of this paper.

[6] F. D. M. Haldane, Scaling theory of the asymmetric Anderson model, *Phys. Rev. Lett.*, vol. 40, p. 416, 1978.

[7] P. W. Anderson, Infrared catastrophe in Fermi gases with local scattering potentials, *Phys. Rev. Lett.*, vol. 18, p. 1049, 1967.

[8] S. Elitzur, Impossibility of spontaneously breaking local symmetries, *Phys. Rev. D*, vol. 12, p. 3978, 1975.

[9] P. Coleman, Large-N as a classical limit ($1/N \approx \hbar$) of mixed valence, *J. Magn. Magn. Mater.*, vol. 47–48, p. 323, 1985.

[10] P. Coleman and N. Andrei, Kondo-stabilized spin liquids and heavy-fermion super-conductivity, *J. Phys.: Condens. Matter.*, vol. 1, p. 4057, 1989.

[11] N. Read, Role of infrared divergences in the $1/N$ expansion of the $U = \infty$ Anderson model, *J. Phys. C: Solid State Phys.*, vol. 18, no. 13, p. 2651, 1985.

[12] E. Witten, Chiral symmetry, the 1/N expansion and the SU(N) Thirring model, *Nucl. Phys. B*, vol. 145, p. 110, 1978.

[13] P. Coleman and N. Andrei, Diagonalisation of the generalised Anderson model, *J. Phys. C: Solid State Phys.*, vol. 19, no. 17, p. 3211, 1986.

[14] P. Nozières and C. De Dominicis, Singularities in the X-ray absorption and emission of metals III. one-body theory exact solution, *Phys. Rev.*, vol. 178, no. 3, p. 1097, 1969.

[15] P. Coleman, J. B. Marston, and A. J. Schofield, Transport anomalies in a simplified model for a heavy electron quantum critical point, *Phys. Rev. B*, vol. 72, p. 245111, 2005.

[16] T. Senthil, M. Vojta, and S. Sachdev, Fractionalized Fermi liquids, *Phys. Rev. Lett.*, vol. 90, p. 216403, 2003.

[17] P. Coleman, J. B. Marston, and A. J. Schofield, Transport anomalies in a simplified model for a heavy-electron quantum critical point, *Phys. Rev.*, vol. 72, p. 245111, 2005.

[18] P. W. Anderson, The resonating valence bond state in La_2CuO_4 and superconductivity, *Science*, vol. 235, p. 1196, 1987.

[19] L. B. Ioffe and A. I. Larkin, Gapless fermions and gauge fields in dielectrics, *Phys. Rev. B*, vol. 39, p. 8988, 1989.

[20] J. W. Allen, B. Batlogg, and P. Wachter, Large low-temperature Hall effect and resistivity in mixed-valent SmB_6, *Phys. Rev. B*, vol. 20, p. 4807, 1979.

[21] J. E. Moore, The birth of topological insulators, *Nature*, vol. 464, no. 7286, p. 194, 2010.

[22] R. B. Laughlin, Quantized Hall conductivity in two-dimensions, *Phys. Rev. B*, vol. 23, no. 10, p. 5632, 1981.

[23] D. J. Thouless, M. Kohmoto, M. P. Nightingale, and M. Den Nijs, Quantized Hall conductance in a two-dimensional periodic potential, *Phys. Rev. Lett.*, vol. 49, no. 6, p. 405, 1982.

[24] F. D. M. Haldane, Model for a quantum Hall effect without Landau levels condensed matter realization of the parity anomaly, *Phys. rev. lett.*, vol. 61, no. 18, p. 2015, 1988.

[25] C. L. Kane and E. J. Mélé, Z_2 topological order and the quantum spin Hall effect, *Phys. Rev. Lett.*, vol. 95, p. 146802, 2005.

[26] B. A. Bernevig, T. L. Hughes, and S.-C. Zhang, Quantum spin Hall effect and topological phase transition in HgTe quantum wells, *Science*, vol. 314, no. 5806, p. 1757, 2006.

[27] R. Roy, Z_2 classification of quantum spin Hall systems; an approach using time-reversal invariance *Phys. Rev. B*, vol. 79, p. 195321, 2009.

[28] L. Fu, C. L. Kane, and E. J. Mélé, Topological insulators in three dimensions, *Phys. Rev. Lett.*, vol. 98, p. 106803, 2007.

[29] J. E. Moore and L. Balents, Topological invariants of time-reversal-invariant band structures, *Phys. Rev. B*, vol. 75, p. 121306(R), 2007.

[30] G.E. Volovik, Fermion zero modes at the boundary of superfluid ^{3}He-B, *JETP Lett.*, vol. 90, no. 5, p. 398, 2009.

[31] G.E. Volovik, Topological invariant for superfluid ^{3}He-B and quantum phase transitions, *JETP Lett.*, vol. 90, no. 8, p. 587, 2009.

[32] M. König, S. Wiedmann, C. Brüne, A. Roth, H. Buhmann, L. W. Molenkamp, X.-L. Qi and S.-C. Zhan, Quantum spin Hall insulator state in HgTe quantum wells, *Science*, vol. 318, p. 766, 2007.

[33] D. Hsieh, D. Qian, L. Wray, Y. Xia, Y. S. Hor, R. J. Cava, and M. Z. Hasan, A topological Dirac insulator in a quantum spin Hall phase, *Nature*, vol. 452, no. 7190, p. 970, 2008.

[34] L. Fu and C. L. Kane, Topological insulators with inversion symmetry, *Phys. Rev. B*, vol. 76, no. 4, p. 45302, 2007.

[35] M. Dzero, K. Sun, V. Galitski, and P. Coleman, Topological Kondo insulators, *Phys. Rev. Lett.*, vol. 104, p. 106408, 2010.

[36] J. C. Cooley, M. C. Aronson, A. Lacerda, Z. Fisk, P. C. Canfield, and R. P. Guertin, High magnetic fields and the correlation gap in SmB_6, *Phys. Rev. B*, vol. 52, p. 7322, 1995.

[37] M. Dzero, K. Sun, P. Coleman, and V. Galitski, Theory of topological Kondo insulators, *Phys. Rev. B*, vol. 85, p. 045130, 2012.

[38] S. Wolgast, Ç. Kurdak, K. Sun, J. W. Allen, D.-J. Kim, and Z. Fisk, Low-temperature surface conduction in the Kondo insulator SmB_6, *Phys. Rev. B*, vol. 88, p. 180405, 2013.

[39] D. J. S. Thomas, T. Grant, J. Botimer, Z. Fisk, and J. Xia, Surface Hall effect and non-local transport in SmB_6: evidence for surface conduction, *Sci. Rep.*, vol. 3, p. 3150, 2014.

[40] N. Xu, P. K. Biswas, J. H. Dil, *et al.* Direct observation of the spin texture in SmB_6 as evidence of the topological Kondo insulator, *Nat. Commun.*, vol. 5, p. 1, 2014.

[41] V. Alexandrov, P. Coleman, and O. Erten, Surface Kondo breakdown and the light surface states in topological Kondo insulators, *Phys. Rev. Lett.*, vol. 114, p. 177202, 2015.

[42] W. Shockley, On the surface states associated with a periodic potential, *Phys. Rev.*, vol. 56, p. 317, 1939.

[43] A. Yu. Kitaev, Unpaired Majorana fermions in quantum wires, *Phys.-Usp (Supplement)*, vol. 44, no. 10S, p. 131, 2001.

[44] V. Alexandrov and P. Coleman, End states in a 1-D topological Kondo insulator, *Phys. Rev. B*, vol. 90, p. 115147, 2014.

[45] V. Alexandrov, M. Dzero, and P. Coleman, Cubic topological Kondo insulators, *Phys. Rev. Lett.*, vol. 111, p. 226403, 2013.

[46] M. B. Maple and D. Wohlleben, Nonmagnetic 4f shell in the high-pressure phase of SmS, *Phys. Rev. Lett.*, vol. 27, no. 8, p. 511, 1971.

[47] T. Takimoto, SmB_6: a promising candidate for a topological insulator, *J. Phys. Soc. Jpn.*, vol. 80, no. 12, p. 123710, 2011.

[48] M. Neupane, N. Alidoust, S. Y. Xu, and T. Kondo, Surface electronic structure of the topological Kondo-insulator candidate correlated electron system SmB_6, *Nat. Commun.*, 4:2991 DOI:0.1038/ncomms3991, 2013.

[49] N. Xu, X. Shi, P. K. Biswas, *et al.*, Surface and bulk electronic structure of the strongly correlated system SmB_6 and implications for a topological Kondo insulator, *Phys. Rev. B*, vol. 88, p. 121102, 2013.

[50] M. Sigrist and K. Ueda, Unconventional superconductivity, *Rev. Mod. Phys.*, vol. 63, p. 239, 1991.

[51] N. Mathur, F. M. Grosche, S. R. Julian, I. R. Walker, D. M. Freye, and R. K. W. Haselwimmer, and G. G. Lonzarich, Magnetically mediated superconductivity in heavy-fermion compounds, *Nature*, vol. 394, p. 39, 1998.

[52] C. Petrovic, P. G. Pagliuso, M. F. Hundley, R. Movshovich, J. L. Sarrao, J. D. Thompson, Z. Fisk, and P. Monthoux, Heavy-fermion superconductivity in $CeCoIn_5$ at 2.3 K, *J. Phys: Condens. Matter*, vol. 13, p. L337, 2001.

[53] N. J. Curro, T. Caldwell, E. D. Bauer et al., Unconventional superconductivity in $PuCoGa_5$, *Nature*, vol. 434, p. 622, 2005.

[54] G. Stewart, Heavy-fermion systems, *Rev. Mod. Phys.*, vol. 73, p. 797, 2001.

[55] G. Stewart, Addendum: Non-Fermi-liquid behavior in d-and f-electron metals, *Rev. Mod. Phys.*, vol. 78, p. 743, 2006.

[56] P. Coleman, C. Pépin, Q. Si, and R. Ramazashvili, How do Fermi liquids get heavy and die?, *J. Phys.: Condens. Matter*, vol. 13, p. 273, 2001.

[57] C. M. Varma, Z. Nussinov, and W. van Saarlos, Singular Fermi liquids, *Phys. Rep.*, vol. 361, p. 267, 2002.

[58] H. von Löhneysen, A. Rosch, M. Vojta, and P. Wölfle, Fermi-liquid instabilities at magnetic quantum phase transitions, *Rev. Mod. Phys.*, vol. 79, p. 1015, Aug 2007.

[59] J. Custers, P. Gegenwart, H. Wilhelm, The break-up of heavy electrons at a quantum critical point, *Nature*, vol. 424, p. 524, 2003.

[60] Y. Matsumoto, S. Nakatsuji, K. Kuga, Y. Karaki, and N. Horie, Quantum criticality without tuning in the mixed valence compound β-YbAlB$_4$, *Science*, vol. 331, p. 316, 2011.

[61] P. Coleman and N. Andrei, Kondo-stabilised spin liquids and heavy-fermion superconductivity, *J. Phys.: Condens. Matter*, vol. 1, no. 26, p. 4057, 1989.

[62] R. Flint, M. Dzero, P. Coleman, and M. Dzero, Heavy electrons and the symplectic symmetry of spin, *Nat. Phys.*, vol. 4, no. 8, p. 643, 2008.

[63] R. Flint and P. Coleman, Tandem pairing in heavy-fermion superconductors, *Phys. Rev. Lett.*, vol. 105, p. 246404, 2010.

[64] R. Flint, A. Nevidomskyy, and P. Coleman, Composite pairing in a mixed-valent two-channel Anderson model, *Phys. Rev. B*, vol. 84, no. 6, p. 064514, 2011.

[65] W. Metzner and D. Vollhardt, Correlated lattice fermions in $d = \infty$ dimensions, *Phys. Rev. Lett.*, vol. 62, p. 324, 1989.

[66] A. Georges and G. Kotliar, Hubbard model in infinite dimensions, *Phys. Rev. B*, vol. 45, p. 6479, 1992.

[67] A. Georges, G. Kotliar, W. Krauth, and M. Rozenberg, Dynamical mean-field theory of strongly correlated fermion systems and the limit of infinite dimensions, *Rev. Mod. Phys.*, vol. 68, p. 13, 1996.

[68] S. R. White, Strongly correlated electron systems and the density matrix renormalization group, *Phys. Rep.*, vol. 301, p. 187, 1998.

[69] K. A. Hallberg, New trends in density matrix renormalization, *Adv. Phys.*, vol. 55, p. 477, 2006.

[70] T. R. Chien, Z. Z. Wang, and N. P. Ong, Effect of Zn impurities on the normal-state Hall angle in single-crystal YBa$_2$Cu$_{3-x}$Zn$_x$ O$_{7-\delta}$, *Phys. Rev. Lett.*, vol. 67, p. 2088, 1991.

[71] J. Paglione, M. A. Tanatar, D. G. Hawthorn, Nonvanishing energy scales at the quantum critical point of CeCoIn$_5$, *Phys. Rev. Lett.*, vol. 97, p. 106606, 2006.

[72] P. W. Anderson, Hall effect in the two-dimensional Luttinger liquid, *Phys. Rev. Lett.*, vol. 67, p. 2092, 1991.

[73] P. Coleman, A. J. Schofield, and A. M. Tsvelik, Phenomenological transport equation for the cuprate metals, *Phys. Rev. Lett.*, vol. 76, p. 1324, 1996.

[74] Y. Nakajima, K. Izawa, Y. Matsuda, Normal-state Hall angle and magnetoresistance in quasi-2D heavy fermion $CeCoIn_5$ near a quantum critical point, *J. Phys. Soc. Jpn.*, vol. 73, p. 5, 2004.

[75] J. M. Drouffe and J. B. Zuber, Strong coupling and mean-field methods in lattice gauge theories, *Phys. Rep.*, vol. 102, p. 1, 1983.

Epilogue: the challenge of the future

On October 19, 2014, as I finished this book, a group of physicists convened at the University of Illinois Urbana-Champaign, home of BCS theory [1], to mark 60 years of progress in strongly correlated electron systems (SCES) and to celebrate the 90th birthday of David Pines. This occasion provided a great opportunity for us to reflect on the dizzying progress of the past 60 years of discovery in many-body physics: a period that spans from the Bohm–Pines plasmon theory of metals [2–4], the discovery of Landau Fermi-liquid theory [5], and BCS theory [1], to the modern era of topological order, strange quantum-critical metals, and high-temperature-superconductivity. This book has only scratched the surface of this period, leaving out some key aspects of the field that I perhaps will go into in a future edition. Yet still, almost everything in this book was unknown 60 years ago. The pioneers of early many-body physics in the 1950s simply could not have imagined the revolution of discovery that has since taken place. David Pines recalls thinking, as a graduate student in the early 1950s, that the physicists 20 years earlier had had all the luck. Yet Pines' thesis work with Bohm marked a beginning of 60 years of tremendous discovery.

You, the reader, might be tempted to think as David Pines did. However, at the "60 years of SCES" meeting in Urbana, many expressed remarkable optimism that the revolution is decidedly unfinished, and that the next 60 years has much in store. Like earthquakes, big scientific discoveries are unpredictable and come at varying intervals. Today, in the early twenty-first century, a large number of unsolved mysteries and problems in condensed matter physics provide the tremors that suggest that there is much yet to discover. I thought it would be interesting, if only as a historical record, to list a few of the challenges we all face:

1 **Wanted: a broader understanding of the classes of emergent order in condensed matter.** Between the many decades of complexity that separate simple elements from primitive life surely lie many layers of emergent material behavior. The possibility of new forms of order, as challenging and surprising as superconductivity, cannot be ruled out. What new ordering processes beyond electron–electron and electron–hole binding are possible? Is it possible, for example, to imagine three-body bound states as a driver of broken symmetry [6]? Another direction we have just brushed upon in this volume is *topological order* and the discovery of topological insulators [7]. Surely much more is possible on this frontier, such as the *quantum time-crystal* in which time translation is spontaneously broken, as proposed by Frank Wilczek [8]?

2 **Wanted: a theory of strange metals.** We now know of many strange metals, such as the optimally doped normal state of the cuprate superconductors and the 115 heavy-electron

metals that exhibit, a T-linear current relaxation rate but a T^2 temperature dependence of the Hall current relaxation rate [9–12]. Many of these phenomena occur close to a quantum critical point, yet despite more than 20 years of observation we still don't have a theory. We don't know whether they are driven by criticality (unstable fixed point) or alternatively signify a new metallic phase (stable fixed point).

3 **Wanted: a theory unifying localized moment magnetism, Mott localization and superconductivity.** This is a key challenge for understanding both heavy-fermion and high-temperature superconductors. How, for instance, can we understand the simultaneous entanglement of local moments in both magnetic and superconducting fluids, as seen in the heavy-electron material $CeRhIn_5$ [13]?

4 **Wanted: new analytic methods for the many-body problem.** First, there is tremendous room for improvement of existing analytic methods. Is it possible, for example, to find a way to extend the mean-field slave boson approach to avoid the false finite-temperature phase transition between strong and weak coupling? Can we find a way to expand the internal symmetry groups of the slave boson to include magnetism? Here, ideas from particle physics such as supersymmetry [14, 15] may be useful. But also, can we develop new types of expansion, akin to the large-N slave boson or perhaps the ϵ expansion of statistical mechanics for approaching the many-body problem? Currently there is great interest in *holographic methods* [16] which strive to map strongly interacting physics onto a higher-dimensional gravity problem, as a possible new tool for the many-body problem. What other novel techniques can be developed that will help us analytically compute and understand many body physics?

5 **Wanted: a new framework for non-equilibrium quantum mechanics** (particularly the non-equilibrium steady state). Our current understanding of the non-equilibrium steady state is based on the application of real-time quantum mechanics, on a case-by-case basis. Can we derive a general theory of the non-equilibrium steady state (e.g. current through a DC-biased quantum dot) that has the same utility as the imaginary-time approach to equilibrium many-body theory?

6 **Wanted: new computational tools.** Here the great unsolved problem is the the minus-sign problem in quantum Monte Carlo, which limits the direct application of Monte Carlo methods in strongly interacting quantum problems, particularly at low temperatures. Although this problem has not yet been solved, amazing progress in the dynamical mean-field theory is currently driving a mini-revolution in computational many-body physics, driven by powerful new methods such as the continuous-time Monte Carlo methods [17]. These new tools will help tremendously in quantifying the search for new materials. At the same time, the codes need to be devleoped to the point where physicists can step back and continue to develop the all-important back-of-the-envelope understanding.

Of course, none of us can tell the future, but it is important to have aspirations. In the near future, I hope this book, and perhaps its future editions, will help launch readers on what is, I believe, a truly exciting and ongoing revolution in the physical sciences.

Piers Coleman, Santa Barbara, California.

References

[1] J. Bardeen, L. N. Cooper, and J. R. Schrieffer, Theory of superconductivity, *Phys. Rev.*, vol. 108, no. 5, p. 1175, 1957.

[2] D. Bohm and D. Pines, A collective description of electron interactions: I. magnetic interactions, *Phys. Rev.*, vol. 82, p. 625, 1951.

[3] D. Pines and D. Bohm, A collective description of electron interactions: II. collective vs individual particle aspects of the interactions, *Phys. Rev.*, vol. 85, p. 338, 1952.

[4] D. Bohm and D. Pines, A collective description of the electron interations: III. Coulomb interactions in a degenerate electron gas, *Phys. Rev.*, vol. 92, p. 609, 1953.

[5] L. D. Landau, The theory of a Fermi liquid, *J. Exp. Theor. Phys.*, vol. 3, p. 920, 1957.

[6] P. Chandra, P. Coleman, and R. Flint, Ising quasiparticles and hidden order in URu_2Si_2, *Philos. Mag.*, vol. 94, no. 32–33, p. 3803, 2014.

[7] J. E. Moore, The birth of topological insulators, *Nature*, vol. 464, no. 7286, p. 194, 2010.

[8] F. Wilczek, Quantum time crystals, *Phys. Rev. Lett.*, vol. 109, p. 160401, 2012.

[9] T. R. Chien, Z. Z. Wang, and N. P. Ong, Effect of Zn impurities on the normal-state Hall angle in single-crystal $YBa_2Cu_{3-x}Zn_x O_{7-\delta}$, *Phys. Rev. Lett.*, vol. 67, p. 2088, 1991.

[10] P. W. Anderson, Hall effect in the two-dimensional Luttinger liquid, *Phys. Rev. Lett.*, vol. 67, p. 2092, 1991.

[11] P. Coleman, A. J. Schofield, and A. M. Tsvelik, Phenomenological transport equation for the cuprate metals, Phys. Rev. Lett., vol. 76, p. 1324, 1996.

[12] Y. Nakajima, K. Izawa, Y. Matsuda, *et al.*, Normal-state Hall angle and magnetoresistance in quasi-2D heavy fermion $CeCoIn_5$ near a quantum critical point, *J. Phys. Soc. Jpn.*, vol. 73, p. 5, 2004.

[13] T. Park, F. Ronning, H. Q. Yuan, M. B. Salamon, R. Movshovich, J. L. Sarrao, and J. D. Thompson, Hidden magnetism and quantum criticality in the heavy-fermion superconductor $CeRhIn_5$, *Nature*, vol. 440, no. 7080, p. 65, 2006.

[14] P. Coleman, C. Pépin, and A. M. Tsvelik, Supersymmetric spin operators, *Phys. Rev. B*, vol. 62, p. 3852, 2000.

[15] S.-S. Lee, Emergence of supersymmetry at a critical point of a lattice model, *Phys. Rev. B*, vol. 76, no. 7, p. 075103, 2007.

[16] S. A. Hartnoll, Lectures on holographic methods for condensed matter physics, *Classical Quantum Gravity*, vol. 26, p. 1, 2009.

[17] E. Gull, A. J. Millis, A. I. Lichtenstein, A. N. Rubtsov, M. Troyer, and P. Werner, Continuous-time Monte Carlo methods for quantum impurity models, *Rev. Mod. Phys.*, vol. 83, p. 349, May 2011.

Author Index

Subject Index